Survey

Part a

Carbon-Centered Radicals I

Part b

Carbon-Centered Radicals II

Part c

Radicals Centered on N, S, P and other Heteroatoms. Nitroxyls

Part d

Oxyl-, Peroxyl-, and Related Radicals

Part e

Proton and Electron Transfer. Biradicals

Übersicht

Teil a

Kohlenstoffzentrierte Radikale I

Teil b

Kohlenstoffzentrierte Radikale II

Teil c

Radikale mit N, S, P und anderen Heteroatomen als Zentralatom. Nitroxylradikale

Teil d

Oxyl-, Peroxyl- und verwandte Radikale

Teil e

Protonen- und Elektronenaustauschreaktionen. Biradikale

LANDOLT-BÖRNSTEIN

Numerical Data and Functional Relationships
in Science and Technology

New Series

Editors in Chief: K.-H. Hellwege · O. Madelung

Group II: Atomic and Molecular Physics

Volume 13
Radical Reaction Rates in Liquids

Subvolume b

Carbon-Centered Radicals II

K.-D. Asmus · M. Bonifačić

Editor: H. Fischer

Springer-Verlag Berlin · Heidelberg · New York · Tokyo 1984

Hans Heinrich

LANDOLT-BÖRNSTEIN

Zahlenwerte und Funktionen aus Naturwissenschaften und Technik

Neue Serie

Gesamtherausgabe: K.-H. Hellwege · O. Madelung

Gruppe II: Atom- und Molekularphysik

Band 13

Kinetische Konstanten von Radikalreaktionen in Flüssigkeiten

Teilband b

Kohlenstoff-Radikale II

K.-D. Asmus · M. Bonifačić

Herausgeber: H. Fischer

Springer-Verlag Berlin · Heidelberg · New York · Tokyo 1984

CIP-Kurztitelaufnahme der Deutschen Bibliothek

Zahlenwerte und Funktionen aus Naturwissenschaften und Technik /Landolt-Börnstein. – Berlin; Heidelberg; New York; Tokyo: Springer. Teilw. mit d. Erscheinungsorten Berlin, Heidelberg, New York. – Parallelt.: Numerical data and functional relationships in science and technology.

NE: Landolt, Hans [Begr.]; PT. N.S./Gesamthrsg.: K.-H. Hellwege; O. Madelung. Gruppe 2, Atom- und Molekularphysik. Bd. 13. Kinetische Konstanten von Radikalreaktionen in Flüssigkeiten. Teilbd. b. Kohlenstoff-Radikale. – 2. K.-D. Asmus; M. Bonifačić. Hrsg.: H. Fischer. – 1984.

ISBN 3-540-13241-4 Berlin, Heidelberg, New York, Tokyo
ISBN 0-387-13241-4 New York, Heidelberg, Berlin, Tokyo

NE: Hellwege, Karl-Heinz [Hrsg.]; Asmus, Klaus-Dieter [Mitverf.]; Fischer, Hanns [Hrsg.]

Printed in Germany

Typesetting: Polyglot Pte. Ltd., Singapore; printing: Druckhaus Langenscheidt KG, Berlin; bookbinding: Lüderitz & Bauer-GmbH, Berlin

2163/3020 – 543210

Vorwort

Die rasche Entwicklung physikalisch-chemischer Meßmethoden, insbesondere solcher auf spektroskopischer Grundlage, hat in den letzten Jahrzehnten das Studium kurzlebiger Zwischenstufen bei chemischen Reaktionen, ihrer Strukturen, sowie ihrer Reaktionen und deren Kinetik sehr begünstigt. Ein äußerst aktiver Zweig ist die Erforschung freier Radikale und ihrer Reaktionen in vielen thermischen, photo- und strahlenchemischen Prozessen von sowohl fundamentaler wie technologischer Bedeutung. Der gewaltige Umfang des heute vorliegenden Datenmaterials ruft nach umfassenden und kritischen Zusammenfassungen, einem Ziel des Landolt-Börnstein. Magnetische Eigenschaften freier Radikale sind in den Bänden II/1 und II/9a–d der neuen Serie bereits tabelliert. Der Band II/13 stellt nun Geschwindigkeitskonstanten und andere kinetische Parameter von Reaktionen freier Radikale in flüssigen Medien vor. Hauptinhalte sind polyatomare organische Radikale unter Einschluß von Biradikalen. Ausgelassen wurden einige freie Radikale in wäßrigen Medien, für die Zusammenstellungen von Geschwindigkeitskonstanten kürzlich in der National Standard Reference Data Reihe des National Bureau of Standards, Washington, D.C., erschienen sind.

Die unerwartet große Datenzahl, verstreut in Tausenden von Publikationen, legte die Verteilung der Aufgaben auf eine ganze Reihe von Autoren nahe, welche ihrem Tätigkeitsbereich entsprechende Kapitel übernahmen. Sie erzwang auch die Aufteilung des Bandes II/13 in fünf Teilbände, von denen die Teile II/13a, II/13c und II/13d bereits erschienen sind. Der hier vorgelegte Teilband II/13b behandelt, anschliessend an II/13a, weitere Reaktionen von kohlenstoffzentrierten Radikalen.

Der Umfang des Gesamtbandes spiegelt den raschen Fortschritt in der Untersuchung der Reaktionskinetik freier Radikale wider, welcher unzweifelhaft auf das Bedürfnis nach gesichertem Zahlenmaterial für die weitere Forschung und die Entwicklung der Technologie gegründet ist. Die noch anwachsende Publikationsrate zeigt, daß das Gebiet noch nicht als einigermaßen abgerundet angesehen werden kann. Dies wird auch dadurch belegt, daß für einige wichtige Reaktionen kinetische Daten bisher fehlen. Es ist dementsprechend geplant, Band II/13 später zu supplementieren, und wir hoffen, daß die hier vorgestellte Zusammenstellung auch die zukünftige Entwicklung des Gebiets befruchtet.

Wir danken allen Autoren des Werks für ihre kompetente, oft mühevolle und zeitraubende Arbeit und die erfreuliche Kooperation. Dank gebührt weiter der Landolt-Börnstein-Redaktion und hier vor allem Frau H. Weise, für die unermüdlich aufmerksame Bearbeitung der Manuskripte und Fahnen. Ferner danken wir dem Springer-Verlag für die sorgfältige Fertigstellung des Bandes, der, wie alle Landolt-Börnstein-Bände, ohne externe finanzielle Unterstützung publiziert wird.

Zürich, im Sommer 1984 **Der Herausgeber**

Preface

The recent enormous advances in physico-chemical technology, such as molecular spectroscopy, have promoted numerous studies of reactive chemical intermediates, the elucidations of their structures, their formation and decay mechanisms and their reaction rates. Free radicals, in particular, key intermediates in many thermal, photo and radiation chemical processes both of fundamental and technological importance have become the subject of active research. The amount of available quantitative data suggests comprehensive and authorative critical compilations as attempted by the Landolt-Börnstein series. Magnetic properties of about 8500 free radical species were collected in volumes II/1 and II/9a–d of the New Series. This volume II/13 presents rate constants and other kinetic data of free radical reactions in liquids. Emphasis is on polyatomic organic free radicals, and biradicals are included. Deliberately omitted were selected species in aqueous solutions for which compilations of rate data have been published in the National Standard Reference Data Series of the National Bureau of Standards, Washington D.C., USA.

The huge amount of available data, widely scattered in several thousand individual publications, required the cooperation of experts who took the charge to prepare individual chapters. It also necessitated the subdivision of the volume into five parts to be published successively of which parts II/13a, II/13c and II/13d have already appeared. The present subvolume II/13b covers further data on reactions of carbon-centered radicals.

The unexpected size of the total volume reflects the rapid progress in free radical reaction kinetics which is undoubtedly stimulated by the requirement of accurate data for future research and technical development. The still continuously increasing number of publications shows that the field has not reached saturation and is corroborated by the lack of data for many important reactions as also evident from this volume. Consequently, we plan future supplementations, and do hope that the present critical evaluations will help to stimulate further work.

We thank all the authors for their enormous and time-consuming efforts and for the most enjoyable cooperation. Thanks are also due to the Landolt-Börnstein office, especially to Frau H. Weise for the untiring careful checking of the manuscripts and galleys, and Springer Verlag for their customary care in the preparation of the volumes, which, as all Landolt-Börnstein volumes, is published without exterior financial support.

Zürich, summer 1984 **The Editor**

Survey

Part a
Carbon-Centered Radicals I

Part b
Carbon-Centered Radicals II

Part c
Radicals Centered on N, S, P and other Heteroatoms. Nitroxyls

Part d
Oxyl-, Peroxyl-, and Related Radicals

Part e
Proton and Electron Transfer. Biradicals

Übersicht

Teil a
Kohlenstoffzentrierte Radikale I

Teil b
Kohlenstoffzentrierte Radikale II

Teil c
Radikale mit N, S, P und anderen Heteroatomen als Zentralatom. Nitroxylradikale

Teil d
Oxyl-, Peroxyl- und verwandte Radikale

Teil e
Protonen- und Elektronenaustauschreaktionen. Biradikale

Table of contents

Radical reaction rates in liquids

Part b: Carbon-centered radicals II

General introduction

H. Fischer, Physikalisch-Chemisches Institut der Universität Zürich, Switzerland

4 Carbon-centered radicals II

K.-D. Asmus, Hahn-Meitner-Institut für Kernforschung,
Bereich Strahlenchemie, Berlin, FRG, and
M. Bonifačić, Ruder Bošković Institute, Zagreb, Yugoslavia

General introduction

A Definitions and coverage

In the following tables rate constants for reactions of free radicals in liquids are collected. The free radical species covered are paramagnetic molecules deriving their paramagnetism from a single unpaired valence electron. They are neutral molecular fragments or ions with positive or negative charges. Emphasis is on polyatomic organic free radicals. Excluded are some small species like the hydrated electron, the hydrogen and other atoms and a variety of polyatomic inorganic radicals. For reaction rates of these in aqueous solutions recent other compilations are available [73Anb, 75Anb, 75Ros, 77Ros, 79Ros, 83Ros]. A table on organic biradicals is included, however, since their reactions are similar to those of monoradicals.

The volume is divided grossly into sections dealing with individual types of free radicals such as carbon-centered radicals, nitrogen-centered radicals, nitroxyl radicals, oxygen-centered radicals and radicals centered on other heteroatoms. These sections deal mainly with irreversible reactions. In addition, there are sections on reversible electron and proton transfer processes and their equilibria and a chapter on biradicals. An index of radicals formulae will facilitate data retrieval.

The reactions covered involve bimolecular processes between like and unlike radicals and between radicals and molecules via atom, group or electron transfer, or addition and unimolecular processes like fragmentations or rearrangements. Within the chapters dealing with special radical types a subdivision according to the types of reaction is applied. In such subdivisions the entries are organized according to the molecular formula of the radical involved, and formulae are in the order of increasing number of C, H atoms and then all other elements (except D, listed with H) in alphabetical order.

The main subject of the volume is the compilation of absolute rate constants for established reactions. In part they were deduced from published relative rate data if the absolute rate constants of the reference reactions were known. Relative rate constants or qualitative data as reaction half-lifes are included occasionally, in particular for important classes of radicals or radical reactions for which absolute data are not yet available. Thus, the tables do not represent a comprehensive compilation of all reactions known to date, yet of all those with sufficiently characterized reaction kinetics.

For details on subdivision into subvolumes, chapters and ordering within chapters, see *Table of contents* and the introductory sections of individual chapters. The literature is generally covered up to 1981, in parts up to 1983.

B Arrangement and contents of tables

As indicated by the general table headings there is one separate entry for each specific reaction or each set of competing reactions. Besides specifying the reaction the entry contains information on the technique of radical generation, the method of rate determination, and experimental conditions such as solvent and temperature. It lists the rate constants, the equilibrium constants and other rate data, such as activation parameters of the reaction, and gives the pertinent reference plus additional references. Further relevant information is given in footnotes. The following explanations apply to the individual parts of the entries.

Reaction: The reaction or the competing reactions are written in stoichiometric form starting with the reacting radical. For reactions between different radicals the radical appearing first in the adopted ordering of substances (see above) is written first and specifies the location of that entry. A cross reference to this reaction is entered at that later position of the tables which corresponds to the order of the second radical. The same principle is obeyed in the ordering of the substrates in competing radical molecule reactions. Where deviations from this scheme occur the reader is referred to the introduction of the individual chapters. Where necessary structural formulae of radicals, reactants and products are written out in full detail. Repeatedly occuring structures are abbreviated by capital bold letters and an entry R = group may specify a substituent within the general structure. Self-evident structures of products are also abbreviated occasionally as OO- or NN-dimers or otherwise. Product structures are not given if they could not be identified from the original literature.

Radical generation: The technique of radical production is outlined in short using abbreviations given in the *List of symbols and abbreviations.*

Method: The methods in use for the determination of reaction rate data are manifold, and a variety of abbreviations had to be introduced (see also *List of symbols and abbreviations*). Whereas earlier literature mostly applied the rather indirect techniques of measurements of product ratios (PR), the consumption of inhibitors (inh. cons.), rates of oxygen absorption (ROA) or consumption of other reactants (RRC) the progress of time resolved spectroscopy is evident more recently, and the most accurate rate data for irreversible processes are now obtained from kinetic absorption spectroscopy (KAS), kinetic electron spin resonance spectroscopy (KESR) or conductometry (cond.) in conjugation with pulsed radical generation. For reversible processes steady-state techniques of absorption spectroscopy (SAS) or electron spin resonance (SESR) or line-shape analyses in magnetic resonance (ESRLA, NMRLA) are common. For details of the methods, the reader is referred to the original literature.

Solvent: Where possible the solvent is given by its molecular formula or name. Special conditions such as pH or solvent composition are indicated.

Temperature T [K]: The temperature of the sample during the rate measurement is given in K. RT stands for an unspecified room temperature. Where activation parameters of rate constants were measured the column temperature indicates the temperature range of measurement.

Rate data: Rate constants of uni- and bimolecular processes are given in their usual dimensions s^{-1} and $M^{-1} s^{-1}$, equilibrium costants in their corresponding appropriate dimensions. The same applies to ratios of rate constants. All rate constants k are defined for product appearance. Consequently, $2k$ governs the rate of radical disappearance in bimolecular self-reactions of radicals. Since the rate of radical disapperance is often measured in these cases, the value of $2k$ is displayed. If available the Arrhenius activation parameters, i.e. the parameters of the equation $k = A \cdot \exp(-E_a/RT)$ are also listed with A given in logarithmic form and E_a in kJ/mol. The column rate data may also give enthalpies (ΔH^{+}), entropies (ΔS^{+}), and volumes (ΔV^{+}) of activation in SI-units. For acid-base equilibrium pK-values are listed. Errors are given in parentheses in units of the last digit displayed for the data.

Reference/additional references: The first entry specifies the reference from which the data were extracted with the first two numbers for the year of appearance (72 = 1972), the following three letters for the family name of the first author and the last number ordering the publications in the year of publication. Additional references contain earlier less reliable work on the same subject, theoretical treatments of rate data or other relevant information.

The following list of symbols and abbreviations is common for all chapters. Additional symbols and abbreviations may appear as necessary in individual chapters. For these and additional information on contents and coverage, on arrangements and ordering and on special data display the reader is referred to the introductory sections of the individual chapters.

C Important monographs, series, compilations

61Jen Jennings, K.R., Cundall, R.B. (eds.): Progress in Reaction Kinetics. Oxford: Pergamon **1961**ff.

63Gol Gold, V., Bethell, D. (eds.): Advances in Physical Organic Chemistry. New York: Acad. Press **1963**ff.

63Taf Taft, R.W. (ed.): Progress in Physical Organic Chemistry. New York: Wiley **1963**ff.

65Kni Knipe, A.C., Watts, W.E. (eds.): Organic Reaction Mechanism. New York: Wiley **1965**ff.

66Pry Pryor, W.A.: Free Radicals. New York: McGraw-Hill **1966**.

68For Forester, A.R., Hay, J.M., Thomson, R.H.: Organic Chemistry of Free Radicals. New York: Acad. Press **1968**.

70Huy Huyser, E.S.: Free Radical Chain Reactions. New York: Interscience **1970**.

70Roz Rozantsev, E.G.: Free Nitroxyl Radicals. New York: Plenum **1970**.

73Anb Anbar, M., Bambenek, M., Ross, A.B.: Selected Specific Rates of Reactions of Transients from Water in Aqueous Solution. 1. Hydrated Electron. Washington: NSRDS-NBS 43 **1973**.

73Buc Buchachenko, A.L., Wasserman, A.L.: Stable Radicals. Weinheim: Verlag Chemie **1973**.

73Koc Kochi, J.K. (ed.): Free Radicals, Vols. I, II. New York: Wiley **1973**.

73Nor Norman, R.O.C., Ayscough, P.B. (eds.): Electron Spin Resonance. Spec. Per. Rept. London, Chemical Society **1973**ff.

74Den Denisov, E.T.: Liquid-Phase Reaction Rate Constants. New York: Plenum **1974**.

74Non Nonhebel, D.C., Walton, J.C.: Free-Radical Chemistry. Cambridge: University Press **1974**.

74Swa Szwarc, M.: Ions and Ion Pairs in Organic Reactions. New York: Wiley **1974**.

75Anb Anbar, M., Ross, A.B., Ross, F.: Selected Specific Rates of Reactions of Transients from Water in Aqueous Solution. II. Hydrogen Atom. Washington: NSRDS-NBS 51 **1975**.

75Ros Ross, A.B.: Selected Specific Rates of Reactions of Transients from Water in Aqueous Solution. Hydrated Electron, Supplemental Data. Washington: NSRDS-NBS 43 – Supplement **1975**.

76Pry Pryor, W.A. (ed.): Free Radicals in Biology. New York: Acad. Press **1976**ff.

77Ros Ross, F., Ross, A.B.: Selected Specific Rates of Reactions of Transients from Water in Aqueous Solution. III. Hydroxyl Radical and Perhydroxyl Radical and Their Radical Ions. Washington: NSRDS-NBS 59 **1977**.

78Jon Jones, M., Jr., Moss, R.A. (eds.): Reactive Intermediates. New York: Wiley **1978**ff.

79Non Nonhebel, D.C., Tedder, J.M., Walton, J.C.: Radicals, Cambridge: Univ. Press **1979**.

79Ros Ross, A.B., Neta, P.: Rate Constants for Reactions of Inorganic Radicals in Aqueous Solution. Washington: NSRDS-NBS 65 **1979**.

83Ros Ross, A.B., Neta, P.: Rate Constants of Aliphatic Carbon Centered Radicals in Aqueous Solution. Washington: NSRDS-NBS, submitted.

D List of symbols and abbreviations

Symbols

$D(\mathrm{R-X})$	bond dissociation energy
E^0, $E^{0\prime}$	standard reduction potentials
G	radiation chemical yield
H_0	Hammett acidity function
k [$\mathrm{s^{-1}, M^{-1}s^{-1}}$]	rate constant
K	equilibrium constant
$2k_t$	rate constant of self-termination
ΔG^{+}	free enthalpy of activation
ΔH^{+}	enthalpy of activation
ΔS^{+}	entropy of activation
ΔV^{+}	volume of activation
$\Delta^0 H$ [$\mathrm{kJ\,mol^{-1}}$]	enthalpy of dissociation
$\Delta^0 S$ [$\mathrm{J\,K^{-1}\,mol^{-1}}$]	entropy of dissociation
η [cP]	viscosity
ε_λ	decadic extinction coefficient at wavelength λ
$\varrho(\sigma)$, $\varrho(\sigma^+)$, $\varrho(\sigma^-)$	Hammett's rho based on σ, σ^+ or σ^- scales
$\tau_{1/2}$ [s, min, h, day]	half-life
V/V	volume by volume mixture
1:1 m	equimolar mixture

Abbreviations

a) General

absorpt.	absorption
Ac	acyl
add.	addition
Ar	aryl
conc.	concentrated, concentration
cons.	consumption
decomp.	decomposition
e	electron
f	foreward
i	iso
mixt.	mixture
MW	microwave
n	normal
phot.	photolysis
prim.	primary
pulse rad.	pulse radiolysis
r	reverse
rad.	radiolysis
reduct.	reduction
RT	room temperature
s, sec.	secondary
soln.	solution
spect.	spectroscopy
t, tert.	tertiary
temp.	temperature
temp. dep.	temperature dependence
therm.	thermolysis

b) Methods

chemil.	chemiluminescence
ch. r.	chain reaction
CIDNP	chemically induced dynamic nuclear polarization
Cond.	conductometry
Co-ox.	co-oxidation

ESRLA	electron spin resonance lineshape analysis
glc	gas liquid chromatography
HPLC	high pressure liquid chromatography
inh. cons.	inhibitor consumption
KAS	kinetic absorption spectroscopy
KESR	kinetic electron spin resonance
MS	mass spectroscopy
NMR	nuclear magnetic resonance
NMRLA	nuclear magnetic resonance lineshape analysis
PR	product ratio
ROA	rate of oxygen consumption
RRC	rate of reactant consumption
RS	rotating sector
SAS	steady-state absorption spectroscopy
SESR	steady-state electron spin resonance
spin trap.	spin trapping
therm. coup.	thermocouple method

c) Substances

ACHN	α,α′-azo-bis-cyclohexanecarbonitrile
AIBN	α,α′-azo-bis-isobutyronitrile
BMP	2,6-di-*t*-butyl-4-methylphenol
BPO	benzoyl peroxide
DBPO	dibenzoyl peroxide
DCP	di-α-cumyl peroxide
DPA	9,10-diphenylanthracene
DPM	diphenylmethanol
DPPH	α,α-diphenyl-β-picryl hydrazyl
DPPH-H	α,α-diphenyl-β-picryl hydrazine
DTBK	di-*t*-butyl ketone
DTBP	di-*t*-butyl peroxide
DTBPO	di-*t*-butyl peroxalate
EDTA	ethylene diamine tetraacetic acid
EN	ethylene diamine
FAD	flavin adenine dinucleotide
FMN	flavin mononucleotide
LTA	lead tetraacetate
MTBP	methyl-*t*-butyl peroxide
NBS	N-bromosuccinimide
NTA	nitrilo triacetate
PAT	phenylazotriphenylmethane
PC	dicyclohexylperoxydicarbonate
PNAP	4-nitroacetophenone
PNBPA	pentaamine(4-nitrobenzoato) cobalt(III)$^{2+}$
THF	tetrahydrofuran
THP	tetrahydropyran

H_2O	water
CH_3OH	methanol
C_2H_4	ethylene
C_2H_5OH	ethanol
C_2H_6	ethane
c-C_3H_6	cyclopropane
C_3H_7	propyl
C_3H_7OH	propanol
C_3H_8	propane
i-C_4H_{10}	isobutane
c-C_5H_{10}	cyclopentane
n-C_5H_{12}	*n*-pentane
C_6H_6	benzene
c-C_6H_{12}	cyclohexane
n-C_6H_{14}	*n*-hexane
n-C_7H_{16}	*n*-heptane
c-C_8H_{16}	cyclooctane
i-C_8H_{18}	isooctane
n-C_8H_{18}	*n*-octane

4 Carbon-centered radicals II

4.1 Rate constants of displacement reactions of carbon-centered radicals with molecules in solutions

4.1.0 Introduction

The absolute and relative rate constants of the displacement reactions of carbon-centered radicals with molecules, which are listed in this chapter have been collected from the literature up to 1982. Considering the large number of relevant publications it has been very helpful to find several data compilations already in print. In particular, we would like to acknowledge the collection of "Rate Constants of Aliphatic Carbon-Centered Radicals in Aqueous Solutions" by A.B. Ross and P. Neta (Radiation Chemistry Data Center, Radiation Laboratory, University of Notre Dame, NSRDS-NBS) and "Liquid Phase Reaction Rate Constants" by E.T. Denisov (IFI/PLENUM, New York, 1974, transl. from Russian, Nauka Press, Moscow, 1971). We would also like to acknowledge the help of Dr. Ch.-H. Fischer in many questions concerning the nomenclature of compounds.

The absolute rate constants listed in this compilation refer to the general equation

$$\dot{R} + AB \xrightarrow{k_a} RB + \dot{A}$$

with $\dot{R}$ being the carbon-centered radical, AB the substrate molecule, RB the product molecule and $\dot{A}$ the product radical. The absolute rate constants have mostly been measured directly using time resolved, mostly spectroscopic methods (for further brief information see introductory part of the following compilation of electron transfer rate constants). Alternatively, they are derived from competition studies, i.e. from rate constant ratios k_a/k_b with k_b being known and referring to a competitive reaction of the general type

$$\dot{R} + X \xrightarrow{k_b} \text{products}$$

About 80% of the rate data extracted from the literature are relative rate constants k_a/k_b in terms of the above two reactions. The second reaction most often also constitutes a displacement process, but may also refer to an addition, electron transfer, radical combination, radical fractionation and radical rearrangement reaction, or simply to a second mechanistic pathway of one particular radical-molecule interaction.

The relative rate constants are usually based on product ratio measurements with k_a/k_b having been identified with the respective yield ratios. A general problem in this connection is of course that the measurable products may not only result from the two competing processes of the radicals but could additionally be formed or consumed in secondary reactions. This would necessarily lead to wrong conclusions if k_a/k_b was identified with the observed product ratios. Unfortunately, but not surprisingly, a number of such cases have become known, the most prominent one in our search concerning the reaction of $\dot{C}Cl_3$ radicals generated from photolysis of $CBrCl_3$. (A corresponding note will be found in the respective $\dot{C}Cl_3$ section.) Data from such systems have of course been omitted. It is anticipated that future research may reveal similar problems in other cases. The relative rate constants should therefore always be viewed in the light of these considerations.

Other possible problems which may arise in connection with the measurement and interpretation of rate constants, such as spin delocalization etc. are briefly dealt with in the introductory part of the following electron transfer rate constant compilation.

Rate constants for displacement reactions are also included in the so-called transfer constants of free radical polymerization. These transfer constants have, however, not been included in this compilation, since a complete collection has already been published in the "Polymer Handbook", Second Edition, J. Brandrup and E.H. Immergut, (eds.), John Wiley and Sons, New York etc., 1975.

Arrangement of rate constants

The rate constant data have been divided into three major sections, namely absolute rate constants, relative rate constants, and isotope effects. The subgrouping is given in the list of contents. It is essentially based on a separation of radicals derived from aliphatic and other nonaromatic compounds from radicals which are formed from substrates containing aromatic and/or heterocyclic constituents. Within these two subgroups it seemed useful to further distinguish between radicals composed only of C and H atoms and radicals containing other atoms as well.

Within each section the radicals are listed in the order of increasing total number of carbon atoms followed by increasing total number of hydrogen atoms and finally increasing total number of other atoms in alphabetical order. Radicals with the same overall stoichiometry are separately grouped according to their structure.

Reactions of one particular radical are listed in the order which is given by application of the above criteria to the substrate molecule. The reactions with organic molecules are however preceded by the reactions with inorganic compounds. Only in a few cases with rather complex substrates it seemed appropriate to list the substrates in alphabetical order.

The same criteria (increasing number of C, H, and other atoms) are finally applied to the second reference substrate if no differentiation is possible on a higher level.

Radicals with "undefined stoichiometry or structure" refer to radicals which either result from radical-molecule reactions in which species with more than one radical site are formed, or the stoichiometry of which is essentially given by a distribution function as in polymer systems. Most of the former result from radiation-chemically induced processes, and the radicals generated in the reaction of $\dot{O}H$ radicals with substrates, for example, are listed as (substrate—OH$^\cdot$). A radical adduct is similarly indicated as substrate $-R^\cdot$, e.g. toluene $-CH_3^\cdot$ denotes a $\dot{C}H_3$ radical adduct to toluene. The other type of radicals is referred to by writing the parent compound in parentheses, e.g. (polyvinylacetate)$^\cdot$.

Similar considerations apply to the products of a particular radical reaction. Whenever their structure or stoichiometry is not exactly defined (or obvious) they are given as "products" or in terms of an overall stoichiometry, e.g. $(C_5H_{11}^\cdot)$ would denote the mixture of radicals which results from hydrogen atom abstraction from pentane. An undefined product radical is also given as substrate minus the abstracted atom, e.g. ethanol $(-H^\cdot)$ or trichloroethylene $(-Cl^\cdot)$ would refer to the radicals left after hydrogen atom abstraction from ethanol and chlorine atom abstraction from trichloroethylene, respectively. Generally, if the unpaired electron can be assigned to a particular atom X it is indicated as $\dot{X}$ (radical dot on top of atom).

The data compiled in the isotope effect section include not only plain isotope effects but also a number of overall relative rate constants which, however, include an isotope effect. Some further data concerning isotope effects are to be found in the section on absolute rate constants. Generally, whenever absolute rate constants were available for a particular reaction or set of reactions other relevant data are also always listed in the "absolute" section.

4.1.1 Absolute rate constants

4.1.1.1 Aliphatic radicals and radicals derived from other non-aromatic compounds

4.1.1.1.1 Radicals containing only C and H atoms

Reaction Radical generation Method	Solvent	T[K]	Rate data	Ref./ add. ref.
$\dot{C}H_3 + [Co(NH_3)_5OOCCH_3]^{2+} \longrightarrow CH_4 + Co^{2+}$ + products				
Phot. of $Co(NH_3)_5O_2CCH_3^{2+}$				71 Kan 1
PR	H_2O	293	$k = 4(1) \cdot 10^3\,M^{-1}s^{-1}$ [1]	
$\dot{C}H_3 + I_2 \longrightarrow CH_3I + \dot{I}$				
^{60}Co γ-irr. of 2,2,4-trimethylpentane				63 Sch 1
PR [2]	2,2,4-trimethylpentane	RT	$k \geqslant 3 \cdot 10^8\,M^{-1}s^{-1}$ [3]	
$\dot{C}H_3 + CH_3OH \longrightarrow CH_4 + \dot{C}H_2OH$				
γ-rad. of $CH_3I + H_2O$				67 Tho 1/
Competition kinetics	H_2O	293	$k = 2.2 \cdot 10^2\,M^{-1}s^{-1}$ [4]	71 Kan 1

[1] Based on $k(\dot{C}H_3 + \dot{C}H_3) = 2 \cdot 10^{10}\,M^{-1}s^{-1}$.
[2] Radiochromatography with ^{131}I.
[3] Based on $k/k(H) = 1.43 \cdot 10^7$ with $k(H)$ referring to $\dot{C}H_3 + (CH_3)_3CCH_2CH(CH_3)_2$.
[4] Based on $k(\dot{C}H_3 + O_2) = 4.7 \cdot 10^9\,M^{-1}s^{-1}$.

Reaction Radical generation Method	Solvent	T[K]	Rate data	Ref./ add. ref.
$\dot{C}H_3 + CH_3SH \longrightarrow CH_4 + CH_3\dot{S}$				
Pulse rad. of $CH_3SH + H_2O$				69 Kar 1
KAS	H_2O, pH = 11	293	$k = 7.4(10) \cdot 10^7\ M^{-1}s^{-1}$ [5]	
$\dot{C}H_3 + CH_3SO_2H \longrightarrow CH_4 + CH_3SO_2^{\cdot}$				
Ti(III) + H_2O_2 in H_2O				75 Gil 2
SESR, flow system	H_2O, pH = 1	RT	$k \approx 10^6\ M^{-1}s^{-1}$ [6]	
$\dot{C}H_3 + ClCH_2COOH \longrightarrow CH_4 + Cl\dot{C}HCOOH$				
$(CH_3)_2SO$ + Ti(III) + H_2O_2 + H_2O				75 Gil 1
SESR [7]	H_2O, pH ≈ 1	293	$k = 3.0 \cdot 10^3\ M^{-1}s^{-1}$	
$\dot{C}H_3 + CH_3CN \longrightarrow CH_4 + \dot{C}H_2CN$				
$(CH_3)_2SO$ + Ti(III) + H_2O_2 + H_2O				75 Gil 1
SESR	H_2O, pH ≈ 1	293	$k < 3 \cdot 10^2\ M^{-1}s^{-1}$	
$\dot{C}H_3 + CH_3COOH \longrightarrow CH_4 + \dot{C}H_2COOH$				
$(CH_3)_2SO$ + Ti(III) + H_2O_2 + H_2O				75 Gil 1
SESR [7]	H_2O, pH ≈ 1	293	$k = 2 \cdot 10^2\ M^{-1}s^{-1}$	
$\dot{C}H_3 + H_2NCH_2COO^- \longrightarrow CH_4 + H_2N\dot{C}HCOO^-$				
Phot. of cumene hydroperoxide + H_2O				70 Mog 1
PR	H_2O, pH = 10	293	$k \approx 1.2 \cdot 10^2\ M^{-1}s^{-1}$ [8]	
$\dot{C}H_3 + H_3N^{(+)}CH_2COO^{(-)} \longrightarrow CH_4 + H_3N^{(+)}\dot{C}HCOO^{(-)}$				
Phot. of cumene hydroperoxide + H_2O				70 Mog 1
PR	H_2O	293	$k \approx 4\ M^{-1}s^{-1}$ [9]	
$\dot{C}H_3 + HOCH_2COOH \longrightarrow CH_4 + HO\dot{C}HCOOH$				
$(CH_3)_2SO$ + Ti(III) + H_2O_2 + H_2O				75 Gil 1
SESR	H_2O	293	$k = 3.6 \cdot 10^3\ M^{-1}s^{-1}$	
$\dot{C}H_3 + CH_3CH_2OH \longrightarrow CH_4 + CH_3\dot{C}HOH$				
γ-rad. of $CH_3I + H_2O$				67 Tho 1
Competition kinetics	H_2O	293	$k = 5.9 \cdot 10^2\ M^{-1}s^{-1}$ [10]	
$\dot{C}H_3 + NCCH_2COOH \longrightarrow CH_4 + NC\dot{C}HCOOH$				
$(CH_3)_2SO$ + Ti(III) + H_2O_2 + H_2O				75 Gil 1
SESR [11]	H_2O, pH ≈ 1	293	$k > 6.6 \cdot 10^3\ M^{-1}s^{-1}$	
$\dot{C}H_3 + CH_3CH_2CN \longrightarrow CH_4 + CH_3\dot{C}HCN$				
$(CH_3)_2SO$ + Ti(III) + H_2O_2 + H_2O				75 Gil 1
SESR [11]	H_2O, pH ≈ 1	293	$k = 1.2 \cdot 10^3\ M^{-1}s^{-1}$	
$\dot{C}H_3 + NCCH_2CH_2OH \longrightarrow CH_4 + NCCH_2\dot{C}HOH$				
$(CH_3)_2SO$ + Ti(III) + H_2O_2 + H_2O				75 Gil 1
SESR	H_2O, pH ≈ 1	293	$k < 1.6 \cdot 10^3\ M^{-1}s^{-1}$	

[5]) Rate constant measured via $RSSR^{\dot{-}}$ formation ($R\dot{S} + RS^- \rightleftharpoons RSSR^{\dot{-}}$).
[6]) Estimated value.
[7]) And competition kinetics.
[8]) Relative to $2k(\dot{C}H_3 + \dot{C}H_3) = 2 \cdot 10^9\ M^{-1}s^{-1}$.
[9]) Based on $k(H)/k(D) = 10.5$; relative to $2k(\dot{C}H_3 + \dot{C}H_3) = 2 \cdot 10^9\ M^{-1}s^{-1}$.
[10]) Based on $k(\dot{C}H_3 + O_2) = 4.7 \cdot 10^9\ M^{-1}s^{-1}$.
[11]) And competition kinetics.

Reaction Radical generation Method	Solvent	T[K]	Rate data	Ref./ add. ref.
$\dot{C}H_3 + CH_3CH_2COOH \longrightarrow CH_4 + CH_3\dot{C}HCOOH$				
$(CH_3)_2SO + Ti(III) + H_2O_2 + H_2O$ SESR [11])	H_2O, pH ≈ 1	293	$k = 3.0 \cdot 10^3\ M^{-1}s^{-1}$	75 Gil 1
$\dot{C}H_3 + CH_3CHOHCOOH \longrightarrow CH_4 + CH_3\dot{C}OHCOOH$				
$(CH_3)_2SO + Ti(III) + H_2O_2 + H_2O$ SESR [11])	H_2O, pH = 1	293	$k = 1.2 \cdot 10^4\ M^{-1}s^{-1}$	75 Gil 1
$\dot{C}H_3 + (CH_3)_2CHOH \longrightarrow CH_4 + (CH_3)_2\dot{C}OH$				
γ-rad. of $CH_3I + H_2O$ Competition kinetics	H_2O	293	$k = 3.4 \cdot 10^3\ M^{-1}s^{-1}$ [12])	67 Tho 1/ 70 Bul 1, 71 Kan 1
$\dot{C}H_3 + (CH_3)_2CHCN \longrightarrow CH_4 + (CH_3)_2\dot{C}CN$				
$(CH_3)_2SO + Ti(III) + H_2O_2 + H_2O$ SESR [11])	H_2O, pH = 1	293	$k = 4.5 \cdot 10^3\ M^{-1}s^{-1}$	75 Gil 1
$\dot{C}H_3 + C_2H_5COOCH_3 \longrightarrow CH_4 + CH_3\dot{C}HCOOCH_3$				
$(CH_3)_2SO + Ti(III) + H_2O_2 + H_2O$ SESR [11])	H_2O, pH = 1	293	$k = 2.9 \cdot 10^3\ M^{-1}s^{-1}$	75 Gil 1
$\dot{C}H_3 + CH_3COOC_2H_5 \longrightarrow CH_4 + CH_3COO\dot{C}HCH_3$				
$(CH_3)_2SO + Ti(III) + H_2O_2 + H_2O$ SESR	H_2O, pH ≈ 1	293	$k < 1.7 \cdot 10^3\ M^{-1}s^{-1}$	75 Gil 1
$\dot{C}H_3 + (CH_3)_2CHCOOH \longrightarrow CH_4 + (CH_3)_2\dot{C}COOH$				
$(CH_3)_2SO + Ti(III) + H_2O_2 + H_2O$ SESR [11])	H_2O, pH = 1	293	$k = 9.0 \cdot 10^3\ M^{-1}s^{-1}$	75 Gil 1
$\dot{C}H_3 + (C_2H_5)_2Se \longrightarrow CH_3\dot{C}H_2 + CH_3SeC_2H_5$				
Phot. of azomethane SESR	[13])	213…253	$k \approx 10^3\ M^{-1}s^{-1}$ [14])	77 Sca 1
$\dot{C}H_3 + C_2H_5COC_2H_5 \longrightarrow CH_4 + CH_3\dot{C}HCOC_2H_5$				
$(CH_3)_2SO + Ti(III) + H_2O_2 + H_2O$ SESR [15])	H_2O, pH = 1	293	$k = 7.4 \cdot 10^4\ M^{-1}s^{-1}$	75 Gil 1
$\dot{C}H_3 + (CH_3)_3CCH_2CH(CH_3)_2 \longrightarrow CH_4$ + 2,2,4-trimethylpentane(−$\dot{H}$)				
^{60}Co γ-irr. of 2,2,4-trimethylpentane PR [16])	2,2,4-trimethylpentane	RT	$k = 20\ M^{-1}s^{-1}$ [17])	63 Sch 1
$\dot{C}H_3 + (CH_3)_3SnSn(CH_3)_3 \longrightarrow (CH_3)_4Sn + (CH_3)_3\dot{S}n$				
Phot. of CH_3Br PR, glc	C_6H_6	296	$k = 8.5 \cdot 10^3\ M^{-1}s^{-1}$ [18])	78 Leh 1/ 68 Car 1
$\dot{C}H_3 + (n\text{-}C_4H_9)_3GeH \longrightarrow CH_4 + (n\text{-}C_4H_9)_3\dot{G}e$				
$\dot{C}H_3$ from CH_3I, phot. of α,α'-azobiscyclohexylnitrile as initiator RS	c-C_6H_{12}	298	$k = 1.4 \cdot 10^5\ M^{-1}s^{-1}$	69 Car 1

[11]) And competition kinetics.
[12]) Based on $k(\dot{C}H_3 + O_2) = 4.7 \cdot 10^9\ M^{-1}s^{-1}$.
[13]) Not given, presumed to be $(C_2H_5)_2Se$.
[14]) Estimate from assumption $k[(C_2H_5)_2Se] = 2k[\dot{C}H_3]^2$.
[15]) And competition kinetics.
[16]) Radiochromatography with ^{131}I.
[17]) Based on $k/k(I) = 1.43 \cdot 10^7$ with $k(I)$ referring to $\dot{C}H_3 + I_2$ reaction.
[18]) Based on $2k_2 = 8.9 \cdot 10^9\ M^{-1}s^{-1}$ for $2\,\dot{C}H_3 \longrightarrow C_2H_6$ [68 Car 1].

Reaction Radical generation Method	Solvent	T[K]	Rate data	Ref./ add. ref.
$\dot{C}H_3 + (n\text{-}C_4H_9)_3SnH \longrightarrow CH_4 + (n\text{-}C_4H_9)_3\dot{S}n$				
Photochem. (α,α′-azobiscyclohexylnitrile as initiator)				68 Car 1
RS [19]	$c\text{-}C_6H_{12}$	298	$k = 5.8 \cdot 10^6\,M^{-1}s^{-1}$ [20]	
$\dot{C}H_3 + (n\text{-}C_4H_9)_3SnH \longrightarrow CH_4 + (n\text{-}C_4H_9)_3\dot{S}n$				
Laser flash phot. of $\{(CH_3)_3CO\}_2 + (CH_3)_3As$ + 2,2,4-trimethylpentane				81 Cha 1
KAS	2,2,4-trimethylpentane	300	$k = 10.6 \cdot 10^6\,M^{-1}s^{-1}$	
		231…317	$\log[A/M^{-1}s^{-1}] = 9.39(28)$ $E_a = 13.5(14)\,kJ\,mol^{-1}$	
$CH_3\dot{C}H_2 + (n\text{-}C_4H_9)_3SnH \longrightarrow C_2H_6 + (n\text{-}C_4H_9)_3\dot{S}n$				
Laser flash phot. of $\{(CH_3)_3CO\}_2 + (C_2H_5)_3As$ + 2,2,4-trimethylpentane				81 Cha 1
KAS	2,2,4-trimethylpentane	300	$k = 2.3 \cdot 10^6\,M^{-1}s^{-1}$	
		248…346	$\log[A/M^{-1}s^{-1}] = 9.14(42)$ $E_a = 15.9(24)\,kJ\,mol^{-1}$	
$CH_2{=}CH\dot{C}H_2 + CH_2{=}CHCH_2I \longrightarrow CH_2{=}CHCH_2I + CH_2{=}CH\dot{C}H_2$				
Decomp. of peroxide				71 Law 1
CIDNP	1,2-dichlorobenzene	373	$k = 3.0 \cdot 10^7\,M^{-1}s^{-1}$	
$CH_3CH_2\dot{C}H_2 + CH_3CH_2CH_2SH \longrightarrow C_3H_8 + CH_3CH_2CH_2\dot{S}$				
$CH_3CH_2CH_2\dot{S} + P(OC_2H_5)_3 \longrightarrow CH_3CH_2\dot{C}H_2 + SP(OC_2H_5)_3$ (AIBN initiated)				69 Bur 1
Potentiometric	C_6H_6	298	$k = 3.0 \cdot 10^4\,M^{-1}s^{-1}$ [21]	
titration	$c\text{-}C_6H_{12}$	298	$k = 2.9 \cdot 10^6\,M^{-1}s^{-1}$ [21]	
$(CH_3)_2\dot{C}H + (n\text{-}C_4H_9)_3SnH \longrightarrow (CH_3)_2CH_2 + (n\text{-}C_4H_9)_3\dot{S}n$				
Laser flash phot. of $((CH_3)_3CO)_2 + ((CH_3)_2CH)_3P$ + 2,2,4-trimethylpentane				81 Cha 1
KAS	2,2,4-trimethylpentane	300	$k = 1.5 \cdot 10^6\,M^{-1}s^{-1}$	
		251…358	$\log[A/M^{-1}s^{-1}] = 8.71(37)$ $E_a = 14.5(21)\,kJ\,mol^{-1}$	
$c\text{-}C_3H_5\dot{C}H_2 + Cu(II)Br_2 \longrightarrow$ products [22]				
Catalytic decomp. of cyclopropylmethylperoxide				71 Jen 1,
PR, glc	CH_3CN	298	$k = 4.3 \cdot 10^9\,M^{-1}s^{-1}$ [23]	72 Jen 1/ 74 Koc 1
$c\text{-}C_3H_5\dot{C}H_2 + Cu(II)Cl_2 \longrightarrow$ products [22]				
Catalytic decomp. of cyclopropylmethylperoxide				71 Jen 1,
PR, glc	CH_3CN	298	$k = 1.1 \cdot 10^9\,M^{-1}s^{-1}$ [23]	72 Jen 1/ 74 Koc 1
$c\text{-}C_3H_5\dot{C}H_2 + Cu(II)(SCN)_2 \longrightarrow$ products [22]				
Catalytic decomp. of cyclopropylmethylperoxide				71 Jen 1,
PR, glc	CH_3CN	298	$k = 3.6 \cdot 10^8\,M^{-1}s^{-1}$ [23]	72 Jen 1/ 74 Koc 1
$CH_3CH_2CH_2\dot{C}H_2 + (n\text{-}C_4H_9)_3SnH \longrightarrow n\text{-}C_4H_{10} + (n\text{-}C_4H_9)_3\dot{S}n$				
Laser flash phot. of $((CH_3)_3CO)_2 + (n\text{-}C_4H_9)_3P$ + 2,2,4-trimethylpentane				81 Cha 1
KAS	2,2,4-trimethylpentane	300	$k = 2.47 \cdot 10^6\,M^{-1}s^{-1}$	
		245…355	$\log[A/M^{-1}s^{-1}] = 9.06(31)$ $E_a = 15.3(17)\,kJ\,mol^{-1}$	

[19]) Determination of rate of temperature increase.

[20]) Based on measured value of $k/(2k_t)^{\frac{1}{2}}$ and $2k_t = 8.9 \cdot 10^9\,M^{-1}s^{-1}$ with k = rate of propagation and $2k_t$ referring to $\dot{C}H_3 + \dot{C}H_3$ reaction in overall chain reaction $CH_3X + (n\text{-}C_4H_9)_3SnH \longrightarrow CH_4 + (n\text{-}C_4H_9)_3SnX$ (X = halide).

[21]) Based on $k/(2k_2)^{\frac{1}{2}} = 2.04$ (in C_6H_6) and 68.8 (in $c\text{-}C_6H_{12}$) with $2k_2 = 2.1 \cdot 10^8\,M^{-1}s^{-1}$ and $1.7 \cdot 10^9\,M^{-1}s^{-1}$, respectively, referring to $2\,CH_3CH_2\dot{C}H_2 \longrightarrow$ products in respective solvents.

[22]) Reaction includes ligand transfer, e^--transfer and addition [74 Koc 1].

[23]) Based on rate constant $k_r = 1 \cdot 10^8\,s^{-1}$ for rearrangement $c\text{-}C_3H_5\dot{C}H_2 \longrightarrow CH_2{=}CHCH_2\dot{C}H_2$.

Reaction Radical generation Method	Solvent	T[K]	Rate data	Ref./ add. ref.
$CH_3CH_2CH_2\dot{C}H_2 + (n\text{-}C_4H_9)_3SnD \longrightarrow n\text{-}C_4H_9D + (n\text{-}C_4H_9)_3\dot{S}n$				
Phot. of $((CH_3)_3CO)_2 + (n\text{-}C_4H_9)_3P$ + 2,2,4-trimethylpentane				81 Cha 1
PR, KAS	2,2,4-trimethylpentane	300	$k(D) = 1.2\cdot10^6\,M^{-1}\,s^{-1}$ $\log[A/M^{-1}\,s^{-1}] = 8.63(106)$ $E_a = 14.6(60)\,kJ\,mol^{-1}$ [24]	
$(CH_3)_3\dot{C} + CCl_4 \longrightarrow (CH_3)_3CCl + \dot{C}Cl_3$				
Phot. of pivalophenone				76 Fri 1
CINDP	PP9 (fluorocarbon solvent)	310 296... 344	$k = 4.9\cdot10^4\,M^{-1}\,s^{-1}$ $\log[A/M^{-1}\,s^{-1}] = 7.02$ $E_a = 13.8(42)\,kJ\,mol^{-1}$	
$(CH_3)_3\dot{C} + CHCl_3$ $\xrightarrow{a}$ $(CH_3)_3CH + \dot{C}Cl_3$ $\xrightarrow{b}$ $(CH_3)_3CCl + \dot{C}HCl_2$				
Phot. of pivalophenone				76 Fri 1
KESR, CIDNP	$CHCl_3$	310	$k_a = 2.54(7)\cdot10^2\,M^{-1}\,s^{-1}$ [25] $k_b = 1.84(7)\cdot10^2\,M^{-1}\,s^{-1}$ [25] $k_{overall} = 5.44(13)\cdot10^3\,s^{-1}$ [26]	
$(CH_3)_3\dot{C} + CHCl_3$ $\xrightarrow{a}$ $(CH_3)_3CH + \dot{C}Cl_3$ $\xrightarrow{b}$ $(CH_3)_3CCl + \dot{C}HCl_2$				
Phot. of DTBK				81 Due 1
KESR	methylcyclopentane	273 283 293 303 313 323	$(k_a + k_b) = 79(2)\,M^{-1}\,s^{-1}$ $= 148(2)$ $= 243(7)$ $= 412(18)$ $= 469(23)$ $= 668(77)$ [27]	
$(CH_3)_3\dot{C} + (CH_3)_3SnH \longrightarrow (CH_3)_3CH + (CH_3)_3\dot{S}n$				
Photochem. (α,α'-azobiscyclohexylnitrile as initiator)				68 Car 1
RS [28]	$c\text{-}C_6H_{12}$	298	$k = 2.9\cdot10^5\,M^{-1}\,s^{-1}$ [29]	
$(CH_3)_3\dot{C} + 3,4\text{-}Cl_2C_6H_3CH_3 \longrightarrow (CH_3)_3CH + 3,4\text{-}Cl_2C_6H_3\dot{C}H_2$				
Phot. of DTBK				82 Due 1
KESR	3,4-dichlorotoluene	321(2)	$k = 37.3(10)\,M^{-1}\,s^{-1}$	
$(CH_3)_3\dot{C} + 3\text{-}ClC_6H_4CH_3 \longrightarrow (CH_3)_3CH + 3\text{-}ClC_6H_4\dot{C}H_2$				
Phot. of DTBK				82 Due 1
KESR	3-chlorotoluene	321(2)	$k = 20.9(7)\,M^{-1}\,s^{-1}$	
$(CH_3)_3\dot{C} + 4\text{-}ClC_6H_4CH_3 \longrightarrow (CH_3)_3CH + 4\text{-}ClC_6H_4\dot{C}H_2$				
Phot. of DTBK				82 Due 1
KESR	4-chlorotoluene	321(2)	$k = 28.5(12)\,M^{-1}\,s^{-1}$	

[24]) $k(H)/k(D) = 2.3$ with $k(H)$ referring to corresponding reaction with $(n\text{-}C_4H_9)_3SnH$.
[25]) CIDNP measurements.
[26]) Overall pseudo-first order rate constant in neat $CHCl_3$ for $(CH_3)_3\dot{C}$ scavenging measured by flash phot.—ESR.
[27]) $\log[k_a + k_b] = 8.41(14) - 34.0(8)/2.303\,RT$($R$ in $kJ\,mol^{-1}\,K^{-1}$).
[28]) Determination of rate of temperature increase.
[29]) Based on measured value of $k/(2k_t)^{\frac{1}{2}}$ and $2k_t = 2.2\cdot10^9\,M^{-1}\,s^{-1}$ with k = rate of propagation and $2k_t$ referring to $2(CH_3)_3\dot{C}$ products in overall chain reaction $(CH_3)_3CX + (CH_3)_3SnH \longrightarrow (CH_3)_3CH + (CH_3)_3SnX$ (X = halide).

Reaction Radical generation Method	Solvent	T[K]	Rate data	Ref./ add. ref.
$(CH_3)_3\dot{C} + 3\text{-}FC_6H_4CH_3 \longrightarrow (CH_3)_3CH + 3\text{-}FC_6H_4\dot{C}H_2$				
Phot. of DTBK				82 Due 1
KESR	3-fluorotoluene	321(2)	$k = 18.0(7)\ M^{-1}\ s^{-1}$	
$(CH_3)_3\dot{C} + 4\text{-}FC_6H_4CH_3 \longrightarrow (CH_3)_3CH + 4\text{-}FC_6H_4\dot{C}H_2$				
Phot. of DTBK				82 Due 1
KESR	4-fluorotoluene	321(2)	$k = 14.1(6)\ M^{-1}\ s^{-1}$	
$(CH_3)_3\dot{C} + C_6H_5CH_3 \longrightarrow (CH_3)_3CH + C_6H_5\dot{C}H_2$				
Phot. of DTBK				82 Due 1
KESR	toluene	321(2)	$k = 14.4(5)\ M^{-1}\ s^{-1}$	
$(CH_3)_3\dot{C} + 4\text{-}CNC_6H_4CH_3 \longrightarrow (CH_3)_3CH + 4\text{-}CNC_6H_4\dot{C}H_2$				
Phot. of DTBK				82 Due 1
KESR	4-cyanotoluene	321(2)	$k = 47.2(17)\ M^{-1}\ s^{-1}$	
$(CH_3)_3\dot{C} + 3\text{-}CH_3C_6H_4CH_3$(*m*-xylene) $\longrightarrow (CH_3)_3CH + CH_3C_6H_4\dot{C}H_2$				
Phot. of DTBK				82 Due 1
KESR	3-methyltoluene (*m*-xylene)	321(2)	$k = 15.0(6)\ M^{-1}\ s^{-1}$ [30]	
$(CH_3)_3\dot{C} + 4\text{-}CH_3C_6H_4CH_3 \longrightarrow (CH_3)_3CH + 4\text{-}CH_3C_6H_4\dot{C}H_2$				
Phot. of DTBK				82 Due 1
KESR	4-methyltoluene	321(2)	$k = 14.2(7)\ M^{-1}\ s^{-1}$ [30]	
$(CH_3)_3\dot{C} + 1,3,5\text{-}(CH_3)_3C_6H_3$(mesitylene) $\longrightarrow (CH_3)_3CH + 3,5\text{-}(CH_3)_2C_6H_3\dot{C}H_2$				
Phot. of DTBK				82 Due 1
KESR	1,3,5-tri- methylbenzene (mesitylene)	321(2)	$k = 14.9(7)\ M^{-1}\ s^{-1}$ [30]	
$(CH_3)_3\dot{C} + 4\text{-}(CH_3)_3CC_6H_4CH_3 \longrightarrow (CH_3)_3CH + 4\text{-}(CH_3)_3CC_6H_4\dot{C}H_2$				
Phot. of DTBK				82 Due 1
KESR	4-*t*-butyltoluene	321(2)	$k = 17.7(12)\ M^{-1}\ s^{-1}$	
$(CH_3)_3\dot{C} + (n\text{-}C_4H_9)_3SnH \longrightarrow (CH_3)_3CH + (n\text{-}C_4H_9)_3\dot{S}n$				
Laser flash phot. of $((CH_3)_3CO)_2 + ((CH_3)_3C)_3P$ + 2,2,4-trimethylpentane				81 Cha 1, 68 Car 1, 68 Car 2
KAS	2,2,4-trimethylpentane	300 263… 351	$k = 1.87 \cdot 10^6\ M^{-1}\ s^{-1}$ $\log[A/M^{-1}\ s^{-1}] = 8.43(14)$ $E_a = 12.4(8)\ kJ\ mol^{-1}$	
	$c\text{-}C_6H_{12}$	298	$k = 3 \cdot 10^5\ M^{-1}\ s^{-1}$ [31] $7.4 \cdot 10^5$ [32]	
$(CH_3)_3\dot{C} + (n\text{-}C_4H_9)_3SnD \longrightarrow (CH_3)_3CD + (n\text{-}C_4H_9)_3\dot{S}n$				
Photochem. (α,α'-azobiscyclohexylnitrile as initiator)				68 Car 1
RS [33]	$c\text{-}C_6H_{12}$	298	$k = 2.7 \cdot 10^5\ M^{-1}\ s^{-1}$ [34]	

30) For single methyl group (statistical correction).
31) From [68 Car 2].
32) From [68 Car 1].
33) Determination of rate of temperature increase.
34) Based on measured value of $k/(2k_t)^{\frac{1}{2}}$ and $2k_t = 2.2 \cdot 10^9\ M^{-1}\ s^{-1}$ with k = rate of propagation and $2k_t$ referring to $2(CH_3)_3\dot{C} \longrightarrow$ products in overall chain reaction $(CH_3)_3CX + (n\text{-}C_4H_9)_3SnD \longrightarrow (CH_3)_3CD + (n\text{-}C_4H_9)_3SnX$ (X = halide).

Reaction Radical generation Method	Solvent	T[K]	Rate data	Ref./ add. ref.
$(CH_3)_3\dot{C} + (C_6H_5)_3SnH \longrightarrow (CH_3)_3CH + (C_6H_5)_3\dot{Sn}$				
Photochem. (α,α'-azobiscyclohexylnitrile as initiator)				68 Car 1,
RS [33]	c-C_6H_{12}	298	$k = 3.1 \cdot 10^6\ M^{-1}s^{-1}$ [35] $5 \cdot 10^6$ [36]	68 Car 2
$(c\text{-}C_5H_7^{\bullet})$ [37] $+ H_2O_2 \longrightarrow$ products				
γ-rad. of c-pentene $+ N_2O + H_2O$				77 Soy 1
[38]	H_2O	293	$k = 3.7 \cdot 10^4\ M^{-1}s^{-1}$ [39]	
$(c\text{-}C_5H_9^{\bullet}) + H_2O_2 \longrightarrow$ products				
γ-rad. of cyclopentane $+ N_2O + H_2O$				77 Soy 1
[38]	H_2O	293	$k = 4.6 \cdot 10^4\ M^{-1}s^{-1}$ [40]	
$(c\text{-}C_5H_9^{\bullet}) + I_2 \longrightarrow c\text{-}C_5H_9I + \dot{I}$				
Pulse rad. of c-pentane				78 Fol 1
KAS	c-C_5H_{10}	296	$k = 1.9(2) \cdot 10^{10}\ M^{-1}s^{-1}$	
$CH_3(CH_2)_3\dot{C}H_2 + CH_3(CH_2)_3CH_2SH \longrightarrow n\text{-}C_5H_{12} + CH_3(CH_2)_3CH_2\dot{S}$				
$CH_3(CH_2)_3CH_2\dot{S} + P(OC_2H_5)_3 \longrightarrow CH_3(CH_2)_3\dot{C}H_2 + SP(OC_2H_5)_3$ (AIBN initiated)				68 Bur 1/
Potentiometric titration	C_6H_6	298	$k = 7.1 \cdot 10^4\ M^{-1}s^{-1}$ [41]	66 Bur 1
$(C_5H_{11}^{\bullet})$ [42] $+ H_2O_2 \longrightarrow$ products				
γ-rad. of n-$C_5H_{12} + N_2O + H_2O$				77 Soy 1
[43]	H_2O	293	$k = 3.4 \cdot 10^4\ M^{-1}s^{-1}$ [44]	
$(c\text{-}C_6H_9^{\bullet})$ [42] $+ CCl_4 \longrightarrow c\text{-}C_6H_9Cl + \dot{C}Cl_3$				
γ-rad. of c-hexene $+ CCl_4$				81 Alf 1
PR	CCl_4	299(1)	$k = 0.77 \ldots 2.30\ M^{-1}s^{-1}$ [45]	
		311(1)	1.02 … 1.31	
		322(1)	1.83 … 3.01	
		331(1)	1.61 … 2.98	
		348(1)	2.17 … 3.20	
		363(1)	2.14 … 6.48	
		393(1)	3.30 … 10.55	
		413(1)	6.35 … 8.98	
$(c\text{-}C_6H_{11}^{\bullet}) + Cl_2 \longrightarrow c\text{-}C_6H_{11}Cl + \dot{C}l$				
Not given				78 Kos 1
Not given	not given	not given	$k = 1.5 \cdot 10^6\ M^{-1}s^{-1}$ [46]	

[33] Determination of rate of temperature increase.
[35] Based on measured value of $k/(2k_t)^{\frac{1}{2}}$ and $2k_t = 2.2 \cdot 10^9\ M^{-1}s^{-1}$ with k = rate of propagation and $2k_t$ referring to $2(CH_3)_3\dot{C} \longrightarrow$ products in overall chain reaction $(CH_3)_3CX + (C_6H_5)_3SnH \longrightarrow (CH_3)_3CH + (C_6H_5)_3SnX$ (X = halide).
[36] From [68 Car 2].
[37] Radical mixture.
[38] Competition kinetics, dose rate effect on H_2O_2 production.
[39] Based on $2k(c\text{-}C_5H_7^{\bullet} + c\text{-}C_5H_7^{\bullet}) = 1.5 \cdot 10^9\ M^{-1}s^{-1}$.
[40] Based on $2k(c\text{-}C_5H_9^{\bullet} + c\text{-}C_5H_9^{\bullet}) = 2 \cdot 10^9\ M^{-1}s^{-1}$.
[41] Based on $k/(2k_2)^{\frac{1}{2}} = 2.24$ measurement with $2k_2 = 1.0 \cdot 10^9\ M^{-1}s^{-1}$ referring to $2\ CH_3(CH_2)_3\dot{C}H_2 \longrightarrow$ products.
[42] Radical mixture.
[43] Competition kinetics, dose rate effect on H_2O_2 production.
[44] Based on $2k(C_5H_{11}^{\bullet} + C_5H_{11}^{\bullet}) = 2 \cdot 10^9\ M^{-1}s^{-1}$.
[45] Based on assumption of $2k(\dot{C}Cl_3 + \dot{C}Cl_3) = k(c\text{-}C_6H_9^{\bullet} + \dot{C}Cl_3)$; rate constants considered to be good within a factor of 2…3.
[46] Based on various experimental data and assuming $2k_2 = 2.7 \cdot 10^9\ M^{-1}s^{-1}$ for $2(c\text{-}C_6H_{11}^{\bullet}) \longrightarrow$ products.

Reaction Radical generation Method	Solvent	T[K]	Rate data	Ref./ add. ref.
$(c\text{-}C_6H_{11}^{\cdot}) + I_2 \longrightarrow c\text{-}C_6H_{11}I + \dot{I}$				
Pulse rad. of $c\text{-}C_6H_{12}$				65 Ebe 1
KAS	$c\text{-}C_6H_{12}$	293... 296	$k = 7 \cdot 10^9\ M^{-1}s^{-1}(\pm 40\%)$	
KAS	$c\text{-}C_6H_{12}$	296	$k = 1.2(1) \cdot 10^{10}\ M^{-1}s^{-1}$	78 Fol 1
$(c\text{-}C_6H_{11}^{\cdot}) + CHCl_2CHCl_2 \longrightarrow c\text{-}C_6H_{11}Cl + \dot{C}HClCHCl_2$				
[47] [47]	[47]	353	$k = 5.75 \cdot 10^2\ M^{-1}s^{-1}$ [47] $\log[A/M^{-1}s^{-1}] = 9.1(1)$ [47] $E_a = 42.7(42)\ kJ\ mol^{-1}$ [47]	77 Kat 1, 75 Kat 1, 75 Kat 2
$(c\text{-}C_6H_{11}^{\cdot}) + (CH_3)_2SO_2Cl \longrightarrow c\text{-}C_6H_{11}Cl + (CH_3)_2SO_2^{\cdot}$				
γ-rad. of $c\text{-}C_6H_{12}$ PR, glc	$c\text{-}C_6H_{12}$		$\log k = 9.07(17) - 20.6(9)/2.303\,RT$ [48]	76 Hor 1/ 73 Hor 1
$(c\text{-}C_6H_{11}^{\cdot}) + (CH_3)_3COCl \longrightarrow c\text{-}C_6H_{11}Cl + (CH_3)_3C\dot{O}$				
Phot. of AIBN (initiator) in $c\text{-}C_6H_{12}$ containing solution				72 Zav 1
PR [49]	$CF_2ClCFCl_2/c\text{-}C_6H_{12}$	313.00(5)	$k = 3.8 \cdot 10^5\ M^{-1}s^{-1}$ [50]	
$(c\text{-}C_6H_{11}^{\cdot}) + C_2H_5N(X)COC_2H_5 \longrightarrow c\text{-}C_6H_{11}X + C_2H_5\dot{N}COC_2H_5$ X = Cl and Br				
$(CH_3)_3CONNOC(CH_3)_3$ as initiator				82 Sut 1
Time-resolved NMR	C_6H_6 [51]	301	$k \geqslant 2 \cdot 10^4\ M^{-1}s^{-1}$	
$(c\text{-}C_6H_{11}^{\cdot}) + c\text{-}C_6H_{11}SH \longrightarrow c\text{-}C_6H_{12} + c\text{-}C_6H_{11}\dot{S}$				
$c\text{-}C_6H_{11}\dot{S} + P(OC_2H_5)_3 \longrightarrow (c\text{-}C_6H_{11}^{\cdot}) + SP(OC_2H_5)_3$				69 Bur 1
Potentiometric	C_6H_6	298	$k = 3.0 \cdot 10^4\ M^{-1}s^{-1}$ [52]	
titration	$c\text{-}C_6H_{12}$	298	$k = 3.9 \cdot 10^5\ M^{-1}s^{-1}$ [52]	
$(c\text{-}C_6H_{11}^{\cdot}) + (n\text{-}C_4H_9)_3SnH \longrightarrow c\text{-}C_6H_{12} + (n\text{-}C_4H_9)_3\dot{S}n$				
Photochem. (α,α'-azobiscyclohexylnitrile as initiator)				68 Car 1
RS [53]	$c\text{-}C_6H_{12}$	298	$k = 1.2 \cdot 10^6\ M^{-1}s^{-1}$ [54]	
Laser flash phot. of $\{(CH_3)_3CO\}_2 + (c\text{-}C_6H_{11})_3P$ + 2,2,4-trimethylpentane				81 Cha 1
KAS	2,2,4-trimethylpentane	300 300... 355	$k = 2.19 \cdot 10^6\ M^{-1}s^{-1}$ $\log[A/M^{-1}s^{-1}] = 9.24(78)$ $E_a = 16.6(48)\ kJ\ mol^{-1}$	
$(c\text{-}C_6H_{11}^{\cdot}) + (n\text{-}C_4H_9)_3SnD \longrightarrow c\text{-}C_6H_{11}D + (n\text{-}C_4H_9)_3\dot{S}n$				
Photochem. (α,α'-azobiscyclohexylnitrile as initiator)				68 Car 1
RS [53]	$c\text{-}C_6H_{12}$	298	$k = 4.4 \cdot 10^5\ M^{-1}s^{-1}$ [55]	

[47]) Estimated values from interpolation of data on Cl-atom abstraction from various $XCCl_3$ and $XCHCl_2$ compounds.
[48]) Based on $\log k/k_c = 0.393(142) + 9.97(30)/2.303\,RT$ with k_c referring to $(c\text{-}C_6H_{11}^{\cdot}) + C_2Cl_4$, and R in $J\,K^{-1}\,mol^{-1}$.
[49]) Photometric determination of $(CH_3)_3COCl$.
[50]) Based on $k/(2k_2)^{\frac{1}{2}} = 11\ M^{-\frac{1}{2}}s^{-\frac{1}{2}}$ with assumed $2k_2 = 1.2 \cdot 10^9\ M^{-1}s^{-1}$ for $2(c\text{-}C_6H_{11}^{\cdot}) \longrightarrow$ products.
[51]) Contains some $CHCl{=}CCl_2$ to suppress possible $\dot{C}l$ or $\dot{B}r$ induced chain reaction.
[52]) Based on $k/(2k_2)^{\frac{1}{2}} = 1.61$ (in C_6H_6) and 15.8 (in $c\text{-}C_6H_{12}$) with $2k_2 = 3.6 \cdot 10^8$ and $6 \cdot 10^8\ M^{-1}s^{-1}$, respectively, referring to $2(c\text{-}C_6H_{11}^{\cdot}) \longrightarrow$ products in respective solvents.
[53]) Determination of rate of temperature increase.
[54]) Based on measured value of $k/(2k_t)^{\frac{1}{2}}$ and $2k_t = 2.2 \cdot 10^9\ M^{-1}s^{-1}$ with k = rate of propagation and $2k_t$ referring to $2(c\text{-}C_6H_{11}^{\cdot}) \longrightarrow$ products in overall chain reaction $c\text{-}C_6H_{11}X + (n\text{-}C_4H_9)_3SnH \longrightarrow c\text{-}C_6H_{12} + (n\text{-}C_4H_9)_3SnX$ (X = halide).
[55]) Based on measured value of $k/(2k_t)^{\frac{1}{2}}$ and $2k_t = 2.2 \cdot 10^9\ M^{-1}s^{-1}$ with k = rate of propagation and $2k_t$ referring to $2(c\text{-}C_6H_{11}^{\cdot}) \longrightarrow$ products in overall chain reaction $c\text{-}C_6H_{11}X + (n\text{-}C_4H_9)_3SnD \longrightarrow c\text{-}C_6H_{11}D + (n\text{-}C_4H_9)_3SnX$ (X = halide).

Reaction Radical generation Method	Solvent	T[K]	Rate data	Ref./ add. ref.
$CH_2{=}CH(CH_2)_3\dot{C}H_2 + Cu(II)Br_2 \longrightarrow$ products [56]				
Catalytic decomp. of 5-hexenylperoxide PR, glc	CH_3CN	298	$k > 2 \cdot 10^8\ M^{-1}s^{-1}$ [57]	71 Jen 1, 72 Jen 1/ 74 Koc 1
$CH_2{=}CH(CH_2)_3\dot{C}H_2 + Cu(II)Cl_2 \longrightarrow$ products [56]				
Catalytic decomp. of 5-hexenylperoxide PR, glc	CH_3CN	298	$k > 2 \cdot 10^8\ M^{-1}s^{-1}$ [57]	71 Jen 1, 72 Jen 1/ 74 Koc 1
$CH_2{=}CH(CH_2)_3\dot{C}H_2 + Cu(II)(SCN)_2 \longrightarrow$ products [56]				
Catalytic decomp. of 5-hexenylperoxide PR, glc	CH_3CN	298	$k = 2.6 \cdot 10^8\ M^{-1}s^{-1}$ [57]	71 Jen 1, 72 Jen 1/ 74 Koc 1
$CH_2{=}CH(CH_2)_3\dot{C}H_2 + (CH_3)_3COOD \longrightarrow CH_2{=}CH(CH_2)_3CH_2D + (CH_3)_3CO\dot{O}$ $+ (CH_3)_3COOH \longrightarrow CH_2{=}CH(CH_2)_3CH_3 + (CH_3)_3CO\dot{O}$				
Decomp. of $(CH_2{=}CH(CH_2)_4)_2O_2$, AIBN iniated PR, glc	$CH_2{=}CH(CH_2)_3CH_3$/ $(CH_3)_3COOD$ or $(CH_3)_3COOH$	323	$k(D) = 8.7 \cdot 10^3\ M^{-1}s^{-1}$ [58] $k(H) \approx 2 \cdot 10^4\ M^{-1}s^{-1}$ [58]	79 How 1/ 79 Sch 1
$CH_3(CH_2)_4\dot{C}H_2 + (n\text{-}C_4H_9)_3SnH \longrightarrow n\text{-}C_6H_{14} + (n\text{-}C_4H_9)_3\dot{S}n$				
Photochem. (α,α'-azobiscyclohexylnitrile as initiator) RS [59]	$c\text{-}C_6H_{12}$	298	$k = 1.0 \cdot 10^6\ M^{-1}s^{-1}$ [60]	68 Car 1
$(n\text{-}C_6H_{13}^{\cdot})$ [61] $+ CHCl_3 \longrightarrow n\text{-}C_6H_{14} + \dot{C}Cl_3$				
γ-rad. of $n\text{-}C_6H_{14} + CHCl_4$ PR	$n\text{-}C_6H_{14}/CHCl_3$	203 263	$k = 3.0 \cdot 10^2\ M^{-1}s^{-1}$ [62] $4.8 \cdot 10^3$ [62]	77 Tua 1
$CH_3(CH_2)_9\dot{C}H_2 + C_6H_5CH_3 \longrightarrow n\text{-}C_{11}H_{24} + C_6H_5\dot{C}H_2$				
Thermal decomp. of lauroylperoxide PR, glc	toluene/C_6H_5Cl	368 373 378	$k(H) = 3.6 \cdot 10^2\ M^{-1}s^{-1}$ [63] $4.2 \cdot 10^2$ [63] $5.0 \cdot 10^2$ [63] $E_a = 40.6(42)\ kJ\ mol^{-1}$	77 Aga 1

[56] Reaction includes ligand transfer, e^--transfer and addition [74 Koc 1].
[57] Based on rate constant $k_r = 1 \cdot 10^5\ s^{-1}$ for rearrangement of 5-hexenyl radical.
[58] Based on $k_r = 2.36 \cdot 10^5\ s^{-1}$ for 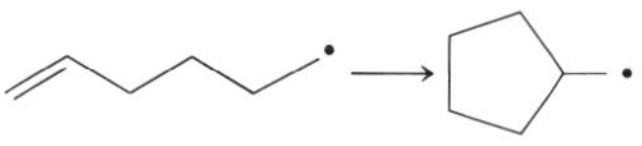 [79 Sch 1], and $k_r/k(D) = 27\ M$ (for deuterated compound).
[59] Determination of rate of temperature increase.
[60] Based on measured value of $k/(2k_t)^{\frac{1}{2}}$ and $2k_t = 2.2 \cdot 10^9\ M^{-1}s^{-1}$ with k = rate of propagation and $2k_t$ referring to $2\,CH_3(CH_2)_4CH_2$ products in overall chain reaction $n\text{-}C_6H_{13}X + (n\text{-}C_4H_9)_3SnH \longrightarrow c\text{-}C_6H_{14} + (n\text{-}C_4H_9)_3SnX$ (X = halide).
[61] Radical mixture.
[62] Values based on some parameter adjustments.
[63] Based on assumed $k_a = 10^8\ M^{-1}s^{-1}$ for $(n\text{-}C_{11}H_{23}^{\cdot}) + O_2 \longrightarrow n\text{-}C_{11}H_{23}\dot{O}_2$ and $k(H)/k_a = 1.07 \cdot 10^{-5}$ (368 K), $1.25 \cdot 10^{-5}$ (373 K) and $1.51 \cdot 10^{-5}$ (378 K).

Reaction Radical generation Method	Solvent	T[K]	Rate data	Ref./ add. ref.
$CH_3(CH_2)_9\dot{C}H_2 + C_6H_5CH_2CH_3 \longrightarrow n\text{-}C_{11}H_{24}$ + ethylbenzene($-\dot{H}$)				
Thermal decomp. of lauroylperoxide				77 Aga 1
PR, glc	ethylbenzene/C_6H_5Cl	373	$k(H) = 1.28 \cdot 10^3\ M^{-1}\ s^{-1}$ [64]	
		383	$1.73 \cdot 10^3$ [64]	
		393	$2.40 \cdot 10^3$ [64]	
			$E_a = 37.3(38)\ kJ\ mol^{-1}$	
$CH_3(CH_2)_9\dot{C}H_2 + C_6H_5CH(CH_3)_2 \longrightarrow n\text{-}C_{11}H_{24} + C_6H_5\dot{C}(CH_3)_2$				
Thermal decomp. of lauroylperoxide				77 Aga 1
PR, glc	cumene/C_6H_5Cl	373	$k(H) = 2.83 \cdot 10^3\ M^{-1}\ s^{-1}$ [65]	
		383	$3.67 \cdot 10^3$ [65]	
		393	$4.72 \cdot 10^3$ [65]	
			$E_a = 31.0(25)\ kJ\ mol^{-1}$	

4.1.1.1.2 Radicals containing C, H, and other atoms

Reaction Radical generation Method	Solvent	T[K]	Rate data	Ref./ add. ref.
$\dot{C}Cl_3 + (CH_3)_2CHOH \longrightarrow CCl_3H + (CH_3)_2\dot{C}OH$				
γ-rad. of 2-propanol + CCl_4 + H_2O				71 Koe 1/
[1]	H_2O	293	$k = 79\ M^{-1}\ s^{-1}$ [2]	75 Wil 1
$\dot{C}Cl_3 + (CH_3)_2CHOH \longrightarrow CCl_3H + (CH_3)_2\dot{C}OH$				
γ-rad. of 2-propanol + CCl_4				70 Rad 1
PR, glc, titration	2-propanol	RT	$k = 3\ M^{-1}\ s^{-1}$ [3]	
$\dot{C}Cl_3$ + THF$\{(CH_2)_4O\} \longrightarrow CCl_3H + (C_4H_7O^{\cdot})$				
γ-rad. of CCl_4 + H_2O				71 Koe 1
[1]	H_2O	293	$k = 57\ M^{-1}\ s^{-1}$ [2]	
$\dot{C}Cl_3 + (CH_3)_3COCl$($t$-butylhypochlorite) $\longrightarrow CCl_4 + (CH_3)_3C\dot{O}$				
Phot. of $(CH_3)_3COCl$				66 Car 1,
RS	CCl_4	297	$k = 1.2(4) \cdot 10^3\ M^{-1}\ s^{-1}$	67 Car 1
$\dot{C}Cl_3 + C_2H_5OC_2H_5 \longrightarrow CCl_3H + CH_3\dot{C}HOC_2H_5$				
γ-rad. of CCl_4 + H_2O				71 Koe 1
PR [1]	H_2O	293	$k = 35\ M^{-1}\ s^{-1}$ [2]	
$\dot{C}Cl_3 + C_6H_5CH_3 \longrightarrow CHCl_3 + C_6H_5\dot{C}H_2$				
Thermal decomp. of acetylperoxide				66 Sch 1 [4]/
PR, glc	chloroform/toluene	373	$k = 28\ M^{-1}\ s^{-1}$	54 Lev 1,
	(ratio 1.44…2.50)	353	12	60 McC 1
		340.5	6.5	
		328	3.4	
			$\log[A/M^{-1}\ s^{-1}] = 8.25$	
			$E_a = 11.6\ kJ\ mol^{-1}$	
$\dot{C}Cl_3 + (n\text{-}C_4H_9)_3GeH \longrightarrow CHCl_3 + (n\text{-}C_4H_9)_3\dot{G}e$				
$\dot{C}Cl_3$ from CCl_4, phot. of α,α'-azobiscyclohexylnitrile as initiator				69 Car 1
RS	$c\text{-}C_6H_{12}$	298	$k = 1.3 \cdot 10^5\ M^{-1}\ s^{-1}$	

[64]) Based on assumed $k_a = 10^8\ M^{-1}\ s^{-1}$ for $(n\text{-}C_{11}H_{23}^{\cdot}) + O_2 \longrightarrow n\text{-}C_{11}H_{23}\dot{O}_2$ and $k(H)/k_a = 2.52 \cdot 10^{-5}$ (373 K), $3.46 \cdot 10^{-5}$ (383 K) and $4.76 \cdot 10^{-5}$ (393 K).

[65]) Based on assumed $k_a = 10^8\ M^{-1}\ s^{-1}$ for $(n\text{-}C_{11}H_{23}^{\cdot}) + O_2$ $n\text{-}C_{11}H_{23}\dot{O}_2$ and $k(H)/k_a = 2.83 \cdot 10^{-5}$ (373 K), $3.67 \cdot 10^{-5}$ (383 K) and $4.72 \cdot 10^{-5}$ (393 K).

[1]) Dose rate dependence of Cl^- yield.

[2]) Based on $2k(\dot{C}Cl_3 + \dot{C}Cl_3) = 10^9\ M^{-1}\ s^{-1}$.

[3]) Based on assumed $2k = 1 \cdot 10^8\ M^{-1}\ s^{-1}$ for $2\ \dot{C}Cl_3 \longrightarrow C_2Cl_6$.

[4]) Based on data in [54 Lev 1] on decomposition of acetylperoxide, and $k = 4 \cdot 10^7\ M^{-1}\ s^{-1}$ for recombination of radicals as from [66 McC 1].

Reaction Radical generation Method	Solvent	T[K]	Rate data	Ref./ add. ref.
$\dot{C}Cl_3$ + cholesta-4-en-3-one ⟶ $CHCl_3$ + cholesta-4-en-3-one(−$\dot{H}$)				
γ-rad. of CCl_4				80 Fel 1/
PR, glc	CCl_4	273	$k = 1.52 \cdot 10^2\ M^{-1}s^{-1}$ [5]	76 Kat 4
		293	$2.34 \cdot 10^2$	
		298	$1.82 \cdot 10^2$	
		306	$3.26 \cdot 10^2$	
		323	$4.66 \cdot 10^2$	
			$\log[A/M^{-1}s^{-1}] = 7.3$	
			$E_a = 28.9\ kJ\ mol^{-1}$	
$\dot{C}Cl_3$ + cholestanic esters [6] ⟶ $CHCl_3$ + cholestanic esters(−$\dot{H}$)				
γ-rad. of CCl_4				80 Fel 1/
PR, glc	CCl_4	298	$k = 96\ M^{-1}s^{-1}$ [7]	76 Kat 4
			$\log[A/M^{-1}s^{-1}] = 7.2$ [8]	
			$E_a = 30.1\ kJ\ mol^{-1}$ [8]	
$\dot{C}Cl_3$ + cholesteric esters [9] ⟶ $CHCl_3$ + cholesteric esters(−$\dot{H}$)				
γ-rad. of CCl_4				80 Fel 1/
PR, glc	CCl_4	298	$k = 6.40 \cdot 10^2\ M^{-1}s^{-1}$ [10]	76 Kat 4
			$\log[A/M^{-1}s^{-1}] = 7.5$ [11]	
			$E_a = 27.2\ kJ\ mol^{-1}$ [11]	
$\dot{C}Cl_3$ + cholestanyl acetate ⟶ $CHCl_3$ + cholestanyl acetate(−$\dot{H}$)				
γ-rad. of CCl_4				80 Fel 1/
PR, glc	CCl_4	273	$k = 5.60 \cdot 10^1\ M^{-1}s^{-1}$ [12]	76 Kat 4
		287	$6.40 \cdot 10^1$ [12]	
		306	$1.12 \cdot 10^2$ [12]	
		328	$1.86 \cdot 10^2$ [12]	

[5]) Based on $k/(2k_2)^{\frac{1}{2}} = 9.1 \cdot 10^{-3}$ with $2k_2 = 4 \cdot 10^8\ M^{-1}s^{-1}$ referring to $2\,\dot{C}Cl_3 \longrightarrow C_2Cl_6$ [76 Kat 4].

[6])

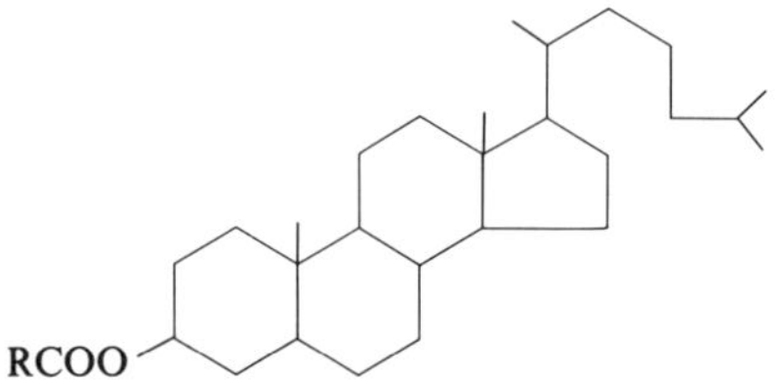

[7]) Based on $k/(2k_2)^{\frac{1}{2}} = 4.8 \cdot 10^{-1}$ with $2k_2 = 4 \cdot 10^8\ M^{-1}s^{-1}$ referring to $2\,\dot{C}Cl_3 \longrightarrow C_2Cl_6(2)$ [76 Kat 4].

[8]) Based on $\log[A/(A_2)^{\frac{1}{2}}] = 1.74$ and $\log[A_2/M^{-1}s^{-1}] = 11$; and $E_a - \frac{1}{2}(E_a)_2 = 23.3\ kJ\ mol^{-1}$ and $(E_a)_2 = 13.8\ kJ\ mol^{-1}$ [76 Kat 4].

[9])

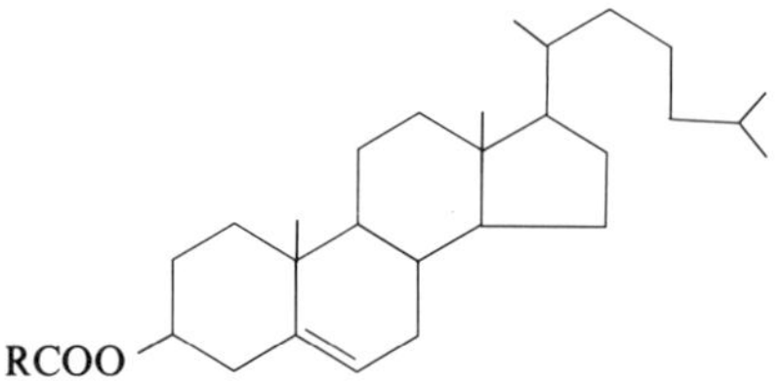

[10]) Based on $k/(2k_2)^{\frac{1}{2}} = 3.2 \cdot 10^{-2}$ with $2k_2 = 4 \cdot 10^8\ M^{-1}s^{-1}$ referring to $2\,\dot{C}Cl_3 \longrightarrow C_2Cl_6(2)$ [76 Kat 4].

[11]) Based on $\log[A/(A_2)^{\frac{1}{2}}] = 2.01$ and $\log[A_2/M^{-1}s^{-1}] = 11$; and $E_a - \frac{1}{2}(E_a)_2 = 20.4\ kJ\ mol^{-1}$ and $(E_a)_2 = 13.8\ kJ\ mol^{-1}$ [76 Kat 4].

[12]) Based on $k/(2k_2)^{\frac{1}{2}}$ values with $2k_2 = 4 \cdot 10^8\ M^{-1}s^{-1}$ referring to $2\,\dot{C}Cl_3 \longrightarrow C_2Cl_6$ [76 Kat 4].

Reaction Radical generation Method	Solvent	T[K]	Rate data	Ref./ add. ref.
$\dot{C}Cl_3$ + cholesteryl isopropylether $\longrightarrow$ $CHCl_3$ + cholesteryl isopropylether(−$\dot{H}$)				
γ-rad. of CCl_4				80 Fel 1/
PR, glc	CCl_4	273	$k = 4.58 \cdot 10^2\,M^{-1}s^{-1}$ [13]	76 Kat 4
		293	$7.80 \cdot 10^2$ [13]	
		298	$8.76 \cdot 10^2$ [13]	
		306	$1.17 \cdot 10^3$ [13]	
		323	$1.82 \cdot 10^3$ [13]	
			$\log[A/M^{-1}s^{-1}] = 7.8$	
			$E_a = 28.1\,kJ\,mol^{-1}$	
$\dot{C}Cl_3$ + cholestanyl nonanoate $\longrightarrow$ $CHCl_3$ + cholestanyl nonanoate(−$\dot{H}$)				
γ-rad. of CCl_4				80 Fel 1/
PR, glc	CCl_4	273	$k = 5.90 \cdot 10^1\,M^{-1}s^{-1}$ [12]	76 Kat 4
		293	$1.00 \cdot 10^2$ [12]	
		323	$1.86 \cdot 10^2$ [12]	
		336	$3.00 \cdot 10^2$ [12]	
$\dot{C}Cl_3$ + cholesteryl acetate $\longrightarrow$ $CHCl_3$ + cholesteryl acetate(−$\dot{H}$)				
γ-rad. of CCl_4				80 Fel 1/
PR, glc	CCl_4	273	$k = 2.84 \cdot 10^2\,M^{-1}s^{-1}$ [12]	76 Kat 4
		287	$4.12 \cdot 10^2$ [12]	
		306	$6.80 \cdot 10^2$ [12]	
		328	$1.12 \cdot 10^3$ [12]	
$\dot{C}Cl_3$ + cholesteryl chloride $\longrightarrow$ $CHCl_3$ + cholesteryl chloride(−$\dot{H}$)				
γ-rad. of CCl_4				80 Fel 1/
PR, glc	CCl_4	298	$k = 6.9 \cdot 10^2\,M^{-1}s^{-1}$ [14]	76 Kat 4
$\dot{C}Cl_3$ + cholesteryl nonanoate $\longrightarrow$ $CHCl_3$ + cholesteryl nonanoate(−$\dot{H}$)				
γ-rad. of CCl_4				80 Fel 1/
PR, glc	CCl_4	273	$k = 3.00 \cdot 10^2\,M^{-1}s^{-1}$ [12]	76 Kat 4
		293	$5.76 \cdot 10^2$ [12]	
		306	$7.10 \cdot 10^2$ [12]	
		323	$1.14 \cdot 10^3$ [12]	
		333	$1.30 \cdot 10^3$ [12]	
		336	$1.38 \cdot 10^3$ [12]	
$\dot{C}F_3 + HCO_2^- \longrightarrow CF_3H + \dot{C}O_2^-$				
Pulse rad. of $CF_3Cl + H_2O$				70 Bul 1
KAS	H_2O, pH = 9…10	293	$k = 3.4(7) \cdot 10^5\,M^{-1}s^{-1}$	
$\dot{C}F_3 + CH_3OH \longrightarrow CF_3H + \dot{C}H_2OH$				
γ-rad. of $CF_3Cl + H_2O$				70 Bul 1
PR, competition kinetics	H_2O, pH = 9…10	293	$k = 8.1(12) \cdot 10^3\,M^{-1}s^{-1}$ [15]	
$\dot{C}F_3 + C_2H_5OH \longrightarrow CF_3H + CH_3\dot{C}HOH$				
γ-rad. of $CF_3Cl + H_2O$				70 Bul 1
PR, competition kinetics	H_2O, pH = 9…10	293	$k = 4.6(5) \cdot 10^4\,M^{-1}s^{-1}$ [15]	

[12]) Based on $k/(2k_2)^{\frac{1}{2}}$ values with $2k_2 = 4 \cdot 10^8\,M^{-1}s^{-1}$ referring to $2\,\dot{C}Cl_3 \longrightarrow C_2Cl_6$ [76 Kat 4].
[13]) Based on $k/(2k_2)^{\frac{1}{2}} = 4.38 \cdot 10^{-2}$ with $2k_2 = 4 \cdot 10^8\,M^{-1}s^{-1}$ referring to $2\,\dot{C}Cl_3 \longrightarrow C_2Cl_6$ [76 Kat 4].
[14]) Based on $k/(2k_2)^{\frac{1}{2}} = 3.45 \cdot 10^{-2}$ with $2k_2 = 4 \cdot 10^8\,M^{-1}s^{-1}$ referring to $2\,\dot{C}Cl_3 \longrightarrow C_2Cl_6$ [76 Kat 4].
[15]) Based on $k(\dot{C}F_3 + HCOO^-) = 3.4 \cdot 10^5\,M^{-1}s^{-1}$.

Reaction Radical generation Method	Solvent	T[K]	Rate data	Ref./ add. ref.
$\dot{C}F_3 + CH_3CH_2CH_2OH \longrightarrow CF_3H + (C_3H_6O^{\cdot})$				
γ-rad. of $CF_3Cl + H_2O$				70 Bul 1
PR, competition kinetics	H_2O, pH = 9...10	293	$k = 4.4(7) \cdot 10^4\ M^{-1}\,s^{-1}$ [15]	
$\dot{C}F_3 + (CH_3)_2CHOH \longrightarrow CF_3H + (CH_3)_2\dot{C}OH$				
γ-rad. of $CF_3Cl + H_2O$				70 Bul 1
PR, competition kinetics	H_2O, pH = 9...10	293	$k = 9.2(9) \cdot 10^4\ M^{-1}\,s^{-1}$ [15]	
$\dot{C}HCl_2 + CHCl_3 \longrightarrow CH_2Cl_2 + \dot{C}Cl_3$				
γ-rad. of $CHCl_3$ (O_2-free)				66 Abr 1
PR, glc	$CHCl_3$	273	$k = 0.74(18) \cdot 10^{-21}$ $cm^3\,molecule^{-1}\,s^{-1}$	
		299	$1.8(2) \cdot 10^{-21}$	
		321	$4.0(6) \cdot 10^{-21}$	
		336	$7.1(17) \cdot 10^{-21}$	
			$\log[A/cm^3\,molecule^{-1}\,s^{-1}] = -15.8$	
			$E_a = 28.1(50)\ kJ\,mol^{-1}$	
γ-irr. of $CHCl_3$				71 Bib 1
PR, glc	$CHCl_3$	298	$k = 1.3(2)\ M^{-1}\,s^{-1}$	
$\dot{C}H_2OH + N_2O \longrightarrow N_2 + \dot{O}CH_2OH$				
γ-rad. of $CH_3OH + H_2O$				78 Rya 1
PR, glc	H_2O/CH_3OH	573	$k = 2.5 \cdot 10^4\ M^{-1}\,s^{-1}$	
$\dot{C}H_2OH + ICH_2COOH \longrightarrow CH_2IOH$ [16] $+ \dot{C}H_2COOH$				
Ti(III) + H_2O_2 + CH_3OH in H_2O				74 Gil 1
SESR	H_2O, pH = 1	293	$k = 2.1(7) \cdot 10^8\ M^{-1}\,s^{-1}$ [17]	
$\dot{C}H_2OH + CH_2(OH)_2 \longrightarrow CH_3OH + \dot{C}H(OH)_2$				
γ-rad. of $CH_3OH + H_2O$				72 Bya 1
[18]	H_2O	293	$k \approx 10^4 \dots 10^5\ M^{-1}\,s^{-1}$	
$\dot{C}H_2OH + HO(CH_2)_2SH \longrightarrow CH_3OH + HO(CH_2)_2\dot{S}$				
Pulse rad. of $CH_3OH + N_2O + H_2O$				69 Kar 1
KAS	H_2O, pH = 10	293	$k = 1.3(2) \cdot 10^8\ M^{-1}\,s^{-1}$ [19]	
$\dot{C}H_2OH + HOCH_2CH_2OH \longrightarrow CH_3OH + HOCH_2\dot{C}HOH$				
γ-rad. of $CH_3OH + H_2O$				72 Bya 1
[18]	H_2O	293	$k \approx 10^3 \dots 10^4\ M^{-1}\,s^{-1}$	
$\dot{C}H_2OH + HSCH_2CH_2NH_2 \longrightarrow CH_3OH + \dot{S}CH_2CH_2NH_2$				
Pulse rad. of $CH_3OH + N_2O + H_2O$				68 Ada 1, 72 Nuc 1
KAS	H_2O	293	$k = 6.8 \cdot 10^7\ M^{-1}\,s^{-1}$ [19] $2.9 \cdot 10^7$ [20]	

[15]) Based on $k(\dot{C}F_3 + HCOO^-) = 3.4 \cdot 10^5\ M^{-1}\,s^{-1}$.
[16]) Decays into HCHO and HI.
[17]) Based on $2k(\dot{C}H_2COOH + \dot{C}H_2COOH) = 1.8 \cdot 10^9\ M^{-1}\,s^{-1}$. $k < 3.5 \cdot 10^8\ M^{-1}\,s^{-1}$ estimated from pulse rad.
[18]) Estimate from dose rate effects on yields.
[19]) Rate constant measured via $RSSR^{\cdot-}$ formation ($R\dot{S} + RS^- \rightleftharpoons RSSR^{\cdot-}$).
[20]) From [72 Nuc 1].

Reaction Radical generation Method	Solvent	T[K]	Rate data	Ref./ add. ref.
$\dot{C}H_2OH + ICH_2CH_2COOH \longrightarrow CH_2IOH$ [21] $+ \dot{C}H_2CH_2COOH$				
$Ti(III) + H_2O_2 + CH_3OH$ in H_2O				74 Gil 1
SESR	H_2O, pH = 1	293	$k = 1.0 \cdot 10^5\ M^{-1}s^{-1}$ [22]	
$\dot{C}H_2OH + HSCH_2CH(NH_2)COOH \longrightarrow CH_3OH + \dot{S}CH_2CH(NH_2)COOH$				
Pulse rad. of $CH_3OH + N_2O + H_2O$				72 Nuc 1/
KAS	H_2O	293	$k = 4.2 \cdot 10^7\ M^{-1}s^{-1}$	69 Mor 1
$\dot{C}H_2OH + HSCH_2CHOHCHOHCH_2SH \longrightarrow CH_3OH + HSCH_2CHOHCHOHCH_2\dot{S}$				
Pulse rad. of $CH_3OH + N_2O + H_2O$				73 Red 1
KAS	H_2O	293	$k = 6.8(6) \cdot 10^7\ M^{-1}s^{-1}$	
$\dot{C}H_2OH$ + penicillamine(RSH) $\longrightarrow CH_3OH + R\dot{S}$				
Pulse rad. of $CH_3OH + H_2O$				73 Pur 1
KAS	H_2O	293	$k = 1.1(1) \cdot 10^8\ M^{-1}s^{-1}$ [23]	
$\dot{C}H_2CN + C_6H_5CH_3 \longrightarrow CH_3CN + C_6H_5\dot{C}H_2$				
Thermal decomp. of acetylperoxide				66 Sch 1/
PR, glc	acetonitrile/toluene	373	$k = 17\ M^{-1}s^{-1}$ [24]	54 Lev 1,
	(ratio 2.4...14.8)	353	7.8	60 McC 1
		340.5	4.7	
		328	2.6	
			$\log[A/M^{-1}s^{-1}] = 7.09$	
			$E_a = 10.0\ kJ\ mol^{-1}$	
$\dot{C}(OH)_2COO^- + HC(OH)_2COO^- \longrightarrow \dot{C}H(OH)COO^-$ + oxalic acid [25]				
Pulse rad. of glyoxylic acid $+ N_2O + H_2O$				72 Seh 1
KAS	H_2O, pH = 6 [26]	RT	$k = 7.5 \cdot 10^5\ M^{-1}s^{-1}$	
$\dot{C}H_2CHO$ + ascorbate ion(AH^-) $\longrightarrow \dot{A}^- + CH_3CHO$				
Pulse rad. of $ClCH_2CH_2OH$ + t-butanol $+ H_2O$				79 Ste 1
KAS	H_2O, pH = 11.5	RT	$k = 8.8 \cdot 10^7\ M^{-1}s^{-1}$	
$\dot{C}H_2CHO$ + catechin $\longrightarrow CH_3CHO$ + products [27]				
Pulse rad. of $CH_2OHCH_2OH + N_2O + H_2O$				82 Ste 1
KAS	H_2O, pH = 13.5	RT	$k = 1.8 \cdot 10^9\ M^{-1}s^{-1}$	
$\dot{C}H_2CHO$ + catechol(R—OH) $\longrightarrow CH_3CHO$ + products (R—$\dot{O}$) [27]				
Pulse rad. of $CH_2OHCH_2OH + N_2O + H_2O$				79 Ste 1
KAS	H_2O, pH = 11.5	RT	$k = 7.4 \cdot 10^8\ M^{-1}s^{-1}$	
$\dot{C}H_2CHO$ + 3,4-dihydroxycinnamate(R—OH) (caffeic acid) $\longrightarrow CH_3CHO$ + R—$\dot{O}$ [27]				
Pulse rad. of $CH_2OHCH_2OH + N_2O + H_2O$				82 Ste 1
KAS	H_2O, pH = 13.5	RT	$k = 2.6 \cdot 10^9\ M^{-1}s^{-1}$	
$\dot{C}H_2CHO$ + 6,7-dihydroxycoumarin(R—OH) (esculetin) $\longrightarrow CH_3CHO$ + R—$\dot{O}$ [27]				
Pulse rad. of $CH_2OHCH_2OH + N_2O + H_2O$				82 Ste 1
KAS	H_2O, pH = 13.5	RT	$k = 2 \cdot 10^9\ M^{-1}s^{-1}$	

[21]) Decays into HCHO and HI.
[22]) Based on $2k(\dot{C}H_2CH_2OH + \dot{C}H_2CH_2OH) = 2.4 \cdot 10^9\ M^{-1}s^{-1}$.
[23]) Rate constant measured via $RSSR^{\dot{-}}$ formation ($R\dot{S} + RS^- \rightleftharpoons RSSR^{\dot{-}}$).
[24]) Based on data in [54 Lev 1] on decomposition of acetylperoxide, and $k = 4 \cdot 10^7\ M^{-1}s^{-1}$ for recombination of radicals as from [60 McC 1].
[25]) Presumed $\dot{O}H$ displacement.
[26]) Reaction of corresp. protonated forms at pH = 1 occurs with $k > 5 \cdot 10^5\ M^{-1}s^{-1}$.
[27]) Possibly involving e^--transfer.

Reaction Radical generation Method	Solvent	T[K]	Rate data	Ref./ add. ref.
$\dot{C}H_2CHO$ + 2,5-dihydroxyphenylacetate(R—OH) ⟶ CH_3CHO + R—$\dot{O}$ [27])				
Pulse rad. of $CH_2OHCH_2OH + N_2O + H_2O$				82 Ste 1
KAS	H_2O, pH = 13.5	RT	$k = 1.7 \cdot 10^9 M^{-1} s^{-1}$	
$\dot{C}H_2CHO$ + DL-β-3,4-dihydrooxyphenylalanine(DOPA) ⟶ CH_3CHO + products [27])				
Pulse rad. of $CH_2OHCH_2OH + N_2O + H_2O$				82 Ste 1
KAS	H_2O, pH = 13.5	RT	$k = 1.4 \cdot 10^9 M^{-1} s^{-1}$	
$\dot{C}H_2CHO$ + p-(N,N-dimethylamino)phenol ⟶ CH_3CHO + products [27])				
Pulse rad. of $CH_2OHCH_2OH + N_2O + H_2O$				82 Ste 1
KAS	H_2O, pH = 13.5	RT	$k = 2.2 \cdot 10^9 M^{-1} s^{-1}$	
$\dot{C}H_2CHO$ + ellagic acid ⟶ CH_3CHO + products [27])				
Pulse rad. of $CH_2OHCH_2OH + N_2O + H_2O$				82 Ste 1
KAS	H_2O, pH = 13.5	RT	$k = 2.4 \cdot 10^9 M^{-1} s^{-1}$	
$\dot{C}H_2CHO$ + hydroquinone ⟶ CH_3CHO + semiquinone [27])				
Pulse rad. of $CH_2OHCH_2OH + N_2O + H_2O$				79 Ste 1
KAS	H_2O, pH = 11.5	RT	$k = 2.2 \cdot 10^9 M^{-1} s^{-1}$	
$\dot{C}H_2CHO$ + 7-hydroxycoumarin(—OH) (umbelliferone) ⟶ CH_3CHO + —$\dot{O}$ [27])				
Pulse rad. of $CH_2OHCH_2OH + N_2O + H_2O$				82 Ste 1
KAS	H_2O, pH = 13.5	RT	$k = 1.3 \cdot 10^9 M^{-1} s^{-1}$	
$\dot{C}H_2CHO$ + 5-hydroxydopamine ⟶ CH_3CHO + products [27])				
Pulse rad. of $CH_2OHCH_2OH + N_2O + H_2O$				82 Ste 1
KAS	H_2O, pH = 13.5	RT	$k = 1.8 \cdot 10^9 M^{-1} s^{-1}$	
$\dot{C}H_2CHO$ + 6-hydroxydopamine ⟶ CH_3CHO + products [27])				
Pulse rad. of $CH_2OHCH_2OH + N_2O + H_2O$				82 Ste 1
KAS	H_2O, pH = 13.5	RT	$k = 1.8 \cdot 10^9 M^{-1} s^{-1}$	
$\dot{C}H_2CHO$ + 5-hydroxyindole ⟶ CH_3CHO + products [27])				
Pulse rad. of $CH_2OHCH_2OH + N_2O + H_2O$				82 Ste 1
KAS	H_2O, pH = 13.5	RT	$k = 1.3 \cdot 10^9 M^{-1} s^{-1}$	
$\dot{C}H_2CHO$ + 6-hydroxy-2,5,7,8-tetramethylchromane-2-carboxylate(HTCC) ⟶ CH_3CHO + products [27])				
Pulse rad. of $CH_2OHCH_2OH + N_2O + H_2O$				82 Ste 1
KAS	H_2O, pH = 13.5	RT	$k = 1.8 \cdot 10^9 M^{-1} s^{-1}$	
$\dot{C}H_2CHO$ + 5-hydroxytryptophan ⟶ CH_3CHO + products [27])				
Pulse rad. of $CH_2OHCH_2OH + N_2O + H_2O$				82 Ste 1
KAS	H_2O, pH = 13.5	RT	$k = 1.3 \cdot 10^9 M^{-1} s^{-1}$	
$\dot{C}H_2CHO$ + 4-methoxyphenol ⟶ CH_3CHO + $CH_3OC_6H_4\dot{O}$ [27])				
Pulse rad. of $CH_2OHCH_2OH + N_2O + H_2O$				82 Ste 1,
KAS	H_2O, pH = 13.5	RT	$k = 8.3 \cdot 10^8 M^{-1} s^{-1}$	79 Ste 1
	pH = 11.5		$9.8 \cdot 10^8$ [28])	
$\dot{C}H_2CHO$ + norepinephrine ⟶ CH_3CHO + products [27])				
Pulse rad. of $CH_2OHCH_2OH + N_2O + H_2O$				82 Ste 1
KAS	H_2O, pH = 13.5	RT	$k = 1.5 \cdot 10^9 M^{-1} s^{-1}$	

[27]) Possibly involving e^--transfer.
[28]) From [79 Ste 1].

Reaction Radical generation Method	Solvent	T[K]	Rate data	Ref./ add. ref.
$\dot{C}H_2CHO$ + 3,3′,4′,5,7-pentahydroxyflavone(quercetin) ⟶ CH_3CHO + products [27])				
Pulse rad. of $CH_2OHCH_2OH + N_2O + H_2O$				82 Ste 1
KAS	H_2O, pH = 13.5	RT	$k = 3.1\cdot10^{9}\,M^{-1}s^{-1}$	
$\dot{C}H_2CHO$ + *o*-phenylenediamine ⟶ CH_3CHO + —$\dot{N}H$ [27])				
Pulse rad. of $CH_2OHCH_2OH + N_2O + H_2O$				82 Ste 1,
KAS	H_2O, pH = 13.5	RT	$k = 7.7\cdot10^{7}\,M^{-1}s^{-1}$	79 Ste 1
	pH = 11.5		$7.3\cdot10^{7}$ [28])	
$\dot{C}H_2CHO$ + *p*-phenylenediamine ⟶ CH_3CHO + —$\dot{N}H$ [27])				
Pulse rad. of $CH_2OHCH_2OH + N_2O + H_2O$				82 Ste 1,
KAS	H_2O, pH = 13.5	RT	$k = 4.6\cdot10^{8}\,M^{-1}s^{-1}$	79 Ste 1
	pH = 11.5		$4.0\cdot10^{8}$ [28])	
$\dot{C}H_2CHO$ + resorcinol (—OH) ⟶ CH_3CHO + products (—$\dot{O}$) [27])				
Pulse rad. of $CH_2OHCH_2OH + N_2O + H_2O$				82 Ste 1,
KAS	H_2O, pH = 13.5	RT	$k = 1.3\cdot10^{9}\,M^{-1}s^{-1}$	79 Ste 1
	pH = 11.5		$1.6\cdot10^{9}$ [28])	
$\dot{C}H_2CHO$ + rutin ⟶ CH_3CHO + products [27])				
Pulse rad. of $CH_2OHCH_2OH + N_2O + H_2O$				82 Ste 1
KAS	H_2O, pH = 13.5	RT	$k = 1.5\cdot10^{9}\,M^{-1}s^{-1}$	
$\dot{C}H_2CHO$ + 1,2,5,8-tetrahydroxyanthraquinone(quinalizarin) ⟶ CH_3CHO + products [27])				
Pulse rad. of $CH_2OHCH_2OH + N_2O + H_2O$				82 Ste 1
KAS	H_2O, pH = 13.5	RT	$k = 2.4\cdot10^{9}\,M^{-1}s^{-1}$	
$\dot{C}H_2CHO$ + N,N,N′,N′-tetramethyl-*p*-phenylenediamine(TMPD) ⟶ CH_3CHO + products [27])				
Pulse rad. of $CH_2OHCH_2OH + N_2O + H_2O$				82 Ste 1,
KAS	H_2O, pH = 13.5	RT	$k = 2.1\cdot10^{9}\,M^{-1}s^{-1}$	79 Ste 1
	pH = 11.5		$2.0\cdot10^{9}$ [28])	
$\dot{C}H_2CHO$ + 3,4,5-trihydroxybenzoate(—OH) (gallate) ⟶ CH_3CHO + —$\dot{O}$ [27])				
Pulse rad. of $CH_2OHCH_2OH + N_2O + H_2O$				82 Ste 1
KAS	H_2O, pH = 13.5	RT	$k = 1.4\cdot10^{9}\,M^{-1}s^{-1}$	
$\dot{C}H_2CHO$ + 2,4,5-trihydroxypyrimidine ⟶ CH_3CHO + products [27])				
Pulse rad. of $CH_2OHCH_2OH + N_2O + H_2O$				82 Ste 1
KAS	H_2O, pH = 13.5	RT	$k = 1.6\cdot10^{9}\,M^{-1}s^{-1}$	
$CH_3\dot{C}HOH + Mn(CO)_5I \longrightarrow Mn(CO)_5^{\cdot} + CH_3CHIOH$				
Pulse rad. of C_2H_5OH				78 Wal 1
KAS	C_2H_5OH	RT	$k = 1.5\cdot10^{8}\,M^{-1}s^{-1}$	
$CH_3\dot{C}HOH + HO(CH_2)_2SH \longrightarrow CH_3CH_2OH + HO(CH_2)_2\dot{S}$				
Pulse rad. of $C_2H_5OH + N_2O + H_2O$				69 Kar 1
KAS	H_2O, pH = 10	293	$k = 2.3(3)\cdot10^{8}\,M^{-1}s^{-1}$ [29])	
$CH_3\dot{C}HOH + HSCH_2CH_2NH_2 \longrightarrow CH_3CH_2OH + \dot{S}CH_2CH_2NH_2$				
Pulse rad. of $C_2H_5OH + N_2O + H_2O$				68 Ada 1
KAS	H_2O	293	$k = 1.4\cdot10^{8}\,M^{-1}s^{-1}$ [29])	

[27]) Possibly involving e^--transfer.
[28]) From [79 Ste 1].
[29]) Rate constant measured via $RSSR^{\dot{-}}$ formation ($R\dot{S} + RS^- \rightleftharpoons RSSR^{\dot{-}}$).

Reaction Radical generation Method	Solvent	T[K]	Rate data	Ref./ add. ref.
$CH_3\dot{C}HOH + HSCH_2CH_2NH_2 \longrightarrow \dot{S}CH_2CH_2NH_2 + CH_3CH_2OH$				
Pulse rad. of $C_2H_5OH + N_2O + H_2O$				82 Wol 1
KAS	H_2O, pH = 4.2	RT	$k = 1.7 \cdot 10^8\ M^{-1}s^{-1}$ [29a]	
$\dot{C}H_2CH_2OH + C_2H_5OH \longrightarrow CH_3CH_2OH + CH_3\dot{C}HOH$				
γ-rad. of $C_2H_5OH + H_2O$				70 Bur 1
[30]	H_2O	293	$k = 16(3)\ M^{-1}s^{-1}$	
$\dot{C}H_2CH_2OH + HO(CH_2)_2SH \longrightarrow CH_3CH_2OH + HO(CH_2)_2\dot{S}$				
Pulse rad. of 2-mercaptoethanol + H_2O				69 Kar 1
KAS	H_2O, pH = 10	293	$k = 4.7(7) \cdot 10^7\ M^{-1}s^{-1}$ [29]	
$\dot{C}H(CO_2^-)_2$ + ascorbate ion(AH^-) $\longrightarrow \dot{A}^- + CH_2(CO_2^-)_2$				
Pulse rad. of malonate + $N_2O + H_2O$				73 Red 2
KAS	H_2O	RT	$k = 1.3(1) \cdot 10^7\ M^{-1}s^{-1}$	
$\dot{C}H(COOH)_2$ [31] + ascorbate(AH^-) $\longrightarrow CH_2(COOH)_2$ [31] + $\dot{A}^-$				
Pulse rad. of malonic acid + $N_2O + H_2O$				73 Red 2
KAS	H_2O, pH = 7.7	RT	$k = 1.3(1) \cdot 10^7\ M^{-1}s^{-1}$	
$\dot{C}H_2COCH_3 + HSCH_2CH_2NH_2 \longrightarrow CH_3COCH_3 + \dot{S}CH_2CH_2NH_2$				
Pulse rad. of acetone + $N_2O + H_2O$				68 Ada 1
KAS	H_2O	293	$k \approx 4 \cdot 10^8\ M^{-1}s^{-1}$ [29]	
$\dot{C}H_2COCH_3 + C_6H_5CH_3 \longrightarrow CH_3COCH_3 + C_6H_5\dot{C}H_2$				
Thermal decomp. of acetylperoxide				66 Sch 1/
PR, glc	acetone/toluene	373	$k = 18\ M^{-1}s^{-1}$ [32]	54 Lev 1,
	(ratio 0.75 ... 4.13)	353	8.7 [32]	60 McC 1
		340.5	5.6 [32]	
		328	3.1 [32]	
			$\log[A/M^{-1}s^{-1}] = 6.71$	
			$E_a = 9.3\ kJ\ mol^{-1}$	
$(CH_3)_2\dot{C}OH + N_2O \longrightarrow N_2 + (CH_3)_2C(\dot{O})OH$				
γ-rad. of 2-propanol + H_2O				77 Rya 1
PR, glc	H_2O/2-propanol	573	$k = 2.9 \cdot 10^6\ M^{-1}s^{-1}$	
$(CH_3)_2\dot{C}OH + HO(CH_2)_2SH \longrightarrow (CH_3)_2CHOH + HO(CH_2)_2\dot{S}$				
Pulse rad. of 2-propanol + $N_2O + H_2O$				69 Kar 1
KAS	H_2O, pH = 10	293	$k = 5.1(8) \cdot 10^8\ M^{-1}s^{-1}$ [33]	
$(CH_3)_2\dot{C}OH + HSCH_2CH_2NH_2 \longrightarrow (CH_3)_2CHOH + \dot{S}CH_2CH_2NH_2$				
Pulse rad. of 2-propanol + $N_2O + H_2O$				68 Ada 1,
KAS	H_2O	293	$k = 4.2 \cdot 10^8\ M^{-1}s^{-1}$	72 Nuc 1
			$2.0 \cdot 10^8$ [34]	

[29]) Rate constant measured via $RSSR^{\dot{-}}$ formation ($R\dot{S} + RS^- \rightleftharpoons RSSR^{\dot{-}}$).
[29a]) Rate constant measured via $ABTS^{\dot{+}}$ formation ($R\dot{S}$ + ABTS $\longrightarrow RS^-$ + $ABTS^{\dot{+}}$) ABTS = 2,2′-azinobis-(3-ethylbenzthiazoline-6-sulfonic acid).
[30]) Observed product yields and assumed mechanism for ox. of C_2H_5OH by H_2O_2.
[31]) Partially ionized at pH = 7.7.
[32]) Based on data in [54 Lev 1] on decomposition of acetylperoxide and $k = 4 \cdot 10^7\ M^{-1}s^{-1}$ for recombination of radicals as from [60 McC 1].
[33]) Rate constant measured via $RSSR^{\dot{-}}$ formation ($R\dot{S} + RS^- \rightleftharpoons RSSR^{\dot{-}}$).
[34]) From [72 Nuc 1].

Reaction Radical generation Method	Solvent	T[K]	Rate data	Ref./ add. ref.
$(CH_3)_2\dot{C}OH + HSCH_2CH_2NH_2 \longrightarrow \dot{S}CH_2CH_2NH_2 + (CH_3)_2CHOH$				
Pulse rad. of 2-propanol + N_2O + H_2O (A) and 2-propanol + acetone + H_2O (B)				82 Wol 1
KAS	H_2O, pH = 4.2	RT	$k = 3.8 \cdot 10^8\,M^{-1}s^{-1}$ (A) [29a] $3.3 \cdot 10^8$ (B)	
$(CH_3)_2\dot{C}OH$ + HSCH₂CHOHCHOHCH₂SH ⟶ $(CH_3)_2CHOH$ + HSCH₂CHOHCHOHCH₂Ṡ				
Pulse rad. of 2-propanol + acetone + H_2O				73 Red 1
KAS	H_2O	293	$k = 2.1(2) \cdot 10^8\,M^{-1}s^{-1}$	
$(CH_3)_2\dot{C}OH$ + ascorbate ion(AH^-) ⟶ $\dot{A}^-$ + $(CH_3)_2CHOH$				
Pulse rad. of 2-propanol + acetone + H_2O				73 Red 2
KAS	H_2O	RT	$k = 1.2(1) \cdot 10^6\,M^{-1}s^{-1}$	
$(CH_3)_2\dot{C}OH$ + 3-bromo-3-deoxy-1,2:5,6-di-O-isopropylidene-α-D-glucofuranose(RBr) ⟶ $\dot{R}$ + HBr + $(CH_3)_2CO$				
γ-rad. of 2-propanol				82 Lem 1
Potentiometric titration, HPLC	2-propanol/acetone	RT	$k \approx 300\,M^{-1}s^{-1}$ [35]	
$(CH_3)_2\dot{C}OH$ + 3-deoxy-3-iodo-1,2:5,6-di-O-isopropylidene-α-D-allofuranose(RI) ⟶ $\dot{R}$ + HI + $(CH_3)_2CO$				
γ-rad. of 2-propanol				82 Lem 1
Potentiometric titration, HPLC	2-propanol/acetone	RT	$k \approx 1200...1500\,M^{-1}s^{-1}$ [35]	
$(CH_3)_2\dot{C}OH$ + glutathione(GSH) ⟶ G$\dot{S}$ + $(CH_3)_2CHOH$				
Pulse rad. of 2-propanol + acetone + H_2O				83 For 1
KAS	H_2O	RT	$k = 1.8 \cdot 10^8\,M^{-1}s^{-1}$	
$\dot{C}H_2CHOHCH_3 + (CH_3)_2CHOH \longrightarrow (CH_3)_2CHOH + (CH_3)_2\dot{C}OH$				
γ-rad. of 2-propanol + H_2O_2 + H_2O				70 Bur 2
[36]	H_2O	293	$k > 53(10)\,M^{-1}s^{-1}$	
$CH_3CH_2CH_2\dot{C}HOH + HSCH_2CH_2NH_2 \longrightarrow CH_3CH_2CH_2CH_2OH + \dot{S}CH_2CH_2NH_2$				
Pulse rad. of 1-butanol + N_2O + H_2O				68 Ada 1
KAS	H_2O	293	$k = 8.2 \cdot 10^7\,M^{-1}s^{-1}$ [33]	
$(CH_3)_2CH\dot{C}HOH + HSCH_2CH_2NH_2 \longrightarrow (CH_3)_2CHCH_2OH + \dot{S}CH_2CH_2NH_2$				
Pulse rad. of 2-methyl-1-propanol + N_2O + H_2O				68 Ada 1
KAS	H_2O	293	$k = 1.4 \cdot 10^8\,M^{-1}s^{-1}$ [33]	
$\dot{C}H_2C(CH_3)_2OH + ICH_2CN \longrightarrow CH_2IC(CH_3)_2OH + \dot{C}H_2CN$				
Ti(III) + H_2O_2 + $(CH_3)_3COH$ in H_2O				74 Gil 1
SESR	H_2O, pH = 1	293	$k = 2.0(5) \cdot 10^7\,M^{-1}s^{-1}$ [37]	
$\dot{C}H_2C(CH_3)_2OH + ICH_2COOH \longrightarrow CH_2IC(CH_3)_2OH + \dot{C}H_2COOH$				
Ti(III) + H_2O_2 + $(CH_3)_3COH$ in H_2O				74 Gil 1
SESR	H_2O, pH = 1	293	$k = 1.3(3) \cdot 10^7\,M^{-1}s^{-1}$ [38]	

[29a]) Rate constant measured via $ABTS^{\dot{+}}$ formation ($R\dot{S}$ + ABTS ⟶ RS^- + $ABTS^{\dot{+}}$) ABTS = 2,2′-azinobis-(3-ethylbenzthiazoline-6-sulfonic acid).
[33]) Rate constant measured via $RSSR^{\dot{-}}$ formation ($R\dot{S}$ + RS^- ⇌ $RSSR^{\dot{-}}$).
[35]) Based on assumption that termination occurs largely by $2(CH_3)_2\dot{C}OH$ ⟶ products.
[36]) Estimate from product yields and rates.
[37]) Based on $2k(\dot{C}H_2CN + \dot{C}H_2CN) = 2 \cdot 10^9\,M^{-1}s^{-1}$.
[38]) Based on $2k(\dot{C}H_2COOH + \dot{C}H_2COOH) = 1.8 \cdot 10^9\,M^{-1}s^{-1}$.

Reaction Radical generation Method	Solvent	T[K]	Rate data	Ref./ add. ref.
$\dot{C}H_2C(CH_3)_2OH + ICH_2CH_2OH \longrightarrow CH_2IC(CH_3)_2OH + \dot{C}H_2CH_2OH$				
Ti(III) + H_2O_2 + $(CH_3)_3COH$ in H_2O				74 Gil 1
SESR	H_2O, pH = 1	293	$k = 1.5 \cdot 10^5\ M^{-1}s^{-1}$ [39]	
$\dot{C}H_2C(CH_3)_2OH + HO(CH_2)_2SH \longrightarrow (CH_3)_3COH + HO(CH_2)_2\dot{S}$				
Pulse rad. of *t*-butanol + N_2O + H_2O				69 Kar 1
KAS	H_2O, pH = 10	293	$k = 8.2(12) \cdot 10^7\ M^{-1}s^{-1}$ [40]	
$\dot{C}H_2C(CH_3)_2OH + HSCH_2CH_2NH_2 \longrightarrow (CH_3)_3COH + \dot{S}CH_2CH_2NH_2$				
Pulse rad. of *t*-butanol + N_2O + H_2O				68 Ada 1
KAS	H_2O	293	$k = 1.8 \cdot 10^7\ M^{-1}s^{-1}$ [40]	
$\dot{C}H_2C(CH_3)_2OH + ICH_2CH_2COOH \longrightarrow CH_2IC(CH_3)_2OH + \dot{C}H_2CH_2COOH$				
Ti(III) + H_2O_2 + $(CH_3)_3COH$ in H_2O				74 Gil 1
SESR	H_2O, pH = 1	293	$k = 1.8 \cdot 10^5\ M^{-1}s^{-1}$ [41]	
$\dot{C}H_2C(CH_3)_2OH + HSCH_2CHOHCHOHCH_2SH \longrightarrow (CH_3)_3COH + HSCH_2CHOHCHOHCH_2\dot{S}$				
Pulse rad. of *t*-butanol + N_2O + H_2O				73 Red 1
KAS	H_2O	293	$k = 6.8(6) \cdot 10^7\ M^{-1}s^{-1}$	
$\dot{C}HOHCH_2CH_2CH_2OH + HSCH_2CH(NH_2)COOH \longrightarrow CH_2OHCH_2CH_2CH_2OH + \dot{S}CH_2CH(NH_2)COOH$				
Pulse rad. of 1,4-butanediol + N_2O + H_2O				68 Ada 1
KAS	H_2O	293	$k = 1.1 \cdot 10^8\ M^{-1}s^{-1}$ [40]	
$CH_2{=}CHCH_2OCH_2\dot{C}H_2$ + (η^5-cyclopentadienyl)tricarbonylhydridovanadate [42] $\longrightarrow CH_2{=}CHCH_2OCH_2CH_3$ + products				
From $CH_2{=}CHCH_2OCH_2CH_2Br$ induced by (η^5-cyclo...)vanadate [42]				78 Kin 1
PR, glc	CH_3CN	298	$k = 2 \cdot 10^7\ M^{-1}s^{-1}$ [43]	
$CH_3(CH_2)_3SCH_2\dot{C}H(CH_2)_2CH_3 + n\text{-}C_4H_9SH \longrightarrow n\text{-}C_4H_9\dot{S} + CH_3(CH_2)_3S(CH_2)_4CH_3$				
Photolytic initiation with AIBN				55 Ony 1/
RS, dilatometric technique	C_6H_6	298	$k = 1.4 \cdot 10^6\ M^{-1}s^{-1}$	59 Siv 1, 54 Bac 1
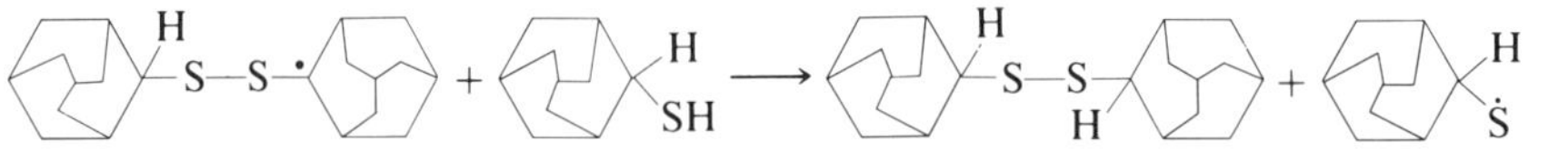				
Phot. (azobiscyclohexylnitrile as initiator), solution containing 2-adamantylthiol and 2-adamantanethione				76 Sca 1
RS, KAS	C_6H_6	323	$k = 4.0 \cdot 10^4\ M^{-1}s^{-1}$	

[39]) Based on $2k(\dot{C}H_2CH_2OH + \dot{C}H_2CH_2OH) = 1.9 \cdot 10^9\ M^{-1}s^{-1}$.
[40]) Rate constant measured via $RSSR^{\dot{-}}$ formation ($R\dot{S} + RS^- \rightleftharpoons RSSR^{\dot{-}}$).
[41]) Based on $2k(\dot{C}H_2CH_2COOH + \dot{C}H_2COOH) = 2.4 \cdot 10^9\ M^{-1}s^{-1}$.
[42])

[43]) Based on $k = 1.2 \cdot 10^6\ s^{-1}$ for cyclization of $CH_2{=}CHCH_2OCH_2\dot{C}H_2$.

4.1.1.2 Aromatic radicals and radicals derived from compounds containing aromatic and heterocyclic constituents

4.1.1.2.1 Radicals containing only C and H atoms

Reaction Radical generation Method	Solvent		T[K]	Rate data	Ref./ add. ref.
$(C_6H_5^{\cdot}) + I_2 \longrightarrow C_6H_5I + \dot{I}$					
Thermal decomp. of PAT					77 Kry 1,
PR, glc	chlorobenzene		318	$k(I) = 1.54 \cdot 10^{10}\,M^{-1}\,s^{-1}$ [1]	78 Lor 1
PR, glc	C_6H_6		318	$k(I) = 1.15 \cdot 10^{10}\,M^{-1}\,s^{-1}$ [2]	77 Kry 1
PR, glc, calc.	chlorobenzene + 20 Vol % mineral oil,		318		77 Kry 1
		η = 0.829 cP		$k = 13.4 \cdot 10^9\,M^{-1}\,s^{-1}$ [3]	
	30	1.03		$13.4 \cdot 10^9$	
	40	1.29		$11.2 \cdot 10^9$	
	50	1.73		$9.05 \cdot 10^9$	
	60	2.41		$6.80 \cdot 10^9$	
	70	3.63		$5.05 \cdot 10^9$	
	75	4.72		$4.20 \cdot 10^9$	
	80	6.32		$2.98 \cdot 10^9$	
	83.3	7.74		$2.72 \cdot 10^9$	
	86.7	9.85		$2.21 \cdot 10^9$	
	90	12.74		$2.05 \cdot 10^9$	
	92.5	15.70		$2.04 \cdot 10^9$	
	95	19.48		$2.10 \cdot 10^9$	
$(C_6H_5^{\cdot}) + CBrCl_3 \longrightarrow C_6H_5Br + \dot{C}Cl_3$					
Thermal decomp. of PAT					78 Lor 1
PR	CCl_4		318	$k = 1.8 \cdot 10^9\,M^{-1}\,s^{-1}$ [4]	
PR, glc	CCl_4		318	$k(Br) = 1.93 \cdot 10^9\,M^{-1}\,s^{-1}$ [5] $1.67 \cdot 10^9$ [6] $1.85 \cdot 10^9$ [7]	77 Kry 1
	C_6H_6		318	$k(Br) = 1.45 \cdot 10^9\,M^{-1}\,s^{-1}$ [8]	
$(C_6H_5^{\cdot}) + CBr_4 \longrightarrow C_6H_5Br + \dot{C}Br_3$					
Thermal decomp. of PAT					77 Kry 1
PR, glc	chlorobenzene		318	$k(Br) = 4.0 \cdot 10^9\,M^{-1}\,s^{-1}$ [9] $5.0 \cdot 10^9$ [10]	
PR, glc (continued)	CCl_4		318	$k(Br) = 5.5 \cdot 10^9\,M^{-1}\,s^{-1}$ [11] $5.0 \cdot 10^9$ [11] [12]	77 Kry 1, 78 Lor 1

[1]) Based on $k(H)(C_6H_5^{\cdot}$ + aliphatic secondary C—H) = $3.3 \cdot 10^5\,M^{-1}\,s^{-1}$ and $k(I)/k(H) = 46700$.
[2]) Based on $k(H)(C_6H_5^{\cdot}$ + aliphatic secondary C—H) = $3.3 \cdot 10^5\,M^{-1}\,s^{-1}$, $k(Br)(C_6H_5^{\cdot} + CBr_4) = 5.5 \cdot 10^9\,M^{-1}\,s^{-1}$ and $k(I)/k(Br) = 2.08$.
[3]) Rate constants calculated from experimental product yields and diffusion equations (Wilke-Chang and Smoluchowski); solutions containing 0.0196 M PAT and 0.050 M I_2.
[4]) Based on $k = 3.3 \cdot 10^5\,M^{-1}\,s^{-1}$ for $(C_6H_5^{\cdot})$ + aliphatic secondary (>CH—H).
[5]) Based on $k(I)(C_6H_5^{\cdot} + (CH_3)_2CHI) = 1.23 \cdot 10^9\,M^{-1}\,s^{-1}$ and $k(Br)/k(I) = 1.57$.
[6]) Rate constant based on $k(Cl)(C_6H_5^{\cdot} + CCl_4) = 3.7 \cdot 10^6\,M^{-1}\,s^{-1}$ and $k(Br)/k(Cl) = 450$.
[7]) Rate constant based on $k(Cl)(C_6H_5^{\cdot} + CCl_4) = 3.7 \cdot 10^6\,M^{-1}\,s^{-1}$ and $k(Br)/k(Cl) = 500$.
[8]) Based on $k(I)(C_6H_5^{\cdot} + (CH_3)_2CHI) = 8.3 \cdot 10^8\,M^{-1}\,s^{-1}$ and $k(Br)/k(I) = 1.75$.
[9]) Rate constant based on $k(H)(C_6H_5^{\cdot}$ + aliphatic secondary C—H) = $3.3 \cdot 10^5\,M^{-1}\,s^{-1}$ and $k(Br)/k(H) = 12100$.
[10]) Rate constant based on $k(H)(C_6H_5^{\cdot}$ + aliphatic secondary C—H) = $3.3 \cdot 10^5\,M^{-1}\,s^{-1}$ and $k(Br)/k(H) = 15200$.
[11]) Based on $k(H)(C_6H_5^{\cdot}$ + aliphatic secondary C—H) = $3.3 \cdot 10^5\,M^{-1}\,s^{-1}$, $k(Cl)(C_6H_5^{\cdot} + CCl_4) = 3.7 \cdot 10^6\,M^{-1}\,s^{-1}$ and $k(Br)/k(Cl) = 1500$.
[12]) From [78 Lor 1].

Reaction Radical generation Method	Solvent	T[K]	Rate data	Ref./ add. ref.
$(C_6H_5^{\cdot} + CBr_4 \longrightarrow C_6H_5Br + \dot{C}Br_3$ (continued)				
Thermal decomp. of PAT				
PR, glc, calc.	chlorobenzene + 20 Vol % mineral oil,	318		77 Kry 1
	η = 0.829 cP		$k = 3.83 \cdot 10^9\ M^{-1}s^{-1}$ [13]	
	30 1.03		$3.37 \cdot 10^9$	
	40 1.29		$3.72 \cdot 10^9$	
	50 1.73		$3.58 \cdot 10^9$	
	60 2.41		$3.18 \cdot 10^9$	
	70 3.63		$3.08 \cdot 10^9$	
	75 4.72		$2.61 \cdot 10^9$	
	80 6.32		$2.26 \cdot 10^9$	
	83.3 7.74		$2.12 \cdot 10^9$	
	86.7 9.85		$1.95 \cdot 10^9$	
	90 12.74		$2.08 \cdot 10^9$	
	92.5 15.70		$1.54 \cdot 10^9$	
	95 19.48		$1.41 \cdot 10^9$	
$(C_6H_5^{\cdot}) + CCl_4 \longrightarrow C_6H_5Cl + \dot{C}Cl_3$				
Thermal decomp. of PAT				77 Kry 1,
PR, glc	CCl_4	318	$k(Cl) = 2.7 \cdot 10^6\ M^{-1}s^{-1}$ [14]	78 Lor 1
			$3.3 \cdot 10^6$ [15]	
			$3.2 \cdot 10^6$ [12]	
PR, glc	C_6H_6/CCl_4	318	$k(Cl) = 5.8 \cdot 10^6\ M^{-1}s^{-1}$ [16]	77 Kry 1
PR, glc	c-C_6H_{12}/CCl_4	318	$k(Cl) = 3.7 \cdot 10^6\ M^{-1}s^{-1}$ [17]	77 Kry 1/ 63 Bri 1
$(C_6H_5^{\cdot}) + CH_3OH \longrightarrow C_6H_6$ + methanol($-\dot{H}$)				
Thermal decomp. of PAT				78 Lor 1
PR	CCl_4	318	$k = 5 \cdot 10^5\ M^{-1}s^{-1}$ [18]	
$(C_6H_5^{\cdot}) + CH_3CH_2OH \longrightarrow C_6H_6 + CH_3\dot{C}HOH$				
Thermal decomp. of PAT				75 Jan 1
KESR	C_2H_5OH	RT	$k = 2.3(1) \cdot 10^5\ M^{-1}s^{-1}$	
$(C_6H_5^{\cdot}) + (CH_3)_2Se \xrightarrow{a} \dot{C}H_3 + C_6H_5SeCH_3$ $+ (C_2H_5)_2Se \xrightarrow{b} CH_3\dot{C}H_2 + C_6H_5SeC_2H_5$ $+ c\text{-}C_5H_{10} \xrightarrow{c} C_6H_6 + (c\text{-}C_5H_9^{\cdot})$				
Phot. of C_6H_5I in presence of hexa-n-butyl tin				77 Sca 1
SESR	[19]	243	$k_a/k_b/k_c = 3.1/70/1.0$	
			$k_a \approx 6 \cdot 10^6\ M^{-1}s^{-1}$ [20]	
			$k_b = 1.4 \cdot 10^8\ M^{-1}s^{-1}$ [20]	
			$k_c = 2 \cdot 10^6\ M^{-1}s^{-1}$ [20]	
		183...263	$\log[A(a)/A(c)] = -0.24(20)$ $E_a(a) - E_a(c) = 2.6(10)\ kJ\ mol^{-1}$	

[12] From [78 Lor 1].
[13] Rate constants calculated from experimental product yields and diffusion equations (Wilke-Chang and Smoluchowski); solutions containing 0.0207 M PAT and 0.050 M CBr_4.
[14] Rate constant based on $k(H)(C_6H_5^{\cdot}$ + aliphatic secondary C—H$) = 3.3 \cdot 10^5\ M^{-1}s^{-1}$, $k(Br)/k(Cl) = 1500$ and $k(Br)(C_6H_5^{\cdot} + CBr_4) = 4.0 \cdot 10^9\ M^{-1}s^{-1}$.
[15] Rate constant based on $k(H)(C_6H_5^{\cdot}$ + aliphatic secondary C—H$) = 3.3 \cdot 10^5\ M^{-1}s^{-1}$, $k(Br)/k(Cl) = 1500$ and $k(Br)(C_6H_5^{\cdot} + CBr_4) = 5.0 \cdot 10^9\ M^{-1}s^{-1}$.
[16] Based on $k(C_6H_5^{\cdot}$ + aliphatic secondary C—H$) = 3.3 \cdot 10^5\ M^{-1}s^{-1}$ and $k(Cl)/k(C_6H_5^{\cdot} + C_6H_6)\cdot$(addition) = 5.6.
[17] Based on $k(H)(C_6H_5^{\cdot}$ + aliphatic secondary C—H$) = 3.3 \cdot 10^5\ M^{-1}s^{-1}$ and $k(H)/k(Cl) = 0.090$ at 333 K [63 Bri 1].
[18] Based on $k = 3.3 \cdot 10^5\ M^{-1}s^{-1}$ for $(C_6H_5^{\cdot})$ + aliphatic secondary (>CH—H).
[19] Not given (presumed: mixture of c-C_5H_{10}, $(CH_3)_2Se$ and $(C_2H_5)_2Se$).
[20] Assumed.

Reaction Radical generation Method	Solvent	T[K]	Rate data	Ref./ add. ref.
$(C_6H_5^{\cdot}) + (CH_3)_2CHI \longrightarrow C_6H_5I + (CH_3)_2\dot{C}H$				
Thermal decomp. of PAT				77 Kry 1
PR, glc	CCl_4	318	$k(I) = 1.19 \cdot 10^9\ M^{-1}\,s^{-1}$ [21] $1.27 \cdot 10^9$ [22]	
	C_6H_6	318	$k(I) = 8.3 \cdot 10^8\ M^{-1}\,s^{-1}$ [23]	
PR	CCl_4	318	$k = 1.1 \cdot 10^9\ M^{-1}\,s^{-1}$ [24]	78 Lor 1
$(C_6H_5^{\cdot}) + (CH_3)_2CHOH \longrightarrow C_6H_6 + (CH_3)_2\dot{C}OH$				
Thermal decomp. of PAT				75 Jan 1
KESR	$(CH_3)_2CHOH$	RT	$k = 4.1(1) \cdot 10^5\ M^{-1}\,s^{-1}$	
$(C_6H_5^{\cdot}) + (CH_3)_3COH \longrightarrow C_6H_6 + \dot{C}H_2C(CH_3)_2OH$				
$Ti(III) + C_6H_5N_2^+BF_4^-$ in H_2O				77 Ash 1
SESR	H_2O, pH 8.0, 9.4	RT	$k > 3 \cdot 10^5\ M^{-1}\,s^{-1}$	
$(C_6H_5^{\cdot}) + C_6H_5SH \longrightarrow C_6H_6 + C_6H_5\dot{S}$				
Thermal decomp. of PAT				77 Kry 1, 78 Lor 1/
PR, glc	CCl_4	318	$k(C_6H_5SH) = 1.9 \cdot 10^9\ M^{-1}\,s^{-1}$ [25]	63 Bri 1
$(C_6H_5^{\cdot})$ + aliphatic primary C—H bond $\longrightarrow C_6H_6$ + products				
Thermal decomp. of PAT				77 Kry 1, 78 Lor 1/
PR, glc	CCl_4	318	$k_{prim} = 3.5 \cdot 10^4\ M^{-1}\,s^{-1}$ [26]	63 Bri 1
$(C_6H_5^{\cdot})$ + aliphatic secondary C—H bond $\longrightarrow C_6H_6$ + products				77 Kry 1, 78 Lor 1/
Thermal decomp. of PAT				
PR, glc, calc.	mineral oil/ chlorobenzene	318	$k(H) = 3.3(7) \cdot 10^5\ M^{-1}\,s^{-1}$ [27]	62 Mac 1, 62 Dun 1, 67 Det 1, 71 Pac 1
PR, glc, calc.	chlorobenzene + Vol % mineral oil, η[cP]	318		77 Kry 1
	20 0.829		$k = 4.57 \cdot 10^5$ [28]; $14.9 \cdot 10^5$ [29] $M^{-1}s^{-1}$	
	30 1.03		$3.81 \cdot 10^5$ $14.2 \cdot 10^5$	
	40 1.29		$3.95 \cdot 10^5$ $10.6 \cdot 10^5$	
	50 1.73		$4.03 \cdot 10^5$ $9.14 \cdot 10^5$	
	60 2.41		$3.77 \cdot 10^5$ $7.41 \cdot 10^5$	
	70 3.63		$3.54 \cdot 10^5$ $5.49 \cdot 10^5$	
	75 4.72		$3.18 \cdot 10^5$ $5.13 \cdot 10^5$	
	80 6.32		$3.78 \cdot 10^5$ $4.62 \cdot 10^5$	
	83.3 7.74		$3.55 \cdot 10^5$ $4.22 \cdot 10^5$	
	86.7 9.85		$3.56 \cdot 10^5$ $3.76 \cdot 10^5$	
	90 12.74		$3.09 \cdot 10^5$ $2.85 \cdot 10^5$	
	92.5 15.70		$2.59 \cdot 10^5$ $3.20 \cdot 10^5$	
(continued)	95 19.48		$2.12 \cdot 10^5$ $2.96 \cdot 10^5$	

21) Based on $k(Br)(C_6H_5^{\cdot} + CBr_4) = 5.0 \cdot 10^9\ M^{-1}\,s^{-1}$ and $k(Br)/k(I) = 4.2$.
22) Based on $k(Cl)(C_6H_5^{\cdot} + CCl_4) = 3.7 \cdot 10^6\ M^{-1}\,s^{-1}$ and $k(I)/k(Cl) = 343$.
23) Based on $k(Br)(C_6H_5^{\cdot} + CBr_4) = 5.0 \cdot 10^9\ M^{-1}\,s^{-1}$ and $k(Br)/k(I) = 6.0$.
24) Based on $k = 3.3 \cdot 10^5\ M^{-1}\,s^{-1}$ for $(C_6H_5^{\cdot})$ + aliphatic secondary (—CH—H).
25) Based on $k(H)(C_6H_5^{\cdot}$ + aliphatic secondary C—H) $= 3.3 \cdot 10^5\ M^{-1}\,s^{-1}$ and $k(C_6H_5SH)/k(H) = 5800$ at 333 K [63 Bri 1].
26) Rate constant per C—H bond. Based on $k(H)(C_6H_5^{\cdot}$ + aliphatic secondary C—H) $= 3.5 \cdot 10^5\ M^{-1}\,s^{-1}$ and $k(H)/k_{prim} = 9.3$ at 333 K [63 Bri 1].
27) Rate constant per C—H bond, calculated on the base of diffusion equations (Wilke-Chang, Smoluchowski) and experimental data.
28) Apparent rate constants calculated from experimental product yields and diffusion equations (Wilke-Chang and Smoluchowski); solutions containing 0.0196 M PAT and 0.050 M I_2.
29) Solutions containing 0.0207 M PAT and 0.050 M CBr_4.

Reaction Radical generation Method	Solvent	T[K]	Rate data	Ref./ add. ref.
$(C_6H_5^{\bullet})$ + aliphatic secondary C—H bond ⟶ C_6H_6 + products (continued)				
Thermal decomp. of PAT				
PR, glc, calc.	chlorobenzene + 10 Vol % mineral oil,	318		77 Kry 1
	η = 0.694 cP		$k = 44.7 \cdot 10^5\ M^{-1}\,s^{-1}$ [30]	
	20 0.829		$22.3 \cdot 10^5$	
	30 1.03		$16.9 \cdot 10^5$	
	50 1.73		$14.0 \cdot 10^5$	
	60 2.41		$9.87 \cdot 10^5$	
	70 3.63		$6.82 \cdot 10^5$	
	80 6.32		$5.41 \cdot 10^5$	
	90 12.74		$4.07 \cdot 10^5$	
$(C_6H_5^{\bullet})$ + aliphatic tertiary C—H bond ⟶ C_6H_6 + products				
Thermal decomp. of PAT				77 Kry 1,
PR, glc	CCl_4	318	$k_{tert} = 1.6 \cdot 10^6\ M^{-1}\,s^{-1}$ [31]	78 Lor 1/ 63 Bri 1
$(C_6H_5^{\bullet})$ + benzylic primary C—H bond ⟶ C_6H_6 + products				
Thermal decomp. of PAT				77 Kry 1,
PR, glc	CCl_4	318	$k_{benz} = 3.3 \cdot 10^5\ M^{-1}\,s^{-1}$ [32]	78 Lor 1/ 63 Bri 1
4-$CH_3C_6H_4^{\bullet}$ + CH_3OH ⟶ $C_6H_5CH_3$ + $\dot{C}H_2OH$				
Pulse rad. of 4-$CH_3C_6H_4N_2^+BF_4^-$ + CH_3OH + H_2O				71 Pac 1,
KAS	H_2O	RT	$k \geqslant 2.8 \cdot 10^5\ M^{-1}\,s^{-1}$ [33]	75 Pac 1
$C_6H_5\dot{C}H_2$ + Cl_2 ⟶ $C_6H_5CH_2Cl$ + $\dot{C}l$				
Not given				78 Kos 1
Not given	not given	not given	$k = 6.7 \cdot 10^5\ M^{-1}\,s^{-1}$ [34]	
$C_6H_5\dot{C}H_2$ + $(CH_2CO)_2NBr$ ⟶ $C_6H_5CH_2Br$ + $(CH_2CO)_2\dot{N}$				
$(CH_2CO)_2\dot{N}$ + $(C_6H_5CH_2)_4Sn$ ⟶ $C_6H_5\dot{C}H_2$ + $(CH_2CO)_2NSn(C_6H_5CH_2)_3$				72 Dav 1
Time-resolved NMR	acetone	308	$k = 5 \cdot 10^5\ M^{-1}\,s^{-1}$	
$C_6H_5\dot{C}H_2$ + $(CH_3)_3COCl$ ⟶ $C_6H_5CH_2Cl$ + $(CH_3)_3C\dot{O}$				
Phot. of AIBN (initiator) in toluene containing soln.				72 Zav 1
PR [35]	CCl_4 or $CF_2ClCFCl_2$/toluene	313.00(5)	$k \approx 6.3 \cdot 10^4\ M^{-1}\,s^{-1}$ [36]	
$C_6H_5\dot{C}H_2$ + $C_6H_5CH_2SH$ ⟶ $C_6H_5CH_3$ + $C_6H_5CH_2\dot{S}$				
$C_6H_5CH_2\dot{S}$ + $P(OC_2H_5)_3$ ⟶ $C_6H_5\dot{C}H_2$ + $SP(OC_2H_5)_3$ (AIBN initiated)				69 Bur 1/
Potentiometric	C_6H_6	298	$k = 2.2 \cdot 10^4\ M^{-1}\,s^{-1}$ [37]	68 Bur 1
titration	c-C_6H_{12}	298	$2.3 \cdot 10^4$ [37]	

[30]) Apparent rate constants calculated from experimental product yields and diffusion equations (Wilke-Chang and Smoluchowski); solution containing 0.026 M PAT and 0.040 M CBr_4.
[31]) Rate constant per C—H bond, based on $k(H)(C_6H_5^{\bullet}$ + aliphatic secondary C—H) = $3.3 \cdot 10^5\ M^{-1}\,s^{-1}$ and $k_{tert}/k(H) = 4.8$ at 333 K [63 Bri 1].
[32]) Rate constant per C—H bond, based on $k(H)(C_6H_5^{\bullet}$ + aliphatic secondary C—H) = $3.3 \cdot 10^5\ M^{-1}\,s^{-1}$ and $k(H)/k_{benz} = 1.0$ at 333 K [63 Bri 1].
[33]) Twice the value from [71 Pac 1] as required in [75 Pac 1].
[34]) Based on various experimental data and assuming $2k_2 = 5.6 \cdot 10^9\ M^{-1}\,s^{-1}$ for 2 $C_6H_5\dot{C}H_2$ ⟶ products.
[35]) Photometric determination of $(CH_3)_3COCl$.
[36]) Based on $k/(2k_2)^{\frac{1}{2}} = 0.82$ ($CF_2ClCFCl_2$ containing soln.) and $1.07\ M^{-1}$ (CCl_4 containing soln.), and assumed $2k_2 = 4 \cdot 10^9\ M^{-1}\,s^{-1}$ for 2 $C_6H_5\dot{C}H_2$ ⟶ products.
[37]) Based on $k/(2k_2)^{\frac{1}{2}} = 0.53$ (in C_6H_6) and 0.51 (in c-C_6H_{12}) measurements with $2k_2 = 1.8 \cdot 10^9$ and $2.0 \cdot 10^9\ M^{-1}\,s^{-1}$ referring to 2 $C_6H_5\dot{C}H_2$ ⟶ products, in respective solvents.

Reaction Radical generation Method	Solvent	T[K]	Rate data	Ref./ add. ref.
$C_6H_5\dot{C}H_2 + CH_3(CH_2)_{11}SH \longrightarrow C_6H_5CH_3 + CH_3(CH_2)_{11}\dot{S}$				
Phot. of $C_6H_5COCH_2C_6H_5$				74 Hei 1
PR, glc	C_6H_6	296(2)	$k = 3.5 \cdot 10^4\ M^{-1}s^{-1}$ [38]	
$C_6H_5\dot{C}H_2 + (n\text{-}C_4H_9)_3SnH \longrightarrow C_6H_5CH_3 + (n\text{-}C_4H_9)_3\dot{S}n$				
Laser flash phot. of $\{(CH_3)_3CO\}_2 + (C_6H_5CH_2)_3P$ + 2,2,4-trimethylpentane				81 Cha 1
KAS	2,2,4-trimethylpentane	RT	$k \leqslant 3 \cdot 10^5\ M^{-1}s^{-1}$	
$C_6H_5\dot{C}(CH_3)_2 + CH_3(CH_2)_{11}SH \longrightarrow C_6H_5CH(CH_3)_2 + CH_3(CH_2)_{11}\dot{S}$				
Phot. of $C_6H_5COC(CH_3)_2C_6H_5$				74 Hei 1
PR, glc	C_6H_6	296(2)	$k = 8.3 \cdot 10^3\ M^{-1}s^{-1}$ [38]	
$(C_6H_5)_3\dot{C} + C_6H_5SH \longrightarrow (C_6H_5)_3CH + C_6H_5\dot{S}$				
Decomp. of 1-diphenylmethylene-4-triphenylmethyl-2-cyclohexadiene				79 Col 1/
KAS [39]	toluene	284.1	$k = 3.08\ M^{-1}s^{-1}$	78 Col 1
		292.4	5.6	
		314.3	15.6	
			$\log[A/M^{-1}s^{-1}] = 7.84$	
			$E_a = 39.9(6)\ kJ\ mol^{-1}$	
$(C_6H_5)_3\dot{C}$ + 2,4,6-$(CH_3)_3C_6H_2SH \longrightarrow (C_6H_5)_3CH$ + 2,4,6-$(CH_3)_3C_6H_2\dot{S}$				
Decomp. of 1-diphenylmethylene-4-triphenylmethyl-2-cyclohexadiene				79 Col 1
KAS [39]	toluene	284.2	$k = 4.4\ M^{-1}s^{-1}$	
		292.8	7.0	
		303.1	11.5	
		315.1	20.8	
			$\log[A/M^{-1}s^{-1}] = 7.50$	
			$E_a = 37.3(5)\ kJ\ mol^{-1}$	
$(C_6H_5)_3\dot{C}$ + bis-(3-chlorobenzoyl)peroxide $\longrightarrow$ 3-$ClC_6H_4COOC(C_6H_5)_3$ + 3-$ClC_6H_4CO\dot{O}$				
Dissociation of $(C_6H_5)_3CC(C_6H_5)_3$				67 Sue 1
KAS	C_6H_6	288	$k = 10.00(22)\ M^{-1}s^{-1}$	
		298	15.3(15)	
			$\Delta H^{\ddagger} = 28.1(172)\ kJ\ mol^{-1}$	
			$\Delta S^{\ddagger} = -130.0(502)\ J\ mol^{-1}K^{-1}$	
$(C_6H_5)_3\dot{C}$ + bis-(4-chlorobenzoyl)peroxide $\longrightarrow$ 4-$ClC_6H_4COOC(C_6H_5)_3$ + 4-$ClC_6H_4CO\dot{O}$				
Dissociation of $(C_6H_5)_3CC(C_6H_5)_3$				67 Sue 1
KAS	C_6H_6	288	$k = 2.900(75)\ M^{-1}s^{-1}$	
		298	5.900(87)	
		308	8.72(69)	
			$\Delta H^{\ddagger} = 38.5(71)\ kJ\ mol^{-1}$	
			$\Delta S^{\ddagger} = -100.5(251)\ J\ mol^{-1}K^{-1}$	
$(C_6H_5)_3\dot{C}$ + bis-(4-fluorobenzoyl)peroxide $\longrightarrow$ 4-$FC_6H_4COOC(C_6H_5)_3$ + 4-$FC_6H_4CO\dot{O}$				
Dissociation of $(C_6H_5)_3CC(C_6H_5)_3$				67 Sue 1
KAS	C_6H_6	288	$k = 1.38(10)\ M^{-1}s^{-1}$	
		298	2.340(87)	
		308	3.66(23)	
			$\Delta H^{\ddagger} = 33.5(84)\ kJ\ mol^{-1}$	
			$\Delta S^{\ddagger} = -125.6(293)\ J\ mol^{-1}K^{-1}$	

[38]) Steady-state calculation.
[39]) Photometric determination of steady-state radical concentration.

Reaction Radical generation Method	Solvent	T[K]	Rate data	Ref./ add. ref.
$(C_6H_5)_3\dot{C}$ + dibenzoylperoxide ⟶ $C_6H_5COOC(C_6H_5)_3 + C_6H_5CO\dot{O}$				
Dissociation of $(C_6H_5)_3CC(C_6H_5)_3$				67 Sue 1
KAS	anisole	288	$k = 1.190(67)\ M^{-1}\,s^{-1}$	
		298	2.26(10)	
		308	3.66(14)	
			$\Delta H^{\ddagger} = 39.4(59)\ kJ\,mol^{-1}$	
			$\Delta S^{\ddagger} = -108.9(209)\ J\,mol^{-1}\,K^{-1}$	
$(C_6H_5)_3\dot{C}$ + dibenzoylperoxide ⟶ $C_6H_5COOC(C_6H_5)_3 + C_6H_5CO\dot{O}$				
Dissociation of $(C_6H_5)_3CC(C_6H_5)_3$				67 Sue 1
KAS	C_6H_6	288	$k = 0.4950(87)\ M^{-1}\,s^{-1}$	
		298	1.160(64)	
		308	1.72(11)	
			$\Delta H^{\ddagger} = 43.5(54)\ kJ\,mol^{-1}$	
			$\Delta S^{\ddagger} = -100.5(209)\ J\,mol^{-1}\,K^{-1}$	
$(C_6H_5)_3\dot{C}$ + dibenzoylperoxide ⟶ $C_6H_5COOC(C_6H_5)_3 + C_6H_5CO\dot{O}$				
Dissociation of $(C_6H_5)_3CC(C_6H_5)_3$				67 Sue 1
KAS	chlorobenzene	288	$k = 0.990(33)\ M^{-1}\,s^{-1}$	
		298	2.400(69)	
		308	3.40(14)	
			$\Delta H^{\ddagger} = 38.9(46)\ kJ\,mol^{-1}$	
			$\Delta S^{\ddagger} = -108.9(168)\ J\,mol^{-1}\,K^{-1}$	
$(C_6H_5)_3\dot{C}$ + dibenzoylperoxide ⟶ $C_6H_5COOC(C_6H_5)_3 + C_6H_5CO\dot{O}$				
Dissociation of $(C_6H_5)_3CC(C_6H_5)_3$				67 Sue 1
KAS	nitrobenzene	288	$k = 4.250(70)\ M^{-1}\,s^{-1}$	
		298	6.58(13)	
		308	8.53(14)	
			$\Delta H^{\ddagger} = 23.4(20)\ kJ\,mol^{-1}$	
			$\Delta S^{\ddagger} = -150.7(84)\ J\,mol^{-1}\,K^{-1}$	
$(C_6H_5)_3\dot{C}$ + bis-(3-methylbenzoyl)peroxide ⟶ $3\text{-}CH_3C_6H_4COOC(C_6H_5)_3 + 3\text{-}CH_3C_6H_4CO\dot{O}$				
Dissociation of $(C_6H_5)_3CC(C_6H_5)_3$				67 Sue 1
KAS	C_6H_6	288	$k = 0.575(4)\ M^{-1}\,s^{-1}$	
		298	0.688(27)	
		308	1.280(39)	
			$\Delta H^{\ddagger} = 27.2(27)\ kJ\,mol^{-1}$	
			$\Delta S^{\ddagger} = -159.0(84)\ J\,mol^{-1}\,K^{-1}$	
$(C_6H_5)_3\dot{C}$ + bis-(4-methylbenzoyl)peroxide ⟶ $4\text{-}CH_3C_6H_4COOC(C_6H_5)_3 + 4\text{-}CH_3C_6H_4CO\dot{O}$				
Dissociation of $(C_6H_5)_3CC(C_6H_5)_3$				67 Sue 1
KAS	C_6H_6	288	$k = 0.195(17)\ M^{-1}\,s^{-1}$	
		298	0.418(34)	
		308	0.656(58)	
			$\Delta H^{\ddagger} = 42.7(105)\ kJ\,mol^{-1}$	
			$\Delta S^{\ddagger} = -113.0(335)\ J\,mol^{-1}\,K^{-1}$	
$(C_6H_5)_3\dot{C}$ + bis-(3-methylbenzoyl)peroxide ⟶ $3\text{-}CH_3C_6H_4COOC(C_6H_5)_3 + 3\text{-}CH_3C_6H_4CO\dot{O}$				
Dissociation of $(C_6H_5)_3CC(C_6H_5)_3$				67 Sue 1
KAS	anisole	298	$k = 1.340(85)\ M^{-1}\,s^{-1}$	
	chlorobenzene	298	$k = 1.770(85)\ M^{-1}\,s^{-1}$	
	nitrobenzene	298	$k = 7.46(23)\ M^{-1}\,s^{-1}$	

Reaction Radical generation Method	Solvent	T[K]	Rate data	Ref./ add. ref.
$(C_6H_5)_3\dot{C}$ + bis-(3-methoxybenzoyl)peroxide $\longrightarrow$ 3-$CH_3OC_6H_4COOC(C_6H_5)_3$ + 3-$CH_3OC_6H_4CO\dot{O}$				
Dissociation of $(C_6H_5)_3CC(C_6H_5)_3$				67 Sue 1
KAS	C_6H_6	288	$k = 1.090(58)\,M^{-1}\,s^{-1}$	
		298	1.480(36)	
		308	2.54(27)	
			$\Delta H^{\ddagger} = 28.9(100)\,kJ\,mol^{-1}$	
			$\Delta S^{\ddagger} = -146.5(335)\,J\,mol^{-1}\,K^{-1}$	
$(C_6H_5)_3\dot{C}$ + bis-(4-methoxybenzoyl)peroxide $\longrightarrow$ $CH_3OC_6H_4COOC(C_6H_5)_3$ + $CH_3OC_6H_4CO\dot{O}$				
Dissociation of $(C_6H_5)_3CC(C_6H_5)_3$				67 Sue 1
KAS	C_6H_6	288	$k = 0.1020(17)\,M^{-1}\,s^{-1}$	
		298	0.1970(64)	
		308	0.312(12)	
			$\Delta H^{\ddagger} = 39.4(36)\,kJ\,mol^{-1}$	
			$\Delta S^{\ddagger} = -125.0(125)\,J\,mol^{-1}\,K^{-1}$	
$(C_6H_5)_3\dot{C}$ + bis-(4-methoxybenzoyl)peroxide $\longrightarrow$ 4-$CH_3OC_6H_4COOC(C_6H_5)_3$ + 4-$CH_3OC_6H_4CO\dot{O}$				
Dissociation of $(C_6H_5)_3CC(C_6H_5)_3$				67 Sue 1
KAS	anisole	298	$k = 0.1560(31)\,M^{-1}\,s^{-1}$	
	chlorobenzene	298	$k = 0.2010(20)\,M^{-1}\,s^{-1}$	
	nitrobenzene	298	$k = 0.7230(70)\,M^{-1}\,s^{-1}$	
$(4\text{-}(CH_3)_3C\text{—}C_6H_4)_3\dot{C}$ + 3-ClC_6H_4SH $\longrightarrow$ $(4\text{-}(CH_3)_3C\text{—}C_6H_4)_3CH$ + 3-$ClC_6H_4\dot{S}$				
React. of tris-*p*-*t*-butylphenylmethylchloride with molecular silver				79 Col 1
KAS [39])	toluene	265.4	$k = 2.00\,M^{-1}\,s^{-1}$	
		272.8	3.45	
		283.0	7.57	
		293.2	13.6	
		304.3	24.1	
		314.3	45.0	
			$\log[A/M^{-1}\,s^{-1}] = 8.77$	
			$E_a = 42.9(5)\,kJ\,mol^{-1}$	
$(4\text{-}(CH_3)_3C\text{—}C_6H_4)_3\dot{C}$ + 4-ClC_6H_4SH $\longrightarrow$ $(4\text{-}(CH_3)_3C\text{—}C_6H_4)_3CH$ + 4-$ClC_6H_4\dot{S}$				
React. of tris-*p*-*t*-butylphenylmethylchloride with molecular silver				79 Col 1
KAS [39])	toluene	264.7	$k = 2.50\,M^{-1}\,s^{-1}$	
		273.5	5.0	
		282.8	10.3	
		303.6	33.0	
		313.1	51.4	
			$\log[A/M^{-1}\,s^{-1}] = 8.96$	
			$E_a = 43.3(7)\,kJ\,mol^{-1}$	
$(4\text{-}(CH_3)_3C\text{—}C_6H_4)_3\dot{C}$ + C_6H_5SH $\longrightarrow$ $(4\text{-}(CH_3)_3C\text{—}C_6H_4)_3CH$ + $C_6H_5\dot{S}$				
React. of tris-*p*-*t*-butylphenylmethylchloride with molecular silver				79 Col 1
KAS [39])	toluene	271.3	$k = 2.70\,M^{-1}\,s^{-1}$	
		285.3	5.74	
		298.2	11.18	
		304.8	20.0	
		314.8	28.5	
			$\log[A/M^{-1}\,s^{-1}] = 7.87$	
			$E_a = 38.7(10)\,kJ\,mol^{-1}$	

[39]) Photometric determination of steady-state radical concentration.

Reaction Radical generation Method	Solvent	T[K]	Rate data	Ref./ add. ref.
$(4\text{-}(CH_3)_3C{-}C_6H_4)_3\dot{C} + 4\text{-}CF_3C_6H_4SH \longrightarrow (4\text{-}(CH_3)_3C{-}C_6H_4)_3CH + 4\text{-}CF_3C_6H_4\dot{S}$				
React. of tris-*p*-*t*-butylphenylmethylchloride with molecular silver				79 Col 1
KAS [39]	toluene	264.1	$k = 0.94\,M^{-1}s^{-1}$	
		273.4	1.90	
		282.7	3.92	
		294.3	9.46	
			$\log[A/M^{-1}s^{-1}] = 9.48$	
			$E_a = 48.1(8)\,kJ\,mol^{-1}$	
$(4\text{-}(CH_3)_3C{-}C_6H_4)_3\dot{C} + 4\text{-}CH_3OC_6H_4SH \longrightarrow (4\text{-}(CH_3)_3C{-}C_6H_4)_3CH + 4\text{-}CH_3OC_6H_4\dot{S}$				
React. of tris-*p*-*t*-butylphenylmethylchloride with molecular silver				79 Col 1
KAS [39]	toluene	273.9	$k = 18.4\,M^{-1}s^{-1}$	
		283.6	28.2	
		293.2	44.4	
		303.7	65.0	
			$\log[A/M^{-1}s^{-1}] = 7.07$	
			$E_a = 30.5(9)\,kJ\,mol^{-1}$	
$(4\text{-}(CH_3)_3C{-}C_6H_4)_3\dot{C} + 4\text{-}(CH_3)_3C{-}C_6H_4SH \longrightarrow (4\text{-}(CH_3)_3C{-}C_6H_4)_3CH + 4\text{-}(CH_3)_3C{-}C_6H_4\dot{S}$				
React. of tris-*p*-*t*-butylphenylmethylchloride with molecular silver				79 Col 1
KAS [39]	toluene	273.8	$k = 2.57\,M^{-1}s^{-1}$	
		283.9	5.02	
		298.4	12.9	
			$\log[A/M^{-1}s^{-1}] = 9.09$	
			$E_a = 45.6(7)\,kJ\,mol^{-1}$	

4.1.1.2.2 Radicals containing C, H, and other atoms

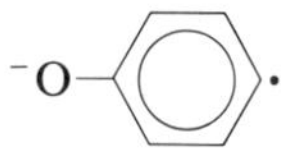 $+ (CH_3)_3COH \longrightarrow C_6H_5O^- + \dot{C}H_2C(CH_3)_2OH$

Rad. of *p*-bromophenol + N_2 + H_2O				76 Sch 1
KAS, Cond.	H_2O, pH = 11.5	RT	$k = 6 \cdot 10^5$	

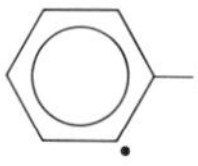 $-OH + (CH_3)_3COH \longrightarrow C_6H_5OH + \dot{C}H_2C(CH_3)_2OH$

Rad. of *o*-bromophenol + N_2 + H_2O				76 Sch 1
KAS, Cond.	H_2O	RT	$k \approx 3 \cdot 10^5\,M^{-1}s^{-1}$	

 $-OH + (CH_3)_2CHOH \longrightarrow C_6H_5OH + (CH_3)_2\dot{C}OH$

γ-rad. of *p*-bromophenol + H_2O				73 Bha 1
PR	H_2O	RT	$k = 3 \cdot 10^7\,M^{-1}s^{-1}$ [40]	

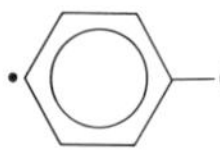 $-COO^- + CH_3OH \longrightarrow C_6H_5COO^- + \dot{C}H_2OH$

Pulse rad. of *p*-bromobenzoate + H_2O				78 Mad 1/
KAS	H_2O	RT	$k = 4.0(8) \cdot 10^5\,M^{-1}s^{-1}$ [41]	75 Jan 1

[39]) Photometric determination of steady-state radical concentration.
[40]) Based on $k(C_6H_4OH^{\cdot} + p\text{-}BrC_6H_4OH) = 7 \cdot 10^7\,M^{-1}s^{-1}$.
[41]) $k = 4.3(4) \cdot 10^5\,M^{-1}s^{-1}$ from product ratio measurements.

Reaction Radical generation Method	Solvent	T[K]	Rate data	Ref./ add. ref.
•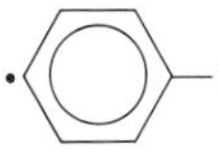$-COO^- + C_2H_5OH \longrightarrow C_6H_5COO^- + CH_3\dot{C}HOH$				
Pulse rad. of p-bromobenzoate + H_2O				78 Mad 1/
KAS	H_2O	RT	$k = 2.0(5) \cdot 10^6\,M^{-1}\,s^{-1}$ [42]	75 Jan 1
•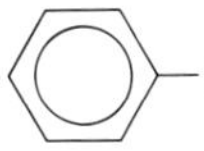$-COO^- + (CH_3)_2CHOH \longrightarrow C_6H_5COO^- + (CH_3)_2\dot{C}OH$				
Pulse rad. of p-bromobenzoate + H_2O				78 Mad 1/
KAS	H_2O	RT	$k = 5.5(10) \cdot 10^6\,M^{-1}\,s^{-1}$ [43]	75 Jan 1
•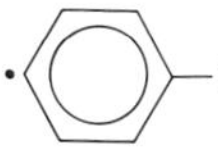$-COO^- + (CH_3)_3COH \longrightarrow C_6H_5COO^- + \dot{C}H_2C(CH_3)_2OH$				
Pulse rad. of p-bromobenzoate + H_2O				78 Mad 1
KAS	H_2O	RT	$k = 3.2(6) \cdot 10^5\,M^{-1}\,s^{-1}$ [44]	
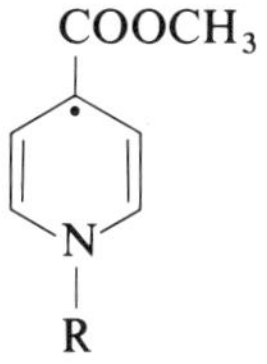 **Ȧ** $+ CH_2BrCl \longrightarrow$ 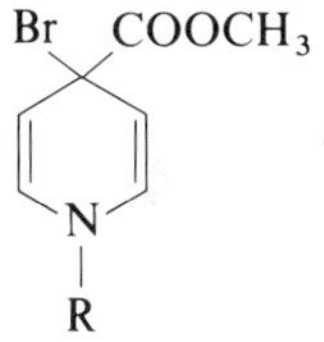 **B** [45] $+ \dot{C}H_2Cl$			$R = CH_3$	
Reduct. of 1-methyl-4-carbomethoxypyridinium iodide by Na-amalgam				78 Kos 2/
KAS	CH_3CN	298	$k = 5.07 \cdot 10^{-5}\,M^{-1}\,s^{-1}$	71 Moh 1
Ȧ $+ CCl_4 \longrightarrow$ 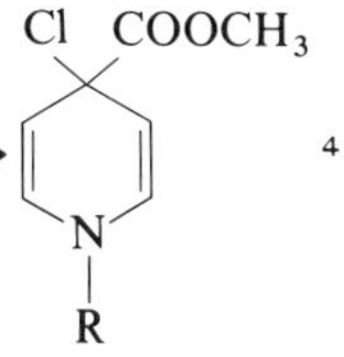 **C** [45] $+ \dot{C}Cl_3$			$R = C_2H_5$	
Reduct. of 1-ethyl-4-carbomethoxypyridinium iodide by Zn				64 Kos 1/
KAS	CH_3CN	298	$k = 3.3\,M^{-1}\,s^{-1}$	71 Moh 1
Ȧ $+ CDCl_3 \longrightarrow$ **C** [45] $+ \dot{C}DCl_2$			$R = C_2H_5$	
Reduct. of 1-ethyl-4-carbomethoxypyridinium iodide by Zn				64 Kos 1/
KAS	CH_3CN	298	$k = 1.72 \cdot 10^{-5}\,M^{-1}\,s^{-1}$	71 Moh 1
Ȧ $+ CHCl_3 \longrightarrow$ **C** [45] $+ \dot{C}HCl_2$			$R = C_2H_5$	
Reduct. of 1-ethyl-4-carbomethoxypyridinium iodide by Zn				74 Kos 1/
KAS	CH_3CN	288	$k = 7.3 \cdot 10^{-6}\,M^{-1}\,s^{-1}$	71 Moh 1
		298	$2.09 \cdot 10^{-5}$	
		308	$5.8 \cdot 10^{-5}$	
			$E_a = 80(4)\,kJ\,mol^{-1}$	

[42]) $k = 2.4(3) \cdot 10^6\,M^{-1}\,s^{-1}$ from product ratio measurements.
[43]) $k = 5.2(5) \cdot 10^6\,M^{-1}\,s^{-1}$ from product ratio measurements, $k = 6.2(12) \cdot 10^6\,M^{-1}\,s^{-1}$ from time-resolved ESR measurements.
[44]) $k = 2.1(4) \cdot 10^5\,M^{-1}\,s^{-1}$ from product ratio measurements.
[45]) Product presumably of ionic character [71 Moh 1].

Reaction Radical generation Method	Solvent	T[K]	Rate data	Ref./ add. ref.
$\dot{\mathbf{A}} + CH_2BrCl \longrightarrow \mathbf{B}$ [45]) $+ \dot{C}H_2Cl$ *)			$R = C_2H_5$	
Reduct. of 1-ethyl-4-carbomethoxypyridinium iodide by Zn				64 Kos 1,
KAS	CH_3CN	298	$k = 5.0 \cdot 10^{-5}\,M^{-1}\,s^{-1}$	78 Kos 2/
		303	$1.04 \cdot 10^{-4}$	64 Kos 2,
		308	$1.56 \cdot 10^{-4}$	71 Moh 1
		298... 308	$E_a = 92(8)\,kJ\,mol^{-1}$	
$\dot{\mathbf{A}} + CH_2Br_2 \longrightarrow \mathbf{B}$ [45]) $+ \dot{C}H_2Br$			$R = C_2H_5$	
Reduct. of 1-ethyl-4-carbomethoxypyridinium iodide by Zn				64 Kos 1/
KAS	CH_3CN	298	$k = 9.8 \cdot 10^{-5}\,M^{-1}\,s^{-1}$	64 Kos 2,
		298... 308	$E_a = 75(8)\,kJ\,mol^{-1}$	71 Moh 1
	CH_2Cl_2	298	$k = 4.8 \cdot 10^{-5}\,M^{-1}\,s^{-1}$	
	C_2H_5OH	298	$k \approx 3.3 \cdot 10^{-5}\,M^{-1}\,s^{-1}$	
	$(CH_3)_2CHOH$	298	$k \approx 2.8 \cdot 10^{-5}\,M^{-1}\,s^{-1}$	
$\dot{\mathbf{A}} + CH_2ClI \longrightarrow$ 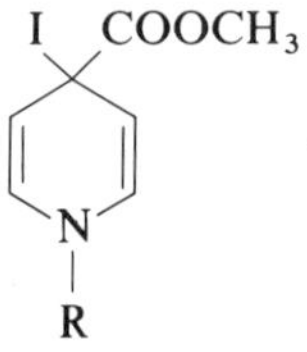 [45]) $+ \dot{C}H_2Cl$ **D**			$R = C_2H_5$	
Reduct. of 1-ethyl-4-carbomethoxypyridinium iodide by Zn				64 Kos 1/
KAS	CH_3CN	298	$k = 1.35 \cdot 10^{-1}\,M^{-1}\,s^{-1}$	71 Moh 1
$\dot{\mathbf{A}} + CH_2Cl_2 \longrightarrow \mathbf{C}$ [45]) $+ \dot{C}H_2Cl$ **)			$R = C_2H_5$	
Reduct. of 1-ethyl-4-carbomethoxypyridinium iodide by Zn				64 Kos 1/
KAS	CH_3CN	298	$k = 2.6(10) \cdot 10^{-8}\,M^{-1}\,s^{-1}$ [46])	64 Kos 2,
		323	$3.1 \cdot 10^{-7}$	71 Moh 1
		333	$5.0 \cdot 10^{-7}$	
		338	$1.1 \cdot 10^{-6}$	
		343	$3.1 \cdot 10^{-6}$	
		348	$2.6 \cdot 10^{-6}$	
		353	$5.0 \cdot 10^{-6}$	
		323... 353	$E_a = 100(8)\,kJ\,mol^{-1}$	
$\dot{\mathbf{A}} + CH_3I \longrightarrow \mathbf{D}$ [45]) $+ \dot{C}H_3$			$R = C_2H_5$	
Reduct. of 1-ethyl-4-carbomethoxypyridinium iodide by Zn				64 Kos 1/
KAS	CH_3CN	293	$k = 2.1 \cdot 10^{-6}\,M^{-1}\,s^{-1}$	71 Moh 1
		298	$4.7 \cdot 10^{-6}$	
		303	$6.6 \cdot 10^{-6}$	
		308	$1.15 \cdot 10^{-5}$	
			$E_a = 80(8)\,kJ\,mol^{-1}$	
$\dot{\mathbf{A}} + 4\text{-}ClC_6H_4CH_2Cl \longrightarrow \mathbf{C}$ [47]) $+ 4\text{-}ClC_6H_4\dot{C}H_2$			$R = C_2H_5$	
Reduct. of 1-ethyl-4-carbomethoxypyridinium iodide by Zn				68 Kos 1/
KAS	CH_3CN	298	$k = 6.5 \cdot 10^{-4}\,M^{-1}s^{-1}$	71 Moh 1

*) For $\dot{\mathbf{A}}$ and **B**, see p. 33.
) For **C, see p. 33.
[45]) Product presumably of ionic character [71 Moh 1].
[46]) Extrapolated value.
[47]) Product presumed to have ionic character (Py^+Cl^-) [71 Moh 1].

Reaction Radical generation Method	Solvent	T[K]	Rate data	Ref./ add. ref.
$\dot{A} + C_6H_5CH_2Br$ B[48]) $+ C_6H_5\dot{C}H_2$ *)			R = C_2H_5	
Reduct. of 1-ethyl-4-carbomethoxypyridinium iodide by Zn				71 Moh 1
KAS	CH_3CN	298	$k = 0.45\,M^{-1}\,s^{-1}$	
	1,2-dimethoxyethane		$k = 1.5(3)\cdot10^{-2}\,M^{-1}\,s^{-1}$	
$\dot{A} + C_6H_5CH_2Cl \longrightarrow$ C[47]) $+ C_6H_5\dot{C}H_2$ *)			R = C_2H_5	
Reduct. of 1-ethyl-4-carbomethoxypyridinium iodide by Zn				68 Kos 1, 71 Moh 1
KAS	CH_3CN	298	$k = 3.31(30)\cdot10^{-4}\,M^{-1}\,s^{-1}$	
	1,2-dimethoxyethane		$k = 5.3(7)\cdot10^{-6}\,M^{-1}\,s^{-1}$	
	CH_3COCH_3		$k = 4.75\cdot10^{-5}\,M^{-1}\,s^{-1}$	
$\dot{A}$ + 4-$CH_3C_6H_4CH_2Cl \longrightarrow$ C[47]) + 4-$CH_3C_6H_4\dot{C}H_2$			R = C_2H_5	
Reduct. of 1-ethyl-4-carbomethoxypyridinium iodide by Zn				68 Kos 1/
KAS	CH_3CN	298	$k = 3.68\cdot10^{-4}\,M^{-1}\,s^{-1}$	71 Moh 1
$\dot{A}$ + 4-$CH_3OC_6H_4CH_2Cl \longrightarrow$ C[47]) + 4-$CH_3OC_6H_4\dot{C}H_2$			R = C_2H_5	
Reduct. of 1-ethyl-4-carbomethoxypyridinium iodide by Zn				68 Kos 1/
KAS	CH_3CN	298	$k = 1.13\cdot10^{-3}\,M^{-1}\,s^{-1}$	71 Moh 1
$\dot{A} + CH_2BrCl \longrightarrow$ B[45]) $+ \dot{C}H_2Cl$			R = $CH(CH_3)_2$	
Reduct. of 1-isopropyl-4-carbomethoxypyridinium iodide by Na-amalgam				78 Kos 2/
KAS	CH_3CN	298	$k = 6.14\cdot10^{-5}\,M^{-1}\,s^{-1}$	71 Moh 1
$\dot{A} + CH_2BrCl \longrightarrow$ B[45]) $+ \dot{C}H_2Cl$			R = $C(CH_3)_3$	
Reduct. of 1-*t*-butyl-4-carbomethoxypyridinium iodide by Na-amalgam				78 Kos 2/
KAS	CH_3CN	298	$k = 7.51\cdot10^{-5}\,M^{-1}\,s^{-1}$	71 Moh 1
$C_6H_5\dot{C}(OH)C(CH_3)_2OCH_3 + c\text{-}C_6H_{12} \longrightarrow C_6H_5CH(OH)C(CH_3)_2OCH_3 + (c\text{-}C_6\dot{H}_{11})$				
Phot. of 1-phenyl-2-methoxy-2-methylpropane in c-C_6H_{12}				80 Eic 1
KAS	c-C_6H_{12}	RT	$k = 8.6(4)\cdot10^{4}\,M^{-1}\,s^{-1}$	
(n-C_4H_9S-styrene˙)[49]) + n-$C_4H_9SH \longrightarrow$ n-$C_4H_9\dot{S}$ + products				
Phot. of azobisisobutyronitrile + n-C_4H_9SH in C_6H_6				59 Siv 1/
Dilatometric measurements	C_6H_6	298	$k = 1.24\cdot10^{3}\,M^{-1}\,s^{-1}$	54 Bac 1
$(C_6H_5)_2\dot{C}OH$ + acrylonitrile $\longrightarrow$ products [50])				
Laser flash phot. of benzophenone				76 Kuh 1
KAS	THF	RT	$k = 3.8(10)\cdot10^{3}\,M^{-1}\,s^{-1}$	
$(C_6H_5)_2\dot{C}OH$ + methylmethacrylate $\longrightarrow$ products [50])				
Laser flash phot. of benzophenone				76 Kuh 1
KAS	THF	RT	$k = 9.0(20)\cdot10^{3}\,M^{-1}\,s^{-1}$	
$(C_6H_5)_2\dot{C}OH$ + vinylacetate $\longrightarrow$ products [50])				
Laser flash phot. of benzophenone				76 Kuh 1
KAS	THF	RT	$k = 5.5(15)\cdot10^{3}\,M^{-1}\,s^{-1}$	

*) For $\dot{A}$, **B** and **C**, see p. 33.
[45]) Product presumably of ionic character [71 Moh 1].
[47]) Product presumed to have ionic character (Py^+Cl^-) [71 Moh 1].
[48]) Product presumed to have ionic character (Py^+Br^-).
[49]) n-$C_4H_9\dot{S}$ adduct to styrene, probably $C_6H_5\dot{C}HCH_2S(CH_2)_3CH_3$.
[50]) $\dot{H}$ atom transfer and addition.

Reaction				
Radical generation				Ref./
Method	Solvent	$T[K]$	Rate data	add. ref.

$(4\text{-}ClC_6H_4)_2CH\dot{C}Cl_2 + (CH_3)_2CHOH \longrightarrow (4\text{-}ClC_6H_4)_2CHCHCl_2 + (CH_3)_2\dot{C}OH$

γ-rad. of $(4\text{-}ClC_6H_4)_2CHCCl_3$ (DDT) + $(CH_3)_2CHOH$				71 Eva 1
HCl titration	2-propanol	RT	$k \leqslant 1\,M^{-1}s^{-1\,51})$	
		(presumed)		

$((C_6H_5)_3CCl)^{\ddot{+}} + (C_6H_5)_3CCl \longrightarrow (C_6H_5)_3C^+ + products$

Pulse rad. of $(C_6H_5)_3CCl + c\text{-}C_6H_{12}$				79 Zad 1,
Time-resolved	$c\text{-}C_6H_{12}$	293	$k = 5(2)\cdot 10^9\,M^{-1}s^{-1}$	69 Cap 1
MW absorpt.				

4.1.1.3 Radicals with undefined stoichiometry and structure

For some explanatory details, see introductory text, and also the text preceding the corresponding section of the electron transfer rate constant compilation, 4.2.1.3

$(\text{Adenosine-5'-monophosphate-OH}^\cdot)^1) + HSCH_2CH(NH_2)COOH \longrightarrow \dot{S}CH_2CH(NH_2)COOH + products$

Pulse rad. of adenosine-5'-monophosphate + N_2O + H_2O				75 Gre 1
KAS	H_2O	RT	$k = 1.5\cdot 10^8\,M^{-1}s^{-1\,2})$	

$(\text{Cytidine-5'-monophosphate-OH}^\cdot)^3) + HSCH_2CH(NH_2)COOH \longrightarrow \dot{S}CH_2CH(NH_2)COOH + products$

Pulse rad. of cytidine-5'-monophosphate + N_2O + H_2O				75 Gre 1
KAS	H_2O	RT	$k = 2.4\cdot 10^8\,M^{-1}s^{-1\,2})$	

$3\text{-Deoxy-3-}\dot{C}\text{-}1,2\text{:}5,6\text{-di-O-isopropylidene-}\alpha\text{-D-glucofuranose}(\dot{R}) + (CH_3)CHOH \longrightarrow$
$3\text{-deoxy-3-CH}\cdots\text{---}(RH) + (CH_3)_2\dot{C}OH$

γ-rad. of 3-deoxy-3-iodo-1,2:5,6-di-O-isopropylidene-α-D-glucofuranose(RI)				82 Lem 1
in alcaline 2-propanol acetone				
Potentiometric	2-propanol/acetone	RT	$k \approx 25\ldots 50\,M^{-1}s^{-1\,4})$	
titration, HPLC	$(10^{-2}\,M\,KOH)$			

$(\text{Dihydrothymine-OH}^\cdot)^5) + HSCH_2CH_2NH_2 \longrightarrow \dot{S}CH_2CH_2NH_2 + products$

Pulse rad. of dihydrothymine + N_2O + H_2O				72 Nuc 1
KAS	H_2O	RT	$k = 1.1\cdot 10^7\,M^{-1}s^{-1\,2})$	

$(\text{Dihydrothymine-OH}^\cdot)^5) + HSCH_2CH(NH_2)COOH \longrightarrow \dot{S}CH_2CH(NH_2)COOH + products$

Pulse rad. of dihydrothymine + N_2O + H_2O				72 Nuc 1
KAS	H_2O	RT	$k = 5.0\cdot 10^7\,M^{-1}s^{-1\,2})$	

$(\text{Glucose-OH}^\cdot)^6) + HSCH_2CH_2NH_2 \longrightarrow RH + \dot{S}CH_2CH_2NH_2$

Pulse rad. of glucose + N_2O + H_2O				68 Ada 1
KAS	H_2O	293	$k = 3.2\cdot 10^7\,M^{-1}s^{-1\,7})$	

---

51) Based on $k/(2k_2)^{\frac{1}{2}} = 2.5\cdot 10^{-5}\,M^{-\frac{1}{2}}s^{-\frac{1}{2}}$ with $2k_2 \leqslant 3\cdot 10^9\,M^{-1}s^{-1}$ referring to $2(4\text{-}ClC_6H_4)_2CH\dot{C}Cl_2 \longrightarrow$ products.
1) Radicals from anedosine-5'-monophosphate + OH reaction.
2) Measured via $RSSR^{\overline{\cdot}}$ absorption $(R\dot{S} + RS^- \rightleftharpoons RSSR^{\overline{\cdot}})$.
3) Radicals from cytidine-5'-monophosphate + $\dot{O}H$ reaction.
4) Based on assumed $2k_2 = (1\ldots 2)\cdot 10^9\,M^{-1}s^{-1}$ for $\dot{R} + \dot{R}$ and $k(RI + (CH_3)_2\dot{C}O^-) \gg k \approx 25\ldots 30\,M^{-1}s^{-1}$.
5) Radicals from dihydrothymine + $\dot{O}H$ reaction.
6) Radicals from glucose + $\dot{O}H$ reaction.
7) Rate constant measured via $RSSR^{\overline{\cdot}}$ formation $(R\dot{S} + RS^- \rightleftharpoons RSSR^{\overline{\cdot}})$.

Reaction Radical generation Method	Solvent	T[K]	Rate data	Ref./ add. ref.
(Guanosine-5′-monophosphate-OH˙)[8]) + cysteine(RSH) RṠ + products				
Pulse rad. of guanosine-5′-monophosphate + N_2O + H_2O				75 Gre 1
KAS	H_2O	RT	$k = 1.8 \cdot 10^8\ M^{-1}\ s^{-1}$ [9])	
R—ĊH($CONH_2$)CH_2—R + Cr^{2+}_{aq} ⟶ Cr(III) + R—CH_2($CONH_2$)CH_2—R [10]) R = polymer chain				
Rad. of polyacrylamide + H_2O [11])				63 Col 1
PR	H_2O	298	$k = 2.8(14) \cdot 10^5\ M^{-1}\ s^{-1}$	
R—ĊH($CONH_2$)CH_2—R + Eu^{2+}_{aq} ⟶ Eu(III) + R—CH_2($CONH_2$)CH_2—R [10]) R = polymer chain				
Rad. of polyacrylamide + H_2O solutions [11])				63 Col 1
PR	H_2O	298	$k = 8(4) \cdot 10^4\ M^{-1}\ s^{-1}$	
R—ĊH($CONH_2$)CH_2—R + Mo^{3+}_{aq} ⟶ Mo(IV) + R—CH_2($CONH_2$)CH_2—R [10]) R = polymer chain				
Rad. of polyacrylamide + H_2O [11])				63 Col 1
PR	H_2O	298	$k = 7.0(7) \cdot 10^3\ M^{-1}\ s^{-1}$	
R—ĊH($CONH_2$)CH_2—R + Ti^{3+}_{aq} ⟶ Ti(IV) + R—CH_2($CONH_2$)CH—R [10]) R = polymer chain				
Rad. of polyacrylamide + H_2O [11])				63 Col 1
PR	H_2O	298	$k = 5.8(3) \cdot 10^2\ M^{-1}\ s^{-1}$	
	D_2O		$k = 8.21(30) \cdot 10^2\ M^{-1}\ s^{-1}$	
R—ĊH($CONH_2$)CH_2—R + V^{2+}_{aq} ⟶ V(III) + R—CH_2($CONH_2$)CH_2—R [10]) R = polymer chain				
Rad. of polyacrylamide + H_2O [11])				63 Col 1
PR	H_2O	298	$k = 1.1(6) \cdot 10^5\ M^{-1}\ s^{-1}$	
R—CH_2ĊHCN [12]) + $CuCl_2$ ⟶ products [13]) R = polymer chain				
AIBN as initiator Inhibition of polymerization	acrylonitrile	333	$k = 2.47 \cdot 10^4\ M^{-1}\ s^{-1}$ [14])	62 Mon 1
R—CH_2ĊHCN [12]) + $FeCl_3$ ⟶ products (R—CH_2CHClCN, R—CH=CHCN, $FeCl_2$, HCl) [13]) R = polymer chain				
AIBN as initiator Inhibition of	acrylonitrile	333	$k = 6.5 \cdot 10^3\ M^{-1}\ s^{-1}$ [14])	57 Bam 1, 62 Bam 1
polymerization	N,N-dimethyl-formamide	333	$k = 4.5 \cdot 10^3\ M^{-1}\ s^{-1}$ [15])	
R—CH_2Ċ(CH_3)CN [16]) + $FeCl_3$ ⟶ products [13]) R = polymer chain				
AIBN as initiator Titration of Fe(II)	N,N-dimethyl-formamide	333	$k = 6.2 \cdot 10^2\ M^{-1}\ s^{-1}$ $3.15 \cdot 10^2$ [15])	57 Bam 1, 62 Bam 1

[8]) Radicals from guanosine-5′-monophosphate + ȮH reaction.
[9]) Measured via $RSSR^{\cdot-}$ absorption (RṠ + RS^- ⇌ $RSSR^{\cdot-}$).
[10]) Reaction suggested to proceed via H-atom transfer from metal ion ligand.
[11]) Solutions 0.8 M in H_2SO_4.
[12]) Polyacrylonitrile radical.
[13]) Ligand transfer assumed.
[14]) Calc. value.
[15]) From [62 Bam 1].
[16]) Polymethacrylonitrile radical.

Reaction Radical generation Method	Solvent	T[K]	Rate data	Ref./ add. ref.
$R—CH_2\dot{C}(CH_3)COOCH_3$ [17]) + $CuCl_2$ ⟶ products [13])			R = polymer chain	
AIBN as initiator Inhibition of polymerization	N,N-dimethyl-formamide	333	$k = 7.7 \cdot 10^5\ M^{-1}s^{-1}$	65 Ben 1
$R—CH_2\dot{C}(CH_3)COOCH_3$ [17]) + $FeCl_3$ ⟶ products [13])			R = polymer chain	
AIBN as initiator Inhibition of polymerization	N,N-dimethyl-formamide	333	$k = 3.05 \cdot 10^3\ M^{-1}s^{-1}$	62 Bam 1
(Polymethylmethacrylate)˙ [18]) + C_2H_5SH ⟶ $C_2H_5\dot{S}$ + products				
Pulse rad. of $N_2O + H_2O$ + polymethylmethacrylate Time-resolved light scattering	H_2O	295	$k = 2.5 \cdot 10^4\ M^{-1}s^{-1}$	77 Bec 1
(Polyriboadenylic acid)˙ [19]) + $CH_2NH_2CH_2SH$ ⟶ $CH_2NH_2CH_2\dot{S}$ + products				
Pulse rad. of $N_2O + H_2O$ + polyriboadenylic acid Time-resolved light scattering	H_2O, pH = 7.8	RT	$k = 3.4 \cdot 10^6\ M^{-1}s^{-1}$	82 Was 1
$R—CH_2\dot{C}(CH_3)C_6H_5$ [20]) + $FeCl_3$ ⟶ products [21])			R = polymer chain	
AIBN as initiator Titration of Fe(II)	N,N-dimethyl-formamide	333	$k = 5.4 \cdot 10^4\ M^{-1}s^{-1}$	57 Bam 1
(Polyvinylacetate)˙ [22]) + 2,2-bis-*p*-hydroxyphenylpropane(R—OH) ⟶ products + R—$\dot{O}$				
AIBN as initiator in vinylacetate + ethylacetate [23])	vinylacetate/ethylacetate	323	$k = 33.4\ M^{-1}s^{-1}$	77 Sim 1/ 75 Sim 1
	vinylacetate/ethyl[D_3]-acetate [24])	323	$k = 18.5\ M^{-1}s^{-1}$	
(Polyvinylacetate)˙ [22]) + 2,2-bis-*p*-hydroxyphenylpropane(R—OH) ⟶ products + R—$\dot{O}$ + (R—OD) ⟶ products-d_1 + R—$\dot{O}$				
AIBN as initiator in vinylacetate [23])	vinylacetate	303	$k(H)$ [25]) = $18.2\ M^{-1}s^{-1}$ $\log[A/M^{-1}s^{-1}] = 6.1(3)$ $E_a = 28(2)\ kJ\ mol^{-1}$ $k(D)$ [25]) = $2.3\ M^{-1}s^{-1}$ $\log[A/M^{-1}s^{-1}] = 5.8(3)$ $E_a = 32(1)\ kJ\ mol^{-1}$	77 Sim 1/ 75 Sim 2, 75 Sim 1, 67 Sim 2
		313	$k(D)$ [25]) = $3.4\ M^{-1}s^{-1}$	
		323	$k(H)$ [25]) = $36\ M^{-1}s^{-1}$ $k(D)$ [25]) = $5.0\ M^{-1}s^{-1}$ $k(H)/k(D) = 7.4$	
		333	$k(D)$ [25]) = $6.8\ M^{-1}s^{-1}$	
		343	$k(H)$ [25]) = $67\ M^{-1}s^{-1}$ $k(D)$ [25]) = $10.3\ M^{-1}s^{-1}$	

[13]) Ligand transfer assumed.
[17]) Polymethylmethacrylate radical.
[18]) Lateral macro radical of polymethylmethacrylate ($\bar{M}_W = (4.6 \ldots 3.8) \cdot 10^6$).
[19]) Macro radical formed from H-atom abstraction by $\dot{O}H$ radicals from 2′-position of the sugar moiety for polyriboadenylic acid ($\bar{M}_W = 6.3 \cdot 10^5$).
[20]) Polystyrene radical.
[21]) Ligand transfer assumed.
[22]) Radical mixture.
[23]) Calc. from inhibition rate of polymerization (dilatometric measurement).
[24]) Deuterated ethylacetate.
[25]) Refers to one reaction center.

Reaction Radical generation Method	Solvent	T[K]	Rate data	Ref./ add. ref.
(Polyvinylacetate)$^{\cdot}$ [22] + 2,2-bis-p-hydrophenylpropane(R—OH) ⟶ products + R—$\dot{O}$ + (R—OD) ⟶ products-d_1 + R—$\dot{O}$				
AIBN as initiator				78 Kar 1
[26]	vinylacetate	323	$k(H) = 66\ M^{-1}s^{-1}$ ($\pm 20\%$)	
	vinylacetate +		$k(H) = 78\ M^{-1}s^{-1}$ ($\pm 20\%$)	
	ethylpivalate		$k(D) = 11.4\ M^{-1}s^{-1}$ ($\pm 20\%$)	
			$k(H)/k(D) = 6.9$ ($\pm 20\%$)	
	trifluoroacetate		$k(H) = 606\ M^{-1}s^{-1}$ ($\pm 40\%$)	
	acetone		$k(H) = 31\ M^{-1}s^{-1}$ ($\pm 20\%$)	
			$k(D) = 8.9\ M^{-1}s^{-1}$ ($\pm 20\%$)	
			$k(H)/k(D) = 3.5$ ($\pm 20\%$)	
	acetone-d_6		$k(H) = 30\ M^{-1}s^{-1}$ ($\pm 20\%$)	
	ethylacetate		$k(H) = 10.5\ M^{-1}s^{-1}$ ($\pm 20\%$)	
	dimethylsulfoxide		$k(D) = 2.5\ M^{-1}s^{-1}$ ($\pm 20\%$)	
	ethylacetate-d_3		$k(D) = 5.9\ M^{-1}s^{-1}$ ($\pm 20\%$)	
(Polyvinylacetate)$^{\cdot}$ [22] + 2,2-bis-p-hydroxyphenylpropane(R—OH) $\xrightarrow{a}$ products + R—$\dot{O}$ + R—OH…M [27] $\xrightarrow{b}$ products + M + R—$\dot{O}$				
AIBN as initiator				76 Sim 1
[26]	vinylacetate/CCl_4	303	$k_a = 39(8)\ M^{-1}s^{-1}$	
			$k_b = 17.5(20)\ M^{-1}s^{-1}$	
		323	$k_a = 109(11)\ M^{-1}s^{-1}$	
			$k_b = 33(4)\ M^{-1}s^{-1}$	
		343	$k_a = 262(50)\ M^{-1}s^{-1}$	
			$k_b = 56(6)\ M^{-1}s^{-1}$	
			$\log[A_a/M^{-1}s^{-1}] = 8.7(10)$	
			$E_a(a) = 41.0(63)\ kJ\ mol^{-1}$	
			$\log[A_b/M^{-1}s^{-1}] = 5.6(5)$	
			$E_a(b) = 25.1(33)\ kJ\ mol^{-1}$	
(Polyvinylacetate)$^{\cdot}$ [22] + 3-bromophenol{3-BrC_6H_4OH} ⟶ products + 3-$BrC_6H_4\dot{O}$ + 3-BrC_6H_4OD ⟶ products-d_1 + 3-$BrC_6H_4\dot{O}$				
AIBN an initiator in vinylacetate				77 Sim 1/
[23]	vinylacetate	323	$k(H) = 12.8\ M^{-1}s^{-1}$ [25]	67 Sim 2
			$k(D) = 3.6\ M^{-1}s^{-1}$ [25]	
			$k(H)/k(D) = 3.6$	
(Polyvinylacetate)$^{\cdot}$ [22] + 3-chlorophenol{3-ClC_6H_4OH} ⟶ products + 3-$ClC_6H_4\dot{O}$ + 3-ClC_6H_4OD ⟶ products-d_1 + 3-$ClC_6H_4\dot{O}$				
AIBN as initiator in vinylacetate				77 Sim 1/
[23]	vinylacetate	323	$k(D) = 4.4\ M^{-1}s^{-1}$ [28]	65 Bir 1,
			$k(H)/k(D) = 2.0$	67 Sim 1,
				71 Sim 1,
				67 Sim 2
(Polyvinylacetate)$^{\cdot}$ [22] + 4-chlorophenol{4-ClC_6H_4OH} ⟶ products + 4-$ClC_6H_4\dot{O}$ + 4-ClC_6H_4OD ⟶ products-d_1 + 4-$ClC_6H_4\dot{O}$				
AIBN as initiator in vinylacetate				77 Sim 1/
[23]	vinylacetate	323	$k(D) = 3.7\ M^{-1}s^{-1}$ [25] [28]	65 Bir 1,
			$k(H)/k(D) = 5.8$	67 Sim 1,
				71 Sim 1,
				67 Sim 2

[22]) Radical mixture.
[23]) Calc. from inhibition rate of polymerization (dilatometric measurement).
[25]) Refers to one reaction center.
[26]) Inhibition of radical polymerization (dilatometry).
[27]) Hydrogen bridging with monomer vinylacetate (M).
[28]) Calculated from $k(H)/k(D)$ values measured at 318 K [65 Bir 1, 67 Sim 1, 71 Sim 1].

Reaction Radical generation Method	Solvent	T[K]	Rate data	Ref./ add. ref.
(Polyvinylacetate)˙ [22]) + 9,10-dihydroanthracene ⟶ products + 9,10-dihydroanthracene(−Ḣ)				
AIBN as initiator				73 Sim 2/
Dilatometry	vinylacetate	303	$k = 478\ M^{-1}s^{-1}$	67 Tüd 1,
		313	710	67 Sim 2,
		323	1050	64 Ber 1
		333	1526	
		343	1638	
			$\log[A/M^{-1}s^{-1}] = 7.40$	
			$E_a = 27.2(33)\ kJ\ mol^{-1}$	
(Polyvinylacetate)˙ [22]) + 2,6-dihydroxyphenol(R—OH) ⟶ products + R—Ȯ + (R—OD) ⟶ products-d_1 + R—Ȯ				
AIBN as initiator in vinylacetate				77 Sim 1/
[23])	vinylacetate	303	$k(H)$ [25]) $= 3810\ M^{-1}s^{-1}$	75 Sim 1,
			$k(D)$ [25]) $= 204\ M^{-1}s^{-1}$	75 Sim 2,
		308	$k(D)$ [25]) $= 241\ M^{-1}s^{-1}$	71 Sim 1,
		313	$k(D)$ [25]) $= 295\ M^{-1}s^{-1}$	67 Sim 2
		318	$k(D)$ [25]) $= 326\ M^{-1}s^{-1}$	
		323	$k(H)$ [25]) $= 5414\ M^{-1}s^{-1}$	
			$k(D)$ [25]) $= 301\ M^{-1}s^{-1}$	
			$k(H)/k(D) = 19.7$ [29])	
		328	$k(D)$ [25]) $= 428\ M^{-1}s^{-1}$	
		333	$k(D)$ [25]) $= 453\ M^{-1}s^{-1}$	
		338	$k(D)$ [25]) $= 490\ M^{-1}s^{-1}$	
		343	$k(H)$ [25]) $= 8320\ M^{-1}s^{-1}$	
			$k(D)$ [25]) $= 637\ M^{-1}s^{-1}$	
			$\log[A/M^{-1}s^{-1}] = 6.5(3)(H)$	
			$E_a = 16.7(21)\ kJ\ mol^{-1}\ (H)$	
			$\log[A/M^{-1}s^{-1}] = 6.2(4)(D)$	
			$E_a = 22.6(25)\ kJ\ mol^{-1}\ (D)$	
(Polyvinylacetate)˙ [22]) + 2,6-dihydroxy-4-t-butylphenol(R—OH) ⟶ products + R—Ȯ + (R—OD) ⟶ products-d_1 + R—Ȯ				
AIBN as initiator in vinylacetate				77 Sim 1/
[23])	vinylacetate	323	$k(H) = 15520\ M^{-1}s^{-1}$ [25])	67 Sim 2,
			$k(D) = 1744\ M^{-1}s^{-1}$ [25])	75 Sim 2
			$k(H)/k(D) = 9.3$	
(Polyvinylacetate)˙ [22]) + 2,6-dihydroxy-4-t-octylphenol(R—OH) ⟶ products + R—Ȯ + (R—OD) ⟶ products-d_1 + R—Ȯ				
AIBN as initiator in vinylacetate				77 Sim 1/
[23])	vinylacetate	323	$k(H) = 9972\ M^{-1}s^{-1}$ [25])	67 Sim 1,
			$k(D) = 528\ M^{-1}s^{-1}$ [25])	75 Sim 2
			$k(H)/k(D) = 20.8$ [29])	
(Polyvinylacetate)˙ [22]) + 2,6-dimethoxy-4-hydroxyphenol(R—OH) ⟶ products + R—Ȯ + (R—OD) ⟶ products-d_1 + R—Ȯ				
AIBN as initiator in vinylacetate				77 Sim 1/
[23])	vinylacetate	323	$k(H) = 3050\ M^{-1}s^{-1}$ [25])	67 Sim 2
			$k(D) = 371\ M^{-1}s^{-1}$ [25])	
			$k(H)/k(D) = 8.5$	

[22]) Radical mixture.
[23]) Calc. from inhibition rate of polymerization (dilatometric measurement).
[25]) Refers to one reaction center.
[29]) $k(H)/k(D) > 10$ indicates tunneling.

Reaction Radical generation Method	Solvent	T[K]	Rate data	Ref./ add. ref.
(Polyvinylacetate)$^{\cdot}$ [22] + 2,6-dimethoxyphenol(R—OH) ⟶ products + R—$\dot{O}$ + (R—OD) ⟶ products-d_1 + R—$\dot{O}$				
AIBN as initiator in vinylacetate [23]	vinylacetate	323	k(H) = 224 $M^{-1}s^{-1}$ [25] k(D) = 25 $M^{-1}s^{-1}$ [25] k(H)/k(D) = 9.4	77 Sim 1/ 75 Sim 2, 67 Sim 2
(Polyvinylacetate)$^{\cdot}$ [22] + 2,2′-dimethyl-2,2-bis-p-hydroxyphenylpropane(R—OH) ⟶ products + R—$\dot{O}$ + (R—OD) ⟶ products-d_1 + R—$\dot{O}$				
AIBN as initiator in vinylacetate [23]	vinylacetate	303	k(H) [25] = 66 $M^{-1}s^{-1}$ k(D) [25] = 4.6 $M^{-1}s^{-1}$	77 Sim 1/ 75 Sim 1, 75 Sim 2, 67 Sim 2
		323	k(H) [25] = 116 $M^{-1}s^{-1}$ k(D) [25] = 10.0 $M^{-1}s^{-1}$ k(H)/k(D) = 12.3 [29]	
		343	k(H) [25] = 271 $M^{-1}s^{-1}$ k(D) [25] = 20.5 $M^{-1}s^{-1}$ $\log[A/M^{-1}s^{-1}]$ = 7.0(5)(H) E_a = 30.6(33) kJ mol^{-1} (H) $\log[A/M^{-1}s^{-1}]$ = 6.2(3)(D) E_a = 32.2(21) kJ mol^{-1} (D)	
(Polyvinylacetate)$^{\cdot}$ [22] + 2,2′-dimethyl-2,2-bis-p-hydroxyphenylpropane(R—OH) ⟶ products + R—$\dot{O}$				
AIBN as initiator [26]	vinylacetate	323	k = 200(±20%) $M^{-1}s^{-1}$	78 Kar 1
	vinylacetate + ethylacetate		250(±20%)	
	ethylacetate-d_3		124(±20%)	
	ethylpivalate		266(±20%)	
(Polyvinylacetate)$^{\cdot}$ [22] + 2,2′-dimethyl-2,2-bis-p-hydroxyphenylpropane(R—OH) $\xrightarrow{a}$ products + R—$\dot{O}$ + R—OH…M [27] $\xrightarrow{b}$ products + M + R—$\dot{O}$				
AIBN as initiator [26]	vinylacetate/CCl_4	323	k_a = 208(21) $M^{-1}s^{-1}$ [30] k_b = 100(10) $M^{-1}s^{-1}$ [30]	76 Sim 1/ 75 Sim 3
(Polyvinylacetate)$^{\cdot}$ [22] + 2,6-dimethyl-2′,6′-di-t-butyl-2,2-bis-p-hydroxyphenylpropane(R—OH) ⟶ products + R—$\dot{O}$				
AIBN as initiator in vinylacetate [23]	vinylacetate	323	k = 467 $M^{-1}s^{-1}$	77 Sim 1/ 67 Sim 2
(Polyvinylacetate)$^{\cdot}$ [22] + 2,2′-dimethyl-6,6′-di-t-butyl-2,2-bis-p-hydroxyphenylpropane(R—OH) ⟶ products + R—$\dot{O}$				
AIBN as initiator [26]	vinylacetate	323	k = 520(±20%) $M^{-1}s^{-1}$	78 Kar 1
	vinylacetate/ ethylpivalate		k = 590(±20%) $M^{-1}s^{-1}$	

[22]) Radical mixture.
[23]) Calc. from inhibition rate of polymerization (dilatometric measurement).
[25]) Refers to one reaction center.
[26]) Inhibition of radical polymerization (dilatometry).
[27]) Hydrogen bridging with monomer vinylacetate (M).
[30]) Corrected values from [75 Sim 3].

Reaction Radical generation Method	Solvent	T[K]	Rate data	Ref./ add. ref.
(Polyvinylacetate)$^\cdot$ [22]) + 2,2′-dimethyl-4,4′-isopropylidenebisphenol(R—OH) $\xrightarrow{a}$ products + R—Ȯ + R—OH...M [27]) $\xrightarrow{b}$ products + M + R—Ȯ				
AIBN as initiator [26])	vinylacetate/CCl_4	323	$k_a = 460(25)\,M^{-1}\,s^{-1}$ $k_b = 186(10)\,M^{-1}\,s^{-1}$	75 Sim 3
(Polyvinylacetate)$^\cdot$ [22]) + 2,2′-dimethyl-4,4′-isopropylidenebisphenol(R—OD) $\xrightarrow{a}$ products-d_1 + R—Ȯ + R—OD...M [27]) $\xrightarrow{b}$ products-d_1 + M + R—Ȯ				
AIBN as initiator [26])	vinylacetate/CCl_4	323	$k_a = 27.8(20)\,M^{-1}\,s^{-1}$ $k_b = 19.5(10)\,M^{-1}\,s^{-1}$ [31])	75 Sim 3
(Polyvinylacetate)$^\cdot$ [22]) + 2,6-dimethyl-4-*t*-butylphenol(R—OH) ⟶ products + (R—Ȯ)				
AIBN as initiator in vinylacetate [23])	vinylacetate	323	$k = 420\,M^{-1}\,s^{-1}$	77 Sim 1/ 67 Sim 2
(Polyvinylacetate)$^\cdot$ [22]) + 2,2′-di-*t*-butyl-2,2-bis-*p*-hydroxyphenylpropane(R—OH) ⟶ products + R—Ȯ + (R—OD) ⟶ products-d_1 + R—Ȯ				
AIBN as initiator in vinylacetate [23])	vinylacetate	323	$k(H) = 66\,M^{-1}\,s^{-1}$ [25]) $k(D) = 5.6\,M^{-1}\,s^{-1}$ [25]) $k(H)/k(D) = 12.6$ [29])	77 Sim 1/ 75 Sim 2, 67 Sim 2
(Polyvinylacetate)$^\cdot$ [22]) + 2,2′-di-*t*-butyl-2,2-bis-*p*-hydroxyphenylpropane(R—OH) $\xrightarrow{a}$ products + R—Ȯ + R—OH...M [27]) $\xrightarrow{b}$ products + M + R—Ȯ				
AIBN as initiator [26])	vinylacetate/CCl_4	323	$k_a = 184(20)\,M^{-1}\,s^{-1}$ $k_b = 68(7)\,M^{-1}\,s^{-1}$	76 Sim 1
(Polyvinylacetate)$^\cdot$ [22]) + 2,6-di-*t*-butyl-2,2-bis-*p*-hydroxyphenylpropane(R—OH) ⟶ products + R—Ȯ				
AIBN as initiator in vinylacetate [23])	vinylacetate	323	$k = 66\,M^{-1}\,s^{-1}$	77 Sim 1/ 67 Sim 2
(Polyvinylacetate)$^\cdot$ [22]) + 2,2′-di-*t*-butyl-6,6′-dimethyl-2,2-bis-*p*-hydroxyphenylpropane(R—OH) ⟶ products + R—Ȯ				
AIBN as initiator in vinylacetate + ethylacetate [23])	vinylacetate/ ethylacetate	323	$k = 219\,M^{-1}\,s^{-1}$	77 Sim 1/ 75 Sim 1
	vinylacetate/ ethyl[D_3]-acetate [32])	323	$k = 161\,M^{-1}\,s^{-1}$	

[22]) Radical mixture.
[23]) Calc. from inhibition rate of polymerization (dilatometric measurement).
[25]) Refers to one reaction center.
[26]) Inhibition of radical polymerization (dilatometry).
[27]) Hydrogen bridging with monomer vinylacetate (M).
[29]) $k(H)/k(D) > 10$ indicates tunneling.
[31]) Isotope effects $k_a(H)/k_a(D) = 17.9(25)$; $k_b(H)/k_b(D) = 10.5(10)$.
[32]) Deuterated ethylacetate.

Reaction Radical generation Method	Solvent	T[K]	Rate data	Ref./ add. ref.
(Polyvinylacetate)$^{\cdot}$ [22]) + 2,2′-di-t-butyl-6,6′-dimethyl-2,2-bis-p-hydroxyphenylpropane(R—OH) $\longrightarrow$ products + R—$\dot{O}$ + (R—OD) $\longrightarrow$ products-d_1 + R—$\dot{O}$				
AIBN as initiator in vinylacetate [23])	vinylacetate	303	$k(H)$ [25]) $= 155\,M^{-1}\,s^{-1}$ $k(D)$ [25]) $= 10\,M^{-1}\,s^{-1}$	77 Sim 1/ 75 Sim 1, 75 Sim 2, 67 Sim 2
		323	$k(H)$ [25]) $= 244\,M^{-1}\,s^{-1}$ $k(D)$ [25]) $= 28\,M^{-1}\,s^{-1}$ $k(H)/k(D) = 9.0$	
		343	$k(H)$ [25]) $= 506\,M^{-1}\,s^{-1}$ $k(D)$ [25]) $= 46.5\,M^{-1}\,s^{-1}$ $\log[A/M^{-1}\,s^{-1}] = 6.5(5)$(H) $E_a = 25.5(33)\,kJ\,mol^{-1}$ (H) $\log[A/M^{-1}\,s^{-1}] = 6.8(5)$(D) $E_a = 33.5(33)\,kJ\,mol^{-1}$ (D)	
(Polyvinylacetate)$^{\cdot}$ [22]) + 2,2′-di-t-butyl-6,6′-dimethyl-2,2-bis-p-hydroxyphenylpropane(R—OH) $\xrightarrow{a}$ products + R—$\dot{O}$ + R—OH...M [27]) $\xrightarrow{b}$ products + M + R—$\dot{O}$				
AIBN as initiator [26])	vinylacetate/CCl_4	303	$k_a = 27(5)\,M^{-1}\,s^{-1}$ $k_b = 166(16)\,M^{-1}\,s^{-1}$	76 Sim 1
		343	$k_a = 210(30)\,M^{-1}\,s^{-1}$ $k_b = 550(60)\,M^{-1}\,s^{-1}$ $\log[A_a/M^{-1}\,s^{-1}] = 9.1(10)$ $E_a(a) = 44.4(63)\,kJ\,mol^{-1}$ $\log[A_b/M^{-1}\,s^{-1}] = 6.6(8)$ $E_a(b) = 26.0(50)\,kJ\,mol^{-1}$	
(Polyvinylacetate)$^{\cdot}$ [22]) + 2,2′-di-t-butyl-6,6′-dimethyl-4,4′-isopropylidenebisphenol(R—OH) $\xrightarrow{a}$ products + R—$\dot{O}$ + R—OH...M [27]) $\xrightarrow{b}$ products + M + R—$\dot{O}$				
AIBN as initiator [26])	vinylacetate/CCl_4	323	$k_a = 80(8)\,M^{-1}\,s^{-1}$ $k_b = 260(26)\,M^{-1}\,s^{-1}$	75 Sim 3
(Polyvinylacetate)$^{\cdot}$ [22]) + 2,2′-di-t-butyl-6,6′-dimethyl-4,4′-isopropylidenebisphenol(R—OD) $\xrightarrow{a}$ products-d_1 + R—$\dot{O}$ + R—OD...M [27]) $\xrightarrow{b}$ products-d_1 + M + R—$\dot{O}$				
AIBN as initiator [26])	vinylacetate/CCl_4	323	$k_a = 31.0(40)\,M^{-1}\,s^{-1}$ $k_b = 27.0(40)\,M^{-1}\,s^{-1}$ [33])	75 Sim 3
(Polyvinylacetate)$^{\cdot}$ [22]) + 2,5-di-t-butyl-4-hydroxyphenol(R—OH) $\longrightarrow$ products + R—$\dot{O}$ + (R—OD) $\longrightarrow$ products-d_1 + R—$\dot{O}$				
AIBN as initiator in vinylacetate [23])	vinylacetate	323	$k(H) = 680\,M^{-1}\,s^{-1}$ [25]) $k(D) = 48\,M^{-1}\,s^{-1}$ [25]) $k(H)/k(D) = 15.2$ [29])	77 Sim 1/ 67 Sim 2

[22]) Radical mixture.
[23]) Calc. from inhibition rate of polymerization (dilatometric measurement).
[25]) Refers to one reaction center.
[26]) Inhibition of radical polymerization (dilatometry).
[27]) Hydrogen bridging with monomer vinylacetate (M).
[29]) $k(H)/k(D) > 10$ indicates tunneling.
[33]) Isotope effects $k_a(H)/k_a(D) = 2.6(5)$; $k_b(H)/k_b(D) = 10.0(20)$.

Reaction Radical generation Method	Solvent	T[K]	Rate data	Ref./ add. ref.
(Polyvinylacetate)˙ [22]) + 2,6-di-*t*-butyl-4-methylphenol(R—OH) ⟶ products + R—Ȯ + (R—OD) ⟶ products-d_1 + R—Ȯ				
AIBN as initiator in vinylacetate				77 Sim 1/
[23])	vinylacetate	323	k(H) = 35 $M^{-1}s^{-1}$	74 Par 1,
			k(D) = 4.0 $M^{-1}s^{-1}$	67 Sim 2
			k(H)/k(D) = 9.2	
(Polyvinylacetate)˙ [22]) + fluorene ⟶ products + fluorene(−Ḣ)				
AIBN as initiator				73 Sim 2/
Dilatometry	vinylacetate	303	k = 133 $M^{-1}s^{-1}$	67 Tüd 1,
		323	234	67 Sim 2,
		343	320	64 Ber 1
			log[A/$M^{-1}s^{-1}$] = 5.66	
			E_a = 20.5(42) kJ mol^{-1}	
(Polyvinylacetate)˙ [22]) + 4-hydroxyphenol(R—OH) ⟶ products + R—Ȯ				
AIBN as initiator				73 Sim 2/
Dilatometry	vinylacetate	303	k = 237 $M^{-1}s^{-1}$	67 Tüd 1,
		313	278	67 Sim 2,
		323	404	64 Ber 1
		333	493	
		343	755	
			log[A/$M^{-1}s^{-1}$] = 6.74	
			E_a = 25.5(25) kJ mol^{-1}	
(Polyvinylacetate)˙ [22]) + 2-methoxy-4-formylphenol(R—OH) ⟶ products + R—Ȯ				
AIBN as initiator in vinylacetate				77 Sim 1/
[23])	vinylacetate	323	k = 11 $M^{-1}s^{-1}$	67 Sim 2
(Polyvinylacetate)˙ [22]) + 4-methoxyphenol(R—OH) ⟶ products + R—Ȯ + (R—OD) ⟶ products-d_1 + R—Ȯ				
AIBN as initiator in vinylacetate				77 Sim 1/
[23])	vinylacetate	323	k(H) = 240 $M^{-1}s^{-1}$ [25])	67 Sim 2
			k(D) = 30.4 $M^{-1}s^{-1}$ [25])	
			k(H)/k(D) = 7.9	
(Polyvinylacetate)˙ [22]) + 2-methyl-2,2-bis-*p*-hydroxyphenylpropane(R—OH) ⟶ products + R—Ȯ				
AIBN as initiator in vinylacetate				77 Sim 1/
[23])	vinylacetate	323	k = 112 $M^{-1}s^{-1}$	67 Sim 2
(Polyvinylacetate)˙ [22]) + 4-methyl-2,6-di-*t*-butylphenol(R—OH) ⟶ products + R—Ȯ				
AIBN as initiator				73 Sim 2/
Dilatometry	vinylacetate	303	k = 42 $M^{-1}s^{-1}$	67 Tüd 1,
		323	57	67 Sim 2,
		343	80	64 Ber 1
			log[A/$M^{-1}s^{-1}$] = 4.08	
			E_a = 14.2(42) kJ mol^{-1}	
(Polyvinylacetate)˙ [22]) + 4-methyl-2,6-di-*t*-butylphenol(R—OH) ⟶ products + R—Ȯ				
AIBN as initiator				74 Par 1/
[34])	vinylacetate	323	k = 160(40) $M^{-1}s^{-1}$ [35])	49 Mat 1
		308...	log[A/$M^{-1}s^{-1}$] = 7.5	
		328	E_a = 31.4 kJ mol^{-1}	

[22]) Radical mixture.
[23]) Calc. from inhibition rate of polymerization (dilatometric measurement).
[25]) Refers to one reaction center.
[34]) Degree of polymerization using viscometry and osmometry.
[35]) Calc. on the basis of k/k_p = 0.020 with k_p = 8(2)·10^3 $M^{-1}s^{-1}$ (propagation rate constant for vinylacetate polymerization) [49 Mat 1].

Reaction Radical generation Method	Solvent	T[K]	Rate data	Ref./ add. ref.
(Polyvinylacetate)$^{\cdot}$ [22]) + 2-methylphenol{$2\text{-}CH_3OC_6H_4OH$} ⟶ products + $2\text{-}CH_3OC_6H_4\dot{O}$ + $2\text{-}CH_3OC_6H_4OD$ ⟶ products-d_1 + $2\text{-}CH_3OC_6H_4\dot{O}$				
AIBN as initiator in vinylacetate [23])	vinylacetate	323	k(H) = 46 $M^{-1}s^{-1}$ [25]) k(D) = 6.4 $M^{-1}s^{-1}$ [25]) k(H)/k(D) = 7.2	77 Sim 1/ 67 Sim 2
(Polyvinylacetate)$^{\cdot}$ [22]) + 3-methylphenol{$3\text{-}CH_3C_6H_4OH$} ⟶ products + $3\text{-}CH_3C_6H_4\dot{O}$ + $3\text{-}CH_3C_6H_4OD$ ⟶ products-d_1 + $3\text{-}CH_3C_6H_4\dot{O}$				
AIBN as initiator in vinylacetate [23])	vinylacetate	323	k(D) = 5.8 $M^{-1}s^{-1}$ [25]) [28]) k(H)/k(D) = 5.9	77 Sim 1/ 65 Bir 1, 67 Sim 1, 71 Sim 1, 67 Sim 2
(Polyvinylacetate)$^{\cdot}$ [22]) + 4-methylphenol{$4\text{-}CH_3C_6H_4OH$} ⟶ products + $4\text{-}CH_3C_6H_4\dot{O}$ + $4\text{-}CH_3C_6H_4OD$ ⟶ products-d_1 + $4\text{-}CH_3C_6H_4\dot{O}$				
AIBN as initiator in vinylacetate [23])	vinylacetate	323	k(D) = 6.4 $M^{-1}s^{-1}$ [25]) [28]) k(H)/k(D) = 6.7	77 Sim 1/ 65 Bir 1, 67 Sim 1, 71 Sim 1, 67 Sim 2
(Polyvinylacetate)$^{\cdot}$ [22]) + pentabromophenol{C_6Br_5OH} ⟶ products + $C_6Br_5\dot{O}$ + C_6Br_5OD ⟶ products-d_1 + $C_6Br_5\dot{O}$				
AIBN as initiator in vinylacetate [23])	vinylacetate	323	k(H) = 36 $M^{-1}s^{-1}$ [25]) k(D) = 5.3 $M^{-1}s^{-1}$ [25]) k(H)/k(D) = 7.0	77 Sim 1/ 75 Sim 2, 67 Sim 2
(Polyvinylacetate)$^{\cdot}$ [22]) + phenol ⟶ products + $C_6H_5\dot{O}$				
AIBN as initiator Dilatometry	vinylacetate	303 323 343	k = 17 $M^{-1}s^{-1}$ 24 38 $\log[A/M^{-1}s^{-1}] = 3.80$ E_a = 18.4(42) kJ mol^{-1}	73 Sim 2/ 67 Tüd 1, 67 Sim 2, 64 Ber 1
(Polyvinylacetate)$^{\cdot}$ [22]) + phenol{C_6H_5OH} ⟶ products + $C_6H_5\dot{O}$ + C_6H_5OD ⟶ products-d_1 + $C_6H_5\dot{O}$				
AIBN as initiator in vinylacetate [23])	vinylacetate	323	k(H) = 12 $M^{-1}s^{-1}$ [25]) k(D) = 2.5 $M^{-1}s^{-1}$ [25]) [28]) k(H)/k(D) = 4.9	77 Sim 1/ 65 Bir 1, 67 Sim 1, 71 Sim 1, 67 Sim 2
(Polyvinylacetate)$^{\cdot}$ [22]) + phenol{C_6H_5OH} $\xrightarrow{a}$ products + $C_6H_5\dot{O}$ + $C_6H_5OH \ldots M$ [27]) $\xrightarrow{b}$ products + M + $C_6H_5\dot{O}$				
AIBN as initiator [26])	vinylacetate/CCl_4	323	k_a = 44(5) $M^{-1}s^{-1}$ k_b = 11(1) $M^{-1}s^{-1}$	76 Sim 1

[22]) Radical mixture.
[23]) Calc. from inhibition rate of polymerization (dilatometric measurement).
[25]) Refers to one reaction center.
[26]) Inhibition of radical polymerization (dilatometry).
[28]) Calculated from k(H)/k(D) values standard at 318 K [65 Bir 1, 67 Sim 1, 71 Sim 1].

Reaction Radical generation Method	Solvent	T[K]	Rate data	Ref./ add. ref.
(Polyvinylacetate)$^{\cdot}$ [22]) + N-phenyl-N-alkylhydroxylamine(R—OH) [36]) ⟶ products + R—Ȯ				
AIBN as initiator				73 Sim 2/
Dilatometry	vinylacetate	303	$k = 1.145 \cdot 10^4 (\pm 10\%)\,M^{-1}\,s^{-1}$	67 Tüd 1,
		313	$1.380 \cdot 10^4 (\pm 10\%)$	67 Sim 2,
		323	$1.885 \cdot 10^4 (\pm 10\%)$	67 Sim 3,
		333	$2.330 \cdot 10^4 (\pm 10\%)$	64 Ber 1
		343	$3.320 \cdot 10^4 (\pm 10\%)$	
			$\log[A/M^{-1}\,s^{-1}] = 7.87$	
			$E_a = 22.2(25)\,kJ\,mol^{-1}$	
(Polyvinylacetate)$^{\cdot}$ [22]) + 2-*t*-butyl-2,2-bis-4-hydroxyphenylpropane(R—OH) ⟶ products + R—Ȯ				
AIBN as initiator in vinylacetate				77 Sim 1/
[23])	vinylacetate	323	$k = 51\,M^{-1}\,s^{-1}$	67 Sim 2
(Polyvinylacetate)$^{\cdot}$ [22]) + 2,2′-*t*-butyl-2,2-bis-4-hydroxyphenylpropane(R—OH) ⟶ R—Ȯ + products + (R—OD) ⟶ R—Ȯ + products-d_1				
Decomp. of AIBN as initiator				75 Sim 2
[37])	vinylacetate	323	$k(H) = 66\,M^{-1}\,s^{-1}$	
			$k(D) = 5.6\,M^{-1}\,s^{-1}$	
			$k(H)/k(D) = 11.9$ [29])	
(Polyvinylacetate)$^{\cdot}$ [22]) + 2,2′-*t*-butyl-6,6′-dimethyl-2,2-bis-*p*-hydroxyphenylpropane(R—OH) ⟶ R—Ȯ + products + (R—OD) ⟶ R—Ȯ + products-d_1				
Decomp. of AIBN as initiator				75 Sim 2
[37])	vinylacetate	323	$k(H) = 244\,M^{-1}\,s^{-1}$	
			$k(D) = 28.0\,M^{-1}\,s^{-1}$	
			$k(H)/k(D) = 8.7$	
(Polyvinylacetate)$^{\cdot}$ [22]) + 2-*t*-butyl-6-methyl-2,2-bis-*p*-hydroxyphenylpropane(R—OH) ⟶ products + R—Ȯ				
AIBN as initiator in vinylacetate				77 Sim 1/
[23])	vinylacetate	323	$k = 254\,M^{-1}\,s^{-1}$	67 Sim 2
(Polyvinylacetate)$^{\cdot}$ [22]) + 2,2′,6,6′-tetraethyl-2,2-bis-*p*-hydroxyphenylpropane(R—OH) $\xrightarrow{a}$ products + R—Ȯ + R—OH...M [27]) $\xrightarrow{b}$ products + M + R—Ȯ				
AIBN as initiator				76 Sim 1
[26])	vinylacetate/CCl_4	323	$k_a = 165(17)\,M^{-1}\,s^{-1}$	
			$k_b = 510(51)\,M^{-1}\,s^{-1}$	
(Polyvinylacetate)$^{\cdot}$ [22]) + 2,2′,6,6′-tetraethyl-2,2-bis-*p*-hydroxyphenylpropane(R—OH) ⟶ products + R—Ȯ + (R—OD) ⟶ products-d_1 + R—Ȯ				
AIBN as initiator in vinylacetate				77 Sim 1/
[23])	vinylacetate	323	$k(H) = 464\,M^{-1}\,s^{-1}$ [25])	67 Sim 2,
			$k(D) = 56\,M^{-1}\,s^{-1}$ [25])	75 Sim 2
			$k(H)/k(D) = 8.6$	

[22]) Radical mixture.
[23]) Calc. from inhibition rate of polymerization (dilatometric measurement).
[25]) Refers to one reaction center.
[26]) Inhibition of radical polymerization (dilatometry).
[27]) Hydrogen bridging with monomer vinylacetate (M).
[29]) $k(H)/k(D) > 10$ indicates tunneling.
[36]) Banfield condensate.
[37]) Inhibition of radical induced polymerization of vinylacetate.

Reaction Radical generation Method	Solvent	T[K]	Rate data	Ref./ add. ref.
(Polyvinylacetate)˙ [22]) + 5,5′,7,7′-tetrahydroxy-1,1,1′,1′-tetramethyl-6,6′-dihydroxy-3,3′-spirobisindane(R—OH) ⟶ products + R—Ȯ + (R—OD) ⟶ products-d_1 + R—Ȯ				
AIBN as initiator in vinylacetate [23])	vinylacetate	323	k(H) = 11820 $M^{-1}s^{-1}$ [25]) k(D) = 1260 $M^{-1}s^{-1}$ [25]) k(H)/k(D) = 9.8	77 Sim 1/ 67 Sim 1
(Polyvinylacetate)˙ [22]) + 2,2′,6,6′-tetraisopropyl-2,2-bis-p-hydroxyphenylpropane(R—OH) $\xrightarrow{a}$ products + R—Ȯ + R—OH...M [27]) $\xrightarrow{b}$ products + M + R—Ȯ				
AIBN as initiator [26])	vinylacetate/CCl_4	323	k_a = 205(21) $M^{-1}s^{-1}$ k_b = 530(53) $M^{-1}s^{-1}$	76 Sim 1
(Polyvinylacetate)˙ [22]) + 2,2′,6,6′-tetraisopropyl-2,2-bis-p-hydroxyphenylpropane(R—OH) ⟶ products + R—Ȯ + (R—OD) ⟶ products-d_1 + R—Ȯ				
AIBN as initiator in vinylacetate [23])	vinylacetate	323	k(H) = 488 $M^{-1}s^{-1}$ [25]) k(D) = 38 $M^{-1}s^{-1}$ [25]) k(H)/k(D) = 13.8 [29])	77 Sim 1/ 75 Sim 2, 67 Sim 2
(Polyvinylacetate)˙ [22]) + 2,2′,6,6′-tetramethyl-2,2-bis-p-hydroxyphenylpropane(R—OH) ⟶ products + R—Ȯ + (R—OD) ⟶ products-d_1 + R—Ȯ				
AIBN as initiator in vinylacetate [23])	vinylacetate	323	k(H) = 460 $M^{-1}s^{-1}$ [25]) k(D) = 33 $M^{-1}s^{-1}$ [25]) k(H)/k(D) = 15.0 [29])	77 Sim 1/ 75 Sim 2, 67 Sim 2
(Polyvinylacetate)˙ [22]) + 1,1,1′,1′-tetramethyl-6,6′-dihydroxy-3,3′-spirobisindane(R—OH) ⟶ products + R—Ȯ + (R—OD) ⟶ products-d_1 + R—Ȯ				
AIBN as initiator in vinylacetate [23])	vinylacetate	323	k(H) = 45 $M^{-1}s^{-1}$ [25]) k(D) = 4.4 $M^{-1}s^{-1}$ [25]) k(H)/k(D) = 10.8 [29])	77 Sim 1/ 67 Sim 2
(Polyvinylacetate)˙ [22]) + 5,5′,7,7′-tetramethyl-6,6′-dihydroxy-1,1,1′,1′-tetramethyl-3,3′-spirobisindane(R—OH) ⟶ products + R—Ȯ + (R—OD) ⟶ products-d_1 + R—Ȯ				
AIBN as initiator [26])	vinylacetate/ ethylacetate	323	k(H) = 1126 $M^{-1}s^{-1}$ (±20%) k(D) = 85.5 $M^{-1}s^{-1}$ (±20%) k(H)/k(D) = 13.2 [29])	78 Kar 1
	vinylacetate/ ethylacetate-d_3		k(D) = 61.6 $M^{-1}s^{-1}$ (±20%)	

[22]) Radical mixture.
[23]) Calc. from inhibition rate of polymerization (dilatometric measurement).
[25]) Refers to one reaction center.
[26]) Inhibition of radical polymerization (dilatometry).
[27]) Hydrogen bridging with monomer vinylacetate (M).
[29]) k(H)/k(D) > 10 indicates tunneling.

Reaction Radical generation Method	Solvent	T[K]	Rate data	Ref./ add. ref.
(Polyvinylacetate)$^{\cdot}$ [22]) + 2,3,5,6-tetramethyl-4-hydroxyphenol(R—OH) ⟶ products + R—Ȯ + (R—OD) ⟶ products-d_1 + R—Ȯ				
AIBN as initiator in vinylacetate [23])	vinylacetate	303	k(H) [25]) = 2110 $M^{-1}s^{-1}$ k(D) [25]) = 160 $M^{-1}s^{-1}$	77 Sim 1/ 75 Sim 1, 75 Sim 2, 67 Sim 2
		313	k(D) [25]) = 228 $M^{-1}s^{-1}$	
		323	k(H) [25]) = 3320 $M^{-1}s^{-1}$ k(D) [25]) = 260 $M^{-1}s^{-1}$ k(H)/k(D) = 13.6 [29])	
		333	k(D) [25]) = 427 $M^{-1}s^{-1}$	
		343	k(H) [25]) = 5060 $M^{-1}s^{-1}$ k(D) [25]) = 579 $M^{-1}s^{-1}$ $\log[A/M^{-1}s^{-1}] = 6.6(3)$(H) E_a = 18.8(21) kJ mol^{-1} (H) $\log[A/M^{-1}s^{-1}] = 6.9(5)$(D) E_a = 27.6(33) kJ mol^{-1} (D)	
(Polyvinylacetate)$^{\cdot}$ [22]) + 2,2′,6,6′-tetramethyl-4,4′-isopropylidenebisphenol(R—OH) $\xrightarrow{a}$ products + R—Ȯ + R—OH…M [27]) $\xrightarrow{b}$ products + M + R—Ȯ				
AIBN as initiator [26])	vinylacetate/CCl_4	323	k_a = 225(30) $M^{-1}s^{-1}$ k_b = 485(50) $M^{-1}s^{-1}$	75 Sim 3
(Polyvinylacetate)$^{\cdot}$ [22]) + 2,2′,6,6′-tetramethyl-4,4′-isopropylidenebisphenol(R—OD) $\xrightarrow{a}$ products-d_1 + R—Ȯ + R—OD…M [27]) $\xrightarrow{b}$ products-d_1 + M + R—Ȯ				
AIBN as initiator [26])	vinylacetate/CCl_4	323	k_a = 36.0(50) $M^{-1}s^{-1}$ k_b = 33.0(40) $M^{-1}s^{-1}$ [38])	75 Sim 3
(Polyvinylacetate)$^{\cdot}$ [22]) + 2,3,4,6-tetramethylphenol(R—OH) ⟶ products + R—Ȯ + (R—OD) ⟶ products-d_1 + R—Ȯ				
AIBN as initiator in vinylacetate [23])	vinylacetate	323	k(H) = 578 $M^{-1}s^{-1}$ [25]) k(D) = 37 $M^{-1}s^{-1}$ [25]) [28]) k(H)/k(D) = 16.7 [29])	77 Sim 1/ 65 Bir 1, 67 Sim 1, 67 Sim 2, 71 Sim 1
(Polyvinylacetate)$^{\cdot}$ [22]) + 5,5′,7,7′-tetramethyl-1,1,1′,1′-tetramethyl-6,6′-dihydroxy-3,3′-spirobisindane(R—OH) ⟶ products + R—Ȯ + (R—OD) ⟶ products-d_1 + R—Ȯ				
AIBN as initiator in vinylacetate [23])	vinylacetate	323	k(H) = 625 $M^{-1}s^{-1}$ [25]) k(D) = 46 $M^{-1}s^{-1}$ [25]) k(H)/k(D) = 14.5 [29])	77 Sim 1/ 67 Sim 2

[22]) Radical mixture.
[23]) Calc. from inhibition rate of polymerization (dilatometric measurement).
[25]) Refers to one reaction center.
[26]) Inhibition of radical polymerization (dilatometry).
[27]) Hydrogen bridging with monomer vinylacetate (M).
[28]) Calculated from k(H)/k(D) values measured at 318 K [65 Bir 1, 67 Sim 1, 71 Sim 1].
[29]) k(H)/k(D) > 10 indicates tunneling.
[38]) Isotope effects k_a(H)/k_a(D) = 6.5(13); k_b(H)/k_b(D) = 16.0(25).

Reaction
  Radical generation
  Method
	Solvent	T[K]	Rate data	Ref./
				add. ref.

(Polyvinylacetate)'[22]) + 2,2',6,6'-tetra-t-butyl-2,2-bis-p-hydroxyphenylpropane(R—OH) $\longrightarrow$
products + R—$\dot{\text{O}}$

				78 Kar 1
AIBN as initiator [26])	vinylacetate	323	$k = 88(\pm 20\%)\,\text{M}^{-1}\,\text{s}^{-1}$	
	vinylacetate + ethylacetate		$k = 92(\pm 20\%)\,\text{M}^{-1}\,\text{s}^{-1}$	
	vinylacetate + ethylacetate-d_3		$k = 64(\pm 20\%)\,\text{M}^{-1}\,\text{s}^{-1}$	

(Polyvinylacetate)'[22]) + 2,2',6,6'-tetra-t-butyl-2,2-bis-p-hydroxyphenylpropane(R—OH) $\longrightarrow$
products + R—$\dot{\text{O}}$
+ (R—OD) $\longrightarrow$ products-d_1 + R—$\dot{\text{O}}$

				77 Sim 1/
AIBN as initiator in vinylacetate [23])	vinylacetate	323	$k(\text{H}) = 36\,\text{M}^{-1}\,\text{s}^{-1}$ [25])	75 Sim 2,
			$k(\text{D}) = 4.0\,\text{M}^{-1}\,\text{s}^{-1}$ [25])	67 Sim 2
			$k(\text{H})/k(\text{D}) = 9.5$	

(Polyvinylacetate)'[22]) + 2,2',6,6'-tetra-t-butyl-2,2-bis-p-hydroxyphenylpropane(R—OH) $\xrightarrow{\text{a}}$
products + R—$\dot{\text{O}}$
+ (R—OH)...M[27]) $\xrightarrow{\text{b}}$ products + M + R—$\dot{\text{O}}$

				76 Sim 1
AIBN as initiator [26])	vinylacetate/CCl$_4$	323	$k_a = 7(1)\,\text{M}^{-1}\,\text{s}^{-1}$	
			$k_b = 44(5)\,\text{M}^{-1}\,\text{s}^{-1}$	

(Polyvinylacetate)'[22]) + 2,4,6-trihydroxyphenol(R—OH) $\longrightarrow$ R—$\dot{\text{O}}$ + products
+ (R—OD) $\longrightarrow$ R—$\dot{\text{O}}$ + products-d_1

				75 Sim 2,
Decomp. of AIBN [37])	vinylacetate	323	$k(\text{H}) = 3.408 \cdot 10^4$	77 Sim 1
			$k(\text{D}) = 4.80 \cdot 10^3$	
			$k(\text{H})/k(\text{D}) = 7.1$	
			7.3 [39])	

(Polyvinylacetate)'[22]) + 2,4,6-trimethylphenol(R—OH) $\longrightarrow$ products + R—$\dot{\text{O}}$

				73 Sim 2,
AIBN as initiator Dilatometry	vinylacetate	303	$k = 124\,\text{M}^{-1}\,\text{s}^{-1}$	77 Sim 1/
		323	280	67 Tüd 1,
			180 [39])	67 Sim 2,
		343	378	64 Ber 1
			$\log[A/\text{M}^{-1}\,\text{s}^{-1}] = 6.38$	
			$E_a = 24.7(25)\,\text{kJ mol}^{-1}$	

(Polyvinylacetate)'[22]) + 2,4,6-trimethylphenol(R—OH) $\longrightarrow$ products + R—$\dot{\text{O}}$

				73 Sim 2/
AIBN as initiator Dilatometry	vinylacetate	303	$k = 544\,\text{M}^{-1}\,\text{s}^{-1}$	67 Tüd 1,
		323	808	67 Sim 2,
		343	1320	64 Ber 1
			$\log[A/\text{M}^{-1}\,\text{s}^{-1}] = 6.64$	
			$E_a = 23.0(25)\,\text{kJ mol}^{-1}$	

[22]) Radical mixture.
[23]) Calc. from inhibition rate of polymerization (dilatometric measurement).
[25]) Refers to one reaction center.
[26]) Inhibition of radical polymerization (dilatometry).
[27]) Hydrogen bridging with monomer vinylacetate (M).
[37]) Inhibition of radical induced polymerization of vinylacetate.
[39]) From [77 Sim 1].

Reaction Radical generation Method	Solvent	T[K]	Rate data	Ref./ add. ref.
(Polyvinylacetate)$^{\cdot}$ [22]) + 2,4,6-trimethylphenol(R—OH) ⟶ products + R—Ȯ + (R—OD) ⟶ products-d_1 + R—Ȯ				
AIBN as initiator in vinylacetate [23])	vinylacetate	323	k(H) = 400 $M^{-1}s^{-1}$ [25]) k(D) = 31 $M^{-1}s^{-1}$ [25]) k(H)/k(D) = 13.6 [29])	77 Sim 1/ 75 Sim 2, 67 Sim 2
(Polyvinylacetate)$^{\cdot}$ [22]) + 2,4,6-trimethylphenol(R—OH) $\xrightarrow{a}$ products + R—Ȯ + (R—OD) $\xrightarrow{b}$ products-d_1 + R—Ȯ				
AIBN as initiator [26])	vinylacetate	323	k(H) = 420 $M^{-1}s^{-1}$ (±20%) k(D) = 32 $M^{-1}s^{-1}$ (±20%)	78 Kar 1
	vinylacetate + ethylpivalate		k(H)/k(D) = 13.1 [29]) k(H) = 421 $M^{-1}s^{-1}$ (±20%) k(D) = 35.5 $M^{-1}s^{-1}$ (±20%) k(H)/k(D) = 11.9 [29])	
	+ trifluoroacetate		k(H) = 2219 $M^{-1}s^{-1}$ (±20%) k(D) = 39.1 $M^{-1}s^{-1}$ (±20%) k(H)/k(D) = 56.8 [29])	
	+ acetone		k(H) = 234 $M^{-1}s^{-1}$ (±20%) k(D) = 28.8 $M^{-1}s^{-1}$ (±20%) k(H)/k(D) = 8.1	
	+ dimethylsulfoxide		k(D) = 6.3 $M^{-1}s^{-1}$ (±20%)	
(Polyvinylacetate)$^{\cdot}$ [22]) + 2,4,6-trimethylphenol(R—OH) $\xrightarrow{a}$ products + R—Ȯ + R—OH...M [27]) $\xrightarrow{b}$ products + M + R—Ȯ				
AIBN as initiator [26])	vinylacetate/CCl_4	323	k_a = 240(40) $M^{-1}s^{-1}$ k_b = 420(60) $M^{-1}s^{-1}$	75 Sim 3
(Polyvinylacetate)$^{\cdot}$ [22]) + 2,4,6-trimethylphenol(R—OD) $\xrightarrow{a}$ products-d_1 + R—Ȯ + R—OD...M [27]) $\xrightarrow{b}$ products-d_1 + M + R—Ȯ				
AIBN as initiator [26])	vinylacetate/CCl_4	323	k_a = 26.0(40) $M^{-1}s^{-1}$ k_b = 32.0(40) $M^{-1}s^{-1}$ [40])	75 Sim 3
(Polyvinylacetate)$^{\cdot}$ [22]) + triphenylmethane{$(C_6H_5)_3CH$} ⟶ products + $(C_6H_5)_3\dot{C}$				
AIBN as initiator Dilatometry	vinylacetate	303 323 343	k = 49 $M^{-1}s^{-1}$ 76 109 log[A/$M^{-1}s^{-1}$] = 4.79 E_a = 18.0(42) kJ mol^{-1}	73 Sim 2/ 67 Tüd 1, 67 Sim 2, 64 Ber 1
(Rhodamine 6G)$^{\overset{+}{\cdot}}$ [41]) + $CH_3CH_2CH_2OH$ ⟶ $CH_3CH_2\dot{C}HOH$ + products				
Flash phot. KAS	n-C_3H_7OH	RT	k = 25 $M^{-1}s^{-1}$	78 Kor 1

[22]) Radical mixture.
[23]) Calc. from inhibition rate of polymerization (dilatometric measurement).
[25]) Refers to one reaction center.
[26]) Inhibition of radical polymerization (dilatometry).
[27]) Hydrogen bridging with monomer vinylacetate (M).
[29]) k(H)/k(D) > 10 indicates tunneling.
[40]) Isotope effects k_a(H)/k_a(D) = 9.6(20); k_b(H)/k_b(D) = 13.9(26).
[41]) At least partially C-centered radical.

Reaction				
Radical generation				Ref./
Method	Solvent	T[K]	Rate data	add. ref.

(Thymine-OH˙)[42]) + $CH_3OH \longrightarrow \dot{C}H_2OH$ + products

Rad. of thymine + N_2O + H_2O				72 Fel 1,
PR, glc	H_2O	RT	$k = 8(2) \cdot 10^3\,M^{-1}\,s^{-1}$ [43])	72 Fel 2/
				68 Dan 1

(Uridine-OH˙)[44]) + $HSCH_2CH(NH_2)COOH \longrightarrow \dot{S}CH_2CH(NH_2)COOH$ + products

Pulse rad. of uridine + N_2O + H_2O				75 Gre 1
KAS	H_2O	RT	$k = 2.0 \cdot 10^8\,M^{-1}\,s^{-1}$ [45])	

(Uridine-5′-monophosphate-OH˙)[46]) + $HSCH_2CH(NH_2)COOH \longrightarrow \dot{S}CH_2CH(NH_2)COOH$ + products

Pulse rad. of uridine-5′-monophosphate + N_2O + H_2O				75 Gre 1
KAS	H_2O	RT	$k = 3.9 \cdot 10^8\,M^{-1}\,s^{-1}$ [45])	

4.1.2 Relative rate constants

4.1.2.1 Aliphatic radicals and radicals derived from other non-aromatic compounds

4.1.2.1.1 Radicals containing only C and H atoms

$\dot{C}D_3 + CH_3OH \xrightarrow{a} CD_3H + \dot{C}H_2OH$
$\quad + CD_3COCD_3 \xrightarrow{b} CD_4 + \dot{C}D_2COCD_3$

Phot. of acetone-d_6				63 Che 1
PR	$CD_3COCD_3/$	303.0(1)	$k_a/k_b = 0.56(5)$	
	CH_3OH			

$\dot{C}H_3 + CBrCl_3 \xrightarrow{a} CH_3Br + \dot{C}Cl_3$
$\quad + CCl_4 \xrightarrow{b} CH_3Cl + \dot{C}Cl_3$

Reduct. of $CH_3HgOCOCH_3$ by $NaBH_4$				79 Gie 1
PR, glc	CCl_4	273	$k_a/k_b = 3400$	
		303	$= 2500$	
		333	$= 1700$ [1])	
		373	$= 1300$	
		403	$= 1000$	
			$\Delta H_a^{\ddagger} - \Delta H_b^{\ddagger} = -8.7\,kJ\,mol^{-1}\,(\pm 10\%)$	
			$\Delta S_a^{\ddagger} - \Delta S_b^{\ddagger} = 36\,J\,mol^{-1}\,K^{-1}\,(\pm 5\%)$	

$\dot{C}H_3 + CBrCl_3 \xrightarrow{a} CH_3Br + \dot{C}Cl_3$
$\quad + C_6H_5CH_3 \xrightarrow{b} CH_4 + C_6H_5\dot{C}H_2$

Thermal decomp. of acetylperoxide				61 Fox 1/
PR, glc	toluene	321.5	$k_a/k_b = 7400$	60 Eva 1
		350.2	$= 7100(400)$	

$\dot{C}H_3 + CCl_4 \xrightarrow{a} CH_3Cl + \dot{C}Cl_3$
$\quad + C_6H_5CH_3 \xrightarrow{b} CH_4 + C_6H_5\dot{C}H_2$

Thermal decomp. of acetylperoxide				61 Fox 1/
PR, glc	toluene	338	$k_a/k_b = 4.2(1)$	60 Eva 1

[42]) Radicals from thymine + $\dot{O}H$ reaction.
[43]) Based on radical-radical termination $2k = 1 \cdot 10^9\,M^{-1}\,s^{-1}$ [68 Dan 1].
[44]) Radicals from uridine + $\dot{O}H$ reaction.
[45]) Measured via RSSR˙⁻ absorption ($R\dot{S} + RS^- \rightleftharpoons RSSR^{\overline{\cdot}}$).
[46]) Radicals from uridine-5′-monophosphate + $\dot{O}H$ reactions.
[1]) Radical generation via therm. of $CH_3CO_3C(CH_3)_3$ (perester).

Reaction Radical generation Method	Solvent	T[K]	Rate data	Ref./ add. ref.
$\dot{C}H_3 + CF_3I \xrightarrow{a} CH_3I + \dot{C}F_3$ $+ C_6H_5CH_3 \xrightarrow{b} CH_4 + C_6H_5\dot{C}H_2$				
Thermal decomp. of acetylperoxide				61 Fox 1/
PR, glc	toluene	338.2	$k_a/k_b = 21500(800)$	60 Eva 1
$\dot{C}H_3 + CHBrCl_2 \xrightarrow{a} CH_3Br + \dot{C}HCl_2$ $+ C_6H_5CH_3 \xrightarrow{b} CH_4 + C_6H_5\dot{C}H_2$				
Thermal decomp. of acetylperoxide				61 Fox 1/
PR, glc	toluene	338.2	$k_a/k_b = 124$	60 Eva 1
		350.1	113	
			$\log[A_a/A_b] = 2.9$	
			$E_a(a) - E_a(b) = -10.9(21)\,kJ\,mol^{-1}$	
Phot. of azomethane				61 Fox 1/
PR, glc	toluene	273	$k_a/k_b = 328$	60 Eva 1
		298.5	227	
		317.6	154	
		338	138	
		367.5	96	
$\dot{C}H_3 + CHBrCl_2 \xrightarrow{a} CH_4 + \dot{C}BrCl_2$ $+ C_6H_5CH_3 \xrightarrow{b} CH_4 + C_6H_5\dot{C}H_2$				
Thermal decomp. of acetylperoxide				61 Fox 1
PR, glc	toluene	338.2	$k_a/k_b = 23$	
		350.1	$= 18$	
Phot. of azomethane				61 Fox 1
PR, glc	toluene	273	$k_a/k_b = 57$	
		298.5	30	
		317.6	29	
		338	30	
		367.5	25	
$\dot{C}H_3 + CHCl_3 \xrightarrow{a} CH_4 + \dot{C}Cl_3$ $+ CCl_4 \xrightarrow{b} CH_3Cl + \dot{C}Cl_3$				
Thermal decomp. of acetylperoxide				50 Edw 1
PR	CCl_4	373	$k_a/k_b = 11.1$	
$\dot{C}H_3 + CH_2BrCl \xrightarrow{a} CH_3Br + \dot{C}H_2Cl$ $+ C_6H_5CH_3 \xrightarrow{b} CH_4 + C_6H_5\dot{C}H_2$				
Thermal decomp. of acetylperoxide				60 Eva 1
PR [2])	toluene	338	$k_a/k_b = 1.4$	
$\dot{C}H_3 + CH_2BrCl \xrightarrow{a} CH_3Br + \dot{C}H_2Cl$ $+ C_6H_5CH_3 \xrightarrow{b} CH_4 + C_6H_5\dot{C}H_2$				
Thermal decomp. of acetylperoxide				61 Fox 1
PR, glc	toluene	321.5	$k_a/k_b = 0.90$	
		350.1	1.75	
$\dot{C}H_3 + CH_2BrCl \xrightarrow{a} CH_4 + \dot{C}HBrCl$ $+ C_6H_5CH_3 \xrightarrow{b} CH_4 + C_6H_5\dot{C}H_2$				
Thermal decomp. of acetylperoxide				61 Fox 1
PR, glc	toluene	321.5	$k_a/k_b = 0.6$ [3])	
		350.1	$= 1.3$ [3])	

[2]) And radiometric methods.
[3]) Per active H-atom.

Reaction Radical generation Method	Solvent	T[K]	Rate data	Ref./ add. ref.
$\dot{C}H_3 + CH_2ClI \xrightarrow{a} CH_3I + \dot{C}H_2Cl$ $+ C_6H_5CH_3 \xrightarrow{b} CH_4 + C_6H_5\dot{C}H_2$				
Thermal decomp. of acetylperoxide				61 Fox 1,
PR, glc	toluene	338.2	$k_a/k_b = 6400(500)$	60 Eva 1
$\dot{C}H_3 + CH_3Br \xrightarrow{a} CH_3Br + \dot{C}H_3$ $+ C_6H_5CH_3 \xrightarrow{b} CH_4 + C_6H_5\dot{C}H_2$				
Thermal decomp. of acetylperoxide				61 Eva 1/
PR, glc	toluene	338	$k_a/k_b = 6 \cdot 10^{-3}$	60 Eva 1
$\dot{C}H_3 + CH_3I \xrightarrow{a} CH_3I + \dot{C}H_3$ $+ C_6H_5CH_3 \xrightarrow{b} CH_4 + C_6H_5\dot{C}H_2$				
Thermal decomp. of acetylperoxide				61 Eva 1,
PR, glc	toluene	338	$k_a/k_b = 45$	60 Eva 1
		328...358	$A_a/A_b = 3.0$ $E_a(a) - E_a(b) = -7.5(21)$ kJ mol^{-1}	
$\dot{C}H_3 + CH_3NO_2 \xrightarrow{a} CH_4 + \dot{C}H_2NO_2$ $+ CH_3COCH_3 \xrightarrow{b} CH_4 + \dot{C}H_2COCH_3$				
Phot. of acetone in H_2O				71 Tra 1
PR, glc	H_2O	295.0(5)	$k_a/k_b = 140(15)$	
$\dot{C}H_3 + CH_3OT \xrightarrow{a} CH_3T + CH_3\dot{O}$ $+ CH_3OH \xrightarrow{b} \dot{C}H_2OH + CH_4$				
Thermal decomp. of acetylperoxide				66 Kel 1
PR, specific activity	T-labelled methanol	333.0(1)	$k_a/k_b = 6.76 \cdot 10^{-2}$	
		343.0(1)	$= 6.14 \cdot 10^{-2}$	
		353.0(1)	$= 5.95 \cdot 10^{-2}$	
		363.0(1)	$= 5.58 \cdot 10^{-2}$	
			$A_a/A_b = 5.3 \cdot 10^{-2}$ $E_a(a) - E_a(b) = -4.2$ kJ mol^{-1}	
$\dot{C}H_3 + CH_3OT \xrightarrow{a} CH_3T + CH_3\dot{O}$ $+ n\text{-}C_7H_{16} \xrightarrow{b} CH_4 + (n\text{-}C_7H_{15}^{\cdot})$				
Thermal decomp. of acetylperoxide				66 Kel 1
PR, specific activity	n-heptane/ T-labelled methanol	333.0(1)	k_a/k_b [4]) $= 5.08 \cdot 10^{-2}$	
		343.0(1)	$5.32 \cdot 10^{-2}$	
		353.0(1)	$5.56 \cdot 10^{-2}$	
		363.0(1)	$5.74 \cdot 10^{-2}$	
			$A_a/A_b = 0.2$ $E_a(a) - E_a(b) = 3.3$ kJ mol^{-1}	
$\dot{C}H_3 + CH_3COOH \xrightarrow{a} CH_4 + \dot{C}H_2COOH$ $+ n\text{-}C_7H_{16} \xrightarrow{b} CH_4 + (n\text{-}C_7H_{15}^{\cdot})$				
Therm. of acetylperoxide				66 Nem 1
PR [5])	$n\text{-}C_7H_{16}$	353	$k_a/k_b = 2.24(12)$	

[4]) Extrapolated to low alcohol concentrations since $k_a/k_b = f$[alcohol] due to hydrogen bonding, and based on $k(\dot{C}H_3 + CH_3OH \longrightarrow CH_4 + CH_3\dot{O})/k_b = 1.3$.
[5]) Specific activity of T-labelled compounds.

Reaction Radical generation Method	Solvent	T[K]	Rate data	Ref./ add. ref.
$\dot{C}H_3 + CH_3COOH \xrightarrow{a} CH_4 + \dot{C}H_2COOH$ $+ n\text{-}C_7H_{16} \xrightarrow{b} CH_4 + (n\text{-}C_7\dot{H}_{15})$				
Therm. of acetylperoxide				66 Dob 1/
PR [5])	CH_3COOH/	333	$k_a/k_b = 1.84(7)$	68 Dob 1
	n-C_7H_{16} (T-labelled)	343	2.05(4)	
		353	2.20(13)	
		363	2.35(17)	
$\dot{C}H_3 + C_2H_5I \xrightarrow{a} CH_3I + CH_3\dot{C}H_2$ $+ C_6H_5CH_3 \xrightarrow{b} CH_4 + C_6H_5\dot{C}H_2$				
Thermal decomp. of acetylperoxide				60 Eva 1
PR [6])	toluene	338	$k_a/k_b = 180$	
			$A_a/A_b = 10$	
			$E_a(a) - E_a(b) = -8.0(42)\ \text{kJ mol}^{-1}$	
$\dot{C}H_3 + C_2H_5I \xrightarrow{a} CH_3I + CH_3\dot{C}H_2$ $+ C_6H_5CH_3 \xrightarrow{b} CH_4 + C_6H_5\dot{C}H_2$				
Thermal decomp. of acetylperoxide				61 Fox 1/
PR, glc	toluene	318	$k_a/k_b = 229(6)$	68 Eac 1
		327.8	$= 198(4)$	
		338.2	$= 181(3)$	
		348.8	$= 166(2)$	
		357.9	$= 159(2)$	
			$A_a/A_b = 14.7$	
			$E_a(a) - E_a(b) = -7.5(21)\ \text{kJ mol}^{-1}$	
$\dot{C}H_3 + CH_3CH{=}CH_2 \xrightarrow{a} CH_4 +$ propylene$(-\dot{H})$ $+ (CH_3)_3CCH_2CH(CH_3)_2 \xrightarrow{b} CH_4 +$ 2,2,4-trimethylpentane$(-\dot{H})$				
Thermal decomp. of acetylperoxide				57 Buc 1/
PR	2,2,4-trimethylpentane	338.1	$k_a/k_b = 1.5$ [7])	54 Lev 1, 55 Lev 1
$\dot{C}H_3 + CH_3COCH_3 \xrightarrow{a} CH_4 + \dot{C}H_2COCH_3$ $2\dot{C}H_3 \xrightarrow{b} C_2H_6$				
Phot. of acetone				55 Pie 1
PR	CH_3COCH_3	249	$k_a/(2k_b)^{\frac{1}{2}} = 0.21 \cdot 10^{-14}$ $\text{cm}^{\frac{3}{2}}\ \text{molecules}^{-\frac{1}{2}}\ \text{s}^{-\frac{1}{2}}$	
		258	$0.35 \cdot 10^{-14}$	
		273	$0.84 \cdot 10^{-14}$	
		279	$1.26 \cdot 10^{-14}$	
		288	$2.24 \cdot 10^{-14}$	
		306	$4.39 \cdot 10^{-14}$	
		313	$6.1 \cdot 10^{-14}$	
		328	$9.5 \cdot 10^{-14}$	
			$E_a(a) - \frac{1}{2}E_a(b) = 33.5\ \text{kJ mol}^{-1}$	

[5]) Specific activity of T-labelled compounds.
[6]) And radiometric methods.
[7]) $k_a/k(\dot{C}H_3$ addition to propylene$) = 0.03$.

Reaction Radical generation Method	Solvent	T[K]	Rate data	Ref./ add. ref.
$\dot{C}H_3 + CH_3COCH_3 \xrightarrow{a} CH_4 + \dot{C}H_2COCH_3$ $2\dot{C}H_3 \xrightarrow{b} C_2H_6$				
Phot. of acetone				61 Doe 1/
PR	perfluorodimethyl-cyclobutane/ CH_3COCH_3	275	$k_a/(2k_b)^{\frac{1}{2}} = 1.2(2)\cdot 10^{-14}$ cm$^{\frac{3}{2}}$ molecule$^{-\frac{1}{2}}$ s$^{-\frac{1}{2}}$	59 Pet 1, 60 Vol 1
		297	$4.0(4)\cdot 10^{-14}$	
		314	$6.0(6)\cdot 10^{-14}$	
			$E_a(a) - \frac{1}{2}E_a(b) = 33.9$ kJ mol^{-1}	
$\dot{C}H_3 + CH_3COCH_3 \xrightarrow{a} CH_4 + \dot{C}H_2COCH_3$ $2\dot{C}H_3 \xrightarrow{b} C_2H_6$				
Phot. of acetone				60 Vol 1/
PR	H_2O/CH_3COCH_3	275	$k_a/(2k_b)^{\frac{1}{2}} = 0.57\cdot 10^{-12}$ cm$^{\frac{3}{2}}$ molecules$^{-\frac{1}{2}}$ s$^{-\frac{1}{2}}$	61 Doe 1, 59 Pet 1
		300	$0.99\cdot 10^{-12}$	
		322	$1.9\cdot 10^{-12}$	
		346	$2.0\cdot 10^{-12}$	
$\dot{C}H_3 + CH_3COCH_3 \xrightarrow{a} CH_4 + \dot{C}H_2COCH_3$ $2\dot{C}H_3 \xrightarrow{b} C_2H_6$				
Phot. of acetone				69 Tah 1/
PR, glc	H_2O	298	$k_a/(2k_b)^{\frac{1}{2}} = 1.0$ cm$^{\frac{3}{2}}$ mol$^{-\frac{1}{2}}$ s$^{-\frac{1}{2}}$	69 Kor 1
		308	1.34	
		318	1.86	
		328	2.56 [8])	
			$E_a(a) = 33.1$ kJ mol^{-1} [9])	
Phot. of acetone in H_2O				71 Tra 1
PR, glc	H_2O	295.0(5)	$k_a/(2k_b)^{\frac{1}{2}} = 2.55(25)\cdot 10^{-2}$ M$^{-\frac{1}{2}}$ s$^{-\frac{1}{2}}$	
$\dot{C}H_3 + CH_3COCH_3 \xrightarrow{a} CH_4 + \dot{C}H_2COCH_3$ $+ CCl_4 \xrightarrow{b} CH_3Cl + \dot{C}Cl_3$				
Thermal decomp. of acetylperoxide				50 Edw 1
PR	CCl_4	373	$k_a/k_b = 0.40$	
$\dot{C}H_3 + CH_3CH_2COOH \xrightarrow{a} CH_4 +$ propionic acid($-\dot{H}$) $+ n\text{-}C_7H_{16} \xrightarrow{b} CH_4 + (n\text{-}C_7\dot{H}_{15})$				
Therm. of acetylperoxide				66 Nem 1
PR [10])	n-C_7H_{16}	353	$k_a/k_b = 16.35(37)$ [11])	
PR [12])	C_2H_5COOH/ n-C_7H_{16} (T-labelled)	333	$k_a/k_b = 27.13(132)$ [13])	66 Dob 1/
			18.82(95) [14])	68 Dob 1
		343	25.08(35) [13])	
			17.79(11) [14])	
		353	22.98(36) [13])	
			16.74(19) [14])	
		363	21.09(92) [13])	
			15.80(64) [14])	

[8]) $\log[k_a/\text{cm}^3\,\text{mol}^{-1}\,\text{s}^{-1}] = (4.47(5) - 1335(18)/T) - 0.5\log k(\dot{R} + \dot{R})$.
[9]) Based on $E_a(b) = 15.1$ kJ mol^{-1} [69 Kor 1].
[10]) Specific activity of T-labelled compounds.
[11]) $k_a/k_b = 8.0$ per α-H.
[12]) Specific activity measurements.
[13]) Extrapolated to 100% C_2H_5COOH.
[14]) Extrapolated to 100% n-C_7H_{16}.

Reaction Radical generation Method	Solvent	T[K]	Rate data	Ref./ add. ref.
$\dot{C}H_3 + CH_3COOCH_3 \xrightarrow{a} CH_4$ + methylacetate(−Ḣ) $+ CCl_4 \xrightarrow{b} CH_3Cl + \dot{C}Cl_3$				
Thermal decomp. of acetylperoxide				50 Edw 1
PR	CCl_4	373	$k_a/k_b = 21$	
$\dot{C}H_3 + CH_3COOCH_3 \xrightarrow{a} CH_4$ + methylacetate(−Ḣ) $+ (CH_3)_3CSD \xrightarrow{b} CH_3D + (CH_3)_3C\dot{S}$				
Thermal decomp. of t-butylperacetate				72 Pry 1
PR [15])	methylacetate/ $(CH_3)_3CSD$ (5:1)	383	$k_a/k_b = 4.02(15) \cdot 10^{-2}$	
$\dot{C}H_3 + (CH_3)_2CHI \xrightarrow{a} CH_3I + (CH_3)_2\dot{C}H$ $+ C_6H_5CH_3 \xrightarrow{b} CH_4 + C_6H_5\dot{C}H_2$				
Thermal decomp. of acetylperoxide				60 Eva 1
PR, radiometric methods	toluene	338	$k_a/k_b = 870$ $A_a/A_b = 11$ $E_a(a) - E_a(b) = -12.1(42)$ kJ mol^{-1}	
$\dot{C}H_3 + (CH_3)_2CHI \xrightarrow{a} CH_3I + (CH_3)_2\dot{C}H$ $+ C_6H_5CH_3 \xrightarrow{b} CH_4 + C_6H_5\dot{C}H_2$				
Thermal decomp. of acetylperoxide				61 Fox 1/ 68 Eac 1
PR, glc	toluene	318 328.1 338.2 348.8 358.2	$k_a/k_b = 1055(50)$ 977(25) 868(20) 686(10) 625(20) $A_a/A_b = 5.4$ $E_a(a) - E_a(b) = -14.2(42)$ kJ mol^{-1}	
$\dot{C}H_3 + (CH_3)_2CHOT \xrightarrow{a} CH_3T + (CH_3)_2CH\dot{O}$ $+ n\text{-}C_7H_{16} \xrightarrow{b} CH_4 + (n\text{-}C_7\dot{H}_{15})$				
Thermal decomp. of acetylperoxide				66 Kel 1
PR [16])	n-heptane/T-labelled 2-propanol	353.0(1)	k_a/k_b [17]) $= 5.7 \cdot 10^{-2}$	
$\dot{C}H_3 + (CH_3)_2CHCOCl \xrightarrow{\alpha} CH_4 + (CH_3)_2\dot{C}COCl$ $\dot{C}H_3 + (CH_3)_2CHCOCl \xrightarrow{\beta} CH_4 + \dot{C}H_2(CH_3)CHCOCl$				
Thermal decomp. of acetylperoxide				53 Pri 1
PR, MS	$(CH_3)_2CHCOCl$	373	$k_\alpha/k_\beta = 12.4(2)$ [18])	
$\dot{C}H_3 + (CH_3)_2CHCOCl \xrightarrow{\beta} CH_4 + \dot{C}H_2(CH_3)CHCOCl$ $+ (CH_3)_2CDCOCl \xrightarrow{\alpha} CH_3D + (CH_3)_2\dot{C}COCl$				
Thermal decomp. of acetylperoxide				53 Pri 1
PR, MS	$(CH_3)_2CHCOCl$/ $(CH_3)_2CDCOCl$ mixt.	373	$k_\beta(H)/k_\alpha(D) = 0.098(10)$	

[15]) MS of CH_4 and CH_3D.
[16]) Specific activity.
[17]) Extrapolated to low alcohol concentration since $k_a/k_b = f$[alcohol] due to hydrogen bonding, and based on $k(\dot{C}H_3 + (CH_3)_2CHOH \longrightarrow CH_4 + (CH_3)_2CH\dot{O})/k_b = 1.3$.
[18]) Derived from experiments with $(CH_3)_2CDCOCl$.

Reaction Radical generation Method	Solvent	T[K]	Rate data	Ref./ add. ref.
$\dot{C}H_3 + CH_2{=}CHCH_2CH_3 \xrightarrow{a} CH_4 + \text{butene-1}(-\dot{H})$ $\xrightarrow{b} (\text{butene-1-}CH_3^{\cdot})$				
Thermal decomp. of acetylperoxide				60 Ste 1/
PR	2,2,4-trimethylpentane	344	$k_a/k_b = 0.28$	60 Fel 1
$\dot{C}H_3 + CH_2{=}CHCH_2CH_3 \xrightarrow{a} CH_4 + \text{butene-1}(-\dot{H})$ $+ (CH_3)_3CCH_2CH(CH_3)_2 \xrightarrow{b} CH_4 + \text{2,2,4-trimethylpentane}(-\dot{H})$				
Thermal decomp. of acetylperoxide				57 Buc 1/
PR	2,2,4-trimethylpentane	338.1	$k_a/k_b = 6.6$	54 Lev 1,
		358.3	5.6 [19])	55 Lev 1
$\dot{C}H_3 + CH_2{=}C(CH_3)_2 \xrightarrow{a} CH_4 + \text{isobutene}(-\dot{H})$ $+ (CH_3)_3CCH_2CH(CH_3)_2 \xrightarrow{b} CH_4 + \text{2,2,4-trimethylpentane}(-\dot{H})$				
Thermal decomp. of acetylperoxide				57 Buc 1/
PR	2,2,4-trimethylpentane	338.1	$k_a/k_b = 2.2$ [20])	54 Lev 1, 55 Lev 1, 56 Buc 1
$\dot{C}H_3 + cis\text{-}CH_3CH{=}CHCH_3 \xrightarrow{a} CH_4 + cis\text{-butene-2}(-\dot{H})$ $+ (CH_3)_3CCH_2CH(CH_3)_2 \xrightarrow{b} CH_4 + \text{2,2,4-trimethylpentane}(-\dot{H})$				
Thermal decomp. of acetylperoxide				57 Buc 1/
PR	2,2,4-trimethylpentane	328.0	$k_a/k_b = 4.7$	54 Lev 1,
		338.1	3.2	55 Lev 1,
		358.3	3.1 [21])	56 Buc 1
$\dot{C}H_3 + trans\text{-}CH_3CH{=}CHCH_3 \xrightarrow{a} CH_4 + trans\text{-butene-2}(-\dot{H})$ $+ (CH_3)_3CCH_2CH(CH_3)_2 \xrightarrow{b} CH_4 + \text{2,2,4-trimethylpentane}(-\dot{H})$				
Thermal decomp. of acetylperoxide				57 Buc 1/
PR	2,2,4-trimethylpentane	328.0	$k_a/k_b = 6.2$	54 Lev 1,
		338.1	4.9	55 Lev 1,
		358.3	4.7 [22])	56 Buc 1
$\dot{C}H_3 + CH_3COCH_2CH_3 \xrightarrow{a} CH_4 + \text{methylethylketone}(-\dot{H})$ $+ (CH_3)_3CCH_2CH(CH_3)_2 \xrightarrow{b} CH_4 + \text{2,2,4-trimethylpentane}(-\dot{H})$				
Thermal decomp. of acetylperoxide				56 Buc 1
PR, glc	2,2,4-trimethylpentane	338	$k_a/k_b = 9$	
$\dot{C}H_3 + CH_3CH_2COCH_3 \xrightarrow{a} CH_4 + \text{methylethylketone}(-\dot{H})$ $+ \text{olefine} \xrightarrow{b} (\text{olefine-}CH_3^{\cdot})$				
Thermal decomp. of acetylperoxide				57 Bad 1
PR	methylethylketone	338	olefine:	[23])/
			diethylmaleate, $k_a/k_b = 4.4 \cdot 10^{-2}$	57 Lea 1
			diethylfumarate, $4.9 \cdot 10^{-3}$	
			maleic anhydride, $2.6 \cdot 10^{-3}$	
			chloromaleic anhydride, $1.8 \cdot 10^{-3}$	
			dichloromaleic anhydride, $\approx 2.5 \cdot 10^{-2}$	
			maleonitrile, $5.7 \cdot 10^{-3}$	
(continued)			fumaronitrile, $5.6 \cdot 10^{-3}$	

[19]) $k_a/k(CH_3$ addition to butene-1) = 0.25 at 338 K.
[20]) $k_a/k(\dot{C}H_3$ addition to isobutene) = 0.06.
[21]) $k_a/k(\dot{C}H_3$ addition to *cis*-butene-2) ≈ 1.0 at 338 K.
[22]) $k_a/k(\dot{C}H_3$ addition to *trans*-butene-2) ≈ 0.75 at 338 K.
[23]) Further data at other temperatures.

Reaction Radical generation Method	Solvent	T[K]	Rate data		Ref./ add. ref.
$\dot{C}H_3 + CH_3CH_2COCH_3 \xrightarrow{a} CH_4$ + methylethylketone($-\dot{H}$) (continued) + olefine $\xrightarrow{b}$ (olefine-$CH_3^{\cdot}$)					
Thermal decomp. of acetylperoxide					
PR	methylethylketone	338	olefine:		55 Lea 1
			styrene,	$k_a/k_b = 1.14 \cdot 10^{-2}$	[24])/
			1,1-diphenyl-ethylene,	$5.56 \cdot 10^{-3}$	55 Szw 1, 60 Fel 1
			diethylmaleate,	$4.46 \cdot 10^{-2}$	
			vinylacetate,	0.26	
Therm. of acetylperoxide					57 Lea 1
PR	methylethylketone	338	olefine:		
			diethylmaleate,	$k_a/k_b = 0.0437$	
			diethylfumarate,	0.00735	
			ethylene glycol-maleic anhydride polyester,	0.00676	
		358	diethylmaleate,	0.0595	
			diethylfumarate,	0.00833	
			ethylene glycol-maleic anhydride polyester,	0.00806	
$\dot{C}H_3 + CH_3CH_2CH_2COOH \xrightarrow{a} CH_4$ + n-butyric acid($-\dot{H}$) + n-$C_7H_{16} \xrightarrow{b} CH_4$ + n-heptane($-\dot{H}$)					
Thermal decomp. of diacetylperoxide					68 Dob 1
PR[25])	n-heptane	353	$k_a/k_b = 18.70$		
		333...363	$A_a/A_b = 2.58$ $E_a(a) - E_a(b) = 5.7\,kJ\,mol^{-1}$		
$\dot{C}H_3 + (CH_3)_2CHCOOH \xrightarrow{a} CH_4 + (CH_3)_2\dot{C}COOH$ + n-$C_7H_{16} \xrightarrow{b} CH_4 + (n\text{-}C_7H_{15}^{\cdot})$					
Therm. of acetylperoxide					66 Nem 1
PR[26])	n-C_7H_{16}	353	$k_a/k_b = 36.71(50)$		
		333...363	$\log[A_a/A_b] = 1.15$ $E_a(a) - E_a(b) = -2.9(2)\,kJ\,mol^{-1}$		
$\dot{C}H_3 + (CH_3)_2CHCOOH \xrightarrow{a} CH_4$ + isobutyric acid($-\dot{H}$) + n-$C_7H_{16} \xrightarrow{b} CH_4$ + n-heptane($-\dot{H}$)					
Thermal decomp. of diacetylperoxide					68 Dob 1
PR[25])	n-heptane	353	$k_a/k_b = 62.93$		
		333...363	$\log[A_a/A_b] = 1.29$ $E_a(a) - E_a(b) = -3\,kJ\,mol^{-1}$		
$\dot{C}H_3$ + dioxan $\xrightarrow{a} CH_4$ + dioxan($-\dot{H}$) $+ (CH_3)_3CSD \xrightarrow{b} CH_3D + (CH_3)_3C\dot{S}$					
Thermal decomp. of t-butylperacetate					72 Pry 1
PR[27])	dioxan/$(CH_3)_3CSD$ (5:1)	383	$k_a/k_b = 4.88(46) \cdot 10^{-2}$		
$\dot{C}H_3 + (CH_3)_3CI \xrightarrow{a} CH_3I + (CH_3)_3\dot{C}$ $+ C_6H_5CH_3 \xrightarrow{b} CH_4 + C_6H_5\dot{C}H_2$					
Thermal decomp. of acetylperoxide					60 Eva 1
PR, radiometric methods	toluene	338	$k_a/k_b = 1680$		

[24]) Also data at 358 K.
[25]) Analysis of T-labelled products.
[26]) Specific activity of T-labelled compounds.
[27]) MS of CH_4 and CH_3D.

Reaction Radical generation Method	Solvent	T[K]	Rate data	Ref./ add. ref.
$\dot{C}H_3 + (CH_3)_3CI \xrightarrow{a} CH_3I + (CH_3)_3\dot{C}$ $+ C_6H_5CH_3 \xrightarrow{b} CH_4 + C_6H_5\dot{C}H_2$				
Thermal decomp. of acetylperoxide				61 Fox 1
PR, glc	toluene	338.2	$k_a/k_b = 1870(100)$	
$\dot{C}H_3 + (CH_3)_3COT \xrightarrow{a} CH_3T + (CH_3)_3C\dot{O}$ $+ n\text{-}C_7H_{16} \xrightarrow{b} CH_4 + (n\text{-}C_7H_{15}^{\cdot})$				
Thermal decomp. of acetylperoxide				66 Kel 1
PR [28])	n-heptane/T-labelled t-butanol	353.0(1)	k_a/k_b [29]) $= 4.0 \cdot 10^{-2}$	
$\dot{C}H_3 + C_2H_5OC_2H_5 \xrightarrow{a} CH_4 + \text{diethylether}(-\dot{H})$ $+ (CH_3)_3CSD \xrightarrow{b} CH_3D + (CH_3)_3C\dot{S}$				
Thermal decomp. of t-butylperacetate				72 Pry 1
PR [30])	diethylether/ $(CH_3)_3CSD$ (5:1)	383	$k_a/k_b = 1.790(181) \cdot 10^{-1}$	
$\dot{C}H_3 + (CH_3)_3COOH$ [31]) $\xrightarrow{a} CH_4$ [31]) $+ (CH_3)_3CO\dot{O}$ $+ n\text{-}C_7H_{16} \xrightarrow{b} CH_4 + n\text{-heptane}(-\dot{H})$				
Thermal decomp. of acetylperoxide				66 Ber 1
PR [32])	n-heptane	333	$k_a/k_b = 3.41$	
		343	$= 3.25$	
		353	$= 3.05$	
		363	$= 2.79$	
			$k_a/k_b = 0.3(5)\, e^{1620(200)/RT}$	
$\dot{C}H_3 + c\text{-pentadiene} \xrightarrow{a} CH_4 + c\text{-pentadiene}(-\dot{H})$ $+ (CH_3)_3CCH_2CH(CH_3)_2 \xrightarrow{b} CH_4 + \text{2,2,4-trimethylpentane}(-\dot{H})$				
Thermal decomp. of acetylperoxide				61 Gre 1
PR, glc	2,2,4-trimethylpentane	338	$k_a/k_b = 30$	
$\dot{C}H_3 + (CH_3)_2C{=}CHCN \xrightarrow{a} CH_4 + \beta,\beta\text{-dimethylacrylonitrile}(-\dot{H})$ $\xrightarrow{b} (\beta,\beta\text{-dimethylacrylonitrile-}CH_3^{\cdot})$				
Thermal decomp. of acetylperoxide				61 Her 1/
PR	2,2,4-trimethylpentane	338	$k_a/k_b = 0.538$	57 Buc 1
$\dot{C}H_3 + c\text{-}C_5H_8 \xrightarrow{a} CH_4 + c\text{-pentene}(-\dot{H})$ $+ (CH_3)_3CCH_2CH(CH_3)_2 \xrightarrow{b} CH_4 + \text{2,2,4-trimethylpentane}(-\dot{H})$				
Thermal decomp. of acetylperoxide				61 Gre 1
PR, glc	2,2,4-trimethylpentane	338	$k_a/k_b = 0.81$	
$\dot{C}H_3 + CH_2{=}CHCH_2CH{=}CH_2 \xrightarrow{a} CH_4 + CH_2{=}CH\dot{C}HCH{=}CH_2$ $\xrightarrow{b} (\text{pentadiene-1,4-}CH_3^{\cdot})$ $+ (CH_3)_3CCH_2CH(CH_3)_2 \xrightarrow{c} CH_4 + \text{2,2,4-trimethylpentane}(-\dot{H})$				
Decomp. of acetylperoxide				59 Raj 1
PR	2,2,4-trimethylpentane	338	$k_a/k_c = 40$ [33]) $k_a/k_b = 0.017$	

[28]) Specific activity.
[29]) Extrapolated to low alcohol concentration since $k_a/k_b = f$[alcohol] due to hydrogen bridging, and based on $k(CH_3 + (CH_3)_3COH \longrightarrow CH_4 + (CH_3)_3C\dot{O})/k_b = 0.9$.
[30]) MS of CH_4 and CH_3D.
[31]) T-labelled compounds ($(CH_3)_3COOT$, CH_3T).
[32]) Analysis of T-labelled products.
[33]) $k_a/k_b = 20$ per C—H bond sec. to two double bonds.

Reaction Radical generation Method	Solvent	T[K]	Rate data	Ref./ add ref.
$\dot{C}H_3 + CH{\equiv}CCH_2CH_2CH_3 \xrightarrow{a} CH_4 + CH{\equiv}C\dot{C}HCH_2CH_3$ $+ (CH_3)_3CCH_2CH(CH_3)_2 \xrightarrow{b} CH_4 +$ 2,2,4-trimethylpentane$(-\dot{H})$				
Thermal decomp. of acetylperoxide				57 Gaz 1
PR	2,2,4-trimethylpentane	338	$k_a/k_b = 8.1$ [34]	
$\dot{C}H_3 + c\text{-}C_5H_8O \xrightarrow{a} CH_4 +$ cyclopentanone$(-\dot{H})$ [35] $+ n\text{-}C_7H_{16}$ [36] $\xrightarrow{b} CH_4$ [36] $+ n$-heptane$(-\dot{H})$ [36]				
Thermal decomp. of acetylperoxide				65 Ber 1
PR [37]	n-heptane [36]	333	$k_a/k_b = 19.8(4)$	
		343	$= 19.5(5)$	
		353	$= 19.5(10)$	
		363	$= 17.5(1)$	
			$E_a(a) - E_a(b) = 0(4)$ kJ mol^{-1}	
$\dot{C}H_3 + CH_2{=}CHCH_2OOCCH_3 \xrightarrow{a} CH_4 +$ allylacetate$(-\dot{H})$ $+ (CH_3)_3CCH_2CH(CH_3)_2 \xrightarrow{b} CH_4 +$ 2,2,4-trimethylpentane$(-\dot{H})$				
Thermal decomp. of acetylperoxide				56 Buc 1
PR, glc	2,2,4-trimethylpentane	338	$k_a/k_b = 1.9$	
		358	1.6	
$\dot{C}H_3 + c\text{-}C_5H_{10} \xrightarrow{a} CH_4 + (c\text{-}C_5\dot{H}_9)$ $+ (CH_3)_3CSD \xrightarrow{b} CH_3D + (CH_3)_3C\dot{S}$				
Thermal decomp. of t-butylperacetate				72 Pry 1
PR [38]	$c\text{-}C_5H_{10}/(CH_3)_3CSD$ (5:1)	383	$k_a/k_b = 2.11(21) \cdot 10^{-2}$	
$\dot{C}H_3 + c\text{-}C_5H_{10} \xrightarrow{a} CH_4 + (c\text{-}C_5\dot{H}_9)$ $+ c\text{-}C_6H_{12} \xrightarrow{b} CH_4 + (c\text{-}C_6\dot{H}_{11})$				
Therm. of acetylperoxide				62 Ber 1
PR [39]	$c\text{-}C_5H_{10}/c\text{-}C_6H_{12}$ (T-labelled)	273	$k_a/k_b = 0.86$	
		353	1.65	
		423	2.4	
			$\log[A_a/A_b] = 1.2$	
			$E_a(a) - E_a(b) = -6.7$ kJ mol^{-1}	
$\dot{C}H_3 + c\text{-}C_5H_{10} \xrightarrow{a} CH_4 + (c\text{-}C_5\dot{H}_9)$ $+ n\text{-}C_7H_{16}$ [40] $\xrightarrow{b} CH_4$ [40] $+ (n\text{-}C_7\dot{H}_{15})$				
Therm. of acetylperoxide				62 Ber 2
PR [41]	$c\text{-}C_5H_{10}/n\text{-}C_7H_{16}$ (T-labelled)	353	$k_a/k_b = 14.2(2)$ [40]	
$\dot{C}H_3 + CH_2{=}CHCH_2CH_2CH_3 \xrightarrow{a} CH_4 +$ pentene-1$(-\dot{H})$ $+ (CH_3)_3CCH_2CH(CH_3)_2 \xrightarrow{b} CH_4 +$ 2,2,4-trimethylpentane$(-\dot{H})$				
Thermal decomp. of acetylperoxide				57 Buc 1/
PR	2,2,4-trimethylpentane	338.1	$k_a/k_b = 8.1$	54 Lev 1,
		358.3	5.8 [42]	55 Lev 1

[34]) Per H-atom for abstraction from CH_2-group α to C≡C bond.
[35]) H-atom abstraction assumed from position α to keto group.
[36]) T-labelled compounds (4T-n-heptane, etc.).
[37]) Analysis of T-labelled product.
[38]) MS of CH_4 and CH_3D.
[39]) Use of T-labelled compounds.
[40]) Refers to one (T-labelled) reaction center.
[41]) Specific activity of T-labelled compounds.
[42]) $k_a/k(\dot{C}H_3$ addition to pentene-1$) = 0.33$ at 338 K.

Reaction Radical generation Method	Solvent	T[K]	Rate data	Ref./ add. ref.
$\dot{C}H_3 + CH_2{=}CHCH(CH_3)CH_3 \xrightarrow{a} CH_4$ + 3-methylbutene-1$(-\dot{H})$ $+ (CH_3)_3CCH_2CH(CH_3)_2 \xrightarrow{b} CH_4$ + 2,2,4-trimethylpentane$(-\dot{H})$				
Thermal decomp. of acetylperoxide				57 Buc 1/
PR	2,2,4-trimethylpentane	338.1	$k_a/k_b = 22.1$ [43]	54 Lev 1, 55 Lev 1
$\dot{C}H_3 + C_2H_5COC_2H_5 \xrightarrow{a} CH_4$ + diethylketone$(-\dot{H})$[44] $+ n\text{-}C_7H_{16}$[45] $\xrightarrow{b} CH_4$[45] + n-heptane$(-\dot{H})$				
Thermal decomp. of acetylperoxide				65 Ber 1
PR[46]	n-heptane[45]	333	$k_a/k_b = 16.3(2)$	
		343	$= 15.35(20)$	
		353	$= 13.9(2)$	
		363	$= 12.6(2)$	
			$E_a(a) - E_a(b) = -13(2)\,\text{kJ mol}^{-1}$	
$\dot{C}H_3 + CH_3(CH_2)_3COOH \xrightarrow{a} CH_4$ + valeric acid$(-\dot{H})$ $+ n\text{-}C_7H_{16} \xrightarrow{b} CH_4$ + n-heptane$(-\dot{H})$				
Thermal decomp. of diacetylperoxide				68 Dob 1
PR[46]	n-heptane	353	$k_a/k_b = 20.00$	
		333...	$A_a/A_b = 5.00$	
		363	$E_a(a) - E_a(b) = -4.1\,\text{kJ mol}^{-1}$	
$\dot{C}H_3 + 2\text{-}NO_2C_6H_4OT \xrightarrow{a} CH_3T + 2\text{-}NO_2C_6H_4\dot{O}$ $+ n\text{-}C_7H_{16} \xrightarrow{b} CH_4 + (C_7H_{15}^{\cdot})$				
Thermal decomp. of acetylperoxide				67 Koe 1
PR[46]	$n\text{-}C_7H_{16}$	353.0(1)	$k_a/k_b = 1.23(5)$	
$\dot{C}H_3 + 4\text{-}NO_2C_6H_4OT \xrightarrow{a} CH_3T + 4\text{-}NO_2C_6H_4\dot{O}$ $+ C_6H_5CH_3 \xrightarrow{b} CH_4 + C_6H_5\dot{C}H_2$				
Thermal decomp. of acetylperoxide				67 Koe 1
PR[46]	toluene	353.0(1)	$k_a/k_b = 4.0(8)$	
$\dot{C}H_3 + C_6H_5OT \xrightarrow{a} CH_3T + C_6H_5\dot{O}$ $+ n\text{-}C_7H_{16} \xrightarrow{b} CH_4 + (n\text{-}C_7H_{15}^{\cdot})$[47]				
Thermal decomp. of acetylperoxide				69 Shi 1
PR[46]	n-heptane/ T-labelled phenol	333.50(5)	k_a/k_b[48] $= 3.68(4)$	
		343.30(5)	3.77(8)	
		353.40(5)	3.92(9)	
		362.95(10)	3.95(4)	
			$A_a/A_b = 9.4(12)$	
			$E_a(a) - E_a(b) = -2.5(4)\,\text{kJ mol}^{-1}$	
$\dot{C}H_3 + C_6H_6 \xrightarrow{a} CH_4 + (C_6H_5^{\cdot})$ $+ CCl_4 \xrightarrow{b} CH_3Cl + \dot{C}Cl_3$				
Thermal decomp. of acetylperoxide				50 Edw 1
PR	CCl_4	373	$k_a/k_b = 0.039$	

[43]) $k_a/k(\dot{C}H_3$ addition to 3-methylbutene-1$) = 0.97$.
[44]) H-atom abstraction assumed from position α to keto group.
[45]) T-labelled compounds (4T-n-heptane, etc.).
[46]) Analysis of T-labelled products.
[47]) Refers to abstraction of sec. hydrogen atom.
[48]) Extrapolated to low phenol concentrations since $k_a/k_b = f$[phenol] due to hydrogen bonding.

Reaction Radical generation Method	Solvent	T[K]	Rate data	Ref./ add. ref.
$\dot{C}H_3 + C_6H_6$ [49] $\xrightarrow{a} CH_4$ [49] $+ (C_6\dot{H}_5)$ $+ n\text{-}C_7H_{16} \xrightarrow{b} CH_4 + (n\text{-}C_7\dot{H}_{15})$				
Therm. of acetylperoxide				62 Ber 2
PR [50]	T-labelled C_6H_6 (extrapol. to 100%)	353	$k_a/k_b = 0.61$ [49]	
	$n\text{-}C_7H_{16}$ (extrapol. to 100%)	353	$k_a/k_b = 0.40(5)$ [49]	
$\dot{C}H_3 + C_6H_5OH \xrightarrow{a} CH_4 + C_6H_5\dot{O}$ $+ n\text{-}C_7H_{16}$ [51] $\xrightarrow{b} CH_4$ [51] $+ (n\text{-}C_7\dot{H}_{15})$ [51]				
Thermal decomp. of acetylperoxide				66 Shi 1
PR, specific activity	phenol/n-heptane (T-labelled in 4-position)	333.50(5)	k_a/k_b [48] $= 443(8)$	
		343.30(5)	398(9)	
		353.40(5)	343(5)	
		362.95(10)	311(7)	
			$A_a/A_b = 5.1(11)$	
			$E_a(a) - E_a(b) = -12.6(6)\,\mathrm{kJ\,mol^{-1}}$	
$\dot{C}H_3$ + 1-cyanocyclopentene $\longrightarrow CH_4$ + 1-cyanocyclopentene($-\dot{H}$) $\longrightarrow$ (1-cyanocyclopentene-$\dot{C}H_3$)				
Thermal decomp. of acetylperoxide				61 Her 1/
PR	2,2,4-trimethylpentane	338	$k_a/k_b = 0.22$	57 Buc 1
$\dot{C}H_3 + c\text{-}C_6H_8\text{-}1,3 \xrightarrow{a} CH_4 + c$-hexadiene-1,3($-\dot{H}$) $+ n\text{-}C_7H_{16}$ [51] $\xrightarrow{b} CH_4$ [51] $+ (n\text{-}C_7\dot{H}_{15})$				
Therm. of acetylperoxide				62 Ber 2
PR [50]	c-hexadiene-1,3/ $n\text{-}C_7H_{16}$(T-labelled)	353	$k_a/k_b = 380(20)$ [49]	
$\dot{C}H_3 + c\text{-}C_6H_8\text{-}1,3 \xrightarrow{a} CH_4 + c$-hexadiene-1,3($-\dot{H}$) $+ (CH_3)_3CCH_2CH(CH_3)_2 \xrightarrow{b} CH_4$ + 2,2,4-trimethylpentane($-\dot{H}$)				
Thermal decomp. of acetylperoxide				61 Gre 1
PR, glc	2,2,4-trimethylpentane	338	$k_a/k_b = 51$	
$\dot{C}H_3 + c\text{-}C_6H_8\text{-}1,4 \xrightarrow{a} CH_4 + c$-hexadiene-1,4($-\dot{H}$) $+ (CH_3)_3CCH_2CH(CH_3)_2 \xrightarrow{b} CH_4$ + 2,2,4-trimethylpentane($-\dot{H}$)				
Thermal decomp. of acetylperoxide				61 Gre 1
PR, glc	2,2,4-trimethylpentane	338	k_a/k_b = high	
$\dot{C}H_3 + c\text{-}C_5H_7CH_3 \xrightarrow{a} CH_4$ + methyl-c-pentene($-\dot{H}$) $+ n\text{-}C_7H_{16}$ [51] $\xrightarrow{b} CH_4$ [51] $+ (n\text{-}C_7\dot{H}_{15})$				
Therm. of acetylperoxide				62 Ber 2
PR [50]	methyl-c-pentene (extrapol. to 100%)	353	$k_a/k_b = 107$ [49]	
	T-labelled $n\text{-}C_7H_{16}$ (extrapol. to 100%)	353	$k_a/k_b = 168(5)$ [49]	

[48]) Extrapolated to low phenol concentrations since $k_a/k_b = f$[phenol] due to hydrogen bonding.
[49]) Refers to one (T-labelled) reaction center.
[50]) Specific activity of T-labelled compounds.
[51]) T-labelled compound.

Reaction Radical generation Method	Solvent	T[K]	Rate data	Ref./ add. ref.
$\dot{C}H_3 + c\text{-}C_6H_{10} \xrightarrow{a} CH_4 + \text{cyclohexene}(-\dot{H})$ $+ (CH_3)_3CSD \xrightarrow{b} CH_3D + (CH_3)_3C\dot{S}$				
Thermal decomp. of *t*-butylperacetate				72 Pry 1
PR [52]	cyclohexene/ $(CH_3)_3CSD$ (5:1)	383	$k_a/k_b = 2.87(16) \cdot 10^{-2}$	
$\dot{C}H_3 + c\text{-}C_6H_{10} \xrightarrow{a} CH_4 + c\text{-hexene}(-\dot{H})$ $+ n\text{-}C_7H_{16}$ [51] $\xrightarrow{b} CH_4$ [51] $+ (n\text{-}C_7\dot{H}_{15})$				
Therm. of acetylperoxide				62 Ber 2
PR [50]	*c*-hexene (extrapol. to 100%)	353	$k_a/k_b = 124$ [49]	
	T-labelled $n\text{-}C_7H_{16}$ (extrapol. to 100%)	353	$k_a/k_b = 144(6)$ [49]	
$\dot{C}H_3 + c\text{-}C_6H_{10} \xrightarrow{a} CH_4 + c\text{-hexene}(-\dot{H})$ $+ (CH_3)_3CCH_2CH(CH_3)_2 \xrightarrow{b} CH_4 + \text{2,2,4-trimethylpentane}(-\dot{H})$				
Thermal decomp. of acetylperoxide				61 Gre 1
PR, glc	2,2,4-trimethylpentane	338	$k_a/k_b = 0.15$	
$\dot{C}H_3 + CH_2{=}CHCH_2CH_2CH{=}CH_2 \xrightarrow{a} CH_4 + CH_2{=}CH\dot{C}HCH_2CH{=}CH_2$ $\xrightarrow{b} (\text{hexadiene-1,5-}\dot{C}H_3)$ $+ (CH_3)_3CCH_2CH(CH_3)_2 \xrightarrow{c} CH_4 + \text{2,2,4-trimethylpentane}(-\dot{H})$				
Decomp. of acetylperoxide				59 Raj 1
PR	2,2,4-trimethylpentane	338	$k_a/k_c = 19.5$ [53] $k_a/k_b = 0.015$	
$\dot{C}H_3 + CH{\equiv}C(CH_2)_3CH_3 \xrightarrow{a} CH_4 + CH{\equiv}C\dot{C}HCH_2CH_2CH_3$ $+ (CH_3)_3CCH_2CH(CH_3)_2 \xrightarrow{b} CH_4 + \text{2,2,4-trimethylpentane}(-\dot{H})$				
Thermal decomp. of acetylperoxide				57 Gaz 1
PR	2,2,4-trimethylpentane	338	$k_a/k_b = 8.5$ [54]	
$\dot{C}H_3 + c\text{-}C_6H_{10}O \xrightarrow{a} CH_4 + \text{cyclohexanone}(-\dot{H})$ [55] $+ n\text{-}C_7H_{16}$ [56] $\xrightarrow{b} CH_4$ [56] $+ n\text{-heptane}(-\dot{H})$ [56]				
Thermal decomp. of acetylperoxide				65 Ber 1
PR [57]	*n*-heptane [56]	333	$k_a/k_b = 16.25(20)$	
		343	$= 16.1(5)$	
		353	$= 14.60(15)$	
		363	$= 13.7(10)$	
			$E_a(a) - E_a(b) = -11.3(21)\,\text{kJ mol}^{-1}$	
$\dot{C}H_3 + (CH_3)_2C{=}CHCOOCH_3 \xrightarrow{a} CH_4 + \text{methyl-}\beta,\beta\text{-dimethylacrylate}(-\dot{H})$ $\xrightarrow{b} (\text{methyl-}\beta,\beta\text{-dimethylacrylate-}\dot{C}H_3)$				
Thermal decomp. of acetylperoxide				61 Her 1/
PR	2,2,4-trimethylpentane	338	$k_a/k_b = 0.65$	57 Buc 1

[49]) Refers to one (T-labelled) reaction center.
[50]) Specific activity of T-labelled compounds.
[51]) T-labelled compound.
[52]) MS of CH_4 and CH_3D.
[53]) $k_a/k_c = 4.9$ per C—H bond sec. to double bond.
[54]) Per H-atom for abstraction from CH_2-group α to C≡C bond.
[55]) H-atom abstraction assumed from position α to keto group.
[56]) T-labelled compounds (4T-*n*-heptane, etc.)
[57]) Analysis of T-labelled products.

Reaction Radical generation Method	Solvent	T[K]	Rate data	Ref./ add. ref.
$\dot{C}H_3 + c\text{-}C_5H_9CH_3 \xrightarrow{a} CH_4$ + methyl-*c*-pentane(−$\dot{H}$) + $n\text{-}C_7H_{16}$ [56]) $\xrightarrow{b} CH_4$ [56]) + ($n\text{-}C_7\dot{H}_{15}$)				
Therm. of acetylperoxide				62 Ber 2
PR [59])	$c\text{-}C_5H_9CH_3$ (extrapol. to 100%)	353	$k_a/k_b = 23.0$ [58])	
	T-labelled $n\text{-}C_7H_{16}$ (extrapol. to 100%)	353	$k_a/k_b = 25.4(2)$ [58])	
$\dot{C}H_3 + c\text{-}C_6H_{12} \xrightarrow{a} CH_4 + (c\text{-}C_6\dot{H}_{11})$ $+ CCl_4 \xrightarrow{b} CH_3Cl + \dot{C}Cl_3$				
Thermal decomp. of acetylperoxide				50 Edw 1
PR	CCl_4	373	$k_a/k_b = 4.8$	
$\dot{C}H_3 + c\text{-}C_6H_{12} \xrightarrow{a} CH_4 + (c\text{-}C_6\dot{H}_{11})$ $+ (CH_3)_3CSD \xrightarrow{b} CH_3D + (CH_3)_3C\dot{S}$				
Thermal decomp. of *t*-butylperacetate				72 Pry 1
PR [60])	cyclohexane/ $(CH_3)_3CSD$ (5:1)	383	$k_a/k_b = 2.09(8) \cdot 10^{-2}$	
$\dot{C}H_3 + c\text{-}C_6H_{12} \xrightarrow{a} CH_4 + (c\text{-}C_6\dot{H}_{11})$ + $n\text{-}C_7H_{16}$ [56]) $\xrightarrow{b} CH_4$ [56]) + ($n\text{-}C_7\dot{H}_{15}$)				
Therm. of acetylperoxide				62 Ber 2
PR [59])	$c\text{-}C_6H_{12}$ (extrapol. to 100%)	353	$k_a/k_b = 11.2$ [58])	
	T-labelled $n\text{-}C_7H_{16}$ (extrapol. to 100%)	353	$k_a/k_b = 10.7(3)$ [58])	
$\dot{C}H_3 + CH_2{=}CH(CH_2)_3CH_3 \xrightarrow{a} CH_4$ + 1-hexene(−$\dot{H}$) $+ (CH_3)_3CSD \xrightarrow{b} CH_3D + (CH_3)_3C\dot{S}$				
Thermal decomp. of *t*-butylperacetate				72 Pry 1
PR [60])	1-hexene/ $(CH_3)_3CSD$ (5:1)	383	$k_a/k_b = 3.45(7) \cdot 10^{-2}$	
$\dot{C}H_3 + CH_2{=}C(CH_3)(CH_2)_2CH_3 \xrightarrow{a} CH_4$ + 2-methyl-1-pentene(−$\dot{H}$) $+ (CH_3)_3CSD \xrightarrow{b} CH_3D + (CH_3)_3C\dot{S}$				
Thermal decomp. of *t*-butylperacetate				72 Pry 1
PR [60])	2-methyl-1-pentene/ $(CH_3)_3CSD$ (5:1)	383	$k_a/k_b = 5.14(31) \cdot 10^{-2}$	
$\dot{C}H_3 + (CH_3)_2C{=}CHCH_2CH_3 \xrightarrow{a} CH_4$ + 2-methyl-2-pentene(−$\dot{H}$) $+ (CH_3)_3CSD \xrightarrow{b} CH_3D + (CH_3)_3C\dot{S}$				
Thermal decomp. of *t*-butylperacetate				72 Pry 1
PR [60])	2-methyl-2-pentene/ $(CH_3)_3CSD$(5:1)	383	$k_a/k_b = 4.21(28) \cdot 10^{-2}$	
$\dot{C}H_3$ + *trans*-$CH_3CH_2CH{=}CHCH_2CH_3 \xrightarrow{a} CH_4$ + *trans*-3-hexene(−$\dot{H}$) $+ (CH_3)_3CSD \xrightarrow{b} CH_3D + (CH_3)_3C\dot{S}$				
Thermal decomp. of *t*-butylperacetate				72 Pry 1
PR [60])	*trans*-3-hexene/ $(CH_3)_3CSD$ (5:1)	383	$k_a/k_b = 4.02(23) \cdot 10^{-2}$	

[56]) T-labelled compounds (4T-*n*-heptane, etc.).
[58]) Refers to one (T-labelled) reaction center.
[59]) Specific activity of T-labelled compound.
[60]) MS of CH_4 and CH_3D.

Reaction Radical generation Method	Solvent	T[K]	Rate data	Ref./ add. ref.
$\dot{C}H_3 + CH_3(CH_2)_4COOH \xrightarrow{a} CH_4 +$ caproic acid$(-\dot{H})$ $+ n\text{-}C_7H_{16} \xrightarrow{b} CH_4 + n$-heptane$(-\dot{H})$				
Thermal decomp. of diacetylperoxide				68 Dob 1
PR [61])	n-heptane	353	$k_a/k_b = 16.38$	
$\dot{C}H_3 + n\text{-}C_6H_{14} \xrightarrow{a} CH_4 +$ hexane$(-\dot{H})$ $+ (CH_3)_3CSD \xrightarrow{b} CH_3D + (CH_3)_3C\dot{S}$				
Thermal decomp. of t-butylperacetate				72 Pry 1
PR [60])	hexane/$(CH_3)_3CSD$ (5:1)	383	$k_a/k_b = 2.07(23) \cdot 10^{-2}$	
$\dot{C}H_3 + (CH_3)_2CHCH(CH_3)_2 \xrightarrow{a} CH_4 +$ 2,3-dimethylbutane$(-\dot{H})$ $+ (CH_3)_3CSD \xrightarrow{b} CH_3D + (CH_3)_3C\dot{S}$				
Thermal decomp. of t-butylperacetate				72 Pry 1
PR [60])	2,3-dimethylbutane/ $(CH_3)_3CSD$ (5:1)	383	$k_a/k_b = 6.22(29) \cdot 10^{-2}$	
$\dot{C}H_3 + (CH_3)_2CHOCH(CH_3)_2 \xrightarrow{a} CH_4 +$ diisopropylether$(-\dot{H})$ $+ (CH_3)_3CSD \xrightarrow{b} CH_3D + (CH_3)_3C\dot{S}$				
Thermal decomp. of t-butylperacetate				72 Pry 1
PR [60])	diisopropylether/ $(CH_3)_3CSD$ (5:1)	383	$k_a/k_b = 2.320(91) \cdot 10^{-1}$	
$\dot{C}H_3 + C_6H_5CH_2Br \xrightarrow{a} CH_3Br + C_6H_5\dot{C}H_2$ $+ C_6H_5CH_3 \xrightarrow{b} CH_4 + C_6H_5\dot{C}H_2$				
Thermal decomp. of acetylperoxide				60 Eva 1
PR, radiometric methods	toluene	338	$k_a/k_b = 6.5$	
$\dot{C}H_3 + C_6H_5CH_2Br \xrightarrow{a} CH_3Br + C_6H_5\dot{C}H_2$ $+ C_6H_5CH_3 \xrightarrow{b} CH_4 + C_6H_5\dot{C}H_2$				
Thermal decomp. of acetylperoxide				61 Fox 1
PR, glc	toluene	338	$k_a/k_b = 7.3(10)$	
$\dot{C}H_3 + 4\text{-}BrC_6H_4CH_3 \xrightarrow{a} CH_4 + 4\text{-}BrC_6H_4\dot{C}H_2$ $+ CCl_4 \xrightarrow{b} CH_3Cl + \dot{C}Cl_3$				
Therm. of acetylperoxide				69 Pry 1
PR, glc	$4\text{-}BrC_6H_4CH_3/CCl_4$	373	$k_a/k_b = 0.28$	
$\dot{C}H_3 + 4\text{-}BrC_6H_4CH_3 \xrightarrow{a} CH_4 + 4\text{-}BrC_6H_4\dot{C}H_2$ $+ C_6H_5CH_3 \xrightarrow{b} CH_4 + C_6H_5\dot{C}H_2$				
Phot. of CH_3HgI				66 Kal 1
PR, glc	toluene	373	$k_a/k_b = 0.9$	
$\dot{C}H_3 + 2\text{-}ClC_6H_4CH_3 \xrightarrow{a} CH_4 + 2\text{-}ClC_6H_4\dot{C}H_2$ $+ C_6H_5CH_3 \xrightarrow{b} CH_4 + C_6H_5\dot{C}H_2$				
Phot. of CH_3HgI				66 Kal 1
PR, glc	toluene	373	$k_a/k_b = 0.5$	
$\dot{C}H_3 + 3\text{-}ClC_6H_4CH_3 \xrightarrow{a} CH_4 + 3\text{-}ClC_6H_4\dot{C}H_2$ $+ CCl_4 \xrightarrow{b} CH_3Cl + \dot{C}Cl_3$				
Therm. of acetylperoxide				69 Pry 1
PR, glc	$3\text{-}ClC_6H_4CH_3/CCl_4$	373	$k_a/k_b = 0.24$	

[60]) MS of CH_4 and CH_3D.

[61]) Analysis of T-labelled products.

Reaction Radical generation Method	Solvent	T[K]	Rate data	Ref./ add. ref.
$\dot{C}H_3 + 4\text{-}ClC_6H_4CH_3 \xrightarrow{a} CH_4 + 4\text{-}ClC_6H_4\dot{C}H_2$ $+ CCl_4 \xrightarrow{b} CH_3Cl + \dot{C}Cl_3$				
Therm. of acetylperoxide				69 Pry 1
PR, glc	$4\text{-}ClC_6H_4CH_3/CCl_4$	373	$k_a/k_b = 0.29$	
$\dot{C}H_3 + 4\text{-}ClC_6H_4CH_3 \xrightarrow{a} CH_4 + 4\text{-}ClC_6H_4\dot{C}H_2$ $+ C_6H_5CH_3 \xrightarrow{b} CH_4 + C_6H_5\dot{C}H_2$				
Phot. of CH_3HgI				66 Kal 1
PR, glc	toluene	373	$k_a/k_b = 1.0$	
$\dot{C}H_3 + C_6H_5CH_2I \xrightarrow{a} CH_3I + C_6H_5\dot{C}H_2$ $+ C_6H_5CH_3 \xrightarrow{b} CH_4 + C_6H_5\dot{C}H_2$				
Thermal decomp. of acetylperoxide				60 Eva 1
PR, radiometric methods	toluene	338	$k_a/k_b = 7560$	
PR, glc	toluene	338	$k_a/k_b = 7630(100)$	61 Fox 1
$\dot{C}H_3 + C_6H_5CH_2T \xrightarrow{a} CH_3T + C_6H_5\dot{C}H_2$ $+ 4\text{-}TC_6H_4CH_3 \xrightarrow{b} CH_3T +$ •  —CH_3				
Therm. of acetylperoxide				61 Ber 1/
PR [62])	T-labelled toluenes	338	$k_a/k_b = 197$	60 Ber 1
		348	174	
		358	156	
		368	143	
$\dot{C}H_3 + 2\text{-}TC_6H_4CH_3 \xrightarrow{a} CH_3T +$ 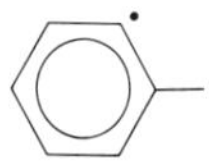 —CH_3 $+ 4\text{-}TC_6H_4CH_3 \xrightarrow{b} CH_3T +$ •  —CH_3				
Therm. of acetylperoxide				61 Ber 1/
PR [62])	T-labelled toluene	348	$k_a/k_b = 0.78$	60 Ber 1
		358	0.76	
		368	0.77	
$\dot{C}H_3 + 3\text{-}TC_6H_4CH_3 \xrightarrow{a} CH_3T +$ 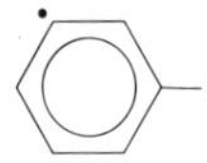 —CH_3 $+ 4\text{-}TC_6H_4CH_3 \xrightarrow{b} CH_3T +$ • 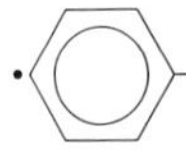 —CH_3				
Therm. of acetylperoxide				61 Ber 1/
PR [62])	T-labelled toluene	338	$k_a/k_b = 0.17$	60 Ber 1
		348	0.20	
		358	0.22	
		368	0.26	

[62]) Specific activity measurements.

Reaction Radical generation Method	Solvent	T[K]	Rate data	Ref./ add. ref.
$\dot{C}H_3 + C_6H_5CH_3 \xrightarrow{a} CH_4 + C_6H_5\dot{C}H_2$ $\dot{C}H_3 + C_6H_5CH_3 \xrightarrow{b} (C_6H_5CH_3{-}CH_3^{\cdot})$				
Therm. of $Pb(OOCCH_3)_4$				68 Hei 1
PR, glc	CH_3COOH	[63]	$k_a/k_b = 2.9$	
$\dot{C}H_3 + C_6H_5CH_3 \xrightarrow{a} CH_4 + C_6H_5\dot{C}H_2$ $+ CCl_4 \xrightarrow{b} CH_3Cl + \dot{C}Cl_3$				
Thermal decomp. of acetylperoxide				50 Edw 1
PR	CCl_4	373	$k_a/k_b = 0.75$	
Therm. of acetylperoxide				69 Pry 1
PR, glc	toluene/CCl_4	373	$k_a/k_b = 0.29$	
$\dot{C}H_3 + C_6H_5CH_3 \xrightarrow{a} CH_4 + C_6H_5\dot{C}H_2$ $+ CH_3COOH \xrightarrow{b} CH_4 + \dot{C}H_2COOH$				
Therm. of $Pb(OOCCH_3)_3$				68 Hei 1
PR, glc	CH_3COOH	[63]	$k_a/k_b = 3.9(5)$	
$\dot{C}H_3 + C_6H_5CH_3 \xrightarrow{a} CH_4 + C_6H_5\dot{C}H_2$ $+ (CH_3)_3CSD \xrightarrow{b} CH_3D + (CH_3)_3C\dot{S}$				
Thermal decomp. of *t*-butylperacetate				72 Pry 1
PR [64]	toluene/$(CH_3)_3CSD$ (5:1)	383	$k_a/k_b = 1.64(10)\cdot 10^{-2}$	
$\dot{C}H_3 + C_6H_5CH_3 \xrightarrow{a} CH_4 + C_6H_5\dot{C}H_2$ $+ (CH_3)_3CCH_2CH(CH_3)_2 \xrightarrow{b} CH_4$ + 2,2,4-trimethylpentane($-\dot{H}$)				
Thermal decomp. of acetylperoxide				56 Buc 1
PR, glc	2,2,4-trimethylpentane	338	$k_a/k_b = 2.5$	
PR	2,2,4-trimethylpentane or toluene	358.2	$k_a/k_b = 3.0(5)$ [65]	55 Lev 1/ 54 Lev 1
$\dot{C}H_3 + C_6H_5CH_3 \xrightarrow{a} CH_4 + C_6H_5\dot{C}H_2$ + diene $\xrightarrow{b}$ (diene-$CH_3^{\cdot}$)				
Decomp. of acetylperoxide				59 Raj 1
PR	$C_6H_5CH_3$	338	1,1,4,4-tetraphenylbutadiene-1,3: $k_a/k_b = 5.26\cdot 10^{-2}$ tetraphenylallene: $k_a/k_b = 5.71\cdot 10^{-2}$	

[63] T at reflux of CH_3COOH solution.
[64] MS of CH_4 and CH_3D.
[65] Based on k_a/k_c and k_b/k_c measurements with k_c referring to $\dot{C}H_3$ addition to various aromatics.

Reaction Radical generation Method	Solvent	T[K]	Rate data	Ref./ add. ref.
$\dot{C}H_3 + C_6H_5CH_3 \xrightarrow{a} CH_4$ + toluene(−$\dot{H}$) + quinone $\xrightarrow{b}$ (quinone-$CH_3^{\cdot}$) [66]				
Thermal decomp. of acetylperoxide				58 Buc 1
PR	toluene	338	quinone:	
			2,5-dimethyl-benzoquinone, $k_a/k_b = 9.7 \cdot 10^{-4}$	
			2,3-dichloro-benzoquinone, $7.5 \cdot 10^{-4}$	
			2,6-dimethyl-benzoquinone, $8.4 \cdot 10^{-4}$	
			2,3-dimethyl-benzoquinone, $9.7 \cdot 10^{-4}$	
			2,5-dichloro-benzoquinone, $1.9 \cdot 10^{-4}$	
			2-chloro-1,4-naphthoquinone, $4.6 \cdot 10^{-4}$	
			6,7-dichloro-1,4-naphtho-quinone, $1.0 \cdot 10^{-3}$	
			2,6-dimethoxy-benzoquinone, $2.9 \cdot 10^{-3}$	
			naphthoquinone, $9.3 \cdot 10^{-4}$	
PR	toluene	338	quinone:	55 Rem 1
			p-benzoquinone, $k_a/k_b = 4.95 \cdot 10^{-4}$	
			toluquinone, $7.25 \cdot 10^{-4}$	
			2-chlorobenzo-quinone, $2.90 \cdot 10^{-4}$	
			1,4-naphthoquinone, $1.54 \cdot 10^{-3}$	
			2-methyl-1,4-naphthoquinone, $2.22 \cdot 10^{-3}$	
			2,7-dimethyl-1,4-naphthoquinone, $1.83 \cdot 10^{-3}$	
			2,3-dimethyl-1,4-naphthoquinone, $1.37 \cdot 10^{-2}$	
			2,5-dimethyl-*p*-benzoquinone, $1.16 \cdot 10^{-3}$	
			2-methoxy-*p*-benzo-quinone, $9.43 \cdot 10^{-4}$	
			2,5-dichloro-*p*-benzoquinone, $1.91 \cdot 10^{-4}$	
			2,6-dichloro-*p*-benzoquinone, $1.95 \cdot 10^{-4}$	
			duroquinone, $9.52 \cdot 10^{-3}$	
			chloranil, $2.50 \cdot 10^{-2}$	
			2,3-dichloronaphtho-quinone, $8.33 \cdot 10^{-2}$	
			1,2-naphthoquinone, $2.25 \cdot 10^{-3}$	
			phenanthraquinone, $1.06 \cdot 10^{-2}$	
			2-*t*-butylanthra-quinone, $8.33 \cdot 10^{-2}$	
$\dot{C}H_3$ + *c*-heptatriene $\xrightarrow{a} CH_4$ + *c*-heptatriene(−$\dot{H}$) + $(CH_3)_3CCH_2CH(CH_3)_2 \xrightarrow{b} CH_4$ + 2,2,4-trimethylpentane(−$\dot{H}$)				
Thermal decomp. of acetylperoxide				61 Gre 1
PR, glc	2,2,4-trimethylpentane	338	$k_a/k_b = 17$	

66) $\dot{C}H_3$ addition to C=C bond.

Reaction Radical generation Method	Solvent	T[K]	Rate data	Ref./ add. ref.
$\dot{C}H_3 + C_6H_5OCH_3 \xrightarrow{a} CH_4$ + anisole$(-\dot{H})$ $+ (CH_3)_3CSD \xrightarrow{b} CH_3D + (CH_3)_3C\dot{S}$				
Thermal decomp. of *t*-butylperacetate PR [67])	anisole/$(CH_3)_3CSD$ (5:1)	383	$k_a/k_b = 2.97(16) \cdot 10^{-2}$	72 Pry 1
$\dot{C}H_3$ + bicyclo[2.2.1]heptene $\xrightarrow{a} CH_4$ + bicyclo[2.2.1]heptene$(-\dot{H})$ $+ (CH_3)_3CCH_2CH(CH_3)_2 \xrightarrow{b} CH_4$ + 2,2,4-trimethylpentane$(-\dot{H})$				
Thermal decomp. of acetylperoxide PR, glc	2,2,4-trimethylpentane	338	$k_a/k_b < 2$	61 Gre 1
$\dot{C}H_3 + c\text{-}C_7H_{12} \xrightarrow{a} CH_4$ + *c*-heptene$(-\dot{H})$ $+ (CH_3)_3CCH_2CH(CH_3)_2 \xrightarrow{b} CH_4$ + 2,2,4-trimethylpentane$(-\dot{H})$				
Thermal decomp. of acetylperoxide PR, glc	2,2,4-trimethylpentane	338	$k_a/k_b = 0.62$	61 Gre 1
$\dot{C}H_3 + c\text{-}C_6H_{11}COOH \xrightarrow{a} CH_4$ + cyclohexane carboxylic acid$(-\dot{H})$ $+ n\text{-}C_7H_{16} \xrightarrow{b} CH_4$ + *n*-heptane$(-\dot{H})$				
Thermal decomp. of diacetylperoxide PR [68])	*n*-heptane	353	$k_a/k_b = 38.59$ $A_a/A_b = 6.57$ $E_a(a) - E_a(b) = 5.3\,\text{kJ mol}^{-1}$	68 Dob 1
$\dot{C}H_3 + c\text{-}C_6H_{11}CH_3 \xrightarrow{a} CH_4$ + methyl-*c*-hexane$(-\dot{H})$ $+ n\text{-}C_7H_{16}$ [68]) $\xrightarrow{b} CH_4$ [68]) $+ (n\text{-}C_7\dot{H}_{15})$				
Therm. of acetylperoxide PR [69])	$c\text{-}C_6H_{11}CH_3$ (extrapol. to 100%)	353	$k_a/k_b = 16.6$ [70])	62 Ber 2
	T-labelled $n\text{-}C_7H_{16}$ (extrapol. to 100%)	353	$k_a/k_b = 18.8(10)$ [70])	
$\dot{C}H_3 + c\text{-}C_7H_{14} \xrightarrow{a} CH_4 + (c\text{-}C_7\dot{H}_{13})$ $+ (CH_3)_3CSD \xrightarrow{b} CH_3D + (CH_3)_3C\dot{S}$				
Thermal decomp. of *t*-butylperacetate PR [67])	cycloheptane/ $(CH_3)_3CSD$ (5:1)	383	$k_a/k_b = 3.64(11) \cdot 10^{-2}$	72 Pry 1
$\dot{C}H_3 + CH_2{=}CH(CH_2)_4CH_3 \xrightarrow{a} CH_4$ + heptene-1$(-\dot{H})$ $+ (CH_3)_3CCH_2CH(CH_3)_2 \xrightarrow{b} CH_4$ + 2,2,4-trimethylpentane$(-\dot{H})$				
Thermal decomp. of acetylperoxide PR	2,2,4-trimethylpentane	338.1	$k_a/k_b = 12.6$ [71])	57 Buc 1/ 54 Lev 1, 55 Lev 1
$\dot{C}H_3 + CH_2{=}CHCH(CH_3)(CH_2)_2CH_3 \xrightarrow{a} CH_4$ + 3-methyl-1-hexene$(-\dot{H})$ $+ (CH_3)_3CSD \xrightarrow{b} CH_3D + (CH_3)_3C\dot{S}$				
Thermal decomp. of *t*-butylperacetate PR [67])	3-methyl-1-hexene/ $(CH_3)_3CSD$ (5:1)	383	$k_a/k_b = 4.97(33) \cdot 10^{-2}$	72 Pry 1

[67]) MS of CH_4 and CH_3D.
[68]) Analysis of T-labelled products.
[69]) Specific activity of T-labelled compounds.
[70]) Refers to one (T-labelled) reaction center.
[71]) $k_a/k(\dot{C}H_3$ addition to heptene-1$) = 0.49$.

Reaction Radical generation Method	Solvent	T[K]	Rate data	Ref./ add. ref.
$\dot{C}H_3 + n\text{-}C_3H_7\text{—CO-}n\text{-}C_3H_7 \xrightarrow{a} CH_4 + \text{di-}n\text{-propylketone}(-\dot{H})$ [72)] $+ n\text{-}C_7H_{16}$ [73)] $\xrightarrow{b} CH_4$ [73)] $+ n\text{-heptane}(-\dot{H})$ [73)]				
Thermal decomp. of acetylperoxide				65 Ber 1
PR [74)]	n-heptane [73)]	333	$k_a/k_b = 15.8(4)$	
		343	14.0(5)	
		353	17.8(1)	
		363	14.4(1)	
			$E_a(a) - E_a(b) = 0.0(63)$ kJ mol^{-1}	
$\dot{C}H_3 + (CH_3)_2CHCOCH(CH_3)_2 \xrightarrow{a} CH_4 + \text{di-2-propylketone}(-\dot{H})$ [72)] $+ n\text{-}C_7H_{16}$ [73)] $\xrightarrow{b} CH_4$ [73)] $+ n\text{-heptane}(-\dot{H})$ [73)]				
Thermal decomp. of acetylperoxide				65 Ber 1
PR [74)]	n-heptane [73)]	343	$k_a/k_b = 34.0(2)$	
		353	$= 30.8(8)$	
		363	$= 27.3(6)$	
			$E_a(a) - E_a(b) = -18.0(21)$ kJ mol^{-1}	
$\dot{C}H_3 + CH_3(CH_2)_5COOH \xrightarrow{a} CH_4 + \text{enanthic acid}(-\dot{H})$ $+ n\text{-}C_7H_{16} \xrightarrow{b} CH_4 + n\text{-heptane}(-\dot{H})$				
Thermal decomp. of diacetylperoxide				68 Dob 1
PR [74)]	n-heptane	353	$k_a/k_b = 22.20$	
$\dot{C}H_3 + CH_2T(CH_2)_5CH_3 \xrightarrow{a} CH_3T + \dot{C}H_2(CH_2)_5CH_3$ $+ CH_3CHT(CH_2)_4CH_3 \xrightarrow{b} CH_3T + CH_3\dot{C}H(CH_2)_4CH_3$				
Therm. of acetylperoxide				60 Ant 1,
PR [75)]	T-labelled n-heptane	358	$k_a/k_b = 0.103(6)$	59 Ant 1
		343	0.098(4)	
		328	0.081(6)	
			$\log[A_a/A_b] = -0.02(30)$	
			$E_a(a) - E_a(b) = 6.6(20)$ kJ mol^{-1}	
$\dot{C}H_3 + CH_3CH_2CHT(CH_2)_3CH_3 \xrightarrow{a} CH_3T + CH_3CH_2\dot{C}H(CH_2)_3CH_3$ $+ CH_2CHT(CH_2)_4CH_3 \xrightarrow{b} CH_3T + CH_3\dot{C}H(CH_2)_4CH_3$				
Therm. of acetylperoxide				60 Ant 1,
PR [75)]	T-labelled n-heptane	358	$k_a/k_b = 0.99(3)$	59 Ant 1
		343	1.05(3)	
		328	1.04(4)	
			$\log[A_a/A_b] = 0.16(11)$	
			$E_a(a) - E_a(b) = 1.1(8)$ kJ mol^{-1}	
$\dot{C}H_3 + n\text{-}C_7H_{16} \xrightarrow{a} CH_4 + (n\text{-}C_7H_{15}^{\cdot})$ $+ c\text{-}C_6H_{12} \xrightarrow{b} CH_4 + (c\text{-}C_6H_{11}^{\cdot})$				
Therm. of acetylperoxide				62 Ber 1
PR [76)]	$n\text{-}C_7H_{16}/c\text{-}C_6H_{12}$	273	$k_a/k_b = 1.75$	
	(T-labelled)	353	$= 1.1$	
		423	$= 0.43$	
			$\log[A_a/A_b] = 0.22(4)$	
			$E_a(a) - E_a(b) = 4.7(5)$	

72) H-atom abstraction assumed from position α to keto group.
73) T-labelled compounds (4T-n-heptane, etc.).
74) Analysis of T-labelled products.
75) Specific activity measurements.
76) Use of T-labelled compounds.

Reaction Radical generation Method	Solvent	T[K]	Rate data	Ref./ add. ref.
$\dot{C}H_3 + (CH_3)_2CH(CH_2)_3CH_3 \xrightarrow{a} CH_4$ + 2-methylhexane($-\dot{H}$) $+ (CH_3)_3CSD \xrightarrow{b} CH_3D + (CH_3)_3C\dot{S}$				
Thermal decomp. of *t*-butylperacetate				72 Pry 1
PR [77]	2-methylhexane/ $(CH_3)_3CSD$ (5:1)	383	$k_a/k_b = 3.79(10) \cdot 10^{-2}$	
$\dot{C}H_3 + CH_3CH_2CH(CH_3)(CH_2)_2CH_3 \xrightarrow{a} CH_4$ + 3-methylhexane($-\dot{H}$) $+ (CH_3)_3CSD \xrightarrow{b} CH_3D + (CH_3)_3C\dot{S}$				
Thermal decomp. of *t*-butylperacetate				72 Pry 1
PR [77]	3-methylhexane/ $(CH_3)_3CSD$ (5:1)	383	$k_a/k_b = 4.38(31) \cdot 10^{-2}$	
$\dot{C}H_3 + (CH_3)_2CHCH(CH_3)CH_2CH_3 \xrightarrow{a} CH_4$ + 2,3-dimethylpentane($-\dot{H}$) $+ (CH_3)_3CSD \xrightarrow{b} CH_3D + (CH_3)_3C\dot{S}$				
Thermal decomp. of *t*-butylperacetate				72 Pry 1
PR [77]	2,3-dimethylpentane/ $(CH_3)_3CSD$ (5:1)	383	$k_a/k_b = 3.86(50) \cdot 10^{-2}$	
$\dot{C}H_3 + (CH_3)_2CHCH_2CH(CH_3)_2 \xrightarrow{a} CH_4$ + 2,4-dimethylpentane($-\dot{H}$) $+ (CH_3)_3CSD \xrightarrow{b} CH_3D + (CH_3)_3C\dot{S}$				
Thermal decomp. of *t*-butylperacetate				72 Pry 1
PR [77]	2,4-dimethylpentane/ $(CH_3)_3CSD$ (5:1)	383	$k_a/k_b = 3.41(8) \cdot 10^{-2}$	
$\dot{C}H_3 + CH_3COC_6H_5 \xrightarrow{a} CH_4$ + acetophenone($-\dot{H}$) [78] $+ n\text{-}C_7H_{16}$ [79] $\xrightarrow{b} CH_4$ [79] + *n*-heptane($-\dot{H}$) [79]				
Thermal decomp. of acetylperoxide				65 Ber 1
PR [80]	*n*-heptane [79]	343 363	$k_a/k_b = 2.45(10)$ $= 2.64(10)$	
$\dot{C}H_3 + C_6H_5COOCH_3 \xrightarrow{a} CH_4$ + methylbenzoate($-\dot{H}$) $+ CCl_4 \xrightarrow{b} CH_3Cl + \dot{C}Cl_3$				
Thermal decomp. of acetylperoxide				50 Edw 1
PR	CCl_4	373	$k_a/k_b = 0.062$	
$\dot{C}H_3 + C_6H_5CH_2CH_3 \xrightarrow{a} CH_4$ + ethylbenzene($-\dot{H}$) $+ CCl_4 \xrightarrow{b} CH_3Cl + \dot{C}Cl_3$				
Therm. of acetylperoxide				69 Pry 1
PR, glc	ethylbenzene/CCl_4	373	$k_a/k_b = 0.77$	
$\dot{C}H_3 + C_6H_5CH_2CH_3 \xrightarrow{a} CH_4$ + ethylbenzene($-\dot{H}$) $+ (CH_3)_3CSD \xrightarrow{b} CH_3D + (CH_3)_3C\dot{S}$				
Thermal decomp. of *t*-butylperacetate				72 Pry 1
PR [77]	ethylbenzene/ $(CH_3)_3CSD$ (5:1)	383	$k_a/k_b = 4.74(42) \cdot 10^{-2}$	

[77]) MS of CH_4 and CH_3D.
[78]) H-atom abstraction assumed from position α to keto group.
[79]) T-labelled compounds (4T-*n*-heptane, etc.).
[80]) Analysis of T-labelled products.

Reaction Radical generation Method	Solvent	T[K]	Rate data	Ref./ add. ref.
$\dot{C}H_3 + C_6H_5CH_2CH_3 \xrightarrow{a} CH_4 + $ ethylbenzene$(-\dot{H})$ $+ C_6H_5CH_3 \xrightarrow{b} CH_4 + C_6H_5\dot{C}H_2$				
Thermal decomp. of acetylperoxide				61 Mey 1
PR, glc	ethylbenzene/toluene	338	$k_a/k_b = 4.14$	
Phot. of CH_3HgI				66 Kal 1
PR, glc	toluene	373	$k_a/k_b = 5.4$	
Phot. of azomethane				68 Eac 1
PR [81])	ethylbenzene/toluene	273	$k_a/k_b = 5.54$	
		298	5.01	
		323	4.63	
		348	4.21	
		368	3.96	
			$A_a/A_b = 1.5$	
			$E_a(a) - E_a(b) = -2.9(8)$ kJ mol^{-1}	
$\dot{C}H_3 + 2\text{-}CH_3C_6H_4CH_3 \xrightarrow{a} CH_4 + 2\text{-}CH_3C_6H_4\dot{C}H_2$ $+ C_6H_5CH_3 \xrightarrow{b} CH_4 + C_6H_5\dot{C}H_2$				
Thermal decomp. of acetylperoxide				61 Mey 1
PR, glc	*o*-xylene/toluene	338	$k_a/k_b = 0.99$	
Phot. of azomethane				68 Eac 1
PR [81])	*o*-xylene/toluene	273	$k_a/k_b = 1.11$	
		298	1.06	
		323	1.03	
		348	1.00	
		368	0.98	
			$A_a/A_b = 0.8$	
			$E_a(a) - E_a(b) = -1.3(8)$ kJ mol^{-1}	
$\dot{C}H_3 + 3\text{-}CH_3C_6H_4CH_3 \xrightarrow{a} CH_4 + 3\text{-}CH_3C_6H_4\dot{C}H_2$ $+ CCl_4 \xrightarrow{b} CH_3Cl + \dot{C}Cl_3$				
Therm. of acetylperoxide				69 Pry 1
PR, glc	$3\text{-}CH_3C_6H_4CH_3/CCl_4$	373	$k_a/k_b = 0.57$	
$\dot{C}H_3 + 3\text{-}CH_3C_6H_4CH_3 \xrightarrow{a} CH_4 + 3\text{-}CH_3C_6H_4\dot{C}H_2$ $+ C_6H_5CH_3 \xrightarrow{b} CH_4 + C_6H_4\dot{C}H_2$				
Thermal decomp. of acetylperoxide				61 Mey 1
PR, glc	*m*-xylene/toluene	338	$k_a/k_b = 1.00$	
Phot. of azomethane				68 Eac 1
PR [81])	*m*-xylene/toluene	273	$k_a/k_b = 1.08$	
		298	1.06	
		323	1.02	
		348	1.00	
		368	0.96	
			$A_a/A_b = 0.8$	
			$E_a(a) - E_a(b) = -0.8(8)$ kJ mol^{-1}	
$\dot{C}H_3 + 4\text{-}CH_3C_6H_4CH_3 \xrightarrow{a} CH_4 + 4\text{-}CH_3C_6H_4\dot{C}H_2$ $+ CCl_4 \xrightarrow{b} CH_3Cl + \dot{C}Cl_3$				
Therm. of acetylperoxide				69 Pry 1
PR, glc	$4\text{-}CH_3C_6H_4CH_3/CCl_4$	373	$k_a/k_b = 0.63$	

[81]) Analysis of T-labelled products from $C_6H_5CH_2T$ reaction.

Reaction Radical generation Method	Solvent	T[K]	Rate data	Ref./ add. ref.
$\dot{C}H_3 + 4\text{-}CH_3C_6H_4CH_3 \xrightarrow{a} CH_4 + 4\text{-}CH_3C_6H_4\dot{C}H_2$ $+ (CH_3)_3CSD \xrightarrow{b} CH_3D + (CH_3)_3C\dot{S}$				
Thermal decomp. of *t*-butylperacetate				72 Pry 1
PR [82])	*p*-xylene/$(CH_3)_3CSD$ (5:1)	383	$k_a/k_b = 3.65(44) \cdot 10^{-2}$	
$\dot{C}H_3 + 4\text{-}CH_3C_6H_4CH_3 \xrightarrow{a} CH_4 + 4\text{-}CH_3C_6H_4\dot{C}H_2$ $+ C_6H_5CH_3 \xrightarrow{b} CH_4 + C_6H_5\dot{C}H_2$				
Thermal decomp. of acetylperoxide				61 Mey 1
PR, glc	*p*-xylene/toluene	338	$k_a/k_b = 1.19$	
Phot. of azomethane				68 Eac 1
PR [81])	*p*-xylene/toluene	273	$k_a/k_b = 1.26$	
		298	1.20	
		323	1.16	
		348	1.11	
		368	1.08	
			$A_a/A_b = 0.8$	
			$E_a(a) - E_a(b) = -1.3(8)$ kJ mol^{-1}	
$\dot{C}H_3 + C_6H_5OC_2H_5 \xrightarrow{a} CH_4 + \text{phenetole}(-\dot{H})$ $+ (CH_3)_3CSD \xrightarrow{b} CH_3D + (CH_3)_3C\dot{S}$				
Thermal decomp. of *t*-butylperacetate				72 Pry 1
PR [82])	phenetole/$(CH_3)_3CSD$ (5:1)	383	$k_a/k_b = 3.40(44) \cdot 10^{-2}$	
$\dot{C}H_3 + 4\text{-}CH_3OC_6H_4CH_3 \xrightarrow{a} CH_4 + \text{4-methoxytoluene}(-\dot{H})$ $+ CCl_4 \xrightarrow{b} CH_3Cl + \dot{C}Cl_3$				
Therm. of acetylperoxide				69 Pry 1
PR, glc	4-methoxytoluene/CCl_4	373	$k_a/k_b = 0.22$	
$\dot{C}H_3 + c\text{-}C_8H_{14} \xrightarrow{a} CH_4 + c\text{-octene}(-\dot{H})$ $+ (CH_3)_3CCH_2CH(CH_3)_2 \xrightarrow{b} CH_4 + \text{2,2,4-trimethylpentane}(-\dot{H})$				
Thermal decomp. of acetylperoxide				61 Gre 1
PR, glc	2,2,4-trimethylpentane	338	$k_a/k_b = 0.40$	
$\dot{C}H_3 + CH_2{=}C(CH_3)CH_2CH_2C(CH_3){=}CH_2 \xrightarrow{a} CH_4 + CH_2{=}C(CH_3)\dot{C}HCH_2C(CH_3){=}CH_2$ $\xrightarrow{b}$ (2,5-dimethyl-hexadiene-1,5-$CH_3^{\cdot}$) $+ (CH_3)_3CCH_2CH(CH_3)_2 \xrightarrow{c} CH_4 + \text{2,2,4-trimethylpentane}(-\dot{H})$				
Decomp. of acetylperoxide				59 Raj 1
PR	2,2,4-trimethylpentane	338	$k_a/k_c = 17$ [83]) $k_a/k_b = 0.013$	
$\dot{C}H_3 + c\text{-}C_8H_{16} \xrightarrow{a} CH_4 + (c\text{-}C_8H_{15}^{\cdot})$ $+ (CH_3)_3CSD \xrightarrow{b} CH_3D + (CH_3)_3C\dot{S}$				
Thermal decomp. of *t*-butylperacetate				72 Pry 1
PR [82])	cyclooctane/$(CH_3)_3CSD$ (5:1)	383	$k_a/k_b = 7.25(29) \cdot 10^{-2}$	

[81]) Analysis of T-labelled products from $C_6H_5CH_2T$ reaction.
[82]) MS of CH_4 and CH_3D.
[83]) $k_a/k_c = 4.2$ per C—H sec. to double bond.

Reaction Radical generation Method	Solvent	T[K]	Rate data	Ref./ add. ref.
$\dot{C}H_3 + CH_2{=}CH(CH_2)_5CH_3 \xrightarrow{a} CH_4$ + octene-1(−Ḣ) $+ CCl_4 \xrightarrow{b} CH_3Cl + \dot{C}Cl_3$				
Thermal decomp. of acetylperoxide				50 Edw 1
PR	CCl_4	373	$k_a/k_b = 3.2$	
$\dot{C}H_3 + CH_2{=}CH(CH_2)_5CH_3 \xrightarrow{a} CH_4$ + 1-octene(−Ḣ) $+ (CH_3)_3CSD \xrightarrow{b} CH_3D + (CH_3)_3C\dot{S}$				
Thermal decomp. of *t*-butylperacetate				72 Pry 1
PR [82])	1-octene/$(CH_3)_3CSD$ (5:1)	383	$k_a/k_b = 3.59(34) \cdot 10^{-2}$	
$\dot{C}H_3 + CH_3CH{=}CH(CH_2)_4CH_3 \xrightarrow{a} CH_4$ + 2-octene(−Ḣ) $+ (CH_3)_3CSD \xrightarrow{b} CH_3D + (CH_3)_3C\dot{S}$				
Thermal decomp. of *t*-butylperacetate				72 Pry 1
PR [82])	2-octene/$(CH_3)_3CSD$ (5:1)	383	$k_a/k_b = 5.42(48) \cdot 10^{-2}$	
$\dot{C}H_3 + n\text{-}C_8H_{18} \xrightarrow{a} CH_4$ + octane(−Ḣ) $+ (CH_3)_3CSD \xrightarrow{b} CH_3D + (CH_3)_3C\dot{S}$				
Thermal decomp. of *t*-butylperacetate				72 Pry 1
PR [82])	octane/$(CH_3)_3CSD$ (5:1)	383	$k_a/k_b = 3.09(29) \cdot 10^{-2}$	
$\dot{C}H_3 + CH_3CH(CH_3)(CH_2)_4CH_3 \xrightarrow{a} CH_4$ + 2-methylheptane(−Ḣ) $+ C_6H_5X \xrightarrow{b} (C_6H_5X\text{-}CH_3^{\cdot})$				
Thermal decomp. of acetylperoxide				57 Hei 1 [84])
PR	2-methylheptane			
	+ 50 mol % C_6H_6	338	X = H $k_a/k_b = 4.35$	
	+ 75 mol % $C_6H_5OCH_3$		OCH_3, 6.67	
	+ 50 mol % C_6H_5F		F, 2.0	
	+ 50 mol % C_6H_5Cl		Cl, 1.1	
	+ 50 mol % C_6H_5Br		Br, 1.2	
	+ 50 mol % C_6H_5CN		CN, 0.29	
	+ 75 mol % $C_6H_5COCH_3$		$COCH_3$, 1.79	
	+ 50 mol % $C_6H_5COOCH_3$		$COOCH_3$, 0.91	
$\dot{C}H_3 + CH_3CH(CH_3)(CH_2)_4CH_3 \xrightarrow{a} CH_4$ + 2-methylheptane(−Ḣ) $+ RCl \xrightarrow{b} (RCl\text{-}CH_3^{\cdot})$				
Thermal decomp. of acetylperoxide				57 Hei 1
PR	2-methylheptane			
	+ 25 mol % 4-ClC_6H_4Cl	338	R = 4-ClC_6H_4, $k_a/k_b = 0.37$	
	+ 25 mol % 3-ClC_6H_4Cl	338	R = 3-ClC_6H_4, $k_a/k_b = 0.36$	
$\dot{C}H_3 + (CH_3)_3CCH_2CH(CH_3)_2 \xrightarrow{a} CH_4$ + 2,2,4-trimethylpentane(−Ḣ) $+ (CH_3)_3CSD \xrightarrow{b} CH_3D + (CH_3)_3C\dot{S}$				
Thermal decomp. of *t*-butylperacetate				72 Pry 1
PR [85])	2,2,4-trimethylpentane/ $(CH_3)_3CSD$ (5:1)	383	$k_a/k_b = 2.21(17) \cdot 10^{-2}$	

[82]) MS of CH_4 and CH_3D.
[84]) Additional data at various other C_6H_5X concentrations, and for $C_6H_5NO_2$.
[85]) MS of CH_4 and CH_3D.

Reaction Radical generation Method	Solvent	T[K]	Rate data		Ref./ add. ref.
$\dot{C}H_3 + (CH_3)_3CCH_2CH(CH_3)_2 \xrightarrow{a} CH_4 + 2,2,4\text{-trimethylpentane}(-\dot{H})$ $+ n\text{-}C_7H_{16}$ [85a]) $\xrightarrow{b} CH_4$ [85a]) $+ (n\text{-}C_7\dot{H}_{15})$					
Therm. of acetylperoxide					62 Ber 2
PR [86])	2,2,4-trimethylpentane (extrapol. to 100%)	353	$k_a/k_b = 4.2$ [87])		
	T-labelled n-C_7H_{16} (extrapol. to 100%)	353	$k_a/k_b = 1.6(1)$ [87])		
$\dot{C}H_3 + (CH_3)_3CCH_2CH(CH_3)_2 \xrightarrow{a} CH_4 + 2,2,4\text{-trimethylpentane}(-\dot{H})$ $+ \text{alkyne} \xrightarrow{b} (\text{alkyne-}\dot{C}H_3)$					
Thermal decomp. of acetylperoxide					57 Gaz 1
PR	2,2,4-trimethylpentane	338	alkyne:	$k_a/k_b =$	[89])
			phenylacetylene,	$5.65 \cdot 10^{-3}$	
			diphenylacetylene,	$7.87 \cdot 10^{-2}$	
			acetylene,	$3.40 \cdot 10^{-2}$	
			methylacetylene,	$9.22 \cdot 10^{-2}$	
			dimethylacetylene,	0.7 [88])	
			pentyne-1,	$7.30 \cdot 10^{-2}$	
			hexyne-1,	$5.71 \cdot 10^{-2}$	
$\dot{C}H_3 + (CH_3)_3CCH_2CH(CH_3)_2 \xrightarrow{a} CH_4 + 2,2,4\text{-trimethylpentane}(-\dot{H})$ $+ \text{aromatic} \xrightarrow{b} (\text{aromatic-}\dot{C}H_3)$					
Thermal decomp. of acetylperoxide					59 Bin 1
PR	2,2,4-trimethylpentane	338	aromatic:		
			anthracene,	$k_a/k_b = 2.54 \cdot 10^{-3}$	
			1-methyl-anthracene,	$2.82 \cdot 10^{-3}$	
			2-methyl-anthracene,	$2.82 \cdot 10^{-3}$	
			2,6-dimethyl-anthracene,	$2.75 \cdot 10^{-3}$	
			9-methyl-anthracene,	$5.26 \cdot 10^{-3}$	
			9,10-dimethyl-anthracene,	$1.64 \cdot 10^{-2}$	
			phenanthrene,	$9.80 \cdot 10^{-2}$	
			2-methyl-phenanthrene,	$7.81 \cdot 10^{-2}$	
			3-methyl-phenanthrene,	$8.06 \cdot 10^{-2}$	
			9,10-dimethyl-phenanthrene,	0.167	
PR	2,2,4-trimethylpentane	358.2	aromatic:		55 Lev 1
			benzene,	$k_a/k_b = 2.56$	54 Lev 1/
			biphenyl,	0.53	
			naphthalene,	0.117	
			pyrene,	0.019	
			stilbene,	0.014	
			anthracene,	0.0030	
			pyridine,	0.83	
			quinoline,	0.089	
			isoquinoline,	0.071	
			acridine,	0.0059	
			benzophenone,	0.23	
			diphenylether,	1.05	
			quinone,	0.00017	

[85a]) T-labelled compound.
[86]) Specific activity of T-labelled compounds.
[87]) Refers to one (T-labelled) reaction center.
[88]) Extrapolated value.
[89]) Further data at other temperatures.

Reaction Radical generation Method	Solvent	T[K]	Rate data		Ref./ add. ref.
$\dot{C}H_3 + (CH_3)_3CCH_2CH(CH_3)_2 \xrightarrow{a} CH_4$ + 2,2,4-trimethylpentane($-\dot{H}$) + diene $\xrightarrow{b}$ (diene-$CH_3^{\cdot}$)					
Decomp. of acetylperoxide					59 Raj 1
PR	2,2,4-trimethylpentane	338	diene:	k_a/k_b =	
			butadiene-1,3,	$4.96(7)\cdot10^{-4}$ [90]	
			isoprene,	$4.78(11)\cdot10^{-4}$ [90]	
			hexadiene-2,4,	$5.56(30)\cdot10^{-3}$ [90]	
			2,5-dimethyl-hexadiene-2,4,	$3.72\cdot10^{-2}$ [90]	
			2,3-dimethyl-butadiene-1,3,	$4.48(14)\cdot10^{-4}$	
			pentadiene-1,3-*cis*,	$8.20(13)\cdot10^{-4}$	
			pentadiene-1,3-*trans*,	$1.19(4)\cdot10^{-3}$	
			4-methyl-pentadiene-1,3,	$1.00(4)\cdot10^{-3}$	
			1-phenyl-butadiene-1,3,	$4.37(10)\cdot10^{-4}$	
			1,4-diphenyl-butadiene-1,3,	$2.63(3)\cdot10^{-3}$	
			chloroprene,	$1.33(8)\cdot10^{-4}$	
			1-methoxy-butadiene-1,3,	$2.00(5)\cdot10^{-3}$	
			allene,	$5.68(6)\cdot10^{-2}$ [90]	
			butadiene-1,2,	$6.76(91)\cdot10^{-2}$ [90]	
			pentadiene-1,2,	$5.21\cdot10^{-2}$	
			pentadiene-2,3,	$7.25\cdot10^{-2}$	
Thermal decomp. of acetylperoxide					57 Raj 1
PR	2,2,4-trimethylpentane	338	diene:	k_a/k_b =	[89]/
			allene,	$5.68\cdot10^{-2}$	60 Fel 1
			butadiene-1,2,	$6.76\cdot10^{-2}$	
			butadiene-1,3,	$4.96\cdot10^{-4}$	
			isoprene,	$4.79\cdot10^{-4}$	
			2,3-dimethyl-butadiene-1,3,	$4.48\cdot10^{-4}$	
			1,4-diphenyl-butadiene-1,3,	$2.65\cdot10^{-3}$	
			2,5-dimethyl-hexadiene-2,4,	$4.69\cdot10^{-2}$ [88]	
			1,1,4,4-tetraphenyl-butadiene-1,3,	$1.7\cdot10^{-2}$	
			hexadiene-1,5	$1.47\cdot10^{-2}$ [88]	
			2,5-dimethyl-hexadiene-1,5,	$1.30\cdot10^{-2}$ [88]	

[88] Extrapolated value.
[89] Further data at other temperatures.
[90] Further data at different temperatures in reference.

Reaction Radical generation Method	Solvent	T[K]	Rate data		Ref./ add. ref.
$\dot{C}H_3 + (CH_3)_3CCH_2CH(CH_3)_2 \xrightarrow{a} CH_4$ + 2,2,4-trimethylpentane($-\dot{H}$) + olefine $\xrightarrow{b}$ (olefine-$CH_3^{\cdot}$)					
Thermal decomp. of acetylperoxide					61 Her 1/
PR	2,2,4-trimethylpentane	338	olefine:		57 Buc 1
			methylvinyl-ketone,	k_a/k_b = $5.3(2)\cdot10^{-4}$	
			methyl-methacrylate,	$6.9(3)\cdot10^{-4}$	
			methylacrylate,	$9.7(3)\cdot10^{-4}$	
			methylcrotonate (95% *trans*),	$1.48(2)\cdot10^{-2}$	
			methyl-β, β-dimethylacrylate,	$8.4\cdot10^{-2}$	
			acrylonitrile,	$5.8(2)\cdot10^{-4}$	
			methacrylonitrile,	$4.7(2)\cdot10^{-4}$	
			crotononitrile (cut 67:33),	$1.383(1)\cdot10^{-2}$	
			crotononitrile (cut 45:55),	$1.344(1)\cdot10^{-2}$	
			β, β-dimethyl-acrylonitrile,	$4.3\cdot10^{-2}$	
			1-cyanocyclo-pentene,	$4.4\cdot10^{-3}$	
			vinylacetylene,	$4.4(2)\cdot10^{-4}$	
PR	2,2,4-trimethylpentane	338.1	olefine:	k_a/k_b =	57 Buc 1/ 54 Lev 1, 55 Lev 1
			ethylene,	0.0294(4) [90]	
			propylene,	0.0457(25) [90]	
			isobutene,	0.0281(8) [90]	
			trans-butene-2,	0.145 [90]	
			cis-butene-2,	0.294 [90]	
			butene-1,	0.0372 [90]	
			pentene-1,	0.0408 [90]	
			heptene-1,	0.0391	
			decene-1,	0.0459	
			hexadecene-1,	0.0400	
			3-methyl-butene-1,	0.0439	
PR	2,2,4-trimethylpentane	338	olefine:		57 Bad 1 [89])
			butene-2-*cis*,	k_a/k_b = 0.29	
			butene-2-*trans*,	0.14	
			di-*t*-butyl-ethylene-*cis*,	0.53	
			di-*t*-butyl-ethylene-*trans*,	≈ 2.5	
			stilbene-*cis*,	= $3.4\cdot10^{-2}$	
			stilbene-*trans*,	$9.6\cdot10^{-3}$	
			diethylmaleate,	$3.0\cdot10^{-3}$	
			diethylfumarate,	$5.0\cdot10^{-4}$	
PR	2,2,4-trimethylpentane	338	olefine:		56 Buc 2 [89])/ 60 Fel 1
			ethylene,	k_a/k_b = $2.93\cdot10^{-2}$	
			tetrafluoro-ethylene,	$2.92\cdot10^{-3}$	
(continued)			tetrachloro-ethylene,	> 3.3	

[89]) Further data at other temperatures.
[90]) Further data at different temperatures in reference.

Reaction Radical generation Method	Solvent	T[K]	Rate data	Ref./ add. ref.
$\dot{C}H_3 + (CH_3)_3CCH_2CH(CH_3)_2 \xrightarrow{a} CH_4$ + 2,2,4-trimethylpentane($-\dot{H}$) (continued) + olefine $\xrightarrow{b}$ (olefine-$CH_3^{\cdot}$)				
Thermal decomp. of acetylperoxide				
PR	2,2,4-trimethylpentane	338	olefine: styrene, $k_a/k_b = 1.26 \cdot 10^{-3}$ α-methyl-styrene, $1.08 \cdot 10^{-3}$ 1,1-diphenyl-ethylene, $6.29 \cdot 10^{-4}$ *trans*-stilbene, $9.52 \cdot 10^{-3}$ 1,1,2-triphenyl-ethylene, $2.17 \cdot 10^{-2}$ diethylmaleate, $3.80 \cdot 10^{-3}$ vinylacetate, $3.22 \cdot 10^{-2}$	55 Lea 1 [91])/ 55 Szw 1, 60 Fel 1, 58 Lea 1
PR	2,2,4-trimethylpentane	344	olefine: propene, $k_a/k_b = 4.24 \cdot 10^{-2}$ butene-1, $4.0 \cdot 10^{-2}$ [88])	60 Ste 1 [89])/ 60 Fel 1
Phot. of azomethane				61 Mat 1
PR, glc, MS	2,2,4-trimethylpentane	323	olefine: styrene, $k_a/k_b = 9.017 \cdot 10^{-4}$ deuterostyrene, $8.258 \cdot 10^{-4}$	
PR, glc, MS	2,2,4-trimethylpentane	338	olefine: $k_a/k_b =$ $CH_2{=}CH_2$, 0.0261 $CD_2{=}CD_2$, 0.0248 $CH_3CH{=}CH_2$, 0.0438 $CH_3CH{=}CD_2$, 0.0392 $CD_3CD{=}CD_2$, 0.0375 $CH_2{=}CHCH{=}CH_2$, 0.00061 $CD_2{=}CDCD{=}CD_2$, 0.00051	62 Fel 1
$\dot{C}H_3 + (CH_3)_3CCH_2CH(CH_3)_2 \xrightarrow{a} CH_4$ + 2,2,4-trimethylpentane($-\dot{H}$) + substrate $\xrightarrow{b}$ (substrate-$CH_3^{\cdot}$)				
Thermal decomp. of acetylperoxide				54 Lev 2
PR	2,2,4-trimethylpentane/ substrate mixt.	358	substrate: benzene, $k_a/k_b = 2.56$ biphenyl, 0.53 naphthalene, 0.12 phenanthrene, $5.95 \cdot 10^{-2}$ pyrene, $2.05 \cdot 10^{-2}$ anthracene, $3.03 \cdot 10^{-3}$ pyridine, 0.83 quinoline, $8.85 \cdot 10^{-2}$ benzophenone, 0.23 diphenylether, 1.11	[92])
(continued)				

[88]) Extrapolated value.
[89]) Further data at other temperatures.
[91]) Also data at 358 K.
[92]) Also data at 338 K.

Reaction Radical generation Method	Solvent	T[K]	Rate data	Ref./ add. ref.
$\dot{C}H_3 + (CH_3)_3CCH_2CH(CH_3)_2 \xrightarrow{a} CH_4$ + 2,2,4-trimethylpentane(−H) (continued) + substrate $\xrightarrow{b}$ (substrate-$CH_3^{\cdot}$)				
Thermal decomp. of acetylperoxide				59 Car 1
PR	2,2,4-trimethylpentane	338	substrate:	
			1-vinyl-naphthalene, $k_a/k_b = 1.23 \cdot 10^{-3}$	
			1-vinyl-anthracene, $7.41 \cdot 10^{-4}$	
			9-vinyl-anthracene, $2.27 \cdot 10^{-3}$	
			vinylmesitylene, $9.62 \cdot 10^{-3}$	
			2-vinylpyridine, $7.35 \cdot 10^{-4}$	
			4-vinylpyridine, $7.35 \cdot 10^{-4}$	
			α-vinyl-thiophene, $4.89 \cdot 10^{-4}$	
			p-di-isopro-penylbenzene, $4.76 \cdot 10^{-4}$	
			β-methylstyrene, (*trans*), $1.08 \cdot 10^{-2}$	
			β-methylstyrene (*cis*), $2.50 \cdot 10^{-2}$	
			α,β,β-trimethyl-styrene, $5.0 \cdot 10^{-2}$	
			indene, $1.04 \cdot 10^{-2}$	
			2-chlorostyrene, $1.0 \cdot 10^{-3}$	
			3-chlorostyrene, $9.62 \cdot 10^{-4}$	
			4-chlorostyrene, $9.80 \cdot 10^{-4}$	
			2,5-dichloro-styrene, $8.26 \cdot 10^{-4}$	
			4-methoxy-styrene, $1.31 \cdot 10^{-3}$	
			dibenzofulvene, $5.0 \cdot 10^{-5}$	
			9-ethylidene-fluorene, $3.13 \cdot 10^{-4}$	
			9-iso-pro-pylidenefluorene, $3.33 \cdot 10^{-3}$	
$\dot{C}H_3 + (CH_3)_3CCH_2CH(CH_3)_2 \xrightarrow{a} CH_4$ + 2,2,4-trimethylpentane(−$\dot{H}$) + $C_6H_5CH_3 \xrightarrow{b} CH_4$ + toluene(−$\dot{H}$) + $CH_3CH_2COCH_3 \xrightarrow{c} CH_4$ + methylethylketone(−$\dot{H}$)				
Thermal decomp. of acetylperoxide				55 Lea 1,
PR	2,2,4-trimethylpentane and toluene and methylethylketone	338	k_a:k_b:k_c = 1:3:9	55 Lev 1/ 55 Szw 1, 60 Fel 1
$\dot{C}H_3$ + indene $\xrightarrow{a} CH_4$ + indene(−$\dot{H}$) + $(CH_3)_3CCH_2CH(CH_3)_2 \xrightarrow{b} CH_4$ + 2,2,4-trimethylpentane(−$\dot{H}$)				
Thermal decomp. of acetylperoxide				59 Car 1
PR, glc	2,2,4-trimethylpentane	338	$k_a/k_b = 47$ [93]	

[93]) k_a per active H-atom.

Reaction Radical generation Method	Solvent	T[K]	Rate data	Ref./ add. ref.
$\dot{C}H_3$ + indan $\xrightarrow{a}$ CH_4 + indan(−$\dot{H}$) + $C_6H_5CH_3$ $\xrightarrow{b}$ CH_4 + $C_6H_5\dot{C}H_2$				
Thermal decomp. of acetylperoxide				61 Mey 1
PR, glc	indan/toluene	338	$k_a/k_b = 8.3$	
Phot. of azomethane				68 Eac 1
PR [94])	indan/toluene	273	$k_a/k_b = 15.0$	
		298	11.4	
		323	9.2	
		348	9.0	
		368	8.0	
			$A_a/A_b = 1.2$	
			$E_a(a) - E_a(b) = -5.4(16)$ kJ mol^{-1}	
$\dot{C}H_3$ + $CH_3CH_2COC_6H_5$ [95]) $\xrightarrow{a}$ CH_4 [95]) + propiophenone(−$\dot{H}$) [95]) [96]) + n-C_7H_{16} $\xrightarrow{b}$ CH_4 + n-heptane(−$\dot{H}$)				
Thermal decomp. of acetylperoxide				65 Ber 1
PR [86])	n-heptane	353	$k_a/k_b = 1.76(4)$ [97])	
$\dot{C}H_3$ + $C_6H_5CH(CH_3)_2$ $\xrightarrow{a}$ CH_4 + $C_6H_5\dot{C}(CH_3)_2$ + CCl_4 $\xrightarrow{b}$ CH_3Cl + $\dot{C}Cl_3$				
Therm. of acetylperoxide				69 Pry 1
PR, glc	cumene/CCl_4	373	$k_a/k_b = 1.29$	
$\dot{C}H_3$ + $C_6H_5CH(CH_3)_2$ $\xrightarrow{a}$ CH_4 + $C_6H_5\dot{C}(CH_3)_2$ + $(CH_3)_3CSD$ $\xrightarrow{b}$ CH_3D + $(CH_3)_3C\dot{S}$				
Thermal decomp. of t-butylperacetate				72 Pry 1
PR [98])	cumene/$(CH_3)_3CSD$ (5:1)	383	$k_a/k_b = 7.64(13) \cdot 10^{-2}$	
$\dot{C}H_3$ + $C_6H_5CH(CH_3)_2$ $\xrightarrow{a}$ CH_4 + $C_6H_5\dot{C}(CH_3)_2$ + $C_6H_5CH_3$ $\xrightarrow{b}$ CH_4 + $C_6H_5\dot{C}H_2$				
Thermal decomp. of acetylperoxide				61 Mey 1
PR, glc	cumene/toluene	338	$k_a/k_b = 12.9$	
Phot. of CH_3HgI				66 Kal 1
PR, glc	toluene	373	$k_a/k_b = 13.5$	
Phot. of azomethane				68 Eac 1
PR [94])	cumene/toluene	273	$k_a/k_b = 19.7$	
		298	17.2	
		323	14.5	
		348	12.8	
		368	11.9	
			$A_a/A_b = 2.7$	
			$E_a(a) - E_a(b) = -4.6(8)$ kJ mol^{-1}	

[86]) Specific activity of T-labelled compounds.
[94]) Analysis of T-labelled products from $C_6H_5CH_2T$ reaction.
[95]) T-labelled compounds.
[96]) H-atom abstraction assumed from position α to keto group.
[97]) Same value at 333 K, 343 K and 363 K.
[98]) MS of CH_4 and CH_3D.

Reaction Radical generation Method	Solvent	T[K]	Rate data	Ref./ add. ref.
$\dot{C}H_3 + 1,3,5\text{-}(CH_3)_3C_6H_3 \xrightarrow{a} CH_4 + \text{mesitylene}(-\dot{H})$ $+ C_6H_5CH_3 \xrightarrow{b} CH_4 + C_6H_5\dot{C}H_2$				
Thermal decomp. of acetylperoxide				61 Mey 1
PR, glc	mesitylene/toluene	338	$k_a/k_b = 0.94$	
Phot. of azomethane				68 Eac 1
PR [1])	mesitylene/toluene	273	$k_a/k_b = 1.18$	
		298	1.09	
		323	1.02	
		348	0.97	
		368	0.93	
			$A_a/A_b = 0.5$	
			$E_a(a) - E_a(b) = -2.1(12)$ kJ mol^{-1}	
$\dot{C}H_3 + n\text{-}C_4H_9\text{—}CO\text{-}n\text{-}C_4H_9 \xrightarrow{a} CH_4 + \text{di-}n\text{-butylketone}(-\dot{H})$ [2]) $+ n\text{-}C_7H_{16}$ [3]) $\xrightarrow{b} CH_4$ [3]) $+ n\text{-heptane}(-\dot{H})$ [3])				
Thermal decomp. of acetylperoxide				65 Ber 1
PR [4])	*n*-heptane [3])	333	$k_a/k_b = 13.6$	
		343	13.9(2)	
		353	12.3(2)	
		363	11.3	
			$E_a(a) - E_a(b) = -8.4(21)$ kJ mol^{-1}	
$\dot{C}H_3 + (CH_3)_2CHCOC_6H_5$ [3]) $\xrightarrow{a} CH_4$ [3]) $+ \text{isobutyrophenone}(-\dot{H})$ $+ n\text{-}C_7H_{16} \xrightarrow{b} CH_4 + n\text{-heptane}(-\dot{H})$				
Thermal decomp. of acetylperoxide				65 Ber 1
PR [4])	*n*-heptane	353	$k_a/k_b = 1.51(3)$ [5])	
$\dot{C}H_3 + \text{vinylphenylacetate} \xrightarrow{a} CH_4 + \text{vinylphenylacetate}(-\dot{H})$ $+ (CH_3)_3CCH_2CH(CH_3)_2 \xrightarrow{b} CH_4 + 2,2,4\text{-trimethylpentane}(-\dot{H})$				
Thermal decomp. of acetylperoxide				56 Buc 1
PR, glc	2,2,4-trimethylpentane	338	$k_a/k_b = 51$ [6])	
$\dot{C}H_3 + \text{tetralin} \xrightarrow{a} CH_4 + \text{tetralin}(-\dot{H})$ [7]) $+ C_6H_5CH_3 \xrightarrow{b} CH_4 + C_6H_5\dot{C}H_2$				
Thermal decomp. of acetylperoxide				61 Mey 1
PR, glc	tetralin/toluene	338	$k_a/k_b = 23$	
Phot. of azomethane				68 Eac 1
PR [8])	tetralin/toluene	273	$k_a/k_b = 50.6$	
		298	33.3	
		323	28.0	
		348	20.8	
		368	18.2	
			$A_a/A_b = 1.1$	
			$E_a(a) - E_a(b) = -8.8(13)$ kJ mol^{-1}	

[1]) Analysis of T-labelled products from $C_6H_5CH_2T$ reaction.
[2]) H-atom abstraction assumed from position α to keto group.
[3]) T-labelled compounds (4T-*n*-heptane, etc).
[4]) Analysis of T-labelled products.
[5]) Same value at 333 K, 343K and 363 K.
[6]) Value should be treated with caution because of the nonlinearity of the competition plot at higher solute concentration.
[7]) Only the two axial H-atoms of tetralin considered to be reactive.
[8]) Analysis of T-labelled products from $C_6H_5CH_2T$ reaction.

Reaction Radical generation Method	Solvent	T[K]	Rate data	Ref./ add. ref.
$\dot{C}H_3$ + *cis*-decalin $\xrightarrow{a}$ CH_4 + *cis*-decalin(−$\dot{H}$) + c-C_6H_{12} [8a]) $\xrightarrow{b}$ CH_4 [8a]) + (c-$C_6\dot{H}_{11}$)				
Therm. of acetylperoxide				62 Ber 1
PR [9])	*cis*-decalin/c-C_6H_{12} (T-labelled)	273 353 423	$k_a/k_b = 7.7$ 2.33 1.05 $A_a/A_b = 2.4(3) \cdot 10^{-2}$ $E_a(a) - E_a(b) = 13.4(4)$ kJ mol^{-1}	
$\dot{C}H_3$ + *cis*-decalin $\xrightarrow{a}$ CH_4 + *cis*-decalin(−$\dot{H}$) + n-C_7H_{16} [8a]) $\xrightarrow{b}$ CH_4 [8a]) + (n-$C_7\dot{H}_{15}$)				
Therm. of acetylperoxide				62 Ber 1
PR [9])	*cis*-decalin/n-C_7H_{16}	353	$k_a/k_b = 2.63$ $A_a/A_b = 0.10(4)$ $E_a(a) - E_a(b) = 9.6(13)$ kJ mol^{-1}	
PR [10])	*cis*-decalin (extrapol. to 100%)	353	$k_a/k_b = 30.0$ [11])	62 Ber 2
	T-labelled n-C_7H_{16} (extrapol. to 100%)	353	$k_a/k_b = 50.0(15)$ [11])	
$\dot{C}H_3$ + *trans*-decalin $\xrightarrow{a}$ CH_4 + *trans*-decalin(−$\dot{H}$) + c-C_6H_{12} [8a]) $\xrightarrow{b}$ CH_4 [8a]) + (c-$C_6\dot{H}_{11}$)				
Therm. of acetylperoxide				62 Ber 1
PR [9])	*trans*-decalin/c-C_6H_{12} (T-labelled)	273 353 423	$k_a/k_b = 8.4$ 1.35 0.43 $A_a/A_b = 2.1(4) \cdot 10^{-3}$ $E_a(a) - E_a(b) = 18.8(4)$ kJ mol^{-1}	
$\dot{C}H_3$ + *trans*-decalin $\xrightarrow{a}$ CH_4 + *trans*-decalin(−$\dot{H}$) + n-C_7H_{16} [8a]) $\xrightarrow{b}$ CH_4 [8a]) + (n-$C_7\dot{H}_{15}$)				
Therm. of acetylperoxide				62 Ber 1
PR [9])	*trans*-decalin/n-C_7H_{16} (T-labelled)	353	$k_a/k_b = 1.38$ $A_a/A_b = 0.9(3) \cdot 10^{-2}$ $E_a(a) - E_a(b) = 14.7(10)$ kJ mol^{-1}	
PR [10])	*trans*-decalin (extrapol. to 100%)	353	$k_a/k_b = 21.0$ [11])	62 Ber 2
	T-labelled n-C_7H_{16} (extrapol. to 100%)	353	$k_a/k_b = 27.5(10)$ [11])	
$\dot{C}H_3$ + $CH_2{=}CH(CH_2)_7CH_3$ $\xrightarrow{a}$ CH_4 + 1-decene(−$\dot{H}$) + $(CH_3)_3CSD$ $\xrightarrow{b}$ CH_3D + $(CH_3)_3C\dot{S}$				
Thermal decomp. of t-butylperacetate				72 Pry 1
PR [12])	1-decene/$(CH_3)_3CSD$ (5:1)	383	$k_a/k_b = 6.12(19) \cdot 10^{-2}$	
$\dot{C}H_3$ + $CH_2{=}CH(CH_2)_7CH_3$ $\xrightarrow{a}$ CH_4 + decene-1(−$\dot{H}$) + $(CH_3)_3CCH_2CH(CH_3)_2$ $\xrightarrow{b}$ CH_4 + 2,2,4-trimethylpentane(−$\dot{H}$)				
Thermal decomp. of acetylperoxide				57 Buc 1/ 54 Lev 1, 55 Lev 1
PR	2,2,4-trimethylpentane	338.1	$k_a/k_b = 12.7$ [13])	

8a) T-labelled compounds.
9) Use of T-labelled compounds.
10) Specific activity of T-labelled compounds.
11) Refers to one (T-labelled) reaction center.
12) MS of CH_4 and CH_3D.
13) $k_a/k(\dot{C}H_3$ addition to decene-1$) = 0.58$.

Reaction Radical generation Method	Solvent	T[K]	Rate data	Ref./ add. ref.
$\dot{C}H_3 + n\text{-}C_{10}H_{22} \xrightarrow{a} CH_4 + \text{decane}(-\dot{H})$ $+ (CH_3)_3CSD \xrightarrow{b} CH_3D + (CH_3)_3C\dot{S}$				
Thermal decomp. of *t*-butylperacetate				72 Pry 1
PR [12])	decane/$(CH_3)_3CSD$ (5:1)	383	$k_a/k_b = 4.06(35) \cdot 10^{-2}$	
$\dot{C}H_3$ + 1-methylnaphthalene $\xrightarrow{a}$ CH_4 + 1-methylnaphthalene($-\dot{H}$) $+ (CH_3)_3CCH_2CH(CH_3)_2 \xrightarrow{b} CH_4$ + 2,2,4-trimethylpentane($-\dot{H}$)				
Thermal decomp. of acetylperoxide				59 Gre 1
PR, glc	2,2,4-trimethylpentane	338	$k_a/k_b = 0.83$	
$\dot{C}H_3$ + 2-methylnaphthalene $\xrightarrow{a}$ CH_4 + 2-methylnaphthalene($-\dot{H}$) $+ (CH_3)_3CCH_2CH(CH_3)_2 \xrightarrow{b} CH_4$ + 2,2,4-trimethylpentane($-\dot{H}$)				
Thermal decomp. of acetylperoxide				59 Gre 1
PR, glc	2,2,4-trimethylpentane	338	$k_a/k_b = 2.05$	
$\dot{C}H_3$ + vinylmesitylene $\xrightarrow{a}$ CH_4 + vinylmesitylene($-\dot{H}$) $+ (CH_3)_3CCH_2CH(CH_3)_2 \xrightarrow{b} CH_4$ + 2,2,4-trimethylpentane($-\dot{H}$)				
Thermal decomp. of acetylperoxide				59 Car 1
PR, glc	2,2,4-trimethylpentane	338	$k_a/k_b = 2.4$ [14])	
$\dot{C}H_3$ + acenaphthene $\xrightarrow{a}$ CH_4 + acenaphthene($-\dot{H}$) $+ (CH_3)_3CCH_2CH(CH_3)_2 \xrightarrow{b} CH_4$ + 2,2,4-trimethylpentane($-\dot{H}$)				
Thermal decomp. of acetylperoxide				59 Gre 1
PR, glc	2,2,4-trimethylpentane	338	$k_a/k_b = 5.15$	
$\dot{C}H_3$ + 1-ethylnaphthalene $\xrightarrow{a}$ CH_4 + 1-ethylnaphthalene($-\dot{H}$) $+ (CH_3)_3CCH_2CH(CH_3)_2 \xrightarrow{b} CH_4$ + 2,2,4-trimethylpentane($-\dot{H}$)				
Thermal decomp. of acetylperoxide				59 Gre 1
PR, glc	2,2,4-trimethylpentane	338	$k_a/k_b = 4.26$	
$\dot{C}H_3$ + 2-ethylnaphthalene $\xrightarrow{a}$ CH_4 + 2-ethylnaphthalene($-\dot{H}$) $+ (CH_3)_3CCH_2CH(CH_3)_2 \xrightarrow{b} CH_4$ + 2,2,4-trimethylpentane($-\dot{H}$)				
Thermal decomp. of acetylperoxide				59 Gre 1
PR, glc	2,2,4-trimethylpentane	338	$k_a/k_b = 4.27$	
$\dot{C}H_3$ + 1,5-dimethylnaphthalene $\xrightarrow{a}$ CH_4 + 1,5-dimethylnaphthalene($-\dot{H}$) $+ (CH_3)_3CCH_2CH(CH_3)_2 \xrightarrow{b} CH_4$ + 2,2,4-trimethylpentane($-\dot{H}$)				
Thermal decomp. of acetylperoxide				59 Gre 1
PR, glc	2,2,4-trimethylpentane	338	$k_a/k_b = 1.09$	
$\dot{C}H_3$ + 2,3-dimethylnaphthalene $\xrightarrow{a}$ CH_4 + 2,3-dimethylnaphthalene($-\dot{H}$) $+ (CH_3)_3CCH_2CH(CH_3)_2 \xrightarrow{b} CH_4$ + 2,2,4-trimethylpentane($-\dot{H}$)				
Thermal decomp. of acetylperoxide				59 Gre 1
PR, glc	2,2,4-trimethylpentane	338	$k_a/k_b = 1.06$	
$\dot{C}H_3$ + 2,6-dimethylnaphthalene $\xrightarrow{a}$ CH_4 + 2,6-dimethylnaphthalene($-\dot{H}$) $+ (CH_3)_3CCH_2CH(CH_3)_2 \xrightarrow{b} CH_4$ + 2,2,4-trimethylpentane($-\dot{H}$)				
Thermal decomp. of acetylperoxide				59 Gre 1
PR, glc	2,2,4-trimethylpentane	338	$k_a/k_b = 1.08$	

[12]) MS of CH_4 and CH_3D.
[14]) k_a per active H-atom.

Reaction Radical generation Method	Solvent	T[K]	Rate data	Ref./ add. ref.
$\dot{C}H_3 + c\text{-}C_6H_{11}C_6H_5 \xrightarrow{a} CH_4 + c$-hexylbenzene($-\dot{H}$) $+ C_6H_5CH_3 \xrightarrow{b} CH_4 + C_6H_5\dot{C}H_2$				
Phot. of azomethane				68 Eac 1
PR [15])	c-hexylbenzene toluene	273	$k_a/k_b = 15.9$	
		298	13.6	
		323	12.3	
		348	11.1	
		368	10.1	
			$A_a/A_b = 3.0$	
			$E_a(a) - E_a(b) = -4.2(8)\,kJ\,mol^{-1}$	
$\dot{C}H_3 + n\text{-}C_{12}H_{26} \xrightarrow{a} CH_4 +$ dodecane($-\dot{H}$) $+ (CH_3)_3CSD \xrightarrow{b} CH_3D + (CH_3)_3C\dot{S}$				
Thermal decomp. of t-butylperacetate				72 Pry 1
PR [16])	dodecane/ $(CH_3)_3CSD$ (5:1)	383	$k_a/k_b = 5.17(35) \cdot 10^{-2}$	
$\dot{C}H_3 + (C_6H_5)_2CH_2 \xrightarrow{a} CH_4 + (C_6H_5)_2\dot{C}H$ $+ C_6H_5CH_3 \xrightarrow{b} CH_4 + C_6H_5\dot{C}H_2$				
Phot. of azomethane				68 Eac 1
PR [15])	diphenylmethane/ toluene	273	$k_a/k_b = 19.3$	
		298	17.2	
		323	16.3	
		348	13.6	
		368	11.7	
			$A_a/A_b = 3.2$	
			$E_a(a) - E_a(b) = -4.2(16)\,kJ\,mol^{-1}$	
$\dot{C}H_3 + 4\text{-}(CH_3)_3CC_6H_4CH(CH_3)_2 \xrightarrow{a} CH_4 + 4\text{-}(CH_3)_3CC_6H_4\dot{C}(CH_3)_2$ $+ C_6H_5CH_3 \xrightarrow{b} CH_4 + C_6H_5\dot{C}H_2$				
Phot. of azomethane				68 Eac 1
PR [15])	p-t-butylcumene/ toluene	273	$k_a/k_b = 24.0$	
		323	16.9	
		368	12.1	
			$A_a/A_b = 2.1$	
			$E_a(a) - E_a(b) = -5.4(13)\,kJ\,mol^{-1}$	
$\dot{C}H_3 + C_6H_5CH_2CH_2C_6H_5 \xrightarrow{a} CH_4 +$ 1,2-diphenylethane($-\dot{H}$) $+ C_6H_5CH_3 \xrightarrow{b} CH_4 + C_6H_5\dot{C}H_2$				
Phot. of azomethane				68 Eac 1
PR [15])	1,2-diphenylethane/ toluene	273	$k_a/k_b = 5.44$	
		323	4.34	
		368	3.74	
			$A_a/A_b = 1.3$	
			$E_a(a) - E_a(b) = -3.3(8)\,kJ\,mol^{-1}$	
$\dot{C}H_3 + n\text{-}C_{14}H_{30} \xrightarrow{a} CH_4 +$ tetradecane($-\dot{H}$) $+ (CH_3)_3CSD \xrightarrow{b} CH_3D + (CH_3)_3C\dot{S}$				
Thermal decomp. of t-butylperacetate				72 Pry 1
PR [16])	tetradecane/ $(CH_3)_3CSD$ (5:1)	383	$k_a/k_b = 6.12(40) \cdot 10^{-2}$	

[15]) Analysis of T-labelled products from $C_6H_5CH_2T$ reaction.
[16]) MS of CH_4 and CH_3D.

Reaction Radical generation Method	Solvent	T[K]	Rate data	Ref./ add. ref.
$\dot{C}H_3 + CH_2{=}CH(CH_2)_{13}CH_3 \xrightarrow{a} CH_4 +$ hexadecene-1$(-\dot{H})$ $+ (CH_3)_3CCH_2CH(CH_3)_2 \xrightarrow{b} CH_4 +$ 2,2,4-trimethylpentane$(-\dot{H})$				
Thermal decomp. of acetylperoxide				57 Buc 1/
PR	2,2,4-trimethylpentane	338.1	$k_a/k_b = 12.8$ [17]	54 Lev 1, 55 Lev 1
$\dot{C}H_3 + n\text{-}C_{16}H_{34} \xrightarrow{a} CH_4 +$ hexadecane$(-\dot{H})$ $+ (CH_3)_3CSD \xrightarrow{b} CH_3D + (CH_3)_3C\dot{S}$				
Thermal decomp. of *t*-butylperacetate				72 Pry 1
PR [16]	hexadecane/ $(CH_3)_3CSD$ (5:1)	383	$k_a/k_b = 6.96(40)\cdot 10^{-2}$	
$\dot{C}H_3 +$ 2,4,6-$((CH_3)_3C)_3C_6H_2OT \xrightarrow{a} CH_3T + R{-}\dot{O}$ $+ n\text{-}C_7H_{16} \xrightarrow{b} CH_4 +$ *n*-heptane$(-\dot{H})$				
Thermal decomp. of acetylperoxide				65 Shi 1
PR [18]	*n*-heptane	333.15(10)	$k_a/k_b = 2.68(10)$	
		343.0(1)	2.46(14)	
		353.3(1)	2.38(2)	
		362.55(10)	2.20(3) [19]	
$\dot{C}H_3 +$ 2,4,6-$\{(CH_3)_3C\}_3C_6H_2OH \xrightarrow{a} CH_4 + R{-}\dot{O}$ $+ CH_3(CH_2)_2CHT(CH_2)_2CH_3 \xrightarrow{b} CH_3T +$ *n*-heptane$(-\dot{H})$				
Thermal decomp. of acetylperoxide				65 Shi 1
PR [20]	*n*-heptane	333.0(1)	$k_a/k_b = 144.2(52)$	
		343.0(1)	121.8(32)	
		348.1(1)	112.4	
		353.35(10)	109.2(25)	
		357.7(1)	110	
		362.9(1)	107.9(33)	
		367.9(1)	108.0(65) [19]	
$\dot{C}H_3 + (CH_3)_3CCH_2CH(CH_3)_2 \xrightarrow{a} CH_4 +$ 2,2,4-trimethylpentane$(-\dot{H})$ $CH_3\dot{C}H_2 + (CH_3)_3CCH_2CH(CH_3)_2 \xrightarrow{b} C_2H_6 +$ 2,2,4-trimethylpentane$(-\dot{H})$				
Thermal decomp. of acetyl and propionyl peroxide				56 Smi 1
PR	2,2,4-trimethylpentane	338	$k_a/k_b = 11 \ldots 13$	
$\dot{C}H_3 + (C_6H_5)_3CH \xrightarrow{a} CH_4 + (C_6H_5)_3\dot{C}$ $+ C_6H_5CH_3 \xrightarrow{b} CH_4 + C_6H_5\dot{C}H_2$				
Phot. of azomethane				68 Eac 1
PR [21]	triphenylmethane/ toluene	273	$k_a/k_b = 170$	
		298	84.3	
		323	77.1	
		348	72.3	
		368	62.4	
			$A_a/A_b = 3.9$	
			$E_a(a) - E_a(b) = -8.4(29)\,kJ\,mol^{-1}$	
$CH_3\dot{C}H_2 + CH_3COC_2H_5 \xrightarrow{a} C_2H_6 +$ methylethylketone$(-\dot{H})$ $2\,CH_3\dot{C}H_2 \xrightarrow{b} C_4H_{10}$				
			See 4.1.2.3, Fig. 1, p. 254	

[16]) MS of CH_4 and CH_3D.
[17]) $k_a/k(\dot{C}H_3$ addition to hexadecene-1$) = 0.51$.
[18]) Analysis of T-labelled products.
[19]) Rate data do not fit simple Arrhenius plot.
[20]) Analysis via T-labelled products.
[21]) Analysis of T-labelled products from $C_6H_5CH_2T$ reaction.

Reaction Radical generation Method	Solvent	T[K]	Rate data	Ref./ add. ref.
$CH_3\dot{C}H_2 + C_2H_5COC_2H_5 \xrightarrow{a} C_2H_6$ + diethylketone(−Ḣ) $2\,CH_3\dot{C}H_2 \xrightarrow{b} C_4H_{10}$				
			See 4.1.2.3, Fig. 1, p. 254	
$CH_3\dot{C}H_2 + CH_3CH_2C(CH_3)_2OCl$ [22]) $\xrightarrow{a} CH_3CH_2Cl + CH_3CH_2C(CH_3)_2\dot{O}$ $+ CBrCl_3 \xrightarrow{b} CH_3CH_2Br + \dot{C}Cl_3$				
Phot.				65 Zav 1
PR, glc	$CBrCl_3$	273	$k_a/k_b = 19.7$	
		343	20.7	
			$A_a/A_b = 26$	
			$E_a(a) - E_a(b) = 0.63\ kJ\ mol^{-1}$	
			$\Delta S_a^\ddagger - \Delta S_b^\ddagger = 6.4\ J\ K^{-1}\ mol^{-1}$	
$CH_3\dot{C}H_2 + n\text{-}C_7H_{16} \xrightarrow{a} C_2H_6 + (n\text{-}C_7H_{15}^{\cdot})$ $+ (C_2H_5)_2N_2 \xrightarrow{b} ((C_2H_5)_3N_2^{\cdot})$				
Phot. of azoethane				66 Kod 1
PR, glc	n-heptane	273	$k_a/k_b = 1.09 \cdot 10^{-2}$ [23])	
$CH_3\dot{C}H_2 + (CH_3)_3CCH_2CH(CH_3)_2 \xrightarrow{a} C_2H_6$ + 2,2,4-trimethylpentane(−Ḣ) + substrate $\xrightarrow{b}$ (substrate-$CH_2CH_3^{\cdot}$)				
Thermal decomp. of propionyl peroxide				56 Smi 1
PR	2,2,4-trimethylpentane	338	substrate:	
			benzene, $k_a/k_b = 0.35$	
			biphenyl, $9.1 \cdot 10^{-2}$	
			naphthalene, $1.0 \cdot 10^{-2}$	
			phenanthrene, $8.1 \cdot 10^{-3}$	
			quinoline, $5.5 \cdot 10^{-3}$	
			trans-stilbene, $6.9 \cdot 10^{-4}$	
			benzophenone, $2.0 \cdot 10^{-2}$	
			vinylacetate, $4.7 \cdot 10^{-3}$	
$(c\text{-}C_3H_5^{\cdot}) + CBrCl_3 \xrightarrow{a} c\text{-}C_3H_5Br + \dot{C}Cl_3$ $+ CCl_4 \xrightarrow{b} c\text{-}C_3H_5Cl + \dot{C}Cl_3$				
Therm. of $c\text{-}C_3H_5COOOC(CH_3)_3$ and red. of $c\text{-}C_3H_5HgX$ by $NaBH_4$				80 Gie 1,
PR, glc	CCl_4	273	$k_a/k_b = 750(\pm 5\%)$	75 Her 1/
		293	$650(\pm 5\%)$	76 Gie 1,
		313	$530(\pm 5\%)$	76 Gie 2,
		343	$323(\pm 5\%)$	76 Gie 3
		383	$278(\pm 5\%)$	
		403	$250(\pm 5\%)$	
			$\Delta H_a^\ddagger - \Delta H_b^\ddagger = -7.9\ kJ\ mol^{-1}$	
			$\Delta S_a^\ddagger - \Delta S_b^\ddagger = 27\ J\ mol^{-1}\ K^{-1}$	

[22]) ... hypochlorite.
[23]) Based on $k_a/(2k_2)^{\frac{1}{2}} = 1.2 \cdot 10^{-4}$ and $k_b/(2k_2)^{\frac{1}{2}} = 1.1 \cdot 10^{-2}\ M^{-\frac{1}{2}}\ s^{-\frac{1}{2}}$ with $2k_2$ referring to $2(C_2\dot{H}_5) \longrightarrow$ products.

Reaction Radical generation Method	Solvent	T[K]	Rate data	Ref./ add. ref.
$(c\text{-}C_3H_5^{\bullet}) + (CH_3)_3CCH_2CH(CH_3)_2 \xrightarrow{a} c\text{-}C_3H_6 + 2,2,4\text{-trimethylpentane}(-\dot{H})$ $+ \text{olefine} \xrightarrow{b} (\text{olefine-}c\text{-}C_3H_5^{\bullet})$				
Therm. of bis-cyclopropaneformylperoxide [24])	2,2,4-trimethylpentane	338	olefine: ethylene, $k_a/k_b = 0.0427$ propylene, 0.0962 *cis*-2-butene, 0.164 *trans*-2-butene, 0.175 isobutylene, 0.152 trimethylethylene, 0.278 tetramethylethylene, 0.588 1-butene, 0.104 1-pentene, 0.141 1-hexene, 0.152 2-octene, 0.238 $CD_2{=}CD_2$, 0.0383 $CH_3CH{=}CD_2$, 0.0885 $CD_3CD{=}CD_2$, 0.0893 vinylchloride, 0.0246 diethylfumarate, 0.00159	70 Ste 1
[24])	2,2,4-trimethylpentane	338	olefine: $k_a/k_b =$ $CH_2{=}CHF$, 0.0621 $CH_2{=}CHCl$, 0.0246 $CH_2{=}CHBr$, 0.0125 $CH_2{=}CHI$, 0.00395 $CHF{=}CHF$, 0.179 $CH_2{=}CF_2$, 0.0893 $CHCl{=}CHCl$(*trans*) 0.556 $CHCl{=}CHCl$(*cis*), 0.154 $CH_2{=}CCl_2$, 0.00251 $CHBr{=}CHBr$, 0.0236 $CF_3CF{=}CFCF_3$, 0.00877 $CH_3OCH{=}CH_2$, 0.0200	71 Ste 1
$CH_3CH_2\dot{C}H_2 + (CH_2CO)_2NBr \xrightarrow{a} C_3H_7Br + (CH_2CO)_2\dot{N}$ $+ (CH_2CO)_2NCl \xrightarrow{b} C_3H_7Cl + (CH_2CO)_2\dot{N}$				
$(CH_2CO)_2\dot{N} + (C_3H_7)_4Sn \longrightarrow CH_3CH_2\dot{C}H_2 + (CH_2CO)_2NSn(C_3H_7)_3$ PR, glc	acetone	308	$k_a/k_b = 7.3$	72 Dav 1
$CH_3CH_2\dot{C}H_2 + (CH_2CO)_2NI \xrightarrow{a} C_3H_7I + (CH_2CO)_2\dot{N}$ $+ (CH_2CO)_2NCl \xrightarrow{b} C_3H_7Cl + (CH_2CO)_2\dot{N}$				
$(CH_2CO)_2\dot{N} + (C_3H_7)_4Sn \longrightarrow CH_3CH_2\dot{C}H_2 + (CH_2CO)_2NSn(C_3H_7)_3$ PR, glc	acetone	308	$k_a/k_b = 22$	72 Dav 1
$CH_3CH_2\dot{C}H_2 + CH_3(CH_2)_2CHO \xrightarrow{a} C_3H_8 + n\text{-butyraldehyde}(-\dot{H})$ $+ Cu(II) \xrightarrow{b}$ products [25])				
Catalytic decomp. of *n*-valerylperoxide PR, glc	CH_3COOH/H_2O (67:33 Vol%)	330	$k_a/k_b = 1.7(1) \cdot 10^{-4}$	65 Koc 1, 65 Koc 2
$CH_3CH_2\dot{C}H_2 + CH_3(CH_2)_3CHO \xrightarrow{a} C_3H_8 + n\text{-valeraldehyde}(-\dot{H})$ $+ Cu(II) \xrightarrow{b}$ products [25])				
Catalytic decomp. of *n*-butyrylperoxide PR, glc	CH_3COOH/H_2O (67:33 Vol%)	330	$k_a/k_b = 1.5 \cdot 10^{-4}$	65 Koc 1, 65 Koc 2

[24]) $c\text{-}C_3H_6$ and CO_2 pressure measurements.
[25]) e^--transfer.

Reaction Radical generation Method	Solvent	T[K]	Rate data	Ref./ add. ref.
$CH_3CH_2\dot{C}H_2 + (CH_3)_3CCH_2CH(CH_3)_2 \xrightarrow{a} C_3H_8$ + 2,2,4-trimethylpentane($-\dot{H}$) + substrate $\xrightarrow{b}$ (substrate-$CH_2CH_2CH_3^{\cdot}$)				
Thermal decomp. of *n*-butyrylperoxide				57 Smi 1
PR	2,2,4-trimethylpentane	338	substrate: benzene, $k_a/k_b = 0.50$ biphenyl, 0.12 benzophenone, $3.0 \cdot 10^{-2}$ naphthalene, $1.46 \cdot 10^{-2}$ phenanthrene, $1.11 \cdot 10^{-2}$ quinoline, $6.85 \cdot 10^{-3}$ vinylacetate, $4.1 \cdot 10^{-3}$ *trans*-stilbene, $1.11 \cdot 10^{-3}$	
$(CH_3)_2\dot{C}H + (CH_3)_2CHCHO \xrightarrow{a} (CH_3)_2CH_2$ + *i*-butyraldehyde($-\dot{H}$) + Cu(II) $\xrightarrow{b}$ products [25]				
Catalytic decomp. of *n*-valerylperoxide				65 Koc 1,
PR, glc	CH_3COOH/H_2O (67:33 Vol %)	330	$k_a/k_b = 2.8 \cdot 10^{-4}$	65 Koc 2
$(CH_3)_2\dot{C}H + (CH_3)_2CHCHO \xrightarrow{a} (CH_3)_2CH_2 + (CH_3)_2CH\dot{C}O$ $+ Cu(II)(NCCH_3)_4^{2+} \xrightarrow{b} CH_3CH{=}CH_2 + Cu(I)(NCCH_3)_4^{+} + H^{+}$ [25]				
Cu(II) catalyzed decomp. of $(CH_3)_2CHOOCH(CH_3)_2$				68 Koc 1
PR, glc	CH_3CN/CH_3COOH (1:1.5)	298.5	$k_a/k_b = 10^{-3}$ [26]	
$(CH_3)_2\dot{C}H + (CH_3)_2CHC(CH_3)_2OCl$ [27] $\xrightarrow{a} (CH_3)_2CHCl + (CH_3)_2CHC(CH_3)_2\dot{O}$ $+ CBrCl_3 \xrightarrow{b} (CH_3)_2CHBr + \dot{C}Cl_3$				
Phot.				65 Zav 1
PR, glc	$CBrCl_3$	273 346	$k_a/k_b = 8.03$ 11.5 $A_a/A_b = 44$ $E_a(a) - E_a(b) = 3.8\ kJ\ mol^{-1}$ $\Delta S_a^{\ddagger} - \Delta S_b^{\ddagger} = 75\ J\ K^{-1}\ mol^{-1}$	
$(CH_3)_2\dot{C}H + 3\text{-}BrC_6H_4CH_3 \xrightarrow{a} (CH_3)_2CH_2 + 3\text{-}BrC_6H_4\dot{C}H_2$ $+ (CH_3)_3CSD \xrightarrow{b} (CH_3)_2CHD + (CH_3)_3C\dot{S}$				
Phot. of azoisopropane				77 Dav 1
PR, MS	$3\text{-}BrC_6H_4CH_3/$ $(CH_3)_3CSD$	303	$k_a/k_b = 6.9 \cdot 10^{-2}$	[28]
$(CH_3)_2\dot{C}H + 3\text{-}BrC_6H_4CH_3 \xrightarrow{a} (CH_3)_2CH_2 + 3\text{-}BrC_6H_4\dot{C}H_2$ $+ C_6H_5SD \xrightarrow{b} (CH_3)_2CHD + C_6H_5\dot{S}$				
Phot. of azoisopropane				77 Dav 1
PR, MS	$3\text{-}BrC_6H_4CH_3/$ C_6H_5SD	303	$k_a/k_b = 4.5 \cdot 10^{-2}$	[28]
$(CH_3)_2\dot{C}H + 4\text{-}BrC_6H_4CH_3 \xrightarrow{a} (CH_3)_2CH_2 + 4\text{-}BrC_6H_4\dot{C}H_2$ $+ (CH_3)_3CSD \xrightarrow{b} (CH_3)_2CHD + (CH_3)_3C\dot{S}$				
Phot. of azoisopropane				77 Dav 1
PR, MS	$4\text{-}BrC_6H_4CH_3/$ $(CH_3)_3CSD$	303	$k_a/k_b = 6.4(4) \cdot 10^{-2}$	[28]

[25] e^--transfer.
[26] Calc. value $k_a = 5 \cdot 10^3\ M^{-1}\ s^{-1}$.
[27] ... hypochlorite.
[28] Data in Supplement to original paper.

Reaction Radical generation Method	Solvent	T[K]	Rate data	Ref./ add. ref.
$(CH_3)_2\dot{C}H + 4\text{-}BrC_6H_4CH_3 \xrightarrow{a} (CH_3)_2CH_2 + 4\text{-}BrC_6H_4\dot{C}H_2$ $+ C_6H_5SD \xrightarrow{b} (CH_3)_2CHD + C_6H_5\dot{S}$				
Phot. of azoisopropane PR, MS	$4\text{-}BrC_6H_4CH_3/$ C_6H_5SD	303	$k_a/k_b = 4.15(40) \cdot 10^{-2}$	77 Dav 1 [28])
$(CH_3)_2\dot{C}H + 3\text{-}ClC_6H_4CH_3 \xrightarrow{a} (CH_3)_2CH_2 + 3\text{-}ClC_6H_4\dot{C}H_2$ $+ (CH_3)_3CSD \xrightarrow{b} (CH_3)_2CHD + (CH_3)_3C\dot{S}$				
Phot. of azoisopropane PR, MS	$3\text{-}ClC_6H_4CH_3/$ $(CH_3)_3CSD$	303	$k_a/k_b = 5.9 \cdot 10^{-2}$	77 Dav 1 [28])
$(CH_3)_2\dot{C}H + 3\text{-}ClC_6H_4CH_3 \xrightarrow{a} (CH_3)_2CH_2 + 3\text{-}ClC_6H_4\dot{C}H_2$ $+ C_6H_5SD \xrightarrow{b} (CH_3)_2CHD + C_6H_5\dot{S}$				
Phot. of azoisopropane PR, MS	$3\text{-}ClC_6H_4CH_3/$ C_6H_5SD	303	$k_a/k_b = 3.95(20) \cdot 10^{-2}$	77 Dav 1 [28])
$(CH_3)_2\dot{C}H + 4\text{-}ClC_6H_4CH_3 \xrightarrow{a} (CH_3)_2CH_2 + 4\text{-}ClC_6H_4\dot{C}H_2$ $+ (CH_3)_3CSD \xrightarrow{b} (CH_3)_3CHD + (CH_3)_3C\dot{S}$				
Phot. of azoisopropane PR, MS	$4\text{-}ClC_6H_4CH_3/$ $(CH_3)_3CSD$	303	$k_a/k_b = 5.5(3) \cdot 10^{-2}$	77 Dav 1 [28])
$(CH_3)_2\dot{C}H + 4\text{-}ClC_6H_4CH_3 \xrightarrow{a} (CH_3)_2CH_2 + 4\text{-}ClC_6H_4\dot{C}H_2$ $+ C_6H_5SD \xrightarrow{b} (CH_3)_2CHD + C_6H_5\dot{S}$				
Phot. of azoisopropane PR, MS	$4\text{-}ClC_6H_4CH_3/$ C_6H_5SD	303	$k_a/k_b = 3.5(3) \cdot 10^{-2}$	77 Dav 1 [28])
$(CH_3)_2\dot{C}H + 3\text{-}FC_6H_4CH_3 \xrightarrow{a} (CH_3)_2CH_2 + 3\text{-}FC_6H_4\dot{C}H_2$ $+ C_6H_5SD \xrightarrow{b} (CH_3)_2CHD + C_6H_5\dot{S}$				
Phot. of azoisopropane PR, MS	$3\text{-}FC_6H_4CH_3/$ C_6H_5SD	303	$k_a/k_b = 4.3 \cdot 10^{-2}$	77 Dav 1 [28])
$(CH_3)_2\dot{C}H + 4\text{-}FC_6H_4CH_3 \xrightarrow{a} (CH_3)_2CH_2 + 4\text{-}FC_6H_4\dot{C}H_2$ $+ C_6H_5SD \xrightarrow{b} (CH_3)_2CHD + C_6H_5\dot{S}$				
Phot. of azoisopropane PR, MS	$4\text{-}FC_6H_4CH_3/$ C_6H_5SD	303	$k_a/k_b = 2.5 \cdot 10^{-2}$	77 Dav 1 [28])
$(CH_3)_2\dot{C}H + 4\text{-}CNC_6H_4CH_3 \xrightarrow{a} (CH_3)_2CH_2 + 4\text{-}CNC_6H_4\dot{C}H_2$ $+ C_6H_5SD \xrightarrow{b} (CH_3)_2CHD + C_6H_5\dot{S}$				
Phot. of azoisopropane PR, MS	$4\text{-}CNC_6H_4CH_3/$ C_6H_5SD	303	$k_a/k_b = 7.0 \cdot 10^{-2}$	77 Dav 1 [28])
$(CH_3)_2\dot{C}H + 3\text{-}CH_3C_6H_4CH_3 \xrightarrow{a} (CH_3)_2CH_2 + 3\text{-}CH_3C_6H_4\dot{C}H_2$ $+ (CH_3)_3CSD \xrightarrow{b} (CH_3)_2CHD + (CH_3)_3C\dot{S}$				
Phot. of azoisopropane PR, MS	$3\text{-}CH_3C_6H_4CH_3/$ $(CH_3)_3CSD$	303	$k_a/k_b = 3.2 \cdot 10^{-2}$ [29])	77 Dav 1 [28])

[28]) Data in Supplement to original paper.
[29]) Statistically corrected to give reactivity per one CH_3-group.

Reaction Radical generation Method	Solvent	T[K]	Rate data	Ref./ add. ref.
$(CH_3)_2\dot{C}H + 3\text{-}CH_3C_6H_4CH_3 \xrightarrow{a} (CH_3)_2CH_2 + 3\text{-}CH_3C_6H_4\dot{C}H_2$ $+ C_6H_5SD \xrightarrow{b} (CH_3)_2CHD + C_6H_5\dot{S}$				
Phot. of azoisopropane				77 Dav 1
PR, MS	$3\text{-}CH_3C_6H_4CH_3/$ C_6H_5SD	303	$k_a/k_b = 1.5(3)\cdot10^{-2}$ [29]	[28]
$(CH_3)_2\dot{C}H + 4\text{-}CH_3C_6H_4CH_3 \xrightarrow{a} (CH_3)_2CH_2 + 4\text{-}CH_3C_6H_4\dot{C}H_2$ $+ (CH_3)_3CSD \xrightarrow{b} (CH_3)_2CHD + (CH_3)_3C\dot{S}$				
Phot. of azoisopropane				77 Dav 1
PR, MS	$4\text{-}CH_3C_6H_4CH_3/$ $(CH_3)_3CSD$	303	$k_a/k_b = 2.7(2)\cdot10^{-2}$ [29]	[28]
$(CH_3)_2\dot{C}H + 4\text{-}CH_3C_6H_4CH_3 \xrightarrow{a} (CH_3)_2CH_2 + 4\text{-}CH_3C_6H_4\dot{C}H_2$ $+ C_6H_5SD \xrightarrow{b} (CH_3)_2CHD + C_6H_5\dot{S}$				
Phot. of azoisopropane				77 Dav 1
PR, MS	$4\text{-}CH_3C_6H_4CH_3/$ C_6H_5SD	303	$k_a/k_b = 1.7(2)\cdot10^{-2}$ [29]	[28]
$(CH_3)_2\dot{C}H + 3,5\text{-}(CH_3)_2C_6H_3CH_3 \xrightarrow{a} (CH_3)_2CH_2 + 3,5\text{-}(CH_3)_2C_6H_3\dot{C}H_2$ $+ C_6H_5SD \xrightarrow{b} (CH_3)_2CHD + C_6H_5\dot{S}$				
Phot. of azoisopropane				77 Dav 1
PR, MS	$3,5\text{-}(CH_3)_2C_6H_3CH_3/$ C_6H_5SD	303	$k_a/k_b = 1.6\cdot10^{-2}$ [29]	[28]
$(c\text{-}C_4H_7^{\cdot}) + CCl_3Br \xrightarrow{a} c\text{-}C_4H_7Br + \dot{C}Cl_3$ $+ CCl_4 \xrightarrow{b} c\text{-}C_4H_7Cl + \dot{C}Cl_3$				
Therm. of $c\text{-}C_4H_7COOOC(CH_3)_3$				75 Her 1/
PR, glc	CCl_4	383	$k_a/k_b = 573$	76 Gie 1, 76 Gie 2, 76 Gie 3
$(CH_3)_2C{=}\dot{C}H + 4\text{-}ClC_6H_4CH_3 \xrightarrow{a} (CH_3)_2C{=}CH_2 + 4\text{-}ClC_6H_4\dot{C}H_2$ $+ CCl_4 \xrightarrow{b} (CH_3)_2C{=}CHCl + \dot{C}Cl_3$				
Therm. of 3-methyl-2-butenoylperoxide				71 Web 1
PR, glc	$4\text{-}ClC_6H_4CH_3/CCl_4$	351	$k_a/k_b = 0.0570(6)$ [30]	
$(CH_3)_2C{=}\dot{C}H + 4\text{-}NO_2C_6H_4CH_3 \xrightarrow{a} (CH_3)_2C{=}CH_2 + 4\text{-}NO_2C_6H_4\dot{C}H_2$ $+ CCl_4 \xrightarrow{b} (CH_3)_2C{=}CHCl + \dot{C}Cl_3$				
Therm. of 3-methyl-2-butenoylperoxide				71 Web 1
PR, glc	$4\text{-}NO_2C_6H_4CH_3/$ CCl_4	351	$k_a/k_b = 0.1210(7)$ [31]	
$(CH_3)_2C{=}\dot{C}H + C_6H_5CH_3 \xrightarrow{a} (CH_3)_2C{=}CH_2 + C_6H_5\dot{C}H_2$ $+ CCl_4 \xrightarrow{b} (CH_3)_2C{=}CHCl + \dot{C}Cl_3$				
Therm. of 3-methyl-2-butenoylperoxide				71 Web 1
PR, glc	$C_6H_5CH_3/CCl_4$	351	$k_a/k_b = 0.0480(9)$ [32]	
$(CH_3)_2C{=}\dot{C}H + C_6H_5OCH_3 \xrightarrow{a} (CH_3)_2C{=}CH_2 + \text{anisole}(-\dot{H})$ $+ CCl_4 \xrightarrow{b} (CH_3)_2C{=}CHCl + \dot{C}Cl_3$				
Therm. of 3-methyl-2-butenoylperoxide				71 Web 1
PR, glc	$C_6H_5OCH_3/CCl_4$	351	$k_a/k_b = 0.0120(3)$	

[28]) Data in Supplement to original paper.
[29]) Statistically corrected to give reactivity per one CH_3-group.
[30]) $k_a/k_b = 0.019$ per α-H.
[31]) $k_a/k_b = 0.040$ per α-H.
[32]) $k_a/k_b = 0.016$ per α-H.

Reaction Radical generation Method	Solvent	T[K]	Rate data	Ref./ add. ref.
$(CH_3)_2C{=}\dot{C}H + C_6H_5CH_2CH_3 \xrightarrow{a} (CH_3)_2C{=}CH_2$ + ethylbenzene(−Ḣ) $+ CCl_4 \xrightarrow{b} (CH_3)_2C{=}CHCl + \dot{C}Cl_3$				
Therm. of 3-methyl-2-butenoylperoxide				71 Web 1
PR, glc	$C_6H_5CH_2CH_3/CCl_4$	351	$k_a/k_b = 0.1270(13)$ [33]	
$(CH_3)_2C{=}\dot{C}H + 4\text{-}CH_3C_6H_4CH_3 \xrightarrow{a} (CH_3)_2C{=}CH_2 + 4\text{-}CH_3C_6H_4\dot{C}H_2$ $+ CCl_4 \xrightarrow{b} (CH_3)_2CH{=}CHCl + \dot{C}Cl_3$				
Therm. of 3-methyl-2-butenoylperoxide				71 Web 1
PR, glc	$4\text{-}CH_3C_6H_4CH_3/CCl_4$	351	$k_a/k_b = 0.1120(12)$ [34]	
$(CH_3)_2C{=}\dot{C}H + 4\text{-}CH_3OC_6H_4CH_3 \xrightarrow{a} (CH_3)_2C{=}CH_2$ + 4-methoxytoluene(−Ḣ) $+ CCl_4 \xrightarrow{b} (CH_3)_2C{=}CHCl + \dot{C}Cl_3$				
Therm. of 3-methyl-2-butenoylperoxide				71 Web 1
PR, glc	$4\text{-}CH_3OC_6H_4CH_3/$ CCl_4	351	$k_a/k_b = 0.0660(22)$ [35]	
$(CH_3)_2C{=}\dot{C}H + C_6H_5CH(CH_3)_2 \xrightarrow{a} (CH_3)_2C{=}CH_2$ + cumene(−Ḣ) $+ CCl_4 \xrightarrow{b} (CH_3)_2C{=}CHCl + \dot{C}Cl_3$				
Therm. of 3-methyl-2-butenoylperoxide				71 Web 1
PR, glc	cumene/CCl_4	351	$k_a/k_b = 0.1430(28)$ [36]	
$(CH_3)_2C{=}\dot{C}H + C_6H_5C(CH_3)_3 \xrightarrow{a} (CH_3)_2C{=}CH_2$ + *t*-butylbenzene(−Ḣ) $+ CCl_4 \xrightarrow{b} (CH_3)_2C{=}CHCl + \dot{C}Cl_3$				
Therm. of 3-methyl-2-butenoylperoxide				71 Web 1
PR, glc	$C_6H_5C(CH_3)_3/CCl_4$	351	$k_a/k_b = 0.0100(15)$	
$CH_3CH_2CH_2\dot{C}H_2 + CHCl_3 \xrightarrow{a} n\text{-}C_4H_{10} + \dot{C}Cl_3$ $+ Cu(II) \xrightarrow{b}$ products [37]				
Catalytic decomp. of *n*-valerylperoxide				65 Koc 1,
PR, glc	CH_3COOH/H_2O (67:33 Vol%)	330	$k_a/k_b = 7 \cdot 10^{-4}$	65 Koc 2
$CH_3CH_2CH_2\dot{C}H_2 + CH_3COOH \xrightarrow{a} n\text{-}C_4H_{10} + \dot{C}H_2COOH$ $+ Cu(II) \xrightarrow{b}$ products [37]				
Catalytic decomp. of *n*-valerylperoxide				65 Koc 1,
PR, glc	CH_3COOH/H_2O (67:33 Vol%)	330	$k_a/k_b = 2.3 \cdot 10^{-6}$	65 Koc 2
$CH_3CH_2CH_2\dot{C}H_2 + CHCl_2COOH \xrightarrow{a} n\text{-}C_4H_{10} + \dot{C}Cl_2COOH$ $+ Cu(II) \xrightarrow{b}$ products [37]				
Catalytic decomp. of *n*-valerylperoxide				65 Koc 1,
PR, glc	CH_3COOH/H_2O (67:33 Vol%)	330	$k_a/k_b = 3.1 \cdot 10^{-4}$	65 Koc 2
	glacial acetic acid	330	$k_a/k_b = 3.4 \cdot 10^{-4}$	
$CH_3CH_2CH_2\dot{C}H_2 + CH_3(CH_2)_2CHO \xrightarrow{a} n\text{-}C_4H_{10}$ + *n*-butyraldehyde(−Ḣ) $+ Cu(II) \xrightarrow{b}$ products [37]				
Catalytic decomp. of *n*-valerylperoxide				65 Koc 1,
PR, glc	CH_3COOH/H_2O (67:33 Vol%)	330	$k_a/k_b = 1.7 \cdot 10^{-4}$	65 Koc 2
	glacial acetic acid	330	$k_a/k_b = 4.4 \cdot 10^{-4}$	

[33]) $k_a/k_b = 0.062$ per α-H.
[34]) $k_a/k_b = 0.019$ per α-H.
[35]) $k_a/k_b = 0.018$ per α-H.
[36]) $k_a/k_b = 0.137$ per α-H.
[37]) e^--transfer.

Reaction Radical generation Method	Solvent	T[K]	Rate data	Ref./ add. ref.
$CH_3CH_2CH_2\dot{C}H_2 + (CH_3)_2CHCHO \xrightarrow{a} n\text{-}C_4H_{10}$ + isobutyraldehyde($-\dot{H}$) $+ Cu(II) \xrightarrow{b}$ products [37])				
Catalytic decomp. of *n*-valerylperoxide				65 Koc 1,
PR, glc	CH_3COOH/H_2O (67:33 Vol%)	330	$k_a/k_b = 4.6 \cdot 10^{-4}$	65 Koc 2
	glacial acetic acid	330	$k_a/k_b = 9.9 \cdot 10^{-4}$	
$CH_3(CH_2)_2\dot{C}H_2 + (CH_3)_2CHCHO \xrightarrow{a} CH_3(CH_2)_2CH_3 + (CH_3)_2CH\dot{C}O$ $+ Cu(II)(NCCH_3)_4^{2+} \xrightarrow{b} Cu(I)(NCCH_3)_4^{+} + CH_3CH_2CH{=}CH_2 + H^+$				
Cu(II) catalyzed decomp. of $n\text{-}C_4H_9OO\text{-}n\text{-}C_4H_9$				68 Koc 1
PR, glc	CH_3CN/CH_3COOH (1:1.5)	298.5	$k_a/k_b = 3.2 \cdot 10^{-3}$ [38])	
$CH_3(CH_2)_2\dot{C}H_2 + (CH_3)_2CHCHO \xrightarrow{a} CH_3(CH_2)_2CH_3 + (CH_3)_2CH\dot{C}O$ + Cu(II)(α,α-bipyridine)$^{2+} \xrightarrow{b} CH_3CH_2CH{=}CH_2 + H^+$ + Cu(I)(α,α-bipyridine)$^+$				
Cu(II) catalyzed decomp. of $n\text{-}C_4H_9OO\text{-}n\text{-}C_4H_9$				68 Koc 1
PR, glc	CH_3CN/CH_3COOH (1:1.5)	298.5	$k_a/k_b = 5.9 \cdot 10^{-4}$ [38])	
$CH_3(CH_2)_2\dot{C}H_2 + CH_3(CH_2)_2CH_2SH \xrightarrow{a} n\text{-}C_4H_{10} + CH_3(CH_2)_2CH_2\dot{S}$ $2\,CH_3(CH_2)_2\dot{C}H_2 \xrightarrow{b}$ products				
$CH_3(CH_2)_2CH_2\dot{S} + P(OC_2H_5) \longrightarrow CH_3(CH_2)_2\dot{C}H_2 + SP(OC_2H_5)$ (AIBN initiated)				69 Bur 2
[39])	C_6H_6	298	$k_a/(2k_b)^{\frac{1}{2}} = 1.96(13)$	
	$c\text{-}C_6H_{12}$	298	$k_a/(2k_b)^{\frac{1}{2}} = 24.9(32)$	
$CH_3CH_2CH_2\dot{C}H_2 + CH_3(CH_2)_3CHO \xrightarrow{a} n\text{-}C_4H_{10}$ + *n*-valeraldehyde($-\dot{H}$) $+ Cu(II) \xrightarrow{b}$ products [37])				
Catalytic decomp. of *i*-valerylperoxide				65 Koc 1,
PR, glc	CH_3COOH/H_2O (67:33 Vol%)	330	$k_a/k_b = 2.0 \cdot 10^{-4}$	65 Koc 2
$CH_3CH_2CH_2\dot{C}H_2 + (CH_3)_2CHCH_2CHO \xrightarrow{a} n\text{-}C_4H_{10}$ + *i*-valeraldehyde($-\dot{H}$) $+ Cu(II) \xrightarrow{b}$ products [37])				
Catalytic decomp. of *n*-valerylperoxide				65 Koc 1,
PR, glc	CH_3COOH/H_2O (67:33 Vol%)	330	$k_a/k_b = 2.0 \cdot 10^{-4}$	65 Koc 2
$CH_3CH_2CH_2\dot{C}H_2 + (CH_3)_3CCHO \xrightarrow{a} n\text{-}C_4H_{10}$ + pivalaldehyde($-\dot{H}$) $+ Cu(II) \xrightarrow{b}$ products [37])				
Catalytic decomp. of *n*-valerylperoxide				65 Koc 1,
PR, glc	CH_3COOH/H_2O (67:33 Vol%)	330	$k_a/k_b = 1.3 \cdot 10^{-4}$	65 Koc 2
$CH_3CH_2CH_2\dot{C}H_2 + CH_2{=}CHCH_2OCH_2CH{=}CH_2 \xrightarrow{a} n\text{-}C_4H_{10}$ + diallylether($-\dot{H}$) $+ Cu(II) \xrightarrow{b}$ products [37])				
Catalytic decomp. of *n*-valerylperoxide				65 Koc 1,
PR	CH_3COOH/H_2O (67:33 Vol%)	330	$k_a/k_b = 9 \cdot 10^{-5}$	65 Koc 2

37) e^--transfer.
38) Assumed value for $k_a = 1 \cdot 10^4\ M^{-1}\,s^{-1}$.
39) Potentiometric titration.

Reaction Radical generation Method	Solvent	T[K]	Rate data	Ref./ add. ref.
$CH_3CH_2CH_2\dot{C}H_2 + (C_2H_5)_2CHCHO \xrightarrow{a} n\text{-}C_4H_{10}$ + 2-ethylbutyraldehyde($-\dot{H}$) $+ Cu(II) \xrightarrow{b}$ products [37])				
Catalytic decomp. of n-valerylperoxide PR, glc	CH_3COOH/H_2O (67:33 Vol%)	330	$k_a/k_b = 3.1 \cdot 10^{-4}$	65 Koc 1, 65 Koc 2
$CH_3CH_2CH_2\dot{C}H_2 + C_6H_5CH_2OH \xrightarrow{a} n\text{-}C_4H_{10} + C_6H_5\dot{C}HOH$ $+ Cu(II) \xrightarrow{b}$ products [37])				
Catalytic decomp. of n-valerylperoxide PR, glc	CH_3COOH/H_2O (67:33 Vol%)	330	$k_a/k_b = 3 \cdot 10^{-5}$	65 Koc 1, 65 Koc 2
$CH_3CH_2CH_2\dot{C}H_2 + C_6H_5CH_2CHO \xrightarrow{a} n\text{-}C_4H_{10}$ + phenylacetaldehyde($-\dot{H}$) $+ Cu(II) \xrightarrow{b}$ products [37])				
Catalytic decomp. of n-valerylperoxide PR, glc	CH_3COOH/H_2O (67:33 Vol%)	330	$k_a/k_b = 6.7 \cdot 10^{-4}$	65 Koc 1, 65 Koc 2
$CH_3CH_2CH_2\dot{C}H_2 + C_6H_5CH_2OCH_2C_6H_5 \xrightarrow{a} n\text{-}C_4H_{10} + C_6H_5\dot{C}HOCH_2C_6H_5$ $+ Cu(II) \xrightarrow{b}$ products [37])				
Catalytic decomp. of n-valerylperoxide PR, glc	CH_3COOH/H_2O (67:33 Vol%)	330	$k_a/k_b = 3 \cdot 10^{-4}$	65 Koc 1, 65 Koc 2
	glacial acetic acid	330	$k_a/k_b = 2.4 \cdot 10^{-4}$	
$(CH_3)_2CH\dot{C}H_2 + CH_3(CH_2)_3CHO \xrightarrow{a} (CH_3)_2CHCH_3$ + n-valeraldehyde($-\dot{H}$) $+ Cu(II) \xrightarrow{b}$ products [37])				
Catalytic decomp. of i-valerylperoxide PR, glc	CH_3COOH/H_2O (67:33 Vol%)	330	$k_a/k_b = 5.3 \cdot 10^{-4}$	65 Koc 1, 65 Koc 2
$(CH_3)_2CH\dot{C}H_2 + CH_3CH_2CH(CH_3)CHO \xrightarrow{a} (CH_3)_2CHCH_3$ + 2-methylbutyraldehyde($-\dot{H}$) $+ Cu(II) \xrightarrow{b}$ products [37])				
Catalytic decomp. of i-valerylperoxide PR, glc	CH_3COOH/H_2O (67:33 Vol%)	330	$k_a/k_b = 7.9 \cdot 10^{-4}$	65 Koc 1, 65 Koc 2
$(CH_3)_2CH\dot{C}H_2 + (CH_3)_2CHCH_2CHO \xrightarrow{a} (CH_3)_2CHCH_3$ + i-valeraldehyde($-\dot{H}$) $+ Cu(II) \xrightarrow{b}$ products [37])				
Catalytic decomp. of n-valerylperoxide PR, glc	CH_3COOH/H_2O (67:33 Vol%)	330	$k_a/k_b = 6.1 \cdot 10^{-4}$	65 Koc 1, 65 Koc 2
$(CH_3)_3\dot{C} + CHCl_3 \xrightarrow{a} (CH_3)_3CH + \dot{C}Cl_3$ $(CH_3)_3\dot{C} + CHCl_3 \xrightarrow{b} (CH_3)_3CCl + \dot{C}HCl_2$				
Phot. of di-t-butylketone PR, NMR	methylcyclopentane	270 303	$k_a/k_b = 1.5(1)$ 1.4(1) $\log[A_a/M^{-1}s^{-1}] = 8.18$ [40]) $E_a(a) = 34\ kJ\ mol^{-1}$ [40]) $\log[A_b/M^{-1}s^{-1}] = 8.03$ [40]) $E_a(b) = 34\ kJ\ mol^{-1}$ [40])	81 Due 1/ 76 Fri 1

[37]) e^--transfer.
[40]) Based on absolute ($k_a + k_b$) measurements.

Reaction Radical generation Method	Solvent	T[K]	Rate data	Ref./ add. ref.
$(CH_3)_3\dot{C} + (C_2H_5)_2NOH \xrightarrow{a} (CH_3)_3CH$ + diethylhydroxylamine$(-\dot{H})$ + isoprene $\xrightarrow{b} (CH_3)_3CH$ + isoprene$(-\dot{H})$				
Phot. of methyl-*t*-butylketone				81 Enc 1
PR, glc	n-C_6H_{14}	293	$k_a/k_b = 1.2$	
$(CH_3)_3\dot{C} + (CH_3)_3CCHO \xrightarrow{a} (CH_3)_3CH$ + pivalaldehyde$(-\dot{H})$ + Cu(II) $\xrightarrow{b}$ products [37]				
Catalytic decomp. of *n*-valerylperoxide				65 Koc 1, 65 Koc 2
PR, glc	CH_3COOH/H_2O (67:33 Vol%)	330	$k_a/k_b \leqslant 9.2 \cdot 10^{-5}$	
$(CH_3)_3\dot{C} + (CH_3)_3CC(CH_3)_2OCl$ [41] $\xrightarrow{a} (CH_3)_3CCl + (CH_3)_3CC(CH_3)_2\dot{O}$ $+ CBrCl_3 \xrightarrow{b} (CH_3)_3CBr + \dot{C}Cl_3$				
Phot.				65 Zav 1
PR, glc	CCl_4	273 333	$k_a/k_b = 4.45$ 13.3 $A_a/A_b = 1990$ $E_a(a) - E_a(b) = 13.9$ kJ mol^{-1} $\Delta S^{\ddagger}_a - \Delta S^{\ddagger}_b = 15.1$ J K^{-1} mol^{-1}	
$(CH_3)_3\dot{C} + CH_3(CH_2)_5SH \xrightarrow{a} (CH_3)_3CH + CH_3(CH_2)_5\dot{S}$ + isoprene $\xrightarrow{b} (CH_3)_3CH$ + isoprene$(-\dot{H})$				
Phot. of methyl-*t*-butylketone				81 Enc 1
PR, glc	n-C_6H_{14}	293	$k_a/k_b = 10.9$	
$(CH_3)_3\dot{C}$ + 3,4-$Cl_2C_6H_3CH_3 \xrightarrow{a} (CH_3)_3CH$ + 3,4-$Cl_2C_6H_3\dot{C}H_2$ $+ C_6H_5CH_3 \xrightarrow{b} (CH_3)_3CH + C_6H_5\dot{C}H_2$				
Phot. of azoisobutane				82 Pry 1
PR, glc	3,4-dichlorotoluene/ toluene	353	$k_a/k_b = 1.55$	
$(CH_3)_3\dot{C}$ + 3-$BrC_6H_4CH_3 \xrightarrow{a} (CH_3)_3CH$ + 3-$BrC_6H_4\dot{C}H_2$ $+ (CH_3)_3CSD \xrightarrow{b} (CH_3)_3CD + (CH_3)_3C\dot{S}$				
Phot. of azoisobutane				77 Dav 1 [42]
PR, MS	3-$BrC_6H_4CH_3$/ $(CH_3)_3CSD$	303	$k_a/k_b = 4.1(2) \cdot 10^{-2}$	
$(CH_3)_3\dot{C}$ + 4-$BrC_6H_4CH_3 \xrightarrow{a} (CH_3)_3CH$ + 4-$BrC_6H_4\dot{C}H_2$ $+ (CH_3)_3CSD \xrightarrow{b} (CH_3)_3CD + (CH_3)_3C\dot{S}$				
Phot. of azoisobutane				77 Dav 1 [42]
PR, MS	4-$BrC_6H_4CH_3$/ $(CH_3)_3CSD$	303	$k_a/k_b = 3.6(3) \cdot 10^{-2}$	
$(CH_3)_3\dot{C}$ + 4-$BrC_6H_4CH_3 \xrightarrow{a} (CH_3)_3CH$ + 4-$BrC_6H_4\dot{C}H_2$ $+ C_6H_5SD \xrightarrow{b} (CH_3)_3CD + C_6H_5\dot{S}$				
Phot. of azoisobutane and therm. of *t*-butylperoxypivalate				77 Dav 1 [42]
PR, MS	4-$BrC_6H_4CH_3$/ C_6H_5SD	303	$k_a/k_b = 3.6(3) \cdot 10^{-2}$	

[37]) e^--transfer.
[41]) ... hypochlorite.
[42]) Data in Supplement to original paper.

Reaction Radical generation Method	Solvent	T[K]	Rate data	Ref./ add. ref.
$(CH_3)_3\dot{C} + 3\text{-}ClC_6H_4CH_3 \xrightarrow{a} (CH_3)_3CH + 3\text{-}ClC_6H_4\dot{C}H_2$ $+ C_6H_5SD \xrightarrow{b} (CH_3)_3CD + C_6H_5\dot{S}$				
Therm. of *t*-butylperoxypivalate PR, MS	$3\text{-}ClC_6H_4CH_3$/ C_6H_5SD	303	$k_a/k_b = 6.2 \cdot 10^{-2}$	77 Dav 1 [42]/ 79 Tan 2
$(CH_3)_3\dot{C} + 3\text{-}ClC_6H_4CH_3 \xrightarrow{a} (CH_3)_3CH$ + 3-chlorotoluene$(-\dot{H})$ $+ C_6H_5CH_3 \xrightarrow{b} (CH_3)_3CH$ + toluene$(-\dot{H})$				
Phot. of azoisobutane PR, glc	3-chlorotoluene/ toluene	353	$k_a/k_b = 1.38$	82 Pry 1/ 79 Tan 2
$(CH_3)_3\dot{C} + 4\text{-}ClC_6H_4CH_3 \xrightarrow{a} (CH_3)_3CH + 4\text{-}ClC_6H_4\dot{C}H_2$ $+ (CH_3)_3CSD \xrightarrow{b} (CH_3)_3CD + (CH_3)_3C\dot{S}$				
Phot. of azoisobutane PR, MS	$4\text{-}ClC_6H_4CH_3$/ $(CH_3)_3CSD$	303	$k_a/k_b = 3.15(140) \cdot 10^{-2}$	77 Dav 1 [42]/ 79 Tan 2
$(CH_3)_3\dot{C} + 4\text{-}ClC_6H_4CH_3 \xrightarrow{a} (CH_3)_3CH + 4\text{-}ClC_6H_4\dot{C}H_2$ $+ C_6H_5SD \xrightarrow{b} (CH_3)_3CD + C_6H_5\dot{S}$				
Phot. of azoisobutane and therm. of *t*-butylperoxypivalate PR, MS	$4\text{-}ClC_6H_4CH_3$/ C_6H_5SD	303	$k_a/k_b = 3.15(30) \cdot 10^{-2}$	77 Dav 1 [42]/ 79 Tan 2
$(CH_3)_3\dot{C} + 4\text{-}ClC_6H_4CH_3 \xrightarrow{a} (CH_3)_3CH$ + 4-chlorotoluene$(-\dot{H})$ $+ C_6H_5CH_3 \xrightarrow{b} (CH_3)_3CH$ + toluene$(-\dot{H})$				
Phot. of azoisobutane PR, glc	4-chlorotoluene/ toluene	353	$k_a/k_b = 1.03$	82 Pry 1/ 79 Tan 2
$(CH_3)_3\dot{C} + 4\text{-}FC_6H_4CH_3 \xrightarrow{a} (CH_3)_3CH + 4\text{-}FC_6H_4\dot{C}H_2$ $+ (CH_3)_3CSD \xrightarrow{b} (CH_3)_3CD + (CH_3)_3C\dot{S}$				
Phot. of azoisobutane PR, MS	$4\text{-}FC_6H_4CH_3$/ $(CH_3)_3CSD$	303	$k_a/k_b = 2.7 \cdot 10^{-2}$	77 Dav 1 [42]
$(CH_3)_3\dot{C} + 4\text{-}FC_6H_4CH_3 \xrightarrow{a} (CH_3)_3CH + 4\text{-}FC_6H_4\dot{C}H_2$ $+ C_6H_5SD \xrightarrow{b} (CH_3)_3CD + C_6H_5\dot{S}$				
Phot. of azoisobutane PR, MS	$4\text{-}FC_6H_4CH_3$/ C_6H_5SD	303	$k_a/k_b = 1.8 \cdot 10^{-2}$	77 Dav 1 [42]
$(CH_3)_3\dot{C} + 3\text{-}NO_2C_6H_4CH_3 \xrightarrow{a} (CH_3)_3CH + 3\text{-}NO_2C_6H_4\dot{C}H_2$ $+ (CH_3)_3CSD \xrightarrow{b} (CH_3)_3CD + (CH_3)_3C\dot{S}$				
Phot. of azoisobutane PR, MS	$3\text{-}NO_2C_6H_4CH_3$/ $(CH_3)_3CSD$	303	$k_a/k_b = 7.0 \cdot 10^{-2}$	77 Dav 1 [42]
$(CH_3)_3\dot{C} + 3\text{-}NO_2C_6H_4CH_3 \xrightarrow{a} (CH_3)_3CH + 3\text{-}NO_2C_6H_4\dot{C}H_2$ $+ C_6H_5SD \xrightarrow{b} (CH_3)_3CD + C_6H_5\dot{S}$				
Phot. of azoisobutane PR, MS	$3\text{-}NO_2C_6H_4CH_3$/ C_6H_5SD	303	$k_a/k_b = 6.9 \cdot 10^{-2}$	77 Dav 1 [42]

[42]) Data in Supplement to original paper.

Reaction Radical generation Method	Solvent	T[K]	Rate data	Ref./ add. ref.
$(CH_3)_3\dot{C} + C_6H_5CH_3 \xrightarrow{a} (CH_3)_3CH + C_6H_5\dot{C}H_2$ $+ (CH_3)_3CSD \xrightarrow{b} (CH_3)_3CD + (CH_3)_3C\dot{S}$				
Phot. of azoisobutane				77 Dav 1
PR, MS	$C_6H_5CH_3$/ $(CH_3)_3CSD$	303	$k_a/k_b = 2.2(3)\cdot 10^{-2}$	[42])/ 79 Tan 2
$(CH_3)_3\dot{C} + C_6H_5CH_3 \xrightarrow{a} (CH_3)_3CH + C_6H_5\dot{C}H_2$ $+ C_6H_5SD \xrightarrow{b} (CH_3)_3CD + C_6H_5\dot{S}$				
Phot. of azoisobutane				77 Dav 1
PR, MS	$C_6H_5CH_3/C_6H_5SD$	303	$k_a/k_b = 1.2(1)\cdot 10^{-2}$	[42])/ 79 Tan 2
$(CH_3)_3\dot{C}$ + 4-$CNC_6H_4CH_3 \xrightarrow{a} (CH_3)_3CH$ + 4-$CNC_6H_4\dot{C}H_2$ $+ C_6H_5SD \xrightarrow{b} (CH_3)_3CD + C_6H_5\dot{S}$				
Phot. of azoisobutane				77 Dav 1
PR, MS	4-$CNC_6H_4CH_3$/ C_6H_5SD	303	$k_a/k_b = 8.4\cdot 10^{-2}$	[42])/ 79 Tan 2
$(CH_3)_3\dot{C}$ + 4-$CNC_6H_4CH_3 \xrightarrow{a} (CH_3)_3CH$ + 4-cyanotoluene($-\dot{H}$) $+ C_6H_5CH_3 \xrightarrow{b} (CH_3)_3CH$ + toluene($-\dot{H}$)				
Phot. of azoisobutane				82 Pry 1/
PR, glc	4-cyanotoluene/ toluene	353	$k_a/k_b = 2.23$	79 Tan 2
$(CH_3)_3\dot{C}$ + 3-$CH_3C_6H_4CH_3 \xrightarrow{a} (CH_3)_3CH$ + 3-$CH_3C_6H_4\dot{C}H_2$ $+ (CH_3)_3CSD \xrightarrow{b} (CH_3)_3CD + (CH_3)_3C\dot{S}$				
Phot. of azoisobutane				77 Dav 1
PR, MS	3-$CH_3C_6H_4CH_3$/ $(CH_3)_3CSD$	303	$k_a/k_b = 1.4(1)\cdot 10^{-2}$ [43])	[42])/ 79 Tan 2
$(CH_3)_3\dot{C}$ + 3-$CH_3C_6H_4CH_3 \xrightarrow{a} (CH_3)_3CH$ + 3-$CH_3C_6H_4\dot{C}H_2$ $+ C_6H_5CH_3 \xrightarrow{b} (CH_3)_3CH$ + toluene($-\dot{H}$)				
Phot. of azoisobutane				82 Pry 1/
PR, glc	1,3-dimethylbenzene/ toluene	353	$k_a/k_b = 0.76$ [44])	79 Tan 2
$(CH_3)_3\dot{C}$ + 4-$CH_3C_6H_4CH_3 \xrightarrow{a} (CH_3)_3CH$ + 4-$CH_3C_6H_4\dot{C}H_2$ $+ (CH_3)_3CSD \xrightarrow{b} (CH_3)_3CD + (CH_3)_3C\dot{S}$				
Phot. of azoisobutane				77 Dav 1
PR, MS	4-$CH_3C_6H_4CH_3$/ $(CH_3)_3CSD$	303	$k_a/k_b = 1.1(1)\cdot 10^{-2}$ [43])	[42])/ 79 Tan 2
$(CH_3)_3\dot{C}$ + 4-$CH_3C_6H_4CH_3 \xrightarrow{a} (CH_3)_3CH$ + 4-$CH_3C_6H_4\dot{C}H_2$ $+ C_6H_5SD \xrightarrow{b} (CH_3)_3CD + C_6H_5\dot{S}$				
Phot. of azoisobutane and therm. of *t*-butylperoxypivalate				77 Dav 1
PR, MS	4-$CH_3C_6H_4CH_3$/ C_6H_5SD	303	$k_a/k_b = 1.25(20)\cdot 10^{-2}$ [43])	[42])/ 79 Tan 2

[42]) Data in Supplement to original paper.
[43]) Statistically corrected to give reactivity per one CH_3-group.
[44]) With statistical correction factor of $\frac{1}{2}$.

Reaction Radical generation Method	Solvent	T[K]	Rate data	Ref./ add. ref.
$(CH_3)_3\dot{C} + 4\text{-}CH_3C_6H_4CH_3 \xrightarrow{a} (CH_3)_3CH + 4\text{-}CH_3C_6H_4\dot{C}H_2$ $+ C_6H_5CH_3 \xrightarrow{b} (CH_3)_3CH + \text{toluene}(-\dot{H})$				
Phot. of azoisobutane				82 Pry 1/
PR, glc	1,4-dimethylbenzene/ toluene	353	$k_a/k_b = 0.73$ [44])	79 Tan 2
$(CH_3)_3\dot{C} + 3,5\text{-}(CH_3)_2C_6H_3CH_3 \xrightarrow{a} (CH_3)_3CH + 3,5\text{-}(CH_3)_2C_6H_3\dot{C}H_2$ $+ (CH_3)_3CSD \xrightarrow{b} (CH_3)_3CD + (CH_3)_3C\dot{S}$				
Phot. of azoisobutane				77 Dav 1
PR, MS	$3,5\text{-}(CH_3)_2C_6H_3CH_3/$ $(CH_3)_3CSD$	303	$k_a/k_b = 1.0 \cdot 10^{-2}$ [43])	[42])
$(CH_3)_3\dot{C} + 3,5\text{-}(CH_3)_2C_6H_3CH_3 \xrightarrow{a} (CH_3)_3CH + 3,5\text{-}(CH_3)_2C_6H_3\dot{C}H_2$ $+ C_6H_5CH_3 \xrightarrow{b} (CH_3)_3CH + \text{toluene}(-\dot{H})$				
Phot. of azoisobutane				82 Pry 1
PR, glc	1,3,5-trimethylbenzene/ toluene	353	$k_a/k_b = 0.80$ [45])	
$(CH_3)_3\dot{C} + 4\text{-}(CH_3)_3CC_6H_4CH_3 \xrightarrow{a} (CH_3)_3CH + 4\text{-}t\text{-butyltoluene}(-\dot{H})$ $+ C_6H_5CH_3 \xrightarrow{b} (CH_3)_3CH + \text{toluene}(-\dot{H})$				
Phot. of azoisobutane				82 Pry 1
PR, glc	4-*t*-butyltoluene/ toluene	353	$k_a/k_b = 0.78$	
$(CH_3)_3\dot{C} + XC_6H_4CH_3$ [46]) $\xrightarrow{a} (CH_3)_3CH + XC_6H_4\dot{C}H_2$ $+ C_6H_5CH_3 \xrightarrow{b} (CH_3)_3CH + C_6H_5\dot{C}H_2$				
			See 4.1.2.3, Fig. 2, p. 254	
$CH_3CH_2\dot{C}HCH_3 + CH_3CH_2CH(CH_3)CHO \xrightarrow{a} n\text{-}C_4H_{10} + \text{2-methylbutyraldehyde}(-\dot{H})$ $+ Cu(II) \xrightarrow{b} \text{products}$ [47])				
Catalytic decomp. of decanoylperoxide (*A*) or *n*-butyrylperoxide (*B*)				65 Koc 1,
PR, glc	CH_3COOH/H_2O (67:33 Vol%)	330	$k_a/k_b = 4.9 \cdot 10^{-4}$ (*A*) $3.9 \cdot 10^{-4}$ (*B*)	65 Koc 2
$(c\text{-}C_5H_7^{\cdot})$ [48]) $+ (CH_3)_3CC(CH_3)_2OCl$ [49]) $\xrightarrow{a} c\text{-}C_5H_7Cl + (CH_3)_3CC(CH_3)_2\dot{O}$ $+ CBrCl_3 \xrightarrow{b} c\text{-}C_5H_7Br + \dot{C}Cl_3$				
Phot. of *c*-pentene containing soln.				65 Zav 1
PR, glc	CCl_4	273 323	$k_a/k_b = 0.095$ 0.013 $A_a/A_b = 2.8 \cdot 10^{-7}$ $E_a(a) - E_a(b) = -28.9\ \text{kJ mol}^{-1}$ $\Delta S^{\ddagger}_a - \Delta S^{\ddagger}_b = -30\ \text{J K}^{-1}\text{mol}^{-1}$	
$(c\text{-}C_5H_9^{\cdot}) + CCl_3Br \xrightarrow{a} c\text{-}C_5H_9Br + \dot{C}Cl_3$ $+ CCl_4 \xrightarrow{b} c\text{-}C_5H_9Cl + \dot{C}Cl_3$				
Therm. of $c\text{-}C_5H_9COOOC(CH_3)_3$				75 Her 1/
PR, glc	CCl_4	383	$k_a/k_b = 662$	76 Gie 1, 76 Gie 2, 76 Gie 3

[42]) Data in Supplement to original paper.
[43]) Statistically corrected to give reactivity per one CH_3-group.
[44]) With statistical correction factor of $\frac{1}{2}$.
[45]) With statistical correction factor of $\frac{1}{3}$.
[46]) Various substituents X.
[47]) e^--transfer.
[48]) *c*-pentenyl radical.
[49]) ... hypochlorite.

Reaction Radical generation Method	Solvent	T[K]	Rate data	Ref./ add. ref.
$(c\text{-}C_5H_9^{\cdot}) + (CH_3)_3COOD \xrightarrow{a} c\text{-}C_5H_9D + (CH_3)_3CO\dot{O}$ $+ O_2 \xrightarrow{b} (c\text{-}C_5H_9O_2^{\cdot})$				
Decomp. of $(c\text{-}C_5H_9)_2O_2$ (AIBN initiated)				79 How 1
PR, glc	$c\text{-}C_5H_{10}/$ $(CH_3)_3COOD$	323	$k_a/k_b \leqslant 10^2$	
$c\text{-}C_4H_7\dot{C}H_2 + (n\text{-}C_4H_9)_3SnH \xrightarrow{a} c\text{-}C_4H_7CH_3 + (n\text{-}C_4H_9)_3\dot{S}n$ $c\text{-}C_4H_7\dot{C}H_2 \xrightarrow{b} CH_2{=}CHCH_2CH_2\dot{C}H_2$				
$(n\text{-}C_4H_9)_3\dot{S}n + c\text{-}C_4H_7CH_2Cl$ reaction (AIBN initiated)				80 Bec 1
PR, glc	decalin	333	$k_a/k_b = 4.26(20)\cdot 10^2\ M^{-1}$	
$(CH_3)_3C\dot{C}H_2 + (CH_3)_2CHCHO \xrightarrow{a} (CH_3)_3CCH_3 + (CH_3)_2CH\dot{C}O$ $+ Cu(II)(NCCH_3)_4^{2+} \xrightarrow{b} Cu(I)\ldots$ + products				
Cu(II) catalyzed decomp. of $(CH_3)_3CCH_2OOCH_2C(CH_3)_3$				68 Koc 1
PR, glc	CH_3CN/CH_3COOH (1:1.5)	298.5	$k_a/k_b = 2.2\cdot 10^{-2}$ [50]	
$(CH_3)_3C\dot{C}H_2 + (CH_3)_2CHCHO \xrightarrow{a} (CH_3)_3CCH_3 + (CH_3)_2CH\dot{C}O$ $+ Cu(II)(\alpha,\alpha\text{-bipyridine})^{2+} \xrightarrow{b} Cu(I)(\alpha,\alpha\text{-bipyridine})^{+}$ + products				
Cu(II) catalyzed decomp. of $(CH_3)_3CCH_2OOCH_2C(CH_3)_3$				68 Koc 1
PR, glc	CH_3CN/CH_3COOH (1:1.5)	298.5	$k_a/k_b = 0.4$ [50]	
$c\text{-}C_4H_7\dot{C}HCH_3 + (n\text{-}C_4H_9)_3SnH \xrightarrow{a} c\text{-}C_4H_7CH_2CH_3 + (n\text{-}C_4H_9)_3\dot{S}n$ $c\text{-}C_4H_7\dot{C}HCH_3 \xrightarrow{b} CH_3CH{=}CHCH_2CH_2\dot{C}H_2$ (*trans* and *cis*)				
$(n\text{-}C_4H_9)_3\dot{S}n + c\text{-}C_4H_7CHClCH_3$ reaction (AIBN initiated)				80 Bec 1
PR, glc	decalin	333	$k_a/k_b(trans) = 6.06(40)\cdot 10^2\ M^{-1}$ $k_a/k_b(cis) = 1.82(40)\cdot 10^3\ M^{-1}$	
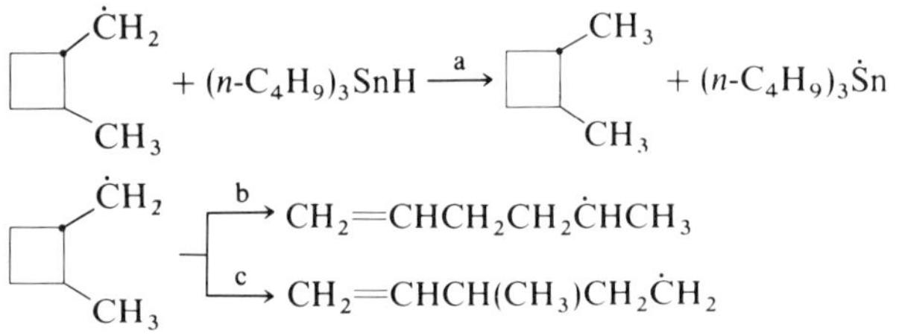				
$(n\text{-}C_4H_9)_3\dot{S}n$ + 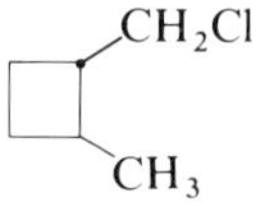 reaction (AIBN initiated)				80 Bec 1
PR, glc	decalin	333	$k_a/k_b = 64.5(8)\ M^{-1}$ $k_a/k_c = 5.26(28)\cdot 10^2\ M^{-1}$	

[50]) Assumed value for $k_a = 1\cdot 10^4\ M^{-1}\,s^{-1}$.

Reaction Radical generation Method	Solvent	T[K]	Rate data	Ref./ add. ref.
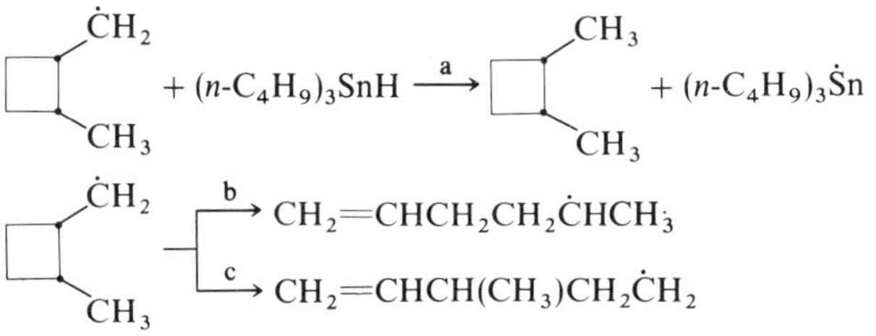				
$(n\text{-}C_4H_9)_3\dot{S}n$ + 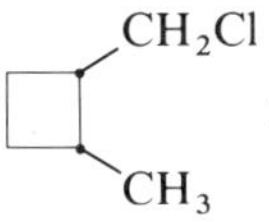 reaction (AIBN initiated)				80 Bec 1
PR, glc	decalin	333	$k_a/k_b = 10.8(1)\,M^{-1}$ $k_a/k_c = 5.0(25)\cdot 10^2\,M^{-1}$	
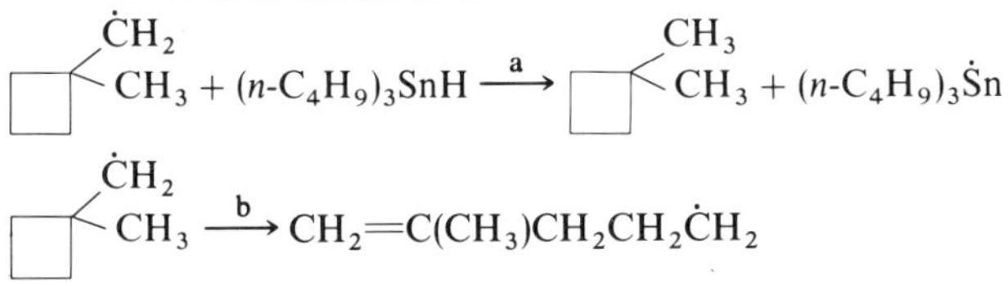				
$(n\text{-}C_4H_9)_3\dot{S}n$ + 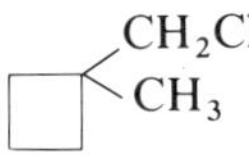 reaction (AIBN initiated)				80 Bec 1
PR, glc	decalin	333	$k_a/k_b = 8.0(3)\cdot 10^2\,M^{-1}$	
$(c\text{-}C_6H_{11}^{\bullet}) + Cl_2 \xrightarrow{a} c\text{-}C_6H_{11}Cl + \dot{C}l$ $(c\text{-}C_6H_{11}^{\bullet}) \xrightarrow{b}$ products [51])				
$\dot{C}l + c\text{-}C_6H_{12}$ reaction (AIBN as initiator) [52])	cyclohexane (containing O_2)	323	$k_a/k_b = 1.20\cdot 10^5\,M^{-1}$	70 Shv 1
$(c\text{-}C_6H_{11}^{\bullet}) + Cl_2 \xrightarrow{a} c\text{-}C_6H_{11}Cl + \dot{C}l$ $2(c\text{-}C_6H_{11}^{\bullet}) \xrightarrow{b}$ products				
$\dot{C}l + c\text{-}C_6H_{12}$ reaction (AIBN as initiator) [52])	cyclohexane	323	$k_a/(k_b)^{\frac{1}{2}} = 19.9\,M^{-\frac{1}{2}}\,s^{-\frac{1}{2}}$	70 Shv 1
$(c\text{-}C_6H_{11}^{\bullet}) + SiHCl_3 \xrightarrow{a} c\text{-}C_6H_{12} + \dot{S}iCl_3$ $+ CCl_4 \xrightarrow{b} c\text{-}C_6H_{11}Cl + \dot{C}Cl_3$				
γ-rad. of $c\text{-}C_6H_{11}Br + CCl_4 + SiHCl_3$				76 Alo 1/
PR, glc	$SiHCl_3$	348	$k_a/k_b = 0.93$	75 Kat 1
		363	0.92	
		373	1.02	
		403	1.05	
		423	0.93	
		443	1.12	
		463	1.0	
			$\log[A_a/A_b] = 0.12(15)$ $E_a(a) - E_a(b) = 1.0(10)\,kJ\,mol^{-1}$ $E_a(a) = 25.6\,kJ\,mol^{-1}$	

[51]) First order termination reaction, likely to be reaction with O_2.
[52]) Cl_2 vapor pressure measurement in gas phase above solution.

Reaction Radical generation Method	Solvent	T[K]	Rate data	Ref./ add. ref.
$(c\text{-}C_6H_{11}^{\cdot}) + CBrCl_3 \xrightarrow{a} c\text{-}C_6H_{11}Br + \dot{C}Cl_3$ $+ CCl_4 \xrightarrow{b} c\text{-}C_6H_{11}Cl + \dot{C}Cl_3$				
Reduct. of $c\text{-}C_6H_{11}HgOCOCH_3$ by $NaBH_4$ and therm. of $c\text{-}C_6H_{11}COOOC(CH_3)_3$				79 Gie 1,
PR, glc	CCl_4	273	$k_a/k_b = 7000$	75 Her 1/
		303	$= 2800$	76 Gie 1,
		323	$= 1800$	76 Gie 2,
		333	$= 1600$	76 Gie 3
		343	$= 1300$	
		383	$= 566$ [53])	
		403	$= 500$ [54])	
			$\Delta H_a^{\ddagger} - \Delta H_b^{\ddagger} = -20\,kJ\,mol^{-1}$ ($\pm 10\%$)	
			$\Delta S_a^{\ddagger} - \Delta S_b^{\ddagger} = 2\,J\,mol^{-1}\,K^{-1}$ ($\pm 5\%$)	
$(c\text{-}C_6H_{11}^{\cdot}) + CCl_3F \xrightarrow{a} c\text{-}C_6H_{11}Cl + \dot{C}Cl_2F$ $+ C_2Cl_4 \xrightarrow{b} c\text{-}C_6H_{11}Cl + (C_2Cl_3^{\cdot})$				
γ-rad. of $c\text{-}C_6H_{12} + CCl_3F + C_2Cl_4$				81 Bar 1/
PR, glc	$c\text{-}C_6H_{12}/CCl_3F/$ C_2Cl_4	363... 453	$\log[k_a/k_b] = 0.328(35) - \frac{2.89(27)}{2.303\,RT}$ [55]) $\log k_a = 9.01 - \frac{33.5}{2.303\,RT}$ [55,56])	73 Hor 1
$(c\text{-}C_6H_{11}^{\cdot}) + CCl_3F \xrightarrow{a} c\text{-}C_6H_{11}Cl + \dot{C}Cl_2F$ $2(c\text{-}C_6H_{11}^{\cdot}) \xrightarrow{b}$ products				
γ-rad. of $c\text{-}C_6H_{12} + CCl_3F$				81 Bar 1/
PR, glc	$c\text{-}C_6H_{12}/CCl_3F$	314... 413	$\log[k_a/(2k_b)^{\frac{1}{2}}] = 2.55(15) - \frac{22.2(50)}{2.303\,RT}$ [55]) $\log k_a = 8.88 - \frac{31.7}{2.303\,RT}$ [55,57])	68 Sau 1, 63 McC 1
$(c\text{-}C_6H_{11}^{\cdot}) + CCl_4 \xrightarrow{a} c\text{-}C_6H_{11}Cl + \dot{C}Cl_3$ $(c\text{-}C_6D_{11}^{\cdot}) + CCl_4 \xrightarrow{b} c\text{-}C_6D_{11}Cl + \dot{C}Cl_3$				
γ-rad. of $c\text{-}C_6H_{12} + c\text{-}C_6D_{12}$				80 Ngu 1
PR, glc	$c\text{-}C_6H_{12}/c\text{-}C_6D_{12}/$ CCl_4	317.7... 357.9	$k_a/k_b = 0.95(6)$	
$(c\text{-}C_6H_{11}^{\cdot}) + CCl_4 \xrightarrow{a} c\text{-}C_6H_{11}Cl + \dot{C}Cl_3$ $+ CCl_2{=}CCl_2 \xrightarrow{b} c\text{-}C_6H_{11}CCl_2\dot{C}Cl_2$				
γ-rad. of $c\text{-}C_6H_{12} + CCl_4 + C_2Cl_4$				75 Kat 1/
PR, glc	$c\text{-}C_6H_{12}/C_2Cl_4/$ CCl_4	333	$k_a/k_b = 55.04$	73 Hor 1
		353	48.20	
		373	43.40	
		393	39.74	
		413	37.85	
		433	32.70	
		453	30.58	
			$\log[A_a/A_b] = 0.72(2)$	
			$E_a(a) - E_a(b) = -5.9(2)\,kJ\,mol^{-1}$	
			$\log[A_a/M^{-1}\,s^{-1}] = 9.40(8)$ [58])	
			$E_a(a) = 24.6(6)\,kJ\,mol^{-1}$ [58])	

[53]) From [75 Her 1], generated by therm. of $c\text{-}C_6H_{11}COOOC(CH_3)_3$.
[54]) Extrapolated value.
[55]) R in $kJ\,mol^{-1}\,K^{-1}$.
[56]) Based on assumed $\log[A_b/M^{-1}\,s^{-1}] = 8.68$ and $E_a(b) = 30.6\,kJ\,mol^{-1}$ for $(c\text{-}C_6H_{11}^{\cdot}) + C_2Cl_4$ reaction [73 Hor 1].
[57]) Based on $k_b = 1 \cdot 10^9\,M^{-1}\,s^{-1}$ [68 Sau 1] at RT and $E_a(b) = 19.0\,kJ\,mol^{-1}$ [63 McC 1] for the radical-radical reaction.
[58]) Based on $\log[A_b/M^{-1}\,s^{-1}] = 8.68(6)$ and $E_a(b) = 30.5(4)$ [73 Hor 1].

Reaction Radical generation Method	Solvent	T[K]	Rate data	Ref./ add. ref.
$(c\text{-}C_6H_{11}^{\cdot}) + CHBr_3 \xrightarrow{a} c\text{-}C_6H_{11}Br + \dot{C}HBr_2$ $+ CCl_3CN \xrightarrow{b} c\text{-}C_6H_{11}Cl + \dot{C}Cl_2CN$				
^{60}Co-γ-rad. of *c*-hexane				76 Gon 1
PR, glc	*c*-hexane	393	$k_a/k_b = 1.78$	
		413	1.82	
		443	1.81	
		473	1.87	
		497	1.88	
			$\log[k_a/k_b] = 0.37(5) - \frac{900(440)}{2.3\,RT}$ [59]	
$(c\text{-}C_6H_{11}^{\cdot}) + CHBr_3 \xrightarrow{a} c\text{-}C_6H_{11}Br + \dot{C}HBr_2$ $+ CHCl_2CN \xrightarrow{b} c\text{-}C_6H_{11}Cl + \dot{C}HClCN$				
γ-rad. of *c*-hexane				77 Gon 1/
PR, glc	*c*-hexane	423	$k_a/k_b = 96.7$	75 Kat 1,
		438	78.9	76 Gon 1
		453	68.2	
		473	55.2	
		493	50.7	
		513	39.2	
			$\log(k_a/k_b) = 0.22(11) + \frac{74.5(40)}{2.303\,RT}$ [60]	
$(c\text{-}C_6H_{11}^{\cdot}) + CHCl_3 \xrightarrow{a} c\text{-}C_6H_{11}Cl + \dot{C}HCl_2$ $+ CCl_2{=}CCl_2 \xrightarrow{b} c\text{-}C_6H_{11}CCl_2\dot{C}Cl_2$				
γ-rad. of $c\text{-}C_6H_{12} + C_2Cl_4 + CHCl_3$				75 Kat 1/
PR, glc	$c\text{-}C_6H_{12}/C_2Cl_4/$ $CHCl_3$	392	$k_a/k_b = 0.1836$	73 Hor 1
		412	0.2098	
		432	0.2677	
		452	0.3020	
		472	0.3397	
		492	0.3841	
			$\log[A_a/A_b] = 0.77(6)$ $E_a(a) - E_a(b) = 12.0(4)\,kJ\,mol^{-1}$ $\log[A_a/M^{-1}s^{-1}] = 9.45(12)$ [61] $E_a(a) = 42.5(5)\,kJ\,mol^{-1}$ [61]	
$(c\text{-}C_6H_{11}^{\cdot}) + CHCl_3 \xrightarrow{a} c\text{-}C_6H_{11}Cl + \dot{C}HCl_2$ $+ CH_2BrCN \xrightarrow{b} c\text{-}C_6H_{11}Br + \dot{C}H_2CN$				
γ-rad. of $CHCl_3 + c\text{-}C_6H_{12} + CH_2BrCN$				81 Gon 1
PR, glc	$CHCl_3/$ $c\text{-}C_6H_{12}$ (0.926 M)/ CH_2BrCN (0.297 M)	453	$k_a/k_b = 3.1$	
$(c\text{-}C_6H_{11}^{\cdot}) + CH_2Cl_2 \xrightarrow{a} c\text{-}C_6H_{11}Cl + \dot{C}H_2Cl$ $+ CCl_2{=}CCl_2 \xrightarrow{b} c\text{-}C_6H_{11}CCl_2\dot{C}Cl_2$				
γ-rad. of $c\text{-}C_6H_{12} + C_2Cl_4 + CH_2Cl_2$				75 Kat 1/
PR, glc	$c\text{-}C_6H_{12}/C_2Cl_4/$ CH_2Cl_2	463	$k_a/k_b = 3.896 \cdot 10^{-3}$	73 Hor 1
		483	$4.728 \cdot 10^{-3}$	
		503	$6.103 \cdot 10^{-3}$	
		523	$8.089 \cdot 10^{-3}$	
		543	$10.035 \cdot 10^{-3}$	
			$\log[A_a/A_b] = 0.56(12)$ $E_a(a) - E_a(b) = 26.7(11)\,kJ\,mol^{-1}$ $\log[A_a/M^{-1}s^{-1}] = 9.24(18)$ [61] $E_a(a) = 57.2(8)\,kJ\,mol^{-1}$ [61]	

[59] In $J\,mol^{-1}$.
[60] R in $kJ\,mol^{-1}K^{-1}$.
[61] Based on $\log[A_b/M^{-1}s^{-1}] = 8.68(6)$ and $E_a(b) = 30.50(4)\,kJ\,mol^{-1}$ [73 Hor 1].

Reaction Radical generation Method	Solvent	T[K]	Rate data	Ref./ add. ref.
$(c\text{-}C_6H_{11}^{\cdot}) + CH_3Br \xrightarrow{a} c\text{-}C_6H_{11}Br + \dot{C}H_3$ $+ CH_2ClCN \xrightarrow{b} c\text{-}C_6H_{11}Cl + \dot{C}H_2CN$				
γ-rad. of *c*-hexane				77 Gon 1/
PR, glc	*c*-hexane	413	$k_a/k_b = 3.91$	75 Kat 1,
		423	4.65	76 Gon 1
		433	3.43	
		473	4.50	
		483	3.37	
			$\log[k_a/k_b] = 0.32(33) + \frac{9.6(14)}{2.303\,RT}$ [60]	
$(c\text{-}C_6H_{11}^{\cdot}) + CCl_3CF_3 \xrightarrow{a} c\text{-}C_6H_{11}Cl + \dot{C}Cl_2CF_3$ $+ CCl_2{=}CCl_2 \xrightarrow{b} c\text{-}C_6H_{11}CCl_2\dot{C}Cl_2$				
γ-rad. of $c\text{-}C_6H_{12} + C_2Cl_4 + CCl_3CF_3$				75 Kat 2/
PR, glc	$c\text{-}C_6H_{12}$	333	$k_a/k_b = 8.75$	73 Hor 1
		353	7.65	
		373	6.98	
		413	6.27	
		423	6.22	
		433	6.04	
		453	5.72	
		473	5.44	
			$\log[A_a/A_b] = 0.29(2)$	
			$E_a(a) - E_a(b) = -4.0(2)\,kJ\,mol^{-1}$	
			$\log[A_a/M^{-1}s^{-1}] = 8.97(9)$ [61]	
			$E_a(b) = 26.5(6)\,kJ\,mol^{-1}$ [61]	
$(c\text{-}C_6H_{11}^{\cdot}) + CF_3CCl_3 \xrightarrow{a} c\text{-}C_6H_{11}Cl + CF_3\dot{C}Cl_2$ $+ CH_2BrCN \xrightarrow{b} c\text{-}C_6H_{11}Br + \dot{C}H_2CN$				
γ-rad. of $CF_3CCl_3 + c\text{-}C_6H_{12} + CH_2BrCN$				81 Gon 1,
PR, glc	CF_3CCl_3 (8.33 M)/ $c\text{-}C_6H_{12}$ (0.926 M)/ CH_2BrCN (0.26 M)	453	$k_a/k_b = 0.75$ [62] 0.59 [63]	77 Gon 2, 75 Kat 1, 75 Kat 2
$(c\text{-}C_6H_{11}^{\cdot}) + CCl_3CN \xrightarrow{a} c\text{-}C_6H_{11}Cl + \dot{C}Cl_2CN$ $+ CCl_2{=}CCl_2 \xrightarrow{b} c\text{-}C_6H_{11}CCl_2\dot{C}Cl_2$				
^{60}Co-γ-rad. of *c*-hexane				76 Gon 1
PR, glc	*c*-hexane	438	$k_a/k_b = 433.6$	
		448	313.5	
		458	281.8	
		473	213.7	
		498	226.3	
		523	151.9	
			$\log[A_a/A_b] = -0.38(47)$	
			$E_a(a) - E_a(b) = -25(4)$	

[60]) R in $kJ\,mol^{-1}\,K^{-1}$.
[61]) Based on $\log[A_b/M^{-1}s^{-1}] = 8.68(6)$ and $E_a(b) = 30.50(4)\,kJ\,mol^{-1}$ [73 Hor 1].
[62]) From [81 Gon 1].
[63]) From [77 Gon 2].

Reaction Radical generation Method	Solvent	T[K]	Rate data	Ref./ add. ref.
$(c\text{-}C_6H_{11}^{\cdot}) + C_2Cl_6 \xrightarrow{a} c\text{-}C_6H_{11}Cl + \dot{C}Cl_2CCl_3$ $+ CCl_2{=}CCl_2 \xrightarrow{b} c\text{-}C_6H_{11}CCl_2\dot{C}Cl_2$				
γ-rad. of $c\text{-}C_6H_{12} + C_2Cl_4 + CCl_3CCl_3$				75 Kat 2/
PR, glc	$c\text{-}C_6H_{12}$	295.5	$k_a/k_b = 56.8$	73 Hor 1
		313	41.5	
		314.5	48.2	
		333	42.5	
		353	39.0	
		393	32.2	
		413	31.0	
		433	26.6	
		463	27.8	
			$\log[A_a/A_b] = 0.88(5)$	
			$E_a(a) - E_a(b) = -4.7(4)$ kJ mol^{-1}	
			$\log[A_a/M^{-1}s^{-1}] = 9.56(11)$ [61]	
			$E_a(a) = 25.8(4)$ kJ mol^{-1} [61]	
$(c\text{-}C_6H_{11}^{\cdot}) + C_2Cl_6 \xrightarrow{a} c\text{-}C_6H_{11}Cl + CCl_3\dot{C}Cl_2$ $+ CH_2BrCN \xrightarrow{b} c\text{-}C_6H_{11}Br + \dot{C}H_2CN$				
γ-rad. of $C_2Cl_6 + c\text{-}C_6H_{12} + CH_2BrCN$				81 Gon 1,
PR, glc	$c\text{-}C_6H_{12}$ (8.53 M)/	453	$k_a/k_b = 2.6$	77 Gon 2,
	C_2Cl_6 (0.5 M)/			75 Kat 1,
	CH_2BrCN (0.297 M)			75 Kat 2
$(c\text{-}C_6H_{11}^{\cdot}) + CCl_3CHCl_2 \xrightarrow{a} c\text{-}C_6H_{11}Cl + \dot{C}Cl_2CHCl_2$ $+ CCl_2{=}CCl_2 \xrightarrow{b} c\text{-}C_6H_{11}CCl_2\dot{C}Cl_2$				
γ-rad. of $c\text{-}C_6H_{12} + C_2Cl_4 + CCl_3CHCl_2$				75 Kat 2/
PR, glc	$c\text{-}C_6H_{12}$	333	$k_a/k_b = 10.03$	73 Hor 1
		353	8.31	
		423	6.12	
		433	5.99	
		453	5.42	
		473	4.99	
			$\log[A_a/A_b] = 0.01(3)$	
			$E_a(a) - E_a(b) = -6.3(2)$ kJ mol^{-1}	
			$\log[A_a/M^{-1}s^{-1}] = 8.69(9)$ [61]	
			$E_a(a) = 24.3(6)$ kJ mol^{-1} [61]	
$(c\text{-}C_6H_{11}^{\cdot}) + CHCl_2CCl_3 \xrightarrow{a} c\text{-}C_6H_{11}Cl + CHCl_2\dot{C}Cl_2$ $+ CHCl_2CCl_3 \xrightarrow{b} c\text{-}C_6H_{12} + \dot{C}Cl_2CCl_3$				
γ-rad. of $c\text{-}C_6H_{12}$				71 Kat 1,
PR, glc	$c\text{-}C_6H_{12}$	323…473	$\log(k_a/k_b) = 1.40(15) - \frac{2.76(96)}{2.303\,RT}$ [60]	75 Kat 2,
			$\log k_a = 8.69(9) - \frac{24.2(6)}{2.303\,RT}$ [60,64]	76 Kat 1
			$\log k_b = 7.29(24) - \frac{21.4(16)}{2.303\,RT}$ [60,65]	

[60] R in kJ mol^{-1} K^{-1}.
[61] Based on $\log[A_b/M^{-1}s^{-1}] = 8.68(6)$ and $E_a(b) = 30.50(4)$ kJ mol^{-1} [73 Hor 1].
[64] From [75 Kat 2].
[65] From [76 Kat 1].

Reaction Radical generation Method	Solvent	T[K]	Rate data	Ref./ add. ref.
$(c\text{-}C_6H_{11}^{\cdot}) + CHCl_2CCl_3 \xrightarrow{a} c\text{-}C_6H_{11}Cl + (C_2HCl_4^{\cdot})$ $+ CH_2BrCN \xrightarrow{b} c\text{-}C_6H_{11}Br + \dot{C}H_2CN$				
γ-rad. of $CHCl_2CCl_3 + c\text{-}C_6H_{12} + CH_2BrCN$				81 Gon 1
PR, glc	$CHCl_2CCl_3$/ $c\text{-}C_6H_{12}$ (0.926 M)/ CH_2BrCN (0.297 M)	453	$k_a/k_b = 41.0$	
$(c\text{-}C_6H_{11}^{\cdot}) + CH_2BrCN \xrightarrow{a} c\text{-}C_6H_{11}Br + \dot{C}H_2CN$ $+ CCl_4 \xrightarrow{b} c\text{-}C_6H_{11}Cl + \dot{C}Cl_3$				
γ-rad. of $c\text{-}C_6H_{12} + CCl_4 + CH_2BrCN$				77 Gon 2/
PR, glc	$c\text{-}C_6H_{12}/CCl_4$/ CH_2BrCN	353... 453	$\log[k_a/k_b] = -0.699(167) + \frac{11.1(13)}{2.303\,RT}$ [66]	75 Kat 1
$(c\text{-}C_6H_{11}^{\cdot}) + CCl_2BrCH_2Cl \xrightarrow{a} c\text{-}C_6H_{11}Br + \dot{C}Cl_2CH_2Cl$ $(c\text{-}C_6H_{11}^{\cdot}) + CCl_2BrCH_2Cl \xrightarrow{b} c\text{-}C_6H_{11}Cl + \dot{C}ClBrCH_2Cl$				
γ-rad. of c-hexane				75 Alo 1
PR, glc	$c\text{-}C_6H_{12}$	423(1)	$k_a/k_b = 756.4$	
		448	535	
		473	334	
			$\log[A_a/A_b] = -0.57(56)$ $E_a(a) - E_a(b) = -28.1(42)\ kJ\ mol^{-1}$	
$(c\text{-}C_6H_{11}^{\cdot}) + CCl_3CH_2Cl \xrightarrow{a} c\text{-}C_6H_{11}Cl + \dot{C}Cl_2CH_2Cl$ $(c\text{-}C_6H_{11}^{\cdot}) + CCl_3CH_2Cl \xrightarrow{b} c\text{-}C_6H_{12} + CCl_3\dot{C}HCl$				
γ-rad. of c-hexane				75 Alo 1
PR, glc	$c\text{-}C_6H_{12}$	379(1)	$k_a/k_b = 55.9$	
		393	35.4	
		423	30.6	
		473	40.3	
		498	44.2	
			$\log[A_a/A_b] = 1.37(24)$ $E_a(a) - E_a(b) = -2.0(20)$	
$(c\text{-}C_6H_{11}^{\cdot}) + CCl_3CH_2Cl \xrightarrow{a} c\text{-}C_6H_{11}Cl + \dot{C}Cl_2CHCl$ $+ CCl_2{=}CCl_2 \xrightarrow{b} c\text{-}C_6H_{11}CCl_2\dot{C}Cl_2$				
γ-rad. of $c\text{-}C_6H_{12} + C_2Cl_4 + CCl_3CH_2Cl$				75 Kat 2/
PR, glc	$c\text{-}C_6H_{12}$	373	$k_a/k_b = 1.85$	73 Hor 1
		393	1.97	
		413	2.07	
		433	2.13	
		463	2.28	
		473	2.37	
		483	2.39	
		493	2.54	
			$\log[A_a/A_b] = 0.78(2)$ $E_a(a) - E_a(b) = 3.7(2)\ kJ\ mol^{-1}$ $\log[A_a/M^{-1}s^{-1}] = 9.46(8)$ $E_a(a) = 34.2(6)\ kJ\ mol^{-1}$ [67]	

[66] R in $kJ\ mol^{-1}\ K^{-1}$. $\log k_a = 8.70(34) - 13.4(25)/2.303\ RT$ based on $\log k_b = 9.4(1) - 24.6(6)/2.303\ RT$ [75 Kat 1].
[67] Based on $\log[A_a(b)/M^{-1}s^{-1}] = 8.68(6)$ and $E_a(b) = 30.5(4)\ kJ\ mol^{-1}$ [73 Hor 1].

Reaction Radical generation Method	Solvent	T[K]	Rate data	Ref./ add. ref.
$(c\text{-}C_6H_{11}^{\cdot}) + CH_2ClCCl_3 \xrightarrow{a} c\text{-}C_6H_{11}Cl + (C_2H_2Cl_3^{\cdot})$ $+ CH_2BrCN \xrightarrow{b} c\text{-}C_6H_{11}Br + \dot{C}H_2CN$				
γ-rad. of $CH_2ClCCl_3 + c\text{-}C_6H_{12} + CH_2BrCN$ PR, glc	CH_2ClCCl_3/ $c\text{-}C_6H_{12}$ (0.926 M)/ CH_2BrCN (0.297 M)	453	$k_a/k_b = 21.7$	81 Gon 1
$(c\text{-}C_6H_{11}^{\cdot}) + CH_2ClCCl_3 \xrightarrow{a} c\text{-}C_6H_{11}Cl + CH_2Cl\dot{C}Cl_2$ $+ CH_2ClCCl_3 \xrightarrow{b} c\text{-}C_6H_{12} + \dot{C}HClCCl_3$				
γ-rad. of $c\text{-}C_6H_{12}$ PR, glc	$c\text{-}C_6H_{12}$	379... 498	$\log(k_a/k_b) = 1.37(24) + \frac{1.96(200)}{2.303\,RT}$ [68]) $\log k_a = 9.46(8) - \frac{34.2(6)}{2.303\,RT}$ [68]) [69]) $\log k_b = 8.09(32) - \frac{36.2(26)}{2.303\,RT}$ [68]) [70])	75 Alo 1, 75 Kat 2, 76 Kat 1
$(c\text{-}C_6H_{11}^{\cdot}) + CHCl_2CHCl_2 \xrightarrow{a} c\text{-}C_6H_{11}Cl + \dot{C}HClCHCl_2$ $+ CHCl_2CHCl_2 \xrightarrow{b} c\text{-}C_6H_{12} + \dot{C}Cl_2CHCl_2$				
γ-rad. of $c\text{-}C_6H_{12}$ PR, glc	$c\text{-}C_6H_{12}$	423... 503	$\log(k_a/k_b) = 1.07(6) - \frac{15.80(50)}{2.303\,RT}$ [68]) $\log k_a = 9.10(30) - \frac{42.6(42)}{2.303\,RT}$ [68]) [71]) $\log k_b = 8.00(40) - \frac{26.8(47)}{2.303\,RT}$ [68]) [70])	71 Kat 1, 77 Kat 1, 76 Kat 1
$(c\text{-}C_6H_{11}^{\cdot}) + CCl_3CH_3 \xrightarrow{a} c\text{-}C_6H_{11}Cl + \dot{C}Cl_2CH_3$ $+ CCl_2{=}CCl_2 \xrightarrow{b} c\text{-}C_6H_{11}CCl_2\dot{C}Cl_2$				
γ-rad. of $c\text{-}C_6H_{12} + C_2Cl_4 + CCl_3CH_3$ PR, glc	$c\text{-}C_6H_{12}$	373	$k_a/k_b = 0.403$	75 Kat 2/ 73 Hor 1
		388	0.424	
		403	0.476	
		420	0.514	
		443	0.593	
		463	0.674	
		473	0.706	
		483	0.747	
		488	0.736	
		493	0.777	
		498	0.765	
			$\log[A_a/A_b] = 0.78(3)$ $E_a(a) - E_a(b) = 8.5(2)\,kJ\,mol^{-1}$ $\log[A_a/M^{-1}s^{-1}] = 9.46(9)$ [67]) $E_a(a) = 39.0(4)\,kJ\,mol^{-1}$ [67])	
$(c\text{-}C_6H_{11}^{\cdot}) + CH_3CCl_3 \xrightarrow{a} c\text{-}C_6H_{11}Cl + CH_3\dot{C}Cl_2$ $+ CH_2BrCN \xrightarrow{b} c\text{-}C_6H_{11}Br + \dot{C}H_2CN$				
γ-rad. of $CH_3CCl_3 + c\text{-}C_6H_{12} + CH_2BrCN$ PR, glc	CH_3CCl_3/ $c\text{-}C_6H_{12}$ (0.926 M)/ CH_2BrCN (0.297 M)	453	$k_a/k_b = 8.7$	81 Gon 1

[67]) Based on $\log[A_a(b)/M^{-1}s^{-1}] = 8.68(6)$ and $E_a(b) = 30.5(4)\,kJ\,mol^{-1}$ [73 Hor 1].
[68]) R in $kJ\,mol^{-1}\,K^{-1}$.
[69]) From [75 Kat 2].
[70]) From [76 Kat 1].
[71]) From [77 Kat 1].

Reaction Radical generation Method	Solvent	T[K]	Rate data	Ref./ add. ref.
$(c\text{-}C_6H_{11}^{\cdot}) + CHCl_2CH_2Cl \xrightarrow{a} c\text{-}C_6H_{11}Cl + \dot{C}HClCH_2Cl$ $+ C_2Cl_4 \xrightarrow{b} (C_2Cl_4\text{-}c\text{-}C_6H_{11}^{\cdot})$				
γ-rad. of $c\text{-}C_6H_{12}$				77 Kat 1/
PR, glc	$c\text{-}C_6H_{12}$	423	$k_a/k_b = 0.0198$	73 Hor 1
		448	0.0242	
		473	0.0345	
		498	0.0370	
		523	0.0481	
			$\log[A_a/A_b] = 0.30(8)$	
			$E_a(a) - E_a(b) = 16.2(7)\,kJ\,mol^{-1}$ [72]	
$(c\text{-}C_6H_{11}^{\cdot}) + CHCl_2CH_2Cl \xrightarrow{a} c\text{-}C_6H_{12} + CHCl_2\dot{C}HCl$ $+ CHCl_2CH_2Cl \xrightarrow{b} c\text{-}C_6H_{11}Cl + \dot{C}HClCH_2Cl$				
γ-rad. of $c\text{-}C_6H_{12}$				76 Kat 2/
PR, glc	$c\text{-}C_6H_{12}$	393	$k_a/k_b = 1.138$	71 Kat 1,
		423	1.118	77 Kat 1,
		473	0.943	76 Kat 1
		523	0.879 [73]	
			$\log[A_a/A_b] = -0.467(37)$	
			$E_a(a) - E_a(b) = -4.06(31)\,kJ\,mol^{-1}$	
$(c\text{-}C_6H_{11}^{\cdot}) + CHCl_2CH_2Cl \xrightarrow{a} \dot{C}Cl_2CH_2Cl + c\text{-}C_6H_{12}$ $(c\text{-}C_6H_{11}^{\cdot}) + CHCl_2CH_2Cl \xrightarrow{b} \dot{C}HClCH_2Cl + c\text{-}C_6H_{11}Cl$				
γ-rad. of $c\text{-}C_6H_{12}$				76 Kat 2/
PR, glc	$c\text{-}C_6H_{12}$	393	$k_a/k_b = 13.39$	75 Alo 1,
		423	10.33	77 Kat 1,
		473	6.34	76 Kat 1
		523	4.73 [74]	
			$\log[A_a/A_b] = -0.752(30)$	
			$E_a(a) - E_a(b) = -14.2(2)\,kJ\,mol^{-1}$	
$(c\text{-}C_6H_{11}^{\cdot}) + CHCl_2CH_3 \xrightarrow{a} c\text{-}C_6H_{11}Cl + \dot{C}HClCH_3$ $+ C_2Cl_4 \xrightarrow{b} (C_2Cl_4\text{-}c\text{-}C_6H_{11}^{\cdot})$				
γ-rad. of $c\text{-}C_6H_{12} + CHCl_2CH_3$				77 Kat 1/
PR, glc	$c\text{-}C_6H_{12}$/ $CHCl_2CH_3$ (1.186 M)	403	$k_a/k_b = 4.46 \cdot 10^{-3}$	73 Hor 1
		423	$4.76 \cdot 10^{-3}$	
		448	$7.14 \cdot 10^{-3}$	
		473	$10.00 \cdot 10^{-3}$	
		498	$12.94 \cdot 10^{-3}$	
		523	$15.85 \cdot 10^{-3}$	
			$\log[A_a/A_b] = 0.13(13)$	
			$E_a(a) - E_a(b) = 19.4(11)\,kJ\,mol^{-1}$ [75]	

72) $\log[A_a/M^{-1}s^{-1}] = 8.98(14)$, $E_a(a) = 46.8(11)\,kJ\,mol^{-1}$ and $\log k_a = 2.07(30)$ at 353 K based on $\log k_b = 8.68(6) - 30.6(4)/2.303\,RT$ [73 Hor 1], R in $kJ\,mol^{-1}\,K^{-1}$.

73) Based on data from [71 Kat 1].

74) Based on data from [75 Alo 1].

75) $\log[A_a/M^{-1}s^{-1}] = 8.81(19)$, $E_a(a) = 50.0(15)\,kJ\,mol^{-1}$, and $\log k_a = 1.42(42)$ at 353 K based on $\log k_b = 8.68(6) - 30.6(4)/2.303\,RT$ [73 Hor 1], R in $kJ\,mol^{-1}\,K^{-1}$.

Reaction Radical generation Method	Solvent	T[K]	Rate data	Ref./ add. ref.
$(c\text{-}C_6H_{11}^{\bullet}) + CH_2ClCH_2Cl \xrightarrow{a} c\text{-}C_6H_{11}Cl + \dot{C}H_2CH_2Cl$ $+ C_2Cl_4 \xrightarrow{b} (C_2Cl_4\text{-}c\text{-}C_6H_{11}^{\bullet})$				
γ-rad. of $c\text{-}C_6H_{12} + CH_2ClCH_2Cl$				77 Kat 1/
PR, glc	$c\text{-}C_6H_{12}$/ CH_2ClCH_2Cl (2.54 M)	423	$k_a/k_b = 3.67 \cdot 10^{-4}$	73 Hor 1,
		448	$6.74 \cdot 10^{-4}$	
		473	$9.94 \cdot 10^{-4}$	
		498	$14.35 \cdot 10^{-4}$	
		523	$21.65 \cdot 10^{-4}$	
			$\log[A_a/A_b] = 0.50(17)$ $E_a(a) - E_a(b) = 31.7(15)\,kJ\,mol^{-1}$ [76])	
$(c\text{-}C_6H_{11}^{\bullet}) + CH_2ClCH_2Cl \xrightarrow{a} c\text{-}C_6H_{12} + \dot{C}HClCH_2Cl$ $+ CH_2ClCH_2Cl \xrightarrow{b} c\text{-}C_6H_{11}Cl + \dot{C}H_2CH_2Cl$				
γ-rad. of $c\text{-}C_6H_{12}$				76 Kat 1,
PR, glc	$c\text{-}C_6H_{12}$	423	$k_a/k_b = 62.9$	77 Kat 1
		448	52.1	
		473	48.2	
		498	36.6	
		523	34.9	
			[77])	
$(c\text{-}C_6H_{11}^{\bullet}) + (CH_3)_3COOD \xrightarrow{a} c\text{-}C_6H_{11}D + (CH_3)_3CO\dot{O}$ $+ O_2 \xrightarrow{b} c\text{-}C_6H_{11}\dot{O}_2$				
Decomp. of $(c\text{-}C_6H_{11})_2O_2$ (AIBN initiated)				79 How 1
PR, glc	$c\text{-}C_6H_{12}$/ $(CH_3)_2COOD$	323	$k_a/k_b \leqslant 10^{-2}$	
$(c\text{-}C_6H_{11}^{\bullet})$ + 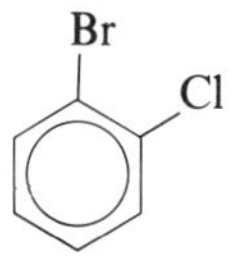 $\xrightarrow{a} c\text{-}C_6H_{11}Br + 2\text{-}Cl\dot{C}_6H_4$ $\xrightarrow{b} c\text{-}C_6H_{11}Cl + 2\text{-}Br\dot{C}_6H_4$				
React. $(CH_3)_3\dot{C} + c\text{-}C_6H_{12}$ after therm. of DTBP				68 She 1
PR, glc	$c\text{-}C_6H_{12}$	378	$k_a/k_b = 2.1$	[78])
$(c\text{-}C_6H_{11}^{\bullet})$ + 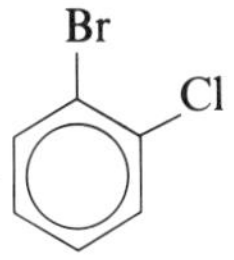 $\xrightarrow{a} c\text{-}C_6H_{11}Br + 2\text{-}Cl\dot{C}_6H_4$ $\xrightarrow{b} c\text{-}C_6H_{12} + (\dot{C}_6H_3BrCl)$				
React. $(CH_3)_3\dot{C} + c\text{-}C_6H_{12}$ after therm. of DTBP				68 She 1
PR, glc	$c\text{-}C_6H_{12}$	378	$k_a/k_b = 2.1$	[78])
$(c\text{-}C_6H_{11}^{\bullet})$ + 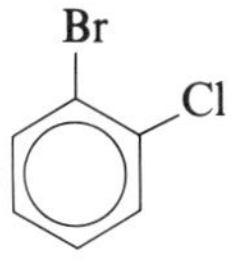 $\xrightarrow{a} c\text{-}C_6H_{11}Cl + 2\text{-}Br\dot{C}_6H_4$ $\xrightarrow{b} c\text{-}C_6H_{12} + (\dot{C}_6H_3BrCl)$				
React. $(CH_3)_3\dot{C} + c\text{-}C_6H_{12}$ after therm. of DTBP				68 She 1
PR, glc	$c\text{-}C_6H_{12}$	378	$k_a/k_b = 1.0$	[78])

[76]) $\log[A_a/M^{-1}s^{-1}] = 9.18(23)$ and $E_a(a) = 62.3(19)\,kJ\,mol^{-1}$ based on $\log k_b = 8.68(6) - 30.6(4)/2.303\,RT$ [73 Hor 1], R in $kJ\,mol^{-1}\,K^{-1}$.

[77]) $\log[k_a/k_b] = 0.394(88) + 11.40(79)/2.303\,RT$, R in $kJ\,mol^{-1}\,K^{-1}$.

[78]) Reaction mechanism proposed to proceed via radical addition.

Reaction				
Radical generation				Ref./
Method	Solvent	$T\,[\mathrm{K}]$	Rate data	add. ref.

$(c\text{-}C_6H_{11}^{\cdot}) +$ $\begin{array}{l} \xrightarrow{a} c\text{-}C_6H_{11}Br + 3\text{-}Cl\dot{C}_6H_4 \\ \xrightarrow{b} c\text{-}C_6H_{11}Cl + 3\text{-}Br\dot{C}_6H_4 \end{array}$

React. $(CH_3)_3\dot{C} + c\text{-}C_6H_{12}$ after therm. of DTBP				68 She 1
PR, glc	$c\text{-}C_6H_{12}$	378	$k_a/k_b = 2.2$	[78])

$(c\text{-}C_6H_{11}^{\cdot}) +$ $\begin{array}{l} \xrightarrow{a} c\text{-}C_6H_{11}Br + 3\text{-}Cl\dot{C}_6H_4 \\ \xrightarrow{b} c\text{-}C_6H_{12} + (\dot{C}_6H_3BrCl) \end{array}$

React. $(CH_3)_3\dot{C} + c\text{-}C_6H_{12}$ after therm. of DTBP				68 She 1
PR, glc	$c\text{-}C_6H_{12}$	378	$k_a/k_b = 1.1$	[78])

$(c\text{-}C_6H_{11}^{\cdot}) +$ $\begin{array}{l} \xrightarrow{a} c\text{-}C_6H_{11}Cl + 3\text{-}Br\dot{C}_6H_4 \\ \xrightarrow{b} c\text{-}C_6H_{12} + (\dot{C}_6H_3BrCl) \end{array}$

React. $(CH_3)_3\dot{C} + c\text{-}C_6H_{12}$ after therm. of DTBP				68 She 1
PR, glc	$c\text{-}C_6H_{12}$	378	$k_a/k_b = 0.4$	[78])

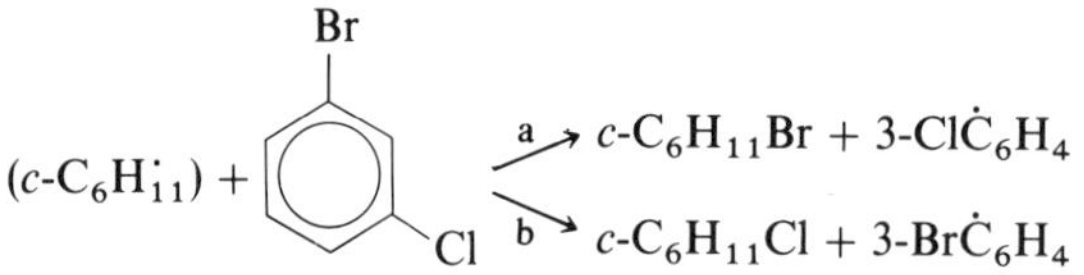

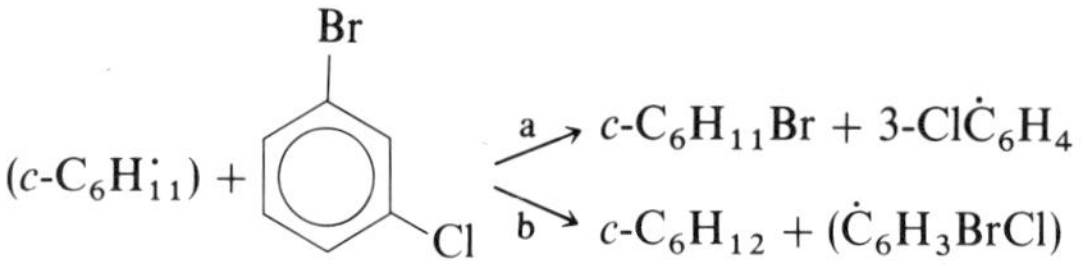

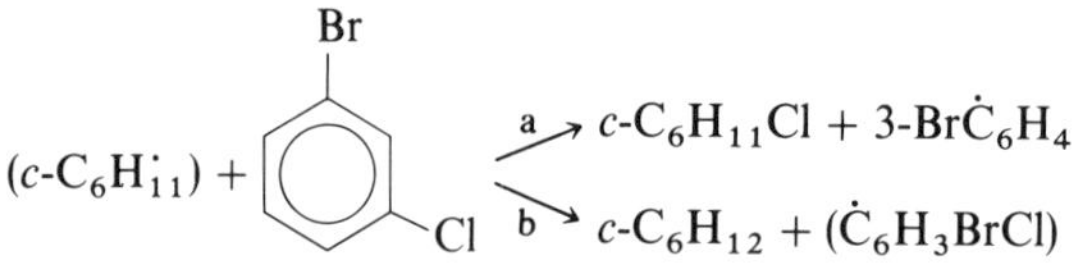

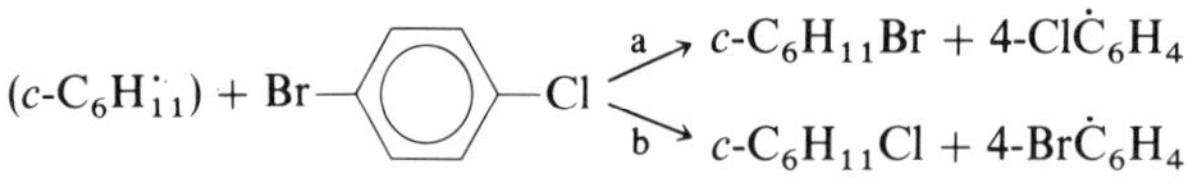

$(c\text{-}C_6H_{11}^{\cdot}) + Br\!-\!$ $\!-\!Cl$ $\begin{array}{l} \xrightarrow{a} c\text{-}C_6H_{11}Br + 4\text{-}Cl\dot{C}_6H_4 \\ \xrightarrow{b} c\text{-}C_6H_{11}Cl + 4\text{-}Br\dot{C}_6H_4 \end{array}$

React. $(CH_3)_3\dot{C} + c\text{-}C_6H_{12}$ after therm. of DTBP				68 She 1
PR, glc	$c\text{-}C_6H_{12}$	378	$k_a/k_b = 2.2$	[78])

$(c\text{-}C_6H_{11}^{\cdot}) + Br\!-\!$ $\!-\!Cl$ $\begin{array}{l} \xrightarrow{a} c\text{-}C_6H_{11}Br + 4\text{-}Cl\dot{C}_6H_4 \\ \xrightarrow{b} c\text{-}C_6H_{12} + (\dot{C}_6H_3BrCl) \end{array}$

React. $(CH_3)_3\dot{C} + c\text{-}C_6H_{12}$ after therm. of DTBP				68 She 1
PR, glc	$c\text{-}C_6H_{12}$	378	$k_a/k_b = 1.1$	[78])

$(c\text{-}C_6H_{11}^{\cdot}) + Br\!-\!$ $\!-\!Cl$ $\begin{array}{l} \xrightarrow{a} c\text{-}C_6H_{11}Cl + 4\text{-}Br\dot{C}_6H_4 \\ \xrightarrow{b} c\text{-}C_6H_{12} + (\dot{C}_6H_3BrCl) \end{array}$

React. $(CH_3)_3\dot{C} + c\text{-}C_6H_{12}$ after therm. of DTBP				68 She 1
PR, glc	$c\text{-}C_6H_{12}$	378	$k_a/k_b = 0.5$	[78])

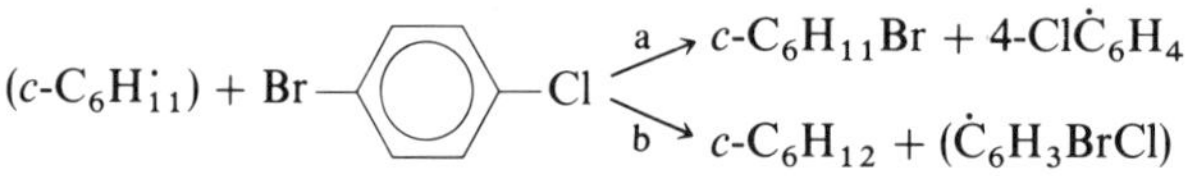

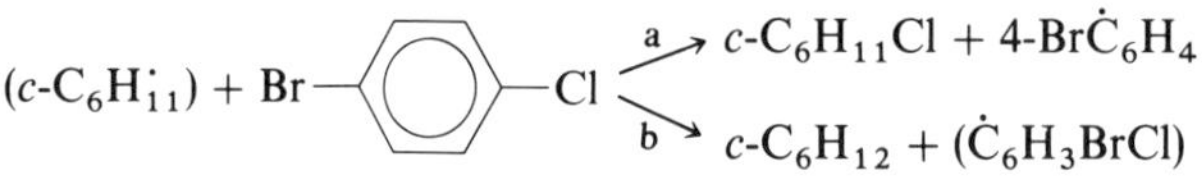

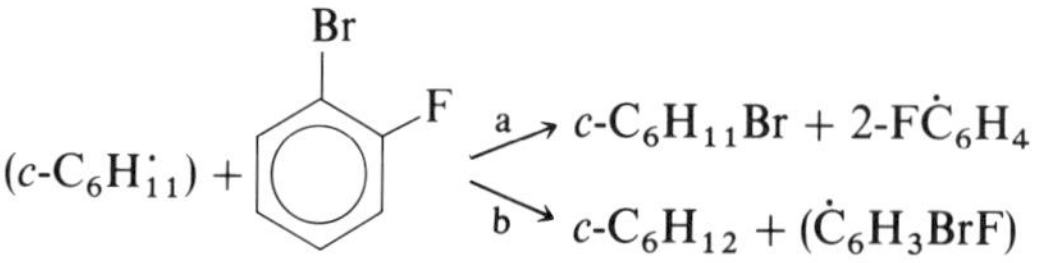

$(c\text{-}C_6H_{11}^{\cdot}) +$ $\begin{array}{l} \xrightarrow{a} c\text{-}C_6H_{11}Br + 2\text{-}F\dot{C}_6H_4 \\ \xrightarrow{b} c\text{-}C_6H_{12} + (\dot{C}_6H_3BrF) \end{array}$

React. $(CH_3)_3\dot{C} + c\text{-}C_6H_{12}$ after therm. of DTBP				68 She 1
PR, glc	$c\text{-}C_6H_{12}$	378	$k_a/k_b = 2.1$	[78])

---

[78]) Reaction mechanism proposed to proceed via radical addition.

Reaction				
Radical generation				Ref./
Method	Solvent	$T[K]$	Rate data	add. ref.

$(c\text{-}C_6\dot{H}_{11}) + $ $\overset{a}{\longrightarrow} c\text{-}C_6H_{11}F + 2\text{-}Br\dot{C}_6H_4$

$\overset{b}{\longrightarrow} c\text{-}C_6H_{11}Br + 2\text{-}F\dot{C}_6H_4$

| React. $(CH_3)_3\dot{C} + c\text{-}C_6H_{12}$ after therm. of DTBP | | | | 68 She 1 |
| PR, glc | $c\text{-}C_6H_{12}$ | 378 | $k_a/k_b = 1.8$ | [78]) |

$(c\text{-}C_6\dot{H}_{11}) + $ $\overset{a}{\longrightarrow} c\text{-}C_6H_{11}F + 2\text{-}Br\dot{C}_6H_4$

$\overset{b}{\longrightarrow} c\text{-}C_6H_{12} + (\dot{C}_6H_3BrF)$

| React. $(CH_3)_3\dot{C} + c\text{-}C_6H_{12}$ after therm. of DTBP | | | | 68 She 1 |
| PR, glc | $c\text{-}C_6H_{12}$ | 378 | $k_a/k_b = 3.8$ | [78]) |

$(c\text{-}C_6\dot{H}_{11}) + $ $\overset{a}{\longrightarrow} c\text{-}C_6H_{11}Br + 3\text{-}F\dot{C}_6H_4$

$\overset{b}{\longrightarrow} c\text{-}C_6H_{12} + (\dot{C}_6H_3BrF)$

| React. $(CH_3)_3\dot{C} + c\text{-}C_6H_{12}$ after therm. of DTBP | | | | 68 She 1 |
| PR, glc | $c\text{-}C_6H_{12}$ | 378 | $k_a/k_b = 0.5$ | [78]) |

$(c\text{-}C_6\dot{H}_{11}) + $ $\overset{a}{\longrightarrow} c\text{-}C_6H_{11}F + 3\text{-}Br\dot{C}_6H_4$

$\overset{b}{\longrightarrow} c\text{-}C_6H_{11}Br + 3\text{-}F\dot{C}_6H_4$

| React. $(CH_3)_3\dot{C} + c\text{-}C_6H_{12}$ after therm. of DTBP | | | | 68 She 1 |
| PR, glc | $c\text{-}C_6H_{12}$ | 378 | $k_a/k_b = 1.1$ | [78]) |

$(c\text{-}C_6\dot{H}_{11}) + $ $\overset{a}{\longrightarrow} c\text{-}C_6H_{11}F + 3\text{-}Br\dot{C}_6H_4$

$\overset{b}{\longrightarrow} c\text{-}C_6H_{12} + (\dot{C}_6H_3BrF)$

| React. $(CH_3)_3\dot{C} + c\text{-}C_6H_{12}$ after therm. of DTBP | | | | 68 She 1 |
| PR, glc | $c\text{-}C_6H_{12}$ | 378 | $k_a/k_b = 0.6$ | [78]) |

$(c\text{-}C_6\dot{H}_{11}) + Br\text{—}$ $\text{—}F \overset{a}{\longrightarrow} c\text{-}C_6H_{11}Br + 4\text{-}F\dot{C}_6H_4$

$\overset{b}{\longrightarrow} c\text{-}C_6H_{12} + (\dot{C}_6H_3BrF)$

| React. $(CH_3)_3\dot{C} + c\text{-}C_6H_{12}$ after therm. of DTBP | | | | 68 She 1 |
| PR, glc | $c\text{-}C_6H_{12}$ | 378 | $k_a/k_b = 0.3$ | [78]) |

$(c\text{-}C_6\dot{H}_{11}) + Br\text{—}$ $\text{—}F \overset{a}{\longrightarrow} c\text{-}C_6H_{11}F + 4\text{-}Br\dot{C}_6H_4$

$\overset{b}{\longrightarrow} c\text{-}C_6H_{11}Br + 4\text{-}F\dot{C}_6H_4$

| React. $(CH_3)_3\dot{C} + c\text{-}C_6H_{12}$ after therm. of DTBP | | | | 68 She 1 |
| PR, glc | $c\text{-}C_6H_{12}$ | 378 | $k_a/k_b = 3.0$ | [78]) |

[78]) Reaction mechanism proposed to proceed via radical addition.

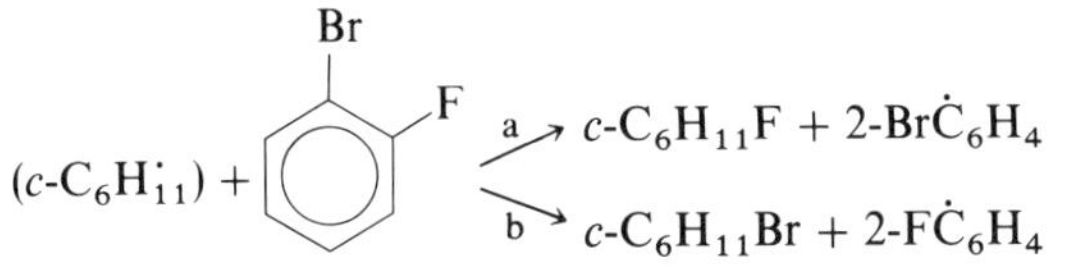

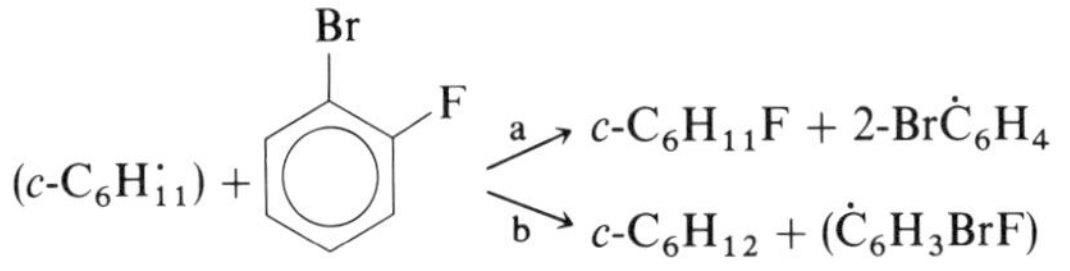

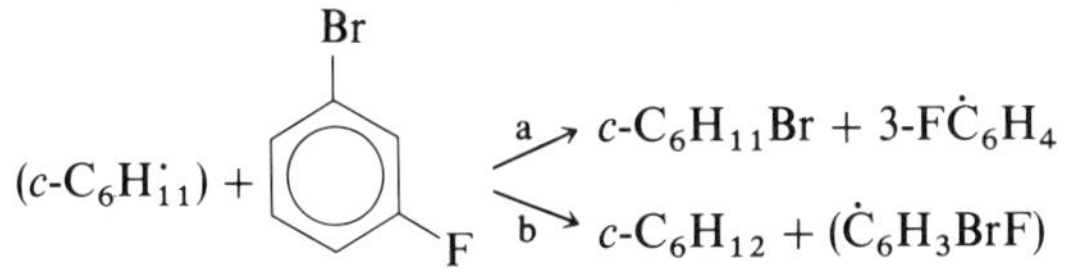

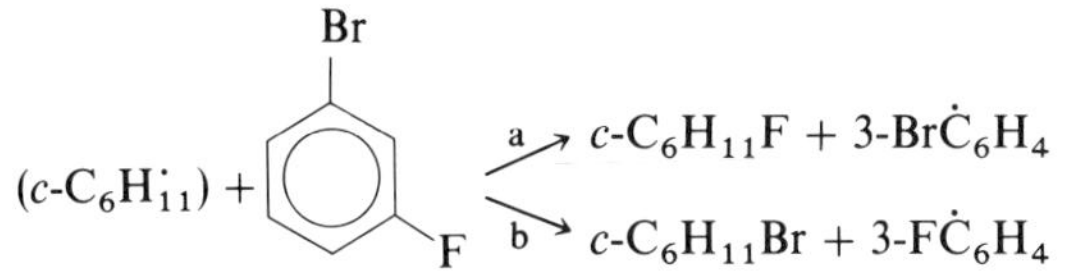

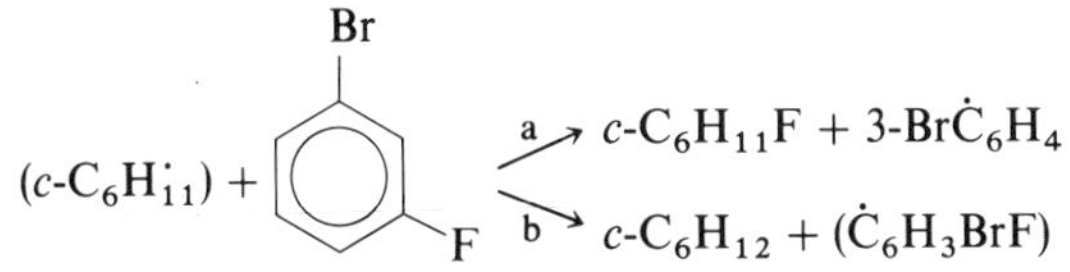

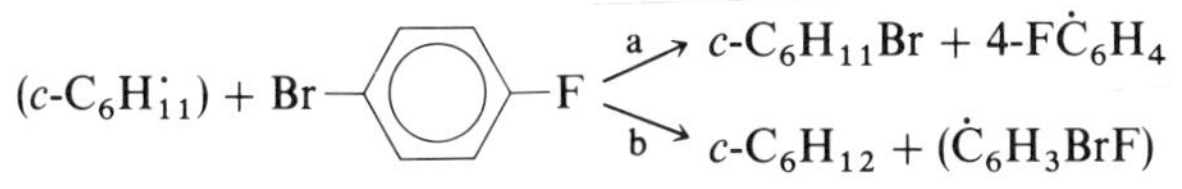

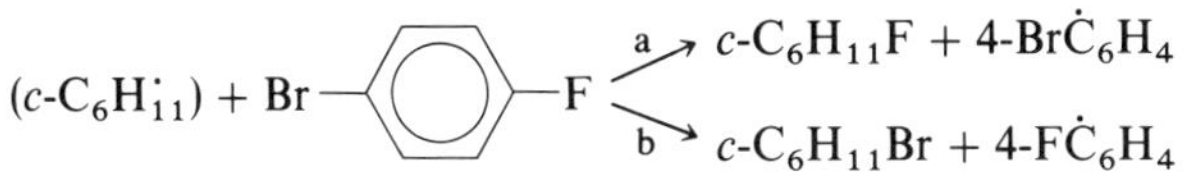

Reaction Radical generation Method	Solvent	T[K]	Rate data	Ref./ add. ref.

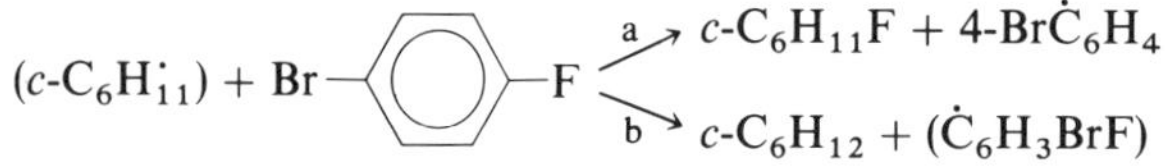

$(c\text{-}C_6H_{11}^{\cdot}) + Br\text{—}\langle\bigcirc\rangle\text{—}F$ $\xrightarrow{a} c\text{-}C_6H_{11}F + 4\text{-}Br\dot{C}_6H_4$ $\xrightarrow{b} c\text{-}C_6H_{12} + (\dot{C}_6H_3BrF)$

| React. $(CH_3)_3\dot{C} + c\text{-}C_6H_{12}$ after therm. of DTBP PR, glc | $c\text{-}C_6H_{12}$ | 378 | $k_a/k_b = 0.8$ | 68 She 1 [78] |

$(c\text{-}C_6H_{11}^{\cdot}) + $ [2-Cl, 1-F benzene] $\xrightarrow{a} c\text{-}C_6H_{11}Cl + 2\text{-}F\dot{C}_6H_4$ $\xrightarrow{b} c\text{-}C_6H_{12} + (\dot{C}_6H_3ClF)$

| React. $(CH_3)_3\dot{C} + c\text{-}C_6H_{12}$ after therm. of DTBP PR, glc | $c\text{-}C_6H_{12}$ | 378 | $k_a/k_b = 1.1$ | 68 She 1 [78] |

$(c\text{-}C_6H_{11}^{\cdot}) + $ [2-Cl, 1-F benzene] $\xrightarrow{a} c\text{-}C_6H_{11}F + 2\text{-}Cl\dot{C}_6H_4$ $\xrightarrow{b} c\text{-}C_6H_{11}Cl + 2\text{-}F\dot{C}_6H_4$

| React. $(CH_3)_3\dot{C} + c\text{-}C_6H_{12}$ after therm. of DTBP PR, glc | $c\text{-}C_6H_{12}$ | 378 | $k_a/k_b = 2.0$ | 68 She 1 [78] |

$(c\text{-}C_6H_{11}^{\cdot}) + $ [2-Cl, 1-F benzene] $\xrightarrow{a} c\text{-}C_6H_{11}F + 2\text{-}Cl\dot{C}_6H_4$ $\xrightarrow{b} c\text{-}C_6H_{12} + (\dot{C}_6H_3ClF)$

| React. $(CH_3)_3\dot{C} + c\text{-}C_6H_{12}$ after therm. of DTBP PR, glc | $c\text{-}C_6H_{12}$ | 378 | $k_a/k_b = 2.3$ | 68 She 1 [78] |

$(c\text{-}C_6H_{11}^{\cdot}) + $ [3-Cl, 1-F benzene] $\xrightarrow{a} c\text{-}C_6H_{11}Cl + 3\text{-}F\dot{C}_6H_4$ $\xrightarrow{b} c\text{-}C_6H_{12} + (\dot{C}_6H_3ClF)$

| React. $(CH_3)_3\dot{C} + c\text{-}C_6H_{12}$ after therm. of DTBP PR, glc | $c\text{-}C_6H_{12}$ | 378 | $k_a/k_b = 0.2$ | 68 She 1 [78] |

$(c\text{-}C_6H_{11}^{\cdot}) + $ [3-Cl, 1-F benzene] $\xrightarrow{a} c\text{-}C_6H_{11}F + 3\text{-}Cl\dot{C}_6H_4$ $\xrightarrow{b} c\text{-}C_6H_{11}Cl + 3\text{-}F\dot{C}_6H_4$

| React. $(CH_3)_3\dot{C} + c\text{-}C_6H_{12}$ after therm. of DTBP PR, glc | $c\text{-}C_6H_{12}$ | 378 | $k_a/k_b = 2.1$ | 68 She 1 [78] |

[78]) Reaction mechanism proposed to proceed via radical addition.

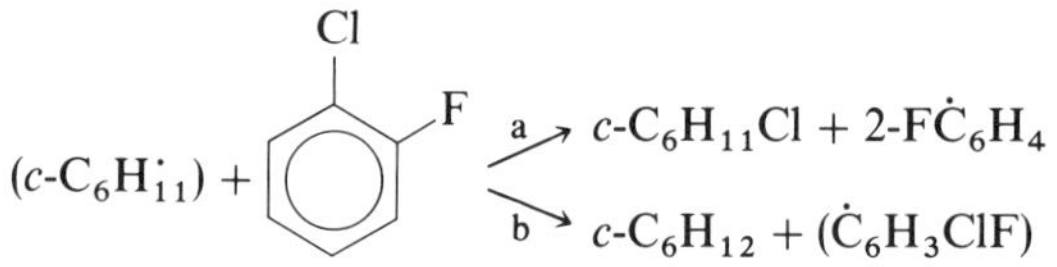

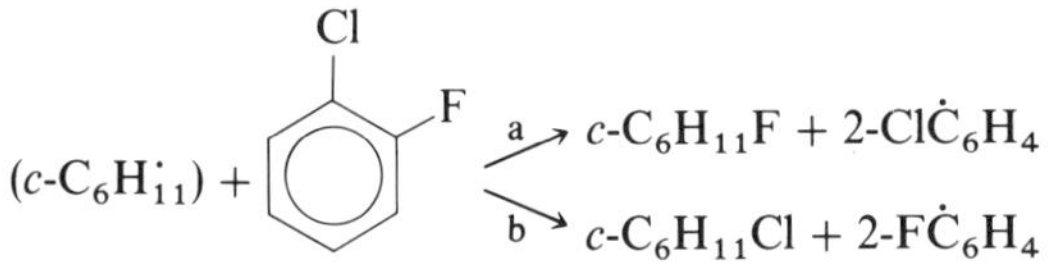

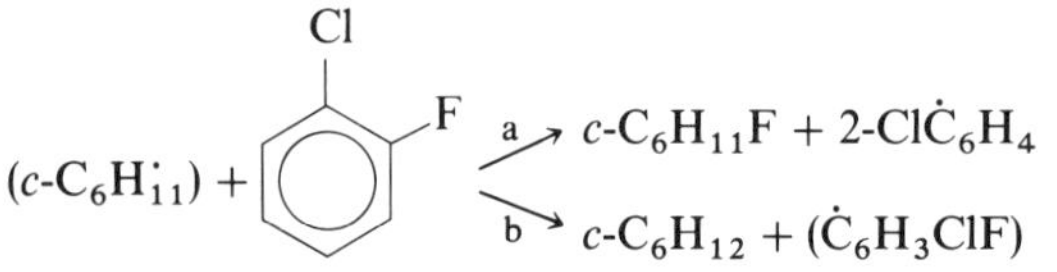

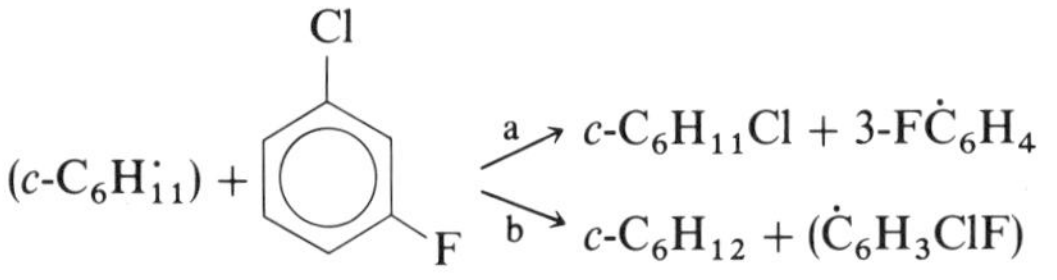

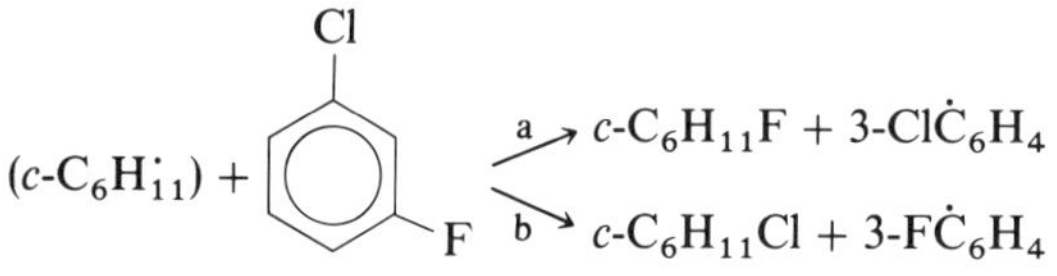

Reaction				
Radical generation				Ref./
Method	Solvent	T[K]	Rate data	add. ref.

$(c\text{-}C_6H_{11}^{\cdot}) +$ [Cl,F-substituted benzene] $\xrightarrow{a}$ $c\text{-}C_6H_{11}F + 3\text{-}ClC_6\dot{H}_4$
$\xrightarrow{b}$ $c\text{-}C_6H_{12} + (\dot{C}_6H_3ClF)$

| React. $(CH_3)_3\dot{C} + c\text{-}C_6H_{12}$ after therm. of DTBP | | | | 68 She 1 |
| PR, glc | $c\text{-}C_6H_{12}$ | 378 | $k_a/k_b = 0.5$ | [78]) |

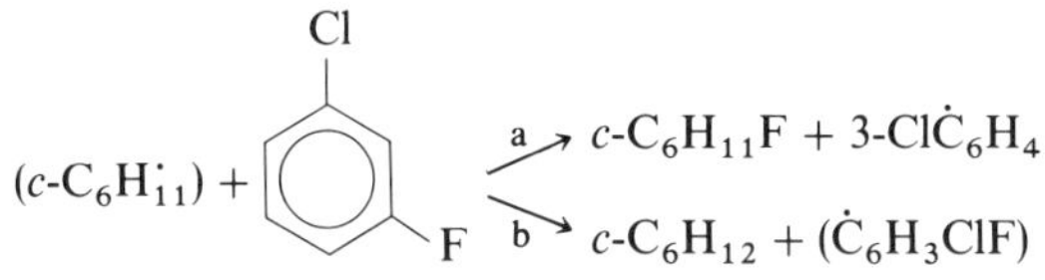

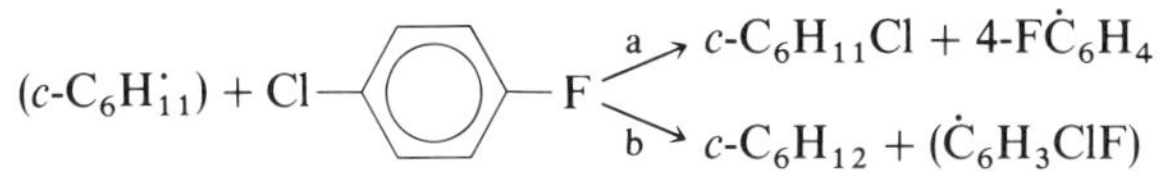

$(c\text{-}C_6H_{11}^{\cdot}) + Cl$—[benzene]—$F$ $\xrightarrow{a}$ $c\text{-}C_6H_{11}Cl + 4\text{-}FC_6\dot{H}_4$
$\xrightarrow{b}$ $c\text{-}C_6H_{12} + (\dot{C}_6H_3ClF)$

| React. $(CH_3)_3\dot{C} + c\text{-}C_6H_{12}$ after therm. of DTBP | | | | 68 She 1 |
| PR, glc | $c\text{-}C_6H_{12}$ | 378 | $k_a/k_b = 0.1$ | [78]) |

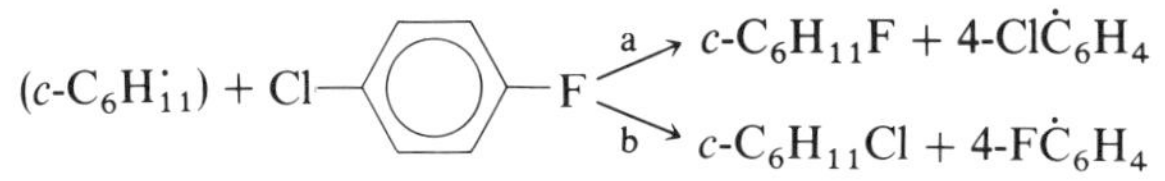

$(c\text{-}C_6H_{11}^{\cdot}) + Cl$—[benzene]—$F$ $\xrightarrow{a}$ $c\text{-}C_6H_{11}F + 4\text{-}ClC_6\dot{H}_4$
$\xrightarrow{b}$ $c\text{-}C_6H_{11}Cl + 4\text{-}FC_6\dot{H}_4$

| React. $(CH_3)_3\dot{C} + c\text{-}C_6H_{12}$ after therm. of DTBP | | | | 68 She 1 |
| PR, glc | $c\text{-}C_6H_{12}$ | 378 | $k_a/k_b = 7.2$ | [78]) |

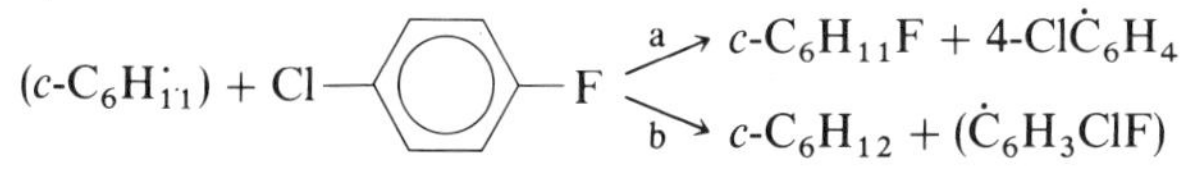

$(c\text{-}C_6H_{11}^{\cdot}) + Cl$—[benzene]—$F$ $\xrightarrow{a}$ $c\text{-}C_6H_{11}F + 4\text{-}ClC_6\dot{H}_4$
$\xrightarrow{b}$ $c\text{-}C_6H_{12} + (\dot{C}_6H_3ClF)$

| React. $(CH_3)_3\dot{C} + c\text{-}C_6H_{12}$ after therm. of DTBP | | | | 68 She 1 |
| PR, glc | $c\text{-}C_6H_{12}$ | 378 | $k_a/k_b = 0.7$ | [78]) |

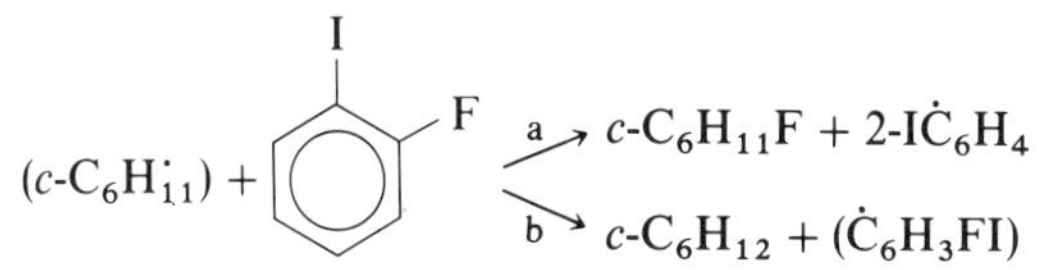

$(c\text{-}C_6H_{11}^{\cdot}) +$ [I,F-substituted benzene] $\xrightarrow{a}$ $c\text{-}C_6H_{11}F + 2\text{-}IC_6\dot{H}_4$
$\xrightarrow{b}$ $c\text{-}C_6H_{12} + (\dot{C}_6H_3FI)$

| React. $(CH_3)_3\dot{C} + c\text{-}C_6H_{12}$ after therm. of DTBP | | | | 68 She 1 |
| PR, glc | $c\text{-}C_6H_{12}$ | 378 | $k_a/k_b = 2.4$ | [78]) |

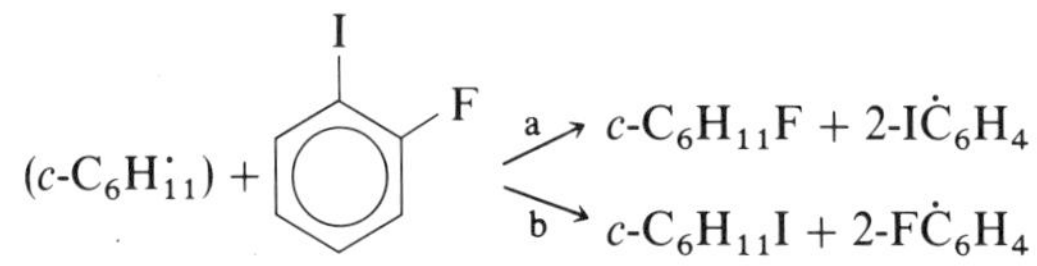

$(c\text{-}C_6H_{11}^{\cdot}) +$ [I,F-substituted benzene] $\xrightarrow{a}$ $c\text{-}C_6H_{11}F + 2\text{-}IC_6\dot{H}_4$
$\xrightarrow{b}$ $c\text{-}C_6H_{11}I + 2\text{-}FC_6\dot{H}_4$

| React. $(CH_3)_3\dot{C} + c\text{-}C_6H_{12}$ after therm. of DTBP | | | | 68 She 1 |
| PR, glc | $c\text{-}C_6H_{12}$ | 378 | $k_a/k_b = 1.1$ | [78]) |

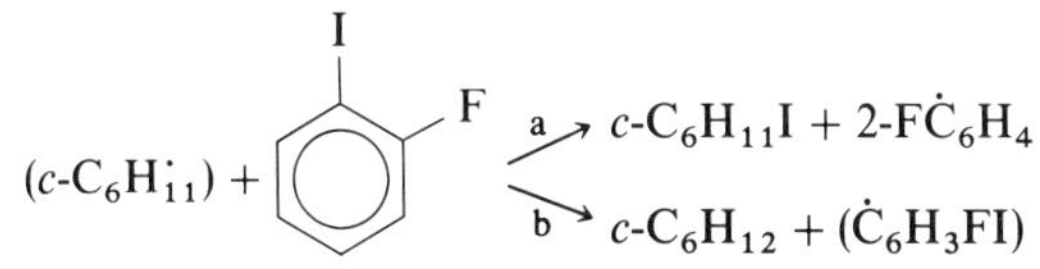

$(c\text{-}C_6H_{11}^{\cdot}) +$ [I,F-substituted benzene] $\xrightarrow{a}$ $c\text{-}C_6H_{11}I + 2\text{-}FC_6\dot{H}_4$
$\xrightarrow{b}$ $c\text{-}C_6H_{12} + (\dot{C}_6H_3FI)$

| React. $(CH_3)_3\dot{C} + c\text{-}C_6H_{12}$ after therm. of DTBP | | | | 68 She 1 |
| PR, glc | $c\text{-}C_6H_{12}$ | 378 | $k_a/k_b = 2.3$ | [78]) |

[78]) Reaction mechanism proposed to proceed via radical addition.

Reaction				
Radical generation				Ref./
Method	Solvent	$T[K]$	Rate data	add. ref.

$(c\text{-}C_6H_{11}^{\cdot}) +$ (1-iodo-3-fluorobenzene) $\begin{array}{l} \xrightarrow{a} c\text{-}C_6H_{11}F + 2\text{-}IC_6H_4 \\ \xrightarrow{b} c\text{-}C_6H_{12} + (\dot{C}_6H_3FI) \end{array}$

| React. $(CH_3)_3\dot{C} + c\text{-}C_6H_{12}$ after therm. of DTBP | | | | 68 She 1 |
| PR, glc | $c\text{-}C_6H_{12}$ | 378 | $k_a/k_b = 1.8$ | [78]) |

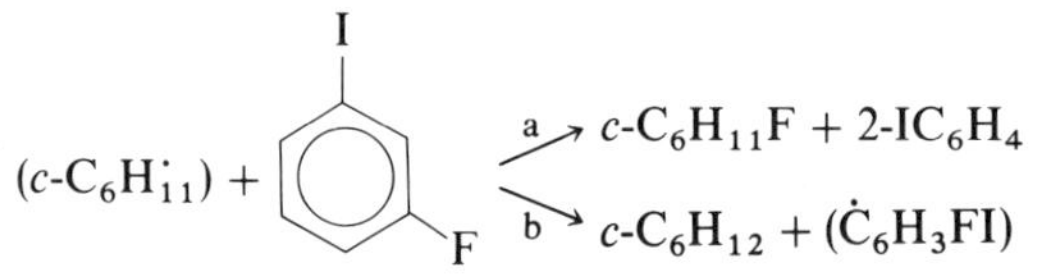

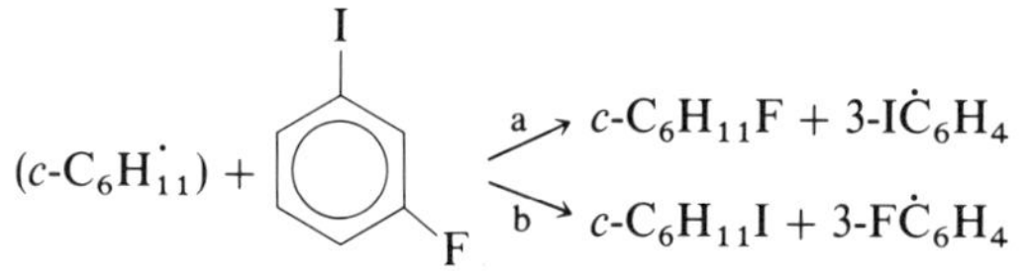

$(c\text{-}C_6H_{11}^{\cdot}) +$ (1-iodo-3-fluorobenzene) $\begin{array}{l} \xrightarrow{a} c\text{-}C_6H_{11}F + 3\text{-}I\dot{C}_6H_4 \\ \xrightarrow{b} c\text{-}C_6H_{11}I + 3\text{-}F\dot{C}_6H_4 \end{array}$

| React. $(CH_3)_3\dot{C} + c\text{-}C_6H_{12}$ after therm. of DTBP | | | | 68 She 1 |
| PR, glc | $c\text{-}C_6H_{12}$ | 378 | $k_a/k_b = 1.9$ | [78]) |

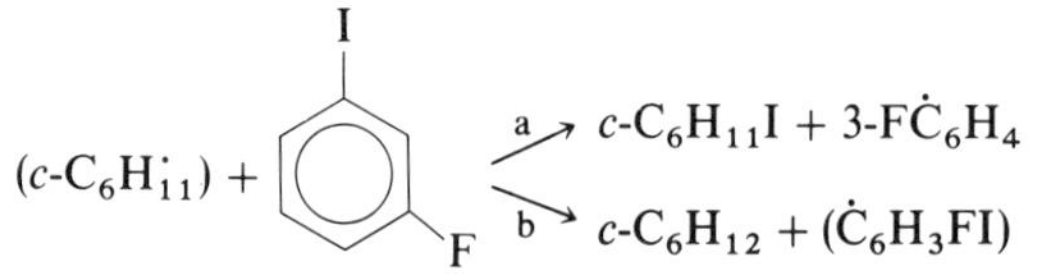

$(c\text{-}C_6H_{11}^{\cdot}) +$ (1-iodo-3-fluorobenzene) $\begin{array}{l} \xrightarrow{a} c\text{-}C_6H_{11}I + 3\text{-}F\dot{C}_6H_4 \\ \xrightarrow{b} c\text{-}C_6H_{12} + (\dot{C}_6H_3FI) \end{array}$

| React. $(CH_3)_3\dot{C} + c\text{-}C_6H_{12}$ after therm. of DTBP | | | | 68 She 1 |
| PR, glc | $c\text{-}C_6H_{12}$ | 378 | $k_a/k_b = 0.9$ | [78]) |

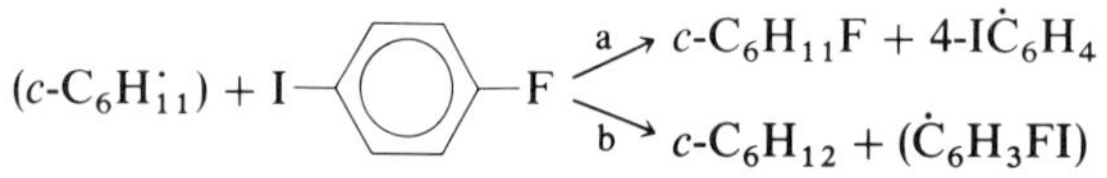

$(c\text{-}C_6H_{11}^{\cdot}) + $ I—(benzene)—F $\begin{array}{l} \xrightarrow{a} c\text{-}C_6H_{11}F + 4\text{-}I\dot{C}_6H_4 \\ \xrightarrow{b} c\text{-}C_6H_{12} + (\dot{C}_6H_3FI) \end{array}$

| React. $(CH_3)_3\dot{C} + c\text{-}C_6H_{12}$ after therm. of DTBP | | | | 68 She 1 |
| PR, glc | $c\text{-}C_6H_{12}$ | 378 | $k_a/k_b = 1.2$ | [78]) |

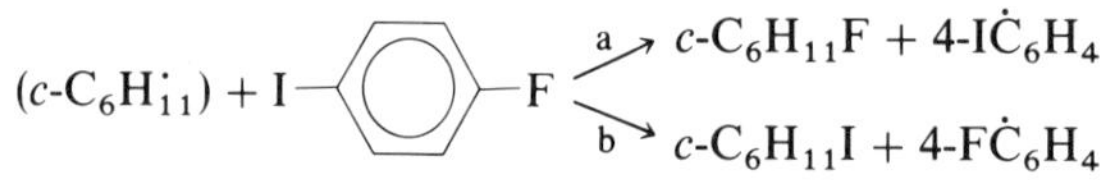

$(c\text{-}C_6H_{11}^{\cdot}) + $ I—(benzene)—F $\begin{array}{l} \xrightarrow{a} c\text{-}C_6H_{11}F + 4\text{-}I\dot{C}_6H_4 \\ \xrightarrow{b} c\text{-}C_6H_{11}I + 4\text{-}F\dot{C}_6H_4 \end{array}$

| React. $(CH_3)_3\dot{C} + c\text{-}C_6H_{12}$ after therm. of DTBP | | | | 68 She 1 |
| PR, glc | $c\text{-}C_6H_{12}$ | 378 | $k_a/k_b = 3.4$ | [78]) |

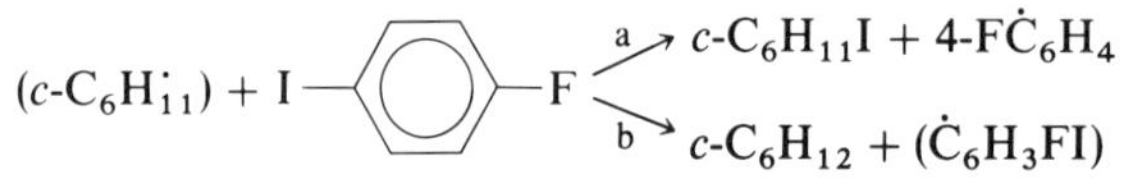

$(c\text{-}C_6H_{11}^{\cdot}) + $ I—(benzene)—F $\begin{array}{l} \xrightarrow{a} c\text{-}C_6H_{11}I + 4\text{-}F\dot{C}_6H_4 \\ \xrightarrow{b} c\text{-}C_6H_{12} + (\dot{C}_6H_3FI) \end{array}$

| React. $(CH_3)_3\dot{C} + c\text{-}C_6H_{12}$ after therm. of DTBP | | | | 68 She 1 |
| PR, glc | $c\text{-}C_6H_{12}$ | 378 | $k_a/k_b = 0.3$ | [78]) |

$(c\text{-}C_6H_{11}^{\cdot}) + (CH_3)_2CHC(CH_3)_2OCl\,^{79)} \xrightarrow{a} c\text{-}C_6H_{11}Cl + (CH_3)_2CHC(CH_3)_2\dot{O}$
$\qquad + CBrCl_3 \xrightarrow{b} c\text{-}C_6H_{11}Br + \dot{C}Cl_3$

Phot. of c-hexane containing soln.				65 Zav 1
PR, glc	$CBrCl_3$	273	$k_a/k_b = 1.22$	
		346	1.90	
			$A_a/A_b = 10$	
			$E_a(a) - E_a(b) = 4.8\,kJ\,mol^{-1}$	
			$\Delta S_a^{\ddagger} - \Delta S_b^{\ddagger} = 4.6\,J\,K^{-1}\,mol^{-1}$	

[78]) Reaction mechanism proposed to proceed via radical addition.
[79]) ...hypochlorite.

Reaction Radical generation Method	Solvent	T[K]	Rate data	Ref./ add. ref.
$CH_2{=}CH(CH_2)_3\dot{C}H_2 + (n\text{-}C_4H_9)_3SnH \xrightarrow{a} CH_2{=}CH(CH_2)_3CH_3 + (n\text{-}C_4H_9)_3\dot{S}n$				
$CH_2{=}CH(CH_2)_3\dot{C}H_2 \xrightarrow{b} c\text{-}C_5H_9\dot{C}H_2$				
$CH_2{=}CH(CH_2)_3\dot{C}H_2 \xrightarrow{c} c\text{-}\dot{C}_6H_{11}$				
$(n\text{-}C_4H_9)_3\dot{S}n + BrCH_2(CH_2)_3CH{=}CH_2$ reaction (AIBN initiated)				74 Bec 1/
PR, glc	C_6H_6	338	$k_a/k_b = 4.59\ M^{-1}$ $k_a/k_c = 2.22 \cdot 10^2\ M^{-1}$ [80])	74 Bec 2
$(n\text{-}C_4H_9)_3\dot{S}n + CH_2{=}CH(CH_2)_4Br$, phot. initiation with AIBN				66 Wal 1
PR, glc	C_6H_6	313	$k_a/k_b = 10\ M^{-1}$ $k_a/k_c > 1000\ M^{-1}$	
$CH_2{=}CH(CH_2)_3\dot{C}H_2 + (n\text{-}C_4H_9)_3SnH \xrightarrow{a} CH_2{=}CH(CH_2)_3CH_3 + (n\text{-}C_4H_9)_3\dot{S}n$ $CH_2{=}CH(CH_2)_3\dot{C}H_2 \xrightarrow{b} c\text{-}C_5H_9\dot{C}H_2$ and $c\text{-}\dot{C}_6H_{11}$				
From $CH_2{=}CH(CH_2)_3CH_2Br$ with AIBN as initiator				72 Wal 1/
PR, glc	C_6H_6	313 343 403	$k_a/k_b = 9.35(52)\ M^{-1}$ $6.54(34)\ M^{-1}$ $3.91(9)\ M^{-1}$	66 Wal 1
$CH_2{=}CH(CH_2)_3\dot{C}H_2 + (C_6H_5)_3SnH \xrightarrow{a} CH_2{=}CH(CH_2)_3CH_3 + (C_6H_5)_3\dot{S}n$ $CH_2{=}CH(CH_2)_3\dot{C}H_2 \xrightarrow{b} c\text{-}C_5H_9\dot{C}H_2$ and $c\text{-}\dot{C}_6H_{11}$				
From $CH_2{=}CH(CH_2)_3CH_2Br$ with AIBN as initiator				72 Wal 1
PR, glc	C_6H_6	343 403	$k_a/k_b = 23.8(6)\ M^{-1}$ $12.8(3)\ M^{-1}$	
$CH_3(CH_2)_4\dot{C}H_2 + CBrCl_3 \xrightarrow{a} CH_3(CH_2)_4CH_2Br + \dot{C}Cl_3$ $+ CCl_4 \xrightarrow{b} CH_3(CH_2)_4CH_2Cl + \dot{C}Cl_3$				
Reduct. of $n\text{-}C_6H_{13}HgOCOCH_3$ by $NaBH_4$				79 Gie 1
PR, glc	CCl_4	273 303 323 343 403	$k_a/k_b = 4500$ 2700 2100 1400 800 [81]) $\Delta H_a^{\ddagger} - \Delta H_b^{\ddagger} = -12\ kJ\,mol^{-1}\ (\pm 10\%)$ $\Delta S_a^{\ddagger} - \Delta S_b^{\ddagger} = 24\ J\,mol^{-1}\,K^{-1}\ (\pm 5\%)$	
$CH_3(CH_2)_4\dot{C}H_2 + CHCl_3 \xrightarrow{a} n\text{-}C_6H_{14} + \dot{C}Cl_3$ $+ CCl_4 \xrightarrow{b} n\text{-}C_6H_{13}Cl + \dot{C}Cl_3$				
Thermal decomp. of n-heptanoylperoxide				60 Det 1
PR, glc	$CHCl_3/CCl_4$	347	$k_a/k_b = 0.40(\pm 25\%)$	
$CH_3(CH_2)_4\dot{C}H_2 + CH_2BrCl \xrightarrow{a} CH_3(CH_2)_4CH_2Br + \dot{C}H_2Cl$ $+ CH_3(CH_2)_2CH_2I \xrightarrow{b} CH_3(CH_2)_4CH_2I + CH_3(CH_2)_2\dot{C}H_2$				
AIBN or BPO as initiator				68 Saf 1
PR, glc	$C_6H_6/CH_2BrCl/$ $n\text{-}C_4H_9I/CH_2{=}CH_2$ mixt.	373	$k_a/k_b = 0.0720(69)$	
$CH_3(CH_2)_4\dot{C}H_2 + CH_3OH \xrightarrow{a} n\text{-}C_6H_{14} + \text{methanol}(-\dot{H})$ $+ CCl_4 \xrightarrow{b} n\text{-}C_6H_{13}Cl + \dot{C}Cl_3$				
Thermal decomp. of n-heptanoylperoxide				60 Det 1
PR, glc	methanol/CCl_4	347	$k_a/k_b = 0.02(\pm 200\ldots 300\%)$	

[80]) $\Delta H^{\ddagger}(b) = -12.3\ kJ\,mol^{-1}$, $\Delta S^{\ddagger}(b) = -23.8\ J\,mol^{-1}\,K^{-1}$; $\Delta H^{\ddagger}(c) = -19.3\ kJ\,mol^{-1}$, $\Delta S^{\ddagger}(c) = -12.1\ J\,mol^{-1}\,K^{-1}$.

[81]) Extrapolated value.

Reaction Radical generation Method	Solvent	T[K]	Rate data	Ref./ add. ref.
$CH_3(CH_2)_4\dot{C}H_2 + CH_2BrCOOH \xrightarrow{a} CH_3(CH_2)_4CH_2Br + \dot{C}H_2COOH$ $+ CH_3(CH_2)_2CH_2I \xrightarrow{b} CH_3(CH_2)_4CH_2I + CH_3(CH_2)_2CH_2$				
AIBN or BPO as initiator				68 Saf 1
PR, glc	$C_6H_6/CH_2BrCOOH/$ n-$C_4H_9I/CH_2{=}CH_2$ mixt.	373	$k_a/k_b = 0.390(81)$	
$CH_3(CH_2)_4\dot{C}H_2 + CH_3CH_2Br \xrightarrow{a} CH_3(CH_2)_4CH_2Br + CH_3\dot{C}H_2$ $+ CH_3(CH_2)_2CH_2I \xrightarrow{b} CH_3(CH_2)_4CH_2I + CH_3(CH_2)_2\dot{C}H_2$				
BPO as initiator				68 Saf 2
PR, glc	C_2H_5Br/n-$C_4H_9I/$ $CH_2{=}CH_2$ mixt.	373	$k_a/k_b = 8.6(24) \cdot 10^{-4}$	
$CH_3(CH_2)_4\dot{C}H_2 + CH_3COCH_3 \xrightarrow{a} n$-$C_6H_{14} + CH_3CO\dot{C}H_2$ $+ CCl_4 \xrightarrow{b} n$-$C_6H_{13}Cl + \dot{C}Cl_3$				
Thermal decomp. of n-heptanoylperoxide				60 Det 1
PR, glc	acetone/CCl_4	347	$k_a/k_b = 0.010(\pm 25\%)$	
$CH_3(CH_2)_4\dot{C}H_2 + CH_2{=}CHCH_2OH \xrightarrow{a} n$-$C_6H_{14}$ + 2-propen-1-ol($-\dot{H}$) $+ CCl_4 \xrightarrow{b} n$-$C_6H_{13}Cl + \dot{C}Cl_3$				
Thermal decomp. of n-heptanoylperoxide				60 Det 1
PR, glc	2-propen-1-ol [82])/CCl_4	347	$k_a/k_b = 0.024(\pm 25\%)$	
$CH_3(CH_2)_4\dot{C}H_2 + (CH_3)_2CHBr \xrightarrow{a} CH_3(CH_2)_4CH_2Br + (CH_3)_2\dot{C}H$ $+ CH_3(CH_2)_2CH_2I \xrightarrow{b} CH_3(CH_2)_4CH_2I + CH_3(CH_2)_2\dot{C}H_2$				
BPO as initiator				68 Saf 2
PR, glc	$(CH_3)_2CHBr/$ n-$C_4H_9I/CH_2{=}CH_2$ mixt.	373	$k_a/k_b = 3.0(4) \cdot 10^{-3}$	
$CH_3(CH_2)_4\dot{C}H_2 + (CH_3)_2CHOH \xrightarrow{a} n$-$C_6H_{14}$ + 2-propanol($-\dot{H}$) $+ CCl_4 \xrightarrow{b} n$-$C_6H_{13}Cl + \dot{C}Cl_3$				
Thermal decomp. of n-heptanoylperoxide				60 Det 1
PR, glc	2-propanol/CCl_4	347	$k_a/k_b = 0.15(\pm 25\%)$	
$CH_3(CH_2)_4\dot{C}H_2 + CHBrClCOOC_2H_5 \xrightarrow{a} CH_3(CH_2)_4CH_2Br + \dot{C}HClCOOC_2H_5$ $+ CH_3(CH_2)_2CH_2I \xrightarrow{b} CH_3(CH_2)_4CH_2I + CH_3(CH_2)_2\dot{C}H_2$				
BPO or AIBN as initiator				68 Saf 1
PR, glc	$C_6H_6/$ $CHBrClCOOC_2H_5/$ n-$C_4H_9I/$ $CH_2{=}CH_2$ mixt.	373	$k_a/k_b = 5.60(75)$	
$CH_3(CH_2)_4\dot{C}H_2$ + THF $\xrightarrow{a} n$-C_6H_{14} + THF($-\dot{H}$) $+ CCl_4 \xrightarrow{b} n$-$C_6H_{13}Cl + \dot{C}Cl_3$				
Thermal decomp. of n-heptanoylperoxide				60 Det 1
PR, glc	THF/CCl_4	347	$k_a/k_b = 0.01(\pm 200 \ldots 300\%)$	
$CH_3(CH_2)_4\dot{C}H_2$ + dioxan $\xrightarrow{a} n$-C_6H_{14} + dioxan($-\dot{H}$) $+ CCl_4 \xrightarrow{b} n$-$C_6H_{13}Cl + \dot{C}Cl_3$				
Thermal decomp. of n-heptanoylperoxide				60 Det 1
PR, glc	dioxan/CCl_4	347	$k_a/k_b = 0.03(\pm 200 \ldots 300\%)$	

[82]) Allyl alcohol.

Reaction Radical generation Method	Solvent	T[K]	Rate data	Ref./ add. ref.
$CH_3(CH_2)_4\dot{C}H_2 + CH_3COOC_2H_5 \xrightarrow{a} n\text{-}C_6H_{14}$ + ethylacetate($-\dot{H}$) $+ CCl_4 \xrightarrow{b} n\text{-}C_6H_{13}Cl + \dot{C}Cl_3$				
Thermal decomp. of n-heptanoylperoxide				60 Det 1
PR, glc	ethylacetate/CCl_4	347	$k_a/k_b = 0.002(\pm 200 \ldots 300\%)$	
$CH_3(CH_2)_4\dot{C}H_2 + CH_3(CH_2)_2CH_2Br \xrightarrow{a} CH_3(CH_2)_4CH_2Br + CH_3(CH_2)_2\dot{C}H_2$ $+ CH_3(CH_2)_2CH_2I \xrightarrow{b} CH_3(CH_2)_4CH_2I + CH_3(CH_2)_2\dot{C}H_2$				
BPO as initiator				68 Saf 2
PR, glc	$n\text{-}C_4H_9Br/n\text{-}C_4H_9I/$ $CH_2{=}CH_2$ mixt.	373	$k_a/k_b = 1.60(23)\cdot 10^{-3}$	
$CH_3(CH_2)_4\dot{C}H_2 + CH_3(CH_2)_2CH_2Cl \xrightarrow{a} n\text{-}C_6H_{14}$ + n-butylchloride($-\dot{H}$) $+ CCl_4 \xrightarrow{b} n\text{-}C_6H_{13}Cl + \dot{C}Cl_3$				
Thermal decomp. of n-heptanoylperoxide				60 Det 1
PR, glc	n-butylchloride/CCl_4	347	$k_a/k_b = 0.006(\pm 200 \ldots 300\%)$	
$CH_3(CH_2)_4\dot{C}H_2 + CH_3CHOHCH_2CH_3 \xrightarrow{a} n\text{-}C_6H_{14}$ + 2-butanol($-\dot{H}$) $+ CCl_4 \xrightarrow{b} n\text{-}C_6H_{13}Cl + \dot{C}Cl_3$				
Thermal decomp. of n-heptanoylperoxide				60 Det 1
PR, glc	2-butanol/CCl_4	347	$k_a/k_b = 0.084(\pm 25\%)$	
$CH_3(CH_2)_4\dot{C}H_2 + (CH_3)_3COH \xrightarrow{a} n\text{-}C_6H_{14}$ + t-butanol($-\dot{H}$) $+ CCl_4 \xrightarrow{b} n\text{-}C_6H_{13}Cl + \dot{C}Cl_3$				
Thermal decomp. of n-heptanoylperoxide				60 Det 1
PR, glc	t-butanol/CCl_4	347	$k_a/k_b = 0.004(\pm 200 \ldots 300\%)$	
$CH_3(CH_2)_4\dot{C}H_2 + c\text{-}C_6H_{10} \xrightarrow{a} n\text{-}C_6H_{14}$ + c-hexene($-\dot{H}$) $+ CCl_4 \xrightarrow{b} n\text{-}C_6H_{13}Cl + \dot{C}Cl_3$				
Thermal decomp. of n-heptanoylperoxide				60 Det 1
PR, glc	c-hexene/CCl_4	347	$k_a/k_b = 0.10(\pm 25\%)$	
$CH_3(CH_2)_4\dot{C}H_2 + c\text{-}C_6H_{11}Cl \xrightarrow{a} n\text{-}C_6H_{14} + (c\text{-}\dot{C}_6H_{10}Cl)$ $+ CCl_4 \xrightarrow{b} n\text{-}C_6H_{13}Cl + \dot{C}Cl_3$				
Thermal decomp. of n-heptanoylperoxide				60 Det 1
PR, glc	c-hexylchloride/CCl_4	343	$k_a/k_b = 0.008(\pm 200 \ldots 300\%)$	
$CH_3(CH_2)_4\dot{C}H_2 + c\text{-}C_6H_{12} \xrightarrow{a} n\text{-}C_6H_{14} + (c\text{-}C_6H_{11}^{\cdot})$ $+ CCl_4 \xrightarrow{b} n\text{-}C_6H_{13}Cl + \dot{C}Cl_3$				
Thermal decomp. of n-heptanoylperoxide				60 Det 1
PR, glc	c-hexane/CCl_4	347	$k_a/k_b = 0.02(\pm 200 \ldots 300\%)$	
$CH_3(CH_2)_4\dot{C}H_2 + (CH_3)_2CHOCH(CH_3)_2 \xrightarrow{a} n\text{-}C_6H_{14}$ + di-(2-propyl)ether($-\dot{H}$) $+ CCl_4 \xrightarrow{b} n\text{-}C_6H_{13}Cl + \dot{C}Cl_3$				
Thermal decomp. of n-heptanoylperoxide				60 Det 1
PR, glc	di-(2-propyl)ether/CCl_4	347	$k_a/k_b = 0.65(\pm 25\%)$	
$CH_3(CH_2)_4\dot{C}H_2 + C_6H_5CH_2Cl \xrightarrow{a} n\text{-}C_6H_{14} + C_6H_5\dot{C}HCl$ $+ CCl_4 \xrightarrow{b} n\text{-}C_6H_{13}Cl + \dot{C}Cl_3$				
Thermal decomp. of n-heptanoylperoxide				60 Det 1
PR, glc	benzylchloride/CCl_4	347	$k_a/k_b = 0.028(\pm 25\%)$	

Reaction / Radical generation / Method	Solvent	T[K]	Rate data	Ref./ add. ref.
$CH_3(CH_2)_4\dot{C}H_2 + C_6H_5CH_3 \xrightarrow{a} n\text{-}C_6H_{14} + C_6H_5\dot{C}H_2$ $+ CCl_4 \xrightarrow{b} n\text{-}C_6H_{13}Cl + \dot{C}Cl_3$				
Thermal decomp. of *n*-heptanoylperoxide				60 Det 1
PR, glc	toluene/CCl_4	347	$k_a/k_b = 0.006(\pm 25\%)$	
$CH_3(CH_2)_4\dot{C}H_2 + C_6H_5COOCH_3 \xrightarrow{a} n\text{-}C_6H_{14}$ + methylbenzoate($-\dot{H}$) $+ CCl_4 \xrightarrow{b} n\text{-}C_6H_{13}Cl + \dot{C}Cl_3$				
Thermal decomp. of *n*-heptanoylperoxide				60 Det 1
PR, glc	methylbenzoate/CCl_4	347	$k_a/k_b = 0.002(\pm 200 \ldots 300\%)$	
$CH_3(CH_2)_4\dot{C}H_2 + CHBr(COOC_2H_5)_2 \xrightarrow{a} CH_3(CH_2)_4CH_2Br + \dot{C}H(COOC_2H_5)_2$ $+ CH_3(CH_2)_2CH_2I \xrightarrow{b} CH_3(CH_2)_4CH_2I + CH_3(CH_2)_2\dot{C}H_2$				
AIBN or BPO as initiator				68 Saf 1
PR, glc	C_6H_6/ $CHBr(COOC_2H_5)_2$/ $n\text{-}C_4H_9I/CH_2{=}CH_2$ mixt.	373	$k_a/k_b = 1.69(29)$	
$CH_3(CH_2)_4\dot{C}H_2 + CH_2{=}CH(CH_2)_5CH_3 \xrightarrow{a} n\text{-}C_6H_{14}$ + 1-octene($-\dot{H}$) $+ CCl_4 \xrightarrow{b} n\text{-}C_6H_{13}Cl + \dot{C}Cl_3$				
Thermal decomp. of *n*-heptanoylperoxide				60 Det 1
PR, glc	1-octene/CCl_4	347	$k_a/k_b = 0.065(\pm 25\%)$	
$CH_3(CH_2)_4\dot{C}H_2 + CH_3CH{=}CH(CH_2)_4CH_3 \xrightarrow{a} n\text{-}C_6H_{14}$ + 2-octene($-\dot{H}$) $+ CCl_4 \xrightarrow{b} n\text{-}C_6H_{13}Cl + \dot{C}Cl_3$				
Thermal decomp. of *n*-heptanoylperoxide				60 Det 1
PR, glc	2-octene/CCl_4	347	$k_a/k_b = 0.051(\pm 25\%)$	
$CH_3(CH_2)_4\dot{C}H_2 + (CH_3)_3CCH_2CH(CH_3)_2 \xrightarrow{a} n\text{-}C_6H_{14}$ + 2,2,4-trimethylpentane($-\dot{H}$) $+ CCl_4 \xrightarrow{b} n\text{-}C_6H_{13}Cl + \dot{C}Cl_3$				
Thermal decomp. of *n*-heptanoylperoxide				60 Det 1
PR, glc	2,2,4-tri-methylpentane/CCl_4	347	$k_a/k_b = 0.013(\pm 200 \ldots 300\%)$	
$CH_3\dot{C}H(CH_2)_3CH_3 + CCl_4 \xrightarrow{a} CH_3CHCl(CH_2)_3CH_3 + \dot{C}Cl_3$ $CH_3CH_2\dot{C}H(CH_2)_2CH_3 + CCl_4 \xrightarrow{b} CH_3CH_2CHCl(CH_2)_2CH_3 + \dot{C}Cl_3$				
γ-rad. of $n\text{-}C_6H_{14} + CCl_4$				78 Tua 1/
PR, glc	$n\text{-}C_6H_{14}/CCl_4$	183 … 383	$k_a/k_b = 0.78(4)$ [83]	77 Tua 1
$CH_3\dot{C}(C_2H_5)_2 + CBrCl_3 \xrightarrow{a} CH_3CBr(C_2H_5)_2 + \dot{C}Cl_3$ $+ CCl_4 \xrightarrow{b} CH_3CCl(C_2H_5)_2 + \dot{C}Cl_3$				
Reduct. of $CH_3C(C_2H_5)_2HgOCOCH_3$ by $NaBH_4$				79 Gie 1
PR, glc	CCl_4	273	$k_a/k_b = 65000$ [84]	
		343	2200	
		373	950	
		383	510	
		403	320	
			$\Delta H_a^\ddagger - \Delta H_b^\ddagger = -38\ \mathrm{kJ\,mol^{-1}}$ $(\pm 10\%)$	
			$\Delta S_a^\ddagger - \Delta S_b^\ddagger = -46\ \mathrm{J\,mol^{-1}\,K^{-1}}$ $(\pm 5\%)$	

[83]) Temperature independent.
[84]) Extrapolated value.

Reaction Radical generation Method	Solvent	$T[K]$	Rate data	Ref./ add. ref.

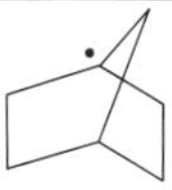

$+ CCl_3Br \xrightarrow{a}$ 1-bromonorbornane + $\dot{C}Cl_3$

$+ CCl_4 \xrightarrow{b}$ 1-chloronorbornane + $\dot{C}Cl_3$

Therm. of 1-norbornyl-COOOC(CH$_3$)$_3$				80 Gie 1,
PR, glc	CCl$_4$	373	$k_a/k_b = 42(\pm 5\%)$	75 Her 1,
		383	$42(\pm 5\%)$	69 Rue 1/
		393	$41(\pm 5\%)$	76 Gie 1,
		403	$39(\pm 5\%)$	76 Gie 2,
			$47^{85)}$	76 Gie 3
			$\Delta H_a^\ddagger - \Delta H_b^\ddagger = -3.2\,\mathrm{kJ\,mol^{-1}}$	
			$\Delta S_a^\ddagger - \Delta S_b^\ddagger = 22\,\mathrm{J\,mol^{-1}\,K^{-1}}$	
			See also 4.1.2.3, Fig. 4, p. 255.	

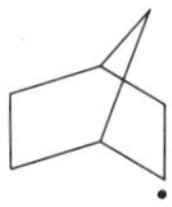

$+ Cl_2 \begin{cases} \xrightarrow{a} exo\text{-2-chloronorbornane} + \dot{C}l \\ \xrightarrow{b} endo\text{-2-chloronorbornane} + \dot{C}l \end{cases}$

$\dot{A}$

Chlorination of norbornane				70 Bar 1
PR	C$_6$H$_6$ or CCl$_3$COOH	353	$k_a/k_b = 2.5$	

$\dot{A} + LiCl \begin{cases} \xrightarrow{a} exo\text{-2-chloronorbornane} + \dot{L}i \\ \xrightarrow{b} endo\text{-2-chloronorbornane} + \dot{L}i \end{cases}$

Initiated by therm. of Pb(OOCCH$_3$)$_4$ in soln. of norbornanecarboxylic acid and LiCl				70 Bar 1
PR	not specified	353	$k_a/k_b = 6$	

$\dot{A} + SO_2Cl_2 \begin{cases} \xrightarrow{a} exo\text{-2-chloronorbornane} + (\dot{S}O_2Cl) \\ \xrightarrow{b} endo\text{-2-chloronorbornane} + (\dot{S}O_2Cl) \end{cases}$

Chlorination of norbornane				70 Bar 1
PR	SO$_2$Cl$_2$	313	$k_a/k_b = 18$	

$\dot{A} + CBrCl_3 \xrightarrow{a}$ 2-bromonorbornane + $\dot{C}Cl_3$

$+ CCl_4 \xrightarrow{b}$ 2-chloronorbornane + $\dot{C}Cl_3$

Therm. of 2-norbornyl-COOOC(CH$_3$)$_3$ $^{86)}$ and				79 Gie 1,
reduct. of 2-norbornyl-HgOCOCH$_3$ by NaBH$_4$				75 Her 1/
PR, glc	CCl$_4$	273	$k_a/k_b = 7100$	76 Gie 1,
		303	3700	76 Gie 2,
		323	1900	76 Gie 3
		333	1700	
		343	1300	
		383	$650^{86)}$	
		403	$470^{87)}$	
			$\Delta H_a^\ddagger - \Delta H_b^\ddagger = -19\,\mathrm{kJ\,mol^{-1}}\,(\pm 10\%)$	
			$\Delta S_a^\ddagger - \Delta S_b^\ddagger = 4\,\mathrm{J\,mol^{-1}\,K^{-1}}\,(\pm 5\%)$	

$^{85)}$ From [75 Her 1, 69 Rue 1].
$^{86)}$ From [75 Her 1].
$^{87)}$ Extrapolated value.

Reaction Radical generation Method	Solvent	T[K]	Rate data	Ref./ add. ref.
$\dot{A} + CCl_4$ $\xrightarrow{a}$ *exo*-2-chloronorbornane + $\dot{C}Cl_3$; $\xrightarrow{b}$ *endo*-2-chloronorbornane + $\dot{C}Cl_3$ *)				
Phot. of 2-azonorbornane + chlorination of norbornane + therm. of 2-norbornylperester				70 Bar 1
PR	CCl_4	353	$k_a/k_b = 42$	
$\dot{A} + CHCl_3$ $\xrightarrow{a}$ *exo*-2-chloronorbornane + $\dot{C}HCl_2$; $\xrightarrow{b}$ *endo*-2-chloronorbornane + $\dot{C}HCl_2$				
Phot. of 2-azonorbornane				70 Bar 1
PR	$CHCl_3$	RT	$k_a/k_b = 14$	
$\dot{A} + CH_2Cl_2$ $\xrightarrow{a}$ *exo*-2-chloronorbornane + $\dot{C}H_2Cl$; $\xrightarrow{b}$ *endo*-2-chloronorbornane + $\dot{C}H_2Cl$				
Phot. of 2-azonorbornane				70 Bar 1
PR	CH_2Cl_2	RT	$k_a/k_b = 16.5$	
$\dot{A} + CCl_3COONa$ $\xrightarrow{a}$ *exo*-2-chloronorbornane + $\dot{C}Cl_2COONa$; $\xrightarrow{b}$ *endo*-2-chloronorbornane + $\dot{C}Cl_2COONa$				
Initiated by therm. of BPO				70 Bar 1
PR	CH_3COOH	353	$k_a/k_b = 28$	
$\dot{A} + (CH_3)_3CCl$ $\xrightarrow{a}$ *exo*-2-chloronorbornane + $(CH_3)_3\dot{C}$; $\xrightarrow{b}$ *endo*-2-chloronorbornane + $(CH_3)_3\dot{C}$				
Therm. of 2-norbornylperester				70 Bar 1
PR	$(CH_3)_3CCl$	353	$k_a/k_b = 2$	
$\dot{A} + (CH_3)_3COCl$ $\xrightarrow{a}$ *exo*-2-chloronorbornane + $(CH_3)_3C\dot{O}$; $\xrightarrow{b}$ *endo*-2-chloronorbornane + $(CH_3)_3C\dot{O}$				
Initiated by therm. of AIBN				70 Bar 1
PR	C_6H_6	313	$k_a/k_b = 7$	
$\dot{A} + C_6H_5CH_2Cl$ $\xrightarrow{a}$ *exo*-2-chloronorbornane + $C_6H_5\dot{C}H_2$; $\xrightarrow{b}$ *endo*-2-chloronorbornane + $C_6H_5\dot{C}H_2$				
Therm. of 2-norbornylperester				70 Bar 1
PR	$C_6H_5CH_2Cl$	353	$k_a/k_b = 6$	
$\dot{A}$ + 4-$CH_3OC_6H_4CH_2Cl$ $\xrightarrow{a}$ *exo*-2-chloronorbornane + 4-$CH_3OC_6H_4\dot{C}H_2$; $\xrightarrow{b}$ *endo*-2-chloronorbornane + 4-$CH_3OC_6H_4\dot{C}H_2$				
Therm. of 2-norbornylperester				70 Bar 1
PR	4-$CH_3OC_6H_4CH_2Cl$	353	$k_a/k_b = 7$	

*) For $\dot{A}$, see p. 117.

Reaction Radical generation Method	Solvent	T[K]	Rate data	Ref./ add. ref.

$\dot{A}$ + *exo*-norbornyldimethylcarbinyl chloride $\xrightarrow{a}$ *exo*-2-chloronorbornane + ...($-\dot{C}l$) *)
$\xrightarrow{b}$ *endo*-2-chloronorbornane + ...($-\dot{C}l$)

Therm. of *exo*-2-norbornyldimethylcarbinyl hypochlorite				70 Bar 1
PR	not specified	353	$k_a/k_b = 6$	

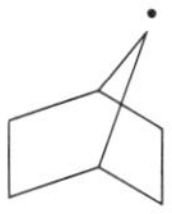
+ $CBrCl_3$ $\xrightarrow{a}$ 7-bromonorbornane + $\dot{C}Cl_3$
+ CCl_4 $\xrightarrow{b}$ 7-chloronorbornane + $\dot{C}Cl_3$

Therm. of 7-norbornyl-$COOOC(CH_3)_3$ and reduct. of 7-norbornyl-HgX by $NaBH_4$				80 Gie 1,
PR, glc	CCl_4	273	$k_a/k_b = 1070(\pm 5\%)$	75 Her 1/
		303	$690(\pm 5\%)$	76 Gie 1,
		323	$510(\pm 5\%)$	76 Gie 2,
		403	$170(\pm 5\%)$	76 Gie 3
			$\Delta H_a^\ddagger - \Delta H_b^\ddagger = -11\ \mathrm{kJ\,mol^{-1}}$	
			$\Delta S_a^\ddagger - \Delta S_b^\ddagger = 15\ \mathrm{J\,mol^{-1}\,K^{-1}}$	

c-$C_4H_7\dot{C}(CH_3)_2 + (n\text{-}C_4H_9)_3SnH \xrightarrow{a} c\text{-}C_4H_7CH(CH_3)_2 + (n\text{-}C_4H_9)_3\dot{S}n$
c-$C_4H_7\dot{C}(CH_3)_2 \xrightarrow{b} (CH_3)_2C{=}CHCH_2CH_2\dot{C}H_2$

$(n\text{-}C_4H_9)_3\dot{S}n$ + c-$C_4H_7CCl(CH_3)_2$ reaction (AIBN initiated)				80 Bec 1
PR, glc	decalin	333	$k_a/k_b = 7.7(30)\,10^2\,M^{-1}$	

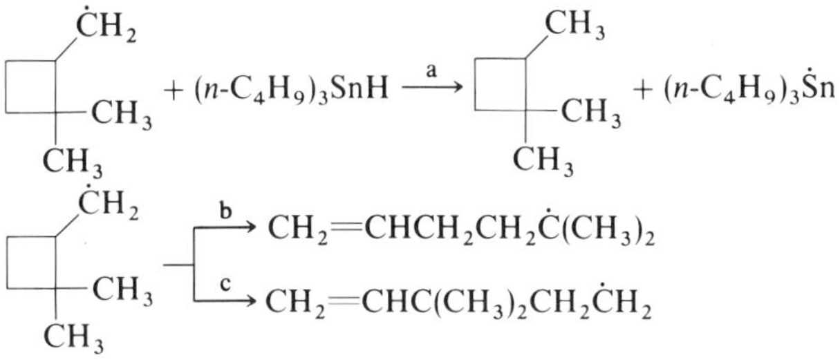

$(n\text{-}C_4H_9)_3\dot{S}n$ + 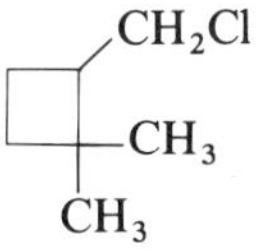

reaction (AIBN initiated) 80 Bec 1

PR, glc	decalin	333	$k_a/k_b = 1.43(4)\,M^{-1}$	
			$k_a/k_c > 2.5 \cdot 10^2\,M^{-1}$	

*) For $\dot{A}$, see p. 117.

Reaction Radical generation			
Method	Solvent	$T[K]$	Rate data

$$\overset{\dot{C}H_2}{\underset{CH_3}{H_3C}} + (n\text{-}C_4H_9)_3SnH \xrightarrow{a} \overset{CH_3}{\underset{CH_3}{H_3C}} + (n\text{-}C_4H_9)_3\dot{S}n$$

$$\overset{\dot{C}H_2}{\underset{CH_3}{H_3C}} \xrightarrow{b} CH_2=CHCH_2C(CH_3)_2\dot{C}H_2$$

$(n\text{-}C_4H_9)_3\dot{S}n + \overset{CH_2Cl}{\underset{CH_3}{H_3C}}$ reaction (AIBN initiated)   80 Bec 1

PR, glc	decalin	333	$k_a/k_b = 7.4(3)\cdot10^2\,M^{-1}$	

$(c\text{-}C_7H_{13}^{\cdot}) + CBrCl_3 \xrightarrow{a} c\text{-}C_7H_{13}Br + \dot{C}Cl_3$
$\quad\quad\quad + CCl_4 \xrightarrow{b} c\text{-}C_7H_{13}Cl + \dot{C}Cl_3$

Therm. of $c\text{-}C_7H_{13}COOOC(CH_3)_3$ [88]) and reduct. of $c\text{-}C_7H_{13}HgOCOCH_3$ by $NaBH_4$ 79 Gie 1,

PR, glc	CCl₄	273	$k_a/k_b = 7900$	75 Her 1/
		303	4000	76 Gie 1,
		323	2400	76 Gie 2,
		343	1500	76 Gie 3
		403	560 [87])	
		383	407 [88])	
			$\Delta H_a^{\ddagger} - \Delta H_b^{\ddagger} = -18\,kJ\,mol^{-1}$	
			$(\pm10\%)$	
			$\Delta S_a^{\ddagger} - \Delta S_b^{\ddagger} = 8\,J\,mol^{-1}\,K^{-1}\,(\pm5\%)$	

$\dot{C}H_2(CH_2)_4CH=CH_2 + (n\text{-}C_4H_9)_3SnH \xrightarrow{a} CH_3(CH_2)_4CH=CH_2 + (n\text{-}C_4H_9)_3\dot{S}n$

$\dot{C}H_2(CH_2)_4CH=CH_2 \xrightarrow[c]{\overset{b}{\nearrow}} \overset{c\text{-}C_6H_{11}\dot{C}H_2}{(c\text{-}C_7H_{13}^{\cdot})}$

$(n\text{-}C_4H_9)_3\dot{S}n + BrCH_2(CH_2)_4CH=CH_2$ reaction (AIBN initiated)   74 Bec 1

PR, glc	not given	338	$k_a/k_b = 1.92\cdot10^2\,M^{-1}$	
			$k_a/k_c = 1.12\cdot10^3\,M^{-1}$ [89])	

$CH_2=C(CH_3)(CH_2)_3\dot{C}H_2 + (n\text{-}C_4H_9)_3SnH \xrightarrow{a} CH_2=C(CH_3)(CH_2)_3CH_3 + (n\text{-}C_4H_9)_3\dot{S}n$

$$CH_2=C(CH_3)(CH_2)_3\dot{C}H_2 \xrightarrow{b} \overset{CH_3}{\underset{}{\bigg\langle}}\overset{\dot{C}H_2}{} \text{ and } \overset{CH_3}{\underset{}{\bigcirc}}^{\cdot}$$

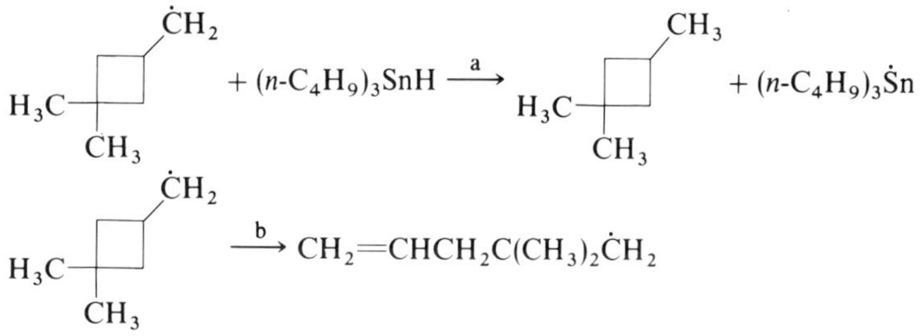

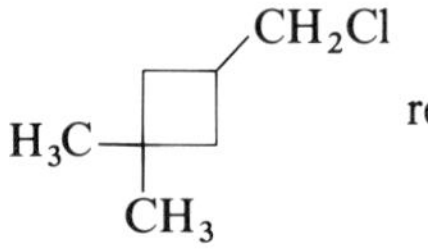

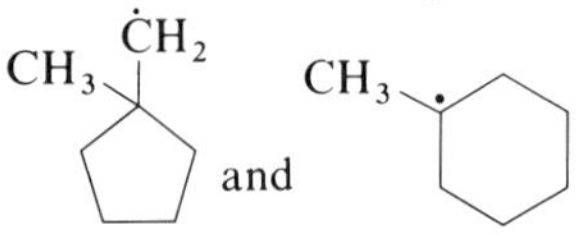

From $CH_2=C(CH_3)(CH_2)_3CH_2Br$ with AIBN (A) and DTBP (B) as initiator   72 Wal 1

PR, glc	C₆H₆	313	$k_a/k_b = 10.1(12)\,M^{-1}\,(A)$	
		343	$7.58(132)\,M^{-1}\,(A)$	
		373	$5.18(110)\,M^{-1}\,(B)$	

---

[87]) Extrapolated value.
[88]) From [75 Her 1].
[89]) $\Delta H_b^{\ddagger} = -17.6\,kJ\,mol^{-1}$, $\Delta S_b^{\ddagger} = -8.4\,J\,mol^{-1}\,K^{-1}$; $\Delta H_c^{\ddagger} = -21.8\,kJ\,mol^{-1}$, $\Delta S_c^{\ddagger} = -5.9\,J\,mol^{-1}\,K^{-1}$.

Reaction Radical generation Method	Solvent	T[K]	Rate data	Ref./ add. ref.
$CH_2{=}C(CH_3)(CH_2)_3\dot{C}H_2 + (n\text{-}C_4H_9)_3SnH \xrightarrow{a} CH_2{=}C(CH_3)(CH_2)_3CH_3 + (n\text{-}C_4H_9)_3\dot{S}n$				
$CH_2{=}C(CH_3)(CH_2)_3\dot{C}H_2$ 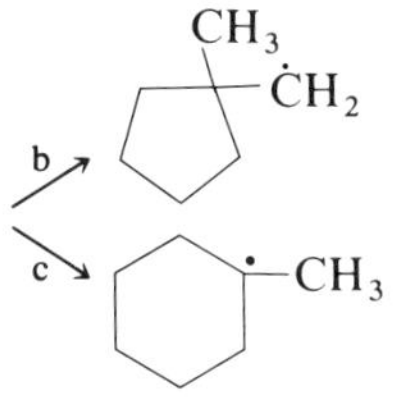				
$(n\text{-}C_4H_9)_3\dot{S}n + CH_2Br(CH_2)_3C(CH_3){=}CH_2$				74 Bec 2
PR, glc	C_6H_6	338	$k_a/k_b = 200\ M^{-1}$ $k_a/k_c = 125\ M^{-1}$	
$CH_2{=}C(CH_3)(CH_2)_3\dot{C}H_2 + (C_6H_5)_3SnH \xrightarrow{a} CH_2{=}C(CH_3)(CH_2)_3CH_3 + (C_6H_5)_3\dot{S}n$				
$CH_2{=}C(CH_3)(CH_2)_3\dot{C}H_2$ 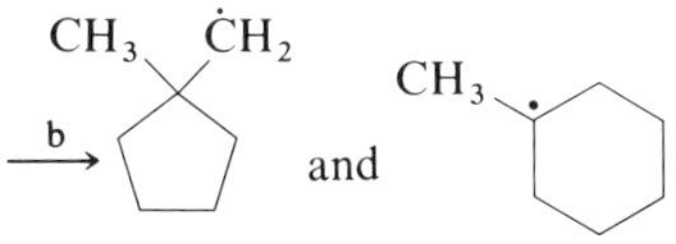				
From $CH_2{=}C(CH_3)(CH_2)_3CH_2Br$ with DTBP as initiator				72 Wal 1
PR, glc	C_6H_6	373	$k_a/k_b = 17.2(3)\ M^{-1}$	
$CH_2{=}CH(CH_2)_3\dot{C}HCH_3 + (n\text{-}C_4H_9)_3SnH \xrightarrow{a} CH_2{=}CH(CH_2)_4CH_3 + (n\text{-}C_4H_9)_3\dot{S}n$				
$CH_2{=}CH(CH_2)_3\dot{C}HCH_3$ 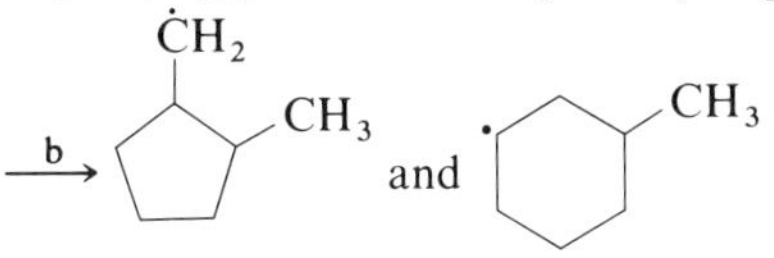				
From $CH_2{=}CH(CH_2)_3CHBrCH_3$ with AIBN (*A*) and DTBP (*B*) as initiators				72 Wal 1
PR, glc	C_6H_6	313 343 373	$k_a/k_b = 7.69(71)\ M^{-1}$ (*A*) $5.62(16)\ M^{-1}$ (*A*) $3.66(13)\ M^{-1}$ (*B*)	
$CH_2{=}CH(CH_2)_3\dot{C}HCH_3 + (n\text{-}C_4H_9)_3SnH \xrightarrow{a} CH_2{=}CH(CH_2)_4CH_3 + (n\text{-}C_4H_9)_3\dot{S}n$				
$CH_2{=}CH(CH_2)_3\dot{C}HCH_3$ 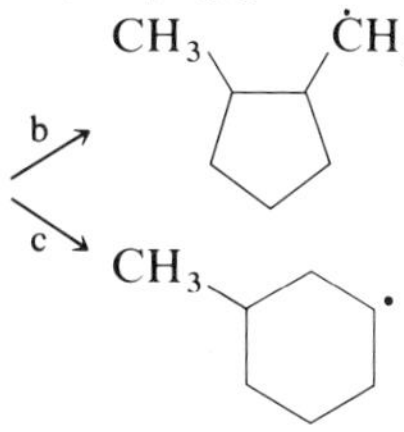				
$(n\text{-}C_4H_9)_3\dot{S}n + CH_3CHBr(CH_2)_3CH{=}CH_2$				74 Bec 2/ 74 Bec 3
PR, glc	C_6H_6	338	$k_a/k_b = 3.85\ M^{-1}$ $k_a/k_c = 303\ M^{-1}$	
$CH_2{=}CH(CH_2)_3\dot{C}HCH_3 + (n\text{-}C_4H_9)_3SnH \xrightarrow{a} CH_2{=}CH(CH_2)_4CH_3 + (n\text{-}C_4H_9)_3\dot{S}n$				
$CH_2{=}CH(CH_2)_3\dot{C}HCH_3$ 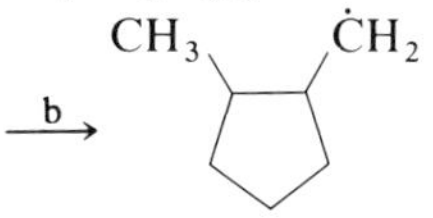				
React. $(n\text{-}C_4H_9)_3\dot{S}n + CH_3CHCl(CH_2)_3CH{=}CH_2$				74 Bec 3
PR, glc	pentane	338	$k_a/k_b = 3.85\ M^{-1}$	

Reaction Radical generation Method	Solvent	T[K]	Rate data	Ref./ add. ref.
$CH_2{=}CH(CH_2)_3\dot{C}HCH_3 + (C_6H_5)_3SnH \xrightarrow{a} CH_2{=}CH(CH_2)_4CH_3 + (C_6H_5)_3\dot{S}n$ $CH_2{=}CH(CH_2)_3\dot{C}HCH_3 \xrightarrow{b}$ 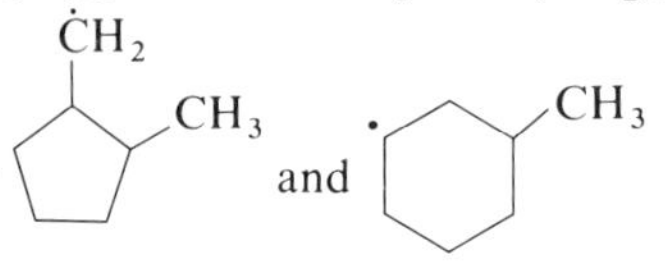				
From $CH_2{=}CH(CH_2)_3CHBrCH_3$ with DTBP as initiator				72 Wal 1
PR, glc	C_6H_6	373	$k_a/k_b = 15.6(10)\,M^{-1}$	
$CH_3(CH_2)_3\dot{C}HCH_2CH_3 + 3\text{-}ClC_6H_4CH_3 \xrightarrow{a} n\text{-}C_7H_{16} + 3\text{-}ClC_6H_4\dot{C}H_2$ $+ CCl_4 \xrightarrow{b} C_4H_9CHClC_2H_5 + \dot{C}Cl_3$				
Thermal decomp. of t-butyl-2-ethylperhexanoate				79 Tan 1
PR, glc	CCl_4/ 3-chlorotoluene mixt.	353	$k_a/k_b = 6.98(19)\cdot 10^{-4}$	
$CH_3(CH_2)_3\dot{C}HCH_2CH_3 + 3\text{-}ClC_6H_4CH_3 \xrightarrow{a} n\text{-}C_7H_{16} + 3\text{-}ClC_6H_4\dot{C}H_2$ $+ C_6H_5CH_3 \xrightarrow{b} n\text{-}C_7H_{16} + C_6H_5\dot{C}H_2$				
			See 4.1.2.3, Fig. 3, p. 255.	
$CH_3(CH_2)_3\dot{C}HCH_2CH_3 + 4\text{-}ClC_6H_4CH_3 \xrightarrow{a} n\text{-}C_7H_{16} + 4\text{-}ClC_6H_4\dot{C}H_2$ $+ CCl_4 \xrightarrow{b} C_4H_9CHClC_2H_5 + \dot{C}Cl_3$				
Thermal decomp. of t-butyl-2-ethylperhexanoate				79 Tan 1
PR, glc	CCl_4/ 4-chlorotoluene mixt.	353	$k_a/k_b = 7.79(9)\cdot 10^{-4}$	
$CH_3(CH_2)_3\dot{C}HCH_2CH_3 + 4\text{-}ClC_6H_4CH_3 \xrightarrow{a} n\text{-}C_7H_{16} + 4\text{-}ClC_6H_4\dot{C}H_2$ $+ C_6H_5CH_3 \xrightarrow{b} n\text{-}C_7H_{16} + C_6H_5\dot{C}H_2$				
			See 4.1.2.3, Fig. 3, p. 255.	
$CH_3(CH_2)_3\dot{C}HCH_2CH_3 + 3\text{-}FC_6H_4CH_3 \xrightarrow{a} n\text{-}C_7H_{16} + 3\text{-}FC_6H_4\dot{C}H_2$ $+ C_6H_5CH_3 \xrightarrow{b} n\text{-}C_7H_{16} + C_6H_5\dot{C}H_2$				
			See 4.1.2.3, Fig. 3, p. 255.	
$CH_3(CH_2)_3\dot{C}HCH_2CH_3 + 4\text{-}FC_6H_4CH_3 \xrightarrow{a} n\text{-}C_7H_{16} + 4\text{-}FC_6H_4\dot{C}H_2$ $+ C_6H_5CH_3 \xrightarrow{b} n\text{-}C_7H_{16} + C_6H_5\dot{C}H_2$				
			See 4.1.2.3, Fig. 3, p. 255.	
$CH_3(CH_2)_3\dot{C}HCH_2CH_3 + C_6H_5CH_3 \xrightarrow{a} n\text{-}C_7H_{16} + C_6H_5\dot{C}H_2$ $+ CCl_4 \xrightarrow{b} C_4H_9CHClC_2H_5 + \dot{C}Cl_3$				
Thermal decomp. of t-butyl-2-ethylperhexanoate				79 Tan 1
PR, glc	CCl_4/toluene mixt.	353	$k_a/k_b = 3.66(6)\cdot 10^{-4}$	
$CH_3(CH_2)_3\dot{C}HCH_2CH_3 + 3\text{-}CNC_6H_4CH_3 \xrightarrow{a} n\text{-}C_7H_{16} + 3\text{-}CNC_6H_4\dot{C}H_2$ $+ C_6H_5CH_3 \xrightarrow{b} n\text{-}C_7H_{16} + C_6H_5\dot{C}H_2$				
			See 4.1.2.3, Fig. 3, p. 255.	
$CH_3(CH_2)_3\dot{C}HCH_2CH_3 + 4\text{-}CNC_6H_4CH_3 \xrightarrow{a} n\text{-}C_7H_{16} + 4\text{-}CNC_6H_4\dot{C}H_2$ $+ CCl_4 \xrightarrow{b} C_4H_9CHClC_2H_5 + \dot{C}Cl_3$				
Thermal decomp. of t-butyl-2-ethylperhexanoate				79 Tan 1
PR, glc	CCl_4/4-cyanotoluene mixt.	353	$k_a/k_b = 1.19(1)\cdot 10^{-3}$	

Reaction Radical generation Method	Solvent	T[K]	Rate data	Ref./ add. ref.
$CH_3(CH_2)_3\dot{C}HCH_2CH_3 + 3\text{-}CH_3C_6H_4CH_3 \xrightarrow{a} n\text{-}C_7H_{16} + 3\text{-}CH_3C_6H_4\dot{C}H_2$ $+ CCl_4 \xrightarrow{b} C_4H_9CHClC_2H_5 + \dot{C}Cl_3$				
Thermal decomp. of *t*-butyl-2-ethylperhexanoate				79 Tan 1
PR, glc	CCl_4/1,3-dimethyl-benzene mixt.	353	$k_a/k_b = 4.05(9) \cdot 10^{-4}$	
$CH_3(CH_2)_3\dot{C}HCH_2CH_3 + 3\text{-}CH_3C_6H_4CH_3 \xrightarrow{a} n\text{-}C_7H_{16} + 3\text{-}CH_3C_6H_4\dot{C}H_2$ $+ C_6H_5CH_3 \xrightarrow{b} n\text{-}C_7H_{16} + C_6H_5\dot{C}H_2$				
			See 4.1.2.3, Fig. 3, p. 255.	
$CH_3(CH_2)_3\dot{C}HCH_2CH_3 + 4\text{-}CH_3C_6H_4CH_3 \xrightarrow{a} n\text{-}C_7H_{16} + 4\text{-}CH_3C_6H_4\dot{C}H_2$ $+ CCl_4 \xrightarrow{b} C_4H_9CHClC_2H_5 + \dot{C}Cl_3$				
Thermal decomp. of *t*-butyl-2-ethylperhexanoate				79 Tan 1
PR, glc	CCl_4/1,4-dimethyl-benzene mixt.	353	$k_a/k_b = 5.18(32) \cdot 10^{-4}$	
$CH_3(CH_2)_3\dot{C}HCH_2CH_3 + 4\text{-}CH_3C_6H_4CH_3 \xrightarrow{a} n\text{-}C_7H_{16} + 4\text{-}CH_3C_6H_4\dot{C}H_2$ $+ C_6H_5CH_3 \xrightarrow{b} n\text{-}C_7H_{16} + C_6H_5\dot{C}H_2$				
			See 4.1.2.3, Fig. 3, p. 255.	
$CH_3(CH_2)_3\dot{C}HCH_2CH_3 + 3,5\text{-}(CH_3)_2C_6H_3CH_3 \xrightarrow{a} n\text{-}C_7H_{16} + 3,5\text{-}(CH_3)_2C_6H_3\dot{C}H_2$ $+ C_6H_5CH_3 \xrightarrow{b} n\text{-}C_7H_{16} + C_6H_5\dot{C}H_2$				
			See 4.1.2.3, Fig. 3, p. 255.	
$(C_7H_{15}^{\bullet})$ [90] $+ Cl_2 \xrightarrow{a} C_7H_{15}Cl + \dot{C}l$ $\xrightarrow{b}$ termination products [91]				
$\dot{C}l + n\text{-}C_7H_{16}$ reaction (AIBN as initiator)				69 Leb 1,
[93]	*n*-heptane	323	$k_a/k_b = 1.05 \cdot 10^5\ M^{-1}$	42 Sta 1
		298	$1.44 \cdot 10^5\ M^{-1}$ [94]	
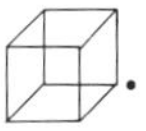 $+ CBrCl_3 \xrightarrow{a}$ bromocubane $+ \dot{C}Cl_3$ $+ CCl_4 \xrightarrow{b}$ chlorocubane $+ \dot{C}Cl_3$				
Thermal decomp. of cubylperester				78 Luh 1/
PR, glc	CCl_4	353	$k_a/k_b = 80$	75 Her 1,
			See also 4.1.2.3, Fig. 4, p. 255.	79 Gie 4
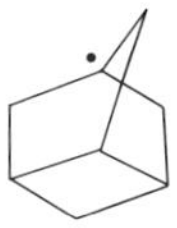 $+ CBrCl_3 \xrightarrow{a}$ 1-bromotricyclo[3,2,1,$0^{3,6}$]octane $+ \dot{C}Cl_3$ $+ CCl_4 \xrightarrow{b}$ 1-chlorotricyclo[3,2,1,$0^{3,6}$]octane $+ \dot{C}Cl_3$				
Thermal decomp. of 1-tricyclo[3,2,1,$0^{3,6}$]octylperester				81 Luh 1
PR, glc	CCl_4	353	$k_a/k_b = 65$	

[90]) Radicals from H-atom abstraction from *n*-heptane by Cl atoms, i.e. likely to be radical mixture.
[91]) First-order termination reaction (possibly reaction with O_2).
[93]) Cl_2 vapor pressure measurement in gas phase above solution.
[94]) From [42 Sta 1].

Reaction				
Radical generation				Ref./
Method	Solvent	T[K]	Rate data	add. ref.

$\quad$ + CBrCl$_3$ $\xrightarrow{a}$ 1-bromobicyclo[3,2,1]octane + $\dot{C}Cl_3$

$\quad$ + CCl$_4$ $\xrightarrow{b}$ 1-chlorobicyclo[3,2,1]octane + $\dot{C}Cl_3$

Therm. of 1-bicyclo[3,2,1]octyl-COOOC(CH$_3$)$_3$				80 Gie 1,
PR, glc	CCl$_4$	353	$k_a/k_b = 43$	75 Her 1/
		363	42	76 Gie 1,
		373	40	76 Gie 2,
		383	39	76 Gie 3
		393	38	
			$\Delta H_a^{\ddagger} - \Delta H_b^{\ddagger} = -4.2\,\text{kJ mol}^{-1}$	
			$\Delta S_a^{\ddagger} - \Delta S_b^{\ddagger} = 20\,\text{J mol}^{-1}\,\text{K}^{-1}$	
		353	$k_a/k_b = 56$ [95]	
		365	54 [95]	
		385	52 [95]	

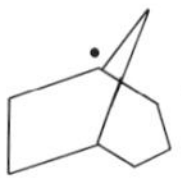

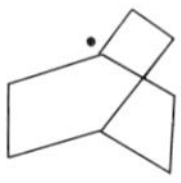

$\quad$ + CBrCl$_3$ $\xrightarrow{a}$ 1-bromobicyclo[2,2,2]octane + $\dot{C}Cl_3$

$\quad$ + CCl$_4$ $\xrightarrow{b}$ 1-chlorobicyclo[2,2,2]octane + $\dot{C}Cl_3$

Therm. of 1-bicyclo[2,2,2]octyl-COOOC(CH$_3$)$_3$				80 Gie 1,
PR, glc	CCl$_4$	343	$k_a/k_b = 40$	75 Her 1,
		363	35	69 Rue 1/
		373	35	76 Gie 1,
		383	33	76 Gie 2,
			$\Delta H_a^{\ddagger} - \Delta H_b^{\ddagger} = -4.9\,\text{kJ mol}^{-1}$	76 Gie 3
			$\Delta S_a^{\ddagger} - \Delta S_b^{\ddagger} = 16\,\text{J mol}^{-1}\,\text{K}^{-1}$	
		353	$k_a/k_b = 59$ [96]	
			See also 4.1.2.3, Fig. 4, p. 255.	

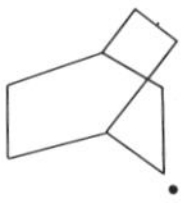

$\quad$ + CBrCl$_3$ $\xrightarrow{a}$ 2-bromobicyclo[2,2,2]octane + $\dot{C}Cl_3$

$\quad$ + CCl$_4$ $\xrightarrow{b}$ 2-chlorobicyclo[2,2,2]octane + $\dot{C}Cl_3$

Reduct. of 2-bicyclo[2,2,2]octyl-HgOCOCH$_3$ by NaBH$_4$ and				79 Gie 1,
therm. of 2-bicyclo[2,2,2]octyl-COOOC(CH$_3$)$_3$ [97]				75 Her 1/
PR, glc	CCl$_4$	273	$k_a/k_b = 6200$	76 Gie 1,
		303	2700	76 Gie 2,
		323	2000	76 Gie 3
		333	1200	
		343	1100	
			$\Delta H_a^{\ddagger} - \Delta H_b^{\ddagger} = -20\,\text{kJ mol}^{-1}\,(\pm 10\%)$	
			$\Delta S_a^{\ddagger} - \Delta S_b^{\ddagger} = 0\,\text{J mol}^{-1}\,\text{K}^{-1}$	
		365	$k_a/k_b = 810$ [97]	
		373	630 [97]	
		383	500 [97]	
		403	310 [97]	

---

[95] From [75 Her 1].
[96] From [75 Her 1, 69 Rue 1].
[97] From [75 Her 1].

Reaction Radical generation Method	Solvent	T[K]	Rate data	Ref./ add. ref.
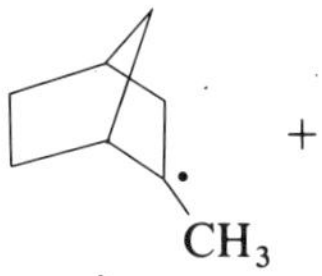 + $CHCl_3$ $\xrightarrow{a}$ *endo*-2-methylnorbornane + $\dot{C}Cl_3$; $\xrightarrow{b}$ *exo*-2-methylnorbornane + $\dot{C}Cl_3$ $\dot{\mathbf{B}}$				
Therm. of 2-methylnorbornylperester				70 Bar 1
PR	$CHCl_3$	343	$k_a/k_b = 3.6$	
$\dot{\mathbf{B}}$ + 4-BrC_6H_4SH $\xrightarrow{a}$ *endo*-2-methylnorbornane + 4-$BrC_6H_4\dot{S}$; $\xrightarrow{b}$ *exo*-2-methylnorbornane + 4-$BrC_6H_4\dot{S}$				
Therm. of 2-methylnorbornylperester				70 Bar 1
PR	C_6H_6	348	$k_a/k_b = 16.2$	
$\dot{\mathbf{B}}$ + 4-$NO_2C_6H_4SH$ $\xrightarrow{a}$ *endo*-2-methylnorbornane + 4-$NO_2C_6H_4\dot{S}$; $\xrightarrow{b}$ *exo*-2-methylnorbornane + 4-$NO_2C_6H_4\dot{S}$				
Therm. of 2-methylnorbornylperester				70 Bar 1
PR	C_6H_6	348	$k_a/k_b = 16.1$	
$\dot{\mathbf{B}}$ + C_6H_5SH $\xrightarrow{a}$ *endo*-2-methylnorbornane + $C_6H_5\dot{S}$; $\xrightarrow{b}$ *exo*-2-methylnorbornane + $C_6H_5\dot{S}$				
Therm. of 2-methylnorbornylperester				70 Bar 1
PR	C_6H_6	348	$k_a/k_b = 16.8$	
$\dot{\mathbf{B}}$ + 4-$NH_2C_6H_4SH$ $\xrightarrow{a}$ *endo*-2-methylnorbornane + 4-$NH_2C_6H_4\dot{S}$; $\xrightarrow{b}$ *exo*-2-methylnorbornane + 4-$NH_2C_6H_4\dot{S}$				
Therm. of 2-methylnorbornylperester				70 Bar 1
PR	C_6H_6	348	$k_a/k_b = 12.8$	
$\dot{\mathbf{B}}$ + 4-$CH_3C_6H_4Br$ $\xrightarrow{a}$ *endo*-2-methylnorbornane + 4-$\dot{C}H_2C_6H_4Br$; $\xrightarrow{b}$ *exo*-2-methylnorbornane + 4-$\dot{C}H_2C_6H_4Br$				
Therm. of 2-methylnorbornylperester				70 Bar 1
PR	C_6H_6	348	$k_a/k_b = 5.7$	
$\dot{\mathbf{B}}$ + 4-$CH_3C_6H_4CN$ $\xrightarrow{a}$ *endo*-2-methylnorbornane + 4-$\dot{C}H_2C_6H_4CN$; $\xrightarrow{b}$ *exo*-2-methylnorbornane + 4-$\dot{C}H_2C_6H_4CN$				
Therm. of 2-methylnorbornylperester				70 Bar 1
PR	C_6H_6	348	$k_a/k_b = 6.6$	
$\dot{\mathbf{B}}$ + 4-$CH_3C_6H_4CH_3$ $\xrightarrow{a}$ *endo*-2-methylnorbornane + 4-$CH_3C_6H_4\dot{C}H_2$; $\xrightarrow{b}$ *exo*-2-methylnorbornane + 4-$CH_3C_6H_4\dot{C}H_2$				
Therm. of 2-methylnorbornylperester				70 Bar 1
PR	4-$CH_3C_6H_4CH_3$	348	$k_a/k_b = 6.2$	

Reaction Radical generation Method	Solvent	T[K]	Rate data	Ref./ add. ref.
$\dot{\mathbf{B}} + 4\text{-}CH_3C_6H_4OCH_3 \xrightarrow{a}$ *endo*-2-methylnorbornane + 4-methoxytoluene(−Ḣ) *) $\xrightarrow{b}$ *exo*-2-methylnorbornane + 4-methoxytoluene(−Ḣ)				
Therm. of 2-methylnorbornylperester (A) and phot. of azo-2-methylnorbornane (B)				70 Bar 1
PR	4-methoxytoluene	343	$k_a/k_b = 7.3(A)$	
		303	$k_a/k_b = 6.9(B)$	
$\dot{\mathbf{B}} + C_8H_{17}SH \xrightarrow{a}$ *endo*-2-methylnorbornane + $C_8H_{17}\dot{S}$ $\xrightarrow{b}$ *exo*-2-methylnorbornane + $C_8H_{17}\dot{S}$				
Therm. of 2-methylnorbornylperester				70 Bar 1
PR	C_6H_6 or cumene or C_6H_5Cl	348	$k_a/k_b = 16.2$	
$\dot{\mathbf{B}} + C_6H_5CH(CH_3)_2 \xrightarrow{a}$ *endo*-2-methylnorbornane + $C_6H_5\dot{C}(CH_3)_2$ $\xrightarrow{b}$ *exo*-2-methylnorbornane + $C_6H_5\dot{C}(CH_3)_2$				
Therm. of 2-methylnorbornylperester (A) and phot. of azo-2-methylnorbornane (B)				70 Bar 1
PR	cumene	343	$k_a/k_b = 7.8(A)$	
		303	$k_a/k_b = 7.1(B)$	
$\dot{\mathbf{B}} + n\text{-}C_{12}H_{25}SH \xrightarrow{a}$ *endo*-2-methylnorbornane + $n\text{-}C_{12}H_{25}\dot{S}$ $\xrightarrow{b}$ *exo*-2-methylnorbornane + $n\text{-}C_{12}H_{25}\dot{S}$				
Therm. of 2-methylnorbornylperester				70 Bar 1
PR	C_6H_5Cl	343	$k_a/k_b = 12.2$	
	C_6H_6	343	$k_a/k_b = 12.4$	
$\dot{\mathbf{B}} + (n\text{-}C_4H_9)_3SnH \xrightarrow{a}$ *endo*-2-methylnorbornane + $(n\text{-}C_4H_9)_3\dot{S}n$ $\xrightarrow{b}$ *exo*-2-methylnorbornane + $(n\text{-}C_4H_9)_3\dot{S}n$				
Therm. of 2-methylnorbornylperester				70 Bar 1
PR	C_6H_6	348	$k_a/k_b = 8.1$	

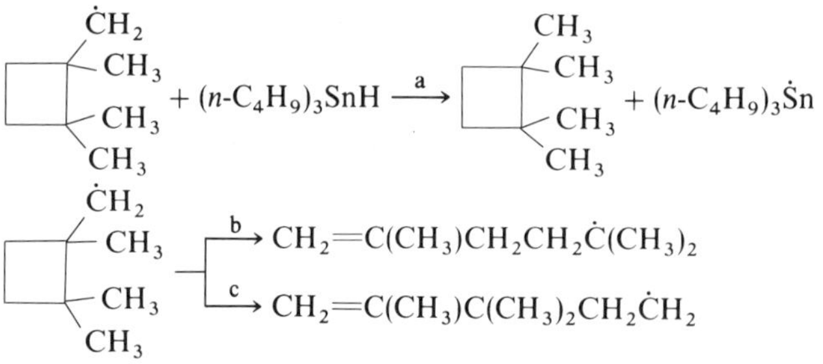

$(n\text{-}C_4H_9)_3\dot{S}n$ + 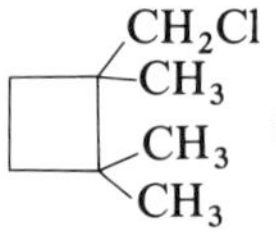 reaction (AIBN initiated) 80 Bec 1

PR, glc	decalin	333	$k_a/k_b = 1.11(3)\,M^{-1}$	
			$k_a/k_c > 10^3\,M^{-1}$	

*) For $\dot{\mathbf{B}}$, see p. 125.

Reaction Radical generation Method	Solvent	T[K]	Rate data	Ref./ add. ref.
$(c\text{-}C_8H_{15}^{\cdot}) + CCl_3Br \xrightarrow{a} c\text{-}C_8H_{15}Br + \dot{C}Cl_3$ $+ CCl_4 \xrightarrow{b} c\text{-}C_8H_{15}Cl + \dot{C}Cl_3$				
Therm. of $c\text{-}C_8H_{15}COOOC(CH_3)_3$ PR, glc	CCl_4	383	$k_a/k_b = 451$	75 Her 1/ 76 Gie 1, 76 Gie 2, 76 Gie 3
$\dot{C}H_2(CH_2)_5CH{=}CH_2 + (n\text{-}C_4H_9)_3SnH \xrightarrow{a} CH_3(CH_2)_5CH{=}CH_2 + (n\text{-}C_4H_9)_3\dot{S}n$ $\dot{C}H_2(CH_2)_5CH{=}CH_2 \xrightarrow{b} c\text{-}C_7H_{13}\dot{C}H_2$ $\dot{C}H_2(CH_2)_5CH{=}CH_2 \xrightarrow{c} (c\text{-}C_8H_{15}^{\cdot})$				
$(n\text{-}C_4H_9)_3\dot{S}n + BrCH_2(CH_2)_5CH{=}CH_2$ reaction (AIBN initiated) PR, glc from 874	not given	338	$k_a/k_b = 6.67 \cdot 10^2\ M^{-1}$ $k_a/k_c > 10^5\ M^{-1}$ [98]	74 Bec 1
$CH_2{=}CHCH_2CH_2C(CH_3)_2\dot{C}H_2 + (n\text{-}C_4H_9)_3SnH \xrightarrow{a} CH_2{=}CHCH_2CH_2C(CH_3)_2CH_3 + (n\text{-}C_4H_9)_3\dot{S}n$ $CH_2{=}CHCH_2CH_2C(CH_3)_2\dot{C}H_2 \xrightarrow{b}$ 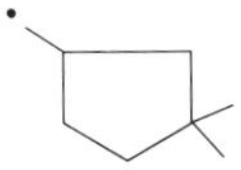				
Decomp. of $CH_2{=}CHCH_2CH_2C(CH_3)_2CH_2Br$ (AIBN initiated) PR, glc, NMR	C_6H_6	303 313 328 353	$k_a/k_b = 0.66\ M^{-1}$ 0.58 0.51 0.43 [99]	79 Bec 1/ 75 Bec 1
$CH_3CH{=}C(CH_3)(CH_2)_3\dot{C}H_2$ [1] $+ (n\text{-}C_4H_9)_3SnH \xrightarrow{a} CH_3CH{=}C(CH_3)(CH_2)_3CH_3$ [1] $+ (n\text{-}C_4H_9)_3\dot{S}n$ $CH_3CH{=}C(CH_3)(CH_2)_3\dot{C}H_2$ (b, c) 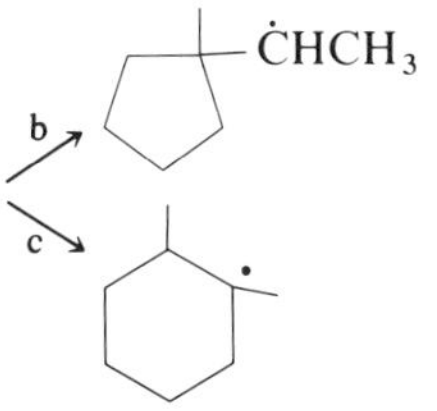				
$(n\text{-}C_4H_9)_3\dot{S}n + CH_2Br(CH_2)_3C(CH_3){=}CHCH_3$ [1]) PR, glc	C_6H_6	338	$k_a/k_b = 41.7\ M^{-1}$ $k_a/k_c = 250\ M^{-1}$	74 Bec 2
trans-$CH_3CH{=}C(CH_3)(CH_2)_3\dot{C}H_2 + (n\text{-}C_4H_9)_3SnH \xrightarrow{a}$ *trans*-$CH_3CH{=}C(CH_3)(CH_2)_3CH_3 + (n\text{-}C_4H_9)_3\dot{S}n$ *trans*-$CH_3CH{=}C(CH_3)(CH_2)_3\dot{C}H_2 \xrightarrow{b}$ 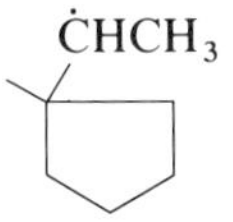				
		353	$k_a/k_b = 5.56\ M^{-1}$ [2])	79 Bec 1/ 75 Jul 1, 75 Bec 1, 74 Bec 2

[98]) $\Delta H_b^{\ddagger} = -24.7\ kJ\ mol^{-1}$, $\Delta S_b^{\ddagger} = -0.4\ J\ mol^{-1}\ K^{-1}$.
[99]) $\Delta H_a^{\ddagger} - \Delta H_b^{\ddagger} = -7.1(4)\ kJ\ mol^{-1}$, $\Delta S_a^{\ddagger} - \Delta S_b^{\ddagger} = -27.6(16)\ J\ mol^{-1}\ K^{-1}$.
[1]) Mixt. of *cis* and *trans*.
[2]) Calculated from experimental data from [75 Jul 1].

Reaction Radical generation Method	Solvent	T[K]	Rate data	Ref./ add. ref.

cis-$CH_3CH{=}C(CH_3)(CH_2)_3\dot{C}H_2$ + $(n\text{-}C_4H_9)_3SnH$ $\xrightarrow{a}$ *cis*-$CH_3CH{=}C(CH_3)(CH_2)_3CH_3$ + $(n\text{-}C_4H_9)_3\dot{S}n$

cis-$CH_3CH{=}C(CH_3)(CH_2)_3\dot{C}H_2$ 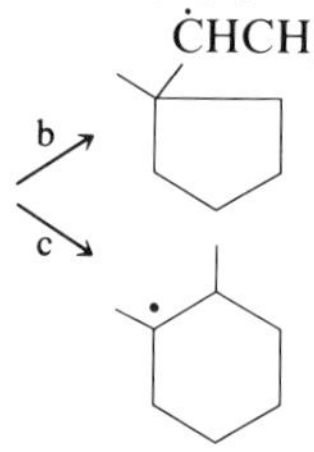

		353	$k_a/k_b = 83.3\ M^{-1}$ [2] $k_a/k_c = 47.6\ M^{-1}$ [2]	79 Bec 1/ 75 Jul 1, 75 Bec 1, 74 Bec 2

$\dot{C}H_2(CH_2)_3CH{=}C(CH_3)_2$ + $(n\text{-}C_4H_9)_3SnH$ $\xrightarrow{a}$ $CH_3(CH_2)_3CH{=}C(CH_3)_2$ + $(n\text{-}C_4H_9)_3\dot{S}n$

$\dot{C}H_2(CH_2)_3CH{=}C(CH_3)_2$ 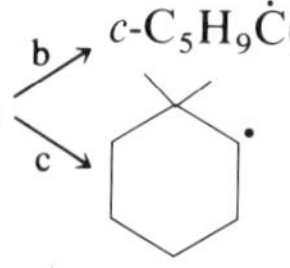

$(n\text{-}C_4H_9)_3\dot{S}n + CH_2Br(CH_2)_3CH{=}C(CH_3)_2$ PR, glc	C_6H_6	338	$k_a/k_b = 1.92\ M^{-1}$ $k_a/k_c > 400\ M^{-1}$	74 Bec 2

$CH_2{=}C(CH_3)(CH_2)_3\dot{C}HCH_3$ + $(n\text{-}C_4H_9)_3SnH$ $\xrightarrow{a}$ $CH_2{=}C(CH_3)(CH_2)_4CH_3$ + $(n\text{-}C_4H_9)_3\dot{S}n$

$CH_2{=}C(CH_3)(CH_2)_3\dot{C}HCH_3$ $\xrightarrow{b}$ 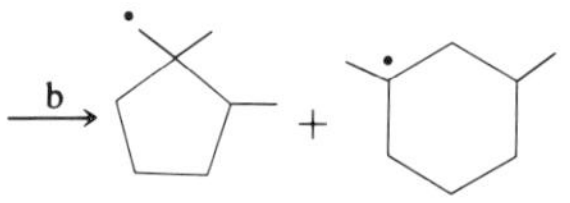

From $CH_2{=}C(CH_3)(CH_2)_3CHBrCH_3$ with AIBN (*A*) and DTBP (*B*) as initiator PR, glc	C_6H_6	313 343 373	$k_a/k_b = 6.67(13)\ M^{-1}$ (*A*) 4.74(61) (*A*) 3.02(27) (*B*)	72 Wal 1

$CH_2{=}C(CH_3)(CH_2)_3\dot{C}HCH_3$ + $(C_6H_5)_3SnH$ $\xrightarrow{a}$ $CH_2{=}C(CH_3)(CH_2)_4CH_3$ + $(C_6H_5)_3\dot{S}n$

$CH_2{=}C(CH_3)(CH_2)_3\dot{C}HCH_3$ $\xrightarrow{b}$ 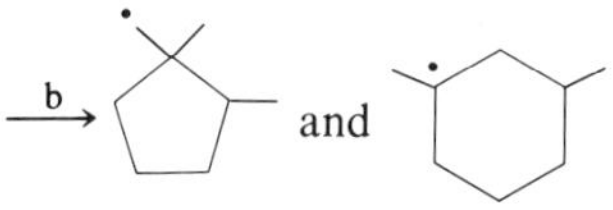

From $CH_2{=}C(CH_3)(CH_2)_3CHBrCH_3$ with DTBP as initiator PR, glc	C_6H_6	373	$k_a/k_b = 10.4(5)\ M^{-1}$	72 Wal 1

$CH_2{=}CH(CH_2)_3\dot{C}(CH_3)_2$ + $(n\text{-}C_4H_9)_3SnH$ $\xrightarrow{a}$ $CH_2{=}CH(CH_2)_3CH(CH_3)_2$ + $(n\text{-}C_4H_9)_3\dot{S}n$

$CH_2{=}CH(CH_2)_3\dot{C}(CH_3)_2$ $\xrightarrow{b}$ 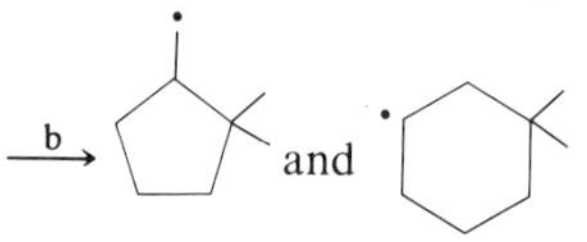

From $CH_2{=}CH(CH_2)_3CBr(CH_3)_2$ with AIBN (*A*) and DTBP (*B*) as initiator PR, glc	C_6H_6	313 343 373	$k_a/k_b = 6.67(40)\ M^{-1}$ (*A*) 4.26(71) (*A*) 1.69(22) (*B*)	72 Wal 1

[2]) Calculated from experimental data from [75 Jul 1].

Reaction Radical generation Method	Solvent	T[K]	Rate data	Ref./ add. ref.
$CH_2{=}CH(CH_2)_3\dot{C}(CH_3)_2 + (n\text{-}C_4H_9)_3SnH \xrightarrow{a} CH_2{=}CH(CH_2)_3CH(CH_3)_2 + (n\text{-}C_4H_9)_3\dot{S}n$				
$CH_2{=}CH(CH_2)_3\dot{C}(CH_3)_2$ 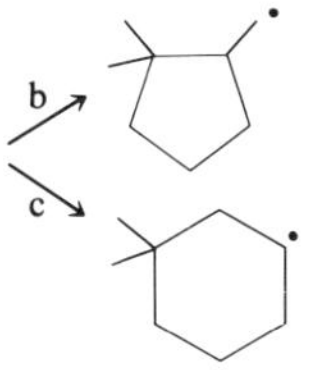				
$(n\text{-}C_4H_9)_3\dot{S}n + (CH_3)_2CBr(CH_2)_3CH{=}CH_2$				74 Bec 2
PR, glc	C_6H_6	338	$k_a/k_b = 2.44\ M^{-1}$ $k_a/k_c = 167\ M^{-1}$	
$CH_2{=}CH(CH_2)_3\dot{C}(CH_3)_2 + (C_6H_5)_3SnH \xrightarrow{a} CH_2{=}CH(CH_2)_3CH(CH_3)_2 + (C_6H_5)_3\dot{S}n$				
$CH_2{=}CH(CH_2)_3\dot{C}(CH_3)_2$ 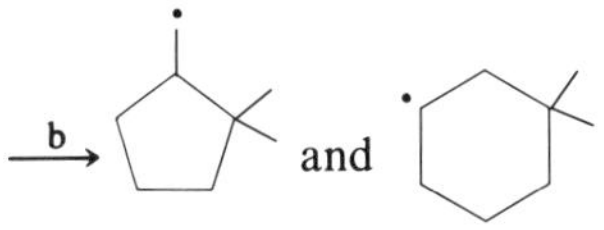				
From $CH_2{=}CH(CH_2)_3CBr(CH_3)_2$ with DTBP as initiator				72 Wal 1
PR, glc	C_6H_6	373	$k_a/k_b = 6.54(77)\ M^{-1}$	
$\dot{C}H_2(CH_2)_6CH_3 + CBrCl_3 \xrightarrow{a} CH_2Br(CH_2)_6CH_3 + \dot{C}Cl_3$ $+ CCl_4 \xrightarrow{b} CH_2Cl(CH_2)_6CH_3 + \dot{C}Cl_3$				
Reduct. of $n\text{-}C_8H_{17}HgOCOCH_3$ by $NaBH_4$ and therm. of $n\text{-}C_8H_{17}COOOC(CH_3)_3$ [3]				79 Gie 1,
PR, glc	CCl_4	273	$k_a/k_b = 4600$	75 Her 1,
		303	2500	69 Rue 1/
		323	1800	76 Gie 1,
		343	1300	76 Gie 2,
			$\Delta H_a^\ddagger - \Delta H_b^\ddagger = -14\ kJ\ mol^{-1}$ ($\pm 10\%$)	76 Gie 3
			$\Delta S_a^\ddagger - \Delta S_b^\ddagger = 19\ J\ mol^{-1}\ K^{-1}$ ($\pm 5\%$)	
		371	$k_a/k_b = 835$ [3]	
		390	722 [3]	
		403	678 [3]	
$CH_3(CH_2)_3C(CH_3)_2\dot{C}H_2 + CCl_3Br \xrightarrow{a} CH_3(CH_2)_3C(CH_3)_2CH_2Br + \dot{C}Cl_3$ $+ CCl_4 \xrightarrow{b} CH_3(CH_2)_3C(CH_3)_2CH_2Cl + \dot{C}Cl_3$				
Therm. of $CH_3(CH_2)_3C(CH_3)_2CH_2COOOC(CH_3)_3$				75 Her 1/
PR, glc	CCl_4	403	$k_a/k_b = 520$	76 Gie 1,
		389	650	76 Gie 2,
		374	890	76 Gie 3, 69 Rue 1

[3]) From [75 Her 1, 69 Rue 1].

Reaction Radical generation Method	Solvent	T[K]	Rate data	Ref./ add. ref.
$CH_3\dot{C}H(CH_2)_5CH_3 + CBrCl_3 \xrightarrow{a} CH_3CHBr(CH_2)_5CH_3 + \dot{C}Cl_3$ $+ CCl_4 \xrightarrow{b} CH_3CHCl(CH_2)_5CH_3 + \dot{C}Cl_3$				
Reduct. of $CH_3(CH_2)_5CH(CH_3)HgOCOCH_3$ by $NaBH_4$ and therm. of $CH_3(CH_2)_5CH(CH_3)COOOC(CH_3)_3$ [4])				79 Gie 1, 75 Her 1/ 76 Gie 1, 76 Gie 2, 76 Gie 3, 69 Rue 1
PR, glc	CCl_4	273	$k_a/k_b = 7100$	
		303	3500	
		323	2400	
		343	1500	
			$\Delta H_a^{\ddagger} - \Delta H_b^{\ddagger} = -19\,kJ\,mol^{-1}$ ($\pm 10\%$)	
			$\Delta S_a^{\ddagger} - \Delta S_b^{\ddagger} = 6\,J\,mol^{-1}\,K^{-1}$ ($\pm 5\%$)	
		366	$k_a/k_b = 805$ [4])	
		373	761 [4])	
		390	655 [4])	
		403	580 [4])	
$CH_3(CH_2)_2\dot{C}H(CH_2)_3CH_3 + CCl_4 \xrightarrow{a} CH_3(CH_2)_2CHCl(CH_2)_3CH_3 + \dot{C}Cl_3$ $+ C_2H_5I \xrightarrow{b} CH_3(CH_2)_2CHI(CH_2)_3CH_3 + CH_3\dot{C}H_2$				
Therm. of BPO as initiator				71 Afa 3
PR, glc	1-hexene/CCl_4/ C_2H_5I	373	$k_a/k_b = 9.4(58)$ [5])	
$CH_3(CH_2)_2\dot{C}H(CH_2)_3CH_3 + CHCl_3 \xrightarrow{a} CH_3(CH_2)_2CHCl(CH_2)_3CH_3 + \dot{C}HCl_2$ $\xrightarrow{b} n\text{-}C_8H_{18} + \dot{C}Cl_3$				
Therm. of BPO as initiator				71 Afa 3
PR, glc	1-hexene/$CHCl_3$/ C_2H_5I	373	$k_a/k_b = 0.23(4)$	
$CH_3(CH_2)_2\dot{C}H(CH_2)_3CH_3 + CHCl_3 \xrightarrow{a} n\text{-}C_8H_{18} + \dot{C}Cl_3$ $+ CCl_4 \xrightarrow{b} CH_3(CH_2)_2CHCl(CH_2)_3CH_3 + \dot{C}Cl_3$				
Therm. of BPO as initiator				71 Afa 3
PR, glc	1-hexene/$CHCl_3$/ CCl_4	373	$k_a/k_b = 1.03(11) \cdot 10^{-2}$	
$CH_3(CH_2)_2\dot{C}H(CH_2)_3CH_3 + CHCl_3 \xrightarrow{a} CH_3(CH_2)_2CHCl(CH_2)_3CH_3 + \dot{C}HCl_2$ $+ C_2H_5I \xrightarrow{b} CH_3(CH_2)_2CHI(CH_2)_3CH_3 + CH_3\dot{C}H_2$				
Therm. of BPO as initiator				71 Afa 3
PR, glc	1-hexene/$CHCl_3$/ C_2H_5I	373	$k_a/k_b = 0.055(12)$	
$CH_3(CH_2)_2\dot{C}H(CH_2)_3CH_3 + CHBrCl_2 \xrightarrow{a} CH_3(CH_2)_2CHBr(CH_2)_3CH_3 + \dot{C}HCl_2$ $+ C_2H_5I \xrightarrow{b} CH_3(CH_2)_2CHI(CH_2)_3CH_3 + CH_3CH_2$				
Therm. of BPO as initiator				71 Afa 4
PR, glc	1-hexene/C_2H_5I/ $CHBrCl_2$	373	$k_a/k_b = 25.8(23)$	
$CH_3(CH_2)_2\dot{C}H(CH_2)_3CH_3 + CH_2BrCl \xrightarrow{a} CH_3(CH_2)_2CHBr(CH_2)_3CH_3 + \dot{C}H_2Cl$ $+ C_2H_5I \xrightarrow{b} CH_3(CH_2)_2CHI(CH_2)_3CH_3 + CH_3\dot{C}H_2$				
Therm. of BPO as initiator				71 Afa 4
PR, glc	1-hexene/C_2H_5I/ CH_2BrCl	373	$k_a/k_b = 0.295(31)$	

[4]) From [75 Her 1].

[5]) $k_a/k_b = 23.3$ from indirect measurement $(k_x/k_b)/(k_x/k_a)$ with k_x referring to $CH_3(CH_2)_2\dot{C}H(CH_2)_3CH_3 + CHCl_3 \longrightarrow n\text{-}C_8H_{18} + \dot{C}Cl_3$.

Reaction Radical generation Method	Solvent	T[K]	Rate data	Ref./ add. ref.
$CH_3(CH_2)_2\dot{C}H(CH_2)_3CH_3 + CH_2Cl_2 \xrightarrow{a} n\text{-}C_8H_{18} + \dot{C}HCl_2$ $+ C_2H_5I \xrightarrow{b} CH_3(CH_2)_2CHI(CH_2)_3CH_3 + CH_3\dot{C}H_2$				
Therm. of BPO as initiator				71 Afa 3
PR, glc	1-hexene/C_2H_5I/ CH_2Cl_2	373	$k_a/k_b = 6.17(97)\cdot 10^{-3}$	
$CH_3(CH_2)_2\dot{C}H(CH_2)_3CH_3 + CH_2BrCOOH \xrightarrow{a} CH_3(CH_2)_2CHBr(CH_2)_3CH_3 + \dot{C}H_2COOH$ $+ C_2H_5I \xrightarrow{b} CH_3(CH_2)_2CHI(CH_2)_3CH_3 + CH_3\dot{C}H_2$				
Therm. of BPO as initiator				71 Afa 4
PR, glc	1-hexene/C_2H_5I/ $CH_2BrCOOH$	373	$k_a/k_b = 2.12(21)$	
$CH_3(CH_2)_2\dot{C}H(CH_2)_3CH_3 + C_2H_5I \xrightarrow{a} n\text{-}C_8H_{18} + CH_3\dot{C}HI$ $CH_3(CH_2)_2\dot{C}H(CH_2)_3CH_3 + C_2H_5I \xrightarrow{b} CH_3(CH_2)_2CHI(CH_2)_3CH_3 + CH_3\dot{C}H_2$				
Therm. of BPO as initiator				71 Afa 3
PR, glc	1-hexene/C_2H_5I	373	$k_a/k_b = 0.019(6)$	
$CH_3(CH_2)_2\dot{C}H(CH_2)_3CH_3 + CH_3COCH_3 \xrightarrow{a} n\text{-}C_8H_{18} + \dot{C}H_2COCH_3$ $+ C_2H_5I \xrightarrow{b} CH_3(CH_2)_2CHI(CH_2)_3CH_3 + CH_3\dot{C}H_2$				
Therm. of BPO as initiator				71 Afa 1
PR, glc	1-hexene/C_2H_5I/ CH_3COCH_3	373	$k_a/k_b = 3.12(55)\cdot 10^{-3}$	
$CH_3(CH_2)_2\dot{C}H(CH_2)_3CH_3 + CH_2{=}CH(CH_2)_3CH_3 \xrightarrow{a} n\text{-}C_8H_{18} + CH_2{=}CH\dot{C}H(CH_2)_2CH_3$ $+ C_2H_5I \xrightarrow{b} CH_3(CH_2)_2CHI(CH_2)_3CH_3 + CH_3\dot{C}H_2$				
Therm. of BPO as initiator				71 Afa 3
PR, glc	1-hexene/C_2H_5I	373	$k_a/k_b = 0.013(1)$	
$CH_3(CH_2)_2\dot{C}H(CH_2)_3CH_3 + C_6H_5CH_2Br \xrightarrow{a} CH_3(CH_2)_2CHBr(CH_2)_3CH_3 + C_6H_5\dot{C}H_2$ $+ C_2H_5I \xrightarrow{b} CH_3(CH_2)_2CHI(CH_2)_3CH_3 + CH_3\dot{C}H_2$				
Therm. of BPO as initiator				71 Afa 4
PR, glc	1-hexene/C_2H_5I/ $C_6H_5CH_2Br$	373	$k_a/k_b = 0.854(55)$	
$CH_3(CH_2)_2\dot{C}H(CH_2)_3CH_3 + C_6H_5CH_3 \xrightarrow{a} n\text{-}C_8H_{18} + C_6H_5\dot{C}H_2$ $+ C_2H_5I \xrightarrow{b} CH_3(CH_2)_2CHI(CH_2)_3CH_3 + CH_3\dot{C}H_2$				
Therm. of BPO as initiator				71 Afa 1
PR, glc	1-hexene/C_2H_5I/ $C_6H_5CH_3$	373	$k_a/k_b = 3.11(45)\cdot 10^{-2}$	
$CH_3(CH_2)_2\dot{C}H(CH_2)_3CH_3 + C_6H_5OCH_3 \xrightarrow{a} n\text{-}C_8H_{18} + \text{anisole}(-\dot{H})$ $+ C_2H_5I \xrightarrow{b} CH_3(CH_2)_2CHI(CH_2)_3CH_3 + CH_3\dot{C}H_2$				
Therm. of BPO as initiator				71 Afa 1
PR, glc	1-hexene/C_2H_5I/ $C_6H_5OCH_3$	373	$k_a/k_b = 3.50(84)\cdot 10^{-3}$	

Reaction Radical generation Method	Solvent	T[K]	Rate data	Ref./ add. ref.
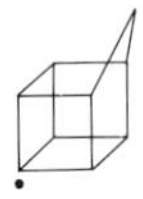 + $CBrCl_3 \xrightarrow{a}$ 4-bromohomocubane + $\dot{C}Cl_3$				
+ $CCl_4 \xrightarrow{b}$ 4-chlorohomocubane + $\dot{C}Cl_3$				
Therm. of 4-homocubyl-$COOOC(CH_3)_3$				80 Gie 1,
PR, glc	CCl_4	353	$k_a/k_b = 68$	75 Her 1/
		367	67	76 Gie 1,
		403	62	76 Gie 2,
			$\Delta H_a^\ddagger - \Delta H_b^\ddagger = -2.1\ \text{kJ mol}^{-1}$	76 Gie 3
			$\Delta S_a^\ddagger - \Delta S_b^\ddagger = 29\ \text{J mol}^{-1}\ \text{K}^{-1}$	
		403	$k_a/k_b = 75$ [6])	
			See also 4.1.2.3, Fig. 4, p. 255.	
$\dot{C}H_2CH_2C(CH_2CH{=}CH_2)_2 + (n\text{-}C_4H_9)_3SnH \xrightarrow{a} CH_3CH_2C(CH_2CH{=}CH_2)_2 + (n\text{-}C_4H_9)_3\dot{S}n$				
$\dot{C}H_2CH_2C(CH_2CH{=}CH_2)_2$ 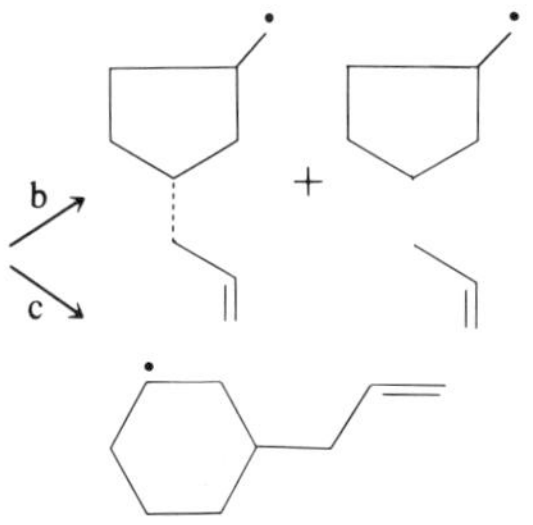				
Decomp. of $BrCH_2CH_2C(CH_2CH{=}CH_2)_2$ (AIBN initiated)				75 Bec 1
PR, glc, NMR	$n\text{-}C_5H_{12}$	303	$k_a/k_b = 0.73\ M^{-1}$	
		333	$k_a/k_b = 0.45\ M^{-1}$	
			$k_a/k_c \approx 20\ M^{-1}$	
		357	$k_a/k_b = 0.33\ M^{-1}$	
			$\Delta H_a^\ddagger - \Delta H_b^\ddagger = -13\ \text{kJ mol}^{-1}$	
			$\Delta S_a^\ddagger - \Delta S_b^\ddagger = -46\ \text{J mol}^{-1}\ \text{K}^{-1}$	
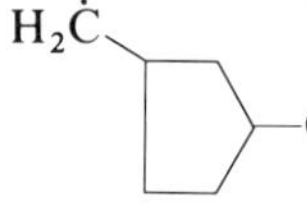—$CH_2CH{=}CH_2 + (n\text{-}C_4H_9)_3SnH \xrightarrow{a}$ 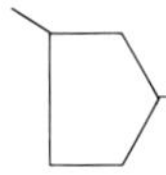—$CH_2CH{=}CH_2 + (n\text{-}C_4H_9)_3\dot{S}n$				
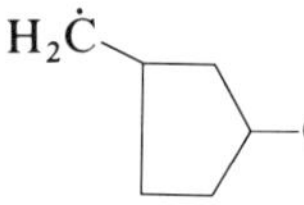—$CH_2CH{=}CH_2$ 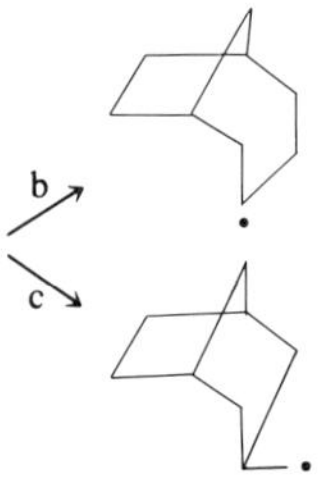				
Decomp. of $BrCH_2CH_2C(CH_2CH{=}CH_2)_2$ (AIBN initiated) and radical rearrangement				75 Bec 1
PR, glc, NMR	$n\text{-}C_5H_{12}$	338	$k_a/k_b = 1.67 \cdot 10^3\ M^{-1}$	
			$k_a/k_c = 1.25 \cdot 10^3\ M^{-1}$	
		357	$k_a/k_b = 1.00 \cdot 10^3\ M^{-1}$	
			$k_a/k_c = 0.91 \cdot 10^3\ M^{-1}$	
		373	$k_a/k_b = 1.00 \cdot 10^3\ M^{-1}$	
			$k_a/k_c = 0.83 \cdot 10^3\ M^{-1}$	

[6]) From [75 Her 1].

Reaction				
Radical generation				Ref./
Method	Solvent	$T[K]$	Rate data	add. ref.

$(CH_2)_3\dot{C}H_2$ ... $(CH_2)_3CH_3$

$+ (n\text{-}C_4H_9)_3SnH \xrightarrow{a}$ ... $+ (n\text{-}C_4H_9)_3\dot{S}n$

$(CH_2)_3\dot{C}H_2$

$\xrightarrow{b}$

$\xrightarrow{c}$

$(CH_2)_3CH_2Br$

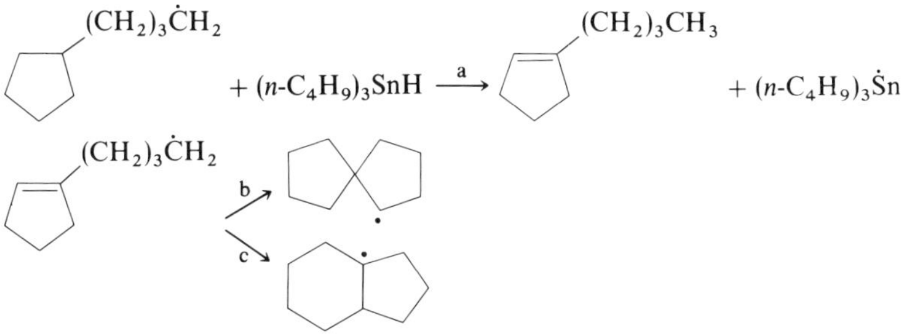

$(n\text{-}C_4H_9)_3\dot{S}n +$ ... reaction   74 Bec 2

PR, glc C_6H_6 338 $k_a/k_b = 40\,M^{-1}$
                                            $k_a/k_c = 167\,M^{-1}$

CH_3 CH_3

$+ LiCl \begin{cases} \xrightarrow{a} exo\text{-}2\text{-chloroapobornane} + Li^{\cdot} \\ \xrightarrow{b} endo\text{-}2\text{-chloroapobornane} + Li^{\cdot} \end{cases}$

$\dot{C}$

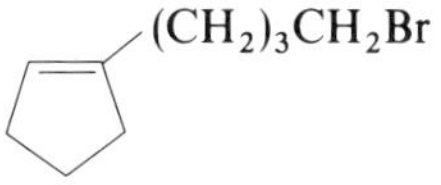

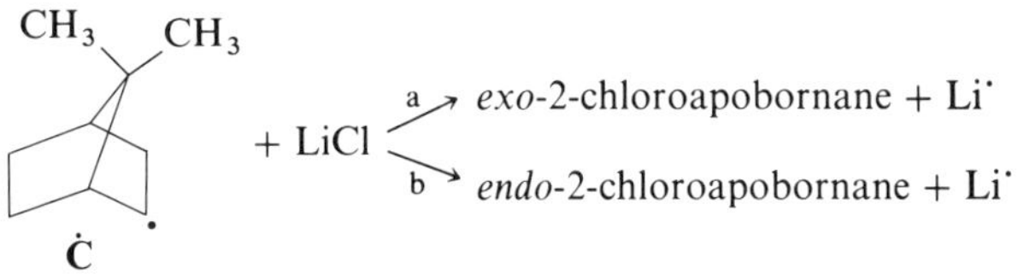

Therm. of $Pb(OOCCH_3)_4$ in presence of $endo$-apobornylcarboxylic acid   70 Bar 1
PR   not specified   353 $k_a/k_b = 2.3$

$\dot{C} + CCl_4 \begin{cases} \xrightarrow{a} exo\text{-}2\text{-chloroapobornane} + \dot{C}Cl_3 \\ \xrightarrow{b} endo\text{-}2\text{-chloroapobornane} + \dot{C}Cl_3 \end{cases}$

Therm. and phot. of apobornylperester   70 Bar 1
PR CCl_4 353 $k_a/k_b = 2.6$
                                 303         $k_a/k_b = 2.6$[7])

$\dot{C} + CHCl_3 \begin{cases} \xrightarrow{a} exo\text{-}2\text{-chloroapobornane} + \dot{C}HCl_2 \\ \xrightarrow{b} endo\text{-}2\text{-chloroapobornane} + \dot{C}HCl_2 \end{cases}$

Initiated by therm. or phot. of apobornylperester   70 Bar 1
PR $CHCl_3$ 353 $k_a/k_b = 1.8$
                                 303         $k_a/k_b = 1.7$[7])

$\dot{C} + CCl_3COONa \begin{cases} \xrightarrow{a} exo\text{-}2\text{-chloroapobornane} + \dot{C}Cl_2COONa \\ \xrightarrow{b} endo\text{-}2\text{-chloroapobornane} + \dot{C}Cl_2COONa \end{cases}$

Therm. of apobornylperester   70 Bar 1
PR CH_3COOH 353 $k_a/k_b = 2.5$

---

[7]) Generation by phot.

Reaction Radical generation Method	Solvent	T[K]	Rate data	Ref./ add. ref.
$\dot{C} + c\text{-}C_3Cl_4$ [8]) → a: *exo*-2-chloroapobornane + $(c\text{-}C_3\dot{Cl}_3)$; b: *endo*-2-chloroapobornane + $(c\text{-}C_3\dot{Cl}_3)$ *)				
Initiated by therm. or phot. of apobornylperester				70 Bar 1
PR	tetrachloro- cyclopropene	353 303	$k_a/k_b = 1.2$ $k_a/k_b = 1.3$ [7])	
$\dot{C} + CH_3COCH_2Cl$ → a: *exo*-2-chloroapobornane + $CH_3CO\dot{C}H_2$; b: *endo*-2-chloroapobornane + $CH_3CO\dot{C}H_2$				
Phot. of apobornylperester				70 Bar 1
PR	CH_3COCH_2Cl	303	$k_a/k_b = 1.8$	
$\dot{C} + (CH_3)_3SiCl$ → a: *exo*-2-chloroapobornane + $(CH_3)_3\dot{Si}$; b: *endo*-2-chloroapobornane + $(CH_3)_3\dot{Si}$				
Phot. of apobornylperester				70 Bar 1
PR	$(CH_3)_3SiCl$	303	$k_a/k_b = 1.9$	
$\dot{C}$ + 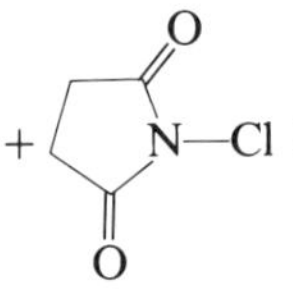 → a: *exo*-2-chloroapobornane + N-chlorosuccinimide($-\dot{H}$); b: *endo*-2-chloroapobornane + N-chlorosuccinimide($-\dot{H}$)				
Therm. of apobornylperester				70 Bar 1
PR	C_6H_6	353	$k_a/k_b = 1.8$	
$\dot{C} + (CH_3)_3CCl$ → a: *exo*-2-chloroapobornane + $(CH_3)_3\dot{C}$; b: *endo*-2-chloroapobornane + $(CH_3)_3\dot{C}$				
Initiated by therm. or phot. of apobornylperester				70 Bar 1
PR	$(CH_3)_3CCl$	353 303	$k_a/k_b = 1.0$ $k_a/k_b = 0.9$ [7])	
$\dot{C} + c\text{-}C_5Cl_6$ [9]) → a: *exo*-2-chloroapobornane + $(c\text{-}C_5\dot{Cl}_5)$; b: *endo*-2-chloroapobornane + $(c\text{-}C_5\dot{Cl}_5)$				
Therm. and phot. of apobornylperester				70 Bar 1
PR	hexachloro- cyclopentadiene	353 303	$k_a/k_b = 3.3$ $k_a/k_b = 3.5$ [7])	
$\dot{C} + 2,4\text{-}(NO_2)_2C_6H_3SCl$ → a: *exo*-2-chloroapobornane + $2,4\text{-}(NO_2)_2C_6H_3\dot{S}$; b: *endo*-2-chloroapobornane + $2,4\text{-}(NO_2)_2C_6H_3\dot{S}$				
Therm. of apobornylperester				70 Bar 1
PR	C_6H_6	353	$k_a/k_b = 3.4$	

*) For $\dot{C}$, see p. 133.
[7]) Generation by phot.
[8]) Tetrachlorocyclopropene.
[9]) Hexachlorocyclopentadiene.

Reaction Radical generation Method	Solvent	T[K]	Rate data	Ref./ add. ref.
$\dot{C} + C_6H_5CH_2Cl$ $\xrightarrow{a}$ *exo*-2-chloroapobornane + $C_6H_5\dot{C}H_2$; $\xrightarrow{b}$ *endo*-2-chloroapobornane + $C_6H_5\dot{C}H_2$ *)				
Initiated by therm. or phot. of apobornylperester				70 Bar 1
PR	$C_6H_5CH_2Cl$	353	$k_a/k_b = 0.6$	
		303	$k_a/k_b = 0.6$ [7])	
$\dot{C}$ + *endo*-apobornyldimethylcarbinyl hypochlorite $\xrightarrow{a}$ *exo*-2-chloroapobornane + *endo*...(−$\dot{H}$); $\xrightarrow{b}$ *endo*-2-chloroapobornane + *endo*...(−$\dot{H}$)				
Phot. of apobornylperester				70 Bar 1
PR	*endo*-apobornyl-dimethylcarbinyl hypochlorite +0.5 M CCl_4	303	$k_a/k_b = 2.2$	
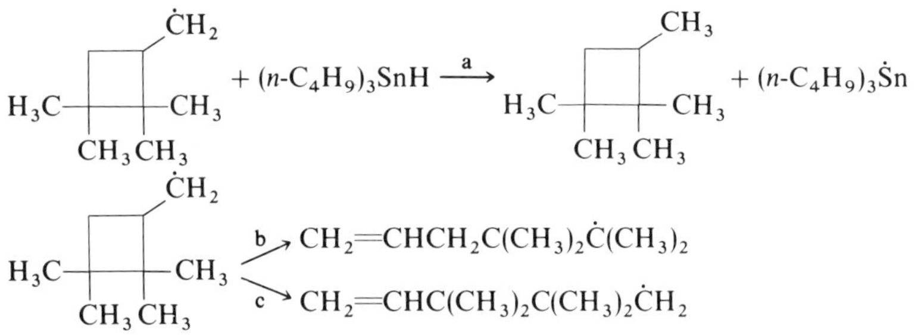				
$(n\text{-}C_4H_9)_3\dot{S}n$ + 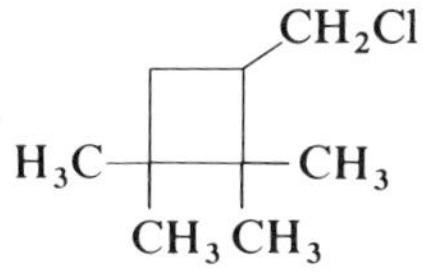 reaction (AIBN initiated)				80 Bec 1
PR, glc	decalin	333	$k_a/k_b = 3.22(10)\ M^{-1}$ $k_a/k_c > 10^3\ M^{-1}$	
$(c\text{-}C_9\dot{H}_{17}) + CCl_3Br \xrightarrow{a} c\text{-}C_9H_{17}Br + \dot{C}Cl_3$ $+ CCl_4 \xrightarrow{b} c\text{-}C_9H_{17}Cl + \dot{C}Cl_3$				
Therm. of $c\text{-}C_9H_{17}COOOC(CH_3)_3$				75 Her 1/
PR, glc	CCl_4	383	$k_a/k_b = 442$	76 Gie 1, 76 Gie 2, 76 Gie 3

*) For $\dot{C}$, see p. 133.
[7]) Generation by phot.

Reaction Radical generation Method	Solvent	T[K]	Rate data	Ref./ add. ref.

$CH_2{=}C(CH_3)CH_2CH_2C(CH_3)_2\dot{C}H_2 + (n\text{-}C_4H_9)_3SnH \xrightarrow{a} CH_2{=}C(CH_3)CH_2CH_2C(CH_3)_2CH_3 + (n\text{-}C_4H_9)_3\dot{S}n$

$CH_2{=}C(CH_3)CH_2CH_2C(CH_3)_2\dot{C}H_2$

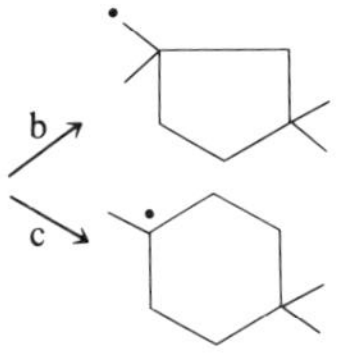

Method	Solvent	T[K]	Rate data	Ref./ add. ref.
Decomp. of $CH_2{=}C(CH_3)CH_2CH_2C(CH_3)_2CH_2Br$ (AIBN initiated)				79 Bec 1/
PR, glc, NMR	C_6H_6	318	$k_a/k_b = 23.8\,M^{-1}$	75 Bec 1
			$k_a/k_c = 55.6\,M^{-1}$	
		353	$k_a/k_b = 15.6\,M^{-1}$	
			$k_a/k_c = 33.3\,M^{-1}$	
		373	$k_a/k_b = 11.8\,M^{-1}$	
			$k_a/k_c = 25.0\,M^{-1}$	
			$\Delta H_a^\ddagger - \Delta H_b^\ddagger = -12.6(4)\,kJ\,mol^{-1}$	
			$\Delta S_a^\ddagger - \Delta S_b^\ddagger = -12.6(13)\,J\,mol^{-1}\,K^{-1}$	
			$\Delta H_a^\ddagger - \Delta H_c^\ddagger = -13.8(5)\,kJ\,mol^{-1}$	
			$\Delta S_a^\ddagger - \Delta S_c^\ddagger = -10.0(8)\,J\,mol^{-1}\,K^{-1}$	

$\dot{C}H_2(CH_2)_3C\{CH(CH_3)_2\}{=}CH_2 + (n\text{-}C_4H_9)_3SnH \xrightarrow{a} CH_3(CH_2)_3C\{CH(CH_3)_2\}{=}CH_2 + (n\text{-}C_4H_9)_3\dot{S}n$

$\dot{C}H_2(CH_2)_3C\{CH(CH_3)_2\}{=}CH_2$

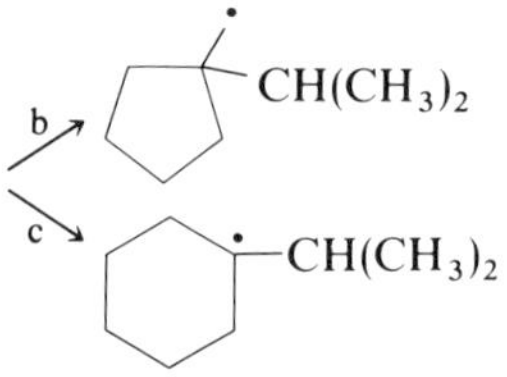

Method	Solvent	T[K]	Rate data	Ref./ add. ref.
$(n\text{-}C_4H_9)_3\dot{S}n + CH_2Br(CH_2)_3C(CH(CH_3)_2){=}CH_2$				74 Bec 2
PR, glc	C_6H_6	338	$k_a/k_b = 200\,M^{-1}$	
			$k_a/k_b = 62.5\,M^{-1}$	

$CH_2{=}C(CH_3)(CH_2)_3\dot{C}(CH_3)_2 + (n\text{-}C_4H_9)_3SnH \xrightarrow{a} CH_2{=}C(CH_3)(CH_2)_3CH(CH_3)_2 + (n\text{-}C_4H_9)_3\dot{S}n$

$CH_2{=}C(CH_3)(CH_2)_3\dot{C}(CH_3)_2$

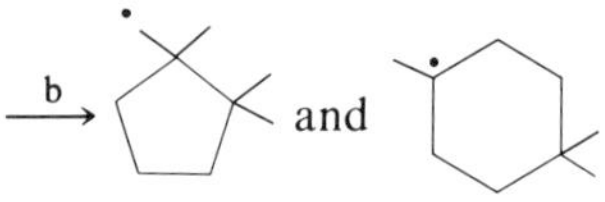

Method	Solvent	T[K]	Rate data	Ref./ add. ref.
From $CH_2{=}C(CH_3)(CH_2)_3CBr(CH_3)_2$ with AIBN (313 K, 343 K) and DTBP (373 K) as initiator				72 Wal 1
PR, glc	C_6H_6	313	$k_a/k_b = 6.25(39)\,M^{-1}$	
		343	3.64(30)	
		373	1.74(40)	

$CH_2{=}C(CH_3)(CH_2)_3\dot{C}(CH_3)_2 + (n\text{-}C_4H_9)_3SnH \xrightarrow{a} CH_2{=}C(CH_3)(CH_2)_3CH(CH_3)_2 + (n\text{-}C_4H_9)_3\dot{S}n$

$CH_2{=}C(CH_3)(CH_2)_3\dot{C}(CH_3)_2$

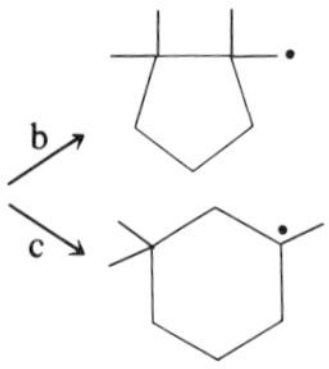

Method	Solvent	T[K]	Rate data	Ref./ add. ref.
$(n\text{-}C_4H_9)_3\dot{S}n + (CH_3)_2CBr(CH_2)_3C(CH_3){=}CH_2$				74 Bec 2
PR, glc	C_6H_6	338	$k_a/k_b > 2\cdot10^4\,M^{-1}$	
			$k_a/k_c = 200\,M^{-1}$	

Reaction Radical generation Method	Solvent	T[K]	Rate data	Ref./ add. ref.
$CH_2{=}C(CH_3)(CH_2)_3\dot{C}(CH_3)_2 + (C_6H_5)_3SnH \xrightarrow{a} CH_2{=}C(CH_3)(CH_2)_3CH(CH_3)_2 + (C_6H_5)_3\dot{S}n$				
$CH_2{=}C(CH_3)(CH_2)_3\dot{C}(CH_3)_2 \xrightarrow{b}$  and 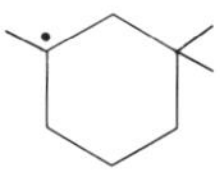				
From $CH_2{=}C(CH_3)(CH_2)_3CBr(CH_3)_2$ with DTBP as initiator				72 Wal 1
PR, glc	C_6H_6	373	$k_a/k_b = 6.41(53)\,M^{-1}$	
$CH_3(CH_2)_2\dot{C}H(CH_2)_3CH{=}CH_2 + (n\text{-}C_4H_9)_3SnH \xrightarrow{a} CH_3(CH_2)_6CH{=}CH_2 + (n\text{-}C_4H_9)_3\dot{S}n$				
$CH_3(CH_2)_2\dot{C}H(CH_2)_3CH{=}CH_2 \xrightarrow{b}$ 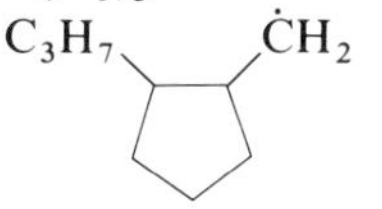				
React. $(n\text{-}C_4H_9)_3\dot{S}n + CH_3(CH_2)_2CHCl(CH_2)_3CH{=}CH_2$				74 Bec 3
PR, glc	pentane	338	$k_a/k_b = 3.33\,M^{-1}$	
$CH_3(CH_2)_2\dot{C}H(CH_2)_3CH{=}CH_2 + (n\text{-}C_4H_9)_3SnH \xrightarrow{a} CH_3(CH_2)_6CH{=}CH_2 + (n\text{-}C_4H_9)_3\dot{S}n$				
$CH_3(CH_2)_2\dot{C}H(CH_2)_3CH{=}CH_2$ $\xrightarrow{b}$ 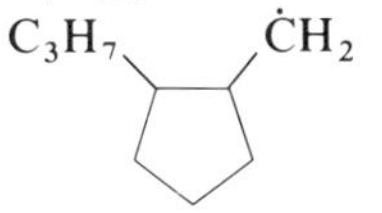 $\xrightarrow{c}$ 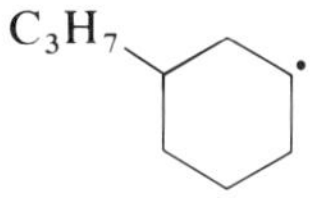				
$(n\text{-}C_4H_9)_3\dot{S}n + CH_3CH_2CH_2CHBr(CH_2)_3CH{=}CH_2$				74 Bec 2/
PR, glc	C_6H_6	338	$k_a/k_b = 3.45\,M^{-1}$ $k_a/k_c = 714\,M^{-1}$	74 Bec 3
$CH_3(CH_2)_7\dot{C}H_2 + CH_2ClCOOH \xrightarrow{a} n\text{-}C_9H_{20} + \dot{C}HClCOOH$ $+ n\text{-}C_7H_{15}I \xrightarrow{b} n\text{-}C_9H_{19}I + (n\text{-}C_7\dot{H}_{15})$				
Not given, see further references				71 Afa 2/
PR, glc	CH_3COOH	373	$k_a/k_b = 0.0123(21)$	68 Saf 1, 68 Saf 2, 70 Saf 1
$CH_3(CH_2)_7\dot{C}H_2 + CH_3CN \xrightarrow{a} n\text{-}C_9H_{20} + \dot{C}H_2CN$ $+ CH_3(CH_2)_5CH_2I \xrightarrow{b} CH_3(CH_2)_7CH_2I + CH_3(CH_2)_5\dot{C}H_2$				
AIBN or BPO as initiator				70 Saf 1
PR, glc	$CH_3CN/n\text{-}C_7H_{15}I/$ $CH_2{=}CH_2$ mixt.	373	$k_a/k_b = 2.86(77)\cdot 10^{-3}$	
$CH_3(CH_2)_7\dot{C}H_2 + CH_3COOH \xrightarrow{a} n\text{-}C_9H_{20} + \dot{C}H_2COOH$ $+ CH_3(CH_2)_5CH_2I \xrightarrow{b} CH_3(CH_2)_7CH_2I + CH_3(CH_2)_5\dot{C}H_2$				
BPO or AIBN as initiator				70 Saf 1
PR, glc	$CH_3COOH/n\text{-}C_7H_{15}I/$ $CH_2{=}CH_2$ mixt.	373	$k_a/k_b = 3.76(49)\cdot 10^{-4}$	
$CH_3(CH_2)_7\dot{C}H_2 + C_2H_5Br \xrightarrow{a} CH_3(CH_2)_7CH_2Br + CH_3\dot{C}H_2$ $+ CH_3(CH_2)_5CH_2I \xrightarrow{b} CH_3(CH_2)_7CH_2I + CH_3(CH_2)_5\dot{C}H_2$				
AIBN or BPO as initiator				70 Afa 1
PR, glc	$C_2H_5Br/n\text{-}C_7H_{15}I/$ $CH_2{=}CH_2$	373	$k_a/k_b = 9.3(16)\cdot 10^{-4}$ $k_a = 17\,M^{-1}s^{-1}$ [10])	

[10]) Not directly measured, based on various assumptions.

Reaction Radical generation Method	Solvent	T[K]	Rate data	Ref./ add. ref.
$CH_3(CH_2)_7\dot{C}H_2 + CH_2(COOH)_2 \xrightarrow{a} n\text{-}C_9H_{20} + HOOC\dot{C}HCH_2COOH$ $+ n\text{-}C_7H_{15}I \xrightarrow{b} n\text{-}C_9H_{19}I + (n\text{-}C_7\dot{H}_{15})$				
Not given, see further references				71 Afa 2/
PR, glc	CH_3COOH	373	$k_a/k_b = 0.0099(37)$	68 Saf 1, 68 Saf 2, 70 Saf 1
$CH_3(CH_2)_7\dot{C}H_2 + CH_3COCH_3 \xrightarrow{a} n\text{-}C_9H_{20} + \dot{C}H_2COCH_3$ $+ CH_3(CH_2)_5CH_2I \xrightarrow{b} CH_3(CH_2)_5CH_2I + CH_3(CH_2)_5\dot{C}H_2$				
AIBN or BPO as initiator				70 Saf 1
PR, glc	CH_3COCH_3/ $n\text{-}C_7H_{15}I/CH_2{=}CH_2$ mixt.	373	$k_a/k_b = 2.98(37) \cdot 10^{-3}$	
$CH_3(CH_2)_7\dot{C}H_2 + CH_3CH_2CH_2Cl \xrightarrow{a} CH_3(CH_2)_7CH_2Cl + CH_3CH_2\dot{C}H_2$ $+ CH_3(CH_2)_5CH_2I \xrightarrow{b} CH_3(CH_2)_7CH_2I + CH_3(CH_2)_5\dot{C}H_2$				
AIBN or BPO as initiator				70 Afa 1
PR, glc	$n\text{-}C_3H_7Cl/n\text{-}C_7H_{15}I$/ $CH_2{=}CH_2$	373	$k_a/k_b = 1.7(1) \cdot 10^{-3}$ $k_a = 31\ M^{-1}s^{-1}$ [10]	
$CH_3(CH_2)_7\dot{C}H_2 + CH_3(CH_2)_2CH_2Br \xrightarrow{a} CH_3(CH_2)_7CH_2Br + CH_3(CH_2)_2\dot{C}H_2$ $+ CH_3(CH_2)_5CH_2I \xrightarrow{b} CH_3(CH_2)_7CH_2I + CH_3(CH_2)_5\dot{C}H_2$				
AIBN or BPO as initiator				70 Afa 1,
PR, glc	$n\text{-}C_4H_9Br/n\text{-}C_7H_{15}I$/ $CH_2{=}CH_2$	373	$k_a/k_b = 2.2(4) \cdot 10^{-3}$ $= 1.80(23) \cdot 10^{-3}$ [11] $k_a = 40\ M^{-1}s^{-1}$ [10]	68 Saf 2
$CH_3(CH_2)_7\dot{C}H_2 + C_6H_5CH_3 \xrightarrow{a} n\text{-}C_9H_{20} + C_6H_5\dot{C}H_2$ $+ CH_3(CH_2)_5CH_2I \xrightarrow{b} CH_3(CH_2)_7CH_2I + CH_3(CH_2)_5\dot{C}H_2$				
AIBN or BPO as initiator				70 Saf 1
PR, glc	$C_6H_5CH_3/n\text{-}C_7H_{15}I$/ $CH_2{=}CH_2$ mixt.	373	$k_a/k_b = 2.56(25) \cdot 10^{-3}$	
$CH_3(CH_2)_7\dot{C}H_2 + CH_3(CH_2)_5CH_2Br \xrightarrow{a} CH_3(CH_2)_7CH_2Br + CH_3(CH_2)_5\dot{C}H_2$ $+ CH_3(CH_2)_5CH_2I \xrightarrow{b} CH_3(CH_2)_7CH_2I + CH_3(CH_2)_5\dot{C}H_2$				
AIBN or BPO as initiator				70 Afa 1,
PR, glc	$n\text{-}C_7H_{15}Br/n\text{-}C_7H_{15}I$/ $CH_2{=}CH_2$	373	$k_a/k_b = 6.8(13) \cdot 10^{-3}$ $= 2.10(46) \cdot 10^{-3}$ [11] $k_a = 122\ M^{-1}s^{-1}$ [10]	68 Saf 2
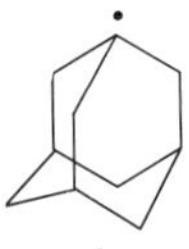 $+ Br_2 \xrightarrow{a}$ 1-bromoadamantane + $\dot{B}r$ $+ CCl_4 \xrightarrow{b}$ 1-chloroadamantane + $\dot{C}Cl_3$ $\dot{D}$				
BPO as initiator				72 Tab 1
PR, glc	CCl_4	368	$k_a/k_b = 0.595$	

[10]) Not directly measured, based on various assumptions.
[11]) From [68 Saf 2].

Reaction Radical generation Method	Solvent	T[K]	Rate data	Ref./ add. ref.
$\dot{D} + CBrCl_3 \xrightarrow{a}$ 1-Br-adamantane + $\dot{C}Cl_3$ *) $+ CCl_4 \xrightarrow{b}$ 1-Cl-adamantane + $\dot{C}Cl_3$				
Therm. of 1-adamantylperester				80 Gie 1
PR, glc	CCl_4	323	$k_a/k_b = 29$	
		333	28	
		343	25	
		353	24	
			$\Delta H^{\ddagger}_a - \Delta H^{\ddagger}_b = -5.5\,kJ\,mol^{-1}$	
			$\Delta S^{\ddagger}_a - \Delta S^{\ddagger}_b = 11\,J\,mol^{-1}\,K^{-1}$	
Therm. of 1-adamantyl-$COOOC(CH_3)_3$				75 Her 1,
PR, glc	CCl_4	363	$k_a/k_b = 24$	78 Luh 1,
		353	29	81 Luh 1/
			30 [12])	76 Gie 1,
		335	25	76 Gie 2,
			See also 4.1.2.3, Fig. 4, p. 255.	76 Gie 3,
				69 Rue 1
$\dot{D} + CH_3COCOCH_3 \xrightarrow{a}$ 1-CH_3CO-adamantane + $CH_3\dot{C}O$ $\dot{D} + CH_3COCOCH_3 \xrightarrow{b}$ adamantane + $\dot{C}H_2COCOCH_3$				
Thermal decomp. of *t*-butyl-1-peroxyadamantanecarboxylate				78 Tab 1
PR, glc, NMR	$CH_3COCOCH_3$	353	$k_a/k_b = 5.7.$	
$\dot{D} + C_6H_5Br \xrightarrow{a}$ adamantylbromobenzene + $\dot{H}$ [13]) $+ C_6H_6 \xrightarrow{b}$ adamantylbenzene + $\dot{H}$ [13])				
Therm. of 1-adamantyl-$COOOC(CH_3)_3$				76 Tes 1
PR, glc	C_6H_5Br/C_6H_6	353	$k_a/k_b = 1.99$ [14])	
$\dot{D} + C_6H_5Cl \xrightarrow{a}$ adamantylchlorobenzene + $\dot{H}$ [13]) $+ C_6H_6 \xrightarrow{b}$ adamantylbenzene + $\dot{H}$ [13])				
Therm. of 1-adamantyl-$COOOC(CH_3)_3$				76 Tes 1
PR, glc	C_6H_5Cl/C_6H_6	353	$k_a/k_b = 1.56$ [14])	
$\dot{D} + C_6H_5F \xrightarrow{a}$ adamantylfluorobenzene + $\dot{H}$ [13]) $+ C_6H_6 \xrightarrow{b}$ adamantylbenzene + $\dot{H}$ [13])				
Therm. of 1-adamantyl-$COOOC(CH_3)_3$				76 Tes 1
PR, glc	C_6H_5F/C_6H_6	353	$k_a/k_b = 1.36$ [14])	
$\dot{D} + C_6H_5CN \xrightarrow{a}$ adamantylcyanobenzene + $\dot{H}$ [13]) $+ C_6H_6 \xrightarrow{b}$ adamantylbenzene + $\dot{H}$ [13])				
Therm. of 1-adamantyl-$COOOC(CH_3)_3$				76 Tes 1
PR, glc	C_6H_5CN/C_6H_6	353	$k_a/k_b = 21.1$ [14])	
$\dot{D} + C_6H_5CH_3 \xrightarrow{a}$ adamantyltoluene + $\dot{H}$ [13]) $+ C_6H_6 \xrightarrow{b}$ adamantylbenzene + $\dot{H}$ [13])				
Therm. of 1-adamantyl-$COOOC(CH_3)_3$				76 Tes 1
PR, glc	$C_6H_5CH_3/C_6H_6$	353	$k_a/k_b = 0.55$ [14])	

*) For $\dot{D}$, see p. 138.
[12]) From [78 Luh 1, 81 Luh 1].
[13]) Reaction assumed to proceed via intermediate adduct.
[14]) Corrected for statistical factor.

Reaction Radical generation Method	Solvent	T[K]	Rate data	Ref./ add. ref.
$\dot{D} + C_6H_5OCH_3 \xrightarrow{a}$ adamantylmethoxybenzene + $\dot{H}$ [13]) *) $+ C_6H_6 \xrightarrow{b}$ adamantylbenzene + $\dot{H}$ [13])				
Therm. of 1-adamantyl-$COOOC(CH_3)_3$				76 Tes 1
PR, glc	$C_6H_5OCH_3/C_6H_6$	353	$k_a/k_b = 0.65$ [14])	
$\dot{D} + C_6H_5COOCH_3 \xrightarrow{a}$ adamantylcarbomethoxybenzene + $\dot{H}$ [13]) $+ C_6H_6 \xrightarrow{b}$ adamantylbenzene + $\dot{H}$ [13])				
Therm. of 1-adamantyl-$COOOC(CH_3)_3$				76 Tes 1
PR, glc	$C_6H_5COOCH_3/C_6H_6$	353	$k_a/k_b = 10.9$ [14])	
$\dot{D} + C_6H_5CH_2CH_3 \xrightarrow{a}$ adamantylethylbenzene + $\dot{H}$ [13]) $+ C_6H_6 \xrightarrow{b}$ adamantylbenzene + $\dot{H}$ [13])				
Therm. of 1-adamantyl-$COOOC(CH_3)_3$				76 Tes 1
PR, glc	$C_6H_5CH_2CH_3/C_6H_6$	353	$k_a/k_b = 0.48$ [14])	
$\dot{D} + C_6H_5C(CH_3)_3 \xrightarrow{a}$ adamantyl-*t*-butylbenzene + $\dot{H}$ [13]) $+ C_6H_6 \xrightarrow{b}$ adamantylbenzene + $\dot{H}$ [13])				
Therm. of 1-adamantyl-$COOOC(CH_3)_3$				76 Tes 1
PR, glc	$C_6H_5C(CH_3)_3/C_6H_6$	353	$k_a/k_b = 0.31$ [14])	
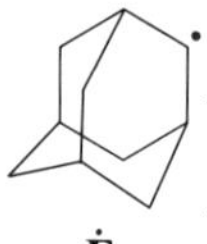 $+ Br_2 \xrightarrow{a}$ 2-bromoadamantane + $\dot{Br}$ $+ CCl_4 \xrightarrow{b}$ 2-chloroadamantane + $\dot{C}Cl_3$ $\dot{E}$				
BPO as initiator				72 Tab 1
PR, glc	CCl_4	368	$k_a/k_b = 9.09$	
$\dot{E} + CH_3COCOCH_3 \xrightarrow{a}$ 2-CH_3CO-adamantane + $CH_3\dot{C}O$ $\xrightarrow{b}$ adamantane + $\dot{C}H_2COCOCH_3$				
Thermal decomp. of *t*-butyl-2-adamantaneperoxycarboxylate				78 Tab 1
PR, glc, NMR	$CH_3COCOCH_3$	353	$k_a/k_b = 0.85$	
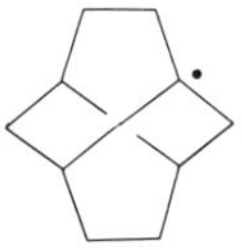 $+ CCl_3Br \xrightarrow{a}$ 1-Br-twistane + $\dot{C}Cl_3$ $+ CCl_4 \xrightarrow{b}$ 1-Cl-twistane + $\dot{C}Cl_3$				
Therm. of 1-twistyl-$COOOC(CH_3)_3$				75 Her 1/
PR, glc	CCl_4	353	$k_a/k_b = 76$	76 Gie 1, 76 Gie 2, 76 Gie 3

*) For $\dot{D}$, see p. 138.

13) Reaction assumed to proceed via intermediate adduct.

14) Corrected for statistical factor.

Reaction Radical generation				Ref./
Method	Solvent	T [K]	Rate data	add. ref.

$\dot{C}(CH_2)_3CH=CH_2 + (n\text{-}C_4H_9)_3SnH \xrightarrow{a} CH(CH_2)_3CH=CH_2 + (n\text{-}C_4H_9)_3\dot{Sn}$

$\dot{C}(CH_2)_3CH=CH_2 \xrightarrow{b} \xrightarrow{c}$ [15])

$(n\text{-}C_4H_9)_3\dot{Sn} + \quad CBr(CH_2)_3CH=CH_2$

| | | | | 74 Bec 2 |
| PR, glc | C_6H_6 | 338 | $k_a/k_b = 3.57\,M^{-1}$ $k_a/k_c = 50\,M^{-1}$ [15]) | |

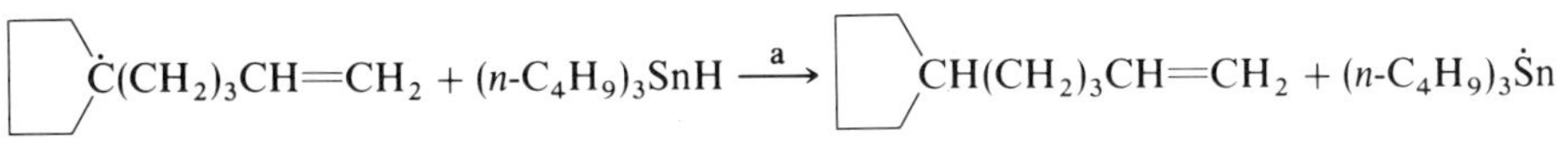

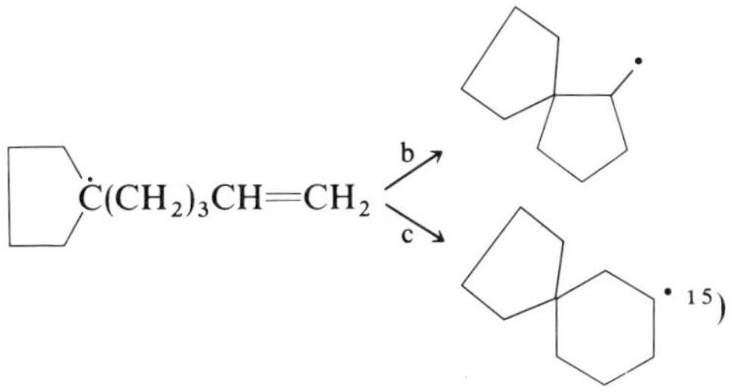

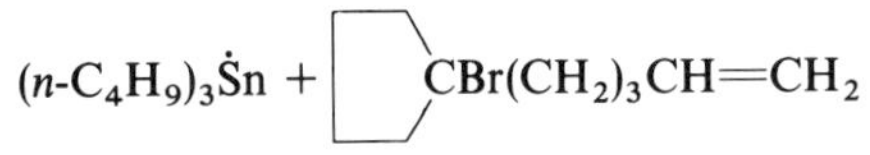

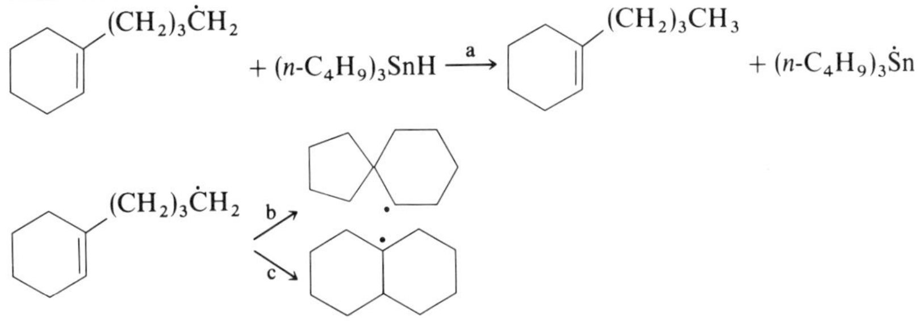

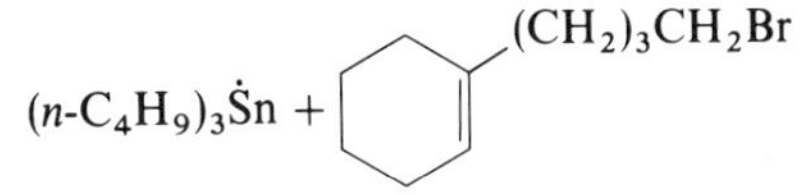

$(CH_2)_3\dot{C}H_2 + (n\text{-}C_4H_9)_3SnH \xrightarrow{a} (CH_2)_3CH_3 + (n\text{-}C_4H_9)_3\dot{Sn}$

$(CH_2)_3\dot{C}H_2 \xrightarrow{b} \xrightarrow{c}$

$(n\text{-}C_4H_9)_3\dot{Sn} + \quad (CH_2)_3CH_2Br \qquad \text{reaction}$

| | | | | 74 Bec 2 |
| PR, glc | C_6H_6 | 338 | $k_a/k_b = 66.7\,M^{-1}$ $k_a/k_c = 83.3\,M^{-1}$ | |

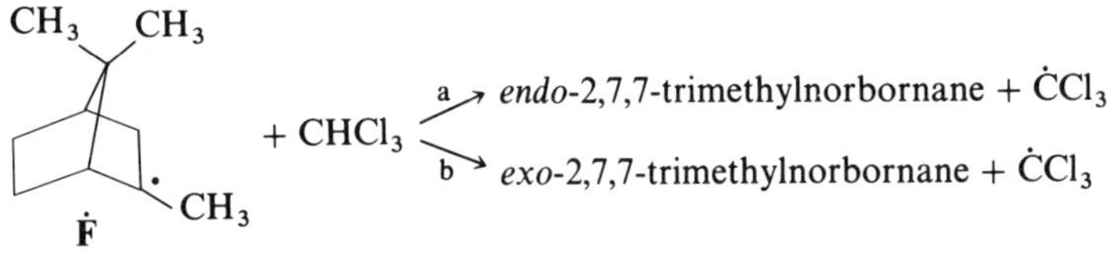

$\dot{F} + CHCl_3 \xrightarrow{a} endo\text{-}2,7,7\text{-trimethylnorbornane} + \dot{C}Cl_3$
$\xrightarrow{b} exo\text{-}2,7,7\text{-trimethylnorbornane} + \dot{C}Cl_3$

| | | | | 70 Bar 1 |
| Therm. of 2,7,7-trimethylnorbornylperester PR | CHCl$_3$ | 353 | $k_a/k_b = 1.0$ | |

$\dot{F} + n\text{-}C_4H_9SH \xrightarrow{a} endo\text{-}2,7,7\text{-trimethylnorbornane} + n\text{-}C_4H_9\dot{S}$
$\xrightarrow{b} exo\text{-}2,7,7\text{-trimethylnorbornane} + n\text{-}C_4H_9\dot{S}$

| | | | | 70 Bar 1 |
| Therm. of 2,7,7-trimethylnorbornylperester PR | $n\text{-}C_4H_9SH$ | 353 | $k_a/k_b = 2.9$ | |

[15]) Mechanism of k_c route not clear, may in part be polar.

Reaction Radical generation Method	Solvent	T[K]	Rate data	Ref./ add. ref.
$\dot{F} + (CH_3)_3CSH \xrightarrow{a}$ *endo*-2,7,7-trimethylnorbornane + $(CH_3)_3\dot{C}$ $\xrightarrow{b}$ *exo*-2,7,7-trimethylnorbornane + $(CH_3)_3\dot{C}$ *)				
Therm. of 2,7,7-trimethylnorbornylperester PR	$(CH_3)_3CSH$	353	$k_a/k_b = 3.1$	70 Bar 1
$\dot{F} + C_6H_5SH \xrightarrow{a}$ *endo*-2,7,7-trimethylnorbornane + $C_6H_5\dot{S}$ $\xrightarrow{b}$ *exo*-2,7,7-trimethylnorbornane + $C_6H_5\dot{S}$				
Therm. of 2,7,7-trimethylnorbornylperester PR	C_6H_5SH	353	$k_a/k_b = 3.0$	70 Bar 1
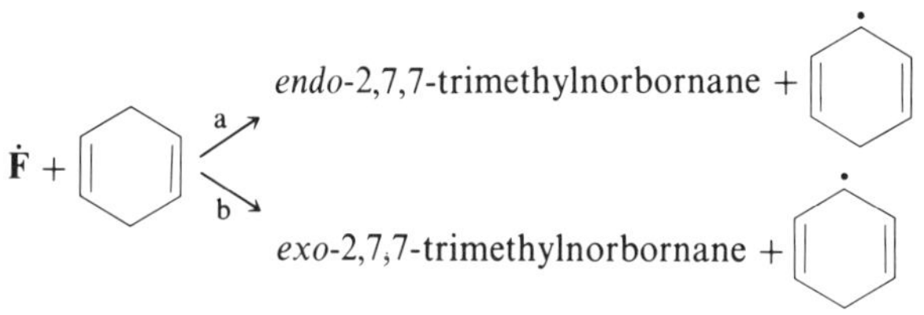				
Therm. of 2,7,7-trimethylnorbornylperester PR	1,4-cyclohexadiene	353	$k_a/k_b = 1.9$	70 Bar 1
$\dot{F} + C_6H_5CH_3 \xrightarrow{a}$ *endo*-2,7,7-trimethylnorbornane + $C_6H_5\dot{C}H_2$ $\xrightarrow{b}$ *exo*-2,7,7-trimethylnorbornane + $C_6H_5\dot{C}H_2$				
Therm. of 2,7,7-trimethylnorbornylperester PR	$C_6H_5CH_3$	353	$k_a/k_b = 1.3$	70 Bar 1
$\dot{F} + C_6H_5C(CH_3)_2 \xrightarrow{a}$ *endo*-2,7,7-trimethylnorbornane + $C_6H_5\dot{C}(CH_3)_2$ $\xrightarrow{b}$ *exo*-2,7,7-trimethylnorbornane + $C_6H_5\dot{C}(CH_3)_2$				
Therm. of 2,7,7-trimethylnorbornylperester PR	cumene	353	$k_a/k_b = 1.5$	70 Bar 1
$\dot{C}H_2(CH_2)_3CH{=}CHCH_2CH_2CH{=}CH_2 + (n\text{-}C_4H_9)_3SnH \xrightarrow{a} CH_3(CH_2)_3CH{=}CHCH_2CH_2CH{=}CH_2 + (n\text{-}C_4H_9)_3\dot{S}n$ $\dot{C}H_2(CH_2)_3CH{=}CHCH_2CH_2CH{=}CH_2$ 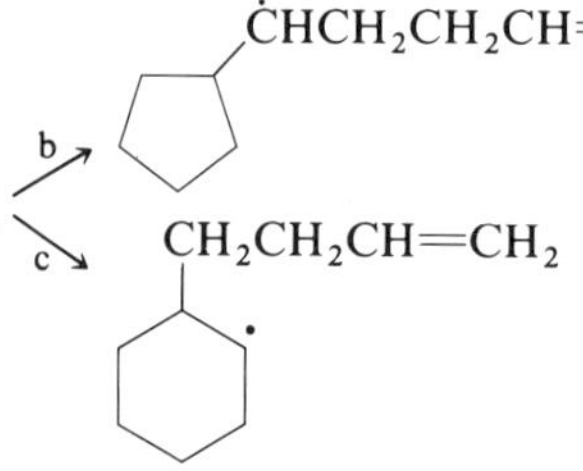				
$(n\text{-}C_4H_9)_3\dot{S}n + CH_2Br(CH_2)_3CH{=}CHCH_2CH_2CH{=}CH_2$ PR, glc	C_6H_6	338	$k_a/k_b = 2.63\ M^{-1}$ [16]) $3.33\ M^{-1}$ [17]) $k_a/k_c > 500\ M^{-1}$ [16]) $> 667\ M^{-1}$ [17])	74 Bec 2/ 73 Bec 1

*) For $\dot{F}$, see p. 141.
[16]) *trans* at $C_5{=}C_6$ double bond.
[17]) *cis* at $C_5{=}C_6$ double bond.

Reaction				Ref./
Radical generation				
Method	Solvent	$T[K]$	Rate data	add. ref.

$(c\text{-}C_{10}H_{19}^{\cdot}) + CCl_3Br \xrightarrow{a} c\text{-}C_{10}H_{19}Br + \dot{C}Cl_3$
$\qquad\qquad + CCl_4 \xrightarrow{b} c\text{-}C_{10}H_{19}Cl + \dot{C}Cl_3$

Therm. of $c\text{-}C_{10}H_{19}COOOC(CH_3)_3$				75 Her 1/
PR, glc	CCl_4	383	$k_a/k_b = 505$	76 Gie 1,
				76 Gie 2,
				76 Gie 3

(3-homoadamantyl)$^{\cdot}$ + $CCl_3Br \xrightarrow{a}$ 3-Br-homoadamantane + $\dot{C}Cl_3$
$\qquad\qquad + CCl_4 \xrightarrow{b}$ 3-Cl-homoadamantane + $\dot{C}Cl_3$

Therm. of 3-homoadamantyl-$COOOC(CH_3)_3$				75 Her 1/
PR, glc	CCl_4	353	$k_a/k_b = 100$	76 Gie 1,
				76 Gie 2,
				76 Gie 3

$\dot{C}H_2(CH_2)_3CH{=}C\langle\hexagon\rangle + (n\text{-}C_4H_9)_3SnH \xrightarrow{a} CH_3(CH_2)_3CH{=}C\langle\hexagon\rangle + (n\text{-}C_4H_9)_3\dot{S}n$

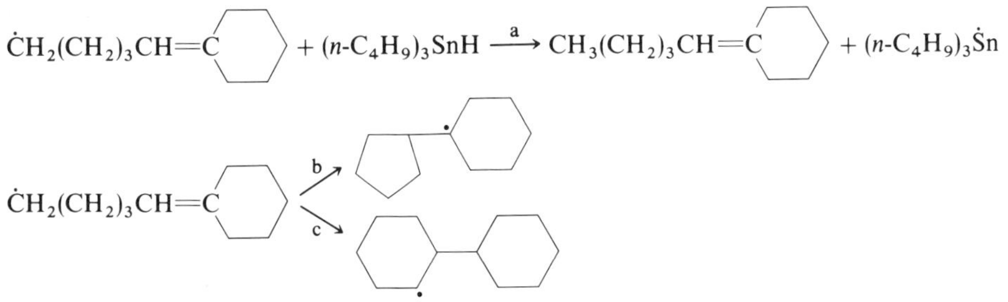

$(n\text{-}C_4H_9)_3\dot{S}n + CH_2Br(CH_2)_3CH{=}C\langle\hexagon\rangle$

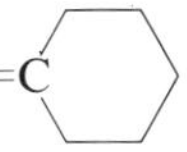

74 Bec 2

| PR, glc | C_6H_6 | 338 | $k_a/k_b = 4.76\,M^{-1}$ | |
| | | | $k_a/k_b > 500\,M^{-1}$ | |

$(c\text{-}C_{11}H_{21}^{\cdot}) + CCl_3Br \xrightarrow{a} c\text{-}C_{11}H_{21}Br + \dot{C}Cl_3$
$\qquad\qquad + CCl_4 \xrightarrow{b} c\text{-}C_{11}H_{21}Cl + \dot{C}Cl_3$

Therm. of $c\text{-}C_{11}H_{21}COOOC(CH_3)_3$				75 Her 1/
PR, glc	CCl_4	383	$k_a/k_b = 568$	76 Gie 1,
				76 Gie 2,
				76 Gie 3

$CH_3(CH_2)_9\dot{C}H_2 + \begin{array}{c}O\\\square\\O\end{array} \xrightarrow{a} n\text{-}C_{11}H_{24} + (C_3H_5O_2^{\cdot})$
$\qquad\qquad + CCl_4 \xrightarrow{b} n\text{-}C_{11}H_{23}Cl + \dot{C}Cl_3$

| Thermal decomp. of lauroylperoxide | | | | 79 Bat 1 |
| PR, glc | /CCl_4 | 353.0(1) | $k_a/k_b = 0.70$ | |

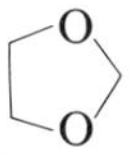

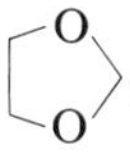

Reaction Radical generation Method	Solvent	T[K]	Rate data	Ref./ add. ref.
$CH_3(CH_2)_9\dot{C}H_2$ + 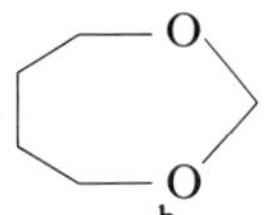 $\xrightarrow{a}$ $n\text{-}C_{11}H_{24} + (C_5H_9\dot{O}_2)$ $+ CCl_4 \xrightarrow{b} n\text{-}C_{11}H_{23}Cl + \dot{C}Cl_3$				
Thermal decomp. of lauroylperoxide				79 Bat 1
PR, glc	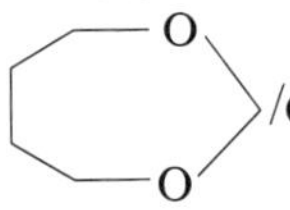 $/CCl_4$	353.0(1)	$k_a/k_b = 0.35$	
$CH_3(CH_2)_9\dot{C}H_2 + CH_3CH_2OCH_2OCH_2CH_3 \xrightarrow{a} n\text{-}C_{11}H_{24} + (C_5H_{11}\dot{O}_2)$ $+ CCl_4 \xrightarrow{b} n\text{-}C_{11}H_{23}Cl + \dot{C}Cl_3$				
Thermal decomp. of lauroylperoxide				79 Bat 1
PR, glc	$CH_3CH_2OCH_2$- OCH_2CH_3/CCl_4	353.0(1)	$k_a/k_b = 0.17$	
$CH_3(CH_2)_9\dot{C}H_2 + C_6H_5Cl \xrightarrow{a}$ products $+ C_6H_5CH_3 \xrightarrow{b}$ products				
Therm. of lauroylperoxide				75 Zav 1
[18])	C_6H_6	353	$k_a/k_b = 0.72(1)$ [19])	
$CH_3(CH_2)_9\dot{C}H_2$ + 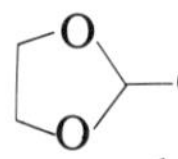 $-C_3H_7 \xrightarrow{a} n\text{-}C_{11}H_{24} + (C_6H_{11}\dot{O}_2)$ $+ CCl_4 \xrightarrow{b} n\text{-}C_{11}H_{23}Cl + \dot{C}Cl_3$				
Thermal decomp. of lauroylperoxide				79 Bat 1
PR, glc	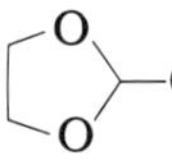 $-C_3H_7/CCl_4$	353.0(1)	$k_a/k_b = 0.98$	
$CH_3(CH_2)_9\dot{C}H_2$ + 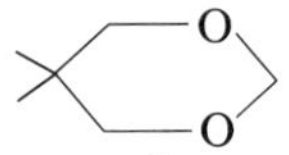 $\xrightarrow{a} n\text{-}C_{11}H_{24} + (C_6H_{11}\dot{O}_2)$ $+ CCl_4 \xrightarrow{b} n\text{-}C_{11}H_{23}Cl + \dot{C}Cl_3$				
Thermal decomp. of lauroylperoxide				79 Bat 1
PR, glc	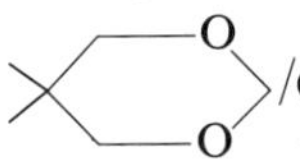 $/CCl_4$	353.0(1)	$k_a/k_b = 0.22$	
$CH_3(CH_2)_9\dot{C}H_2 + CH_3CH_2OCH(CH_3)OCH_2CH_3 \xrightarrow{a} n\text{-}C_{11}H_{24} + (C_6H_{13}\dot{O}_2)$ $+ CCl_4 \xrightarrow{b} n\text{-}C_{11}H_{23}Cl + \dot{C}Cl_3$				
Thermal decomp. of lauroylperoxide				79 Bat 1
PR, glc	$CH_3CH_2OCH(CH_3)$- OCH_2CH_3/CCl_4	353.0(1)	$k_a/k_b = 0.37$	
$CH_3(CH_2)_9\dot{C}H_2 + C_6H_5CN \xrightarrow{a}$ products $+ C_6H_5CH_3 \xrightarrow{b}$ products				
Therm. of lauroylperoxide				75 Zav 1
[18])	C_6H_6	354	$k_a/k_b = 1.78(2)$ [19])	

[18]) Disappearance of C_6H_5Cl and $C_6H_5CH_3$.
[19]) Refers to total reaction (includes possible radical addition to aromatic ring besides H-atom abstraction).

Reaction Radical generation Method	Solvent	T[K]	Rate data	Ref./ add. ref.
$CH_3(CH_2)_9\dot{C}H_2 + 3\text{-}BrC_6H_4CH_3 \xrightarrow{a} n\text{-}C_{11}H_{24} + 3\text{-}BrC_6H_4\dot{C}H_2$ $+ CCl_4 \xrightarrow{b} n\text{-}C_{11}H_{23}Cl + \dot{C}Cl_3$				
Thermal decomp. of *n*-lauroylperoxide PR, glc	3-bromotoluene/ CCl_4 (ratio 0... ≈ 15)	353	$k_a/k_b = 0.017$	74 Hen 1/ 74 Pry 1 [20]
$CH_3(CH_2)_9\dot{C}H_2 + 3\text{-}BrC_6H_4CH_3 \xrightarrow{a} n\text{-}C_{11}H_{24} + 3\text{-}BrC_6H_4\dot{C}H_2$ $+ C_6H_5CH_3 \xrightarrow{b} n\text{-}C_{11}H_{24} + C_6H_5\dot{C}H_2$			See 4.1.2.3, Fig. 5, p. 256.	
$CH_3(CH_2)_9\dot{C}H_2 + 3\text{-}ClC_6H_4CH_3 \xrightarrow{a} n\text{-}C_{11}H_{24} + 3\text{-}ClC_6H_4\dot{C}H_2$ $+ CCl_4 \xrightarrow{b} n\text{-}C_{11}H_{23}Cl + \dot{C}Cl_3$				
Thermal decomp. of *n*-lauroylperoxide PR, glc	3-chlorotoluene/ CCl_4 (ratio 0... ≈ 15)	353	$k_a/k_b = 0.023$	74 Hen 1/ 74 Pry 1 [20]
$CH_3(CH_2)_9\dot{C}H_2 + 3\text{-}ClC_6H_4CH_3 \xrightarrow{a} n\text{-}C_{11}H_{24} + 3\text{-}ClC_6H_4\dot{C}H_2$ $+ C_6H_5CH_3 \xrightarrow{b} n\text{-}C_{11}H_{24} + C_6H_5\dot{C}H_2$				
Therm. of lauroylperoxide [21]	C_6H_6	354	$k_a/k_b = 1.24(3)$ [19] See also 4.1.2.3, Fig. 5, p. 256.	75 Zav 1
$CH_3(CH_2)_9\dot{C}H_2 + 4\text{-}ClC_6H_4CH_3 \xrightarrow{a} n\text{-}C_{11}H_{24} + 4\text{-}ClC_6H_4\dot{C}H_2$ $+ CCl_4 \xrightarrow{b} n\text{-}C_{11}H_{23}Cl + \dot{C}Cl_3$				
Thermal decomp. of *n*-lauroylperoxide PR, glc	4-chlorotoluene/ CCl_4 (ratio 0... ≈ 15)	353	$k_a/k_b = 0.018$	74 Hen 1/ 74 Pry 1 [20]
$CH_3(CH_2)_9\dot{C}H_2 + 4\text{-}ClC_6H_4CH_3 \xrightarrow{a} n\text{-}C_{11}H_{24} + 4\text{-}ClC_6H_4\dot{C}H_2$ $+ C_6H_5CH_3 \xrightarrow{b} n\text{-}C_{11}H_{24} + C_6H_5\dot{C}H_2$				
Therm. of lauroylperoxide [21]	C_6H_6	354	$k_a/k_b = 1.09(1)$ [19] See also 4.1.2.3, Fig. 5, p. 256.	75 Zav 1
$CH_3(CH_2)_9\dot{C}H_2 + 3\text{-}FC_6H_4CH_3 \xrightarrow{a} n\text{-}C_{11}H_{24} + 3\text{-}FC_6H_4\dot{C}H_2$ $+ CCl_4 \xrightarrow{b} n\text{-}C_{11}H_{23}Cl + \dot{C}Cl_3$				
Thermal decomp. of *n*-lauroylperoxide PR, glc	3-fluorotoluene/ CCl_4 (ratio 0... ≈ 15)	353	$k_a/k_b = 0.021$	74 Hen 1/ 74 Pry 1 [20]
$CH_3(CH_2)_9\dot{C}H_2 + 3\text{-}FC_6H_4CH_3 \xrightarrow{a} n\text{-}C_{11}H_{24} + 3\text{-}FC_6H_4\dot{C}H_2$ $+ C_6H_5CH_3 \xrightarrow{b} n\text{-}C_{11}H_{24} + C_6H_5\dot{C}H_2$				
Therm. of lauroylperoxide [21]	C_6H_6	354	$k_a/k_b = 1.16(3)$ [19] See also 4.1.2.3, Fig. 5, p. 256.	75 Zav 1
$CH_3(CH_2)_9\dot{C}H_2 + 4\text{-}FC_6H_4CH_3 \xrightarrow{a} n\text{-}C_{11}H_{24} + 4\text{-}FC_6H_4\dot{C}H_2$ $+ C_6H_5CH_3 \xrightarrow{b} n\text{-}C_{11}H_{24} + C_6H_5\dot{C}H_2$				
Therm. of lauroylperoxide [21]	C_6H_6	354	$k_a/k_b = 1.04(3)$ [19] See also 4.1.2.3, Fig. 5, p. 256.	75 Zav 1

[19]) Refers to total reaction (includes possible radical addition to aromatic ring besides H-atom abstraction).
[20]) Graphic presentation of data in *t*-butylbenzene as solvent.
[21]) Disappearance of toluenes.

Reaction Radical generation Method	Solvent	T[K]	Rate data	Ref./ add. ref.
$CH_3(CH_2)_9\dot{C}H_2 + 3\text{-}NO_2C_6H_4CH_3 \xrightarrow{a} n\text{-}C_{11}H_{24} + 3\text{-}NO_2C_6H_4\dot{C}H_2$ $+ CCl_4 \xrightarrow{b} n\text{-}C_{11}H_{23}Cl + \dot{C}Cl_3$				
Thermal decomp. of *n*-lauroylperoxide PR, glc	3-nitrotoluene/ CCl_4 (ratio 0… ≈ 15)	353	$k_a/k_b = 0.028$	74 Hen 1/ 74 Pry 1 [20]
$CH_3(CH_2)_9\dot{C}H_2 + C_6H_5CH_3 \xrightarrow{a} n\text{-}C_{11}H_{24} + C_6H_5\dot{C}H_2$ $+ CCl_4 \xrightarrow{b} n\text{-}C_{11}H_{23}Cl + \dot{C}Cl_3$				
Thermal decomp. of *n*-lauroylperoxide PR, glc	toluene/CCl_4 (ratio 0… ≈ 15)	353	$k_a/k_b = 0.011$	74 Hen 1/ 74 Pry 1 [20]
$CH_3(CH_2)_9\dot{C}H_2 + C_6H_5OCH_3 \xrightarrow{a}$ products $+ C_6H_5CH_3 \xrightarrow{b}$ products				
Therm. of lauroylperoxide [22]	C_6H_6	354	$k_a/k_b = 0.45(4)$ [19]	75 Zav 1
$CH_3(CH_2)_9\dot{C}H_2 + 3\text{-}CNC_6H_4CH_3 \xrightarrow{a} n\text{-}C_{11}H_{24} + 3\text{-}CNC_6H_4\dot{C}H_2$ $+ CCl_4 \xrightarrow{b} n\text{-}C_{11}H_{23}Cl + \dot{C}Cl_3$				
Thermal decomp. of *n*-lauroylperoxide PR, glc	3-cyanotoluene/CCl_4 (ratio 0… ≈ 15)	353	$k_a/k_b = 0.019$	74 Hen 1/ 74 Pry 1 [20]
$CH_3(CH_2)_9\dot{C}H_2 + 3\text{-}CNC_6H_4CH_3 \xrightarrow{a} n\text{-}C_{11}H_{24} + 3\text{-}CNC_6H_4\dot{C}H_2$ $+ C_6H_5CH_3 \xrightarrow{b} n\text{-}C_{11}H_{24} + C_6H_5\dot{C}H_2$				
Therm. of lauroylperoxide [21]	C_6H_6	354	$k_a/k_b = 1.66(2)$ [19] See also 4.1.2.3, Fig. 5, p. 256	75 Zav 1
$CH_3(CH_2)_9\dot{C}H_2 + 4\text{-}CNC_6H_4CH_3 \xrightarrow{a}$ products $+ C_6H_5CH_3 \xrightarrow{b}$ products				
Therm. of lauroylperoxide [21]	C_6H_6	354	$k_a/k_b = 2.03(2)$ [19]	75 Zav 1
$CH_3(CH_2)_9\dot{C}H_2 + 3\text{-}CH_3C_6H_4CH_3 \xrightarrow{a} C_{11}H_{24} +$ 3-methyltoluene($-\dot{H}$) $+ CCl_4 \xrightarrow{b} C_{11}H_{23}Cl + \dot{C}Cl_3$				
Therm. of *n*-lauroylperoxide PR, glc	3-methyltoluene/ CCl_4	353	$k_a/k_b = 0.021$ [23]	74 Hen 1/ 74 Pry 1 [20]
$CH_3(CH_2)_9\dot{C}H_2 + 3\text{-}CH_3C_6H_4CH_3 \xrightarrow{a} n\text{-}C_{11}H_{24} + 3\text{-}CH_3C_6H_4\dot{C}H_2$ $+ C_6H_5CH_3 \xrightarrow{b} n\text{-}C_{11}H_{24} + C_6H_5\dot{C}H_2$				
Therm. of lauroylperoxide [21]	C_6H_6	354	$k_a/k_b = 0.97(1)$ [19] [23] See also 4.1.2.3, Fig. 5, p. 256	75 Zav 1
$CH_3(CH_2)_9\dot{C}H_2 + 4\text{-}CH_3C_6H_4CH_3 \xrightarrow{a} C_{11}H_{24} +$ 4-methyltoluene($-\dot{H}$) $+ CCl_4 \xrightarrow{b} C_{11}H_{23}Cl + \dot{C}Cl_3$				
Therm. of *n*-lauroylperoxide PR, glc	4-methyltoluene/ CCl_4	353	$k_a/k_b = 0.023$ [23]	74 Hen 1/ 74 Pry 1 [20]

[19]) Refers to total reaction (includes possible radical addition to aromatic ring besides H-atom abstraction).
[20]) Graphic presentation of data in *t*-butylbenzene as solvent.
[21]) Disappearance of toluenes.
[22]) Disappearance of anisole and toluene.
[23]) With statistical correction by a factor of 2.

Reaction Radical generation Method	Solvent	T[K]	Rate data	Ref./ add. ref.
$CH_3(CH_2)_9\dot{C}H_2 + 4\text{-}CH_3C_6H_4CH_3 \xrightarrow{a} n\text{-}C_{11}H_{24} + 4\text{-}CH_3C_6H_4\dot{C}H_2$ $+ C_6H_5CH_3 \xrightarrow{b} n\text{-}C_{11}H_{24} + C_6H_5\dot{C}H_2$				
Therm. of lauroylperoxide [21]	C_6H_6	354	$k_a/k_b = 0.89(2)$ [19] [23] See also 4.1.2.3, Fig. 5, p. 256.	75 Zav 1
$CH_3(CH_2)_9\dot{C}H_2 + 3\text{-}CH_3OC_6H_4CH_3 \xrightarrow{a} n\text{-}C_{11}H_{24} + 3\text{-}CH_3OC_6H_4\dot{C}H_2$ $+ C_6H_5CH_3 \xrightarrow{b} n\text{-}C_{11}H_{24} + C_6H_5\dot{C}H_2$				
Therm. of lauroylperoxide [21]	C_6H_6	354	$k_a/k_b = 1.12(1)$ [19] See also 4.1.2.3, Fig. 5, p. 256.	75 Zav 1
$CH_3(CH_2)_9\dot{C}H_2 + 4\text{-}CH_3OC_6H_4CH_3 \xrightarrow{a} n\text{-}C_{11}H_{24}$ + 4-methoxytoluene($-\dot{H}$) $+ C_6H_5CH_3 \xrightarrow{b} n\text{-}C_{11}H_{24} + C_6H_5\dot{C}H_2$				
Therm. of lauroylperoxide [21]	C_6H_6	354	$k_a/k_b = 1.24(1)$ [19] See also 4.1.2.3, Fig. 5, p. 256.	75 Zav 1
$CH_3(CH_2)_9\dot{C}H_2$ + 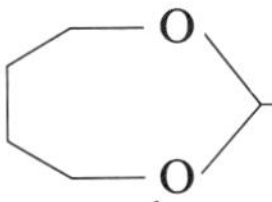 $-C_3H_7 \xrightarrow{a} n\text{-}C_{11}H_{24} + (C_8H_{15}O_2^{\bullet})$ $+ CCl_4 \xrightarrow{b} n\text{-}C_{11}H_{23}Cl + \dot{C}Cl_3$				
Thermal decomp. of lauroylperoxide				79 Bat 1
PR, glc	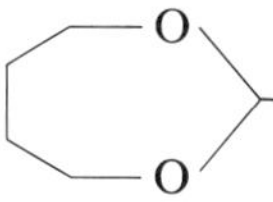 $-C_3H_7/CCl_4$	353.0(1)	$k_a/k_b = 0.54$	
$CH_3(CH_2)_9\dot{C}H_2$ + chroman $\xrightarrow{a} n\text{-}C_{11}H_{24}$ + chroman($-\dot{H}$) $+ CCl_4 \xrightarrow{b} n\text{-}C_{11}H_{23}Cl + \dot{C}Cl_3$				
Thermal decomp. of lauroylperoxide PR, glc	chroman/CCl_4	353	$k_a/k_b = 0.04$	79 Zlo 1
$CH_3(CH_2)_9\dot{C}H_2$ + 2-methyl-2,3-dihydrobenzofuran $\xrightarrow{a} n\text{-}C_{11}H_{24}$ + 2-methyl-2,3-dihydrobenzofuran($-\dot{H}$) $+ CCl_4 \xrightarrow{b} n\text{-}C_{11}H_{23}Cl + \dot{C}Cl_3$				
Thermal decomp. of lauroylperoxide PR, glc	2-methyl-2,3-dihydro-benzofuran/CCl_4	353	$k_a/k_b = 0.32$	79 Zlo 1
$CH_3(CH_2)_9\dot{C}H_2 + 1,3,5\text{-}(CH_3)_3C_6H_3 \xrightarrow{a} C_{11}H_{24}$ + 1,3,5-trimethylbenzene($-\dot{H}$) $+ CCl_4 \xrightarrow{b} C_{11}H_{23}Cl + \dot{C}Cl_3$				
Therm. of *n*-lauroylperoxide PR, glc	1,3,5-trimethylbenzene/CCl_4	353	$k_a/k_b = 0.039$ [24]	74 Hen 1/ 74 Pry 1 [25]

[19]) Refers to total reaction (includes possible radical addition to aromatic ring besides H-atom abstraction).
[21]) Disappearance of toluenes.
[23]) With statistical correction by a factor of 2.
[24]) With statistical correction by a factor of 3.
[25]) Graphic presentation in *t*-butylbenzene as solvent.

Reaction Radical generation Method	Solvent	T[K]	Rate data	Ref./ add. ref.
$CH_3(CH_2)_9\dot{C}H_2$ + 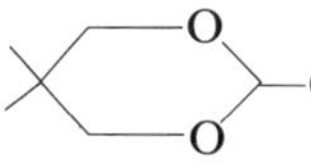 $-C_3H_7 \xrightarrow{a} n\text{-}C_{11}H_{24} + (C_{10}H_{17}O_2^{\bullet})$ $+ CCl_4 \xrightarrow{b} n\text{-}C_{11}H_{23}Cl + \dot{C}Cl_3$				
Thermal decomp. of lauroylperoxide				79 Bat 1
PR, glc	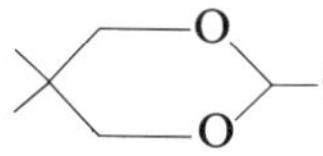 $-C_3H_7$/ CCl_4	353.0(1)	$k_a/k_b = 0.35$	
$CH_3(CH_2)_9\dot{C}H_2$ + tetralin $\xrightarrow{a} n\text{-}C_{11}H_{24}$ + tetralin(−Ḣ) $+ CCl_4 \xrightarrow{b} n\text{-}C_{11}H_{23}Cl + \dot{C}Cl_3$				
Thermal decomp. of lauroylperoxide				79 Zlo 1
PR, glc	tetralin/CCl_4	353	$k_a/k_b = 0.09$	
$CH_3(CH_2)_9\dot{C}H_2$ + 2,2-dimethyl-2,3-dihydrobenzofuran $\xrightarrow{a}$ $n\text{-}C_{11}H_{24}$ + 2,2-dimethyl-2,3-dihydrobenzofuran(−Ḣ) $+ CCl_4 \xrightarrow{b} n\text{-}C_{11}H_{23}Cl + \dot{C}Cl_3$				
Thermal decomp. of lauroylperoxide				79 Zlo 1
PR, glc	2,2-dimethyl- 2,3-dihydro- benzofuran/CCl_4	353	$k_a/k_b = 0.28$	
$CH_3(CH_2)_9\dot{C}H_2 + 4\text{-}(CH_3)_3CC_6H_4CH_3 \xrightarrow{a} n\text{-}C_{11}H_{24}$ + 4-t-butyl-toluene(−Ḣ) $+ C_6H_5CH_3 \xrightarrow{b} n\text{-}C_{11}H_{24} + C_6H_5\dot{C}H_2$				
			See also 4.1.2.3, Fig. 5, p. 256.	
$CH_3(CH_2)_7\dot{C}(CH_3)_2 + CBrCl_3 \xrightarrow{a} n\text{-}C_8H_{17}CBr(CH_3)_2 + \dot{C}Cl_3$ $+ CCl_4 \xrightarrow{b} n\text{-}C_8H_{17}CCl(CH_3)_2 + \dot{C}Cl_3$				
Reduct. of $C_8H_{17}C(CH_3)_2HgOCOCH_3$ by $NaBH_4$				79 Gie 1
PR, glc	CCl_4	273	k_a/k_b = 25000 [26]	
		333	2400	
		343	1600	
		373	620	
		403	300	
			$\Delta H_a^{\ddagger} - \Delta H_b^{\ddagger} = -33\ \text{kJ mol}^{-1}$ (±10%)	
			$\Delta S_a^{\ddagger} - \Delta S_b^{\ddagger} = -35\ \text{J mol}^{-1}\text{K}^{-1}$ (±5%)	
$(c\text{-}C_{12}H_{23}^{\bullet}) + CCl_3Br \xrightarrow{a} c\text{-}C_{12}H_{23}Br + \dot{C}Cl_3$ $+ CCl_4 \xrightarrow{b} c\text{-}C_{12}H_{23}Cl + \dot{C}Cl_3$				
Therm. of $c\text{-}C_{12}H_{23}COOOC(CH_3)_3$				75 Her 1/
PR, glc	CCl_4	383	$k_a/k_b = 583$	76 Gie 1, 76 Gie 2, 76 Gie 3

[26]) Extrapolated value.

Reaction Radical generation Method	Solvent	T[K]	Rate data	Ref./ add. ref.

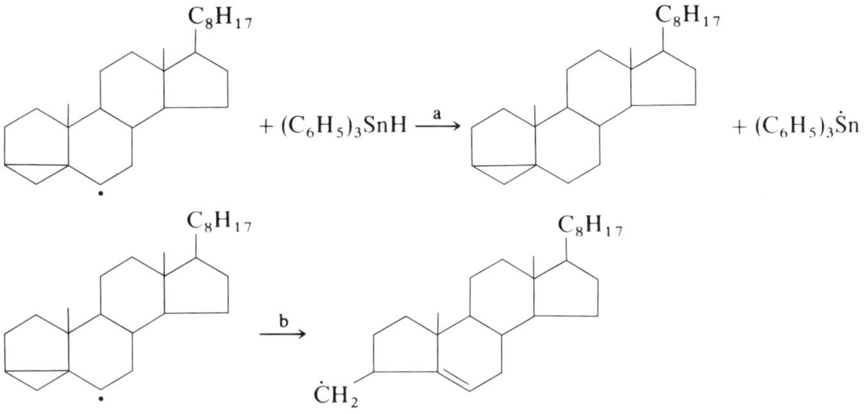

$(C_6H_5)_3\dot{S}n$ + 6β-chloro-3β,5-cyclo-5β-cholestane (AIBN as initiator)				76 Bec 1
PR, glc	pentane	298	$k_a/k_b = 0.133\ M^{-1}$	

4.1.2.1.2 Radicals containing C, H, and other atoms

$\dot{C}Cl_3 + Cl_3SiH \xrightarrow{a} CHCl_3 + Cl_3\dot{S}i$ $+ (C_2H_5)_3SiH \xrightarrow{b} CHCl_3 + (C_2H_5)_3\dot{S}i$				
Therm. of BPO as initiator				71 Nag 1
PR, glc	CCl_4	353	$k_a/k_b = 0.093$	
$\dot{C}Cl_3 + CH_3CHOH \xrightarrow{a} CHCl_3$ + ethanol(−Ḣ) $2\dot{C}Cl_3 \xrightarrow{b} C_2Cl_6$				
γ-rad. of $CH_3CH_2OH + CCl_4$				82 Fel 1/
PR, glc	C_2H_5OH	299	$k_a/(2k_b)^{\frac{1}{2}} = 8.1(8)\ M^{-\frac{1}{2}}s^{-\frac{1}{2}}$ [1])	55 Wat 1,
		301	9.3(10)	79 Pau 1
		325	22.3(10)	
		338	37.2(42)	
		357	66.5(45)	
		361	88.2(74)	
		387	168.4(27)	
		393	157.6(114)	
		403	255.8(16)	
		404	190.0(79)	
		433	403.7(71)	
		443	630.0(102)	
$\dot{C}Cl_3 + C_2H_5Cl_2SiH \xrightarrow{a} CHCl_3 + C_2H_5Cl_2\dot{S}i$ $+ (C_2H_5)_3SiH \xrightarrow{b} CHCl_3 + (C_2H_5)_3\dot{S}i$				
Therm. of BPO as initiator				71 Nag 1
PR, glc	CCl_4	353	$k_a/k_b = 0.24$	
$\dot{C}Cl_3 + (CH_3)_2CHOH \xrightarrow{a} CHCl_3 + (CH_3)_2\dot{C}OH$ $2\dot{C}Cl_3 \xrightarrow{b} C_2Cl_6$				
Phot.				78 Van 1
PR, glc	$CCl_4/(CH_3)_2CHOH$	RT	$k_a/(2k_b)^{\frac{1}{2}} = 0.34 \cdot 10^{-2}\ M^{-\frac{1}{2}}s^{-\frac{1}{2}}$	
	CCl_4/C_2H_5OH		$0.10 \cdot 10^{-2}$	
	$(CCl_3)_2/(CH_3)_2CHOH$		$1.34 \cdot 10^{-2}$	
	$(CCl_3)_2/C_2H_5OH$		$0.23 \cdot 10^{-2}$	

[1]) $\log k_a = 8.22 - 39.1/2.303\,RT$ (R in $kJ\,mol^{-1}K^{-1}$) based on $\log[k_a/(2k_b)^{\frac{1}{2}}] = 2.51(12) - 32.2(8)/2.303\,RT$, and $E_a = 13.8\ kJ\,mol^{-1}$ [55 Wat 1] and $\log[A/M^{-1}s^{-1}] = 11.42$ [79 Pau 1] for $\dot{C}Cl_3 + \dot{C}Cl_3$ reaction, respectively.

Reaction Radical generation Method	Solvent	T[K]	Rate data	Ref./ add. ref.
$\dot{C}Cl_3 + (CH_3)_3SiH \xrightarrow{a} CHCl_3 + (CH_3)_3\dot{S}i$ $+ c\text{-}C_6H_{12} \xrightarrow{b} CHCl_3 + (c\text{-}C_6H_{11}^{\cdot})$				
γ-rad. of CCl_4 + c-C_6H_{12} solutions				80 Bar 1/
PR, glc	$CCl_4/c\text{-}C_6H_{12}$	323.0(3)	$k_a/k_b = 37.3$	76 Kat 3,
		348.0(3)	29.0	78 Kat 1
		373.0(3)	24.3	
		398.0(3)	20.7	
		423.0(3)	14.7	
			$\log[A_a/A_b] = -0.304(150)$ [2]	
			$E_a(a) - E_a(b) = -10.0(11)$ kJ mol^{-1} [2]	
$\dot{C}Cl_3 + (C_2H_5)_2ClSiH \xrightarrow{a} CHCl_3 + (C_2H_5)_2Cl\dot{S}i$ $+ (C_2H_5)_3SiH \xrightarrow{b} CHCl_3 + (C_2H_5)_3\dot{S}i$				
Therm. of BPO as initiator				71 Nag 1
PR, glc	CCl_4	353	$k_a/k_b = 0.37$	
$\dot{C}Cl_3 + (CH_3)_2CHOH \xrightarrow{a} CHCl_3 + (CH_3)_2\dot{C}OH$ $2\dot{C}Cl_3 \xrightarrow{b} C_2Cl_6$				
γ-rad. of $CCl_4 + (CH_3)_2CHOH$				81 Fel 1/
PR, glc	CCl_4	306(1)	$k_a/(2k_b)^{\frac{1}{2}} = 3.75(37)\cdot 10^{-3}\,M^{-\frac{1}{2}}s^{-\frac{1}{2}}$	55 Wat 1,
		339(1)	$12.4(7)\cdot 10^{-3}$	79 Pau 1
		347(1)	$16.6(9)\cdot 10^{-3}$	
		363(1)	$28.8(13)\cdot 10^{-3}$	
		377(1)	$34.6(1)\cdot 10^{-3}$	
		385(1)	$53.3(10)\cdot 10^{-3}$	
		400(1)	$65.7(6)\cdot 10^{-3}$	
		417(1)	$107(1)\cdot 10^{-3}$	
			$\log A_a - \frac{1}{2}\log A_b/cm^{\frac{3}{2}}mol^{-\frac{1}{2}}s^{-\frac{1}{2}} = 2.99(13)$	
			$E_a(a) - \frac{1}{2}E_a(b) = 31.9(9)$ kJ mol^{-1}	
$\dot{C}Cl_3 + CH_3(CH_2)_3CH_2OH \xrightarrow{\alpha} CHCl_3 + CH_3(CH_2)_3\dot{C}HOH$ $2\dot{C}Cl_3 \xrightarrow{b} C_2Cl_6$				
γ-rad. of n-pentanol + CCl_4				82 Fel 1/
PR, glc	n-pentanol	317	$k_\alpha/(2k_b)^{\frac{1}{2}} = 15.0\,M^{-\frac{1}{2}}s^{-\frac{1}{2}}$ [3]	81 Alf 2,
		323	23.0	55 Wat 1
		348	43.0	
		356	61.0	
		385	164.0	
		408	247.0	
		428	767.0	
			[4]	
$\dot{C}Cl_3 + 3\text{-}ClC_6H_4SiH_3 \xrightarrow{a} CHCl_3 + 3\text{-}ClC_6H_4\dot{S}iH_2$ $+ C_6H_5SiH_3 \xrightarrow{b} CHCl_3 + C_6H_5\dot{S}iH_2$				
Therm. of BPO as initiator				71 Nag 2
PR, glc	$CCl_4/3\text{-}ClC_6H_4SiH_3/C_6H_5SiH_3$	353	$k_a/k_b = 0.71$	

[2]) $\log[A_a/M^{-1}s^{-1}] = 8.49$ and $E_a(a) = 36.4$ kJ mol^{-1} assuming $\log[A_b/M^{-1}s^{-1}] = 8.79$ and $E_a(b) = 46.4$ kJ mol^{-1} [76 Kat 3, 78 Kat 1].

[3]) Calc. from overall H-atom abstraction rates by correction for abstraction of non-α-H-atoms using $k_a/(2k_b)^{\frac{1}{2}} = \exp(8.12 - 4.6/T)$ [81 Alf 2].

[4]) $\log k_\alpha = 8.94 - 44.0/2.303\,RT$ (R in kJ mol^{-1} K^{-1}), and $E_a = 13.8$ kJ mol^{-1} [55 Wat 1] and $\log[A/M^{-1}s^{-1}] = 11.42$ [79 Pau 1] for $\dot{C}Cl_3 + \dot{C}Cl_3$ reaction, respectively.

Reaction Radical generation Method	Solvent	T[K]	Rate data	Ref./ add. ref.
$\dot{C}Cl_3 + 4\text{-}ClC_6H_4SiH_3 \xrightarrow{a} CHCl_3 + 4\text{-}ClC_6H_4\dot{S}iH_2$ $+ C_6H_5SiH_3 \xrightarrow{b} CHCl_3 + C_6H_5\dot{S}iH_2$				
Therm. of BPO as initiator				71 Nag 2
PR, glc	$CCl_4/4\text{-}ClC_6H_4SiH_3/$ $C_6H_5SiH_3$	353	$k_a/k_b = 0.84$	
$\dot{C}Cl_3 + c\text{-}C_6H_{10} \xrightarrow{a} CHCl_3 + (c\text{-}C_6H_9^{\cdot})$ $2\,\dot{C}Cl_3 \xrightarrow{b} C_2Cl_6$				
γ-rad. of CCl_4 + c-hexene				81 Alf 1/
PR, glc	CCl_4	299(1)	$k_a/(2k_b)^{\frac{1}{2}} = 3.0 \cdot 10^{-2}\,M^{-\frac{1}{2}}s^{-\frac{1}{2}}$ [5]	55 Wat 1,
		311(1)	$4.2 \cdot 10^{-2}$	79 Pau 1
		322(1)	$7.0 \cdot 10^{-2}$	
		331(1)	$6.4 \cdot 10^{-2}$	
		348(1)	$11.1 \cdot 10^{-2}$	
		358(1)	$13.6 \cdot 10^{-2}$	
		363(1)	$11.8 \cdot 10^{-2}$	
		393(1)	$17.7 \cdot 10^{-2}$	
		398(1)	$24.4 \cdot 10^{-2}$	
		413(1)	$29.3 \cdot 10^{-2}$	
			[6]	
$\dot{C}Cl_3 + c\text{-}C_6H_{12} \xrightarrow{a} CHCl_3 + (c\text{-}C_6H_{11}^{\cdot})$ $2\,\dot{C}Cl_3 \xrightarrow{b} C_2Cl_6$				
γ-rad. of $c\text{-}C_6H_{12} + CCl_4$				76 Kat 3
PR, glc	$c\text{-}C_6H_{12}$	303	$k_a/(k_b)^{\frac{1}{2}} = 2.54 \cdot 10^{-2}$	
		327	$= 7.75 \cdot 10^{-2}$	
		353	$= 23.3 \cdot 10^{-2}$	
		383	$= 52.2 \cdot 10^{-2}$	
			$\log[A_a/(A_b)^{\frac{1}{2}}/cm^{\frac{3}{2}}\,mol^{-\frac{1}{2}}\,s^{-\frac{1}{2}}] = 4.78(8)$	
			$E_a(a) - \frac{1}{2}E_a(b) = 36.9(5)\,kJ\,mol^{-1}$	
$\dot{C}Cl_3 + c\text{-}C_6H_{12} \xrightarrow{a} CHCl_3 + (c\text{-}C_6H_{11}^{\cdot})$ $2\,\dot{C}Cl_3 \xrightarrow{b} C_2Cl_6$				
γ-rad. of $CCl_4 + c\text{-}C_6H_{12}$				81 Alf 2/
PR, glc	CCl_4	300(1)	$k_a/(2k_b)^{\frac{1}{2}} = 10.1(35) \cdot 10^{-4}\,M^{-\frac{1}{2}}s^{-\frac{1}{2}}$	55 Wat 1,
		317(1)	$18.7(35) \cdot 10^{-4}$	79 Pau 1
		340(1)	$37.0(42) \cdot 10^{-4}$	
		361(1)	$75.5(32) \cdot 10^{-4}$	
		367(1)	$77.4(70) \cdot 10^{-4}$	
		398(1)	$354.3(65) \cdot 10^{-4}$	
		428(1)	$995.3(249) \cdot 10^{-4}$	
		463(1)	$1747.3(321) \cdot 10^{-4}$	
			[7]	

[5]) $\log[k_a/(2k_b)^{\frac{1}{2}}] = 1.93(18) - 19.7(12)/2.303\,RT (R = kJ\,mol^{-1}\,K^{-1})$.

[6]) $\log k_a = 7.64 - 26.6/2.303\,RT$ based on $E_a = 13.8\,kJ\,mol^{-1}$ [55 Wat 1] and $\log[A/M^{-1}s^{-1}] = 11.42$ [79 Pau 1] for $\dot{C}Cl_3 + \dot{C}Cl_3$ reaction.

[7]) $\log[A_a/M^{-1}s^{-1}] = 9.24(34)$ and $E_a(a) = 45.4(14)\,kJ\,mol^{-1}$ based on $\log[k_a/(2k_b)^{\frac{1}{2}}] = 3.53(34) - 38.5(21)/2.303\,RT$ (R in $kJ\,mol^{-1}\,K^{-1}$), and $E_a = 13.8\,kJ\,mol^{-1}$ [55 Wat 1] and $\log[A/M^{-1}s^{-1}] = 11.42$ [79 Pau 1] for $\dot{C}Cl_3 + \dot{C}Cl_3$ reaction, respectively.

Reaction Radical generation Method	Solvent	T[K]	Rate data	Ref./ add. ref.
$\dot{C}Cl_3 + c\text{-}C_6H_{12} \xrightarrow{a} CHCl_3 + (c\text{-}C_6H_{11}^{\cdot})$ $+ C_2Cl_4 \xrightarrow{b} (C_2Cl_4\text{-}\dot{C}Cl_3^{\cdot})$				
γ-rad. of $c\text{-}C_6H_{12} + CCl_4 + C_2Cl_4$				79 Hor 1/
PR, glc	$c\text{-}C_6H_{12}/CCl_4/$	363	$k_a/k_b = 1.855$	76 Kat 3,
	C_2Cl_4 mixt.	373	1.792	78 Kat 1
		393	2.101	
		413	2.500	
		433	2.392	
		448	2.882	
			$\log[A_a/A_b] = 1.21(10)$	
			$E_a(a) - E_a(b) = 6.7(10)\,kJ\,mol^{-1}$	
$\dot{C}Cl_3 + c\text{-}C_6H_{12} \xrightarrow{a} CHCl_3 + (c\text{-}C_6H_{11}^{\cdot})$ $+ n\text{-}C_6H_{14} \xrightarrow{b} CHCl_3 + (n\text{-}C_6H_{13}^{\cdot})$				
γ-rad. of $c\text{-}C_6H_{12} + n\text{-}C_6H_{14} + CCl_4$				78 Kat 1
PR, glc	$c\text{-}C_6H_{12}/n\text{-}C_6H_{14}/$	296	$k_a/k_b = 1.67$	
	CCl_4 mixt.	333	1.71	
		373	1.72	
		413	1.79	
			$\log[A_a/A_b] = 0.32(4)$	
			$E_a(a) - E_a(b) = 0.6(3)\,kJ\,mol^{-1}$	
$\dot{C}Cl_3 + c\text{-}C_6H_{12} \xrightarrow{a} CHCl_3 + (c\text{-}C_6H_{11}^{\cdot})$ $+ (CH_3)_3CCH_2CH(CH_3)_2 \xrightarrow{b} CHCl_3 +$ 2,2,4-trimethylpentane(−Ḣ)				
Di-n-butyryl peroxide as initiator				77 Nug 1
PR, glc	CCl_4	368	$k_a/k_b = 0.93$ [8])	
$\dot{C}Cl_3 + c\text{-}C_6H_{11}OH \xrightarrow{a} CHCl_3 + (c\text{-}C_6H_{10}OH^{\cdot})$ $2\dot{C}Cl_3 \xrightarrow{b} C_2Cl_6$				
γ-rad. of $CCl_4 + c\text{-}C_6H_{11}OH$				81 Fel 1/
PR, glc	CCl_4	303(1)	$k_a/(2k_b)^{\frac{1}{2}} = 5.25(66)\cdot 10^{-3}\,M^{-\frac{1}{2}}s^{-\frac{1}{2}}$	55 Wat 1,
		323(1)	$10.6(2)\cdot 10^{-3}$	79 Pau 1
		335(1)	$13.2(2)\cdot 10^{-3}$	
		346(1)	$17.6(10)\cdot 10^{-3}$	
		348(1)	$18.0(9)\cdot 10^{-3}$	
		363(1)	$27.6(18)\cdot 10^{-3}$	
		385(1)	$62.9(16)\cdot 10^{-3}$	
		417(1)	$97.7(20)\cdot 10^{-3}$	
		428(1)	$203(6)\cdot 10^{-3}$	
			$\log A_a - \frac{1}{2}\log A_b/cm^{\frac{3}{2}}mol^{-\frac{1}{2}}s^{-\frac{1}{2}} = 2.85(26)$	
			$E_a(a) - \frac{1}{2}E_a(b) = 30.4(18)\,kJ\,mol^{-1}$	
$\dot{C}Cl_3 + n\text{-}C_6H_{14} \xrightarrow{a} CHCl_3 + CH_3\dot{C}H(CH_2)_3CH_3$ $\dot{C}Cl_3 + n\text{-}C_6H_{14} \xrightarrow{b} CHCl_3 + CH_3CH_2\dot{C}H(CH_2)_2CH_3$				
γ-rad. of $n\text{-}C_6H_{14} + CCl_4$				78 Tua 1/
PR, glc	$n\text{-}C_6H_{12}/CCl_4$	183...383	$k_a/k_b = 1.53(6)$ [9])	77 Tua 1

[8]) Per reactive H-atom. Overall relative reactivity $k_a/k_b = 11.2$.
[9]) Temperature independent from 293...383 K.

Reaction Radical generation Method	Solvent	T[K]	Rate data	Ref./ add. ref.
$\dot{C}Cl_3 + n\text{-}C_6H_{14} \xrightarrow{a} CHCl_3 + (n\text{-}C_6\dot{H}_{13})$ $2\dot{C}Cl_3 \xrightarrow{b} C_2Cl_6$				
γ-rad. of $CCl_4 + n\text{-}C_6H_{14}$				76 Kat 4/
PR, glc	$n\text{-}C_6H_{14}/CCl_4$	297.5	$k_a/(2k_b)^{\frac{1}{2}} = 1.28 \cdot 10^{-2}$ $mol^{-\frac{1}{2}} s^{-\frac{1}{2}} cm^{\frac{3}{2}}$	74 Gri 1
		299	$1.51 \cdot 10^{-2}$	
		302	$1.62 \cdot 10^{-2}$	
		313	$2.89 \cdot 10^{-2}$	
		333	$6.94 \cdot 10^{-2}$	
		353	$1.71 \cdot 10^{-1}$	
		373	$3.63 \cdot 10^{-1}$	
			$\log[A_a/M^{-1}s^{-1}] = 11.2(3)$ [10]	
			$E_a(a) = 42.3(4)\,kJ\,mol^{-1}$ [10]	
$\dot{C}Cl_3 + (CH_3)_2CHCH(CH_3)_2 \xrightarrow{a} CHCl_3 +$ 2,3-dimethylbutane$(-\dot{H})$ $2\dot{C}Cl_3 \xrightarrow{b} C_2Cl_6$				
γ-rad. of CCl_4 + 2,3-dimethylbutane				80 Alf 1/
PR, glc	CCl_4	305(1)	$k_a/(2k_b)^{\frac{1}{2}} = 2.7(3) \cdot 10^{-3}\,M^{-\frac{1}{2}} s^{-\frac{1}{2}}$	55 Wat 1,
		327(1)	$6.7(9) \cdot 10^{-3}$	79 Pau 1
		331(1)	$6.8(4) \cdot 10^{-3}$	
		352(1)	$14.2(3) \cdot 10^{-3}$	
		385(1)	$32.0(10) \cdot 10^{-3}$	
		391(1)	$32.1(15) \cdot 10^{-3}$	
			[11]	
$\dot{C}Cl_3 + (C_2H_5)_3SiH \xrightarrow{a} CHCl_3 + (C_2H_5)_3\dot{S}i$ $+ c\text{-}C_6H_{12} \xrightarrow{b} CHCl_3 + (c\text{-}C_6\dot{H}_{11})$				
γ-rad. of $CCl_4 + c\text{-}C_6H_{12}$				80 Bar 1/
PR, glc	$CCl_4/c\text{-}C_6H_{12}$	335.0(3)	$k_a/k_b = 58.8$	76 Kat 3,
		348.0(3)	53.7	78 Kat 1
		373.0(3)	43.3	
		388.0(3)	35.2	
		413.0(3)	27.1	
		423.0(3)	22.3	
			$\log[A_a/A_b] = -0.17(25)$ [12]	
			$E_a(a) - E_a(b) = -12.6(10)\,kJ\,mol^{-1}$ [12]	
$\dot{C}Cl_3 + 3\text{-}BrC_6H_4CH_3 \xrightarrow{a} CHCl_3 + 3\text{-}BrC_6H_4\dot{C}H_2$ $+ C_6H_5CH_3 \xrightarrow{b} CHCl_3 + C_6H_5\dot{C}H_2$				
Phot. of CCl_3Br				74 Tan 1
PR, glc	toluene (containing 20% ethylene oxide)	323.0(1)	$k_a/k_b = 0.57(2)$	
$\dot{C}Cl_3 + 4\text{-}BrC_6H_4CH_3 \xrightarrow{a} CHCl_3 + 4\text{-}BrC_6H_4\dot{C}H_2$ $+ C_6H_5CH_3 \xrightarrow{b} CHCl_3 + C_6H_5\dot{C}H_2$				
Phot. of CCl_3Br				74 Tan 1
PR, glc	toluene (containing 20% ethylene oxide)	323.0(1)	$k_a/k_b = 0.75(3)$	

[10]) Based on $2k_b = 8(6) \cdot 10^{11}\,mol^{-1}\,s^{-1}\,cm^3$ at RT [74 Gri 1].

[11]) $\log[A_a/M^{-1}s^{-1}] = 8{,}1$ and $E_a(a) = 36\,kJ\,mol^{-1}$ based on $\log[k_a/(2k_b)^{\frac{1}{2}}] = 2.40(17) - 29.1(11)/2.303\,RT$ (R in $kJ\,mol^{-1}\,K^{-1}$), and $E_a = 13.8\,kJ\,mol^{-1}$ [55 Wat 1] and $\log[A/M^{-1}s^{-1}] = 11.42$ [79 Pau 1] for $\dot{C}Cl_3 + \dot{C}Cl_3$ reaction, respectively.

[12]) $\log[A_a/M^{-1}s^{-1}] = 8.62$ and $E_a(a) = 33.8\,kJ\,mol^{-1}$ assuming $\log[A_b/M^{-1}s^{-1}] = 8.79$ and $E_a(b) = 46.4\,kJ\,mol^{-1}$.

Reaction Radical generation Method	Solvent	T[K]	Rate data	Ref./ add. ref.
$\dot{C}Cl_3$ + 4-$ClC_6H_4CH_3$ $\xrightarrow{a}$ $CHCl_3$ + 4-$ClC_6H_4\dot{C}H_2$ + $C_6H_5CH_3$ $\xrightarrow{b}$ $CHCl_3$ + $C_6H_5\dot{C}H_2$				
Phot. of CCl_3Br				74 Tan 1
PR, glc	toluene (containing 20% ethylene oxide)	323.0(1)	$k_a/k_b = 0.92(2)$	
$\dot{C}Cl_3$ + 3-$ClC_6H_4Si(CH_3)H_2$ $\xrightarrow{a}$ $CHCl_3$ + 3-$ClC_6H_4\dot{S}i(CH_3)H$ + $C_6H_5Si(CH_3)H_2$ $\xrightarrow{b}$ $CHCl_3$ + $C_6H_5\dot{S}i(CH_3)H$				
Therm. of BPO as initiator				71 Nag 2
PR, glc	CCl_4/ 3-$ClC_6H_4Si(CH_3)H_2$/ $C_6H_5Si(CH_3)H_2$	353	$k_a/k_b = 0.78$	
$\dot{C}Cl_3$ + 4-$ClC_6H_4Si(CH_3)H_2$ $\xrightarrow{a}$ $CHCl_3$ + 4-$ClC_6H_4\dot{S}i(CH_3)H$ + $C_6H_5Si(CH_3)H_2$ $\xrightarrow{b}$ $CHCl_3$ + $C_6H_5\dot{S}i(CH_3)H$				
Therm. of BPO as initiator				71 Nag 2
PR, glc	CCl_4/ 4-$ClC_6H_4Si(CH_3)H_2$/ $C_6H_5Si(CH_3)H_2$	353	$k_a/k_b = 0.89$	
$\dot{C}Cl_3$ + 3-$CH_3C_6H_4SiH_3$ $\xrightarrow{a}$ $CHCl_3$ + 3-$CH_3C_6H_4\dot{S}iH_2$ + $C_6H_5SiH_3$ $\xrightarrow{b}$ $CHCl_3$ + $C_6H_5\dot{S}iH_2$				
Therm. of BPO as initiator				71 Nag 2
PR, glc	CCl_4/ 3-$CH_3C_6H_4SiH_3$/ $C_6H_5SiH_3$	353	$k_a/k_b = 1.08$	
$\dot{C}Cl_3$ + 4-$CH_3C_6H_4SiH_3$ $\xrightarrow{a}$ $CHCl_3$ + 4-$CH_3C_6H_4\dot{S}iH_2$ + $C_6H_5SiH_3$ $\xrightarrow{b}$ $CHCl_3$ + $C_6H_5\dot{S}iH_2$				
Therm. of BPO as initiator				71 Nag 2
PR, glc	CCl_4/ 4-$CH_3C_6H_4SiH_3$/ $C_6H_5SiH_3$	353	$k_a/k_b = 1.33$	
$\dot{C}Cl_3$ + $CH_3CH_2HgCH_2C(CH_3)_3$ $\xrightarrow{a}$ $CHCl_3$ + ethylneopentyl-Hg(−$\dot{H}$) + c-C_6H_{12} $\xrightarrow{b}$ $CHCl_3$ + (c-$C_6\dot{H}_{11}$)				
Therm. and phot. of ethylneopentylmercury in CCl_4				77 Nug 1
PR, glc	CCl_4	368	$k_a/k_b = 4.4$ [13])	
$\dot{C}Cl_3$ + $C_6H_5CH_2CH_3$ $\xrightarrow{a}$ $CHCl_3$ + ethylbenzene(−$\dot{H}$) + $C_6H_5CH_3$ $\xrightarrow{b}$ $CHCl_3$ + $C_6H_5\dot{C}H_2$				
Phot. of $CBrCl_3$				80 Tan 1
PR	$C_6H_5Br/CBrCl_3$ (molar ratio 2:3)	343	$k_a/k_b = 4.70(4)$ [14])	
PR, glc	Freon 113	313.0(1)		75 Tan 1
	+ K_2CO_3		$k_a/k_b = 6.76(49)$	
	+ ethylene oxide		$k_a/k_b = 6.96(57)$	

[13]) Per reactive hydrogen atom. Overall relative reactivity $k_a/k_b = 1.09$.
[14]) Solutions contained ethylene oxide as HBr scavenger.

Reaction Radical generation Method	Solvent	T[K]	Rate data	Ref./ add. ref.
$\dot{C}Cl_3 + 3\text{-}CH_3C_6H_4CH_3 \xrightarrow{a} CHCl_3 + 3\text{-}CH_3C_6H_4\dot{C}H_2$ [15]) $+ C_6H_5CH_3 \xrightarrow{b} CHCl_3 + C_6H_5\dot{C}H_2$				
Phot. of CCl_3Br				74 Tan 1
PR, glc	toluene (containing 20% ethylene oxide)	323.0(1)	$k_a/k_b = 1.23(2)$ [15])	
$\dot{C}Cl_3 + 4\text{-}CH_3C_6H_4CH_3 \xrightarrow{a} CHCl_3 + 4\text{-}CH_3C_6H_4\dot{C}H_2$ [15]) $+ C_6H_5CH_3 \xrightarrow{b} CHCl_3 + C_6H_5\dot{C}H_2$				
Phot. of CCl_3Br				74 Tan 1
PR, glc	toluene (containing 20% ethylene oxide)	323.0(1)	$k_a/k_b = 1.71(1)$ [15])	
$\dot{C}Cl_3 + 3\text{-}CH_3OC_6H_4CH_3 \xrightarrow{a} CHCl_3 + 3\text{-}CH_3OC_6H_4\dot{C}H_2$ $+ C_6H_5CH_3 \xrightarrow{b} CHCl_3 + C_6H_5\dot{C}H_2$				
Phot. of CCl_3Br				74 Tan 1
PR, glc	toluene (containing 20% ethylene oxide)	323.0(1)	$k_a/k_b = 0.97(8)$	
$\dot{C}Cl_3 + 4\text{-}CH_3OC_6H_4CH_3 \xrightarrow{a} CHCl_3 + 4\text{-}CH_3OC_6H_4\dot{C}H_2$ $+ C_6H_5CH_3 \xrightarrow{b} CHCl_3 + C_6H_5\dot{C}H_2$				
Phot. of CCl_3Br				74 Tan 1
PR, glc	toluene (containing 20% ethylene oxide)	323.0(1)	$k_a/k_b = 3.59(3)$	
$\dot{C}Cl_3 + 3\text{-}ClC_6H_4Si(CH_3)_2H \xrightarrow{a} CHCl_3 + 3\text{-}ClC_6H_4\dot{S}i(CH_3)_2$ $+ C_6H_5Si(CH_3)_2H \xrightarrow{b} CHCl_3 + C_6H_5\dot{S}i(CH_3)_2$				
Therm. of BPO as initiator				71 Nag 2
PR, glc	CCl_4/ $3\text{-}ClC_6H_4Si(CH_3)_2H$/ $C_6H_5Si(CH_3)_2H$	353	$k_a/k_b = 0.71$	
$\dot{C}Cl_3 + 4\text{-}ClC_6H_4Si(CH_3)_2H \xrightarrow{a} CHCl_3 + 4\text{-}ClC_6H_4\dot{S}i(CH_3)_2$ $+ C_6H_5Si(CH_3)_2H \xrightarrow{b} CHCl_3 + C_6H_5\dot{S}i(CH_3)_2$				
Therm. of BPO as initiator				71 Nag 2
PR, glc	CCl_4/ $4\text{-}ClC_6H_4Si(CH_3)_2H$/ $C_6H_5Si(CH_3)_2H$	353	$k_a/k_b = 0.83$	
$\dot{C}Cl_3 + C_6H_5(CH_3)_2SiH \xrightarrow{a} CHCl_3 + C_6H_5(CH_3)_2\dot{S}i$ $+ (C_2H_5)_3SiH \xrightarrow{b} CHCl_3 + (C_2H_5)_3\dot{S}i$				
Therm. of BPO as initiator				71 Nag 1
PR, glc	CCl_4	353	$k_a/k_b = 0.84$	
$\dot{C}Cl_3 + 3\text{-}CH_3C_6H_4Si(CH_3)H_2 \xrightarrow{a} CHCl_3 + 3\text{-}CH_3C_6H_4\dot{S}i(CH_3)H$ $+ C_6H_5Si(CH_3)H_2 \xrightarrow{b} CHCl_3 + C_6H_5\dot{S}i(CH_3)H$				
Therm. of BPO as initiator				71Nag 2
PR, glc	CCl_4/ $3\text{-}CH_3C_6H_4Si(CH_3)H_2$/ $C_6H_5Si(CH_3)H_2$	353	$k_a/k_b = 1.13$	

[15]) Corrected by statistical factor of 2.

Reaction Radical generation Method	Solvent	T[K]	Rate data	Ref./ add. ref.
$\dot{C}Cl_3 + 4\text{-}CH_3C_6H_4Si(CH_3)H_2 \xrightarrow{a} CHCl_3 + 4\text{-}CH_3C_6H_4\dot{S}i(CH_3)H$ $+ C_6H_5Si(CH_3)H_2 \xrightarrow{b} CHCl_3 + C_6H_5\dot{S}i(CH_3)H$				
Therm. of BPO as initiator				71 Nag 2
PR, glc	CCl_4/ $4\text{-}CH_3C_6H_4Si(CH_3)H_2$/ $C_6H_5Si(CH_3)H_2$	353	$k_a/k_b = 1.28$	
$\dot{C}Cl_3 + c\text{-}C_8H_{16} \xrightarrow{a} CHCl_3 + (c\text{-}C_8\dot{H}_{15})$ $2\dot{C}Cl_3 \xrightarrow{b} C_2Cl_6$				
γ-rad. of CCl_4 + $c\text{-}C_8H_{16}$				81 Alf 2/ 55 Wat 1, 79 Pau 1
PR, glc	CCl_4	321(1)	$k_a/(2k_b)^{\frac{1}{2}} = 1.254(71) \cdot 10^{-2}$ $M^{-\frac{1}{2}} s^{-\frac{1}{2}}$ [16])	
		335(1)	$1.895(68) \cdot 10^{-2}$	
		340(1)	$2.526(92) \cdot 10^{-2}$	
		354(1)	$3.647(26) \cdot 10^{-2}$	
		367(1)	$4.080(85) \cdot 10^{-2}$	
		408(1)	$14.449(249) \cdot 10^{-2}$	
		425(1)	$26.299(195) \cdot 10^{-2}$ [16])	
$\dot{C}Cl_3 + CH_3CH_2CH_2HgCH_2C(CH_3)_3 \xrightarrow{a} CHCl_3$ + n-propylneopentyl-Hg($-\dot{H}$) $+ c\text{-}C_6H_{12} \xrightarrow{b} CHCl_3 + (c\text{-}C_6\dot{H}_{11})$				
Therm. and phot. of n-propylneopentylmercury in CCl_4				77 Nug 1
PR, glc	CCl_4	368	$k_a/k_b = 95.5$ [17])	
$\dot{C}Cl_3 + C_6H_5CH(CH_3)_2 \xrightarrow{a} CHCl_3 + C_6H_5\dot{C}(CH_3)_2$ $+ C_6H_5CH_3 \xrightarrow{b} CHCl_3 + C_6H_5\dot{C}H_2$				
Phot. of $CBrCl_3$				80 Tan 1
PR	$C_6H_5Br/CBrCl_3$ (molar ratio 2:3)	343	$k_a/k_b = 5.00(5)$ [14])	
PR, glc	Freon 113	313.0(1)		75 Tan 1
	+ K_2CO_3		$k_a/k_b = 9.97(123)$	
	+ ethylene oxide		$k_a/k_b = 9.26(240)$	
$\dot{C}Cl_3 + C_6H_5CH(CH_3)_2 \xrightarrow{a} CHCl_3 + C_6H_5\dot{C}(CH_3)_2$ $+ C_6H_5CH_2CH_3 \xrightarrow{b} CHCl_3$ + ethylbenzene($-\dot{H}$)				
Phot. of CCl_3Br				75 Tan 1
PR, glc	Freon 113	313.0(1)		
	+ K_2CO_3		$k_a/k_b = 1.47(7)$	
	+ ethylene oxide		$k_a/k_b = 1.33(22)$	
$\dot{C}Cl_3 + 3\text{-}CH_3C_6H_4Si(CH_3)_2H \xrightarrow{a} CHCl_3 + 3\text{-}CH_3C_6H_4\dot{S}i(CH_3)_2$ $+ C_6H_5Si(CH_3)_2H \xrightarrow{b} CHCl_3 + C_6H_5\dot{S}i(CH_3)_2$				
Therm. of BPO as initiator				71 Nag 2
PR, glc	CCl_4/ $3\text{-}CH_3C_6H_4Si(CH_3)_2H$/ $C_6H_5Si(CH_3)_2H$	353	$k_a/k_b = 1.16$	

[14]) Solutions contained ethylene oxide as HBr scavenger.

[16]) $\log[A_a/M^{-1}s^{-1}] = 9.13(13)$ and $E_a(a) = 40.0(8)\,kJ\,mol^{-1}$ based $\log[k_a/(2k_b)^{\frac{1}{2}}] = 3.42(14) - 33.1(8)/2.303\,RT$ (R in $kJ\,mol^{-1}K^{-1}$), and $E_a = 13.8\,kJ\,mol^{-1}$ [55 Wat 1] and $\log[A/M^{-1}s^{-1}] = 11.42$ [79 Pau 1] for $\dot{C}Cl_3 + \dot{C}Cl_3$ reaction, respectively.

[17]) Per reactive hydrogen atom. Overall relative reactivity $k_a/k_b = 16$.

Reaction Radical generation Method	Solvent	T[K]	Rate data	Ref./ add. ref.
$\dot{C}Cl_3 + 4\text{-}CH_3C_6H_4Si(CH_3)_2H \xrightarrow{a} CHCl_3 + 4\text{-}CH_3C_6H_4\dot{S}i(CH_3)_2$ $+ C_6H_5Si(CH_3)_2H \xrightarrow{b} CHCl_3 + C_6H_5\dot{S}i(CH_3)_2$				
Therm. of BPO as initiator				71 Nag 2
PR, glc	CCl_4/ $4\text{-}CH_3C_6H_4Si(CH_3)_2H$/ $C_6H_5Si(CH_3)_2H$	353	$k_a/k_b = 1.25$	
$\dot{C}Cl_3 + (CH_3)_2CHCH_2HgCH_2C(CH_3)_3 \xrightarrow{a} CHCl_3$ + isobutylneopentyl-Hg(−Ḣ) $+ c\text{-}C_6H_{12} \xrightarrow{b} CHCl_3 + (c\text{-}C_6\dot{H}_{11})$				
Therm. and phot. of isobutylneopentylmercury in CCl_4				77 Nug 1
PR, glc	CCl_4	368	$k_a/k_b = 605$ [18])	
$\dot{C}Cl_3 + (CH_3)_2CHCH_2HgCH_2C(CH_3)_3 \xrightarrow{a} CHCl_3$ + isobutylneopentyl-Hg(−Ḣ) $+ (CH_3)_3CCH_2CH(CH_3)_2 \xrightarrow{b} CHCl_3$ + 2,2,4-trimethylpentane(−Ḣ)				
Therm. and phot. of isobutylneopentylmercury in CCl_4				77 Nug 1
PR, glc	CCl_4	368	$k_a/k_b = 577$ [19])	
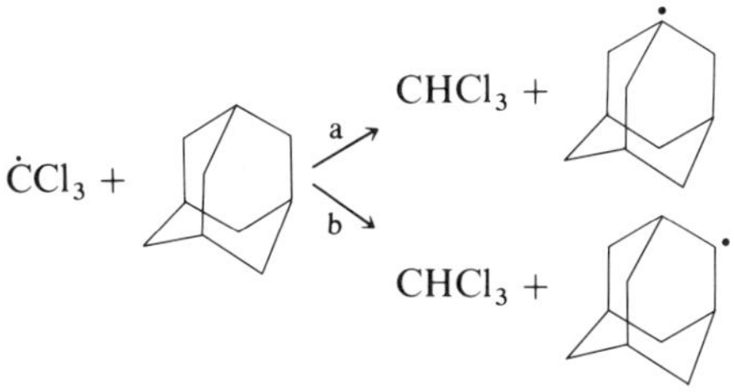				
BPO as initiator				72 Tab 1
PR, glc	CCl_4	368	$k_a/k_b = 25.5(15)$	
$\dot{C}Cl_3 + 4\text{-}(CH_3)_3CC_6H_4CH_3 \xrightarrow{a} CHCl_3 + 4\text{-}(CH_3)_3CC_6H_4\dot{C}H_2$ $+ C_6H_5CH_3 \xrightarrow{b} CHCl_3 + C_6H_5\dot{C}H_2$				
Phot. of CCl_3Br				74 Tan 1
PR, glc	toluene (containing 20% ethylene oxide)	323.0(1)	$k_a/k_b = 1.82(6)$	
$\dot{C}Cl_3 + c\text{-}C_{12}H_{24} \xrightarrow{a} CHCl_3 + (c\text{-}C_{12}\dot{H}_{23})$ $2\dot{C}Cl_3 \xrightarrow{b} C_2Cl_6$				
γ-rad. of $c\text{-}C_{12}H_{24} + CCl_4$				81 Alf 2/
PR, glc	CCl_4	321(1)	$k_a/(2k_b)^{\frac{1}{2}} = 0.636(36)\cdot10^{-2}$ $M^{-\frac{1}{2}}s^{-\frac{1}{2}}$ [20])	55 Wat 1, 79 Pau 1
		335(1)	$1.085(51)\cdot10^{-2}$	
		340(1)	$1.281(43)\cdot10^{-2}$	
		354(1)	$2.271(74)\cdot10^{-2}$	
		367(1)	$2.417(16)\cdot10^{-2}$	
		408(1)	$5.529(328)\cdot10^{-2}$	
		425(1)	$20.386(175)\cdot10^{-2}$ [20])	

[18]) Per reactive hydrogen atom. Overall relative reactivity $k_a/k_b = 50.5$.

[19]) Per reactive hydrogen atom.

[20]) $\log[A_a/M^{-1}s^{-1}] = 9.04(49)$ and $E_a(a) = 37.9(33)\,kJ\,mol^{-1}$ based on $\log[k_a/(2k_b)^{\frac{1}{2}}] = 3.30(49) - 34.3(33)/2.303\,RT$($R$ in $kJ\,mol^{-1}K^{-1}$), and $E_a = 13.8\,kJ\,mol^{-1}$ [55 Wat 1] and $\log[A/M^{-1}s^{-1}] = 11.42$ [79 Pau 1] for $\dot{C}Cl_3 + \dot{C}Cl_3$ reaction, respectively.

Reaction Radical generation Method	Solvent	T[K]	Rate data	Ref./ add. ref.
$\dot{C}Cl_3$ + fluorene $\xrightarrow{a}$ $CHCl_3$ + fluorene$(-\dot{H})$ $+ C_6H_5CH_3 \xrightarrow{b} CHCl_3 + C_6H_5\dot{C}H_2$				
Phot. of $CBrCl_3$				80 Tan 1
PR	$C_6H_5Br/CBrCl_3$/ (molar ratio 2:3)	343	$k_a/k_b = 8.80(5)$ [21]	
$\dot{C}Cl_3$ + 10-bromo-9-methylanthracene $\xrightarrow{a}$ $CHCl_3$ + 10-bromo-9-methylanthracene$(-\dot{H})$ + fluorene $\xrightarrow{b}$ $CHCl_3$ + fluorene$(-\dot{H})$				
Phot. of $CBrCl_3$				80 Tan 1
PR, NMR	$C_6H_5Br/CBrCl_3$ (molar ratio 2:3)		$k_a/k_b = 2.88(5)$ [21]	
$\dot{C}Cl_3$ + 10-chloro-9-methylanthracene $\xrightarrow{a}$ $CHCl_3$ + 10-chloro-9-methylanthracene$(-\dot{H})$ + fluorene $\xrightarrow{b}$ $CHCl_3$ + fluorene$(-\dot{H})$				
Phot. of $CBrCl_3$				80 Tan 1
PR, NMR	$C_6H_5Br/CBrCl_3$ (molar ratio 2:3)	343	$k_a/k_b = 2.83(8)$ [21]	
$\dot{C}Cl_3$ + 9-methylanthracene $\xrightarrow{a}$ $CHCl_3$ + 9-methylanthracene$(-\dot{H})$ + fluorene $\xrightarrow{b}$ $CHCl_3$ + fluorene$(-\dot{H})$				
Phot. of $CBrCl_3$				80 Tan 1/
PR, NMR	$C_6H_5Br/CBrCl_3$ (molar ratio 2:3)	343	$k_a/k_b = 2.83(8)$ [21] [22]	80 Nol 1
$\dot{C}Cl_3$ + 10-cyano-9-methylanthracene $\xrightarrow{a}$ $CHCl_3$ + 10-cyano-9-methylanthracene$(-\dot{H})$ + fluorene $\xrightarrow{b}$ $CHCl_3$ + fluorene$(-\dot{H})$				
Phot. of $CBrCl_3$				80 Tan 1
PR, NMR	$C_6H_5Br/CBrCl_3$ (molar ratio 2:3)	343	$k_a/k_b = 2.77(17)$ [21] $= 2.39(18)$ [23]	
$\dot{C}Cl_3$ + 9,10-dimethylanthracene $\xrightarrow{a}$ $CHCl_3$ + 9,10-dimethylanthracene$(-\dot{H})$ $+ C_6H_5CH_3 \xrightarrow{b} CHCl_3 + C_6H_5\dot{C}H_2$				
Phot. of $CBrCl_3$				80 Tan 1
PR	$C_6H_5Br/CBrCl_3$ (molar ratio 2:3)	343	$k_a/k_b = 47.5(7)$ [21]	
$\dot{C}Cl_3$ + 9,10-dimethylanthracene $\xrightarrow{a}$ $CHCl_3$ + 9,10-dimethylanthracene$(-\dot{H})$ + fluorene $\xrightarrow{b}$ $CHCl_3$ + fluorene$(-\dot{H})$				
Phot. of $CBrCl_3$				80 Tan 1
PR, NMR	$C_6H_5Br/CBrCl_3$ (molar ratio 2:3)	343	$k_a/k_b = 2.71(5)$ [21] [24]	
$\dot{C}Cl_3$ + 10-methoxy-9-methylanthracene $\xrightarrow{a}$ $CHCl_3$ + 10-methoxy-9-methylanthracene$(-\dot{H})$ $+ C_6H_5CH_3 \xrightarrow{b} CHCl_3 + C_6H_5\dot{C}H_2$				
Phot. of $CBrCl_3$				80 Tan 1
PR	$C_6H_5Br/CBrCl_3$ (molar ratio 2:3)	343	$k_a/k_b = 27.30(35)$ [21]	

[21]) Solution contained ethylene oxide as HBr scavenger.
[22]) Corrected for ring substitution [80 Nol 1].
[23]) Solution contained K_2CO_3 as HBr scavenger.
[24]) Statistically corrected.

Reaction Radical generation Method	Solvent	T[K]	Rate data	Ref./ add. ref.
$\dot{C}Cl_3$ + 10-methoxy-9-methylanthracene $\xrightarrow{a}$ $CHCl_3$ + 10-methoxy-9-methylanthracene(−$\dot{H}$) + fluorene $\xrightarrow{b}$ $CHCl_3$ + fluorene(−$\dot{H}$)				
Phot. of $CBrCl_3$				80 Tan 1
PR, NMR	$C_6H_5Br/CBrCl_3$ (molar ratio 2:3)	343	k_a/k_b = 3.05(35) [21]) = 3.08(26) [23])	
$\dot{C}Cl_3$ + 10-carbomethoxy-9-methylanthracene $\xrightarrow{a}$ $CHCl_3$ + 10-carbomethoxy-9-methylanthracene(−$\dot{H}$) + fluorene $\xrightarrow{b}$ $CHCl_3$ + fluorene(−$\dot{H}$)				
Phot. of $CBrCl_3$				80 Tan 1
PR, NMR	$C_6H_5Br/CBrCl_3$ (molar ratio 2:3)	343	k_a/k_b = 2.85(15) [21])	

Further data on relative rate constants for reactions of $\dot{C}Cl_3$ radicals are to be found in references 51 Mel 1; 60 Huy 1; 61 Huy 1; 63 Huy 1; 63 Rus 1; 64 Hua 1; 64 Mar 1,2; 66 Car 1; 66 Hua 1; 67 Car 1; 68 Gle 1; 68 Lee 1; 68 Owe 1; 69 Cha 1; 69 Tot 1; 69 Unr 1; 71 Koc 1; 71 Unr 1; 73 Lee 1; 73 Won 1; 74 Gle 1; 74 New 1; 75 Cha 1; 77 Che 1; 80 Nol 1. They have all been evaluated from systems where $\dot{C}Cl_3$ was generated via photolysis of $CBrCl_3$. Such systems have been shown to be complicated by secondary reactions of HBr with $\dot{C}Cl_3$ and possible reversible hydrogen transfer between Br atoms and primary product radicals (see [80 Tan 1] and references cited therein). Since all data in the above cited papers may be affected by such complications they have been omitted from this compilation. This does not mean that the published rate constant ratios are necessarily incorrect, but in all cases it would need additional experiments with appropriate solutions (presence of HBr scavengers, e.g. ethylene oxide, K_2CO_3 etc.) to evaluate the possible influence of secondary reactions (see also introductory text to this rate constant compilation).

Reaction Radical generation Method	Solvent	T[K]	Rate data	Ref./ add. ref.
$\dot{C}F_3$ + CH_2=$CHCH_2CH_3$ $\xrightarrow{a}$ CHF_3 + butene-1(−$\dot{H}$) $\xrightarrow{b}$ (butene-1-$CF_3^{\cdot}$)				
Phot. of hexafluoroazomethane				61 Ste 1
PR, glc	2,2,4-trimethylpentane	338	k_a/k_b = 0.213	
$\dot{C}F_3$ + CH_2=$CHCH_2CH_3$ $\xrightarrow{a}$ CHF_3 + butene-1(−$\dot{H}$) + $(CH_3)_3CCH_2CH(CH_3)_2$ $\xrightarrow{b}$ CHF_3 + 2,2,4-trimethylpentane(−$\dot{H}$)				
Phot. of hexafluoroazomethane				61 Ste 1
PR, glc	2,2,4-trimethylpentane	338	k_a/k_b = 146.5 73.2 [25])	
$\dot{C}F_3$ + *cis*-CH_3CH=$CHCH_3$ $\xrightarrow{a}$ CHF_3 + *cis*-butene-2(−$\dot{H}$) $\xrightarrow{b}$ (*cis*-butene-2-$CF_3^{\cdot}$)				
Phot. of hexafluoroazomethane				61 Ste 1
PR, glc	2,2,4-trimethylpentane	338	k_a/k_b = 0.315	
$\dot{C}F_3$ + *cis*-CH_3CH=$CHCH_3$ $\xrightarrow{a}$ CHF_3 + *cis*-butene-2(−$\dot{H}$) + $(CH_3)_3CCH_2CH(CH_3)_2$ $\xrightarrow{b}$ CHF_3 + 2,2,4-trimethylpentane(−$\dot{H}$)				
Phot. of hexafluoroazomethane				61 Ste 1
PR, glc	2,2,4-trimethylpentane	338	k_a/k_b = 183 30.4 [25])	
$\dot{C}F_3$ + *trans*-CH_3CH=$CHCH_3$ $\xrightarrow{a}$ CHF_3 + *trans*-butene-2(−$\dot{H}$) $\xrightarrow{b}$ (*trans*-butene-2-$CF_3^{\cdot}$)				
Phot. of hexafluoroazomethane				61 Ste 1
PR, glc	2,2,4-trimethylpentane	338	k_a/k_b = 0.332	

[21]) Solution contained ethylene oxide as HBr scavenger.
[23]) Solution contained K_2CO_3 as HBr scavenger.
[25]) Per reactive H atom.

Reaction Radical generation Method	Solvent	T[K]	Rate data	Ref./ add. ref.
$\dot{C}F_3$ + *trans*-$CH_3CH{=}CHCH_3$ $\xrightarrow{a}$ CHF_3 + *trans*-butene-2(−$\dot{H}$) + $(CH_3)_3CCH_2CH(CH_3)_2$ $\xrightarrow{b}$ CHF_3 + 2,2,4-trimethylpentane(−$\dot{H}$)				
Phot. of hexafluoroazomethane PR, glc	2,2,4-trimethylpentane	338	$k_a/k_b = 194$ 32.3 [25])	61 Ste 1
$\dot{C}F_3$ + $CH_3CH{=}C{=}CHCH_3$ $\xrightarrow{a}$ CHF_3 + 2,3-pentadiene(−$\dot{H}$) $\xrightarrow{b}$ (2,3-pentadiene-$CF_3^{\bullet}$)				
Phot. of hexafluoroazomethane PR, glc	2,2,4-trimethylpentane	338	$k_a/k_b = 0.121$	61 Ste 1
$\dot{C}F_3$ + $CH_3CH{=}C{=}CHCH_3$ $\xrightarrow{a}$ CHF_3 + 2,3-pentadiene(−$\dot{H}$) + $(CH_3)_3CCH_2CH(CH_3)_2$ $\xrightarrow{b}$ CHF_3 + 2,2,4-trimethylpentane(−$\dot{H}$)				
Phot. of hexafluoroazomethane PR, glc	2,2,4-trimethylpentane	338	$k_a/k_b = 50$ 8.3 [25])	61 Ste 1
$\dot{C}F_3$ + $CH_2{=}CHCH(CH_3)_2$ $\xrightarrow{a}$ CHF_3 + 3-methylbutene-1(−$\dot{H}$) $\xrightarrow{b}$ (3-methylbutene-1-$CF_3^{\bullet}$)				
Phot. of hexafluoroazomethane PR, glc	2,2,4-trimethylpentane	338	$k_a/k_b = 0.098$	61 Ste 1
$\dot{C}F_3$ + $CH_2{=}CHCH(CH_3)_2$ $\xrightarrow{a}$ CHF_3 + 3-methylbutene-1(−$\dot{H}$) + $(CH_3)_3CCH_2CH(CH_3)_2$ $\xrightarrow{b}$ CHF_3 + 2,2,4-trimethylpentane(−$\dot{H}$)				
Phot. of hexafluoroazomethane PR, glc	2,2,4-trimethylpentane	338	$k_a/k_b = 69.2$ [25])	61 Ste 1
$\dot{C}F_3$ + c-C_6H_{12} $\xrightarrow{a}$ CF_3H + (c-$C_6H_{11}^{\bullet}$) + I_2 $\xrightarrow{b}$ CF_3I + $\dot{I}$				
γ-rad. of CF_3Br or CF_3Cl in c-C_6H_{12} PR, glc, using $^{131}I_2$	c-C_6H_{12}	RT	$k_a/k_b = 3.15 \cdot 10^{-5}$	69 Inf 1
$\dot{C}F_3$ + $(CH_3)_2C{=}C(CH_3)_2$ $\xrightarrow{a}$ CF_3H + tetramethylethylene(−$\dot{H}$) $\xrightarrow{b}$ $CF_3(CH_3)_2C\dot{C}(CH_3)_2$				
Phot. of hexafluoroazomethane PR, glc	tetramethylethylene	338	$k_a/k_b = 5.0 \cdot 10^{-3}$	63 Kom 1
$\dot{C}F_3$ + $(CH_3)_2C{=}C(CH_3)_2$ $\xrightarrow{a}$ CF_3H + tetramethylethylene(−$\dot{H}$) + $CH_3C(CH_3)_2CH_2CH(CH_3)CH_3$ $\xrightarrow{b}$ CF_3H + 2,2,4-trimethylpentane(−$\dot{H}$)				
Phot. of hexafluoroazomethane PR, glc	tetramethylethylene and 2,2,4-trimethylpentane	338	$k_a/k_b = 2.7$ [26])	63 Kom 1
$\dot{C}F_3$ + $(CH_3)_2CHCH(CH_3)_2$ $\xrightarrow{a}$ CF_3H + 2,3-dimethylbutane(−$\dot{H}$) + *trans*-$CHCl{=}CHCl$ $\xrightarrow{b}$ $CF_3CHCl\dot{C}HCl$				
Phot. of hexafluoroazomethane PR, glc	2,3-dimethylbutane	338	$k_a/k_b = 0.34$	63 Kom 1

[25]) Per reactive H atom.
[26]) Derived from individual experiments in either solvent.

Reaction Radical generation Method	Solvent	T[K]	Rate data	Ref./ add. ref.
$\dot{C}F_3 + (CH_3)_2CHCH(CH_3)_2 \xrightarrow{a} CF_3H + 2,3\text{-dimethylbutane}(-\dot{H})$ $+ (CH_3)_3CCH_2CH(CH_3)_2 \xrightarrow{b} CF_3H + 2,2,4\text{-trimethylpentane}(-\dot{H})$				
Phot. of hexafluoroazomethane				64 Owe 1,
PR	2,3-dimethylbutane and 2,2,4-trimethylpentane	338	k_a/k_b [26]) = 4.53 4.7 [27])	63 Kom 1
$\dot{C}F_3 + (CH_3)_2CHCH(CH_3)_2 \xrightarrow{a} CF_3H + 2,3\text{-dimethylbutane}(-\dot{H})$ $+ \text{olefine} \xrightarrow{b} (\text{olefine-}CF_3^{\bullet})$				
Phot. of hexafluoroazomethane				63 Dix 1,
PR, glc	2,3-dimethylbutane	338	olefine:	63 Kom 1,
			propylene, $k_a/k_b = 9.21 \cdot 10^{-3}$	64 Owe 1
			$7.46 \cdot 10^{-3}$ [28])	[29])
			isobutene, $2.92 \cdot 10^{-3}$	
			$2.65 \cdot 10^{-3}$ [28])	
			tetramethyl-ethylene, $8.16 \cdot 10^{-3}$	
			$8.13 \cdot 10^{-3}$ [28])	
			1,3-butadiene, $1.15 \cdot 10^{-3}$	
			cyclopentene, $1.66 \cdot 10^{-2}$	
			benzene, 0.6	
			vinylfluoride, $6.45 \cdot 10^{-2}$	
			2-fluoropropylene, $1.93 \cdot 10^{-2}$	
			vinylchloride, $1.50 \cdot 10^{-2}$	
			$1.58 \cdot 10^{-2}$ [28])	
			ethylene-dichloride (*trans*), 0.345 [27])	
			ethylene, $1.03 \cdot 10^{-2}$ [28])	
			butene-1, $8.43 \cdot 10^{-3}$ [28])	
			butadiene, $1.16 \cdot 10^{-3}$ [28])	
			isoprene, $7.12 \cdot 10^{-4}$ [28])	
			styrene, $2.16 \cdot 10^{-3}$ [28])	
			α-methyl-styrene, $1.24 \cdot 10^{-3}$ [28])	
$\dot{C}F_3 + C_6H_5CH_3 \xrightarrow{a} CHF_3 + C_6H_5\dot{C}H_2$ $\dot{C}F_3 + C_6H_5CH_3 \xrightarrow{b} (\text{toluene-}CF_3^{\bullet})$				
Phot. of hexafluoroazomethane				62 Whi 1
PR, glc	2,2,4-trimethylpentane	338	$k_a/k_b = 0.030$	
$\dot{C}F_3 + C_6H_5CH_3 \xrightarrow{a} CHF_3 + C_6H_5\dot{C}H_2$ $+ (CH_3)_3CCH_2CH(CH_3)_2 \xrightarrow{b} CHF_3 + 2,2,4\text{-trimethylpentane}(-\dot{H})$				
Phot. of hexafluoroazomethane				62 Whi 1
PR, glc	2,2,4-trimethylpentane	338	$k_a/k_b = 0.12(2)$ [30])	
$\dot{C}F_3 + C_6H_5CH_2CH_3 \xrightarrow{a} CHF_3 + \text{ethylbenzene}(-\dot{H})$ $\dot{C}F_3 + C_6H_5CH_2CH_3 \xrightarrow{b} (\text{ethylbenzene-}CF_3^{\bullet})$				
Phot. of hexafluoroazomethane				62 Whi 1
PR, glc	2,2,4-trimethylpentane	338	$k_a/k_b = 0.15$	

[26]) Derived from individual experiments in either solvent.
[27]) From [63 Kom 1].
[28]) From [64 Owe 1].
[29]) Further data at other temperatures.
[30]) Per reactive H atom.

Reaction Radical generation Method	Solvent	T[K]	Rate data	Ref./ add. ref.
$\dot{C}F_3 + C_6H_5CH_2CH_3 \xrightarrow{a} CHF_3$ + ethylbenzene($-\dot{H}$) $+ (CH_3)_3CCH_2CH(CH_3)_2 \xrightarrow{b} CHF_3$ + 2,2,4-trimethylpentane($-\dot{H}$)				
Phot. of hexafluoroazomethane				62 Whi 1
PR, glc	2,2,4-trimethylpentane	338	$k_a/k_b = 0.83(5)$ [30]	
$\dot{C}F_3 + (CH_3)_3CCH_2CH(CH_3)_2 \xrightarrow{a} CHF_3$ + 2,2,4-trimethylpentane($-\dot{H}$) + aromatic $\xrightarrow{b}$ (aromatic-$CF_3^{\cdot}$)				
Phot. of hexafluoroazomethane				62 Ste 1
PR	2,2,4-trimethylpentane	338	aromatic: benzene, $k_a/k_b = 0.141$ biphenyl, 0.0629 naphthalene, 0.00974 phenanthrene, 0.0101 pyrene, 0.00278 anthracene, 0.000623	
$\dot{C}F_3 + (CH_3)_3CCH_2CH(CH_3)_2 \xrightarrow{a} CF_3H$ + 2,2,4-trimethylpentane($-\dot{H}$) + olefine $\xrightarrow{b}$ (olefine-$CF_3^{\cdot}$)				
Phot. of hexafluoroazomethane				61 Ste 1,
PR	2,2,4-trimethylpentane	338	olefine: $k_a/k_b =$ ethylene, $2.19 \cdot 10^{-3}$ propylene, $1.61 \cdot 10^{-3}$ isobutene, $5.81 \cdot 10^{-4}$ styrene, $4.72 \cdot 10^{-4}$ [28] butadiene, $1.94 \cdot 10^{-4}$ [28] tetramethyl-ethylene, $1.86 \cdot 10^{-3}$ [28] vinylchloride, $3.66 \cdot 10^{-3}$ [28] vinylfluoride, $1.36 \cdot 10^{-2}$ [28] ethylene-dichloride (*trans*), $8.0 \cdot 10^{-2}$ [28]	64 Owe 1
PR, glc, MS	2,2,4-trimethylpentane	338	olefine: $k_a/k_b =$ $CH_2{=}CH_2$, 0.00238 $CD_2{=}CD_2$, 0.00222 $CH_3CH{=}CH_2$, 0.00157 $CH_3CH{=}CD_2$, 0.00146 $CD_3CD{=}CD_2$, 0.00145 $CH_2{=}CHCH{=}CH_2$, 0.000195 $CD_2{=}CDCD{=}CD_2$, 0.000178 $C_6H_5CH{=}CH_2$, 0.000468 $C_6H_5CD{=}CD_2$, 0.000426	62 Fel 1
PR, glc	2,2,4-trimethylpentane	338	olefine: $k_a/k_b =$ *cis*-dichloroethylene, 0.154 *trans*-dichloro-ethylene, $8.0 \cdot 10^{-2}$ trichloroethylene, $7.81 \cdot 10^{-2}$ tetrachloroethylene, 0.769 diethylmaleate, $3.75 \cdot 10^{-2}$ diethylfumarate, $1.18 \cdot 10^{-2}$ tetramethylethylene, $1.86 \cdot 10^{-3}$	63 Kom 1

[28]) From [64 Owe 1].
[30]) Per reactive H atom.

Reaction Radical generation Method	Solvent	T[K]	Rate data	Ref./ add. ref.
$\dot{C}F_3 + C_6H_5CH(CH_3)_2 \xrightarrow{a} CHF_3 + C_6H_5\dot{C}(CH_3)_2$ $\xrightarrow{b}$ (cumene-$CF_3^{\cdot}$)				
Phot. of hexafluoroazomethane PR, glc	2,2,4-trimethylpentane	338	$k_a/k_b = 0.26$	62 Whi 1
$\dot{C}F_3 + C_6H_5CH(CH_3)_2 \xrightarrow{a} CHF_3 + C_6H_5\dot{C}(CH_3)_2$ $+ (CH_3)_3CCH_2CH(CH_3)_2 \xrightarrow{b} CHF_3 +$ 2,2,4-trimethylpentane($-\dot{H}$)				
Phot. of hexafluoroazomethane PR, glc	2,2,4-trimethylpentane	338	$k_a/k_b = 2.98(18)$ [30]	62 Whi 1
$\dot{C}F_3 + 1,3,5\text{-}(CH_3)_3C_6H_3 \xrightarrow{a} CHF_3 +$ mesitylene($-\dot{H}$) $\xrightarrow{b}$ (mesitylene—$CF_3^{\cdot}$)				
Phot. of hexafluoroazomethane PR, glc	2,2,4-trimethylpentane	338	$k_a/k_b = 0.031$	62 Whi 1
$\dot{C}F_3 + 1,3,5\text{-}(CH_3)_3C_6H_3 \xrightarrow{a} CHF_3 +$ mesitylene($-\dot{H}$) $+ (CH_3)_3CCH_2CH(CH_3)_2 \xrightarrow{b} CHF_3 +$ 2,2,4-trimethylpentane($-\dot{H}$)				
Phot. of hexafluoroazomethane PR, glc	2,2,4-trimethylpentane	338	$k_a/k_b = 0.14(1)$ [30]	62 Whi 1
$\dot{C}HCl_2 + Cl_3SiH \xrightarrow{a} CH_2Cl_2 + Cl_3\dot{S}i$ $+ (C_2H_5)_3SiH \xrightarrow{b} CH_2Cl_2 + (C_2H_5)_3\dot{S}i$				
Therm. of BPO as initiator PR, glc	$CHCl_3$	353	$k_a/k_b = 0.19$	71 Nag 1
$\dot{C}HCl_2 + C_2H_5Cl_2SiH \xrightarrow{a} CH_2Cl_2 + C_2H_5Cl_2\dot{S}i$ $+ (C_2H_5)_3SiH \xrightarrow{b} CH_2Cl_2 + (C_2H_5)_3\dot{S}i$				
Therm. of BPO as initiator PR, glc	$CHCl_3$	353	$k_a/k_b = 0.37$	71 Nag 1
$\dot{C}HCl_2 + (C_2H_5)_2ClSiH \xrightarrow{a} CH_2Cl_2 + (C_2H_5)_2Cl\dot{S}i$ $+ (C_2H_5)_3SiH \xrightarrow{b} CH_2Cl_2 + (C_2H_5)_3\dot{S}i$				
Therm. of BPO as initiator PR, glc	$CHCl_3$	353	$k_a/k_b = 0.53$	71 Nag 1
$\dot{C}HCl_2 +$ 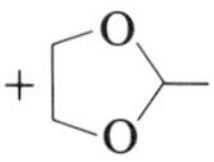 $-(CH_2)_3CH_3 \xrightarrow{a} CH_2Cl_2 +$ 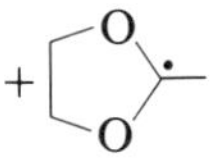 $-(CH_2)_3CH_3$ $2\,\dot{C}HCl_2 \xrightarrow{b} CHCl_2CHCl_2$				
Initiated by thermal decomp. of BPO in 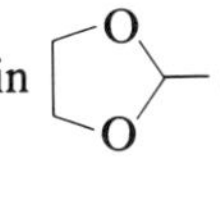 $-(CH_2)_3CH_3 + CHCl_3 + C_6H_6$				79 Sam 1
PR, glc	$C_6H_6/CHCl_3$ (1.5 M)/ 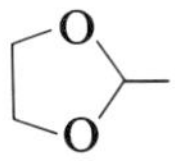 $-(CH_2)_3CH_3$ (3 M)	353	$k_a/(2k_b)^{\frac{1}{2}} = 2.5 \cdot 10^{-3}\ M^{-\frac{1}{2}}\ s^{-\frac{1}{2}}$	

[30]) Per reactive H atom.

Reaction Radical generation Method	Solvent	$T[K]$	Rate data	Ref./ add. ref.

$\dot{C}HCl_2 + $ $CH_2CH(CH_3)_2 \xrightarrow{a} CH_2Cl_2 + $ $CH_2CH(CH_3)_2$

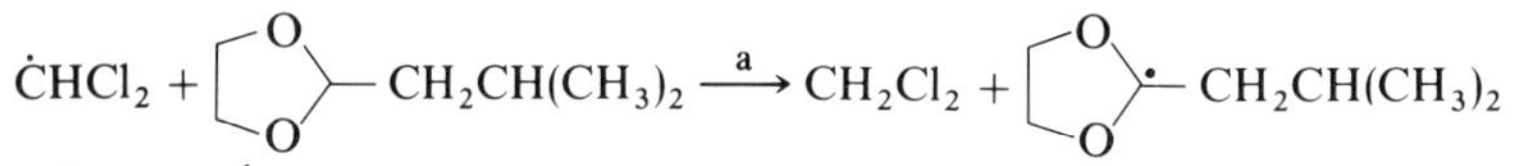

$2\,\dot{C}HCl_2 \xrightarrow{b} CHCl_2CHCl_2$

Initiated by thermal decomp. of BPO in $CH_2CH(CH_3)_2 + CHCl_3 + C_6H_6$

| PR, glc | $C_6H_6/CHCl_3\,(1.5\,M)/$ | 353 | $k_a/(2k_b)^{\frac{1}{2}} = 1.6\cdot 10^{-3}\,M^{-\frac{1}{2}}s^{-\frac{1}{2}}$ | |

$CH_2CH(CH_3)_2$

(3 M)

$\dot{C}HCl_2 + C_6H_5(CH_3)_2SiH \xrightarrow{a} CH_2Cl_2 + C_6H_5(CH_3)_2\dot{S}i$
$\quad + (C_2H_5)_3SiH \xrightarrow{b} CH_2Cl_2 + (C_2H_5)_3\dot{S}i$

Therm. of BPO as initiator

| PR, glc | $CHCl_3$ | 353 | $k_a/k_b = 0.89$ | 71 Nag 1 |

$\dot{C}Cl_2CN + c\text{-}C_5H_{10} \xrightarrow{a} CHCl_2CN + (c\text{-}C_5H_9^{\cdot})$
$\dot{C}Cl_2CN \xrightarrow{b} (CCl_2CN)_2$

$^{60}Co\text{-}\gamma\text{-rad. of }c\text{-pentane} + CCl_3CN$

PR, glc	c-pentane	369	$k_a/(2k_b)^{\frac{1}{2}} = 0.70\cdot 10^{-4}$ $dm^{\frac{3}{2}}mol^{-\frac{1}{2}}s^{-\frac{1}{2}}$ [31]	76 Gon 1
		383	$1.22\cdot 10^{-4}$	
		393	$1.43\cdot 10^{-4}$	
		403	$2.07\cdot 10^{-4}$	
		413	$3.51\cdot 10^{-4}$	
		423	$5.24\cdot 10^{-4}$	
		433	$5.83\cdot 10^{-4}$	
		443	$8.77\cdot 10^{-4}$	
		450	$9.43\cdot 10^{-4}$	

$\dot{C}Cl_2CN + c\text{-}C_6H_{12} \xrightarrow{a} CHCl_2CN + (c\text{-}C_6H_{11}^{\cdot})$
$2\,\dot{C}Cl_2CN \xrightarrow{b} (CCl_2CN)_2$

$^{60}Co\text{-}\gamma\text{-irr. of }c\text{-hexane} + CCl_3CN$

PR, glc	c-hexane	373	$k_a/(2k_b)^{\frac{1}{2}} = 0.5\cdot 10^{-4}$ $dm^{\frac{3}{2}}mol^{-\frac{1}{2}}s^{-\frac{1}{2}}$ [32] $k_a/k_b^{\frac{1}{2}} = 0.64\cdot 10^{-4}\,dm^{\frac{3}{2}}mol^{-\frac{1}{2}}s^{-\frac{1}{2}}$	76 Gon 1
		383	$1.04\cdot 10^{-4}$	
		398	$1.83\cdot 10^{-4}$	
		413	$k_a/(2k_b)^{\frac{1}{2}} = 3.09\cdot 10^{-4}\,dm^{\frac{3}{2}}mol^{-\frac{1}{2}}s^{-\frac{1}{2}}$ $k_a/k_b^{\frac{1}{2}} = 3.67\cdot 10^{-4}\,dm^{\frac{3}{2}}mol^{-\frac{1}{2}}s^{-\frac{1}{2}}$	
		423	$5.28\cdot 10^{-4}$	
		443	$k_a/(2k_b)^{\frac{1}{2}} = 6.99\cdot 10^{-4}\,dm^{\frac{3}{2}}mol^{-\frac{1}{2}}s^{-\frac{1}{2}}$ $k_a/k_b^{\frac{1}{2}} = 8.93\cdot 10^{-4}\,dm^{\frac{3}{2}}mol^{-\frac{1}{2}}s^{-\frac{1}{2}}$	
		463	$k_a/(2k_b)^{\frac{1}{2}} = 19.49\cdot 10^{-4}$ $dm^{\frac{3}{2}}mol^{-\frac{1}{2}}s^{-\frac{1}{2}}$ $k_a/k_b^{\frac{1}{2}} = 20.17\cdot 10^{-4}\,dm^{\frac{3}{2}}mol^{-\frac{1}{2}}s^{-\frac{1}{2}}$	
		478	$30.22\cdot 10^{-4}$	
		493	$k_a(2k_b)^{\frac{1}{2}} = 55.65\cdot 10^{-4}$ $dm^{\frac{3}{2}}mol^{-\frac{1}{2}}s^{-\frac{1}{2}}$ $k_a/k_b^{\frac{1}{2}} = 73.89\cdot 10^{-4}\,dm^{\frac{3}{2}}mol^{-\frac{1}{2}}s^{-\frac{1}{2}}$	
		508	$75.77\cdot 10^{-4}$	
		523	$k_a/(2k_b)^{\frac{1}{2}} = 125.53\cdot 10^{-4}$ $dm^{\frac{3}{2}}mol^{-\frac{1}{2}}s^{-\frac{1}{2}}$ $k_a/k_b^{\frac{1}{2}} = 143.56\cdot 10^{-4}\,dm^{\frac{3}{2}}mol^{-\frac{1}{2}}s^{-\frac{1}{2}}$	

[31] $\log[k_a/(2k_b)^{\frac{1}{2}}/dm^{\frac{3}{2}}mol^{-\frac{1}{2}}s^{-\frac{1}{2}}] = 3.22(36) - 53110(2890)/2.3\,RT$, in $J\,mol^{-1}$.
[32] $\log[k_a/(2k_b)^{\frac{1}{2}}/dm^{\frac{3}{2}}mol^{-\frac{1}{2}}s^{-\frac{1}{2}}] = 3.80(30) - 57380(2590)/2.3\,RT$, in $J\,mol^{-1}$.

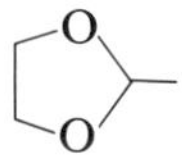

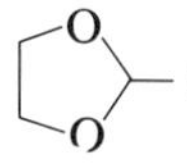

Reaction Radical generation Method	Solvent	T[K]	Rate data	Ref./ add. ref.
$CCl_3\dot{C}Cl_2 + CH_3OH \xrightarrow{a} CCl_3CHCl_2 + \dot{C}H_2OH$ $CCl_3\dot{C}Cl_2 \xrightarrow{b} C_2Cl_4 + \dot{C}l$				
γ-rad. of $C_2Cl_6 + CH_3OH$ PR, glc	CH_3OH	RT (presumed)	$k_a/k_b = 0.02\ M^{-1}$	78 Saw 1
$CCl_3\dot{C}Cl_2 + C_2H_5OH \xrightarrow{a} CCl_3CHCl_2 + CH_3\dot{C}HOH$ $CCl_3\dot{C}Cl_2 \xrightarrow{b} C_2Cl_4 + \dot{C}l$				
γ-rad. of $C_2Cl_6 + C_2H_5OH$ PR, glc	C_2H_5OH	RT (presumed)	$k_a/k_b = 0.59\ M^{-1}$	78 Saw 1
$CCl_3\dot{C}Cl_2 + (CH_3)_2CHOH \xrightarrow{a} CCl_3CHCl_2 + (CH_3)_2\dot{C}OH$ $CCl_3\dot{C}Cl_2 \xrightarrow{b} C_2Cl_4 + \dot{C}l$				
γ-rad. of $C_2Cl_6 + (CH_3)_2CHOH$ PR, glc	$(CH_3)_2CHOH$	RT (presumed)	$k_a/k_b = 3.1\ M^{-1}$	78 Saw 1
$CCl_3\dot{C}Cl_2 + c\text{-}C_6H_{12} \xrightarrow{a} CCl_3CHCl_2 + (c\text{-}C_6\dot{H}_{11})$ $CCl_3\dot{C}Cl_2 \xrightarrow{b} C_2Cl_4 + \dot{C}l$				
γ-rad. of $C_2Cl_6 + c\text{-}C_6H_{12}$ PR, glc	$c\text{-}C_6H_{12}$	297...373	$\log[A_a/A_b/M^{-1}] = 5.9(2)$ $E_a(a) - E_a(b) = 29.3(8)\ kJ\ mol^{-1}$	70 Hor 1
$CCl_3\dot{C}Cl_2 + (C_2H_5)_3SiH \xrightarrow{a} C_2HCl_5 + (C_2H_5)_3\dot{S}i$ $CCl_3\dot{C}Cl_2 \xrightarrow{b} C_2Cl_4 + \dot{C}l$				
γ-rad. of triethylsilane + hexachloroethane [33]) PR, glc	$(C_2H_5)_3SiH/C_2Cl_6$	323	$k_a/k_b = 4.76\ M^{-1}$	81 Alo 1/ 76 Kat 2
		333	2.94	
		348	1.47	
		363	1.00	
		373	0.60	
		398	0.28	
		423	0.167	
		447	0.091	
		463	0.066	
		471	0.044	
			$\log[A_a/M^{-1}s^{-1}] = 8.80(30)$ [34]) $E_a(a) = 35.7(17)\ kJ\ mol^{-1}$ [35])	
$CF_3\dot{C}F_2 + c\text{-}C_6H_{12} \xrightarrow{a} CF_3CF_2H + (c\text{-}C_6\dot{H}_{11})$ $+ I_2 \xrightarrow{b} CF_3CF_2I + \dot{I}$				
γ-rad. of $C_2F_5Br + c\text{-}C_6H_{12}$ PR, glc, using $^{131}I_2$	$c\text{-}C_6H_{12}$	RT	$k_a/k_b = 1.63 \cdot 10^{-5}$	69 Inf 1

33) No change of results if $n\text{-}C_5H_{11}Br$ is added.
34) Based on $\log[A_b/s^{-1}] = 14.3$ and $\log[A_b/A_a/M] = 5.50(6)$.
35) Based on $E_a(b) = 73.7\ kJ\ mol^{-1}$ and $E_a(b) - E_a(a) = 38.0(46)\ kJ\ mol^{-1}$.

Reaction Radical generation Method	Solvent	T[K]	Rate data	Ref./ add. ref.
$CHCl_2\dot{C}Cl_2 + (C_2H_5)_3SiH \xrightarrow{a} CHCl_2CHCl_2 + (C_2H_5)_3\dot{S}i$				
$CHCl_2\dot{C}Cl_2 \xrightarrow{b} CHCl{=}CCl_2 + \dot{C}l$				
Rad. of triethylsilane + pentachloroethane				81 Alo 1/ 76 Kat 2
PR, glc	$(C_2H_5)_3SiH/C_2Cl_5H$	348	$k_a/k_b = 9.09\ M^{-1}$ [36]	
		373	3.35 [36]	
		398	2.27 [36]	
			1.75	
		423	0.787	
		448	0.476	
			$\log[A_a/M^{-1}s^{-1}] = 8.60(60)$ [37]	
			$E_a(a) = 34.8(33)\ kJ\ mol^{-1}$ [38]	
$(C_2Cl_4H^{\cdot})$ [39] $+ c\text{-}C_6H_{12} \xrightarrow{a} C_2Cl_4H_2 + (c\text{-}C_6\dot{H}_{11})$				
$(C_2Cl_4\dot{H}^{\cdot}) \xrightarrow{b} C_2Cl_3H + \dot{C}l$				
^{60}Co-γ-irr. of *c*-hexane + C_2Cl_5H				71 Kat 1
PR, glc	*c*-hexane	323 ... 473	$\log[k_b/k_a] = 5.63(7) - \frac{8100(100)}{4.576 \cdot T}$	
			$\log[A_b/A_a] = 5.6(1)$	
			$E_a(b) - E_a(a) = 33.9(4)\ kJ\ mol^{-1}$	
$\dot{C}Cl_2CH_2Cl + c\text{-}C_6H_{12} \xrightarrow{a} CHCl_2CH_2Cl + (c\text{-}C_6\dot{H}_{11})$				
$\dot{C}Cl_2CH_2Cl \xrightarrow{b} CCl_2{=}CH_2 + \dot{C}l$				
γ-rad. of $c\text{-}C_6H_{12} + CCl_3CH_2Cl$				75 Alo 1
PR, glc	$c\text{-}C_6H_{12}$	379(1)	$k_a/k_b = 146.3 \cdot 10^{-2}\ M^{-1}$	
		393	$77.1 \cdot 10^{-2}$	
		423	$25.3 \cdot 10^{-2}$	
		473	$7.0 \cdot 10^{-2}$	
		498	$4.3 \cdot 10^{-2}$	
			$\log[A_a/A_b/M^{-1}] = -5.49(9)$	
			$E_a(a) - E_a(b) = -39.6(8)\ kJ\ mol^{-1}$	
γ-rad. of $c\text{-}C_6H_{12} + CCl_2BrCH_2Cl$				75 Alo 1
PR, glc	$c\text{-}C_6H_{12}$	373	$k_a/k_b = 119.7 \cdot 10^{-2}\ M^{-1}$	
		398	$65 \cdot 10^{-2}$	
		423	$23 \cdot 10^{-2}$	
		448	$15.2 \cdot 10^{-2}$	
		473	$8.5 \cdot 10^{-2}$	
		513	$4.2 \cdot 10^{-2}$	
			$\log[A_a/A_b/M^{-1}] = -5.40(12)$	
			$E_a(a) - E_a(b) = -38.6(11)\ kJ\ mol^{-1}$	
$CHCl_2\dot{C}HCl + (C_2H_5)_3SiH \xrightarrow{a} CHCl_2CH_2Cl + (C_2H_5)_3\dot{S}i$				
$CHCl_2\dot{C}HCl \xrightarrow{b} CHCl{=}CHCl + \dot{C}l$				
Rad. of triethylsilane + 1,1,2,2-tetrachloroethane				81 Alo 1/ 76 Kat 2
PR, glc	$(C_2H_5)_3SiH/$ $CHCl_2CHCl_2$	323	$k_a/k_b = 1.33\ M^{-1}$	
			1.37 [36]	
		333	1.11	
		345	0.794	
			1.20 [36]	
		373	0.339	
		383	0.267	
		398	0.248	
		406	0.241	
		413	0.091	
			$\log[A_a/M^{-1}s^{-1}] = 8.80(60)$ [40]	
			$E_a(a) = 37.3(33)\ kJ\ mol^{-1}$ [41]	

[36]) $n\text{-}C_5H_{11}Br$ added.
[37]) Based on $\log[A_b/s^{-1}] = 14.1$ and $\log[A_b/A_a/M] = 4.88(27)$.
[38]) Based on $E_a(b) = 78.3\ kJ\ mol^{-1}$ and $E_a(b) - E_a(a) = 43.5(21)\ kJ\ mol^{-1}$.
[39]) Presumably radical mixture of $CCl_3\dot{C}HCl$ and $\dot{C}Cl_2CHCl_2$.
[40]) Based on $\log[A_b/s^{-1}] = 14.3$ and $\log[A_b/A_a/M] = 4.22(34)$.
[41]) Based on $E_a(b) = 73.4\ kJ\ mol^{-1}$ and $E_a(b) - E_a(a) = 35.6(23)\ kJ\ mol^{-1}$.

Reaction Radical generation Method	Solvent	$T[K]$	Rate data	Ref./ add. ref.

$\dot{C}ClHCCl_2H + c\text{-}C_6H_{12} \xrightarrow{\text{a}} CClH_2CCl_2H + (c\text{-}C_6H_{11}^{\cdot})$
$\dot{C}ClHCCl_2H \xrightarrow{\text{b}} C_2Cl_2H_2 + \dot{C}l$

^{60}Co-γ-rad. of c-hexane + 1,1,2,2-tetrachloroethane PR, glc	c-hexane	423... 503	$\log(k_b/k_a) = 5.91(21) - \dfrac{6900(400)}{4.576\,T}$ $\log[A_b/A_a/M] = 5.9(2)$ $E_a(b) - E_a(a) = 28.9(17)\,kJ\,mol^{-1}$	71 Kat 1

$\dot{C}H_2CN + CHCl_3 \underset{\text{b}}{\overset{\text{a}}{\lessgtr}} \begin{array}{l} CH_3CN + \dot{C}Cl_3 \\ CH_2ClCN + \dot{C}HCl_2 \end{array}$

γ-rad. of CHCl$_3$ + CH$_2$BrCN PR, glc	CHCl$_3$/ CH$_2$BrCN (0.37 M)	453	$k_a/k_b = 0.21$	81 Gon 1

$\dot{C}H_2CN + CHCl_2CCl_3 \underset{\text{b}}{\overset{\text{a}}{\lessgtr}} \begin{array}{l} CH_3CN + \dot{C}Cl_2CCl_3 \\ CH_2ClCN + (C_2HCl_4^{\cdot}) \end{array}$

γ-rad. of CHCl$_2$CCl$_3$ + CH$_2$BrCN PR, glc	CHCl$_2$CCl$_3$/ CH$_2$BrCN (0.37 M)	453	$k_a/k_b = 0.23$	81 Gon 1

$\dot{C}H_2CN + CH_2ClCCl_3 \underset{\text{b}}{\overset{\text{a}}{\lessgtr}} \begin{array}{l} CH_3CN + \dot{C}HClCCl_3 \\ CH_2ClCN + (C_2H_2Cl_3^{\cdot}) \end{array}$

γ-rad. of CH$_2$ClCCl$_3$ + CH$_2$BrCN PR, glc	CH$_2$ClCCl$_3$/ CH$_2$BrCN(0.37 M)	453	$k_a/k_b = 0.06$	81 Gon 1

$\dot{C}H_2CN + CH_3CCl_3 \underset{\text{b}}{\overset{\text{a}}{\lessgtr}} \begin{array}{l} CH_3CN + \dot{C}H_2CCl_3 \\ CH_2ClCN + CH_3\dot{C}Cl_2 \end{array}$

γ-rad. of CH$_3$CCl$_3$ + CH$_2$BrCN PR, glc	CH$_3$CCl$_3$/ CH$_2$BrCN (0.37 M)	453	$k_a/k_b = 0.085$	81 Gon 1

$\dot{C}H_2CN + c\text{-}C_6H_{12} \xrightarrow{\text{a}} CH_3CN + (c\text{-}C_6H_{11}^{\cdot})$
$\quad\quad + CCl_4 \xrightarrow{\text{b}} CH_2ClCN + \dot{C}Cl_3$

γ-rad. of CCl$_4$ + c-C$_6$H$_{12}$ + CH$_2$BrCN PR, glc	CCl$_4$	373 393 423 438 453	$k_a/k_b = 84.5$ 76.4 41.9 34.1 33.9 $E_a(b) = 73.7\,kJ\,mol^{-1}$ [42])	81 Gon 1/ 77 Gon 2

$\dot{C}H_2CN + c\text{-}C_6H_{12} \xrightarrow{\text{a}} CH_3CN + (c\text{-}C_6H_{11}^{\cdot})$
$\quad\quad + CHCl_3 \xrightarrow{\text{b}} CH_2ClCN + \dot{C}HCl_2$

γ-rad. of CHCl$_3$ + c-C$_6$H$_{12}$ + CH$_2$BrCN PR, glc	CHCl$_3$/ c-C$_6$H$_{12}$ (0.926 M)/ CH$_2$BrCN (0.297 M)	453	$k_a/k_b = 129.2$	81 Gon 1

[42]) Based on $\log[k_b/k_a] = -0.11(20) - 3.34(39)/2.303\,RT$ (RT in k cal mol^{-1}) and $E_a(a) = 59.9(25)\,kJ\,mol^{-1}$ [77 Gon 2].

Reaction Radical generation Method	Solvent	T[K]	Rate data	Ref./ add. ref.
$\dot{C}H_2CN + c\text{-}C_6H_{12} \xrightarrow{a} CH_3CN + (c\text{-}C_6H_{11}^{\cdot})$ $+ CF_3CCl_3 \xrightarrow{b} CH_2ClCN + CF_3\dot{C}Cl_2$				
γ-rad. of $CF_3CCl_3 + c\text{-}C_6H_{12} + CH_2BrCN$				81 Gon 1
PR, glc	CF_3CCl_3 (8.33 M)/ $c\text{-}C_6H_{12}$(0.926 M)/ CH_2BrCN(0.26 M)	453	$k_a/k_b = 124.5$	
$\dot{C}H_2CN + c\text{-}C_6H_{12} \xrightarrow{a} CH_3CN + (c\text{-}C_6H_{11}^{\cdot})$ $+ CCl_3CN \xrightarrow{b} CH_2ClCN + \dot{C}Cl_2CN$				
γ-rad. of $CCl_3CN + c\text{-}C_6H_{12} + CH_2BrCN$				81 Gon 1
PR, glc	$c\text{-}C_6H_{12}$ (7.22 M)/ CCl_3CN (2.0 M)/ CH_2BrCN (0.26 M)	453	$k_a/k_b = 0.05$	
$\dot{C}H_2CN + c\text{-}C_6H_{12} \xrightarrow{a} CH_3CN + (c\text{-}C_6H_{11}^{\cdot})$ $+ C_2Cl_6 \xrightarrow{b} \dot{C}H_2ClCN + CCl_3\dot{C}Cl_2$				
γ-rad. of $C_2Cl_6 + c\text{-}C_6H_{12} + CH_2BrCN$				81 Gon 1
PR, glc	$c\text{-}C_6H_{12}$ (8.53 M)/ C_2Cl_6 (0.5 M)/ CH_2BrCN (0.297 M)	453	$k_a/k_b = 1.33$	
$\dot{C}H_2CN + c\text{-}C_6H_{12} \xrightarrow{a} CH_3CN + (c\text{-}C_6H_{11}^{\cdot})$ $+ CHCl_2CCl_3 \xrightarrow{b} CH_2ClCN + (C_2HCl_4^{\cdot})$				
γ-rad. of $CHCl_2CCl_3 + c\text{-}C_6H_{12} + CH_2BrCN$				81 Gon 1
PR, glc	$CHCl_2CCl_3$/ $c\text{-}C_6H_{12}$ (0.926 M)/ CH_2BrCN (0.297 M)	453	$k_a/k_b = 23.2$	
$\dot{C}H_2CN + c\text{-}C_6H_{12} \xrightarrow{a} CH_3CN + (c\text{-}C_6H_{11}^{\cdot})$ $+ CH_2ClCCl_3 \xrightarrow{b} CH_2ClCN + (C_2H_2Cl_3^{\cdot})$				
γ-rad. of $CH_2ClCCl_3 + c\text{-}C_6H_{12} + CH_2BrCN$				81 Gon 1
PR, glc	CH_2ClCCl_3/ $c\text{-}C_6H_{12}$ (0.926 M)/ CH_2BrCN (0.297 M)	453	$k_a/k_b = 11.4$	
$\dot{C}H_2CN + c\text{-}C_6H_{12} \xrightarrow{a} CH_3CN + (c\text{-}C_6H_{11}^{\cdot})$ $2\dot{C}H_2CN \xrightarrow{b} (CH_2CN)_2$				
γ-rad. of $c\text{-}C_6H_{12} + CH_2BrCN$				77 Gon 2/
PR, glc	$c\text{-}C_6H_{12}/CH_2BrCN$	333... 443	$\log[k_a/(2k_b)^{\frac{1}{2}}] = 4.07(35) - \frac{50.1(28)}{2.303\,RT}$ [43] [44]	63 McC 1
$\dot{C}H_2CN + c\text{-}C_6H_{12} \xrightarrow{a} CH_3CN + (c\text{-}C_6H_{11}^{\cdot})$ $+ CH_3CCl_3 \xrightarrow{b} CH_2ClCN + CH_3\dot{C}Cl_2$				
γ-rad. of $CH_3CCl_3 + c\text{-}C_6H_{12} + CH_2BrCN$				81 Gon 1
PR, glc	CH_3CCl_3/ $c\text{-}C_6H_{12}$ (0.926 M)/ CH_2BrCN (0.297 M)	453	$k_a/k_b = 23.3$	

[43]) R in $kJ\,mol^{-1}\,K^{-1}$.

[44]) $E_a(a) = 59.9(26)\,kJ\,mol^{-1}$ based on assumed $E_a = 19\,kJ\,mol^{-1}$ for diffusion controlled $\dot{C}H_2CN + \dot{C}H_2CN$ reactions [63 McC 1].

Reaction Radical generation Method	Solvent	T[K]	Rate data	Ref./ add. ref.
$\dot{C}HClCH_2Cl + c\text{-}C_6H_{12} \xrightarrow{a} CH_2ClCH_2Cl + (c\text{-}C_6\dot{H}_{11})$ $\dot{C}HClCH_2Cl \xrightarrow{b} CHCl{=}CH_2 + \dot{C}l$				
γ-rad. of $c\text{-}C_6H_{12} + CHCl_2CH_2Cl$				76 Kat 2
PR, glc; titration methods	$c\text{-}C_6H_{12}$	393 423	$k_a/k_b = 0.086\,M^{-1}$ $= 0.039\,M^{-1}$ $\log[A_a/A_b/M^{-1}] = -5.74(29)$ $E_a(a) - E_a(b) = -35.3(22)\,kJ\,mol^{-1}$	
$\dot{C}H_2CHO + CH_3CHO \xrightarrow{a} CH_3CHO + CH_3\dot{C}O$ $+ 4\text{-}CH_3C_6H_5N_2^+ \xrightarrow{b} (4\text{-}CH_3C_6H_4N_2CH_2CHO^{\cdot})^+$				
γ-rad. of $4\text{-}CH_3C_6H_4N_2^+BF_4^- + CH_3CHO + H_2O$				75 Pac 1
PR [45])	H_2O	RT	$k_a/k_b = 5.1 \cdot 10^{-3}$	
$\dot{C}H_2COOH + 3\text{-}BrC_6H_4CH_3 \xrightarrow{a} CH_3COOH + 3\text{-}BrC_6H_4\dot{C}H_2$ $+ C_6H_5CH_3 \xrightarrow{b} CH_3COOH + C_6H_5\dot{C}H_2$				
Therm. of $Mn(OOCCH_3)_3$				69 Hei 1
PR, glc	glacial acetic acid	403(1)	$k_a/k_b = 0.58$	
$\dot{C}H_2COOH + 4\text{-}BrC_6H_4CH_3 \xrightarrow{a} CH_3COOH + 4\text{-}BrC_6H_4\dot{C}H_2$ $+ C_6H_5CH_3 \xrightarrow{b} CH_3COOH + C_6H_5\dot{C}H_2$				
Therm. of $Mn(OOCCH_3)_3$				69 Hei 1
PR, glc	glacial acetic acid	403(1)	$k_a/k_b = 0.87$	
$\dot{C}H_2COOH + 4\text{-}ClC_6H_4CH_3 \xrightarrow{a} CH_3COOH + 4\text{-}ClC_6H_4\dot{C}H_2$ $+ C_6H_5CH_3 \xrightarrow{b} CH_3COOH + C_6H_5\dot{C}H_2$				
Therm. of $Mn(OOCCH_3)_3$				69 Hei 1
PR, glc	glacial acetic acid	403(1)	$k_a/k_b = 0.89$	
$\dot{C}H_2COOH + 4\text{-}FC_6H_4CH_3 \xrightarrow{a} CH_3COOH + 4\text{-}FC_6H_4\dot{C}H_2$ $+ C_6H_5CH_3 \xrightarrow{b} CH_3COOH + C_6H_5\dot{C}H_2$				
Therm. of $Mn(OOCCH_3)_3$				69 Hei 1
PR, glc	glacial acetic acid	403(1)	$k_a/k_b = 1.11$	
$\dot{C}H_2COOH + C_6H_5CH_3 \xrightarrow{a} CH_3COOH + C_6H_5\dot{C}H_2$ $\dot{C}H_2COOH + C_6H_5CH_3 \xrightarrow{b} (C_6H_5CH_3{-}CH_2COOH^{\cdot})$				
Therm. of $Pb(OOCCH_3)_4$ and $\dot{C}H_3 + CH_3COOH$ reaction				68 Hei 1
PR, glc	CH_3COOH	[46])	$k_a/k_b = 0.27$	
$\dot{C}H_2COOH + 3\text{-}CH_3C_6H_4CH_3 \xrightarrow{a} CH_3COOH + 3\text{-}CH_3C_6H_4\dot{C}H_2$ $+ C_6H_5CH_3 \xrightarrow{b} CH_3COOH + C_6H_5\dot{C}H_2$				
Therm. of $Mn(OOCCH_3)_3$				69 Hei 1
PR, glc	glacial acetic acid	403(1)	$k_a/k_b = 1.19$	
$\dot{C}H_2COOH + 4\text{-}CH_3C_6H_4CH_3 \xrightarrow{a} CH_3COOH + 4\text{-}CH_3C_6H_4\dot{C}H_2$ $+ C_6H_5CH_3 \xrightarrow{b} CH_3COOH + C_6H_5\dot{C}H_2$				
Therm. of $Mn(OOCCH_3)_3$				69 Hei 1
PR, glc	glacial acetic acid	403(1)	$k_a/k_b = 1.66$	
$\dot{C}H_2COOH + 3\text{-}CH_3OC_6H_4CH_3 \xrightarrow{a} CH_3COOH + 3\text{-methoxytoluene}(-\dot{H})$ $+ C_6H_5CH_3 \xrightarrow{b} CH_3COOH + C_6H_5\dot{C}H_2$				
Therm. of $Mn(OOCCH_3)_3$				69 Hei 1
PR, glc	acetic acid anhydride	403(1)	$k_a/k_b = 0.92$	

[45]) Spectrophotometric determination of $4\text{-}CH_3C_6H_4N_2^+$.

[46]) T at reflux of CH_3COOH solution.

Reaction Radical generation Method	Solvent	T[K]	Rate data	Ref./ add. ref.
$\dot{C}H_2COOH + 4\text{-}CH_3OC_6H_4CH_3 \xrightarrow{a} CH_3COOH + 4\text{-methoxytoluene}(-\dot{H})$ $+ C_6H_5CH_3 \xrightarrow{b} CH_3COOH + C_6H_5\dot{C}H_2$				
Therm. of $Mn(OOCCH_3)_3$				69 Hei 1
PR, glc	acetic acid anhydride	403(1)	$k_a/k_b = 3.64$	
$\dot{C}H_2COOH + (C_6H_5)_2CH_2 \xrightarrow{a} CH_3COOH + (C_6H_5)_2\dot{C}H$ $+ C_6H_5CH_3 \xrightarrow{b} CH_3COOH + C_6H_5\dot{C}H_2$				
Therm. of $Mn(OOCCH_3)_3$				69 Hei 1
PR, glc	glacial acetic acid	403(1)	$k_a/k_b = 11.7$	
$\dot{C}H_2COOH + 4\text{-}C_6H_5C_6H_4CH_3 \xrightarrow{a} CH_3COOH + 4\text{-}C_6H_5C_6H_4\dot{C}H_2$ $+ C_6H_5CH_3 \xrightarrow{b} CH_3COOH + C_6H_5\dot{C}H_2$				
Therm. of $Mn(OOCCH_3)_3$				69 Hei 1
PR, glc	acetic acid anhydride	403(1)	$k_a/k_b = 1.60$	
$\dot{C}H_2COOH + (C_6H_5)_3CH \xrightarrow{a} CH_3COOH + (C_6H_5)_3\dot{C}$ $+ C_6H_5CH_3 \xrightarrow{b} CH_3COOH + C_6H_5\dot{C}H_2$				
Therm. of $Mn(OOCCH_3)_3$				69 Hei 1
PR, glc	glacial acetic acid	403(1)	$k_a/k_b = 39$	
$\dot{C}H_2CH_2Cl + c\text{-}C_6H_{12} \xrightarrow{a} CH_3CH_2Cl + (c\text{-}C_6H_{11}^{\cdot})$ $\dot{C}H_2CH_2Cl \xrightarrow{b} CH_2{=}CH_2 + \dot{C}l$				
γ-rad. of $c\text{-}C_6H_{12} + CH_2ClCH_2Cl$				76 Kat 1
PR, glc; titration methods	$c\text{-}C_6H_{12}$	473 523	$k_a/k_b = 3.03 \cdot 10^{-2}\,M^{-1}$ $1.72 \cdot 10^{-2}$ $\log[A_a/A_b/M^{-1}] = -5.4(4)$ $E_a(a) - E_a(b) = -35.5(60)\,kJ\,mol^{-1}$	
$\dot{C}H_2CH_2OH + C_2H_5OH \xrightarrow{a} C_2H_5OH + CH_3\dot{C}HOH$ $+ 4\text{-}CH_3C_6H_4N_2^+ \xrightarrow{b} (4\text{-}CH_3C_6H_4N_2CH_2CH_2OH^{\cdot})^+$				
γ-rad. of $4\text{-}CH_3C_6H_4N_2^+BF_4^- + C_2H_5OH + H_2O$				75 Pac 1
PR [45])	H_2O	RT	$k_a/k_b = 2.9 \cdot 10^{-4}$	
$CCl_3CH{=}\dot{C}H + CHCl_3 \xrightarrow{a} CCl_3CH{=}CH_2 + \dot{C}Cl_3$ $+ CCl_3Br \xrightarrow{b} CCl_3CH{=}CHBr + \dot{C}Cl_3$				
$\dot{C}Cl_3 + CH{\equiv}CH \longrightarrow CCl_3CH{=}\dot{C}H$ reaction after dimethylaniline catalyzed decomp. of BPO as initiator				73 Afa 1/ 72 Afa 2
PR, glc	$CHCl_3$	293	$k_a/k_b = 3.14(40) \cdot 10^{-3}$	
$CCl_3CH{=}\dot{C}H + CH_2Cl_2 \xrightarrow{a} CCl_3CH{=}CH_2 + \dot{C}HCl_2$ $+ CCl_3Br \xrightarrow{b} CCl_3CH{=}CHBr + \dot{C}Cl_3$				
$\dot{C}Cl_3 + CH{\equiv}CH \longrightarrow CCl_3CH{=}\dot{C}H$ reaction after dimethylaniline catalyzed decomp. of BPO as initiator				73 Afa 1/ 72 Afa 2
PR, glc	CH_2Cl_2	293	$k_a/k_b = 7.86(100) \cdot 10^{-4}$	
$CCl_3CH{=}\dot{C}H + CH_3OH \xrightarrow{a} CCl_3CH{=}CH_2 + \dot{C}H_2OH$ $+ CCl_3Br \xrightarrow{b} CCl_3CH{=}CHBr + \dot{C}Cl_3$				
$\dot{C}Cl_3 + CH{\equiv}CH \longrightarrow CCl_3CH{=}\dot{C}H$ reaction after dimethylaniline catalyzed decomp. of BPO as initiator				73 Afa 1/ 72 Afa 1
PR, glc	CH_3OH	293	$k_a/k_b = 1.56(15) \cdot 10^{-3}$	

[45]) Spectrophotometric determination of $4\text{-}CH_3C_6H_4N_2^+$.

Reaction Radical generation Method	Solvent	T[K]	Rate data	Ref./ add. ref.
$CCl_3CH{=}\dot{C}H + CH_3CN \xrightarrow{a} CCl_3CH{=}CH_2 + \dot{C}H_2CN$ $+ CCl_3Br \xrightarrow{b} CCl_3CH{=}CHBr + \dot{C}Cl_3$				
$\dot{C}Cl_3 + CH{\equiv}CH \longrightarrow CCl_3CH{=}\dot{C}H$ reaction after dimethylaniline catalyzed decomp. of BPO as initiator				73 Afa 1/ 72 Afa 2
PR, glc	CH_3CN	293	$k_a/k_b = 3.92(48) \cdot 10^{-4}$	
$CCl_3CH{=}\dot{C}H + CH_3CHO \xrightarrow{a} CCl_3CH{=}CH_2 + CH_3\dot{C}O$ $+ CCl_3Br \xrightarrow{b} CCl_3CH{=}CHBr + \dot{C}Cl_3$				
$\dot{C}Cl_3 + CH{\equiv}CH \longrightarrow CCl_3CH{=}\dot{C}H$ reaction after dimethylaniline catalyzed decomp. of BPO as initiator				73 Afa 1/ 72 Afa 1
PR, glc	CH_3CHO	293	$k_a/k_b = 1.346(250) \cdot 10^{-2}$	
$CCl_3CH{=}\dot{C}H + CH_3COCH_3 \xrightarrow{a} CCl_3CH{=}CH_2 + \dot{C}H_2COCH_3$ $+ CCl_3Br \xrightarrow{b} CCl_3CH{=}CHBr + \dot{C}Cl_3$				
$\dot{C}Cl_3 + CH{\equiv}CH \longrightarrow CCl_3CH{=}\dot{C}H$ reaction after dimethylaniline catalyzed decomp. of BPO as initiator				73 Afa 1/ 72 Afa 2
PR, glc	CH_3COCH_3	293	$k_a/k_b = 1.19(27) \cdot 10^{-3}$	
$CCl_3CH{=}\dot{C}H + C_6H_5CH_3 \xrightarrow{a} CCl_3CH{=}CH_2 + C_6H_5\dot{C}H_2$ $+ CCl_3Br \xrightarrow{b} CCl_3CH{=}CHBr + \dot{C}Cl_3$				
$\dot{C}Cl_3 + CH{\equiv}CH \longrightarrow CCl_3CH{=}\dot{C}H$ reaction after dimethylaniline catalyzed decomp. of BPO as initiator				73 Afa 1/ 72 Afa 1
PR, glc	$C_6H_5CH_3$	293	$k_a/k_b = 5.17(82) \cdot 10^{-3}$	
$CCl_3CH{=}\dot{C}H + C_6H_5OCH_3 \xrightarrow{a} CCl_3CH{=}CH_2 + C_6H_5O\dot{C}H_2$ $+ CCl_3Br \xrightarrow{b} CCl_3CH{=}CHBr + \dot{C}Cl_3$				
$\dot{C}Cl_3 + CH{\equiv}CH \longrightarrow CCl_3CH{=}\dot{C}H$ reaction after dimethylaniline catalyzed decomp. of BPO as initiator				73 Afa 1/ 72 Afa 1
PR, glc	$C_6H_5OCH_3$	293	$k_a/k_b = 1.96(24) \cdot 10^{-3}$	
$\dot{C}D_2CH(CD_3)OH + (CD_3)_2CHOH \xrightarrow{a} CD_2HCH(CD_3)OH + (CD_3)_2\dot{C}OH$ $+ 4\text{-}CH_3C_6H_4N_2^+ \xrightarrow{b} (4\text{-}CH_3C_6H_4N_2CD_2CH(CD_3)OH^{\cdot})^+$				
γ-rad. of $4\text{-}CH_3C_6H_4N_2^+BF_4^- + (CD_3)_2CHOH + H_2O$				75 Pac 1
PR [45])	H_2O	RT	$k_a/k_b = 1.4 \cdot 10^{-3}$	
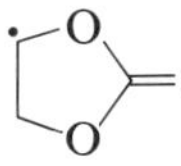 $=O + Cl_2 \xrightarrow{a}$ products $+ \dot{C}l$ $\dot{C}l$ + ethylene carbonate $\xrightarrow{b}$ products + ethylene carbonate$(-\dot{H}) \xrightarrow{c}$ products				
$\dot{C}l$ + ethylene carbonate (AIBN as initiator)				70 Shv 1
[47])	ethylene carbonate	333	$(k_a \cdot k_b)^{\frac{1}{2}}/k_c^{\frac{1}{2}} = 3.95\,M^{-\frac{1}{2}}s^{-\frac{1}{2}}$	
$ICH_2CH_2\dot{C}H_2 + CCl_4 \xrightarrow{a} ICH_2CH_2CH_2Cl + \dot{C}Cl_3$ $ICH_2CH_2\dot{C}H_2 \xrightarrow{b} c\text{-}C_3H_6 + \dot{I}$				
Thermal decomp. of BPO and $(C_6H_5^{\cdot}) + ICH_2CH_2CH_2I$ reaction				73 Dru 1/
PR, glc, NMR	C_6H_6/CCl_4	384	$k_a/k_b = 0.25 \ldots 0.45\,M^{-1}$	72 Dru 1
$\dot{C}H_2CH(CH_3)OH + (CH_3)_2CHOH \xrightarrow{a} (CH_3)_2CHOH + (CH_3)_2\dot{C}OH$ $+ 4\text{-}CH_3C_6H_4N_2^+ \xrightarrow{b} (4\text{-}CH_3C_6H_4N_2CH_2CH(CH_3)OH^{\cdot})^+$				
γ-rad. of $4\text{-}CH_3C_6H_4N_2^+BF_4^- + (CH_3)_2CHOH + H_2O$				75 Pac 1
PR [45])	H_2O	RT	$k_a/k_b = 1.3 \cdot 10^{-3}$	

[45]) Spectrophotometric determination of $4\text{-}CH_3C_6H_4N_2^+$.
[47]) Cl_2 vapor pressure measurement in gas phase above solution.

Reaction Radical generation Method	Solvent	T[K]	Rate data	Ref./ add. ref.
$CD_3CD_2CD_2\dot{C}DCl + HBr_3 \xrightarrow{a} CD_3CD_2CD_2CHDCl + \dot{B}r_3$ $+ HBr \xrightarrow{b} CD_3CD_2CD_2CHDCl + \dot{B}r$				
Phot. of 1-ClC_4D_9 + Freon 113 + HBr + Br_2				77 Tan 1
PR, glc, MS, NMR	Freon 113	296.0(1)	$k_a/k_b = 57.6(12)$	
$CD_3CD_2\dot{C}DCD_2Cl + HBr_3 \xrightarrow{a} CD_3CD_2CHDCD_2Cl + \dot{B}r_3$ $+ HBr \xrightarrow{b} CD_3CD_2CHDCD_2Cl + \dot{B}r$				
Phot. of 1-ClC_4D_9 + Freon 113 + HBr + Br_2				77 Tan 1
PR, glc, MS, NMR	Freon 113	296.0(1)	$k_a/k_b = 13.0(4)$	
$CD_3\dot{C}DCD_2CD_2Cl + HBr_3 \xrightarrow{a} CD_3CHDCD_2CD_2Cl + \dot{B}r_3$ $+ HBr \xrightarrow{b} CD_3CHDCD_2CD_2Cl + \dot{B}r$				
Phot. of 1-ClC_4D_9, + Freon 113 + HBr + Br_2				77 Tan 1
PR, glc	Freon 113	296.0(1)	$k_a/k_b = 26.8(18)$	
$(CH_3)_2\dot{C}CN + (C_2H_5)_3N \xrightarrow{a} (CH_3)_2CHCN + ((C_2H_5)_2NC_2H_4^{\cdot})$ $2\,(CH_3)_2\dot{C}CN \xrightarrow{b} (CH_3)_2(CN)CC(CN)(CH_3)_2$				
Therm. of AIBN				68 Tro 1/
PR, glc	$(C_2H_5)_3N$	353 368 383	$k_a/(2k_b)^{\frac{1}{2}} = 2.2\cdot10^{-4}\,M^{-\frac{1}{2}}s^{-\frac{1}{2}}$ $3.7\cdot10^{-4}$ $6.5\cdot10^{-4}$ $\log[A_a/A_b^{\frac{1}{2}}/M^{-\frac{1}{2}}s^{-\frac{1}{2}}] = 2$ $E_a(a) - \frac{1}{2}E_a(b) = 37.7\,kJ\,mol^{-1}$ $\log[A_a/M^{-1}s^{-1}] = 6.15$ [48]) $E_a(a) = 37.7\,kJ\,mol^{-1}$ [48])	64 Vic 1
$CH_3\dot{C}HCH_2OCH_2CH{=}CH_2 + (n\text{-}C_4H_9)_3SnH \xrightarrow{a} CH_3(CH_2)_2OCH_2CH{=}CH_2 + (n\text{-}C_4H_9)_3\dot{S}n$ $CH_3\dot{C}HCH_2OCH_2CH{=}CH_2 \xrightarrow{b}$ 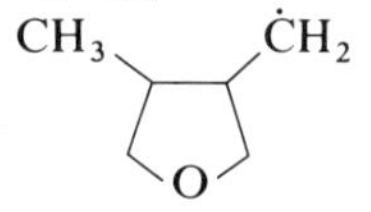				
React. $(n\text{-}C_4H_9)_3\dot{S}n + CH_3CHClCH_2OCH_2CH{=}CH_2$				74 Bec 3
PR	pentane	338	$k_a/k_b = 0.33\,M^{-1}$	
$\left.\begin{array}{l}\dot{C}H_2CH_2OOC(CH_3)_3\\ \dot{C}H_2CH(CH_3)OOC(CH_3)_3\\ \dot{C}H_2C(CH_3)_2OOC(CH_3)_3\end{array}\right\}$ [49]) $+ (n\text{-}C_4H_9)_3SnH \xrightarrow{a,b,c} \left\{\begin{array}{l}C_2H_5OOC(CH_3)_3\\ CH_3CH(CH_3)OOC(CH_3)_3\\ (CH_3)_3COOC(CH_3)_3\end{array}\right\} + (n\text{-}C_4H_9)_3\dot{S}n$				
$(n\text{-}C_4H_9)_3\dot{S}n$ + corresp. 1-bromoperoxide (DTBH initiated)				74 Blo 1
PR	C_6H_6	298	$k_a{:}k_b{:}k_c = 1{:}20{:}350$	
$\left.\begin{array}{l}\dot{C}H_2CH_2OOC(CH_3)_3\\ \dot{C}H_2CH(CH_3)OOC(CH_3)_3\\ \dot{C}H_2C(CH_3)_2OOC(CH_3)_3\end{array}\right\}$ [49]) $+ (C_6H_5)_3SnH \xrightarrow{a,b,c} \left\{\begin{array}{l}C_2H_5OOC(CH_3)_3\\ CH_3CH(CH_3)OOC(CH_3)_3\\ (CH_3)_3COOC(CH_3)_3\end{array}\right\} + (C_6H_5)_3\dot{S}n$				
$(C_6H_5)_3\dot{S}n$ + corresp. 1-bromoperoxide (DTBH initiated)				74 Blo 1
PR	C_6D_6	298	$k_a{:}k_b{:}k_c = 1{:}20{:}350$	

[48]) Assuming $2k_b = 2\cdot10^8\,M^{-1}s^{-1}$ and $E_a(b) = 0$ [64 Vic 1].
[49]) Radical mixture.

Reaction				
Radical generation				Ref./
Method	Solvent	T[K]	Rate data	add. ref.

$^{50})$ + $(n\text{-}C_4H_9)_3SnH$ $\xrightarrow{a}$ + $(n\text{-}C_4H_9)_3\dot{S}n$

+ $NaBH_4$ $\xrightarrow{b}$ + $Na\dot{B}H_3$

Br-abstract. from 7,7-dibromonorcarane via radical chain mechanism   77 Gro 1
PR, glc C_2H_5OH RT $k_a/k_b = 2.7 \cdot 10^3$

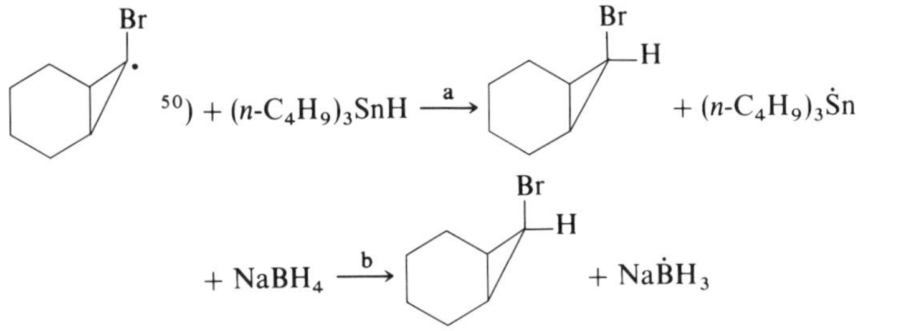

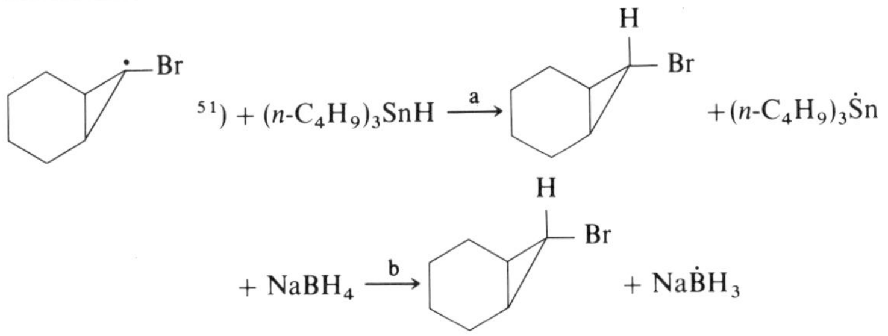

$^{51})$ + $(n\text{-}C_4H_9)_3SnH$ $\xrightarrow{a}$ + $(n\text{-}C_4H_9)_3\dot{S}n$

+ $NaBH_4$ $\xrightarrow{b}$ + $Na\dot{B}H_3$

Br-abstract. from 7,7-dibromonorcarane via radical chain mechanism   77 Gro 1
PR, glc C_2H_5OH RT $k_a/k_b = 3.8 \cdot 10^2$

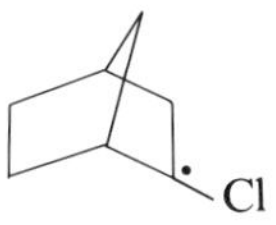

+ $CHCl_3$ $\begin{cases} \xrightarrow{a} endo\text{-2-chloronorbornane} + \dot{C}Cl_3 \\ \xrightarrow{b} exo\text{-2-chloronorbornane} + \dot{C}Cl_3 \end{cases}$

$\dot{G}$

Therm. of 2-chloronorborylperester   70 Bar 1
PR $CHCl_3$ 333... $k_a/k_b = 3$
                                        353

$\dot{G} + CH_2Cl_2$ $\begin{cases} \xrightarrow{a} endo\text{-2-chloronorbornane} + \dot{C}HCl_2 \\ \xrightarrow{b} exo\text{-2-chloronorbornane} + \dot{C}HCl_2 \end{cases}$

Therm. of 2-chloronorbornylperester   70 Bar 1
PR CH_2Cl_2 333... $k_a/k_b = 2$
                                        353

$\dot{G} + C_2H_5SH$ $\begin{cases} \xrightarrow{a} endo\text{-2-chloronorbornane} + C_2H_5\dot{S} \\ \xrightarrow{b} exo\text{-2-chloronorbornane} + C_2H_5\dot{S} \end{cases}$

Therm. of 2-chloronorbornylperester   70 Bar 1
PR C_2H_5SH 333... $k_a/k_b = 12$
                                        353

$^{50})$ *cis*-bromo-*c*-propyl radical.
$^{51})$ *trans*-bromo-*c*-propyl radical.

Reaction				
Radical generation				Ref./
Method	Solvent	T[K]	Rate data	add. ref.

$\dot{G}$ + (cyclohexadiene) $\xrightarrow{a}$ *endo*-2-chloronorbornane + (radical) *)
$\xrightarrow{b}$ *exo*-2-chloronorbornane + (radical)

Therm. of 2-chloronorbornylperester				70 Bar 1
PR	1,4-cyclohexadiene	333... 353	$k_a/k_b = 5$	

$\dot{G}$ + $C_6H_5CH_3$ $\xrightarrow{a}$ *endo*-2-chloronorbornane + $C_6H_5\dot{C}H_2$
$\xrightarrow{b}$ *exo*-2-chloronorbornane + $C_6H_5\dot{C}H_2$

Therm. of 2-chloronorbornylperester				70 Bar 1
PR	$C_6H_5CH_3$	333... 353	$k_a/k_b = 5$	

$\dot{G}$ + $C_6H_5CH(CH_3)_2$ $\xrightarrow{a}$ *endo*-2-chloronorbornane + $C_6H_5\dot{C}(CH_3)_2$
$\xrightarrow{b}$ *exo*-2-chloronorbornane + $C_6H_5\dot{C}(CH_3)_2$

Therm. of 2-chloronorbornylperester				70 Bar 1
PR	cumene	333... 353	$k_a/k_b = 7$	

$CH_3(CH_2)_3\dot{C}{=}CHCCl_3 + CCl_4 \xrightarrow{a} CH_3(CH_2)_3CCl{=}CHCCl_3 + \dot{C}Cl_3$
$CH_3(CH_2)_3\dot{C}{=}CHCCl_3 \xrightarrow{1,5}\;^{52)} \dot{C}H_2(CH_2)_3CH{=}CHCCl_3$

Therm. of BPO as initiator				67 Hei 1
PR, glc	1-hexyne/CCl_4	350	$k_a/k_{1,5} = 3.33\,M^{-1}$	

(cyclohexyl-CN radical) $\dot{H}$ + O_2N—⟨ ⟩—$SH \xrightarrow{a} c$-$C_6H_{11}CN + O_2N$—⟨ ⟩—$\dot{S}$
+ 4-$ClC_6H_4SH \xrightarrow{b} c$-$C_6H_{11}CN + 4$-$ClC_6H_4\dot{S}$

Thermal decomp. of 1,1'-azobis-1-cyanocyclohexane				57 Sch 1
PR, redox titration	toluene	383	$k_a/k_b = 0.45(5)$	

$\dot{H}$ + $C_6H_5SH \xrightarrow{a} c$-$C_6H_{11}CN + C_6H_5\dot{S}$
+ 4-$ClC_6H_4SH \xrightarrow{b} c$-$C_6H_{11}CN + 4$-$ClC_6H_4\dot{S}$

Thermal decomp. of 1,1'-azobis-1-cyanocyclohexane				57 Sch 1
PR, redox titration	toluene	383	$k_a/k_b = 1.20(7)$	

$\dot{H}$ + CH_3O—⟨ ⟩—$SH \xrightarrow{a} c$-$C_6H_{11}CN + CH_3O$—⟨ ⟩—$\dot{S}$
+ 4-$ClC_6H_4SH \xrightarrow{b} c$-$C_6H_{11}CN + 4$-$ClC_6H_4\dot{S}$

Thermal decomp. of 1,1'-azobis-1-cyanocyclohexane				57 Sch 1
PR, redox titration	toluene	383	$k_a/k_b = 1.75(5)$	

---

*) For $\dot{G}$, see p. 173.
[52]) Intramolecular 1,5-hydrogen shift.

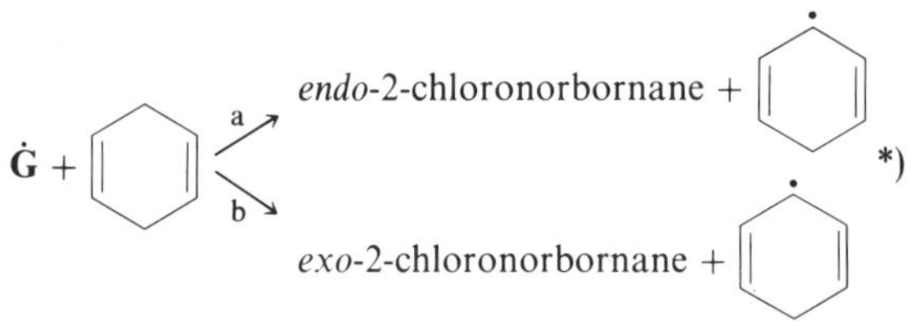

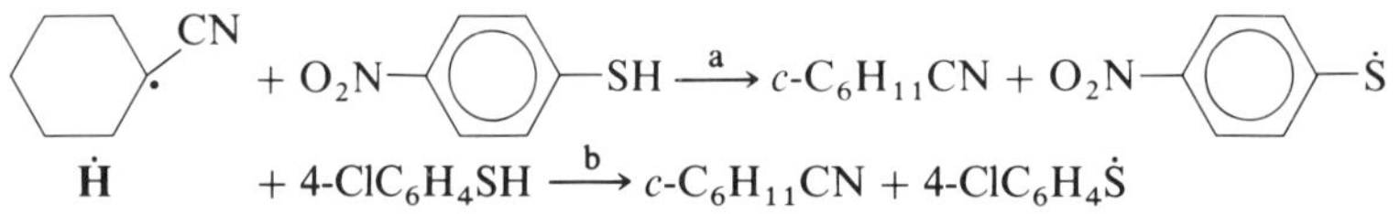

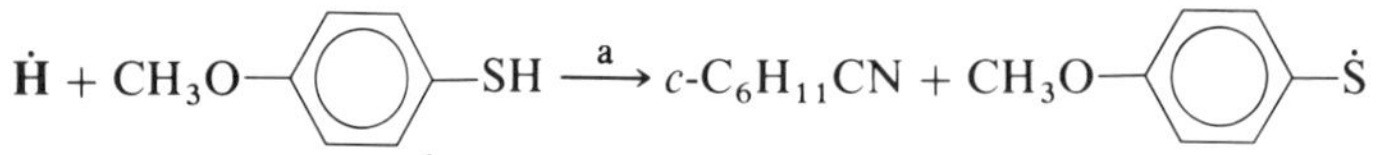

Reaction				
Radical generation				Ref./
Method	Solvent	T[K]	Rate data	add. ref.

$\dot{H} + CH_3CH_2$—⬡—$SH \xrightarrow{a} c\text{-}C_6H_{11}CN + CH_3CH_2$—⬡—$\dot{S}$ *)

$+ 4\text{-}ClC_6H_4SH \xrightarrow{b} c\text{-}C_6H_{11}CN + 4\text{-}ClC_6H_4\dot{S}$

| Thermal decomp. of 1,1′-azobis-1-cyanocyclohexane | | | | 57 Sch 1 |
| PR, redox titration | toluene | 383 | $k_a/k_b = 1.35(5)$ | |

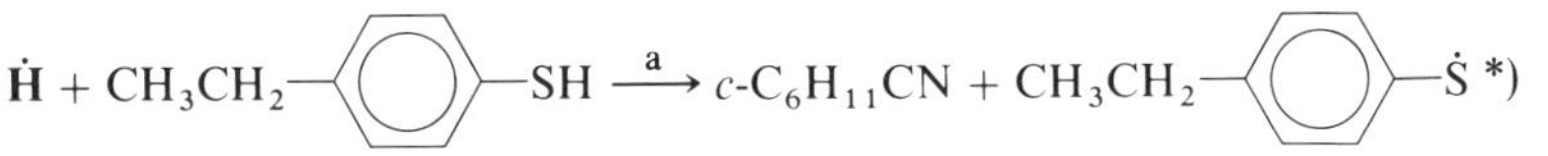

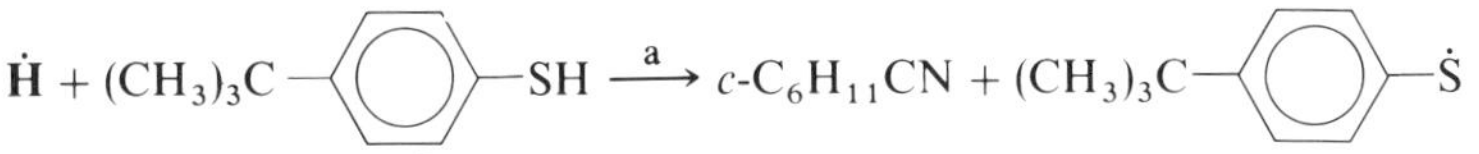

$\dot{H} + (CH_3)_3C$—⬡—$SH \xrightarrow{a} c\text{-}C_6H_{11}CN + (CH_3)_3C$—⬡—$\dot{S}$

$+ 4\text{-}ClC_6H_4SH \xrightarrow{b} c\text{-}C_6H_{11}CN + 4\text{-}ClC_6H_4\dot{S}$

| Thermal decomp. of 1,1′-azobis-1-cyanocyclohexane | | | | 57 Sch 1 |
| PR, redox titration | toluene | 383 | $k_a/k_b = 1.40(10)$ | |

$\dot{H} + 2,3\text{-benzothiophenol}(-SH) \xrightarrow{a} c\text{-}C_6H_{11}CN + (-\dot{S})$
$+ 4\text{-}ClC_6H_4SH \xrightarrow{b} c\text{-}C_6H_{11}CN + 4\text{-}ClC_6H_4\dot{S}$

| Thermal decomp. of 1,1′-azobis-1-cyanocyclohexane | | | | 57 Sch 1 |
| PR, redox titration | toluene | 383 | $k_a/k_b = 0.80$ | |

$\dot{H} + 3,4\text{-benzothiophenol}(-SH) \xrightarrow{a} c\text{-}C_6H_{11}CN + (-\dot{S})$
$+ 4\text{-}ClC_6H_4SH \xrightarrow{b} c\text{-}C_6H_{11}CN + 4\text{-}ClC_6H_4\dot{S}$

| Thermal decomp. of 1,1′-azobis-1-cyanocyclohexane | | | | 57 Sch 1 |
| PR, redox titration | toluene | 383 | $k_a/k_b = 1.15(5)$ | |

$\dot{C}H_2CH(CH_3)OOC(CH_3)_3 + (n\text{-}C_4H_9)_3SnH \longrightarrow CH_3CH(CH_3)OOC(CH_3)_3 + (n\text{-}C_4H_9)_3\dot{S}n$

| | See corresp. reaction of | 74 Blo 1 |
| | $\dot{C}H_2CH_2OOC(CH_3)_3$ | |

$\dot{C}H_2CH(CH_3)OOC(CH_3)_3 + (C_6H_5)_3SnH \longrightarrow CH_3CH(CH_3)OOC(CH_3)_3 + (C_6H_5)_3\dot{S}n$

| | See corresp. reaction of | 74 Blo 1 |
| | $\dot{C}H_2CH_2OOC(CH_3)_3$ | |

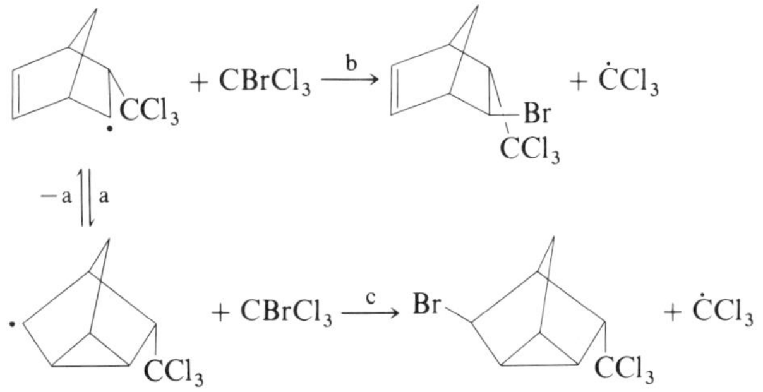

endo add. of $\dot{C}Cl_3$ radical to norbornadiene. $\dot{C}Cl_3$ by phot. of $CBrCl_3$ (at 273 and 313 K)
or AIBN as initiator (at 353 K)

PR, glc, NMR	CH_2Cl_2	273(1)	$k_b/k_a = 43.2 \cdot 10^{-2}$	79 Gie 3
			$k_b \cdot k_{-a}/k_c \cdot k_a = 99.7 \cdot 10^{-2}$	
	CH_2Cl_2	313(1)	$k_b/k_a = 16.1 \cdot 10^{-2}$	
			$k_b \cdot k_{-a}/k_c \cdot k_a = 46.0 \cdot 10^{-2}$	
	C_6H_5Cl	353.0(5)	$k_b/k_a = 7.45 \cdot 10^{-2}$	
			$k_b \cdot k_{-a}/k_c \cdot k_a = 37.7 \cdot 10^{-2}$	
			$\Delta H_a^\ddagger - \Delta H_b^\ddagger = 17.6(8)\,kJ\,mol^{-1}$	
			$\Delta H_a^\ddagger - \Delta H_c^\ddagger = 7.5(25)\,kJ\,mol^{-1}$	
			$\Delta S_a^\ddagger - \Delta S_b^\ddagger = 71(42)\,J\,mol^{-1}\,K^{-1}$	
			$\Delta S_a^\ddagger - \Delta S_c^\ddagger = 33.5(84)\,J\,mol^{-1}\,K^{-1}$	

*) For $\dot{H}$, see p. 174.

Reaction Radical generation Method	Solvent	T[K]	Rate data	Ref./ add. ref.
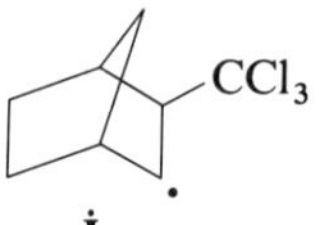 İ + $CBrCl_3$ $\xrightarrow{a}$ *endo*-2-bromo-3-trichloromethylnorbornane + $\dot{C}Cl_3$ $\xrightarrow{b}$ *exo*-2-bromo-3-trichloromethylnorbornane + $\dot{C}Cl_3$			See 4.1.2.3, Fig. 6, p. 256.	
İ + CCl_4 $\xrightarrow{a}$ *endo*-2-chloro-3-trichloromethylnorbornane + $\dot{C}Cl_3$ $\xrightarrow{b}$ *exo*-2-chloro-3-trichloromethylnorbornane + $\dot{C}Cl_3$			See 4.1.2.3, Fig. 6, p. 256.	
(c-$C_6H_{11}C_2HCl_3^{\cdot}$)[53] + c-C_6H_{12} $\xrightarrow{a}$ c-$C_6H_{11}C_2H_2Cl_3$ + (c-$C_6H_{11}^{\cdot}$) (c-$C_6H_{11}C_2HCl_3^{\cdot}$)[53] $\xrightarrow{b}$ c-$C_6H_{11}C_2HCl_2$ + $\dot{C}l$				
γ-rad. of c-C_6H_{12} + CCl_2=CHCl				73 Hor 1
PR, glc	c-C_6H_{12}	348	$k_a/k_b = 2.86\cdot10^{-2}\ M^{-1}$	
		363	$1.90\cdot10^{-2}$	
		378	$1.19\cdot10^{-2}$	
		393	$8.40\cdot10^{-3}$	
		408	$5.65\cdot10^{-3}$	
		423	$3.82\cdot10^{-3}$	
			$\log[A_a/A_b/M^{-1}] = -6.48(6)$	
			$E_a(a) - E_a(b) = -33.0(4)$ kJ mol^{-1}	
$CH_3(CH_2)_4\dot{C}$=$CHCCl_3$ + CCl_4 $\xrightarrow{a}$ $CH_3(CH_2)_4CCl$=$CHCCl_3$ + $\dot{C}Cl_3$ $CH_3(CH_2)_4\dot{C}$=$CHCCl_3$ $\longrightarrow$ [52]) $CH_3\dot{C}H(CH_2)_3CH$=$CHCCl_3$				
Therm. of BPO as initiator				67 Hei 1
PR, glc	1-heptyne/CCl_4	350	$k_a/k_b = 0.233\ M^{-1}$	
		334…373	$\log[A_a/A_b/M^{-1}] = -2.4$ $E_a(a) - E_a(b) = -11.3(21)$ kJ mol^{-1}	
$CH_3(CH_2)_4\dot{C}$=$CHCCl_3$ + $CHCl_3$ $\xrightarrow{a}$ $CH_3(CH_2)_4CH$=$CHCCl_3$ + $\dot{C}Cl_3$ + CCl_4 $\xrightarrow{b}$ $CH_3(CH_2)_4CCl$=$CHCCl_3$ + $\dot{C}Cl_3$				
Therm. of BPO as initiator				67 Hei 1
PR, glc	CCl_4/$CHCl_3$/ 1-heptyne	350	$k_a/k_b = 0.9$	
c-$C_6H_{11}CHCl\dot{C}HCl$ + c-C_6H_{12} $\xrightarrow{a}$ c-$C_6H_{11}CHClCH_2Cl$ + (c-$C_6H_{11}^{\cdot}$) c-$C_6H_{11}CHCl\dot{C}HCl$ $\xrightarrow{b}$ c-$C_6H_{11}CH$=CHCl + $\dot{C}l$				
γ-rad. of c-C_6H_{12} + *trans*-CHCl=CHCl				73 Hor 1
PR, glc	c-C_6H_{12}	348	$k_a/k_b = 1.78\cdot10^{-2}\ M^{-1}$	
		363	$1.28\cdot10^{-2}$	
		378	$7.75\cdot10^{-3}$	
		393	$5.78\cdot10^{-3}$	
		403	$3.86\cdot10^{-3}$	
		423	$2.46\cdot10^{-3}$	
			$\log[A_a/A_b/M^{-1}] = -6.5(3)$	
			$E_a(a) - E_a(b) = -32.2(17)$ kJ mol^{-1}	

[52]) Intramolecular 1,5-hydrogen shift.
[53]) Radicals from (c-$C_6H_{11}^{\cdot}$) addition to CCl_2=CHCl (mixture of c-$C_6H_{11}CHCl\dot{C}Cl_2$ and possibly c-$C_6H_{11}CCl_2\dot{C}HCl$).

Reaction Radical generation Method	Solvent	T[K]	Rate data	Ref./ add. ref.
$\dot{C}H_2C(CH_3)_2OOC(CH_3)_3 + (n\text{-}C_4H_9)_3SnH \xrightarrow{a} (CH_3)_3COOC(CH_3)_3 + (n\text{-}C_4H_9)_3\dot{S}n$				
$\dot{C}H_2C(CH_3)_2OOC(CH_3)_3 \xrightarrow{b}$ 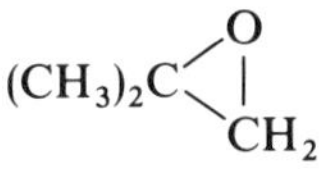 $+ (CH_3)_3C\dot{O}$				
$(n\text{-}C_4H_9)_3\dot{S}n + BrCH_2C(CH_3)_2OOC(CH_3)_3$ reaction (DTBH initiated)				74 Blo 1
PR, glc	C_6H_6	298	$k_a/k_b = 1.37\ M^{-1}$ [54])	
$\dot{C}H_2C(CH_3)_2OOC(CH_3)_3 + (C_6H_5)_3SnH \xrightarrow{a} (CH_3)_3COOC(CH_3)_3 + (C_6H_5)_3\dot{S}n$				
$\dot{C}H_2C(CH_3)_2OOC(CH_3)_3 \xrightarrow{b}$ 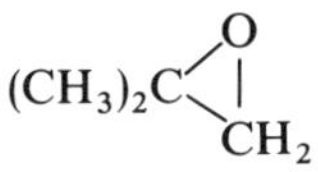 $+ (CH_3)_3C\dot{O}$				
$(C_6H_5)_3\dot{S}n + BrC(CH_3)_2OOC(CH_3)_3$ reaction (DTBH initiated)				74 Blo 1
PR, NMR	C_6D_6	298	$k_a/k_b = 7.14\ M^{-1}$ [55])	
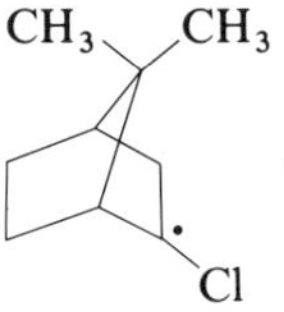 K̇ $+ CHCl_3$ $\xrightarrow{a}$ *endo*-2-chloroapobornane $+ \dot{C}Cl_3$; $\xrightarrow{b}$ *exo*-2-chloroapobornane $+ \dot{C}Cl_3$				
Therm. of 2-chloroapobornylperester				70 Bar 1
PR	$CHCl_3$	333...353	$k_a/k_b = 1.0$	
K̇ $+ C_2H_5SH$ $\xrightarrow{a}$ *endo*-2-chloroapobornane $+ C_2H_5\dot{S}$; $\xrightarrow{b}$ *exo*-2-chloroapobornane $+ C_2H_5\dot{S}$				
Therm. of 2-chloroapobornylperester				70 Bar 1
PR	C_2H_5SH	333...353	$k_a/k_b = 2.5$	
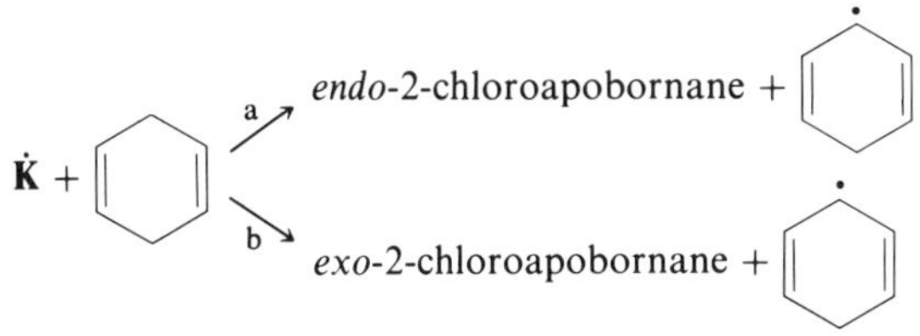				
Therm. of 2-chloroapobornylperester				70 Bar 1
PR	1,4-cyclohexadiene	333...353	$k_a/k_b = 2.0$	
K̇ $+ C_6H_5CH_3$ $\xrightarrow{a}$ *endo*-2-chloroapobornane $+ C_6H_5\dot{C}H_2$; $\xrightarrow{b}$ *exo*-2-chloroapobornane $+ C_6H_5\dot{C}H_2$				
Therm. of 2-chloroapobornylperester				70 Bar 1
PR	$C_6H_5CH_3$	333...353	$k_a/k_b = 1.2$	

[54]) See also corresp. k_a reaction of $\dot{C}H_2CH_2OOC(CH_3)_3$.
[55]) See also corresp. k_a reaction of $\dot{C}H_2CH_2OOC(CH_3)_3$.

Reaction Radical generation Method	Solvent	T[K]	Rate data	Ref./ add. ref.
$\dot{\mathbf{K}} + C_6H_5CH(CH_3)_2 \xrightarrow{a}$ *endo*-2-chloroapobornane + $C_6H_5\dot{C}(CH_3)_2$ $\dot{\mathbf{K}} + C_6H_5CH(CH_3)_2 \xrightarrow{b}$ *exo*-2-chloroapobornane + $C_6H_5\dot{C}(CH_3)_2$ *)				
Therm. of 2-chloroapobornylperester				70 Bar 1
PR	cumene	333... 353	$k_a/k_b = 1.5$	
$(CH_3)_2CH(CH_2)_3\dot{C}{=}CHCCl_3 + CCl_4 \xrightarrow{a} (CH_3)_2CH(CH_2)_3CCl{=}CHCCl_3 + \dot{C}Cl_3$ $(CH_3)_2CH(CH_2)_3\dot{C}{=}CHCCl_3 \xrightarrow{b}$ [56] $(CH_3)_2\dot{C}(CH_2)_3CH{=}CHCCl_3$				
Therm. of BPO as initiator				67 Hei 1
PR, glc	6-methylheptyne-1/ CCl_4	350	$k_a/k_b = 0.0154\ M^{-1}$	
		334... 373	$\log[A_a/A_b/M^{-1}] = -1.4$ $E_a(a) - E_a(b) = 5.9(4)\ kJ\ mol^{-1}$	
n-$C_6H_{13}\dot{C}HCH_2CH_2COCH_3 + CH_3COCH_3 \xrightarrow{a} CH_3CO\dot{C}H_2$ + products + Ce(IV)-acetate $\xrightarrow{b}$ Ce(III)... + products				
Ox. of CH_3COCH_3 by Ce(IV)-acetate and $CH_3CO\dot{C}H_2$ add. to $C_6H_{13}CH{=}CH_2$ in glacial acetic acid				71 Hei 1
PR	glacial acetic acid/ 10% $NaOOCCH_3$	340	$k_a/k_b = 1.85 \cdot 10^{-3}$	
		318	$1.58 \cdot 10^{-3}$	
n-$C_6H_{13}\dot{C}HCH_2CH_2COCH_3 + CH_3COCH_3 \xrightarrow{a} C_6H_{13}(CH_2)_3COCH_3 + CH_3CO\dot{C}H_2$ + $Mn(III)(OOCCH_3)_3 \xrightarrow{b}$ Mn(II)... + products				
$Mn(III)(OOCCH_3)_3$ induced ox. of CH_3COCH_3 and $CH_3CO\dot{C}H_2$ add. to $C_6H_{13}CHCH_2$ in glacial acetic acid				71 Hei 1
PR	glacial acetic acid/ 10% $NaOOCCH_3$	343	$k_a/k_b = 2.20 \cdot 10^{-2}$	
		318	$2.62 \cdot 10^{-2}$	
(Polyvinylacetate)$^{\cdot}$ [57] + 2,2′-dimethyl-6,6′-di-t-butyl-4,4′-isopropylidene-bisphenol(R—OH) $\xrightarrow{a}$ products + R—$\dot{O}$ + 4,4′-isopropylidene-bisphenol(R—OH) $\xrightarrow{b}$ products + R—$\dot{O}$				
Therm. of AIBN as initiator				73 Sim 1/ 67 Sim 2
[58]	vinylacetate	323	$k_a/k_b = 7.17$	
(Polyvinylacetate)$^{\cdot}$ [57] + 2,6-dimethyl-2′,6′-di-t-butyl-4,4′-isopropylidene-bisphenol(R—OH) $\xrightarrow{a}$ products + R—$\dot{O}$ + 4,4′-isopropylidene-bisphenol(R—OH) $\xrightarrow{b}$ products + R—$\dot{O}$				
Therm. of AIBN as initiator				73 Sim 1/ 67 Sim 2
[58]	vinylacetate	323	$k_a/k_b = 6.74$	
(Polyvinylacetate)$^{\cdot}$ [57] + 2,2′-dimethyl-4,4′-isopropylidene-bisphenol(R—OH) $\xrightarrow{a}$ products + R—$\dot{O}$ + 4,4′-isopropylidene-bisphenol(R—OH) $\xrightarrow{b}$ products + R—$\dot{O}$				
Therm. of AIBN as initiator				73 Sim 1/ 67 Sim 2
[58]	vinylacetate	323	$k_a/k_b = 3.80$	
(Polyvinylacetate)$^{\cdot}$ [57] + 2,6-dimethyl-4,4′-isopropylidene-bisphenol(R—OH) $\xrightarrow{a}$ products + R—$\dot{O}$ + 4,4′-isopropylidene-bisphenol (R—OH) $\xrightarrow{b}$ products + R—$\dot{O}$				
Therm. of AIBN as initiator				73 Sim 1/ 67 Sim 2
[58]	vinylacetate	323	$k_a/k_b = 5.86$	

*) For $\dot{\mathbf{K}}$, see p. 177.
[56]) Intramolecular 1,5-hydrogen shift.
[57]) Radical mixture.
[58]) Inhibition of radical polymerization.

Reaction Radical generation Method	Solvent	T[K]	Rate data	Ref./ add. ref.
(Polyvinylacetate)˙ [57]) + 2,2′-di-*t*-butyl-4,4′-isopropylidene-bisphenol(R—OH) $\xrightarrow{a}$ products + R—Ȯ + 4,4′-isopropylidene-bisphenol(R—OH) $\xrightarrow{b}$ products + R—Ȯ				
Therm. of AIBN as initiator [58])	vinylacetate	323	$k_a/k_b = 2.01$	73 Sim 1/ 67 Sim 2
(Polyvinylacetate)˙ [57]) + 2,6-di-*t*-butyl-4,4′-isopropylidene-bisphenol(R—OH) $\xrightarrow{a}$ products + R—Ȯ + 4,4′-isopropylidene-bisphenol(R—OH) $\xrightarrow{b}$ products + R—Ȯ				
Therm. of AIBN as initiator [58])	vinylacetate	323	$k_a/k_b = 1.35$	73 Sim 1/ 67 Sim 2
(Polyvinylacetate)˙ [57]) + 2-methyl-4,4′-isopropylidene-bisphenol(R—OH) $\xrightarrow{a}$ products + R—Ȯ + 4,4′-isopropylidene-bisphenol(R—OH) $\xrightarrow{b}$ products + R—Ȯ				
Therm. of AIBN as initiator [58])	vinylacetate	323	$k_a/k_b = 2.08$	73 Sim 1/ 67 Sim 2
(Polyvinylacetate)˙ [57]) + 2-methyl-6-*t*-butyl-4,4′-isopropylidene-bisphenol(R—OH) $\xrightarrow{a}$ products + R—Ȯ + 4,4′-isopropylidene-bisphenol(R—OH) $\xrightarrow{b}$ products + R—Ȯ				
Therm. of AIBN as initiator [58])	vinylacetate	323	$k_a/k_b = 3.96$	73 Sim 1/ 67 Sim 2
(Polyvinylacetate)˙ [57]) + 2-*t*-butyl-4,4′-isopropylidene-bisphenol(R—OH) $\xrightarrow{a}$ products + R—Ȯ + 4,4′-isopropylidene-bisphenol(R—OH) $\xrightarrow{b}$ products + R—Ȯ				
Therm. of AIBN as initiator [58])	vinylacetate	323	$k_a/k_b = 1.59$	73 Sim 1/ 67 Sim 2
(Polyvinylacetate)˙ [57]) + 2,2′,6,6′-tetramethyl-4,4′-isopropylidene-bisphenol(R—OH) $\xrightarrow{a}$ products + R—Ȯ + 4,4′-isopropylide e-bisphenol(R—OH) $\xrightarrow{b}$ products + R—Ȯ				
Therm. of AIBN as initiator [58])	vinylacetate	323	$k_a/k_b = 11.62$	73 Sim 1/ 67 Sim 2
(Polyvinylacetate)˙ [57]) + 2,2′,6,6′-tetra-*t*-butyl-4,4′-isopropylidene-bisphenol(R—OH) $\xrightarrow{a}$ products + R—Ȯ + 4,4′-isopropylidene-bisphenol(R—OH) $\xrightarrow{b}$ products + R—Ȯ				
Therm. of AIBN as initiator [58])	vinylacetate	323	$k_a/k_b = 1.27$	73 Sim 1/ 67 Sim 2

4.1.2.2 Aromatic radicals and radicals derived from compounds containing aromatic and heterocyclic constituents

4.1.2.2.1 Radicals containing only C and H atoms

Reaction Radical generation Method	Solvent	T[K]	Rate data	Ref./ add. ref.
$(C_6H_5^{\cdot}) + I_2 \xrightarrow{a} C_6H_5I + \dot{I}$ $+ O_2 \xrightarrow{b}$ products				
Phot. of C_6H_5I PR [1])	methylcyclohexane	297(1)	$k_a/k_b = 2.2(3)$	71 Lev 1
$(C_6H_5^{\cdot}) + I_2 \xrightarrow{a} C_6H_5I + \dot{I}$ $+ C_6H_5Cl \xrightarrow{b}$ products				
Phot. of C_6H_5I PR [1])	C_6H_5Cl	RT	$k_a/k_b = 1.3 \cdot 10^4$	73 Lev 1

[57]) Radical mixture.
[58]) Inhibition of radical polymerization.
[1]) ^{131}I radioactivity measurements.

Reaction Radical generation Method	Solvent	T[K]	Rate data	Ref./ add. ref.
$(C_6H_5^{\cdot}) + I_2 \xrightarrow{a} C_6H_5I + \dot{I}$ $+ C_6H_5I \xrightarrow{b}$ products				
Phot. of C_6H_5I PR [1])	C_6H_5I	RT	$k_a/k_b = 9 \cdot 10^3$	73 Lev 1/ 74 Lev 1
$(C_6H_5^{\cdot}) + I_2 \xrightarrow{a} C_6H_5I + \dot{I}$ $+ C_6H_6 \xrightarrow{b}$ products				
Phot. of C_6H_5I PR [1])	C_6H_6	RT	$k_a/k_b = 2.0 \cdot 10^4$	73 Lev 1
$(C_6H_5^{\cdot}) + I_2 \xrightarrow{a} C_6H_5I + \dot{I}$ $+ C_6H_5CN \xrightarrow{b}$ products				
Phot. of C_6H_5I PR [1])	C_6H_5CN	RT	$k_a/k_b = 5.1 \cdot 10^3$	73 Lev 1/ 74 Lev 1
$(C_6H_5^{\cdot}) + I_2 \xrightarrow{a} C_6H_5I + \dot{I}$ $+ C_6H_5CH_3 \xrightarrow{b}$ products				
Phot. of C_6H_5I PR [1])	$C_6H_5CH_3$	RT	$k_a/k_b = 4.0 \cdot 10^4$	73 Lev 1
$(C_6H_5^{\cdot})$ + iodoferrocene $\xrightarrow{a} C_6H_5I$ + (ferrocene)$^{\cdot}$ $+ CCl_4 \xrightarrow{b} C_6H_5Cl + \dot{C}Cl_3$				
Thermal decomp. of PAT PR, glc	CCl_4	333.0(1)	$k_a/k_b = 33(\pm 5\%)$	73 Dan 1
$(C_6H_5^{\cdot}) + CBrCl_3 \xrightarrow{a} C_6H_5Br + \dot{C}Cl_3$ $+ CCl_4 \xrightarrow{b} C_6H_5Cl + \dot{C}Cl_3$				
Therm. of $C_6H_5COOOC(CH_3)_3$ or PAT PR, glc	CCl_4	403 393 385 353 343	$k_a/k_b = 144$ 157 165 234 257 [2]) 305 [2])	75 Her 1, 69 Rue 1/ 76 Gie 1, 76 Gie 2, 76 Gie 3
$(C_6H_5^{\cdot}) + CBrCl_3 \xrightarrow{a} C_6H_5Br + \dot{C}Cl_3$ $+ CCl_4 \xrightarrow{b} C_6H_5Cl + \dot{C}Cl_3$				
Thermal decomp. of PAT PR, glc	CCl_4	333	$k_a/k_b = 265$	79 Mig 1
$(C_6H_5^{\cdot}) + CCl_4 \xrightarrow{a} C_6H_5Cl + \dot{C}Cl_3$ $+ I_2 \xrightarrow{b} C_6H_5I + \dot{I}$				
Thermal decomp. of PAT PR, glc	CCl_4	333	$k_a/k_b = 4.62 \cdot 10^{-4}$	79 Mig 1
$(C_6H_5^{\cdot}) + CCl_4 \xrightarrow{a} C_6H_5Cl + \dot{C}Cl_3$ $+ O_2 \xrightarrow{b} (C_6H_5O_2^{\cdot})$				
Thermal decomp. of PAT PR, glc	CCl_4	333(2)	$k_a/k_b = 8.33 \cdot 10^{-4}$	63 Rus 2

[1]) ^{131}I radioactivity measurements.
[2]) Generation by therm. of phenylazotriphenyl methane (PAT).

Reaction Radical generation Method	Solvent	T[K]	Rate data	Ref./ add. ref.
$(C_6H_5^{\cdot}) + CCl_4 \xrightarrow{a} C_6H_5Cl + \dot{C}Cl_3$ $+ c\text{-}C_6H_{12} \xrightarrow{b} C_6H_6 + (c\text{-}C_6H_{11}^{\cdot})$				
Thermal decomp. of PAT				63 Rus 2
PR	CCl_4 (4.54 M)/ $c\text{-}C_6H_{12}$ (5.22 M)	333(2)	$k_a/k_b = 1.02$	
$(C_6H_5^{\cdot})$[3]) $+ CCl_4 \xrightarrow{a} C_6H_5Cl + \dot{C}Cl_3$ $+$ aromatic $\xrightarrow{b}$ (aromatic-$C_6H_5^{\cdot}$)				
Therm. of D-labelled BPO				60 Bag 1
PR[4])	CCl_4/aromatic	373	aromatic: benzene, $k_a/k_b = 4.26$ nitrobenzene, 1.00 naphthalene, 0.192	
$(C_6H_5^{\cdot}) + CCl_4 \xrightarrow{a} C_6H_5Cl + \dot{C}Cl_3$ $+$ substrate $\xrightarrow{b}$ (substrate-$C_6H_5^{\cdot}$)				
Therm. of PAT				64 Baz 1
PR, glc	CCl_4	333	substrate: phenol, $k_a/k_b = 2.50$ aniline, 1.67 anisole, 2.78	
PR, glc	CCl_4/substrate mixt.	333	substrate: benzene $k_a/k_b = 5.0$ toluene, 3.33 t-butylbenzene, 5.0 2,2-diphenyl-propane, 2.0 2,3-diphenyl-2,3-dimethyl-butane, 2.5	64 Baz 2
$(C_6H_5^{\cdot}) + CHCl_3 \xrightarrow{a} C_6H_6 + \dot{C}Cl_3$ $+ CCl_4 \xrightarrow{b} C_6H_5Cl + \dot{C}Cl_3$				
Thermal decomp. of PAT				63 Bri 1
PR, glc	$CCl_4/CHCl_3$	333	$k_a/k_b = 3.2$	
$(C_6H_5^{\cdot}) + CH_2Cl_2 \xrightarrow{a} C_6H_6 + \dot{C}HCl_2$ $+ CCl_4 \xrightarrow{b} C_6H_5Cl + \dot{C}Cl_3$				
Thermal decomp. of PAT				63 Bri 1
PR, glc	CCl_4/CH_2Cl_2	333	$k_a/k_b = 0.48$	
$(C_6H_5^{\cdot}) + CH_3Cl \xrightarrow{a} C_6H_6 + \dot{C}H_2Cl$ $+ CCl_4 \xrightarrow{b} C_6H_5Cl + \dot{C}Cl_3$				
Thermal decomp. of PAT				63 Bri 1
PR, glc	CCl_4/CH_3Cl	333	$k_a/k_b = 0.80$	
$(C_6H_5^{\cdot}) + CH_3I \xrightarrow{a} C_6H_5I + \dot{C}H_3$ $+ CBrCl_3 \xrightarrow{b} C_6H_5Br + \dot{C}Cl_3$				
Thermal decomp. of PAT				71 Dan 2
PR, glc	$CBrCl_3/CH_3I$ mixt.	333	$k_a/k_b = 0.17$	

[3]) D-labelled.
[4]) Measurement of D-labelled compounds.

Reaction Radical generation Method	Solvent	T[K]	Rate data	Ref./ add. ref.
$(C_6H_5^{\cdot}) + CH_3NO_2 \xrightarrow{a} C_6H_6 + \dot{C}H_2NO_2$ $+ CCl_4 \xrightarrow{b} C_6H_5Cl + \dot{C}Cl_3$				
Thermal decomp. of PAT PR, glc	CCl_4/CH_3NO_2	333	$k_a/k_b = 0.047$	63 Bri 1
$(C_6H_5^{\cdot}) + CH_3OH$ [5]) $\xrightarrow{a} C_6H_6 + \dot{C}H_2OH$ $\xrightarrow{b} C_6H_6 + CH_3\dot{O}$				
Decomp. of nitrosoacetanilide, or PAT, or DBPO PR, MS	methanol [5])	RT (presumed)	$k_a/k_b = 3$ [6])	72 Koe 1
$(C_6H_5^{\cdot}) + CH_3OH \xrightarrow{a} C_6H_6 +$ methanol$(-\dot{H})$ $+ CCl_4 \xrightarrow{b} C_6H_5Cl + \dot{C}Cl_3$				
Thermal decomp. of PAT PR, glc	CCl_4/CH_3OH	333	$k_a/k_b = 0.13$	63 Bri 1
$(C_6H_5^{\cdot}) + CH_3OH \xrightarrow{a} C_6H_6 + \dot{C}H_2OH$ $+ C_6H_5CH{=}N(O)C(CH_3)_3 \xrightarrow{b} (C_6H_5)_2CHN(\dot{O})C(CH_3)_3$				
Thermal decomp. of PAT PR, SESR, spin trap	CH_3OH	297	$k_a/k_b = 1.16 \cdot 10^{-2}$	75 Jan 2
$(C_6H_5^{\cdot}) + CF_3CH_2I \xrightarrow{a} C_6H_5I + CF_3\dot{C}H_2$ $+ CBrCl_3 \xrightarrow{b} C_6H_5Br + \dot{C}Cl_3$				
Thermal decomp. of PAT PR, glc	$CBrCl_3/CF_3CH_2I$ mixt.	333	$k_a/k_b = 0.92$	71 Dan 2
$(C_6H_5^{\cdot}) + ICH_2COOH \xrightarrow{a} C_6H_5I + \dot{C}H_2COOH$ $+ CBrCl_3 \xrightarrow{b} C_6H_5Br + \dot{C}Cl_3$				
Thermal decomp. of PAT PR, glc	$CBrCl_3/ICH_2COOH$ mixt.	333	$k_a/k_b = 1.73$	71 Dan 2
$(C_6H_5^{\cdot}) + CH_3CN \xrightarrow{a} C_6H_6 + \dot{C}H_2CN$ $+ CCl_4 \xrightarrow{b} C_6H_5Cl + \dot{C}Cl_3$				
Thermal decomp. of PAT PR, glc	CCl_4/CH_3CN	333	$k_a/k_b = 0.090$	63 Bri 1
$(C_6H_5^{\cdot}) + BrCH_2CH_2I \xrightarrow{a} C_6H_5I + BrCH_2\dot{C}H_2$ $+ CBrCl_3 \xrightarrow{b} C_6H_5Br + \dot{C}Cl_3$				
Thermal decomp. of PAT PR, glc	$CBrCl_3/BrCH_2CH_2I$ mixt.	333	$k_a/k_b = 0.86$	71 Dan 2
$(C_6H_5^{\cdot}) + BrCH_2CH_2Br \xrightarrow{a} C_6H_5Br + BrCH_2\dot{C}H_2$ $+ CCl_4 \xrightarrow{b} C_6H_5Cl + \dot{C}Cl_3$				
Therm. of PAT PR, glc	CH_2BrCH_2Br/CCl_4	333.0(1)	$k_a/k_b = 0.37$ [7]) $(\pm 0.5\%)$	75 Dan 1

[5]) Differently deuterated.
[6]) Statistically corrected to give value per C—H.
[7]) Corrected by statistical factor of 2.

Reaction Radical generation Method	Solvent	T[K]	Rate data	Ref./ add. ref.
$(C_6H_5^{\cdot}) + ClCH_2CH_2I \xrightarrow{a} C_6H_5I + ClCH_2\dot{C}H_2$ $+ CBrCl_3 \xrightarrow{b} C_6H_5Br + \dot{C}Cl_3$				
Thermal decomp. of PAT				71 Dan 2
PR, glc	$CBrCl_3/ClCH_2CH_2I$	333	$k_a/k_b = 0.52$	
$(C_6H_5^{\cdot}) + CH_3CHO \xrightarrow{a} C_6H_6$ + acetaldehyde$(-\dot{H})$ $+ CCl_4 \xrightarrow{b} C_6H_5Cl + \dot{C}Cl_3$				
Thermal decomp. of PAT				65 Baz 2
PR, glc	CCl_4/CH_3CHO	333	$k_a/k_b = 2.4$	
$(C_6H_5^{\cdot}) + CH_3COOH \xrightarrow{a} C_6H_6$ + acetic acid$(-\dot{H})$ $+ CCl_4 \xrightarrow{b} C_6H_5Cl + \dot{C}Cl_3$				
Thermal decomp. of PAT				63 Bri 1
PR, glc	CCl_4/CH_3COOH	333	$k_a/k_b = 0.086$	
$(C_6H_5^{\cdot}) + CH_3CH_2Br \xrightarrow{a} C_6H_5Br + CH_3\dot{C}H_2$ $+ CCl_4 \xrightarrow{b} C_6H_5Cl + \dot{C}Cl_3$				
Therm. of PAT				75 Dan 1
PR, glc	CH_3CH_2Br/CCl_4	333.0(1)	$k_a/k_b = 0.076(\pm 0.5\%)$	
$(C_6H_5^{\cdot}) + CH_3CH_2I \xrightarrow{a} C_6H_5I + CH_3\dot{C}H_2$ $+ CBrCl_3 \xrightarrow{b} C_6H_5Br + \dot{C}Cl_3$				
Thermal decomp. of PAT				71 Dan 2
PR, glc	$CBrCl_3/CH_3CH_2I$ mixt.	333	$k_a/k_b = 0.33$	
$(C_6H_5^{\cdot}) + CH_3CH_2NO_2 \xrightarrow{a} C_6H_6$ + nitroethane$(-\dot{H})$ $+ CCl_4 \xrightarrow{b} C_6H_5Cl + \dot{C}Cl_3$				
Thermal decomp. of PAT				63 Bri 1
PR, glc	$CCl_4/CH_3CH_2NO_2$	333	$k_a/k_b = 0.19$	
$(C_6H_5^{\cdot}) + CH_3OCH_3 \xrightarrow{a} C_6H_6 + \dot{C}H_2OCH_3$ $+ CCl_4 \xrightarrow{b} C_6H_5Cl + \dot{C}Cl_3$				
Thermal decomp. of PAT				63 Bri 1
PR, glc	CCl_4/CH_3OCH_3	333	$k_a/k_b = 0.28$	
$(C_6H_5^{\cdot}) + (CH_3)_2SO \xrightarrow{a} C_6H_6 + \dot{C}H_2SOCH_3$ $+ CCl_4 \xrightarrow{b} C_6H_5Cl + \dot{C}Cl_3$				
Thermal decomp. of PAT				63 Bri 1
PR, glc	$CCl_4/(CH_3)_2SO$	333	$k_a/k_b = 0.039$	
$(C_6H_5^{\cdot}) + CH_3SSCH_3 \xrightarrow{a} C_6H_5SCH_3 + CH_3\dot{S}$ $(C_6H_5^{\cdot}) + CH_3SSCH_3 \xrightarrow{b} C_6H_6 + \dot{C}H_2SSCH_3$				
Thermal decomp. of PAT				63 Pry 1
PR, glc	dimethyldisulfide	333	$k_a/k_b = 55.3(11)$	
$(C_6H_5^{\cdot}) + CH_3SSCH_3 \xrightarrow{a} C_6H_6 + \dot{C}H_2SSCH_3$ $+ CCl_4 \xrightarrow{b} C_6H_5Cl + \dot{C}Cl_3$				
Thermal decomp. of PAT				64 Pry 1, 70 Pry 1
PR	CCl_4/CH_3SSCH_3 mixt.	333	$k_a/k_b = 0.57$ 0.50 [8])	

[8]) From [70 Pry 1].

Reaction Radical generation Method	Solvent	T[K]	Rate data	Ref./ add. ref.
$(C_6H_5^{\cdot}) + CH_3SSCH_3 \xrightarrow{a} C_6H_5SCH_3 + CH_3\dot{S}$ $+ CCl_4 \xrightarrow{b} C_6H_5Cl + \dot{C}Cl_3$ Thermal decomp. of PAT PR	CCl_4/CH_3SSCH_3 mixt.	333	$k_a/k_b = 31.0$ 17.9 [8])	64 Pry 1, 70 Pry 1
$(C_6H_5^{\cdot}) + (CH_3)_2Se \xrightarrow{a} \dot{C}H_3 + C_6H_5SeCH_3$ $+ (C_2H_5)_2Se \xrightarrow{b} CH_3\dot{C}H_2 + C_6H_5SeC_2H_5$ Phot. of C_6H_5I in presence of hexa-*n*-butyl tin SESR	[9])	243	$k_a/k_b = 0.044$	77 Sca 1
$(C_6H_5^{\cdot}) + (CH_3)_2Se \xrightarrow{a} \dot{C}H_3 + C_6H_5SeCH_3$ $+ c\text{-}C_5H_{10} \xrightarrow{b} C_6H_6 + (c\text{-}C_5H_9^{\cdot})$ Phot. of C_6H_5I in presence of hexa-*n*-butyl tin SESR	[9])	243	$k_a/k_b = 3.1$	77 Sca 1
$(C_6H_5^{\cdot}) + CD_3COCD_3 \xrightarrow{a} C_6H_5D + \dot{C}D_2COCD_3$ $+ CCl_4 \xrightarrow{b} C_6H_5Cl + \dot{C}Cl_3$ Thermal decomp. of PAT PR, glc	CCl_4/CD_3COCD_3	333	$k_a/k_b = 0.041$	63 Bri 1
$(C_6H_5^{\cdot}) + CH{\equiv}CCH_3 \xrightarrow{a} C_6H_6 + (C_3H_3^{\cdot})$ $+ CCl_4 \xrightarrow{b} C_6H_5Cl + \dot{C}Cl_3$ Thermal decomp. of PAT PR, glc	$CCl_4/CH{\equiv}CCH_3$	333	$k_a/k_b = 0.3$	63 Bri 1
$(C_6H_5^{\cdot}) + c\text{-}trans\text{-}C_3H_4Br_2 \xrightarrow{a} C_6H_5Br + (c\text{-}trans\text{-}C_3H_4Br^{\cdot})$ $+ CCl_4 \xrightarrow{b} C_6H_5Cl + \dot{C}Cl_3$ Therm. of PAT PR, glc	$c\text{-}trans\text{-}C_3H_4Br_2/CCl_4$	333.0(1)	$k_a/k_b = 0.31$ [10]) $(\pm 0.5\%)$	75 Dan 1
$(C_6H_5^{\cdot}) + c\text{-}C_3H_5Br \xrightarrow{a} C_6H_5Br + (c\text{-}C_3H_5^{\cdot})$ $+ CCl_4 \xrightarrow{b} C_6H_5Cl + \dot{C}Cl_3$ Therm. of PAT PR, glc	$c\text{-}C_3H_5Br/CCl_4$	333.0(1)	$k_a/k_b = 0.035(\pm 0.5\%)$	75 Dan 1
$(C_6H_5^{\cdot}) + ICH_2CH_2COOH \xrightarrow{a} C_6H_5I + \dot{C}H_2CH_2COOH$ $+ CBrCl_3 \xrightarrow{b} C_6H_5Br + \dot{C}Cl_3$ Thermal decomp. of PAT PR, glc	$CBrCl_3/$ ICH_2CH_2COOH mixt.	333	$k_a/k_b = 0.46$	71 Dan 2
$(C_6H_5^{\cdot}) + CH_3CH_2CN \xrightarrow{a} C_6H_6 + (C_2H_4CN^{\cdot})$ $+ CCl_4 \xrightarrow{b} C_6H_5Cl + \dot{C}Cl_3$ Thermal decomp. of PAT PR, glc	CCl_4/CH_3CH_2CN	333	$k_a/k_b = 0.37$	63 Bri 1
$(C_6H_5^{\cdot}) + CH_3CH_2CN \xrightarrow{a} C_6H_6 + (C_2H_4CN^{\cdot})$ $+ (CH_3)_2CHCOOH \xrightarrow{b} C_6H_6 + (C_4H_7O_2^{\cdot})$ $Ti(III) + C_6H_5N_2^+BF_4^-$ in H_2O PR, SESR	H_2O, pH = 4	RT	$k_a/k_b = 0.17(\pm 15\%)$	77 Ash 1

8) From [70 Pry 1].
9) Not given (presumed: mixture of $c\text{-}C_5H_{10}$, $(CH_3)_2Se$ and $(C_2H_5)_2Se$).
10) Corrected by statistical factor of 2.

Reaction Radical generation Method	Solvent	T[K]	Rate data	Ref./ add. ref.
$(C_6H_5^{\cdot}) + ICH_2CH_2CH_2I \begin{cases} \xrightarrow{a} c\text{-}C_3H_6 + C_6H_5I + \dot{I} \\ \xrightarrow{b} ICH_2CH_2\dot{C}H_2 + C_6H_5I \end{cases}$				
Thermal decomp. of BPO PR, glc, NMR	C_6H_6/CCl_4	384	$k_a/k_b < 0.17$	73 Dru 1/ 72 Dru 1
$(C_6H_5^{\cdot}) + ICH_2CH_2CH_2I \xrightarrow{a} C_6H_5I + \dot{C}H_2CH_2CH_2I$ $+ CBrCl_3 \xrightarrow{b} C_6H_5Br + \dot{C}Cl_3$				
Thermal decomp. of PAT PR, glc	$CBrCl_3/$ $ICH_2CH_2CH_2I$ mixt.	333	$k_a/k_b = 0.47$ [11])	71 Dan 2
$(C_6H_5^{\cdot}) + CH_3COCH_3 \xrightarrow{a} C_6H_6 + \dot{C}H_2COCH_3$ $+ CCl_4 \xrightarrow{b} C_6H_5Cl + \dot{C}Cl_3$				
Thermal decomp. of PAT PR, glc	CCl_4/CH_3COCH_3	333	$k_a/k_b = 0.17$	63 Bri 1, 66 Pry 1
$(C_6H_5^{\cdot}) + CH_3CH_2COOH \xrightarrow{a} C_6H_6 + (C_3H_5O_2^{\cdot})$ $+ (CH_3)_2CHCOOH \xrightarrow{b} C_6H_6 + (C_4H_7O_2^{\cdot})$				
Ti(III) + $C_6H_5N_2^+BF_4^-$ in H_2O PR, SESR	H_2O, pH = 4	RT	$k_a/k_b = 0.28(\pm 15\%)$	77 Ash 1
$(C_6H_5^{\cdot}) + CH_3COOCH_3 \xrightarrow{a} C_6H_6$ + acetic acid methylester$(-\dot{H})$ $+ CCl_4 \xrightarrow{b} C_6H_5Cl + \dot{C}Cl_3$				
Thermal decomp. of PAT PR, glc	CCl_4/CH_3COOCH_3	333	$k_a/k_b = 0.086$	63 Bri 1, 66 Pry 1
$(C_6H_5^{\cdot}) + (CH_3)_2CHI \xrightarrow{a} C_6H_5I + (CH_3)_2\dot{C}H$ $+ CBrCl_3 \xrightarrow{b} C_6H_5Br + \dot{C}Cl_3$				
Thermal decomp. of PAT PR, glc	$CBrCl_3/(CH_3)_2CHI$ mixt.	333	$k_a/k_b = 0.58$	71 Dan 2
$(C_6H_5^{\cdot}) + (CH_3)_2CHNO_2 \xrightarrow{a} C_6H_6$ + 2-nitropropane$(-\dot{H})$ $+ CCl_4 \xrightarrow{b} C_6H_5Cl + \dot{C}Cl_3$				
Thermal decomp. of PAT PR, glc	$CCl_4/(CH_3)_2CHNO_2$	333	$k_a/k_b = 0.25$	63 Bri 1
$(C_6H_5^{\cdot}) + CH_3CH_2CH_2SH \xrightarrow{a} C_6H_6 + CH_3CH_2CH_2\dot{S}$ $+ CH_3CH_2CH_2SSCH_2CH_2CH_3 \xrightarrow{b} \begin{cases} C_6H_6 + RSSR(-\dot{H}) \\ C_6H_5SCH_2CH_2CH_3 + R\dot{S} \end{cases}$				
Thermal decomp. of PAT PR, glc	di-*n*-propyl-disulfide	333	$k_a/k_b = 8.5$	63 Pry 1
$(C_6H_5^{\cdot}) + (CH_3)_2CHSH \xrightarrow{a} C_6H_6 + (CH_3)_2CH\dot{S}$ $+ (CH_3)_2CHSSCH(CH_3)_2 \xrightarrow{b} \begin{cases} C_6H_6 + RSSR(-\dot{H}) \\ C_6H_5SCH(CH_3)_2 + R\dot{S} \end{cases}$				
Thermal decomp. of PAT PR, glc	di-isopropyl-disulfide	333	$k_a/k_b = 23$	63 Pry 1

[11]) Ratio per I atom.

Reaction Radical generation Method	Solvent	T[K]	Rate data	Ref./ add. ref.
$(C_6H_5^{\cdot}) + (CH_3)_3N \xrightarrow{a} C_6H_6 + \dot{C}H_2N(CH_3)_2$ $+ CCl_4 \quad C_6H_5Cl + \dot{C}Cl_3$				
Thermal decomp. of PAT				63 Bri 1
PR, glc	$CCl_4/(CH_3)_3N$	333	$k_a/k_b = 3.3$	
$(C_6H_5^{\cdot})$ + 2-iodothiophene $\xrightarrow{a} C_6H_5I +$ 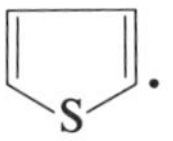 $+ CCl_4 \xrightarrow{b} C_6H_5Cl + \dot{C}Cl_3$				
Thermal decomp. of PAT				74 Dan 1
PR, glc	CCl_4/2-iodothiophene	333.0(1)	$k_a/k_b = 4.6$	
$(C_6H_5^{\cdot})$ + 3-iodothiophene $\xrightarrow{a} C_6H_5I +$ 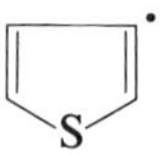 $+ CCl_4 \xrightarrow{b} C_6H_5Cl + \dot{C}Cl_3$				
Thermal decomp. of PAT				74 Dan1
PR, glc	CCl_4/ 3-iodothiophene	333.0(1)	$k_a/k_b = 4.0$	
$(C_6H_5^{\cdot}) + CH{\equiv}C{-}CH_2CH_3 \xrightarrow{a} C_6H_6 + (C_4H_5^{\cdot})$ $+ CCl_4 \xrightarrow{b} C_6H_5Cl + \dot{C}Cl_3$				
Thermal decomp. of PAT				63 Bri 1
PR, glc	$CCl_4/CH{\equiv}CCH_2CH_3$	333	$k_a/k_b = 1.02$	
$(C_6H_5^{\cdot}) + CH_3C{\equiv}CCH_3 \xrightarrow{a} C_6H_6 + (C_4H_5^{\cdot})$ $+ CCl_4 \xrightarrow{b} C_6H_5Cl + \dot{C}Cl_3$				
Thermal decomp. of PAT				63 Bri 1
PR, glc	$CCl_4/CH_3C{\equiv}CCH_3$	333	$k_a/k_b = 0.87$	
$(C_6H_5^{\cdot}) + c\text{-}C_4H_7Br \xrightarrow{a} C_6H_5Br + (c\text{-}C_4H_7^{\cdot})$ $+ CCl_4 \xrightarrow{b} C_6H_5Cl + \dot{C}Cl_3$				
Therm. of PAT				75 Dan 1
PR, glc	$c\text{-}C_4H_7Br/CCl_4$	333.0(1)	$k_a/k_b = 0.18(\pm 0.5\%)$	
$(C_6H_5^{\cdot}) + ICH_2COOC_2H_5 \xrightarrow{a} C_6H_5I + \dot{C}H_2COOC_2H_5$ $+ CBrCl_3 \xrightarrow{b} C_6H_5Br + \dot{C}Cl_3$				
Thermal decomp. of PAT				71 Dan 2
PR, glc	$CBrCl_3$/ $ICH_2COOC_2H_5$ mixt.	333	$k_a/k_b = 1.34$	
$(C_6H_5^{\cdot}) + (CH_3)_2CHCN \xrightarrow{a} C_6H_6$ + isobutyronitrile$(-\dot{H})$ $+ CCl_4 \xrightarrow{b} C_6H_5Cl + \dot{C}Cl_3$				
Thermal decomp. of PAT				63 Bri 1
PR, glc	$CCl_4/(CH_3)_2CHCN$	333	$k_a/k_b = 0.73$	
$(C_6H_5^{\cdot}) + (CH_3)_2CHCOO^- \xrightarrow{a} C_6H_6 + (C_4H_6O_2^{\cdot})^-$ $+ CH_3CH_2COO^- \xrightarrow{b} C_6H_6 + (C_3H_4O_2^{\cdot})^-$				
Ti(III) + $C_6H_5N_2^+BF_4^-$ in H_2O				77 Ash 1
PR, SESR	H_2O, pH = 8	RT	$k_a/k_b = 1.29$	

Reaction Radical generation Method	Solvent	T[K]	Rate data	Ref./ add. ref.
$(C_6H_5^{\cdot}) + CH_2{=}CHCH_2CH_3 \xrightarrow{a} C_6H_6 + (C_4H_7^{\cdot})$ $+ CCl_4 \xrightarrow{b} C_6H_5Cl + \dot{C}Cl_3$				
Thermal decomp. of PAT				63 Bri 1
PR, glc	CCl_4/ $CH_2{=}CHCH_2CH_3$	333	$k_a/k_b = 0.66$	
$(C_6H_5^{\cdot})$ + *cis*-$CH_3CH{=}CHCH_3 \xrightarrow{a} C_6H_6 + (C_4H_7^{\cdot})$ $+ CCl_4 \xrightarrow{b} C_6H_5Cl + \dot{C}Cl_3$				
Thermal decomp. of PAT				63 Bri 1
PR, glc	CCl_4/ *cis*-$CH_3CH{=}CHCH_3$	333	$k_a/k_b = 0.88$	
$(C_6H_5^{\cdot})$ + *trans*-$CH_3CH{=}CHCH_3 \xrightarrow{a} C_6H_6 + (C_4H_7^{\cdot})$ $+ CCl_4 \xrightarrow{b} C_6H_5Cl + \dot{C}Cl_3$				
Thermal decomp. of PAT				63 Bri 1
PR, glc	CCl_4/ *trans*-$CH_3CH{=}CHCH_3$	333	$k_a/k_b = 0.63$	
$(C_6H_5^{\cdot})$ + *meso*-$CH_3CHBrCHBrCH_3 \xrightarrow{a} C_6H_5Br + CH_3CHBr\dot{C}HCH_3$ $+ CCl_4 \xrightarrow{b} C_6H_5Cl + \dot{C}Cl_3$				
Therm. of PAT				75 Dan 1
PR, glc	*meso*- $CH_3CHBrCHBrCH_3$/ CCl_4	333.0(1)	$k_a/k_b = 1.49$ [12] $(\pm 0.5\%)$	
$(C_6H_5^{\cdot})$ + *d,l*-$CH_3CHBrCHBrCH_3 \xrightarrow{a} C_6H_5Br + CH_3CHBr\dot{C}HCH_3$ $+ CCl_4 \xrightarrow{b} C_6H_5Cl + \dot{C}Cl_3$				
Therm. of PAT				75 Dan 1
PR, glc	*d,l*- $CH_3CHBrCHBrCH_3$/ CCl_4	333.0(1)	$k_a/k_b = 1.22$ [12] $(\pm 0.5\%)$	
$(C_6H_5^{\cdot}) + CH_2{=}CHCH_2SCH_3 \xrightarrow{a} C_6H_5CH_2CH{=}CH_2 + CH_3\dot{S}$ $+ CCl_4 \xrightarrow{b} C_6H_5Cl + \dot{C}Cl_3$				
Thermal decomp. of PAT				79 Mig 1
PR, glc	CCl_4/ $CH_2{=}CHCH_2SCH_3$ mixt.	333	$k_a/k_b = 3.0$	
$(C_6H_5^{\cdot}) + CH_3CH_2CHBrCH_3 \xrightarrow{a} C_6H_5Br + CH_3CH_2\dot{C}HCH_3$ $+ CCl_4 \xrightarrow{b} C_6H_5Cl + \dot{C}Cl_3$				
Therm. of PAT				75 Dan 1
PR, glc	$CH_3CH_2CHBrCH_3$/ CCl_4	333.0(1)	$k_a/k_b = 0.24(\pm 0.5\%)$	
$(C_6H_5^{\cdot}) + CH_3CH_2CH_2CH_2I \xrightarrow{a} C_6H_5I + CH_3CH_2CH_2\dot{C}H_2$ $+ CBrCl_3 \xrightarrow{b} C_6H_5Br + \dot{C}Cl_3$				
Thermal decomp. of PAT				71 Dan 2
PR, glc	$CBrCl_3$/ $CH_3CH_2CH_2CH_2I$ mixt.	333	$k_a/k_b = 0.31$	

[12]) Corrected by statistical factor of 2.

Reaction Radical generation Method	Solvent	T[K]	Rate data	Ref./ add. ref.
$(C_6H_5^{\bullet}) + CH_3CH_2CH(CH_3)I \xrightarrow{a} C_6H_5I + CH_3CH_2\dot{C}HCH_3$ $+ CBrCl_3 \xrightarrow{b} C_6H_5Br + \dot{C}Cl_3$				
Thermal decomp. of PAT PR, glc	$CBrCl_3/$ $CH_3CH_2CH(CH_3)I$ mixt.	333	$k_a/k_b = 0.54$	71 Dan 2
$(C_6H_5^{\bullet}) + (CH_3)_2CHCH_2I \xrightarrow{a} C_6H_5I + (CH_3)_2CH\dot{C}H_2$ $+ CBrCl_3 \xrightarrow{b} C_6H_5Br + \dot{C}Cl_3$				
Thermal decomp. of PAT PR, glc	$CBrCl_3/$ $(CH_3)_2CHCH_2I$ mixt.	333	$k_a/k_b = 0.27$	71 Dan 2
$(C_6H_5^{\bullet}) + (C_2H_5)_2SO_2 \xrightarrow{a} C_6H_6 + (C_2H_5SO_2C_2H_4^{\bullet})$ $+ (CH_3)_2CHCOOH \xrightarrow{b} C_6H_6 + (C_4H_7O_2^{\bullet})$				
Ti(III) + $C_6H_5N_2^+BF_4^-$ in H_2O PR, SESR	H_2O, pH = 4	RT	$k_a/k_b = 0.20(\pm 15\%)$	77 Ash 1
$(C_6H_5^{\bullet}) + C_2H_5SSC_2H_5 \xrightarrow{a} C_6H_6 +$ diethyldisulfide$(-\dot{H})$ $+ CCl_4 \xrightarrow{b} C_6H_5Cl + \dot{C}Cl_3$				
Thermal decomp. of PAT PR	$CCl_4/C_2H_5SSC_2H_5$ mixt.	333	$k_a/k_b = 1.3$	64 Pry 1/ 70 Pry 1
$(C_6H_5^{\bullet}) + C_2H_5SSC_2H_5 \xrightarrow{a} C_6H_5SC_2H_5 + C_2H_5\dot{S}$ $+ CCl_4 \xrightarrow{b} C_6H_5Cl + \dot{C}Cl_3$				
Thermal decomp. of PAT PR	$CCl_4/C_2H_5SSC_2H_5$ mixt.	333	$k_a/k_b = 17.4$	64 Pry 1/ 70 Pry 1
$(C_6H_5^{\bullet}) + (C_2H_5)_2Se \xrightarrow{a} CH_3\dot{C}H_2 + C_6H_5SeC_2H_5$ $+ c\text{-}C_5H_{10} \xrightarrow{b} C_6H_6 + (c\text{-}C_5H_9^{\bullet})$				
Phot. of C_6H_5I in presence of hexa-*n*-butyl tin SESR	[13])	243	$k_a/k_b = 70$	77 Sca 1
$(C_6H_5^{\bullet}) + (CH_3)_4Si \xrightarrow{a} C_6H_6 + \dot{C}H_2Si(CH_3)_3$ $+ CCl_4 \xrightarrow{b} C_6H_5Cl + \dot{C}Cl_3$				
Thermal decomp. of PAT PR, glc	$CCl_4/(CH_3)_4Si$	333	$k_a/k_b = 0.29$	63 Bri 1
$(C_6H_5^{\bullet})$ + 2-iodopyridine $\xrightarrow{a} C_6H_5I +$ 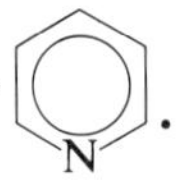 $+ CCl_4 \xrightarrow{b} C_6H_5Cl + \dot{C}Cl_3$				
Thermal decomp. of PAT PR, glc	CCl_4	333.0(1)	$k_a/k_b = 10.6(\pm 5\%)$	74 Dan 1

[13]) Not given (presumed: mixture of $c\text{-}C_5H_{10}$, $(CH_3)_2Se$ and $(C_2H_5)_2Se$).

Reaction Radical generation Method	Solvent	T[K]	Rate data	Ref./ add. ref.
$(C_6H_5^{\cdot})$ + 3-iodopyridine $\xrightarrow{a}$ C_6H_5I + 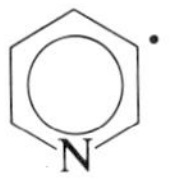 + CCl_4 $\xrightarrow{b}$ C_6H_5Cl + $\dot{C}Cl_3$				
Thermal decomp. of PAT				74 Dan 1
PR, glc	CCl_4	333.0(1)	$k_a/k_b = 25.6(\pm 5\%)$	
$(C_6H_5^{\cdot})$ + 4-iodopyridine $\xrightarrow{a}$ C_6H_5I +  + CCl_4 $\xrightarrow{b}$ C_6H_5Cl + $\dot{C}Cl_3$				
Thermal decomp. of PAT				74 Dan 1
PR, glc	CCl_4	333.0(1)	$k_a/k_b = 31.2(\pm 5\%)$	
$(C_6H_5^{\cdot})$ + 2-methylfuran $\xrightarrow{a}$ C_6H_6 + 2-methylfuran($-\dot{H}$) + CCl_4 $\xrightarrow{b}$ C_6H_5Cl + $\dot{C}Cl_3$				
Thermal decomp. of PAT				63 Bri 1
PR, glc	CCl_4/2-methylfuran	333	$k_a/k_b = 0.25$	
$(C_6H_5^{\cdot})$ + 2-methylfuran $\xrightarrow{a}$ C_6H_6 + 2-methylfuran($-\dot{H}$) + CCl_4 $\xrightarrow{b}$ C_6H_5Cl + $\dot{C}Cl_3$				
Thermal decomp. of PAT				66 Str 1
PR, glc	CCl_4/2-methylfuran	333	$k_a/k_b = 8.6 \cdot 10^{-2}$	
$(C_6H_5^{\cdot})$ + 3-methylfuran $\xrightarrow{a}$ C_6H_6 + 3-methylfuran($-\dot{H}$) + CCl_4 $\xrightarrow{b}$ C_6H_5Cl + $\dot{C}Cl_3$				
Thermal decomp. of PAT				66 Str 1
PR, glc	CCl_4/3-methylfuran	333	$k_a/k_b = 2.5 \cdot 10^{-2}$	
$(C_6H_5^{\cdot})$ + 2-methylthiophene $\xrightarrow{a}$ C_6H_6 + 2-methylthiophene($-\dot{H}$) + CCl_4 $\xrightarrow{b}$ C_6H_5Cl + $\dot{C}Cl_3$				
Thermal decomp. of PAT				63 Bri 1
PR, glc	CCl_4/ 2-methylthiophene	333	$k_a/k_b = 0.44$	
$(C_6H_5^{\cdot})$ + 3-methylthiophene $\xrightarrow{a}$ C_6H_6 + 3-methylthiophene($-\dot{H}$) + CCl_4 $\xrightarrow{b}$ C_6H_5Cl + $\dot{C}Cl_3$				
Thermal decomp. of PAT				63 Bri 1
PR, glc	CCl_4/ 3-methylthiophene	333	$k_a/k_b = 0.24$	
$(C_6H_5^{\cdot})$ + N-methylpyrrole $\xrightarrow{a}$ C_6H_6 + N-methylpyrrole($-\dot{H}$) + CCl_4 $\xrightarrow{b}$ C_6H_5Cl + $\dot{C}Cl_3$				
Thermal decomp. of PAT				63 Bri 1
PR, glc	CCl_4/ N-methylpyrrole	333	$k_a/k_b = 0.3$	
$(C_6H_5^{\cdot})$ + c-C_5H_8 $\xrightarrow{a}$ C_6H_6 + (c-$C_5H_7^{\cdot}$) + CCl_4 $\xrightarrow{b}$ C_6H_5Cl + $\dot{C}Cl_3$				
Thermal decomp. of PAT				63 Bri 1
PR, glc	CCl_4/c-C_5H_8	333	$k_a/k_b = 3.7$	

Reaction Radical generation Method	Solvent	T[K]	Rate data	Ref./ add. ref.
$(C_6H_5^{\cdot}) + c\text{-}trans\text{-}C_5H_8Br_2 \xrightarrow{a} C_6H_5Br + (c\text{-}trans\text{-}C_5H_8Br^{\cdot})$ $+ CCl_4 \xrightarrow{b} C_6H_5Cl + \dot{C}Cl_3$				
Therm. of PAT				75 Dan 1
PR, glc	$c\text{-}trans\text{-}C_5H_8Br_2/CCl_4$	333.0(1)	$k_a/k_b = 1.48$ [14] $(\pm 0.5\%)$	
$(C_6H_5^{\cdot}) + CH_3COCH_2COCH_3 \xrightarrow{a} C_6H_6 + $ 2,4-pentadione$(-\dot{H})$ $+ CCl_4 \xrightarrow{b} C_6H_5Cl + \dot{C}Cl_3$				
Thermal decomp. of PAT				63 Bri 1
PR, glc	CCl_4/2,4-pentadione	333	$k_a/k_b = 0.48$	
$(C_6H_5^{\cdot}) + CH_3OOCCH_2COOCH_3 \xrightarrow{a} C_6H_6 + $ malonic acid dimethylester$(-\dot{H})$ $+ CCl_4 \xrightarrow{b} C_6H_5Cl + \dot{C}Cl_3$				
Thermal decomp. of PAT				63 Bri 1
PR, glc	$CCl_4/CH_3OOCCH_2COOCH_3$	333	$k_a/k_b = 0.18$	
$(C_6H_5^{\cdot}) + c\text{-}C_5H_9Br \xrightarrow{a} C_6H_5Br + (c\text{-}C_5H_9^{\cdot})$ $+ CCl_4 \xrightarrow{b} C_6H_5Cl + \dot{C}Cl_3$				
Therm. of PAT				75 Dan 1
PR, glc	$c\text{-}C_5H_9Br/CCl_4$	333.0(1)	$k_a/k_b = 0.26(\pm 0.5\%)$	
$(C_6H_5^{\cdot}) + c\text{-}C_5H_{10} \xrightarrow{a} C_6H_6 + (c\text{-}C_5H_9^{\cdot})$ $+ CCl_4 \xrightarrow{b} C_6H_5Cl + \dot{C}Cl_3$				
Thermal decomp. of PAT				63 Bri 1
PR, glc	$CCl_4/c\text{-}C_6H_{10}$	333	$k_a/k_b = 1.04$	
$(C_6H_5^{\cdot}) + c\text{-}C_5H_{10} \xrightarrow{a} C_6H_6 + (c\text{-}C_5H_9^{\cdot})$ $+ n\text{-}C_6H_{14} \xrightarrow{b} C_6H_6 + (n\text{-}C_6H_{13}^{\cdot})$				
Therm. of BPO				74 Bun 1
PR, glc	CH_3CN	347.6	$k_a/k_b = 1.731(19)$	
$(C_6H_5^{\cdot}) + CH_2{=}CHCH_2CH_2CH_3 \xrightarrow{a} C_6H_6 + (C_5H_9^{\cdot})$ $+ CCl_4 \xrightarrow{b} C_6H_5Cl + \dot{C}Cl_3$				
Thermal decomp. of PAT				63 Bri 1
PR, glc	$CCl_4/CH_2{=}CHCH_2CH_2CH_3$	333	$k_a/k_b = 0.73$	
$(C_6H_5^{\cdot}) + CH_2{=}C(CH_3)CH_2CH_3 \xrightarrow{a} C_6H_6 + (C_5H_9^{\cdot})$ $+ CCl_4 \xrightarrow{b} C_6H_5Cl + \dot{C}Cl_3$				
Thermal decomp. of PAT				63 Bri 1
PR, glc	$CCl_4/CH_2{=}C(CH_3)CH_2CH_3$	333	$k_a/k_b = 0.93$	
$(C_6H_5^{\cdot}) + CH_2{=}CHCH(CH_3)_2 \xrightarrow{a} C_6H_6 + (C_5H_9^{\cdot})$ $+ CCl_4 \xrightarrow{b} C_6H_5Cl + \dot{C}Cl_3$				
Thermal decomp. of PAT				63 Bri 1
PR, glc	$CCl_4/CH_2{=}CHCH(CH_3)_2$	333	$k_a/k_b = 1.47$	

[14]) Corrected by statistical factor of 2.

Reaction Radical generation Method	Solvent	$T[K]$	Rate data	Ref./ add. ref.
$(C_6H_5^{\cdot}) + CH_3CH{=}CHCH_2CH_3 \xrightarrow{a} C_6H_6 + (C_5H_9^{\cdot})$ $+ CCl_4 \xrightarrow{b} C_6H_5Cl + \dot{C}Cl_3$				
Thermal decomp. of PAT PR, glc	$CCl_4/$ $CH_3CH{=}CHCH_2CH_3$	333	$k_a/k_b = 1.29$	63 Bri 1
$(C_6H_5^{\cdot}) + CH_3C(CH_3){=}CHCH_3 \xrightarrow{a} C_6H_6 + (C_5H_9^{\cdot})$ $+ CCl_4 \xrightarrow{b} C_6H_5Cl + \dot{C}Cl_3$				
Thermal decomp. of PAT PR, glc	$CCl_4/$ $CH_3C(CH_3){=}CHCH_3$	333	$k_a/k_b = 1.24$	63 Bri 1
$(C_6H_5^{\cdot}) + C_2H_5COC_2H_5 \xrightarrow{a} C_6H_6 + \text{diethylketone}(-\dot{H})$ $+ CCl_4 \xrightarrow{b} C_6H_5Cl + \dot{C}Cl_3$				
Thermal decomp. of PAT PR, glc	$CCl_4/C_2H_5COC_2H_5$	333	$k_a/k_b = 1.30$	63 Bri 1
$(C_6H_5^{\cdot}) + C_2H_5COC_2H_5 \xrightarrow{a} C_6H_6 + (C_5H_9O^{\cdot})$ $+ (CH_3)_2CHCOOH \xrightarrow{b} C_6H_6 + (C_4H_7O_2^{\cdot})$				
Ti(III) $+ C_6H_5N_2^+ BF_4^-$ in H_2O PR, SESR	$H_2O, pH = 4$	RT	$k_a/k_b = 0.79(\pm 15\%)$	77 Ash 1
$(C_6H_5^{\cdot}) + (CH_3)_3CCH_2I \xrightarrow{a} C_6H_5I + (CH_3)_3C\dot{C}H_2$ $+ CBrCl_3 \xrightarrow{b} C_6H_5Br + \dot{C}Cl_3$				
Thermal decomp. of PAT PR, glc	$CBrCl_3/(CH_3)_3CCH_2I$ mixt.	333	$k_a/k_b = 0.24$	71 Dan 2
$(C_6H_5^{\cdot}) + n\text{-}C_5H_{12} \xrightarrow{a} C_6H_6 + (C_5H_{11}^{\cdot})$ $+ CCl_4 \xrightarrow{b} C_6H_5Cl + \dot{C}Cl_3$				
Thermal decomp. of PAT PR, glc	$CCl_4/n\text{-}C_5H_{12}$	333	$k_a/k_b = 0.60$	63 Bri 1
$(C_6H_5^{\cdot}) + CH_3C(CH_3)_2CH_3 \xrightarrow{a} C_6H_6 + (C_5H_{11}^{\cdot})$ $+ CCl_4 \xrightarrow{b} C_6H_5Cl + \dot{C}Cl_3$				
Thermal decomp. of PAT PR, glc	$CCl_4/$ $CH_3C(CH_3)_2CH_3$	333	$k_a/k_b = 0.14$	63 Bri 1
$(C_6H_5^{\cdot}) + 2\text{-}BrC_6H_4I \xrightarrow{a} C_6H_5I + \langle\!\langle\bigcirc\rangle\!\rangle{-}Br$ $+ CCl_4 \xrightarrow{b} C_6H_5Cl + \dot{C}Cl_3$				
Thermal decomp. of PAT PR, glc	CCl_4	333.0(1)	$k_a/k_b = 50(\pm 5\%)$	73 Dan 1
$(C_6H_5^{\cdot}) + 3\text{-}BrC_6H_4I \xrightarrow{a} C_6H_5I + Br{-}\langle\!\langle\bigcirc\rangle\!\rangle$ [15]$)$ $+ CCl_4 \xrightarrow{b} C_6H_5Cl + \dot{C}Cl_3$				
Therm. of PAT PR, glc	$3\text{-}BrC_6H_4I/CCl_4$	333	$k_a/k_b = 1.51(0)$ See also 4.1.2.3, Fig. 7, p. 257.	82 Tan 1/ 69 Dan 1

[15]) Reaction proceeds via intermediate adduct.

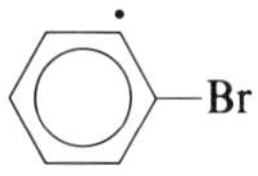

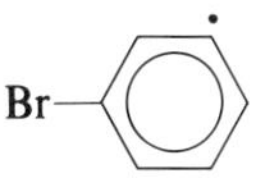

Reaction Radical generation Method	Solvent	T[K]	Rate data	Ref./ add. ref.
$(C_6H_5^{\cdot}) + 4\text{-}BrC_6H_4I \xrightarrow{a} C_6H_5I + Br$— 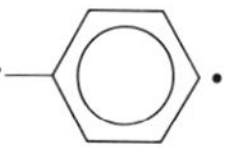 [15] $+ CCl_4 \xrightarrow{b} C_6H_5Cl + \dot{C}Cl_3$				
Therm. of PAT				82 Tan 1/
PR, glc	$4\text{-}BrC_6H_4I/CCl_4$	333	$k_a/k_b = 1.33(0)$ See also 4.1.2.3, Fig. 7, p. 257.	69 Dan 1
$(C_6H_5^{\cdot}) + 2\text{-}ClC_6H_4I \xrightarrow{a} C_6H_5I +$ —Cl $+ CCl_4 \xrightarrow{b} C_6H_5Cl + \dot{C}Cl_3$				
Thermal decomp. of PAT				73 Dan 1
PR, glc	CCl_4	333.0(1)	$k_a/k_b = 25.9(\pm 5\%)$	
$(C_6H_5^{\cdot}) + 2\text{-}FC_6H_4I \xrightarrow{a} C_6H_5I +$ 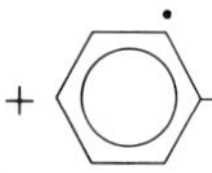—F $+ CCl_4 \xrightarrow{b} C_6H_5Cl + \dot{C}Cl_3$				
Thermal decomp. of PAT				73 Dan 1
PR, glc	$2\text{-}FC_6H_4I/CCl_4$	333.0(1)	$k_a/k_b = 14.5(\pm 5\%)$	
$(C_6H_5^{\cdot}) + 2\text{-}NO_2C_6H_4I \xrightarrow{a} C_6H_5I +$ 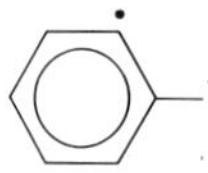—NO_2 $+ CCl_4 \xrightarrow{b} C_6H_5Cl + \dot{C}Cl_3$				
Thermal decomp. of PAT				73 Dan 1
PR, glc	CCl_4	333.0(1)	$k_a/k_b = 186(\pm 5\%)$	
$(C_6H_5^{\cdot}) + 3\text{-}NO_2C_6H_4I \xrightarrow{a} C_6H_5I + O_2N$— 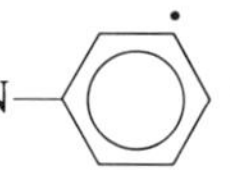 [15] $+ CCl_4 \xrightarrow{b} C_6H_5Cl + \dot{C}Cl_3$				
Therm. of PAT				82 Tan 1/
PR, glc	$3\text{-}NO_2C_6H_4I/CCl_4$	333	$k_a/k_b = 1.63(1)$ See also 4.1.2.3, Fig. 7, p. 257.	69 Dan 1
$(C_6H_5^{\cdot}) + 4\text{-}NO_2C_6H_4I \xrightarrow{a} C_6H_5I + O_2N$—  [15] $+ CCl_4 \xrightarrow{b} C_6H_5Cl + \dot{C}Cl_3$				
Therm. of PAT				82 Tan 1/
PR, glc	$4\text{-}NO_2C_6H_4I/CCl_4$	333	$k_a/k_b = 1.36(1)$ See also 4.1.2.3, Fig. 7, p. 257.	69 Dan 1
$(C_6H_5^{\cdot}) + 2\text{-}IC_6H_4I \xrightarrow{a} C_6H_5I +$ 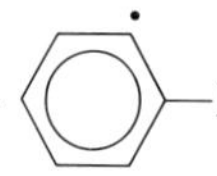—I $+ CCl_4 \xrightarrow{b} C_6H_5Cl + \dot{C}Cl_3$				
Thermal decomp. of PAT				73 Dan 1
PR, glc	CCl_4	333.0(1)	$k_a/k_b = 77$ [16] $(\pm 5\%)$	

[15]) Reaction proceeds via intermediate adduct. [16]) Statistically corrected.

Reaction Radical generation Method	Solvent	T[K]	Rate data	Ref./ add. ref.
$(C_6H_5^{\bullet}) + C_6H_5I \xrightarrow{a} C_6H_5I + (C_6H_5^{\bullet})$ $+ CCl_4 \xrightarrow{b} C_6H_5Cl + \dot{C}Cl_3$				
Thermal decomp. of PAT PR, glc	CCl_4	333.0(1)	$k_a/k_b = 16 \dots 18$ [17])	74 Dan 1/ 69 Dan 1
$(C_6H_5^{\bullet})$ [18]) $+ C_6H_5NO_2 \xrightarrow{a} C_6H_6 + (C_6H_4NO_2^{\bullet})$ $+ CCl_4 \xrightarrow{b} C_6H_5Cl + \dot{C}Cl_3$				
Therm. of D-labelled BPO PR [19])	$CCl_4/C_6H_5NO_2$	373	$k_a/k_b = 0.11$	60 Bag 1
$(C_6H_5^{\bullet})$ + 4-nitrophenol $\xrightarrow{a} C_6H_6$ + 4-nitrophenol$(-\dot{H})$ $+ CCl_4 \xrightarrow{b} C_6H_5Cl + \dot{C}Cl_3$				
Thermal decomp. of PAT PR, glc	CCl_4/4-nitrophenol	333	$k_a/k_b = 0.6$	65 Baz 2
$(C_6H_5^{\bullet}) + 2\text{-}NH_2C_6H_4I \xrightarrow{a} C_6H_5I +$ 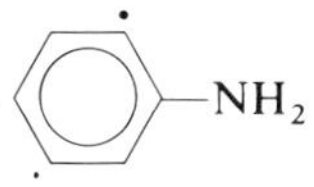 $+ CCl_4 \xrightarrow{b} C_6H_5Cl + \dot{C}Cl_3$				
Thermal decomp. of PAT PR, glc	$2\text{-}NH_2C_6H_4I/CCl_4$	333.0(1)	$k_a/k_b = 7.6(\pm 5\%)$	73 Dan 1
$(C_6H_5^{\bullet}) + 3\text{-}NH_2C_6H_4I \xrightarrow{a} C_6H_5I +$ 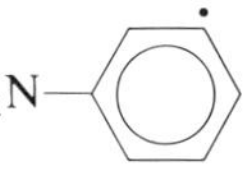 [20]) $+ CCl_4 \xrightarrow{b} C_6H_5Cl + \dot{C}Cl_3$				
Therm. of PAT PR, glc	$3\text{-}NH_2C_6H_4I/CCl_4$	333	$k_a/k_b = 1.25(1)$	82 Tan 1
$(C_6H_5^{\bullet}) + 4\text{-}NH_2C_6H_4I \xrightarrow{a} C_6H_5I +$ 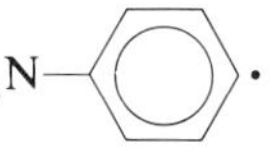 [20]) $+ CCl_4 \xrightarrow{b} C_6H_5Cl + \dot{C}Cl_3$				
Therm. of PAT PR, glc	$4\text{-}NH_2C_6H_4I/CCl_4$	333	$k_a/k_b = 0.92(1)$	82 Tan 1
$(C_6H_5^{\bullet}) + C_6H_5OH \xrightarrow{a} C_6H_6$ + phenol$(-\dot{H})$ $+ CCl_4 \xrightarrow{b} C_6H_5Cl + \dot{C}Cl_3$				
Thermal decomp. of PAT PR, glc	CCl_4/C_6H_5OH	333	$k_a/k_b = 8.7$	65 Baz 2, 64 Baz 1
$(C_6H_5^{\bullet}) + C_6H_5SH \xrightarrow{a} C_6H_6 + C_6H_5\dot{S}$ $+ CCl_4 \xrightarrow{b} C_6H_5Cl + \dot{C}Cl_3$				
Thermal decomp. of PAT PR, glc	CCl_4	333	$k_a/k_b = 518$	66 Pry 1
$(C_6H_5^{\bullet})$ + 3-methylpyridine $\xrightarrow{a} C_6H_6$ + 3-methylpyridine$(-\dot{H})$ $+ CCl_4 \xrightarrow{b} C_6H_5Cl + \dot{C}Cl_3$				
Thermal decomp. of PAT PR, glc	CCl_4/ 3-methylpyridine	333	$k_a/k_b = 0.12$ 0.17 [21])	63 Bri 1, 63 Rus 3

[17]) Estimated from data in [69 Dan 1].
[18]) D-labelled.
[19]) Measurement of D-labelled compounds.
[20]) Reaction proceeds via intermediate adduct.
[21]) From [63 Rus 3].

Reaction Radical generation Method	Solvent	T[K]	Rate data	Ref./ add. ref.
$(C_6H_5^{\bullet})$ + 4-methylpyridine $\xrightarrow{a}$ C_6H_6 + 4-methylpyridine$(-\dot{H})$ $+ CCl_4 \xrightarrow{b} C_6H_5Cl + \dot{C}Cl_3$				
Thermal decomp. of PAT				63 Bri 1
PR, glc	CCl_4/ 4-methylpyridine	333	$k_a/k_b = 0.12$	
$(C_6H_5^{\bullet}) + C_6H_5NH_2 \xrightarrow{a} C_6H_6$ + aniline$(-\dot{H})$ $+ CCl_4 \xrightarrow{b} C_6H_5Cl + \dot{C}Cl_3$				
Thermal decomp. of PAT				65 Baz 2,
PR, glc	$CCl_4/C_6H_5NH_2$	333	$k_a/k_b = 1.7$ [22])	64 Baz 1
$(C_6H_5^{\bullet})$ + 2,5-dimethylpyrrole $\xrightarrow{a}$ C_6H_6 + 2,5-dimethylpyrrole$(-\dot{H})$ $+ CCl_4 \xrightarrow{b} C_6H_5Cl + \dot{C}Cl_3$				
Thermal decomp. of PAT				63 Bri 1
PR, glc	CCl_4/ 2,5-dimethylpyrrole	333	$k_a/k_b = 1.43$	
$(C_6H_5^{\bullet})$ + 2,5-dimethylpyrazine $\xrightarrow{a}$ C_6H_6 + 2,5-dimethylpyrazine$(-\dot{H})$ $+ CCl_4 \xrightarrow{b} C_6H_5Cl + \dot{C}Cl_3$				
Thermal decomp. of PAT				63 Bri 1
PR, glc	CCl_4/ 2,5-dimethylpyrazine	333	$k_a/k_b = 0.33$	
$(C_6H_5^{\bullet}) + c\text{-}C_6H_{10} \xrightarrow{a} C_6H_6 + (c\text{-}C_6H_9^{\bullet})$ $+ CCl_4 \xrightarrow{b} C_6H_5Cl + \dot{C}Cl_3$				
Thermal decomp. of PAT				63 Bri 1
PR, glc	$CCl_4/c\text{-}C_6H_{10}$	333	$k_a/k_b = 4.4$	
$(C_6H_5^{\bullet}) + c\text{-}trans\text{-}C_6H_{10}Br_2 \xrightarrow{a} C_6H_5Br + (c\text{-}trans\text{-}C_6H_{10}Br^{\bullet})$ $+ CCl_4 \xrightarrow{b} C_6H_5Cl + \dot{C}Cl_3$				
Therm. of PAT				75 Dan 1
PR, glc	$c\text{-}trans\text{-}C_6H_{10}Br_2$/ CCl_4	333.0(1)	$k_a/k_b = 1.32$ [23]) $(\pm 0.5\%)$	
$(C_6H_5^{\bullet}) + c\text{-}C_6H_{11}Br \xrightarrow{a} C_6H_5Br + (c\text{-}C_6H_{11}^{\bullet})$ $+ CCl_4 \xrightarrow{b} C_6H_5Cl + \dot{C}Cl_3$				
Therm. of PAT				75 Dan 1
PR, glc	$c\text{-}C_6H_{11}Br/CCl_4$	333.0(1)	$k_a/k_b = 0.16(\pm 0.5\%)$	
$(C_6H_5^{\bullet}) + c\text{-}C_6H_{11}I \xrightarrow{a} C_6H_5I + (c\text{-}C_6H_{11}^{\bullet})$ $+ CBrCl_3 \xrightarrow{b} C_6H_5Br + \dot{C}Cl_3$				
Thermal decomp. of PAT				71 Dan 2
PR, glc	$CBrCl_3/c\text{-}C_6H_{11}I$ mixt.	333	$k_a/k_b = 0.49$	
$(C_6H_5^{\bullet}) + c\text{-}C_5H_9CH_3 \xrightarrow{a} C_6H_6$ + methylcyclopentane$(-\dot{H})$ $+ CCl_4 \xrightarrow{b} C_6H_5Cl + \dot{C}Cl_3$				
Thermal decomp. of PAT				63 Bri 1
PR, glc	CCl_4/ methylcyclopentane	333	$k_a/k_b = 1.56$	

[22]) $k_a/k_b = 0.85$ per active atom.
[23]) Corrected by statistical factor of 2.

Reaction Radical generation Method	Solvent	T[K]	Rate data	Ref./ add. ref.
$(C_6H_5^{\cdot}) + c\text{-}C_6H_{12} \xrightarrow{a} C_6H_6 + (c\text{-}C_6H_{11}^{\cdot})$ $+ O_2 \xrightarrow{b} (C_6H_5O_2^{\cdot})$				
Thermal decomp. of PAT				63 Rus 2
PR	$c\text{-}C_6H_{12}$	333(2)	$k_a/k_b = 8.33 \cdot 10^{-4}$	
$(C_6H_5^{\cdot}) + c\text{-}C_6H_{12} \xrightarrow{a} C_6H_6 + (c\text{-}C_6H_{11}^{\cdot})$ $+ CCl_4 \xrightarrow{b} C_6H_5Cl + \dot{C}Cl_3$				
Thermal decomp. of PAT				63 Bri 1
PR, glc	$CCl_4/c\text{-}C_6H_{12}$	333	$k_a/k_b = 1.08$	
		333	$k_a/k_b = 1.14$	
$(C_6H_5^{\cdot}) + c\text{-}C_6H_{12} \xrightarrow{a} C_6H_6 + (c\text{-}C_6H_{11}^{\cdot})$ $+ n\text{-}C_6H_{14} \xrightarrow{b} C_6H_6 + (n\text{-}C_6H_{13}^{\cdot})$				
Therm. of BPO				76 Lew 2
PR, glc	CH_3CN	347.6	$k_a/k_b = 0.739(24)$	
$(C_6H_5^{\cdot}) + CH_3CH{=}CHCH(CH_3)_2 \xrightarrow{a} C_6H_6 + (C_6H_{11}^{\cdot})$ $+ CCl_4 \xrightarrow{b} C_6H_5Cl + \dot{C}Cl_3$				
Thermal decomp. of PAT				63 Bri 1
PR, glc	$CCl_4/CH_3CH{=}CH\text{-}CH(CH_3)_2$	333	$k_a/k_b = 1.45$	
$(C_6H_5^{\cdot}) + CH_3C(CH_3){=}C(CH_3)_2 \xrightarrow{a} C_6H_6 + (C_6H_{11}^{\cdot})$ $+ CCl_4 \xrightarrow{b} C_6H_5Cl + \dot{C}Cl_3$				
Thermal decomp. of PAT				63 Bri 1
PR, glc	$CCl_4/CH_3C(CH_3){=}C(CH_3)_2$	333	$k_a/k_b = 2.3$	
$(C_6H_5^{\cdot}) + n\text{-}C_6H_{14} \xrightarrow{a} C_6H_6 + (C_6H_{13}^{\cdot})$ $+ CCl_4 \xrightarrow{b} C_6H_5Cl + \dot{C}Cl_3$				
Thermal decomp. of PAT				63 Bri 1,
PR, glc	$CCl_4/n\text{-}C_6H_{14}$	333	$k_a/k_b = 0.80$	66 Pry 1
$(C_6H_5^{\cdot}) + n\text{-}C_6H_{14} \xrightarrow{a} C_6H_6 + n\text{-hexane}(-\dot{H})$ $+ CCl_4 \xrightarrow{b} C_6H_5Cl + \dot{C}Cl_3$				
Thermal decomp. of PAT				63 Rus 3
PR, glc	$n\text{-}C_6H_{14}/CCl_4$	333	$k_a/k_b = 1.1$	
$(C_6H_5^{\cdot}) + (CH_3)_2CHCH_2CH_2CH_3 \xrightarrow{a} C_6H_6 + (C_6H_{13}^{\cdot})$ $+ CCl_4 \xrightarrow{b} C_6H_5Cl + \dot{C}Cl_3$				
Thermal decomp. of PAT				63 Bri 1
PR, glc	$CCl_4/(CH_3)_2CHCH_2CH_2CH_3$	333	$k_a/k_b = 0.86$	
$(C_6H_5^{\cdot}) + CH_3CH_2CH(CH_3)CH_2CH_3 \xrightarrow{a} C_6H_6 + (C_6H_{13}^{\cdot})$ $+ CCl_4 \xrightarrow{b} C_6H_5Cl + \dot{C}Cl_3$				
Thermal decomp. of PAT				63 Bri 1
PR, glc	$CCl_4/CH_3CH_2CH(CH_3)\text{-}CH_2CH_3$	333	$k_a/k_b = 0.90$	

Reaction Radical generation Method	Solvent	T[K]	Rate data	Ref./ add. ref.
$(C_6H_5^{\bullet}) + (CH_3)_3CCH_2CH_3 \xrightarrow{a} C_6H_6 + (C_6H_{13}^{\bullet})$ $+ CCl_4 \xrightarrow{b} C_6H_5Cl + \dot{C}Cl_3$				
Thermal decomp. of PAT				63 Bri 1
PR, glc	CCl_4/ $(CH_3)_3CCH_2CH_3$	333	$k_a/k_b = 0.31$	
$(C_6H_5^{\bullet}) + (CH_3)_2CHCH(CH_3)_2 \xrightarrow{a} C_6H_6 + (C_6H_{13}^{\bullet})$ $+ CCl_4 \xrightarrow{b} C_6H_5Cl + \dot{C}Cl_3$				
Thermal decomp. of PAT				63 Bri 1
PR, glc	CCl_4/ $(CH_3)_2CHCH(CH_3)_2$	333	$k_a/k_b = 1.19$	
$(C_6H_5^{\bullet}) + CH_3CH_2CH_2SSCH_2CH_2CH_3$ $\xrightarrow{a} C_6H_5SCH_2CH_2CH_3 + n\text{-}C_3H_7\dot{S}$ $\xrightarrow{b} C_6H_6 + RSSR(-\dot{H})$				
Thermal decomp. of PAT				63 Pry 1
PR, glc	di-n-propyl-disulfide	333	$k_a/k_b = 9.3(3)$	
$(C_6H_5^{\bullet}) + CH_3CH_2CH_2SSCH_2CH_2CH_3 \xrightarrow{a} C_6H_6 + RSSR(-\dot{H})$ $+ CCl_4 \xrightarrow{b} C_6H_5Cl + \dot{C}Cl_3$				
Thermal decomp. of PAT				64 Pry 1/
PR	CCl_4/ $n\text{-}C_3H_7SS\text{-}n\text{-}C_3H_7$ mixt.	333	$k_a/k_b = 1.8$	70 Pry 1
$(C_6H_5^{\bullet}) + CH_3CH_2CH_2SSCH_2CH_2CH_3 \xrightarrow{a} C_6H_5SCH_2CH_2CH_3 + n\text{-}C_3H_7\dot{S}$ $+ CCl_4 \xrightarrow{b} C_6H_5Cl + \dot{C}Cl_3$				
Thermal decomp. of PAT				64 Pry 1/
PR	CCl_4/ $n\text{-}C_3H_7SS\text{-}n\text{-}C_3H_7$ mixt.	333	$k_a/k_b = 16.0$	70 Pry 1
$(C_6H_5^{\bullet}) + (CH_3)_2CHSSCH(CH_3)_2$ $\xrightarrow{a} C_6H_5SCH(CH_3)_2 + (CH_3)_2CH\dot{S}$ $\xrightarrow{b} C_6H_6 + RSSR(-\dot{H})$				
Thermal decomp. of PAT				63 Pry 1
PR, glc	di-isopropyl-disulfide	333	$k_a/k_b = 1.79(1)$	
$(C_6H_5^{\bullet}) + (CH_3)_2CHSSCH(CH_3)_2 \xrightarrow{a} C_6H_6 +$ di-2-propyl-disulfide$(-\dot{H})$ $+ CCl_4 \xrightarrow{b} C_6H_5Cl + \dot{C}Cl_3$				
Thermal decomp. of PAT				64 Pry 1,
PR	CCl_4/di-2-propyl- disulfide mixt.	333	$k_a/k_b = 1.90$ 1.81 [24])	70 Pry 1
$(C_6H_5^{\bullet}) + (CH_3)_2CHSSCH(CH_3)_2 \xrightarrow{a} C_6H_5SCH(CH_3)_2 + (CH_3)_2CH\dot{S}$ $+ CCl_4 \xrightarrow{b} C_6H_5Cl + \dot{C}Cl_3$				
Thermal decomp. of PAT				64 Pry 1,
PR	CCl_4/di-2-propyl- disulfide mixt.	333	$k_a/k_b = 3.30$ 3.52 [24])	70 Pry 1

[24]) From [70 Pry 1].

Reaction Radical generation Method	Solvent	T[K]	Rate data	Ref./ add. ref.
$(C_6H_5^{\cdot}) + 2\text{-}CF_3C_6H_4I \xrightarrow{a} C_6H_5I +$ 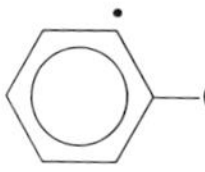 $-CF_3$ $+ CCl_4 \xrightarrow{b} C_6H_5Cl + \dot{C}Cl_3$				
Thermal decomp. of PAT PR, glc	CCl_4	333.0(1)	$k_a/k_b = 103(\pm 5\%)$	73 Dan 1
$(C_6H_5^{\cdot}) + 3\text{-}CF_3C_6H_4I \xrightarrow{a} C_6H_5I + CF_3-$ 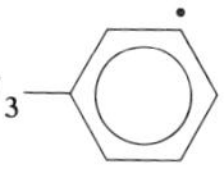 [25] $+ CCl_4 \xrightarrow{b} C_6H_5Cl + \dot{C}Cl_3$				
Therm. of PAT PR, glc	$3\text{-}CF_3C_6H_4I/CCl_4$	333	$k_a/k_b = 1.45(0)$	82 Tan 1
$(C_6H_5^{\cdot}) + 3\text{-}CNC_6H_4I \xrightarrow{a} C_6H_5I + NC-$ 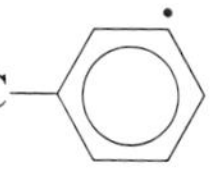 [25] $+ CCl_4 \xrightarrow{b} C_6H_5Cl + \dot{C}Cl_3$				
Therm. of PAT PR, glc	$3\text{-}CNC_6H_4I/CCl_4$	333	$k_a/k_b = 1.58(1)$	82 Tan 1
$(C_6H_5^{\cdot}) + 4\text{-}CNC_6H_4I \xrightarrow{a} C_6H_5I + NC-$ 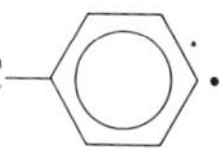 [25] $+ CCl_4 \xrightarrow{b} C_6H_5Cl + \dot{C}Cl_3$				
Therm. of PAT PR, glc	$4\text{-}CNC_6H_4I/CCl_4$	333	$k_a/k_b = 1.52(0)$ See also 4.1.2.3, Fig. 7, p. 257	82 Tan 1/ 69 Dan 1
$(C_6H_5^{\cdot}) + C_6H_5CD_3 \xrightarrow{a} C_6H_5D + C_6H_5\dot{C}D_2$ $+ CCl_4 \xrightarrow{b} C_6H_5Cl + \dot{C}Cl_3$				
Thermal decomp. of PAT PR, glc	$CCl_4/C_6H_5CD_3$	333	$k_a/k_b = 0.06$ 0.09 [26]	63 Bri 1, 63 Rus 3
$(C_6H_5^{\cdot}) + 3\text{-}CHOC_6H_4I \xrightarrow{a} C_6H_5I + OCH-$ 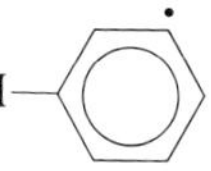 [25] $+ CCl_4 \xrightarrow{b} C_6H_5Cl + \dot{C}Cl_3$				
Therm. of PAT PR, glc	$3\text{-}CHOC_6H_4I/CCl_4$	333	$k_a/k_b = 1.44(1)$	82 Tan 1
$(C_6H_5^{\cdot}) + 4\text{-}CHOC_6H_4I \xrightarrow{a} C_6H_5I + OCH-$ 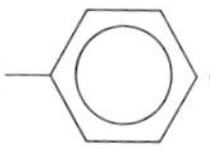 [25] $+ CCl_4 \xrightarrow{b} C_6H_5Cl + \dot{C}Cl_3$				
Therm. of PAT PR, glc	$4\text{-}CHOC_6H_4I/CCl_4$	333	$k_a/k_b = 1.45(0)$	82 Tan 1

[25]) Reaction proceeds via intermediate adduct.
[26]) From [63 Rus 3].

Reaction Radical generation Method	Solvent	T[K]	Rate data	Ref./ add. ref.
$(C_6H_5^{\bullet}) + 3\text{-}NO_2C_6H_4CHO \xrightarrow{a} C_6H_6 +$ 3-nitrobenzaldehyde$(-\dot{H})$ $+ CCl_4 \xrightarrow{b} C_6H_5Cl + \dot{C}Cl_3$				
Thermal decomp. of PAT				65 Baz 2
PR, glc	CCl_4/ 3-nitrobenzaldehyde	333	$k_a/k_b = 3.6$	
$(C_6H_5^{\bullet}) + 4\text{-}NO_2C_6H_4CHO \xrightarrow{a} C_6H_6 +$ 4-nitrobenzaldehyde$(-\dot{H})$ $+ CCl_4 \xrightarrow{b} C_6H_5Cl + \dot{C}Cl_3$				
Thermal decomp. of PAT				65 Baz 2
PR, glc	CCl_4/ 4-nitrobenzaldehyde	333	$k_a/k_b = 4.5$	
$(C_6H_5^{\bullet}) + 3\text{-}ClC_6H_4CH_2Br \xrightarrow{a} C_6H_5Br + 3\text{-}ClC_6H_4\dot{C}H_2$ $+ CCl_4 \xrightarrow{b} C_6H_5Cl + \dot{C}Cl_3$				
Therm. of PAT				75 Mig 1
PR, glc	$3\text{-}ClC_6H_4CH_2Br$/ CCl_4	333	$k_a/k_b = 4.17(16)$	
$(C_6H_5^{\bullet}) + 4\text{-}ClC_6H_4CH_2Br \xrightarrow{a} C_6H_5Br + 4\text{-}ClC_6H_4\dot{C}H_2$ $+ CCl_4 \xrightarrow{b} C_6H_5Cl + \dot{C}Cl_3$				
Therm. of PAT				75 Mig 1
PR, glc	$4\text{-}ClC_6H_4CH_2Br$/ CCl_4	333	$k_a/k_b = 5.79(25)$	
$(C_6H_5^{\bullet}) + 3\text{-}NO_2C_6H_4CH_2Br \xrightarrow{a} C_6H_5Br + 3\text{-}NO_2C_6H_4\dot{C}H_2$ $+ CCl_4 \xrightarrow{b} C_6H_5Cl + \dot{C}Cl_3$				
Therm. of PAT				75 Mig 1
PR, glc	$3\text{-}NO_2C_6H_4CH_2Br$/ CCl_4	333	$k_a/k_b = 4.50(30)$	
$(C_6H_5^{\bullet}) + 4\text{-}NO_2C_6H_4CH_2Br \xrightarrow{a} C_6H_5Br + 4\text{-}NO_2C_6H_4\dot{C}H_2$ $+ CCl_4 \xrightarrow{b} C_6H_5Cl + \dot{C}Cl_3$				
Therm. of PAT				75 Mig 1
PR, glc	$4\text{-}NO_2C_6H_4CH_2Br$/ CCl_4	333	$k_a/k_b = 6.97(50)$	
$(C_6H_5^{\bullet}) + C_6H_5CHO \xrightarrow{a} C_6H_6 +$ benzaldehyde$(-\dot{H})$ $+ CCl_4 \xrightarrow{b} C_6H_5Cl + \dot{C}Cl_3$				
Thermal decomp. of PAT				65 Baz 2
PR, glc	CCl_4/C_6H_5CHO	333	$k_a/k_b = 2.3$	
$(C_6H_5^{\bullet}) + C_6H_5CH_2Br \xrightarrow{a} C_6H_5Br + C_6H_5\dot{C}H_2$ $+ CCl_4 \xrightarrow{b} C_6H_5Cl + \dot{C}Cl_3$				
Thermal decomp. of PAT				79 Mig 1, 75 Mig 1
PR, glc	$CCl_4/C_6H_5CH_2Br$ mixt.	333	$k_a/k_b = 3.57$ 3.71(25) [27])	
$(C_6H_5^{\bullet}) + 3\text{-}BrC_6H_4CH_3 \xrightarrow{a} C_6H_6 + 3\text{-}BrC_6H_4\dot{C}H_2$ $+ CCl_4 \xrightarrow{b} C_6H_5Cl + \dot{C}Cl_3$				
Thermal decomp. of PAT				66 Pry 1
PR, glc	CCl_4/3-bromotoluene	333	$k_a/k_b = 0.26$	

[27]) From [75 Mig 1].

Reaction Radical generation Method	Solvent	T[K]	Rate data	Ref./ add. ref.
$(C_6H_5^{\cdot}) + 3\text{-}ClC_6H_4CH_3 \xrightarrow{a} C_6H_6 + 3\text{-}ClC_6H_4\dot{C}H_2$ $+ CCl_4 \xrightarrow{b} C_6H_5Cl + \dot{C}Cl_3$				
Thermal decomp. of PAT				63 Bri 1
PR, glc	CCl_4/3-chlorotoluene	333	$k_a/k_b = 0.24$	
$(C_6H_5^{\cdot}) + 3\text{-}ClC_6H_4CH_3 \xrightarrow{a} C_6H_6 + 3\text{-}ClC_6H_4\dot{C}H_2$ $+ CCl_4 \xrightarrow{b} C_6H_5Cl + \dot{C}Cl_3$				
Therm. of BPO				77 Sue 1
PR, glc	$3\text{-}ClC_6H_4CH_3/CCl_4$	353	$k_a/k_b = 0.288(55)$	
$(C_6H_5^{\cdot}) + 4\text{-}ClC_6H_4CH_3 \xrightarrow{a} C_6H_6 + 4\text{-}ClC_6H_4\dot{C}H_2$ $+ CCl_4 \xrightarrow{b} C_6H_5Cl + \dot{C}Cl_3$				
Therm. of BPO				77 Sue 1
PR, glc	$4\text{-}ClC_6H_4CH_3/CCl_4$	353	$k_a/k_b = 0.277(54)$	
$(C_6H_5^{\cdot}) + 4\text{-}ClC_6H_4CH_3 \xrightarrow{a} C_6H_6 + 4\text{-}ClC_6H_4\dot{C}H_2$ $+ CCl_4 \xrightarrow{b} C_6H_5Cl + \dot{C}Cl_3$				
Thermal decomp. of PAT and BPO [28])				63 Bri 1, 64 Tro 1
PR, glc	CCl_4/4-chlorotoluene	333 373	$k_a/k_b = 0.29$ 0.36 [28])	
$(C_6H_5^{\cdot}) + 2\text{-}CH_3C_6H_4I \xrightarrow{a} C_6H_5I +$ 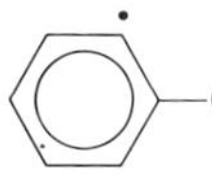 $-CH_3$ $+ CCl_4 \xrightarrow{b} C_6H_5Cl + \dot{C}Cl_3$				
Thermal decomp. of PAT				73 Dan 1
PR, glc	CCl_4	333.0(1)	$k_a/k_b = 16.9(\pm 5\%)$	
$(C_6H_5^{\cdot}) + 3\text{-}CH_3C_6H_4I \xrightarrow{a} C_6H_5I + CH_3-$ 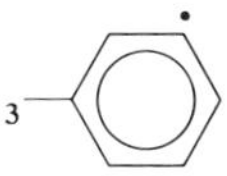 [29]) $+ CCl_4 \xrightarrow{b} C_6H_5Cl + \dot{C}Cl_3$				
Therm. of PAT				82 Tan 1/
PR, glc	$3\text{-}CH_3C_6H_4I/CCl_4$	333	$k_a/k_b = 1.29(0)$ See also 4.1.2.3, Fig. 7, p. 257.	69 Dan 1
$(C_6H_5^{\cdot}) + 4\text{-}CH_3C_6H_4I \xrightarrow{a} C_6H_5I + CH_3-$ 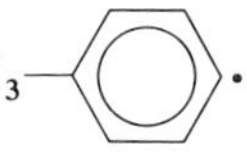 [29]) $+ CCl_4 \xrightarrow{b} C_6H_5Cl + \dot{C}Cl_3$				
Therm. of PAT				82 Tan 1/
PR, glc	$4\text{-}CH_3C_6H_4I/CCl_4$	333	$k_a/k_b = 1.13(0)$ See also 4.1.2.3, Fig. 7, p. 257.	69 Dan 1
$(C_6H_5^{\cdot}) + 2\text{-}CH_3OC_6H_4I \xrightarrow{a} C_6H_5I +$ 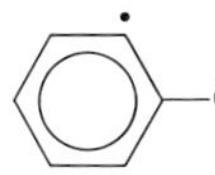 $-OCH_3$ $+ CCl_4 \xrightarrow{b} C_6H_5Cl + \dot{C}Cl_3$				
Thermal decomp. of PAT				73 Dan 1
PR, glc	$2\text{-}CH_3OC_6H_4I/CCl_4$	333.0(1)	$k_a/k_b = 8.7(\pm 5\%)$	

[28]) From [64 Tro 1].
[29]) Reaction proceeds via intermediate adduct.

Reaction
   Radical generation
   Method
	Solvent	T[K]	Rate data	Ref./
				add. ref.

$(C_6H_5^{\cdot}) + 3\text{-}CH_3OC_6H_4I \xrightarrow{\text{a}} C_6H_5I + CH_3O\text{—}\langle\bigcirc\rangle^{\cdot}$ [29])

$+ CCl_4 \xrightarrow{\text{b}} C_6H_5Cl + \dot{C}Cl_3$

Therm. of PAT PR, glc	3-CH$_3$OC$_6$H$_4$I/CCl$_4$	333	$k_a/k_b = 1.33(0)$ See also 4.1.2.3, Fig. 7, p. 257.	82 Tan 1/ 69 Dan 1

$(C_6H_5^{\cdot}) + 4\text{-}CH_3OC_6H_4I \xrightarrow{\text{a}} C_6H_5I + CH_3O\text{—}\langle\bigcirc\rangle\cdot$ [29])

$+ CCl_4 \xrightarrow{\text{b}} C_6H_5Cl + \dot{C}Cl_3$

Therm. of PAT PR, glc	4-CH$_3$OC$_6$H$_4$I/CCl$_4$	333	$k_a/k_b = 1.00(0)$ See also 4.1.2.3, Fig. 7, p. 257.	82 Tan 1/ 69 Dan 1

$(C_6H_5^{\cdot}) + 3\text{-}NO_2C_6H_4CH_3 \xrightarrow{\text{a}} C_6H_6 + 3\text{-}NO_2C_6H_4\dot{C}H_2$
$+ CCl_4 \xrightarrow{\text{b}} C_6H_5Cl + \dot{C}Cl_3$

Therm. of BPO PR, glc	3-NO$_2$C$_6$H$_4$CH$_3$/CCl$_4$	353	$k_a/k_b = 0.332(75)$	77 Sue 1

$(C_6H_5^{\cdot}) + 3\text{-}NO_2C_6H_4CH_3 \xrightarrow{\text{a}} C_6H_6 + 3\text{-}NO_2C_6H_4\dot{C}H_2$
$+ C_6H_5CH_3 \xrightarrow{\text{b}} C_6H_6 + C_6H_5\dot{C}H_2$

Thermal decomp. of BPO PR, glc	toluene	373	$k_a/k_b = 0.475$	64 Tro 1

$(C_6H_5^{\cdot}) + 4\text{-}NO_2C_6H_4CH_3 \xrightarrow{\text{a}} C_6H_6 + 4\text{-}NO_2C_6H_4\dot{C}H_2$
$+ CCl_4 \xrightarrow{\text{b}} C_6H_5Cl + \dot{C}Cl_3$

Thermal decomp. of PAT PR, glc	CCl$_4$/4-nitrotoluene	333	$k_a/k_b = 0.22$	63 Bri 1

$(C_6H_5^{\cdot}) + 4\text{-}NO_2C_6H_4CH_3 \xrightarrow{\text{a}} C_6H_6 + 4\text{-}NO_2C_6H_4\dot{C}H_2$
$+ CCl_4 \xrightarrow{\text{b}} C_6H_5Cl + \dot{C}Cl_3$

Therm. of BPO PR, glc	4-NO$_2$C$_6$H$_4$CH$_3$/CCl$_4$	353	$k_a/k_b = 0.646(110)$	77 Sue 1

$(C_6H_5^{\cdot}) + 4\text{-}NO_2C_6H_4CH_3 \xrightarrow{\text{a}} C_6H_6 + 4\text{-}NO_2C_6H_4\dot{C}H_2$
$+ C_6H_5CH_3 \xrightarrow{\text{b}} C_6H_6 + C_6H_5\dot{C}H_2$

Thermal decomp. of BPO PR, glc	toluene	373	$k_a/k_b = 0.80$	64 Tro 1

$(C_6H_5^{\cdot}) + C_6H_5CH_3 \xrightarrow{\text{a}} C_6H_6 + C_6H_5\dot{C}H_2$
$+ CCl_4 \xrightarrow{\text{b}} C_6H_5Cl + \dot{C}Cl_3$

Therm. of PAT PR, glc	C$_6$H$_5$CH$_3$/CCl$_4$	333	$k_a/k_b = 0.31$	63 Rus 3
PR, glc	C$_6$H$_5$CH$_3$/CCl$_4$	333	$k_a/k_b = 0.32$	64 Baz 2
Thermal decomp. of PAT and BPO [31]) PR, glc	CCl$_4$/C$_6$H$_5$CH$_3$	333 333 373	$k_a/k_b = 0.27$ 0.36 [30]) 0.32 [31])	63 Bri 1, 66 Pry 1, 64 Tro 1
Therm. of BPO PR, glc	C$_6$H$_5$CH$_3$/CCl$_4$	353	$k_a/k_b = 0.210(55)$	77 Sue 1

[29]) Reaction proceeds via intermediate adduct.   [30]) From [66 Pry 1].   [31]) From [64 Tro 1].

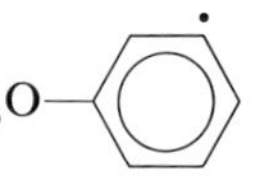

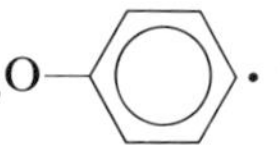

Reaction Radical generation Method	Solvent	T[K]	Rate data	Ref./ add. ref.
$(C_6H_5^{\cdot}) + C_6H_5OCH_3 \xrightarrow{a} C_6H_6 + \text{anisole}(-\dot{H})$ $+ CCl_4 \xrightarrow{b} C_6H_5Cl + \dot{C}Cl_3$				
Therm. of PAT				63 Rus 3
PR, glc	anisole/CCl_4	333	$k_a/k_b = 0.13$	
$(C_6H_5^{\cdot}) + C_6H_5OCH_3 \xrightarrow{a} C_6H_6 + \text{methoxybenzene}(-\dot{H})$ $+ CCl_4 \xrightarrow{b} C_6H_5Cl + \dot{C}Cl_3$				
Thermal decomp. of PAT				63 Bri 1
PR, glc	$CCl_4/C_6H_5OCH_3$	333	$k_a/k_b = 0.094$	
$(C_6H_5^{\cdot}) + C_6H_5OCH_3 \xrightarrow{a} C_6H_6 + C_6H_5O\dot{C}H_2$ $+ CCl_4 \xrightarrow{b} C_6H_5Cl + \dot{C}Cl_3$				
Therm. of PAT				64 Baz 1
PR, glc	CCl_4	333	$k_a/k_b = 0.09$ [32]	
$(C_6H_5^{\cdot}) + 4\text{-}CH_3OC_6H_4OH \xrightarrow{a} C_6H_6 + \text{4-methoxyphenol}(-\dot{H})$ $+ CCl_4 \xrightarrow{b} C_6H_5Cl + \dot{C}Cl_3$				
Thermal decomp. of PAT				65 Baz 2
PR, glc	CCl_4/ 4-methoxyphenol	333	$k_a/k_b = 52$	
$(C_6H_5^{\cdot}) + C_6H_5SCH_3 \xrightarrow{a} C_6H_6 + \text{thioanisole}(-\dot{H})$ $+ CCl_4 \xrightarrow{b} C_6H_5Cl + \dot{C}Cl_3$				
Therm. of PAT				63 Rus 3
PR, glc	$C_6H_5SCH_3/CCl_4$	333	$k_a/k_b = 0.46$	
$(C_6H_5^{\cdot}) + C_6H_5SCH_3 \xrightarrow{a} C_6H_6 + \text{thioanisole}(-\dot{H})$ $+ CCl_4 \xrightarrow{b} C_6H_5Cl + \dot{C}Cl_3$				
Thermal decomp. of PAT				63 Bri 1
PR, glc	$CCl_4/C_6H_5SCH_3$	333	$k_a/k_b = 0.39$	
$(C_6H_5^{\cdot}) + \text{1-iodobicyclo[2.2.1]heptane} \xrightarrow{a} C_6H_5I + (\text{1-bicyclo[2.2.1]heptyl}^{\cdot})$ $+ CBrCl_3 \xrightarrow{b} C_6H_5Br + \dot{C}Cl_3$				
Thermal decomp. of PAT				71 Dan 1
PR, glc	$CBrCl_3$/1-iodobicyclo- [2.2.1]heptane mixt.	333	$k_a/k_b = 0.17$	
$(C_6H_5^{\cdot}) + c\text{-}trans\text{-}C_7H_{12}Br_2 \xrightarrow{a} C_6H_5Br + (c\text{-}trans\text{-}C_7H_{12}Br^{\cdot})$ $+ CCl_4 \xrightarrow{b} C_6H_5Cl + \dot{C}Cl_3$				
Therm. of PAT				75 Dan 1
PR, glc	$c\text{-}trans\text{-}C_7H_{12}Br_2$/ CCl_4	333.0(1)	$k_a/k_b = 1.52$ [33] $(\pm 0.5\%)$	
$(C_6H_5^{\cdot}) + c\text{-}C_7H_{13}Br \xrightarrow{a} C_6H_5Br + (c\text{-}C_7H_{13}^{\cdot})$ $+ CCl_4 \xrightarrow{b} C_6H_5Cl + \dot{C}Cl_3$				
Therm. of PAT				75 Dan 1
PR, glc	$c\text{-}C_7H_{13}Br/CCl_4$	333.0(1)	$k_a/k_b = 0.34(\pm 0.5\%)$	
$(C_6H_5^{\cdot}) + c\text{-}C_6H_{11}CH_3 \xrightarrow{a} C_6H_6 + \text{methylcyclohexane}(-\dot{H})$ $+ CCl_4 \xrightarrow{b} C_6H_5Cl + \dot{C}Cl_3$				
Thermal decomp. of PAT				63 Bri 1
PR, glc	CCl_4/methylcyclo- hexane	333	$k_a/k_b = 1.55$	

[32] $k_a/k_b = 0.03$ per active H-atom.

[33] Corrected by statistical factor of 2.

Reaction Radical generation Method	Solvent	T[K]	Rate data	Ref./ add. ref.
$(C_6H_5^{\bullet}) + c\text{-}C_7H_{14} \xrightarrow{a} C_6H_6 + (c\text{-}C_7H_{13}^{\bullet})$ $+ CCl_4 \xrightarrow{b} C_6H_5Cl + \dot{C}Cl_3$				
Thermal decomp. of PAT				63 Bri 1
PR, glc	$CCl_4/c\text{-}C_7H_{14}$	333	$k_a/k_b = 2.3$	
$(C_6H_5^{\bullet}) + c\text{-}C_7H_{14} \xrightarrow{a} C_6H_6 + (c\text{-}C_7H_{13}^{\bullet})$ $+ n\text{-}C_8H_{18} \xrightarrow{b} C_6H_6 + (n\text{-}C_8H_{17}^{\bullet})$				
Therm. of BPO				74 Bun 1
PR, glc	CH_3CN	347.6	$k_a/k_b = 2.63(5)$	
$(C_6H_5^{\bullet}) + (CH_3)_2CHCOCH(CH_3)_2 \xrightarrow{a} C_6H_6$ + di-2-propylketone$(-\dot{H})$ $+ CCl_4 \xrightarrow{b} C_6H_5Cl + \dot{C}Cl_3$				
Thermal decomp. of PAT				63 Bri 1
PR, glc	CCl_4/di-2-propyl-ketone	333	$k_a/k_b = 1.30$	
$(C_6H_5^{\bullet}) + n\text{-}C_7H_{16} \xrightarrow{a} C_6H_6 + (C_7H_{15}^{\bullet})$ $+ CCl_4 \xrightarrow{b} C_6H_5Cl + \dot{C}Cl_3$				
Thermal decomp. of PAT				63 Bri 1
PR, glc	$CCl_4/n\text{-}C_7H_{16}$	333	$k_a/k_b = 0.96$	
$(C_6H_5^{\bullet}) + CH_3CH_2CH(CH_3)CH_2CH_2CH_3 \xrightarrow{a} C_6H_6 + (C_7H_{15}^{\bullet})$ $+ CCl_4 \xrightarrow{b} C_6H_5Cl + \dot{C}Cl_3$				
Thermal decomp. of PAT				63 Bri 1
PR, glc	$CCl_4/$ $CH_3CH_2CH(CH_3)$- $CH_2CH_2CH_3$	333	$k_a/k_b = 1.07$	
$(C_6H_5^{\bullet}) + (CH_3)_2CHCH_2CH(CH_3)_2 \xrightarrow{a} C_6H_6 + (C_7H_{15}^{\bullet})$ $+ CCl_4 \xrightarrow{b} C_6H_5Cl + \dot{C}Cl_3$				
Thermal decomp. of PAT				63 Bri 1
PR, glc	$CCl_4/(CH_3)_2CHCH_2$- $CH(CH_3)CH_3$	333	$k_a/k_b = 0.67$	
$(C_6H_5^{\bullet}) + (CH_3)_3CCH(CH_3)_2 \xrightarrow{a} C_6H_6 + (C_7H_{15}^{\bullet})$ $+ CCl_4 \xrightarrow{b} C_6H_5Cl + \dot{C}Cl_3$				
Thermal decomp. of PAT				63 Bri 1
PR, glc	$CCl_4/$ $(CH_3)_3CCH(CH_3)_2$	333	$k_a/k_b = 0.71$	
$(C_6H_5^{\bullet}) + 3\text{-}CH_3COC_6H_4I \xrightarrow{a} C_6H_5I + CH_3CO$– 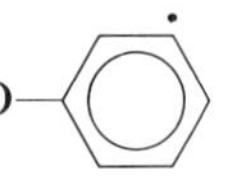 [34]) $+ CCl_4 \xrightarrow{b} C_6H_5Cl + \dot{C}Cl_3$				
Therm. of PAT				82 Tan 1/
PR, glc	$3\text{-}CH_3COC_6H_4I/CCl_4$	333	$k_a/k_b = 1.43(4)$ See also 4.1.2.3, Fig. 7, p. 257.	69 Dan 1

[34]) Reaction proceeds via intermediate adduct.

Reaction Radical generation Method	Solvent	T[K]	Rate data	Ref./ add. ref.
$(C_6H_5^{\cdot}) + 4\text{-}CH_3COC_6H_4I \xrightarrow{a} C_6H_5I + CH_3CO$– [34] $+ CCl_4 \xrightarrow{b} C_6H_5Cl + \dot{C}Cl_3$				
Therm. of PAT PR, glc	$4\text{-}CH_3COC_6H_4I/CCl_4$	333	$k_a/k_b = 1.26(0)$	82 Tan 1
$(C_6H_5^{\cdot}) + 3\text{-}CH_3OOCC_6H_4I \xrightarrow{a} C_6H_5I + CH_3OOC$–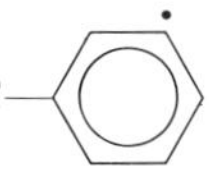 [34] $+ CCl_4 \xrightarrow{b} C_6H_5Cl + \dot{C}Cl_3$				
Therm. of PAT PR, glc	$3\text{-}CH_3OOCC_6H_4I/$ CCl_4	333	$k_a/k_b = 1.40(0)$	82 Tan 1
$(C_6H_5^{\cdot}) + 4\text{-}CH_3OOCC_6H_4I \xrightarrow{a} C_6H_5I + CH_3OOC$–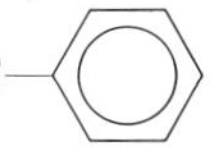 [34] $+ CCl_4 \xrightarrow{b} C_6H_5Cl + \dot{C}Cl_3$				
Therm. of PAT PR, glc	$4\text{-}CH_3OOCC_6H_4I/$ CCl_4	333	$k_a/k_b = 1.28(1)$	82 Tan 1
$(C_6H_5^{\cdot}) + 3\text{-}CNC_6H_4CH_3 \xrightarrow{a} C_6H_6 + 3\text{-}CNC_6H_4\dot{C}H_2$ $+ CCl_4 \xrightarrow{b} C_6H_5Cl + \dot{C}Cl_3$				
Therm. of BPO PR, glc	$3\text{-}CNC_6H_4CH_3/CCl_4$	353	$k_a/k_b = 0.220(42)$	77 Sue 1
$(C_6H_5^{\cdot}) + 4\text{-}CNC_6H_4CH_3 \xrightarrow{a} C_6H_6 + 4\text{-}CNC_6H_4\dot{C}H_2$ $+ CCl_4 \xrightarrow{b} C_6H_5Cl + \dot{C}Cl_3$				
Therm. of BPO PR, glc	$4\text{-}CNC_6H_4CH_3/CCl_4$	353	$k_a/k_b = 0.265(51)$	77 Sue 1
$(C_6H_5^{\cdot}) + 4\text{-}CNC_6H_4CH_3 \xrightarrow{a} C_6H_6 + 4\text{-}CNC_6H_4\dot{C}H_2$ $+ CCl_4 \xrightarrow{b} C_6H_5Cl + \dot{C}Cl_3$				
Thermal decomp. of BPO PR, glc	CCl_4/4-cyanotoluene	373	$k_a/k_b = 0.29$	64 Tro 1
$(C_6H_5^{\cdot}) + C_6H_5COCH_3 \xrightarrow{a} C_6H_6$ + acetophenone$(-\dot{H})$ $+ CCl_4 \xrightarrow{b} C_6H_5Cl + \dot{C}Cl_3$				
Thermal decomp. of PAT PR, glc	$CCl_4/C_6H_5COCH_3$	333	$k_a/k_b = 0.17$	65 Baz 2
$(C_6H_5^{\cdot}) + C_6H_5COOCH_3 \xrightarrow{a} C_6H_6$ + benzoic acid methylester$(-\dot{H})$ $+ CCl_4 \xrightarrow{b} C_6H_5Cl + \dot{C}Cl_3$				
Thermal decomp. of PAT PR, glc	$CCl_4/C_6H_5COOCH_3$	333	$k_a/k_b = 0.01$	63 Bri 1
$(C_6H_5^{\cdot}) + 3\text{-}CH_3C_6H_4CH_2Br \xrightarrow{a} C_6H_5Br + 3\text{-}CH_3C_6H_4\dot{C}H_2$ $+ CCl_4 \xrightarrow{b} C_6H_5Cl + \dot{C}Cl_3$				
Therm. of PAT PR, glc	$3\text{-}CH_3C_6H_4CH_2Br/$ CCl_4	333	$k_a/k_b = 3.54(10)$	75 Mig 1

[34]) Reaction proceeds via intermediate adduct.

Reaction Radical generation Method	Solvent	T[K]	Rate data	Ref./ add. ref.
$(C_6H_5^{\cdot}) + 4\text{-}CH_3C_6H_4CH_2Br \xrightarrow{a} C_6H_5Br + 4\text{-}CH_3C_6H_4\dot{C}H_2$ $+ CCl_4 \xrightarrow{b} C_6H_5Cl + \dot{C}Cl_3$				
Therm. of PAT				75 Mig 1
PR, glc	$4\text{-}CH_3C_6H_4CH_2Br/$ CCl_4	333	$k_a/k_b = 6.17(45)$	
$(C_6H_5^{\cdot}) + 4\text{-}CH_3OC_6H_4CH_2Br \xrightarrow{a} C_6H_5Br + 4\text{-}CH_3OC_6H_4\dot{C}H_2$ $+ CCl_4 \xrightarrow{b} C_6H_5Cl + \dot{C}Cl_3$				
Therm. of PAT				75 Mig 1
PR, glc	$4\text{-}CH_3OC_6H_4CH_2Br/$ CCl_4	333	$k_a/k_b = 4.78(32)$	
$(C_6H_5^{\cdot}) + C_6H_5CH_2CH_2I \xrightarrow{a} C_6H_5I + C_6H_5CH_2\dot{C}H_2$ $+ CBrCl_3 \xrightarrow{b} C_6H_5Br + \dot{C}Cl_3$				
Thermal decomp. of PAT				71 Dan 2
PR, glc	$CBrCl_3/$ $C_6H_5CH_2CH_2I$ mixt.	333	$k_a/k_b = 0.37$	
$(C_6H_5^{\cdot}) + 2,6\text{-}(CH_3)_2C_6H_3I \xrightarrow{a} C_6H_5I +$ 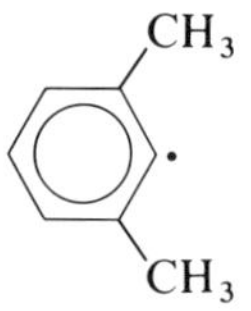 $+ CCl_4 \xrightarrow{b} C_6H_5Cl + \dot{C}Cl_3$				
Thermal decomp. of PAT				73 Dan 1
PR, glc	$2,6\text{-}(CH_3)_2C_6H_3I/$ CCl_4	333.0(1)	$k_a/k_b = 14.6(\pm 5\%)$	
$(C_6H_5^{\cdot}) + 2\text{-}C_2H_5C_6H_4I \xrightarrow{a} C_6H_5I +$ 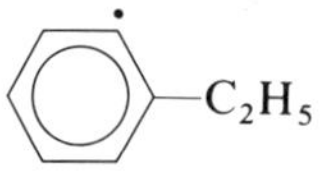 $+ CCl_4 \xrightarrow{b} C_6H_5Cl + \dot{C}Cl_3$				
Thermal decomp. of PAT				73 Dan 1
PR, glc	CCl_4	333.0(1)	$k_a/k_b = 19.7(\pm 5\%)$	
$(C_6H_5^{\cdot}) + 2\text{-}C_2H_5OC_6H_4I \xrightarrow{a} C_6H_5I +$ 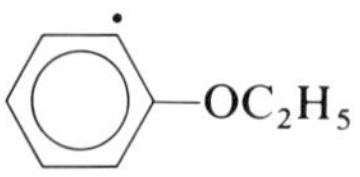 $+ CCl_4 \xrightarrow{b} C_6H_5Cl + \dot{C}Cl_3$				
Thermal decomp. of PAT				73 Dan 1
PR, glc	$2\text{-}C_2H_5OC_6H_4I/CCl_4$	333.0(1)	$k_a/k_b = 8.5(\pm 5\%)$ See also 4.1.2.3, Fig. 7, p. 257.	
$(C_6H_5^{\cdot}) + C_6H_5CH_2CH_3 \xrightarrow{a} C_6H_6 +$ ethylbenzene$(-\dot{H})$ $+ CCl_4 \xrightarrow{b} C_6H_5Cl + \dot{C}Cl_3$				
Thermal decomp. of PAT				63 Bri 1
PR, glc	CCl_4/ethylbenzene	333	$k_a/k_b = 0.84$	
$(C_6H_5^{\cdot}) + C_6H_5CH_2CH_3 \xrightarrow{a} C_6H_6 +$ ethylbenzene$(-\dot{H})$ $+ CCl_4 \xrightarrow{b} C_6H_5Cl + \dot{C}Cl_3$				
Therm. of PAT				63 Rus 3
PR, glc	$C_6H_5C_2H_5/CCl_4$	333	$k_a/k_b = 0.91$	

Reaction Radical generation Method	Solvent	T[K]	Rate data	Ref./ add. ref.
$(C_6H_5^{\cdot}) + C_6H_5CH_2CH_3 \xrightarrow{a} C_6H_6$ + ethylbenzene$(-\dot{H})$ $+ C_6H_5CH_3 \xrightarrow{b} C_6H_6 + C_6H_5\dot{C}H_2$				
Therm. of PAT				63 Rus 4
PR, glc	$C_6H_5CH_2CH_3$ or $C_6H_5CH_3/CCl_4$	333	$k_a/k_b = 4.4$	
$(C_6H_5^{\cdot})$ + 3-$CH_3C_6H_4CH_3 \xrightarrow{a} C_6H_6$ + 3-$CH_3C_6H_4\dot{C}H_2$ $+ CCl_4 \xrightarrow{b} C_6H_5Cl + \dot{C}Cl_3$				
Therm. of BPO				77 Sue 1
PR, glc	3-$CH_3C_6H_4CH_3/CCl_4$	353	$k_a/k_b = 0.214(46)$	
$(C_6H_5^{\cdot})$ + 3-$CH_3C_6H_4CH_3$ [35] $\xrightarrow{a} C_6H_6$ + 3-$CH_3C_6H_4\dot{C}H_2$ $+ CCl_4 \xrightarrow{b} C_6H_5Cl + \dot{C}Cl_3$				
Thermal decomp. of BPO				64 Tro 1
PR, glc	CCl_4/3-methyltoluene	373	$k_a/k_b = 0.70$	
$(C_6H_5^{\cdot})$ + 4-$CH_3C_6H_4CH_3 \xrightarrow{a} C_6H_6$ + 4-$CH_3C_6H_4\dot{C}H_2$ $+ CCl_4 \xrightarrow{b} C_6H_5Cl + \dot{C}Cl_3$				
Therm. of PAT				63 Rus 3
PR, glc	4-$CH_3C_6H_4CH_3/CCl_4$	333	$k_a/k_b = 0.84$	
$(C_6H_5^{\cdot})$ + 4-$CH_3C_6H_4CH_3$ [36] $\xrightarrow{a} C_6H_6$ + 4-$CH_3C_6H_4\dot{C}H_2$ $+ CCl_4 \xrightarrow{b} C_6H_5Cl + \dot{C}Cl_3$				
Thermal decomp. of PAT [37] [38] and BPO [39]				63 Bri 1, 66 Pry 1, 64 Tro 1
PR, glc	CCl_4/4-methyltoluene	333 333 373	$k_a/k_b = 0.79$ [37] 1.02 [38] 0.95 [39]	
$(C_6H_5^{\cdot})$ + 4-$CH_3C_6H_4CH_3 \xrightarrow{a} C_6H_6$ + 4-$CH_3C_6H_4\dot{C}H_2$ $+ CCl_4 \xrightarrow{b} C_6H_5Cl + \dot{C}Cl_3$				
Therm. of BPO				77 Sue 1
PR, glc	4-$CH_3C_6H_4CH_3/CCl_4$	353	$k_a/k_b = 0.270(35)$	
$(C_6H_5^{\cdot})$ + 4-$CH_3C_6H_4CH_3$ [40] $\xrightarrow{a} C_6H_6$ + 4-$CH_3C_6H_4\dot{C}H_2$ $+ C_6H_5CH_3 \xrightarrow{b} C_6H_6 + C_6H_5\dot{C}H_2$				
Thermal decomp. of BPO				64 Tro 1
PR, glc	toluene	373	$k_a/k_b = 3.3$	
$(C_6H_5^{\cdot})$ + 4-$CH_3OC_6H_4CH_3 \xrightarrow{a} C_6H_6$ + 4-methoxytoluene$(-\dot{H})$ $+ CCl_4 \xrightarrow{b} C_6H_5Cl + \dot{C}Cl_3$				
Thermal decomp. of PAT				66 Pry 1
PR, glc	CCl_4/4-methoxytoluene	333	$k_a/k_b = 0.57$	

[35]) *m*-xylene, 1,3-dimethylbenzene.
[36]) *p*-xylene, 1,4-dimethylbenzene.
[37]) From [63 Bri 1].
[38]) From [66 Pry 1].
[39]) From [64 Tro 1].
[40]) *p*-xylene, 1,4-dimethylbenzene.

Reaction Radical generation Method	Solvent	T[K]	Rate data	Ref./ add. ref.
$(C_6H_5^{\cdot}) + 4\text{-}CH_3OC_6H_4CH_3 \xrightarrow{a} C_6H_6 + 4\text{-}CH_3OC_6H_4\dot{C}H_2$ $+ CCl_4 \xrightarrow{b} C_6H_5Cl + \dot{C}Cl_3$				
Therm. of BPO				77 Sue 1
PR, glc	$4\text{-}CH_3OC_6H_4CH_3/CCl_4$	353	$k_a/k_b = 0.389(90)$	
$(C_6H_5^{\cdot})$ + 2,4,6-trimethylpyridine $\xrightarrow{a} C_6H_6$ + 2,4,6-trimethylpyridine$(-\dot{H})$ $+ CCl_4 \xrightarrow{b} C_6H_5Cl + \dot{C}Cl_3$				
Thermal decomp. of PAT				63 Bri 1, 63 Rus 3
PR, glc	CCl_4/2,4,6-trimethylpyridine	333	$k_a/k_b = 0.37$ 0.42 [41]	
$(C_6H_5^{\cdot}) + C_6H_5N(CH_3)_2 \xrightarrow{a} C_6H_6$ + dimethylphenylamine$(-\dot{H})$ $+ CCl_4 \xrightarrow{b} C_6H_5Cl + \dot{C}Cl_3$				
Thermal decomp. of PAT				63 Bri 1
PR, glc	$CCl_4/C_6H_5N(CH_3)_2$	333	$k_a/k_b = 2.8$	
$(C_6H_5^{\cdot}) + C_6H_5N(CH_3)_2 \xrightarrow{a} C_6H_6$ + dimethylphenylamine$(-\dot{H})$ $+ CCl_4 \xrightarrow{b} C_6H_5Cl + \dot{C}Cl_3$				
Thermal decomp. of PAT				65 Baz 1
PR, glc	$CCl_4/C_6H_5N(CH_3)_2$	333	$k_a/k_b = 5.4$	
$(C_6H_5^{\cdot})$ + 1-iodobicyclo[2,2,2]octane $\xrightarrow{a} C_6H_5I$ + (1-bicyclo[2.2.2]octyl)$^{\cdot}$ $+ CBrCl_3 \xrightarrow{b} C_6H_5Br + \dot{C}Cl_3$				
Thermal decomp. of PAT				71 Dan 1
PR, glc	$CBrCl_3$/1-iodobicyclo-[2.2.2]octane mixt.	333	$k_a/k_b = 0.55$	
$(C_6H_5^{\cdot}) + (CH_3)_2C{=}CHCH{=}C(CH_3)_2 \xrightarrow{a} C_6H_6 + (C_8H_{13}^{\cdot})$ $+ CCl_4 \xrightarrow{b} C_6H_5Cl + \dot{C}Cl_3$				
Thermal decomp. of PAT				63 Bri 1
PR, glc	$CCl_4/(CH_3)_2C{=}CH\text{-}CH{=}C(CH_3)_2$	333	$k_a/k_b = 2.9$	
$(C_6H_5^{\cdot})$ + *c-trans*-$C_8H_{14}Br_2 \xrightarrow{a} C_6H_5Br$ + (*c-trans*-$C_8H_{14}Br^{\cdot}$) $+ CCl_4 \xrightarrow{b} C_6H_5Cl + \dot{C}Cl_3$				
Therm. of PAT				75 Dan 1
PR, glc	*c-trans*-$C_8H_{14}Br_2/CCl_4$	333.0(1)	$k_a/k_b = 1.36$ [42] $(\pm 0.5\%)$	
$(C_6H_5^{\cdot})$ + *c*-$C_8H_{15}Br \xrightarrow{a} C_6H_5Br$ + (*c*-$C_8H_{15}^{\cdot}$) $+ CCl_4 \xrightarrow{b} C_6H_5Cl + \dot{C}Cl_3$				
Therm. of PAT				75 Dan 1
PR, glc	*c*-$C_8H_{15}Br/CCl_4$	333.0(1)	$k_a/k_b = 0.58(\pm 0.5\%)$	
$(C_6H_5^{\cdot})$ + *c*-$C_8H_{16} \xrightarrow{a} C_6H_6$ + (*c*-$C_8H_{15}^{\cdot}$) $+ CCl_4 \xrightarrow{b} C_6H_5Cl + \dot{C}Cl_3$				
Thermal decomp. of PAT				63 Bri 1
PR, glc	CCl_4/*c*-C_8H_{16}	333	$k_a/k_b = 2.9$	

[41]) From [63 Rus 3].
[42]) Corrected by statistical factor of 2.

Reaction Radical generation Method	Solvent	T[K]	Rate data	Ref./ add. ref.
$(C_6H_5^{\cdot}) + c\text{-}C_8H_{16} \xrightarrow{a} C_6H_6 + (c\text{-}C_8H_{15}^{\cdot})$ $+ n\text{-}C_8H_{18} \xrightarrow{b} C_6H_6 + (n\text{-}C_8H_{17}^{\cdot})$				
Therm. of BPO PR, glc	CH_3CN	347.6	$k_a/k_b = 4.20(12)$	75 Bun 1
$(C_6H_5^{\cdot}) + c\text{-}C_8H_{16} \xrightarrow{a} C_6H_6 + (c\text{-}C_8H_{15}^{\cdot})$ $+ n\text{-}C_{10}H_{22} \xrightarrow{b} C_6H_6 + (n\text{-}C_{10}H_{21}^{\cdot})$				
Therm. of BPO PR, glc	CH_3CN	347.6	$k_a/k_b = 4.11(12)$	74 Bun 1
$(C_6H_5^{\cdot}) + CH_2{=}CH(CH_2)_5CH_3 \xrightarrow{a} C_6H_6 + (C_8H_{15}^{\cdot})$ $+ CCl_4 \xrightarrow{b} C_6H_5Cl + \dot{C}Cl_3$				
Thermal decomp. of PAT PR, glc	$CCl_4/$ $CH_2{=}CH(CH_2)_5CH_3$	333	$k_a/k_b = 1.38$	63 Bri 1
$(C_6H_5^{\cdot}) + CH_3CH{=}CH(CH_2)_4CH_3 \xrightarrow{a} C_6H_6 + (C_8H_{15}^{\cdot})$ $+ CCl_4 \xrightarrow{b} C_6H_5Cl + \dot{C}Cl_3$				
Thermal decomp. of PAT PR, glc	$CCl_4/CH_3CH{=}CH$- $(CH_2)_4CH_3$	333	$k_a/k_b = 1.60$	63 Bri 1
$(C_6H_5^{\cdot}) + n\text{-}C_8H_{18} \xrightarrow{a} C_6H_6 + (C_8H_{17}^{\cdot})$ $+ CCl_4 \xrightarrow{b} C_6H_5Cl + \dot{C}Cl_3$				
Thermal decomp. of PAT PR, glc	$CCl_4/n\text{-}C_8H_{18}$	333	$k_a/k_b = 1.15$	63 Bri 1
$(C_6H_5^{\cdot}) + (CH_3)_2CH(CH_2)_2CH(CH_3)_2 \xrightarrow{a} C_6H_6 + (C_8H_{17}^{\cdot})$ $+ CCl_4 \xrightarrow{b} C_6H_5Cl + \dot{C}Cl_3$				
Thermal decomp. of PAT PR, glc	$CCl_4/(CH_3)_2CH$- $(CH_2)_2CH(CH_3)_2$	333	$k_a/k_b = 1.27$	63 Bri 1
$(C_6H_5^{\cdot}) + (CH_3)_2CHCH(CH_3)CH(CH_3)_2 \xrightarrow{a} C_6H_6 + (C_8H_{17}^{\cdot})$ $+ CCl_4 \xrightarrow{b} C_6H_5Cl + \dot{C}Cl_3$				
Thermal decomp. of PAT PR, glc	$CCl_4/(CH_3)_2CH$- $CH(CH_3)CH(CH_3)_2$	333	$k_a/k_b = 0.97$	63 Bri 1
$(C_6H_5^{\cdot}) + (CH_3)_3CCH_2CH(CH_3)_2 \xrightarrow{a} C_6H_6 + (C_8H_{17}^{\cdot})$ $+ CCl_4 \xrightarrow{b} C_6H_5Cl + \dot{C}Cl_3$				
Thermal decomp. of PAT PR, glc	$CCl_4/(CH_3)_3CCH_2$- $CH(CH_3)_2$	333	$k_a/k_b = 0.34$	63 Bri 1
$(C_6H_5^{\cdot}) + (CH_3)_3CC(CH_3)_3 \xrightarrow{a} C_6H_6 + $ 2,2,3,3-tetramethylbutane$(-\dot{H})$ $+ CCl_4 \xrightarrow{b} C_6H_5Cl + \dot{C}Cl_3$				
Therm. of PAT PR, glc	$(CH_3)_3CC(CH_3)_3/$ CCl_4	333	$k_a/k_b = 0.18$	64 Baz 2

Reaction Radical generation Method	Solvent	T[K]	Rate data	Ref./ add. ref.
$(C_6H_5^\bullet) + (CH_3)_3CC(CH_3)_3 \xrightarrow{a} C_6H_6 + (C_8H_{17}^\bullet)$ $+ CCl_4 \xrightarrow{b} C_6H_5Cl + \dot{C}Cl_3$				
Thermal decomp. of PAT PR, glc	CCl_4/ $(CH_3)_3CC(CH_3)_3$	333	$k_a/k_b = 0.21$	63 Bri 1
$(C_6H_5^\bullet) + CH_3(CH_2)_3SS(CH_2)_3CH_3 \xrightarrow{a} C_6H_5S(CH_2)_3CH_3 + n\text{-}C_4H_9\dot{S}$ $\xrightarrow{b} C_6H_6 + RSSR(-\dot{H})$				
Thermal decomp. of PAT PR, glc	di-*n*-butyldisulfide	333	$k_a/k_b = 6.90(9)$	63 Pry 1
$(C_6H_5^\bullet) + CH_3CH_2CH(CH_3)SSCH(CH_3)CH_2CH_3 \xrightarrow{a} C_6H_5SCH(CH_3)CH_2CH_3 + CH_3CH_2CH(CH_3)\dot{S}$ $\xrightarrow{b} C_6H_6 + RSSR(-\dot{H})$				
Thermal decomp. of PAT PR, glc	di-*s*-butyldisulfide	333	$k_a/k_b = 1.65(1)$	63 Pry 1
$(C_6H_5^\bullet) + (CH_3)_2CHCH_2SSCH_2CH(CH_3)_2 \xrightarrow{a} C_6H_5SCH_2CH(CH_3)_2 + (CH_3)_2CHCH_2\dot{S}$ $\xrightarrow{b} C_6H_6 + RSSR(-\dot{H})$				
Thermal decomp. of PAT PR, glc	di-iso-butyldisulfide	333	$k_a/k_b = 6.72(11)$	63 Pry 1
$(C_6H_5^\bullet) + (CH_3)_3CSSC(CH_3)_3 \xrightarrow{a} C_6H_5SC(CH_3)_3 + (CH_3)_3C\dot{S}$ $\xrightarrow{b} C_6H_6 + RSSR(-\dot{H})$				
Thermal decomp. of PAT PR, glc	di-*t*-butyldisulfide	333	$k_a/k_b = 0.95(1)$	63 Pry 1
$(C_6H_5^\bullet) + (CH_3)_3CSSC(CH_3)_3 \xrightarrow{a} C_6H_6 +$ di-*t*-butyldisulfide$(-\dot{H})$ $+ CCl_4 \xrightarrow{b} C_6H_5Cl + \dot{C}Cl_3$				
Thermal decomp. of PAT PR	CCl_4/ di-*t*-butyldisulfide mixt.	333	$k_a/k_b = 0.25$ 0.28 [43])	64 Pry 1, 70 Pry 1
$(C_6H_5^\bullet) + (CH_3)_3CSSC(CH_3)_3 \xrightarrow{a} C_6H_5SC(CH_3)_3 + (CH_3)_3C\dot{S}$ $+ CCl_4 \xrightarrow{b} C_6H_5Cl + \dot{C}Cl_3$				
Thermal decomp. of PAT PR	CCl_4/ di-*t*-butyldisulfide mixt.	333	$k_a/k_b = 0.23$ $= 0.16$ [43])	64 Pry 1, 70 Pry 1
$(C_6H_5^\bullet) +$ 2-methylbenzofuran $\xrightarrow{a} C_6H_6 +$ 2-methylbenzofuran$(-\dot{H})$ $+ CCl_4 \xrightarrow{b} C_6H_5Cl + \dot{C}Cl_3$				
Thermal decomp. of PAT PR, glc	CCl_4/ 2-methylbenzofuran	333	$k_a/k_b = 1.93 \cdot 10^{-1}$	66 Str 1

[43]) From [70 Pry 1].

Reaction Radical generation Method	Solvent	T[K]	Rate data	Ref./ add. ref.
$(C_6H_5^{\cdot})$ + 3-methylbenzofuran $\xrightarrow{a}$ C_6H_6 + 3-methylbenzofuran$(-\dot{H})$ + CCl_4 $\xrightarrow{b}$ $C_6H_5Cl + \dot{C}Cl_3$				
Thermal decomp. of PAT PR, glc	CCl_4/ 3-methylbenzofuran	333	$k_a/k_b \approx 1.7 \cdot 10^{-2}$	66 Str 1
$(C_6H_5^{\cdot})$ + 2-methylbenzothiophene $\xrightarrow{a}$ C_6H_6 + 2-methylbenzothiophene$(-\dot{H})$ + CCl_4 $\xrightarrow{b}$ $C_6H_5Cl + \dot{C}Cl_3$				
Thermal decomp. of PAT PR, glc	CCl_4/ 2-methylbenzothiophene	333	$k_a/k_b = 1.37 \cdot 10^{-1}$	66 Str 1
$(C_6H_5^{\cdot})$ + 3-methylbenzothiophene $\xrightarrow{a}$ C_6H_6 + 3-methylbenzothiophene$(-\dot{H})$ + CCl_4 $\xrightarrow{b}$ $C_6H_5Cl + \dot{C}Cl_3$				
Thermal decomp. of PAT PR, glc	CCl_4/3-methyl- benzothiophene	333	$k_a/k_b = 7.9 \cdot 10^{-2}$	66 Str 1
$(C_6H_5^{\cdot})$ + di-2-thienylmethane $\xrightarrow{a}$ C_6H_6 + di-2-thienylmethane$(-\dot{H})$ + CCl_4 $\xrightarrow{b}$ $C_6H_5Cl + \dot{C}Cl_3$				
Thermal decomp. of PAT PR, glc	CCl_4/ di-2-thienylmethane	333	$k_a/k_b = 4.2$	63 Bri 1
$(C_6H_5^{\cdot}) + CH_2{=}CHCH_2C_6H_5 \xrightarrow{a} C_6H_6 + (C_9H_9^{\cdot})$ + CCl_4 $\xrightarrow{b}$ $C_6H_5Cl + \dot{C}Cl_3$				
Thermal decomp. of PAT PR, glc	CCl_4/ $CH_2{=}CHCH_2C_6H_5$	333	$k_a/k_b = 1.82$	63 Bri 1
$(C_6H_5^{\cdot})$ + indan $\xrightarrow{a}$ C_6H_6 + indan$(-\dot{H})$ + CCl_4 $\xrightarrow{b}$ $C_6H_5Cl + \dot{C}Cl_3$				
Thermal decomp. of PAT PR, glc	CCl_4/indan	333	$k_a/k_b = 3.1$	63 Bri 1
$(C_6H_5^{\cdot})$ + 2-methyl-2,3-dihydrobenzofuran $\xrightarrow{a}$ C_6H_6 + 2-methyl-2,3-dihydrobenzofuran$(-\dot{H})$ + CCl_4 $\xrightarrow{b}$ $C_6H_5Cl + \dot{C}Cl_3$				
Thermal decomp. of *t*-butylperbenzoate PR, glc	2-methyl-2,3-dihydro- benzofuran/CCl_4	373	$k_a/k_b = 3.4$	79 Zlo 1
$(C_6H_5^{\cdot})$ + chroman $\xrightarrow{a}$ C_6H_6 + chroman$(-\dot{H})$ + CCl_4 $\xrightarrow{b}$ $C_6H_5Cl + \dot{C}Cl_3$				
Thermal decomp. of *t*-butylperbenzoate PR, glc	chroman/CCl_4	373	$k_a/k_b = 2.4$	79 Zlo 1
$(C_6H_5^{\cdot}) + CH_2{=}CHCH_2SC_6H_5 \xrightarrow{a} C_6H_5CH_2CH{=}CH_2 + C_6H_5\dot{S}$ + CCl_4 $\xrightarrow{b}$ $C_6H_5Cl + \dot{C}Cl_3$				
Thermal decomp. of PAT PR, glc	CCl_4/ $CH_2{=}CHCH_2SC_6H_5$ mixt.	333	$k_a/k_b = 3.7$	79 Mig 1

Reaction Radical generation Method	Solvent	T[K]	Rate data	Ref./ add. ref.
$(C_6H_5^{\cdot}) + C_6H_5CO\underset{b}{O}N(CH_3)_2 \xrightarrow{a} C_6H_6$ + N,N-dimethylbenzoic acid amide(−$\dot{H}$) $+ CCl_4 \xrightarrow{b} C_6H_5Cl + \dot{C}Cl_3$				
Thermal decomp. of PAT PR, glc	$CCl_4/$ $C_6H_5COON(CH_3)_2$	333	$k_a/k_b = 0.80$	65 Baz 1
$(C_6H_5^{\cdot})$[44]$) + C_6H_5C\underset{b}{H}(CH_3)_2 \xrightarrow{a} C_6H_6 + C_6H_5\dot{C}(CH_3)_2$ $+ CCl_4 \xrightarrow{b} C_6H_5Cl + \dot{C}Cl_3$				
Therm. of D-labelled BPO PR [45])	cumene/CCl_4	373	$k_a/k_b = 1.1$	60 Bag 1
$(C_6H_5^{\cdot}) + C_6H_5C\underset{b}{H}(CH_3)_2 \xrightarrow{a} C_6H_6 + C_6H_5\dot{C}(CH_3)_2$ $+ CCl_4 \xrightarrow{b} C_6H_5Cl + \dot{C}Cl_3$				
Thermal decomp. of PAT PR, glc	$CCl_4/$ $C_6H_5CH(CH_3)_2$	333	$k_a/k_b = 0.93$	63 Bri 1
$(C_6H_5^{\cdot}) + C_6H_5C\underset{b}{H}(CH_3)_2 \xrightarrow{a} C_6H_6 + C_6H_5\dot{C}(CH_3)_2$ $+ CCl_4 \xrightarrow{b} C_6H_5Cl + \dot{C}Cl_3$				
Therm. of PAT PR, glc	cumene/CCl_4	333	$k_a/k_b = 1.0$	63 Rus 3
$(C_6H_5^{\cdot}) + C_6H_5CH(CH_3)_2 \xrightarrow{a} C_6H_6 + C_6H_5\dot{C}(CH_3)_2$ $+ C_6H_5CH_3 \xrightarrow{b} C_6H_6 + C_6H_5\dot{C}H_2$				
Therm. of PAT PR, glc	$C_6H_5CH(CH_3)_2$ or $C_6H_5CH_3/CCl_4$	333	$k_a/k_b = 9.7$	63 Rus 4
$(C_6H_5^{\cdot})$ + 1,3,5-$(C\underset{b}{H}_3)_3C_6H_3 \xrightarrow{a} C_6H_6$ + mesitylene(−$\dot{H}$) $+ CCl_4 \xrightarrow{b} C_6H_5Cl + \dot{C}Cl_3$				
Thermal decomp. of PAT PR, glc	CCl_4/mesitylene	333	$k_a/k_b = 0.85$ 1.14 [46]) 0.91 [47])	63 Bri 1, 66 Pry 1, 63 Rus 3
$(C_6H_5^{\cdot}) + C_6H_5C(CH_3)_2OH \xrightarrow{a} C_6H_6$ + 2-hydroxy-2-phenylpropane(−$\dot{H}$) $+ CCl_4 \xrightarrow{b} C_6H_5Cl + \dot{C}Cl_3$				
Thermal decomp. of PAT PR, glc	$CCl_4/$ $C_6H_5C(CH_3)_2OH$	333	$k_a/k_b = 0.14$	65 Baz 2
$(C_6H_5^{\cdot}) + C_6H_5C(CH_3)_2NH_2 \xrightarrow{a} C_6H_6$ + 2-amino-2-phenylpropane(−$\dot{H}$) $+ CCl_4 \xrightarrow{b} C_6H_5Cl + \dot{C}Cl_3$				
Thermal decomp. of PAT PR, glc	$CCl_4/$ $C_6H_5C(CH_3)_2NH_2$	333	$k_a/k_b = 0.38$	65 Baz 2
$(C_6H_5^{\cdot}) + C_6H_5Si(C\underset{b}{H}_3)_3 \xrightarrow{a} C_6H_6$ + phenyltrimethylsilane(−$\dot{H}$) $+ CCl_4 \xrightarrow{b} C_6H_5Cl + \dot{C}Cl_3$				
Thermal decomp. of PAT PR, glc	$CCl_4/C_6H_5Si(CH_3)_3$	333	$k_a/k_b = 0.21$	63 Bri 1

[44]) D-labelled.
[45]) Measurement of D-labelled compounds.
[46]) From [66 Pry 1].
[47]) From [63 Rus 3].

Reaction Radical generation Method	Solvent	T[K]	Rate data	Ref./ add. ref.
$(C_6H_5^{\cdot}) + (CH_3)_3CCH_2CH_2CH(CH_3)_2 \xrightarrow{a} C_6H_6 + (C_9H_{19}^{\cdot})$ $+ CCl_4 \xrightarrow{b} C_6H_5Cl + \dot{C}Cl_3$				
Thermal decomp. of PAT PR, glc	$CCl_4/(CH_3)_3CCH_2$- $CH_2CH(CH_3)_2$	333	$k_a/k_b = 0.81$	63 Bri 1
$(C_6H_5^{\cdot})$ + 1-iodonaphthalene $\xrightarrow{a} C_6H_5I +$ 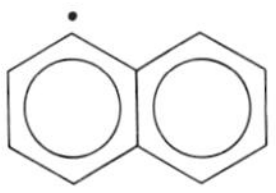 $+ CCl_4 \xrightarrow{b} C_6H_5Cl + \dot{C}Cl_3$				
Thermal decomp. of PAT PR, glc	CCl_4	333.0(1)	$k_a/k_b = 22.0(\pm 5\%)$	74 Dan 1
$(C_6H_5^{\cdot})$ + 2-iodonaphthalene $\xrightarrow{a} C_6H_5I +$ 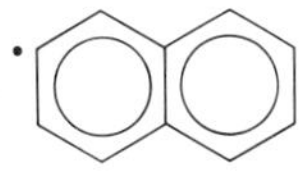 $+ CCl_4 \xrightarrow{b} C_6H_5Cl + \dot{C}Cl_3$				
Thermal decomp. of PAT PR, glc	CCl_4	333.0(1)	$k_a/k_b = 17.4(\pm 5\%)$	74 Dan 1
$(C_6H_5^{\cdot})$ + tetralin $\xrightarrow{a} C_6H_6$ + tetralin$(-\dot{H})$ $+ CCl_4 \xrightarrow{b} C_6H_5Cl + \dot{C}Cl_3$				
Thermal decomp. of PAT PR, glc	CCl_4/tetralin	333	$k_a/k_b = 4.8$	63 Bri 1
Thermal decomp. of *t*-butylperbenzoate PR, glc	tetralin/CCl_4	373	$k_a/k_b = 4.8$	79 Zlo 1
$(C_6H_5^{\cdot}) + C_6H_5C(CH_3)_2CHO \xrightarrow{a} C_6H_6$ + 2-methyl-2-phenylpropanal$(-\dot{H})$ $+ CCl_4 \xrightarrow{b} C_6H_5Cl + \dot{C}Cl_3$				
Thermal decomp. of PAT PR, glc	CCl_4/ $C_6H_5C(CH_3)_2CHO$	333	$k_a/k_b = 8.6$	65 Baz 2
$(C_6H_5^{\cdot})$ + 2,2-dimethyl-2,3-dihydrobenzofuran $\xrightarrow{a} C_6H_5Cl$ + 2,2-dimethyl-2,3-dihydrobenzofuran$(-\dot{H})$ $+ CCl_4 \xrightarrow{b} C_6H_5Cl + \dot{C}Cl_3$				
Thermal decomp. of *t*-butylperbenzoate PR, glc	2,2-dimethyl- 2,3-dihydro- benzofuran/CCl_4	373	$k_a/k_b = 1.6$	79 Zlo 1
$(C_6H_5^{\cdot})$ + 3-carboethoxytoluene $\xrightarrow{a} C_6H_5$ + 3-carboethoxytoluene$(-\dot{H})$ $+ CCl_4 \xrightarrow{b} C_6H_5Cl + \dot{C}Cl_3$				
Thermal decomp. of BPO PR, glc	CCl_4/ 3-carboethoxytoluene	373	$k_a/k_b = 0.28$	64 Tro 1
$(C_6H_5^{\cdot})$ + 4-carboethoxytoluene $\xrightarrow{a} C_6H_6$ + 4-carboethoxytoluene$(-\dot{H})$ $+ CCl_4 \xrightarrow{b} C_6H_5Cl + \dot{C}Cl_3$				
Thermal decomp. of BPO PR, glc	CCl_4/ 4-carboethoxytoluene	373	$k_a/k_b = 0.24$	64 Tro 1

Reaction Radical generation Method	Solvent	T[K]	Rate data	Ref./ add. ref.
$(C_6H_5^{\cdot}) + C_6H_5C(CH_3)_3 \xrightarrow{a} C_6H_6$ + 2-methyl-2-phenylpropane($-\dot{H}$) $+ CCl_4 \xrightarrow{b} C_6H_5Cl + \dot{C}Cl_3$				
Thermal decomp. of PAT PR, glc	$CCl_4/C_6H_5C(CH_3)_3$	333	$k_a/k_b = 0.11$	63 Bri 1, 64 Baz 2
$(C_6H_5^{\cdot}) + C_6H_5C(CH_3)_2OCH_3 \xrightarrow{a} C_6H_6$ + 2-methoxy-2-phenylpropane($-\dot{H}$) $+ CCl_4 \xrightarrow{b} C_6H_5Cl + \dot{C}Cl_3$				
Thermal decomp. of PAT PR, glc	$CCl_4/$ $C_6H_5C(CH_3)_2OCH_3$	333	$k_a/k_b = 0.09$	65 Baz 1
$(C_6H_5^{\cdot})$ + 1-iodoadamantane $\xrightarrow{a} C_6H_5I$ + (1-adamantyl)$^{\cdot}$ $+ CBrCl_3 \xrightarrow{b} C_6H_5Br + \dot{C}Cl_3$				
Thermal decomp. of PAT PR, glc	$CBrCl_3/$ 1-iodoadamantane mixt.	333	$k_a/k_b = 0.66$	71 Dan 1
$(C_6H_5^{\cdot}) + c\text{-}C_{10}H_{20} \xrightarrow{a} C_6H_6 + (c\text{-}C_{10}H_{19}^{\cdot})$ $+ n\text{-}C_{10}H_{22} \xrightarrow{b} C_6H_6 + (n\text{-}C_{10}H_{21}^{\cdot})$				
Therm. of BPO PR, glc	CH_3CN	347.6	$k_a/k_b = 4.21(17)$	74 Bun 1
$(C_6H_5^{\cdot}) + C_6H_5C(CH_3)_2COCH_3 \xrightarrow{a} C_6H_6$ + methyl-(2-phenyl)-2-propylketone($-\dot{H}$) $+ CCl_4 \xrightarrow{b} C_6H_5Cl + \dot{C}Cl_3$				
Thermal decomp. of PAT PR, glc	$CCl_4/$ $C_6H_5C(CH_3)_2COCH_3$	333	$k_a/k_b = 0.18$	65 Baz 1
$(C_6H_5^{\cdot}) + C_6H_5C(CH_3)_2N(CH_3)_2 \xrightarrow{a} C_6H_6$ + 2-dimethylamino-2-phenylpropane($-\dot{H}$) $+ CCl_4 \xrightarrow{b} C_6H_5Cl + \dot{C}Cl_3$				
Thermal decomp. of PAT PR, glc	$CCl_4/$ $C_6H_5C(CH_3)_2N(CH_3)_2$	333	$k_a/k_b = 5.3$	65 Baz 1
$(C_6H_5^{\cdot}) + 2\text{-}C_6H_5C_6H_4I \xrightarrow{a} C_6H_5I +$ 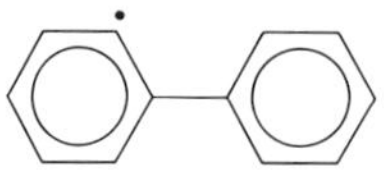 $+ CCl_4 \xrightarrow{b} C_6H_5Cl + \dot{C}Cl_3$				
Thermal decomp. of PAT PR, glc	CCl_4	333.0(1)	$k_a/k_b = 34(\pm 5\%)$	73 Dan 1
$(C_6H_5^{\cdot}) + 4\text{-}C_6H_5C_6H_4I \xrightarrow{a} C_6H_5I + C_6H_5$  [48]) $+ CCl_4 \xrightarrow{b} C_6H_5Cl + \dot{C}Cl_3$				
Therm. of PAT PR, glc	$4\text{-}C_6H_5C_6H_4I/CCl_4$	333	$k_a/k_b = 1.14(2)$ See also 4.1.2.3, Fig. 7, p. 257.	82 Tan 1/ 69 Dan 1

[48]) Reaction proceeds via intermediate adduct.

Reaction Radical generation Method	Solvent	T[K]	Rate data	Ref./ add. ref.
$(C_6H_5^{\bullet}) + (C_6H_5)_2NH \xrightarrow{a} C_6H_6$ + diphenylamine($-\dot{H}$) $+ CCl_4 \xrightarrow{b} C_6H_5Cl + \dot{C}Cl_3$				
Thermal decomp. of PAT PR, glc	$CCl_4/(C_6H_5)_2NH$	333	$k_a/k_b = 2.3$ 2.8 [49])	63 Bri 1, 63 Rus 3
$(C_6H_5^{\bullet}) + (C_6H_5)_2PH \xrightarrow{a} C_6H_6$ + diphenylphosphine($-\dot{H}$) $+ CCl_4 \xrightarrow{b} C_6H_5Cl + \dot{C}Cl_3$				
Thermal decomp. of PAT PR, glc	CCl_4	333	$k_a/k_b = 60$ 52 [49])	63 Bri 1, 63 Rus 3
$(C_6H_5^{\bullet}) + (C_6H_5)_2SiH_2 \xrightarrow{a} C_6H_6$ + diphenylsilane($-\dot{H}$) $+ CCl_4 \xrightarrow{b} C_6H_5Cl + \dot{C}Cl_3$				
Thermal decomp. of PAT PR, glc	$CCl_4/(C_6H_5)_2SiH_2$	333	$k_a/k_b = 7.7$ 8.1 [49])	63 Bri 1, 63 Rus 3
$(C_6H_5^{\bullet})$ + hexamethylbenzene $\xrightarrow{a} C_6H_6$ + hexamethylbenzene($-\dot{H}$) $+ CCl_4 \xrightarrow{b} C_6H_5Cl + \dot{C}Cl_3$				
Thermal decomp. of PAT PR, glc	CCl_4/ hexamethylbenzene	333	$k_a/k_b = 3.1$ 3.2 [49])	63 Bri 1, 63 Rus 3
$(C_6H_5^{\bullet}) + c\text{-}C_{12}H_{24} \xrightarrow{a} C_6H_6 + (c\text{-}C_{12}H_{23}^{\bullet})$ $+ c\text{-}C_{10}H_{20} \xrightarrow{b} C_6H_6 + (c\text{-}C_{10}H_{19}^{\bullet})$				
Therm. of BPO PR, glc	CH_3CN	347.6	$k_a/k_b = 0.363(11)$	74 Bun 1
$(C_6H_5^{\bullet})$ + 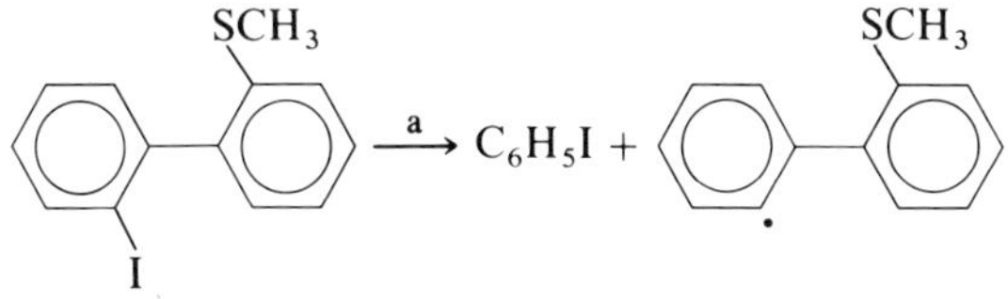 $+ CCl_4 \xrightarrow{b} C_6H_5Cl + \dot{C}Cl_3$				
Thermal decomp. of PAT PR, glc	CCl_4	333.0(1)	$k_a/k_b = 24.9(\pm 5\%)$	73 Dan 1
$(C_6H_5^{\bullet}) + (C_6H_5)_2CH_2 \xrightarrow{a} C_6H_6 + (C_6H_5)_2\dot{C}H$ $+ CCl_4 \xrightarrow{b} C_6H_5Cl + \dot{C}Cl_3$				
Thermal decomp. of PAT PR, glc	$CCl_4/(C_6H_5)_2CH_2$	333	$k_a/k_b = 1.4$ 1.5 [49])	63 Bri 1, 63 Rus 3
$(C_6H_5^{\bullet}) + (C_6H_5)_2CH_2 \xrightarrow{a} C_6H_6 + (C_6H_5)_2\dot{C}H$ $+ C_6H_5CH_3 \xrightarrow{b} C_6H_6 + C_6H_5\dot{C}H_2$				
Therm. of PAT PR, glc	$(C_6H_5)_2CH_2$ or $C_6H_5CH_3/CCl_4$	333	$k_a/k_b = 7.5$	63 Rus 4

[49]) From [63 Rus 3].

Reaction Radical generation Method	Solvent	T[K]	Rate data	Ref./ add. ref.
$(C_6H_5^{\cdot}) + 4\text{-}C_6H_5OC_6H_4CH_3 \xrightarrow{a} C_6H_6 + 4\text{-}C_6H_5OC_6H_4\dot{C}H_2$ $+ CCl_4 \xrightarrow{b} C_6H_5Cl + \dot{C}Cl_3$				
Thermal decomp. of PAT				63 Bri 1
PR, glc	CCl_4/ 4-phenoxytoluene	333	$k_a/k_b = 0.26$	
$(C_6H_5^{\cdot}) + (C_6H_5)_2C(CH_3)_2 \xrightarrow{a} C_6H_6 +$ 2,2-diphenylpropane$(-\dot{H})$ $+ CCl_4 \xrightarrow{b} C_6H_5Cl + \dot{C}Cl_3$				
Therm. of PAT				64 Baz 2
PR, glc	2,2-diphenylpropane/ CCl_4	333	$k_a/k_b = 0.07$	
$(C_6H_5^{\cdot}) + n\text{-}C_{16}H_{34} \xrightarrow{a} C_6H_6 + (C_{16}H_{33}^{\cdot})$ $+ CCl_4 \xrightarrow{b} C_6H_5Cl + \dot{C}Cl_3$				
Thermal. decomp. of PAT				63 Bri 1
PR, glc	$CCl_4/n\text{-}C_{16}H_{34}$	333	$k_a/k_b = 2.5$	
$(C_6H_5^{\cdot}) + (C_6H_5)_3SiH \xrightarrow{a} C_6H_6 +$ triphenylsilane$(-\dot{H})$ $+ CCl_4 \xrightarrow{b} C_6H_5Cl + \dot{C}Cl_3$				
Thermal decomp. of PAT				63 Bri 1
PR, glc	$CCl_4/(C_6H_5)_3SiH$	333	$k_a/k_b = 4.8$	
$(C_6H_5^{\cdot}) + (C_6H_5OH)_3$ [50]) $\xrightarrow{a} C_6H_6 + C_6H_5\dot{O} + 2\,C_6H_5OH$ $+ CCl_4 \xrightarrow{b} C_6H_5Cl + \dot{C}Cl_3$				
Therm. of PAT				64 Baz 1
PR, glc	CCl_4	333	$k_a/k_b = 0.86$	
$(C_6H_5^{\cdot}) + (CH_3)_2C(C_6H_5)C(C_6H_5)(CH_3)_2 \xrightarrow{a} C_6H_6 +$ 2,3-diphenyl-2,3-dimethylbutane$(-\dot{H})$ $+ CCl_4 \xrightarrow{b} C_6H_5Cl + \dot{C}Cl_3$				
Therm. of PAT				64 Baz 2
PR, glc	2,3-diphenyl- 2,3-dimethylbutane/ CCl_4	333	$k_a/k_b = 0.15$	
$(C_6H_5^{\cdot}) + (C_6H_5)_3CH \xrightarrow{a} C_6H_6 + (C_6H_5)_3\dot{C}$ $+ CCl_4 \xrightarrow{b} C_6H_5Cl + \dot{C}Cl_3$				
Thermal decomp. of PAT				63 Bri 1, 63 Rus 3
PR, glc	$CCl_4/(C_6H_5)_3CH$	333	$k_a/k_b = 3.5$ 3.7 [51])	
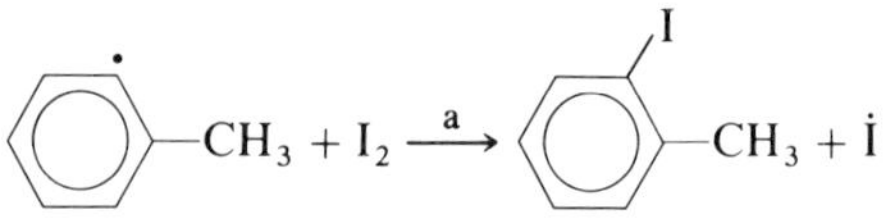 $+ O_2 \xrightarrow{b}$ products				
Phot. of $2\text{-}IC_6H_4CH_3$				73 Lev 1
PR [52])	methylcyclohexane	RT	$k_a/k_b = 2.13$	

[50]) Associated trimer.
[51]) From [63 Rus 3].
[52]) ^{131}I radioactivity measurements.

Reaction					
Radical generation					Ref./
Method	Solvent		$T[K]$	Rate data	add. ref.

$\cdot C_6H_4$—$CH_3 + CCl_3Br \xrightarrow{a} 2\text{-}CH_3C_6H_4Br + \dot{C}Cl_3$

$+ CCl_4 \xrightarrow{b} 2\text{-}CH_3C_6H_4Cl + \dot{C}Cl_3$

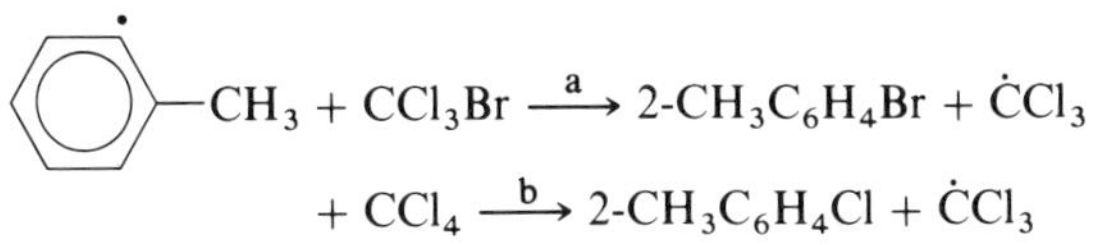

Therm. of 2-CH$_3$C$_6$H$_4$COOOC(CH$_3$)$_3$				75 Her 1,
PR, glc	CCl$_4$	403	$k_a/k_b = 81$	69 Rue 1/
				76 Gie 1,
				76 Gie 2,
				76 Gie 3

—$CH_3 + I_2 \xrightarrow{a}$ I—C_6H_4—$CH_3 + \dot{I}$

$+ O_2 \xrightarrow{b}$ products

Phot. of 3-IC$_6$H$_4$CH$_3$				73 Lev 1
PR [52]	methylcyclohexane	RT	$k_a/k_b = 1.82$	

—$CH_3 + I_2 \xrightarrow{a}$ I—C_6H_4—$CH_3 + \dot{I}$

$+ O_2 \xrightarrow{b}$ products

Phot. of 4-IC$_6$H$_4$CH$_3$				73 Lev 1
PR [52]	methylcyclohexane	RT	$k_a/k_b = 1.96$	

—$CH_3 + CBrCl_3 \xrightarrow{a} 4\text{-}CH_3C_6H_4Br + \dot{C}Cl_3$

$+ CCl_4 \xrightarrow{b} 4\text{-}CH_3C_6H_4Cl + \dot{C}Cl_3$

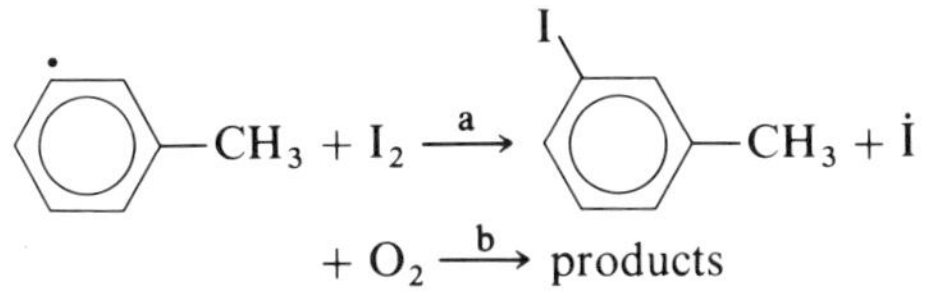

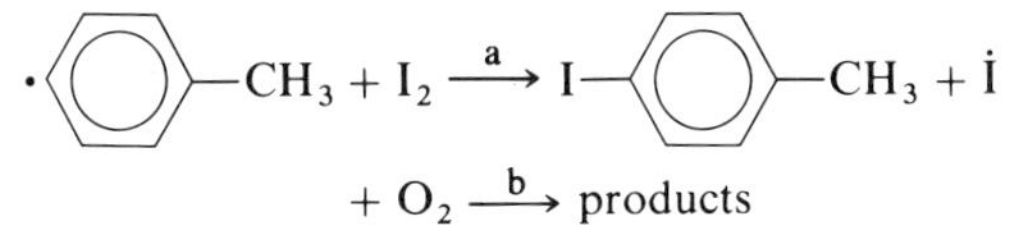

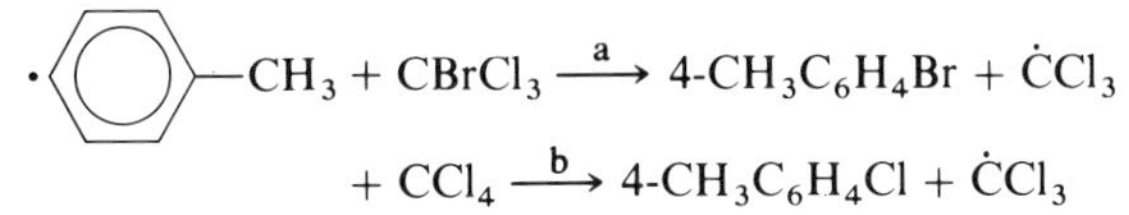

Therm. of 4-CH$_3$C$_6$H$_4$COOOC(CH$_3$)$_3$				75 Her 1,
PR, glc	CCl$_4$	403	$k_a/k_b = 150$	69 Rue 1/
				76 Gie 1,
				76 Gie 2,
				76 Gie 3

—$CH_3 + HCOO^- \xrightarrow{a} C_6H_5CH_3 + \dot{C}O_2^-$

$+ 4\text{-}CH_3C_6H_4N_2^+ \xrightarrow{b} (4\text{-}CH_3C_6H_4)_2\dot{N}_2^+$

γ-rad. of 4-CH$_3$C$_6$H$_4$N$_2^+$BF$_4^-$ + HCOO$^-$ + H$_2$O				74 Pac 1,
PR [53]	H$_2$O	RT	$k_a/k_b = 3.6$ [54]	75 Pac 1

—$CH_3 + HCOOH \xrightarrow{a} C_6H_5CH_3 + \dot{C}O_2H$

$+ 4\text{-}CH_3C_6H_4N_2^+ \xrightarrow{b} (4\text{-}CH_3C_6H_4)_2\dot{N}_2^+$

γ-rad. of 4-CH$_3$C$_6$H$_4$N$_2^+$BF$_4^-$ + HCOOH + H$_2$O				74 Pac 1,
PR [53]	H$_2$O (pH = 0)	RT	$k_a/k_b = 6.0 \cdot 10^{-3}$ [54]	75 Pac 1

---

[52] ^{131}I radioactivity measurements.
[53] Spectrophotometric determination of 4-CH$_3$C$_6$H$_4$N$_2^+$.
[54] Twice the value of [74 Pac 1] as required in [75 Pac 1].

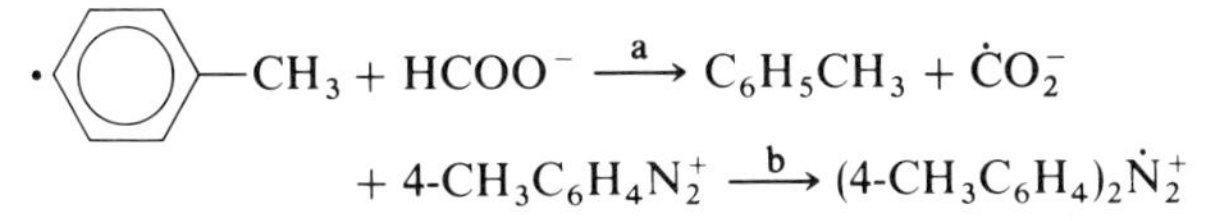

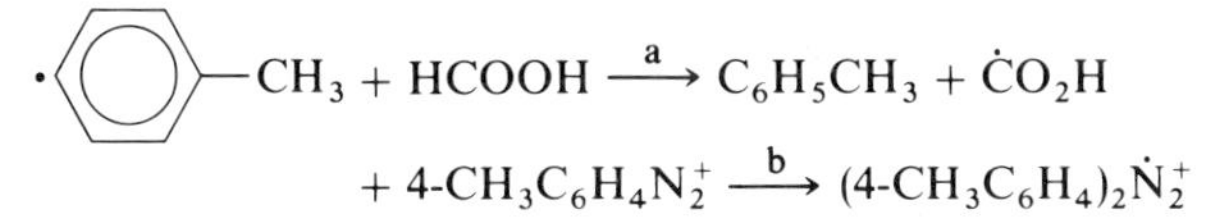

Reaction Radical generation Method	Solvent	T[K]	Rate data	Ref./ add. ref.
$\cdot$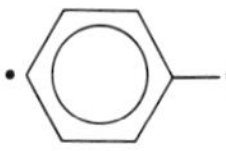$-CH_3 + CH_3OH \xrightarrow{a} C_6H_5CH_3 + \dot{C}H_2OH$ $+ 4\text{-}CH_3C_6H_4N_2^+ \xrightarrow{b} (4\text{-}CH_3C_6H_4)_2\dot{N}_2^+$				
γ-rad. of $4\text{-}CH_3C_6H_4N_2^+BF_4^- + CH_3OH + H_2O$				71 Pac 1,
PR [53]	H_2O	RT	$k_a/k_b = 6.7 \cdot 10^{-2}$ [54]	75 Pac 1
$\cdot$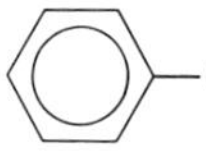$-CH_3 + CH_3CHO \xrightarrow{\alpha} C_6H_5CH_3 + CH_3\dot{C}O$ $\xrightarrow{\beta} C_6H_5CH_3 + \dot{C}H_2CHO$				
γ-rad. of $4\text{-}CH_3C_6H_5N_2^+BF_4^- + CH_3CHO + H_2O$				75 Pac 1
PR	H_2O	RT	$k_\alpha/k_\beta = 40.6$ [55]	
$\cdot$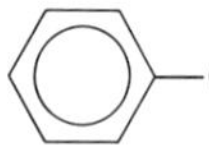$-CH_3 + C_2H_5OH \xrightarrow{\alpha} C_6H_5CH_3 + CH_3\dot{C}HOH$ $\xrightarrow{\beta} C_6H_5CH_3 + \dot{C}H_2CH_2OH$				
γ-rad. of $4\text{-}CH_3C_6H_4N_2^+BF_4^- + C_2H_5OH + H_2O$				75 Pac 1/
PR	H_2O	RT	$k_\alpha/k_\beta = 55.4$ [56]	71 Bur 1, 69 Dra 1
$\cdot$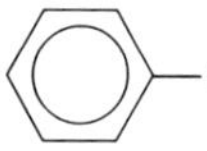$-CH_3 + CD_3COCD_3 \xrightarrow{a} 4\text{-}DC_6H_4CH_3 + \dot{C}D_2COCD_3$ $+ CCl_4 \xrightarrow{b} 4\text{-}CH_3C_6H_4Cl + \dot{C}Cl_3$				
Thermal decomp. of 4-methylphenylazotriphenylmethane				66 Pry 1
PR, glc	CCl_4/CD_3COCD_3	333	$k_a/k_b = 0.040$	
$\cdot$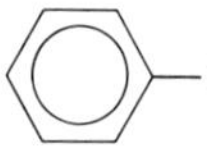$-CH_3 + (CD_3)_2CHOH \xrightarrow{\alpha} (CD_3)_2\dot{C}OH + C_6H_5CH_3$ $\xrightarrow{\beta} \dot{C}D_2CH(CD_3)OH + C_6H_5CH_2D$				
γ-rad. of $4\text{-}CH_3C_6H_4N_2^+BF_4^- + (CD_3)_2CHOH + H_2O$				75 Pac 1
PR	H_2O	RT	$k_\alpha/k_\beta = 210$	
$\cdot$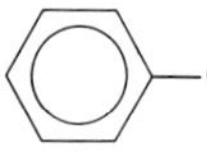$-CH_3 + CH_3COCH_3 \xrightarrow{a} C_6H_5CH_3 + \dot{C}H_2COCH_3$ $+ CCl_4 \xrightarrow{b} 4\text{-}CH_3C_6H_4Cl + \dot{C}Cl_3$				
Thermal decomp. of 4-methylphenylazotriphenylmethane				66 Pry 1
PR, glc	CCl_4/CH_3COCH_3	333	$k_a/k_b = 0.14$	
$\cdot$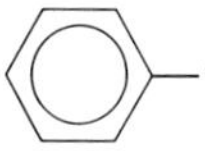$-CH_3 + CH_3COOCH_3 \xrightarrow{a} C_6H_5CH_3$ + acetic acid methylester($-\dot{H}$) $+ CCl_4 \xrightarrow{b} 4\text{-}CH_3C_6H_4Cl + \dot{C}Cl_3$				
Thermal decomp. of 4-methylphenylazotriphenylmethane				66 Pry 1
PR, glc	CCl_4/CH_3COOCH_3	333	$k_a/k_b = 0.075$	

[53]) Spectrophotometric determination of $4\text{-}CH_3C_6H_4N_2^+$.
[54]) Twice the value of [74 Pac 1] as required in [75 Pac 1].
[55]) Based on assumed $k_\alpha/k_\beta = 6$ for $\dot{O}H + CH_3CHO$ and $k_\alpha/k_\beta = 10.5$ for $\dot{H} + CH_3CHO$.
[56]) Based on $k_\alpha/k_\beta = 6$ for $\dot{O}H + C_2H_5OH$ [71 Bur 1] and $k_\alpha/k_\beta = 10.5$ for $\dot{H} + C_2H_5OH$ [69 Dra 1].

Reaction Radical generation Method	Solvent	T[K]	Rate data	Ref./ add. ref.

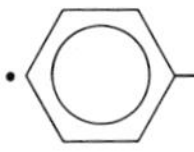 $-CH_3 + (CH_3)_2CHOH \xrightarrow{\alpha} C_6H_5CH_3 + (CH_3)_2\dot{C}OH$
$\xrightarrow{\beta} C_6H_5CH_3 + \dot{C}H_2CH(CH_3)OH$

γ-rad. of $4\text{-}CH_3C_6H_4N_2^+BF_4^- + (CH_3)_2CHOH + H_2O$				75 Pac 1/
PR	H_2O	RT	$k_\alpha/k_\beta = 57.3$ [57])	71 Bur 1, 69 Dra 1

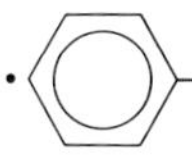 $-CH_3 + C_6H_5SH \xrightarrow{a} C_6H_5CH_3 + C_6H_5\dot{S}$
$+ CCl_4 \xrightarrow{b} 4\text{-}CH_3C_6H_4Cl + \dot{C}Cl_3$

Thermal decomp. of 4-methylphenylazotriphenylmethane				66 Pry 1
PR, glc	CCl_4	333	$k_a/k_b = 438$	

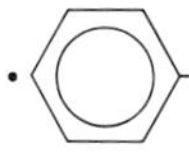 $-CH_3 + c\text{-}C_6H_{10} \xrightarrow{a} C_6H_5CH_3 + (c\text{-}C_6H_9^{\cdot})$
$+ CCl_4 \xrightarrow{b} 4\text{-}CH_3C_6H_4Cl + \dot{C}Cl_3$

Thermal decomp. of 4-methylphenylazotriphenylmethane				66 Pry 1
PR, glc	$CCl_4/c\text{-}C_6H_{10}$	333	$k_a/k_b = 7.1$	

 $-CH_3 + c\text{-}C_6H_{12} \xrightarrow{a} C_6H_5CH_3 + (c\text{-}C_6H_{11}^{\cdot})$
$+ CCl_4 \xrightarrow{b} 4\text{-}CH_3C_6H_4Cl + \dot{C}Cl_3$

Thermal decomp. of 4-methylphenylazotriphenylmethane				66 Pry 1
PR, glc	$CCl_4/c\text{-}C_6H_{12}$	333	$k_a/k_b = 1.11$	

 $-CH_3 + n\text{-}C_6H_{14} \xrightarrow{a} C_6H_5CH_3 + (C_6H_{13}^{\cdot})$
$+ CCl_4 \xrightarrow{b} 4\text{-}CH_3C_6H_4Cl + \dot{C}Cl_3$

Thermal decomp. of 4-methylphenylazotriphenylmethane				66 Pry 1
PR, glc	$CCl_4/n\text{-}C_6H_{14}$	333	$k_a/k_b = 0.89$	

 $-CH_3 + 3\text{-}BrC_6H_4CH_3 \xrightarrow{a} C_6H_5CH_3 +$ 3-bromotoluene$(-\dot{H})$
$+ CCl_4 \xrightarrow{b} 4\text{-}CH_3C_6H_4Cl + \dot{C}Cl_3$

Thermal decomp. of 4-methylphenylazotriphenylmethane				66 Pry 1
PR, glc	CCl_4/3-bromotoluene	333	$k_a/k_b = 0.29$	

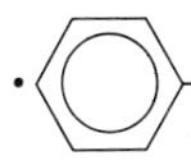 $-CH_3 + 4\text{-}BrC_6H_4CH_3 \xrightarrow{a} C_6H_5CH_3 +$ 4-bromotoluene$(-\dot{H})$
$+ CCl_4 \xrightarrow{b} 4\text{-}CH_3C_6H_4Cl + \dot{C}Cl_3$

Thermal decomp. of 4-methylphenylazotriphenylmethane				66 Pry 1
PR, glc	CCl_4/4-bromotoluene	333	$k_a/k_b = 0.28$	

[57]) Based on $k_\alpha/k_\beta = 5.2$ for $\dot{O}H + (CH_3)_2CHOH$ [71 Bur 1] and $k_\alpha/k_\beta = 20$ for $\dot{H} + (CH_3)_2CHOH$ [69 Dra 1].

Reaction				
Radical generation				Ref./
Method	Solvent	$T[K]$	Rate data	add. ref.

$\cdot\langle\bigcirc\rangle$—$CH_3$ + 4-$FC_6H_4CH_3$ $\xrightarrow{a}$ $C_6H_5CH_3$ + 4-fluorotoluene($-\dot{H}$)

$\qquad$ + CCl_4 $\xrightarrow{b}$ 4-$CH_3C_6H_4Cl$ + $\dot{C}Cl_3$

Thermal decomp. of 4-methylphenylazotriphenylmethane				66 Pry 1
PR, glc	CCl_4/4-fluorotoluene	333	$k_a/k_b = 0.30$	

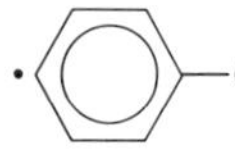

$\cdot\langle\bigcirc\rangle$—$CH_3$ + 4-$NO_2C_6H_4CH_3$ $\xrightarrow{a}$ $C_6H_5CH_3$ + 4-nitrotoluene($-\dot{H}$)

$\qquad$ + CCl_4 $\xrightarrow{b}$ 4-$CH_3C_6H_4Cl$ + $\dot{C}Cl_3$

Thermal decomp. of 4-methylphenylazotriphenylmethane				66 Pry 1
PR, glc	CCl_4/4-nitrotoluene	333	$k_a/k_b = 0.26$	

$\cdot\langle\bigcirc\rangle$—$CH_3$ + 3-$CH_3C_6H_4CH_3$ $\xrightarrow{a}$ $C_6H_5CH_3$ + 3-methyltoluene($-\dot{H}$)

$\qquad$ + CCl_4 $\xrightarrow{b}$ 4-$CH_3C_6H_4Cl$ + $\dot{C}Cl_3$

Thermal decomp. of 4-methylphenylazotriphenylmethane				66 Pry 1
PR, glc	CCl_4/3-methyltoluene	333	$k_a/k_b = 0.64$	

$\cdot\langle\bigcirc\rangle$—$CH_3$ + 4-$CH_3OC_6H_4CH_3$ $\xrightarrow{a}$ $C_6H_5CH_3$ + 4-methoxytoluene($-\dot{H}$)

$\qquad$ + CCl_4 $\xrightarrow{b}$ 4-$CH_3C_6H_4Cl$ + $\dot{C}Cl_3$

Thermal decomp. of 4-methylphenylazotriphenylmethane				66 Pry 1
PR, glc	CCl_4/4-methoxytoluene	333	$k_a/k_b = 0.35$	

$C_6H_5\dot{C}H_2$ + Cl_2 $\xrightarrow{a}$ $C_6H_5CH_2Cl$ + $\dot{C}l$
2 $C_6H_5\dot{C}H_2$ $\xrightarrow{b}$ products

$\dot{C}l$ + $C_6H_5CH_3$ reaction (AIBN as initiator)				70 Shv 1
[58])	toluene	323	$k_a/(k_b)^{\frac{1}{2}} = 13.0\ M^{-\frac{1}{2}}s^{-\frac{1}{2}}$	

$C_6H_5\dot{C}H_2$ + Cl_2 $\xrightarrow{a}$ $C_6H_5CH_2Cl$ + $\dot{C}l$
$C_6H_5\dot{C}H_2$ $\xrightarrow{b}$ products [59])

$\dot{C}l$ + $C_6H_5CH_3$ reaction (AIBN as initiator)				70 Shv 1
[58])	toluene (containing O_2)	323	$k_a/k_b = 1.22 \cdot 10^5\ M^{-1}$	

$C_6H_5\dot{C}H_2$ + $CBrCl_3$ $\xrightarrow{a}$ $C_6H_5CH_2Br$ + $\dot{C}Cl_3$
$\qquad$ + CCl_4 $\xrightarrow{b}$ $C_6H_5CH_2Cl$ + $\dot{C}Cl_3$

Therm. of $C_6H_5CH_2COOOC(CH_3)_3$				75 Her 1/
PR, glc	CCl_4	354	$k_a/k_b = 1700(\pm 15\%)$	76 Gie 1, 76 Gie 2, 76 Gie 3

$C_6H_5\dot{C}H_2$ + $(CH_2CO)_2NBr$ $\xrightarrow{a}$ $C_6H_5CH_2Br$ + $(CH_2CO)_2\dot{N}$
$\qquad$ + $(CH_2CO)_2NCl$ $\xrightarrow{b}$ $C_6H_5CH_2Cl$ + $(CH_2CO)_2\dot{N}$

$(CH_2CO)_2\dot{N}$ + $(C_6H_5CH_2)_4Sn$ $\longrightarrow$ $C_6H_5\dot{C}H_2$ + $(CH_2CO)_2NSn(C_6H_5CH_2)_3$				72 Dav 1
PR, glc	acetone	308	$k_a/k_b = 7$	

---

[58]) Cl_2 vapor pressure measurement in gas phase above solution.
[59]) First order termination reaction, likely to be reaction with O_2.

Reaction Radical generation Method	Solvent	T[K]	Rate data	Ref./ add. ref.
$C_6H_5\dot{C}H_2 + (CH_2CO)_2NI \xrightarrow{a} C_6H_5CH_2I + (CH_2CO)_2\dot{N}$ $+ (CH_2CO)_2NCl \xrightarrow{b} C_6H_5CH_2Cl + (CH_2CO)_2\dot{N}$ $(CH_2CO)_2\dot{N} + (C_6H_5CH_2)_4Sn \longrightarrow (CH_2CO)_2NSn(C_6H_5CH_2)_3 + C_6H_5\dot{C}H_2$				72 Dav 1
PR, glc	acetone	308	$k_a/k_b = 22$	
$C_6H_5\dot{C}HCH_3 + CHBr_3 \xrightarrow{a} C_6H_5CHBrCH_3 + \dot{C}HBr_2$ $C_6H_5\dot{C}HCH_3 + CHBr_3 \xrightarrow{b} C_6H_5CH_2CH_3 + \dot{C}Br_3$				
Therm. of azobis-α-phenylethane				77 Tho 1
PR, glc	$CHBr_3$	353	$k_a/k_b = 5.7(17)$ [60]	
$C_6H_5\dot{C}HCH_3 + CH_2Br_2 \xrightarrow{a} C_6H_5CHBrCH_3 + \dot{C}H_2Br$ $C_6H_5\dot{C}HCH_3 + CH_2Br_2 \xrightarrow{b} C_6H_5CH_2CH_3 + \dot{C}HBr_2$				
Therm. of azobis-α-phenylethane				77 Tho 1
PR, glc	CH_2Br_2	368	$k_a/k_b = 1.1(4)$ [60]	
$C_6H_5\dot{C}HCH_3 + CH_3Br \xrightarrow{a} C_6H_5CHBrCH_3 + \dot{C}H_3$ $C_6H_5\dot{C}HCH_3 + CH_3Br \xrightarrow{b} C_6H_5CH_2CH_3 + \dot{C}H_2Br$				
Therm. of azobis-α-phenylmethane				77 Tho 1
PR, glc	CH_3Br	368	$k_a/k_b = 84(30)$ [60]	
$C_6H_5CH_2\dot{C}H_2 + (CH_3)_2CHCHO \xrightarrow{a} C_6H_5CH_2CH_3 + (CH_3)_2CH\dot{C}O$ $+ Cu(II)(NCCH_3)_4^{2+} \xrightarrow{b} C_6H_5CH{=}CH_2 + H^+ + Cu(I)(NCCH_3)_4^+$				
Cu(II) catalyzed decomp. of $C_6H_5CH_2CH_2OOCH_2CH_2C_6H_5$				68 Koc 1
PR, glc	CH_3CN/CH_3COOH (1:1.5)	298.5	$k_a/k_b = 6.25 \cdot 10^{-3}$ [61]	
$C_6H_5CH_2\dot{C}H_2 + (CH_3)_2CHCHO \xrightarrow{a} C_6H_5CH_2CH_3 + (CH_3)_2CH\dot{C}O$ $+ Cu(II)(bipyridine)^{2+} \xrightarrow{b} C_6H_5CH{=}CH_2 + H^+ + Cu(I)(bipyridine)^+$				
Cu(II) catalyzed decomp. of $C_6H_5CH_2CH_2OOCH_2CH_2C_6H_5$				68 Koc 1
PR, glc	CH_3CN/CH_3COOH (1:1.5)	298.5	$k_a/k_b = 7.1 \cdot 10^{-4}$ [61]	
Ethylbenzene($-\dot{H}$)[62] + $C_6H_5OH \xrightarrow{a} C_6H_5C_2H_5$ + phenol($-\dot{H}$) 2 ethylbenzene($-\dot{H}$)[62] $\xrightarrow{b}$ products				
AIBN as initiator				77 Bel 1
Chemil.	$C_6H_5C_2H_5$	333	$k_a/(2k_b)^{\frac{1}{2}} = 0.088\ M^{-\frac{1}{2}}s^{-\frac{1}{2}}$	
Ethylbenzene($-\dot{H}$)[62] + 4-$CH_3OC_6H_4OH \xrightarrow{a} C_6H_5C_2H_5$ + 4-methoxyphenol($-\dot{H}$) 2 ethylbenzene($-\dot{H}$)[62] $\xrightarrow{b}$ products				
PC (313 K) or AIBN (333 K)				77 Bel 1
Chemil.	$C_6H_5C_2H_5$	313 333	$k_a/(2k_b)^{\frac{1}{2}} = 0.094\ M^{-\frac{1}{2}}s^{-\frac{1}{2}}$ 0.062	
Ethylbenzene($-\dot{H}$)[62] + 2,4,6-$(CH_3)_3C_6H_2OH \xrightarrow{a} C_6H_5C_2H_5$ + 2,4,6-trimethylphenol($-\dot{H}$) 2 ethylbenzene($-\dot{H}$)[62] $\xrightarrow{b}$ products				
PC as initiator				77 Bel 1
Chemil.	$C_6H_5C_2H_5$	313	$k_a/(2k_b)^{\frac{1}{2}} = 0.175\ M^{-\frac{1}{2}}s^{-\frac{1}{2}}$	

[60]) Per abstractable atom.
[61]) Assumed value for $k_a = 1 \cdot 10^4\ M^{-1}s^{-1}$.
[62]) Possibly radical mixture.

Reaction Radical generation Method	Solvent	T[K]	Rate data	Ref./ add. ref.
Ethylbenzene($-\dot{H}$)[62]) + α-naphthol $\xrightarrow{a}$ $C_6H_5C_2H_5$ + α-naphthol($-\dot{H}$)				
2 ethylbenzene($-\dot{H}$)[62]) $\xrightarrow{b}$ products				
PC (313 K) and AIBN (333 K) as initiator				77 Bel 1
Chemil.	$C_6H_5C_2H_5$	313	$k_a/(2k_b)^{\frac{1}{2}} = 0.094\ M^{-\frac{1}{2}}s^{-\frac{1}{2}}$	
		333	0.047	
Ethylbenzene($-\dot{H}$)[62]) + α-naphthylamine $\xrightarrow{a}$ $C_6H_5C_2H_5$ + α-naphthylamine($-\dot{H}$)				
2 ethylbenzene($-\dot{H}$)[62]) $\xrightarrow{b}$ products				
AIBN as initiator				77 Bel 1
Chemil.	$C_6H_5C_2H_5$	333	$k_a/(2k_b)^{\frac{1}{2}} = 0.040\ M^{-\frac{1}{2}}s^{-\frac{1}{2}}$	
Ethylbenzene($-\dot{H}$)[62]) + 4-$(CH_3)_3CC_6H_4OH$ $\xrightarrow{a}$ $C_6H_5C_2H_5$ + 4-*t*-butylphenol($-\dot{H}$)				
2 ethylbenzene($-\dot{H}$)[62]) $\xrightarrow{b}$ products				
AIBN as initiator				77 Bel 1
Chemil.	$C_6H_5C_2H_5$	333	$k_a/(2k_b)^{\frac{1}{2}} = 0.088\ M^{-\frac{1}{2}}s^{-\frac{1}{2}}$	
Ethylbenzene($-\dot{H}$)[62]) + 4-*t*-butylpyrocatechol $\xrightarrow{a}$ $C_6H_5C_2H_5$ + 4-*t*-butylpyrocatechol($-\dot{H}$)				
2 ethylbenzene($-\dot{H}$)[62]) $\xrightarrow{b}$ products				
AIBN as initiator				77 Bel 1
Chemil.	$C_6H_5C_2H_5$	333	$k_a/(2k_b)^{\frac{1}{2}} = 0.075\ M^{-\frac{1}{2}}s^{-\frac{1}{2}}$	
Ethylbenzene($-\dot{H}$)[62]) + $(C_6H_5)_2NH$ $\xrightarrow{a}$ $C_6H_5C_2H_5$ + $(C_6H_5)_2\dot{N}$				
2 ethylbenzene($-\dot{H}$)[62]) $\xrightarrow{b}$ products				
AIBN as initiator				77 Bel 1
Chemil.	$C_6H_5C_2H_5$	333	$k_a/(2k_b)^{\frac{1}{2}} = 0.088\ M^{-\frac{1}{2}}s^{-\frac{1}{2}}$	
Ethylbenzene($-\dot{H}$)[62]) + dimethyl-*p*-phenylaminophenoxysilane $\xrightarrow{a}$ $C_6H_5C_2H_5$ + dimethyl-*p*-... silane($-\dot{H}$)				
2 ethylbenzene($-\dot{H}$)[62]) $\xrightarrow{b}$ products				
AIBN as initiator				77 Bel 1
Chemil.	$C_6H_5C_2H_5$	333	$k_a/(2k_b)^{\frac{1}{2}} = 0.12\ M^{-\frac{1}{2}}s^{-\frac{1}{2}}$	
Ethylbenzene($-\dot{H}$)[62]) + 2,6-$((CH_3)_3C)_2C_6H_3OH$ $\xrightarrow{a}$ $C_6H_5C_2H_5$ + 2,6-di-*t*-butylphenol($-\dot{H}$)				
2 ethylbenzene($-\dot{H}$)[62]) $\xrightarrow{b}$ products				
PC as initiator				77 Bel 1
Chemil.	$C_6H_5C_2H_5$	313	$k_a/(2k_b)^{\frac{1}{2}} = 0.044\ M^{-\frac{1}{2}}s^{-\frac{1}{2}}$	
Ethylbenzene($-\dot{H}$)[62]) + N-phenyl-N′-isopropyl-*p*-phenylenediamine $\xrightarrow{a}$ $C_6H_5C_2H_5$ + N-phenyl-... diamine($-\dot{H}$)				
2 ethylbenzene($-\dot{H}$)[62]) $\xrightarrow{b}$ products				
AIBN as initiator				77 Bel 1
Chemil.	$C_6H_5C_2H_5$	333	$k_a/(2k_b)^{\frac{1}{2}} = 0.075\ M^{-\frac{1}{2}}s^{-\frac{1}{2}}$	
Ethylbenzene($-\dot{H}$)[62]) + 2,6-$((CH_3)_3C)_2$-4-CH_3-C_6H_2OH $\xrightarrow{a}$ $C_6H_5C_2H_5$ + 2,6-di-*t*-butyl-4-methylphenol($-\dot{H}$)				
2 ethylbenzene($-\dot{H}$)[62]) $\xrightarrow{b}$ products				
PC (313 K) and AIBN (333 K) as initiator				77 Bel 1
Chemil.	$C_6H_5C_2H_5$	313	$k_a/(2k_b)^{\frac{1}{2}} = 0.094\ M^{-\frac{1}{2}}s^{-\frac{1}{2}}$	
		333	0.131	
Ethylbenzene($-\dot{H}$)[62]) + N-phenyl-β-naphthylamine $\xrightarrow{a}$ $C_6H_5C_2H_5$ + N-phenyl-β-naphthylamine($-\dot{H}$)				
2 ethylbenzene($-\dot{H}$)[62]) $\xrightarrow{b}$ products				
AIBN as initiator				77 Bel 1
Chemil.	$C_6H_5C_2H_5$	333	$k_a/(2k_b)^{\frac{1}{2}} = 0.094\ M^{-\frac{1}{2}}s^{-\frac{1}{2}}$	

[62]) Possibly radical mixture.

Reaction Radical generation Method	Solvent	T[K]	Rate data	Ref./ add. ref.
Ethylbenzene$(-\dot{H})$[62] + 2,6-$(c\text{-}C_6H_{11})_2C_6H_3OH \xrightarrow{a} C_6H_5CH_2CH_3$ + 2,6-di-c-hexylphenol$(-\dot{H})$ 2 ethylbenzene$(-\dot{H})$[62] $\xrightarrow{b}$ products				
PC as initiator Chemil.	$C_6H_5C_2H_5$	313	$k_a/(2k_b)^{\frac{1}{2}} = 0.131\ M^{-\frac{1}{2}}s^{-\frac{1}{2}}$	77 Bel 1
Ethylbenzene$(-\dot{H})$[62] + 2,4,6-$((CH_3)_3C)_3C_6H_2OH \xrightarrow{a} C_6H_5C_2H_5$ + 2,4,6-tri-t-butylphenol$(-\dot{H})$ 2 ethylbenzene$(-\dot{H})$[62] $\xrightarrow{b}$ products				
PC as initiator Chemil.	$C_6H_5C_2H_5$	313	$k_a/(2k_b)^{\frac{1}{2}} = 0.175\ M^{-\frac{1}{2}}s^{-\frac{1}{2}}$	77 Bel 1
Ethylbenzene$(-\dot{H})$[62] + bis(2-hydroxy-5-methyl-3-t-butylphenyl)sulfide $\xrightarrow{a}$ $C_6H_5C_2H_5$ + bis(2-...)sulfide$(-\dot{H})$ 2 ethylbenzene$(-\dot{H})$[62] $\xrightarrow{b}$ products				
PC as initiator Chemil.	$C_6H_5C_2H_5$	313	$k_a/(2k_b)^{\frac{1}{2}} = 0.22\ M^{-\frac{1}{2}}s^{-\frac{1}{2}}$	77 Bel 1
Ethylbenzene$(-\dot{H})$[62] + bis(2-hydroxy-5-methyl-3-t-butylphenyl)sulfoxide $\xrightarrow{a}$ $C_6H_5C_2H_5$ + bis(2-...)sulfoxide$(-\dot{H})$ 2 ethylbenzene$(-\dot{H})$[62] $\xrightarrow{b}$ products				
PC as initiator Chemil.	$C_6H_5C_2H_5$	313	$k_a/(2k_b)^{\frac{1}{2}} = 0.076\ M^{-\frac{1}{2}}s^{-\frac{1}{2}}$	77 Bel 1
Ethylbenzene$(-\dot{H})$[62] + bis(5-methyl-3-t-butyl-2-hydroxyphenyl)methane $\xrightarrow{a}$ $C_6H_5C_2H_5$ + bis(5-methyl...)methane$(-\dot{H})$ 2 ethylbenzene$(-\dot{H})$[62] $\xrightarrow{b}$ products				
PC as initiator Chemil.	$C_6H_5C_2H_5$	313	$k_a/(2k_b)^{\frac{1}{2}} = 0.094\ M^{-\frac{1}{2}}s^{-\frac{1}{2}}$	77 Bel 1
CH_3–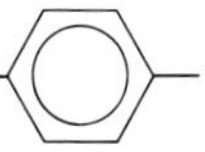–$\dot{C}H_2 + CBrCl_3 \xrightarrow{a}$ 4-$CH_3C_6H_4CH_2Br + \dot{C}Cl_3$ $+ CCl_4 \xrightarrow{b}$ 4-$CH_3C_6H_4CH_2Cl + \dot{C}Cl_3$				
Therm. of 4-$CH_3C_6H_4CH_2COOOC(CH_3)_3$ PR, glc	CCl_4	355	$k_a/k_b = 1870(\pm 15\%)$	75 Her 1/ 76 Gie 1, 76 Gie 2, 76 Gie 3
$C_6H_5\dot{C}(CH_3)_2 + (CH_3)_3COOD \xrightarrow{a} C_6H_5CD(CH_3)_2 + (CH_3)_3CO\dot{O}$ $+ O_2 \xrightarrow{b} C_6H_5C(CH_3)_2O_2^{\cdot}$				
Decomp. of $\{C_6H_5C(CH_3)_2\}_2O_2$, AIBN initiated PR, glc	$C_6H_5CH(CH_3)_2$/ $(CH_3)_3COOD$	303	$k_a/k_b \approx 2\cdot 10^{-4}$	79 How 1
4-$CH_3C_6H_4CH_2\dot{C}H_2 + (CH_3)_2CHCHO \xrightarrow{a}$ 4-$CH_3C_6H_4C_2H_5 + (CH_3)_2CH\dot{C}O$ $+ Cu(II)(NCCH_3)_4^{2+} \xrightarrow{b}$ 4-$CH_3C_6H_4CH{=}CH_2 + H^+ + Cu(I)(NCCH_3)_4^+$				
Cu(II) catalyzed decomp. of 4-$CH_3C_6H_4CH_2CH_2OOCH_2CH_2C_6H_4CH_3$ PR, glc	CH_3CN/CH_3COOH (1:1.5)	298.5	$k_a/k_b = 10^{-2}$[63]	68 Koc 1

[62]) Possibly radical mixture.
[63]) Assumed value for $k_a = 1\cdot 10^4\ M^{-1}s^{-1}$.

Reaction Radical generation Method	Solvent	T[K]	Rate data	Ref./ add. ref.
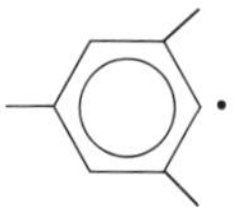 $\cdot + CBrCl_3 \xrightarrow{a} 2,4,6\text{-}(CH_3)_3C_6H_2Br + \dot{C}Cl_3$ $+ CCl_4 \xrightarrow{b} 2,4,6\text{-}(CH_3)_3C_6H_2Cl + \dot{C}Cl_3$				
Therm. of $2,4,6\text{-}(CH_3)_3C_6H_2COOOC(CH_3)_3$ PR, glc	CCl_4	403	$k_a/k_b = 86$	75 Her 1, 69 Rue 1/ 76 Gie 1, 76 Gie 2, 76 Gie 3
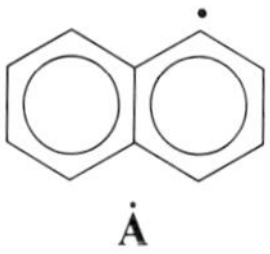 $+ CBrCl_3 \xrightarrow{a} 1\text{-}C_{10}H_7Br + \dot{C}Cl_3$ Ȧ $+ CCl_4 \xrightarrow{b} 1\text{-}C_{10}H_7Cl + \dot{C}Cl_3$				
Therm. of $1\text{-}C_{10}H_7COOOC(CH_3)_3$ PR, glc	CCl_4	403	$k_a/k_b = 103$	75 Her 1/ 76 Gie 1, 76 Gie 2, 76 Gie 3
Ȧ $+ CH_3SOCH_3 \xrightarrow{a} C_{10}H_8 + CH_3SO\dot{C}H_2$ $+ CD_3SOCD_3 \xrightarrow{a} C_{10}H_7D + CD_3SO\dot{C}D_2$ 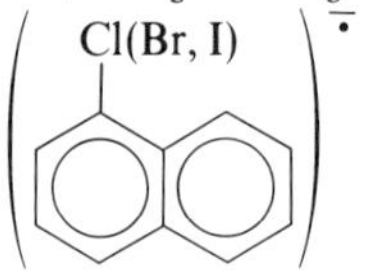 $\xrightarrow{b}$ Ȧ $+ Cl^-(Br^-, I^-)$				
Electrochem. reduct. of 1-chloronaphthalene, 1-bromonaphthalene, 1-iodonaphthalene MS, electrochem.	CH_3SOCH_3 (CD_3SOCD_3)/ $H_2O(D_2O)$ (9:1)	[64]	$k_a(H)/k_b = 1.96\ M^{-1}$ (Cl) $k_a(D)/k_b = 0.244$ $k_a(H)/k_b = 0.51\ M^{-1}$ (Br) $k_a(D)/k_b = 0.0704$ $k_a(H)/k_b = 0.208\ M^{-1}$ (I) $k_a(D)/k_b = 0.0153$	80 M'Ha 1
Ȧ $+ c\text{-}C_6H_{12} \xrightarrow{a} C_{10}H_8 + (c\text{-}C_6\dot{H}_{11})$ $+ CCl_4 \xrightarrow{b}$ 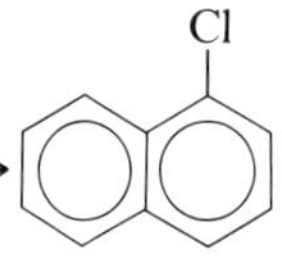 $+ \dot{C}Cl_3$				
Decomp. of (α-naphthyl)N=NC$(C_6H_5)_3$ PR, glc	$c\text{-}C_6H_{12}/CCl_4$	333	$k_a/k_b = 1.03$	76 Lew 2/ 63 Bri 1

[64]) Not given (presumably RT).

Reaction Radical generation				
Method	Solvent	T[K]	Rate data	Ref./ add. ref.

$+ CBrCl_3 \xrightarrow{a} 2\text{-}C_{10}H_7Br + \dot{C}Cl_3$

$+ CCl_4 \xrightarrow{b} 2\text{-}C_{10}H_7Cl + \dot{C}Cl_3$

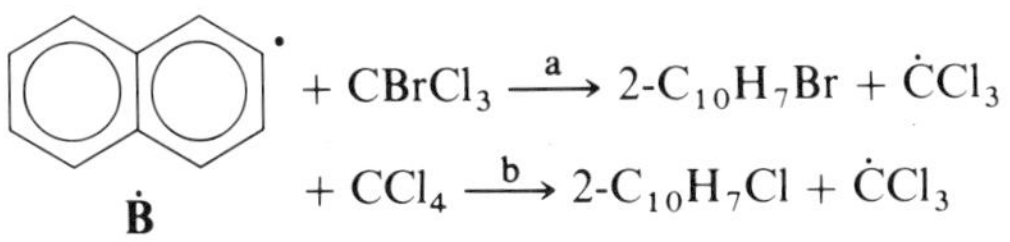

Therm. of 2-$C_{10}H_7$COOOC(CH$_3$)$_3$				75 Her 1/
PR, glc	CCl$_4$	403	$k_a/k_b = 107$	76 Gie 1,
				76 Gie 2,
				76 Gie 3

$\dot{B} + c\text{-}C_6H_{12} \xrightarrow{a} C_{10}H_8 + (c\text{-}C_6\dot{H}_{11})$

$+ CCl_4 \xrightarrow{b}$ ⟨structure⟩ $+ \dot{C}Cl_3$

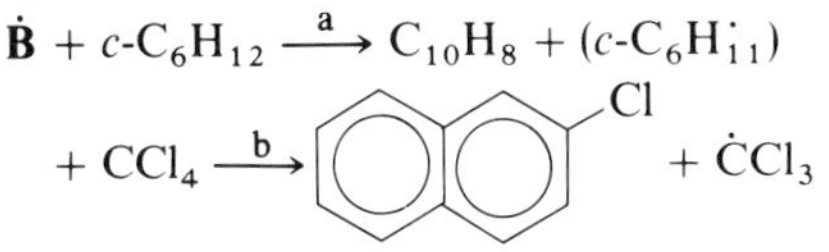

Decomp. of (β-naphthyl)N=NC(C$_6$H$_5$)$_3$				76 Lew 2/
PR, glc	c-C$_6$H$_{12}$/CCl$_4$	333	$k_a/k_b = 1.05$	63 Bri 1

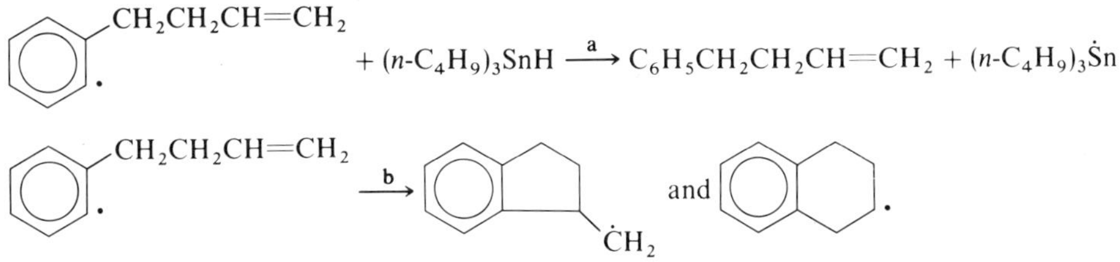

$+ (n\text{-}C_4H_9)_3SnH \xrightarrow{a} C_6H_5CH_2CH_2CH{=}CH_2 + (n\text{-}C_4H_9)_3\dot{Sn}$

Decomp. of ⟨structure⟩ (AIBN initiated)				75 Bec 2
PR, glc	C$_6$H$_6$	403(3)	$k_a/k_b = 2\,M^{-1}$	

$-C(CH_3)_3 + CBrCl_3 \xrightarrow{a} 2\text{-}(CH_3)_3CC_6H_4Br + \dot{C}Cl_3$

$+ CCl_4 \xrightarrow{b} 2\text{-}(CH_3)_3CC_6H_4Cl + \dot{C}Cl_3$

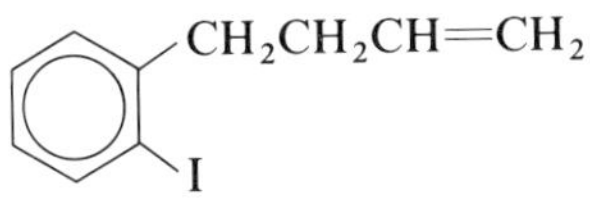

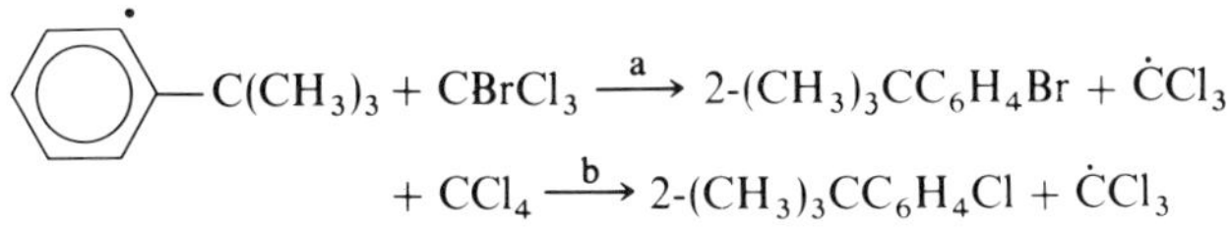

Therm. of 2-(CH$_3$)$_3$CC$_6$H$_4$COOOC(CH$_3$)$_3$				75 Her 1/
PR, glc	CCl$_4$	403	$k_a/k_b = $ 75	76 Gie 1,
		396	87	76 Gie 2,
		384	104	76 Gie 3
		373	139	

$(CH_3)_3C$—⟨structure⟩$\cdot + CBrCl_3 \xrightarrow{a} 4\text{-}(CH_3)_3CC_6H_4Br + \dot{C}Cl_3$

$+ CCl_4 \xrightarrow{b} 4\text{-}(CH_3)_3CC_6H_4Cl + \dot{C}Cl_3$

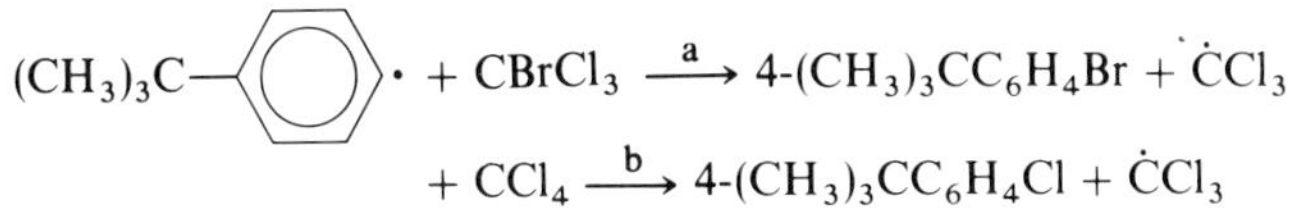

Therm. of 4-(CH$_3$)$_3$CC$_6$H$_4$COOOC(CH$_3$)$_3$				75 Her 1,
PR, glc	CCl$_4$	403	$k_a/k_b = 150$	69 Rue 1/
				76 Gie 1,
				76 Gie 2,
				76 Gie 3

Reaction Radical generation Method	Solvent	T[K]	Rate data	Ref./ add. ref.

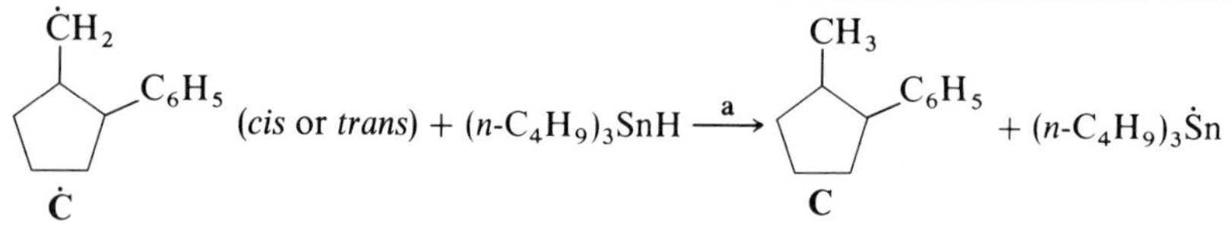

$\dot{C} \xrightarrow{b} CH_2{=}CH(CH_2)_3\dot{C}HC_6H_5$

From *cis*-2-phenylcyclopentylmethylbromide with AIBN or DTBP as initiator				72 Wal 2
PR, glc	C_6H_6	343	$k_a/k_b = 47.6\ M^{-1}$ (*cis*)	65)
From $CH_2{=}CH(CH_2)_3CHBrC_6H_5$ or *trans*-2-phenylcyclopentylmethylbromide with AIBN or DTBP as initiator				72 Wal 2 65)
PR, glc	C_6H_6	343 373	$k_a/k_b = 16.7\ M^{-1}$ (*trans*) $30.3\ M^{-1}$ (*trans*)	

$\dot{C} + (n\text{-}C_4H_9)_3SnH \xrightarrow{a} C + (n\text{-}C_4H_9)_3\dot{S}n$

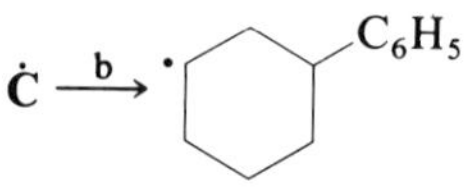

From $CH_2{=}CH(CH_2)_3CHBrC_6H_5$ or *trans*-2-phenylcyclopentylmethylbromide with AIBN or DTBP as initiator				72 Wal 2
PR, glc	C_6H_6	343 373	$k_a/k_b = 83.3\ M^{-1}$ 50.0	

$CH_2{=}CH(CH_2)_3\dot{C}HC_6H_5 + (n\text{-}C_4H_9)_3SnH \xrightarrow{a} CH_2{=}CH(CH_2)_4C_6H_5 + (n\text{-}C_4H_9)_3\dot{S}n$

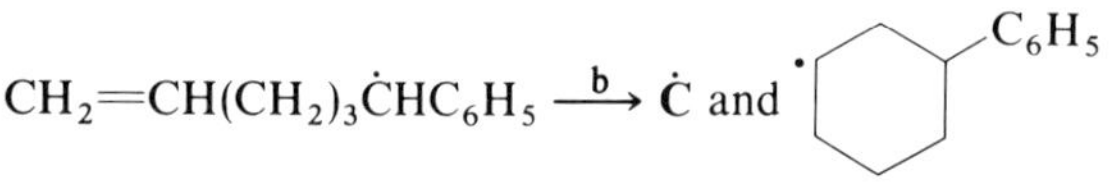

From $CH_2{=}CH(CH_2)_3CBrC_6H_5$ with AIBN or DTBP as initiator				72 Wal 2
PR, glc	C_6H_6	343 373	$k_a/k_b = 1.11\ M^{-1}$ 0.71	65)

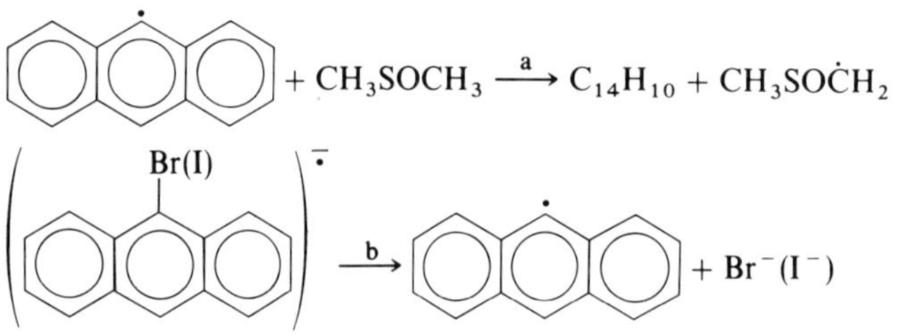

Electrochem. reduct. of 9-bromoanthracene and 9-iodoanthracene				80 M'Ha 1
MS, electrochem.	CH_3SOCH_3/H_2O (9:1)	66)	$k_a/k_b = 119$ (Br) 12.5 (I)	

65) Rate constant $k_a \approx 20$ times slower for $(C_2H_5)_3SnH$ as substrate.
66) Not given (presumably RT).

Reaction Radical generation Method	Solvent	T[K]	Rate data	Ref./ add. ref.
(9-triptycyl)˙ [67]) + $CBrCl_3 \xrightarrow{a}$ 9-Br-triptycene + $\dot{C}Cl_3$ + $CCl_4 \xrightarrow{b}$ 9-Cl-triptycene + $\dot{C}Cl_3$				
Therm. of 9-triptycyl-$COOOC(CH_3)_3$				75 Her 1/
PR, glc	CCl_4	403	$k_a/k_b = 32$	76 Gie 1, 76 Gie 2, 76 Gie 3
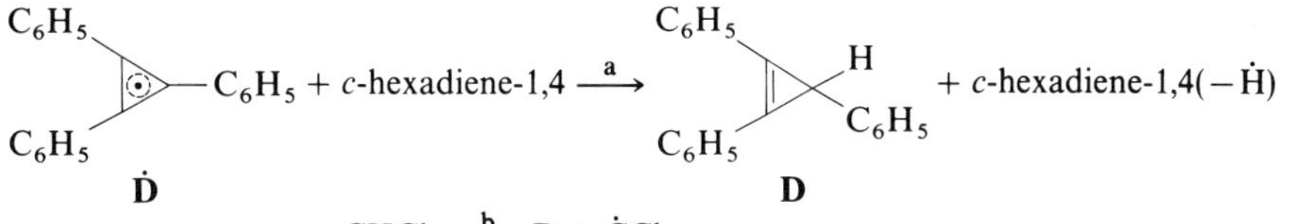 + $CHCl_3 \xrightarrow{b}$ **D** + $\dot{C}Cl_3$				
Electrolytic reduct. of triphenyl-*c*-propenium bromide				70 Sho 1
PR	CH_3CN	RT	$k_a/k_b = 1.34$	
$\dot{\mathbf{D}}$ + *c*-$C_6H_{10} \xrightarrow{a}$ **D** + *c*-hexene(−Ḣ) + $CHCl_3 \xrightarrow{b}$ **D** + $\dot{C}Cl_3$				
Electrolytic reduct. of triphenyl-*c*-propenium bromide				70 Sho 1
PR	CH_3CN	RT	$k_a/k_b = 1.22$	

4.1.2.2.2 Radicals containing C, H, and other atoms

$(4\text{-}BrC_6F_4^{\bullet}) + CHCl_3 \xrightarrow{a} C_6HBrF_4 + \dot{C}Cl_3$ $\xrightarrow{b} C_6BrClF_4 + \dot{C}HCl_2$ $+ CBrCl_3 \xrightarrow{a} C_6Br_2F_4 + \dot{C}Cl_3$ $\xrightarrow{b} C_6BrClF_4 + \dot{C}BrCl_2$			See 4.1.2.3, Fig. 8, p. 257.	
$(C_6F_5^{\bullet}) + (CH_3)_2Se \xrightarrow{a} \dot{C}H_3 + C_6F_5SeCH_3$ $+ (C_2H_5)_2Se \xrightarrow{b} CH_3\dot{C}H_2 + C_6F_5SeC_2H_5$ $+ c\text{-}C_5H_{10} \xrightarrow{c} C_6F_5H + (c\text{-}C_5H_9^{\bullet})$				
Phot. of C_6F_5Br in presence of hexa-*n*-butyltin				77 Sca 1
SESR	[1])	243	$k_a/k_b/k_c = 2.1/11/1.0$	
$(C_6F_5^{\bullet}) + CHCl_3 \xrightarrow{a} C_6HF_5 + \dot{C}Cl_3$ $\xrightarrow{b} C_6ClF_5 + \dot{C}HCl_2$ $+ CBrCl_3 \xrightarrow{a} C_6BrF_5 + \dot{C}Cl_3$ $\xrightarrow{b} C_6ClF_5 + \dot{C}BrCl_2$			See 4.1.2.3, Fig. 8, p. 257.	

[67])
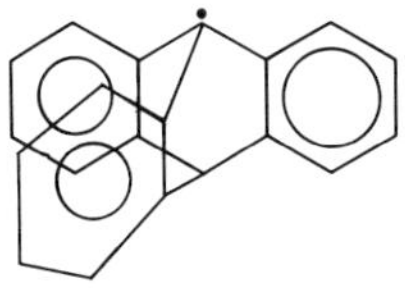

[1]) Not given (presumed: mixture of *c*-C_5H_{10}, $(CH_3)_2Se$ and $(C_2H_5)_2Se$).

Reaction Radical generation Method	Solvent	T[K]	Rate data	Ref./ add. ref.

$(4\text{-}HC_6F_4^{\bullet}) + CHCl_3 \xrightarrow{a} C_6H_2F_4 + \dot{C}Cl_3$
$\xrightarrow{b} C_6HClF_4 + \dot{C}HCl_2$

$+ CBrCl_3 \xrightarrow{a} C_6HBrF_4 + \dot{C}Cl_3$
$\xrightarrow{b} C_6HClF_4 + \dot{C}BrCl_2$

See 4.1.2.3, Fig. 8, p. 257.

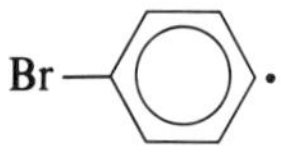 $\cdot + CD_3COCD_3 \xrightarrow{a} 4\text{-}DC_6H_4Br + \dot{C}D_2COCD_3$

$+ CCl_4 \xrightarrow{b} 4\text{-}BrC_6H_4Cl + \dot{C}Cl_3$

Thermal decomp. of 4-bromophenylazotriphenylmethane				66 Pry 1
PR, glc	CCl_4/CD_3COCD_3	333	$k_a/k_b = 0.084$	

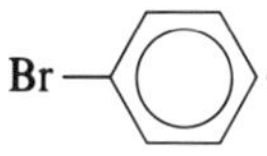 $\cdot + CH_3COCH_3 \xrightarrow{a} C_6H_5Br + \dot{C}H_2COCH_3$

$+ CCl_4 \xrightarrow{b} 4\text{-}BrC_6H_4Cl + \dot{C}Cl_3$

Thermal decomp. of 4-bromophenylazotriphenylmethane				66 Pry 1
PR, glc	CCl_4/CH_3COCH_3	333	$k_a/k_b = 0.35$	

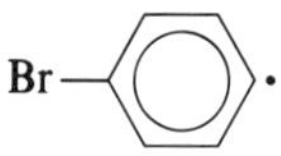 $\cdot + CH_3COOCH_3 \xrightarrow{a} C_6H_5Br +$ acetic acid methylester$(-\dot{H})$

$+ CCl_4 \xrightarrow{b} 4\text{-}BrC_6H_4Cl + \dot{C}Cl_3$

Thermal decomp. of 4-bromophenylazotriphenylmethane				66 Pry 1
PR, glc	CCl_4/CH_3COOCH_3	333	$k_a/k_b = 0.22$	

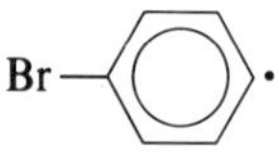 $\cdot + (CH_3)_2CHCH_2CH_3 \xrightarrow{a} C_6H_5Br + (C_5H_{11}^{\bullet})$

$+ CCl_4 \xrightarrow{b} 4\text{-}BrC_6H_4Cl + \dot{C}Cl_3$

Thermal decomp. of 4-bromophenylazotriphenylmethane				66 Pry 1
PR, glc	$CCl_4/$ $(CH_3)_2CHCH_2CH_3$	333	$k_a/k_b = 0.64$	

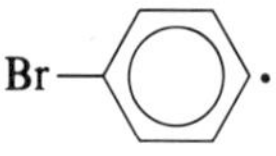 $\cdot + C_6H_5SH \xrightarrow{a} C_6H_5Br + C_6H_5\dot{S}$

$+ CCl_4 \xrightarrow{b} 4\text{-}BrC_6H_4Cl + \dot{C}Cl_3$

Thermal decomp. of 4-bromophenylazotriphenylmethane				66 Pry 1
PR, glc	CCl_4	333	$k_a/k_b = 569$	

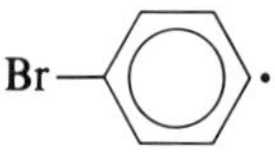 $\cdot + c\text{-}C_6H_{10} \xrightarrow{a} C_6H_5Br + (c\text{-}C_6H_9^{\bullet})$

$+ CCl_4 \xrightarrow{b} 4\text{-}BrC_6H_4Cl + \dot{C}Cl_3$

Thermal decomp. of 4-bromophenylazotriphenylmethane				66 Pry 1
PR, glc	$CCl_4/c\text{-}C_6H_{10}$	333	$k_a/k_b = 14.9$	

Reaction Radical generation Method	Solvent	T[K]	Rate data	Ref./ add. ref.
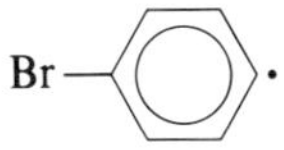 $\cdot + c\text{-}C_6H_{12} \xrightarrow{a} C_6H_5Br + (c\text{-}C_6\dot{H}_{11})$ $+ CCl_4 \xrightarrow{b} 4\text{-}BrC_6H_4Cl + \dot{C}Cl_3$				
Thermal decomp. of 4-bromophenylazotriphenylmethane				66 Pry 1
PR, glc	$CCl_4/c\text{-}C_6H_{12}$	333	$k_a/k_b = 3.5$	
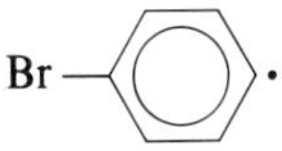 $\cdot + n\text{-}C_6H_{14} \xrightarrow{a} C_6H_5Br + (C_6\dot{H}_{13})$ $+ CCl_4 \xrightarrow{b} 4\text{-}BrC_6H_4Cl + \dot{C}Cl_3$				
Thermal decomp. of 4-bromophenylazotriphenylmethane				66 Pry 1
PR, glc	$CCl_4/n\text{-}C_6H_{14}$	333	$k_a/k_b = 2.76$	
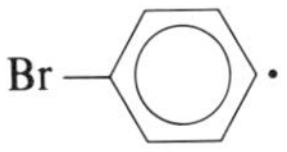 $\cdot + 4\text{-}FC_6H_4CH_3 \xrightarrow{a} C_6H_5Br + 4\text{-}FC_6H_4\dot{C}H_2$ $+ CCl_4 \xrightarrow{b} 4\text{-}BrC_6H_4Cl + \dot{C}Cl_3$				
Thermal decomp. of 4-bromophenyltriphenylmethane				66 Pry 1
PR, glc	CCl_4/4-fluorotoluene	333	$k_a/k_b = 0.61$	
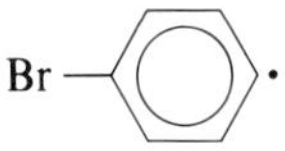 $\cdot + 4\text{-}NO_2C_6H_4CH_3 \xrightarrow{a} C_6H_5Br + 4\text{-}NO_2C_6H_4\dot{C}H_2$ $+ CCl_4 \xrightarrow{b} 4\text{-}BrC_6H_4Cl + \dot{C}Cl_3$				
Thermal decomp. of 4-bromophenylazotriphenylmethane				66 Pry 1
PR, glc	CCl_4/4-nitrotoluene	333	$k_a/k_b = 0.54$	
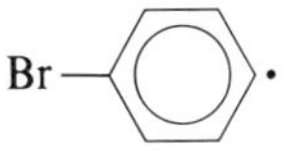 $\cdot + C_6H_5CH_3 \xrightarrow{a} C_6H_5Br + C_6H_5\dot{C}H_2$ $+ CCl_4 \xrightarrow{b} 4\text{-}BrC_6H_4Cl + \dot{C}Cl_3$				
Thermal decomp. of 4-bromophenylazotriphenylmethane				66 Pry 1
PR, glc	CCl_4/toluene	333	$k_a/k_b = 0.79$	
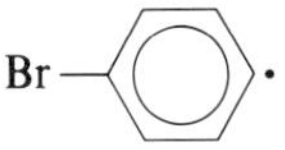 $\cdot + (CH_3)_2CHCH_2CH(CH_3)_2 \xrightarrow{a} C_6H_5Br + (C_7\dot{H}_{15})$ $+ CCl_4 \xrightarrow{b} 4\text{-}BrC_6H_4Cl + \dot{C}Cl_3$				
Thermal decomp. of 4-bromophenylazotriphenylmethane				66 Pry 1
PR, glc	$CCl_4/$ $(CH_3)_2CHCH_2$- $CH(CH_3)_2$	333	$k_a/k_b = 4.1$	
 $\cdot + C_6H_5CH_2CH_3 \xrightarrow{a} C_6H_5Br + \text{ethylbenzene}(-\dot{H})$ $+ CCl_4 \xrightarrow{b} 4\text{-}BrC_6H_4Cl + \dot{C}Cl_3$				
Thermal decomp. of 4-bromophenylazotriphenylmethane				66 Pry 1
PR, glc	$CCl_4/C_6H_5CH_2CH_3$	333	$k_a/k_b = 2.49$	

Reaction Radical generation Method	Solvent	T[K]	Rate data	Ref./ add. ref.

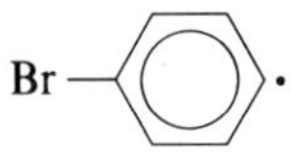 $+ 3\text{-}CH_3C_6H_4CH_3 \xrightarrow{a} C_6H_5Br + 3\text{-}CH_3C_6H_4\dot{C}H_2$

$+ CCl_4 \xrightarrow{b} 4\text{-}BrC_6H_4Cl + \dot{C}Cl_3$

Thermal decomp. of 4-bromophenylazotriphenylmethane				66 Pry 1
PR, glc	CCl_4/ 3-methyltoluene [2])	333	$k_a/k_b = 1.66$	

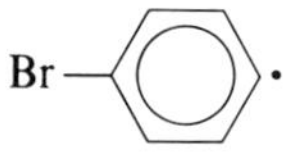 $+ 4\text{-}CH_3C_6H_4CH_3 \xrightarrow{a} C_6H_5Br + 4\text{-}CH_3C_6H_4\dot{C}H_2$

$+ CCl_4 \xrightarrow{b} 4\text{-}BrC_6H_4Cl + \dot{C}Cl_3$

Thermal decomp. of 4-bromophenylazotriphenylmethane				66 Pry 1
PR, glc	CCl_4/ 4-methyltoluene [3])	333	$k_a/k_b = 2.16$	

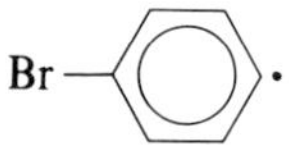 $+ 4\text{-}CH_3OC_6H_4CH_3 \xrightarrow{a} C_6H_5Br +$ 4-methoxytoluene$(-\dot{H})$

$+ CCl_4 \xrightarrow{b} 4\text{-}BrC_6H_4Cl + \dot{C}Cl_3$

Thermal decomp. of 4-bromophenylazotriphenylmethane				66 Pry 1
PR, glc	CCl_4/ 4-methoxytoluene	333	$k_a/k_b = 1.28$	

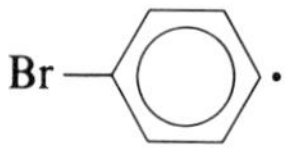 $+ n\text{-}C_8H_{18} \xrightarrow{a} C_6H_5Br + (C_8H_{17}^{\bullet})$

$+ CCl_4 \xrightarrow{b} 4\text{-}BrC_6H_4Cl + \dot{C}Cl_3$

Thermal decomp. of 4-bromophenylazotriphenylmethane				66 Pry 1
PR, glc	$CCl_4/c\text{-}C_8H_{18}$	333	$k_a/k_b = 5.1$	

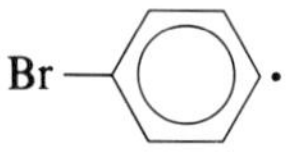 $+ (CH_3)_3CCH_2CH(CH_3)_2 \xrightarrow{a} C_6H_5Br + (C_8H_{17}^{\bullet})$

$+ CCl_4 \xrightarrow{b} 4\text{-}BrC_6H_4Cl + \dot{C}Cl_3$

Thermal decomp. of 4-bromophenylazotriphenylmethane				66 Pry 1
PR, glc	CCl_4/ $(CH_3)_2CCH_2CH(CH_3)_2$	333	$k_a/k_b = 0.82$	

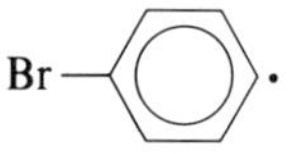 $+ n\text{-}C_3H_7C_6H_5 \xrightarrow{a} C_6H_5Br +$ n-propylbenzene$(-\dot{H})$

$+ CCl_4 \xrightarrow{b} 4\text{-}BrC_6H_4Cl + \dot{C}Cl_3$

Thermal decomp. of 4-bromophenylazotriphenylmethane				66 Pry 1
PR, glc	CCl_4/n-propylbenzene	333	$k_a/k_b = 2.11$	

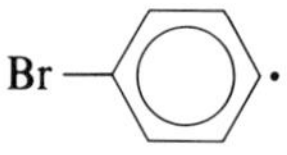 $+ C_6H_5CH(CH_3)_2$ [4]) $\xrightarrow{a} C_6H_5Br + C_6H_5\dot{C}(CH_3)_2$

$+ CCl_4 \xrightarrow{b} 4\text{-}BrC_6H_4Cl + \dot{C}Cl_3$

Thermal decomp. of 4-bromophenylazotriphenylmethane				66 Pry 1
PR, glc	CCl_4/ $C_6H_5CH(CH_3)_2$ [4])	333	$k_a/k_b = 3.75$	

[2]) m-xylene. [3]) p-xylene. [4]) Cumene.

Reaction Radical generation Method	Solvent	T[K]	Rate data	Ref./ add. ref.
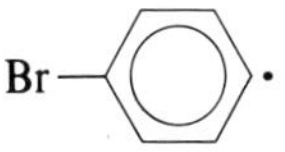 $+ 1,3,5\text{-}(CH_3)_3C_6H_3 \xrightarrow{a} C_6H_5Br + \text{mesitylene}(-\dot{H})$ $+ CCl_4 \xrightarrow{b} 4\text{-}BrC_6H_4Cl + \dot{C}Cl_3$				
Thermal decomp. of 4-bromophenylazotriphenylmethane				66 Pry 1
PR, glc	CCl_4/mesitylene	333	$k_a/k_b = 2.84$	
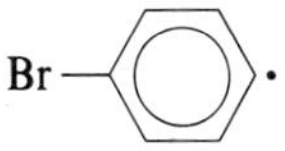 $+ C_6H_5C(CH_3)_3 \xrightarrow{a} C_6H_5Br + \text{2-methyl-2-phenylpropane}(-\dot{H})$ $+ CCl_4 \xrightarrow{b} 4\text{-}BrC_6H_4Cl + \dot{C}Cl_3$				
Thermal decomp. of 4-bromophenylazotriphenylmethane				66 Pry 1
PR, glc	$CCl_4/C_6H_5C(CH_3)_3$	333	$k_a/k_b = 0.21$	
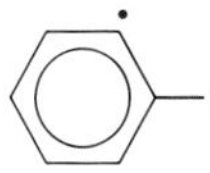 $-Cl + CBrCl_3 \xrightarrow{a} 2\text{-}ClC_6H_4Br + \dot{C}Cl_3$ $+ CCl_4 \xrightarrow{b} 2\text{-}ClC_6H_4Cl + \dot{C}Cl_3$				
Therm. of $2\text{-}ClC_6H_4COOOC(CH_3)_3$				75 Her 1,
PR, glc	CCl_4	403	$k_a/k_b = 128$	69 Rue 1/ 76 Gie 1, 76 Gie 2, 76 Gie 3
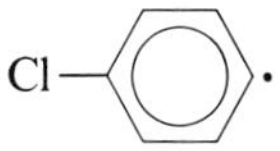 $+ CBrCl_3 \xrightarrow{a} 4\text{-}ClC_6H_4Br + \dot{C}Cl_3$ $+ CCl_4 \xrightarrow{b} 4\text{-}ClC_6H_4Cl + \dot{C}Cl_3$				
Thermal decomp. of 4-chlorophenylazotriphenylmethane				79 Mig 1
PR, glc	CCl_4	333	$k_a/k_b = 288$	
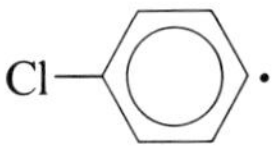 $+ CBrCl_3 \xrightarrow{a} 4\text{-}ClC_6H_4Br + \dot{C}Cl_3$ $+ CCl_4 \xrightarrow{b} 4\text{-}ClC_6H_4Cl + \dot{C}Cl_3$				
Therm. of $4\text{-}ClC_6H_4COOOC(CH_3)_3$ (403 K) or 4-chlorophenylazotriphenylmethane				75 Her 1,
(353 K and 341 K)				69 Rue 1/
PR, glc	CCl_4	403	$k_a/k_b = 124$	76 Gie 1,
		353	220	76 Gie 2,
		341	270	76 Gie 3
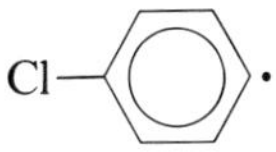 $+ CCl_4 \xrightarrow{a} 4\text{-}ClC_6H_4Cl + \dot{C}Cl_3$ $+ I_2 \xrightarrow{b} 4\text{-}ClC_6H_4I + \dot{I}$				
Thermal decomp. of 4-chlorophenylazotriphenylmethane				79 Mig 1
PR, glc	CCl_4	333	$k_a/k_b = 3.48 \cdot 10^{-4}$	

Reaction Radical generation Method	Solvent	T[K]	Rate data	Ref./ add. ref.
 $\cdot + CH_2{=}CHCH_2SCH_3 \xrightarrow{a} 4\text{-}ClC_6H_4CH_2CH{=}CH_2 + CH_3\dot{S}$ $+ CCl_4 \xrightarrow{b} 4\text{-}ClC_6H_4Cl + \dot{C}Cl_3$				
Thermal decomp. of 4-chlorophenylazotriphenylmethane				79 Mig 1
PR, glc	CCl_4/ $CH_2{=}CHCH_2SCH_3$ mixt.	333	$k_a/k_b = 7.0$	
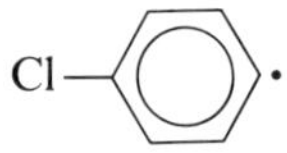 $\cdot + 3\text{-}ClC_6H_4CH_2Br \xrightarrow{a} 4\text{-}ClC_6H_4Br + 3\text{-}ClC_6H_4\dot{C}H_2$ $+ CCl_4 \xrightarrow{b} 4\text{-}ClC_6H_4Cl + \dot{C}Cl_3$				
Therm. of 4-chlorophenylazotriphenylmethane				75 Mig 1
PR, glc	$3\text{-}ClC_6H_4CH_2Br/CCl_4$	333	$k_a/k_b = 8.98(31)$	
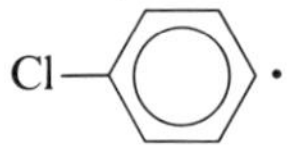 $\cdot + 4\text{-}ClC_6H_4CH_2Br \xrightarrow{a} 4\text{-}ClC_6H_4Br + 4\text{-}ClC_6H_4\dot{C}H_2$ $+ CCl_4 \xrightarrow{b} 4\text{-}ClC_6H_4Cl + \dot{C}Cl_3$				
Therm. of 4-chlorophenylazotriphenylmethane				75 Mig 1
PR, glc	$4\text{-}ClC_6H_4CH_2Br/CCl_4$	333	$k_a/k_b = 10.5(20)$	
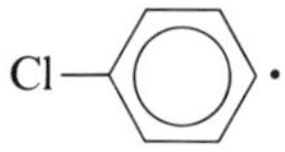 $\cdot + 3\text{-}NO_2C_6H_4CH_2Br \xrightarrow{a} 4\text{-}ClC_6H_4Br + 3\text{-}NO_2C_6H_4\dot{C}H_2$ $+ CCl_4 \xrightarrow{b} 4\text{-}ClC_6H_4Cl + \dot{C}Cl_3$				
Therm. of 4-chlorophenylazotriphenylmethane				75 Mig 1
PR, glc	$3\text{-}NO_2C_6H_4CH_2Br$/ CCl_4	333	$k_a/k_b = 8.82(74)$	
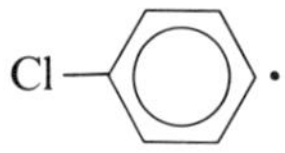 $\cdot + 4\text{-}NO_2C_6H_4CH_2Br \xrightarrow{a} 4\text{-}ClC_6H_4Br + 4\text{-}NO_2C_6H_4\dot{C}H_2$ $+ CCl_4 \xrightarrow{b} 4\text{-}ClC_6H_4Cl + \dot{C}Cl_3$				
Therm. of 4-chlorophenylazotriphenylmethane				75 Mig 1
PR, glc	$4\text{-}NO_2C_6H_4CH_2Br$/ CCl_4	333	$k_a/k_b = 10.7(10)$	
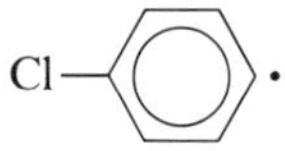 $\cdot + C_6H_5CH_2Br \xrightarrow{a} 4\text{-}ClC_6H_4Br + C_6H_5\dot{C}H_2$ $+ CCl_4 \xrightarrow{b} 4\text{-}ClC_6H_4Cl + \dot{C}Cl_3$				
Thermal decomp. of 4-chlorophenylazotriphenylmethane				79 Mig 1/
PR, glc	$CCl_4/C_6H_5CH_2Br$ mixt.	333	$k_a/k_b = 8.13(50)$	75 Mig 1
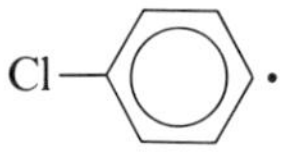 $\cdot + 3\text{-}CH_3C_6H_4CH_2Br \xrightarrow{a} 4\text{-}ClC_6H_4Br + 3\text{-}CH_3C_6H_4\dot{C}H$ $+ CCl_4 \xrightarrow{b} 4\text{-}ClC_6H_4Cl + \dot{C}Cl_3$				
Therm. of 4-chlorophenylazotriphenylmethane				75 Mig 1
PR, glc	$3\text{-}CH_3C_6H_4CH_2Br$/ CCl_4	333	$k_a/k_b = 8.14(30)$	

Reaction Radical generation Method	Solvent	T[K]	Rate data	Ref./ add. ref.

Cl—• + $4\text{-}CH_3C_6H_4CH_2Br \xrightarrow{a} 4\text{-}ClC_6H_4Br + 4\text{-}ClC_6H_4\dot{C}H_2$

+ $CCl_4 \xrightarrow{b} 4\text{-}ClC_6H_4Cl + \dot{C}Cl_3$

Therm. of 4-chlorophenylazotriphenylmethane				75 Mig 1
PR, glc	$4\text{-}ClC_6H_4CH_2Br/CCl_4$	333	$k_a/k_b = 9.56(92)$	

Cl—• + $4\text{-}CH_3OC_6H_4CH_2Br \xrightarrow{a} 4\text{-}ClC_6H_4Br + 4\text{-}CH_3OC_6H_4\dot{C}H_2$

+ $CCl_4 \xrightarrow{b} 4\text{-}ClC_6H_4Cl + \dot{C}Cl_3$

Therm. of 4-chlorophenylazotriphenylmethane				75 Mig 1
PR, glc	$4\text{-}CH_3OC_6H_4CH_2Br/$ CCl_4	333	$k_a/k_b = 11.1(40)$	

Cl—• + $CH_2{=}CHCH_2SC_6H_5 \xrightarrow{a} 4\text{-}ClC_6H_4CH_2CH{=}CH_2 + C_6H_5\dot{S}$

+ $CCl_4 \xrightarrow{b} 4\text{-}ClC_6H_4Cl + \dot{C}Cl_3$

Thermal decomp. of 4-chlorophenylazotriphenylmethane				79 Mig 1
PR, glc	$CCl_4/$ $CH_2{=}CHCH_2SC_6H_5$ mixt.	333	$k_a/k_b = 14.3$	

F—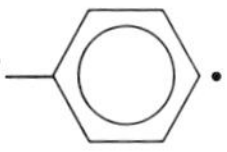• + $CBrCl_3 \xrightarrow{a} 4\text{-}FC_6H_4Br + \dot{C}Cl_3$

+ $CCl_4 \xrightarrow{b} 4\text{-}FC_6H_4Cl + \dot{C}Cl_3$

Therm. of $4\text{-}FC_6H_4COOOC(CH_3)_3$				75 Her 1,
PR, glc	CCl_4	403	$k_a/k_b = 123$	69 Rue 1/ 76 Gie 1, 76 Gie 2, 76 Gie 3

•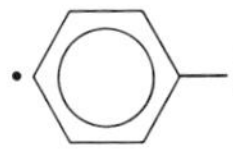—NO_2 + $CBrCl_3 \xrightarrow{a} 4\text{-}NO_2C_6H_4Br + \dot{C}Cl_3$

+ $CCl_4 \xrightarrow{b} 4\text{-}NO_2C_6H_4Cl + \dot{C}Cl_3$

Thermal decomp. of 4-nitrophenylazotriphenylmethane				79 Mig 1
PR, glc	CCl_4	333	$k_a/k_b = 330$	

•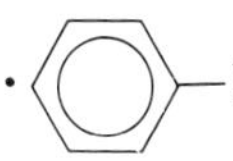—NO_2 + $CCl_4 \xrightarrow{a} 4\text{-}NO_2C_6H_4Cl + \dot{C}Cl_3$

+ $I_2 \xrightarrow{b} 4\text{-}NO_2C_6H_4I + \dot{I}$

Thermal decomp. of 4-nitrophenylazotriphenylmethane				79 Mig 1
PR, glc	CCl_4	333	$k_a/k_b = 1.23 \cdot 10^{-4}$	

•—NO_2 + prim. alkyl C—H bond $\xrightarrow{a} C_6H_5NO_2 + {—}\dot{C}$

+ $C_6H_5CH_3 \xrightarrow{b} C_6H_5NO_2 + C_6H_5\dot{C}H_2$

Therm. of 4-nitrophenylazotriphenylmethane				72 Pry 2
PR, glc	C—H substrate/ toluene	333	$k_a/k_b = 0.17$	

Reaction Radical generation Method	Solvent	T[K]	Rate data	Ref./ add. ref.
$-NO_2$ + sec. alkyl C—H bond $\xrightarrow{a}$ $C_6H_5NO_2$ + $-\dot{C}$ + $C_6H_5CH_3$ $\xrightarrow{b}$ $C_6H_5NO_2$ + $C_6H_5\dot{C}H_2$				
Therm. of 4-nitrophenylazotriphenylmethane				72 Pry 2
PR, glc	C—H substrate/ toluene	333	$k_a/k_b = 1.8$	
$-NO_2$ + tert. alkyl C—H bond $\xrightarrow{a}$ $C_6H_5NO_2$ + $-\dot{C}$ + $C_6H_5CH_3$ $\xrightarrow{b}$ $C_6H_5NO_2$ + $C_6H_5\dot{C}H_2$				
Therm. of 4-nitrophenylazotriphenylmethane				72 Pry 2
PR, glc	C—H substrate/ toluene	333	$k_a/k_b = 8.3$	
$-NO_2$ + prim. allylic C—H bond $\xrightarrow{a}$ $C_6H_5NO_2$ + $-\dot{C}$ + $C_6H_5CH_3$ $\xrightarrow{b}$ $C_6H_5NO_2$ + $C_6H_5\dot{C}H_2$				
Therm. of 4-nitrophenylazotriphenylmethane				72 Pry 2
PR, glc	C—H substrate/ toluene	333	$k_a/k_b = 2.8$	
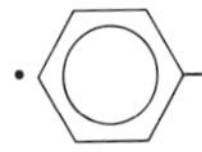$-NO_2$ + sec. allylic C—H bond $\xrightarrow{a}$ $C_6H_5NO_2$ + $-\dot{C}$ + $C_6H_5CH_3$ $\xrightarrow{b}$ $C_6H_5NO_2$ + $C_6H_5\dot{C}H_2$				
Therm. of 4-nitrophenylazotriphenylmethane				72 Pry 2
PR, glc	C—H substrate/ toluene	333	$k_a/k_b = 6.0$	
$-NO_2 + CH_2Cl_2 \xrightarrow{a} C_6H_5NO_2 + \dot{C}HCl_2$ $+ CCl_4 \xrightarrow{b}$ 4-$ClC_6H_4NO_2 + \dot{C}Cl_3$				
Therm. of 4-nitrophenylazotriphenylmethane				72 Pry 2
PR, glc	CH_2Cl_2/CCl_4	333	$k_a/k_b = 0.94(1)$	
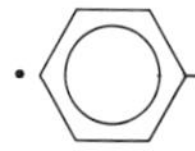$-NO_2 + CH_3NO_2 \xrightarrow{a} C_6H_5NO_2 + \dot{C}H_2NO_2$ $+ CCl_4 \xrightarrow{b}$ 4-$ClC_6H_4NO_2 + \dot{C}Cl_3$				
Therm. of 4-nitrophenylazotriphenylmethane				72 Pry 2
PR, glc	CH_3NO_2/CCl_4	333	$k_a/k_b = 0.05$	
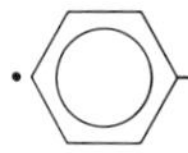$-NO_2 + CH_3OH \xrightarrow{a} C_6H_5NO_2$ + methanol$(-\dot{H})$ $+ CCl_4 \xrightarrow{b}$ 4-$ClC_6H_4NO_2 + \dot{C}Cl_3$				
Therm. of 4-nitrophenylazotriphenylmethane				72 Pry 2
PR, glc	CH_3OH/CCl_4	333	$k_a/k_b = 1.4$	

Reaction Radical generation Method	Solvent	T[K]	Rate data	Ref./ add. ref.

• 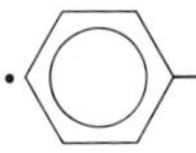 $-NO_2 + CH_3CN \xrightarrow{a} C_6H_5NO_2 + \dot{C}H_2CN$

$+ CCl_4 \xrightarrow{b} 4\text{-}ClC_6H_4NO_2 + \dot{C}Cl_3$

Therm. of 4-nitrophenylazotriphenylmethane				72 Pry 2
PR, glc	CH_3CN/CCl_4	333	$k_a/k_b = 0.23$	

• 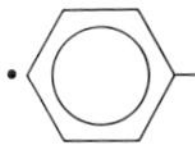 $-NO_2 + CH_3SCN \xrightarrow{a} C_6H_5NO_2 + \dot{C}H_2SCN$

$+ CCl_4 \xrightarrow{b} 4\text{-}ClC_6H_4NO_2 + \dot{C}Cl_3$

Therm. of 4-nitrophenylazotriphenylmethane				72 Pry 2
PR, glc	CH_3SCN/CCl_4	333	$k_a/k_b = 0.16$	

• 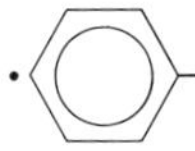 $-NO_2 + CH_2BrCH_3 \xrightarrow{a} C_6H_5NO_2 +$ bromoethane$(-\dot{H})$

$+ CCl_4 \xrightarrow{b} 4\text{-}ClC_6H_4NO_2 + \dot{C}Cl_3$

Therm. of 4-nitrophenylazotriphenylmethane				72 Pry 2
PR, glc	bromoethane/CCl_4	333	$k_a/k_b = 0.75$	

• 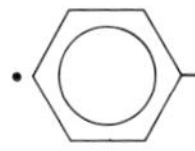 $-NO_2 + C_2H_5NO_2 \xrightarrow{a} C_6H_5NO_2 +$ nitroethane$(-\dot{H})$

$+ CCl_4 \xrightarrow{b} 4\text{-}ClC_6H_4NO_2 + \dot{C}Cl_3$

Therm. of 4-nitrophenylazotriphenylmethane				72 Pry 2
PR, glc	$C_2H_5NO_2/CCl_4$	333	$k_a/k_b = 0.43(2)$	

• 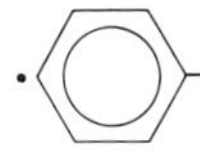 $-NO_2 + CD_3COCD_3 \xrightarrow{a} 4\text{-}DC_6H_4NO_2 + \dot{C}D_2COCD_3$

$+ CCl_4 \xrightarrow{b} 4\text{-}NO_2C_6H_4Cl + \dot{C}Cl_3$

Thermal decomp. of 4-nitrophenylazotriphenylmethane				66 Pry 1
PR, glc	CCl_4/CD_3COCD_3	333	$k_a/k_b = 0.123$	

• 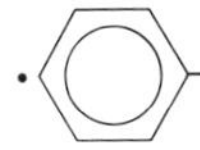 $-NO_2 + c\text{-}C_3H_5Br \xrightarrow{a} C_6H_5NO_2 +$ bromocyclopropane$(-\dot{H})$

$+ CCl_4 \xrightarrow{b} 4\text{-}ClC_6H_4NO_2 + \dot{C}Cl_3$

Therm. of 4-nitrophenylazotriphenylmethane				72 Pry 2
PR, glc	bromocyclopropane/ CCl_4	333	$k_a/k_b = 0.32$	

•  $-NO_2 + CH_3COCH_3 \xrightarrow{a} C_6H_5NO_2 + \dot{C}H_2COCH_3$

$+ CCl_4 \xrightarrow{b} 4\text{-}NO_2C_6H_4Cl + \dot{C}Cl_3$

Thermal decomp. of 4-nitrophenylazotriphenylmethane				66 Pry 1
PR, glc	CCl_4/CH_3COCH_3	333	$k_a/k_b = 0.61$	

Reaction Radical generation Method	Solvent	T[K]	Rate data	Ref./ add. ref.

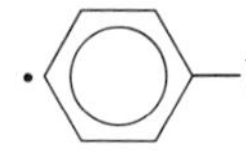 $-NO_2 + CH_3COOCH_3 \xrightarrow{a} C_6H_5NO_2$ + methyl acetate(− $\dot{H}$)

$+ CCl_4 \xrightarrow{b} 4\text{-}ClC_6H_4NO_2 + \dot{C}Cl_3$

Therm. of 4-nitrophenylazotriphenylmethane				72 Pry 2
PR, glc	CH_3COOCH_3/CCl_4	333	$k_a/k_b = 0.4$	

 $-NO_2 + CH_2BrCH_2CH_3 \xrightarrow{a} C_6H_5NO_2$ + 1-bromopropane(− $\dot{H}$)

$+ CCl_4 \xrightarrow{b} 4\text{-}ClC_6H_4NO_2 + \dot{C}Cl_3$

Therm. of 4-nitrophenylazotriphenylmethane				72 Pry 2
PR, glc	1-bromopropane/CCl_4	333	$k_a/k_b = 1.93$	

 $-NO_2 + CH_3CHBrCH_3 \xrightarrow{a} C_6H_5NO_2$ + 2-bromopropane(− $\dot{H}$)

$+ CCl_4 \xrightarrow{b} 4\text{-}ClC_6H_4NO_2 + \dot{C}Cl_3$

Therm. of 4-nitrophenylazotriphenylmethane				72 Pry 2
PR, glc	2-bromopropane/CCl_4	333	$k_a/k_b = 1.67(5)$	

 $-NO_2 + CH_3CH(NO_2)CH_3 \xrightarrow{a} C_6H_5NO_2$ + 2-nitropropane(− $\dot{H}$)

$+ CCl_4 \xrightarrow{b} 4\text{-}ClC_6H_4NO_2 + \dot{C}Cl_3$

Therm. of 4-nitrophenylazotriphenylmethane				72 Pry 2
PR, glc	2-nitropropane/CCl_4	333	$k_a/k_b = 0.6$	

 $-NO_2$ +  $\xrightarrow{a} C_6H_5NO_2$ + THF(− $\dot{H}$)

$+ CCl_4 \xrightarrow{b} 4\text{-}ClC_6H_4NO_2 + \dot{C}Cl_3$

Therm. of 4-nitrophenylazotriphenylmethane				72 Pry 2
PR, glc	THF/CCl_4	333	$k_a/k_b = 19.2(45)$	

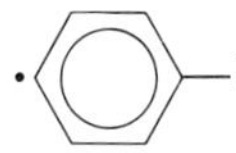 $-NO_2 + CH_3COOC_2H_5 \xrightarrow{a} C_6H_5NO_2$ + ethyl acetate(− $\dot{H}$)

$+ CCl_4 \xrightarrow{b} 4\text{-}ClC_6H_4NO_2 + \dot{C}Cl_3$

Therm. of 4-nitrophenylazotriphenylmethane				72 Pry 2
PR, glc	$CH_3COOC_2H_5/CCl_4$	333	$k_a/k_b = 0.8$	

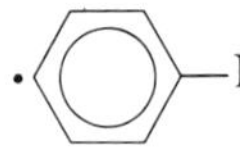 $-NO_2$ +  $\xrightarrow{a} C_6H_5NO_2$ + tetrahydrothiophene(− $\dot{H}$)

$+ CCl_4 \xrightarrow{b} 4\text{-}ClC_6H_4NO_2 + \dot{C}Cl_3$

Therm. of 4-nitrophenylazotriphenylmethane				72 Pry 2
PR, glc	tetrahydrothiophene/ CCl_4	333	$k_a/k_b = 32.1$	

Reaction Radical generation Method	Solvent	T[K]	Rate data	Ref./ add. ref.
 $-NO_2 + CH_2{=}CHCH_2SCH_3 \xrightarrow{a} 4\text{-}NO_2C_6H_4CH_2CH{=}CH_2 + CH_3\dot{S}$ $+ CCl_4 \xrightarrow{b} 4\text{-}NO_2C_6H_4Cl + \dot{C}Cl_3$				
Thermal decomp. of 4-nitrophenylazotriphenylmethane				79 Mig 1
PR, glc	CCl_4/ $CH_2{=}CHCH_2SCH_3$ mixt.	333	$k_a/k_b = 32$	
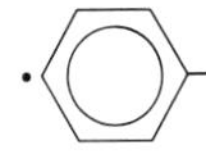 $-NO_2 + CH_2BrCH_2CH_2CH_3 \xrightarrow{a} C_6H_5NO_2 +$ 1-bromobutane$(-\dot{H})$ $+ CCl_4 \xrightarrow{b} 4\text{-}ClC_6H_4NO_2 + \dot{C}Cl_3$				
Therm. of 4-nitrophenylazotriphenylmethane				72 Pry 2
PR, glc	1-bromobutane/CCl_4	333	$k_a/k_b = 2.53(12)$	
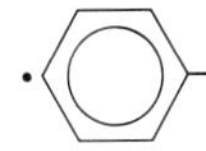 $-NO_2 + (CH_3)_4Si \xrightarrow{a} \dot{C}H_2(CH_3)_3Si + C_6H_5NO_2$ $+ CCl_4 \xrightarrow{b} 4\text{-}ClC_6H_4NO_2 + \dot{C}Cl_3$				
Therm. of 4-nitrophenylazotriphenylmethane				72 Pry 2
PR, glc	$(CH_3)_4Si/CCl_4$	333	$k_a/k_b = 0.8$	
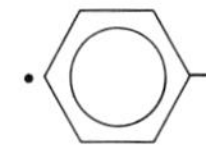 $-NO_2 +$ 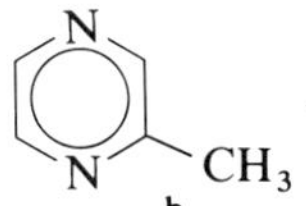 $\xrightarrow{a} C_6H_5NO_2 +$ 2-methylpyrazine$(-\dot{H})$ $+ CCl_4 \xrightarrow{b} 4\text{-}ClC_6H_4NO_2 + \dot{C}Cl_3$				
Therm. of 4-nitrophenylazotriphenylmethane				72 Pry 2
PR, glc	2-methylpyrazine/ CCl_4	333	$k_a/k_b = 0.32$	
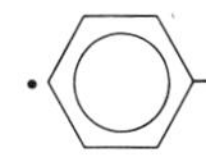 $-NO_2 +$ 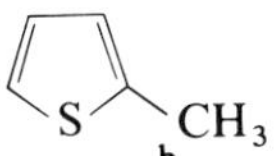 $\xrightarrow{a} C_6H_5NO_2 +$ 2-methylthiophene$(-\dot{H})$ $+ CCl_4 \xrightarrow{b} 4\text{-}ClC_6H_4NO_2 + \dot{C}Cl_3$				
Therm. of 4-nitrophenylazotriphenylmethane				72 Pry 2
PR, glc	2-methylthiophene/ CCl_4	333	$k_a/k_b = 1.5$	
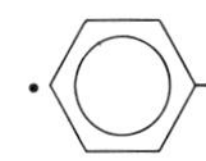 $-NO_2 +$ 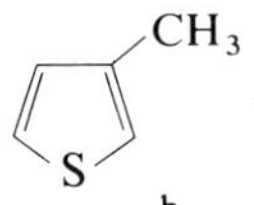 $\xrightarrow{a} C_6H_5NO_2 +$ 3-methylthiophene$(-\dot{H})$ $+ CCl_4 \xrightarrow{b} 4\text{-}ClC_6H_4NO_2 + \dot{C}Cl_3$				
Therm. of 4-nitrophenylazotriphenylmethane				72 Pry 2
PR, glc	3-methylthiophene/ CCl_4	333	$k_a/k_b = 0.4$	

Reaction Radical generation Method	Solvent	T[K]	Rate data	Ref./ add. ref.
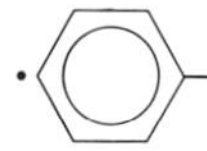 $-NO_2 + c\text{-}C_5H_8 \xrightarrow{a} C_6H_5NO_2 + c\text{-pentene}(-\dot{H})$ $+ CCl_4 \xrightarrow{b} 4\text{-}ClC_6H_4NO_2 + \dot{C}Cl_3$				
Therm. of 4-nitrophenylazotriphenylmethane PR, glc	cyclopentene/CCl_4	333	$k_a/k_b = 20.7$	72 Pry 2
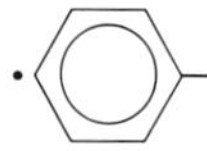 $-NO_2 + c\text{-}C_5H_8 \xrightarrow{a} C_6H_5NO_2 + c\text{-pentene}(-\alpha\text{-}\dot{H})$ $+ C_6H_5CH_3 \xrightarrow{b} C_6H_5NO_2 + C_6H_5\dot{C}H_2$				
Therm. of 4-nitrophenylazotriphenylmethane PR, glc	cyclopentene/toluene	333	$k_a/k_b = 10.4$	72 Pry 2
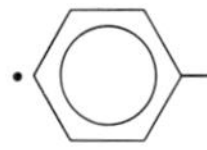 $-NO_2 + c\text{-}C_5H_{10} \xrightarrow{a} C_6H_5NO_2 + (c\text{-}C_5H_9^{\cdot})$ $+ CCl_4 \xrightarrow{b} 4\text{-}ClC_6H_4NO_2 + \dot{C}Cl_3$				
Therm. of 4-nitrophenylazotriphenylmethane PR, glc	cyclopentane/CCl_4	333	$k_a/k_b = 8.85(5)$	72 Pry 2
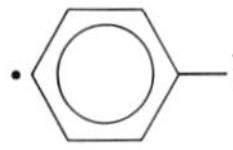 $-NO_2 + c\text{-}C_5H_{10} \xrightarrow{a} C_6H_5NO_2 + (c\text{-}C_5H_9^{\cdot})$ $+ C_6H_5CH_3 \xrightarrow{b} C_6H_5NO_2 + C_6H_5\dot{C}H_2$				
Therm. of 4-nitrophenylazotriphenylmethane PR, glc	$c\text{-}C_5H_{10}$/toluene	333	$k_a/k_b = 2.1$	72 Pry 2
 $-NO_2 + (CH_3)_2C{=}CHCH_3 \xrightarrow{a} C_6H_5NO_2 + \text{2-methylbutene}(-\dot{H})$ $+ CCl_4 \xrightarrow{b} 4\text{-}ClC_6H_4NO_2 + \dot{C}Cl_3$				
Therm. of 4-nitrophenylazotriphenylmethane PR, glc	2-methyl-2-butene/ CCl_4	333	$k_a/k_b = 9.1$	72 Pry 2
 $-NO_2 + CH_3CH_2COCH_2CH_3 \xrightarrow{a} C_6H_5NO_2 + \text{3-pentanone}(-\dot{H})$ $+ CCl_4 \xrightarrow{b} 4\text{-}ClC_6H_4NO_2 + \dot{C}Cl_3$				
Therm. of 4-nitrophenylazotriphenylmethane PR, glc	3-pentanone/CCl_4	333	$k_a/k_b = 5.2$	72 Pry 2
 $-NO_2 + (CH_3)_2CHCH_2CH_3 \xrightarrow{a} C_6H_5NO_2 + (C_5H_{11}^{\cdot})$ $+ CCl_4 \xrightarrow{b} 4\text{-}NO_2C_6H_4Cl + \dot{C}Cl_3$				
Thermal decomp. of 4-nitrophenylazotriphenylmethane PR, glc	CCl_4/ $(CH_3)_2CHCH_2CH_3$	333	$k_a/k_b = 2.2$	66 Pry 1

Reaction Radical generation Method	Solvent	T[K]	Rate data	Ref./ add. ref.
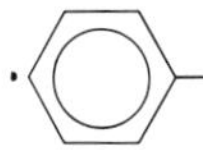 $-NO_2 + (CH_3)_3COCH_3 \xrightarrow{a} C_6H_5NO_2 + t\text{-butylmethyl ether}(-\dot{H})$ $+ CCl_4 \xrightarrow{b} 4\text{-}ClC_6H_4NO_2 + \dot{C}Cl_3$				
Therm. of 4-nitrophenylazotriphenylmethane				72 Pry 2
PR, glc	$(CH_3)_3COCH_3/CCl_4$	333	$k_a/k_b = 2.5$	
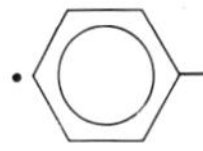 $-NO_2 + 4\text{-}BrC_6H_4I \xrightarrow{a} 4\text{-}NO_2C_6H_4I + (4\text{-}BrC_6\dot{H}_4)$ $+ CCl_4 \xrightarrow{b} 4\text{-}NO_2C_6H_4Cl + \dot{C}Cl_3$				
Therm. of 4-nitrophenylazotriphenylmethane				77 Pry 1
PR, glc	$CCl_4/4\text{-}BrC_6H_4I$	333	$k_a/k_b = 47(2)$	
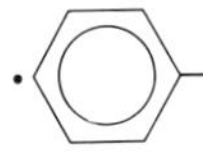 $-NO_2 + 3\text{-}FC_6H_4I \xrightarrow{a} 4\text{-}NO_2C_6H_4I + (3\text{-}FC_6\dot{H}_4)$ $+ CCl_4 \xrightarrow{b} 4\text{-}NO_2C_6H_4Cl + \dot{C}Cl_3$				
Therm. of 4-nitrophenylazotriphenylmethane				77 Pry 1
PR, glc	$CCl_4/3\text{-}FC_6H_4I$	333	$k_a/k_b = 64(1)$	
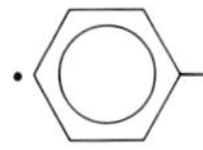 $-NO_2 + 4\text{-}FC_6H_4I \xrightarrow{a} 4\text{-}NO_2C_6H_4I + (4\text{-}FC_6\dot{H}_4)$ $+ CCl_4 \xrightarrow{b} 4\text{-}NO_2C_6H_4Cl + \dot{C}Cl_3$				
Therm. of 4-nitrophenylazotriphenylmethane				77 Pry 1
PR, glc	$CCl_4/4\text{-}FC_6H_4I$	333	$k_a/k_b = 48(2)$	
 $-NO_2 + C_6H_5I \xrightarrow{a} 4\text{-}NO_2C_6H_4I + (C_6\dot{H}_5)$ $+ CCl_4 \xrightarrow{b} 4\text{-}NO_2C_6H_4Cl + \dot{C}Cl_3$				
Therm. of 4-nitrophenylazotriphenylmethane				77 Pry 1
PR, glc	CCl_4/C_6H_5I	333	$k_a/k_b = 57.3(9)$	
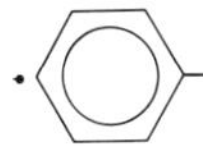 $-NO_2 + C_6H_5SH \xrightarrow{a} C_6H_5NO_2 + C_6H_5\dot{S}$ $+ CCl_4 \xrightarrow{b} 4\text{-}NO_2C_6H_4Cl + \dot{C}Cl_3$				
Thermal decomp. of 4-nitrophenylazotriphenylmethane				66 Pry 1
PR, glc	CCl_4	333	$k_a/k_b = 869$	
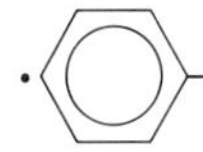 $-NO_2 + c\text{-}C_6H_{10} \xrightarrow{a} C_6H_5NO_2 + c\text{-hexene}(-\dot{H})$ $+ CCl_4 \xrightarrow{b} 4\text{-}ClC_6H_4NO_2 + \dot{C}Cl_3$				
Therm. of 4-nitrophenylazotriphenylmethane				72 Pry 2/
PR, glc	cyclohexene/CCl_4	333	$k_a/k_b = 30.3(6)$	66 Pry 1

Reaction Radical generation Method	Solvent	T[K]	Rate data	Ref./ add. ref.
•$-NO_2 + c\text{-}C_6H_{10} \xrightarrow{a} C_6H_5NO_2 + c\text{-hexene}(-\alpha\text{-}\dot{H})$ $+ C_6H_5CH_3 \xrightarrow{b} C_6H_5NO_2 + C_6H_5\dot{C}H_2$				
Therm. of 4-nitrophenylazotriphenylmethane PR, glc	cyclohexene/toluene	333	$k_a/k_b = 15.4$	72 Pry 2
•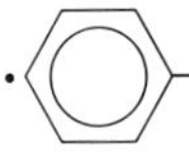$-NO_2 + c\text{-}C_6H_{12} \xrightarrow{a} C_6H_5NO_2 + (c\text{-}C_6H_{11}^{\cdot})$ $+ CCl_4 \xrightarrow{b} 4\text{-}ClC_6H_4NO_2 + \dot{C}Cl_3$				
Therm. of 4-nitrophenylazotriphenylmethane PR, glc	cyclohexane/CCl_4	333	$k_a/k_b = 9.14(74)$	72 Pry 2/ 66 Pry 1
Decomp. of 4-nitrophenylazotriphenylmethane PR, glc	$c\text{-}C_6H_{12}/CCl_4$	333	$k_a/k_b = 9.2$	76 Lew 2/ 63 Bri 1
•$-NO_2 + c\text{-}C_6H_{12} \xrightarrow{a} C_6H_5NO_2 + (c\text{-}C_6H_{11}^{\cdot})$ $+ C_6H_5CH_3 \xrightarrow{b} C_6H_5NO_2 + C_6H_5\dot{C}H_2$				
Therm. of 4-nitrophenylazotriphenylmethane PR, glc	$c\text{-}C_6H_{12}$/toluene	333	$k_a/k_b = 1.7$	72 Pry 2
•$-NO_2 + CH_2{=}CH(CH_2)_3CH_3 \xrightarrow{a} C_6H_5NO_2 + 1\text{-hexene}(-\dot{H})$ $+ CCl_4 \xrightarrow{b} 4\text{-}NO_2C_6H_4Cl + \dot{C}Cl_3$				
Therm. of 4-nitrophenylazotriphenylmethane PR, glc	1-hexene/CCl_4	333	$k_a/k_b = 4.2(2)$	72 Pry 2
•$-NO_2 + CH_2{=}CHCH_2CH(CH_3)_2 \xrightarrow{a} C_6H_5NO_2 + 4\text{-methyl-1-pentene}(-\dot{H})$ $+ CCl_4 \xrightarrow{b} 4\text{-}ClC_6H_4NO_2 + \dot{C}Cl_3$				
Therm. of 4-nitrophenylazotriphenylmethane PR, glc	4-methyl-1-pentene/ CCl_4	333	$k_a/k_b = 6.4$	72 Pry 2
•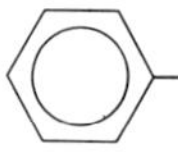$-NO_2 + CH_3CH{=}CH(CH_2)_2CH_3 \xrightarrow{a} C_6H_5NO_2 + 2\text{-hexene}(-\dot{H})$ $+ CCl_4 \xrightarrow{b} 4\text{-}ClC_6H_4NO_2 + \dot{C}Cl_3$				
Therm. of 4-nitrophenylazotriphenylmethane PR, glc	2-hexene/CCl_4	333	$k_a/k_b = 14.5$	72 Pry 2
•$-NO_2 + CH_3CH{=}CHCH(CH_3)_2 \xrightarrow{a} C_6H_5NO_2 + 4\text{-methyl-2-pentene}(-\dot{H})$ $+ CCl_4 \xrightarrow{b} 4\text{-}ClC_6H_4NO_2 + \dot{C}Cl_3$				
Therm. of 4-nitrophenylazotriphenylmethane PR, glc	4-methyl-2-pentene/ CCl_4	333	$k_a/k_b = 5.75(15)$	72 Pry 2

Reaction Radical generation Method	Solvent	T[K]	Rate data	Ref./ add. ref.

$-NO_2 + CH_3CH_2CH{=}CHCH_2CH_3 \xrightarrow{a} C_6H_5NO_2$ + 3-hexene($-\dot{H}$)

$+ CCl_4 \xrightarrow{b}$ 4-$ClC_6H_4NO_2 + \dot{C}Cl_3$

Therm. of 4-nitrophenylazotriphenylmethane				72 Pry 2
PR, glc	3-hexene/CCl_4	333	$k_a/k_b = 18.0(7)$	

$-NO_2 + (CH_3)_2C{=}CHCH_2CH_3 \xrightarrow{a} C_6H_5NO_2$ + 2-methyl-2-pentene($-\dot{H}$)

$+ CCl_4 \xrightarrow{b}$ 4-$ClC_6H_4NO_2 + \dot{C}Cl_3$

Therm. of 4-nitrophenylazotriphenylmethane				72 Pry 2
PR, glc	2-methyl-2-pentene/ CCl_4	333	$k_a/k_b = 14.2$	

$-NO_2 + (CH_3)_2C{=}C(CH_3)_2 \xrightarrow{a} C_6H_5NO_2$ + 2,3-dimethylbutene($-\dot{H}$)

$+ CCl_4 \xrightarrow{b}$ 4-$ClC_6H_4NO_2 + \dot{C}Cl_3$

Therm. of 4-nitrophenylazotriphenylmethane				72 Pry 2
PR, glc	2,3-dimethylbutene/ CCl_4	333	$k_a/k_b = 22.3$	

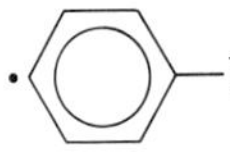$-NO_2 + CH_2Br(CH_2)_4CH_3 \xrightarrow{a} C_6H_5NO_2$ + 1-bromohexane($-\dot{H}$)

$+ CCl_4 \xrightarrow{b}$ 4-$NO_2C_6H_4Cl + \dot{C}Cl_3$

Thermal decomp. of 4-nitrophenylazotriphenylmethane				66 Pry 1
PR, glc	CCl_4/1-bromohexane	333	$k_a/k_b = 5.4$	

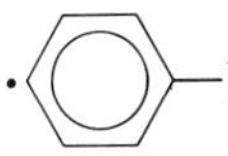$-NO_2 + n\text{-}C_6H_{14} \xrightarrow{a} C_6H_5NO_2$ + n-hexane($-\dot{H}$)

$+ CCl_4 \xrightarrow{b}$ 4-$ClC_6H_4NO_2 + \dot{C}Cl_3$

Therm. of 4-nitrophenylazotriphenylmethane				72 Pry 2/
PR, glc	n-C_6H_{14}/CCl_4	333	$k_a/k_b = 7.6(5)$	66 Pry 1

$-NO_2 + (CH_3)_2CHCH_2CH_2CH_3 \xrightarrow{a} C_6H_5NO_2$ + 2-methylpentane($-\dot{H}$)

$+ CCl_4 \xrightarrow{b}$ 4-$ClC_6H_4NO_2 + \dot{C}Cl_3$

Therm. of 4-nitrophenylazotriphenylmethane				72 Pry 2
PR, glc	2-methylpentane/ CCl_4	333	$k_a/k_b = 7.7$	

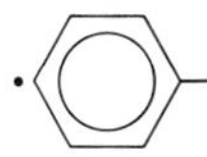$-NO_2 + CH_3CH_2CH(CH_3)CH_2CH_3 \xrightarrow{a} C_6H_5NO_2$ + 3-methylpentane($-\dot{H}$)

$+ CCl_4 \xrightarrow{b}$ 4-$ClC_6H_4NO_2 + \dot{C}Cl_3$

Therm. of 4-nitrophenylazotriphenylmethane				72 Pry 2
PR, glc	3-methylpentane/ CCl_4	333	$k_a/k_b = 8.0$	

Reaction Radical generation Method	Solvent	T[K]	Rate data	Ref./ add. ref.
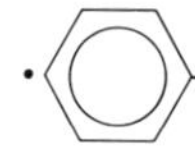$-NO_2 + (CH_3)_2CHCH(CH_3)_2 \xrightarrow{a} C_6H_5NO_2 +$ 2,3-dimethylbutane($-\dot{H}$) $+ CCl_4 \xrightarrow{b}$ 4-$ClC_6H_4NO_2 + \dot{C}Cl_3$				
Therm. of 4-nitrophenylazotriphenylmethane				72 Pry 2
PR, glc	2,3-dimethylbutane/ CCl_4	333	$k_a/k_b = 8.9(13)$	
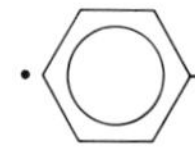$-NO_2 +$ 3-$CF_3C_6H_4I \xrightarrow{a}$ 4-$NO_2C_6H_4I$ + (3-$CF_3C_6\dot{H}_4$) $+ CCl_4 \xrightarrow{b}$ 4-$NO_2C_6H_4Cl + \dot{C}Cl_3$				
Therm. of 4-nitrophenylazotriphenylmethane				77 Pry 1
PR, glc	CCl_4/3-$CF_3C_6H_4I$	333	$k_a/k_b = 61(2)$	
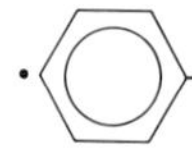$-NO_2 +$ 4-$CF_3C_6H_4I \xrightarrow{a}$ 4-$NO_2C_6H_4I$ + (4-$CF_3C_6\dot{H}_4$) $+ CCl_4 \xrightarrow{b}$ 4-$NO_2C_6H_4Cl + \dot{C}Cl_3$				
Therm. of 4-nitrophenylazotriphenylmethane				77 Pry 1
PR, glc	CCl_4/4-$CF_3C_6H_4I$	333	$k_a/k_b = 48(1)$	
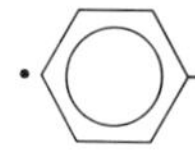$-NO_2 +$ 3-$ClC_6H_4CH_2Br \xrightarrow{a}$ 4-$NO_2C_6H_4Br$ + 3-$ClC_6H_4\dot{C}H_2$ $+ CCl_4 \xrightarrow{b}$ 4-$NO_2C_6H_4Cl + \dot{C}Cl_3$				
Therm. of 4-nitrophenylazotriphenylmethane				75 Mig 1
PR, glc	3-$ClC_6H_4CH_2Br$/ CCl_4	333	$k_a/k_b = 26.1(7)$	
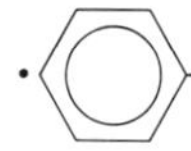$-NO_2 +$ 4-$ClC_6H_4CH_2Br \xrightarrow{a}$ 4-$NO_2C_6H_4Br$ + 4-$ClC_6H_4\dot{C}H_2$ $+ CCl_4 \xrightarrow{b}$ 4-$NO_2C_6H_4Cl + \dot{C}Cl_3$				
Therm. of 4-nitrophenylazotriphenylmethane				75 Mig 1
PR, glc	4-$ClC_6H_4CH_2Br$/ CCl_4	333	$k_a/k_b = 27.5(5)$	
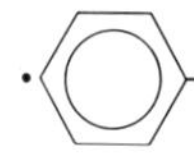$-NO_2 +$ 3-$NO_2C_6H_4CH_2Br \xrightarrow{a}$ 4-$NO_2C_6H_4Br$ + 3-$NO_2C_6H_4\dot{C}H_2$ $+ CCl_4 \xrightarrow{b}$ 4-$NO_2C_6H_4Cl + \dot{C}Cl_3$				
Therm. of 4-nitrophenylazotriphenylmethane				75 Mig 1
PR, glc	3-$NO_2C_6H_4CH_2Br$/ CCl_4	333	$k_a/k_b = 22.4(9)$	
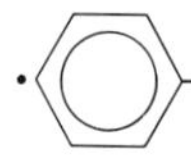$-NO_2 +$ 4-$NO_2C_6H_4CH_2Br \xrightarrow{a}$ 4-$NO_2C_6H_4Br$ + 4-$NO_2C_6H_4\dot{C}H_2$ $+ CCl_4 \xrightarrow{b}$ 4-$NO_2C_6H_4Cl + \dot{C}Cl_3$				
Therm. of 4-nitrophenylazotriphenylmethane				75 Mig 1
PR, glc	4-$NO_2C_6H_4CH_2Br$/ CCl_4	333	$k_a/k_b = 15.7(7)$	

Reaction Radical generation Method	Solvent	T[K]	Rate data	Ref./ add. ref.

 $-NO_2 + C_6H_5CH_2Br \xrightarrow{a} 4\text{-}NO_2C_6H_4Br + C_6H_5\dot{C}H_2$

$+ CCl_4 \xrightarrow{b} 4\text{-}NO_2C_6H_4Cl + \dot{C}Cl_3$

Therm. of 4-nitrophenylazotriphenylmethane				75 Mig 1/
PR, glc	$C_6H_5CH_2Br/CCl_4$	333	$k_a/k_b = 25.9(18)$	79 Mig 1

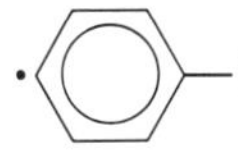 $-NO_2 + 3\text{-}BrC_6H_4CH_3 \xrightarrow{a} C_6H_5NO_2 + 3\text{-}BrC_6H_4\dot{C}H_2$

$+ CCl_4 \xrightarrow{b} 4\text{-}NO_2C_6H_4Cl + \dot{C}Cl_3$

Thermal decomp. of 4-nitrophenylazotriphenylmethane				66 Pry 1
PR, glc	CCl_4/3-bromotoluene	333	$k_a/k_b = 1.06$	

 $-NO_2 + 4\text{-}BrC_6H_4CH_3 \xrightarrow{a} C_6H_5NO_2 + 4\text{-}BrC_6H_4\dot{C}H_2$

$+ CCl_4 \xrightarrow{b} 4\text{-}NO_2C_6H_4Cl + \dot{C}Cl_3$

Thermal decomp. of 4-nitrophenylazotriphenylmethane				66 Pry 1
PR, glc	CCl_4/4-bromotoluene	333	$k_a/k_b = 1.33$	

 $-NO_2 + 3\text{-}ClC_6H_4CH_3 \xrightarrow{a} C_6H_5NO_2 + 3\text{-}ClC_6H_4\dot{C}H_2$

$+ CCl_4 \xrightarrow{b} 4\text{-}NO_2C_6H_4Cl + \dot{C}Cl_3$

Thermal decomp. of 4-nitrophenylazotriphenylmethane				66 Pry 1
PR, glc	CCl_4/3-chlorotoluene	333	$k_a/k_b = 0.972$	

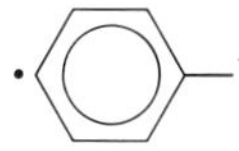 $-NO_2 + 4\text{-}ClC_6H_4CH_3 \xrightarrow{a} C_6H_5NO_2 + 4\text{-}ClC_6H_4\dot{C}H_2$

$+ CCl_4 \xrightarrow{b} 4\text{-}NO_2C_6H_4Cl + \dot{C}Cl_3$

Thermal decomp. of 4-nitrophenylazotriphenylmethane				66 Pry 1
PR, glc	CCl_4/4-chlorotoluene	333	$k_a/k_b = 1.28$	

 $-NO_2 + 4\text{-}FC_6H_4CH_3 \xrightarrow{a} C_6H_5NO_2 + 4\text{-}FC_6H_4\dot{C}H_2$

$+ CCl_4 \xrightarrow{b} 4\text{-}NO_2C_6H_4Cl + \dot{C}Cl_3$

Thermal decomp. of 4-nitrophenylazotriphenylmethane				66 Pry 1
PR, glc	CCl_4/4-fluorotoluene	333	$k_a/k_b = 1.24$	

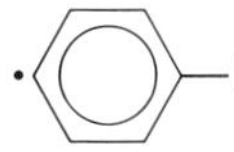 $-NO_2 + 3\text{-}CH_3C_6H_4I \xrightarrow{a} 4\text{-}NO_2C_6H_4I + (3\text{-}CH_3C_6H_4^{\cdot})$

$+ CCl_4 \xrightarrow{b} 4\text{-}NO_2C_6H_4Cl + \dot{C}Cl_3$

Therm. of 4-nitrophenylazotriphenylmethane				77 Pry 1
PR, glc	$CCl_4/3\text{-}CH_3C_6H_4I$	333	$k_a/k_b = 60(10)$	

Reaction Radical generation Method	Solvent	T[K]	Rate data	Ref./ add. ref.
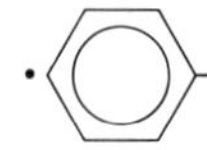 $-NO_2 + 4\text{-}CH_3C_6H_4I \xrightarrow{a} 4\text{-}NO_2C_6H_4I + (4\text{-}CH_3C_6\dot{H}_4)$ $+ CCl_4 \xrightarrow{b} 4\text{-}NO_2C_6H_4Cl + \dot{C}Cl_3$				
Therm. of 4-nitrophenylazotriphenylmethane				77 Pry 1
PR, glc	$CCl_4/4\text{-}CH_3C_6H_4I$	333	$k_a/k_b = 54.0(1)$	
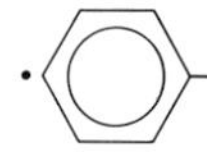 $-NO_2 + 4\text{-}CH_3OC_6H_4I \xrightarrow{a} 4\text{-}NO_2C_6H_4I + (4\text{-}CH_3OC_6\dot{H}_4)$ $+ CCl_4 \xrightarrow{b} 4\text{-}NO_2C_6H_4Cl + \dot{C}Cl_3$				
Therm. of 4-nitrophenylazotriphenylmethane				77 Pry 1
PR, glc	$CCl_4/4\text{-}CH_3OC_6H_4I$	333	$k_a/k_b = 36(2)$	
 $-NO_2 +$ 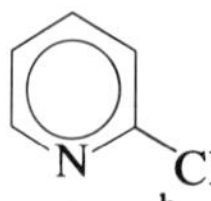 $\xrightarrow{a} C_6H_5NO_2 +$ 2-methylpyridine($-\dot{H}$) $+ CCl_4 \xrightarrow{b} 4\text{-}ClC_6H_4NO_2 + \dot{C}Cl_3$				
Therm. of 4-nitrophenylazotriphenylmethane				72 Pry 2
PR, glc	2-methylpyridine/ CCl_4	333	$k_a/k_b = 0.93(18)$	
 $-NO_2 + 4\text{-}NO_2C_6H_4CH_3 \xrightarrow{a} C_6H_5NO_2 + 4\text{-}NO_2C_6H_4\dot{C}H_2$ $+ CCl_4 \xrightarrow{b} 4\text{-}NO_2C_6H_4Cl + \dot{C}Cl_3$				
Thermal decomp. of 4-nitrophenylazotriphenylmethane				66 Pry 1
PR, glc	CCl_4/4-nitrotoluene	333	$k_a/k_b = 0.68$	
 $-NO_2 + C_6H_5CH_3 \xrightarrow{a} C_6H_5NO_2 + C_6H_5\dot{C}H_2$ $+ CCl_4 \xrightarrow{b} 4\text{-}NO_2C_6H_4Cl + \dot{C}Cl_3$				
Thermal decomp. of 4-nitrophenylazotriphenylmethane				66 Pry 1
PR, glc	CCl_4/toluene	333	$k_a/k_b = 1.67$	
 $-NO_2 + C_6H_5OCH_3 \xrightarrow{a} C_6H_5NO_2 +$ anisole($-\dot{H}$) $+ CCl_4 \xrightarrow{b} 4\text{-}NO_2C_6H_4Cl + \dot{C}Cl_3$				
Thermal decomp. of 4-nitrophenylazotriphenylmethane				66 Pry 1
PR, glc	CCl_4/anisole	333	$k_a/k_b = 0.14$	
 $-NO_2 + CH_3SC_6H_5 \xrightarrow{a} C_6H_5NO_2 +$ methyl phenyl sulfide($-\dot{H}$) $+ CCl_4 \xrightarrow{b} 4\text{-}ClC_6H_4NO_2 + \dot{C}Cl_3$				
Therm. of 4-nitrophenylazotriphenylmethane				72 Pry 2
PR, glc	$CH_3SC_6H_5/CCl_4$	333	$k_a/k_b = 0.97$	

Reaction Radical generation Method	Solvent	T[K]	Rate data	Ref./ add. ref.
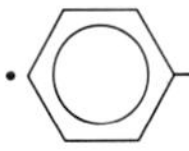 $-NO_2 + (c\text{-}C_3H_5)_2CO \xrightarrow{a} C_6H_5NO_2 +$ dicyclopropylketone$(-\dot{H})$ $+ CCl_4 \xrightarrow{b} 4\text{-}ClC_6H_4NO_2 + \dot{C}Cl_3$				
Therm. of 4-nitrophenylazotriphenylmethane PR, glc	$(c\text{-}C_3H_5)_2CO/CCl_4$	333	$k_a/k_b = 1.1$	72 Pry 2
 $-NO_2 + c\text{-}C_7H_{14} \xrightarrow{a} C_6H_5NO_2 + (c\text{-}C_7\dot{H}_{13})$ $+ CCl_4 \xrightarrow{b} 4\text{-}ClC_6H_4NO_2 + \dot{C}Cl_3$				
Therm. of 4-nitrophenylazotriphenylmethane PR, glc	cycloheptane/CCl_4	333	$k_a/k_b = 18.35(74)$	72 Pry 2
 $-NO_2 + c\text{-}C_7H_{14} \xrightarrow{a} C_6H_5NO_2 + (c\text{-}C_7\dot{H}_{13})$ $+ C_6H_5CH_3 \xrightarrow{b} C_6H_5NO_2 + C_6H_5\dot{C}H_2$				
Therm. of 4-nitrophenylazotriphenylmethane PR, glc	$c\text{-}C_7H_{14}$/toluene	333	$k_a/k_b = 2.9$	72 Pry 2
 $-NO_2 + CH_3C_6H_{11} \xrightarrow{a} C_6H_5NO_2 +$ methylcyclohexane$(-\dot{H})$ $+ CCl_4 \xrightarrow{b} 4\text{-}ClC_6H_4NO_2 + \dot{C}Cl_3$				
Therm. of 4-nitrophenylazotriphenylmethane PR, glc	methylcyclohexane/ CCl_4	333	$k_a/k_b = 13.9(9)$	72 Pry 2
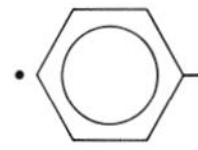 $-NO_2 + CH_2{=}CH(CH_2)_4CH_3 \xrightarrow{a} C_6H_5NO_2 +$ 1-heptene$(-\dot{H})$ $+ CCl_4 \xrightarrow{b} 4\text{-}ClC_6H_4NO_2 + \dot{C}Cl_3$				
Therm. of 4-nitrophenylazotriphenylmethane PR, glc	1-heptene/CCl_4	333	$k_a/k_b = 10.35(25)$	72 Pry 2
 $-NO_2 + CH_3CH{=}CH(CH_2)_3CH_3 \xrightarrow{a} C_6H_5NO_2 +$ 2-heptene$(-\dot{H})$ $+ CCl_4 \xrightarrow{b} 4\text{-}ClC_6H_4NO_2 + \dot{C}Cl_3$				
Therm. of-4-nitrophenylazotriphenylmethane PR, glc	2-heptene/CCl_4	333	$k_a/k_b = 15.15(55)$	72 Pry 2
 $-NO_2 + n\text{-}C_7H_{16} \xrightarrow{a} C_6H_5NO_2 + (C_7\dot{H}_{15})$ $+ CCl_4 \xrightarrow{b} 4\text{-}NO_2C_6H_4Cl + \dot{C}Cl_3$				
Thermal decomp. of 4-nitrophenylazotriphenylmethane PR, glc	$CCl_4/n\text{-}C_7H_{16}$	333	$k_a/k_b = 11.1$ 8.4(4) [5])	66 Pry 1, 72 Pry 2

[5]) From [72 Pry 2].

Reaction Radical generation Method	Solvent	T[K]	Rate data	Ref./ add. ref.

 $-NO_2 + (CH_3)_2CH(CH_2)_3CH_3 \xrightarrow{a} C_6H_5NO_2 + \text{2-methylhexane}(-\dot{H})$

$+ CCl_4 \xrightarrow{b} \text{4-ClC}_6H_4NO_2 + \dot{C}Cl_3$

Therm. of 4-nitrophenylazotriphenylmethane				72 Pry 2
PR, glc	2-methylhexane/ CCl_4	333	$k_a/k_b = 8.6(3)$	

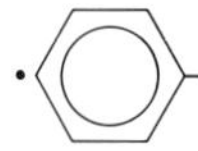 $-NO_2 + CH_3CH_2CH(CH_3)CH_2CH_2CH_3 \xrightarrow{a} C_6H_5NO_2 + \text{3-methylhexane}(-\dot{H})$

$+ CCl_4 \xrightarrow{b} \text{4-ClC}_6H_4NO_2 + \dot{C}Cl_3$

Therm. of 4-nitrophenylazotriphenylmethane				72 Pry 2
PR, glc	3-methylhexane/ CCl_4	333	$k_a/k_b = 8.8$	

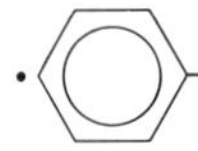 $-NO_2 + (CH_3)_2CHCH_2CH(CH_3)_2 \xrightarrow{a} C_6H_5NO_2 + \text{2,4-dimethylpentane}(-\dot{H})$

$+ CCl_4 \xrightarrow{b} \text{4-ClC}_6H_4NO_2 + \dot{C}Cl_3$

Therm. of 4-nitrophenylazotriphenylmethane				72 Pry 2/
PR, glc	2,4-dimethylpentane/ CCl_4	333	$k_a/k_b = 5.2(2)$	66 Pry 1

 $-NO_2 + (CH_3)_3CCH(CH_3)_2 \xrightarrow{a} C_6H_5NO_2 + \text{2,2,3-trimethylbutane}(-\dot{H})$

$+ CCl_4 \xrightarrow{b} \text{4-ClC}_6H_4NO_2 + \dot{C}Cl_3$

Therm. of 4-nitrophenylazotriphenylmethane				72 Pry 2
PR, glc	2,2,3-tri-methylbutane/CCl_4	333	$k_a/k_b = 6.2$	

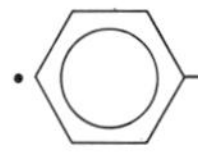 $-NO_2 + \text{3-}CH_3C_6H_4CH_2Br \xrightarrow{a} \text{4-}NO_2C_6H_4Br + \text{3-}CH_3C_6H_4\dot{C}H_2$

$+ CCl_4 \xrightarrow{b} \text{4-}NO_2C_6H_4Cl + \dot{C}Cl_3$

Therm. of 4-nitrophenylazotriphenylmethane				75 Mig 1
PR, glc	3-$CH_3C_6H_4CH_2Br$/ CCl_4	333	$k_a/k_b = 24.3(3)$	

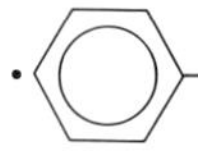 $-NO_2 + \text{4-}CH_3C_6H_4CH_2Br \xrightarrow{a} \text{4-}NO_2C_6H_4Br + \text{4-}CH_3C_6H_4\dot{C}H_2$

$+ CCl_4 \xrightarrow{b} \text{4-}NO_2C_6H_4Cl + \dot{C}Cl_3$

Therm. of 4-nitrophenylazotriphenylmethane				75 Mig 1
PR, glc	4-$CH_3C_6H_4CH_2Br$/ CCl_4	333	$k_a/k_b = 26.5(9)$	

Reaction Radical generation Method	Solvent	T[K]	Rate data	Ref./ add. ref.
 $-NO_2 + 4\text{-}CH_3OC_6H_4CH_2Br \xrightarrow{a} 4\text{-}NO_2C_6H_4Br + 4\text{-}CH_3OC_6H_4\dot{C}H_2$ $+ CCl_4 \xrightarrow{b} 4\text{-}NO_2C_6H_4Cl + \dot{C}Cl_3$				
Therm. of 4-nitrophenylazotriphenylmethane				75 Mig 1
PR, glc	$4\text{-}CH_3OC_6H_4CH_2Br/$ CCl_4	333	$k_a/k_b = 40.2(20)$	
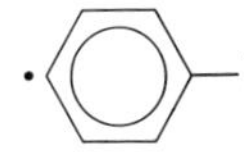 $-NO_2 + C_6H_5CH_2CH_3 \xrightarrow{a} C_6H_5NO_2 +$ ethylbenzene$(-\dot{H})$ $+ CCl_4 \xrightarrow{b} 4\text{-}NO_2C_6H_4Cl + \dot{C}Cl_3$				
Thermal decomp. of 4-nitrophenylazotriphenylmethane				66 Pry 1
PR, glc	$CCl_4/C_6H_5CH_2CH_3$	333	$k_a/k_b = 7.0$	
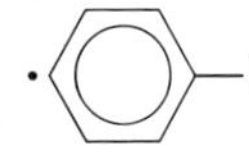 $-NO_2 + C_6H_5CH_2CH_3 \xrightarrow{a} C_6H_5NO_2 +$ ethylbenzene$(-\dot{H})$ $+ C_6H_5CH_3 \xrightarrow{b} C_6H_5NO_2 + C_6H_5\dot{C}H_2$				
Therm. of 4-nitrophenylazotriphenylmethane				72 Pry 2
PR, glc	ethylbenzene/toluene	333	$k_a/k_b = 4.7$	
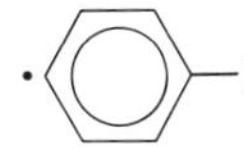 $-NO_2 + 3\text{-}CH_3C_6H_4CH_3 \xrightarrow{a} C_6H_5NO_2 + 3\text{-}CH_3C_6H_4\dot{C}H_2$ $+ CCl_4 \xrightarrow{b} 4\text{-}NO_2C_6H_4Cl + \dot{C}Cl_3$				
Thermal decomp. of 4-nitrophenylazotriphenylmethane				66 Pry 1
PR, glc	CCl_4/3-methyl- toluene [6])	333	$k_a/k_b = 3.66$	
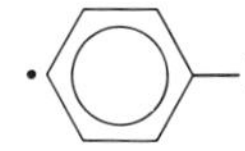 $-NO_2 + 4\text{-}CH_3C_6H_4CH_3 \xrightarrow{a} C_6H_5NO_2 + 4\text{-}CH_3C_6H_4\dot{C}H_2$ $+ CCl_4 \xrightarrow{b} 4\text{-}NO_2C_6H_4Cl + \dot{C}Cl_3$				
Thermal decomp. of 4-nitrophenylazotriphenylmethane				66 Pry 1
PR, glc	CCl_4/4-methyl- toluene [7])	333	$k_a/k_b = 4.86$	
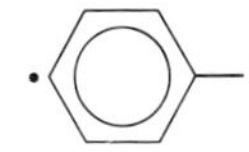 $-NO_2 + 4\text{-}CH_3OC_6H_4CH_3 \xrightarrow{a} C_6H_5NO_2 +$ 4-methoxytoluene$(-\dot{H})$ $+ CCl_4 \xrightarrow{b} 4\text{-}NO_2C_6H_4Cl + \dot{C}Cl_3$				
Thermal decomp. of 4-nitrophenylazotriphenylmethane				66 Pry 1
PR, glc	CCl_4/4-methoxy- toluene	333	$k_a/k_b = 2.98$	
 $-NO_2 + CH_3CH_2SC_6H_5 \xrightarrow{a} C_6H_5NO_2 +$ ethyl phenyl sulfide$(-\dot{H})$ $+ CCl_4 \xrightarrow{b} 4\text{-}ClC_6H_4NO_2 + \dot{C}Cl_3$				
Therm. of 4-nitrophenylazotriphenylmethane				72 Pry 2
PR, glc	$CH_3CH_2SC_6H_5/$ CCl_4	333	$k_a/k_b = 5.1$	

[6]) *m*-xylene. [7]) *p*-xylene.

Reaction Radical generation Method	Solvent	T[K]	Rate data	Ref./ add. ref.
 $-NO_2 + (CH_3)_2C{=}CHCH{=}C(CH_3)_2 \xrightarrow{a} C_6H_5NO_2$ + 2,5-dimethyl-2,4-hexadiene($-\dot{H}$) $+ CCl_4 \xrightarrow{b}$ 4-$ClC_6H_4NO_2 + \dot{C}Cl_3$				
Therm. of 4-nitrophenylazotriphenylmethane PR, glc	2,5-dimethyl- 2,4-hexadiene/CCl_4	333	$k_a/k_b = 19.4$	72 Pry 2
 $-NO_2 + c\text{-}C_8H_{16} \xrightarrow{a} C_6H_5NO_2 + (c\text{-}C_8\dot{H}_{15})$ $+ CCl_4 \xrightarrow{b}$ 4-$ClC_6H_4NO_2 + \dot{C}Cl_3$				
Therm. of 4-nitrophenylazotriphenylmethane PR, glc	cyclooctane/CCl_4	333	$k_a/k_b = 24.62(155)$	72 Pry 2
 $-NO_2 + c\text{-}C_8H_{16} \xrightarrow{a} C_6H_5NO_2 + (c\text{-}C_8\dot{H}_{15})$ $+ C_6H_5CH_3 \xrightarrow{b} C_6H_5NO_2 + C_6H_5\dot{C}H_2$				
Therm. of 4-nitrophenylazotriphenylmethane PR, glc	$c\text{-}C_8H_{16}$/toluene	333	$k_a/k_b = 4.6$	72 Pry 2
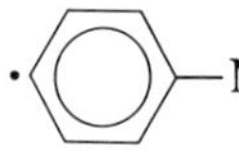 $-NO_2 + CH_2{=}CH(CH_2)_5CH_3 \xrightarrow{a} C_6H_5NO_2$ + 1-octene($-\dot{H}$) $+ CCl_4 \xrightarrow{b}$ 4-$ClC_6H_4NO_2 + \dot{C}Cl_3$				
Therm. of 4-nitrophenylazotriphenylmethane PR, glc	1-octene/CCl_4	333	$k_a/k_b = 9.35(35)$	72 Pry 2
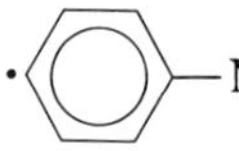 $-NO_2 + CH_3CH{=}CH(CH_2)_4CH_3 \xrightarrow{a} C_6H_5NO_2$ + 2-octene($-\dot{H}$) $+ CCl_4 \xrightarrow{b}$ 4-$ClC_6H_4NO_2 + \dot{C}Cl_3$				
Therm. of 4-nitrophenylazotriphenylmethane PR, glc	2-octene/CCl_4	333	$k_a/k_b = 15.6$	72 Pry 2
 $-NO_2 + n\text{-}C_8H_{18} \xrightarrow{a} C_6H_5NO_2$ + n-octane($-\dot{H}$) $+ CCl_4 \xrightarrow{b}$ 4-$ClC_6H_4NO_2 + \dot{C}Cl_3$				
Therm. of 4-nitrophenylazotriphenylmethane PR, glc	$n\text{-}C_8H_{18}/CCl_4$	333	$k_a/k_b = 10.3(3)$	72 Pry 2/ 66 Pry 1
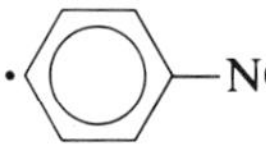 $-NO_2 + (CH_3)_2CHCH_2CH_2CH(CH_3)_2 \xrightarrow{a} C_6H_5NO_2$ + 2,5-dimethylhexane($-\dot{H}$) $+ CCl_4 \xrightarrow{b}$ 4-$ClC_6H_4NO_2 + \dot{C}Cl_3$				
Therm. of 4-nitrophenylazotriphenylmethane PR, glc	2,5-dimethylhexane/ CCl_4	333	$k_a/k_b = 9.7$	72 Pry 2

Reaction Radical generation Method	Solvent	T[K]	Rate data	Ref./ add. ref.

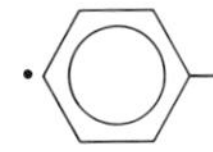 $NO_2 + (CH_3)_2CHCH(CH_3)CH(CH_3)_2 \xrightarrow{a} C_6H_5NO_2 + $ 2,3,4-trimethylpentane$(-\dot{H})$

$+ CCl_4 \xrightarrow{b}$ 4-$ClC_6H_4NO_2 + \dot{C}Cl_3$

Therm. of 4-nitrophenylazotriphenylmethane				72 Pry 2
PR, glc	2,3,4-trimethylpentane/ CCl_4	333	$k_a/k_b = 7.27(48)$	

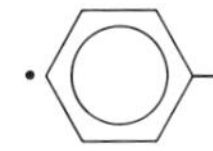 $NO_2 + (CH_3)_3CCH_2CH(CH_3)_2 \xrightarrow{a} C_6H_5NO_2 + $ 2,2,4-trimethylpentane$(-\dot{H})$

$+ CCl_4 \xrightarrow{b}$ 4-$ClC_6H_4NO_2 + \dot{C}Cl_3$

Therm. of 4-nitrophenylazotriphenylmethane				72 Pry 2/
PR, glc	2,2,4-trimethylpentane/ CCl_4	333	$k_a/k_b = 2.77(53)$	66 Pry 1

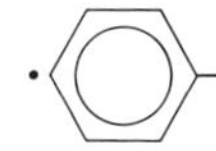 $NO_2 + (CH_3)_3CC(CH_3)_3 \xrightarrow{a} C_6H_5NO_2 + $ 2,2,3,3-tetramethylbutane$(-\dot{H})$

$+ CCl_4 \xrightarrow{b}$ 4-$ClC_6H_4NO_2 + \dot{C}Cl_3$

Therm. of 4-nitrophenylazotriphenylmethane				72 Pry 2
PR, glc	2,2,3,3-tetra-methylbutane/CCl_4	333	$k_a/k_b = 1.33(47)$	

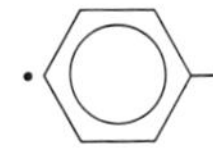 $NO_2 + $ indan $\xrightarrow{a} C_6H_5NO_2 + $ indan$(-\dot{H})$

$+ CCl_4 \xrightarrow{b}$ 4-$ClC_6H_4NO_2 + \dot{C}Cl_3$

Therm. of 4-nitrophenylazotriphenylmethane				72 Pry 2
PR, glc	indan/CCl_4	333	$k_a/k_b = 23.5$	

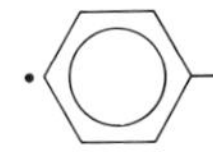 $NO_2 + $ indan $\xrightarrow{a} C_6H_5NO_2 + $ indan$(-\dot{H})$

$+ C_6H_5CH_3 \xrightarrow{b} C_6H_5NO_2 + C_6H_5\dot{C}H_2$

Therm. of 4-nitrophenylazotriphenylmethane				72 Pry 2
PR, glc	indan/toluene	333	$k_a/k_b = 11.7$	

 $NO_2 + CH_2{=}CHCH_2SC_6H_5 \xrightarrow{a}$ 4-$NO_2C_6H_4CH_2CH{=}CH_2 + C_6H_5\dot{S}$

$+ CCl_4 \xrightarrow{b}$ 4-$NO_2C_6H_4Cl + \dot{C}Cl_3$

Thermal decomp. of 4-nitrophenylazotriphenylmethane				79 Mig 1
PR, glc	CCl_4/ $CH_2{=}CHCH_2SC_6H_5$ mixt.	333	$k_a/k_b = 57$	

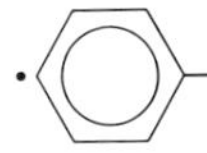 $NO_2 + n$-$C_3H_7C_6H_5 \xrightarrow{a} C_6H_5NO_2 + n$-propylbenzene$(-\dot{H})$

$+ CCl_4 \xrightarrow{b}$ 4-$NO_2C_6H_4Cl + \dot{C}Cl_3$

Thermal decomp. of 4-nitrophenylazotriphenylmethane				66 Pry 1
PR, glc	CCl_4/ n-propylbenzene	333	$k_a/k_b = 14.4$	

Reaction / Radical generation / Method	Solvent	T[K]	Rate data	Ref./ add. ref.

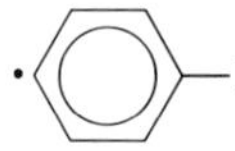

$-NO_2 + C_6H_5CH(CH_3)_2 \xrightarrow{a} C_6H_5NO_2 + C_6H_5\dot{C}(CH_3)_2$

$+ CCl_4 \xrightarrow{b} 4\text{-}NO_2C_6H_4Cl + \dot{C}Cl_3$

Thermal decomp. of 4-nitrophenylazotriphenylmethane				66 Pry 1
PR, glc	CCl_4/ $C_6H_5(CH_3)_2CH$	333	$k_a/k_b = 14.6$	

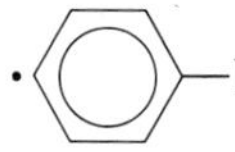

$-NO_2 + C_6H_5CH(CH_3)_2 \xrightarrow{a} C_6H_5NO_2 + C_6H_5\dot{C}(CH_3)_2$

$+ C_6H_5CH_3 \xrightarrow{b} C_6H_5NO_2 + C_6H_5\dot{C}H_2$

Therm. of 4-nitrophenylazotriphenylmethane				72 Pry 2
PR, glc	cumene/toluene	333	$k_a/k_b = 28.5$	

$-NO_2 + 1,3,5\text{-}(CH_3)_3C_6H_3 \xrightarrow{a} C_6H_5NO_2 + \text{mesitylene}(-\dot{H})$

$+ CCl_4 \xrightarrow{b} 4\text{-}NO_2C_6H_4Cl + \dot{C}Cl_3$

Thermal decomp. of 4-nitrophenylazotriphenylmethane				66 Pry 1
PR, glc	CCl_4/mesitylene [8])	333	$k_a/k_b = 5.90$	

$-NO_2 + (CH_3)_2CHSC_6H_5 \xrightarrow{a} C_6H_5NO_2 + \text{2-propyl phenyl sulfide}(-\dot{H})$

$+ CCl_4 \xrightarrow{b} 4\text{-}ClC_6H_4NO_2 + \dot{C}Cl_3$

Therm. of 4-nitrophenylazotriphenylmethane				72 Pry 2
PR, glc	$(CH_3)_2CHSC_6H_5$/ CCl_4	333	$k_a/k_b = 8.0$	

$-NO_2 + C_6H_5C(CH_3)_3 \xrightarrow{a} C_6H_5NO_2 + \text{2-methyl-2-phenylpropane}(-\dot{H})$

$+ CCl_4 \xrightarrow{b} 4\text{-}NO_2C_6H_4Cl + \dot{C}Cl_3$

Thermal decomp. of 4-nitrophenylazotriphenylmethane				66 Pry 1
PR, glc	$CCl_4/C_6H_5C(CH_3)_3$	333	$k_a/k_b = 0.19$	

$-NO_2 + n\text{-}C_{10}H_{22} \xrightarrow{a} C_6H_5NO_2 + n\text{-decane}(-\dot{H})$

$+ CCl_4 \xrightarrow{b} 4\text{-}ClC_6H_4NO_2 + \dot{C}Cl_3$

Therm. of 4-nitrophenylazotriphenylmethane				72 Pry 2
PR, glc	$n\text{-}C_{10}H_{22}/CCl_4$	333	$k_a/k_b = 12.8(6)$	

$-NO_2 + 4\text{-}C_6H_5C_6H_4I \xrightarrow{a} 4\text{-}NO_2C_6H_4I + (4\text{-}C_6H_5C_6H_4^{\cdot})$

$+ CCl_4 \xrightarrow{b} 4\text{-}NO_2C_6H_4Cl + \dot{C}Cl_3$

Therm. of 4-nitrophenylazotriphenylmethane				77 Pry 1
PR, glc	$CCl_4/4\text{-}C_6H_5C_6H_4I$	333	$k_a/k_b = 40.0(9)$	

[8]) 1,3,5-trimethylbenzene.

Reaction Radical generation Method	Solvent	T[K]	Rate data	Ref./ add. ref.

 $—NO_2 + C_6H_5NHNHC_6H_5 \xrightarrow{a} C_6H_5NO_2$ + hydrazobenzene$(-\dot{H})$

$+ CCl_4 \xrightarrow{b}$ 4-$ClC_6H_4NO_2 + \dot{C}Cl_3$

Therm. of 4-nitrophenylazotriphenylmethane				72 Pry 2
PR, glc	hydrazobenzene/CCl_4	333	$k_a/k_b = 36.1$	

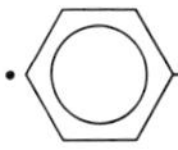 $—NO_2 + C_6(CH_3)_6 \xrightarrow{a} C_6H_5NO_2$ + hexamethylbenzene$(-\dot{H})$

$+ CCl_4 \xrightarrow{b}$ 4-$ClC_6H_4NO_2 + \dot{C}Cl_3$

Therm. of 4-nitrophenylazotriphenylmethane				72 Pry 2
PR, glc	hexamethylbenzene/ CCl_4	333	$k_a/k_b = 38.7(9)$	

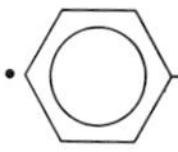 $—NO_2 + (C_6H_5)_2CH_2 \xrightarrow{a} C_6H_5NO_2 + (C_6H_5)_2\dot{C}H$

$+ CCl_4 \xrightarrow{b}$ 4-$ClC_6H_4NO_2 + \dot{C}Cl_3$

Therm. of 4-nitrophenylazotriphenylmethane				72 Pry 2
PR, glc	diphenylmethane/ CCl_4	333	$k_a/k_b = 6.7$	

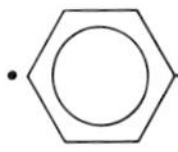 $—NO_2 + (C_6H_5)_2CH_2 \xrightarrow{a} C_6H_5NO_2 + (C_6H_5)_2\dot{C}H$

$+ C_6H_5CH_3 \xrightarrow{b} C_6H_5NO_2 + C_6H_5\dot{C}H_2$

Therm. of 4-nitrophenylazotriphenylmethane				72 Pry 2
PR, glc	$(C_6H_5)_2CH_2$/toluene	333	$k_a/k_b = 7.6$	

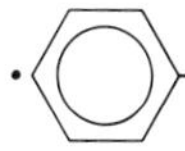 $—NO_2$ + 4-$(CH_3)_3CC_6H_4C(CH_3)_3 \xrightarrow{a} C_6H_5NO_2$ + 1,4-di-*t*-butylbenzene$(-\dot{H})$

$+ CCl_4 \xrightarrow{b}$ 4-$ClC_6H_4NO_2 + \dot{C}Cl_3$

Therm. of 4-nitrophenylazotriphenylmethane				72 Pry 2
PR, glc	1,4-di-*t*-butylbenzene/ CCl_4	333	$k_a/k_b = 1.0$	

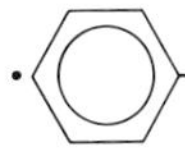 $—NO_2 + C_6(C_2H_5)_6 \xrightarrow{a} C_6H_5NO_2$ + hexaethylbenzene$(-\dot{H})$

$+ CCl_4 \xrightarrow{b}$ 4-$ClC_6H_4NO_2 + \dot{C}Cl_3$

Therm. of 4-nitrophenylazotriphenylmethane				72 Pry 2
PR, glc	hexaethylbenzene/ CCl_4	333	$k_a/k_b = 4.3(1)$	

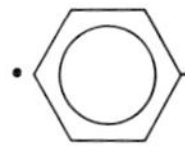 $—NO_2 + (C_6H_5)_3CH \xrightarrow{a} C_6H_5NO_2 + (C_6H_5)_3\dot{C}$

$+ CCl_4 \xrightarrow{b}$ 4-$ClC_6H_4NO_2 + \dot{C}Cl_3$

Therm. of 4-nitrophenylazotriphenylmethane				72 Pry 2
PR, glc	triphenylmethane/ CCl_4	333	$k_a/k_b = 16.1$	

Reaction Radical generation Method	Solvent	T[K]	Rate data	Ref./ add. ref.

$$(4\text{-}CH_3OC_6F_4^{\cdot}) + CHCl_3 \underset{b}{\overset{a}{<}} \begin{array}{l} CH_3OC_6HF_4 + \dot{C}Cl_3 \\ CH_3OC_6ClF_4 + \dot{C}HCl_2 \end{array}$$

$$(4\text{-}CH_3OC_6F_4^{\cdot}) + CBrCl_3 \underset{b}{\overset{a}{<}} \begin{array}{l} CH_3OC_6BrF_4 + \dot{C}Cl_3 \\ CH_3OC_6ClF_4 + \dot{C}BrCl_2 \end{array}$$

See 4.1.2.3, Fig. 8, p. 257.

$$\cdot\bigcirc\!\!-CN + CBrCl_3 \xrightarrow{a} 4\text{-}NCC_6H_4Br + \dot{C}Cl_3$$

$$+ CCl_4 \xrightarrow{b} 4\text{-}NCC_6H_4Cl + \dot{C}Cl_3$$

Therm. of 4-NCC$_6$H$_4$COOOC(CH$_3$)$_3$ PR, glc	CCl$_4$	403	$k_a/k_b = 123$	75 Her 1/ 76 Gie 1, 76 Gie 2, 76 Gie 3

$$\cdot\bigcirc\!\!-CN + CH_3CN \xrightarrow{a} C_6H_5CN + \dot{C}H_2CN$$

$$+ CD_3CN \xrightarrow{a} 4\text{-}DC_6H_4CN + \dot{C}D_2CN$$

$$\left\{(I,Br)Cl\!\!-\!\bigcirc\!\!-CN\right\}^{\overline{\cdot}} \xrightarrow{b} \cdot\bigcirc\!\!-CN + Cl^-(Br^-,I^-)$$

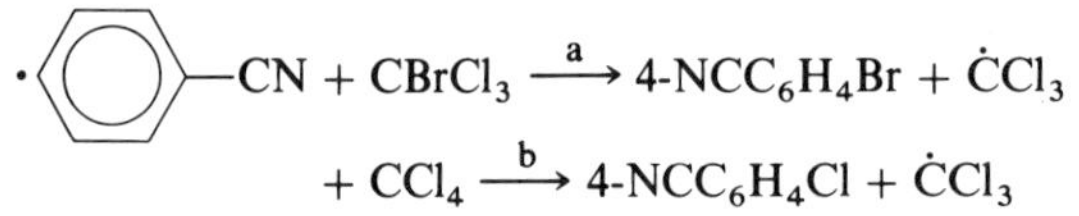

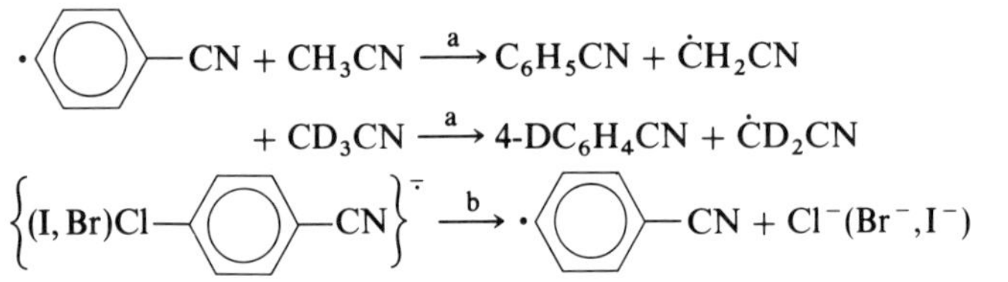

Electrochem. reduct. of 4-chlorobenzonitrile, 4-bromobenzonitrile and 4-iodobenzonitrile				80 M'Ha 1
MS, electrochem.	CH$_3$CN/H$_2$O(D$_2$O) (9:1)	[9])	$k_a(H)/k_b = 0.345\,M^{-1}$ (Cl) $k_a(D)/k_b = 2.44\cdot10^{-2}\,M^{-1}$ $k_a(H)/k_b = 0.0213\,M^{-1}$ (Br) $k_a(D)/k_b = 1.74\cdot10^{-3}\,M^{-1}$ $k_a(H)/k_b = 0.00386\,M^{-1}$ (I) $k_a(D)/k_b = 2.32\cdot10^{-4}\,M^{-1}$	

$$Cl\!\!-\!\bigcirc\!\!-\dot{C}H_2 + CBrCl_3 \xrightarrow{a} 4\text{-}ClC_6H_4CH_2Br + \dot{C}Cl_3$$

$$+ CCl_4 \xrightarrow{b} 4\text{-}ClC_6H_4CH_2Cl + \dot{C}Cl_3$$

Therm. of 4-ClC$_6$H$_4$CH$_2$COOOC(CH$_3$)$_3$ PR, glc	CCl$_4$	354	$k_a/k_b = 1500(\pm15\%)$	75 Her 1/ 76 Gie 1, 76 Gie 2, 76 Gie 3

$$\cdot\bigcirc\!\!-OCH_3 + CBrCl_3 \xrightarrow{a} 4\text{-}CH_3OC_6H_4Br + \dot{C}Cl_3$$

$$+ CCl_4 \xrightarrow{b} 4\text{-}CH_3OC_6H_4Cl + \dot{C}Cl_3$$

Therm. of 4-CH$_3$OC$_6$H$_4$COOOC(CH$_3$)$_3$ PR, glc	CCl$_4$	403	$k_a/k_b = 131$	75 Her 1/ 76 Gie 1, 76 Gie 2, 76 Gie 3

---

[9]) Not given (presumably RT).

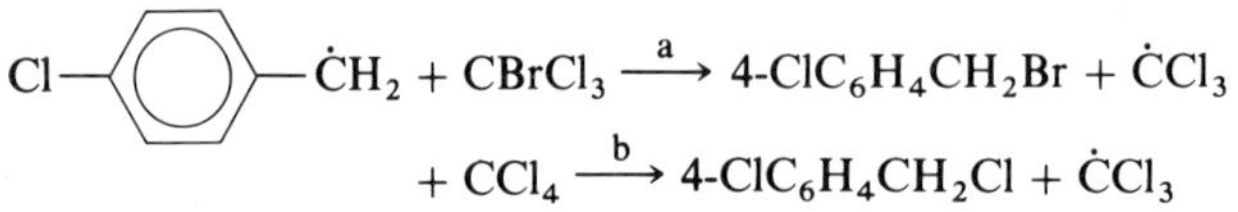

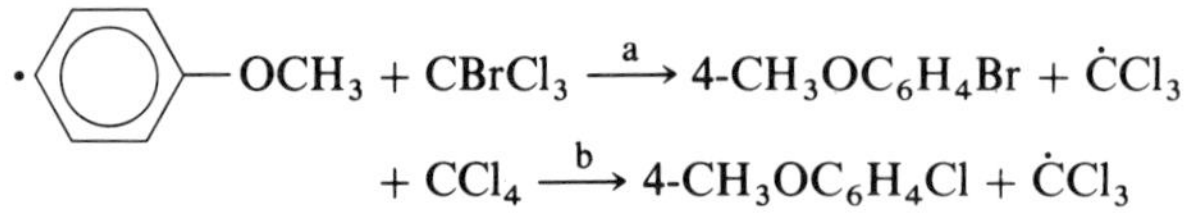

Reaction Radical generation Method	Solvent	T[K]	Rate data	Ref./ add. ref.
 $+ CH_3OH \xrightarrow{a} C_6H_5OCH_3 + \dot{C}H_2OH$ $+ 4\text{-}CH_3OC_6H_4N_2^+ \xrightarrow{b} (4\text{-}CH_3OC_6H_4)_2\dot{N}_2^+$				
γ-rad. of $4\text{-}CH_3OC_6H_4N_2^+BF_4^- + CH_3OH + H_2O$				74 Pac 1,
PR [10])	H_2O	RT	$k_a/k_b = 0.128$ [11])	75 Pac 1
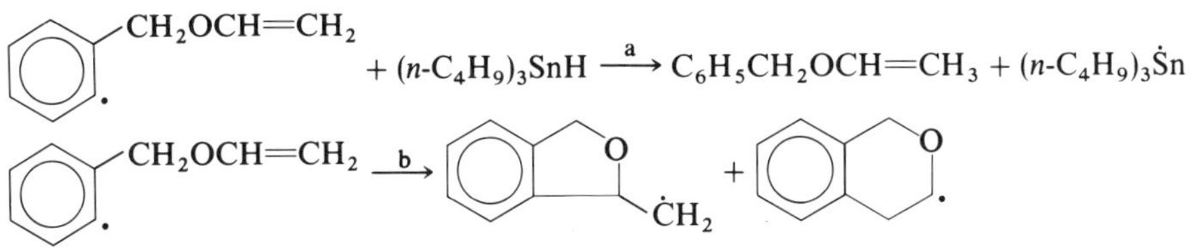				
Decomp. of 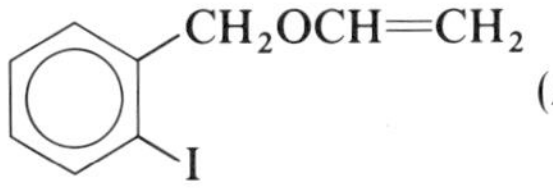			(AIBN initiated)	75 Bec 2
PR, glc	C_6H_6	403(3)	$k_a/k_b = 5.26\ M^{-1}$	
$4\text{-}CH_3OC_6H_4CH_2\dot{C}H_2 + (CH_3)_2CHCHO \xrightarrow{a} 4\text{-}CH_3OC_6H_4C_2H_5 + (CH_3)_2CH\dot{C}O$ $+ Cu(II)(NCCH_3)_4^{2+} \xrightarrow{b} 4\text{-}CH_3OC_6H_4CH{=}CH_2 + H^+ + Cu(I)(NCCH_3)_4^+$				
Cu(II) catalyzed decomp. of $4\text{-}CH_3OC_6H_4CH_2CH_2OOCH_2CH_2C_6H_4OCH_3$				68 Koc 1
PR, glc	CH_3CN/CH_3COOH (1:1.5)	298.5	$k_a/k_b = 0.48$ [12])	
$4\text{-}CH_3OC_6H_4CH_2\dot{C}H_2 + (CH_3)_2CHCHO \xrightarrow{a} 4\text{-}CH_3OC_6H_4C_2H_5 + (CH_3)_2CH\dot{C}O$ $+ Cu(II)(bipyridine)^{2+} \xrightarrow{b} 4\text{-}CH_3OC_6H_4CH{=}CH_2 + H^+ + Cu(I)(bipyridine)^+$				
Cu(II) catalyzed decomp. of $4\text{-}CH_3OC_6H_4CH_2CH_2OOCH_2CH_2C_6H_4OCH_3$				68 Koc 1
PR, glc	CH_3CN/CH_3COOH (1:1.5)	298.5	$k_a/k_b = 3.3 \cdot 10^{-3}$ [12])	
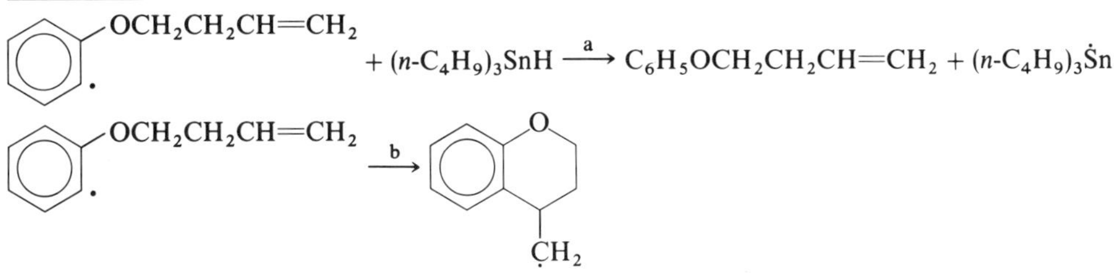				
Decomp. of 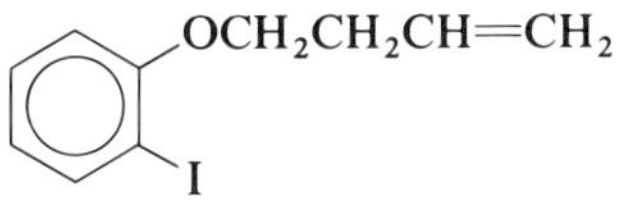			(AIBN initiated)	75 Bec 2
PR, glc	C_6H_6	403(3)	$k_a/k_b = 0.91\ M^{-1}$	

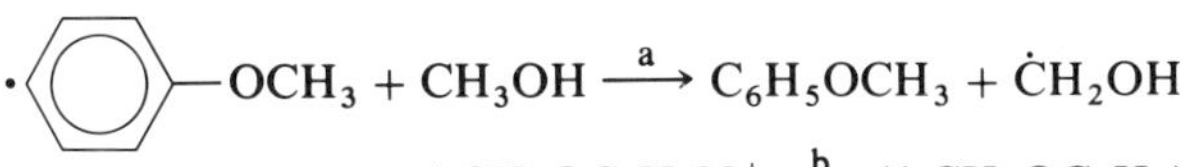

[10]) Spectrophotometric determination of $4\text{-}CH_3OC_6H_4N_2^+$.
[11]) Twice the value from [74 Pac 1] as required in [75 Pac 1].
[12]) Assumed value for $k_a = 1 \cdot 10^4\ M^{-1}\ s^{-1}$.

Reaction				
Radical generation Method	Solvent	T[K]	Rate data	Ref./ add. ref.

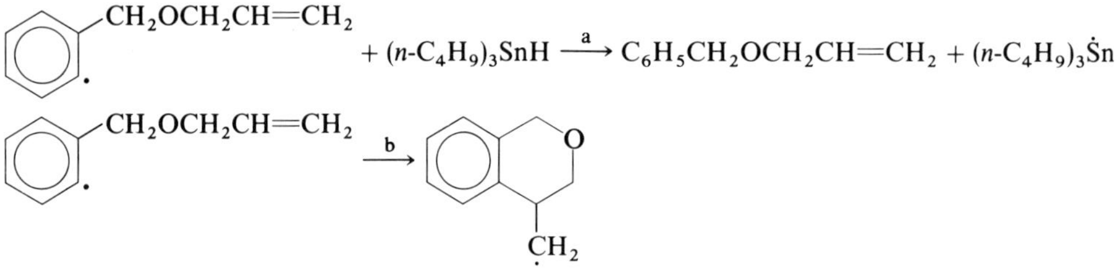

$$+\ (n\text{-}C_4H_9)_3SnH \xrightarrow{a} C_6H_5CH_2OCH_2CH{=}CH_2 + (n\text{-}C_4H_9)_3\dot{S}n$$

Decomp. of		(AIBN initiated)		75 Bec 2
PR, glc	C_6H_6	403(3)	$k_a/k_b = 6.67\,M^{-1}$	

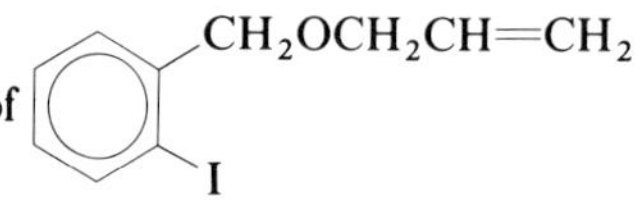

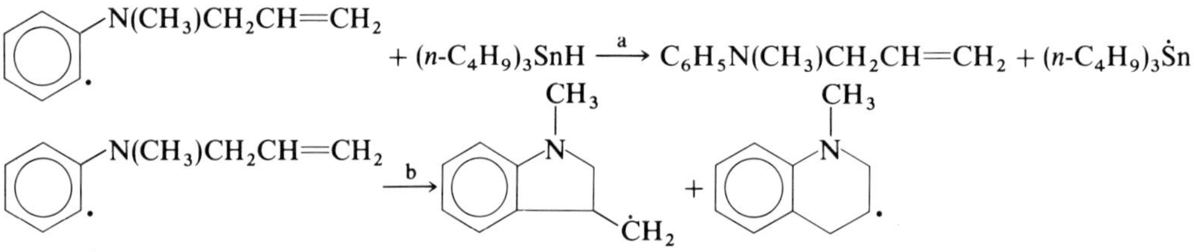

$$+\ (n\text{-}C_4H_9)_3SnH \xrightarrow{a} C_6H_5N(CH_3)CH_2CH{=}CH_2 + (n\text{-}C_4H_9)_3\dot{S}n$$

Decomp. of		(AIBN initiated)		75 Bec 2
PR, glc	C_6H_6	403(3)	$k_a/k_b = 0.25\,M^{-1}$	

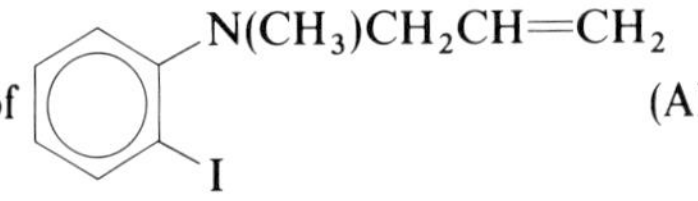

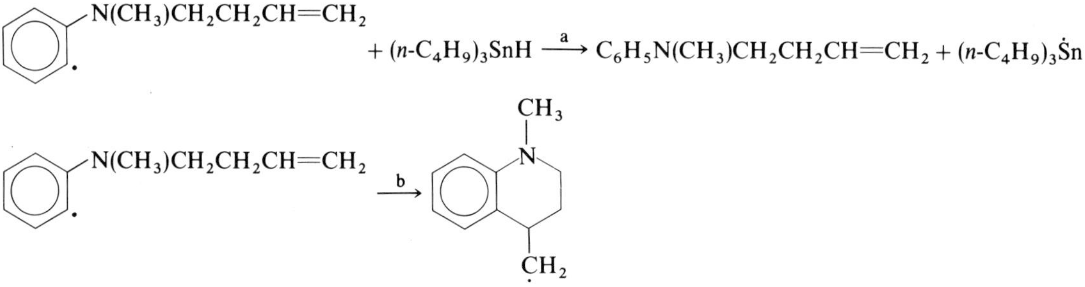

$$+\ (n\text{-}C_4H_9)_3SnH \xrightarrow{a} C_6H_5N(CH_3)CH_2CH_2CH{=}CH_2 + (n\text{-}C_4H_9)_3\dot{S}n$$

Decomp. of		(AIBN initiated)		75 Bec 2
PR, glc	C_6H_6	403(3)	$k_a/k_b = 0.77\,M^{-1}$	

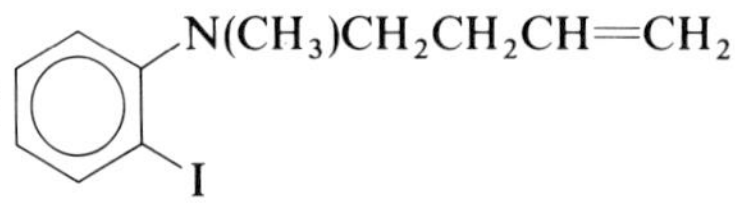

Reaction Radical generation Method	Solvent	T[K]	Rate data	Ref./ add. ref.

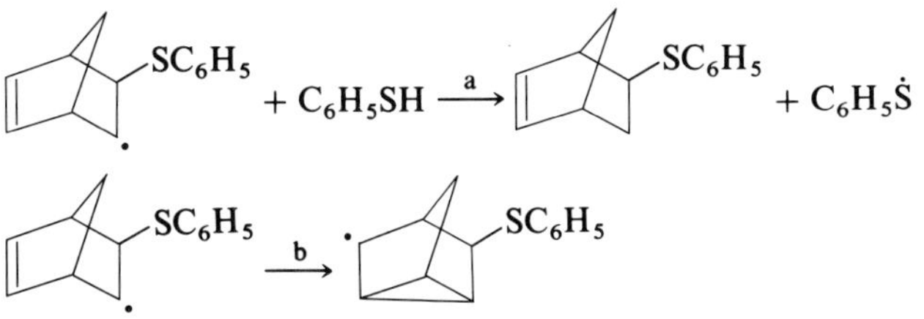

$C_6H_5\dot{S}$ add. to 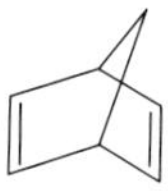 (norbornadiene) — 79 Gie 2

Method	Solvent	T[K]	Rate data	Ref./add. ref.
PR	CH_2Cl_2	233	$k_a/k_b = 2.5(1)\,M^{-1}$	
		243	1.8(1)	
		253	1.20(5)	
			$\Delta H^{\ddagger}_a - \Delta H^{\ddagger}_b = -18.0(13)\,kJ\,mol^{-1}$	
			$\Delta S^{\ddagger}_a - \Delta S^{\ddagger}_b = -71.2(84)\,J\,mol^{-1}\,K^{-1}$	

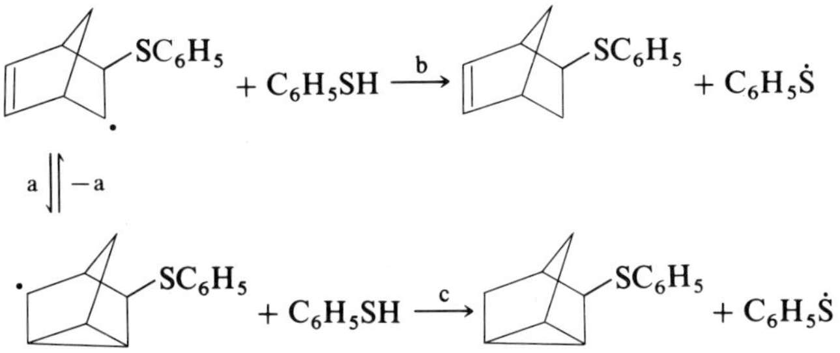

$C_6H_5\dot{S}$ add. to 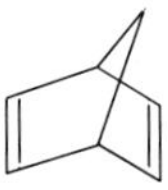 (norbornadiene) — 79 Gie 2

Method	Solvent	T[K]	Rate data	Ref./add. ref.
PR, glc	CH_2Cl_2	233	$\frac{k_b k_{-a}}{k_a k_c} = 0.28(15)$	
		243	0.24(15)	
		253	0.18(7)	
			$\Delta H^{\ddagger}_a - \Delta H^{\ddagger}_c = 6.7(33)\,kJ\,mol^{-1}$	
			$\Delta S^{\ddagger}_a - \Delta S^{\ddagger}_c = 10.5(84)\,J\,mol^{-1}\,K^{-1}$	

Reaction Radical generation Method	Solvent	T[K]	Rate data	Ref./ add. ref.

4.1.2.3 Graphical data

$CH_3\dot{C}H_2 + C_2H_5COC_2H_5 \xrightarrow{a} C_2H_6$ + diethylketone($-\dot{H}$)
$CH_3\dot{C}H_2 + CH_3COC_2H_5 \xrightarrow{a'} C_2H_6$ + methyl ethyl ketone($-\dot{H}$)
$2\,CH_3\dot{C}H_2 \xrightarrow{b} C_4H_{10}$

Phot. PR	diethyl ketone and methyl ethyl ketone		$E_a(a) - \frac{1}{2}E_a(b) = 21$ kJ mol^{-1} ($\Delta T = 238 \ldots 367$) $E_a(a') - \frac{1}{2}E_a(b) = 21$ kJ mol^{-1} ($\Delta T = 278 \ldots 348$)	58 Aus 1

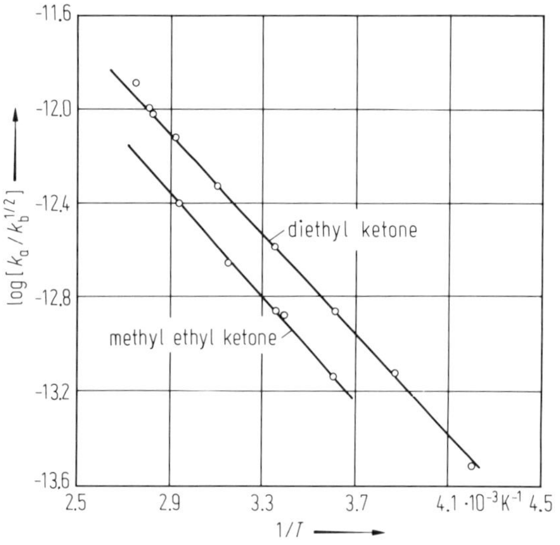

Fig. 1. Plot of log $k_a/k_b^{\frac{1}{2}}$ against $1/T$ for the phot. of diethyl ketone and methyl ethyl ketone. [58 Aus 1].

$(CH_3)_3\dot{C} + ArCH_3 \longrightarrow (CH_3)_3CH + Ar\dot{C}H_2$
Ar = substituted toluenes, not individually specified

Phot. of 2,2′-azoisobutane or t-butyl peroxypivalate PR, MS	Mixt. of $ArCH_3$ and thiophenol-d or t-butyl mercaptan-d	303		73 Pry 2

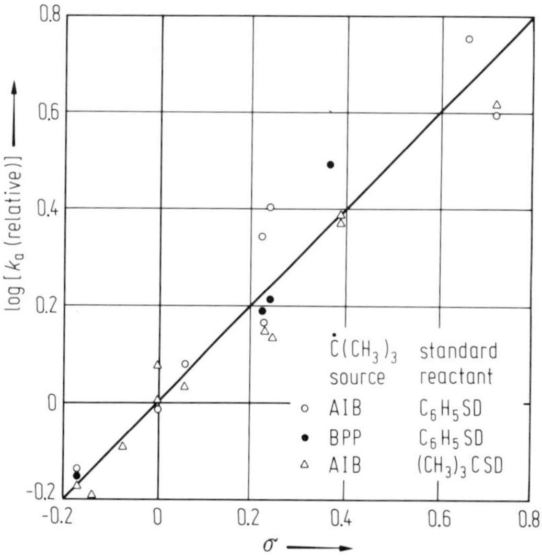

Fig. 2. Hammett equation plot of relative k_a values vs. σ constants. For each system, k_a/k_a^0 is plotted, where k_a^0 is the intercept of the least-squares line for that system: system I, 0.1 M azoisobutane (AIB) and thiophenol-d; system II, 0.1 M t-butyl peroxypivalate (BPP) and thiophenol-d; system III, 0.1 M AIB and t-butyl mercaptan-d [73 Pry 2].

Reaction Radical generation Method	Solvent	T[K]	Rate data	Ref./ add. ref.
$CH_3CH_2\dot{C}HCH_2CH_2CH_2CH_3 + ArCH_3 \longrightarrow n\text{-}C_7H_{16} + Ar\dot{C}H_2$ Ar = 4-$CH_3C_6H_4$(*1*); 3,5-$(CH_3)_2C_6H_3$(*2*); 3-$CH_3C_6H_4$(*3*); C_6H_5(*4*); 4-FC_6H_4(*5*); 4-ClC_6H_4(*6*); 3-FC_6H_4(*7*); 3-ClC_6H_4(*8*); 3-CNC_6H_4(*9*)				
Thermal decomp. of *t*-butyl 2-ethylperhexanoate				75 Hen 1
PR, glc	Mixt. of toluene and CCl_4	353		

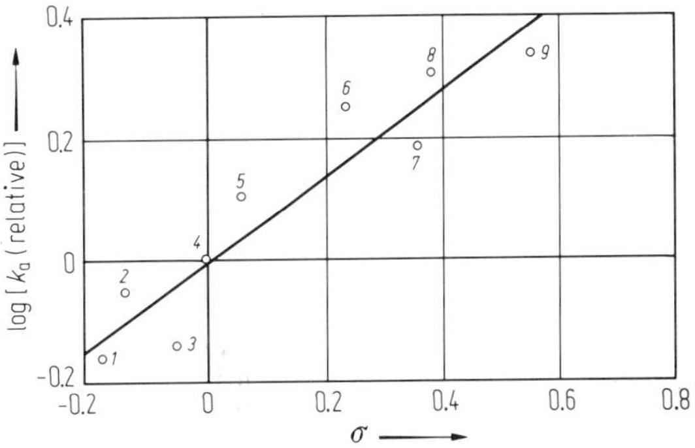

Fig. 3. Hammett plot of relative k_a values for the 3-heptyl radical vs. σ substituent constants. *1*, 4-xylene; *2*, mesitylene; *3*, 3-xylene; *4*, toluene; *5*, 4-fluorotoluene; *6*, 4-chlorotoluene; *7*, 3-fluorotoluene; *8*, 3-chlorotoluene; *9*, 3-tolunitrile. The values for mesitylene and the xylenes have been statistically corrected [75 Hen 1].

$$\dot{R} + CBrCl_3 \xrightarrow{a} RBr + \dot{C}Cl_3$$
$$+ CCl_4 \xrightarrow{b} RCl + \dot{C}Cl_3$$

R =

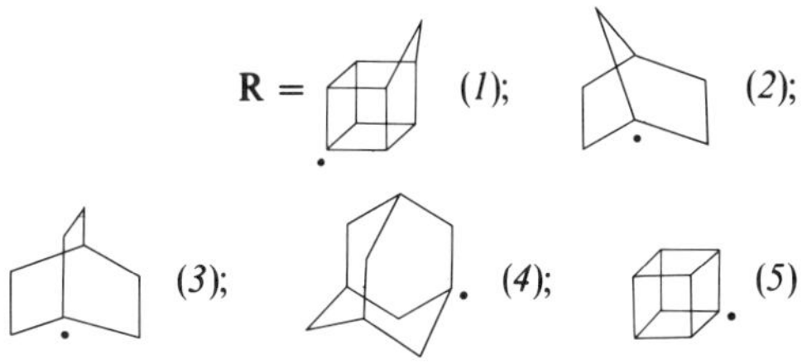

Thermal decomp. of peresters				79 Gie 4/
PR	CCl_4	323...413	$\dot{R}$ = *1*: $\Delta H_b^{\ddagger} - \Delta H_a^{\ddagger} = 2.1(2)$ kJ mol^{-1} $\Delta S_b^{\ddagger} - \Delta S_a^{\ddagger} = -29(3)$ J mol^{-1} K^{-1} R = *2*: $\Delta H_b^{\ddagger} - \Delta H_a^{\ddagger} = 4.2(4)$ kJ mol^{-1} $\Delta S_b^{\ddagger} - \Delta S_a^{\ddagger} = -20(3)$ J mol^{-1} K^{-1} $\dot{R}$ = *3*: $\Delta H_b^{\ddagger} - \Delta H_a^{\ddagger} = 4.8(5)$ kJ mol^{-1} $\Delta S_b^{\ddagger} - \Delta S_a^{\ddagger} = -16(3)$ J mol^{-1} K^{-1} $\dot{R}$ = *4*: $\Delta H_b^{\ddagger} - \Delta H_a^{\ddagger} = 5.5(6)$ kJ mol^{-1} $\Delta S_b^{\ddagger} - \Delta S_a^{\ddagger} = -11(3)$ J mol^{-1} K^{-1}	75 Her 1, 78 Luh 1

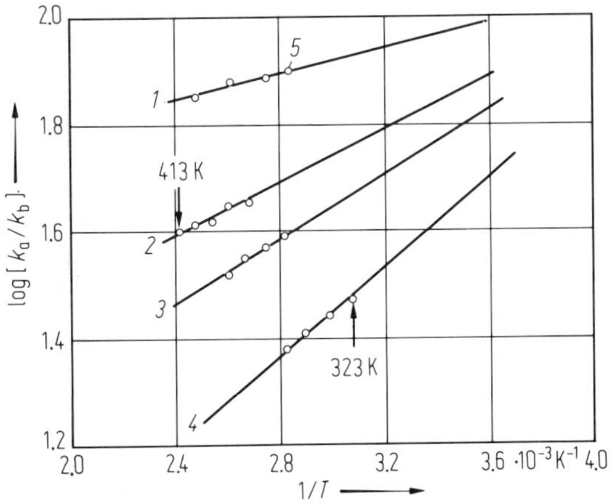

Fig. 4. Temperature dependence of selectivities log[k_a/k_b] of bridgehead radicals using competition system $BrCCl_3/CCl_4$ [79 Gie 4]. *1*, homocubyl; *2*, norbonyl; *3*, bicyclooctyl; *4*, adamantyl; *5*, cubyl. The experimental value of the cubyl radical at 353 K, cited from [78 Luh 1], fits onto the selectivity line of the homocubyl radical.

Reaction Radical generation Method	Solvent	T[K]	Rate data	Ref./ add. ref.

$CH_3(CH_2)_9\dot{C}H_2 + ArCH_3 \longrightarrow n\text{-}C_{11}H_{24} + Ar\dot{C}H_2$

Ar = $4\text{-}CH_3C_6H_4$(*1*); $3\text{-}CH_3C_6H_4$(*2*); C_6H_5(*3*); $4\text{-}FC_6H_4$(*4*); $4\text{-}ClC_6H_4$(*5*); $3\text{-}FC_6H_4$(*6*); $3\text{-}ClC_6H_4$(*7*); $3\text{-}CNC_6H_4$(*8*); $4\text{-}CH_3OC_6H_4$(*9*); $4\text{-}C(CH_3)_3C_6H_4$(*10*); $3\text{-}CH_3OC_6H_4$(*11*); $3\text{-}BrC_6H_4$(*12*)

Thermal decomp. of lauroyl peroxide				74 Pry 1
PR, NMR	*t*-butylbenzene	354		

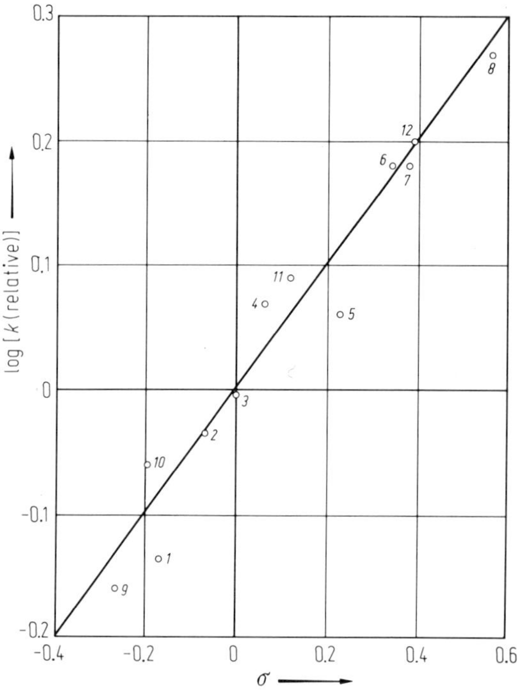

Fig. 5. Hammett equation plot of relative k values vs. σ for hydrogen abstraction from substituted toluenes by the undecyl radical at 354 K. The least-squares treatment gives $\rho = 0.50(2)$ ($r = 0.97$) [74 Pry 1]. *1*, 4-methyltoluene; *2*, 3-methyltoluene; *3*, toluene; *4*, 4-fluorotoluene; *5*, 4-chlorotoluene; *6*, 3-fluorotoluene; *7*, 3-chlorotoluene; *8*, 3-cyanotoluene; *9*, 4-methoxytoluene; *10*, 4-*t*-butyltoluene; *11*, 3-methoxytoluene; *12*, 3-bromotoluene.

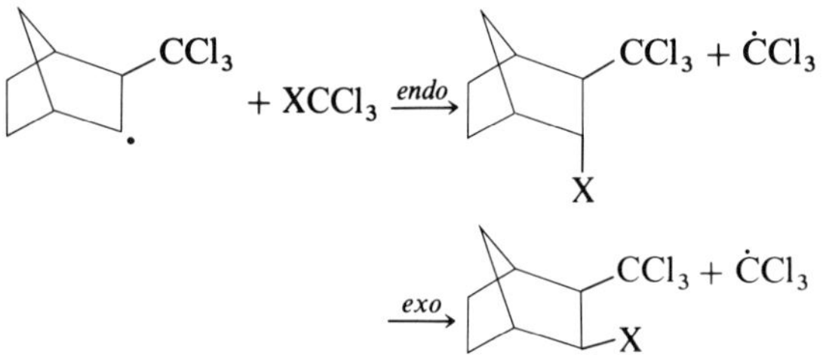

not specified				77 Gie 1/
not specified	not specified	273…400	X = Br: $\Delta H^{\ddagger}_{exo} - \Delta H^{\ddagger}_{endo} = 2.10(17)\,\text{kJ mol}^{-1}$ $\Delta S^{\ddagger}_{exo} - \Delta S^{\ddagger}_{endo} = -17(2)\,\text{J K}^{-1}\,\text{mol}^{-1}$ X = Cl: $\Delta H^{\ddagger}_{exo} - \Delta H^{\ddagger}_{endo} = 8.40(42)\,\text{kJ mol}^{-1}$ $\Delta S^{\ddagger}_{exo} - \Delta S^{\ddagger}_{endo} = -0.4(20)\,\text{J K}^{-1}\,\text{mol}^{-1}$	79 Gie 3

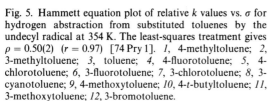

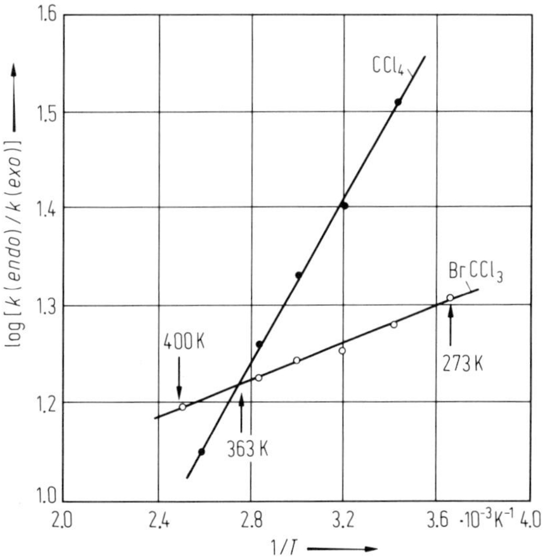

Fig. 6. Temperature dependence of stereoselectives log[k(*endo*)/k(*exo*)] of *exo*-trichloromethylnorbonyl radical for halogen abstraction from $XCCl_3$ (X = Br, Cl) [77 Gie 1].

Reaction Radical generation Method	Solvent	T[K]	Rate data	Ref./ add. ref.

$(C_6H_5^{\bullet}) + ArI \xrightarrow{a} C_6H_5I + \dot{A}r$

$+ CCl_4 \xrightarrow{b} C_6H_5Cl + \dot{C}Cl_3$

Ar = 4-$CH_3C_6H_4$(*1*); 3-$CH_3C_6H_4$(*2*); 4-CNC_6H_4(*3*); 4-$CH_3OC_6H_4$(*4*); 3-$CH_3OC_6H_4$(*5*); 3-BrC_6H_4(*6*); 4-$C_2H_5OC_6H_4$(*7*); 4-$C_6H_5C_6H_4$(*8*); 4-BrC_6H_4(*9*); 3-$CH_3COC_6H_4$(*10*); 3-$NO_2C_6H_4$(*11*); 4-$NO_2C_6H_4$(*12*)

Thermal decomp. of PAT				69 Dan 1
not specified	CCl_4	333		

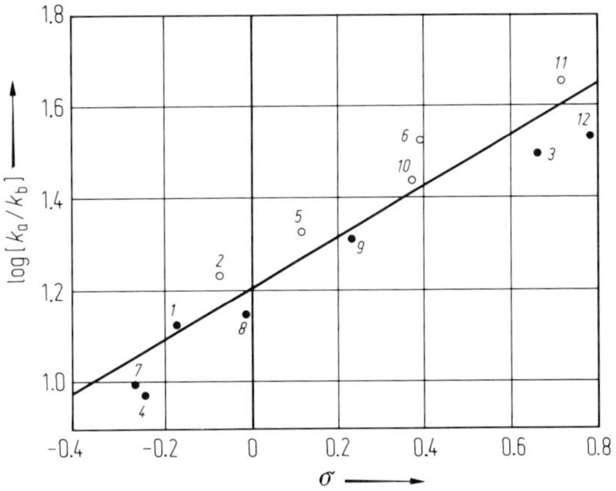

Fig. 7. Plot of log[k_a/k_b] at 333 K vs. Hammett σ constants. Each point represents the average of at least three independent experiments [69 Dan 1]. *1*, 4-methyliodobenzene; *2*, 3-methyliodobenzene; *3*, 4-cyanoiodobenzene; *4*, 4-methoxyiodobenzene; *5*, 3-methoxyiodobenzene; *6*, 3-bromoiodobenzene; *7*, 4-ethoxyiodobenzene; *8*, 4-phenyliodobenzene; *9*, 4-bromoiodobenzene; *10*, 3-acetyliodobenzene; *11*, 3-nitroiodobenzene; *12*, 4-nitroiodobenzene.

$\dot{A}r + HCCl_3 \xrightarrow{a} ArH + \dot{C}Cl_3$

$+ BrCCl_3 \xrightarrow{b} ArBr + \dot{C}Cl_3$

$+ X\text{-}CCl_3 \xrightarrow{c} ArCl + \dot{C}Cl_2X$

Ar = C_6F_5; 4-HC_6F_4; 4-BrC_6F_4; 4-$CH_3OC_6F_4$; X = Br or H

Decomp. of pentafluoroaniline; 2,3,5,6-tetrafluoroaniline; 4-bromo-2,3,5,6-tetrafluoroaniline, and 4-methoxy-2,3,5,6-tetrafluoroaniline by pentylnitrite				75 Bol 1
PR, glc	$CHCl_3$ or $BrCCl_3$	reflux		

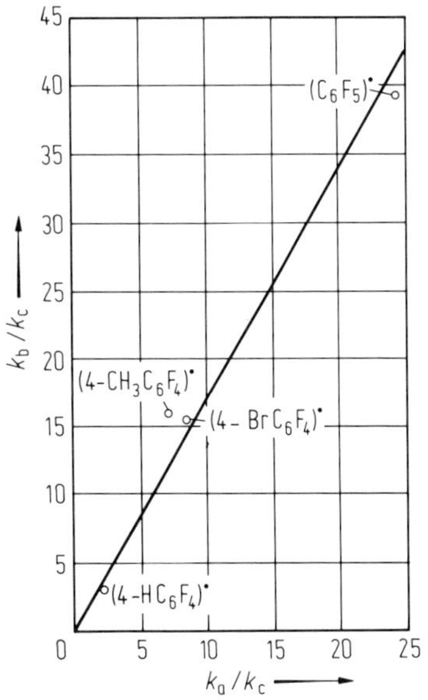

Fig. 8. Relative rates of abstraction reactions of some aryl radicals, as measured by product yields [75 Bol 1].

Reaction Radical generation Method	Solvent	T[K]	Rate data	Ref./ add. ref.

4.1.3 Isotope effects

For some explanatory comments, see introductory text.

4.1.3.1 Aliphatic radicals and radicals derived from other non-aromatic compounds

Further data on isotope effects are to be found in the section on absolute rate constants for the reactions of

$(\text{Polyvinylacetate})^{\cdot} + \text{Substrates}$

$R{-}\dot{C}H(CONH_2)CH_2{-}R + Ti^{3+}_{aq}$

$CH_2{=}CH(CH_2)_3\dot{C}H_2 + (CH_3)_3COOD/(CH_3)_3COOH$

$(c\text{-}C_6H^{\cdot}_{11}) + (n\text{-}C_4H_9)_3SnD/(n\text{-}C_4H_9)_3SnH$

$(CH_3)_3\dot{C} + (n\text{-}C_4H_9)_3SnD/(n\text{-}C_4H_9)_3SnH$

$\dot{C}H_3 + H_3N^+CH_2COO^-$

4.1.3.1.1 Radicals containing only C, H, and D atoms

$$\dot{C}D_3 + CH_3OH \xrightarrow{a} CD_3H + \dot{C}H_2OH$$

$$+ CD_3OH \xrightarrow{b} CD_4 + \dot{C}D_2OH$$

$$+ CD_3OH \xrightarrow{c} CD_3H + CD_3\dot{O}$$

Phot. of acetone-d_6				63 Che 1
PR	$CD_3COCD_3/$ CD_3OH/CH_3OH	303.0(1)	$k_a/k_b = 8.7(28)$ $k_a/k_c = 45(12)$	

$$\dot{C}H_3 + CH_3CH_2COOH \xrightarrow{a} CH_4 + \text{propionic acid}(-\dot{H})$$

$$+ CH_3CHTCOOH \xrightarrow{b} CH_3T + CH_3\dot{C}HCOOH$$

Therm. of acetylperoxide or t-butylperoxide + H_2O_2				66 Nem 1/ 60 Ant 1, 59 Ant 1
PR [1])	decane	423	$k_a/k_b = 27.1(55)$	
	H_2O [2])	423	27.5(6)	
	propionic acid	423	26.9(2)	
		413	28.8(4)	
		403	31.0(2)	
		393	33.0(4)	
			$\log[A_a/A_b] = -0.07$	
			$E_a(a) - E_a(b) = -9.6(4)\,kJ\,mol^{-1}$	

$$\dot{C}H_3 + (CH_3)_2CHCOCl \xrightarrow{a} CH_4 + (CH_3)_2\dot{C}COCl$$

$$+ (CH_3)_2CDCOCl \xrightarrow{b} CH_3D + (CH_3)_2\dot{C}COCl$$

Thermal decomp. of acetylperoxide				53 Pri 1
PR, MS	$(CH_3)_2CHCOCl/$ $(CH_3)_2CDCOCl$ mixt.	373	$k_a/k_b = 1.22(20)$	

$$\dot{C}H_3 + (CH_3)_3CSH \xrightarrow{a} CH_4 + (CH_3)_3C\dot{S}$$

$$+ (CH_3)_3CSD \xrightarrow{b} CH_3D + (CH_3)_3C\dot{S}$$

Not given				71 Pry 1
not given	not given	298 [3])	$k_a/k_b = 2.71$ $A_a/A_b = 0.59$ $E_a(a) - E_a(b) = -3.8\,kJ\,mol^{-1}$	

[1]) Specific activity measurements.
[2]) Generation by therm. of t-butylperoxide + H_2O_2.
[3]) Extrapolated to 298 K from Arrhenius plots.

Reaction Radical generation Method	Solvent	T[K]	Rate data	Ref./ add. ref.
$\dot{C}H_3 + C_6H_6 \xrightarrow{a} CH_4 + (C_6H_5^{\cdot})$				
$+ C_6H_5T \xrightarrow{b} CH_3T + (C_6H_5^{\cdot})$				
Therm. of acetylperoxide				60 Ant 2
PR [4])	C_6H_6	358.00(5)	$k_a/k_b = 6.0(1)$	
	(T-labelled)	343.00(5)	6.63(13)	
		328.00(5)	7.36(9)	
			$E_a(a) - E_a(b) = -6.6(4)\ kJ\ mol^{-1}$	
$\dot{C}H_3 + C_6H_5OH \xrightarrow{a} CH_4 + C_6H_5\dot{O}$				
$+ C_6H_5OT \xrightarrow{b} CH_3T + C_6H_5\dot{O}$				
Thermal decomp. of acetylperoxide				69 Shi 1/
PR [1])	n-heptane/	333.50(5)	k_a/k_b [5]) = 38.5	66 Shi 1
	T-labelled phenol	343.30(5)	32.7	
		353.40(5)	27.1	
		362.95(10)	22.8	
			$\log[A_a/A_b] = 0.06(1)$	
			$E_a(a) - E_a(b) = -18.0(17)\ kJ\ mol^{-1}$	
$\dot{C}H_3 + C_6H_5CH_3 \xrightarrow{a} CH_4 + \text{toluene}(-\dot{H})$				
$+ C_6H_4TCH_3 \xrightarrow{b} CH_3T + (C_6H_4CH_3^{\cdot})$				
Therm. of acetylperoxide				60 Ant 2
PR [4])	toluene (T-labelled)	358	$k_a/k_b = 12.1(1)$	
		343	14.14(20)	
		328	15.9(3)	
			$E_a(a) - E_a(b) = -9.2(4)\ kJ\ mol^{-1}$	
$\dot{C}H_3 + (CH_3)_3CCH_2CH(CH_3)_2 \xrightarrow{a} CH_4 + \text{2,2,4-trimethylpentane}(-H)$				
$+ (CH_3)_3CCHTCH(CH_3)_2 \xrightarrow{b} CH_3T + (CH_3)_3C\dot{C}HCH(CH_3)_2$				
Therm. of acetylperoxide				64 Ber 2
PR [4])	2,2,4-trimethylpentane	333.2	$k_a/k_b = 21.55$	
		343.2	16.33	
		353.2	14.01	
		363.2	11.36	
			$E_a(a) - E_a(b) = -20.1(8)\ kJ\ mol^{-1}$	
$\dot{C}H_3 + (CH_3)_3CCH_2CH(CH_3)_2 \xrightarrow{a} CH_4 + \text{2,2,4-trimethylpentane}(-\dot{H})$				
$+ (CH_3)_3CCH_2CT(CH_3)_2 \xrightarrow{b} CH_3T + (CH_3)_3CCH_2\dot{C}(CH_3)_2$				
Therm. of acetylperoxide				64 Ber 2
PR [4])	2,2,4-trimethylpentane	333.2	$k_a/k_b = 37.31$	
		343.2	28.41	
		353.2	25.58	
		363.2	21.98	
			$E_a(a) - E_a(b) = -15.5(8)\ kJ\ mol^{-1}$	
$\dot{C}H_3 + (n\text{-}C_4H_9)_3SnH \xrightarrow{a} CH_4 + (n\text{-}C_4H_9)_3\dot{S}n$				
$+ (n\text{-}C_4H_9)_3SnD \xrightarrow{b} CH_3D + (n\text{-}C_4H_9)_3\dot{S}n$				
Phot. of $\{(CH_3)_3CO\}_2 + (CH_3)_3As$				81 Cha 1
PR, KAS	2,2,4-trimethylpentane	300	$k_a/k_b = 2.3$	
$CH_3\dot{C}H_2 + (n\text{-}C_4H_9)_3SnH \xrightarrow{a} C_2H_6 + (n\text{-}C_4H_9)_3\dot{S}n$				
$+ (n\text{-}C_4H_9)_3SnD \xrightarrow{b} C_2H_5D + (n\text{-}C_4H_9)_3\dot{S}n$				
Phot. of $\{(CH_3)_3CO\}_2 + (C_2H_5)_3As$				81 Cha 1
PR, KAS	2,2,4-trimethylpentane	300	$k_a/k_b = 1.9$	

[1]) Specific activity measurements.
[4]) Specific activity of T-labelled compounds.
[5]) Based on k_a/k [66 Shi 1] and k_b/k [69 Shi 1] with k referring to $\dot{C}H_3 + n\text{-}C_7H_{16} \longrightarrow CH_4 + (n\text{-}C_7H_{15}^{\cdot})$.

Reaction				
Radical generation				Ref./
Method	Solvent	T[K]	Rate data	add. ref.

$(c\text{-}C_5H_9^{\boldsymbol{\cdot}}) + (C_4H_9)_3SnH \xrightarrow{\text{a}} c\text{-}C_5H_{10} + (C_4H_9)_3\dot{S}n$

$\phantom{(c\text{-}C_5H_9^{\boldsymbol{\cdot}})} + (C_4H_9)_3SnT \xrightarrow{\text{b}} c\text{-}C_5H_9T + (C_4H_9)_3\dot{S}n$

Thermal decomp. of AIBN as initiator				76 Koz 1
PR [6])	C_6H_6 or 1,2-dimethoxyethane	353	$k_a/k_b = 2.71$	

$(c\text{-}C_6H_{11}^{\boldsymbol{\cdot}}) + (CH_3)_3CSH \xrightarrow{\text{a}} c\text{-}C_6H_{12} + (CH_3)_3C\dot{S}$

$\phantom{(c\text{-}C_6H_{11}^{\boldsymbol{\cdot}})} + (CH_3)_3CSD \xrightarrow{\text{b}} c\text{-}C_6H_{11}D + (CH_3)_3C\dot{S}$

Not given				71 Pry 1/
not given	not given	298 [7])	k_a/k_b [8]) $= 2.96$	58 Swa 1
			$A_a/A_b = 0.21$	
			$E_a(a) - E_a(b) = -6.5\,\text{kJ mol}^{-1}$	

$(c\text{-}C_6H_{11}^{\boldsymbol{\cdot}}) + (CH_3)_3CSH\,[9]) \xrightarrow{\text{a}} c\text{-}C_6H_{12}\,[9]) + (CH_3)_3C\dot{S}$

$\phantom{(c\text{-}C_6H_{11}^{\boldsymbol{\cdot}})} + (CH_3)_3CSD\,[9]) \xrightarrow{\text{b}} c\text{-}C_6H_{11}D\,[9]) + (CH_3)_3C\dot{S}$

γ-rad. of $c\text{-}C_6H_{12}$ or decomp. of $C_6H_5COOOC_4H_9$				73 Pry 1
PR [10])	$c\text{-}C_6H_{12}$	298	$k_a/k_b = 2.7$	
	$(CH_3)_3CSH$		$k_a/k_b = 2.9$ [11])	
	or $(CH_3)_3CSD$ [11])			

$(c\text{-}C_6H_{11}^{\boldsymbol{\cdot}}) + (C_4H_9)_3SnH \xrightarrow{\text{a}} c\text{-}C_6H_{12} + (C_4H_9)_3\dot{S}n$

$\phantom{(c\text{-}C_6H_{11}^{\boldsymbol{\cdot}})} + (C_4H_9)_3SnT \xrightarrow{\text{b}} c\text{-}C_6H_{11}T + (C_4H_9)_3\dot{S}n$

Photoinitiated process				76 Koz 1
PR [12])	toluene	298	$k_a/k_b = 2.60$	
		353	$= 2.38$	

$CH_3(CH_2)_4\dot{C}H_2 + (C_4H_9)_3SnH \xrightarrow{\text{a}} n\text{-}C_6H_{14} + (C_4H_9)_3\dot{S}n$

$\phantom{CH_3(CH_2)_4\dot{C}H_2} + (C_4H_9)_3SnT \xrightarrow{\text{b}} n\text{-}C_6H_{13}T + (C_4H_9)_3\dot{S}n$

Thermal decomp. of AIBN as initiator and phot. of $n\text{-}C_6H_{13}Br$ solutions				76 Koz 1
PR [12])	C_6H_6 or	354	$k_a/k_b = 2.56$	
	1,2-dimethoxyethane	335	$k_a/k_b = 2.71$	
		295	$k_a/k_b = 2.95$	
		280	$k_a/k_b = 3.06$	
		298	$k_a/k_b = 3.07$ [13])	

Thermal decomp. of AIBN as initiator in solutions containing $n\text{-}C_6H_{11}Br$,				76 Koz 1
$n\text{-}C_6H_{11}Cl$, $n\text{-}C_6H_{11}I$				
PR [12])	C_6H_6 or	353	$k_a/k_b = 2.65$ [14])	
	1,2-dimethoxyethane		2.96 [15])	
			2.70 [16])	

Thermal decomp. of AIBN as initiator in $n\text{-}C_6H_{13}Br$ containing solutions				76 Koz 1
PR [12])	n-decane	353	$k_a/k_b = 2.55$	

---

[6]) Analysis of T-labelled products.
[7]) Extrapolated to 298 K from Arrhenius plot.
[8]) Based on $k(H)/k(T)$ measurements with T-labelled thiol and equation for $k(D)/k(T)$ ratio in [58 Swa 1].
[9]) T-labelled.
[10]) Specific activity of T-labelled products.
[11]) Generation by decomp. of $C_6H_5COOC_4H_9$.
[12]) Analysis of T-labelled products.
[13]) From phot. of $n\text{-}C_6H_{13}Cl$.
[14]) In solution containing $n\text{-}C_6H_{11}Br$.
[15]) In solution containing $n\text{-}C_6H_{11}Cl$.
[16]) In solution containing $n\text{-}C_6H_{11}I$.

Reaction Radical generation Method	Solvent	T[K]	Rate data	Ref./ add. ref.
$CH_3(CH_2)_3\dot{C}HCH_3 + (C_4H_9)_3SnH \xrightarrow{a} n\text{-}C_6H_{14} + (C_4H_9)_3\dot{S}n$ $+ (C_4H_9)_3SnT \xrightarrow{b} n\text{-}C_6H_{13}T + (C_4H_9)_3\dot{S}n$				
Thermal decomp. of AIBN as initiator PR [12]	C_6H_6 or 1,2-dimethoxyethane	353	$k_a/k_b = 2.72$	76 Koz 1
$CH_3(CH_2)_3\dot{C}HCH_3 + (C_4H_9)_3SnH \xrightarrow{a} n\text{-}C_6H_{14} + (C_4H_9)_3\dot{S}n$ $+ (C_4H_9)_3SnT \xrightarrow{b} CH_3(CH_2)_3CHTCH_3 + (C_4H_9)_3\dot{S}n$				
Thermal decomp. of AIBN as initiator PR [12]	n-decane	353	$k_a/k_b = 2.30$	76 Koz 1
$CH_3CH_2CH_2\dot{C}(CH_3)_2 + (C_4H_9)_3SnH \xrightarrow{a} CH_3CH_2CH_2CH(CH_3)_2 + (C_4H_9)_3\dot{S}n$ $+ (C_4H_9)_3SnT \xrightarrow{b} CH_3CH_2CH_2CT(CH_3)_2 + (C_4H_9)_3\dot{S}n$				
Thermal decomp. of AIBN as initiator PR [12]	C_6H_6 or 1,2-dimethoxyethane	353	$k_a/k_b = 2.53$	76 Koz 1
$CH_3CH_2CH_2\dot{C}(CH_3)_2 + (C_4H_9)_3SnH \xrightarrow{a} CH_3CH_2CH_2CH(CH_3)_2 + (C_4H_9)_3\dot{S}n$ $+ (C_4H_9)_3SnT \xrightarrow{b} CH_3CH_2CH_2CT(CH_3)_2 + (C_4H_9)_3\dot{S}n$				
Thermal decomp. of AIBN as initiator PR [12]	n-decane	353	$k_a/k_b = 2.14$	76 Koz 1
$CH_3CH_2\dot{C}H(CH_2)_3CH_3 + (CH_3)_3CSH \xrightarrow{a} n\text{-}C_7H_{16} + (CH_3)_3C\dot{S}$ $+ (CH_3)_3CSD \xrightarrow{b} CH_3CH_2CHD(CH_2)_3CH_3 + (CH_3)_3C\dot{S}$				
Not given not given	not given	298 [17]	k_a/k_b [18] $= 4.40$ $A_a/A_b = 0.87$ $E_a(a) - E_a(b) = -4.0\ \text{kJ mol}^{-1}$	71 Pry 1/ 58 Swa 1
$(C_2H_5)_3\dot{C} + (CH_3)_3CSH \xrightarrow{a} (C_2H_5)_3CH + (CH_3)_3C\dot{S}$ $+ (CH_3)_3CSD \xrightarrow{b} (C_2H_5)_3CD + (CH_3)_3C\dot{S}$				
Not given not given	not given	298 [17]	k_a/k_b [18] $= 5.35$ $A_a/A_b = 1.14$ $E_a(a) - E_a(b) = -3.4\ \text{kJ mol}^{-1}$	71 Pry 1/ 58 Swa 1
$CH_3(CH_2)_7\dot{C}H_2 + (CH_3)_3CSH \xrightarrow{a} n\text{-}C_9H_{20} + (CH_3)_3C\dot{S}$ $+ (CH_3)_3CSD \xrightarrow{b} n\text{-}C_9H_{19}D + (CH_3)_3C\dot{S}$				
Not given not given	not given	298 [17]	k_a/k_b [18] $= 3.40$ $A_a/A_b = 0.99$ $E_a(a) - E_a(b) = -3.1\ \text{kJ mol}^{-1}$	71 Pry 1/ 58 Swa 1
(1-adamantyl)$^{\cdot}$ $+ (CH_3)_3CSH \xrightarrow{a}$ adamantane $+ (CH_3)_3C\dot{S}$ $+ (CH_3)_3CSD \xrightarrow{b}$ adamantane-d_1 $+ (CH_3)_3C\dot{S}$				
Not given not given	not given	298 [17]	k_a/k_b [18] $= 1.89$ $A_a/A_b = 0.53$ $E_a(a) - E_a(b) = -3.1\ \text{kJ mol}^{-1}$	71 Pry 1/ 58 Swa 1

[12]) Analysis of T-labelled products.
[17]) Extrapolated to 298 K from Arrhenius plot.
[18]) Based on $k(H)/k(T)$ measurements with T-labelled thiol and equation for $k(D)/k(T)$ ratio in [58 Swa 1].

Reaction Radical generation Method	Solvent	T[K]	Rate data	Ref./ add. ref.
4.1.3.1.2 Radicals containing C, H, and other atoms				
$\dot{C}Cl_3 + c\text{-}C_6H_{12} \xrightarrow{a} CHCl_3 + (c\text{-}C_6\dot{H}_{11})$ $+ c\text{-}C_6D_{12} \xrightarrow{b} CDCl_3 + (c\text{-}C_6\dot{D}_{11})$				
γ-rad. of $c\text{-}C_6H_{12} + c\text{-}C_6D_{12} + CCl_4$				80 Ngu 1
PR, glc	$c\text{-}C_6H_{12}$/ $c\text{-}C_6D_{12}$/ CCl_4	301.7(5) 317.7(5) 336.2(5) 357.2(5) 384.2(5) 413.2(5)	k_a/k_b = 23.5(2) 19.9(3) 16.4(3) 13.1(1) 10.0(1) 7.9(1) $\log[A_a/A_b] = -0.38(4)$ $E_a(a) - E_a(b) = 10.1(3)\,kJ\,mol^{-1}$	
$Br\dot{C}HCH_2Br + HBr \xrightarrow{a} BrCH_2CH_2Br + \dot{B}r$ $+ TBr \xrightarrow{b} BrCHTCH_2Br + \dot{B}r$				
Phot. of AIBN as initiator				73 Lew 1
[19])	diethylether	273	$k_a/k_b = 2.83$	
$\dot{C}H_2COOH + C_6H_5CH_3 \xrightarrow{a} CH_3COOH + C_6H_5\dot{C}H_2$ $+ C_6H_5CD_3 \xrightarrow{b} CH_2DCOOH + C_6H_5\dot{C}D_2$				
Therm. of $Mn(OOCCH_3)_3$				69 Hei 1
PR, glc	glacial acetic acid	403(1)	$k_a/k_b = 5.46$	
$\dot{C}H_2CH_2Br + HBr \xrightarrow{a} CH_3CH_2Br + \dot{B}r$ $+ TBr \xrightarrow{b} CH_2TCH_2Br + \dot{B}r$				
Phot. of AIBN as initiator				73 Lew 1
[19])	diethylether	273	$k_a/k_b = 4.20$	
$\dot{C}H_2CH_2OH + CH_3CH_2OH \xrightarrow{a} CH_3CH_2OH + CH_3\dot{C}HOH$ $+ CH_3CD_2OH \xrightarrow{b} CH_2DCH_2OH + CH_3\dot{C}DOH$				
γ-rad. of ethanol + H_2O				71 Bur 2
PR	H_2O	RT	$k_a/k_b = 2.9(3)$	
$\dot{C}H_2CH(CH_3)OH + (CH_3)_2CHOH \xrightarrow{a} (CH_3)_2CHOH + (CH_3)_2\dot{C}OH$ $+ (CH_3)_2CDOH \xrightarrow{b} CH_2D(CH_3)CHOH + (CH_3)_2\dot{C}OH$				
γ-rad. of 2-propanol + H_2O				71 Bur 2
PR	H_2O	RT	$k_a/k_b = 7.3(15)$	
$(CH_3)_2\dot{C}CN + C_6H_5SD \xrightarrow{a} (CH_3)_2CDCN + C_6H_5\dot{S}$ $+ C_6H_5ST \xrightarrow{b} (CH_3)_2CTCN + C_6H_5\dot{S}$				
Decomp. of $(CH_3)_2C(CN)N{=}N(CN)C(CH_3)_2$				76 Lew 2/
PR [20])	thiophenol (89% C_6H_5SD)	376(1)	$k_a/k_b = 1.45(\pm 3\%)$	58 Swa 1 [21])
$(CH_3)_2\dot{C}COOCH_3 + C_6H_5SD \xrightarrow{a} (CH_3)_2CDCOOCH_3 + C_6H_5\dot{S}$ $+ C_6H_5ST \xrightarrow{b} (CH_3)_2CTCOOCH_3 + C_6H_5\dot{S}$				
Decomp. of $(CH_3)_2C(COOCH_3)N{=}N(CH_3OOC)C(CH_3)_2$				76 Lew 2/
PR [20])	thiophenol (89% C_6H_5SD)	376(1)	$k_a/k_b = 1.24(\pm 3\%)$	58 Swa 1 [22])

[19]) Measurement of T-labelled compounds.
[20]) Specific activity, NMR.
[21]) $k(H)/k(T) = 3.36$ calculated from $k(H)/k(T) = (k(D)/k(T))^{3.262}$.
[22]) $k(H)/k(T) = 3.14$ calculated from $k(H)/k(T) = (k(D)/k(T))^{3.262}$.

Reaction Radical generation Method	Solvent	T[K]	Rate data	Ref./ add. ref.
$NCCH_2\dot{C}HCH_2PS(OCH_3)_2 + HPS(OCH_3)_2 \xrightarrow{a} NC(CH_2)_3PS(OCH_3)_2 + \dot{P}S(OCH_3)_2$ $+ TPS(OCH_3)_2 \xrightarrow{b} NCCH_2CHTCH_2PS(OCH_3)_2 + \dot{P}S(OCH_3)_2$				
$\dot{P}S(OCH_3)_2$ add. to $NCCH_2CH{=}CH_2$ (AIBN catalyzed)				76 Lew 3
PR [23])	$HPS(OCH_3)_2$	353.0(1)	$k_a/k_b = 8.3(\pm 5\%)$	
$(CH_3)_3C\dot{C}HCH_2Br + HBr \xrightarrow{a} (CH_3)_3CCH_2CH_2Br + \dot{B}r$ $+ TBr \xrightarrow{b} (CH_3)_3CCHTCH_2Br + \dot{B}r$				
Phot. of AIBN as initiator				73 Lew 1
[26])	diethylether	273	$k_a/k_b = 1.99$	
$(CH_3)_2N\dot{C}HCOOC_2H_5 + (CH_3)_3CSH$ [24]) $\xrightarrow{a} (CH_3)_2NCH_2COOC_2H_5$ [24]) $+ (CH_3)_3C\dot{S}$ $+ (CH_3)_3CSD \xrightarrow{b} (CH_3)_2NCHDCOOC_2H_5$ [24]) $+ (CH_3)_3C\dot{S}$				
γ-rad. of $(CH_3)_2NCH_2COOC_2H_5$				73 Pry 1
PR [25])	$(CH_3)_2NCH_2$- $COOC_2H_5$	298	$k_a/k_b = 4.7$	
$CH_3OCO\dot{C}HCH_2PS(OCH_3)_2 + HPS(OCH_3)_2 \xrightarrow{a} CH_3OCOCH_2CH_2PS(OCH_3)_2 + \dot{P}S(OCH_3)_2$ $+ TPS(OCH_3)_2 \xrightarrow{b} CH_3OCOCHTCH_2PS(OCH_3)_2 + \dot{P}S(OCH_3)_2$				
$\dot{P}S(OCH_3)_2$ add. to $CH_3OCOCH{=}CH_2$ (AIBN catalyzed)				76 Lew 3
PR [23])	$HPS(OCH_3)_2$	353.0(1)	$k_a/k_b = 6.7(\pm 5\%)$	
$CH_3COO\dot{C}HCH_2PS(OCH_3)_2 + HPS(OCH_3)_2 \xrightarrow{a} CH_3COOCH_2CH_2PS(OCH_3)_2 + \dot{P}S(OCH_3)_2$ $+ TPS(OCH_3)_2 \xrightarrow{b} CH_3COOCHTCH_2PS(OCH_3)_2 + \dot{P}S(OCH_3)_2$				
$\dot{P}S(OCH_3)_2$ add. to $CH_3COOCH{=}CH_2$ (AIBN catalyzed)				76 Lew 3
PR [23])	$HPS(OCH_3)_2$	353.0(1)	$k_a/k_b = 8.1(\pm 5\%)$	
$C_2H_5O\dot{C}HCH_2PS(OCH_3)_2 + HPS(OCH_3)_2 \xrightarrow{a} C_2H_5OCH_2CH_2PS(OCH_3)_2 + \dot{P}S(OCH_3)_2$ $+ TPS(OCH_3)_2 \xrightarrow{b} C_2H_5OCHTCH_2PS(OCH_3)_2 + \dot{P}S(OCH_3)_2$				
$\dot{P}S(OCH_3)_2$ add. to $C_2H_5OCH{=}CH_2$ (AIBN catalyzed)				76 Lew 3
PR [23])	$HPS(OCH_3)_2$	353.0(1)	$k_a/k_b = 8.6(\pm 5\%)$	
$C_2H_5O\dot{C}HCH_2PO(OCH_3)_2 + HPO(OCH_3)_2 \xrightarrow{a} C_2H_5OCH_2CH_2PO(OCH_3)_2 + \dot{P}O(OCH_3)_2$ $+ TPO(OCH_3)_2 \xrightarrow{b} C_2H_5OCHTCH_2PO(OCH_3)_2 + \dot{P}O(OCH_3)_2$				
$\dot{P}O(OCH_3)_2$ add. to $C_2H_5OCH{=}CH_2$ (PPO catalyzed)				76 Lew 3
PR [23])	$HPO(OCH_3)_2$	363.0(1)	$k_a/k_b = 6.1(\pm 5\%)$	
$CH_3COO\dot{C}HCH_2PO(OCH_3)_2 + HPO(OCH_3)_2 \xrightarrow{a} CH_3COOCH_2CH_2PO(OCH_3)_2 + \dot{P}O(OCH_3)_2$ $+ TPO(OCH_3)_2 \xrightarrow{b} CH_3COOCHTCH_2PO(OCH_3)_2 + \dot{P}O(OCH_3)_2$				
$\dot{P}O(OCH_3)_2$ add. to $CH_3COOCH{=}CH_2$ (BPO catalyzed)				76 Lew 3
PR [23])	$HPO(OCH_3)_2$	363.0(1)	$k_a/k_b = 12.4(\pm 5\%)$	
$\underline{CH_2(CH_2)_2\dot{C}H}CHPO(OCH_3)_2 + HPO(OCH_3)_2 \xrightarrow{a} \underline{CH_2(CH_2)_3C}HPO(OCH_3)_2 + \dot{P}O(OCH_3)_2$ $+ TPO(OCH_3)_2 \xrightarrow{b} \underline{CH_2(CH_2)_2CHTC}HPO(OCH_3)_2 + \dot{P}O(OCH_3)_2$				
$\dot{P}O(OCH_3)_2$ add. to $\underline{CH_2(CH_2)_2CH{=}CH}$ (BPO catalyzed)				76 Lew 3
PR [23])	$HPO(OCH_3)_2$	363.0(1)	$k_a/k_b = 7.5(\pm 5\%)$	

[23]) Specific activity.
[24]) T-labelled.
[25]) Specific activity of T-labelled products.
[26]) Measurement of T-labelled compounds.

Reaction Radical generation Method	Solvent	T[K]	Rate data	Ref./ add. ref.
$CH_3COOCH_2\dot{C}HCH_2PS(OCH_3)_2 + HPS(OCH_3)_2 \xrightarrow{a} CH_3COO(CH_2)_3PS(OCH_3)_2 + \dot{P}S(OCH_3)_2$ $+ TPS(OCH_3)_2 \xrightarrow{b} CH_3COOCH_2CHTCH_2PS(OCH_3)_2 + \dot{P}S(OCH_3)_2$				
$\dot{P}S(OCH_3)_2$ add. to $CH_3COOCH_2CH{=}CH_2$ (AIBN catalyzed)				76 Lew 3
PR [23]	$HPS(OCH_3)_2$	353.0(1)	$k_a/k_b = 5.6(\pm 5\%)$	
$CH_3COO\dot{C}(CH_3)CH_2PS(OCH_3)_2 + HPS(OCH_3)_2 \xrightarrow{a} CH_3COOCH(CH_3)CH_2PS(OCH_3)_2 + \dot{P}S(OCH_3)_2$ $+ TPS(OCH_3)_2 \xrightarrow{b} CH_3COOCT(CH_3)CH_2PS(OCH_3)_2 + \dot{P}S(OCH_3)_2$				
$\dot{P}S(OCH_3)_2$ add. to $CH_3COOC(CH_3){=}CH_2$ (AIBN catalyzed)				76 Lew 3
PR [23]	$HPS(OCH_3)_2$	353.0(1)	$k_a/k_b = 7.2(\pm 5\%)$	
$CH_3COOCH_2\dot{C}HCH_2PO(OCH_3)_2 + HPO(OCH_3)_2 \xrightarrow{a} CH_3COO(CH_2)_3PO(OCH_3)_2 + \dot{P}O(OCH_3)_2$ $+ TPO(OCH_3)_2 \xrightarrow{b}$ $CH_3COOCH_2CHTCH_2PO(OCH_3)_2 + \dot{P}O(OCH_3)_2$				
$\dot{P}O(OCH_3)_2$ add. to $CH_3COOCH_2CH{=}CH_2$ (BPO catalyzed)				76 Lew 3
PR [23]	$HPO(OCH_3)_2$	363.0(1)	$k_a/k_b = 4.9(\pm 5\%)$	
$CH_3COO\dot{C}(CH_3)CH_2PO(OCH_3)_2 + HPO(OCH_3)_2 \xrightarrow{a}$ $CH_3COOCH(CH_3)CH_2PO(OCH_3)_2 + \dot{P}O(OCH_3)_2$ $+ TPO(OCH_3)_2 \xrightarrow{b}$ $CH_3COOCT(CH_3)CH_2PO(OCH_3)_2 + \dot{P}O(OCH_3)_2$				
$PO(OCH_3)_2$ add. to $CH_3COOC(CH_3){=}CH_2$ (BPO catalyzed)				76 Lew 3
PR [23]	$HPO(OCH_3)_2$	363.0(1)	$k_a/k_b = 10.0(\pm 5\%)$	
$CH_3(CH_2)_5\dot{C}HCH_2Br + HBr \xrightarrow{a} n\text{-}C_8H_{17}Br + \dot{B}r$ $+ TBr \xrightarrow{b} CH_3(CH_2)_5CHTCH_2Br + \dot{B}r$				
Phot. of AIBN as initiator				73 Lew 1
[26]	diethylether	273	$k_a/k_b = 2.07$	
$\underline{CH_2(CH_2)_3\dot{C}HC}HPO(OCH_3)_2 + HPO(OCH_3)_2 \xrightarrow{a} \underline{CH_2(CH_2)_4C}HPO(OCH_3)_2 + \dot{P}O(OCH_3)_2$ $+ TPO(OCH_3)_2 \xrightarrow{b} \underline{CH_2(CH_2)_3CHTC}HPO(OCH_3)_2 + \dot{P}O(OCH_3)_2$				
$\dot{P}O(OCH_3)_2$ add. to $\underline{CH_2(CH_2)_3CH{=}CH}$ (BPO catalyzed)				73 Lew 3
PR [23]	$HPO(OCH_3)_2$	363.0(1)	$k_a/k_b = 9.1(\pm 5\%)$	
$CH_3(CH_2)_3\dot{C}{=}CHPO(OCH_3)_2 + HPO(OCH_3)_2 \xrightarrow{a} n\text{-}C_4H_9CH{=}CHPO(OCH_3)_2 + \dot{P}O(OCH_3)_2$ $+ TPO(OCH_3)_2 \xrightarrow{b} n\text{-}C_4H_9CT{=}CHPO(OCH_3)_2 + \dot{P}O(OCH_3)_2$				
$\dot{P}O(OCH_3)_2$ add. to $n\text{-}C_4H_9C{\equiv}CH$ (BPO catalyzed)				76 Lew 3
PR [23]	$HPO(OCH_3)_2$	363.0(1)	$k_a/k_b = 6.4(\pm 5\%)$	
$CH_3COOCH_2\dot{C}(CH_3)CH_2PS(OCH_3)_2 + HPS(OCH_3)_2 \xrightarrow{a}$ $CH_3COOCH_2CH(CH_3)CH_2PS(OCH_3)_2 + \dot{P}S(OCH_3)_2$ $+ TPS(OCH_3)_2 \xrightarrow{b}$ $CH_3COOCH_2CT(CH_3)CH_2PS(OCH_3)_2 + \dot{P}S(OCH_3)_2$				
$\dot{P}S(OCH_3)_2$ add. to $CH_3COOCH_2C(CH_3){=}CH_2$ (AIBN catalyzed)				76 Lew 3
PR [23]	$HPS(OCH_3)_2$	353.0(1)	$k_a/k_b = 7.6(\pm 5\%)$	
$CH_3COOCH_2\dot{C}(CH_3)CH_2PO(OCH_3)_2 + HPO(OCH_3)_2 \xrightarrow{a}$ $CH_3COOCH_2CH(CH_3)CH_2PO(OCH_3)_2 + \dot{P}O(OCH_3)_2$ $+ TPO(OCH_3)_2 \xrightarrow{b}$ $CH_3COOCH_2CT(CH_3)CH_2PO(OCH_3)_2 + \dot{P}O(OCH_3)_2$				
$\dot{P}O(OCH_3)_2$ add. to $CH_3COOCH_2C(CH_3){=}CH_2$ (BPO catalyzed)				76 Lew 3
PR [23]	$HPO(OCH_3)_2$	363.0(1)	$k_a/k_b = 6.0(\pm 5\%)$	

[23] Specific activity.
[26] Measurement of T-labelled compounds.

Reaction Radical generation Method	Solvent	T[K]	Rate data	Ref./ add. ref.
$(CH_3)_3C\dot{C}HCH_2PS(OCH_3)_2 + HPS(OCH_3)_2 \xrightarrow{a} (CH_3)_3CCH_2CH_2PS(OCH_3)_2 + \dot{P}S(OCH_3)_2$ $+ TPS(OCH_3)_2 \xrightarrow{b} (CH_3)_3CCHTCH_2PS(OCH_3)_2 + \dot{P}S(OCH_3)_2$				
$\dot{P}S(OCH_3)_2$ add. to $(CH_3)_3CCH{=}CH_2$ (AIBN catalyzed)				76 Lew 3
PR [23]	$HPS(OCH_3)_2$	353.0(1)	$k_a/k_b = 6.4(\pm 5\%)$	
$(CH_3)_3C\dot{C}HCH_2PO(OCH_3)_2 + HPO(OCH_3)_2 \xrightarrow{a} (CH_3)_3CCH_2CH_2PO(OCH_3)_2 + \dot{P}O(OCH_3)_2$ $+ TPO(OCH_3)_2 \xrightarrow{b} (CH_3)_3CCHTCH_2PO(OCH_3)_2 + \dot{P}O(OCH_3)_2$				
$\dot{P}O(OCH_3)_2$ add. to $(CH_3)_3CCH{=}CH_2$ (BPO catalyzed)				76 Lew 3
PR [23]	$HPO(OCH_3)_2$	363.0(1)	$k_a/k_b = 6.1(\pm 5\%)$	
$(CH_3)_3SiCH_2\dot{C}HCH_2PS(OCH_3)_2 + HPS(OCH_3)_2 \xrightarrow{a} (CH_3)_3Si(CH_2)_3PS(OCH_3)_2 + \dot{P}S(OCH_3)_2$ $+ TPS(OCH_3)_2 \xrightarrow{b} (CH_3)_3SiCH_2CHTCH_2PS(OCH_3)_2 + \dot{P}S(OCH_3)_2$				
$\dot{P}S(OCH_3)_2$ add. to $(CH_3)_3SiCH_2CH{=}CH_2$ (AIBN catalyzed)				76 Lew 3
PR [23]	$HPS(OCH_3)_2$	353.0(1)	$k_a/k_b = 9.6(\pm 5\%)$	
$(CH_3)_3SiCH_2\dot{C}HCH_2PO(OCH_3)_2 + HPO(OCH_3)_2 \xrightarrow{a} (CH_3)_3Si(CH_2)_3PO(OCH_3)_2 + \dot{P}O(OCH_3)_2$ $+ TPO(OCH_3)_2 \xrightarrow{b} (CH_3)_3SiCH_2CHTCH_2PO(OCH_3)_2 + \dot{P}O(OCH_3)_2$				
$\dot{P}O(OCH_3)_2$ add. to $(CH_3)_3SiCH_2CH{=}CH_2$ (BPO catalyzed)				76 Lew 3
PR [23]	$HPO(OCH_3)_2$	363.0(1)	$k_a/k_b = 11.1(\pm 5\%)$	
$\underline{CH_2(CH_2)_5\dot{C}HC}HPS(OCH_3)_2 + HPS(OCH_3)_2 \xrightarrow{a} \underline{CH_2(CH_2)_6C}HPS(OCH_3)_2 + \dot{P}S(OCH_3)_2$ $+ TPS(OCH_3)_2 \xrightarrow{b} \underline{CH_2(CH_2)_5CHTC}HPS(OCH_3)_2 + \dot{P}S(OCH_3)_2$				
$\dot{P}S(OCH_3)_2$ add. to $\underline{CH_2(CH_2)_5CH{=}CH}$ (AIBN catalyzed)				76 Lew 3
PR [23]	$HPS(OCH_3)_2$	353.0(1)	$k_a/k_b = 7.3(\pm 5\%)$	
$\underline{CH_2(CH_2)_5\dot{C}HC}HPO(OCH_3)_2 + HPO(OCH_3)_2 \xrightarrow{a} \underline{CH_2(CH_2)_6C}HPO(OCH_3)_2 + \dot{P}O(OCH_3)_2$ $+ TPO(OCH_3)_2 \xrightarrow{b} \underline{CH_2(CH_2)_5CHTC}HPO(OCH_3)_2 + \dot{P}O(OCH_3)_2$				
$\dot{P}O(OCH_3)_2$ add. to $\underline{CH_2(CH_2)_5CH{=}CH}$ (BPO catalyzed)				76 Lew 3
PR [23]	$HPO(OCH_3)_2$	363.0(1)	$k_a/k_b = 8.4(\pm 5\%)$	
$CH_3(CH_2)_5\dot{C}HCH_2PS(OCH_3)_2 + HPS(OCH_3)_2 \xrightarrow{a} n\text{-}C_8H_{17}PS(OCH_3)_2 + \dot{P}S(OCH_3)_2$ $+ TPS(OCH_3)_2 \xrightarrow{b} n\text{-}C_6H_{13}CHTCH_2PS(OCH_3)_2 + \dot{P}S(OCH_3)_2$				
$\dot{P}S(OCH_3)_2$ add. to $n\text{-}C_6H_{13}CH{=}CH_2$ (AIBN catalyzed)				76 Lew 3
PR [23]	$HPS(OCH_3)_2$	353.0(1)	$k_a/k_b = 6.8(\pm 5\%)$	
$CH_3(CH_2)_5\dot{C}HCH_2PO(OCH_3)_2 + HPO(OCH_3)_2 \xrightarrow{a} n\text{-}C_8H_{17}PO(OCH_3)_2 + \dot{P}O(OCH_3)_2$ $+ TPO(OCH_3)_2 \xrightarrow{b} n\text{-}C_6H_{13}CHTCH_2PO(OCH_3)_2 + \dot{P}O(OCH_3)_2$				
$\dot{P}O(OCH_3)_2$ add. to $n\text{-}C_6H_{13}CH{=}CH_2$ (BPO catalyzed)				76 Lew 3
PR [23]	$HPO(OCH_3)_2$	363.0(1)	$k_a/k_b = 8.7(\pm 5\%)$	
(Polyvinylacetate)$^{\cdot}$ [27] + 1,2-dihydroxybenzene(R—OH) $\xrightarrow{a}$ products + R—$\dot{O}$ + (R—OD) $\xrightarrow{b}$ products-d_1 + R—$\dot{O}$				
AIBN as initiator in vinylacetate				77 Sim 1/
[28]	vinylacetate	323	$k_a/k_b = 13.0$ [29]	67 Sim 2

[23]) Specific activity.
[27]) Radical mixture.
[28]) Calculated from inhibition rate of polymerization (dilatometric measurement).
[29]) $k_a/k_b > 10$ indicates tunneling.

Reaction Radical generation Method	Solvent	$T[K]$	Rate data	Ref./ add. ref.

$(\text{Polyvinylacetate})^{\cdot\,27)} + \text{1,4-dihydroxybenzene}(\text{R}\!-\!\text{OH}) \xrightarrow{a} \text{products} + \text{R}\!-\!\dot{\text{O}}$

$\qquad\qquad + (\text{R}\!-\!\text{OD}) \xrightarrow{b} \text{products-}d_1 + \text{R}\!-\!\dot{\text{O}}$

AIBN as initiator in vinylacetate $^{28})$	vinylacetate	323	$k_a/k_b = 9.5$	77 Sim 1/ 67 Sim 2

4.1.3.2 Aromatic radicals and radicals derived from compounds containing aromatic and heterocyclic constituents

4.1.3.2.1 Radicals containing only C and H atoms

$(\text{C}_6\text{H}_5^{\cdot}) + (\text{CH}_3)_3\text{CSH} \xrightarrow{a} \text{C}_6\text{H}_6 + (\text{CH}_3)_3\text{C\dot{S}}$

$\qquad\qquad + (\text{CH}_3)_3\text{CSD} \xrightarrow{b} \text{C}_6\text{H}_5\text{D} + (\text{CH}_3)_3\text{C\dot{S}}$

Not given not given	not given	$298\,^{30})$	$k_a/k_b\,^{31}) = 1.88$ $A_a/A_b = 0.42$ $E_a(a) - E_a(b) = -3.8\,\text{kJ mol}^{-1}$	71 Pry 1/ 58 Swa 1

$(\text{C}_6\text{H}_5^{\cdot}) + \text{4-ClC}_6\text{H}_4\text{SH} \xrightarrow{a} \text{C}_6\text{H}_6 + \text{4-ClC}_6\text{H}_4\dot{\text{S}}$

$\qquad\qquad + \text{4-ClC}_6\text{H}_4\text{ST} \xrightarrow{b} \text{C}_6\text{H}_5\text{T} + \text{4-ClC}_6\text{H}_4\dot{\text{S}}$

Decomp. of PAT PR $^{32})$	4-ClC$_6$H$_4$SH	333(1)	$k_a/k_b = 1.84$	76 Lew 2

$(\text{C}_6\text{H}_5^{\cdot}) + \text{C}_6\text{H}_5\text{SH} \xrightarrow{a} \text{C}_6\text{H}_6 + \text{C}_6\text{H}_5\dot{\text{S}}$

$\qquad\qquad + \text{C}_6\text{H}_5\text{ST} \xrightarrow{b} \text{C}_6\text{H}_5\text{T} + \text{C}_6\text{H}_5\dot{\text{S}}$

Decomp. of PAT PR $^{32})$	C$_6$H$_5$SH	333(1)	$k_a/k_b = 1.66$	76 Lew 2

$(\text{C}_6\text{H}_5^{\cdot}) + \text{C}_6\text{H}_5\text{CH}_3 \xrightarrow{a} \text{C}_6\text{H}_6 + \text{C}_6\text{H}_5\dot{\text{C}}\text{H}_2$

$\qquad\qquad + \text{C}_6\text{H}_5\text{CD}_3 \xrightarrow{b} \text{C}_6\text{H}_5\text{D} + \text{C}_6\text{H}_5\dot{\text{C}}\text{D}_2$

Therm. of PAT PR, glc	C$_6$H$_5$CH$_3$ or C$_6$H$_5$CD$_3$/CCl$_4$	333	$k_a/k_b = 3.4\,^{33})$	63 Rus 3

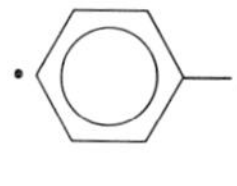

$\cdot\langle\!\bigcirc\!\rangle\!-\!\text{CH}_3 + (\text{CH}_3)_2\text{CHOH} \xrightarrow{a} \text{C}_6\text{H}_5\text{CH}_3 + \dot{\text{C}}\text{H}_2\text{CH}(\text{CH}_3)\text{OH}$

$\qquad\qquad + (\text{CD}_3)_2\text{CHOH} \xrightarrow{b} \text{C}_6\text{H}_5\text{CH}_2\text{D} + \dot{\text{C}}\text{D}_2\text{CH}(\text{CD}_3)\text{OH}$

γ-rad. of 4-CH$_3$C$_6$H$_4$N$_2^+$BF$_4^-$ + H$_2$O PR	H$_2$O	RT	$k_a/k_b = 3.7$	75 Pac 1 $^{34})$

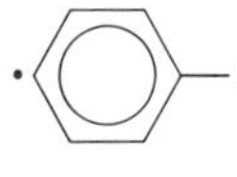

$\cdot\langle\!\bigcirc\!\rangle\!-\!\text{CH}_3 + \text{C}_6\text{H}_5\text{SH} \xrightarrow{a} \text{C}_6\text{H}_5\text{CH}_3 + \text{C}_6\text{H}_5\dot{\text{S}}$

$\qquad\qquad + \text{C}_6\text{H}_5\text{ST} \xrightarrow{b} \text{4-TC}_6\text{H}_4\text{CH}_3 + \text{C}_6\text{H}_5\dot{\text{S}}$

Decomp. of 4-CH$_3$C$_6$H$_4$N=NC(C$_6$H$_5$)$_3$ PR $^{32})$	C$_6$H$_5$SH	333(1)	$k_a/k_b = 1.63$	76 Lew 2

$^{27})$ Radical mixture.
$^{28})$ Calculated from inhibition rate of polymerization (dilatometric measurement).
$^{30})$ Extrapolated to 298 K from Arrhenius plot.
$^{31})$ Based on $k(\text{H})/k(\text{T})$ measurements with T-labelled thiol and equation for $k(\text{D})/k(\text{T})$ ratio in [58 Swa 1].
$^{32})$ Specific activity.
$^{33})$ Calculated from $k(\text{D})/k(\text{Cl})$ and $k(\text{H})/k(\text{Cl})$ with $k(\text{Cl})$ referring to $(\text{C}_6\text{H}_5^{\cdot}) + \text{CCl}_4 \longrightarrow \text{C}_6\text{H}_5\text{Cl} + \dot{\text{C}}\text{Cl}_3$.
$^{34})$ β-abstraction accounts only for $\approx 2\%$ H and $\approx 0.5\%$ D of total reaction.

Reaction Radical generation Method	Solvent	T[K]	Rate data	Ref./ add. ref.
$C_6H_5\dot{C}H_2 + (CH_3)_3CSH \xrightarrow{a} C_6H_5CH_3 + (CH_3)_3C\dot{S}$ $+ (CH_3)_3CSD \xrightarrow{b} C_6H_5CH_2D + (CH_3)_3C\dot{S}$				
Not given not given	not given	298 [30]	k_a/k_b [31] $= 6.35$ $A_a/A_b = 0.89$ $E_a(a) - E_a(b) = -4.9\,kJ\,mol^{-1}$	71 Pry 1/ 58 Swa 1
$C_6H_5\dot{C}H_2 + C_6H_5SH \xrightarrow{a} C_6H_5CH_3 + C_6H_5\dot{S}$ $+ C_6H_5SD \xrightarrow{b} C_6H_5CH_2D + C_6H_5\dot{S}$				
Not given not given	not given	298 [30]	k_a/k_b [31] $= 3.90$ $A_a/A_b = 0.61$ $E_a(a) - E_a(b) = -4.6\,kJ\,mol^{-1}$	71 Pry 1/ 58 Swa 1
$C_6H_5\dot{C}H_2 + 3\text{-}DC_6H_4CH_3 \xrightarrow{a} C_6H_5CH_3 + 3\text{-}DC_6H_4\dot{C}H_2$ $+ C_6D_5CD_3 \xrightarrow{b} C_6H_5CH_2D + C_6D_5\dot{C}D_2$				
Therm. of $(C_6H_5CH_2)_2Hg$ PR, MS	3-D-toluene	441 428	$k_a/k_b = 6.75$ 7.65 for k_a: $\log[A/M^{-1}s^{-1}] = 10.5(37)$ $E_a = 83.3(88)\,kJ\,mol^{-1}$	69 Jac 1 [35]
$C_6H_5\dot{C}H_2 + (C_4H_9)_3SnH \xrightarrow{a} C_6H_5CH_3 + (C_4H_9)_3\dot{S}n$ $+ (C_4H_9)_3SnT \xrightarrow{b} C_6H_5CH_2T + (C_4H_9)_3\dot{S}n$				
Thermal decomp. of AIBN as initiator and phot. of $C_6H_5CH_2Cl$ containing solutions				76 Koz 1
PR [36]	C_6H_6 or 1,2-dimethoxyethane	352 332 294 277	$k_a/k_b = 3.9$ 4.32 5.69 6.32	
$C_6H_5\dot{C}H_2 + (C_4H_9)_3SnH \xrightarrow{a} C_6H_5CH_3 + (C_4H_9)_3\dot{S}n$ $+ (C_4H_9)_3SnT \xrightarrow{b} C_6H_5CH_2T + (C_4H_9)_3\dot{S}n$				
Thermal decomp. of AIBN as initiator and benzyl radical from $C_6H_5CH_2Cl$, $C_6H_5CH_2Br$ and $C_6H_5CH_2I$				76 Koz 1
PR [36]	C_6H_6 or 1,2-dimethoxyethane	353	$k_a/k_b = 4.12$ [37] 4.01 [38] 3.86 [39]	
$3\text{-}CH_3C_6H_4\dot{C}H_2 + (C_4H_9)_3SnH \xrightarrow{a} 3\text{-}CH_3C_6H_4CH_3 + (C_4H_9)_3\dot{S}n$ $+ (C_4H_9)_3SnT \xrightarrow{b} 3\text{-}CH_3C_6H_4CH_2T + (C_4H_9)_3\dot{S}n$				
Thermal decomp. of AIBN as initiator PR [36]	C_6H_6 or 1,2-dimethoxyethane	353	$k_a/k_b = 3.82$	76 Koz 1
$4\text{-}CH_3C_6H_4\dot{C}H_2 + (C_4H_9)_3SnH \xrightarrow{a} 4\text{-}CH_3C_6H_4CH_3 + (C_4H_9)_3\dot{S}n$ $+ (C_4H_9)_3SnT \xrightarrow{b} 4\text{-}CH_3C_6H_4CH_2T + (C_4H_9)_3\dot{S}n$				
Thermal decomp. of AIBN as initiator PR [36]	C_6H_6 or 1,2-dimethoxyethane	353	$k_a/k_b = 3.92$	76 Koz 1

[30]) Extrapolated to 298 K from Arrhenius plot.
[31]) Based on $k(H)/k(T)$ measurements with T-labelled thiol and equation for $k(D)/k(T)$ ratio in [58 Swa 1].
[35]) Tunnelling mechanism assumed.
[36]) Analysis of T-labelled products.
[37]) Benzyl radical from $C_6H_5CH_2Cl$.
[38]) Benzyl radical from $C_6H_5CH_2Br$.
[39]) Benzyl radical from $C_6H_5CH_2I$.

Reaction Radical generation Method	Solvent	T[K]	Rate data	Ref./ add. ref.
$(CH_3)_2\dot{C}C_6H_5 + C_6H_5SD \xrightarrow{a} (CH_3)_2CDC_6H_5 + C_6H_5\dot{S}$ $+ C_6H_5ST \xrightarrow{b} (CH_3)_2CTC_6H_5 + C_6H_5\dot{S}$				
Decomp. of $(CH_3)_2C(C_6H_5)N{=}N(C_6H_5)C(CH_3)_2$				76 Lew 2/
PR [40]	thiophenol (80% C_6H_5SD)	376(1)	$k_a/k_b = 1.50(\pm 3\%)$	58 Swa 1 [41]
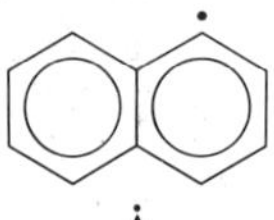 $+ CH_3SOCH_3 \xrightarrow{a} C_{10}H_8 + CH_3SO\dot{C}H_2$ Ȧ $+ CD_3SOCD_3 \xrightarrow{b} C_{10}H_7D + CD_3SO\dot{C}D_2$				
Electrochem. reduct. of 1-iodonaphthalene(A), 1-bromonaphthalene(B), 1-chloronaphthalene(C)				80 M'Ha 1
MS, electrochem.	CH_3SOCH_3 or CD_3SOCD_3/H_2O or D_2O (9:1)	[42]	$k_a/k_b = 13(A)$ [43] 7(B) 8(C)	
$\dot{A} + (CH_3)_3CSH \xrightarrow{a} C_{10}H_8 + (CH_3)_3C\dot{S}$ $+ (CH_3)_3CST \xrightarrow{b}$ α-T-naphthalene $+ (CH_3)_3C\dot{S}$				
Decomp. of (α-naphthyl)N=NC$(C_6H_5)_3$				76 Lew 2
PR [44]	$(CH_3)_3CSH$	333(1)	$k_a/k_b = 2.75$	
$\dot{A}$ + 4-Cl$C_6H_4SH \xrightarrow{a} C_{10}H_8$ + 4-Cl$C_6H_4\dot{S}$ + 4-Cl$C_6H_4ST \xrightarrow{b}$ α-T-naphthalene + 4-Cl$C_6H_4\dot{S}$				
Decomp. of (α-naphthyl)N=NC$(C_6H_5)_3$				76 Lew 2
PR [44]	4-ClC_6H_4SH	333(1)	$k_a/k_b = 1.63$	
$\dot{A} + C_6H_5SH \xrightarrow{a} C_{10}H_8 + C_6H_5\dot{S}$ $+ C_6H_5ST \xrightarrow{b}$ α-T-naphthalene $+ C_6H_5\dot{S}$				
Decomp. of (α-naphthyl)N=NC$(C_6H_5)_3$				76 Lew 2
PR [44]	C_6H_5SH C_6H_6	333(1) 333(1)	$k_a/k_b = 1.42$ $k_a/k_b = 1.56$	
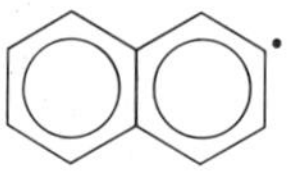 $+ C_6H_5SH \xrightarrow{a} C_{10}H_8 + C_6H_5\dot{S}$ $+ C_6H_5ST \xrightarrow{b}$ β-T-naphthalene $+ C_6H_5\dot{S}$				
Decomp. of (β-naphthyl)N=NC$(C_6H_5)_3$				76 Lew 2
PR [44]	C_6H_5SH	333(1)	$k_a/k_b = 1.41$	
$(C_6H_5)_2\dot{C}H + (CH_3)_3CSH \xrightarrow{a} (C_6H_5)_2CH_2 + (CH_3)_3C\dot{S}$ $+ (CH_3)_3CSD \xrightarrow{b} (C_6H_5)_2CHD + (CH_3)_3C\dot{S}$				
Not given				71 Pry 1/
not given	not given	298 [45]	k_a/k_b [46] $= 6.59$ $A_a/A_b = 1.92$ $E_a(a) - E_a(b) = -3.1\ \mathrm{kJ\,mol^{-1}}$	58 Swa 1

[40]) Specific activity, NMR.
[41]) $k(H)/k(T) = 3.75$ calculated from $k(H)/k(T) = (k(D)/k(T))^{3.262}$.
[42]) Not given (presumably RT).
[43]) A, B, C based on measured k/k_a and k/k_b ratios with k referring to $[C_{10}H_7Cl(Br, I)^{\dot{-}}] \longrightarrow$ 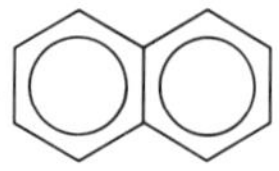 $+ Cl^-(Br^-, I^-)$.
[44]) Specific activity.
[45]) Extrapolated to 298 K from Arrhenius plot.
[46]) Based on $k(H)/k(T)$ measurements with T-labelled thiol and equation for $k(D)/k(T)$ ratio in [58 Swa 1].

Reaction Radical generation Method	Solvent	T[K]	Rate data	Ref./ add. ref.
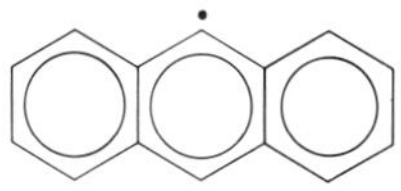 $+ CH_3SOCH_3 \xrightarrow{a} C_{14}H_{10} + CH_3SO\dot{C}H_2$ $+ CD_3SOCD_3 \xrightarrow{b} C_{14}H_9D + CD_3SO\dot{C}D_2$				
Electrochem. reduct. of 9-chloroanthracene, 9-bromoanthracene, 9-iodoanthracene				80 M'Ha 1
PR, MS, electrochem.	CH_3SOCH_3 or CD_3SOCD_3/H_2O or D_2O (9:1)	[47])	$k_a/k_b = 6$	
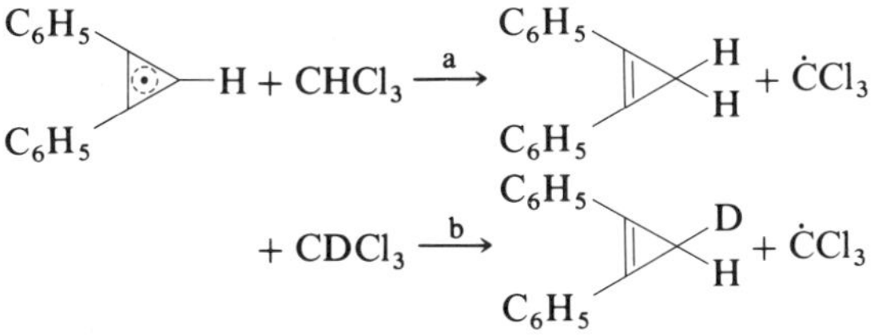				
Electrolytic reduct. of diphenyl-*c*-propenium bromide				70 Sho 1
PR, MS	CH_3CN	RT	$k_a/k_b = 2.8(1)$	
$(C_6H_5)_3\dot{C} + C_6H_5SH \xrightarrow{a} (C_6H_5)_3CH + C_6H_5\dot{S}$ $+ C_6H_5ST \xrightarrow{b} (C_6H_5)_3CT + C_6H_5\dot{S}$				
Decomp. of 1-diphenylmethylene-4-triphenylmethyl-2,5-cyclohexadiene				71 Lew 1 [49])
PR [48])	toluene	248	$k_a/k_b = 27.4(3)$	
		258	27.4(3)	
		273	22.3(3)	
		298	14.9(3)	
		313	12.1(3)	
		333	1.9(3)	
		273... 313	$\log[A_a/A_b] = 0.187$ $E_a(a) - E_a(b) = -10.8\ kJ\ mol^{-1}$	
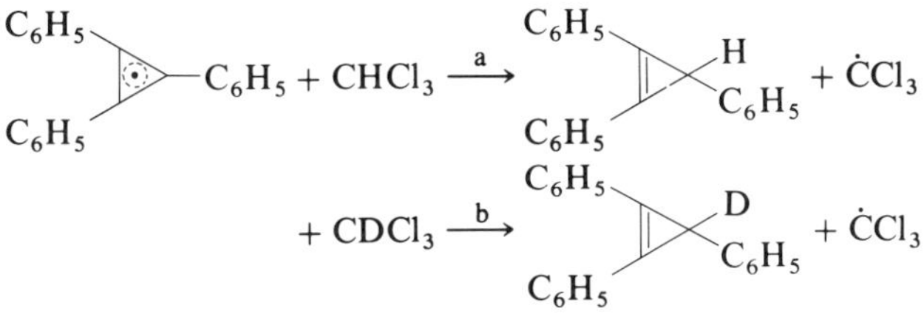				
Electrolytic reduct. of triphenyl-*c*-propenium bromide				70 Sho 1
PR, MS	CH_3CN	RT	$k_a/k_b = 2.30(16)$	
$4\text{-}(C_6H_5)_3CC_6H_4\dot{C}(C_6H_5)_2 + C_6H_5SH \xrightarrow{a} 4\text{-}(C_6H_5)_3CC_6H_4CH(C_6H_5)_2 + C_6H_5\dot{S}$ $+ C_6H_5ST \xrightarrow{b} 4\text{-}(C_6H_5)_3CC_6H_4CT(C_6H_5)_2 + C_6H_5\dot{S}$				
From $C_6H_5\dot{S}$ + 1-diphenylmethylene-4-triphenylmethyl-2,5-cyclohexadiene react.				71 Lew 1
PR [48])	toluene	248	$k_a/k_b = 27.9$	
		258	27.4	

[47]) Not given (presumably RT).
[48]) Specific activity measurement.
[49]) $k_a \approx 1\ M^{-1}\ s^{-1}$ (very rough estimate).

4.1.3.2.2 Radicals containing C, H, and other atoms

Reaction Radical generation Method	Solvent	T[K]	Rate data	Ref./ add. ref.
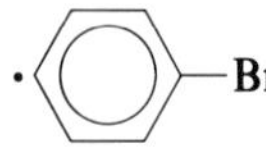 —Br + $C_6H_5SH \xrightarrow{a} C_6H_5Br + C_6H_5\dot{S}$ $+ C_6H_5ST \xrightarrow{b}$ 4-$TC_6H_4Br + C_6H_5\dot{S}$				
Decomp. of 4-$BrC_6H_5N{=}NC(C_6H_5)_3$ PR [50]	C_6H_5SH	333(1)	$k_a/k_b = 1.74$	76 Lew 2
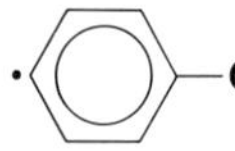 —Cl + $(CH_3)_3CSH \xrightarrow{a} C_6H_5Cl + (CH_3)_3C\dot{S}$ $+ (CH_3)_3CST \xrightarrow{b}$ 4-$TC_6H_4Cl + (CH_3)_3C\dot{S}$				
Decomp. of 4-$ClC_6H_4N{=}NC(C_6H_5)_3$ PR [50]	$(CH_3)_3CSH$	333(1)	$k_a/k_b = 3.03$	76 Lew 2
 —Cl + 4-$ClC_6H_4SH \xrightarrow{a} C_6H_5Cl$ + 4-$ClC_6H_4\dot{S}$ + 4-$ClC_6H_4ST \xrightarrow{b}$ 4-TC_6H_4Cl + 4-$ClC_6H_4\dot{S}$				
Decomp. of 4-$ClC_6H_4N{=}NC(C_6H_5)_3$ PR [50]	4-ClC_6H_4SH	333(1)	$k_a/k_b = 1.94$	76 Lew 2
 —Cl + $C_6H_5SH \xrightarrow{a} C_6H_5Cl + C_6H_5\dot{S}$ $+ C_6H_5ST \xrightarrow{b}$ 4-$TC_6H_4Cl + C_6H_5\dot{S}$				
Decomp. of 4-$ClC_6H_4N{=}NC(C_6H_5)_3$ PR [50]	C_6H_5SH	333(1)	$k_a/k_b = 1.76$	76 Lew 2
 —$NO_2 + (CH_3)_3CSH \xrightarrow{a} C_6H_5NO_2 + (CH_3)_3C\dot{S}$ $+ (CH_3)_3CSD \xrightarrow{b}$ 4-$DC_6H_4NO_2 + (CH_3)_3C\dot{S}$				
Not given not given	not given	298 [51]	k_a/k_b [52] = 2.49 $A_a/A_b = 1.10$ $E_a(a) - E_a(b) = -1.9\ \mathrm{kJ\,mol^{-1}}$	71 Pry 1/ 58 Swa 1
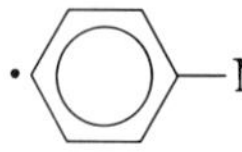 —$NO_2 + (CH_3)_3CSH \xrightarrow{a} C_6H_5NO_2 + (CH_3)_3C\dot{S}$ $+ (CH_3)_3CST \xrightarrow{b}$ 4-$TC_6H_4NO_2 + (CH_3)_3C\dot{S}$				
Decomp. of 4-$NO_2C_6H_4N{=}NC(C_6H_5)_3$ PR [50]	$(CH_3)_3CSH$	333(1)	$k_a/k_b = 3.08$	76 Lew 2
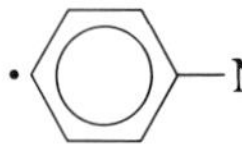 —NO_2 + 4-$ClC_6H_4SH \xrightarrow{a} C_6H_5NO_2$ + 4-$ClC_6H_4\dot{S}$ + 4-$ClC_6H_4ST \xrightarrow{b}$ 4-$TC_6H_4NO_2$ + 4-$ClC_6H_4\dot{S}$				
Decomp. of 4-$NO_2C_6H_4N{=}NC(C_6H_5)_3$ PR [50]	4-ClC_6H_4SH	333(1)	$k_a/k_b = 2.08$	76 Lew 2

[50]) Specific activity.
[51]) Extrapolated to 258 K from Arrhenius plot.
[52]) Based on k(H)/k(T) measurements with T-labelled thiol and equation for k(D)/k(T) ratio in [58 Swa 1].

Reaction Radical generation Method	Solvent	T[K]	Rate data	Ref./ add. ref.
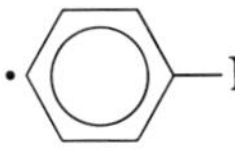 $-NO_2 + C_6H_5SH \xrightarrow{a} C_6H_5NO_2 + C_6H_5\dot{S}$ $+ C_6H_5ST \xrightarrow{b} 4\text{-}TC_6H_4NO_2 + C_6H_5\dot{S}$				
Decomp. of $4\text{-}NO_2C_6H_4N{=}NC(C_6H_5)_3$				76 Lew 2
PR [50]	C_6H_5SH	333(1)	$k_a/k_b = 1.92$	
	C_6H_6	333(1)	$k_a/k_b = 1.99$	
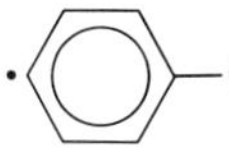 $-CN + CH_3CN \xrightarrow{a} C_6H_5CN + \dot{C}H_2CN$ $+ CD_3CN \xrightarrow{b} 4\text{-}DC_6H_4CN + \dot{C}D_2CN$				
Electrochem. reduct. of 4-chlorobenzonitrile(*A*), 4-bromobenzonitrile(*B*), 4-iodobenzonitrile(*C*)				80 M'Ha 1
MS, electrochem.	CH_3CN or CD_3CN/ H_2O or D_2O (9:1)	[54]	$k_a/k_b = 14(A)$ [55] 12(*B*) 16(*C*)	
$3\text{-}ClC_6H_4\dot{C}H_2 + (C_4H_9)_3SnH \xrightarrow{a} 3\text{-}ClC_6H_4CH_3 + (C_4H_9)_3\dot{S}n$ $+ (C_4H_9)_3SnT \xrightarrow{b} 3\text{-}ClC_6H_4CH_2T + (C_4H_9)_3\dot{S}n$				
Thermal decomp. of AIBN as initiator				76 Koz 1
PR [56]	C_6H_6 or 1,2-dimethoxyethane	353	$k_a/k_b = 3.68$	
$4\text{-}ClC_6H_4\dot{C}H_2 + (C_4H_9)_3SnH \xrightarrow{a} 4\text{-}ClC_6H_4CH_3 + (C_4H_9)_3\dot{S}n$ $+ (C_4H_9)_3SnT \xrightarrow{b} 4\text{-}ClC_6H_4CH_2T + (C_4H_9)_3\dot{S}n$				
Thermal decomp. of AIBN as initiator				76 Koz 1
PR [56]	C_6H_6 or 1,2-dimethoxyethane	353	$k_a/k_b = 3.76$	
$4\text{-}ClC_6H_4\dot{C}HCH_2Br + HBr \xrightarrow{a} ClC_6H_4CH_2CH_2Br + \dot{B}r$ $+ TBr \xrightarrow{b} ClC_6H_4CHTCH_2Br + \dot{B}r$				
Phot. of AIBN as initiator				73 Lew 1
[57]	diethylether	273	$k_a/k_b = 1.59$	
$C_6H_5\dot{C}HCH_2Br + HBr \xrightarrow{a} C_6H_5CH_2CH_2Br + \dot{B}r$ $+ TBr \xrightarrow{b} C_6H_5CHTCH_2Br + \dot{B}r$				
Phot. of AIBN as initiator				73 Lew 1
[57]	diethylether	273	$k_a/k_b = 1.47$	
$C_6H_5SCH_2\dot{C}HCN + C_6H_5SH \xrightarrow{a} C_6H_5SCH_2CH_2CN + C_6H_5\dot{S}$ $+ C_6H_5ST \xrightarrow{b} C_6H_5SCH_2CHTCN + C_6H_5\dot{S}$				
Phot.				76 Lew 1/
[58]	C_6H_5SH	343.0(1)	$k_a/k_b = 4.69(\pm 1.5\%)$ [59]	71 Lew 2

[50]) Specific activity.
[54]) Not given (presumably RT).
[55]) *A*, *B*, *C* based on measured k/k_a and k/k_b ratios with k referring to

$[4\text{-}(I, Br)Cl\text{-}C_6H_4CN^{\overline{\cdot}}] \longrightarrow$  $-CN + Cl^-(Br^-, I^-)$.

[56]) Analysis of T-labelled products.
[57]) Measurement of T-labelled compounds.
[58]) Specific activity.
[59]) Mechanism probably ionic.

Reaction Radical generation Method	Solvent	T[K]	Rate data	Ref./ add. ref.
$4\text{-}CH_3C_6H_4\dot{C}HCH_2Br + HBr \xrightarrow{a} CH_3C_6H_4CH_2CH_2Br + \dot{B}r$ $+ TBr \xrightarrow{b} CH_3C_6H_4CHTCH_2Br + \dot{B}r$				
Phot. of AIBN as initiator				73 Lew 1
[57]	diethylether	273	$k_a/k_b = 1.04$	
$(CH_3)_2\dot{C}SC_6H_5 + C_6H_5SD \xrightarrow{a} (CH_3)_2CDSC_6H_5 + C_6H_5\dot{S}$ $+ C_6H_5ST \xrightarrow{b} (CH_3)_2CTSC_6H_5 + C_6H_5\dot{S}$				
Decomp. of $(CH_3)_2C(SC_6H_5)N{=}N(C_6H_5S)C(CH_3)_2$				76 Lew 2/
PR [60]	thiophenol (89% C_6H_5SD)	376(1)	$k_a/k_b = 1.28(\pm 3\%)$	58 Swa 1 [61]
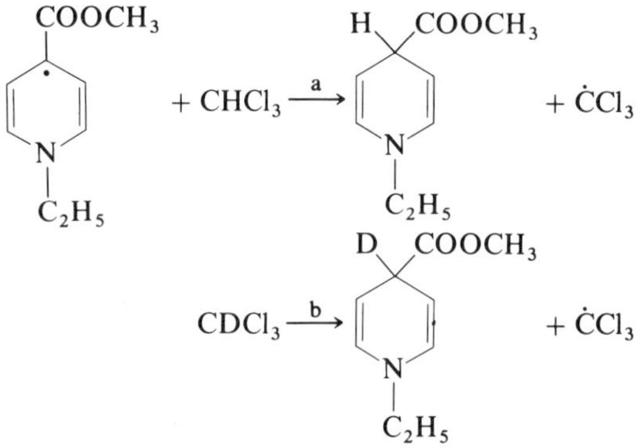				
Reduct. of 1-ethyl-4-carbomethoxypyridinium iodide by Zn				64 Kos 1
KAS	CH_3CN	298	$k_a/k_b = 1.21$	
$C_6H_5CH_2SCH_2\dot{C}HCN + C_6H_5SH \xrightarrow{a} C_6H_5CH_2SCH_2CH_2CN + C_6H_5\dot{S}$ $+ C_6H_5ST \xrightarrow{b} C_6H_5CH_2SCH_2CHTCN + C_6H_5\dot{S}$				
Phot.				76 Lew 1/
[58]	C_6H_5SH	343	$k_a/k_b = 7.75(\pm 1.5\%)$ [59]	71 Lew 2
$C_6H_5SCH_2\dot{C}HOCOCH_3 + C_6H_5SH \xrightarrow{a} C_6H_5SCH_2CH_2OCOCH_3 + C_6H_5\dot{S}$ $+ C_6H_5ST \xrightarrow{b} C_6H_5SCH_2CHTOCOCH_3 + C_6H_5\dot{S}$				
Phot.				76 Lew 1/
[58]	C_6H_5SH	343.0(1)	$k_a/k_b = 2.67(\pm 1.5\%)$	71 Lew 2
$C_6H_5SCH_2\dot{C}HCOOCH_3 + C_6H_5SH \xrightarrow{a} C_6H_5SCH_2CH_2COOCH_3 + C_6H_5\dot{S}$ $+ C_6H_5ST \xrightarrow{b} C_6H_5SCH_2CHTCOOCH_3 + C_6H_5\dot{S}$				
Phot.				76 Lew 1/
[58]	C_6H_5SH	343.0(1)	$k_a/k_b = 3.89(\pm 1.5\%)$ [62]	71 Lew 2
$C_6H_5\dot{C}{=}CHPS(OCH_3)_2 + HPS(OCH_3)_2 \xrightarrow{a} C_6H_5CH{=}CHPS(OCH_3)_2 + \dot{P}S(OCH_3)_2$ $+ TPS(OCH_3)_2 \xrightarrow{b} C_6H_5CT{=}CHPS(OCH_3)_2 + \dot{P}S(OCH_3)_2$				
$\dot{P}S(OCH_3)_2$ add. to $C_6H_5C{\equiv}CH$ (AIBN catalyzed)				76 Lew 3
PR [58]	$HPS(OCH_3)_2$	353.0(1)	$k_a/k_b = 6.5(\pm 5\%)$	

[57]) Measurement of T-labelled compounds.
[58]) Specific activity.
[59]) Mechanism probably ionic.
[60]) Specific activity, NMR.
[61]) $k(H)/k(T) = 2.24$ calculated from $k(H)/k(T) = (k(D)/k(T))^{3.262}$.
[62]) Possibility of ionic mechanism.

Reaction Radical generation Method	Solvent	T[K]	Rate data	Ref./ add. ref.
$C_6H_5\dot{C}HCH_2PS(OCH_3)_2 + HPS(OCH_3)_2 \xrightarrow{a} C_6H_5CH_2CH_2PS(OCH_3)_2 + \dot{P}S(OCH_3)_2$ $+ TPS(OCH_3)_2 \xrightarrow{b} C_6H_5CHTCH_2PS(OCH_3)_2 + \dot{P}S(OCH_3)_2$				
$\dot{P}S(OCH_3)_2$ add. to $C_6H_5CH{=}CH_2$ (AIBN catalyzed)				76 Lew 3
PR [58])	$HPS(OCH_3)_2$	353.0(1)	$k_a/k_b = 3.8(\pm 5\%)$	
$C_6H_5CH_2SCH_2\dot{C}HOCOCH_3 + C_6H_5SH \xrightarrow{a} C_6H_5CH_2SCH_2CH_2OCOCH_3 + C_6H_5\dot{S}$ $+ C_6H_5ST \xrightarrow{b} C_6H_5CH_2SCH_2CHTOCOCH_3 + C_6H_5\dot{S}$				
Phot. [58])	C_6H_5SH	343	$k_a/k_b = 5.75(\pm 1.5\%)$	76 Lew 1/ 71 Lew 2
$C_6H_5CH_2SCH_2\dot{C}HCOOCH_3 + C_6H_5SH \xrightarrow{a} C_6H_5CH_2SCH_2CH_2COOCH_3 + C_6H_5\dot{S}$ $+ C_6H_5ST \xrightarrow{b} C_6H_5CH_2SCH_2CHTCOOCH_3 + C_6H_5\dot{S}$				
Phot. [58])	C_6H_5SH	343	$k_a/k_b = 8.04(\pm 1.5\%)$ [62])	76 Lew 1/ 71 Lew 2
$C_6H_5SCH_2\dot{C}(CH_3)COOCH_3 + C_6H_5SH \xrightarrow{a} C_6H_5SCH_2CH(CH_3)COOCH_3 + C_6H_5\dot{S}$ $+ C_6H_5ST \xrightarrow{b} C_6H_5SCH_2CT(CH_3)COOCH_3 + C_6H_5\dot{S}$				
Phot. [58])	C_6H_5SH	343.0(1)	$k_a/k_b = 4.42(\pm 1.5\%)$ [62])	76 Lew 1/ 71 Lew 2
$C_6H_5CH_2\dot{C}HCH_2PS(OCH_3)_2 + HPS(OCH_3)_2 \xrightarrow{a} C_6H_5(CH_2)_3PS(OCH_3)_2 + \dot{P}S(OCH_3)_2$ $+ TPS(OCH_3)_2 \xrightarrow{b} C_6H_5CH_2CHTCH_2PS(OCH_3)_2 + \dot{P}S(OCH_3)_2$				
$\dot{P}S(OCH_3)_2$ add. to $C_6H_5CH_2CH{=}CH_2$ (AIBN catalyzed)				76 Lew 3
PR [58])	$HPS(OCH_3)_2$	353.0(1)	$k_a/k_b = 8.1(\pm 5\%)$	
$C_6H_5CH_2\dot{C}HCH_2PO(OCH_3)_2 + HPO(OCH_3)_2 \xrightarrow{a} C_6H_5(CH_2)_3PO(OCH_3)_2 + \dot{P}O(OCH_3)_2$ $+ TPO(OCH_3)_2 \xrightarrow{b} C_6H_5CH_2CHTCH_2PO(OCH_3)_2 + \dot{P}O(OCH_3)_2$				
$\dot{P}O(OCH_3)_2$ add. to $C_6H_5CH_2CH{=}CH_2$ (BPO catalyzed)				76 Lew 3
PR [58])	$HPO(OCH_3)_2$	363.0(1)	$k_a/k_b = 11.9(\pm 5\%)$	
$C_6H_5OCH_2\dot{C}HCH_2PS(OCH_3)_2 + HPS(OCH_3)_2 \xrightarrow{a} C_6H_5O(CH_2)_3PS(OCH_3)_2 + \dot{P}S(OCH_3)_2$ $+ TPS(OCH_3)_2 \xrightarrow{b} C_6H_5OCH_2CHTCH_2PS(OCH_3)_2 + \dot{P}S(OCH_3)_2$				
$\dot{P}S(OCH_3)_2$ add. to $C_6H_5OCH_2CH{=}CH_2$ (AIBN catalyzed)				76 Lew 3
PR [58])	$HPS(OCH_3)_2$	353.0(1)	$k_a/k_b = 8.4(\pm 5\%)$	
$2,4,6\text{-}(CH_3)_3C_6H_2SCH_2\dot{C}HCN + C_6H_5SH \xrightarrow{a} 2,4,6\text{-}(CH_3)_3C_6H_2SCH_2CH_2CN + C_6H_5\dot{S}$ $+ C_6H_5ST \xrightarrow{b} 2,4,6\text{-}(CH_3)_3C_6H_2SCH_2CHTCN + C_6H_5\dot{S}$				
Phot. [58])	C_6H_5SH	343	$k_a/k_b = 10.72(\pm 1.5\%)$ [62])	76 Lew 1/ 71 Lew 2
$C_6H_5CH_2SCH_2\dot{C}(CH_3)COOCH_3 + C_6H_5SH \xrightarrow{a} C_6H_5CH_2SCH_2CH(CH_3)COOCH_3 + C_6H_5\dot{S}$ $+ C_6H_5ST \xrightarrow{b} C_6H_5CH_2SCH_2CT(CH_3)COOCH_3 + C_6H_5\dot{S}$				
Phot. [58])	C_6H_5SH	343	$k_a/k_b = 9.75(\pm 1.5\%)$ [62])	76 Lew 1/ 71 Lew 2
$C_6H_5CH_2\dot{C}(CH_3)CH_2PS(OCH_3)_2 + HPS(OCH_3)_2 \xrightarrow{a}$ $C_6H_5CH_2CH(CH_3)CH_2PO(OCH_3)_2 + \dot{P}S(OCH_3)_2$ $+ TPS(OCH_3)_2 \xrightarrow{b} C_6H_5CH_2CT(CH_3)CH_2PS(OCH_3)_2 + \dot{P}S(OCH_3)_2$				
$\dot{P}S(OCH_3)_2$ add. to $C_6H_5CH_2C(CH_3){=}CH_2$ (AIBN catalyzed)				76 Lew 3
PR [58])	$HPS(OCH_3)_2$	353.0(1)	$k_a/k_b = 7.0(\pm 5\%)$	

[58]) Specific activity.
[62]) Possibility of ionic mechanism.

Reaction Radical generation Method	Solvent	T [K]	Rate data	Ref./ add. ref.

$C_6H_5CH_2\dot{C}(CH_3)CH_2PO(OCH_3)_2 + HPO(OCH_3)_2 \xrightarrow{a}$

$\qquad\qquad\qquad\qquad\qquad\qquad\qquad\qquad C_6H_5CH_2CH(CH_3)CH_2PO(OCH_3)_2 + \dot{P}O(OCH_3)_2$

$\qquad\qquad\qquad\qquad + TPO(OCH_3)_2 \xrightarrow{b}$

$\qquad\qquad\qquad\qquad\qquad\qquad\qquad\qquad C_6H_5CH_2CT(CH_3)CH_2PO(OCH_3)_2 + \dot{P}O(OCH_3)_2$

| $\dot{P}O(OCH_3)_2$ add. to $C_6H_5CH_2C(CH_3){=}CH_2$ (BPO catalyzed) | | | | 76 Lew 3 |
| PR [58]) | HPO(OCH$_3$)$_2$ | 363.0(1) | $k_a/k_b = 10.0(\pm5\%)$ | |

$2,4,6\text{-}(CH_3)_3C_6H_2SCH_2\dot{C}HCOOCH_3 + C_6H_5SH \xrightarrow{a} 2,4,6\text{-}(CH_3)_3C_6H_2SCH_2CH_2COOCH_3 + C_6H_5\dot{S}$

$\qquad\qquad\qquad\qquad\qquad + C_6H_5ST \xrightarrow{b} 2,4,6\text{-}(CH_3)_3C_6H_2SCH_2CHTCOOCH_3 + C_6H_5\dot{S}$

| Phot. [58]) | C_6H_5SH | 343 | $k_a/k_b = 3.73(\pm1.5\%)$ [62]) | 76 Lew 1/ 71 Lew 2 |

$C_6H_5SCH_2\dot{C}H-\!\!\left\langle\bigcirc\right\rangle\!\!-Cl + C_6H_5SH \xrightarrow{a} C_6H_5SCH_2CH_2-\!\!\left\langle\bigcirc\right\rangle\!\!-Cl + C_6H_5\dot{S}$

$\qquad\qquad\qquad\qquad + C_6H_5ST \xrightarrow{b} C_6H_5SCH_2CHT-\!\!\left\langle\bigcirc\right\rangle\!\!-Cl + C_6H_5\dot{S}$

| Phot. [58]) | C_6H_5SH | 343.0(1) | $k_a/k_b = 7.13(\pm1.5\%)$ | 76 Lew 1/ 71 Lew 2 |

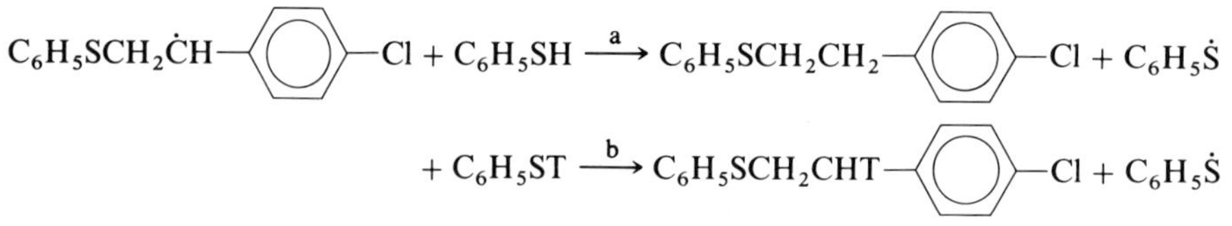

$C_6H_5SCH_2\dot{C}H-\!\!\left\langle\bigcirc\right\rangle\!\!\overset{NO_2}{} + C_6H_5SH \xrightarrow{a} C_6H_5SCH_2CH_2-\!\!\left\langle\bigcirc\right\rangle\!\!\overset{NO_2}{} + C_6H_5\dot{S}$

$\qquad\qquad\qquad\qquad + C_6H_5ST \xrightarrow{b} C_6H_5SCH_2CHT-\!\!\left\langle\bigcirc\right\rangle\!\!\overset{NO_2}{} + C_6H_5\dot{S}$

| Phot. [58]) | C_6H_5SH | 343(1) | $k_a/k_b = 8.33(\pm1.5\%)$ | 76 Lew 1/ 71 Lew 2 |

$C_6H_5SCH_2\dot{C}HC_6H_5 + C_6H_5SH \xrightarrow{a} C_6H_5SCH_2CH_2C_6H_5 + C_6H_5\dot{S}$

$\qquad\qquad\qquad\qquad\qquad + C_6H_5ST \xrightarrow{b} C_6H_5SCH_2CHTC_6H_5 + C_6H_5\dot{S}$

Phot. [58])	C_6H_5SH	343(1)	$k_a/k_b = 6.74(\pm1.5\%)$	76 Lew 1/ 71 Lew 2
		333(1)	$= 7.20(\pm1.5\%)$	
		313(1)	$= 8.40(\pm1.5\%)$	
		293(1)	$= 9.9(\pm1.5\%)$	
		273(1)	$= 12.1(\pm1.5\%)$	
			$k_a/k_b = 0.738\exp(1512/RT)$	

$2,4,6\text{-}(CH_3)_3C_6H_2SCH_2\dot{C}(CH_3)COOCH_3 + C_6H_5SH \xrightarrow{a}$

$\qquad\qquad\qquad\qquad\qquad\qquad\qquad 2,4,6\text{-}(CH_3)_3C_6H_2SCH_2CH(CH_3)COOCH_3 + C_6H_5\dot{S}$

$\qquad\qquad\qquad\qquad + C_6H_5ST \xrightarrow{b}$

$\qquad\qquad\qquad\qquad\qquad\qquad\qquad 2,4,6\text{-}(CH_3)_3C_6H_2SCH_2CT(CH_3)COOCH_3 + C_6H_5\dot{S}$

| Phot. [58]) | C_6H_5SH | 343 | $k_a/k_b = 6.72(\pm1.5\%)$ [62]) | 76 Lew 1/ 71 Lew 2 |

---

[58]) Specific activity.
[62]) Possibility of ionic mechanism.

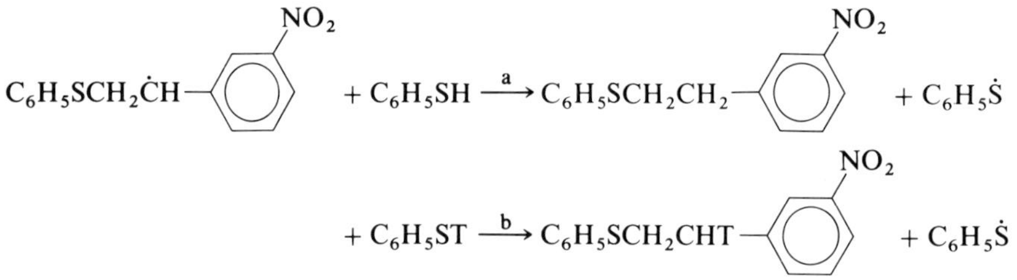

Reaction				
Radical generation				Ref./
Method	Solvent	T [K]	Rate data	add. ref.

$C_6H_5SCH_2\dot{C}H(CH_2)_5CH_3 + C_6H_5SH \xrightarrow{a} C_6H_5S(CH_2)_7CH_3 + C_6H_5\dot{S}$
$\qquad\qquad\qquad + C_6H_5ST \xrightarrow{b} C_6H_5SCH_2CHT(CH_2)_5CH_3 + C_6H_5\dot{S}$

Phot.				76 Lew 1/
[58])	C_6H_5SH	273.0(1)	$k_a/k_b = 4.46(\pm 1.5\%)$	71 Lew 2

$C_6H_5CH_2SCH_2\dot{C}H\!-\!\!\bigcirc\!\!-\!Cl + C_6H_5SH \xrightarrow{a} C_6H_5CH_2SCH_2CH_2\!-\!\!\bigcirc\!\!-\!Cl + C_6H_5\dot{S}$

$\qquad\qquad\qquad + C_6H_5ST \xrightarrow{b} C_6H_5CH_2SCH_2CHT\!-\!\!\bigcirc\!\!-\!Cl + C_6H_5\dot{S}$

Phot.				76 Lew 1/
[58])	C_6H_5SH	343	$k_a/k_b = 11.3(\pm 1.5\%)$	71 Lew 2

$C_6H_5CH_2SCH_2\dot{C}H\!-\!\!\bigcirc\!\!\overset{NO_2}{} + C_6H_5SH \xrightarrow{a} C_6H_5CH_2SCH_2CH_2\!-\!\!\bigcirc\!\!\overset{NO_2}{} + C_6H_5\dot{S}$

$\qquad\qquad\qquad + C_6H_5ST \xrightarrow{b} C_6H_5CH_2SCH_2CHT\!-\!\!\bigcirc\!\!\overset{NO_2}{} + C_6H_5\dot{S}$

Phot.				76 Lew 1/
[58])	C_6H_5SH	343	$k_a/k_b = 11.3(\pm 1.5\%)$	71 Lew 2

$C_6H_5SCH_2\dot{C}H\!-\!\!\bigcirc\!\!-\!OCH_3 + C_6H_5SH \xrightarrow{a} C_6H_5SCH_2CH_2\!-\!\!\bigcirc\!\!-\!OCH_3 + C_6H_5\dot{S}$

$\qquad\qquad\qquad + C_6H_5ST \xrightarrow{b} C_6H_5SCH_2CHT\!-\!\!\bigcirc\!\!-\!OCH_3 + C_6H_5\dot{S}$

Phot.				76 Lew 1/
[58])	C_6H_5SH	343.0(1)	$k_a/k_b = 6.59(\pm 1.5\%)$	71 Lew 2

$C_6H_5CH_2SCH_2\dot{C}HCHC_6H_5 + C_6H_5SH \xrightarrow{a} C_6H_5CH_2SCH_2CH_2C_6H_5 + C_6H_5\dot{S}$
$\qquad\qquad\qquad + C_6H_5ST \xrightarrow{b} C_6H_5CH_2SCH_2CHTC_6H_5 + C_6H_5\dot{S}$

Phot.				76 Lew 1/
[58])	C_6H_5SH	343	$k_a/k_b = 10.40(\pm 1.5\%)$	71 Lew 2
			$k_a/k_b = 0.55 \exp(2000/RT)$	

$C_6H_5SCH_2\dot{C}HCH_2C_6H_5 + C_6H_5SH \xrightarrow{a} C_6H_5SCH_2CH_2CH_2C_6H_5 + C_6H_5\dot{S}$
$\qquad\qquad\qquad + C_6H_5ST \xrightarrow{b} C_6H_5SCH_2CHTCH_2C_6H_5 + C_6H_5\dot{S}$

Phot.				76 Lew 1/
[58])	C_6H_5SH	343.0(1)	$k_a/k_b = 2.61(\pm 1.5\%)$	71 Lew 2

$C_6H_5CH_2SCH_2\dot{C}H(CH_2)_5CH_3 + C_6H_5SH \xrightarrow{a} C_6H_5CH_2S(CH_2)_7CH_3 + C_6H_5\dot{S}$
$\qquad\qquad\qquad + C_6H_5ST \xrightarrow{b} C_6H_5CH_2SCH_2CHT(CH_2)_5CH_3 + C_6H_5\dot{S}$

Phot.				76 Lew 1/
[58])	C_6H_5SH	343	$k_a/k_b = 6.73(\pm 1.5\%)$	71 Lew 2

[58]) Specific activity.

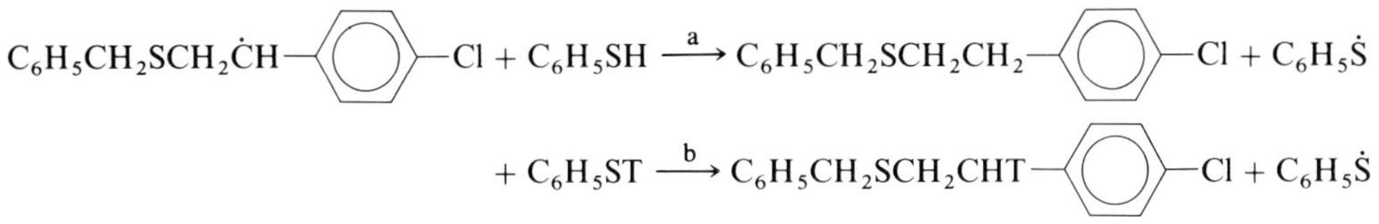

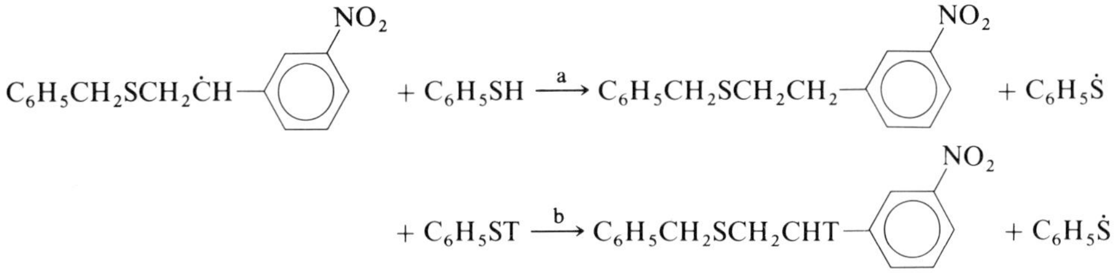

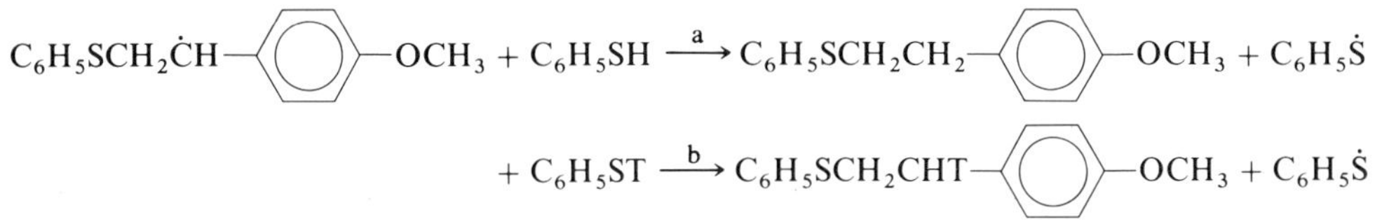

Reaction				
Radical generation				Ref./
Method	Solvent	$T\,[K]$	Rate data	add. ref.

$C_6H_5CH_2SCH_2\dot{C}H$—⬡—$OCH_3 + C_6H_5SH \xrightarrow{a} C_6H_5CH_2SCH_2CH_2$—⬡—$OCH_3 + C_6H_5\dot{S}$

$+ C_6H_5ST \xrightarrow{b} C_6H_5CH_2SCH_2CHT$—⬡—$OCH_3 + C_6H_5\dot{S}$

| Phot. | | | | 76 Lew 1/ |
| [58]) | C_6H_5SH | 343 | $k_a/k_b = 9.43\,(\pm 1.5\%)$ | 71 Lew 2 |

$C_6H_5CH_2SCH_2\dot{C}HCH_2C_6H_5 + C_6H_5SH \xrightarrow{a} C_6H_5CH_2S(CH_2)_3C_6H_5 + C_6H_5\dot{S}$
$+ C_6H_5ST \xrightarrow{b} C_6H_5CH_2SCH_2CHTCH_2C_6H_5 + C_6H_5\dot{S}$

| Phot. | | | | 76 Lew 1/ |
| [58]) | C_6H_5SH | 343 | $k_a/k_b = 4.81\,(\pm 1.5\%)$ | 71 Lew 2 |

$2,4,6\text{-}(CH_3)_3C_6H_2SCH_2\dot{C}HC_6H_5 + C_6H_5SH \xrightarrow{a} 2,4,6\text{-}(CH_3)_3C_6H_2SCH_2CH_2C_6H_5 + C_6H_5\dot{S}$
$+ C_6H_5ST \xrightarrow{b} 2,4,6\text{-}(CH_3)_3C_6H_2SCH_2CHTC_6H_5 + C_6H_5\dot{S}$

| Phot. | | | | 76 Lew 1/ |
| [58]) | C_6H_5SH | 343 | $k_a/k_b = 7.15\,(\pm 1.5\%)$ | 71 Lew 2 |

$2,4,6\text{-}(CH_3)_3C_6H_2SCH_2\dot{C}H$—⬡—$OCH_3$

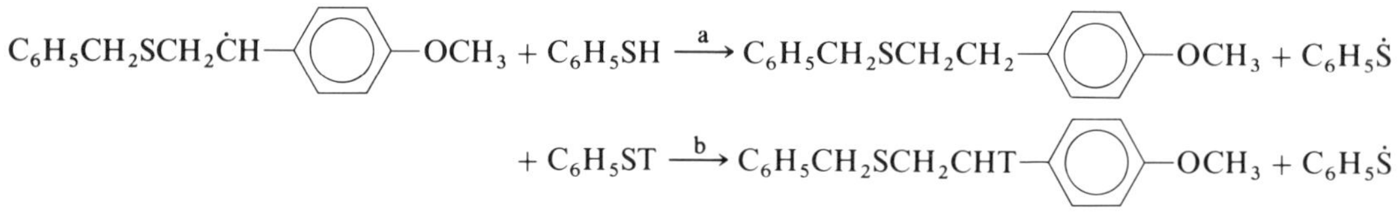

$+ C_6H_5SH \xrightarrow{a} 2,4,6\text{-}(CH_3)_3C_6H_2SCH_2CH_2C_6H_4OCH_3 + C_6H_5\dot{S}$
$+ C_6H_5ST \xrightarrow{b} 2,4,6\text{-}(CH_3)_3C_6H_2SCH_2CHTC_6H_4OCH_3 + C_6H_5\dot{S}$

| Phot. | | | | 76 Lew 1/ |
| [58]) | C_6H_5SH | 343 | $k_a/k_b = 7.02\,(\pm 1.5\%)$ | 71 Lew 2 |

$C_6H_5SCH_2\dot{C}\left(\text{—⬡—}Cl\right)_2 + C_6H_5SH \xrightarrow{a} C_6H_5SCH_2CH\left(\text{—⬡—}Cl\right)_2 + C_6H_5\dot{S}$

$+ C_6H_5ST \xrightarrow{b} C_6H_5SCH_2CT\left(\text{—⬡—}Cl\right)_2 + C_6H_5\dot{S}$

| Phot. | | | | 76 Lew 1/ |
| [58]) | C_6H_5SH | 343.0(1) | $k_a/k_b = 8.93\,(\pm 1.5\%)$ | 71 Lew 2 |

$C_6H_5SCH_2\dot{C}H$—N⟨carbazolyl⟩ $+ C_6H_5SH \xrightarrow{a} C_6H_5SCH_2CH_2(\text{-N-carbazolyl}) + C_6H_5\dot{S}$

$+ C_6H_5ST \xrightarrow{b} C_6H_5SCH_2CHT(\text{-N-carbazolyl}) + C_6H_5\dot{S}$

| Phot. | | | | 76 Lew 1/ |
| [58]) | C_6H_5SH | 343.0(1) | $k_a/k_b = 2.68\,(\pm 1.5\%)$ | 71 Lew 2 |

[58]) Specific activity.

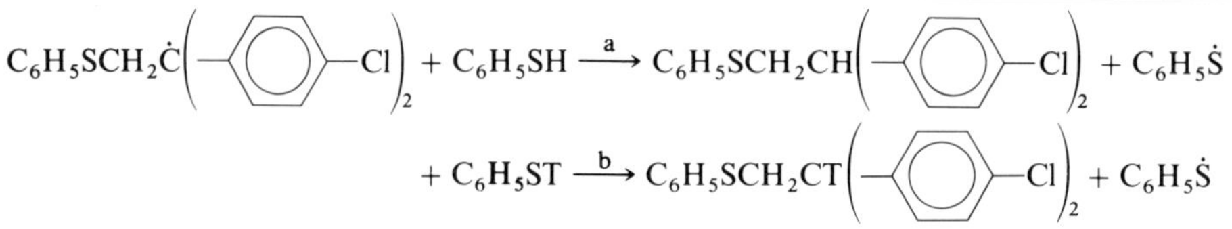

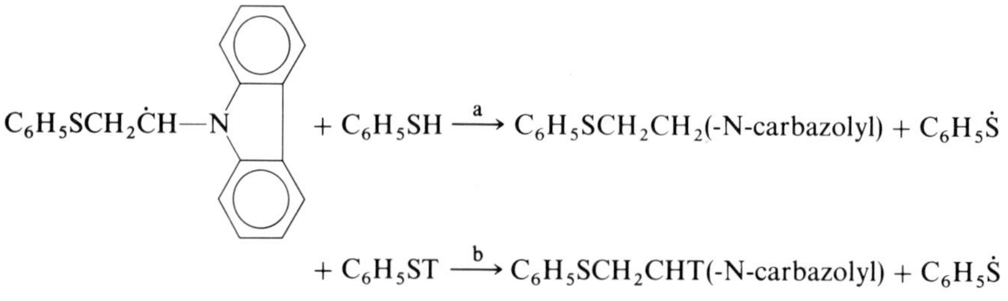

Reaction Radical generation Method	Solvent	T [K]	Rate data	Ref./ add. ref.

$C_6H_5SCH_2\dot{C}(C_6H_5)_2 + C_6H_5SH \xrightarrow{a} C_6H_5SCH_2CH(C_6H_5)_2 + C_6H_5\dot{S}$
$\phantom{C_6H_5SCH_2\dot{C}(C_6H_5)_2} + C_6H_5ST \xrightarrow{b} C_6H_5SCH_2CT(C_6H_5)_2 + C_6H_5\dot{S}$

| Phot. [58]) | C_6H_5SH | 343.0(1) | $k_a/k_b = 10.89(\pm 1.5\%)$ | 76 Lew 1/ 71 Lew 2 |

$C_6H_5CH_2SCH_2\dot{C}\left(-\text{⟨○⟩}-Cl\right)_2 + C_6H_5SH \xrightarrow{a} C_6H_5CH_2SCH_2CH\left(-\text{⟨○⟩}-Cl\right)_2 + C_6H_5\dot{S}$

$ + C_6H_5ST \xrightarrow{b} C_6H_5CH_2SCH_2CT\left(-\text{⟨○⟩}-Cl\right)_2 + C_6H_5\dot{S}$

| Phot. [58]) | C_6H_5SH | 343 | $k_a/k_b = 8.13(\pm 1.5\%)$ | 76 Lew 1/ 71 Lew 2 |

$C_6H_5CH_2SCH_2\dot{C}H-N\text{(carbazolyl)} + C_6H_5SH \xrightarrow{a} C_6H_5CH_2SCH_2CH_2(\text{-N-carbazolyl}) + C_6H_5\dot{S}$

$ + C_6H_5ST \xrightarrow{b} C_6H_5CH_2SCH_2CHT(\text{-N-carbazolyl}) + C_6H_5\dot{S}$

| Phot. [58]) | C_6H_5SH | 343 | $k_a/k_b = 4.44(\pm 1.5\%)$ | 76 Lew 1/ 71 Lew 2 |

$C_6H_5CH_2SCH_2\dot{C}(C_6H_5)_2 + C_6H_5SH \xrightarrow{a} C_6H_5CH_2SCH_2CH(C_6H_5)_2 + C_6H_5\dot{S}$
$\phantom{C_6H_5CH_2SCH_2\dot{C}(C_6H_5)_2} + C_6H_5ST \xrightarrow{b} C_6H_5CH_2SCH_2CT(C_6H_5)_2 + C_6H_5\dot{S}$

| Phot. [58]) | C_6H_5SH | 343 | $k_a/k_b \geqslant 6.70$ | 76 Lew 1/ 71 Lew 2 |

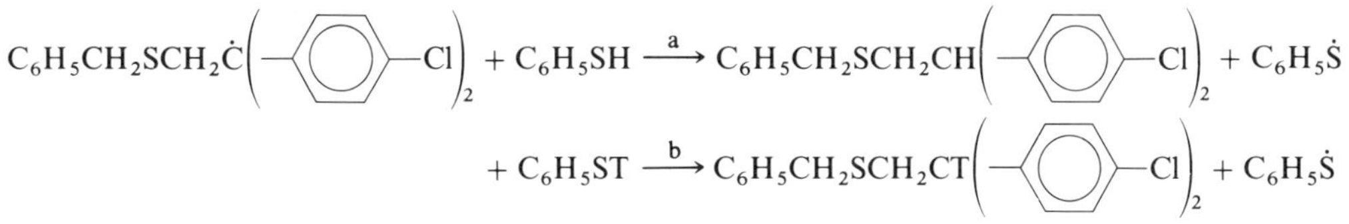

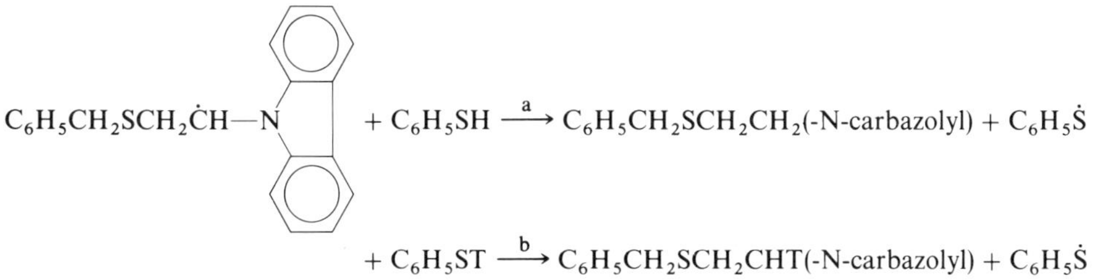

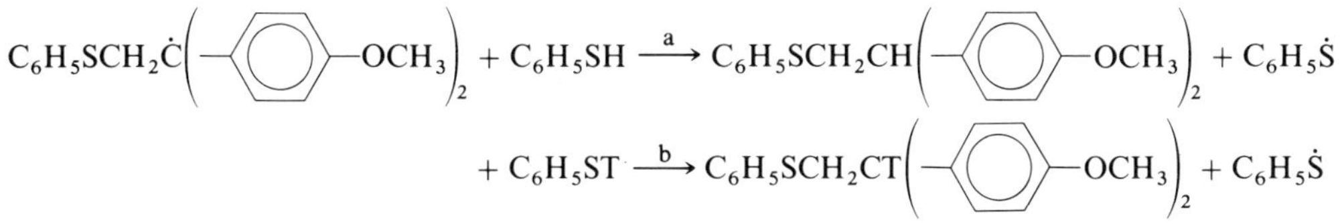

$C_6H_5SCH_2\dot{C}\left(-\text{⟨○⟩}-OCH_3\right)_2 + C_6H_5SH \xrightarrow{a} C_6H_5SCH_2CH\left(-\text{⟨○⟩}-OCH_3\right)_2 + C_6H_5\dot{S}$

$ + C_6H_5ST \xrightarrow{b} C_6H_5SCH_2CT\left(-\text{⟨○⟩}-OCH_3\right)_2 + C_6H_5\dot{S}$

| Phot. [58]) | C_6H_5SH | 343.0(1) | $k_a/k_b = 14.10(\pm 1.5\%)$ | 76 Lew 1/ 71 Lew 2 |

$2,4,6\text{-}(CH_3)_3C_6H_2SCH_2\dot{C}\left(-\text{⟨○⟩}-Cl\right)_2 + C_6H_5SH \xrightarrow{a} 2,4,6\text{-}(CH_3)_3C_6H_2SCH_2CH(C_6H_4Cl)_2 + C_6H_5\dot{S}$

$ + C_6H_5ST \xrightarrow{b} 2,4,6\text{-}(CH_3)_3C_6H_2SCH_2CT(C_6H_4Cl)_2 + C_6H_5\dot{S}$

| Phot. [58]) | C_6H_5SH | 343 | $k_a/k_b \geqslant 13.05$ | 76 Lew 1/ 71 Lew 2 |

[58]) Specific activity.

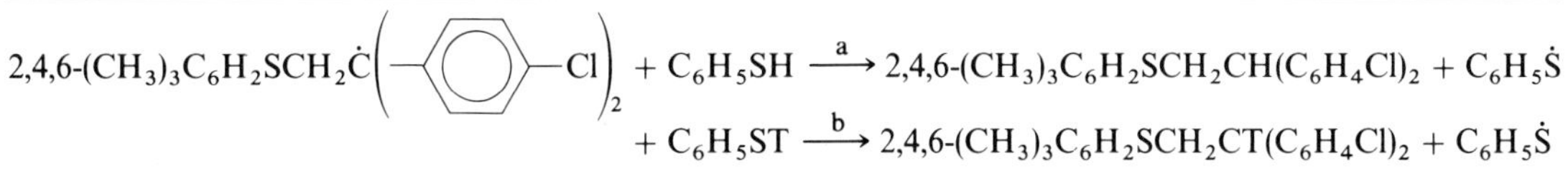

References for 4.1

42 Sta 1	Stauff, I., Schumacher, H.I.: Z. Elektrochem. **48** (1942) 271.
49 Mat 1	Matheson, M.S., Auer, E.E., Bevilacqua, E.B., Hart, E.J.: J. Am. Chem. Soc. **71** (1949) 2610.
50 Edw 1	Edwards, F.G., Mayo, F.R.: J. Am. Chem. Soc. **72** (1950) 1265.
51 Mel 1	Melville, H.W., Robb, J.C., Tutton, R.C.: Disc. Faraday Soc. **10** (1951) 154.
53 Pri 1	Price, C.C., Morita, H.: J. Am. Chem. Soc. **75** (1953) 3686.
54 Bac 1	Back, R., Trick, G., McDonald, C., Sivertz, C.: Can. J. Chem. **32** (1954) 1078.
54 Lev 1	Levy, M., Steinberg, M., Szwarc, M.: J. Am. Chem. Soc. **76** (1954) 5978.
54 Lev 2	Levy, M., Szwarc, M.: J. Chem. Phys. **22** (1954) 1621.
55 Lea 1	Leavitt, F., Levy, M., Szwarc, M., Stannett, V.: J. Am. Chem. Soc. **77** (1955) 5493.
55 Lev 1	Levy, M., Szwarc, M.: J. Am. Chem. Soc. **77** (1955) 1949.
55 Ony 1	Onyszchuk, M., Sivertz, C.: Can. J. Chem. **33** (1955) 1034.
55 Pie 1	Pieck, R., Steacie, E.W.R.: Can. J. Chem. **33** (1955) 1304.
55 Rem 1	Rembaum, A., Szwarc, M.: J. Am. Chem. Soc. **77** (1955) 4468.
55 Szw 1	Szwarc, M.: J. Polym. Sci **16** (1955) 367.
55 Wat 1	Watts, H., Adler B.J., Hildebrand, J.H.: J. Chem. Phys. **23** (1955) 659.
56 Buc 1	Buckley, R.P., Leavitt, F., Szwarc, M.: J. Am. Chem. Soc. **78** (1956) 5557.
56 Buc 2	Buckley, R.P., Szwarc, M.: J. Am. Chem. Soc. **78** (1956) 5696.
56 Smi 1	Smid, J., Szwarc, M.: J. Am. Chem. Soc. **78** (1956) 3322.
57 Bad 1	Bader, A.R., Buckley, R.P., Leavitt, F., Szwarc, M.: J. Am. Chem. Soc. **79** (1957) 5621.
57 Bam 1	Bamford, C.H., Jenkins, A.D., Johnston, R.: Proc. Roy. Soc. (London) Ser. A **239** (1957) 214.
57 Buc 1	Buckley, R.P., Szwarc, M.: Proc. Roy. Soc. (London) Ser. A **240** (1957) 396.
57 Gaz 1	Gazith, M., Szwarc, M.: J. Am. Chem. Soc. **79** (1957) 3339.
57 Hei 1	Heilman, W.J., Rembaum, A., Szwarc, M.: J. Chem. Soc. **1957**, 1127.
57 Lea 1	Leavitt, F., Stannett, V., Szwarz, M.: Chem. Ind. (London) **28** (1957) 985.
57 Raj 1	Rajbenbach, A., Szwarc, M.: J. Am. Chem. Soc. **79** (1957) 6343.
57 Sch 1	Schaafsma, Y., Bickel, A.F., Kooyman, E.C.: Rec. Trav. Chim Pays-Bas **76** (1957) 180.
57 Smi 1	Smid, J., Szwarc, M.: J. Am. Chem. Soc. **79** (1957) 1534.
58 Aus 1	Ausloos, P.: Can. J. Chem. **36** (1958) 400.
58 Buc 1	Buckley, R.P., Rembaum, A., Szwarc, M.: J. Chem. Soc. **1958**, 3442.
58 Lea 1	Leavitt, F., Stannett, V., Szwarc, M.: J. Polym. Sci. **31** (1958) 193.
58 McD 1	McDaniel, D.H., Brown, H.C.: J. Org. Chem. **23** (1958) 420.
58 Swa 1	Swain, C.G., Stivers, E.G., Reuver, J.F., Schaad, L.J.: J. Am. Chem. Soc. **80** (1958) 5885.
59 Ant 1	Antonovskii, V.L., Berezin, I.V.: Dokl. Akad. Nauk SSSR **127** (1959) 124.
59 Bin 1	Binks, J.H., Szwarc, M.: J. Chem. Phys. **30** (1959) 1494.
59 Car 1	Carrock, F., Szwarc, M.: J. Am. Chem. Soc. **81** (1959) 4138.
59 Gre 1	Gresser, J., Binks, J.H., Szwarc, M.: J. Am. Chem. Soc. **81** (1959) 5004.
59 Pet 1	Peterson, D.B., Mains, G.J.: J. Am. Chem. Soc. **81** (1959) 3510.
59 Raj 1	Rajbenbach, A., Szwarc, M.: Proc. Roy. Soc. (London) Ser. A **251** (1959) 1226.
59 Siv 1	Sivertz, C.: J. Phys. Chem. **63** (1959) 34.
60 Ant 1	Antonovskii, V.L., Berezin, I.V.: Russ. J. Phys. Chem. (English Transl.) **34** (1960) 614.
60 Ant 2	Antonovskii, V.L., Berezin, I.V.: Dokl. Akad. Nauk SSSR **134** (1960) 860.
60 Bag 1	Bagdasar'yan, Kh.S., Milyutinskaya, R.I.: Zh. Fiz. Khim. **34** (1960) 234.
60 Ber 1	Berezin, I.V., Kazanskaya, N.F., Martinek, K.: Zh. Obshch. Khim. **30** (1960) 4092.
60 DeT 1	DeTar, D.F., Wells, D.V.: J. Am. Chem. Soc. **82** (1960) 5839.
60 Eva 1	Evans, F.W., Fox, R.J., Szwarc, M.: J. Am. Chem. Soc. **82** (1960) 6414.
60 Fel 1	Feld, M., Szwarc, M.: J. Am. Chem. Soc. **82** (1960) 3791.
60 Huy 1	Huyser, E.S.: J. Am. Chem. Soc. **82** (1960) 394.
60 McC 1	McCarthy, R.L., MacLachlan, A.: Trans. Faraday Soc. **56** (1960) 1187.
60 Ste 1	Steel, C., Szwarc, M.: J. Chem. Phys. **33** (1960) 1677.
60 Vol 1	Volman, D.H., Swanson, L.W.: J. Am. Chem. Soc. **82** (1960) 4141.
61 Ber 1	Berezin, I.V., Kazanskaya, N.F., Martinek, K.: Zh. Fiz. Khim. **35** (1961) 2039.
61 Doe 1	Doepker, R., Mains, G.J.: J. Am. Chem. Soc. **83** (1961) 294.
61 Eva 1	Evans, F.W., Szwarc, M.: Trans. Faraday Soc. **57** (1961) 1905.
61 Fox 1	Fox, R.J., Evans, F.W., Szwarc, M.: Trans. Faraday Soc. **57** (1961) 1915.
61 Gre 1	Gresser, J., Rajbenbach, A., Szwarc, M.: J. Am. Chem. Soc. **83** (1961) 3005.
61 Her 1	Herk, L., Stefani, A., Szwarc, M.: J. Am. Chem. Soc. **83** (1961) 3008.
61 Huy 1	Huyser, E.S.: J. Org. Chem. **26** (1961) 3261.
61 Mat 1	Matsuoka, M., Szwarc, M.: J. Am. Chem. Soc. **83** (1961) 1260.
61 Mey 1	Meyer, J.A., Stannett, V., Szwarc, M.: J. Am. Chem. Soc. **83** (1961) 25.
61 Ste 1	Stefani, A.P., Herk, L., Szwarc, M.: J. Am. Chem. Soc. **83** (1961) 4732.

62 Bam 1	Bamford, C., Jenkins, A., Johnston, R.: Trans. Faraday Soc. **58** (1962) 1212.
62 Ber 1	Berezin, I.V., Dobish, O.: Proc. Acad. Sci. USSR (English Transl.) **142** (1962) 1.
62 Ber 2	Berezin, I.V., Dobish, O.: Proc. Acad. Sci. USSR (English Transl.) **144** (1962) 382.
62 Dun 1	Duncan, F.J., Trotman-Dickenson, A.F.: J. Chem. Soc. **1962** 4672.
62 Fel 1	Feld, M., Stefani, A.P., Szwarc, M.: J. Am. Chem. Soc. **84** (1962) 4451.
62 Mac 1	MacLachlan, A., McCarthy, R.L.: J. Am. Chem. Soc. **84** (1962) 2519.
62 Mon 1	Monteiro, H.: J. Chim. Phys. **59** (1962) 9.
62 Ste 1	Stefani, A.P., Szwarc, M.: J. Am. Chem. Soc. **84** (1962) 3661.
62 Whi 1	Whittemore, I.M., Stefani, A.P., Szwarc, M.: J. Am. Chem. Soc. **84** (1962) 3799.
63 Bri 1	Bridger, R.F., Russell, G.A.: J. Am. Chem. Soc. **85** (1963) 3754.
63 Che 1	Cher, M.: J. Phys. Chem. **67** (1963) 605.
63 Col 1	Collinson, E., Dainton, F.S., Mile, B., Tazuke, S., Schmith, D.R.: Nature (London) **198** (1963) 26.
63 Dix 1	Dixon, P.S., Szwarc, M.: Trans. Faraday Soc. **59** (1963) 112.
63 Huy 1	Huyser, E.S., Schimke, H., Burham, R.L.: J. Org. Chem. **28** (1963) 2141.
63 Kom 1	Komazawa, H., Stefani, A.P., Szwarc, M.: J. Am. Chem. Soc. **85** (1963) 2043.
63 McC 1	McCall, D.W., Douglass, D.C., Anderson, E.W.: Ber. Bunsenges. Phys. Chem. **67** (1963) 336.
63 Pry 1	Pryor, W.A., Platt, P.K.: J. Am. Chem. Soc. **85** (1963) 1496.
63 Rus 1	Russell, G.A., DeBoer, C.: J. Am. Chem. Soc. **85** (1963) 3136.
63 Rus 2	Russell, G.A., Bridger, R.F.: J. Am. Chem. Soc. **85** (1963) 3765.
63 Rus 3	Russell, G.A., Bridger, R.F.: Tetrahedron Lett. **1963**, 737.
63 Rus 4	Russell, G.A., DeBoer, C., Desmond, K.M.: J. Am. Chem. Soc. **85** (1963) 365.
63 Sch 1	Schuler, R.H., Kuntz, R.R.: J. Phys. Chem. **67** (1963) 1004.
64 Baz 1	Bazilevskii, M.V., Bagdasar'yan, Kh. S.: Kinet. Katal. **5** (1964) 215.
64 Baz 2	Bazilevskii, M.V.: Zh. Fiz. Khim. **38** (1964) 225.
64 Ber 1	Bereznikh-Földes, T., Tüdös, F.: Vysokomol. Soedin. **6** (1964) 1523.
64 Ber 2	Berezin, I.V., Kazanskaya, N.F., Pentin, Yu.A.: Zh. Fiz. Khim. **38** (1964) 125.
64 Hua 1	Huang, R.L., Lee, K.H.: J. Chem. Soc. **1964**, 5963.
64 Kos 1	Kosower, E.M., Schwager, I.: J. Am. Chem. Soc. **86** (1964) 5528.
64 Kos 2	Kosower, E.M., Schwager, I.: J. Am. Chem. Soc. **86** (1964) 4493.
64 Mar 1	Martin, M.M., Gleicher, G.J.: J. Am. Chem. Soc. **86** (1964) 233.
64 Mar 2	Martin, M.M., Gleicher, G.J.: J. Am. Chem. Soc. **86** (1964) 238.
64 Owe 1	Owen, Jr., G.E., Pearson, J.M., Szwarc, M.: Trans. Faraday Soc. **60** (1964) 564.
64 Pry 1	Pryor, W.A., Guard, H.: J. Am. Chem. Soc. **86** (1964) 1150.
64 Tro 1	Trosman, E.A., Bagdasar'yan, Kh. S.: Zh. Fiz. Khim. **38** (1964) 2698.
64 Vic 1	Vichutinskii, A.A., Prokof'ev, A.I., Shabalkin, V.A.: Zh. Fiz. Khim. **38** (1964) 983.
65 Baz 1	Bazilevskii, M.V., Piskun, N.I.: Zh. Fiz. Khim. **39** (1965) 762.
65 Baz 2	Bazilevskii, M.V., Piskun, N.I.: Zh. Fiz. Khim. **39** (1965) 951.
65 Ben 1	Bengough, W.I., Fairservice, W.H.: Trans. Faraday Soc. **61** (1965) 1206.
65 Ber 1	Berezin, I.V., Ivanov, V.L., Kazanskaya, N.F., Ugarova, N.N.: Zh. Fiz. Khim. **39** (1965) 3011.
65 Bir 1	Bird, R.A., Russell, K.E.: Can. J. Chem. **43** (1965) 2123.
65 Ebe 1	Ebert, M., Keene, J.P., Land, E.J., Swallow, A.J.: Proc. Roy. Soc. (London) Ser. A **287** (1965) 1.
65 Koc 1	Kochi, J.K., Subramanian, R.V.: Inorg. Chem. **4** (1965) 1527.
65 Koc 2	Kochi, J.K., Subramanian, R.V.: J. Am. Chem. Soc. **87** (1965) 4855.
65 Shi 1	Shishkina, L.N., Berezin, I.V.: Zh. Fiz. Khim. **39** (1965) 2547.
65 Zav 1	Zavitsas, A.A., Ehrenson, S.: J. Am. Chem. Soc. **87** (1965) 2841.
66 Abr 1	Abramson, F.P., Firestone, R.F.: J. Phys. Chem. **70** (1966) 3596.
66 Ber 1	Berezin, I.V., Kazanskaya, N.F., Ugarova, N.N.: Zh. Fiz. Khim. **40** (1966) 766.
66 Bur 1	Burkhart, R.D.: J. Phys. Chem. **70** (1966) 605.
66 Car 1	Carlsson, D.J., Howard, J.A., Ingold, K.U.: J. Am. Chem. Soc. **88** (1966) 4726.
66 Dob 1	Dobis, O., Nemes, I., Kerepes, R.: Zh. Fiz. Khim. **40** (1966) 328.
66 Hua 1	Huang, R.L., Lee, K.H.: J. Chem. Soc. C **1966**, 932.
66 Kal 1	Kalatzis, E., Williams, G.H.: J. Chem. Soc. B **1966**, 1112.
66 Kel 1	Keler, V., Kazanskaya, N.F., Berezin, I.V.: Vestn. Mosk. Gos. Univ. **2** (1966) 29.
66 Kod 1	Kodama, S., Fujita, S., Takeishi, J., Tayama, O.: Bull. Chem. Soc. Jpn. **39** (1966) 1009.
66 Nem 1	Nemes, I., Ugarova, N.N., Dobis, O.: Russ. J. Phys. Chem. (English Transl.) **40** (1966) 249.
66 Pry 1	Pryor, W.A., Echols, Jr., J.T., Smith, K.: J. Am. Chem. Soc. **88** (1966) 1189.
66 Sch 1	Schwetlick, K., Kelm, S.: Tetrahedron **22** (1966) 793.
66 Shi 1	Shishkina, L.N., Berezin, I.V.: Vetsn. Mosk. Gos. Univ. **2** (1966) 13.
66 Str 1	Strom, E.T., Russell, G.A., Schoeb, J.H.: J. Am. Chem. Soc. **88** (1966) 2004.
66 Wal 1	Walling, C., Cooley, J.H., Ponaras, A.A., Racah, E.J.: J. Am. Chem. Soc. **88** (1966) 5361.
67 Car 1	Carlsson, D.J., Ingold, K.U.: J. Am. Chem. Soc. **89** (1967) 4891.
67 DeT 1	DeTar, D.F.: J. Am. Chem. Soc. **90** (1967) 4058.
67 Hei 1	Heiba, E.I., Dessau, R.M.: J. Am. Chem. Soc. **89** (1967) 3772.

67 Koe 1	Koehler, W., Kazanskaya, N.F., Nagler, L.G., Berezin, I.V.: Ber. Buns. Phys. Chem. **71** (1967) 736.
67 Sim 1	Simonyi, M., Tüdös, F., Holly, S., Pospišil, J.: Eur. Polymer J. **3** (1967) 559.
67 Sim 2	Simonyi, M., Tüdös, F., Pospišil, J.: Eur. Polymer J. **3** (1967) 101.
67 Sim 3	Simonyi. M., Tüdös, F., Heidt, J.: Acta Chim. Acad. Sci. Hung. **53** (1967) 43.
67 Sue 1	Suehiro, T., Kanoya, A., Hara, H., Nakahama, T., Omori, M., Komori, T.: Bull. Chem. Soc. Jpn. **40** (1967) 668.
67 Tho 1	Thomas, J.K.: J. Phys. Chem. **71** (1967) 1919.
67 Tud 1	Tüdös, F., Berezhnikh-Földes, T., Simonyi, M.: Vysokomol. Soedin. **9** (1967) 2284.
68 Ada 1	Adams, G.E., McNaughton, G.S., Michael, B.D.: Trans. Faraday Soc. **64** (1968) 902.
68 Bur 1	Burkhart, R.D.: J. Am. Chem. Soc. **90** (1968) 273.
68 Car 1	Carlsson, D.J., Ingold, K.U.: J. Am. Chem. Soc. **90** (1968) 7047.
68 Car 2	Carlsson, D.J., Ingold. K.U.: J. Am. Chem. Soc. **90** (1968) 1055.
68 Dan 1	Danziger, R.M., Hayon, E., Langmuir, M.: J. Phys. Chem. **12** (1968) 3842.
68 Dob 1	Dobis, O., Nemes, I., Kerepes, R.: Acta Chim. Acad. Sci. Hung. **55** (1968) 215.
68 Eac 1	Eachus, A.E., Meyer, J.A., Pearson, J., Szwarc, M.: J. Am. Chem. Soc. **90** (1968) 3646.
68 Gle 1	Gleicher, G.J.: J. Org. Chem. **33** (1968) 332.
68 Hei 1	Heiba, E.I., Dessau, R.M., Koehl, Jr., W.J.: J. Am. Chem. Soc. **90** (1968) 1082.
68 Koc 1	Kochi, J.K., Bemis, A., Jenkins, C.L.: J. Am. Chem. Soc. **90** (1968) 4616.
68 Kos 1	Kosower, E.M., Mohammad, M.: J. Am. Chem. Soc. **90** (1968) 3271.
68 Lee 1	Lee, K.H.: Tetrahedron **24** (1968) 4793.
68 Owe 1	Owens, P.H., Gleicher, G.J., Smith, L.H.: J. Am. Chem. Soc. **90** (1968) 4122.
68 Saf 1	Safronenko, E.D., Afanas'ev, I.B.: Zh. Org. Khim. **4** (1968) 2086.
68 Saf 2	Safronenko, E.D., Afanas'ev, I.B.: Zh. Org. Khim. **4** (1968) 2092.
68 Sau 1	Sauer, M.C., Mani, M.: J. Phys. Chem. **72** (1968) 3856.
68 She 1	Shelton, J.R., Uzelmeier, C.W.: Rec. Trav. Chim. Pays-Bas **87** (1968) 1211.
68 Tro 1	Trosman, E.A., Bazilevskii, M.V.: Kinet. Katal. **9** (1968) 684.
69 Bur 1	Burkhart, R.D.: J. Phys. Chem. **73** (1969) 2703.
69 Bur 2	Burkhart, R.D., Merrill, J.C.: J. Phys. Chem. **73** (1969) 2699.
69 Cap 1	Capellos, C., Allen, A.O.: J. Phys. Chem. **73** (1969) 3264.
69 Car 1	Carlsson, D.J., Ingold, K.U., Bray, L.C.: Int. J. Chem. Kinet. **1** (1969) 315.
69 Cha 1	Chang, E.P., Huang, R.L., Lee, K.H.: J. Chem. Soc. B **1969**, 878.
69 Dan 1	Danen, W.C., Saunders, D.G.: J. Am. Chem. Soc. **91** (1969) 5924.
69 Dra 1	Draganić, I.G., Nenadović, M.T., Draganić, Z.D.: J. Phys. Chem. **73** (1969) 2564.
69 Hei 1	Heiba, E.I., Dessau, R.M., Koehl, W.J.: J. Am. Chem. Soc. **91** (1969) 138.
69 Inf 1	Infelta, P.P., Schuler, R.H.: J. Phys. Chem. **73** (1969) 2083.
69 Jac 1	Jackson, R.A., O'Neill, D.W.: Chem. Commun. **1969**, 1210.
69 Kar 1	Karmann, W., Granzow, A., Meissner, G., Henglein, A.: Int. J. Radiat. Phys. Chem. **1** (1969) 395.
69 Kor 1	Korson, L., Drost-Hanson, W., Millero, F.J.: J. Phys. Chem. **73** (1969) 34.
69 Leb 1	Lebedev, N.N., Shvets, V.F., Karimov, Kh. Sh., Romashkin, A.V.: Kinet. Katal. **10** (1969) 255.
69 Mor 1	Morita, M., Tajima, M., Fujimaki, M.: Agric. Biol. Chem. **33** (1969) 250.
69 Pry 1	Pryor, W.A., Tonellato, U., Fuller, D.L., Jumonville, S.: J. Org. Chem. **34** (1969) 2018.
69 Rue 1	Ruechardt, C., Herwig, K., Eichler, S.: Tetrahedron Lett. **1969**, 421.
69 Shi 1	Shishkina, L.N., Berezin, I.V.: Zh. Fiz. Khim. **43** (1969) 912.
69 Tah 1	Taha, I.A.I., Kuntz, R.R.: J. Phys. Chem. **73** (1969) 4406.
69 Tot 1	Totherow, W.D., Gleicher, G.J.: J. Am. Chem. Soc. **91** (1969) 7150.
69 Unr 1	Unruh, J.D., Gleicher, G.J.: J. Am. Chem. Soc. **91** (1969) 6211.
70 Afa 1	Afanas'ev, I.B., Safronenko, E.D.: Zh. Org. Khim. **6** (1970) 1537.
70 Bar 1	Bartlett, P.D., Fickes, G.N., Haupt, F.C., Helgeson, R.: Acc. Chem. Res. **3** (1970) 177.
70 Bul 1	Bullock, G., Cooper, R.: Trans. Faraday Soc. **66** (1970) 2055.
70 Bur 1	Burchill, C.E., Ginns, I.S.: Can. J. Chem. **48** (1970) 2628.
70 Bur 2	Burchill, C.E., Ginns, I.S.: Can. J. Chem. **48** (1970) 1232.
70 Hor 1	Horowitz, A., Rajbenbach, L.A.: J. Phys. Chem. **74** (1970) 678.
70 Mog 1	Moger, G., Dobis, O.: Magy. Kem. Foly. **76** (1970) 328.
70 Pry 1	Pryor, W.A., Smith, K.: J. Am. Chem. Soc. **92** (1970) 2731.
70 Rad 1	Radlowski, C., Sherman, W.V.: J. Phys. Chem. **74** (1970) 3043.
70 Saf 1	Safronenko, E.D., Afanas'ev, I.B.: Zh. Org. Khim. **6** (1970) 209.
70 Sho 1	Shono, T., Toda, T., Oda, R.: Tetrahedron Lett. **1970**, 369.
70 Shv 1	Shvets, V.F., Lebedev, N.N., Karimov, Kh. Sh., Zuev, A.V., Turikova, T.V.: Kinet. Katal. **11** (1970) 57.
70 Ste 1	Stefani, A.P., Chuang, L-Y.Y., Todd, H.E.: J. Am. Chem. Soc. **92** (1970) 4168.
71 Afa 1	Afanas'ev, I.B., Momontova, I.V.: Zh. Org. Khim. 7 (1971) 682.
71 Afa 2	Afanas'ev, I.B., Safronenko, E.D.: Zh. Org. Kim. **7** (1971) 453.

71 Afa 3 Afanas'ev, I.B., Momontova, I.V., Samokhvalov, G.I.: Zh. Org. Khim. **7** (1971) 457.
71 Afa 4 Afanas'ev, I.B., Momontova, I.V.: Zh. Org. Khim. **7** (1971) 678.
71 Bib 1 Bibler, N.E.: J. Phys. Chem. **75** (1971) 2436.
71 Bur 1 Burchill, C.E., Perron, K.M.: Can. J. Chem. **49** (1971) 2382.
71 Bur 2 Burchill, C.E., Thompson, G.F.: Can. J. Chem. **49** (1971) 1305.
71 Dan 1 Danen, W.C., Tipton, T.J., Saunders, D.G.: J. Am. Chem. Soc. **93** (1971) 5186.
71 Dan 2 Danen, W.C., Winter, R.L.: J. Am. Chem. Soc. **93** (1971) 716.
71 Eva 1 Evans, R., Nesyto, E., Radlowski, C., Sherman, W.V.: J. Phys. Chem. **75** (1971) 2762.
71 Hei 1 Heiba, E.J., Dessau, R.M.: J. Am. Chem. Soc. **93** (1971) 524.
71 Jen 1 Jenkins, C.L., Kochi, J.K.: J. Org. Chem. **36** (1971) 3103.
71 Kan 1 Kantrowitz, E.R., Hoffman, M.Z., Endicott, J.F.: J. Phys. Chem. **75** (1971) 1914.
71 Kat 1 Katz, M.G., Horowitz, A., Rajbenbach, L.A.: Trans. Faraday Soc. **67** (1971) 2354.
71 Koc 1 Koch, V.R., Gleicher, G.J.: J. Am. Chem. Soc. **93** (1971) 1657.
71 Koe 1 Koester, R., Asmus, K.-D.: Z. Naturforsch. **26b** (1971) 1104.
71 Law 1 Lawler, R.G., Ward, H.R., Allen, R.B., Ellenlogen, P.E.: J. Am. Chem. Soc. **93** (1971) 789.
71 Lev 1 Levy, A., Meyerstein, D., Ottolenghi, M.: J. Phys. Chem. **75** (1971) 3350.
71 Lew 1 Lewis, E.S., Butler, M.M.: J. Org. Chem. **36** (1971) 2582.
71 Lew 2 Lewis, E.S., Butler, M.M.: Chem. Commun. **1971**, 941.
71 Moh 1 Mohammad, M., Kosower, E.M.: J. Am. Chem. Soc. **93** (1971) 2709.
71 Nag 1 Nagai, Y., Matsumoto, H., Hayashi, M., Tajima, E., Watanabe, H.: Bull. Chem. Soc. Jpn. **44** (1971) 3113.
71 Nag 2 Nagai, Y., Matsumoto, H., Hayashi, M., Tajima, E., Ohtsuki, M., Sekikawa, N.: J. Organomet. Chem. **29** (1971) 209.
71 Pac 1 Packer, J.E., House, D.B., Rasburn, E.J.: J. Chem. Soc. B **1971**, 1574.
71 Pry 1 Pryor, W.A., Kneipp, K.G.: J. Am. Chem. Soc. **93** (1971) 5584.
71 Sim 1 Simonyi, M., Tüdös, F.: Adv. Phys. Org. Chem. **9** (1971) 127.
71 Ste 1 Stefani, A.P., Todd, H.E.: J. Am. Chem. Soc. **93** (1971) 2982.
71 Tra 1 Tran-Dinh-Son, Sutton, J.: J. Phys. Chem. **75** (1971) 851.
71 Unr 1 Unruh, J.D., Gleicher, G.J.: J. Am. Chem. Soc. **93** (1971) 2008.
71 Web 1 Webb, P.G., Kampmeier, J.A.: J. Am. Chem. Soc. **93** (1971) 3730.
72 Afa 1 Afanas'ev, I.B., Baranova, N.G., Samokhvalov, G.I.: Zh. Org. Khim. **8** (1972) 2449.
72 Afa 2 Afanas'ev, I.B., Baranova, N.G., Samokhvalov, G.I.: Zh. Org. Khim. **8** (1972) 1113.
72 Bya 1 Byakov, V.M., Zhiltsova, L.V., Kalyazin, E.P., Kovalyova, E.P., Petryaev, E.P.: Proc. Third Tihany Symp. Radiat. Chem., Vol. 2, J. Dobo and P. Hedvig (eds.), Budapest: Académiai Kiado, **1972**, p. 1301.
72 Dav 1 Davies, A.G., Roberts, B.P., Smith, J.M.: J. Chem. Soc., Perkin Trans. II **1972**, 2221.
72 Dru 1 Drury, R.F., Kaplan, L.: J. Am. Chem. Soc. **94** (1972) 3982.
72 Fel 1 Fel', N.S., Zaozerskaya, L.A., Dolin, P.I.: High Energy Chem. (USSR) (English Transl.) **6** (1972) 418.
72 Fel 2 Fel', N.S., Zaozerskaya, L.A., Dolin, P.I.: Radiat. Eff. **15** (1972) 57.
72 Jen 1 Jenkins, C.L., Kochi, J.K.: J. Am. Chem. Soc. **94** (1972) 856.
72 Koe 1 Koenig, E., Musso, H., Zahorszky, U.I.: Angew. Chem. **84** (1972) 33.
72 Nuc 1 Nucifora, G., Smaller, B., Remko, R., Avery, E.C.: Radiat. Res. **49** (1972) 96.
72 Pry 1 Pryor, W.A., Fuller, D.L., Stanley, J.P.: J. Am. Chem. Soc. **94** (1972) 1632.
72 Pry 2 Pryor, W.A., Smith, K., Echols, Jr., J.T., Fuller, D.L.: J. Org. Chem. **37** (1972) 1753.
72 Seh 1 Sehested, K., Marković, V.: Proc. Third Tihany Symp. Radiat. Chem., J. Dobo and P. Hedvig (eds.), Budapest: Académiai Kiado, **1972**, p. 1255.
72 Tab 1 Tabushi, I., Aoyama, Y., Kojo, S., Hamuro, J., Yoshida, Z.: J. Am. Chem. Soc. **94** (1972) 1177.
72 Wal 1 Walling, C., Cioffari, A.: J. Am. Chem. Soc. **94** (1972) 6059.
72 Wal 2 Walling, C., Cioffari, A.: J. Am. Chem. Soc. **94** (1972) 6064.
72 Zav 1 Zavitsas, A.A., Blank, J.D.: J. Am. Chem. Soc. **94** (1972) 4603.
73 Afa 1 Afanas'ev, I.B., Baranova, N.G., Samokhvalov, G.I.: Int. J. Chem. Kinet. **5** (1973) 477.
73 Bec 1 Beckwith, A.L.J., Phillipou, G.: J. Chem. Soc. Chem. Commun. **1973**, 280.
73 Bha 1 Bhatia, K., Schuler, R.H.: J. Phys. Chem. **77** (1973) 1356.
73 Dan 1 Danen, W.C., Saunders, D.G., Rose, K.A.: J. Am. Chem. Soc. **95** (1973) 1612.
73 Dru 1 Drury, R.F., Kaplan, L.: J. Am. Chem. Soc. **95** (1973) 2217.
73 Hor 1 Horowitz, A., Rajbenbach, L.A.: J. Am. Chem. Soc. **95** (1973) 6308.
73 Lee 1 Lee, K.H.: J. Chem. Soc., Perkin Trans. II **1973**, 693.
73 Lev 1 Levy, A., Meyerstein, D., Ottolenghi, M.: J. Phys. Chem. **77** (1973) 3044.
73 Lew 1 Lewis, E.S., Kozuka, S.: J. Am. Chem. Soc. **95** (1973) 282.
73 Pry 1 Pryor, W.A., Kneipp, K.G., Morkved, E.H., Stanley, J.P.: Radiat. Res. **53** (1973) 181.
73 Pry 2 Pryor, W.A., Davis, Jr., W.H., Stanley, J.P.: J. Am. Chem. Soc. **95** (1973) 4754.
73 Pur 1 Purdie, J.W., Gillis, H.A., Klassen, N.V.: Can. J. Chem. **51** (1973) 3132.

73 Red 1 Redpath, J.L.: Radiat. Res. **54** (1973) 364.
73 Red 2 Redpath, J.L., Willson, R.L.: Int. J. Radiat. Biol. **23** (1973) 51.
73 Sim 1 Simonyi, M., Kardos, J., Tüdös, F., Pospišil, J.: J. Polymer Sci., Part C; Polymer Symp. **40** (1973) 163.
73 Sim 2 Simonyi, M., Tüdös, F.: Acta Chim. Acad. Sci. Hung. **77** (1973) 315.
73 Won 1 Wong, R.H.W., Gleicher, G.J.: J. Org. Chem. **38** (1973) 1957.
74 Bec 1 Beckwith, A.L.J., Moad, G.: J. Chem. Soc. Chem. Commun. **1974**, 472.
74 Bec 2 Beckwith, A.L.J., Blair, I.A., Phillipou, G.: Tetrahedron Lett. **1974**, 2251.
74 Bec 3 Beckwith, A.L.J., Blair, I.A., Phillipou, G.: J. Am. Chem. Soc. **96** (1974) 1613.
74 Blo 1 Bloodworth, A.J., Davies, A.G., Griffin, I.M., Muggleton, B., Roberts, B.P.: J. Am. Chem. Soc. **96** (1974) 7599.
74 Bun 1 Bunce, N.J., Hadley, M.: J. Org. Chem. **39** (1974) 2271.
74 Dan 1 Danen, W.D., Saunders, D.G., Rose, K.A.: J. Am. Chem. Soc. **96** (1974) 4558.
74 Gil 1 Gilbert, B.C., Norman, R.O.C., Sealy, R.C.: J. Chem. Soc., Perkin Trans. II **1974**, 1435.
74 Gle 1 Gleicher, G.J.: Tetrahedron **30** (1974) 935.
74 Gri 1 Griller, D., Ingold, K.U.: Int. J. Chem. Kinet. **6** (1974) 453.
74 Hei 1 Heine, H.-G., Hartmann, W., Kory, D.R., Magyar, J.G., Hoyle, C.E., McVey, J.K., Lewis, F.D.: J. Org. Chem. **39** (1974) 691.
74 Hen 1 Henderson, R.W., Ward, Jr., R.D.: J. Am. Chem. Soc. **96** (1974) 7556.
74 Hen 2 Hendry, D.G., Mill, T., Piszkiewicz, L., Howard, J.A., Eigenmann, H.K.: J. Phys. Chem. Ref. Data **3** (1974) 934.
74 Koc 1 Kochi, J.K.: Acc. Chem. Res. **7** (1974) 351.
74 Lev 1 Levy, A., Meyerstein, D., Ottolenghi, M.: Int. J. Appl. Radiat. Isot. **25** (1974) 9.
74 New 1 Newkirk, D.D., Gleicher, G.J.: J. Am. Chem. Soc. **96** (1974) 3543.
74 Pac 1 Packer, J.E., Richardson, R.K., Soole, P.J., Webster, D.R.: J. Chem. Soc., Perkin Trans. II **1974**, 1472.
74 Par 1 Parnell, R.D., Russell, K.E.: J. Polymer Sci., Part A1, Polymer Chem. **12** (1974) 347.
74 Pry 1 Pryor, W.A., Davis, W.H.: J. Am. Chem. Soc. **96** (1974) 7557.
74 Tan 1 Tanner, D.D., Arhart, R.J., Blackburn, E.V., Das, N.C., Wada, N.: J. Am. Chem. Soc. **96** (1974) 829.
75 Alo 1 Aloni, R., Katz, M.G., Rajbenbach, L.A.: Int. J. Chem. Kinet. **7** (1975) 699.
75 Bec 1 Beckwith, A.L.J., Moad, G.: J. Chem. Soc., Perkin Trans. II **1975**, 1726.
75 Bec 2 Beckwith, A.L.J., Gara, W.B.: J. Chem. Soc., Perkin Trans. II **1975**, 795.
75 Bol 1 Bolton, R., Seabrooke, J.M., Williams, G.H.: J. Fluorine Chem. **5** (1975) 1.
75 Cha 1 Chang, S.C., Gleicher, G.J.: J. Org. Chem. **40** (1975) 3800.
75 Dan 1 Danen, W.C., Rose, K.A.: J. Org. Chem. **40** (1975) 619.
75 Gil 1 Gilbert, B.C., Norman, R.O.C., Placucci, G., Sealy, R.C.: J. Chem. Soc., Perkin Trans. II **1975**, 885.
75 Gil 2 Gilbert, B.C., Norman, R.O.C., Selay, R.C.: J. Chem. Soc., Perkin Trans. II **1975**, 303.
75 Gre 1 Greenstock, C.L., Dunlop, I.: Fast Processes in Radiation Chemistry and Biology, G.E. Adams, E.M. Fielden and B.D. Michael, (eds.), London: Wiley, **1975**, p. 247.
75 Hen 1 Henderson, R.W.: J. Am. Chem. Soc. **97** (1975) 213.
75 Her 1 Herwig, K., Lorenz, P., Rüchardt, C.: Chem. Ber **108** (1975) 1421.
75 Jan 1 Janzen, E.G., Nutler, D.E., Evans, C.A.: J. Phys. Chem. **79** (1975) 1983.
75 Jan 2 Janzen, E.G., Evans, C.A.: J. Am. Chem. Soc. **97** (1975) 205.
75 Jul 1 Julia, M., Descoins, C., Baillarge, M., Jacquet, B., Uguen, D., Groeger, F.A.: Tetrahedron **31** (1975) 1737.
75 Kat 1 Katz, M.G., Horowitz, A., Rajbenbach, L.A.: Int. J. Chem. Kinet. **7** (1975) 183.
75 Kat 2 Katz, M.G., Rajbenbach, L.A.: Int. J. Chem. Kinet. **7** (1975) 785.
75 Mig 1 Migita, T., Nagai, T., Abe, Y.: Chem. Lett. (Tokyo) **1975**, 543.
75 Pac 1 Packer, J.E., Richardson, R.K.: J. Chem. Soc., Perkin Trans. II **1975**, 751.
75 Sim 1 Simonyi, M., Kardos, J., Fitos, I., Kovács, I., Pospišil, J.: J. Chem. Soc. Chem. Commun. **1975**, 15.
75 Sim 2 Simonyi, M., Fitos, I., Kardos, J., Lukovits, I.: J. Chem. Soc. Chem. Commun. **1975**, 252.
75 Sim 3 Simonyi, M., Kardos, J., Fitos, I., Kovacs, I., Holly, S., Pospišil, J.: Tetrahedron Lett. **1975**, 565.
75 Tan 1 Tanner, D.D., Wada, N.: J. Am. Chem. Soc. **97** (1975) 2190.
75 Wil 1 Willson, R.L., Slater, T.F.: Fast Processes in Radiation Chemistry and Biology; Adams, G.E.: Fielden, E.M.; Michael, B.D., (eds). New York: John Wiley & Sons, **1975**, pp. 147.
75 Zav 1 Zavitsas, A.A., Hanna, G.M.: J. Org. Chem. **40** (1975) 3782.
76 Alo 1 Aloni, R., Horowitz, A., Rajbenbach, L.A.: Int. J. Chem. Kinet. **8** (1976) 673.
76 Bec 1 Beckwith, A.L.J., Phillipou, G.: Austral. J. Chem. **20** (1976) 123.
76 Fri 1 Frith, P.G., McLauchlan, K.A.: J. Chem. Soc., Faraday Trans. II **72** (1976) 87.
76 Gie 1 Giese, B.: Angew. Chem. **88** (1976) 159.
76 Gie 2 Giese, B.: Angew. Chem. **88** (1976) 161.
76 Gie 3 Giese, B.: Angew. Chem. **88** (1976) 723.
76 Gon 1 Gonen, Y., Horowitz, A., Rajbenbach, L.A.: J. Chem. Soc., Faraday Trans. I **72** (1976) 901.

76 Hor 1 Horowitz, A.: Int. J. Chem. Kinet. **8** (1976) 709.
76 Kat 1 Katz, M.G., Baruch, G., Rajbenbach, L.A.: J. Chem. Soc., Faraday Trans. I **72** (1976) 2462.
76 Kat 2 Katz, M.G., Baruch, G., Rajbenbach, L.A.: J. Chem. Soc., Faraday Trans. I **72** (1976) 1903.
76 Kat 3 Katz, M.G., Baruch, G., Rajbenbach, L.A.: Int. J. Chem. Kinet. **8** (1976) 131.
76 Kat 4 Katz, M.G., Baruch, G., Rajbenbach, L.A.: Int. J. Chem. Kinet. **8** (1976) 599.
76 Koz 1 Kozuka, S., Lewis, E.S.: J. Am. Chem. Soc. **98** (1976) 2254.
76 Kuh 1 Kuhlmann, R., Schnabel, W.: Polymer **17** (1976) 419.
76 Lew 1 Lewis, E.S., Butler, M.M.: J. Am. Chem. Soc. **98** (1976) 2257.
76 Lew 2 Lewis, E.S., Ogino, K.: J. Am. Chem. Soc. **98** (1976) 2260.
76 Lew 3 Lewis, E.S., Nieh, E.C.: J. Am. Chem. Soc. **98** (1976) 2268.
76 Sca 1 Scaiano, J.C., Tremblay, J.P.-A., Ingold, K.U.: Can. J. Chem. **54** (1976) 3407.
76 Sch 1 Schuler, R.H., Neta, P., Zemel, H., Fessenden, R.W.: J. Am. Chem. Soc. **98** (1976) 3825.
76 Sim 1 Simonyi, M., Kardos, Y., Fitos, I., Kovács, I., Pospišil, J.: J. Chem. Soc., Perkin Trans. II **1976**, 1913.
76 Tes 1 Testaferri, L., Tiecco, M., Spagnolo, P., Martelli, G.: J. Chem. Soc., Perkin Trans. II **1976**, 662.
77 Aga 1 Agabekov, V.E., Budeiko, N.L., Denisov, E.T., Mitskevich, N.I.: React. Kinet. Catal. Lett. **7** (1977) 437.
77 Ash 1 Ashworth, B., Gilbert, B.C., Norman, R.O.C.: J. Chem. Res. (M) **1977**, 1101.
77 Bec 1 Beck, G., Lindenau, D., Schnabel, W.: Macromolecules **10** (1977) 135.
77 Bel 1 Belova, L.I., Karpukhina, G.V.: Bull. Acad. Sci. USSR (English Transl.) **26** (1977) 1605.
77 Che 1 Chess, E.K., Schatz, B.S., Gleicher, G.J.: J. Org. Chem. **42** (1977) 752.
77 Dav 1 Davis, Jr., W.H., Pryor, W.A.: J. Am. Chem. Soc. **99** (1977) 6365.
77 Gie 1 Giese, B., Joy, K.: Angew. Chem. **89** (1977) 482.
77 Gon 1 Gonen, Y., Horowitz, A., Rajbenbach, L.A.: J. Chem. Soc., Faraday Trans. I **73** (1977) 866.
77 Gon 2 Gonen, Y., Rajbenbach, L.A., Horowitz, A.: Int. J. Chem. Kinet. **9** (1977) 361.
77 Gro 1 Groves, J.T., Kittisopikul, S.: Tetrahedron Lett. **1977**, 4291.
77 Kat 1 Katz, M.G., Baruch, G., Rajbenbach, L.A.: Int. J. Chem. Kinet. **9** (1977) 55.
77 Kry 1 Kryger, R.G., Lorand, J.P., Stevens, N.R., Herron, N.R.: J. Am. Chem. Soc. **99** (1977) 7589.
77 Nug 1 Nugent, W.A., Kochi, J.K.: J. Organomet. Chem. **124** (1977) 327.
77 Pry 1 Pryor, W.A., Gleaton, J.H., Davis, W.H.: J. Org. Chem. **42** (1977) 7.
77 Rya 1 Ryan, T.G., Freeman, G.R.: J. Phys. Chem. **81** (1977) 1455.
77 Sca 1 Scaiano, J.C., Ingold, K.U.: J. Am. Chem. Soc. **99** (1977) 2079.
77 Sim 1 Simonyi, M., Fitos, I., Kardos, J., Kovács, I., Lukovits, I., Pospišil, J.: J. Chem. Soc., Faraday Trans. I **73** (1977) 1286.
77 Soy 1 Soylemez, T., Balkas, T.I.; Proc. Fourth Tihany Symp. Radiat. Chem., P. Hedvig and R. Schiller (eds.), Budapest: Akad. Kiado, **1977**, p. 853.
77 Sue 1 Suehiro, T., Suzuki, A., Tsuchida, Y., Yamazaki, J.: Bull. Chem. Soc. Jpn. **50** (1977) 3324.
77 Tan 1 Tanner, D.D., Ochiai, T., Rowe, J., Pace, T., Takiguchi, H., Samal, P.W.: Can. J. Chem. **55** (1977) 3536.
77 Tho 1 Thomson, R.A.M., Manolis, C.S.: Chem. Ind. (London) **1977**, 274.
77 Tua 1 Tuan, N.Q., Gäumann, T.: Radiat. Phys. Chem. **10** (1977) 263.
78 Col 1 Colle, T.H., Glaspie, P.S., Lewis, E.S.: J. Org. Chem. **43** (1978) 2722.
78 Fol 1 Foldiak, G., Schuler, R.H.: J. Phys. Chem. **82** (1978) 2756.
78 Kar 1 Kardos, J., Fitos, I., Kovács, I., Szammer, J., Simonyi, M.: J. Chem. Soc., Perkin Trans. II **1978**, 405.
78 Kat 1 Katz, M.G., Baruch, G., Rajbenbach, L.A.: Int. J. Chem. Kinet. **10** (1978) 905.
78 Kin 1 Kinney, R.J., Jones, W.D., Bergman, R.G.: J.Am. Chem. Soc. **100** (1978) 7902.
78 Kor 1 Korobov, V.E., Chibisov, A.K.: J. Photochem. **9** (1978) 411.
78 Kos 1 Kosorotov, V.I., Dzhagatspanyan, R.V.: Kinet. Katal. **19** (1978) 1123.
78 Kos 2 Kosower, E.M., Waits, H.P., Teuerstein, A., Butler, L.C.: J. Org. Chem. **43** (1978) 800.
78 Leh 1 Lehnig, M., Neumann, W.P., Seifert, P.: J. Organometal. Chem. **162** (1978) 145.
78 Lor 1 Lorand, J.P., Kryger, R.G., Herron, N.R.: Colloqu. Int. CNRS (1977) Radicaux Libres Org. **1978**, 463.
78 Luh 1 Luh, T.-Y., Stock, L.M.: J. Org. Chem. **43** (1978) 3271.
78 Mad 1 Madhavan, V., Schuler, R.H., Fessenden, R.W.: J. Am. Chem. Soc. **100** (1978) 888.
78 Rya 1 Ryan, T.G., Sambrook, T.E.M., Freeman, G.R.: J. Phys. Chem. **82** (1978) 26.
78 Saw 1 Sawai, T., Ohara, N., Shimokawa, T.: Bull. Chem. Soc. Jpn. **51** (1978) 1300.
78 Tab 1 Tabushi, I., Kojo, S., Fukunishi, K.: J. Org. Chem. **43** (1978) 2370.
78 Tua 1 Tuan, N.Q., Gäumann, T.: Radiat. Phys. Chem. **11** (1978) 183.
78 Van 1 Van Beek, H.C.A., van der Stoep, H.J.: Recl. Trav. Chim. Pays-Bas **97** (1978) 279.
78 Wal 1 Waltz, W.L., Hachelberg, O., Dorfman, L.M., Wojcicki, A.: J. Am. Chem. Soc. **100** (1978) 7259.
79 Bat 1 Batyrbaev, N.A., Zorin, V.V., Imashev, U.B., Zlotskii, S.S., Karakhanov, R.A., Rakhmankulov, D.L.: Arm. Khim. Zh. **32** (1979) 822.

79 Bec 1	Beckwith, A.L.J., Lawrence, T.: J. Chem. Soc. Perkin Trans. II **1979**, 1535.
79 Col 1	Colle, T.H., Lewis, E.S.: J. Am. Chem. Soc. **101** (1979) 1810.
79 Gie 1	Giese, B., Keller, K.: Chem. Ber. **112** (1979) 1743.
79 Gie 2	Giese, B., Jay, K.: Chem. Ber. **112** (1979) 304.
79 Gie 3	Giese, B., Jay, K.: Chem. Ber. **112** (1979) 298.
79 Gie 4	Giese, B., Stellmach, J.: Tetrahedron Lett. **1979**, 857.
79 Hor 1	Horowitz, A., Baruch, G.: Int. J. Chem. Kinet. **11** (1979) 1263.
79 How 1	Howard, J.A., Tong, S.B.: Can. J. Chem. **57** (1979) 2755.
79 Mig 1	Migita, T., Takayama, K., Abe, Y., Kosugi, M.: J. Chem. Soc., Perkin Trans. II **1979**, 1137.
79 Pau 1	Paul, H.: Int. J. Chem. Kinet. **11** (1979) 495.
79 Sam 1	Samirkhanov, Sh. M., Zlotskii, S.S., Rakhmankulov, D.L.: Zh. Org. Khim. **15** (1979) 1815.
79 Sch 1	Schmid, P., Griller, D., Ingold, K.U.: Int. J. Chem. Kinet. **11** (1979) 333.
79 Ste 1	Steenken, S.: J. Phys. Chem. **83** (1979) 595.
79 Tan 1	Tanner, D.D., Henriquez, R., Reed, D.W.: Can. J. Chem. **57** (1979) 2578.
79 Tan 2	Tanner, D.D., Samal, P.W., Ruo, T.C.-S., Henriquez, R.: J. Am. Chem. Soc. **101** (1979) 1168.
79 Zad 1	Zador, E., Warman, J.M., Hummel, A.: J. Chem. Soc., Faraday Trans. I **75** (1979) 914.
79 Zlo 1	Zlotskii, S.S., Rakhmankulov, D.L., Borodina, L.N., Karakhanov, E.A., Vestn. Mosk. Univ., Ser. 2: Khim. **20** (1979) 164.
80 Alf 1	Alfassi, Z.B., Feldman, L.: Int. J. Chem. Kinet. **12** (1980) 379.
80 Bar 1	Baruch, G., Horowitz, A.: J. Phys. Chem. **84** (1980) 2535.
80 Bec 1	Beckwith, A.L.J., Moad, G.: J. Chem. Soc., Perkin Trans. II **1980**, 1083.
80 Eic 1	Eichler, J., Herz, C.P., Schnabel, W.: Angew. Makromol. Chem. **91** (1980) 39.
80 Fel 1	Feldman, L., Alfassi, Z.B.: Radiat. Phys. Chem. **15** (1980) 687.
80 Gie 1	Giese, B., Stellmach, J.: Chem. Ber. **113** (1980) 3294.
80 M'Ha 1	M'Halla, F., Pinson, J., Savéant, J.M.: J. Am. Chem. Soc. **102** (1980) 4120.
80 Ngu 1	Nguyen, T.Q., Dang, T.M., Gäumann, T.: Radiat. Phys. Chem. **15** (1980) 223.
80 Nol 1	Nolan, G.S., Gleicher, G.J., Schatz, B., Cordova, R.: J. Org. Chem. **45** (1980) 444.
80 Tan 1	Tanner, D.D., Blackburn, E.V., Reed, D.W., Setiloane, B.P.: J. Org. Chem. **45** (1980) 5183.
81 Alf 1	Alfassi, Z.B., Feldman, L.: Int. J. Chem. Kinet. **13** (1981) 771.
81 Alf 2	Alfassi, Z.B., Feldman, L.: Int. J. Chem. Kinet. **13** (1981) 517.
81 Alo 1	Aloni, R., Rajbenbach, L.A., Horowitz, A.: Int. J. Chem. Kinet. **13** (1981) 23.
81 Bar 1	Baruch, G., Rajbenbach, L.A., Horowitz, A.: Int. J. Chem. Kinet. **13** (1981) 473.
81 Cha 1	Chatgilialoglu, C., Ingold, K.U., Scaiano, J.C.: J. Am. Chem. Soc. **103** (1981) 7739.
81 Due 1	Duetsch, H.-R., Fischer, H.: Int. J. Chem. Kinet. **13** (1981) 527.
81 Enc 1	Encinas, M.V., Lissi, E.A., Soto, H.: J. Photochem. **16** (1981) 43.
81 Fel 1	Feldman, L., Alfassi, Z.B.: J. Phys. Chem. **85** (1981) 3060.
81 Gon 1	Gonen, Y., Horowitz, A., Rajbenbach, L.A.: Int. J. Chem. Kinet. **13** (1981) 219.
81 Luh 1	Luh, T.-Y., Lei, K.L.: J. Org. Chem. **46** (1981) 5328.
82 Due 1	Duetsch, H.R., Fischer, H.: Int. J. Chem. Kinet. **14** (1982) 195.
82 Fel 1	Feldman, L., Alfassi, Z.B.: Int. J. Chem. Kinet. **14** (1982) 659.
82 Lem 1	Lemmes, R., von Sonntag, C.: Carbohydr. Res. **105** (1982) 276.
82 Pry 1	Pryor, W.A., Tang, F.Y., Tang, R.H., Church, D.F.: J. Am. Chem. Soc. **104** (1982) 2885.
82 Ste 1	Steenken, S., Neta, P.: J. Phys. Chem. **86** (1982) 3661.
82 Sut 1	Sutcliffe, R., Anpo, M., Stolow, A., Ingold, K.U.: J. Am. Chem. Soc. **104** (1982) 6064.
82 Tan 1	Tanner, D.D., Reed, D.W., Setiloane, B.P.: J. Am. Chem. Soc. **104** (1982) 3917.
82 Was 1	Washino, K., Schnabel, W.: Makromol. Chem., Rapid Commun. **3** (1982) 427.
82 Wol 1	Wolfenden, B.S., Willson, R.L.: J. Chem. Soc., Perkin Trans. II **1982**, 805.
83 For 1	Forni, L.G., Mönig, J., Mora-Arellano, V.O., Willson, R.L.: J. Chem. Soc., Perkin Trans. II **1983**, 961.

4.2 Rate constants of electron transfer reactions of carbon-centered radicals with molecules in solutions

4.2.0 Introduction

4.2.0.1 General remarks

The rate constants of electron transfer reactions of carbon-centered radicals with molecules have been collected from the literature up to the end of 1981. Several thousand papers and references had to be screened and cross checked, and it is hoped that omissions have been kept to a minimum. It was very helpful to find some compilations of data in the recent literature and we would like to acknowledge in particular the collection on "Rate Constants of Aliphatic Carbon-Centered Radicals in Aqueous Solutions" by A.B. Ross and P. Neta, Radiation Chemistry Data Center, Radiation Laboratory, University of Notre Dame, NSRDS-NBS series and an article by A.J. Swallow on "Reactions of Free Radicals Produced from Organic Compounds in Aqueous Solutions by Means of Radiation" which appeared in Progr. Reaction Kinetics Vol. 9, No 3/4, 1978, pp 195–366. Equally useful particularly for spotting the relevant original literature were the books by M. Szwarc "Ions and Ion Pairs in Organic Reactions", J. Wiley, New York 1974 and by J.K. Kochi "Free Radicals", J. Wiley, New York 1973. We also like to acknowledge the help of Dr. Ch.-H. Fischer concerning the nomenclature of many compounds.

The majority of the electron transfer rate constants are based on direct measurements, namely time resolved observations of the reaction

$$\dot{R} + S \xrightarrow{k_2} P_i$$

where $\dot{R}$ = radical, S = substrate, P_i = products and k_2 = bimolecular rate constant. Observed parameters are usually physical properties of either the reactant radical $\dot{R}$ or products, radical or molecular, such as optical absorption, conductivity etc. "Mixing" of the reactants is generally achieved by *in situ* generation of the radicals $\dot{R}$ on application of a short (compared with the lifetime of $\dot{R}$ in its reaction with S) and intense pulse of energy. The latter is provided in particular by photons (flash photolysis) and high energy electrons (pulse radiolysis). Generation of $\dot{R}$ occurs in the majority of cases not directly but indirectly by reaction of a primary radical with a molecule, e.g. $\dot{O}H + RH \longrightarrow \dot{R} + H_2O$. In such cases experimental conditions are chosen which provide for fast formation of $\dot{R}$, i.e. within the duration of the pulse.

An obvious problem resulting from this method of radical generation is that the reaction of the primary radical to produce $\dot{R}$ may not be regiospecific, with the result that more than one type of R radical is produced. This consideration applies in particular to the radiation chemical technique of pulse radiolysis.

Another and general problem is partial spin delocalization which is often the case for radicals containing heteroatoms. A typical example is the $\dot{C}H_2CHO$ radical which has oxidizing properties. Its spin is predominantly localized at carbon ($\approx 85\%$) and thus this radical will be listed as carbon-centered. Its oxidizing action can be anticipated to occur however through its mesomeric form $CH_2{=}CH\dot{O}$, i.e. an oxygen-centered radical. A similar situation usually pertains to heteroaromatic radicals, particularly radical anions and cations. The reactions of all these species have been included in the listing of carbon-centered species if spin localization is either known to be mainly on a carbon atom or if a reasonably high probability of spin localization on carbon can be anticipated. Spin and charge delocalization is also apparent in most radical anions and cations of substrates composed of only C and H. While their classification as carbon-centered radicals is hardly questionable, the radical reaction site cannot necessarily be assigned to a particular carbon atom.

The rate constants in this collection include both electron transfer from the radical $\dot{R}$ to the substrate S as well as transfer in the opposite direction. The latter constitutes an oxidative action of the radical $\dot{R}$. For cationic radicals this is usually referred to as charge transfer. In irradiated solvents, particularly of non-polar nature, the primary radical cation often exhibits the property of a highly mobile hole. Reactions of such species which involve electron transfer from a substrate molecule to such a hole have also been included.

In some cases an ambiguity exists as to whether a particular reaction is an H-atom transfer (which would have to be listed as a displacement reaction) or an electron transfer followed by immediate protonation. Generally all these and other cases of questionable assignment have been listed in the electron transfer section but may possibly appear in other sections by other authors as well. The total number of such ambiguous cases is however relatively small.

4.2.0.2 Arrangement of the rate constants

Since more than half of the electron transfer rate constants refer to aqueous solutions it seems reasonable to divide the data into an aqueous and a non-aqueous section. Within both sections the radicals derived from aliphatic substrates are separated from those derived from aromatic compounds. A third group of rate constants in the aqueous section covers radicals with undefined or questionable stoichiometry and structure or systems where more than one radical is simultaneously involved in an overall electron transfer process. Some further and more detailed explanations immediately precede the listing of these particular rate constants.

Subgrouping of the stoichiometrically and structurally defined radicals lists species together containing the same kind of atoms, e.g. aliphatic radicals composed of C, H and O atoms or aromatic radicals composed of C and H atoms. Within any such subgroup the radicals are listed with increasing number of C-atoms. If this is the same the listing is defined by the alphabetical order and number of the other atoms, e.g. $CH_3\dot{C}HOH$ radicals will be listed under C_2H_5O, or $CH_3SCH_2CH_2\dot{C}HNH_3^+$ radical cations under $C_4H_{11}NS$. For nonaqueous systems each subgroup is further divided into neutral, anionic and cationic radical sections because of the particular interest of organic chemists in ionic reactions in non-aqueous, especially nonpolar, liquids.

Rate constants for a particular radical are divided into reactions with inorganic and metal containing substrates (including e.g. cytochrome-III-c, etc.) followed by the reactions with organic substrates. These are listed alphabetically according to symbols (inorganic substrates) and names (organic substrates), i.e. reactions of a particular radical with Ag substrates precede those with Co, Fe etc. and reactions with acetophenone are followed by the reactions with benzoquinone, 1,1'-dimethyl-4,4'-bipyridinium etc.

For aqueous solutions the pH of the solution is included whenever specifically mentioned in the reference paper. In all other cases the pH of the solutions can generally be assumed to be between 5 and 9, i.e. near neutrality.

Error limits have been listed whenever given in the literature. If not given the experimental error limit for pulse radiolysis data is usually considered to be $\pm 10\%$.

4.2.1 Reactions in aqueous solutions

4.2.1.1 Aliphatic radicals and radicals derived from other non-aromatic compounds

4.2.1.1.1. Radicals containing only C and H, and C, H, and Cl atoms

Reaction / Radical generation / Method	Solvent	T[K]	Rate data	Ref./ add. ref.
$\dot{C}Cl_3 + IrCl_6^{2-} \longrightarrow Ir(III) + products$				
Pulse rad. of $CCl_4 + H_2O$				82 Ste 1 [1])
KAS	H_2O	295	$k = 2.8 \cdot 10^7\,M^{-1}s^{-1}$	
$\dot{C}Cl_3 + MnO_4^- \longrightarrow Mn(VI) + products$				
Pulse rad. of $CCl_4 + H_2O$				82 Ste 1 [2])
KAS	H_2O	295	$k = 4 \cdot 10^8\,M^{-1}s^{-1}$	
$\dot{C}HCl_2 + Fe(CN)_6^{3-} \longrightarrow Fe(II) + products$				
Pulse rad. of $CHCl_3 + H_2O$				82 Ste 1 [2])
KAS	H_2O	295	$k < 5 \cdot 10^5\,M^{-1}s^{-1}$	
$\dot{C}HCl_2 + IrCl_6^{2-} \longrightarrow Ir(III) + products$				
Pulse rad. of $CHCl_3 + H_2O$				82 Ste 1 [1])
KAS	H_2O	295	$k \approx 5 \cdot 10^8\,M^{-1}s^{-1}$	
$\dot{C}HCl_2 + MnO_4^- \longrightarrow Mn(VI) + products$				
Pulse rad. of $CHCl_3 + H_2O$				82 Ste 1 [2])
KAS	H_2O	295	$k \approx 1 \cdot 10^9\,M^{-1}s^{-1}$	

[1]) Mechanism discussed as e^-- or $\dot{C}l$-transfer.
[2]) Mechanism discussed as e^--transfer.

Reaction Radical generation Method	Solvent	T[K]	Rate data	Ref./ add. ref.
$\dot{C}H_2Cl + Fe(CN)_6^{3-} \longrightarrow Fe(II)$ + products				
Pulse rad. of $CH_2Cl_2 + H_2O$				82 Ste 1 [2])
KAS	H_2O	295	$k < 5 \cdot 10^5 M^{-1} s^{-1}$	
$\dot{C}H_2Cl + IrCl_6^{2-} \longrightarrow Ir(III)$ + products				
Pulse rad. of $CH_2Cl_2 + H_2O$				82 Ste 1 [1])
KAS	H_2O	295	$k \approx 1 \cdot 10^9 M^{-1} s^{-1}$	
$\dot{C}H_2Cl + MnO_4^- \longrightarrow Mn(VI)$ + products				
Pulse rad. of $CH_2Cl_2 + H_2O$				82 Ste 1 [2])
KAS	H_2O	295	$k \approx 1 \cdot 10^9 M^{-1} s^{-1}$	
$\dot{C}H_3 + Fe(CN)_6^{3-} \longrightarrow Fe(II)$ + products				
Pulse rad. of $(CH_3)_2SO + N_2O + H_2O$ and $CH_3Cl + H_2O$				82 Ste 1 [2])
KAS	H_2O	295	$k = 5(1) \cdot 10^6 M^{-1} s^{-1}$	
$\dot{C}H_3 + IrCl_6^{2-} \longrightarrow Ir(III)$ + products				
Pulse rad. of $(CH_3)_2SO + N_2O + H_2O$ and $CH_3Cl + H_2O$				82 Ste 1 [1])
KAS, Cond.	H_2O	295	$k = 1.15 \cdot 10^9 M^{-1} s^{-1}$	
$\dot{C}H_3 + MnO_4^- \longrightarrow Mn(VI)$ + products				
Pulse rad. of $(CH_3)_2SO + N_2O + H_2O$ and $CH_3Cl + H_2O$				82 Ste 1 [2])
KAS	H_2O	295	$k = 1.05 \cdot 10^9 M^{-1} s^{-1}$	
$\dot{C}H_2CH_2Cl + IrCl_6^{2-} \longrightarrow Ir(III)$ + products				
Pulse rad. of $ClCH_2CH_2Cl + H_2O$				82 Ste 1 [1])
KAS	H_2O	295	$k \approx 1 \cdot 10^9 M^{-1} s^{-1}$	
$\dot{C}H_2CH_3 + Fe(CN)_6^{3-} \longrightarrow Fe(II)$ + products				
Pulse rad. of $(C_2H_5)_2SO + N_2O + H_2O$ and $C_2H_5Cl + H_2O$				82 Ste 1 [2])
KAS	H_2O	293	$k = 5.0 \cdot 10^7 M^{-1} s^{-1}$	
$\dot{C}H_2CH_3 + IrCl_6^{2-} \longrightarrow Ir(III)$ + products				
Pulse rad. of $(C_2H_5)_2SO + N_2O + H_2O$ and $C_2H_5Cl + H_2O$				82 Ste 1 [1])
KAS, Cond.	H_2O	295	$k = 3.1 \cdot 10^9 M^{-1} s^{-1}$	
$\dot{C}H_2CH_3 + MnO_4^- \longrightarrow Mn(VI)$ + products				
Pulse rad. of $(C_2H_5)_2SO + N_2O + H_2O$ and $C_2H_5Cl + H_2O$				82 Ste 1 [2])
KAS	H_2O	295	$k \approx 2 \cdot 10^9 M^{-1} s^{-1}$	
$\dot{C}H(CH_3)_2 + Fe(CN)_6^{3-} \longrightarrow Fe(II)$ + products				
Pulse rad. of $((CH_3)_2CH)_2SO + N_2O + H_2O$ and $(CH_3)_2CHCl + H_2O$				82 Ste 1 [2])
KAS	H_2O	295	$k = 1.25 \cdot 10^9 M^{-1} s^{-1}$	
$\dot{C}H(CH_3)_2 + IrCl_6^{2-} \longrightarrow Ir(III)$ + products				
Pulse rad. of $((CH_3)_2CH)_2SO + N_2O + H_2O$ and $(CH_3)_2CHCl + H_2O$				82 Ste 1 [3])
KAS, Cond.	H_2O	295	$k = 3.6 \cdot 10^9 M^{-1} s^{-1}$	
$(CH_3)_3\dot{C} + Fe(CN)_6^{3-} \longrightarrow Fe(II)$ + products				
Pulse rad. of $((CH_3)_3C)_2SO + N_2O + H_2O$ and $(CH_3)_3CH + N_2O + H_2O$				82 Ste 1 [2])
KAS	H_2O	295	$k = 3.6 \cdot 10^9 M^{-1} s^{-1}$	

[1]) Mechanism discussed as e^-- or $\dot{C}l$-transfer.
[2]) Mechanism discussed as e^--transfer.
[3]) Mechanism discussed as e^-- and $\dot{C}l$-transfer.

Reaction Radical generation Method	Solvent	T[K]	Rate data	Ref./ add. ref.
$(CH_3)_3\dot{C} + IrCl_6^{2-} \longrightarrow Ir(III) + products$				
Pulse rad. of $((CH_3)_3C)_2SO + N_2O + H_2O$ and $(CH_3)_3CH + N_2O + H_2O$				82 Ste 1 [3]
KAS, Cond.	H_2O	295	$k = 3.8 \cdot 10^9\ M^{-1}\,s^{-1}$	
$CH_2{=}CHCH_2CH_2CH_2\dot{C}H_2 + Cu(II) \longrightarrow Cu(I) + H^+ + CH_2{=}CHCH_2CH_2CH{=}CH_2$				
Phot. of hex-5-enyl(aquo)cobaloxime				81 Gol 1/
PR	H_2O	RT	$k = 7.7 \cdot 10^6\ M^{-1}\,s^{-1}$	72 Jen 1
4.2.1.1.2 Radicals containing only C, H, and O atoms				
$\dot{C}O_2^- + Co(III)(NH_3)_6^{3+} \longrightarrow CO_2 + Co(NH_3)_6^{2+}$				
Pulse rad. of formate $+ H_2O + N_2O$				72 Coh 1/
KAS	H_2O, pH = 4.8	RT	$k = 4.0 \cdot 10^7\ M^{-1}\,s^{-1}$	73 Hof 1
$\dot{C}O_2^- + (Co(III)(NH_3)_5Cl)^{2+} \longrightarrow$ products				
Pulse rad. of formate $+ N_2O + H_2O$				73 Hof 1
PR, KAS, competition kinetics	H_2O	RT	$k = 1.5(3) \cdot 10^8\ M^{-1}\,s^{-1}$ [4]	
$\dot{C}O_2^- + Co(III)(NH_3)_5OH^{2+} \longrightarrow Co(II)$ + products				
Pulse rad. of formate $+ H_2O + N_2O$				73 Hof 1
PR, KAS, competition kinetics	H_2O, pH = 7.8	RT	$k = 3.0(15) \cdot 10^7\ M^{-1}\,s^{-1}$ [4]	
$\dot{C}O_2^- + Co(III)(NH_3)_5OH_2^{3+} \longrightarrow Co(II)$ + products				
Pulse rad. of formate $+ H_2O + N_2O$				73 Hof 1
PR, KAS, competition kinetics	H_2O, pH = 5.2	RT	$k = 1.7(3) \cdot 10^8\ M^{-1}\,s^{-1}$ [4]	
$\dot{C}O_2^- + (Co(III)(NH_3)_5NO_3)^{2+} \longrightarrow$ products				
Pulse rad. of formate $+ N_2O + H_2O$				73 Hof 1
PR, KAS, competition kinetics	H_2O	RT	$k = 2.1(3) \cdot 10^8\ M^{-1}\,s^{-1}$ [4]	
$\dot{C}O_2^- + (Co(III)(NH_3)_5O_2CC_6H_5)^{2+}$ [5] $\longrightarrow$ products				
Pulse rad. of formate $+ N_2O + H_2O$				73 Hof 1
PR, KAS, competition kinetics	H_2O	RT	$k = 4.5(20) \cdot 10^7\ M^{-1}\,s^{-1}$ [4]	
$\dot{C}O_2^- + (Co(III)(NH_3)_5O_2CC_6H_4CN)^{2+}$ [6] $\longrightarrow$ products				
Pulse rad. of formate $+ N_2O + H_2O$				73 Hof 1
PR, KAS, competition kinetics	H_2O	RT	$k = 4.6(20) \cdot 10^7\ M^{-1}\,s^{-1}$ [4]	

[3]) Mechanism discussed as e^-- and $\dot{C}l$-transfer.
[4]) Based on competition kinetics with $k(\dot{C}O_2^- + \text{PNBPA}) = 1.9 \cdot 10^9\ M^{-1}\,s^{-1}$.
[5]) Pentaammine(benzoato)cobalt(III)$^{2+}$ ion.
[6]) Pentaammine(4-cyanobenzoato)cobalt(III)$^{2+}$ ion.

Reaction Radical generation Method	Solvent	T[K]	Rate data	Ref./ add. ref.
$\dot{C}O_2^- +$ pentaammine(2,4-dinitrobenzoato)cobalt(III)$^{2+} \longrightarrow (Co(III)(NH_3)_2O_2CC_6H_3NO_2(\dot{N}O_2^-))^+ + CO_2$				
Pulse rad. of formate + N_2O + H_2O				77 Sim 1
KAS	H_2O	RT	$k = 7.5 \cdot 10^9\ M^{-1}s^{-1}$	
$\dot{C}O_2^- +$ pentaammine(3,5-dinitrobenzoato)cobalt(III)$^{2+} \longrightarrow (Co(III)(NH_3)_5O_2CC_6H_3NO_2(\dot{N}O_2^-))^+ + CO_2$				
Pulse rad. of formate + N_2O + H_2O				77 Sim 1
KAS	H_2O	RT	$k = 8.1 \cdot 10^9\ M^{-1}s^{-1}$	
$\dot{C}O_2^- +$ pentaammine(2-nitrobenzoato)cobalt(III)$^{2+} \longrightarrow (Co(III)(NH_3)_5O_2CC_6H_4(\dot{N}O_2^-))^+ + CO_2$				
Pulse rad. of formate + N_2O + H_2O				77 Sim 1/
KAS	H_2O	RT	$k = 2.0 \cdot 10^9\ M^{-1}s^{-1}$	72 Hof 1
$\dot{C}O_2^- +$ pentaammine(3-nitrobenzoato)cobalt(III)$^{2+} \longrightarrow (Co(III)(NH_3)_5O_2CC_6H_4(\dot{N}O_2^-))^+ + CO_2$				
Pulse rad. of formate + N_2O + H_2O				77 Sim 1/
KAS	H_2O	RT	$k = 1.5 \cdot 10^9\ M^{-1}s^{-1}$	72 Hof 1
$\dot{C}O_2^- + (Co(III)(NH_3)_5O_2CCH_3)^{2+} \longrightarrow$ products				
Pulse rad. of formate + N_2O + H_2O				73 Hof 1
PR, KAS, competition kinetics	H_2O	RT	$k = 1.1(3) \cdot 10^8\ M^{-1}s^{-1}$ [4]	
$\dot{C}O_2^- + (Co(III)(NH_3)_5O_2CCH_2C_6H_4NO_2)^{2+}$ [7] $\longrightarrow$ products				
Pulse rad. of formate + N_2O + H_2O				73 Hof 1
PR, KAS, competition kinetics	H_2O	RT	$k = 1.2(1) \cdot 10^9\ M^{-1}s^{-1}$ [4]	
$\dot{C}O_2^- + (Co(III)(NH_3)_5O_2CCH_2C_6H_5)^{2+}$ [8] $\longrightarrow$ products				
Pulse rad. of formate + N_2O + H_2O				73 Hof 1
PR, KAS, competition kinetics	H_2O	RT	$k = 7.0(20) \cdot 10^7\ M^{-1}s^{-1}$ [4]	
$\dot{C}O_2^- + (Co(III)(NH_3)_5C_5H_5N)^{3+}$ [9] $\longrightarrow$ products				
Pulse rad. of formate + N_2O + H_2O				73 Hof 1
PR, KAS, competition kinetics	H_2O	RT	$k = 3.3(4) \cdot 10^8\ M^{-1}s^{-1}$ [4]	
$\dot{C}O_2^- + (Co(III)(NH_3)_5O_2CC_5H_4N)^{2+}$ [10] $\longrightarrow$ products				
Pulse rad. of formate + N_2O + H_2O				73 Hof 1
PR, KAS, competition kinetics	H_2O	RT	$k = 5.1(20) \cdot 10^7\ M^{-1}s^{-1}$ [4]	
$\dot{C}O_2^- + [(NH_3)_4Co(III)(\mu O_2, \mu NH_2)Co(III)(NH_3)_4]^{4+} \longrightarrow CO_2$ + products [11]				
Pulse rad. of formate + N_2O + H_2O				81 Nat 1
KAS	H_2O, pH = 5	RT	$k = 5.4 \cdot 10^9\ M^{-1}s^{-1}$	

[4]) Based on competition kinetics with $k(\dot{C}O_2^- + \text{PNBPA}) = 1.9 \cdot 10^9\ M^{-1}s^{-1}$.
[7]) Pentaammine(*p*-nitrophenylacetato-O)cobalt(III)$^{2+}$ ion.
[8]) Pentaammine(phenylacetato-O)cobalt(III)$^{2+}$ ion.
[9]) Pentaammine(pyridine-N)cobalt(III)$^{3+}$ ion.
[10]) Pentaammine(pyridinecarboxylato-O)cobalt(III)$^{2+}$ ion.
[11]) e^--transfer assumed to occur at dioxygen center.
μ defines a bridging group.

Reaction Radical generation Method	Solvent	T[K]	Rate data	Ref./ add. ref.
$\dot{C}O_2^-$ + [$(EN)_2$Co(III)(μO_2, μNH_2)Co(III)$(EN)_2]^{4+}$ ⟶ CO_2 + products [11]				
Pulse rad. of formate + N_2O + H_2O				81 Nat 1
KAS	H_2O, pH = 5	RT	$k = 5.7 \cdot 10^9\ M^{-1}s^{-1}$	
$\dot{C}O_2^-$ + vitamin B12r(cobalt(II)amine) ⟶ CO_2 + B12s(Co(I))				
Pulse rad. of formate + N_2O + H_2O				74 Bla 1
KAS	H_2O, pH = 9.2	RT	$k = 8.2 \cdot 10^8\ M^{-1}s^{-1}$	
$\dot{C}O_2^-$ + diaqua(2,3,9,10-tetramethyl-1,4,8,11-tetraazacyclotetradeca-1,3,8,10-tetraene-N,N′,N″,N‴)-cobalt(II) ion ⟶ Co(I)... + products				
Pulse rad. of formate + the corresp. ion + H_2O				76 Tai 1
KAS	H_2O	RT	$k = 4.7 \cdot 10^9\ M^{-1}s^{-1}$	
$\dot{C}O_2^-$ + diaqua(5,7,7,12,14,14-hexamethyl-1,4,8,11-tetraazacyclotetradeca-4,11-diene-N,N′,N″,N‴)-cobalt(III) ion ⟶ Co(II)... + products				
Pulse rad. of formate + H_2O + N_2O				76 Tai 2
KAS	H_2O, pH = 2.5	RT	$k = 8.1 \cdot 10^8\ M^{-1}s^{-1}$	
$\dot{C}O_2^-$ + diaqua(2,3,9,10-tetramethyl-1,4,8,11-tetraazacyclotetradeca-1,3,8,10-tetraene-N,N′,N″, N‴)-cobalt(III) ion ⟶ Co(II)... + products				
Pulse rad. of formate + H_2O + N_2O				76 Tai 2
KAS	H_2O, pH = 2.5	RT	$k = 6.4 \cdot 10^9\ M^{-1}s^{-1}$	
$\dot{C}O_2^-$ + tris(2,2′-bipyridine-N,N′)cobalt(III) ⟶ CO_2 + tris(2,2′-bipyridine-N,N′)cobalt(II)				
Pulse rad. of formate + H_2O + N_2O				73 Hof 1
PR, KAS, competition kinetics	H_2O	RT	$k = 7.6(2) \cdot 10^9\ M^{-1}s^{-1}$ [12]	
$\dot{C}O_2^-$ + dichloro(5,7,7,12,14,14-hexamethyl-1,4,8,11-tetraazacyclotetradeca-4,11-diene-N,N′,N″,N‴)-cobalt(III) ion ⟶ Co(II)... + products				
Pulse rad. of formate + H_2O + N_2O				76 Tai 2
KAS	H_2O, pH = 2.5	RT	$k = 1.1 \cdot 10^9\ M^{-1}s^{-1}$	
$\dot{C}O_2^-$ + [$(CN)_5$Co(III)(μO_2)Co(III)$(CN)_5]^{5-}$ ⟶ CO_2 + products [11]				
Pulse rad. of formate + N_2O + H_2O				81 Nat 1
KAS	H_2O, pH = 5	RT	$k = 1.7 \cdot 10^7\ M^{-1}s^{-1}$	
$\dot{C}O_2^-$ + vitamin B12(Co) ⟶ products				
Pulse rad. of CO_2 + t-butanol + H_2O				73 Far 2,
KAS	H_2O	RT	$k = 1.2 \cdot 10^9\ M^{-1}s^{-1}$ no reaction [13]	74 Bla 1
$\dot{C}O_2^-$ + vitamin B12a(Co) ⟶ products				
Pulse rad. of CO_2 + t-butanol + H_2O				74 Bla 1
KAS	H_2O, pH = 9.2	RT	$k = 1.45 \cdot 10^9\ M^{-1}s^{-1}$	
$\dot{C}O_2^-$ + Cu^{2+} ⟶ CO_2 + Cu^+				
Pulse rad. of formate + H_2O + N_2O				78 Ila 1
KAS	H_2O	RT	$k = 1.5(3) \cdot 10^8\ M^{-1}s^{-1}$	
$\dot{C}O_2^-$ + β-alanylhistidine copper(II) complex ⟶ Cu(I)... + CO_2				
Pulse rad. of formate + H_2O + N_2O				77 Bet 1
KAS	H_2O, pH = 7.5...11	RT	$k = 3.5(4) \cdot 10^8\ M^{-1}s^{-1}$	

[11]) e^--transfer assumed to occur at dioxygen center. μ defines a bridging group.
[12]) Based on $k(\dot{C}O_2^- + \text{PNBPA}) = 1.9 \cdot 10^9\ M^{-1}s^{-1}$.
[13]) From [74 Bla 1].

Reaction Radical generation Method	Solvent	T[K]	Rate data	Ref./ add. ref.
$\dot{C}O_2^-$ + oxidized gluthathione copper(II) complex ⟶ Cu(I)... + CO_2				
Pulse rad. of formate + H_2O + N_2O				76 Far 1
KAS	H_2O, pH = 11	RT	$k = 1.0(2) \cdot 10^8\ M^{-1}s^{-1}$	
$\dot{C}O_2^-$ + glycylglycylglycinato copper(II) complex ⟶ Cu(I)... + CO_2				
Pulse rad. of formate + H_2O + N_2O				76 Far 1
KAS	H_2O, pH = 9.1	RT	$k = 2.8(3) \cdot 10^8\ M^{-1}s^{-1}$	
$\dot{C}O_2^-$ + glycylhistidine copper(II) complex ⟶ Cu(I)... + CO_2				
Pulse rad. of formate + H_2O + N_2O				77 Bet 1
KAS	H_2O, pH = 6.6	RT	$k = 4.5(3) \cdot 10^8\ M^{-1}s^{-1}$	
	pH = 11		$k = 1.6(2) \cdot 10^7\ M^{-1}s^{-1}$	
$\dot{C}O_2^-$ + (5,7,7,12,14,14-hexamethyl-1,4,8,11-tetraazacyclotetradeca-4,11-diene-N,N',N'',N''') copper(II) ⟶ Cu(I)... + products				
Pulse rad. of formate + H_2O + N_2O				76 Tai 3
KAS	H_2O	RT	$k = 2.3 \cdot 10^9\ M^{-1}s^{-1}$	
$\dot{C}O_2^-$ + histidine copper(II) complex ⟶ Cu(I)... + CO_2				
Pulse rad. of formate + H_2O + N_2O				77 Bet 1
KAS	H_2O, pH = 11	RT	$k = 4.1(4) \cdot 10^8\ M^{-1}s^{-1}$	
$\dot{C}O_2^-$ + $Fe(CN)_6^{3-}$ ⟶ CO_2 + $Fe(CN)_6^{4-}$				
Pulse rad. of formate + H_2O				69 Ada 1
KAS	H_2O	RT	$k = 1.06 \cdot 10^9\ M^{-1}s^{-1}$	
$\dot{C}O_2^-$ + $Fe(CN)_5NO^{2-}$ ⟶ CO_2 + $Fe(CN)_5NO^{3-}$				
Pulse rad. of formate + H_2O + N_2O				69 Bux 1,
KAS	H_2O	RT	$k = 3.7 \cdot 10^8\ M^{-1}s^{-1}$	77 Che 1
			$4.0 \cdot 10^8$ [14])	
$\dot{C}O_2^-$ + Fe(III) cytochrome c ⟶ Fe(II) cytochrome c + CO_2				
Pulse rad. of formate + H_2O + N_2O				76 Sek 1/
KAS	H_2O, pH = 8.7	RT	$k = 6.3 \cdot 10^8\ M^{-1}s^{-1}$	71 Lan 1,
			$\log[A/M^{-1}s^{-1}] = 11$	75 Sim 1,
			$E_a = 12.3\ kJ\ mol^{-1}$	75 Wil 1,
	pH = 6.2	RT	$k = 1.0 \cdot 10^9\ M^{-1}s^{-1}$	77 Sha 1,
		277...	$\log[A/M^{-1}s^{-1}] = 11$	77 Sek 1,
		312	$E_a = 11.1\ kJ\ mol^{-1}$	78 Sim 1,
				79 Ila 1,
				78 Fav 1
$\dot{C}O_2^-$ + cytochrome-c-3(Fe(III)) ⟶ products				
Pulse rad. of formate + H_2O + N_2O				78 Fav 1
KAS	H_2O	RT	$k = 2.1 \cdot 10^8\ M^{-1}s^{-1}$	
$\dot{C}O_2^-$ + cytochrome c(acetylated)(Fe(III)) ⟶ products				
Pulse rad. of formate + H_2O + N_2O				79 Ila 1
KAS	H_2O	RT	$k = 1.5 \cdot 10^9\ M^{-1}s^{-1}$	
$\dot{C}O_2^-$ + cytochrome c(carboxymethylated)(Fe(III)) ⟶ products				
Pulse rad. of formate + H_2O + N_2O				78 Sim 1
KAS	H_2O	RT	$k = 1.4 \cdot 10^8\ M^{-1}s^{-1}$	

[14]) From [77 Che 1].

Reaction Radical generation Method	Solvent	T[K]	Rate data	Ref./ add. ref.
$\dot{C}O_2^- +$ cytochrome c(dicarboxymethyl)(Fe(III)) $\longrightarrow$ products				
Pulse rad. of formate + H_2O + N_2O				79 Ila 1
KAS	H_2O	RT	$k = 1.3 \cdot 10^8\ M^{-1} s^{-1}$	
$\dot{C}O_2^- +$ cytochrome-c(succinylated)(Fe(III)) $\longrightarrow$ products				
Pulse rad. of formate + H_2O + N_2O				79 Ila 1
KAS	H_2O	RT	$k = 4.0 \cdot 10^9\ M^{-1} s^{-1}$	
$\dot{C}O_2^- +$ hematoporphyrine(Fe(III)) $\longrightarrow$ products				
Pulse rad. of formate + H_2O + N_2O				74 Har 1
KAS	H_2O, pH = 13	RT	$k = 4(1) \cdot 10^7\ M^{-1} s^{-1}$	
$\dot{C}O_2^- +$ hemin c(Fe(III)) $\longrightarrow CO_2 +$ hemin c(Fe(II))				75 Gof 1
Pulse rad. of formate + H_2O + N_2O				
KAS	H_2O	RT	$k = 1.3(3) \cdot 10^9\ M^{-1} s^{-1}$	
$\dot{C}O_2^- +$ methemerythrin(Fe(III)) $\longrightarrow$ products				
Pulse rad. of formate + H_2O + N_2O				79 Har 1
KAS	H_2O	RT	$k = 6.8 \cdot 10^7\ M^{-1} s^{-1}$	
$\dot{C}O_2^- +$ metmyoglobin(Fe(III)) $\longrightarrow CO_2 +$ metmyoglobin(Fe(II))				
Pulse rad. of formate + H_2O + N_2O				78 Sim 1,
KAS	H_2O, pH = 7	RT	$k = 2.0(4) \cdot 10^9\ M^{-1} s^{-1}$	79 Har 1/
	pH = 8.2		$k = 2.9 \cdot 10^9\ M^{-1} s^{-1}$ [15])	76 Ila 2
$\dot{C}O_2^- +$ ferridoxin $\longrightarrow$ products				
Pulse rad. of formate + H_2O + N_2O				73 Hof 1
KAS	H_2O	RT	$k = 8.0(7) \cdot 10^7\ M^{-1} s^{-1}$	
$\dot{C}O_2^- + Hg(CN)_2 \longrightarrow CO_2 + Hg(I)\ldots$				
Pulse rad. of formate + H_2O + N_2O				75 Fuj 1
KAS	H_2O	RT	$k = 3.4(2) \cdot 10^9\ M^{-1} s^{-1}$	
$\dot{C}O_2^- + HgBr_2 \xrightarrow{a} CO_2 + Hg(I)\ldots$ $\dot{C}H_2OH + HgBr_2 \xrightarrow{b} HgBr + Br^- + H^+ + HCHO$				
Pulse rad. of formate + CH_3OH + H_2O + N_2O				76 Fuj 1
PR	H_2O	RT	$k_b/k_a = 0.63$ [16])	
$\dot{C}O_2^- + HgBr_2 \xrightarrow{a} CO_2 + Hg(I)\ldots$ $(CH_3)_2\dot{C}OH + HgBr_2 \xrightarrow{b} HgBr + Br^- + H^+ + (CH_3)_2CO$				
Pulse rad. of formate + 2-propanol + H_2O + N_2O				76 Fuj 1
PR	H_2O	RT	$k_b/k_a = 0.89$	
$\dot{C}O_2^- + HgBr_2 \xrightarrow{a} CO_2 + Hg(I)\ldots$ $CH_3\dot{C}HOH + HgBr_2 \xrightarrow{b} HgBr + Br^- + H^+ + CH_3CHO$				
Pulse rad. of formate + ethanol + H_2O + N_2O				76 Fuj 1
PR	H_2O	RT	$k_b/k_a = 0.87$	
$\dot{C}O_2^- + HgI_2 \longrightarrow$ products				
Pulse rad. of formate + H_2O + N_2O				78 Fuj 1
KAS	H_2O	RT	$k = 3.0(10) \cdot 10^9\ M^{-1} s^{-1}$	

[15]) From [79 Har 1].
[16]) Upper limit for k_a estimated to be $9 \cdot 10^9\ M^{-1} s^{-1}$.

Reaction Radical generation Method	Solvent	T[K]	Rate data	Ref./ add. ref.
$\dot{C}O_2^- + IrCl_6^{2-} \longrightarrow$ Ir(III) + products				
Pulse rad. of formate + N_2O + H_2O				82 Ste 1
KAS	H_2O	295	$k = 1.7 \cdot 10^9\ M^{-1}s^{-1}$	
$\dot{C}O_2^- + Ni(CN)_4^{2-} \longrightarrow CO_2 + Ni(CN)_4^{3-}$				
Pulse rad. of formate + H_2O + N_2O				74 Mul 1
KAS	H_2O	RT	$k = 1.2(1) \cdot 10^9\ M^{-1}s^{-1}$	
$\dot{C}O_2^-$ + (5,7,7,12,12,14-hexamethyl-1,4,8,11-tetraazacyclotetradecane-N,N′,N″,N‴)nickel(II) ion $\longrightarrow$ Ni(I)... + CO_2				
Pulse rad. of formate + H_2O + Ar				76 Tai 3
KAS	H_2O	RT	$k = 5.7 \cdot 10^9\ M^{-1}s^{-1}$	
$\dot{C}O_2^-$ + (5,7,7,12,12,14-hexamethyl-1,4,8,11-tetraazacyclotetradeca-4,11-diene-N,N′,N″,N‴)-nickel(II) ion $\longrightarrow$ Ni(I)... + CO_2				
Pulse rad. of formate + H_2O + Ar				76 Tai 3
KAS	H_2O	RT	$k = 6.7 \cdot 10^9\ M^{-1}s^{-1}$	
$\dot{C}O_2^- + O_2 \longrightarrow CO_2 + \dot{O}_2^-$				
Pulse rad. of formate + H_2O				76 Bux 1,
KAS	H_2O, pH = 8	RT	$k = 2.0(4) \cdot 10^9\ M^{-1}s^{-1}$	76 Ila 1/
	pH = 6.8		$k = 4.2(4) \cdot 10^9\ M^{-1}s^{-1}$ [17])	69 Bax 1,
				69 Ada 1,
				78 Sut 1
$\dot{C}O_2^-$ + Rh(III)(2,2′-bipyridine)$_3^{3+}$ $\longrightarrow$ Rh(II)(2,2′-bipyridine)$_3^{2+}$ + CO_2				
Pulse rad. of formate + N_2O + H_2O				81 Mul 1,
KAS	H_2O	RT	$k = 6.2(6) \cdot 10^9\ M^{-1}s^{-1}$	74 Mul 2
$\dot{C}O_2^- + Ru(NH_3)_6^{3+} \longrightarrow Ru(NH_3)_6^{2+} + CO_2$				
Pulse rad. of formate + H_2O + N_2O				72 Coh 1
KAS	H_2O, pH = 4.8	RT	$k = 2.0(6) \cdot 10^9\ M^{-1}s^{-1}$	
$\dot{C}O_2^- + Ru(NH_3)_5NO^{3+} \longrightarrow Ru(NH_3)_5NO^{2+} + CO_2$				
Pulse rad. of formate + H_2O + N_2O				75 Arm 1
KAS	H_2O	RT	$k = 3.1 \cdot 10^9\ M^{-1}s^{-1}$	
$\dot{C}O_2^- + SO_2 \longrightarrow SO_2^- + CO_2$				
Pulse rad. of formate + H_2O				75 Eri 1
KAS	H_2O, pH = 3.1	RT	$k = 7.6(10) \cdot 10^8\ M^{-1}s^{-1}$	
$\dot{C}O_2^- + S_4O_6^{2-} \longrightarrow CO_2 + S_4O_6^{2\dot{-}}$				
Pulse rad. of formate + H_2O + N_2O				73 Sch 1
KAS	H_2O	RT	$k = 5.8 \cdot 10^7\ M^{-1}s^{-1}$	
$\dot{C}O_2^-$ + Ti(III) $\longrightarrow$ Ti(IV) + products				
Pulse rad. of formate + H_2O				73 Ell 1
KAS	H_2O, pH = 1.4	RT	$k \approx 5 \cdot 10^6\ M^{-1}s^{-1}$	
$\dot{C}O_2^- + 2Tl^+ \longrightarrow Tl_2^+ + CO_2$				
Pulse rad. of formate + H_2O				80 But 1
KAS	H_2O	RT	$k = 2.3 \cdot 10^6\ M^{-1}s^{-1}$	

[17]) From [76 Ila 1].

Reaction Radical generation Method	Solvent	T[K]	Rate data	Ref./ add. ref.
$\dot{C}O_2^- + Zn^+ + H_2O \longrightarrow Zn^{2+} + HCO_2^- + OH^-$				
Pulse rad. of formate + Zn^{2+} + H_2O				77 Rab 1
KAS	H_2O	RT	$k \approx 4 \cdot 10^9\ M^{-1}\,s^{-1}$	
$\dot{C}O_2^-$ + Zn(II)-insulin $\longrightarrow CO_2$ + products				
Pulse rad. of formate + N_2O + H_2O				80 Ell 1
KAS	H_2O	RT	$k = 6 \cdot 10^8\ M^{-1}\,s^{-1}$ [18])	
$\dot{C}O_2^-$ + acetophenone ($C_6H_5COCH_3$) $\longrightarrow CO_2 + C_6H_5\dot{C}(O^-)CH_3$				
Pulse rad. of formate + H_2O				68 Ada 1
KAS	H_2O, pH = 12	RT	$k = 1 \cdot 10^7\ M^{-1}\,s^{-1}$	
$\dot{C}O_2^-$ + acridine $\longrightarrow$ products				
Pulse rad. of formate + H_2O + N_2O				79 Net 1
KAS	H_2O	RT	$k \approx 3 \cdot 10^8\ M^{-1}\,s^{-1}$	
$\dot{C}O_2^-$ + acriflavin [19]) $\longrightarrow$ products				
Pulse rad. of formate + H_2O + Ar				70 Pru 1
KAS	H_2O	RT	$k = 3.7(4) \cdot 10^8\ M^{-1}\,s^{-1}$	
$\dot{C}O_2^-$ + alloxane(A) + $H^+ \longrightarrow CO_2 + AH^{\cdot}$				
Pulse rad. of formate + N_2O + H_2O				80 Hou 1/
KAS	H_2O	RT	$k = 3.7(11) \cdot 10^7\ M^{-1}\,s^{-1}$	79 Hou 1
$\dot{C}O_2^-$ + 2-amino-4-[2-(formylamino)phenyl]-4-oxobutanoic acid [20]) $\longrightarrow$ products				
Pulse rad. of formate + N_2O + H_2O				75 Wal 1
KAS	H_2O	RT	$k > 3 \cdot 10^7\ M^{-1}\,s^{-1}$	
$\dot{C}O_2^-$ + 9,10-anthraquinone-2,6-disulfonate ion $\longrightarrow$ ...semiquinone + CO_2				
Pulse rad. of formate + H_2O + N_2O				73 Rao 1
KAS	H_2O	RT	$k = 2.4(2) \cdot 10^9\ M^{-1}\,s^{-1}$	
$\dot{C}O_2^-$ + 9,10-anthraquinone-1-sulfonate ion $\longrightarrow$ ...semiquinone + CO_2				
Pulse rad. of formate + H_2O				72 Hul 1
KAS	H_2O, pH = 3	RT	$k = 1.0 \cdot 10^9\ M^{-1}\,s^{-1}$	
	pH = 7		$k = 3.3 \cdot 10^9\ M^{-1}\,s^{-1}$	
$\dot{C}O_2^-$ + 9,10-anthraquinone-2-sulfonate ion $\longrightarrow$ ...semiquinone + CO_2				
Pulse rad. of formate + H_2O + N_2O				72 Hul 1,
KAS	H_2O	RT	$k = 3.1 \cdot 10^9\ M^{-1}\,s^{-1}$	73 Rao 1
			$k = 1.6(2) \cdot 10^9\ M^{-1}\,s^{-1}$ [21])	
$\dot{C}O_2^-$ + benzoquinone $\longrightarrow CO_2$ + ...semiquinone				
Pulse rad. of formate + H_2O + N_2O				71 Wil 1/
KAS	H_2O	RT	$k = 6.6(7) \cdot 10^9\ M^{-1}\,s^{-1}$	73 Sim 1,
				73 Rao 1

[18]) From single pulse experiments; k decreases to $2 \cdot 10^8\ M^{-1}\,s^{-1}$ at fourth successive pulse.

[19]) 3,10-diamino-10-methylacridinium chloride.

[20]) N-formylkynurenine

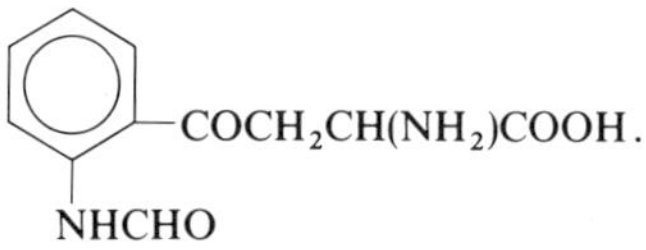

[21]) From [73 Rao 1].

Reaction Radical generation Method	Solvent	T[K]	Rate data	Ref./ add. ref.
$\dot{C}O_2^- + $ 1,1'-dibenzyl-4,4'-bipyridinium(BV^{2+}) [22] $\longrightarrow CO_2 + BV^{\dot{+}}$				
Pulse rad. of formate + N_2O + H_2O				76 And 1,
KAS	H_2O	RT	$k = 6.7 \cdot 10^9\ M^{-1}s^{-1}$ $1.7 \cdot 10^{10}$ [23])	78 Far 1
$\dot{C}O_2^- + $ 2,2'-bipyridine [24]) $\longrightarrow$ products				
Pulse rad. of formate + H_2O + N_2O				79 Mul 1
KAS	H_2O, pH = 4.4	RT	$k = 5.0 \cdot 10^8\ M^{-1}s^{-1}$ [25])	
$\dot{C}O_2^- + $ 5-bromouracil $\longrightarrow$ [Br-uracil-$\dot{C}O_2^-$ intermediate] $\longrightarrow CO_2 + Br^- +$ (uracil)$^{\cdot}$				
Pulse rad. of formate + H_2O + N_2O				69 Zim 1
KAS	H_2O	RT	$k > 1 \cdot 10^8\ M^{-1}s^{-1}$	
$\dot{C}O_2^- + $ 2-*t*-butyl-2,3-diazabicyclo[2.2.2]octane($R_3N_2^+$) $\longrightarrow CO_2 + R_3N_2^{\cdot}$				
Pulse rad. of formate + N_2O + H_2O				80 Nel 1
KAS	H_2O	RT	$k \approx 5 \cdot 10^8\ M^{-1}s^{-1}$	
$\dot{C}O_2^- + $ 3-carbamoyl-1-methylpyridinium ion $\longrightarrow$ products				
Pulse rad. of formate + H_2O + N_2O				68 Lan 1
KAS	H_2O, pH = 8.5	RT	$k = 4.6 \cdot 10^9\ M^{-1}s^{-1}$	
$\dot{C}O_2^- + $ 1-chloro-4-nitrobenzene $\longrightarrow$ products				
Pulse rad. of formate + H_2O				77 Bia 1
KAS	H_2O	RT	$k = 3 \cdot 10^8\ M^{-1}s^{-1}$	
$\dot{C}O_2^- + $ crystal violet $\longrightarrow CO_2 +$ products				
Pulse rad. of formate + H_2O				73 Rao 2
KAS	H_2O	RT	$k = 1.5 \cdot 10^9\ M^{-1}s^{-1}$	
$\dot{C}O_2^- + $ (1,1'-dicarboxyethyl-4,4'-bipyridinium)$^{2+} \longrightarrow CO_2 +$ (1,1-dicarboxyethyl-4,4'-bipyridinium)$^{\dot{+}}$				
Pulse rad. of formate + H_2O + N_2O				76 And 1
KAS	H_2O	RT	$k = 2.0 \cdot 10^9\ M^{-1}s^{-1}$	
$\dot{C}O_2^- + $ dichloroindophenol $\longrightarrow$ products				
Pulse rad. of formate + H_2O + N_2O				73 Rao 2
KAS	H_2O	RT	$k = 3.5 \cdot 10^9\ M^{-1}s^{-1}$	
$\dot{C}O_2^- + $ (1,1'-di(4-cyanophenyl)-4,4'-bipyridinium)$^{2+} \longrightarrow CO_2 +$ (1,1'-di...)$^{\dot{+}}$				
Pulse rad. of formate + H_2O + N_2O				78 Far 1
KAS	H_2O	RT	$k = 1.4 \cdot 10^{10}\ M^{-1}s^{-1}$	
$\dot{C}O_2^- + NBT^{2+}$ [26]) $\longrightarrow CO_2 + NBT^{\dot{+}}$				
Pulse rad. of formate + N_2O + H_2O				80 Bie 1
KAS	H_2O, pH = 10	297	$k = 6.4(2) \cdot 10^9\ M^{-1}s^{-1}$	
$\dot{C}O_2^- + $ (1,1'-dimethyl-4,4'-bipyridinium)$^{2+}$ [27]) $\longrightarrow CO_2 +$ (1,1'-di...)$^{\dot{+}}$				
Pulse rad. of formate + H_2O + N_2O				73 Far 1
KAS	H_2O	RT	$k = 1.5 \cdot 10^{10}\ M^{-1}s^{-1}$	

[22]) Benzylviologen.
[23]) From [78 Far 1].
[24]) Protonated form (bipy H^+).
[25]) No observable reaction with deprotonated 2,2'-bipyridine.
[26]) 3,3'-(3,3'-dimethoxy[1,1'-diphenyl]-4,4'-diyl)bis[2-(4-nitrophenyl)-5-phenyl]-2H-tetrazolium ion.
[27]) Paraquat, methylviologen.

Reaction Radical generation Method	Solvent	T[K]	Rate data	Ref./ add. ref.
$\dot{C}O_2^- + \text{dimethylfumerate}(CH_3OOCCH{=}CHCOOCH_3) \longrightarrow$ products [28]				
Pulse rad. of formate + H_2O + N_2O				73 Hay 2
KAS	H_2O	RT	$k = 9 \cdot 10^8\ M^{-1}\,s^{-1}$	
$\dot{C}O_2^- + (\text{2,4-dinitrobenzoate})^- \longrightarrow CO_2 + (\text{2,4-dinitrobenzoate})^{2-\cdot}$ [29]				
Pulse rad. of formate + H_2O + N_2O				76 Net 1
KAS	H_2O	RT	$k = 1.8 \cdot 10^9\ M^{-1}\,s^{-1}$	
$\dot{C}O_2^- + (\text{2,5-dinitrobenzoate})^- \longrightarrow CO_2 + (\text{2,5-dinitrobenzoate})^{2-\cdot}$ [29]				
Pulse rad. of formate + H_2O + N_2O				76 Net 2
KAS	H_2O	RT	$k = 1.9 \cdot 10^9\ M^{-1}\,s^{-1}$	
$\dot{C}O_2^- + (\text{3,4-dinitrobenzoate})^- \longrightarrow CO_2 + (\text{3,4-dinitrobenzoate})^{2-\cdot}$ [29]				
Pulse rad. of formate + H_2O + N_2O				76 Net 2
KAS	H_2O	RT	$k = 1.8 \cdot 10^9\ M^{-1}\,s^{-1}$	
$\dot{C}O_2^- + (\text{3,5-dinitrobenzoate})^- \longrightarrow CO_2 + (\text{3,5-dinitrobenzoate})^{2-\cdot}$ [29]				
Pulse rad. of formate + H_2O + N_2O				76 Net 2
KAS	H_2O	RT	$k = 2.5 \cdot 10^9\ M^{-1}\,s^{-1}$	
$\dot{C}O_2^- + (\text{1-(2,4-dinitrophenyl)pyridinium})^+ \longrightarrow CO_2 + (\text{1-(2,4-dinitrophenyl)pyridinium})^{\pm\cdot}$ [29]				
Pulse rad. of formate + H_2O + N_2O				77 Bia 1
KAS	H_2O	RT	$k = 4.0 \cdot 10^8\ M^{-1}\,s^{-1}$	
$\dot{C}O_2^- + (\text{1,1}'\text{-diphenyl-4,4}'\text{-bipyridinium})^{2+} \longrightarrow CO_2 + (\text{1,1}'\text{-di}\ldots)^{+\cdot}$				
Pulse rad. of formate + H_2O + N_2O				78 Far 1
KAS	H_2O	RT	$k = 1.3 \cdot 10^{10}\ M^{-1}\,s^{-1}$	
$\dot{C}O_2^- + \text{eosin} \longrightarrow CO_2 + (\text{eosin})^{-\cdot}$				
Pulse rad. of formate + H_2O_2 + H_2O				67 Chr 1
KAS	H_2O, pH = 8.5…9.0	RT	$k = 2.5(5) \cdot 10^8\ M^{-1}\,s^{-1}$	
$\dot{C}O_2^- + (\text{1,1}'\text{-ethylene-2,2}'\text{-bipyridinium})^{2+} \longrightarrow CO_2 + (\text{1,1}'\text{-ethylene}\ldots)^{+\cdot}$				
Pulse rad. of formate + H_2O + N_2O				76 And 1,
KAS	H_2O	RT	$k = 4.0 \cdot 10^9\ M^{-1}\,s^{-1}$ $1.2 \cdot 10^{10}$ [30]	78 Far 1
$\dot{C}O_2^- + \text{N-ethylmaleimide} \longrightarrow CO_2 + (\text{N-ethylmaleimide})^{-\cdot}$				
Pulse rad. of formate + H_2O + N_2O				72 Hay 1
KAS	H_2O	RT	$k = 5.4 \cdot 10^9\ M^{-1}\,s^{-1}$	
$\dot{C}O_2^- + \text{fluorescein} \longrightarrow CO_2 + (\text{fluorescein})^{-\cdot}$ [31]				
Pulse rad. of formate + H_2O + N_2O				68 Cor 1
KAS	H_2O, pH = 10.4	RT	$k = 2.6(9) \cdot 10^7\ M^{-1}\,s^{-1}$	
$\dot{C}O_2^- + \text{fumaric acid}\ (HOOCCH{=}CHCOO^-) \longrightarrow$ products [32]				
Pulse rad. of formate + H_2O + N_2O				73 Hay 2
KAS	H_2O, pH = 5.2	RT	$k = 1.1 \cdot 10^8\ M^{-1}\,s^{-1}$	

[28]) $\geq 80\%$ e^-- transfer.
[29]) e^--transfer to nitro groups.
[30]) From [78 Far 1].
[31]) Semiquinone form.
[32]) $\geq 65\%$ e^--transfer at pH = 5.2; no e^--transfer to dianion at pH = 10.5.

Reaction Radical generation Method	Solvent	T[K]	Rate data	Ref./ add. ref.
$\dot{C}O_2^-$ + 1,1′-bis(2-hydroxyethyl)-4,4′-bipyridinium ion ⟶ CO_2 + $(1,1'\text{-bis}\ldots)^{\dot{+}}$				
Pulse rad. of formate + H_2O + N_2O				78 Far 1
KAS	H_2O	RT	$k = 1.9 \cdot 10^{10}\,M^{-1}\,s^{-1}$	
$\dot{C}O_2^-$ + 1-(2-hydroxyethyl)-2-methyl-5-nitroimidazole ⟶ products				
Pulse rad. of formate + H_2O + N_2O				74 Wil 1
KAS	H_2O	293	$k = 8 \cdot 10^{8}\,M^{-1}\,s^{-1}$	
$\dot{C}O_2^-$ + 2-hydroxy-1,4-naphthoquinone ⟶ CO_2 + ...-semiquinone				
Pulse rad. of formate + H_2O + N_2O				73 Rao 1
KAS	H_2O	RT	$k = 1.95(20) \cdot 10^{9}\,M^{-1}\,s^{-1}$	
$\dot{C}O_2^-$ + 6-hydroxy-5-nitrothymine [33]) ⟶ CO_2 + $(6\text{-hydroxy}\ldots)^{\dot{-}}$ [34])				
Pulse rad. of formate + N_2O + H_2O				80 Eri 1
KAS	H_2O, pH = 2	RT	$k = 1.7(2) \cdot 10^{8}\,M^{-1}\,s^{-1}$	
	pH = 6.5		$9(2) \cdot 10^{7}$	
$\dot{C}O_2^-$ + indigo disulfonate ⟶ CO_2 + products				
Pulse rad. of formate + H_2O + N_2O				73 Rao 2
KAS	H_2O	RT	$k = 2.0 \cdot 10^{9}\,M^{-1}\,s^{-1}$	
$\dot{C}O_2^-$ + indophenol ⟶ products				
Pulse rad. of formate + H_2O + N_2O				73 Rao 2
KAS	H_2O, pH = 9	RT	$k = 2.8 \cdot 10^{9}\,M^{-1}\,s^{-1}$	
$\dot{C}O_2^-$ + 3-iodotyrosine ⟶ I^- + CO_2 + (tyrosine)$^{\cdot}$				
γ-rad. of formate + H_2O				72 Ste 1
[35])	H_2O	RT	$k = 1.3(1) \cdot 10^{5}\,M^{-1}\,s^{-1}$	
$\dot{C}O_2^-$ + lipoate ion(—S—S—) ⟶ CO_2 + $—S\overset{(-)}{\therefore}S—$				
Pulse rad. of formate + H_2O + N_2O				70 Wil 1,
KAS	H_2O, pH = 7	RT	$k = 5.5 \cdot 10^{8}\,M^{-1}\,s^{-1}$	75 Far 1
	pH = 3		$9 \cdot 10^{8}$ [36])	
$\dot{C}O_2^-$ + lysozyme(—S—S—) ⟶ CO_2 + $—S\overset{(-)}{\therefore}S—$				
Pulse rad. of formate + H_2O + N_2O				75 Hof 1
KAS	H_2O, pH = 6	RT	$k \approx 5 \cdot 10^{8}\,M^{-1}\,s^{-1}$ [37])	
$\dot{C}O_2^-$ + methylene blue (MB^+) ⟶ CO_2 + $MB^{\cdot}$				
Pulse rad. of formate + H_2O + N_2O				65 Kee 1
KAS	H_2O, pH ≈ 9	RT	$k = 5.6 \cdot 10^{9}\,M^{-1}\,s^{-1}$	
$\dot{C}O_2^-$ + 2-methyl-1,4-naphthoquinone ⟶ ...semiquinone + CO_2				
Pulse rad. of formate + H_2O + N_2O				73 Rao 3/
KAS	H_2O	RT	$k = 4.8(5) \cdot 10^{9}\,M^{-1}\,s^{-1}$	73 Rao 1,
				72 Sim 1
$\dot{C}O_2^-$ + nicotinamide adenine dinucleotide (NAD^+) ⟶ CO_2 + $NA\dot{D}$				
Pulse rad. of formate + H_2O + N_2O				68 Lan 1
KAS	H_2O	RT	$k = 1.6 \cdot 10^{9}\,M^{-1}\,s^{-1}$	

[33]) Deprotonated form at pH = 6.5.
[34]) Possibly protonated at lower pH.
[35]) Estimated from dependence of tyrosine yields on irr. time assuming $2k(\dot{R} + \dot{R}) = 5.0 \cdot 10^{8}\,M^{-1}\,s^{-1}$.
[36]) From [75 Far 1].
[37]) Rate constant pH dependent.

Reaction Radical generation Method	Solvent	T[K]	Rate data	Ref./ add. ref.
$\dot{C}O_2^-$ + 4-nitroacetophenone(PNAP) ⟶ $PNAP^{\dot{+}}$ + CO_2				
Pulse rad. of formate + N_2O + H_2O				77 Bia 1,
KAS	H_2O, pH = 7	RT	$k = 7 \cdot 10^8\ M^{-1}s^{-1}$	73 Ada 1
	pH = 10		$1.0(1) \cdot 10^9$ [38]	
$\dot{C}O_2^-$ + nitrobenzene ($C_6H_5NO_2$) ⟶ $C_6H_5\dot{N}O_2^-$ + CO_2				
Pulse rad. of formate + N_2O + H_2O				70 Foj 1,
KAS	H_2O, pH = 6...7	RT	$k = 1.0 \cdot 10^9\ M^{-1}s^{-1}$	73 Bux 1
	pH = 3		$5.6 \cdot 10^8$ [39]	
	pH = 9.4		$5.8 \cdot 10^8$ [40]	
	pH = 2.5		$7.5 \cdot 10^8$ [39] [40]	
	pH = 0		$4.6 \cdot 10^8$ [39] [40]	
$\dot{C}O_2^-$ + nitrosobenzene + $H^+(H_2O)$ ⟶ $C_6H_5\dot{N}OH$ + $CO_2(+OH^-)$				
Pulse rad. of formate + N_2O + H_2O				66 Asm 2
KAS	H_2O	RT	$k = 4.0 \cdot 10^9\ M^{-1}s^{-1}$	
$\dot{C}O_2^-$ + 2-nitrobenzoate ⟶ $^-OOCC_6H_4\dot{N}O_2^-$ + CO_2				
Pulse rad. of formate + N_2O + H_2O				76 Net 2
KAS	H_2O, pH = 7 and 0.8	RT	$k = 2.4 \cdot 10^8\ M^{-1}s^{-1}$	
$\dot{C}O_2^-$ + 3-nitrobenzoate ⟶ $^-OOCC_6H_4\dot{N}O_2^-$ + CO_2				
Pulse rad. of formate + N_2O + H_2O				76 Net 2
KAS	H_2O, pH = 7 and 0.8	RT	$k = 6.3 \cdot 10^8\ M^{-1}s^{-1}$	
$\dot{C}O_2^-$ + 4-nitrobenzoate ⟶ $^-OOCC_6H_4\dot{N}O_2^-$ + CO_2				
Pulse rad. of formate + N_2O + H_2O				76 Net 2
KAS	H_2O, pH = 7 and 0.8	RT	$k = 8.0 \cdot 10^8\ M^{-1}s^{-1}$	
$\dot{C}O_2^-$ + *anti*-5-nitro-2-furaldoxime [41] ⟶ $(anti\text{-}\ldots)^{\dot{-}}$ + CO_2				
Pulse rad. of CO_2 + *t*-butanol or formate + H_2O				73 Gre 1
KAS	H_2O	RT	$k = 2.7 \cdot 10^9\ M^{-1}s^{-1}$	
$\dot{C}O_2^-$ + pterin ⟶ products [42]				
Pulse rad. of formate + N_2O + H_2O				76 Moo 1
KAS	H_2O, pH = 7	RT	$k = 4.6 \cdot 10^8\ M^{-1}s^{-1}$ [43]	
	pH > 9.5		$\leqslant 10^7$ [44]	
$\dot{C}O_2^-$ + ^{3}(pyrene) [45] ⟶ $(pyrene)^{\dot{-}}$ + CO_2				
Combined pulse rad. and phot. of formate + N_2O + micellar solutions				76 Fra 1
KAS	micellar solution (H_2O, CTAB [46]) pH = 13	RT	$k = 5 \cdot 10^9\ M^{-1}s^{-1}$	
$\dot{C}O_2^-$ + rhodamine B ⟶ CO_2 + products				
Pulse rad. of formate + N_2O + H_2O				67 Pru 1
KAS	H_2O	RT	$k = 1.8(5) \cdot 10^8\ M^{-1}s^{-1}$	

[38]) From [73 Ada 1].
[39]) Reaction product at pH = 0 and partially at pH = 2.5 and 3: $C_6H_5\dot{N}O_2H$.
[40]) From [73 Bux 1].
[41]) Nifuroxime.
[42]) 100% e^--transfer.
[43]) Reaction with neutral form of pterin (PtH).
[44]) Reaction with anionic form of pterin (Pt^-).
[45]) Pyrene triplet.
[46]) $5 \cdot 10^{-3}$ M hexadecyl trimethyl ammonium bromide (CTAB).

Reaction Radical generation Method	Solvent	T[K]	Rate data	Ref./ add. ref.
$\dot{C}O_2^-$ + riboflavin ⟶ products				
Pulse rad. of formate + N_2O + H_2O				69 Lan 1,
KAS	H_2O, pH = 3 and 5.9	RT	$k = 3.6 \cdot 10^9\ M^{-1}s^{-1}$	73 Rao 1
	pH = 7		$1.7 \cdot 10^9$ [47])	
	pH ≈ 11.5		$1.4 \cdot 10^9$	
$\dot{C}O_2^-$ + tetramethyldiazenedicarboxamide ($(CH_3)_2NCON{=}NCON(CH_3)_2$) ⟶ products				
Pulse rad. of formate + H_2O + N_2O				75 Whi 1
KAS	H_2O	RT	$k \approx 2.5 \cdot 10^9\ M^{-1}s^{-1}$	
$\dot{C}O_2^-$ + 1,1′-tetramethylene-2,2′-bipyridinium$^{(2+)}$ ⟶ (1,1′-tetra...)$^{\dot{+}}$ + CO_2				
Pulse rad. of formate + N_2O + H_2O				76 And 1,
KAS	H_2O	RT	$k = 9 \cdot 10^9\ M^{-1}s^{-1}$	78 Far 1
			$7 \cdot 10^9$ [48])	
$\dot{C}O_2^-$ + 2,2,6,6-tetramethyl-4-oxo-1-piperidinyloxy(TAN) ⟶ products				
Pulse rad. of formate + N_2O + H_2O				71 Wil 2
KAS	H_2O	RT	$k = 7.0 \cdot 10^8\ M^{-1}s^{-1}$	
$\dot{C}O_2^-$ + tetranitromethane($C(NO_2)_4$) ⟶ CO_2 + $C(NO_2)_3^-$ + NO_2				
Pulse rad. of formate + N_2O + H_2O				70 Foj 1
KAS	H_2O	RT	$k = 4(1) \cdot 10^9\ M^{-1}s^{-1}$	
$\dot{C}O_2^-$ + 1,1′-trimethylene-2,2′-bipyridinium$^{(2+)}$ ⟶ (1,1′-tri...)$^{\dot{+}}$ + CO_2				
Pulse rad. of formate + N_2O + H_2O				76 And 1,
KAS	H_2O	293	$k = 7.5 \cdot 10^9\ M^{-1}s^{-1}$	78 Far 1
			$1.1 \cdot 10^{10}$ [48])	
$\dot{C}O_2^-$ + 2,4,6-trinitrobenzoate ion ⟶ CO_2 + $^-OOCC_6H_2(NO_2)_2\dot{N}O_2^-$				
Pulse rad. of formate + N_2O + H_2O				76 Net 2
KAS	H_2O	RT	$k = 3.4 \cdot 10^9\ M^{-1}s^{-1}$	
$\dot{C}O_2H$ + methylene blue (MB) [49]) ⟶ CO_2 + [49])				
Pulse rad. of formate + H_2O				65 Kee 1
KAS	H_2O, pH = 1.75	RT	$k \approx 2 \cdot 10^9\ M^{-1}s^{-1}$	
	$H_0 = -0.8$		$\approx 10^9$	
	$H_0 = -5.7$		$\approx 7 \cdot 10^8$	
$\dot{C}H_2O^-$ + $Co(NH_3)_6^{3+}$ ⟶ HCHO + $Co(NH_3)_6^{2+}$				
Pulse rad. of CH_3OH + H_2O + N_2O				72 Coh 1
KAS	H_2O, pH = 12	RT	$k = 9.0(14) \cdot 10^9\ M^{-1}s^{-1}$	
$\dot{C}H_2O^-$ + (5,7,7,12,14,14-hexamethyl-1,4,8,11-tetraazacyclotetradeca-4,11-diene-N,N′,N″,N‴)-copper(II) ion ⟶ Cu(I)... + products				
Pulse rad. of CH_3OH + H_2O + N_2O				76 Tai 3
KAS	H_2O, pH = 12	RT	$k = 9.0 \cdot 10^8\ M^{-1}s^{-1}$ [50])	
$\dot{C}H_2O^-$ + $Fe(CN)_6^{3-}$ ⟶ HCHO + $Fe(CN)_6^{4-}$				
Pulse rad. of CH_3OH + H_2O				68 Ada 1
KAS	H_2O, pH = 13	RT	$k = 3.1 \cdot 10^9\ M^{-1}s^{-1}$	

[47]) From [73 Rao 1].
[48]) From [78 Far 1].
[49]) At pH = 1.75 MB^+ ⟶ $MB^{\cdot}$; at $H_0 = -0.8$ MBH^{2+} ⟶ $MBH^{2\dot{+}}$; at $H_0 = -5.7$ MBH_2^{3+} ⟶ $MBH_2^{2\dot{+}}$.
[50]) No reaction in neutral solution.

Reaction Radical generation Method	Solvent	T[K]	Rate data	Ref./ add. ref.
$\dot{C}H_2O^- +$ hematoporphyrine $\longrightarrow$ products				
Pulse rad. of $CH_3OH + H_2O + N_2O$				74 Har 1
KAS	H_2O, pH = 13	RT	$k = 3.3(5) \cdot 10^8\ M^{-1}s^{-1}$ [51]	
$\dot{C}H_2O^- + Ga(OH)_6^{4-} \longrightarrow$ products				
Pulse rad. of $CH_3OH + Ga^{3+} + H_2O$				79 Suk 1
KAS	H_2O, pH = 12	RT	$k = 7.8 \cdot 10^8\ M^{-1}s^{-1}$	
$\dot{C}H_2O^- +$ Ru(II)(2,2'-bipyridine-N,N')$_3^{2+} \longrightarrow$ Ru(I)(2,2'-bipyridine-N,N')$_3^+$ + HCHO				
Pulse rad. of $CH_3OH + H_2O$				78 Mul 1
KAS	H_2O, pH = 11...13	RT	$k = 2.9 \cdot 10^9\ M^{-1}s^{-1}$	
$\dot{C}H_2O^- +$ Ru(1,10-phenanthroline)$_3^{2+} \longrightarrow$ Ru(1,10-phenanthroline)$_3^+$ + HCHO				
Pulse rad. of $CH_3OH + N_2O + H_2O$				80 Ven 1
KAS	H_2O, pH = 11	RT	$k = 5.2 \cdot 10^9\ M^{-1}s^{-1}$	
$\dot{C}H_2O^- +$ azobenzene $\longrightarrow$ HCHO + (azobenzene)$^{\dot{-}}$				
Pulse rad. of $CH_3OH + H_2O + N_2O$				77 Net 1
KAS	H_2O, pH = 14	RT	$k = 1 \cdot 10^9\ M^{-1}s^{-1}$	
$\dot{C}H_2O^- +$ benzophenone ($(C_6H_5)_2CO$) $\longrightarrow$ HCHO + $(C_6H_5)_2\dot{C}O^-$				
Pulse rad. of $CH_3OH + H_2O + N_2O$				75 Bre 1
KAS	H_2O, pH = 13	RT	$k = 1.2 \cdot 10^8\ M^{-1}s^{-1}$	
$\dot{C}H_2O^- +$ 4-carboxy-1-methylpyridinium ion $\longrightarrow$ products				
Pulse rad. of $CH_3OH + H_2O + N_2O$				79 Ste 1
KAS	H_2O, pH = 12.7	RT	$k = 3.8 \cdot 10^9\ M^{-1}s^{-1}$	
$\dot{C}H_2O^- +$ 2-methyl-1,4-naphthoquinone $\longrightarrow$ ...-semiquinone + products [52]				
Pulse rad. of $CH_3OH + H_2O + N_2O$				73 Rao 3
KAS	H_2O, pH = 12.4	RT	$k = 4.4(4) \cdot 10^9\ M^{-1}s^{-1}$	
$\dot{C}H_2O^- +$ 4-nitroacetophenone(PNAP) $\longrightarrow$ PNAP$^{\dot{-}}$ + HCHO				
Pulse rad. of $CH_3OH + N_2O + H_2O$				73 Ada 1
KAS	H_2O, pH = 13	RT	$k = 4.7(5) \cdot 10^9\ M^{-1}s^{-1}$	
$\dot{C}H_2O^- +$ nitrobenzene($C_6H_5NO_2$) $\longrightarrow C_6H_5\dot{N}O_2^-$ + HCHO				
Pulse rad. of $CH_3OH + N_2O + H_2O$				66 Asm 1
KAS	H_2O, pH = 13	RT	$k = 2.7 \cdot 10^9\ M^{-1}s^{-1}$	
$\dot{C}H_2O^- +$ nitrosobenzene(C_6H_5NO) $\longrightarrow C_6H_5\dot{N}O^-$ + HCHO				
Pulse rad. of $CH_3OH + N_2O + H_2O$				66 Asm 2
KAS	H_2O, pH = 13	RT	$k = 6.8 \cdot 10^9\ M^{-1}s^{-1}$	
$\dot{C}H_2O^- +$ 1,10-phenanthroline $\longrightarrow$ (1,10-phenanthroline)$^{\dot{-}}$ + HCHO				
Pulse rad. of alkaline $CH_3OH + N_2O + H_2O$				80 Tep 1
KAS	H_2O	RT	$k < 10^7\ M^{-1}s^{-1}$	
$\dot{C}H_2O^- +$ pterin [53] $\longrightarrow$ products				
Pulse rad. of $CH_3OH + N_2O + H_2O$				76 Moo 1
KAS	H_2O, pH = 13	RT	$k = 6.0 \cdot 10^8\ M^{-1}s^{-1}$	

[51]) $k \leqslant 10^7$ for reaction of $\dot{C}H_2OH$ at pH = 7.
[52]) 92% e$^-$-transfer.
[53]) Reaction with anionic form of pterin; 100% e$^-$-transfer.

Reaction Radical generation Method	Solvent	T[K]	Rate data	Ref./ add. ref.
$\dot{C}H_2O^- + {}^3$(pyrene) [54]) $\longrightarrow$ (pyrene)$^{\dot{-}}$ + HCHO				
Combined phot. and pulse rad. of micellar solutions containing $CH_3OH + N_2O$				76 Fra 1
KAS	micellar solution (H_2O, CTAB [55])), pH = 13	RT	$k = 1.8 \cdot 10^{10}\,M^{-1}s^{-1}$	
$\dot{C}H(OH)O^-$ + nitrobenzene($C_6H_5NO_2$) $\longrightarrow C_6H_5\dot{N}O_2^- + H^+ + HCO_2^-$				
Pulse rad. of HCHO + $N_2O + H_2O$				71 Sto 1
KAS	H_2O, pH = 12	RT	$k = 4.5 \cdot 10^9\,M^{-1}s^{-1}$	
$\dot{C}H_2OH + Co(NH_3)_6^{3+} \longrightarrow HCHO + H^+ + Co(NH_3)_6^{2+}$				
Pulse rad. of $CH_3OH + N_2O + H_2O$				72 Coh 1, 77 Coh 1
KAS	H_2O, pH = 5.75	RT	$k = 1.4(2) \cdot 10^8\,M^{-1}s^{-1}$	
	pH = 4.9		$6.0(9) \cdot 10^7$	
	pH = 4.5		$4.5(7) \cdot 10^7$	
	pH = 3.5		$k < 10^7\,M^{-1}s^{-1}$	
$\dot{C}H_2OH + (Co(III)(NH_3)_5Br)^{2+} \longrightarrow$ products				
Pulse rad. of $CH_3OH + N_2O + H_2O$				77 Coh 1
KAS	H_2O, pH = 6.1	RT	$k = 9.0(14) \cdot 10^7\,M^{-1}s^{-1}$	
	pH = 4.9		$2.5(4) \cdot 10^7$	
	pH = 4.5		$2.0(3) \cdot 10^7$	
	pH = 3.5...1.0		$1.8(3) \cdot 10^7$	
$\dot{C}H_2OH + (Co(III)(NH_3)_5Cl)^{2+} \longrightarrow$ products				
Pulse rad. of $CH_3OH + N_2O + H_2O$				77 Coh 1
KAS	H_2O, pH = 3.5...4.0	RT	$k = 3(1) \cdot 10^6\,M^{-1}s^{-1}$ [56])	
$\dot{C}H_2OH + (Co(III)(NH_3)_5F)^{2+} \longrightarrow$ products				
Pulse rad. of $CH_3OH + N_2O + H_2O$				77 Coh 1
PR, KAS	H_2O	RT	$k = 5.5(20) \cdot 10^5\,M^{-1}s^{-1}$ [56])	
$\dot{C}H_2OH + Co(NH_3)_5OH_2^{3+} \longrightarrow$ Co(II) + products				
γ-rad. of $CH_3OH + H_2O$				77 Coh 1
[56a])	H_2O, pH = 3.5...4	RT	$k = 1.5(5) \cdot 10^6\,M^{-1}s^{-1}$ [57])	
$\dot{C}H_2OH + [(NH_3)_4Co(III)(\mu O_2, \mu NH_2)Co(III)(NH_3)_4]^{4+} \longrightarrow H^+ + HCHO$ + products [58])				
Pulse rad. of $CH_3OH + N_2O + H_2O$				81 Nat 1
KAS	H_2O, pH = 5	RT	$k = 1.2 \cdot 10^8\,M^{-1}s^{-1}$	
$\dot{C}H_2OH + [(CN)_5Co(III)(\mu O_2)Co(III)(CN)_5]^{5-} \longrightarrow H^+ + HCHO$ + products [58])				
Pulse rad. of $CH_3OH + N_2O + H_2O$				81 Nat 1
KAS	H_2O, pH = 5.0	RT	$k = 2.8 \cdot 10^8\,M^{-1}s^{-1}$	
$\dot{C}H_2OH + [(EN)_2Co(III)(\mu O_2, \mu NH_2)Co(III)(EN)_2]^{4+} \longrightarrow H^+ + HCHO$ + products [58])				
Pulse rad. of $CH_3OH + N_2O + H_2O$				81 Nat 1
KAS	H_2O, pH = 5	RT	$k = 4 \cdot 10^7\,M^{-1}s^{-1}$	

[54]) Pyrene triplet.
[55]) $5 \cdot 10^{-3}$ M hexadecyl trimethyl ammonium bromide (CTAB).
[56]) Estimated from effect of complex conc. on yield of Co^{2+} and assuming $2k(\dot{C}H_2OH + \dot{C}H_2OH) = 2.4 \cdot 10^9\,M^{-1}s^{-1}$.
[56a]) Competition kinetics and Co^{2+} yield.
[57]) Based on competition with $2k(\dot{C}H_2OH + \dot{C}H_2OH) = 2.4 \cdot 10^9\,M^{-1}s^{-1}$.
[58]) e^--transfer assumed to occur at dioxygen center. μ defines a bridging group.

Reaction Radical generation Method	Solvent	T[K]	Rate data	Ref./ add. ref.
$\dot{C}H_2OH$ + *cis*-(aquachlorobis(1,2-ethanediamine-N,N')cobalt(III))$^{2+}$ ⟶ products				
Pulse rad. of $CH_3OH + N_2O + H_2O$ KAS	H_2O, pH = 3.5...4.0	RT	$k = 1.8(5)\cdot 10^6\ M^{-1}s^{-1}$	77 Coh 1
$\dot{C}H_2OH$ + *trans*-(dibromobis(1,2-ethanediamine-N,N')cobalt(III))$^+$ ⟶ products				
Pulse rad. of $CH_3OH + N_2O + H_2O$ KAS	H_2O, pH = 3.5...4.0	RT	$k = 2.6\cdot 10^8\ M^{-1}s^{-1}$	77 Coh 1
$\dot{C}H_2OH$ + *trans*-(dichlorobis(1,2-ethanediamine-N,N')cobalt(III))$^+$ ⟶ products				
Pulse rad. of $CH_3OH + N_2O + H_2O$ KAS	H_2O, pH = 3.5...4.0	RT	$k = 8(2)\cdot 10^6\ M^{-1}s^{-1}$	77 Coh 1
$\dot{C}H_2OH$ + tris(2,2'-bipyridine-N,N')cobalt(III)$^{3+}$ ⟶ tris(2,2'-bipyridine-N,N')cobalt(II)$^{2+}$ + H^+ + HCHO				
Pulse rad. of $CH_3OH + H_2O + N_2O$ KAS	H_2O, pH = 1.7	RT	$k = 2\cdot 10^8\ M^{-1}s^{-1}$	79 Sim 1
$\dot{C}H_2OH$ + [{Co(pts)}$_2$]$^{8-}$ [59]) ⟶ [{Co(pts)(pts)$^{\cdot}$}$_2$]$^{9-}$ [60]) + HCHO + H^+				
Pulse rad. of $CH_3OH + N_2O + H_2O$ KAS	H_2O	RT	$k = 2.2\cdot 10^8\ M^{-1}s^{-1}$	80 Fer 1
$\dot{C}H_2OH + Cu^{2+}$ ⟶ Cu^+ + products				
Pulse rad. of $CH_3OH + H_2O + N_2O$ KAS	H_2O, pH = 5...6 pH = 2...5 pH ⩽ 3	RT	$k = 1.1(2)\cdot 10^8\ M^{-1}s^{-1}$ $1.6(3)\cdot 10^8$ [61]) [62]) $1.9(4)\cdot 10^8$ [62])	72 Coh 1, 78 Bux 1
$\dot{C}H_2OH + Fe^{3+}$ ⟶ $Fe^{2+} + H^+$ + HCHO				
γ-rad. of $CH_3OH + H_2O$ PR, competition kinetics	H_2O, pH ≈ 1	RT	$k = 1.0\cdot 10^8\ M^{-1}s^{-1}$ [63])	77 Ber 1
$\dot{C}H_2OH + Fe(CN)_6^{3-}$ ⟶ HCHO + H^+ + $Fe(CN)_6^{4-}$				
Pulse rad. of $CH_3OH + H_2O + N_2O$ KAS and time resolved cond.	H_2O	RT	$k = 4.2(4)\cdot 10^9\ M^{-1}s^{-1}$	70 Bar 1/ 68 Ada 1, 69 Ada 1
$\dot{C}H_2OH + Fe(CN)_5NO^{2-}$ ⟶ HCHO + H^+ + $Fe(CN)_5NO^{3-}$				
Pulse rad. of $CH_3OH + H_2O + N_2O$ KAS	H_2O, pH = 8.5	RT	$k = 6.7\cdot 10^8\ M^{-1}s^{-1}$	77 Che 1
$\dot{C}H_2OH$ + Fe(III)cytochrome c ⟶ HCHO + H^+ + Fe(II)cytochrome c				
Pulse rad. of $CH_3OH + H_2O + N_2O$ KAS	H_2O	RT	$k = 3.0(7)\cdot 10^7\ M^{-1}s^{-1}$	79 Lee 1
$\dot{C}H_2OH$ + hemoglobin(Fe(III)) ⟶ HCHO + H^+ + hemoglobin(Fe(II))				
Pulse rad. of $CH_3OH + H_2O + N_2O$ KAS	H_2O	RT	$k = 9.5(15)\cdot 10^6\ M^{-1}s^{-1}$	79 Lee 1
$\dot{C}H_2OH$ + metmyoglobin(Fe(III)) ⟶ HCHO + H^+ + metmyoglobin(Fe(II))				
Pulse rad. of $CH_3OH + H_2O + N_2O$ KAS	H_2O	RT	$k = 2.4(5)\cdot 10^7\ M^{-1}s^{-1}$	79 Lee 1

[59]) Dimeric Co(II)-sulfophthalocyanine.
[60]) (pts) = phthalocyanine-3,10,17,24-tetrasulfonate hexa anion; (pts)$^{\cdot}$ = one-electron reduction product of (pts).
[61]) Based on formation kinetics of $Cu(I)CH_2CHCONH_2$ in presence of acrylamide.
[62]) From [78 Bux 1].
[63]) Based on $k(\dot{C}H_2OH + C(NO_2)_4) = 5\cdot 10^9\ M^{-1}s^{-1}$.

Reaction Radical generation Method	Solvent	T[K]	Rate data	Ref./ add. ref.
$\dot{C}H_2OH + Ga^{2+}(Ga(OH)^+) \longrightarrow HCHO + Ga^+ + H^+(H_2O)$				
Pulse rad. of $CH_3OH + Ga^{3+} + H_2O$				79 Suk 1
KAS	H_2O	RT	$k = 1\cdot 10^9\ M^{-1}\,s^{-1}$	
$\dot{C}H_2OH + H_2O_2 \longrightarrow HCHO + H_2O + \dot{O}H$				
γ-rad. of $CH_3OH + H_2O$ and $Ti(III) + H_2O_2 + CH_3OH$				70 Bur 1,
PR and ESR	H_2O	RT	$k = 4.0(4)\cdot 10^4\ M^{-1}\,s^{-1}$ [64] $2.3(8)\cdot 10^4$ [65]	74 Gil 1
$\dot{C}H_2OH + HgBr_2 \xrightarrow{a} HgBr + Br^- + H^+ + HCHO$ $\dot{C}O_2^- + HgBr_2 \xrightarrow{b} CO_2 + Hg(I)\ldots$				
Pulse rad. of formate $+ CH_3OH + N_2O + H_2O$				76 Fuj 1
PR	H_2O	RT	$k_a/k_b = 1.59$	
$\dot{C}H_2OH + IrCl_6^{2-} \longrightarrow Ir(III) +$ products				
Pulse rad. of $CH_3OH + N_2O + H_2O$				82 Ste 1
KAS, Cond.	H_2O	295	$k = 6.0\cdot 10^9\ M^{-1}\,s^{-1}$	
$\dot{C}H_2OH$ + *trans*-dichlorobisethylene diamine platinum(IV) ion $\longrightarrow$ Pt(III)... + products				
Pulse rad. of $CH_3OH + H_2O + N_2O$				75 Sto 1
KAS	H_2O	RT	$k = 6.9(10)\cdot 10^8\ M^{-1}\,s^{-1}$	
$\dot{C}H_2OH$ + tris(2,2′-bipyridine-N,N′)rhodium(III) ion $\longrightarrow$ products				
Pulse rad. of $CH_3OH + H_2O$				74 Mul 2
KAS	H_2O	RT	$k = 2.2(2)\cdot 10^8\ M^{-1}\,s^{-1}$	
$\dot{C}H_2OH + Ru(NH_3)_6^{3+} \longrightarrow Ru(NH_3)_6^{2+} + H^+ + HCHO$				
Pulse rad. of $CH_3OH + H_2O + N_2O$				72 Coh 1
KAS	H_2O	RT	$k = 4.1(6)\cdot 10^7\ M^{-1}\,s^{-1}$	
$\dot{C}H_2OH + (Ru(III)(NH_3)_5Cl)^{2+} \longrightarrow$ products				
Pulse rad. of $CH_3OH + N_2O + H_2O$				77 Coh 1
KAS	H_2O, pH = 3.5...4.0	RT	$k = 1.2(2)\cdot 10^8\ M^{-1}\,s^{-1}$	
$\dot{C}H_2OH + Zn^+ + H_2O \longrightarrow Zn^{2+} + CH_3OH + OH^-$				
Pulse rad. of $CH_3OH + Zn^{2+} + H_2O$				77 Rab 1
KAS	H_2O	RT	$k = 2.5(3)\cdot 10^9\ M^{-1}\,s^{-1}$ [66]	
$\dot{C}H_2OH$ + acridine $(C_{13}H_9NH^+) \longrightarrow$ products [67]				
Pulse rad. of $CH_3OH + H_2O$				74 Moo 1
KAS	H_2O, pH = 2	RT	$k = 5.0\cdot 10^8\ M^{-1}\,s^{-1}$ [68]	
$\dot{C}H_2OH$ + benzoquinone $\longrightarrow$ HCHO + ...semiquinone				
Pulse rad. of $CH_3OH + H_2O + N_2O$				71 Wil 1,
KAS	H_2O	RT	$k = 6.1\cdot 10^9\ M^{-1}\,s^{-1}$ $4.8\cdot 10^9$ [69]	73 Sim 1

[64]) Based on $2k(\dot{C}H_2OH + \dot{C}H_2OH) = 2.4\cdot 10^9\ M^{-1}\,s^{-1}$ (γ-rad. of $CH_3OH + H_2O$) [70 Bur 1].
[65]) Based on $2k(\dot{C}H_2OH + \dot{C}H_2OH) = 2\cdot 10^9\ M^{-1}\,s^{-1}$ ($Ti(III) + H_2O_2 + CH_3OH$) [74 Gil 1].
[66]) Based on $k(Zn^+ + Zn^+) = 4.5\cdot 10^8\ M^{-1}\,s^{-1}$; $k(Zn^+ + H_2O_2) = 2.4\cdot 10^9\ M^{-1}\,s^{-1}$; $k(Zn^+ + N_2O) = 1.6\cdot 10^7\ M^{-1}\,s^{-1}$; $k(\dot{C}H_2OH + \dot{C}H_2OH) = 1.5\cdot 10^9\ M^{-1}\,s^{-1}$.
[67]) Estimated 30% e^--transfer.
[68]) No reaction with deprotonated acridine.
[69]) From [73 Sim 1].

Reaction Radical generation Method	Solvent	T[K]	Rate data	Ref./ add. ref.
$\dot{C}H_2OH$ + 2,3-butanedione($CH_3COCOCH_3$) ⟶ HCHO + H^+ + $(CH_3COCOCH_3)^{\overline{\cdot}}$				
Pulse rad. of $CH_3OH + H_2O + N_2O$ KAS	H_2O	RT	$k = 6.5(10) \cdot 10^7\,M^{-1}s^{-1}$ $1.1 \cdot 10^8$ [70]	72 Coh 1, 68 Lil 1
$\dot{C}H_2OH$ + 4-chlorobenzenediazonium ($ClC_6H_4N_2^+$) ⟶ H^+ + HCHO + N_2 + Cl— 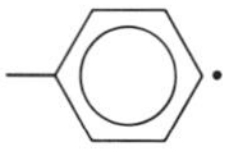				
Pulse rad. of $CH_3OH + N_2O + H_2O$ Cond. (time resolved)	H_2O	RT	$k = 4.2 \cdot 10^9\,M^{-1}s^{-1}$	81 Pac 1
$\dot{C}H_2OH$ + dichloroindophenol ⟶ products				
Pulse rad. of $CH_3OH + H_2O + N_2O$ KAS	H_2O	RT	$k = 3.2 \cdot 10^9\,M^{-1}s^{-1}$	73 Rao 2
$\dot{C}H_2OH$ + (1,1′-dimethyl-4,4′-bipyridinium)$^{2+}$ [71]) ⟶ HCHO + H^+ + $(1,1'\text{-di}\ldots)^{\dot{+}}$				
Pulse rad. of $CH_3OH + H_2O + N_2O$ KAS	H_2O/CH_3OH (97%:3%)	RT	$k = 3 \cdot 10^8\,M^{-1}s^{-1}$	77 Pat 1
$\dot{C}H_2OH$ + N-ethylmaleimide ⟶ products [72])				
Pulse rad. of $CH_3OH + H_2O + N_2O$ KAS	H_2O	RT	$k = 2.4 \cdot 10^9\,M^{-1}s^{-1}$	72 Hay 1
$\dot{C}H_2OH$ + [1-(2-hydroxyethyl)-2-methyl-5-nitroimidazole] [73]) ⟶ products				
Pulse rad. of $CH_3OH + H_2O + N_2O$ KAS	H_2O	293	$k = 10^8\,M^{-1}s^{-1}$	74 Wil 1
$\dot{C}H_2OH$ + indigo disulfonate ⟶ products [74])				
Pulse rad. of $CH_3OH + H_2O + N_2O$ KAS	H_2O	RT	$k = 2.0 \cdot 10^9\,M^{-1}s^{-1}$	73 Rao 2
$\dot{C}H_2OH$ + indigo tetrasulfonate ⟶ products [75])				
Pulse rad. of $CH_3OH + H_2O + N_2O$ KAS	H_2O	RT	$k = 3.0 \cdot 10^9\,M^{-1}s^{-1}$	73 Rao 2
$\dot{C}H_2OH$ + indophenol ⟶ products				
Pulse rad. of $CH_3OH + H_2O + N_2O$ KAS	H_2O, pH = 9	RT	$k = 3.1 \cdot 10^9\,M^{-1}s^{-1}$	73 Rao 2
$\dot{C}H_2OH$ + 4-methoxybenzenediazonium ($CH_3OC_6H_4N_2^+$) ⟶ products				
Pulse rad. of $CH_3OH + N_2O + H_2O$ KAS	H_2O	RT	$k = 1.9 \cdot 10^9\,M^{-1}s^{-1}$	81 Pac 1
$\dot{C}H_2OH$ + methylene blue ⟶ products				
Pulse rad. of $CH_3OH + H_2O + N_2O$ KAS	H_2O	RT	$k = 3.4 \cdot 10^9\,M^{-1}s^{-1}$	73 Rao 2

[70]) From [68 Lil 1].
[71]) Paraquat; methylviologen.
[72]) 15% e^--transfer.
[73]) Metronidazole.
[74]) 75% e^--transfer at pH = 7; 62% e^--transfer at pH = 9.
[75]) 80% e^--transfer.

Reaction Radical generation Method	Solvent	T[K]	Rate data	Ref./ add. ref.
$\dot{C}H_2OH$ + 2-methyl-1,4-naphthoquinone ⟶ ...-semiquinone + products [76]				
Pulse rad. of $CH_3OH + H_2O + N_2O$ KAS	H_2O	RT	$k = 3.7(4) \cdot 10^9\ M^{-1}\,s^{-1}$	73 Rao 3
$\dot{C}H_2OH$ + 3-methylpterin ⟶ products [77]				
Pulse rad. of $CH_3OH + H_2O + N_2O$ KAS	H_2O, pH = 0.8	RT	$k = 6 \cdot 10^7\ M^{-1}\,s^{-1}$	76 Moo 1
$\dot{C}H_2OH$ + nicotinamide adenine dinucleotide(NAD^+) ⟶ products				
Pulse rad. of $CH_3OH + H_2O + N_2O$ KAS	H_2O	RT	$k = 1.0 \cdot 10^9\ M^{-1}\,s^{-1}$	73 Rao 1
$\dot{C}H_2OH$ + 4-nitroacetophenone (PNAP) ⟶ $PNA\dot{P}^{\cdot -} + H^+$ + HCHO				
Pulse rad. of $CH_3OH + N_2O + H_2O$ KAS	H_2O	RT	$k = 1 \cdot 10^7\ M^{-1}\,s^{-1}$	73 Ada 1
$\dot{C}H_2OH$ + nitrobenzene ($C_6H_5NO_2$) ⟶ $C_6H_5\dot{N}O_2^- + H^+$ + HCHO				
Pulse rad. of $CH_3OH + N_2O + H_2O$ KAS	H_2O	RT	$k = 6.0(9) \cdot 10^7\ M^{-1}\,s^{-1}$	72 Coh 1/ 66 Asm 1
$\dot{C}H_2OH$ + 4-nitrobenzene diazonium ($O_2NC_6H_4N_2^+$) ⟶ H^+ + HCHO + N_2 + O_2N—  ·				
Pulse rad. of $CH_3OH + N_2O + H_2O$ Cond. (time resolved)	H_2O	RT	$k = 5.2 \cdot 10^9\ M^{-1}\,s^{-1}$	81 Pac 1
$\dot{C}H_2OH$ + *anti*-5-nitro-2-furaldoxime [78] ⟶ products [79]				
Pulse rad. of $CH_3OH + N_2O + H_2O$ KAS	H_2O	RT	$k = 7.2 \cdot 10^8\ M^{-1}\,s^{-1}$	73 Gre 1
$\dot{C}H_2OH$ + 4-nitroperoxybenzoic acid ⟶ products [80]				
Pulse rad. of $CH_3OH + N_2O + H_2O$ KAS, Cond.	H_2O	RT	$k = 2 \cdot 10^8\ M^{-1}\,s^{-1}$	74 Lil 1
$\dot{C}H_2OH$ + nitrosobenzene (C_6H_5NO) ⟶ $C_6H_5\dot{N}OH$ + HCHO				
Pulse rad. of $CH_3OH + N_2O + H_2O$ KAS	H_2O	RT	$k = 3.2 \cdot 10^9\ M^{-1}\,s^{-1}$	66 Asm 2
$\dot{C}H_2OH$ + phenosafranine ⟶ products [81]				
Pulse rad. of $CH_3OH + N_2O + H_2O$ KAS	H_2O	RT	$k = 1.2 \cdot 10^9\ M^{-1}\,s^{-1}$	73 Rao 2
$\dot{C}H_2OH$ + pterin ⟶ products				
Pulse rad. of $CH_3OH + N_2O + H_2O$ KAS	H_2O, pH = 0.8 pH = 7	RT	$k = 9.0 \cdot 10^7\ M^{-1}\,s^{-1}$ [82] $k \leqslant 10^7\ M^{-1}\,s^{-1}$ [83]	76 Moo 1

[76] 88% e^--transfer.
[77] ≈45% e^--transfer.
[78] Nifuroxime.
[79] 90% e^--transfer.
[80] 20% e^--transfer (based on Cond.).
[81] 22% e^--transfer.
[82] Reaction with protonated form of pterin; 40% e^--transfer.
[83] Reaction with neutral form of pterin.

Reaction Radical generation Method	Solvent	T[K]	Rate data	Ref./ add. ref.
$\dot{C}H_2OH$ + tetranitromethane ($C(NO_2)_4$) $\longrightarrow$ $C(NO_2)_3^-$ + NO_2 + H^+ + HCHO				
Pulse rad. of $CH_3OH + N_2O + H_2O$				64 Asm 1
KAS	H_2O	RT	$k = 5.0(10) \cdot 10^9\ M^{-1}s^{-1}$	
$\dot{C}H_2OH$ + thionine $\longrightarrow$ products [84]				
Pulse rad. of $CH_3OH + N_2O + H_2O$				73 Rao 2
KAS	H_2O	RT	$k = 2.6 \cdot 10^9\ M^{-1}s^{-1}$	
$\dot{C}H_2OH$ + 4-toluene diazonium ($CH_3C_6H_4N_2^+$) $\longrightarrow$ H^+ + HCHO + N_2 + CH_3– 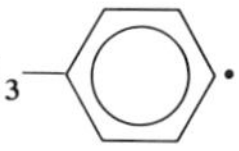 ·				
Pulse rad. of $CH_3OH + N_2O + H_2O$				81 Pac 1
Cond. (time resolved)	H_2O	RT	$k = 1.8 \cdot 10^9\ M^{-1}s^{-1}$	
$\dot{C}H(OH)_2$ + nitrobenzene ($C_6H_5NO_2$) $\longrightarrow$ $C_6H_5\dot{N}O_2^-$ + $2H^+$ + HCO_2^-				
Pulse rad. of HCHO + $N_2O + H_2O$				71 Sto 1
KAS	H_2O	RT	$k = 1.9 \cdot 10^9\ M^{-1}s^{-1}$	
$\dot{C}H(O^-)COO^-$ + $Fe(CN)_6^{3-}$ $\longrightarrow$ Fe(II) + products				
Pulse rad. of glycolate($CH_2(O^-)COO^-$) + $N_2O + H_2O$				82 Ste 1 [85]
KAS	H_2O, basic pH	295	$k = 7.5 \cdot 10^8\ M^{-1}s^{-1}$	
$\dot{C}H(O^-)COO^-$ + $IrCl_6^{2-}$ $\longrightarrow$ Ir(III) + products				
Pulse rad. of glycolate($CH_2(O^-)COO^-$) + $N_2O + H_2O$				82 Ste 1 [85]
KAS	H_2O, basic pH	295	$k = 1.8 \cdot 10^9\ M^{-1}s^{-1}$	
$\dot{C}H(O^-)COO^-$ + 2-methyl-1,4-naphthoquinone $\longrightarrow$ ...-semiquinone + products [86]				
Pulse rad. of glycolate($CH_2(O^-)COO^-$) + $H_2O + N_2O$				73 Rao 3
KAS	H_2O, pH = 10.6	RT	$k = 1.6(2) \cdot 10^9\ M^{-1}s^{-1}$	
$\dot{C}H_2COO^-$ + $IrCl_6^{2-}$ $\longrightarrow$ Ir(III) + products				
Pulse rad. of CH_3COO^- + $N_2O + H_2O$				82 Ste 1
KAS	H_2O	295	$k = 4.2 \cdot 10^8\ M^{-1}s^{-1}$	
$\dot{C}HOHCOO^-$ + $Fe(CN)_6^{3-}$ $\longrightarrow$ $Fe(CN)_6^{4-}$ + products				
Pulse rad. of glycolate($CHOHCOO^-$) + H_2O				69 Ada 1
KAS	H_2O, basic pH	RT	$k = 5 \cdot 10^8\ M^{-1}s^{-1}$	
$\dot{C}HOHCOO^-$ + $IrCl_6^{2-}$ $\longrightarrow$ Ir(III) + products				
Pulse rad. of glycolate($CHOHCOO^-$) + $N_2O + H_2O$				82 Ste 1 [87]
KAS	H_2O, basic pH	295	$k = 2 \cdot 10^9\ M^{-1}s^{-1}$	
$\dot{C}HOHCOO^-$ + 9,10-anthraquinone-2-sulfonate $\longrightarrow$ ...-semiquinone + products				
Pulse rad. of glycolate($CHOHCOO^-$) + $H_2O + N_2O$				73 Rao 1
KAS	H_2O	RT	$k = 7.1(7) \cdot 10^8\ M^{-1}s^{-1}$	
$\dot{C}HOHCOO^-$ + benzoquinone $\longrightarrow$ benzosemiquinone + products				
Pulse rad. of glycolate($CHOHCOO^-$) + $H_2O + N_2O$				73 Hay 1/ 73 Rao 1
KAS	H_2O	RT	$k = 2.2 \cdot 10^9\ M^{-1}s^{-1}$	

[84]) 86% e^--transfer.
[85]) Mechanism discussed as e^--transfer.
[86]) 77% e^--transfer.
[87]) Mechanism discussed as e^--transfer.

Reaction Radical generation Method	Solvent	T[K]	Rate data	Ref./ add. ref.
$\dot{C}HOHCOO^-$ + 2-hydroxy-1,4-naphthoquinone ⟶ ...-semiquinone + products				
Pulse rad. of glycolate($CHOHCOO^-$) + H_2O + N_2O				73 Rao 1
KAS	H_2O	RT	$k = 9.1(9) \cdot 10^8\ M^{-1}\,s^{-1}$	
$\dot{C}HOHCOO^-$ + 2-methyl-1,4-naphthoquinone ⟶ ...-semiquinone + products [88]				
Pulse rad. of glycolate($CHOHCOO^-$) + H_2O + N_2O				73 Rao 3/
KAS	H_2O	RT	$k = 1.5(2) \cdot 10^9\ M^{-1}\,s^{-1}$	73 Rao 1
$\dot{C}HOHCOO^-$ + 1,4-naphthoquinone-2-sulfonate ion ⟶ ...-semiquinone + products				
Pulse rad. of glycolate($CHOHCOO^-$) + H_2O + N_2O				73 Rao 1
KAS	H_2O	RT	$k = 1.7(2) \cdot 10^9\ M^{-1}\,s^{-1}$	
$\dot{C}HOHCOO^-$ + riboflavin ⟶ products				
Pulse rad. of glycolate($CHOHCOO^-$) + N_2O + H_2O				73 Rao 1
KAS	H_2O	RT	$k = 9.3(9) \cdot 10^8\ M^{-1}\,s^{-1}$	
$\dot{C}H_2CHO$ [89] + Co(II)-tetra(4-sulfonatophenyl)porphyrine ⟶ Co(III)-tetra... + products				
Pulse rad. of ethylene glycol + N_2O + H_2O				81 Net 2
KAS	H_2O, pH = 12	RT	$k = 2.0 \cdot 10^9\ M^{-1}\,s^{-1}$	
$\dot{C}H_2CHO$ [89] + Fe(II) ⟶ Fe(III) + $^-CH_2CHO$ [90]				
Pulse rad. of ethylene glycol + H_2O				73 Gil 1
KAS	H_2O, pH = 0.7	RT	$k = 4.5 \cdot 10^5\ M^{-1}\,s^{-1}$	
$\dot{C}H_2CHO$ [89] + $IrCl_6^{2-}$ ⟶ Ir(III) + products				
Pulse rad. of 2-chloroethanol + N_2O + H_2O				82 Ste 1
KAS	H_2O	295	$k = 1.7 \cdot 10^9\ M^{-1}\,s^{-1}$	
$\dot{C}H_2CHO$ [89] + Ti(III) ⟶ Ti(IV) + $^-CH_2CHO$ [90]				
Ti(III)/H_2O_2 flow expt. and pulse rad. of ethylene glycol + H_2O				73 Gil 1
SESR, KAS	H_2O, pH = 0.7	RT	$k \leqslant 10^5\ M^{-1}\,s^{-1}$ (pulse rad.) $\leqslant 6 \cdot 10^5$ (ESR)	
$\dot{C}H_2CHO$ [89] + Ti(III)-EDTA ⟶ Ti(IV)-EDTA + $^-CH_2CHO$ [90]				
Pulse rad. of ethylene glycol + H_2O				73 Gil 1
KAS	H_2O, pH = 9	RT	$k = 6 \cdot 10^7\ M^{-1}\,s^{-1}$ [91]	
$\dot{C}H_2CHO$ [89] + Zn-tetra(4-sulfonatophenyl)porphyrine ⟶ $(\text{Zn-tetra...})^{\dot{+}}$ + products				
Pulse rad. of ethylene glycol + N_2O + H_2O				81 Net 2
KAS	H_2O, pH = 12	RT	$k \approx 1.5 \cdot 10^8\ M^{-1}\,s^{-1}$	
$\dot{C}H_2CHO$ [89] + 4-aminophenoxide ($NH_2C_6H_4O^-$) ⟶ CH_3CHO + $NH_2C_6H_4\dot{O}$				
Pulse rad. of ethylene glycol + H_2O + N_2O				79 Ste 1
KAS	H_2O, pH ≈ 11.5	RT	$k = 2.1 \cdot 10^9\ M^{-1}\,s^{-1}$	
$\dot{C}H_2CHO$ [89] + 1,2-diaminobenzene [92] ⟶ CH_3CHO + $H_2NC_6H_4\dot{N}H$				
Pulse rad. of ethylene glycol + N_2O + H_2O				79 Ste 1
KAS	H_2O, pH = 11.5	RT	$k = 7.3 \cdot 10^7\ M^{-1}\,s^{-1}$	

[88]) 69% e^--transfer.
[89]) Mainly C-centered radical, oxidizing action, however, likely to occur through mesomeric O-centered radical.
[90]) Immediate protonation.
[91]) $k > 10^7\ M^{-1}\,s^{-1}$ from ESR experiment.
[92]) *o*-Phenylenediamine.

Reaction Radical generation Method	Solvent	T[K]	Rate data	Ref./ add. ref.
$\dot{C}H_2CHO$ [89]) + 1,4-diaminobenzene [93]) $\longrightarrow$ $CH_3CHO + H_2NC_6H_4\dot{N}H$				
Pulse rad. of ethylene glycol + $N_2O + H_2O$				79 Ste 1
KAS	H_2O	RT	$k = 4.0 \cdot 10^8\ M^{-1}s^{-1}$	
$\dot{C}H_2CHO$ [89]) + 1,2-dihydroxybenzene + OH^- $\longrightarrow$ $CH_3CHO + {}^-OC_6H_4\dot{O} + H_2O$				
Pulse rad. of ethylene glycol + $N_2O + H_2O$				79 Ste 1
KAS	H_2O, pH ≈ 11.5	RT	$k = 7.5 \cdot 10^8\ M^{-1}s^{-1}$	
$\dot{C}H_2CHO$ [89]) + 1,3-dihydroxybenzene + OH^- $\longrightarrow$ $CH_3CHO + {}^-OC_6H_4\dot{O} + H_2O$				
Pulse rad. of ethylene glycol + $N_2O + H_2O$				79 Ste 1
KAS	H_2O, pH ≈ 11.5	RT	$k = 1.6 \cdot 10^9\ M^{-1}s^{-1}$	
$\dot{C}H_2CHO$ [89]) + 4-ethylphenoxide ($C_2H_5C_6H_4O^-$) + H_2O $\longrightarrow$ $CH_3CHO + OH^- + C_2H_5C_6H_4\dot{O}$				
Pulse rad. of ethylene glycol + $H_2O + N_2O$				79 Ste 1
KAS	H_2O, pH = 11.5	RT	$k = 7.0 \cdot 10^7\ M^{-1}s^{-1}$	
$\dot{C}H_2CHO$ [89]) + hydroquinone ($HOC_6H_4O^-$) $\longrightarrow$ $CH_3CHO + {}^-OC_6H_4\dot{O}$				
Pulse rad. of ethylene glycol + $N_2O + H_2O$				79 Ste 1
KAS	H_2O, pH ≈ 11.5	RT	$k = 2.2(1) \cdot 10^9\ M^{-1}s^{-1}$	
	pH = 7.2		$k \leqslant 2 \cdot 10^6\ M^{-1}s^{-1}$	
$\dot{C}H_2CHO$ [89]) + 4-methoxyphenoxide ($CH_3OC_6H_4O^-$) + H_2O $\longrightarrow$ $CH_3CHO + CH_3OC_6H_4\dot{O} + OH^-$				
Pulse rad. of ethylene glycol + $H_2O + N_2O$				79 Ste 1
KAS	H_2O, pH = 11.5	RT	$k = 9.8 \cdot 10^8\ M^{-1}s^{-1}$	
$\dot{C}H_2CHO$ [89]) + 4-methylphenol $\longrightarrow$ $CH_3CHO + CH_3C_6H_4\dot{O}$				
Pulse rad. of ethylene glycol + $N_2O + H_2O$				79 Ste 1
KAS	H_2O, pH ≈ 11.5	RT	$k = 9.0 \cdot 10^7\ M^{-1}s^{-1}$	
$\dot{C}H_2CHO$ [89]) + metiazinic acid(MZ^-) $\longrightarrow$ $MZ^{\dot{\pm}}$ + products				
Pulse rad. of ethylene glycol + $N_2O + H_2O$				81 Bah 1
KAS	H_2O, pH = 8...11	RT	$k = 2 \cdot 10^8\ M^{-1}s^{-1}$	
$\dot{C}H_2CHO$ [89]) + phenolate($C_6H_5O^-$) + H_2O $\longrightarrow$ $CH_3CHO + C_6H_5\dot{O} + OH^-$				
Pulse rad. of ethylene glycol + $N_2O + H_2O$				79 Ste 1
KAS	H_2O, pH = 11.5	RT	$k = 4.3 \cdot 10^6\ M^{-1}s^{-1}$	
$\dot{C}H_2CHO$ [89]) + N,N,N′,N′-tetramethyl-4-phenylenediamine $\longrightarrow$ products				
Pulse rad. of ethylene glycol + $N_2O + H_2O$				79 Ste 1
KAS	H_2O	RT	$k = 2.0 \cdot 10^9\ M^{-1}s^{-1}$	
$\dot{C}H_2COOH + Fe(CN)_6^{3-}$ $\longrightarrow$ Fe(II) + products				
Pulse rad. of $CH_3COOH + N_2O + H_2O$				82 Ste 1
KAS	H_2O	295	$k = 2 \cdot 10^6\ M^{-1}s^{-1}$	[94])
$\dot{C}H_2COOH + IrCl_6^{2-}$ $\longrightarrow$ Ir(III) + products				
Pulse rad. of $CH_3COOH + N_2O + H_2O$				82 Ste 1
KAS	H_2O	295	$k = 1.4 \cdot 10^9\ M^{-1}s^{-1}$	[95])
$\dot{C}H_2COOH + (Ru(III)(NH_3)_5Br)^{2+}$ $\longrightarrow$ products				
Pulse rad. of $CH_3COOH + N_2O + H_2O$				77 Coh 1
KAS	H_2O, pH = 3.9	RT	$k = 4.6(7) \cdot 10^8\ M^{-1}s^{-1}$	

[89]) Mainly C-centered radical, oxidizing action, however, likely to occur through mesomeric O-centered radical.
[93]) *p*-Phenylenediamine.
[94]) Mechanism discussed as e^--transfer.
[95]) Mechanism discussed as e^-- or $Cl^{\cdot}$-transfer.

Reaction Radical generation Method	Solvent	T[K]	Rate data	Ref./ add. ref.
$\dot{C}H_2COOH + (Ru(III)(NH_3)_5Cl)^{2+} \longrightarrow$ products				
Pulse rad. of $CH_3COOH + N_2O + H_2O$				77Coh 1
KAS	H_2O, pH = 3.9	RT	$k = 4.0(6) \cdot 10^7\,M^{-1}s^{-1}$	
$\dot{C}HOHCOOH + Fe(CN)_6^{3-} \longrightarrow Fe(II)$ + products				
Pulse rad. of glycolic acid + $N_2O + H_2O$				82 Ste 1
KAS	H_2O	295	$k = 1.0 \cdot 10^8\,M^{-1}s^{-1}$	[94])
$\dot{C}HOHCOOH + IrCl_6^{2-} \longrightarrow Ir(III)$ + products				
Pulse rad. of glycolic acid + $N_2O + H_2O$				82 Ste 1
KAS	H_2O	295	$k = 2.3 \cdot 10^9\,M^{-1}s^{-1}$	[94])
$\dot{C}HOHCOOH$ + 2-methyl-1,4-naphthoquinone $\longrightarrow$ ...-semiquinone + products [96])				
Pulse rad. of glycolic acid + $H_2O + N_2O$				73 Rao 3
KAS	H_2O, pH = 3.2	RT	$k = 9.2(9) \cdot 10^8\,M^{-1}s^{-1}$	
$CH_3\dot{C}HO^- + BrO_3^- \longrightarrow$ products				
Pulse rad. of $CH_3CH_2OH + BrO_3^- + H_2O + N_2O$				72 Coh 1
KAS	H_2O, pH = 11.8	RT	$k = 3.0(5) \cdot 10^7\,M^{-1}s^{-1}$ [97])	
$CH_3\dot{C}HO^- + Co(NH_3)_6^{3+} \longrightarrow CH_3CHO + Co(NH_3)_6^{2+}$				
Pulse rad. of $C_2H_5OH + H_2O + N_2O$				72 Coh 1
KAS	H_2O, pH = 12	RT	$k = 8.5(13) \cdot 10^9\,M^{-1}s^{-1}$	
$CH_3\dot{C}HO^-$ + hematoporphyrine $\longrightarrow$ products				
Pulse rad. of $C_2H_5OH + H_2O + N_2O$				74 Har 1
KAS	H_2O, pH = 13	RT	$k = 7.0(10) \cdot 10^8\,M^{-1}s^{-1}$	
$CH_3\dot{C}HO^-$ + hemin(Fe(III)) $\longrightarrow CH_3CHO$ + hemin(Fe(II))				
Pulse rad. of $C_2H_5OH + H_2O$				74 Har 1
KAS	H_2O/C_2H_5OH (70:30%), pH = 13	RT	$k = 9.0(10) \cdot 10^8\,M^{-1}s^{-1}$	
$CH_3\dot{C}HO^- + Ga(OH)_6^{4-} \longrightarrow$ products				
Pulse rad. of $C_2H_5OH + Ga^{3+} + H_2O$				79 Suk 1
KAS	H_2O, pH = 12	RT	$k = 1.2 \cdot 10^9\,M^{-1}s^{-1}$	
$CH_3\dot{C}HO^- + IO_3^- \longrightarrow$ products				
Pulse rad. of $C_2H_5OH + H_2O + N_2O$				72 Coh 1
KAS	H_2O, pH = 11.8	RT	$k = 7.5(2) \cdot 10^8\,M^{-1}s^{-1}$ [98])	
$CH_3\dot{C}HO^-$ + tris(2,2'-bipyridine-N,N')ruthenium(II) ion $\longrightarrow$ products				
Pulse rad. of $C_2H_5OH + H_2O$				78 Mul 1
KAS	H_2O, pH = 11...13	RT	$k = 7.0 \cdot 10^9\,M^{-1}s^{-1}$	
$CH_3\dot{C}HO^-$ + Ru(1,10-phenanthroline)$_3^{2+}$ $\longrightarrow$ Ru(1,10-phenanthroline)$_3^+$ + CH_3CHO				
Pulse rad. of $C_2H_5OH + N_2O + H_2O$				80 Ven 1
KAS	H_2O, pH = 12	RT	$k = 5.9 \cdot 10^9\,M^{-1}s^{-1}$	
$CH_3\dot{C}HO^- + 2Tl^+ \longrightarrow Tl_2^+ + CH_3CHO$				
Pulse rad. of $C_2H_5OH + H_2O$				80 But 1
KAS	H_2O, pH = 13	RT	$k = 1.5 \cdot 10^9\,M^{-1}s^{-1}$	

[94]) Mechanism discussed as e^--transfer.
[96]) 13% e^--transfer.
[97]) Protonated radical $CH_3\dot{C}HOH$ reacts with $k < 5 \cdot 10^6\,M^{-1}s^{-1}$ at pH = 6.
[98]) Protonated radical $CH_3\dot{C}HOH$ reacts with $k < 5 \cdot 10^6\,M^{-1}s^{-1}$.

Reaction Radical generation Method	Solvent	T[K]	Rate data	Ref./ add. ref.
$CH_3\dot{C}HO^- $ + acetophenone ($C_6H_5COCH_3$) ⟶ CH_3CHO + $(C_6H_5COCH_3)^{\dot{-}}$				
Pulse rad. of $C_2H_5OH + H_2O + N_2O$				73 Ada 1
KAS	H_2O	RT	$k = 1.1 \cdot 10^9\ M^{-1}\,s^{-1}$ [99])	
$CH_3\dot{C}HO^-$ + benzophenone ($(C_6H_5)_2CO$) ⟶ CH_3CHO + $(C_6H_5)_2\dot{C}O^-$				
Pulse rad. of $C_2H_5OH + H_2O + N_2O$				75 Bre 1,
KAS	H_2O, pH = 13	RT	$k = 2.6 \cdot 10^8\ M^{-1}\,s^{-1}$ $1 \cdot 10^9$ [1])	74 Mic 1
		291 ... 348	$E_a = 6\ kJ\,mol^{-1}$	
$CH_3\dot{C}HO^-$ + 4-carboxy-1-methylpyridinium ion ⟶ products				
Pulse rad. of $C_2H_5OH + H_2O + N_2O$				79 Ste 1
KAS	H_2O, pH = 12.7	RT	$k = 3.8 \cdot 10^9\ M^{-1}\,s^{-1}$	
$CH_3\dot{C}HO^-$ + 2-methyl-1,4-naphthoquinone ⟶ ...-semiquinone + products				
Pulse rad. of $C_2H_5OH + H_2O + N_2O$				73 Rao 3
KAS	H_2O, pH = 12.5	RT	$k = 4.2(4) \cdot 10^9\ M^{-1}\,s^{-1}$	
$CH_3\dot{C}HO^-$ + 9-methylpurine ⟶ CH_3CHO + (9-methylpurine)$^{\dot{-}}$				
Pulse rad. of $C_2H_5OH + H_2O + N_2O$				76 Moo 1
KAS	H_2O, pH = 13.6	RT	$k = 5.1 \cdot 10^8\ M^{-1}\,s^{-1}$ [2])	
$CH_3\dot{C}HO^-$ + nitrobenzene ($C_6H_5NO_2$) ⟶ $C_6H_5\dot{N}O_2^-$ + CH_3CHO				
Pulse rad. of $C_2H_5OH + N_2O + H_2O$				66 Asm 1,
KAS	H_2O, pH = 13	293	$k = 3.1 \cdot 10^9\ M^{-1}\,s^{-1}$	74 Mic 1
		291 ... 348	$E_a = 14\ kJ\,mol^{-1}$ [1])	
$CH_3\dot{C}HO^-$ + nitrosobenzene (C_6H_5NO) ⟶ $C_6H_5\dot{N}O^-$ + CH_3CHO				
Pulse rad. of $C_2H_5OH + N_2O + H_2O$				66 Asm 2
KAS	H_2O, pH = 13	RT	$k = 6.4 \cdot 10^9\ M^{-1}\,s^{-1}$	
$CH_3\dot{C}HO^-$ + 1,10-phenanthroline ⟶ (1,10-phenanthroline)$^{\dot{-}}$ + CH_3CHO				
Pulse rad. of alkaline $C_2H_5OH + N_2O + H_2O$				80 Tep 1
KAS	H_2O, basic pH	RT	$k = 8(2) \cdot 10^8\ M^{-1}\,s^{-1}$	
$CH_3\dot{C}HO^-$ + pterin ⟶ products [3])				
Pulse rad. of $C_2H_5OH + N_2O + H_2O$				76 Moo 1
KAS	H_2O, pH = 13	RT	$k = 1.2 \cdot 10^9\ M^{-1}\,s^{-1}$	
$CH_3\dot{C}HO^-$ + pyrene ⟶ (pyrene)$^{\dot{-}}$ + CH_3CHO				
Pulse rad. of C_2H_5OH + micellar (CTAB) solutions				76 Fra 1
KAS	micellar solution (H_2O, CTAB [4])), pH = 13	RT	$k = 1.8 \cdot 10^8\ M^{-1}\,s^{-1}$	
$CH_3\dot{C}HO^-$ + ^{3}(pyrene) [5]) ⟶ (pyrene)$^{\dot{-}}$ + CH_3CHO				
Combined pulse rad. and phot. of C_2H_5OH + micellar (CTAB) solutions				76 Fra 1
KAS	micellar solution (H_2O, CTAB[4])), pH = 13	RT	$k = 8 \cdot 10^9\ M^{-1}\,s^{-1}$	

[99]) k is pH dependent and extrapolated to pH = 14.
[1]) From [74 Mic 1].
[2]) $k < 2 \cdot 10^7$ at pH = 8.2 for conjugated acid $CH_3\dot{C}HOH$.
[3]) Reaction with anionic form of pterin; 100% e^--transfer.
[4]) $5 \cdot 10^{-3}$ M hexadecyltrimethyl ammonium bromide (CTAB).
[5]) Pyrene triplet.

Reaction Radical generation Method	Solvent	T[K]	Rate data	Ref./ add. ref.
$CH_3\dot{C}HOH + Ag_2^+ \longrightarrow CH_3CHO + H^+ + Ag_2$				
Pulse rad. of $C_2H_5OH + AgClO_4 + H_2O$				78 Tau 1
Time resolved conductivity	H_2O	RT	$k = 1.0 \cdot 10^9\ M^{-1}s^{-1}$ [6])	
$CH_3\dot{C}HOH + Co(NH_3)_6^{3+} \longrightarrow CH_3CHO + H^+ + Co(NH_3)_6^{2+}$				
Pulse rad. of $C_2H_5OH + H_2O + N_2O$				72 Coh 1
KAS	H_2O, pH = 5...6	RT	$k = 5.2(8) \cdot 10^7\ M^{-1}s^{-1}$	
$CH_3\dot{C}HOH + (Co(III)(NH_3)_5Br)^{2+} \longrightarrow$ products				
Pulse rad. of $C_2H_5OH + N_2O + H_2O$				77 Coh 1
KAS	H_2O, pH = 3.5...4	RT	$k = 1.5(2) \cdot 10^8\ M^{-1}s^{-1}$	
$CH_3\dot{C}HOH + (Co(III)(NH_3)_5Cl)^{2+} \longrightarrow$ products				
Pulse rad. of $C_2H_5OH + N_2O + H_2O$				77 Coh 1
KAS	H_2O, pH = 3.5...4.0	RT	$k = 3.0(4) \cdot 10^6\ M^{-1}s^{-1}$	
$CH_3\dot{C}HOH + [(NH_3)_4Co(III)(\mu O_2, \mu NH_2)Co(III)(NH_3)_4]^{4+} \longrightarrow H^+ + CH_3CHO$ + products [7])				
Pulse rad. of $C_2H_5OH + N_2O + H_2O$				81 Nat 1
KAS	H_2O, pH = 5	RT	$k = 1.0 \cdot 10^8\ M^{-1}s^{-1}$	
$CH_3\dot{C}HOH + [(EN)_2Co(III)(\mu O_2, \mu NH_2)Co(III)(EN)_2]^{4+} \longrightarrow H^+ + CH_3CHO$ + products [7])				
Pulse rad. of $C_2H_5OH + N_2O + H_2O$				81 Nat 1
KAS	H_2O, pH = 5	RT	$k = 2 \cdot 10^7\ M^{-1}s^{-1}$	
$CH_3\dot{C}HOH$ + *cis*-(ammine chlorobis(1,2-ethanediamine-N,N')cobalt(III))$^{2+} \longrightarrow$ products				
Pulse rad. of $C_2H_5OH + N_2O + H_2O$				77 Coh 1
KAS	H_2O, pH = 3.5...4.0	RT	$k = 4.2 \cdot 10^6\ M^{-1}s^{-1}$	
$CH_3\dot{C}HOH$ + *trans*-(dibromobis(1,2-ethanediamine-N,N')cobalt(III))$^+ \longrightarrow$ products				
Pulse rad. of $C_2H_5OH + N_2O + H_2O$				77 Coh 1
KAS	H_2O, pH = 3.5...4.0	RT	$k = 5.7 \cdot 10^8\ M^{-1}s^{-1}$	
$CH_3\dot{C}HOH$ + *cis*-(bromobis(1,2-ethanediamine-N,N')fluorocobalt(III))$^+ \longrightarrow$ products				
Pulse rad. of $C_2H_5OH + N_2O + H_2O$				77 Coh 1
KAS	H_2O, pH = 3.5...4.0	RT	$k = 2.8 \cdot 10^7\ M^{-1}s^{-1}$	
$CH_3\dot{C}HOH$ + *cis*-(aquachlorobis(1,2-ethanediamine-N,N')cobalt(III))$^{2+} \longrightarrow$ products				
Pulse rad. of $C_2H_5OH + N_2O + H_2O$				77 Coh 1
KAS	H_2O, pH = 3.5...4.0	RT	$k = 2.0 \cdot 10^7\ M^{-1}s^{-1}$	
$CH_3\dot{C}HOH$ + *cis*-(dichlorobis(1,2-ethanediamine-N,N')cobalt(III))$^+ \longrightarrow$ products				
Pulse rad. of $C_2H_5OH + N_2O + H_2O$				77 Coh 1
KAS	H_2O, pH = 3.5...4.0	RT	$k = 3.8. \cdot 10^7\ M^{-1}s^{-1}$	
$CH_3\dot{C}HOH$ + *trans*-(dichlorobis(1,2-ethanediamine-N,N')cobalt(III))$^+ \longrightarrow$ products				
Pulse rad. of $C_2H_5OH + N_2O + H_2O$				77 Coh 1
KAS	H_2O, pH = 3.5...4.0	RT	$k = 1.5 \cdot 10^8\ M^{-1}s^{-1}$	
$CH_3\dot{C}HOH + [(CN)_5Co(III)(\mu O_2)Co(III)(CN)_5]^{5-} \longrightarrow H^+ + CH_3CHO$ + products [7])				
Pulse rad. of $C_2H_5OH + N_2O + H_2O$				81 Nat 1
KAS	H_2O, pH = 5	RT	$k = 1.2 \cdot 10^8\ M^{-1}s^{-1}$	

[6]) Based on $2k(CH_3\dot{C}HOH + CH_3\dot{C}HOH) = 2.3 \cdot 10^9\ M^{-1}s^{-1}$ and assumed $[Ag_2^+]$, $[CH_3\dot{C}HOH]$ and $[H_2O_2]$ concentrations.
[7]) e^--transfer assumed to occur at dioxygen center. μ defines a bridging group.

Reaction Radical generation Method	Solvent	T[K]	Rate data	Ref./ add. ref.
$CH_3\dot{C}HOH + Cu^{2+} \longrightarrow Cu^{+}$ + products				
Pulse rad. of $C_2H_5OH + H_2O + N_2O$				72 Coh 1,
KAS	H_2O, pH = 5...6	RT	$k = 7.4(11)\cdot10^{7}\,M^{-1}s^{-1}$	78 Bux 1
	pH = 2...5		$9.4(19)\cdot10^{7}$ [8])	
$CH_3\dot{C}HOH + Fe^{3+} \longrightarrow Fe^{2+} + H^{+} + CH_3CHO$				
γ-rad. of $C_2H_5OH + H_2O$				77 Ber 1
PR, competition kinetics	H_2O, pH ≈ 1	RT	$k = 2.7\cdot10^{8}\,M^{-1}s^{-1}$ [9])	
$CH_3\dot{C}HOH + Fe(CN)_6^{3-} \longrightarrow CH_3CHO + H^{+} + Fe(CN)_6^{4-}$				
Pulse rad. of $C_2H_5OH + H_2O$				69 Ada 1,
KAS	H_2O	RT	$k = 5.3\cdot10^{9}\,M^{-1}s^{-1}$	79 Alm 1
			$4.0\cdot10^{9}$ [10])	
$CH_3\dot{C}HOH$ + cytochrome c(Fe(III)) $\longrightarrow CH_3CHO + H^{+}$ + cytochrome c(Fe(II))				
Pulse rad. of $C_2H_5OH + H_2O + N_2O$				74 Sha 1/
KAS	H_2O	RT	$k = 1.8(2)\cdot10^{8}\,M^{-1}s^{-1}$	77 Sha 1, 79 Lee 1, 79 Ila 1
$CH_3\dot{C}HOH$ + cytochrome c(acetylated)(Fe(III)) $\longrightarrow$ products				
Pulse rad. of $C_2H_5OH + H_2O + N_2O$				79 Ila 1
KAS	H_2O		$k = 2.5\cdot10^{8}\,M^{-1}s^{-1}$	
$CH_3\dot{C}HOH$ + cytochrome c(dicarboxymethyl)(Fe(III)) $\longrightarrow$ products				
Pulse rad. of $C_2H_5OH + H_2O + N_2O$				79 Ila 1
KAS	H_2O	RT	$k = 3.5\cdot10^{8}\,M^{-1}s^{-1}$	
$CH_3\dot{C}HOH$ + cytochrome c(succinylated)(Fe(III)) $\longrightarrow$ products				
Pulse rad. of $C_2H_5OH + H_2O + N_2O$				79 Ila 1
KAS	H_2O	RT	$k = 1.8\cdot10^{9}\,M^{-1}s^{-1}$	
$CH_3\dot{C}HOH$ + hemin(Fe(III)) $\longrightarrow CH_3CHO + H^{+}$ + hemin(Fe(II))				
Pulse rad. of $C_2H_5OH + H_2O + N_2O$				78 Eve 1
KAS	H_2O (micellar solution) [11]),	RT		
	pH = 9.2		$k = 5.6(6)\cdot10^{8}\,M^{-1}s^{-1}$	
	pH = 4		$1.6(6)\cdot10^{9}$	
$CH_3\dot{C}HOH$ + hemoglobin(Fe(III)) $\longrightarrow CH_3CHO + H^{+}$ + hemoglobin(Fe(II))				
Pulse rad. of $C_2H_5OH + H_2O + N_2O$				79 Lee 1
KAS	H_2O	RT	$k = 4.0(4)\cdot10^{7}\,M^{-1}s^{-1}$	
$CH_3\dot{C}HOH$ + metmyoglobin(Fe(III)) $\longrightarrow CH_3CHO + H^{+}$ + metmyoglobin(Fe(II))				
Pulse rad. of $C_2H_5OH + H_2O + N_2O$				79 Lee 1
KAS	H_2O	RT	$k = 5.5(5)\cdot10^{7}\,M^{-1}s^{-1}$	
$CH_3\dot{C}HOH + H_2O_2 \longrightarrow CH_3CHO + H_2O + \dot{O}H$				
γ-rad. of $C_2H_5OH + H_2O$				67 Sed 1
PR	H_2O	RT	$k = 1.5\cdot10^{5}\,M^{-1}s^{-1}$ [12])	

[8]) Based on formation kinetics of Cu(I)$CH_2CHCONH_2$ in presence of acrylamide [78 Bux 1].
[9]) Based on $k(CH_3\dot{C}HOH + C(NO_2)_4) = 5\cdot10^{9}\,M^{-1}s^{-1}$.
[10]) From [79 Alm 1].
[11]) In presence of 0.2 M sodium dodecyl sulfate (micelles).
[12]) Based on $2k(CH_3\dot{C}HOH + CH_3\dot{C}HOH) = 2.0\cdot10^{9}\,M^{-1}s^{-1}$.

Reaction Radical generation Method	Solvent	T[K]	Rate data	Ref./ add. ref.
$CH_3\dot{C}HOH + HgI_2 \longrightarrow HgI + I^- + H^+ + CH_3CHO$				
Pulse rad. of $C_2H_5OH + H_2O + N_2O$				78 Fuj 1
KAS	H_2O	RT	$k = 7.0(25) \cdot 10^8\ M^{-1} s^{-1}$	
$CH_3\dot{C}HOH + HgBr_2 \xrightarrow{a} HgBr + Br^- + H^+ + CH_3CHO$ $\dot{C}O_2^- + HgBr_2 \xrightarrow{b} Hg(I)\ldots + CO_2$				
Pulse rad. of formate + $C_2H_5OH + N_2O + H_2O$				76 Fuj 1
PR	H_2O	RT	$k_a/k_b = 1.15$	
$CH_3\dot{C}HOH + IrCl_6^{2-} \longrightarrow$ Ir(III) + products				
Pulse rad. of $C_2H_5OH + N_2O + H_2O$				82 Ste 1
KAS	H_2O	295	$k = 4.5 \cdot 10^9\ M^{-1} s^{-1}$	
$CH_3\dot{C}HOH + Ru(NH_3)_6^{3+} \longrightarrow Ru(NH_3)_6^{2+} + H^+ + CH_3CHO$				
Pulse rad. of $C_2H_5OH + H_2O + N_2O$				72 Coh 1
KAS	H_2O	RT	$k = 5.5(16) \cdot 10^8\ M^{-1} s^{-1}$	
$CH_3\dot{C}HOH + (Ru(III)(NH_3)_5Cl)^{2+} \longrightarrow$ products				
Pulse rad. of $C_2H_5OH + N_2O + H_2O$				77 Coh 1
KAS	H_2O, pH = 3.5…4.0	RT	$k = 8.0(12) \cdot 10^8\ M^{-1} s^{-1}$	
$CH_3\dot{C}HOH$ + 1,4-benzoquinone $\longrightarrow CH_3CHO$ + …semiquinone				
Pulse rad. of $C_2H_5OH + H_2O + N_2O$				71 Wil 1
KAS	H_2O	RT	$k = 4.5 \cdot 10^9\ M^{-1} s^{-1}$	
$CH_3\dot{C}HOH$ + 2,3-butanedione $(CH_3COCOCH_3) \longrightarrow CH_3CHO + H^+ + (CH_3COCOCH_3)^{\dot{-}}$				
Pulse rad. of $C_2H_5OH + H_2O + N_2O$				68 Lil 1
KAS	H_2O	RT	$k = 5.6 \cdot 10^8\ M^{-1} s^{-1}$	
$CH_3\dot{C}HOH$ + eosin $\longrightarrow CH_3CHO + H^+$ + (eosin)$^{\dot{-}}$				
Pulse rad. of $C_2H_5OH + H_2O_2 + H_2O$				67 Chr 1
KAS	H_2O, pH = 9	RT	$k = 1.1(2) \cdot 10^9\ M^{-1} s^{-1}$	
$CH_3\dot{C}HOH$ + fluorescein $\longrightarrow CH_3CHO + H^+$ + (fluorescein)$^{\dot{-}}$ [13])				
Pulse rad. of $C_2H_5OH + H_2O + N_2O$				73 Rao 2
KAS	H_2O, pH = 10.8	RT	$k = 4.5 \cdot 10^8\ M^{-1} s^{-1}$	
$CH_3\dot{C}HOH$ + lipoate ion(—S—S—) $\longrightarrow CH_3CHO + H^+ + -S\overset{(\dot{-})}{-}S-$				
Pulse rad. of $C_2H_5OH + H_2O + N_2O$				70 Wil 1
KAS	H_2O	RT	$k = 1.0 \cdot 10^8\ M^{-1} s^{-1}$	
$CH_3\dot{C}HOH$ + 2-methyl-1,4-naphthoquinone $\longrightarrow$ …-semiquinone + products				
Pulse rad. of $C_2H_5OH + H_2O + N_2O$				73 Rao 3
KAS	H_2O	RT	$k = 3.8(4) \cdot 10^9\ M^{-1} s^{-1}$	
$CH_3\dot{C}HOH$ + 3-methylpterin $\longrightarrow$ products [14])				
Pulse rad. of $C_2H_5OH + H_2O + N_2O$				76 Moo 1
KAS	H_2O	RT	$k = 3.2 \cdot 10^7\ M^{-1} s^{-1}$	
$CH_3\dot{C}HOH$ + 4-nitroacetophenone(PNAP) $\longrightarrow PNAP^{\dot{-}} + H^+ + CH_3CHO$				
Pulse rad. of $C_2H_5OH + N_2O + H_2O$				73 Ada 1
KAS	H_2O, pH = 11	RT	$k = 8(2) \cdot 10^8\ M^{-1} s^{-1}$	

[13]) Semiquinone form.
[14]) 55% e^--transfer.

Reaction Radical generation Method	Solvent	T[K]	Rate data	Ref./ add. ref.
$CH_3\dot{C}HOH$ + nitrobenzene ($C_6H_5NO_2$) ⟶ $C_6H_5\dot{N}O_2^-$ + H^+ + CH_3CHO				
Pulse rad. of $C_2H_5OH + H_2O + N_2O$ KAS	H_2O, pH = 7 pH = 5...6	RT	$k = 3.3 \cdot 10^8\ M^{-1}s^{-1}$ $k = 2.4(7) \cdot 10^8\ M^{-1}s^{-1}$ [15]	66 Asm 1, 72 Coh 1
$CH_3\dot{C}HOH$ + *anti*-5-nitro-2-furaldoxime [16] ⟶ products [17]				
Pulse rad. of $C_2H_5OH + N_2O + H_2O$ KAS	H_2O	RT	$k > 1.1 \cdot 10^9\ M^{-1}s^{-1}$	73 Gre 1
$CH_3\dot{C}HOH$ + nitrosobenzene (C_6H_5NO) ⟶ $C_6H_5\dot{N}OH$ + CH_3CHO				
Pulse rad. of $C_2H_5OH + N_2O + H_2O$ KAS	H_2O	RT	$k = 3.9 \cdot 10^9\ M^{-1}s^{-1}$	66 Asm 2
$CH_3\dot{C}HOH$ + pterin ⟶ products [18]				
Pulse rad. of $C_2H_5OH + N_2O + H_2O$ KAS	H_2O	RT	$k = 3.7 \cdot 10^7\ M^{-1}s^{-1}$	76 Moo 1
$CH_3\dot{C}HOH$ + quinoxaline(Qx) ⟶ $(QxH)^\cdot$ + CH_3CHO [19]				
Pulse rad. of $C_2H_5OH + N_2O + H_2O$ KAS	H_2O	RT	$k = 6.5 \cdot 10^7\ M^{-1}s^{-1}$	74 Moo 1
$CH_3\dot{C}HOH$ + tetrachlorobenzoquinone (chloranil) ⟶ products				
Pulse rad. of C_2H_5OH + micellar solutions KAS	micellar solution (H_2O, CTAB or NaLS) [20]	RT	$k = (2.9 \ldots 3.4) \cdot 10^9\ M^{-1}s^{-1}$	76 Fra 2
$CH_3\dot{C}HOH$ + 1,2,4,5-tetracyanobenzene ⟶ $C_6H_2(CN)_4^{\cdot-}$ + H^+ + CH_3CHO				
Pulse rad. of C_2H_5OH + micellar solutions KAS	micellar solution (H_2O, CTAB or NaLS) [20]	RT	$k = (3.3 \ldots 4.1) \cdot 10^9\ M^{-1}s^{-1}$	76 Fra 2
$CH_3\dot{C}HOH$ + 2,2,6,6-tetramethyl-4-hydroxy-1-piperidinyloxy(TMPN) ⟶ products				
Pulse rad. of $C_2H_5OH + N_2O + H_2O$ Cond. (time resolved)	H_2O, pH = 3...5	RT	$k = 4.9(5) \cdot 10^8\ M^{-1}s^{-1}$	76 Asm 1
$CH_3\dot{C}HOH$ + 2,2,6,6-tetramethyl-4-oxo-1-piperidinyloxy(TAN) ⟶ products				
Pulse rad. of $C_2H_5OH + N_2O + H_2O$ Cond. (time resolved)	H_2O, pH = 3...5	RT	$k = 4.0(4) \cdot 10^8\ M^{-1}s^{-1}$	76 Asm 1/ 71 Wil 2
$CH_3\dot{C}HOH$ + 2,2,5,5-tetramethyl-1-pyrrolidinyloxy-3-carboxamide ⟶ products				
Pulse rad. of $C_2H_5OH + N_2O + H_2O$ Cond. (time resolved)	H_2O, pH = 3...5	RT	$k = 4.3(4) \cdot 10^8\ M^{-1}s^{-1}$	76 Nig 1

[15]) From [72 Coh 1].
[16]) Nifuroxime.
[17]) > 75% e^--transfer.
[18]) Reaction with neutral form of pterin; 40% e^--transfer.
[19]) 70% e^--transfer.
[20]) Hexadecyltetramethyl ammonium bromide (CTAB) and sodium laurylsulfate (NaLS).

Reaction Radical generation Method	Solvent	T[K]	Rate data	Ref./ add. ref.
$CH_3\dot{C}HOH$ + 2,2,5,5-tetramethyl-3-pyrrolin-1-yloxy-3-carboxamide ⟶ products				
Pulse rad. of $C_2H_5OH + N_2O + H_2O$				76 Nig 1
Cond. (time resolved)	H_2O, pH = 3…5	RT	$k = 6.2(6) \cdot 10^8\ M^{-1}s^{-1}$	
$CH_3\dot{C}HOH$ + tetranitromethane($C(NO_2)_4$) ⟶ $C(NO_2)_3^- + NO_2 + H^+ + CH_3CHO$				
Pulse rad. of $C_2H_5OH + H_2O$				65 Rab 1,
KAS	H_2O, pH ≈ 1	RT	$k = 5.6 \cdot 10^9\ M^{-1}s^{-1}$	76 Fra 2
			$\approx 3.5 \cdot 10^9$ [21])	
$\dot{C}H_2CH_2OH + Cu^{2+}$ ⟶ Cu^+ + products				
Pulse rad. of $C_2H_5OH + H_2O + N_2O$				78 Bux 1
KAS	H_2O, pH = 4.5	RT	$k = 1.9(4) \cdot 10^7\ M^{-1}s^{-1}$	
	pH = 2		$k = 2.2(4) \cdot 10^7\ M^{-1}s^{-1}$	
$\dot{C}H_2CH_2OH + CuC_2H_4^+$ ⟶ $2C_2H_4 + Cu^{2+} + OH^-$				
Pulse rad. of $C_2H_4 + H_2O$				78 Bux 1
KAS	H_2O, pH = 4.5	RT	$k = 7.8(25) \cdot 10^7\ M^{-1}s^{-1}$ [22])	
$\dot{C}H_2CH_2OH + IrCl_6^{2-}$ ⟶ Ir(III) + products				
Pulse rad. of $C_2H_4 + N_2O + H_2O$				82 Ste 1 [23])
KAS	H_2O	295	$k \approx 2 \cdot 10^9\ M^{-1}s^{-1}$	
$\dot{C}H_2OCH_3 + Fe(CN)_6^{3-}$ ⟶ Fe(II) + products				
Pulse rad. of $CH_3OCH_3 + N_2O + H_2O$				82 Ste 1
KAS	H_2O	295	$k = 4.3 \cdot 10^9\ M^{-1}s^{-1}$	
$\dot{C}H_2OCH_3 + IrCl_6^{2-}$ ⟶ Ir(III) + products				
Pulse rad. of $CH_3OCH_3 + N_2O + H_2O$				82 Ste 1
KAS	H_2O	295	$k = 6.5 \cdot 10^9\ M^{-1}s^{-1}$	
$\dot{C}HOHCH_2OH + Fe(CN)_6^{3-}$ ⟶ $Fe(CN)_6^{4-}$ + products				
Pulse rad. of ethylene glycol + H_2O				69 Ada 1
KAS	H_2O	RT	$k = 3.6 \cdot 10^9\ M^{-1}s^{-1}$	
$\dot{C}O^-(COO^-)_2$ + 2-methyl-1,4-naphthoquinone ⟶ …-semiquinone + products				
Pulse rad. of ketomalonate + t-butanol + H_2O				73 Rao 3
KAS	H_2O, pH = 9.2	RT	$k = 2.5(3) \cdot 10^9\ M^{-1}s^{-1}$	
$CH_3\dot{C}(O^-)COO^- + Co(NH_3)_6^{3+}$ ⟶ $Co(NH_3)_6^{2+}$ + …				
Pulse rad. of lactate + $H_2O + N_2O$				72 Coh 1
KAS	H_2O, pH = 12	RT	$k = 2.4 \cdot 10^{10}\ M^{-1}s^{-1}$	
$CH_3\dot{C}(O^-)COO^-$ + 9,10-anthraquinone-2,6-disulfonate ⟶ …semiquinone + products				
Pulse rad. of acetoacetate + t-butanol + H_2O + Ar				73 Rao 1
KAS	H_2O, pH = 9.2	RT	$k = 7.2(7) \cdot 10^8\ M^{-1}s^{-1}$	
$CH_3\dot{C}(O^-)COO^-$ + 9,10-anthraquinone-2-sulfonate ⟶ …semiquinone + products				
Pulse rad. of acetoacetate + t-butanol + H_2O + Ar				73 Rao 1
KAS	H_2O, pH = 9.2	RT	$k = 2.1 \cdot 10^9\ M^{-1}s^{-1}$	

[21]) Solutions contained also 0.1 M sodium dodecyl sulfate or 0.02 M dodecyltrimethylammonium chloride [76 Fra 2].
[22]) Based on effect of C_2H_4 concentration on formation and decay of $CuC_2H_4^+$.
[23]) Mechanism discussed as e^-- or $Cl^\cdot$-transfer.

Reaction Radical generation Method	Solvent	T[K]	Rate data	Ref./ add. ref.
$CH_3\dot{C}(O^-)COO^-$ + 2-methyl-1,4-naphthoquinone ⟶ ...semiquinone + products [24])				
Pulse rad. of lactate + H_2O + N_2O				73 Rao 3
KAS	H_2O, pH = 10.6	RT	$k = 1.9(2) \cdot 10^9\ M^{-1}s^{-1}$	
$\dot{C}H(COOH)_2$ + Ti(III) ⟶ Ti(IV) + products				
In situ rad. of malonic acid + H_2SO_4 + H_2O				73 Beh 1
PR, ESR	H_2O, pH ≈ 0	RT	$k = 8 \cdot 10^6\ M^{-1}s^{-1}$	
$CH_3\dot{C}(OH)COO^-$ + $Co(NH_3)_6^{3+}$ ⟶ $Co(NH_3)_6^{2+}$ + products				
Pulse rad. of lactate + H_2O + N_2O				72 Coh 1
KAS	H_2O, pH = 6	RT	$k = 7.0 \cdot 10^6\ M^{-1}s^{-1}$	
$CH_3\dot{C}(OH)COO^-$ + $Fe(CN)_6^{3-}$ ⟶ $Fe(CN)_6^{4-}$ + products				
Pulse rad. of lactate + H_2O				69 Ada 1
KAS	H_2O	RT	$k = 1.5 \cdot 10^9\ M^{-1}s^{-1}$	
$CH_3\dot{C}(OH)COO^-$ + Fe(III)cytochrome c ⟶ Fe(II)cytochrome c + products				
Pulse rad. of lactate + H_2O + N_2O				74 Sha 1/
KAS	H_2O	RT	$k = 2.4(2) \cdot 10^8\ M^{-1}s^{-1}$	75 Sim 1
$CH_3\dot{C}(OH)COO^-$ + hemin c(Fe(III)) ⟶ hemin c(Fe(II)) + products				
Pulse rad. of lactate + H_2O + N_2O				75 Gof 1
KAS	H_2O	RT	$k = 5.6(11) \cdot 10^8\ M^{-1}s^{-1}$	
$CH_3\dot{C}(OH)COO^-$ + $Ru(NH_3)_6^{3+}$ ⟶ $Ru(NH_3)_6^{2+}$ + products				
Pulse rad. of lactate + N_2O + H_2O				72 Coh 1
KAS	H_2O	RT	$k = 2.5(4) \cdot 10^9\ M^{-1}s^{-1}$	
$CH_3\dot{C}(OH)COO^-$ + 1,4-benzoquinone ⟶ 1,4-benzosemiquinone + products				
Pulse rad. of lactate + H_2O + N_2O				73 Hay 1
KAS	H_2O	RT	$k = 6.5 \cdot 10^9\ M^{-1}s^{-1}$	
$CH_3\dot{C}(OH)COO^-$ + 2,3-butanedione ($CH_3COCOCH_3$) ⟶ products				
Pulse rad. of lactate + H_2O + N_2O				72 Coh 1
KAS	H_2O	RT	$k = 2.8 \cdot 10^7\ M^{-1}s^{-1}$	
$CH_3\dot{C}(OH)COO^-$ + 2-methyl-1,4-naphthoquinone ⟶ ...-semiquinone + products [25])				
Pulse rad. of lactate + H_2O + N_2O				73 Rao 3
KAS	H_2O	RT	$k = 1.4(1) \cdot 10^9\ M^{-1}s^{-1}$	
$CH_3\dot{C}(OH)COO^-$ + *anti*-5-nitro-2-furaldoxime [26]) ⟶ (*anti*-...)$^{\dot{-}}$ + products				
Pulse rad. of lactate + N_2O + H_2O				73 Gre 1
KAS	H_2O	293	$k = 1.5 \cdot 10^9\ M^{-1}s^{-1}$	
$\dot{C}H_2COCH_3$ + Fe^{2+} ⟶ Fe^{3+} + $^-CH_2COCH_3$ [27])				
Fe(II)/H_2O_2 in acetone + H_2O				73 Wal 1
PR	H_2O	RT	$k = 1 \cdot 10^7\ M^{-1}s^{-1}$ [28])	
$\dot{C}H_2COCH_3$ + Fe(II) ⟶ Fe(III) + $^-CH_2COCH_3$ [27])				
Fe(II)/H_2O_2 flow expts. in acetone + H_2O				73 Gil 1
SESR	H_2O, pH = 1	RT	$k \leqslant 10^5\ M^{-1}s^{-1}$	

[24]) 72% e^--transfer.
[25]) 55% e^--transfer.
[26]) Nifuroxime.
[27]) Immediate protonation.
[28]) Calculated on the basis of various assumptions.

Reaction Radical generation Method	Solvent	T[K]	Rate data	Ref./ add. ref.
$\dot{C}H_2COCH_3 + Ti(III) \longrightarrow Ti(IV) + {}^-CH_2COCH_3$ [27]				
$Ti(III)/H_2O_2$ flow expts. with acetone + H_2O				73 Gil 1
SESR	H_2O, pH = 1	RT	$k \leqslant 10^6\,M^{-1}\,s^{-1}$	
$\dot{C}H_2COCH_3$ + Ti(III)-EDTA $\longrightarrow$ Ti(IV)-EDTA + $^-CH_2COCH_3$ [27]				
$Ti(III)/H_2O_2$ flow expts. with acetone + H_2O				73 Gil 1
SESR	H_2O, pH = 7	RT	$k = 1.4 \cdot 10^7\,M^{-1}\,s^{-1}$ [29]	
$\left.\begin{matrix} CH_3\dot{C}HCHO \\ CH_3CO\dot{C}H_2 \end{matrix}\right\}$ [30] + hydroquinone $\longrightarrow {}^-OC_6H_4\dot{O} + \left\{\begin{matrix} CH_3CH_2CHO \\ CH_3COCH_3 \end{matrix}\right.$				
Pulse rad. of 1,2-propanediol + N_2O + H_2O				79 Ste 1
KAS	H_2O	RT	$k = 1.2(1) \cdot 10^9\,M^{-1}\,s^{-1}$	
$HOCH_2\dot{C}HCHO$ [31] + hydroquinone $\longrightarrow HOCH_2CH_2CHO + {}^-OC_6H_4\dot{O}$				
Pulse rad. of glycerol 2-phosphate + N_2O + H_2O				79 Ste 1
KAS	H_2O, pH ≈ 11.5	RT	$k = 1.7(2) \cdot 10^9\,M^{-1}\,s^{-1}$	
$\left.\begin{matrix} HOCH_2\dot{C}HCHO \\ HOCH_2CO\dot{C}H_2 \end{matrix}\right\}$ [30] + hydroquinone $\longrightarrow {}^-OC_6H_4\dot{O} + \left\{\begin{matrix} HOCH_2CH_2CHO \\ HOCH_2COCH_3 \end{matrix}\right.$				
Pulse rad. of 2,3-epoxypropanol (A) or glycerol (B) + N_2O + H_2O				79 Ste 1
KAS	H_2O, pH ≈ 11.5	RT	$k = 1.3(1) \cdot 10^9\,M^{-1}\,s^{-1}$ (A) $1.5(1) \cdot 10^9$ (B)	
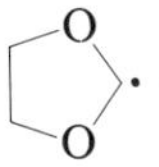 $\cdot + H_2O_2 \longrightarrow$ products				
$Ti(III)/H_2O_2$ flow expt. with dioxolan + H_2O				74 Gil 1
SESR	H_2O	RT	$k > 6 \cdot 10^4\,M^{-1}\,s^{-1}$ [32]	
$CH_3\dot{C}OHCOOH$ + 9,10-anthraquinone-2,6-disulfonate ion $\longrightarrow$ ...semiquinone + products [33]				
Pulse rad. of lactate + H_2O + N_2O				75 Rao 1
KAS	H_2O	RT	$k = 3.0 \cdot 10^9\,M^{-1}\,s^{-1}$	
$CH_3CH_2\dot{C}HO^-$ + nitrobenzene ($C_6H_5NO_2$) $\longrightarrow C_6H_5\dot{N}O_2^- + CH_3CH_2CHO$				
Pulse rad. of 1-propanol + N_2O + H_2O				66 Asm 1
KAS	H_2O, pH = 13	RT	$k = 3.1 \cdot 10^9\,M^{-1}\,s^{-1}$	
$(CH_3)_2\dot{C}O^- + Co(NH_3)_6^{3+} \longrightarrow (CH_3)_2CO + Co(NH_3)_6^{2+}$				
Pulse rad. of 2-propanol + H_2O + N_2O				72 Coh 1
KAS	H_2O, pH = 12	RT	$k = 5.0(8) \cdot 10^9\,M^{-1}\,s^{-1}$	
$(CH_3)_2\dot{C}O^-$ + Co(III)-tetra(4-N-methylpyridyl)porphyrine $\longrightarrow (CH_3)_2CO$ + Co(II)-tetra...				
Pulse rad. of 2-propanol + N_2O + H_2O				81 Net 2
KAS	H_2O, pH = 13	RT	$k = 7.0 \cdot 10^9\,M^{-1}\,s^{-1}$	
$(CH_3)_2\dot{C}O^-$ + Co(III)-tetra(4-sulfonatophenyl)porphyrine $\longrightarrow (CH_3)_2CO$ + Co(II)-tetra...				
Pulse rad. of 2-propanol + N_2O + H_2O				81 Net 2
KAS	H_2O, pH = 13	RT	$k = 1.1 \cdot 10^9\,M^{-1}\,s^{-1}$	

[27]) Immediate protonation.
[29]) $k > 4 \cdot 10^6\,M^{-1}\,s^{-1}$ from pulse rad. expts.
[30]) Mixt. of C-centered radicals, oxidation action likely to occur through mesomeric O-centered radical.
[31]) Mainly C-centered radical, oxidation likely to occur through mesomeric O-centered radical.
[32]) Calculated rate constant assuming $2k(\dot{R} + \dot{R}) = 3 \cdot 10^9\,M^{-1}\,s^{-1}$.
[33]) 58% e^--transfer.

Reaction Radical generation Method	Solvent	T[K]	Rate data	Ref./ add. ref.
$(CH_3)_2\dot{C}O^- + $ Co(III)-tetra(4-(N,N,N-trimethylamino)phenyl)porphyrine $\longrightarrow (CH_3)_2CO + $ Co(II)-tetra...				
Pulse rad. of 2-propanol + $N_2O + H_2O$				81 Net 2
KAS	H_2O, pH = 13	RT	$k = 3.4 \cdot 10^9\ M^{-1}s^{-1}$	
$(CH_3)_2\dot{C}O^- + $ (5,7,7,12,14,14-hexamethyl-1,4,8,11-tetraazacyclotetradeca-4,11-diene-N,N′,N″,N‴)copper(II) ion $\longrightarrow$ Cu(I)...				
Pulse rad. of 2-propanol + $H_2O + N_2O$				76 Tai 3
KAS	H_2O, pH = 12.5	RT	$k = 9.0 \cdot 10^8\ M^{-1}s^{-1}$ [34])	
$(CH_3)_2\dot{C}O^- + $ deuterohemin(DPFe(III)) [35]) $\longrightarrow$ DPFe(II) + $(CH_3)_2CO$				
Pulse rad. of 2-propanol + NaOH + H_2O				81 Bra 1
KAS	H_2O/2-propanol (1:1) mixt. [36])	RT	$k = 9(1) \cdot 10^8\ M^{-1}s^{-1}$	
$(CH_3)_2\dot{C}O^- + $ hematoporphyrine(Fe(III)) $\longrightarrow$ products				
Pulse rad. of 2-propanol + $H_2O + N_2O$				74 Har 1
KAS	H_2O, pH = 13	RT	$k = 1.1(2) \cdot 10^9\ M^{-1}s^{-1}$	
$(CH_3)_2\dot{C}O^- + Ga(OH)_6^{4-} \longrightarrow$ products				
Pulse rad. of 2-propanol + $Ga^{3+} + H_2O$				79 Suk 1
KAS	H_2O	RT	$k = 1.7 \cdot 10^9\ M^{-1}s^{-1}$	
$(CH_3)_2\dot{C}O^- + N_2O \longrightarrow (CH_3)_2CO + $ products				
Rad. of 2-propanol + $N_2O + H_2O$				72 Bur 1
PR	H_2O	RT	$k = 3.8(4) \cdot 10^4\ M^{-1}s^{-1}$	
$(CH_3)_2\dot{C}O^- + $ (tris-(2,2′-bipyridine-N,N′)ruthenium(II))$^{2+} \longrightarrow$ (tris-(2,2-bipyridine-N,N′)ruthenium(I))$^+ + (CH_3)_2CO$				
Pulse rad. of 2-propanol + H_2O				78 Mul 1
KAS	H_2O, pH = 11...13	RT	$k = 4.9 \cdot 10^9\ M^{-1}s^{-1}$	
$(CH_3)_2\dot{C}O^- + $ Ru(1,10-phenanthroline)$_3^{2+} \longrightarrow$ Ru(1,10-phenanthroline)$_3^+ + (CH_3)_2CO$				
Pulse rad. of 2-propanol + $N_2O + H_2O$				80 Ven 1
KAS	H_2O, pH = 13	RT	$k = 3.7 \cdot 10^8\ M^{-1}s^{-1}$	
$(CH_3)_2\dot{C}O^- + 2Tl^+ \longrightarrow Tl_2^+ + (CH_3)_2CO$				
Pulse rad. of 2-propanol + H_2O				80 But 1
KAS	H_2O, pH = 13	RT	$k = 3.0 \cdot 10^9\ M^{-1}s^{-1}$	
$(CH_3)_2\dot{C}O^- + $ zinc hematoporphyrine $\longrightarrow$ products				
Pulse rad. of 2-propanol + $H_2O + N_2O$				74 Har 1
KAS	H_2O, pH = 13	RT	$k = 1.0(1) \cdot 10^9\ M^{-1}s^{-1}$	
$(CH_3)_2\dot{C}O^- + $ Zn-tetra(4-N-methylpyridyl)porphyrine $\longrightarrow (CH_3)_2CO + $ (Zn-tetra...)$^{\dot{-}}$				
Pulse rad. of 2-propanol + $N_2O + H_2O$				81 Net 2
KAS	H_2O, pH = 13	RT	$k = 6.7 \cdot 10^9\ M^{-1}s^{-1}$	
$(CH_3)_2\dot{C}O^- + $ Zn-tetra(4-sulfonatophenyl)porphyrine $\longrightarrow (CH_3)_2CO + $ (Zn-tetra...)$^{\dot{-}}$				
Pulse rad. of 2-propanol + $N_2O + H_2O$				81 Net 2
KAS	H_2O, pH = 13	RT	$k = 1.0 \cdot 10^9\ M^{-1}s^{-1}$	

[34]) No reaction in neutral solution.
[35]) Ferrideuteroporphyrine(IX)chloride.
[36]) 0.1 M in NaOH.

Reaction Radical generation Method	Solvent	T[K]	Rate data	Ref./ add. ref.
$(CH_3)_2\dot{C}O^- +$ Zn-tetra(4-(N,N,N-trimethylamino)phenyl)porphyrine $\longrightarrow (CH_3)_2CO +$ (Zn-tetra...)$^{\dot{-}}$				
Pulse rad. of 2-propanol + N_2O + H_2O				81 Net 2
KAS	H_2O, pH = 13	RT	$k = 3.5 \cdot 10^9\ M^{-1}\,s^{-1}$	
$(CH_3)_2\dot{C}O^- +$ acetophenone $(C_6H_5COCH_3) \longrightarrow (CH_3)_2CO + (C_6H_5COCH_3)^{\dot{-}}$				
Pulse rad. of 2-propanol + H_2O + N_2O				73 Ada 1/
KAS	H_2O	RT	$k = 9 \cdot 10^8\ M^{-1}\,s^{-1}$ [37]	67 Ada 1
$(CH_3)_2\dot{C}O^- +$ acridine $(C_{13}H_9N) \longrightarrow (CH_3)_2CO +$ (acridine)$^{\dot{-}}$				
Pulse rad. of 2-propanol + H_2O + N_2O				79 Net 1
KAS	H_2O, pH = 13	RT	$k = 3 \cdot 10^9\ M^{-1}\,s^{-1}$	
$(CH_3)_2\dot{C}O^- +$ azobenzene $\longrightarrow (CH_3)_2CO +$ (azobenzene)$^{\dot{-}}$				
Pulse rad. of 2-propanol + H_2O + N_2O				77 Net 1
KAS	H_2O, pH = 14	RT	$k = 2 \cdot 10^9\ M^{-1}\,s^{-1}$	
$(CH_3)_2\dot{C}O^- +$ benzophenone $((C_6H_5)_2CO) \longrightarrow (CH_3)_2CO + (C_6H_5)_2\dot{C}O^-$				
Pulse rad. of 2-propanol + H_2O + N_2O				68 Ada 1,
KAS	H_2O, pH = 12	RT	$k = 1.2 \cdot 10^9\ M^{-1}\,s^{-1}$	75 Bre 1/
	pH = 13		$7.0 \cdot 10^8$ [38]	72 Nel 1
$(CH_3)_2\dot{C}O^- +$ 2-benzoylpyridine $(C_6H_5COC_5H_4N) \longrightarrow (CH_3)_2CO + C_6H_5\dot{C}(O^-)C_5H_4N$				
Pulse rad. of 2-propanol + H_2O + N_2O				72 Nel 1
KAS	H_2O, pH = 13.2	RT	$k = 2.3(2) \cdot 10^9\ M^{-1}\,s^{-1}$	
$(CH_3)_2\dot{C}O^- +$ 3-benzoylpyridine $(C_6H_5COC_5H_4N) \longrightarrow (CH_3)_2CO + C_6H_5\dot{C}(O^-)C_5H_4N$				
Pulse rad. of 2-propanol + H_2O + N_2O				72 Nel 1
KAS	H_2O, pH = 13	RT	$k = 2.0(2) \cdot 10^9\ M^{-1}\,s^{-1}$	
$(CH_3)_2\dot{C}O^- +$ 4-benzoylpyridine $(C_6H_5COC_5H_4N) \longrightarrow (CH_3)_2CO + C_6H_5\dot{C}(O^-)C_5H_4N$				
Pulse rad. of 2-propanol + H_2O + N_2O				72 Nel 1
KAS	H_2O, pH = 13.2	RT	$k = 2.5(2) \cdot 10^9\ M^{-1}\,s^{-1}$	
$(CH_3)_2\dot{C}O^- +$ 2,2′-bipyridine $\longrightarrow (CH_3)_2CO +$ (2,2′-bipyridine)$^{\dot{-}}$				
Pulse rad. of 2-propanol + H_2O + N_2O				79 Mul 1
KAS	H_2O, pH = 13	RT	$k = 1.3 \cdot 10^8\ M^{-1}\,s^{-1}$	
$(CH_3)_2\dot{C}O^- +$ deuteroporphyrinedimethylester(DP) $\longrightarrow DP^{\dot{-}} + (CH_3)_2CO$				
Pulse rad. of 2-propanol + NaOH + H_2O				81 Bra 1
KAS	H_2O/2-propanol (1:1) mixt. [39]	RT	$k = 6(1) \cdot 10^8\ M^{-1}\,s^{-1}$	
$(CH_3)_2\dot{C}O^- +$ methyliodide$(CH_3I) \longrightarrow \dot{C}H_3 + I^- + (CH_3)_2CO$				
Pulse rad. of 2-propanol + acetone + NaOH + H_2O				81 Bra 1
[40]	H_2O [39]	RT	$k = 1.1(1) \cdot 10^8\ M^{-1}\,s^{-1}$	[41]
$(CH_3)_2\dot{C}O^- +$ 2-methyl-1,4-naphthoquinone $\longrightarrow$ ...-semiquinone + products				
Pulse rad. of 2-propanol + H_2O + N_2O				73 Rao 3
KAS	H_2O, pH = 12.4	RT	$k = 4.2(4) \cdot 10^9\ M^{-1}\,s^{-1}$	

[37]) k is pH dependent and extrapolated to pH = 14.
[38]) From [75 Bre 1].
[39]) 0.1 M in NaOH.
[40]) Pulse rad. competition kinetics relative to $(CH_3)_2\dot{C}O^- +$ p-nitroacetophenone.
[41]) For $(CH_3)_2\dot{C}OH + CH_3I$: $k \leqslant 10^5\ M^{-1}\,s^{-1}$.

Reaction Radical generation Method	Solvent	T[K]	Rate data	Ref./ add. ref.
$(CH_3)_2\dot{C}O^- + 9\text{-methylpurine} \longrightarrow (9\text{-methylpurine})^{\dot{-}} + (CH_3)_2CO$				
Pulse rad. of 2-propanol + H_2O + N_2O				76 Moo 1
KAS	H_2O, pH = 13.6	RT	$k = 8.7 \cdot 10^8\ M^{-1}\,s^{-1}$	
$(CH_3)_2\dot{C}O^- + \text{nicotinic acid } (NC_5H_4COO^-) \longrightarrow \text{products}$ [42]				
Pulse rad. of 2-propanol + N_2O + H_2O				74 Net 1
KAS	H_2O, pH = 13.1	RT	$k \geq 10^8\ M^{-1}\,s^{-1}$	
$(CH_3)_2\dot{C}O^- + \text{nitrobenzene } (C_6H_5NO_2) \longrightarrow C_6H_5\dot{N}O_2^- + (CH_3)_2CO$				
Pulse rad. of 2-propanol + N_2O + H_2O				66 Asm 1
KAS	H_2O, pH = 13	RT	$k = 3.0 \cdot 10^9\ M^{-1}\,s^{-1}$	
$(CH_3)_2\dot{C}O^- + 2\text{-nitrophenol} \longrightarrow {}^-OC_6H_4\dot{N}O_2^- + (CH_3)_2CO$				
Pulse rad. of 2-propanol + H_2O				69 Gru 1
KAS	H_2O, pH = 13	RT	$k = 1.4 \cdot 10^9\ M^{-1}\,s^{-1}$	
$(CH_3)_2\dot{C}O^- + \text{nitrosobenzene } (C_6H_5NO) \longrightarrow C_6H_5\dot{N}O^- + (CH_3)_2CO$				
Pulse rad. of 2-propanol + N_2O + H_2O				66 Asm 2
KAS	H_2O, pH = 13	RT	$k = 7.0 \cdot 10^9\ M^{-1}\,s^{-1}$	
$(CH_3)_2\dot{C}O^- + 1{,}10\text{-phenanthroline}(+H_2O) \longrightarrow (\text{phen}\ldots\text{-H})^{\cdot} + (CH_3)_2CO(+OH^-)$				
Pulse rad. of 2-propanol + acetone + H_2O				79 Mul 1,
KAS	H_2O	RT	$k = 6.0 \cdot 10^8\ M^{-1}\,s^{-1}$	79 Net 1/
			$5.7 \cdot 10^8$ [43]	80 Tep 1
$(CH_3)_2\dot{C}O^- + 1{,}10\text{-phenanthroline} \longrightarrow (1{,}10\text{-phen}\ldots)^{\dot{-}} + (CH_3)_2CO$				
Pulse rad. of alkaline 2-propanol + N_2O + H_2O				80 Tep 1/
KAS	H_2O, basic pH	RT	$k = 3.0(5) \cdot 10^9\ M^{-1}\,s^{-1}$	79 Mul 1,
				79 Net 1
$(CH_3)_2\dot{C}O^- + \text{phenazine (PZ)}\,(+H_2O) \longrightarrow (\text{PzH})^{\cdot} + (CH_3)_2CO\,(+OH^-)$				
Pulse rad. of 2-propanol + acetone + H_2O				79 Net 1
KAS	H_2O, pH = 13	RT	$k = 3.0 \cdot 10^9\ M^{-1}\,s^{-1}$	
$(CH_3)_2\dot{C}O^- + \text{pterin} \longrightarrow \text{products}$ [44]				
Pulse rad. of 2-propanol + N_2O + H_2O				76 Moo 1
KAS	H_2O, pH = 13	RT	$k = 1.5 \cdot 10^9\ M^{-1}\,s^{-1}$	
$(CH_3)_2\dot{C}O^- + \text{pyrazine} \longrightarrow \text{products}$				
Pulse rad. of 2-propanol + N_2O + H_2O				74 Moo 1
KAS	H_2O, pH = 13.6	RT	$k = 1.7 \cdot 10^9\ M^{-1}\,s^{-1}$	
$(CH_3)_2\dot{C}O^- + \text{pyrene} \longrightarrow (\text{pyrene})^{\dot{-}} + (CH_3)_2CO$				
Pulse rad. of 2-propanol + acetone + micellar (CTAB) solutions				76 Fra 1
KAS	micellar solution (H_2O, CTAB [45]), pH = 13	RT	$k = 2.3 \cdot 10^9\ M^{-1}\,s^{-1}$	
$(CH_3)_2\dot{C}O^- + {}^3(\text{pyrene})$ [46] $\longrightarrow (\text{pyrene})^{\dot{-}} + (CH_3)_2CO$				
Combined pulse rad. and phot. of 2-propanol + acetone + micellar (CTAB) solutions				76 Fra 1
KAS	micellar solution (H_2O, CTAB [45]), pH = 13	RT	$k = 2.3 \cdot 10^9\ M^{-1}\,s^{-1}$	

[42]) Pyridinyl radical.
[43]) From [79 Net 1].
[44]) Reaction with anionic form of pterin; 100% e^--transfer.
[45]) $5 \cdot 10^{-3}$ M hexadecyltrimethylammonium bromide (CTAB).
[46]) Pyrene triplet.

Reaction Radical generation Method	Solvent	T[K]	Rate data	Ref./ add. ref.
$(CH_3)_2\dot{C}O^- + \text{pyridazine (Pdz)} (+H_2O) \longrightarrow (PdzH)^{\cdot} + (CH_3)_2CO (+OH^-)$				
Pulse rad. of 2-propanol + $N_2O + H_2O$				74 Moo 1
KAS	H_2O, pH = 13.6	RT	$k = 2.1 \cdot 10^9\ M^{-1}s^{-1}$	
$(CH_3)_2\dot{C}O^- + \text{4-pyridinecarboxaldoxime} \longrightarrow \text{products}$				
Pulse rad. of 2-propanol + $N_2O + H_2O$				76 Net 3
KAS	H_2O, pH = 13.6	RT	$k = 1.7 \cdot 10^8\ M^{-1}s^{-1}$	
$(CH_3)_2\dot{C}O^- + \text{pyridoxal-5-phosphate} \longrightarrow \text{products}$				
Pulse rad. of 2-propanol + $N_2O + H_2O$				75 Moo 3
KAS	H_2O, pH = 13.3	RT	$k = 2.9 \cdot 10^8\ M^{-1}s^{-1}$	
$(CH_3)_2\dot{C}O^- + \text{trichloroacetate}(CCl_3COO^-) \longrightarrow \text{products}$				
Pulse rad. of 2-propanol + acetone + H_2O				75 Wil 2
Competition kinetics	H_2O	293	$k = 3 \cdot 10^8\ M^{-1}s^{-1}$ [47])	
$CH_3CH_2\dot{C}HOH + Fe(CN)_6^{3-} \longrightarrow Fe(CN)_6^{4-} + H^+ + CH_3CH_2CHO$				
Pulse rad. of 1-propanol + H_2O				69 Ada 1
KAS	H_2O	RT	$k = 3.7 \cdot 10^9\ M^{-1}s^{-1}$	
$CH_3CH_2\dot{C}HOH + \text{2,3-butanedione} (CH_3COCOCH_3) \longrightarrow CH_3CH_2CHO + H^+ + (CH_3COCOCH_3)^{\dot{-}}$				
Pulse rad. of 1-propanol + $H_2O + N_2O$				68 Lil 1
KAS	H_2O	RT	$k = 6.8 \cdot 10^8\ M^{-1}s^{-1}$	
$CH_3CH_2\dot{C}HOH + \text{4-nitroacetophenone(PNAP)} \longrightarrow \text{products}$ [48])				
Pulse rad. of 1-propanol + $N_2O + H_2O$				73 Gre 2
KAS	H_2O	RT	$k = 1.7 \cdot 10^9\ M^{-1}s^{-1}$	
$CH_3CH_2\dot{C}HOH + \text{nitrobenzene} (C_6H_5NO_2) \longrightarrow C_6H_5\dot{N}O_2^- + H^+ + CH_3CH_2CHO$				
Pulse rad. of 1-propanol + $N_2O + H_2O$				66 Asm 1,
KAS	H_2O	RT	$k = 3.5 \cdot 10^8\ M^{-1}s^{-1}$ [49]) $7.5 \cdot 10^8$ [50])	73 Gre 2
$CH_3CH_2\dot{C}HOH$ + *anti*-5-nitro-2-furaldoxime [51]) $\longrightarrow$ products [52])				
Pulse rad. of 1-propanol + $N_2O + H_2O$				73 Gre 1/
KAS	H_2O	RT	$k = 3.1 \cdot 10^9\ M^{-1}s^{-1}$	73 Gre 2
$CH_3CH_2\dot{C}HOH + \text{nitrosobenzene} (C_6H_5NO) \longrightarrow C_6H_5\dot{N}OH + CH_3CH_2CHO$				
Pulse rad. of 1-propanol + $N_2O + H_2O$				66 Asm 2
KAS	H_2O	RT	$k = 4.0 \cdot 10^9\ M^{-1}s^{-1}$	
$CH_3CH_2\dot{C}HOH + \text{tetranitromethane} (C(NO_2)_4) \longrightarrow C(NO_2)_3^- + NO_2 + H^+ + CH_3CH_2CHO$				
Pulse rad. of 1-propanol + $N_2O + H_2O$				64 Asm 1
KAS	H_2O	RT	$k = 4.7(10) \cdot 10^9\ M^{-1}s^{-1}$	
$(CH_3)_2\dot{C}OH + Ag_2^+ \longrightarrow (CH_3)_2CO + H^+ + Ag_2$				
Pulse rad. of 2-propanol + $AgClO_4 + H_2O$				78 Tau 1
Time-resolved Cond.	H_2O	RT	$k = 2.5 \cdot 10^9\ M^{-1}s^{-1}$ [53])	

[47]) Relative to $k((CH_3)_2\dot{C}O^- + PNAP) = 3.8 \cdot 10^9\ M^{-1}s^{-1}$.
[48]) 63% e^--transfer.
[49]) 68% e^--transfer [66 Asm 1].
[50]) 84% e^--transfer [73 Gre 2].
[51]) Nifuroxime.
[52]) 95% e^--transfer.
[53]) Based on $2k((CH_3)_2\dot{C}OH + (CH_3)_2\dot{C}OH) = 1.4 \cdot 10^9\ M^{-1}s^{-1}$ and assumed Ag_2^+, $(CH_3)_2\dot{C}OH$ and H_2O_2 concentrations.

Reaction Radical generation Method	Solvent	T[K]	Rate data	Ref./ add. ref.
$(CH_3)_2\dot{C}OH$ + Ag(II)-tetrakis(4-sulfonatophenyl)porphyrine $\longrightarrow H^+ + (CH_3)_2CO$ + products				
Pulse rad. of 2-propanol + $N_2O + H_2O$				81 Kum 1
KAS	H_2O, pH = 8.9	RT	$k = 6(1) \cdot 10^8\ M^{-1}s^{-1}$	
$(CH_3)_2\dot{C}OH + Co(NH_3)_6^{3+} \longrightarrow (CH_3)_2CO + H^+ + Co(NH_3)_6^{2+}$				
Pulse rad. of 2-propanol + $H_2O + N_2O$				72 Coh 1
KAS	H_2O, pH = 5…6	RT	$k = 1.3(2) \cdot 10^7\ M^{-1}s^{-1}$	
$(CH_3)_2\dot{C}OH + Co(III)(NH_3)_6 \ldots NO_2$ [54]) $\longrightarrow (CH_3)_2CO + H^+ + Co(III)(NH_3)_6 \ldots \dot{N}O_2^-$				
Pulse rad. of 2-propanol + $H_2O + N_2O$				78 Wie 1
KAS	H_2O, pH = 0.7…6.0	RT	$k = 2.0(4) \cdot 10^9\ M^{-1}s^{-1}$	
$(CH_3)_2\dot{C}OH + Co(III)(NH_3)_6 \ldots N$ [55]) $\longrightarrow (CH_3)_2CO + Co(III)(NH_3)_6 \ldots \dot{N}H$				
Pulse rad. of 2-propanol + H_2O				78 Wie 2
KAS	H_2O, pH = 0…6	RT	$k = 4.2(4) \cdot 10^8\ M^{-1}s^{-1}$	
$(CH_3)_2\dot{C}OH + (Co(III)(NH_3)_5Br)^{2+} \longrightarrow$ products				
Pulse rad. of 2-propanol + $N_2O + H_2O$				77 Coh 1
KAS	H_2O, pH = 3.5…4.0	RT	$k = 3.00(45) \cdot 10^8\ M^{-1}s^{-1}$	
$(CH_3)_2\dot{C}OH + (Co(III)(NH_3)_5Cl)^{2+} \longrightarrow$ products				
Pulse rad. of 2-propanol + $N_2O + H_2O$				77 Coh 1
KAS	H_2O, pH = 3.5…4.0	RT	$k = 4.0(6) \cdot 10^7\ M^{-1}s^{-1}$	
$(CH_3)_2\dot{C}OH$ + pentaammine(2,4-dinitrobenzoato)cobalt(III)$^{2+} \longrightarrow$ $(Co(III)(NH_3)_5O_2CC_6H_3NO_2(\dot{N}O_2^-))^+ + H^+ + (CH_3)_2CO$				
Pulse rad. of 2-propanol + $N_2O + H_2O$				77 Sim 1
KAS	H_2O	RT	$k = 4.3 \cdot 10^9\ M^{-1}s^{-1}$	
$(CH_3)_2\dot{C}OH$ + pentaammine(3,5-dinitrobenzoato)cobalt(III)$^{2+} \longrightarrow$ $(Co(III)(NH_3))_5O_2CC_6H_3NO_2(\dot{N}O_2^-))^+ + H^+ + (CH_3)_2CO$				
Pulse rad. of 2-propanol + $N_2O + H_2O$				77 Sim 1
KAS	H_2O	RT	$k = 2.9 \cdot 10^9\ M^{-1}s^{-1}$	
$(CH_3)_2\dot{C}OH$ + pentaammine(2-nitrobenzoato)cobalt(III)$^{2+} \longrightarrow$ $(Co(III)(NH_3)_5O_2CC_6H_4(\dot{N}O_2^-))^+ + H^+ + (CH_3)_2CO$				
Pulse rad. of 2-propanol + $N_2O + H_2O$				77 Sim 1
KAS	H_2O	RT	$k = 1.7 \cdot 10^9\ M^{-1}s^{-1}$	
$(CH_3)_2\dot{C}OH$ + pentaammine(3-nitrobenzoato)cobalt(III)$^{2+} \longrightarrow$ $(Co(III)(NH_3)_5O_2CC_6H_4(\dot{N}O_2^-))^+ + H^+ + (CH_3)_2CO$				
Pulse rad. of 2-propanol + $N_2O + H_2O$				77 Sim 1
KAS	H_2O	RT	$k = 1.5 \cdot 10^9\ M^{-1}s^{-1}$	
$(CH_3)_2\dot{C}OH$ + pentaammine(4-nitrobenzoato)cobalt(III)$^{2+} \longrightarrow$ $(Co(III)(NH_3)_5O_2CC_6H_4(\dot{N}O_2^-))^+ + H^+ + (CH_3)_2CO$				
Pulse rad. of 2-propanol + $N_2O + H_2O$				74 Sim 1,
KAS, time resolved Cond.	H_2O, pH = 4.5 and 7	RT	$k = 2.6 \cdot 10^9\ M^{-1}s^{-1}$	77 Sim 1
$(CH_3)_2\dot{C}OH + Co(NH_3)_5C_5H_5N^{3+} \longrightarrow Co^{2+} + 5NH_3 + C_5H_5N + (CH_3)_2CO + H^+$				
γ-rad. of 2-propanol + $H_2O + N_2O$				79 Hof 1
PR	H_2O	RT	$k \approx 10^9\ M^{-1}s^{-1}$ [56])	

[54]) Hexaaminebis(μ-hydroxy)[μ-(4-nitrobenzoato-O,O')]dicobalt(III) ion.
[55]) Hexaaminebis(μ-hydroxy)[μ-(pyrazinecarboxylato-O,O')]dicobalt(III) ion.
[56]) Estimated from Co^{2+} yield.

Reaction Radical generation Method	Solvent	T[K]	Rate data	Ref./ add. ref.
$(CH_3)_2\dot{C}OH + Co(NH_3)_5O_2CC_4H_3N_2^{2+} \longrightarrow$ products				
Pulse rad. of 2-propanol + H_2O				78 Wie 2
KAS	H_2O, pH = 0...5	RT	$k = 4.0(6) \cdot 10^8\ M^{-1}s^{-1}$	
$(CH_3)_2\dot{C}OH + Co(NH_3)_4O_2CC_4H_3N_2^{2+} \longrightarrow$ products				
Pulse rad. of 2-propanol + H_2O				78 Wie 2
KAS	H_2O	RT	$k = 9(2) \cdot 10^8\ M^{-1}s^{-1}$	
$(CH_3)_2\dot{C}OH + ((NH_3)_4Co(III)(\mu O_2, \mu NH_2)Co(III)(NH_3)_4)^{4+} \longrightarrow H^+ + (CH_3)_2CO$ + products [57])				
Pulse rad. of 2-propanol + N_2O + H_2O				81 Nat 1
KAS	H_2O, pH = 5	RT	$k = 1.49 \cdot 10^9\ M^{-1}s^{-1}$	
$(CH_3)_2\dot{C}OH$ + *cis*-(amminechlorobis(1,2-ethanediamine-N,N')cobalt(III))$^{2+} \longrightarrow$ products				
Pulse rad. of 2-propanol + N_2O + H_2O				77 Coh 1
KAS	H_2O, pH = 3.5...4.0	RT	$k = 2.2 \cdot 10^7\ M^{-1}s^{-1}$	
$(CH_3)_2\dot{C}OH$ + tris(2,2'-bipyridine-N,N')cobalt(III)$^{3+} \longrightarrow$ tris(2,2'-bipyridine-N,N')cobalt(II)$^{2+}$ + H^+ + $(CH_3)_2CO$				
Pulse rad. of 2-propanol + H_2O + N_2O				79 Sim 1
KAS	H_2O, pH = 0.5 and 7.8	RT	$k = 2.5(3) \cdot 10^9\ M^{-1}s^{-1}$	
$(CH_3)_2\dot{C}OH$ + *trans*-(dibromobis(1,2-ethanediamine-N,N')cobalt(III))$^+ \longrightarrow$ products				
Pulse rad. of 2-propanol + N_2O + H_2O				77 Coh 1
KAS	H_2O, pH = 3.5...4.0	RT	$k = 6.8 \cdot 10^8\ M^{-1}s^{-1}$	
$(CH_3)_2\dot{C}OH$ + *cis*-(bromobis(1,2-ethanediamine-N,N')fluorocobalt(III))$^+ \longrightarrow$ products				
Pulse rad. of 2-propanol + N_2O + H_2O				77 Coh 1
KAS	H_2O, pH = 3.5...4.0	RT	$k = 1.1 \cdot 10^8\ M^{-1}s^{-1}$	
$(CH_3)_2\dot{C}OH$ + *cis*-(aquachlorobis(1,2-ethanediamine-N,N')cobalt(III))$^{2+} \longrightarrow$ products				
Pulse rad. of 2-propanol + N_2O + H_2O				77 Coh 1
KAS	H_2O, pH = 3.5...4.0	RT	$k = 8.2 \cdot 10^7\ M^{-1}s^{-1}$	
$(CH_3)_2\dot{C}OH$ + *cis*-(dichlorobis(1,2-ethanediamine-N,N')cobalt(III))$^+ \longrightarrow$ products				
Pulse rad. of 2-propanol + N_2O + H_2O				77 Coh 1
KAS	H_2O, pH = 3.5...4.0	RT	$k = 1.0 \cdot 10^8\ M^{-1}s^{-1}$	
$(CH_3)_2\dot{C}OH$ + *trans*-(dichlorobis(1,2-ethanediamine-N,N')cobalt(III))$^+ \longrightarrow$ products				
Pulse rad. of 2-propanol + N_2O + H_2O				77 Coh 1
KAS	H_2O, pH = 3.5...4.0	RT	$k = 3.8 \cdot 10^8\ M^{-1}s^{-1}$	
$(CH_3)_2\dot{C}OH + ((EN)_2Co(III)(\mu O_2, \mu NH_2)Co(III)(EN)_2)^{4+} \longrightarrow H^+ + (CH_3)_2CO$ + products [57])				
Pulse rad. of 2-propanol + N_2O + H_2O				81 Nat 1
KAS	H_2O, pH = 5	RT	$k = 1.26 \cdot 10^9\ M^{-1}s^{-1}$	
$(CH_3)_2\dot{C}OH + ((CN)_5Co(III)(\mu O_2)Co(III)(CN)_5)^{5-} \longrightarrow H^+ + (CH_3)_2CO$ + products [57])				
Pulse rad. of 2-propanol + N_2O + H_2O				81 Nat 1
KAS	H_2O, pH = 5	RT	$k = 2.6 \cdot 10^8\ M^{-1}s^{-1}$	
$(CH_3)_2\dot{C}OH$ + Co(III)(5,6-dimethyl-1,10-phenanthroline)$_3^{3+} \longrightarrow$ $(CH_3)_2CO$ + H^+ + Co(II)(5,6-dimethyl-1,10-phenanthroline)$_3^{2+}$				
Pulse rad. of 2-propanol + N_2O + H_2O				80 Ven 1
KAS	H_2O	RT	$k = 3.2 \cdot 10^9\ M^{-1}s^{-1}$	

[57]) e^--transfer assumed to occur both at dioxygen center and at Co(III). μ defines a bridging group.

Reaction Radical generation Method	Solvent	T[K]	Rate data	Ref./ add. ref.
$(CH_3)_2\dot{C}OH$ + (5,7,7,12,14,14-hexamethyl-1,4,8,11-tetraazacyclotetradeca-4,11-diene-N,N′,N″,N‴)dihydroxy-cobalt(III) ion ⟶ Co(II)... + H^+ + $(CH_3)_2CO$				
Pulse rad. of 2-propanol + H_2O + N_2O				76 Tai 2
KAS	H_2O, pH = 10	RT	$k = 1.1 \cdot 10^8\ M^{-1}s^{-1}$	
$(CH_3)_2\dot{C}OH$ + aqua(5,7,7,12,14,14-hexamethyl-1,4,8,11-tetraazacyclotetradeca-4,11-diene-N,N′,N″,N‴)-hydroxycobalt(III) ion ⟶ Co(II)... + H^+ + $(CH_3)_2CO$				
Pulse rad. of 2-propanol + H_2O + N_2O				76 Tai 2
KAS	H_2O, pH = 6	RT	$k = 1.1 \cdot 10^8\ M^{-1}s^{-1}$	
$(CH_3)_2\dot{C}OH$ + diaqua(5,7,7,12,14,14-hexamethyl-1,4,8,11-tetraazacyclotetradeca-4,11-diene-N,N′,N″,N‴)-cobalt(III) ion ⟶ Co(II)... + H^+ + $(CH_3)_2CO$				
Pulse rad. of 2-propanol + H_2O + N_2O				76 Tai 2
KAS	H_2O, pH = 2.0	RT	$k = 2.0 \cdot 10^8\ M^{-1}s^{-1}$	
$(CH_3)_2\dot{C}OH$ + dichloro(5,7,7,12,14,14-hexamethyl-1,4,8,11-tetraazacyclotetradeca-4,11-diene-N,N′,N″,N‴)-cobalt(III) ion ⟶ Co(II)... + H^+ + $(CH_3)_2CO$				
Pulse rad. of 2-propanol + H_2O + N_2O				76 Tai 2
KAS	H_2O, pH = 1.0	RT	$k = 7.0 \cdot 10^8\ M^{-1}s^{-1}$	
$(CH_3)_2\dot{C}OH$ + Co(1,10-phenanthroline)$_3^{3+}$ ⟶ Co(1,10-phenanthroline)$_3^{2+}$ + $(CH_3)_2CO$ + H^+				
Pulse rad. of 2-propanol + H_2O + N_2O				79 Sim 1
KAS	H_2O	RT	$k = 4.6 \cdot 10^9\ M^{-1}s^{-1}$	
$(CH_3)_2\dot{C}OH$ + ($\{Co(pts)\}_2)^{8-}$ [58]) ⟶ ($\{Co(pts)(pts)^{\cdot}\}_2)^{9-}$ [59]) + $(CH_3)_2CO$ + H^+				
Pulse rad. of 2-propanol + N_2O + H_2O				80 Fer 1
KAS	H_2O	RT	$k = 1.5 \cdot 10^9\ M^{-1}s^{-1}$	
$(CH_3)_2\dot{C}OH$ + Co(III)-tetra(4-N-methylpyridyl)porphyrine ⟶ H^+ + $(CH_3)_2CO$ + Co(II)-tetra...				
Pulse rad. of 2-propanol + N_2O + H_2O				81 Net 2
KAS	H_2O, pH = 8	RT	$k = 1.8 \cdot 10^9\ M^{-1}s^{-1}$	
$(CH_3)_2\dot{C}OH$ + Co(III)-tetra(4-sulfonatophenyl)porphyrine ⟶ H^+ + $(CH_3)_2CO$ + Co(II)-tetra...				
Pulse rad. of 2-propanol + N_2O + H_2O				81 Net 2
KAS	H_2O, pH = 8	RT	$k = 8 \cdot 10^8\ M^{-1}s^{-1}$	
$(CH_3)_2\dot{C}OH$ + dihydroxy(2,3,9,10-tetramethyl-1,4,8,11-tetraazacyclotetradeca-1,3,8,10-tetraene-N,N′,N″,N‴)-cobalt(III) ion ⟶ Co(II)... + H^+ + $(CH_3)_2CO$				
Pulse rad. of 2-propanol + H_2O + N_2O				76 Tai 2
KAS	H_2O, pH = 9	RT	$k = 3.3 \cdot 10^8\ M^{-1}s^{-1}$	
$(CH_3)_2\dot{C}OH$ + diaqua(2,3,9,10-tetramethyl-1,4,8,11-tetraazacyclotetradeca-1,3,8,10-tetraene-N,N′,N″,N‴)-cobalt(III) ion ⟶ Co(II)... + H^+ + $(CH_3)_2CO$				
Pulse rad. of 2-propanol + H_2O + N_2O				76 Tai 2
KAS	H_2O, pH = 1.0	RT	$k = 1.9 \cdot 10^9\ M^{-1}s^{-1}$	
$(CH_3)_2\dot{C}OH$ + aquahydroxy(2,3,9,10-tetramethyl-1,4,8,11-tetraazacyclotetradeca-1,3,8,10-tetraene-N,N′,N″,N‴)cobalt(III) ion ⟶ Co(II)... + H^+ + $(CH_3)_2CO$				
Pulse rad. of 2-propanol + H_2O + N_2O				76 Tai 2
KAS	H_2O, pH = 5	RT	$k = 5.5 \cdot 10^8\ M^{-1}s^{-1}$	

[58]) Dimeric Co(II)-sulfophthalocyanine.
[59]) (pts) = phthalocyanine-3,10,17,24-tetrasulfonate hexa anion; (pts)$^{\cdot}$ = one-electron reduction product of (pts).

Reaction Radical generation Method	Solvent	T[K]	Rate data	Ref./ add. ref.
$(CH_3)_2\dot{C}OH$ + diaqua(2,3,9,10-tetramethyl-1,4,8,11-tetraazacyclotetradeca-1,3,8,10-tetraene-N,N′,N″,N‴)-Co(II) ⟶ Co(I)... + H^+ + $(CH_3)_2CO$				
Pulse rad. of 2-propanol + N_2O + H_2O				76 Tai 1
KAS	H_2O, pH = 1.25 and 6.5	RT	$k = 5.5 \cdot 10^9\,M^{-1}s^{-1}$	
$(CH_3)_2\dot{C}OH$ + Cr(III)(5-bromo-1,10-phenanthroline)$_3^{3+}$ ⟶ Cr(II)(5-bromo-1,10-phenanthroline)$_3^{2+}$ + $(CH_3)_2CO$ + H^+				
Pulse rad. of 2-propanol + H_2O				81 Ser 1
KAS	H_2O [60]	295...297	$k = 3.8 \cdot 10^9\,M^{-1}s^{-1}$	
$(CH_3)_2\dot{C}OH$ + Cr(III)(5-chloro-1,10-phenanthroline)$_3^{3+}$ ⟶ Cr(II)(5-chloro-1,10-phenanthroline)$_3^{2+}$ + H^+ + $(CH_3)_2CO$				
Pulse rad. of 2-propanol + H_2O				81 Ser 1
KAS	H_2O [60]	295...297	$k = 2.8 \cdot 10^9\,M^{-1}s^{-1}$	
$(CH_3)_2\dot{C}OH$ + Cr(III)(4,4′-dimethyl-2,2′-bipyridine)$_3^{3+}$ ⟶ Cr(II)(4,4′-dimethyl-2,2′-bipyridine)$_3^{2+}$ + H^+ + $(CH_3)_2CO$				
Pulse rad. of 2-propanol + H_2O				81 Ser 1
KAS	H_2O [60]	295...297	$k = 2 \cdot 10^9\,M^{-1}s^{-1}$	
$(CH_3)_2\dot{C}OH$ + Cr(III)(5,6-dimethyl-1,10-phenanthroline)$_3^{3+}$ ⟶ Cr(II)(5,6-dimethyl-1,10-phenanthroline)$_3^{2+}$ + H^+ + $(CH_3)_2CO$				
Pulse rad. of 2-propanol + H_2O				81 Ser 1
KAS	H_2O [60]	295...297	$k = 3.6 \cdot 10^9\,M^{-1}s^{-1}$	
$(CH_3)_2\dot{C}OH$ + Cr(III)(5-methyl-1,10-phenanthroline)$_3^{3+}$ ⟶ Cr(II)(5-methyl-1,10-phenanthroline)$_3^{2+}$ + H^+ + $(CH_3)_2CO$				
Pulse rad. of 2-propanol + H_2O				81 Ser 1
KAS	H_2O [60]	295...297	$k = 3.1 \cdot 10^9\,M^{-1}s^{-1}$	
$(CH_3)_2\dot{C}OH$ + Cr(III)(1,10-phenanthroline)$_3^{3+}$ ⟶ Cr(II)(1,10-phenanthroline)$_3^{2+}$ + H^+ + $(CH_3)_2CO$				
Pulse rad. of 2-propanol + H_2O				81 Ser 1
KAS	H_2O [60]	295...297	$k = 4.1 \cdot 10^9\,M^{-1}s^{-1}$	
$(CH_3)_2\dot{C}OH$ + Cu^{2+} ⟶ Cu^+ + H^+ + $(CH_3)_2CO$				
Pulse rad. of 2-propanol + H_2O + N_2O				72 Coh 1,
KAS	H_2O, pH = 5...6 pH = 2...5	RT	$k = 4.5(7) \cdot 10^7\,M^{-1}s^{-1}$ $k = 5.2(10) \cdot 10^7\,M^{-1}s^{-1}$ [61]	78 Bux 1
$(CH_3)_2\dot{C}OH$ + Fe^{3+} ⟶ Fe^{2+} + H^+ + $(CH_3)_2CO$				
Pulse rad. of 2-propanol + H_2O				74 But 1/
KAS	H_2O, pH ≈ 1	RT	$k = 4.5(4) \cdot 10^8\,M^{-1}s^{-1}$	77 Ber 1
$(CH_3)_2\dot{C}OH$ + $Fe(CN)_6^{3-}$ ⟶ $Fe(CN)_6^{4-}$ + H^+ + $(CH_3)_2CO$				
Pulse rad. of 2-propanol + H_2O				69 Ada 1,
KAS	H_2O	RT	$k = 4.7 \cdot 10^9\,M^{-1}s^{-1}$ $5.6(6) \cdot 10^9$ [62]	73 Rao 1

[60]) 1 M in HCl.
[61]) Based on formation kinetics of $Cu(I)CH_2CHCONH_2$ in presence of acrylamide [78 Bux 1].
[62]) From [73 Rao 1].

Reaction Radical generation Method	Solvent	T[K]	Rate data	Ref./ add. ref.
$(CH_3)_2\dot{C}OH + Fe(CN)_5NO^{2-} \longrightarrow (CH_3)_2CO + H^+ + Fe(CN)_5NO^{3-}$				
Pulse rad. of 2-propanol + H_2O + N_2O				77 Che 1
KAS	H_2O	RT	$k = 2.9 \cdot 10^9\,M^{-1}s^{-1}$	
$(CH_3)_2\dot{C}OH$ + Fe(III)cytochrome c $\longrightarrow (CH_3)_2CO + H^+$ + Fe(II)cytochrome c				
Pulse rad. of 2-propanol + H_2O + N_2O				75 Sim 1
KAS	H_2O, pH = 7 pH = 9.3	RT	$k = 3.8 \cdot 10^8\,M^{-1}s^{-1}$ $1.6 \cdot 10^8$	
$(CH_3)_2\dot{C}OH$ + hematoporphyrine(Fe(III)) $\longrightarrow$ products				
Pulse rad. of 2-propanol + H_2O + N_2O				74 Har 1
KAS	H_2O	RT	$k = 2.4(4) \cdot 10^8\,M^{-1}s^{-1}$	
$(CH_3)_2\dot{C}OH$ + hemin c(Fe(III)) $\longrightarrow (CH_3)_2CO + H^+$ + hemin c(Fe(II))				
Pulse rad. of 2-propanol + H_2O + N_2O				75 Gof 1
KAS	H_2O	RT	$k = 2.8 \cdot 10^9\,M^{-1}s^{-1}$	
$(CH_3)_2\dot{C}OH$ + deutero-hemin(Fe(III)) $\longrightarrow$ products				
Pulse rad. of 2-propanol + acetone + H_2O				80 Bra 1
KAS	H_2O/2-propanol (60:40%)	RT	$k = 3.7 \cdot 10^8\,M^{-1}s^{-1}$	
$(CH_3)_2\dot{C}OH$ + metmyoglobin(Fe(III)) $\longrightarrow$ products				
Pulse rad. of 2-propanol + H_2O				79 Shi 1
KAS	H_2O	RT	$k \approx 10^8\,M^{-1}s^{-1}$	
$(CH_3)_2\dot{C}OH + H_2O_2 \longrightarrow (CH_3)_2CO + H_2O + \dot{O}H$				
Photochem., react. of ^{3}acetone + 2-propanol + H_2O				71 Bur 1
PR, ESR	H_2O, pH = 1.7 and neutral sol.	RT	$k = 5 \cdot 10^5\,M^{-1}s^{-1}$ [63]	
$(CH_3)_2\dot{C}OH + HgBr_2 \longrightarrow HgBr + Br^- + H^+ + (CH_3)_2CO$				
Pulse rad. of 2-propanol + H_2O + N_2O				76 Jun 1
KAS	H_2O	RT	$k = 2.4(6) \cdot 10^9\,M^{-1}s^{-1}$	
$(CH_3)_2\dot{C}OH + HgCl_2 \longrightarrow HgCl + Cl^- + H^+ + (CH_3)_2CO$				
Pulse rad. of 2-propanol + H_2O + N_2O				73 Naz 1/
KAS	H_2O	RT	$k = 2.0(2) \cdot 10^9\,M^{-1}s^{-1}$	76 Jun 1
$(CH_3)_2\dot{C}OH + HgI_2 \longrightarrow HgI + I^- + H^+ + (CH_3)_2CO$				
Pulse rad. of 2-propanol + H_2O + N_2O				76 Jun 1,
KAS	H_2O	RT	$k = 2.0(5) \cdot 10^9\,M^{-1}s^{-1}$ $1.0(5) \cdot 10^9$ [64]	78 Fuj 1
$(CH_3)_2\dot{C}OH + Hg(SCN)_2 \longrightarrow HgSCN + SCN^- + H^+ + (CH_3)_2CO$				
Pulse rad. of 2-propanol + H_2O + N_2O				76 Jun 1
KAS	H_2O	RT	$k = 2.2(5) \cdot 10^9\,M^{-1}s^{-1}$	
$(CH_3)_2\dot{C}OH + HgBr_2 \xrightarrow{a} HgBr + Br^- + H^+ + (CH_3)_2CO$ $\dot{C}O_2^- + HgBr_2 \xrightarrow{b} CO_2 + Hg(I)\ldots$				
Pulse rad. of formate + 2-propanol + N_2O + H_2O				76 Fuj 1/
PR	H_2O	RT	$k_a/k_b = 1.12$	76 Jun 1

[63]) Based on $2k((CH_3)_2\dot{C}OH + (CH_3)_2\dot{C}OH) = 1.4 \cdot 10^9\,M^{-1}s^{-1}$.
[64]) From [78 Fuj 1].

Reaction Radical generation Method	Solvent	T[K]	Rate data	Ref./ add. ref.
$(CH_3)_2\dot{C}OH + IrCl_6^{2-} \longrightarrow Ir(III) + products$				
Pulse rad. of 2-propanol + $N_2O + H_2O$				82 Ste 1
KAS	H_2O	295	$k = 4.7 \cdot 10^9\ M^{-1}s^{-1}$	
$(CH_3)_2\dot{C}OH + MnO_4^- \longrightarrow products$				
Pulse rad. of 2-propanol + $H_2O + N_2O$				73 Rao 1
KAS	H_2O	RT	$k = 4.2(4) \cdot 10^9\ M^{-1}s^{-1}$	
$(CH_3)_2\dot{C}OH + Pb^{2+} \longrightarrow (CH_3)_2CO + H^+ + Pb^+$				
Pulse rad. of 2-propanol + H_2O				76 Bre 1
KAS	H_2O, pH = 5	RT	$k = 3.0 \cdot 10^4\ M^{-1}s^{-1}$	
$(CH_3)_2\dot{C}OH$ + *trans*-dichlorobisethylenediamineplatinum(IV) ion $\longrightarrow$ Pt(III)... + products				
Pulse rad. of 2-propanol + H_2O				75 Sto 1
KAS	H_2O	RT	$k = 8.1(11) \cdot 10^8\ M^{-1}s^{-1}$	
$(CH_3)_2\dot{C}OH$ + Rh(III)(2,2′-bipyridine)$_3^{3+}$ $\longrightarrow$ Rh(II)(2,2′-bipyridine)$_3^{2+}$ + $(CH_3)_2CO + H^+$				
Pulse rad. of 2-propanol + $N_2O + H_2O$				81 Mul 1/
KAS	H_2O, pH = 1...10	RT	$k = 1.8(2) \cdot 10^9\ M^{-1}s^{-1}$	74 Mul 2
$(CH_3)_2\dot{C}OH$ + Rh(III)(1,10-phenanthroline)$_3^{3+}$ $\longrightarrow$ $(CH_3)_2CO + H^+$ + Rh(II)(1,10-phenanthroline)$_3^{2+}$				
Pulse rad. of 2-propanol + $N_2O + H_2O$				80 Ven 1
KAS	H_2O	RT	$k = 3.2 \cdot 10^9\ M^{-1}s^{-1}$	
$(CH_3)_2\dot{C}OH + Ru(NH_3)_6^{3+} \longrightarrow Ru(NH_3)_6^{2+} + H^+ + (CH_3)_2CO$				
Pulse rad. of 2-propanol + $H_2O + N_2O$				72 Coh 1
KAS	H_2O	RT	$k = 9.2(14) \cdot 10^8\ M^{-1}s^{-1}$	
$(CH_3)_2\dot{C}OH + (Ru(III)(NH_3)_5Cl)^{2+} \longrightarrow products$				
Pulse rad. of 2-propanol + $N_2O + H_2O$				77 Coh 1
KAS	H_2O, pH = 3.5...4.0	RT	$k = 1.3(2) \cdot 10^9\ M^{-1}s^{-1}$	
$(CH_3)_2\dot{C}OH + Ru(NH_3)_5NO^{3+} \longrightarrow Ru(NH_3)_5NO^{2+} + H^+ + (CH_3)_2CO$				
Pulse rad. of 2-propanol + $H_2O + N_2O$				75 Arm 1
KAS	H_2O, pH = 5	RT	$k = 5.5 \cdot 10^8\ M^{-1}s^{-1}$	
$(CH_3)_2\dot{C}OH + Zn^+ + H_2O \longrightarrow Zn^{2+} + (CH_3)_2CHOH + OH^-$				
Pulse rad. of 2-propanol + Zn^{2+} + H_2O				77 Rab 1
KAS	H_2O	RT	$k = 1.30(25) \cdot 10^9\ M^{-1}s^{-1}$ [65])	
$(CH_3)_2\dot{C}OH$ + Zn-tetra(4-N-methylpyridyl)porphyrine $\longrightarrow$ $H^+ + (CH_3)_2CO$ + (Zn-tetra...)$^{\dot{-}}$				
Pulse rad. of 2-propanol + $N_2O + H_2O$				81 Net 2
KAS	H_2O, pH = 8	RT	$k = 2.3 \cdot 10^9\ M^{-1}s^{-1}$	
$(CH_3)_2\dot{C}OH$ + Zn-tetra(4-sulfonatophenyl)porphyrine $\longrightarrow$ $H^+ + (CH_3)_2CO$ + (Zn-tetra...)$^{\dot{-}}$				
Pulse rad. of 2-propanol + $N_2O + H_2O$				81 Net 2
KAS	H_2O, pH = 8	RT	$k \approx 8 \cdot 10^7\ M^{-1}s^{-1}$	
$(CH_3)_2\dot{C}OH$ + Zn-tetra(4-(N,N,N-trimethylamine)phenyl)porphyrine $\longrightarrow$ $H^+ + (CH_3)_2CO$ + (Zn-tetra...)$^{\dot{-}}$				
Pulse rad. of 2-propanol + $N_2O + H_2O$				81 Net 2
KAS	H_2O, pH = 8	RT	$k = 2.4 \cdot 10^8\ M^{-1}s^{-1}$	

[65]) Based on $k(Zn^+ + Zn^+) = 4.5 \cdot 10^8\ M^{-1}s^{-1}$.

Reaction Radical generation Method	Solvent	T[K]	Rate data	Ref./ add. ref.
$(CH_3)_2\dot{C}OH$ + 3-acetylpyridine ($CH_3COC_5H_4NH^+$) ⟶ products				
Pulse rad. of 2-propanol + H_2O				74 Bru 1
KAS	H_2O, pH = 0.6	RT	$k = 8.6(17) \cdot 10^9\ M^{-1} s^{-1}$ [66]	
$(CH_3)_2\dot{C}OH$ + acridine ($C_{13}H_9N$) ⟶ products [67]				
Pulse rad. of 2-propanol + H_2O + N_2O				74 Moo 1/
KAS	H_2O	RT	$k = 3.0 \cdot 10^8\ M^{-1} s^{-1}$	79 Net 1
$(CH_3)_2\dot{C}OH$ + acridine ($C_{13}H_9NH^+$) ⟶ $(CH_3)_2CO$ + H^+ + $(C_{13}H_9NH)^{\cdot}$ [68]				
Pulse rad. of 2-propanol + H_2O + N_2O				74 Moo 1/
KAS	H_2O, pH = 2.0	RT	$k = 3.7 \cdot 10^9\ M^{-1} s^{-1}$	79 Net 1
$(CH_3)_2\dot{C}OH$ + adenosine ⟶ products				
Pulse rad. of 2-propanol				75 Moo 1
KAS	H_2O, pH = 2.2	RT	$k = 4.6 \cdot 10^7\ M^{-1} s^{-1}$	
	pH = 7 and 13.6		$k < 10^6\ M^{-1} s^{-1}$	
$(CH_3)_2\dot{C}OH$ + 2-amino-5-nitrothiazole ⟶ $(CH_3)_2CO$ + H^+ + (2-amino-5-nitrothiazole)$^{\dot{-}}$				
Pulse rad. of 2-propanol + H_2O				76 Gre 1
KAS	H_2O	RT	$k = 2.0 \cdot 10^9\ M^{-1} s^{-1}$	
$(CH_3)_2\dot{C}OH$ + 9,10-anthraquinone ⟶ ...semiquinone + $(CH_3)_2CO$				
Pulse rad. of 2-propanol + H_2O + N_2O				73 Rao 1
KAS	H_2O	RT	$k = 1.6(1) \cdot 10^9\ M^{-1} s^{-1}$	
$(CH_3)_2\dot{C}OH$ + 9,10-anthraquinone-2,6-disulfonate ion ⟶ ...semiquinone + H^+ + $(CH_3)_2CO$				
Pulse rad. of 2-propanol + H_2O + N_2O				73 Rao 1
KAS	H_2O	RT	$k = 4.6(5) \cdot 10^9\ M^{-1} s^{-1}$	
$(CH_3)_2\dot{C}OH$ + 9,10-anthraquinone-2-sulfonate ⟶ ...semiquinone + H^+ + $(CH_3)_2CO$				
Pulse rad. of 2-propanol + H_2O + N_2O				73 Rao 1,
KAS	H_2O	RT	$k = 3.0(3) \cdot 10^9\ M^{-1} s^{-1}$	76 War 1
			$5.6(6) \cdot 10^9$ [69]	
$(CH_3)_2\dot{C}OH$ + azobenzene ⟶ $(CH_3)_2CO$ + H^+ + (azobenzene)$^{\dot{-}}$				
Pulse rad. of 2-propanol + H_2O + N_2O				77 Net 1
KAS	H_2O	RT	$k = 4 \cdot 10^8\ M^{-1} s^{-1}$	
$(CH_3)_2\dot{C}OH$ + benzoquinone ⟶ $(CH_3)_2CO$ + benzosemiquinone				
Pulse rad. of 2-propanol + H_2O + N_2O				71 Wil 1/
KAS	H_2O	RT	$k = 5.0 \cdot 10^9\ M^{-1} s^{-1}$	73 Rao 1,
				73 Sim 1,
				73 Pat 1
$(CH_3)_2\dot{C}OH$ + 3-benzoyl-1-methylpyridinium ion ⟶ products				
Pulse rad. of 2-propanol + H_2O + N_2O				72 Nel 1
KAS	H_2O, pH = 1 and 5	RT	$k = 2.3(3) \cdot 10^9\ M^{-1} s^{-1}$	
$(CH_3)_2\dot{C}OH$ + 2-benzoylpyridine ($C_6H_5COC_5H_4N$) ⟶ $(CH_3)_2CO$ + $C_6H_5\dot{C}OHC_5H_4N$				
Pulse rad. of 2-propanol + H_2O + N_2O				72 Nel 1
KAS	H_2O, pH = 5.8	RT	$k = 1.5(2) \cdot 10^8\ M^{-1} s^{-1}$	

[66] No reaction with deprotonated form of 3-acetylpyridine.
[67] ≈40...50% e^--transfer.
[68] ≈90% e^--transfer.
[69] From [76 War 1].

Reaction Radical generation Method	Solvent	T[K]	Rate data	Ref./ add. ref.
$(CH_3)_2\dot{C}OH$ + 2-benzoylpyridine (protonated form, $C_6H_5COC_5H_4NH^+$) ⟶ $(CH_3)_2CO + C_6H_5\dot{C}OHC_5H_4NH^+$				
Pulse rad. of 2-propanol + H_2O + N_2O				72 Nel 1
KAS	H_2O, pH = 1	RT	$k = 3.0(3) \cdot 10^9\ M^{-1}s^{-1}$	
$(CH_3)_2\dot{C}OH$ + 3-benzoylpyridine ($C_6H_5COC_5H_4N$) ⟶ $(CH_3)_2CO + C_6H_5\dot{C}OHC_5H_4N$				
Pulse rad. of 2-propanol + H_2O + N_2O				72 Nel 1
KAS	H_2O, pH = 5.1	RT	$k = 1.0(2) \cdot 10^8\ M^{-1}s^{-1}$	
$(CH_3)_2\dot{C}OH$ + 3-benzoylpyridine (protonated form, $C_6H_5COC_5H_4NH^+$) ⟶ $(CH_3)_2CO + C_6H_5\dot{C}OHC_5H_4NH^+$				
Pulse rad. of 2-propanol + H_2O + N_2O				72 Nel 1
KAS	H_2O, pH = 0.9	RT	$k = 1.7(2) \cdot 10^9\ M^{-1}s^{-1}$	
$(CH_3)_2\dot{C}OH$ + 4-benzoylpyridine ($C_6H_5COC_5H_4N$) ⟶ $(CH_3)_2CO + C_6H_5\dot{C}OHC_5H_4N$				
Pulse rad. of 2-propanol + H_2O + N_2O				72 Nel 1
KAS	H_2O, pH = 7.7	RT	$k = 2.4(4) \cdot 10^8\ M^{-1}s^{-1}$	
$(CH_3)_2\dot{C}OH$ + 4-benzoylpyridine (protonated form, $C_6H_5COC_5H_4NH^+$) ⟶ $(CH_3)_2CO + C_6H_5\dot{C}OHC_5H_4NH^+$				
Pulse rad. of 2-propanol + H_2O + N_2O				72 Nel 1
KAS	H_2O, pH = 0.7	RT	$k = 2.5(2) \cdot 10^9\ M^{-1}s^{-1}$	
$(CH_3)_2\dot{C}OH$ + 2,2'-bipyridine [70]) ⟶ $(CH_3)_2CO$ + bipy $\overset{+}{H_2^{\cdot}}$				
Pulse rad. of 2-propanol + H_2O + N_2O				79 Mul 1
KAS	H_2O, pH ⩽ 3.7	RT	$k = 3.5 \cdot 10^8\ M^{-1}s^{-1}$ [71])	
$(CH_3)_2\dot{C}OH$ + 2-bromo-5-nitrothiazole ⟶ $(CH_3)_2CO + H^+$ + (2-bromo...)$^{\cdot-}$				
Pulse rad. of 2-propanol + H_2O + N_2O				76 Gre 1
KAS	H_2O	RT	$k = 3.0 \cdot 10^9\ M^{-1}s^{-1}$	
$(CH_3)_2\dot{C}OH$ + α-bromo-4-nitrotoluene ($BrCH_2C_6H_4NO_2$) ⟶ ($BrCH_2C_6H_4\dot{N}O_2^-$) + H^+ + $(CH_3)_2CO$				
Pulse rad. of 2-propanol + N_2O + H_2O				80 Net 1
KAS	H_2O, pH < 2.3	RT	$k = 2.6 \cdot 10^9\ M^{-1}s^{-1}$	
$(CH_3)_2\dot{C}OH$ + 5-bromouracil ⟶ products				
Pulse rad. of 2-propanol + H_2O + N_2O				75 Wil 2
PR, KAS	H_2O	RT	$k = 2 \cdot 10^7\ M^{-1}s^{-1}$ [72])	
$(CH_3)_2\dot{C}OH$ + 2,3-butanedione ($CH_3COCOCH_3$) ⟶ $(CH_3)_2CO$ + $(CH_3COCOCH_3)^{\cdot-}$ + H^+				
Pulse rad. of 2-propanol + H_2O + N_2O				68 Lil 1,
KAS	H_2O	RT	$k = 8.6 \cdot 10^8\ M^{-1}s^{-1}$ $6.0 \cdot 10^8$ [73])	72 Coh 1
$(CH_3)_2\dot{C}OH$ + 2-*t*-butyl-2,3-diazabicyclo[2.2.2]octane($R_3N_2^+$) ⟶ $R_3\dot{N}_2$ + H^+ + $(CH_3)_2CO$				
Pulse rad. of 2-propanol + N_2O + H_2O				80 Nel 1
KAS	H_2O	RT	$k \approx 3 \cdot 10^8\ M^{-1}s^{-1}$	
$(CH_3)_2\dot{C}OH$ + 3-carbamoyl-1-methylpyridinium ion ⟶ products				
Pulse rad. of 2-propanol + H_2O + N_2O				74 Bru 1
KAS	H_2O, pH = 9.5	RT	$k = 3.6(7) \cdot 10^8\ M^{-1}s^{-1}$	

[70]) Protonated form (bipy H^+).
[71]) No observable reaction with deprotonated 2,2'-bipyridine.
[72]) Based on $k((CH_3)_2\dot{C}OH$ + *p*-nitroacetophenone) = $3.8 \cdot 10^9\ M^{-1}s^{-1}$.
[73]) From [72 Coh 1].

Reaction Radical generation Method	Solvent	T[K]	Rate data	Ref./ add. ref.
$(CH_3)_2\dot{C}OH$ + carbontetrachloride (CCl_4) ⟶ $\dot{C}Cl_3 + Cl^- + H^+ + (CH_3)_2CO$				
Pulse rad. of 2-propanol + H_2O + N_2O PR, KAS, time resolved Cond.	H_2O	RT	$k = 1.0 \cdot 10^8\,M^{-1}s^{-1}$ [74] $7 \cdot 10^8$ [75]	71 Koe 1, 75 Wil 2/ 73 Wil 1
$(CH_3)_2\dot{C}OH$ + 3-carboxy-1-methylpyridinium ion ⟶ products (pyridinyl radical etc.)				
Pulse rad. of 2-propanol + H_2O + N_2O KAS	H_2O, pH = 9.2	RT	$k = 1.0 \cdot 10^8\,M^{-1}s^{-1}$	74 Net 1
$(CH_3)_2\dot{C}OH$ + 4-carboxy-1-methylpyridinium ion ⟶ products				
Pulse rad. of 2-propanol + H_2O + N_2O KAS	H_2O, pH = 8.6	RT	$k = 1.5 \cdot 10^9\,M^{-1}s^{-1}$	79 Ste 1
$(CH_3)_2\dot{C}OH$ + 6-carboxyuracil (anionic form) [76] ⟶ products				
Pulse rad. of 2-propanol + acetone + H_2O KAS	H_2O	RT	$k = 1.0(5) \cdot 10^8\,M^{-1}s^{-1}$	73 Wil 1
$(CH_3)_2\dot{C}OH$ + crystal violet ⟶ products				
Pulse rad. of 2-propanol + H_2O + N_2O KAS	H_2O	RT	$k = 2.3 \cdot 10^9\,M^{-1}s^{-1}$	73 Rao 2
$(CH_3)_2\dot{C}OH$ + 1,1'-dibenzyl-4,4-bipyridinium$^{(2+)}$ [77] ⟶ (1,1'-dibenzyl...)$^{\dot{+}}$ + $H^+ + (CH_3)_2CO$				
Pulse rad. of 2-propanol + N_2O + H_2O KAS	H_2O	RT	$k = 3.0(1) \cdot 10^9\,M^{-1}s^{-1}$	76 War 1
$(CH_3)_2\dot{C}OH$ + dichloroindophenol ⟶ products				
Pulse rad. of 2-propanol + H_2O + N_2O KAS	H_2O	RT	$k = 4.4 \cdot 10^9\,M^{-1}s^{-1}$	73 Rao 2
$(CH_3)_2\dot{C}OH$ + 1,3-dihydroxy-2-nitrobenzene ($(OH)_2C_6H_3NO_2$) ⟶ $(CH_3)_2CO + H^+ + (OH)_2C_6H_3\dot{N}O_2^-$				
Pulse rad. of 2-propanol + H_2O + N_2O KAS	H_2O	RT	$k = 6.8 \cdot 10^8\,M^{-1}s^{-1}$	76 Net 1
$(CH_3)_2\dot{C}OH$ + 2,3-dimethylbenzoquinone ⟶ $(CH_3)_2CO + H^+$ + ... semiquinone				
Pulse rad. of 2-propanol + acetone + H_2O KAS	H_2O	RT	$k = 3.5 \cdot 10^9\,M^{-1}s^{-1}$	73 Pat 1
$(CH_3)_2\dot{C}OH$ + 2,5-dimethylbenzoquinone ⟶ $(CH_3)_2CO + H^+$ + ... semiquinone				
Pulse rad. of 2-propanol + acetone + H_2O KAS	H_2O	RT	$k = 3.9 \cdot 10^9\,M^{-1}s^{-1}$	73 Pat 1
$(CH_3)_2\dot{C}OH$ + 2,6-dimethylbenzoquinone ⟶ $(CH_3)_2CO + H^+$ + ... semiquinone				
Pulse rad. of 2-propanol + acetone + H_2O KAS	H_2O	RT	$k = 4.2 \cdot 10^9\,M^{-1}s^{-1}$	73 Pat 1
$(CH_3)_2\dot{C}OH$ + dimethylfumerate ($CH_3OOCCH{=}CHCOOCH_3$) ⟶ products [78]				
Pulse rad. of 2-propanol + H_2O + N_2O KAS	H_2O	RT	$k = 4 \cdot 10^9\,M^{-1}s^{-1}$	73 Hay 2

[74] Build-up of H^+/Cl^- conductivity [71 Koe 1].
[75] Competition kinetics relative to $k((CH_3)_2\dot{C}OH$ + 4-nitroacetophenone $= 3.8 \cdot 10^9\,M^{-1}s^{-1}$ [75 Wil 1].
[76] Orotate.
[77] Benzylviologen.
[78] 12% e^--transfer.

Reaction Radical generation Method	Solvent	T[K]	Rate data	Ref./ add. ref.
$(CH_3)_2\dot{C}OH$ + 2,3-dimethylnaphthoquinone ⟶ $(CH_3)_2CO + H^+$ + ... semiquinone				
Pulse rad. of 2-propanol + acetone + H_2O				73 Pat 1
KAS	H_2O	RT	$k = 3.9 \cdot 10^9\ M^{-1}s^{-1}$	
$(CH_3)_2\dot{C}OH$ + 3,5-dinitroanisole ⟶ (3,5-dinitroanisole)$^{\dot{-}}$ + H^+ + $(CH_3)_2CO$ [79]				
Pulse rad. of 2-propanol + H_2O				79 Tam 1
KAS	H_2O	RT	$k = 2.5 \cdot 10^9\ M^{-1}s^{-1}$	
$(CH_3)_2\dot{C}OH$ + 1,2-dinitrobenzene ⟶ $(CH_3)_2CO + H^+$ + (1,2-dinitrobenzene)$^{\dot{-}}$ [79]				
Pulse rad. of 2-propanol + H_2O + N_2O				76 Net 2
KAS	H_2O	RT	$k = 2.9 \cdot 10^9\ M^{-1}s^{-1}$	
$(CH_3)_2\dot{C}OH$ + 1,3-dinitrobenzene ⟶ $(CH_3)_2CO + H^+$ + (1,3-dinitrobenzene)$^{\dot{-}}$ [79]				
Pulse rad. of 2-propanol + H_2O + N_2O				76 Net 2
KAS	H_2O	RT	$k = 3.6 \cdot 10^9\ M^{-1}s^{-1}$	
$(CH_3)_2\dot{C}OH$ + 1,4-dinitrobenzene ⟶ $(CH_3)_2CO + H^+$ + (1,4-dinitrobenzene)$^{\dot{-}}$ [79]				
Pulse rad. of 2-propanol + H_2O + N_2O				76 Net 2
KAS	H_2O	RT	$k = 3.2 \cdot 10^9\ M^{-1}s^{-1}$	
$(CH_3)_2\dot{C}OH$ + (2,4-dinitrobenzoate)$^-$ ⟶ $(CH_3)_2\dot{C}OH$ + (2,4-dinitrobenzoate)$^{2\dot{-}}$ [79]				
Pulse rad. of 2-propanol + H_2O + N_2O				76 Net 2
KAS	H_2O	RT	$k = 2.9 \cdot 10^9\ M^{-1}s^{-1}$	
$(CH_3)_2\dot{C}OH$ + (2,5-dinitrobenzoate)$^-$ ⟶ $(CH_3)_2CO + H^+$ + (2,5-dinitrobenzoate)$^{2\dot{-}}$ [79]				
Pulse rad. of 2-propanol + H_2O + N_2O				76 Net 2
KAS	H_2O	RT	$k = 3.3 \cdot 10^9\ M^{-1}s^{-1}$	
$(CH_3)_2\dot{C}OH$ + (3,4-dinitrobenzoate)$^-$ ⟶ $(CH_3)_2CO + H^+$ + (3,4-dinitrobenzoate)$^{2\dot{-}}$ [79]				
Pulse rad. of 2-propanol + H_2O + N_2O				76 Net 2
KAS	H_2O	RT	$k = 3.2 \cdot 10^9\ M^{-1}s^{-1}$	
$(CH_3)_2\dot{C}OH$ + (3,5-dinitrobenzoate)$^-$ ⟶ $(CH_3)_2CO + H^+$ + (3,5-dinitrobenzoate)$^{2\dot{-}}$ [79]				
Pulse rad. of 2-propanol + H_2O + N_2O				76 Net 2
KAS	H_2O	RT	$k = 3.1 \cdot 10^9\ M^{-1}s^{-1}$	
$(CH_3)_2\dot{C}OH$ + N-ethylmaleimide ⟶ products [80]				
Pulse rad. of 2-propanol + H_2O + N_2O				72 Hay 1
KAS	H_2O	RT	$k = 5.0 \cdot 10^9\ M^{-1}s^{-1}$	
$(CH_3)_2\dot{C}OH$ + folic acid ⟶ products				
Pulse rad. of 2-propanol + H_2O				76 Moo 1
KAS	H_2O, pH = 0.5	RT	$k = 1.1 \cdot 10^9\ M^{-1}s^{-1}$	
	pH = 6		$4.0 \cdot 10^8$	
$(CH_3)_2\dot{C}OH$ + fumaric acid (HOOCCH=CHCOOH) ⟶ products [81]				
Pulse rad. of 2-propanol + N_2O + H_2O				73 Hay 2
KAS	H_2O, pH = 0.5	RT	$k = 9.0 \cdot 10^8\ M^{-1}s^{-1}$	
$(CH_3)_2\dot{C}OH$ + 1-(2-hydroxyethyl)-2-methyl-5-nitroimidazole [82] ⟶ ...-$\dot{N}O_2^-$ + H^+ + $(CH_3)_2CO$ [83]				
Pulse rad. of 2-propanol + H_2O + N_2O				74 Wil 1/
KAS	H_2O	RT	$k = 7 \cdot 10^8\ M^{-1}s^{-1}$	75 Ays 1

[79] e^--transfer to nitro groups.
[80] 47% e^--transfer.
[81] 14% e^--transfer. No e^--transfer to dianion $^-$OOCCH=CHCOO$^-$ at pH = 9.
[82] Metronidazole.
[83] ESR indicates e^--transfer in basic solutions, but addition in neutral solutions [75 Ays 1].

Reaction Radical generation Method	Solvent	T[K]	Rate data	Ref./ add. ref.
$(CH_3)_2\dot{C}OH$ + 1-(2-hydroxy-3-methoxypropyl)-2-nitroimidazole [84]) ⟶ ...-$\dot{N}O_2^-$ + H^+ + $(CH_3)_2CO$				
Pulse rad. of 2-propanol + H_2O + N_2O				76 War 1
KAS	H_2O	RT	$k = 2.8(4) \cdot 10^9\ M^{-1}s^{-1}$	
$(CH_3)_2\dot{C}OH$ + 2-hydroxy-1,4-naphthoquinone ⟶ ...-semiquinone + products				
Pulse rad. of 2-propanol + H_2O + N_2O				73 Rao 1
KAS	H_2O	RT	$k = 3.4(3) \cdot 10^9\ M^{-1}s^{-1}$	
$(CH_3)_2\dot{C}OH$ + indigo disulfonate ⟶ products				
Pulse rad. of 2-propanol + H_2O + N_2O				73 Rao 2
KAS	H_2O	RT	$k = 4.0 \cdot 10^9\ M^{-1}s^{-1}$	
$(CH_3)_2\dot{C}OH$ + indigo tetrasulfonate ⟶ products				
Pulse rad. of 2-propanol + H_2O + N_2O				73 Rao 2
KAS	H_2O	RT	$k = 4.2 \cdot 10^9\ M^{-1}s^{-1}$	
$(CH_3)_2\dot{C}OH$ + indophenol ⟶ products				
Pulse rad. of 2-propanol + H_2O + N_2O				73 Rao 2
KAS	H_2O, pH = 9	RT	$k = 4.0 \cdot 10^9\ M^{-1}s^{-1}$	
$(CH_3)_2\dot{C}OH$ + iodoacetamide (ICH_2CONH_2) ⟶ products				
Pulse rad. of 2-propanol + acetone + H_2O				75 Wil 2
PR, KAS, competition kinetics	H_2O	RT	$k = 4 \cdot 10^8\ M^{-1}s^{-1}$ [85])	
$(CH_3)_2\dot{C}OH$ + iodoacetate (ICH_2COO^-) ⟶ products				
Pulse rad. of 2-propanol + acetone + H_2O				75 Wil 2
PR, KAS, competition kinetics	H_2O	RT	$k = 7 \cdot 10^7\ M^{-1}s^{-1}$ [85])	
$(CH_3)_2\dot{C}OH$ + (isonicotinamide)$^+$ ⟶ $(CH_3)_2CO$ + H^+ + (isonicotinamide)$^{\cdot}$ [86])				
Pulse rad. of 2-propanol + H_2O + N_2O				74 Bru 1
KAS	H_2O, pH = 0.7	RT	$k = 3.1(6) \cdot 10^9\ M^{-1}s^{-1}$	
$(CH_3)_2\dot{C}OH$ + isonicotinic acid [87]) [88]) [89]) ⟶ products [86])				
Pulse rad. of 2-propanol + H_2O + N_2O				74 Net 1
KAS	H_2O, pH = 0.4	RT	$k = 2.0 \cdot 10^9\ M^{-1}s^{-1}$ [87])	
	pH = 3.2		$k = 8.5 \cdot 10^8\ M^{-1}s^{-1}$ [88])	
	pH = 9		$k < 10^6\ M^{-1}s^{-1}$ [89])	
$(CH_3)_2\dot{C}OH$ + lipoate ion(—S—S—) ⟶ $(CH_3)_2CO$ + H^+ + —$\overset{(-)}{S \therefore S}$—				
Pulse rad. of 2-propanol + acetone + H_2O				70 Wil 1
KAS	H_2O	RT	$k = 1.8 \cdot 10^8\ M^{-1}s^{-1}$	
$(CH_3)_2\dot{C}OH$ + lumazine [90]) [91]) [92]) (LH_2-L^{2-}) ⟶ $(CH_3)_2CO$ + (lumazine)$^{\cdot}$ [90]) [91]) [92])				
Pulse rad. of 2-propanol + H_2O + N_2O				75 Moo 2
KAS	H_2O, pH = 0.8...5.1	RT	$k = 1.3(1) \cdot 10^9\ M^{-1}s^{-1}$ [90])	
	pH = 9.5		$k = 1.7(2) \cdot 10^8\ M^{-1}s^{-1}$ [91])	
	pH = 14		$k = 1.0(2) \cdot 10^9\ M^{-1}s^{-1}$ [92])	

[84]) Misonidazole.
[85]) Based on $k((CH_3)_2\dot{C}OH$ + 4-nitroacetophenone$) = 3.8 \cdot 10^9\ M^{-1}s^{-1}$.
[86]) Pyridinyl radical.
[87]) $HN^+C_5H_4COOH$.
[88]) $HN^+C_5H_4COO^-$.
[89]) $NC_5H_4COO^-$.
[90]) $LH_2 \longrightarrow LH_3^{\cdot}$.
[91]) $LH^- \longrightarrow LH_2^{\dot{-}}$.
[92]) $L^{2-} \longrightarrow LH^{2\dot{-}}$.

Reaction Radical generation Method	Solvent	T[K]	Rate data	Ref./ add. ref.
$(CH_3)_2\dot{C}OH$ + maleic acid ⟶ products [93]				
Pulse rad. of 2-propanol + N_2O + H_2O				73 Hay 2
KAS	H_2O, pH = 0.5	RT	$k = 2.2 \cdot 10^8\ M^{-1}s^{-1}$	
$(CH_3)_2\dot{C}OH$ + 4-methoxybenzenediazonium ($CH_3OC_6H_4N_2^+$) ⟶ products				
Pulse rad. of 2-propanol + N_2O + H_2O				81 Pac 1
KAS	H_2O	RT	$k = 3.2 \cdot 10^9\ M^{-1}s^{-1}$	
$(CH_3)_2\dot{C}OH$ + 3-methoxy-2-nitrotoluene($ArNO_2$) ⟶ $(CH_3)_2CO + H^+ + Ar\dot{N}O_2^-$				
Pulse rad. of 2-propanol + H_2O + N_2O				76 Net 1
KAS	H_2O	RT	$k = 2.3 \cdot 10^8\ M^{-1}s^{-1}$	
$(CH_3)_2\dot{C}OH$ + 2-methylbenzoquinone ⟶ $(CH_3)_2CO + H^+$ + ...-semiquinone				
Pulse rad. of 2-propanol + acetone + H_2O				73 Pat 1
KAS	H_2O	RT	$k = 3.5 \cdot 10^9\ M^{-1}s^{-1}$	
$(CH_3)_2\dot{C}OH$ + methylene blue ⟶ products				
Pulse rad. of 2-propanol + H_2O + N_2O				73 Rao 2/
KAS	H_2O	RT	$k = 4.4 \cdot 10^9\ M^{-1}s^{-1}$	73 Wil 1
$(CH_3)_2\dot{C}OH$ + 1-methylguanosine(MG^+) ⟶ $(CH_3)_2CO + H^+ + MG^{\cdot}$				
Pulse rad. of 2-propanol + H_2O				75 Moo 1
KAS	H_2O, pH = 0.5	RT	$k = 8.0 \cdot 10^7\ M^{-1}s^{-1}$	
$(CH_3)_2\dot{C}OH$ + 2-methyl-1,4-naphthoquinone ⟶ ...-semiquinone + products				
Pulse rad. of 2-propanol + H_2O + N_2O				72 Sim 1,
KAS	H_2O	RT	$k = 6.2 \cdot 10^9\ M^{-1}s^{-1}$	73 Pat 1,
			$k = 4.2(4) \cdot 10^9\ M^{-1}s^{-1}$ [94]	73 Rao 3,
			$k = 4.8(5) \cdot 10^9\ M^{-1}s^{-1}$ [95]	73 Rao 1
$(CH_3)_2\dot{C}OH$ + 3-methyl-2-nitrobenzoate ion ⟶ ...-$\dot{N}O_2^-$ + H^+ + $(CH_3)_2CO$				
Pulse rad. of 2-propanol + H_2O + N_2O				76 Net 1
KAS	H_2O	RT	$k = 1.9 \cdot 10^8\ M^{-1}s^{-1}$	
$(CH_3)_2\dot{C}OH$ + 2-methyl-5-nitroimidazole ⟶ ...-$\dot{N}O_2^-$ + H^+ + $(CH_3)_2CO$				
Pulse rad. of 2-propanol + H_2O				76 Gre 1
KAS	H_2O	RT	$k = 2.5 \cdot 10^9\ M^{-1}s^{-1}$	
$(CH_3)_2\dot{C}OH$ + 3-methyl-2-nitrophenol ⟶ $CH_3(OH)C_6H_3\dot{N}O_2^- + H^+ + (CH_3)_2CO$				
Pulse rad. of 2-propanol + H_2O + N_2O				76 Net 1
KAS	H_2O	RT	$k = 2.9 \cdot 10^8\ M^{-1}s^{-1}$	
$(CH_3)_2\dot{C}OH$ + 2-methyl-3-phytyl-1,4-naphthoquinone [96] ⟶ ...semiquinone + H^+ + $(CH_3)_2CO$				
Pulse rad. of 2-propanol + acetone + H_2O				73 Pat 1
KAS	H_2O	RT	$k = 1.7 \cdot 10^9\ M^{-1}s^{-1}$	
$(CH_3)_2\dot{C}OH$ + 3-methylpterin ⟶ products				
Pulse rad. of 2-propanol + H_2O				76 Moo 1
KAS	H_2O, pH = 0.8	RT	$k = 1.9 \cdot 10^9\ M^{-1}s^{-1}$	
	pH = 7.0		$2.9 \cdot 10^8$	

[93]) 18% e^--transfer. No e^--transfer to dianion at pH = 10.
[94]) From [73 Pat 1] and [73 Rao 3].
[95]) From [73 Rao 1].
[96]) Vitamin K_1.

Reaction Radical generation Method	Solvent	T[K]	Rate data	Ref./ add. ref.
$(CH_3)_2\dot{C}OH$ + 9-methylpurine(MP) ⟶ products [97]				
Pulse rad. of 2-propanol + H_2O + N_2O				76 Moo 1
KAS	H_2O, pH = 0	RT	$k = 1.9 \cdot 10^9\ M^{-1}s^{-1}$	
	pH = 8.6		$1.7 \cdot 10^8$	
$(CH_3)_2\dot{C}OH$ + 1,4-naphthoquinone ⟶ ...semiquinone + H^+ + $(CH_3)_2CO$				
Pulse rad. of 2-propanol + H_2O + acetone				73 Pat 1
KAS	H_2O	RT	$k = 3.6 \cdot 10^9\ M^{-1}s^{-1}$	
$(CH_3)_2\dot{C}OH$ + nicotinamide(NH^+) ⟶ $NH^{\cdot}$ + H^+ + $(CH_3)_2CO$ [98]				
Pulse rad. of 2-propanol + H_2O + N_2O				74 Bru 1,
KAS	H_2O, pH = 0.9	RT	$k = 2.1(4) \cdot 10^8\ M^{-1}s^{-1}$	74 Net 1
	pH = 1.9		$4.0 \cdot 10^8$ [99]	
$(CH_3)_2\dot{C}OH$ + nicotinamide-adenine dinucleotide(NAD^+) ⟶ $NAD^{\cdot}$ + H^+ + $(CH_3)_2CO$				
Pulse rad. of 2-propanol + H_2O + acetone				70 Wil 2/
KAS	H_2O	RT	$k = 1.0 \cdot 10^9\ M^{-1}s^{-1}$	73 Rao 1
$(CH_3)_2\dot{C}OH$ + nicotinic acid [1] ⟶ products [2]				
Pulse rad. of 2-propanol + H_2O + N_2O				74 Net 1
KAS	H_2O, pH = 0	RT	$k = 3.5 \cdot 10^8\ M^{-1}s^{-1}$	
	pH = 3.4		$k = 1.8 \cdot 10^8\ M^{-1}s^{-1}$	
	pH = 8.2		$k < 10^6\ M^{-1}s^{-1}$	
$(CH_3)_2\dot{C}OH$ + 4-nitroacetophenone(PNAP) ⟶ $PNAP^{\cdot -}$ + H^+ + $(CH_3)_2CO$				
Pulse rad. of 2-propanol + N_2O + H_2O				73 Ada 1
KAS	H_2O, pH = 11	RT	$k = 3.8(4) \cdot 10^9\ M^{-1}s^{-1}$	
$(CH_3)_2\dot{C}OH$ + 2-nitroaniline ⟶ $NH_2C_6H_4\dot{N}O_2^-$ + H^+ + $(CH_3)_2CO$				
Pulse rad. of 2-propanol + N_2O + H_2O				76 Net 1
KAS	H_2O	RT	$k = 9.2 \cdot 10^8\ M^{-1}s^{-1}$	
$(CH_3)_2\dot{C}OH$ + 3-nitroaniline ⟶ $NH_2C_6H_4\dot{N}O_2^-$ + H^+ + $(CH_3)_2CO$				
Pulse rad. of 2-propanol + N_2O + H_2O				76 Net 1
KAS	H_2O	RT	$k = 1.5 \cdot 10^9\ M^{-1}s^{-1}$	
$(CH_3)_2\dot{C}OH$ + 4-nitroaniline ⟶ $NH_2C_6H_4\dot{N}O_2^-$ + H^+ + $(CH_3)_2CO$				
Pulse rad. of 2-propanol + N_2O + H_2O				76 Net 1,
KAS	H_2O, pH = 7	RT	$k = 7.2 \cdot 10^8\ M^{-1}s^{-1}$	77 Lin 1
	pH = 10.7		$1.9(2) \cdot 10^9$ [3]	
$(CH_3)_2\dot{C}OH$ + nitrobenzene ($C_6H_5NO_2$) ⟶ $C_6H_5\dot{N}O_2^-$ + H^+ + $(CH_3)_2CO$				
Pulse rad. of 2-propanol + N_2O + H_2O				66 Asm 1
KAS	H_2O	RT	$k = 1.6 \cdot 10^9\ M^{-1}s^{-1}$	
$(CH_3)_2\dot{C}OH$ + 2-nitrobenzoate ⟶ $^-OOCC_6H_5\dot{N}O_2^-$ + H^+ + $(CH_3)_2CO$				
Pulse rad. of 2-propanol + N_2O + H_2O				76 Net 2
KAS	H_2O, pH = 7	RT	$k = 5.4 \cdot 10^8\ M^{-1}s^{-1}$ [4]	

[97] MPH_3^{2+} at pH = 0; $MPH^{\cdot}$ at pH = 8.6.
[98] No e^--transfer in neutral solution.
[99] From [74 Net 1].
[1] $^+HNC_5H_4COOH$ at pH = 0; $^+HNC_5H_4COO^-$ at pH = 3.4; $NC_5H_4COO^-$ at pH = 8.2.
[2] Pyridinyl radical.
[3] From [77 Lin 1].
[4] At pH = 0.8 $k \approx (1.0 \ldots 1.5) \cdot 10^8\ M^{-1}s^{-1}$.

Reaction Radical generation Method	Solvent	T[K]	Rate data	Ref./ add. ref.
$(CH_3)_2\dot{C}OH$ + 3-nitrobenzoate $\longrightarrow$ $^-OOCC_6H_5\dot{N}O_2^-$ + H^+ + $(CH_3)_2CO$				
Pulse rad. of 2-propanol + N_2O + H_2O				76 Net 2
KAS	H_2O, pH = 7	RT	$k = 9.0 \cdot 10^8\ M^{-1}s^{-1}$ [5])	
$(CH_3)_2\dot{C}OH$ + 4-nitrobenzoate $\longrightarrow$ $^-OOCC_6H_5\dot{N}O_2^-$ + $(CH_3)_2CO$ + H^+				
Pulse rad. of 2-propanol + H_2O + N_2O				76 Net 2
KAS	H_2O, pH = 7	RT	$k = 2.1 \cdot 10^9\ M^{-1}s^{-1}$ [6])	
$(CH_3)_2\dot{C}OH$ + *anti*-5-nitro-2-furaldoxime [7]) $\longrightarrow$ (*anti*-...)$^{\dot{-}}$ + H^+ + $(CH_3)_2CO$				
Pulse rad. of 2-propanol + N_2O + H_2O				73 Gre 1,
KAS	H_2O	RT	$k = 3.3 \cdot 10^9\ M^{-1}s^{-1}$ $3.5 \cdot 10^9$ [8])	76 Gre 1
$(CH_3)_2\dot{C}OH$ + 5-nitrofuroate ion $\longrightarrow$ ...$\dot{N}O_2^-$ + H^+ + $(CH_3)_2CO$				
Pulse rad. of 2-propanol + N_2O + H_2O				73 Gre 3
KAS	H_2O	RT	$k = 1.5 \cdot 10^9\ M^{-1}s^{-1}$	
$(CH_3)_2\dot{C}OH$ + 2-nitroimidazole $\longrightarrow$ (2-nitroimidazole)$^{\dot{-}}$ + H^+ + $(CH_3)_2CO$				
Pulse rad. of 2-propanol + N_2O + H_2O				76 Gre 1
KAS	H_2O	RT	$k = 3.5 \cdot 10^9\ M^{-1}s^{-1}$	
$(CH_3)_2\dot{C}OH$ + 4-nitroimidazole $\longrightarrow$ (4-nitroimidazole)$^{\dot{-}}$ + H^+ + $(CH_3)_2CO$				
Pulse rad. of 2-propanol + N_2O + H_2O				76 Gre 1
KAS	H_2O	RT	$k = 3.5 \cdot 10^9\ M^{-1}s^{-1}$	
$(CH_3)_2\dot{C}OH$ + 2-nitroisophthalate ion $\longrightarrow$ $(^-OOC)_2C_6H_3\dot{N}O_2^-$ + H^+ + $(CH_3)_2CO$				
Pulse rad. of 2-propanol + N_2O + H_2O				76 Net 1
KAS	H_2O	RT	$k = 1.5 \cdot 10^8\ M^{-1}s^{-1}$	
$(CH_3)_2\dot{C}OH$ + 4-nitroperoxybenzoic acid $\longrightarrow$ $(HOO)OCC_6H_4\dot{N}O_2^-$ + H^+ + $(CH_3)_2CO$				
Pulse rad. of 2-propanol + N_2O + H_2O				74 Lil 1
KAS	H_2O	RT	$k = 3.3 \cdot 10^9\ M^{-1}s^{-1}$	
$(CH_3)_2\dot{C}OH$ + 2-nitrophenol $\longrightarrow$ $HOC_6H_4\dot{N}O_2H$ + $(CH_3)_2CO$				
Pulse rad. of 2-propanol + H_2O				69 Gru 1
KAS	H_2O, pH = 1	RT	$k = 2.6 \cdot 10^9\ M^{-1}s^{-1}$	
$(CH_3)_2\dot{C}OH$ + 2-nitropyrrole $\longrightarrow$ (2-nitropyrrole)$^{\dot{-}}$ + H^+ + $(CH_3)_2CO$				
Pulse rad. of 2-propanol + N_2O + H_2O				76 Gre 1
KAS	H_2O	RT	$k = 2.0 \cdot 10^9\ M^{-1}s^{-1}$	
$(CH_3)_2\dot{C}OH$ + 3-nitropyrrole $\longrightarrow$ (3-nitropyrrole)$^{\dot{-}}$ + H^+ + $(CH_3)_2CO$				
Pulse rad. of 2-propanol + N_2O + H_2O				76 Gre 1
KAS	H_2O	RT	$k = 2.0 \cdot 10^9\ M^{-1}s^{-1}$	
$(CH_3)_2\dot{C}OH$ + nitrosobenzene (C_6H_5NO) $\longrightarrow$ $C_6H_5\dot{N}OH$ + $(CH_3)_2CO$				
Pulse rad. of 2-propanol + N_2O + H_2O				66 Asm 2
KAS	H_2O	RT	$k = 5.0 \cdot 10^9\ M^{-1}s^{-1}$	
$(CH_3)_2\dot{C}OH$ + 2-nitrothiophene $\longrightarrow$ (2-nitrothiophene)$^{\dot{-}}$ + H^+ + $(CH_3)_2CO$				
Pulse rad. of 2-propanol + N_2O + H_2O				76 Gre 1
KAS	H_2O	RT	$k = 3.0 \cdot 10^9\ M^{-1}s^{-1}$	

[5]) At pH = 0.8 $k \approx (1.8...2.7) \cdot 10^8\ M^{-1}s^{-1}$.
[6]) At pH = 0.8 $k \approx (4...6) \cdot 10^8\ M^{-1}s^{-1}$.
[7]) Nifuroxime.
[8]) From [76 Gre 1].

Reaction Radical generation Method	Solvent	T[K]	Rate data	Ref./ add. ref.
$(CH_3)_2\dot{C}OH$ + 3-nitrothiophene ⟶ (3-nitrothiophene)$^{\overline{\cdot}}$ + H^+ + $(CH_3)_2CO$				
Pulse rad. of 2-propanol + N_2O + H_2O				76 Gre 1
KAS	H_2O	RT	$k = 2.0\cdot10^9\ M^{-1}s^{-1}$	
$(CH_3)_2\dot{C}OH$ + 2-nitrotoluene ⟶ $CH_3C_6H_4\dot{N}O_2^-$ + H^+ + $(CH_3)_2CO$				
Pulse rad. of 2-propanol + N_2O + H_2O				76 Net 1
KAS	H_2O	RT	$k = 4.8\cdot10^8\ M^{-1}s^{-1}$	
$(CH_3)_2\dot{C}OH$ + 5-nitrouracil ⟶ (5-nitrouracil)$^{\overline{\cdot}}$ + H^+ + $(CH_3)_2CO$				
Pulse rad. of 2-propanol + $(CH_3)_2CO$ + H_2O				73 Wil 1
KAS	H_2O	RT	$k = 7.0(35)\cdot10^8\ M^{-1}s^{-1}$	
$(CH_3)_2\dot{C}OH$ + norpseudopelletierine-N-oxyl ⟶ products				
Pulse rad. of 2-propanol + H_2O				71 Fie 1
KAS	H_2O	RT	$k = 8.1(8)\cdot10^8\ M^{-1}s^{-1}$	
$(CH_3)_2\dot{C}OH$ + 1,10-phenanthroline ⟶ products				
Pulse rad. of 2-propanol + acetone + H_2O				79 Mul 1/
KAS	H_2O, pH = 7	RT	$k = 1.2\cdot10^7\ M^{-1}s^{-1}$ [9]	80 Tep 1
	pH = 3		$1.9\cdot10^9$ [10]	
$(CH_3)_2\dot{C}OH$ + 1,10-phenanthroline [11] ⟶ (1,10-phenanthroline)$^{\overline{\cdot}}$ [12] + H^+ + $(CH_3)_2CO$				
Pulse rad. of 2-propanol + N_2O + H_2O				80 Tep 1/
KAS	H_2O, pH = 1	RT	$k = 3.2(2)\cdot10^9\ M^{-1}s^{-1}$	79 Mul 1
	pH = 5.5		$1\cdot10^8$	
	pH = 8		$< 10^7$	
$(CH_3)_2\dot{C}OH$ + phenosafranine ⟶ products [13]				
Pulse rad. of 2-propanol + N_2O + H_2O				73 Rao 2
KAS	H_2O	RT	$k = 3.2\cdot10^9\ M^{-1}s^{-1}$	
$(CH_3)_2\dot{C}OH$ + pterin ⟶ products [14]				
Pulse rad. of 2-propanol + N_2O + H_2O				76 Moo 1
KAS	H_2O, pH = 0.8	RT	$k = 2.0\cdot10^9\ M^{-1}s^{-1}$ [15]	
	pH = 7		$4.5\cdot10^8$ [16]	
	pH = 9.4		$\leqslant 10^7$ [17]	
$(CH_3)_2\dot{C}OH$ + purine ⟶ products				
Pulse rad. of 2-propanol + N_2O + H_2O				75 Moo 1
KAS	H_2O, pH = 0	RT	$k = 2.7\cdot10^9\ M^{-1}s^{-1}$ [18]	
	pH = 6 and 13		$< 10^7$ [19]	
$(CH_3)_2\dot{C}OH$ + pyrazine ⟶ products [20]				
Pulse rad. of 2-propanol + N_2O + H_2O				74 Moo 1
KAS	H_2O, pH = 0	RT	$k = 2.8\cdot10^9\ M^{-1}s^{-1}$	
	in 70% $HClO_4$		$5\cdot10^9$	

[9] 1,10-phenanthroline neutral form.
[10] 1,10-phenanthroline protonated form.
[11] Fully protonated at pH = 1, partially protonated at pH = 5.5.
[12] Possibly protonated at lower pH.
[13] 82% e^--transfer.
[14] 100% e^--transfer.
[15] Reaction with protonated form of pterin.
[16] Reaction with neutral form of pterin.
[17] Reaction with anionic form of pterin.
[18] Reaction with protonated form of purine.
[19] Reaction with neutral and anionic form of purine, respectively.
[20] Reaction with protonated form of pyrazine; $k < 10^7\ M^{-1}s^{-1}$ at pH = 5 and 11.

Reaction Radical generation Method	Solvent	T[K]	Rate data	Ref./ add. ref.
$(CH_3)_2\dot{C}OH$ + pyrazinecarboxylic acid $\longrightarrow$ products				
Pulse rad. of 2-propanol + N_2O + H_2O				78 Wie 2
KAS	H_2O	RT	$k \approx 8.5 \cdot 10^8\ M^{-1}s^{-1}$	
$(CH_3)_2\dot{C}OH$ + pyridazine $\longrightarrow$ products				
Pulse rad. of 2-propanol + N_2O + H_2O				74 Moo 1
KAS	H_2O, pH = 0	RT	$k = 2.6 \cdot 10^9\ M^{-1}s^{-1}$ [21])	
	pH = 5, 11		$< 10^7$ [22])	
$(CH_3)_2\dot{C}OH$ + 4-pyridinecarboxaldoxime $\longrightarrow$ products				
Pulse rad. of 2-propanol + N_2O + H_2O				76 Net 3
KAS	H_2O, acid pH	RT	$k = 1.7 \cdot 10^9\ M^{-1}s^{-1}$	
	pH = 7		$< 10^7$	
$(CH_3)_2\dot{C}OH$ + 2-pyridinecarboxaldoxime methochloride $\longrightarrow$ products				
Pulse rad. of 2-propanol + N_2O + H_2O				76 Net 3
KAS	H_2O	RT	$k = 6 \cdot 10^8\ M^{-1}s^{-1}$	
$(CH_3)_2\dot{C}OH$ + pyridoxal-5-phosphate $\longrightarrow$ products				
Pulse rad. of 2-propanol + N_2O + H_2O				75 Moo 3
KAS	H_2O, pH = 1	RT	$k = 5.8 \cdot 10^8\ M^{-1}s^{-1}$	
	pH = 5.6		$1.3 \cdot 10^8$	
$(CH_3)_2\dot{C}OH$ + pyrimidine $\longrightarrow$ products				
Pulse rad. of 2-propanol + N_2O + H_2O				74 Moo 1
KAS	H_2O, pH = 0	RT	$k = 2.2 \cdot 10^9\ M^{-1}s^{-1}$ [23])	
	pH = 5, 11, 13.6		$< 10^7$ [24])	
$(CH_3)_2\dot{C}OH$ + quinoxaline $\longrightarrow$ products				
Pulse rad. of 2-propanol + N_2O + H_2O				74 Moo 1
KAS	H_2O, pH = 6.5	RT	$k = 1.6 \cdot 10^8\ M^{-1}s^{-1}$ [25])	
	pH = 0		$3.7 \cdot 10^9$ [26])	
	in 70% $HClO_4$		$7.0 \cdot 10^8$	
$(CH_3)_2\dot{C}OH$ + riboflavin $\longrightarrow$ products				
Pulse rad. of 2-propanol + N_2O + H_2O				73 Rao 1
KAS	H_2O	RT	$k = 2.3(2) \cdot 10^9\ M^{-1}s^{-1}$	
$(CH_3)_2\dot{C}OH$ + riboflavin-5-phosphate [27]) $\longrightarrow$ products				
Pulse rad. of 2-propanol + acetone + H_2O				75 Wil 2
KAS	H_2O	RT	$k = 1.0(5) \cdot 10^9\ M^{-1}s^{-1}$	
$(CH_3)_2\dot{C}OH$ + safranine T $\longrightarrow$ products [28])				
Pulse rad. of 2-propanol + N_2O + H_2O				73 Rao 2
KAS	H_2O	RT	$k = 2.8 \cdot 10^9\ M^{-1}s^{-1}$	
$(CH_3)_2\dot{C}OH$ + *meso*-tetra(4-carboxyphenyl)porphyrine (H_2TCPP) $\longrightarrow$ ($H_2TCPP^{\dot{-}}$) + H^+ + $(CH_3)_2CO$				
Pulse rad. of 2-propanol + N_2O + H_2O				79 Net 2
KAS	H_2O, pH = 7...11	RT	$k = 9(1) \cdot 10^8\ M^{-1}s^{-1}$	

[21]) Reaction with protonated form of pyridazine.
[22]) Reaction with neutral form of pyridazine.
[23]) Reaction with protonated form of pyrimidine.
[24]) Reaction with neutral form of pyrimidine.
[25]) Reaction with neutral form of quinoxaline.
[26]) Reaction with protonated form of quinoxaline.
[27]) Flavin mononucleotide.
[28]) 86% e^--transfer.

Reaction Radical generation Method	Solvent	T[K]	Rate data	Ref./ add. ref.
$(CH_3)_2\dot{C}OH$ + tetramethylbenzoquinone [29]) $\longrightarrow$ $(CH_3)_2CO + H^+ + \ldots$semiquinone				
Pulse rad. of 2-propanol + acetone + H_2O				73 Pat 1
KAS	H_2O	RT	$k = 4.0 \cdot 10^9\ M^{-1}s^{-1}$	
$(CH_3)_2\dot{C}OH$ + tetramethyldiazenedicarboxamide ($(CH_3)_2NCON{=}NCON(CH_3)_2$) $\longrightarrow$ products				
Pulse rad. of 2-propanol + H_2O + N_2O				75 Whi 1
KAS	H_2O	RT	$k \approx 2.5 \cdot 10^9\ M^{-1}s^{-1}$	
$(CH_3)_2\dot{C}OH$ + 2,2,6,6-tetramethyl-4-hydroxy-1-piperidinyloxy(TMPN) $\longrightarrow$ products				
Pulse rad. of 2-propanol + N_2O + H_2O				76 Asm 1
Cond. (time resolved)	H_2O, pH = 3...5	RT	$k = 3.6(4) \cdot 10^8\ M^{-1}s^{-1}$	
$(CH_3)_2\dot{C}OH$ + 2,2,6,6-tetramethyl-4-oxo-1-piperidinyloxy(TAN) $\longrightarrow$ products				
Pulse rad. of 2-propanol + N_2O + H_2O				76 Asm 1/ 71 Wil 2
Cond. (time resolved)	H_2O, pH = 3...5	RT	$k = 4.3(4) \cdot 10^8\ M^{-1}s^{-1}$	
$(CH_3)_2\dot{C}OH$ + 2,2,5,5-tetramethyl-3-pyrroline-1-yloxy-3-carboxamide $\longrightarrow$ products				
Pulse rad. of 2-propanol + N_2O + H_2O				76 Nig 1
Cond. (time resolved)	H_2O, pH = 3...5	RT	$k = 3.4(4) \cdot 10^8\ M^{-1}s^{-1}$	
$(CH_3)_2\dot{C}OH$ + 2,2,5,5-tetramethyl-1-pyrrolidinyloxy-3-carboxamide $\longrightarrow$ products				
Pulse rad. of 2-propanol + N_2O + H_2O				76 Nig 1
Cond. (time resolved)	H_2O, pH = 3...5	RT	$k = 3.3(3) \cdot 10^8\ M^{-1}s^{-1}$	
$(CH_3)_2\dot{C}OH$ + tetranitromethane ($C(NO_2)_4$) $\longrightarrow$ $C(NO_2)_3^- + NO_2 + H^+ + (CH_3)_2CO$				
Pulse rad. of 2-propanol + N_2O + H_2O				64 Asm 1
KAS	H_2O	RT	$k = 5.0(10) \cdot 10^9\ M^{-1}s^{-1}$	
$(CH_3)_2\dot{C}OH$ + thiamine $\longrightarrow$ products				
Pulse rad. of 2-propanol + H_2O				77 Moo 1
KAS	H_2O, pH = 0.5 pH = 6.6	RT	$k = 2.2 \cdot 10^8\ M^{-1}s^{-1}$ [30]) $1.9 \cdot 10^8$ [31])	
$(CH_3)_2\dot{C}OH$ + thiazole ($C_3H_3SNH^+$) $\longrightarrow$ $(CH_3)_2CO + C_3H_3SNH_2^{\dot{+}}$				
Pulse rad. of 2-propanol + H_2O				77 Moo 1
KAS	H_2O	RT	$k = 6.2 \cdot 10^8\ M^{-1}s^{-1}$ [32])	
$(CH_3)_2\dot{C}OH$ + thionine $\longrightarrow$ products				
Pulse rad. of 2-propanol + N_2O + H_2O				73 Rao 2
KAS	H_2O	RT	$k = 4.2 \cdot 10^9\ M^{-1}s^{-1}$ [33])	
$(CH_3)_2\dot{C}OH$ + trichloroacetate (CCl_3COO^-) $\longrightarrow$ products				
Pulse rad. of 2-propanol + acetone + H_2O				75 Wil 2
Competition kinetics	H_2O	RT	$k = 5 \cdot 10^6\ M^{-1}s^{-1}$ [34])	

[29]) Duroquinone.
[30]) Reaction with protonated form of thiamine.
[31]) Reaction with neutral form of thiamine.
[32]) No reaction at high pH with neutral form of thiazole.
[33]) 88% e^--transfer.
[34]) Relative to $k((CH_3)_2\dot{C}OH$ + 4-nitroacetophenone$) = 3.8 \cdot 10^9\ M^{-1}s^{-1}$.

Reaction Radical generation Method	Solvent	T[K]	Rate data	Ref./ add. ref.
$(CH_3)_2\dot{C}OH$ + trichloroacetaldehyde hydrate $(CCl_3CH(OH)_2) \longrightarrow Cl^- +$ products $+ H^+ + (CH_3)_2CO$				
Pulse rad. of 2-propanol + $H_2O + N_2O$				75 Wil 2/
PR, KAS, Competition kinetics	H_2O	RT	$k = 1 \cdot 10^6\ M^{-1}s^{-1}$ [35]	73 Wil 1
$(CH_3)_2\dot{C}OH$ + 2,3,5-trimethylbenzoquinone $\longrightarrow$ ...semiquinone $+ (CH_3)_2CO + H^+$				
Pulse rad. of 2-propanol + acetone + H_2O				73 Pat 1
KAS	H_2O	RT	$k = 3.6 \cdot 10^9\ M^{-1}s^{-1}$	
$(CH_3)_2\dot{C}OH$ + 2,4,6-trinitrobenzoate ion $\longrightarrow (CH_3)_2CO + H^+ + {}^-OOCC_6H_2(NO_2)_2\dot{N}O_2^-$				
Pulse rad. of 2-propanol + $N_2O + H_2O$				76 Net 2
KAS	H_2O	RT	$k = 3.9 \cdot 10^9\ M^{-1}s^{-1}$	
$(CH_3)_2\dot{C}OH$ + trypan blue $\longrightarrow$ products				
Pulse rad. of 2-propanol + acetone + H_2O				73 Wil 1
KAS	H_2O	RT	$k = 3.0(15) \cdot 10^9\ M^{-1}s^{-1}$	
$(CH_3)_2\dot{C}OH$ + ubiquinone [36] $\longrightarrow$ ubisemiquinone $+ H^+ + (CH_3)_2CO$				
Pulse rad. of 2-propanol + acetone + H_2O				73 Pat 1
KAS	H_2O	RT	$k = 1.9 \cdot 10^9\ M^{-1}s^{-1}$	
$\dot{C}H(OCH_3)_2 + H_2O_2 \longrightarrow$ products				
H_2O + Ti(III) + H_2O_2 + dimethoxymethane				74 Gil 1
PR, ESR	H_2O	RT	$k = 1.1(3) \cdot 10^6\ M^{-1}s^{-1}$ [37]	
$\dot{C}HOHCHOHCH_2OH + Fe(CN)_6^{3-} \longrightarrow Fe(CN)_6^{4-}$ + products				
Pulse rad. of glycerol + H_2O				69 Ada 1
KAS	H_2O	RT	$k = 3.3 \cdot 10^9\ M^{-1}s^{-1}$	
$\dot{C}HOHCHOHCH_2OH$ + Fe(III)cytochrome c $\longrightarrow$ Fe(II)cytochrome c + products				
Pulse rad. of glycerol + $H_2O + N_2O$				75 Sim 1
KAS	H_2O	RT	$k = 2.5 \cdot 10^6\ M^{-1}s^{-1}$	
$\dot{C}HOHCHOHCH_2OH$ + *anti*-5-nitro-2-furaldoxime [38] $\longrightarrow$ products [39]				
Pulse rad. of glycerol + $N_2O + H_2O$				73 Gre 1
KAS	H_2O	RT	$k = 3.0 \cdot 10^8\ M^{-1}s^{-1}$	
$CH_2OH\dot{C}OHCH_2OH$ + hemin c(Fe(III)) $\longrightarrow$ hemin c(Fe(II)) + products				
Pulse rad. of glycerol + $H_2O + N_2O$				75 Gof 1
KAS	H_2O	RT	$k = 1.3(3) \cdot 10^9\ M^{-1}s^{-1}$	
${}^-OOC\dot{C}HCH_2COO^- + IrCl_6^{2-} \longrightarrow$ Ir(III) + products				
Pulse rad. of $(CH_2COO^-)_2 + N_2O + H_2O$				82 Ste 1 [40]
KAS	H_2O	295	$k = 1.1 \cdot 10^8\ M^{-1}s^{-1}$	

[35] Based on competition kinetics with $k((CH_3)_2\dot{C}OH$ + 4-nitroacetophenone$) = 3.8 \cdot 10^9\ M^{-1}s^{-1}$.

[36]

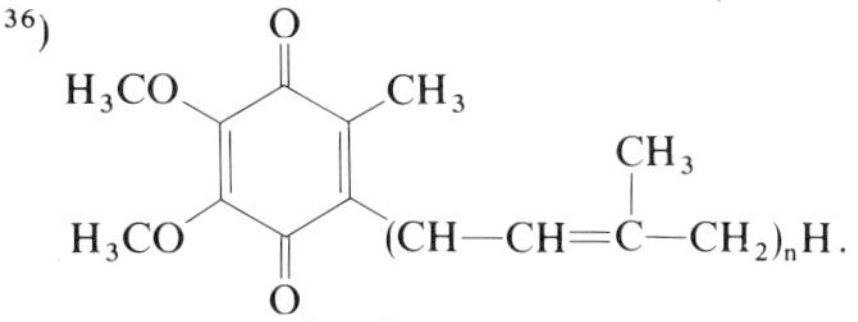

[37] Based on $2k(\dot{R} + \dot{R}) = 2 \cdot 10^9\ M^{-1}s^{-1}$.

[38] Nifuroxime.

[39] 20% e^--transfer.

[40] Mechanism discussed as e^-- or $Cl^{\cdot}$-transfer.

Reaction Radical generation Method	Solvent	T[K]	Rate data	Ref./ add. ref.
$^-OOCCH_2\dot{C}OHCOO^-$ + 2-methyl-1,4-naphthoquinone ⟶ ...semiquinone + products				
Pulse rad. of malate + H_2O + N_2O [41]) and of oxaloacetate + t-butanol + H_2O [42])				73 Rao 3
KAS	H_2O, pH = 7	RT	$k = 1.1(1) \cdot 10^9\ M^{-1}s^{-1}$ [41])	
	pH = 6.2		$k = 3.1(3) \cdot 10^9\ M^{-1}s^{-1}$ [42])	
$^-OOCCH_2\dot{C}OHCOO^-$ + *anti*-5-nitro-2-furaldoxime [43]) ⟶ products [44])				
Pulse rad. of malate + N_2O + H_2O				73 Gre 1
KAS	H_2O	RT	$k = 3.6 \cdot 10^8\ M^{-1}s^{-1}$	
$^-OOC\dot{C}OHCHOHCOO^-$ + Fe(III)cytochrome c ⟶ Fe(II)cytochrome c + products				
Pulse rad. of tartrate + H_2O + N_2O				75 Sim 1/
KAS	H_2O	RT	$k = 1.7 \cdot 10^8\ M^{-1}s^{-1}$	78 Sim 1
$^-OOC\dot{C}OHCHOHCOO^-$ + cytochrome c (carboxymethylated) ⟶ products				
Pulse rad. of tartrate + H_2O + N_2O				78 Sim 1
KAS	H_2O	RT	$k = 2.8 \cdot 10^7\ M^{-1}s^{-1}$	
$^-OOC\dot{C}OHCHOHCOO^-$ + hemin c(Fe(III)) ⟶ hemin c(Fe(II)) + products				
Pulse rad. of tartrate + H_2O + N_2O				75 Gof 1
KAS	H_2O	RT	$k = 8.1(16) \cdot 10^7\ M^{-1}s^{-1}$	
$^-OOC\dot{C}OHCHOHCOO^-$ + metmyoglobin(Fe(III)) ⟶ metmyoglobin(Fe(II)) + products				
Pulse rad. of tartrate + H_2O + N_2O				78 Sim 1
KAS	H_2O	RT	$k = 3.5(7) \cdot 10^7\ M^{-1}s^{-1}$	
$^-OOC\dot{C}OHCHOHCOO^-$ + 2-methyl-1,4-naphthoquinone ⟶ semiquinone + products [45])				
Pulse rad. of tartrate + H_2O + N_2O				73 Rao 3
KAS	H_2O, pH = 11.0	RT	$k = 7.0(7) \cdot 10^8\ M^{-1}s^{-1}$	
$CH_3\dot{C}(O^-)CH_2COO^-$ + $Fe(CN)_6^{3-}$ ⟶ products				
Pulse rad. of acetoacetate + t-butanol + H_2O				73 Rao 1
KAS	H_2O	RT	$k = 7.3(7) \cdot 10^8\ M^{-1}s^{-1}$	
$CH_3\dot{C}(O^-)CH_2COO^-$ + MnO_4^- ⟶ products				
Pulse rad. of acetoacetate + t-butanol + H_2O + N_2O				73 Rao 1
KAS	H_2O, pH = 9.2	RT	$k = 4.8(5) \cdot 10^9\ M^{-1}s^{-1}$	
$CH_3\dot{C}(O^-)CH_2COO^-$ + 9,10-anthraquinone ⟶ ...semiquinone + products				
Pulse rad. of acetoacetate + t-butanol + H_2O + Ar				73 Rao 1
KAS	H_2O, pH = 9.2	RT	$k = 6.7(7) \cdot 10^8\ M^{-1}s^{-1}$	
$CH_3\dot{C}(O^-)CH_2COO^-$ + 2,5-dimethylbenzoquinone ⟶ ...semiquinone + products				
Pulse rad. of acetoacetate + t-butanol + H_2O + Ar				73 Rao 1
KAS	H_2O, pH = 9.2	RT	$k = 3.3 \cdot 10^9\ M^{-1}s^{-1}$	
$CH_3\dot{C}(O^-)CH_2COO^-$ + 2-hydroxy-1,4-naphthoquinone ⟶ ...semiquinone + products				
Pulse rad. of acetoacetate + t-butanol + H_2O				73 Rao 1
KAS	H_2O, pH = 9.2	RT	$k = 1.5(2) \cdot 10^9\ M^{-1}s^{-1}$	

[41]) 47% e^--transfer.
[42]) 92% e^--transfer.
[43]) Nifuroxime.
[44]) 30% e^--transfer.
[45]) 69% e^--transfer.

Reaction Radical generation Method	Solvent	T[K]	Rate data	Ref./ add. ref.
$CH_3\dot{C}(O^-)CH_2COO^-$ + 2-methyl-1,4-naphthoquinone ⟶ ...semiquinone + products				
Pulse rad. of acetoacetate + t-butanol + H_2O				73 Rao 3/
KAS	H_2O	RT	$k = 3.7(4)\cdot 10^9\ M^{-1}s^{-1}$	73 Rao 1
$HOOC\dot{C}HCH_2COOH + IrCl_6^{2-}$ ⟶ Ir(III) + products				
Pulse rad. of $(CH_2COOH)_2 + N_2O + H_2O$				82 Ste 1 [46]
KAS	H_2O	295	$k = 4.6\cdot 10^8\ M^{-1}s^{-1}$	
$HOOC\dot{C}OHCHOHCOOH$ + 2-methyl-1,4-naphthoquinone ⟶ ...semiquinone + products [47]				
Pulse rad. of tartrate + $H_2O + N_2O$				73 Rao 3
KAS	H_2O, pH = 3.2	RT	$k = 7.0(7)\cdot 10^8\ M^{-1}s^{-1}$	
$CH_3\dot{C}(O^-)COOCH_3$ + 2-methyl-1,4-naphthoquinone ⟶ ...semiquinone + products [48]				
Pulse rad. of methyl lactate + $H_2O + N_2O$				73 Rao 3
KAS	H_2O, pH = 10.4	RT	$k = 2.3(2)\cdot 10^9\ M^{-1}s^{-1}$	
$CH_3CO\dot{C}HCH_3$ [49] + hydroquinone ⟶ $CH_3COCH_2CH_3 + {}^-OC_6H_4\dot{O}$				
Pulse rad. of 2,3-butanediol + $N_2O + H_2O$				79 Ste 1
KAS	H_2O, pH ≈ 11.5	RT	$k = 5.6(5)\cdot 10^8\ M^{-1}s^{-1}$	
 + H_2O_2 ⟶ products $\dot{A}$				
H_2O + Ti(III) + H_2O_2 + THF				74 Gil 1
PR, ESR	H_2O	RT	$k = 3.0\cdot 10^4\ M^{-1}s^{-1}$ [50]	
$\dot{A}$ + carbontetrachloride (CCl_4) ⟶ Cl^- + products				
Pulse rad. of THF + $H_2O + N_2O$				71 Koe 1
Cond., time resolved	H_2O	RT	$k = 2.0\cdot 10^7\ M^{-1}s^{-1}$	
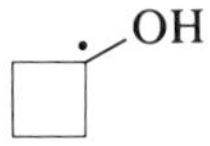 + $Fe(CN)_6^{3-}$ ⟶ 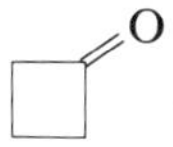 + H^+ + $Fe(CN)_6^{4-}$ **$\dot{B}$** **C**				
Pulse rad. of cyclobutanone + t-butanol + H_2O				76 Tof 1
KAS	H_2O	275 337	$k = 2.6\cdot 10^9\ M^{-1}s^{-1}$ $5.3\cdot 10^9$ $E_a = 8.8\ kJ\,mol^{-1}$	
$\dot{B}$ + nitrobenzene ($C_6H_5NO_2$) ⟶ **C** + H^+ + $C_6H_5\dot{N}O_2^-$				
Pulse rad. of cyclobutanone + t-butanol + H_2O				76 Tof 1
KAS	H_2O	275 337	$k = 1.5\cdot 10^9\ M^{-1}s^{-1}$ $3.4\cdot 10^9$ $E_a = 9.6\ kJ\,mol^{-1}$	
$CH_3CO\dot{C}OHCH_3 + Ru(NH_3)_6^{3+}$ ⟶ products				
Pulse rad. of biacetyl + $H_2O + N_2O$				72 Coh 1
KAS	H_2O, pH = 5.5	RT	$k = 2.0\cdot 10^9\ M^{-1}s^{-1}$	

[46] Mechanism discussed as e^-- or $Cl^{\cdot}$-transfer.
[47] 14% e^--transfer.
[48] 74% e^--transfer.
[49] Mainly C-centered radical, oxidizing action, however, likely to occur through mesomeric O-centered radical.
[50] Based on $2k(\dot{R} + \dot{R}) = 3\cdot 10^9\ M^{-1}s^{-1}$.

Reaction Radical generation Method	Solvent	T[K]	Rate data	Ref./ add. ref.
$HOCH_2CH_2CO\dot{C}H_2$ [51]) + hydroquinone ⟶ $HOCH_2CH_2COCH_3$ + $^{-}OC_6H_4\dot{O}$				
Pulse rad. of 2-hydroxyfuran + N_2O + H_2O				79 Ste 1
KAS	H_2O, pH ≈ 11.5	RT	$k = 6.2(8) \cdot 10^8\ M^{-1}s^{-1}$	
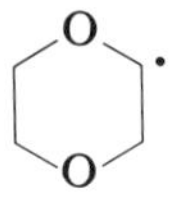 + *trans*-(dibromobis(1,2-ethanediamine-N,N')cobalt(III))$^+$ ⟶ products Ḋ				
Pulse rad. of dioxan + N_2O + H_2O				77 Coh 1
KAS	H_2O, pH = 3.5...4.0	RT	$k = 4.4 \cdot 10^8\ M^{-1}s^{-1}$	
Ḋ + *trans*-(dichlorobis(1,2-ethanediamine-N,N')cobalt(III))$^+$ ⟶ products				
Pulse rad. of dioxan + N_2O + H_2O				77 Coh 1
KAS	H_2O, pH = 3.5...4.0	RT	$k = 3.5 \cdot 10^7\ M^{-1}s^{-1}$	
Ḋ + H_2O_2 ⟶ products				
H_2O + Ti(III) + H_2O_2 + dioxan				74 Gil 1
PR, ESR	H_2O	RT	$k = 3.0(6) \cdot 10^4\ M^{-1}s^{-1}$ [52])	
Ḋ + $IrCl_6^{2-}$ ⟶ Ir(III) + products				
Pulse rad. of dioxan + N_2O + H_2O				82 Ste 1
KAS, Cond.	H_2O	295	$k = 5.4 \cdot 10^9\ M^{-1}s^{-1}$	
Ḋ + MnO_4^- ⟶ Mn(VI) + products				
Pulse rad. of dioxan + N_2O + H_2O				82 Ste 1
KAS	H_2O	295	$k = 6.5 \cdot 10^9\ M^{-1}s^{-1}$	
Ḋ + $Ru(NH_3)_6^{3+}$ ⟶ products				
Pulse rad. of dioxan + H_2O + N_2O				77 Coh 1
KAS	H_2O, pH = 3.5...4.0	RT	$k = 5.00(75) \cdot 10^6\ M^{-1}s^{-1}$	
Ḋ + $(Ru(III)(NH_3)_5Br)^{2+}$ ⟶ products				
Pulse rad. of dioxan + N_2O + H_2O				77 Coh 1
KAS	H_2O, pH = 3.5...4.0	RT	$k = 2.7(4) \cdot 10^8\ M^{-1}s^{-1}$	
Ḋ + $(Ru(III)(NH_3)_5Cl)^{2+}$ ⟶ products				
Pulse rad. of dioxan + N_2O + H_2O				77 Coh 1
KAS	H_2O	RT	$k = 8.3(12) \cdot 10^7\ M^{-1}s^{-1}$	
$CH_3CH_2CH_2\dot{C}HO^-$ + nitrobenzene ($C_6H_5NO_2$) ⟶ $C_6H_5\dot{N}O_2^-$ + $CH_3CH_2CH_2CHO$				
Pulse rad. of 1-butanol + N_2O + H_2O				66 Asm 1
KAS	H_2O, pH = 13	RT	$k = 3.1 \cdot 10^9\ M^{-1}s^{-1}$	
$(CH_3)_2CH\dot{C}HO^-$ + nitrobenzene ($C_6H_5NO_2$) ⟶ $C_6H_5\dot{N}O_2^-$ + $(CH_3)_2CHCHO$				
Pulse rad. of 2-methyl-1-propanol + N_2O + H_2O				66 Asm 1
KAS	H_2O, pH = 13	RT	$k = 2.9 \cdot 10^9\ M^{-1}s^{-1}$	
$CH_3CH_2CH_2\dot{C}HOH$ + 2-methyl-1,4-naphthoquinone ⟶ ...semiquinone + products [53])				
Pulse rad. of 1-butanol + H_2O + N_2O				73 Rao 3
KAS	H_2O	RT	$k = 4.1(4) \cdot 10^9\ M^{-1}s^{-1}$	

[51]) Mainly C-centered radical, oxidizing action, however, to occur through mesomeric O-centered radical.
[52]) Based on $2k(\dot{R} + \dot{R}) = 2 \cdot 10^9\ M^{-1}s^{-1}$.
[53]) 32% e^--transfer.

Reaction Radical generation Method	Solvent	T[K]	Rate data	Ref./ add. ref.
$CH_3CH_2CH_2\dot{C}HOH$ + 4-nitroacetophenone(PNAP) ⟶ products [54]				
Pulse rad. of 1-butanol + N_2O + H_2O				73 Gre 2
KAS	H_2O	RT	$k = 6 \cdot 10^8\ M^{-1}s^{-1}$	
$CH_3CH_2CH_2\dot{C}HOH$ + nitrobenzene ($C_6H_5NO_2$) ⟶ $C_6H_5\dot{N}O_2^-$ + $CH_3CH_2CH_2CHO$				
Pulse rad. of 1-butanol + N_2O + H_2O				66 Asm 1,
KAS	H_2O	RT	$k = 4.0 \cdot 10^8\ M^{-1}s^{-1}$ [55] $5.0 \cdot 10^8$ [56]	73 Gre 2
$CH_3CH_2CH_2\dot{C}HOH$ + *anti*-5-nitro-2-furaldoxime [57] ⟶ products [58]				
Pulse rad. of 1-butanol + N_2O + H_2O				73 Gre 1
KAS	H_2O	RT	$k = 3.8 \cdot 10^9\ M^{-1}s^{-1}$	
$CH_3CH_2CH_2\dot{C}HOH$ + nitrosobenzene (C_6H_5NO) ⟶ $C_6H_5\dot{N}OH$ + $CH_3CH_2CH_2CHO$				
Pulse rad. of 1-butanol + N_2O + H_2O				66 Asm 2
KAS	H_2O	RT	$k = 4.0 \cdot 10^9\ M^{-1}s^{-1}$	
$CH_3\dot{C}OHCH_2CH_3$ + $Fe(CN)_6^{3-}$ ⟶ $Fe(CN)_6^{4-}$ + H^+ + $CH_3COCH_2CH_3$				
Pulse rad. of 2-butanol + H_2O				69 Ada 1
KAS	H_2O	RT	$k = 4.8 \cdot 10^9\ M^{-1}s^{-1}$	
$CH_3\dot{C}OHCH_2CH_3$ + 2,3-butanedione ($CH_3COCOCH_3$) ⟶ $CH_3COCH_2CH_3$ + H^+ + $(CH_3COCOCH_3)^{\dot{-}}$				
Pulse rad. of 2-butanol + H_2O + N_2O				68 Lil 1
KAS	H_2O	RT	$k = 7.2 \cdot 10^8\ M^{-1}s^{-1}$	
$(CH_3)_2CH\dot{C}HOH$ + $Fe(CN)_6^{3-}$ ⟶ $Fe(CN)_6^{4-}$ + H^+ + $(CH_3)_2CHCHO$				
Pulse rad. of 2-methyl-1-propanol + H_2O				69 Ada 1
KAS	H_2O	RT	$k = 3.0 \cdot 10^9\ M^{-1}s^{-1}$	
$(CH_3)_2CH\dot{C}HOH$ + nitrobenzene ($C_6H_5NO_2$) ⟶ $C_6H_5\dot{N}O_2^-$ + H^+ + $(CH_3)_2CHCHO$ [59]				
Pulse rad. of 2-methyl-1-propanol + N_2O + H_2O				66 Asm 1
KAS	H_2O	RT	$k = 3.9 \cdot 10^8\ M^{-1}s^{-1}$	
$(CH_3)_2CH\dot{C}HOH$ + nitrosobenzene (C_6H_5NO) ⟶ $C_6H_5\dot{N}OH$ + $(CH_3)_2CHCHO$				
Pulse rad. of 2-methyl-1-propanol + N_2O + H_2O				66 Asm 2
KAS	H_2O	RT	$k = 4.0 \cdot 10^9\ M^{-1}s^{-1}$	
$\dot{C}H_2C(CH_3)_2OH$ + Ag_2^+ ⟶ 2 Ag^+ + OH^- + $CH_2{=}C(CH_3)_2$				
Pulse rad. of *t*-butanol + $AgClO_4$ + H_2O				78 Tau 1
Cond., time resolved	H_2O	RT	$k = 2.0 \cdot 10^9\ M^{-1}s^{-1}$ [60]	
$\dot{C}H_2C(CH_3)_2OH$ + Cd^+ ⟶ Cd^{2+} + $(CH_3)_2C{=}CH_2$ + OH^-				
Pulse rad. of *t*-butanol + Cd^{2+} + H_2O				75 Kel 1
KAS, and time- resolved Cond.	H_2O	RT	$k \approx 1 \cdot 10^9\ M^{-1}s^{-1}$ [61]	

[54]) 40% e^--transfer.
[55]) 35% e^--transfer [66 Asm 1].
[56]) 64% e^--transfer [73 Gre 2].
[57]) Nifuroxime.
[58]) 75% e^--transfer.
[59]) 39% e^--transfer.
[60]) Based on $2k(\dot{C}H_2C(CH_3)_2OH + \dot{C}H_2C(CH_3)_2OH) = 1.4 \cdot 10^9\ M^{-1}s^{-1}$.
[61]) Based on assumed values for various competing reactions; mechanism not clear.

Reaction Radical generation Method	Solvent	T[K]	Rate data	Ref./ add. ref.
$\dot{C}H_2C(CH_3)_2OH + Cd_2^{2+} \longrightarrow OH^- + (CH_3)_2C{=}CH_2 + Cd^+ + Cd^{2+}$				
Pulse rad. of t-butanol + Cd^{2+} + H_2O				75 Kel 1
KAS, and time-resolved Cond.	H_2O	RT	$k \approx 1.10^9\ M^{-1}s^{-1}$ [61]	
$\dot{C}H_2C(CH_3)_2OH + Cu^{2+} \longrightarrow Cu^+$ + products				
Pulse rad. of t-butanol + H_2O + N_2O				78 Bux 1
KAS	H_2O, pH = 4.5	RT	$k = 2.7(5)\cdot 10^6\ M^{-1}s^{-1}$ [62]	
	pH = 3		$k = 3.2(6)\cdot 10^6\ M^{-1}s^{-1}$ [62]	
$\dot{C}H_2C(CH_3)_2OH + CuC_2H_4^+ \longrightarrow C_2H_4 + CH_2{=}C(CH_3)_2 + Cu^{2+} + OH^-$				
Pulse rad. of C_2H_4 + H_2O + t-butanol				78 Bux 1
KAS	H_2O, pH = 4.5	RT	$k = 5.3(16)\cdot 10^7\ M^{-1}s^{-1}$ [63]	
$\dot{C}H_2C(CH_3)_2OH$ + tris-1,10-phenanthroline-iron(III) ion $\longrightarrow$ Fe(II)... + products				
Pulse rad. of acid t-butanol + H_2O				79 Jan 1
KAS	H_2O	RT	$k \approx 10^7\ M^{-1}s^{-1}$	
$\dot{C}H_2C(CH_3)_2OH + IrCl_6^{2-} \longrightarrow$ Ir(III) + products				
Pulse rad. of $(CH_3)_3COH$ + N_2O + H_2O				82 Ste 1
KAS, Cond.	H_2O	295	$k = 1.2\cdot 10^9\ M^{-1}s^{-1}$	[64]
$\dot{C}H_2C(CH_3)_2OH + Ni^+ \longrightarrow Ni^{2+} + CH_2{=}C(CH_3)_2 + OH^-$				
Pulse rad. of t-butanol + Ni^{2+} + H_2O				74 Kel 1
KAS	H_2O	RT	$k = 3\cdot 10^9\ M^{-1}s^{-1}$ [65]	
$\dot{C}H_2C(CH_3)_2OH$ + Ru(2,2'-bipyridine)$_3^{3+}$ $\longrightarrow$ Ru(II)... + products				
Pulse rad. of t-butanol + H_2O + N_2O				72 Mar 1,
KAS	H_2O, acid pH	RT	$k = 1.3\cdot 10^8\ M^{-1}s^{-1}$	78 Jon 1
	pH = 4.6		$k = 1.9(2)\cdot 10^8\ M^{-1}s^{-1}$ [66]	
$\dot{C}H_2C(CH_3)_2OH + Zn^+ + H_2O \longrightarrow Zn^{2+} + (CH_3)_3COH + OH^-$				
Pulse rad. of t-butanol + Zn^{2+} + H_2O				77 Rab 1
KAS	H_2O	RT	$k = 1.0(3)\cdot 10^9\ M^{-1}s^{-1}$ [67]	
$CH_3\dot{C}HOC_2H_5 + (Co(III)(NH_3)_5Br)^{2+} \longrightarrow$ products				
Pulse rad. of $C_2H_5OC_2H_5$ + N_2O + H_2O				77 Coh 1
KAS	H_2O, pH = 3.5...4.0	RT	$k = 1.6(2)\cdot 10^8\ M^{-1}s^{-1}$	
$CH_3\dot{C}HOC_2H_5 + (Co(III)(NH_3)_5Cl)^{2+} \longrightarrow$ products				
Pulse rad. of $C_2H_5OC_2H_5$ + N_2O + H_2O				77 Coh 1
KAS	H_2O, pH = 3.5...4.0	RT	$k = 1.4(2)\cdot 10^7\ M^{-1}s^{-1}$	
$CH_3\dot{C}HOC_2H_5$ + *cis*-(amminechlorobis(1,2-ethanediamine-N,N')cobalt(III))$^{2+}$ $\longrightarrow$ products				
Pulse rad. of $C_2H_5OC_2H_5$ + N_2O + H_2O				77 Coh 1
KAS	H_2O, pH = 3.5...4.0	RT	$k = 4.6\cdot 10^6\ M^{-1}s^{-1}$	
$CH_3\dot{C}HOC_2H_5$ + *cis*-(aquachlorobis(1,2-ethanediamine-N,N')cobalt(III))$^{2+}$ $\longrightarrow$ products				
Pulse rad. of $C_2H_5OC_2H_5$ + N_2O + H_2O				77 Coh 1
KAS	H_2O, pH = 3.5...4.0	RT	$k = 3.5\cdot 10^7\ M^{-1}s^{-1}$	

[61]) Based on assumed values for various competing reactions; mechanism not clear.
[62]) Based on formation kinetics of $Cu(I)CH_2CHCONH_2$ in presence of acrylamide.
[63]) Based on effect of C_2H_4 concentration on formation and decay of $CuC_2H_4^+$.
[64]) Mechanism discussed as e^-- or Cl'-transfer.
[65]) Based on equal concentrations of Ni^+ and $\dot{C}H_2C(CH_3)_2OH$ (2.7 species per 100 eV absorbed energy).
[66]) From [78 Jon 1].
[67]) Based on $k(\dot{H} + \dot{H}) = 1.3\cdot 10^{10}\ M^{-1}s^{-1}$; $k(Zn^+ + Zn^+) = 4\cdot 10^8\ M^{-1}s^{-1}$; $k(\dot{R} + \dot{R}) = 6.5\cdot 10^8\ M^{-1}s^{-1}$; $k(Zn^+ + \dot{H}) = 2.8\cdot 10^9\ M^{-1}s^{-1}$.

Reaction Radical generation Method	Solvent	T[K]	Rate data	Ref./ add. ref.
$CH_3\dot{C}HOC_2H_5$ + *trans*-(dibromobis(1,2-ethanediamine-N,N')cobalt(III))$^+$ ⟶ products				
Pulse rad. of $C_2H_5OC_2H_5 + N_2O + H_2O$				77 Coh 1
KAS	H_2O, pH = 3.5...4.0	RT	$k = 6.5 \cdot 10^8\ M^{-1}s^{-1}$	
$CH_3\dot{C}HOC_2H_5$ + *cis*-(bromobis(1,2-ethanediamine-N,N')fluorocobalt(III))$^+$ ⟶ products				
Pulse rad. of $C_2H_5OC_2H_5 + N_2O + H_2O$				77 Coh 1
KAS	H_2O, pH = 3.5...4.0	RT	$k = 4.8 \cdot 10^7\ M^{-1}s^{-1}$	
$CH_3\dot{C}HOC_2H_5$ + *trans*-(dichlorobis(1,2-ethanediamine-N,N')cobalt(III))$^+$ ⟶ products				
Pulse rad. of $C_2H_5OC_2H_5 + N_2O + H_2O$				77 Coh 1
KAS	H_2O, pH = 3.5...4.0	RT	$k = 1.5 \cdot 10^8\ M^{-1}s^{-1}$	
$CH_3\dot{C}HOC_2H_5 + H_2O_2$ ⟶ products				
H_2O + Ti(III) + $H_2O_2 + C_2H_5OC_2H_5$ solutions				74 Gil 1
PR, ESR	H_2O	RT	$k = 5.5(11) \cdot 10^4\ M^{-1}s^{-1}$ [68]	
$CH_3\dot{C}HOC_2H_5 + IrCl_6^{2-}$ ⟶ Ir(III) + products				
Pulse rad. of $C_2H_5OC_2H_5 + N_2O + H_2O$				82 Ste 1
KAS	H_2O	295	$k = 5.7 \cdot 10^9\ M^{-1}s^{-1}$	
$CH_3\dot{C}HOC_2H_5 + Ru(NH_3)_6^{3+}$ ⟶ products				
Pulse rad. of $C_2H_5OC_2H_5 + H_2O + N_2O$				77 Coh 1
KAS	H_2O, pH = 3.5...4.0	RT	$k = 1.00(15) \cdot 10^8\ M^{-1}s^{-1}$	
$CH_3\dot{C}HOC_2H_5 + (Ru(III)(NH_3)_5Br)^{2+}$ ⟶ products				
Pulse rad. of $C_2H_5OC_2H_5 + N_2O + H_2O$				77 Coh 1
KAS	H_2O, pH = 3.5...4.0	RT	$k = 5.8(9) \cdot 10^8\ M^{-1}s^{-1}$	
$CH_3\dot{C}HOC_2H_5 + (Ru(III)(NH_3)_5Cl)^{2+}$ ⟶ products				
Pulse rad. of $C_2H_5OC_2H_5 + N_2O + H_2O$				77 Coh 1
KAS	H_2O, pH = 3.5...4.0	RT	$k = 2.6(4) \cdot 10^8\ M^{-1}s^{-1}$	
$CH_3\dot{C}HOC_2H_5$ + carbontetrachloride (CCl_4) ⟶ Cl^- + products				
Pulse rad. of $C_2H_5OC_2H_5 + H_2O + N_2O$				71 Koe 1
Cond., time resolved	H_2O	RT	$k = 2.5 \cdot 10^7\ M^{-1}s^{-1}$	
$CH_3CHOH\dot{C}HCH_2OH$ + *anti*-5-nitro-2-furaldoxime [69] ⟶ products [70]				
Pulse rad. of crotyl alcohol + $N_2O + H_2O$				73 Gre 1
KAS	H_2O	RT	$k = 1.4 \cdot 10^9\ M^{-1}s^{-1}$	
$^-OOC(CH_2)_2\dot{C}(O^-)COO^-$ + 2-methyl-1,4-naphthoquinone ⟶ ...semiquinone + products				
Pulse rad. of α-ketoglutarate + H_2O + *t*-butanol				73 Rao 3
KAS	H_2O, pH = 10.2	RT	$k = 3.8(4) \cdot 10^9\ M^{-1}s^{-1}$	
$^-OOCCH_2CH_2\dot{C}OHCOOH$ + Fe(III)cytochrome c ⟶ Fe(II)cytochrome c + products				
Pulse rad. of malate + $N_2O + H_2O$				74 Sha 1
KAS	H_2O	RT [71]	$k = 8.5(8) \cdot 10^7\ M^{-1}s^{-1}$ $E_a = 12\ kJ\ mol^{-1}$	

[68] Based on $2k(CH_3\dot{C}HOC_2H_5 + CH_3\dot{C}HOC_2H_5) = 3 \cdot 10^9\ M^{-1}s^{-1}$.
[69] Nifuroxime.
[70] 24% e^--transfer.
[71] No temp. range given.

Reaction Radical generation Method	Solvent	T[K]	Rate data	Ref./ add. ref.
$CH_3\dot{C}HCOC_2H_5 + Ti(III) \longrightarrow Ti(IV) + CH_3^{(-)}CHCOC_2H_5$ [72]				
$Ti(III)/H_2O_2$ flow expt. with $C_2H_5COC_2H_5 + H_2O$				73 Gil 1
SESR	H_2O, pH = 1	RT	$k \leqslant 6 \cdot 10^5\,M^{-1}s^{-1}$	
$CH_3\dot{C}HCOC_2H_5 + Ti(III)\text{-}EDTA \longrightarrow Ti(IV)\text{-}EDTA + CH_3^{(-)}CHCOC_2H_5$ [72]				
$Ti(III)/H_2O_2$ flow expt. with $C_2H_5COC_2H_5 + H_2O$				73 Gil 1
SESR	H_2O, pH = 7	RT	$k \leqslant 1.2 \cdot 10^6\,M^{-1}s^{-1}$	
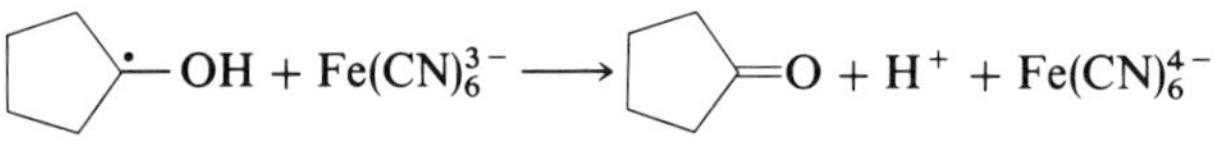 Ė F				
Pulse rad. of cyclopentanone + *t*-butanol + H_2O				76 Tof 1
KAS	H_2O	275	$k = 2.3 \cdot 10^9\,M^{-1}s^{-1}$	
		337	$5.4 \cdot 10^9$	
			$E_a = 10.9\,kJ\,mol^{-1}$	
Ė + nitrobenzene ($C_6H_5NO_2$) ⟶ **F** + $H^+ + C_6H_5\dot{N}O_2^-$				
Pulse rad. of cyclopentanone + *t*-butanol + H_2O				76 Tof 1
KAS	H_2O	275	$k = 1.4 \cdot 10^9\,M^{-1}s^{-1}$	
		337	$5.2 \cdot 10^9$	
			$E_a = 17\,kJ\,mol^{-1}$	
$HO(CH_2)_3\dot{C}HCHO$ [73] + hydroquinone ⟶ $HO(CH_2)_4CHO + {}^-OC_6H_4\dot{O}$				
Pulse rad. of 2-(hydroxymethyl)furan + $N_2O + H_2O$				79 Ste 1
KAS	H_2O, pH ≈ 11.5	RT	$k = 8.6(12) \cdot 10^8\,M^{-1}s^{-1}$	
$CH_3(CH_2)_3\dot{C}HOH + Fe(CN)_6^{3-} \longrightarrow Fe(CN)_6^{4-} + H^+ + CH_3(CH_2)_3CHO$				
Pulse rad. of 1-pentanol + H_2O				79 Alm 1
KAS	H_2O	RT	$k = 9(2) \cdot 10^8\,M^{-1}s^{-1}$	
$\dot{C}HOHC(CH_2OH)_3$ + Fe(III)cytochrome c ⟶ Fe(II)cytochrome c + products				
Pulse rad. of pentaerythritol				75 Sim 1
KAS	H_2O, pH = 9...10	RT	$k = 1.5 \cdot 10^8\,M^{-1}s^{-1}$	
	pH = 5.6		$k < 10^6\,M^{-1}s^{-1}$	
$\dot{C}HOHC(CH_2OH)_3$ + hemin c(Fe(III)) ⟶ hemin c(Fe(II)) + products				
Pulse rad. of pentaerythritol + $H_2O + N_2O$				75 Gof 1
KAS	H_2O, pH = 7	RT	$k = 3.0(6) \cdot 10^8\,M^{-1}s^{-1}$	
	pH = 11.8		$k = 2.8(6) \cdot 10^8\,M^{-1}s^{-1}$	
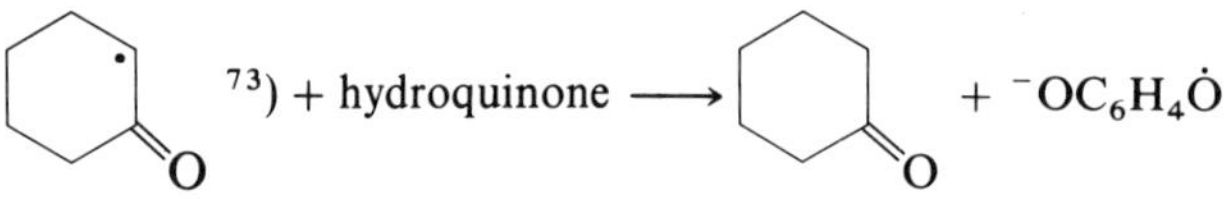				
Pulse rad. of 1,2-cyclohexanediol + $N_2O + H_2O$				79 Ste 1
KAS	H_2O, pH ≈ 11.5	RT	$k = 5.5(19) \cdot 10^8\,M^{-1}s^{-1}$	
$\cdot\overline{CH(CHOH)_4CO}$ [73] + hydroquinone ⟶ products + $^-OC_6H_4\dot{O}$				
Pulse rad. of *meso*-inositol + $N_2O + H_2O$				79 Ste 1
KAS	H_2O, pH ≈ 11.5	RT	$k = 6.4(11) \cdot 10^8\,M^{-1}s^{-1}$	

[72]) Immediate protonation.
[73]) Mainly C-centered radical, oxidizing action, however, likely to occur through mesomeric O-centered radical.

| Reaction | | | | Ref./ |
| Radical generation | | | | add. ref. |
Method	Solvent	$T[K]$	Rate data	

$$\dot{G}\text{—OH} + Fe(CN)_6^{3-} \longrightarrow \text{=O} + H^+ + Fe(CN)_6^{4-} \quad H$$

Pulse rad. of cyclohexanone + t-butanol + H_2O				76 Tof 1/
KAS	H_2O	275	$k = 1.8 \cdot 10^9 \, M^{-1} s^{-1}$	79 Alm 1
		337	$5.0 \cdot 10^9$	
			$E_a = 12.1 \, kJ \, mol^{-1}$	

$\dot{G}$ + nitrobenzene $\longrightarrow$ **H** + H^+ + $C_6H_5\dot{N}O_2^-$

Pulse rad. of cyclohexanone + t-butanol + H_2O				76 Tof 1
KAS	H_2O	275	$k = 7 \cdot 10^8 \, M^{-1} s^{-1}$	
		337	$2.0 \cdot 10^9$	
			$E_a = 13 \, kJ \, mol^{-1}$	

$HO(CH_2)_4\dot{C}HCHO$ [73]) + hydroquinone $\longrightarrow HO(CH_2)_5CHO + {}^-OC_6H_4\dot{O}$

Pulse rad. of 2-(hydroxymethyl)pyran + N_2O + H_2O				79 Ste 1
KAS	$H_2O, pH \approx 11.5$	RT	$k = 5.2(9) \cdot 10^8 \, M^{-1} s^{-1}$	

$(CH_3)_2\dot{C}OCH(CH_3)_2 + IrCl_6^{2-} \longrightarrow Ir(III) + products$

Pulse rad. of $(CH_3)_2CHOCH(CH_3)_2$ + N_2O + H_2O				82 Ste 1
KAS	H_2O	295	$k = 3.6 \cdot 10^9 \, M^{-1} s^{-1}$	

$c\text{-}C_6H_{11}\dot{C}HOH$ + 2,3-butanedione $(CH_3COCOCH_3) \longrightarrow c\text{-}C_6H_{11}CHO + H^+ + (CH_3COCOCH_3)^{\overline{\cdot}}$

Pulse rad. of cyclohexanemethanol + H_2O + N_2O				68 Lil 1
KAS	H_2O	RT	$k = 2.6 \cdot 10^8 \, M^{-1} s^{-1}$	

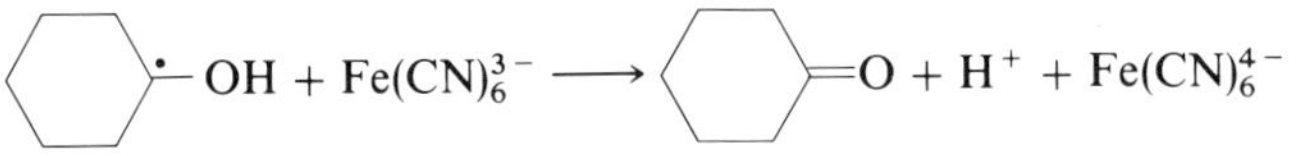

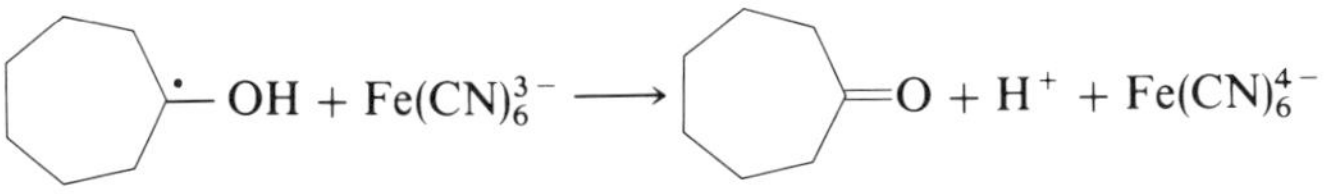

$$\dot{I}\text{—OH} + Fe(CN)_6^{3-} \longrightarrow \text{=O} + H^+ + Fe(CN)_6^{4-} \quad K$$

Pulse rad. of cycloheptanone + t-butanol + H_2O				76 Tof 1
KAS	H_2O	275	$k = 2.1 \cdot 10^9 \, M^{-1} s^{-1}$	
		337	$5.4 \cdot 10^9$	
			$E_a = 11.7 \, kJ \, mol^{-1}$	

$\dot{I}$ + nitrobenzene $(C_6H_5NO_2) \longrightarrow$ **K** + H^+ + $C_6H_5\dot{N}O_2^-$

Pulse rad. of cycloheptanone + t-butanol + H_2O				76 Tof 1
KAS	H_2O	275	$k = 1.4 \cdot 10^9 \, M^{-1} s^{-1}$	
		337	$4.9 \cdot 10^9$	
			$E_a = 16 \, kJ \, mol^{-1}$	

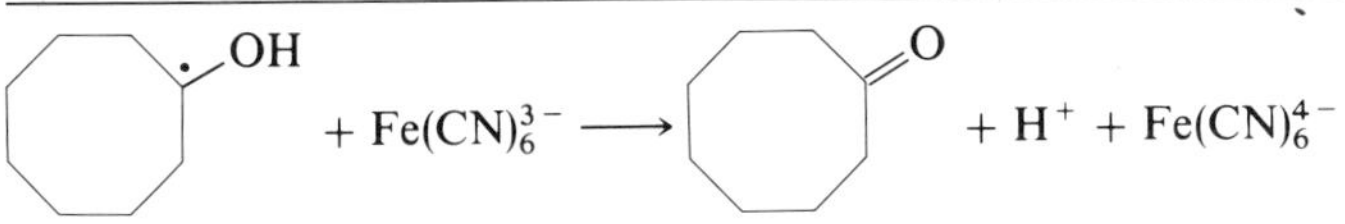

$$\dot{L}\underset{}{\overset{\dot{}\text{OH}}{}} + Fe(CN)_6^{3-} \longrightarrow \overset{O}{} + H^+ + Fe(CN)_6^{4-} \quad M$$

Pulse rad. of cyclooctanone + t-butanol + H_2O				76 Tof 1
KAS	H_2O	275	$k = 2.2 \cdot 10^9 \, M^{-1} s^{-1}$	
		337	$5.4 \cdot 10^9$	
			$E_a = 11.7 \, kJ \, mol^{-1}$	

---

[73]) Mainly C-centered radical, oxidizing action, however, likely to occur through mesomeric O-centered radical.

Reaction Radical generation Method	Solvent	T[K]	Rate data	Ref./ add. ref.
$\dot{L}$ + nitrobenzene ($C_6H_5NO_2$) ⟶ M + H^+ + $C_6H_5\dot{N}O_2^-$ *)				
Pulse rad. of cyclooctanone + *t*-butanol + H_2O				76 Tof 1
KAS	H_2O	275	$k = 1.6 \cdot 10^9\ M^{-1}s^{-1}$	
		337	$5.2 \cdot 10^9$	
			$E_a = 15.5\ kJ\ mol^{-1}$	

4.2.1.1.3 Radicals containing C, H, N, and other atoms

Reaction Radical generation Method	Solvent	T[K]	Rate data	Ref./ add. ref.
$\dot{C}H(O^-)CONH_2$ + 2-methyl-1,4-naphthoquinone ⟶ ...semiquinone + products [74])				
Pulse rad. of glycolamide + H_2O + N_2O				73 Rao 3
KAS	H_2O	RT	$k = 2.3(2) \cdot 10^9\ M^{-1}s^{-1}$	
$NH_2\dot{C}HCOO^-$ + $Fe(CN)_6^{3-}$ ⟶ $Fe(CN)_6^{4-}$ + products				
Pulse rad. of glycine + N_2O + H_2O				69 Ada 1
KAS	H_2O	RT	$k = 1 \cdot 10^9\ M^{-1}s^{-1}$	
$NH_2\dot{C}HCOO^-$ + 9,10-anthraquinone-2,6-disulfonate ion ⟶ ... semiquinone + products				
Pulse rad. of glycine + H_2O + N_2O				73 Rao 1
KAS	H_2O, pH = 8	RT	$k = 2.6(3) \cdot 10^9\ M^{-1}s^{-1}$	
$NH_2\dot{C}HCOO^-$ + 9,10-anthraquinone-2-sulfonate ion ⟶ ...semiquinone + products				
Pulse rad. of glycine + H_2O + N_2O				73 Rao 1
KAS	H_2O, pH = 8	RT	$k = 2.2(2) \cdot 10^9\ M^{-1}s^{-1}$	
$NH_2\dot{C}HCOO^-$ + benzoquinone ⟶ benzosemiquinone + products				
Pulse rad. of glycine + H_2O + N_2O				73 Rao 1
KAS	H_2O, pH = 8	RT	$k = 3.9(4) \cdot 10^9\ M^{-1}s^{-1}$	
$NH_2\dot{C}HCOO^-$ + crystal violet ⟶ products [75])				
Pulse rad. of glycine + H_2O + N_2O				73 Rao 2
KAS	H_2O	RT	$k = 1.2 \cdot 10^9\ M^{-1}s^{-1}$	
$NH_2\dot{C}HCOO^-$ + dichloroindophenol ⟶ products				
Pulse rad. of glycine + H_2O + N_2O				73 Rao 2
KAS	H_2O	RT	$k = 3.6 \cdot 10^9\ M^{-1}s^{-1}$	
$NH_2\dot{C}HCOO^-$ + 2-hydroxy-1,4-naphthoquinone ⟶ ...semiquinone + products				
Pulse rad. of glycine + H_2O + N_2O				73 Rao 1
KAS	H_2O, pH = 8	RT	$k = 3.1(3) \cdot 10^9\ M^{-1}s^{-1}$	
$NH_2\dot{C}HCOO^-$ + indigo disulfonate ⟶ products [76])				
Pulse rad. of glycine + H_2O + N_2O				73 Rao 2
KAS	H_2O, pH = 9	RT	$k = 2.8 \cdot 10^9\ M^{-1}s^{-1}$	
$NH_2\dot{C}HCOO^-$ + indigo tetrasulfonate ⟶ products [77])				
Pulse rad. of glycine + H_2O + N_2O				73 Rao 2
KAS	H_2O	RT	$k = 2.6 \cdot 10^9\ M^{-1}s^{-1}$	
$NH_2\dot{C}HCOO^-$ + methylene blue ⟶ products [78])				
Pulse rad. of glycine + H_2O + N_2O				73 Rao 2
KAS	H_2O	RT	$k = 3.7 \cdot 10^9\ M^{-1}s^{-1}$	

*) For $\dot{L}$ and M, see p. 347.
[74]) 48% e^--transfer.
[75]) 56% e^--transfer.
[76]) 71% e^--transfer.
[77]) 78% e^--transfer.
[78]) 86% e^--transfer.

Reaction Radical generation Method	Solvent	T[K]	Rate data	Ref./ add. ref.
$NH_2\dot{C}HCOO^-$ + 2-methyl-1,4-naphthoquinone ⟶ ...semiquinone + products [79]				
Pulse rad. of glycine + H_2O + N_2O KAS	H_2O, pH = 8	RT	$k = 4.0(4) \cdot 10^9\ M^{-1}s^{-1}$	73 Rao 3/ 72 Sim 1
$NH_2\dot{C}HCOO^-$ + 1,4-naphthoquinone-2-sulfonate ion ⟶ ...semiquinone + products				
Pulse rad. of glycine + H_2O + N_2O KAS	H_2O, pH = 8	RT	$k = 3.3(3) \cdot 10^9\ M^{-1}s^{-1}$	73 Rao 1
$NH_2\dot{C}HCOO^-$ + nicotinamide adenine dinucleotide(NAD^+) ⟶ products				
Pulse rad. of glycine + H_2O + N_2O KAS	H_2O	RT	$k = 1.5 \cdot 10^9\ M^{-1}s^{-1}$	73 Rao 1
$NH_2\dot{C}HCOO^-$ + phenosafranine ⟶ products [80]				
Pulse rad. of glycine + N_2O + H_2O KAS	H_2O	RT	$k = 1.9 \cdot 10^9\ M^{-1}s^{-1}$	73 Rao 2
$NH_2\dot{C}HCOO^-$ + riboflavin ⟶ products				
Pulse rad. of glycine + N_2O + H_2O KAS	H_2O, pH = 8	RT	$k = 2.7(3) \cdot 10^9\ M^{-1}s^{-1}$	73 Rao 1
$NH_2\dot{C}HCOO^-$ + safranine T ⟶ products [81]				
Pulse rad. of glycine + N_2O + H_2O KAS	H_2O	RT	$k = 1.6 \cdot 10^9\ M^{-1}s^{-1}$	73 Rao 2
$NH_2\dot{C}HCOO^-$ + thionine ⟶ products [82]				
Pulse rad. of glycine + N_2O + H_2O KAS	H_2O	RT	$k = 3.2 \cdot 10^9\ M^{-1}s^{-1}$	73 Rao 2
$\dot{C}H_2CONH_2$ [83] + 2-methyl-1,4-naphthoquinone ⟶ ...semiquinone + products [84]				
Pulse rad. of acetamide + H_2O + N_2O KAS	H_2O	RT	$k = 1.1 \cdot 10^9\ M^{-1}s^{-1}$	73 Rao 3
$NH_3^+\dot{C}HCOO^-$ + $Ru(NH_3)_6^{3+}$ ⟶ products				
Pulse rad. of glycine + H_2O + N_2O KAS	H_2O	RT	$k = 4.0(6) \cdot 10^8\ M^{-1}s^{-1}$	72 Coh 1
$NH_3^+\dot{C}HCO_2^-$ + 2-methyl-1,4-naphthoquinone ⟶ ...semiquinone + products [85]				
Pulse rad. of glycine + H_2O + N_2O KAS	H_2O	RT	$k = 5.5(6) \cdot 10^9\ M^{-1}s^{-1}$	73 Rao 3/ 72 Sim 1
$CH_3\dot{C}HOSO_3^-$ + $Fe(CN)_6^{3-}$ ⟶ $Fe(CN)_6^{4-}$ + products				
Pulse rad. of ethylsulfate + H_2O + N_2O KAS	H_2O	RT	$k \approx 2 \cdot 10^8\ M^{-1}s^{-1}$	79 Alm 1
$NH_2\dot{C}HCONH_2$ + 2-methyl-1,4-naphthoquinone ⟶ ...semiquinone + products [86]				
Pulse rad. of glycinamide + H_2O + N_2O KAS	H_2O, pH = 10.4	RT	$k = 5.4(5) \cdot 10^9\ M^{-1}s^{-1}$	73 Rao 3/ 72 Sim 1
$CH_3\dot{C}HNH_2$ + 2-methyl-1,4-naphthoquinone ⟶ ...semiquinone + products [87]				
Pulse rad. of ethylamine + H_2O + N_2O KAS	H_2O, pH = 11.6	RT	$k = 3.3(3) \cdot 10^9\ M^{-1}s^{-1}$	73 Rao 3

[79]) 79% e^--transfer.
[80]) 66% e^--transfer.
[81]) 60% e^--transfer.
[82]) 85% e^--transfer.
[83]) Possibly also $CH_3CO\dot{N}H$.
[84]) 17% e^--transfer.
[85]) 71% e^--transfer.
[86]) 41% e^--transfer.
[87]) 34% e^--transfer.

Reaction Radical generation Method	Solvent	T[K]	Rate data	Ref./ add. ref.
$H_3N^+\dot{C}HCONH_2$ + 2-methyl-1,4-naphthoquinone ⟶ ...semiquinone + products [88]				
Pulse rad. of glycinamide + H_2O + N_2O				73 Rao 3
KAS	H_2O, pH = 3.2	RT	$k = 3.6(4)\cdot10^9\ M^{-1}s^{-1}$	
$CH_3N^-\dot{C}HCO_2^-$ + 2-methyl-1,4-naphthoquinone ⟶ ...semiquinone + products [89]				
Pulse rad. of sarcosine + H_2O + N_2O				73 Rao 3
KAS	H_2O, pH = 12.5	RT	$k = 1.7(2)\cdot10^9\ M^{-1}s^{-1}$	
$(CH_2{=}CHCONH_2)^{\dot{-}}$ [90] + Ag^+ ⟶ $\dot{A}g$ + $CH_2{=}CHCONH_2$				
Pulse rad. of $CH_2{=}CHCONH_2$ + H_2O				70 Cha 1
KAS	H_2O, pH ≈ 6	RT	$k = 1.1(4)\cdot10^8\ M^{-1}s^{-1}$	
$\dot{C}H(OH)CH(NH_2)COO^-$ + $Fe(CN)_6^{3-}$ ⟶ $Fe(CN)_6^{4-}$ + products				
Pulse rad. of serine + H_2O				69 Ada 1
KAS	H_2O	RT	$k = 3.2\cdot10^9\ M^{-1}s^{-1}$	
$CH_3CONH\dot{C}H_2$ + 2-methyl-1,4-naphthoquinone ⟶ ...semiquinone + products [91]				
Pulse rad. of N-methylacetamide + H_2O + N_2O				73 Rao 3
KAS	H_2O	RT	$k = 2.0(2)\cdot10^9\ M^{-1}s^{-1}$	
$CH_3\dot{C}OHCONH_2$ + 1,4-benzoquinone ⟶ benzosemiquinone + products				
Pulse rad. of lactamide + H_2O + N_2O				73 Hay 1
KAS	H_2O, pH = 5	RT	$k = 2.0\cdot10^9\ M^{-1}s^{-1}$	
	pH = 7.3		$3.6\cdot10^9$	
$CH_3NH\dot{C}HCOOH$ + 2-methyl-1,4-naphthoquinone ⟶ ...semiquinone + products [92]				
Pulse rad. of sarcosine + H_2O + N_2O				73 Rao 3
KAS	H_2O	RT	$k = 1.1(1)\cdot10^9\ M^{-1}s^{-1}$	
$(CH_3)_2\dot{C}NH_2$ + 2-methyl-1,4-naphthoquinone ⟶ ...semiquinone + products [93]				
Pulse rad. of isopropylamine + H_2O + N_2O				73 Rao 3
KAS	H_2O, pH = 11.4	RT	$k = 3.6(4)\cdot10^9\ M^{-1}s^{-1}$	
$\{(^-O_2CCH_2)NH\dot{C}HCO_2^-\}Ni(II)$ + O_2 ⟶ products				
Pulse rad. of Ni(II)-iminodiacetic acid + N_2O + H_2O				81 Bha 1
KAS	H_2O	RT	$k = 3\cdot10^4\ M^{-1}s^{-1}$	
$(^-OOCCH_2)NH\dot{C}HCOO^-$ + $Fe(CN)_6^{3-}$ ⟶ $Fe(CN)_6^{4-}$ + products				
Pulse rad. of iminodiacetic acid + N_2O + H_2O				81 Bha 1
KAS	H_2O	RT	$k = 5.0\cdot10^8\ M^{-1}s^{-1}$	
$(^-OOCCH_2)NH\dot{C}HCOO^-$ + O_2 ⟶ products				
Pulse rad. of iminodiacetic acid + N_2O + H_2O				81 Bha 1
KAS	H_2O	RT	$k = 8\cdot10^8\ M^{-1}s^{-1}$	

[88] 24% e^--transfer.
[89] 64% e^--transfer.
[90] Electron adduct to $CH_2{=}CHCONH_2$, presumably $>\dot{C}O^-$.
[91] 19% e^--transfer.
[92] 33% e^--transfer.
[93] 41% e^--transfer.

Reaction Radical generation Method	Solvent	T[K]	Rate data	Ref./ add. ref.
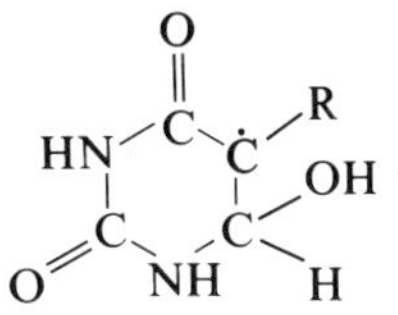 Ȧ [94]) + tetramethylphenylenediamine (TMPD) $\longrightarrow$ $(TMPD)^{\dot{+}}$ + products			R = H	
Pulse rad. of uracil + N_2O + H_2O KAS	H_2O, pH = 8	293(2)	$k = 1.7(1)\cdot 10^9\ M^{-1}s^{-1}$ [95])	81 Fuj 1
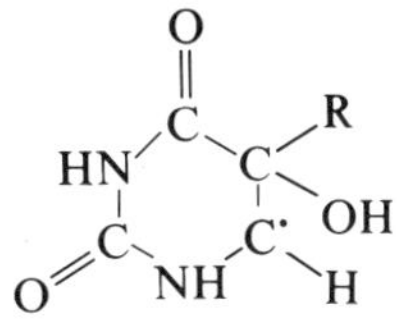 Ḃ + tetranitromethane ($C(NO_2)_4$) $\longrightarrow$ $C(NO_2)_3^-$ + NO_2 + H^+ + products			R = H	
Pulse rad. of uracil + N_2O + H_2O KAS	H_2O	293(2)	$k = 1.9\cdot 10^9\ M^{-1}s^{-1}$	81 Fuj 1
$CH_3CH_2CH_2\dot{C}HNH_2$ + 4-nitroacetophenone(PNAP) $\longrightarrow$ $PNAP^{\dot{-}}$ + $CH_3CH_2CH_2CHNH$ + H^+				
Pulse rad. of *n*-butylamine + N_2O + H_2O KAS	H_2O, pH = 10.8	RT	$k = 4.1(4)\cdot 10^9\ M^{-1}s^{-1}$	83 Hil 1
$CH_3SCH_2CH_2\dot{C}HNH_2$ + $Fe(CN)_6^{3-}$ $\longrightarrow$ $Fe(CN)_6^{4-}$ + H^+ + $CH_3SCH_2CH_2CHNH$				
Pulse rad. of methionine + N_2O + H_2O [96]) KAS	H_2O, pH = 5.8	RT	$k = 3.5(4)\cdot 10^9\ M^{-1}s^{-1}$ [97])	83 Hil 1/ 79 Hil 1
$CH_3SCH_2CH_2\dot{C}HNH_2$ + cytochrome(III)-c(Fe(III)) $\longrightarrow$ $CH_3SCH_2CH_2CHNH$ + H^+ + products				
Pulse rad. of methionine + N_2O + H_2O [96]) KAS	H_2O, pH = 5.0	RT	$k = 6.6(6)\cdot 10^8\ M^{-1}s^{-1}$ [97])	83 Hil 1/ 79 Hil 1
$CH_3SCH_2CH_2\dot{C}HNH_2$ + O_2 $\longrightarrow$ $CH_3SCH_2CH_2CHNH$ + H^+ + $\dot{O}_2^-$				
Pulse rad. of methionine + N_2O + H_2O [96]) KAS	H_2O, pH = 5.5	RT	$k = 1.8(4)\cdot 10^9\ M^{-1}s^{-1}$ [97])	83 Hil 1/ 79 Hil 1
$CH_3SCH_2CH_2\dot{C}HNH_2$ + 1,1′-dimethyl-4,4′-bipyridinium [98]) (MV^{2+}) $\longrightarrow$ $CH_3SCH_2CH_2CHNH$ + H^+ + $MV^{\dot{+}}$				
Pulse rad. of methionine + N_2O + H_2O [96]) KAS	H_2O, pH = 7.0	RT	$k = 3.6(3)\cdot 10^9\ M^{-1}s^{-1}$ [97])	83 Hil 1/ 81 Hil 1
$CH_3SCH_2CH_2\dot{C}HNH_2$ + lipoate (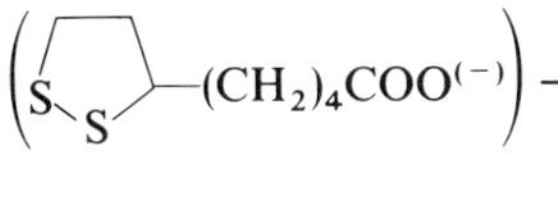) $\longrightarrow$ $CH_3SCH_2CH_2CHNH$ + H^+ + 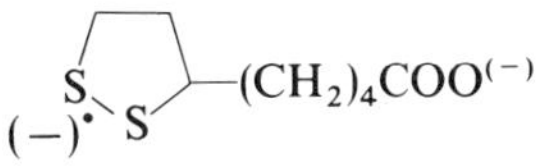				
Pulse rad. of methionine + N_2O + H_2O [96]) KAS	H_2O, pH = 8.2	RT	$k = (1\ldots2)\cdot 10^8\ M^{-1}s^{-1}$	83 Hil 1/ 79 Hil 1

[94]) Mesomeric form between C- and O-centered radical, oxidizing action likely through O-centered radical.
[95]) Possible contribution also by radicals formed from ȮH addition to C-5.
[96]) $CH_3SCH_2CH_2\dot{C}HNH_2$ radical formed via ȮH radical induced decarboxylation of methionine.
[97]) Possibly includes up to 20% contribution of $\dot{C}H_2SCH_2CH_2CHNH_3^+COO^-$ and $CH_3S\dot{C}HCH_2CHNH_3^+COO^-$ radicals.
[98]) Methylviologen.

Reaction Radical generation Method	Solvent	T[K]	Rate data	Ref./ add. ref.
$CH_3SCH_2CH_2\dot{C}HNH_2$ + nicotinamide adenine dinucleotide(NAD^+) ⟶ $CH_3SCH_2CH_2CHNH + H^+ + NAD^\cdot$				
Pulse rad. of methionine + $N_2O + H_2O$ [96] KAS	H_2O, pH = 6.6	RT	$k = 8.5(3) \cdot 10^8\ M^{-1} s^{-1}$ [97]	83 Hil 1/ 79 Hil 1
$CH_3SCH_2CH_2\dot{C}HNH_2$ + 4-nitroacetophenone(PNAP) ⟶ $CH_3SCH_2CH_2CHNH + H^+ + PNAP^{\cdot-}$				
Pulse rad. of methionine + $N_2O + H_2O$ [96] KAS	H_2O, pH = 4.3...11.0	RT	$k = 3.9(4) \cdot 10^9\ M^{-1} s^{-1}$ [97]	83 Hil 1/ 79 Hil 1
$CH_3SCH_2CH_2\dot{C}HNH_2$ + tetramethyl-piperidino-N-oxyl(TMPN) ⟶ $CH_3SCH_2CH_2CHNH + H^+$ + products				
Pulse rad. of methionine + $N_2O + H_2O$ [96] KAS, Cond., time resolved	H_2O, pH = 5.0	RT	$k = 5.4(5) \cdot 10^8\ M^{-1} s^{-1}$ [97]	83 Hil 1/ 79 Hil 1
$CH_3SCH_2CH_2\dot{C}HNH_2$ + tetranitromethane ($C(NO_2)_4$) ⟶ $C(NO_2)_3^- + NO_2 + H^+ + CH_3SCH_2CH_2CHNH$				
Pulse rad. of methionine + $N_2O + H_2O$ [96] KAS	H_2O, pH = 4.5	RT	$k = 4.2(5) \cdot 10^9\ M^{-1} s^{-1}$	83 Hil 1/ 79 Hil 1
$CH_3SCH_2CH_2\dot{C}HNH_3^+$ + 1,1'-dimethyl-4,4'-bipyridinium [98] (MV^{2+}) ⟶ $CH_3SCH_2CH_2CHNH/2H^+ + MV^{\cdot+}$				
Pulse rad. of methionine + $N_2O + H_2O$ [96] KAS	H_2O, pH < 3	RT	$k = 1.0(2) \cdot 10^7\ M^{-1} s^{-1}$ [97] [99]	83 Hil 1
$CH_3SCH_2CH_2\dot{C}HNH_3^+$ + tetramethyl-piperidino-N-oxyl(TMPN) ⟶ $CH_3SCH_2CH_2CHNH/2H^+$ + products				
Pulse rad. of methionine + $N_2O + H_2O$ [96] KAS	H_2O, pH < 3	RT	$k = 2.3(3) \cdot 10^8\ M^{-1} s^{-1}$ [97] [99]	83 Hil 1/ 79 Hil 1
Ȧ *) [1]) + tetramethylphenylenediamine(TMPD) ⟶ $(TMPD)^{\cdot+}$ + products				R = COOH
Pulse rad. of 5-carboxyuracil [2]) + $N_2O + H_2O$ KAS	H_2O, pH = 8	293(2)	$k = 1.8(2) \cdot 10^9\ M^{-1} s^{-1}$ [3])	81 Fuj 1
Ḃ *) + tetranitromethane ($C(NO_2)_4$) ⟶ $C(NO_2)_3^- + NO_2 + H^+$ + products				R = COOH
Pulse rad. of 5-carboxyuracil [2]) + $N_2O + H_2O$ KAS	H_2O	293(2)	$k = 1.7 \cdot 10^9\ M^{-1} s^{-1}$	81 Fuj 1

[96]) $CH_3SCH_2CH_2\dot{C}HNH_2$ radical formed via $\dot{O}H$ radical induced decarboxylation of methionine.
[97]) Possibly includes up to 20% contribution of $\dot{C}H_2SCH_2CH_2CHNH_3^+COO^-$ and $CH_3S\dot{C}HCH_2CHNH_3^+COO^-$ radicals.
[98]) Methylviologen.
[99]) Extrapolated value (computer evaluation) from pH dependence of MV^{2+} reduction based on pK = 3.85 for $CH_3SCH_2CH_2\dot{C}HNH_3^+ \rightleftharpoons CH_3SCH_2CH_2\dot{C}HNH_2 + H^+$ equilibrium.
*) For **Ȧ** and **Ḃ**, see p. 351.
[1]) Mesomeric form between C- and O-centered radical, oxidizing action likely through O-centered radical.
[2]) Iso-orotic acid.
[3]) Possible contribution also by radicals formed from $\dot{O}H$ addition to C-5.

Reaction Radical generation Method	Solvent	T[K]	Rate data	Ref./ add. ref.
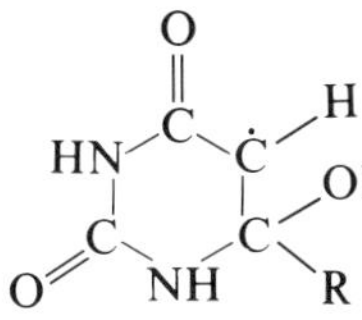 Ċ [1]) + tetramethylphenylenediamine (TMPD) ⟶ $(TMPD)^{+\cdot}$ + products R = COOH				
Pulse rad. of 6-carboxyuracil [4]) + N_2O + H_2O				81 Fuj 1
KAS	H_2O	293(2)	$k \approx 10^9\ M^{-1}s^{-1}$ [3])	
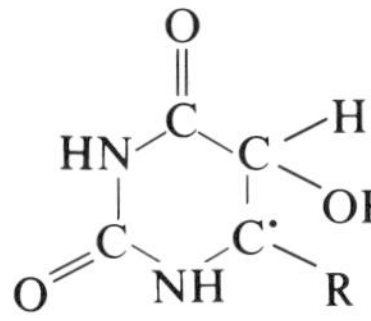 Ḋ + tetranitromethane ($C(NO_2)_4$) ⟶ $C(NO_2)_3^-$ + NO_2 + H^+ + products R = COOH				
Pulse rad. of 6-carboxyuracil [4]) + N_2O + H_2O				81 Fuj 1
KAS	H_2O	293(2)	$k = 2 \cdot 10^8\ M^{-1}s^{-1}$	
$CH_3CON(CH_3)\dot{C}HCOO^-$ + 2-methyl-1,4-naphthoquinone ⟶ ...semiquinone + products [5])				
Pulse rad. of acetylsarcosine + H_2O + N_2O				73 Rao 3
KAS	H_2O, pH = 7	RT	$k = 1.3(1) \cdot 10^9\ M^{-1}s^{-1}$	
	pH = 12.5		$1.0(1) \cdot 10^9$	
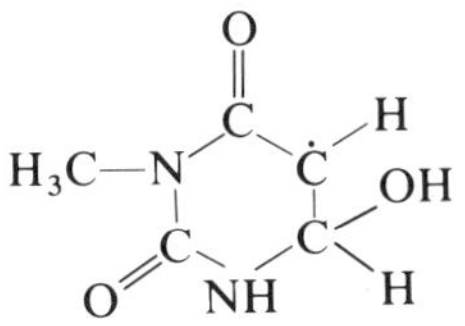 [1]) + tetramethylphenylenediamine (TMPD) ⟶ $(TMPD)^{+\cdot}$ + products				
Pulse rad. of 3-methyluracil + N_2O + H_2O				81 Fuj 1
KAS	H_2O, pH = 8	293(2)	$k = 2.3(3) \cdot 10^9\ M^{-1}s^{-1}$ [3])	
Ȧ *) [1]) + tetramethylphenylenediamine (TMPD) ⟶ $(TMPD)^{+\cdot}$ + products R = CH_3				
Pulse rad. of 5-methyluracil [6]) + N_2O + H_2O				81 Fuj 1
KAS	H_2O, pH = 8	293(2)	$k = 1.3(1) \cdot 10^9\ M^{-1}s^{-1}$ [3])	
Ḃ *) + tetranitromethane ($C(NO_2)_4$) ⟶ $C(NO_2)_3^-$ + NO_2 + H^+ + products R = CH_3				
Pulse rad. of 5-methyluracil [6]) + N_2O + H_2O				81 Fuj 1
KAS	H_2O	293(2)	$k = 1.5 \cdot 10^9\ M^{-1}s^{-1}$	
Ċ [1]) + tetramethylphenylenediamine (TMPD) ⟶ $(TMPD)^{+\cdot}$ + products R = CH_3				
Pulse rad. of 6-methyluracil + N_2O + H_2O				81 Fuj 1
KAS	H_2O, pH = 8	293(2)	$k = 1.1(2) \cdot 10^9\ M^{-1}s^{-1}$ [3])	

*) For **Ȧ** and **Ḃ**, see p. 351.
[1]) Mesomeric form between C- and O-centered radical, oxidizing action likely through O-centered radical.
[3]) Possible contribution also by radicals formed from ȮH addition to C-5.
[4]) Orotic acid.
[5]) 39% e^--transfer.
[6]) Thymine.

Reaction
Radical generation
Method | Solvent | T[K] | Rate data | Ref./ add. ref.

$\dot{\textbf{D}}$ *) + tetranitromethane ($C(NO_2)_4$) $\longrightarrow$ $C(NO_2)_3^-$ + NO_2 + H^+ + products R = CH_3

Pulse rad. of 6-methyluracil + N_2O + H_2O — 81 Fuj 1
KAS | H_2O | 293(2) | $k = 2.7 \cdot 10^9\ M^{-1}s^{-1}$

$\dot{C}H_2SCH_2CH_2CHNH_3^+(COO^-)$ / $CH_3S\dot{C}HCH_2CHNH_3^+(COO^-)$ [7] + $Fe(CN)_6^{3-}$ $\longrightarrow$ $Fe(CN)_6^{4-}$ + products

Pulse rad. of methionine + N_2O + Tl^+ + H_2O — 81 Hil 2
KAS | H_2O | RT | $k = 4.7(5) \cdot 10^9\ M^{-1}s^{-1}$

$CH_3SCH_2CH_2\dot{C}HNHCOCH_3$ + 1,1′-dimethyl-4,4′-bipyridinium(MV^{2+}) [8] $\longrightarrow$ $(CH_3SCH_2CH_2CHNHCOCH_3)^+$ + $MV^{\dot{+}}$

Pulse rad. of N-acetylmethionine + N_2O + H_2O [9] — 83 Hil 1
KAS | H_2O, pH = 4.7 | RT | $k = (1\ldots2) \cdot 10^7\ M^{-1}s^{-1}$

$CH_3\dot{C}HN(C_2H_5)_2$ + 2-methyl-1,4-naphthoquinone $\longrightarrow$ ...semiquinone + products [10]

Pulse rad. of triethylamine + H_2O + N_2O — 73 Rao 3
KAS | H_2O, pH = 11.6 | RT | $k = 4.6(5) \cdot 10^9\ M^{-1}s^{-1}$

$[\{HN(CH_2CO_2^-)_2\}Ni(II)\{HN(CH_2CO_2^-)\dot{C}HCO_2^-\}]$ + $Fe(CN)_6^{3-}$ $\longrightarrow$ $Fe(CN)_6^{4-}$ + products

Pulse rad. of Ni(II)-iminodiacetic acid complex + N_2O + H_2O — 81 Bha 1
KAS | H_2O | RT | $k = 7 \cdot 10^4\ M^{-1}s^{-1}$

$[\{HN(CH_2CO_2^-)_2\}Ni(II)\{HN(CH_2CO_2^-)\dot{C}HCO_2^-\}]$ + O_2 $\longrightarrow$ products

Pulse rad. of Ni(II)-iminodiacetic acid complex + N_2O + H_2O — 81 Bha 1
KAS | H_2O | RT | $k = 5 \cdot 10^3\ M^{-1}s^{-1}$

4.2.1.2 Aromatic radicals and radicals derived from compounds containing aromatic and heterocyclic constituents

4.2.1.2.1 Radicals containing only C and H atoms

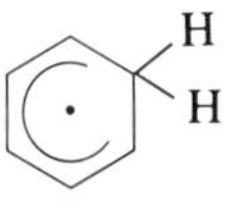
+ $Ce^{4+} \cdot nH_2O$ $\longrightarrow$ $Ce^{3+} \cdot nH_2O$ + products

$\dot{\textbf{E}}$

Pulse rad. of C_6H_6 + t-butanol + N_2O + H_2O — 79 And 1
KAS | H_2O, pH = 0...1 | RT | $k = 1.5(1) \cdot 10^8\ M^{-1}s^{-1}$

$\dot{\textbf{E}}$ + $Cu(H_2O)_6^{2+}$ $\longrightarrow$ $Cu(H_2O)_6^+$ + products

Pulse rad. of benzene + t-butanol + N_2O + H_2O — 79 And 1
KAS | H_2O, pH = 0...1 | RT | $k = 3.6(6) \cdot 10^6\ M^{-1}s^{-1}$

$\dot{\textbf{E}}$ + $Fe(H_2O)_6^{3+}$ $\longrightarrow$ $Fe(H_2O)_6^{2+}$ + products

Pulse rad. of benzene + t-butanol + N_2O + H_2O — 79 And 1
KAS | H_2O, pH = 0...1 | RT | $k = 2.5(4) \cdot 10^7\ M^{-1}s^{-1}$
| | 281...348 | $\log[A/M^{-1}s^{-1}] = 11.5(2)$
| | | $E_a = 22.3(11)\ kJ\ mol^{-1}$

---

*) For $\dot{\textbf{D}}$, see p. 353.
[7]) Radical mixture.
[8]) Methylviologen.
[9]) $CH_3SCH_2CH_2\dot{C}HNHCOCH_3$ radical formed via $\dot{O}H$ radical induced decarboxylation of N-acetylmethionine.
[10]) 37% e^--transfer.

Reaction Radical generation Method	Solvent	T[K]	Rate data	Ref./ add. ref.
$\dot{E}$ *) + $Fe(C_2O_4)_3^{3-} \longrightarrow Fe(C_2O_4)_3^{4-}$ + products				
Pulse rad. of benzene + t-butanol + N_2O + H_2O				79 And 1
KAS	H_2O, pH = 0...1	RT	$k = 1.30(5) \cdot 10^6\ M^{-1}s^{-1}$	

4.2.1.2.2 Radicals containing only C, H, and O atoms

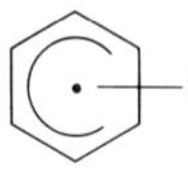

OH [1]) + $Fe^{3+} \longrightarrow C_6H_5OH + Fe^{2+} + H^+$

$\dot{F}$

γ-rad. of C_6H_6 + H_2O				69 Chr 1,
PR	H_2O	RT	$k \approx 7 \cdot 10^3\ M^{-1}s^{-1}$ $9 \cdot 10^3$ [2])	53 Bax 1
$\dot{F}$ [1]) + $Fe(CN)_6^{3-} \longrightarrow Fe(CN)_6^{4-} + H^+ + C_6H_5OH$				
Pulse rad. of C_6H_6 + N_2O + H_2O				80 Mad 1
KAS, Cond.	H_2O, pH = 4...11	RT	$k = 1.8 \cdot 10^7\ M^{-1}s^{-1}$	
$\dot{F}$ [1]) + $IrCl_6^{2-} \longrightarrow$ Ir(III) + products				
[3])				80 Sel 1
[3])	H_2O	RT	$k = 2.7 \cdot 10^9\ M^{-1}s^{-1}$	

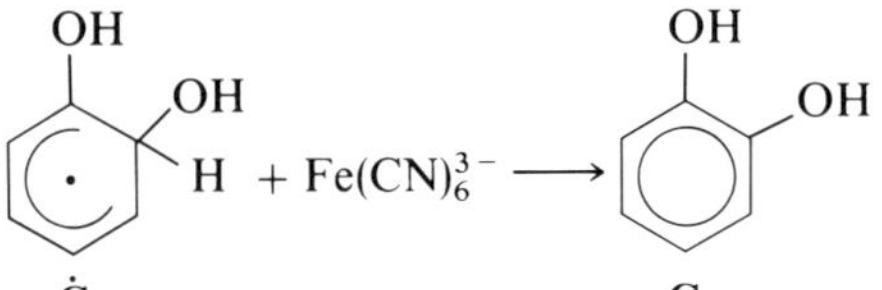

$\dot{G}$ + $Fe(CN)_6^{3-} \longrightarrow$ G + $Fe(CN)_6^{4-} + H^+$

Pulse rad. of phenol + N_2O + H_2O				80 Rag 1
KAS	H_2O	RT	$k = 3.6 \cdot 10^9\ M^{-1}s^{-1}$	
$\dot{G}$ + anthraquinone 2-sulfonate $\longrightarrow$ G + products				
Pulse rad. of phenol + N_2O + H_2O				80 Rag 1
KAS	H_2O	RT	$k \leqslant 2 \cdot 10^7\ M^{-1}s^{-1}$	
$\dot{G}$ + 1,4-benzoquinone $\longrightarrow$ G + 1,4-benzosemiquinone				
Pulse rad. of phenol + N_2O + H_2O				80 Rag 1
KAS	H_2O	RT	$k = 3.7 \cdot 10^9\ M^{-1}s^{-1}$	
$\dot{G}$ + 2,5-dimethylbenzoquinone $\longrightarrow$ G + 2,5-dimethylbenzosemiquinone				
Pulse rad. of phenol + N_2O + H_2O				80 Rag 1
KAS	H_2O	RT	$k \approx 1 \cdot 10^9\ M^{-1}s^{-1}$	
$\dot{G}$ + 2,6-dimethylbenzoquinone $\longrightarrow$ G + 2,6-dimethylbenzosemiquinone				
Pulse rad. of phenol + N_2O + H_2O				80 Rag 1
KAS	H_2O	RT	$k \approx 1 \cdot 10^9\ M^{-1}s^{-1}$	

*) For $\dot{E}$, see p. 354.
[1]) $\dot{O}H$ radical adduct to benzene.
[2]) From [53 Bax 1].
[3]) Conditions not specified; reference to unpublished results.

Reaction Radical generation Method	Solvent	T[K]	Rate data	Ref./ add. ref.
Ġ *) + 2-methylbenzoquinone ⟶ **G** + 2-methylbenzosemiquinone				
Pulse rad. of phenol + N_2O + H_2O				80 Rag 1
KAS	H_2O	RT	$k = 2.2 \cdot 10^9\ M^{-1}s^{-1}$	
Ġ + 4-nitrobenzoate ⟶ **G** + $^{-}OOCC_6H_4\dot{N}O_2^{-}$ + H^+				
Pulse rad. of phenol + N_2O + H_2O				80 Rag 1
KAS	H_2O	RT	$k \leqslant 2 \cdot 10^7\ M^{-1}s^{-1}$	
Ġ + tetramethylbenzoquinone [4]) ⟶ **G** + tetramethylbenzosemiquinone				
Pulse rad. of phenol + N_2O + H_2O				80 Rag 1
KAS	H_2O	RT	$k \leqslant 2 \cdot 10^7\ M^{-1}s^{-1}$	
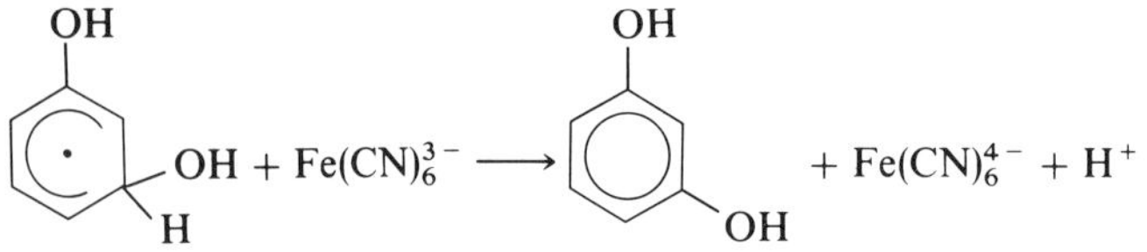 **Ḣ** **H**				
γ-irr. of phenol + N_2O + H_2O				80 Rag 1
PR	H_2O	RT	$k = (1 \ldots 2) \cdot 10^7\ M^{-1}s^{-1}$ [5])	
Ḣ + anthraquinone 2-sulfonate ⟶ **H** + products				
γ-rad. of phenol + N_2O + H_2O				80 Rag 1
PR	H_2O	RT	$k \leqslant 2 \cdot 10^5\ M^{-1}s^{-1}$ [5])	
Ḣ + *p*-benzoquinone ⟶ **H** + *p*-benzosemiquinone				
γ-rad. of phenol + N_2O + H_2O				80 Rag 1
PR	H_2O	RT	$k \approx 2 \cdot 10^6\ M^{-1}s^{-1}$ [5])	
Ḣ + 2,5-dimethylbenzoquinone ⟶ **H** + 2,5-dimethylbenzosemiquinone				
γ-rad. of phenol + N_2O + H_2O				80 Rag 1
PR	H_2O	RT	$k \approx 1 \cdot 10^6\ M^{-1}s^{-1}$ [5])	
Ḣ + 2,6-dimethylbenzoquinone ⟶ **H** + 2,6-dimethylbenzosemiquinone				
γ-rad. of phenol + N_2O + H_2O				80 Rag 1
PR	H_2O	RT	$k \approx 1 \cdot 10^6\ M^{-1}s^{-1}$ [5])	
Ḣ + 4-nitrobenzoate ⟶ **H** + $^{-}OOCC_6H_4\dot{N}O_2^{-}$ + H^+				
Pulse rad. of phenol + N_2O + H_2O				80 Rag 1
KAS	H_2O	RT	$k \leqslant 3 \cdot 10^7\ M^{-1}s^{-1}$	
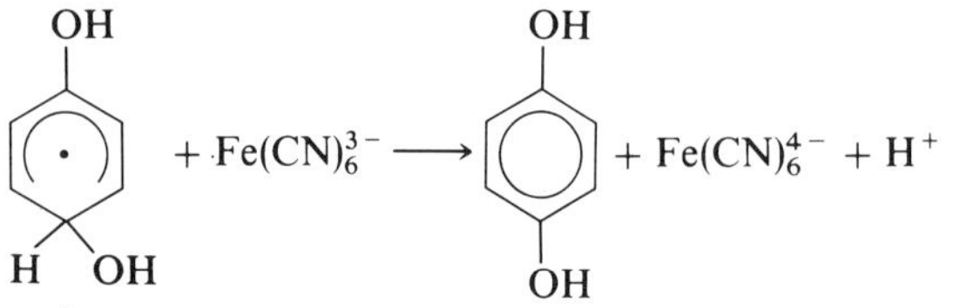 **İ** **I**				
Pulse rad. of phenol + N_2O + H_2O				80 Rag 1
KAS	H_2O	RT	$k = 3.6 \cdot 10^9\ M^{-1}s^{-1}$ [6])	

*) For **Ġ**, see p. 355.

[4]) Duroquinone.

[5]) Based on $k = 10^3\ s^{-1}$ for dehydration of ȮH adduct.

[6]) $k = 4.0 \cdot 10^9$ from competition experiments.

Reaction Radical generation Method	Solvent	T[K]	Rate data	Ref./ add. ref.
$\dot{\mathbf{I}}$ *) + anthraquinone 2-sulfonate ⟶ **I** + products				
Pulse rad. of phenol + N_2O + H_2O				80 Rag 1
KAS	H_2O	RT	$k = 2.1 \cdot 10^9\,M^{-1}\,s^{-1}$	
$\dot{\mathbf{I}}$ + *p*-benzoquinone ⟶ **I** + *p*-benzosemiquinone				
Pulse rad. of phenol + N_2O + H_2O				80 Rag 1
KAS	H_2O	RT	$k = 3.7 \cdot 10^9\,M^{-1}\,s^{-1}$	
$\dot{\mathbf{I}}$ + 2,5-dimethylbenzoquinone ⟶ **I** + 2,5-dimethylbenzosemiquinone				
Pulse rad. of phenol + N_2O + H_2O				80 Rag 1
KAS	H_2O	RT	$k \approx 2 \cdot 10^9\,M^{-1}\,s^{-1}$	
$\dot{\mathbf{I}}$ + 2,6-dimethylbenzoquinone ⟶ **I** + 2,6-dimethylbenzosemiquinone				
Pulse rad. of phenol + N_2O + H_2O				80 Rag 1
KAS	H_2O	RT	$k \approx 2 \cdot 10^9\,M^{-1}\,s^{-1}$	
$\dot{\mathbf{I}}$ + 2-methylbenzoquinone ⟶ **I** + 2-methylbenzosemiquinone				
Pulse rad. of phenol + N_2O + H_2O				80 Rag 1
KAS	H_2O	RT	$k = 2.2 \cdot 10^9\,M^{-1}\,s^{-1}$	
$\dot{\mathbf{I}}$ + 4-nitrobenzoate ⟶ **I** + $^{-}OOCC_6H_4\dot{N}O_2^{-}$ + H^+				
Pulse rad. of phenol + N_2O + H_2O				80 Rag 1
KAS	H_2O	RT	$k \leqslant 3 \cdot 10^7\,M^{-1}\,s^{-1}$	
$\dot{\mathbf{I}}$ + tetramethylbenzoquinone [7]) ⟶ **I** + tetramethylbenzosemiquinone				
Pulse rad. of phenol + N_2O + H_2O				80 Rag 1
KAS	H_2O	RT	$k = 1.8 \cdot 10^9\,M^{-1}\,s^{-1}$	
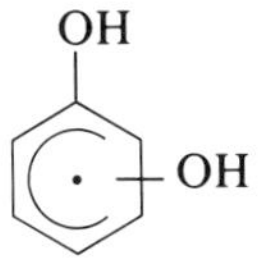 [8]) + 2-methyl-1,4-naphthoquinone ⟶ products [9])				
Pulse rad. of phenol + N_2O + H_2O				73 Rao 3
KAS	H_2O, pH = 5.8 and 7.1	RT	$k = 3.3 \cdot 10^9\,M^{-1}\,s^{-1}$	
	pH = 3		$3.8 \cdot 10^9$	
$C_6H_5\dot{C}HOH$ + 4-chlorobenzenediazonium($ClC_6H_4N_2^+$) ⟶ products				
γ-rad. of benzylalcohol + N_2O + H_2O				80 Pac 1
[10])	H_2O	RT	$k = 4.1 \cdot 10^5\,M^{-1}\,s^{-1}$ [11])	
$C_6H_5\dot{C}HOH$ + 4-methoxybenzenediazonium ($CH_3OC_6H_4N_2^+$) ⟶ products				
γ-rad. of benzylalcohol + N_2O + H_2O				80 Pac 1
[10])	H_2O	RT	$k = 1.5 \cdot 10^5\,M^{-1}\,s^{-1}$ [11])	
$C_6H_5\dot{C}HOH$ + 4-nitrobenzenediazonium ($O_2NC_6H_4N_2^+$) ⟶ products				
γ-rad. of benzylalcohol + N_2O + H_2O				80 Pac 1
[10])	H_2O	RT	$k = 1.5 \cdot 10^6\,M^{-1}\,s^{-1}$ [11])	

*) For $\dot{\mathbf{I}}$, see p. 356.
[7]) Duroquinone.
[8]) Radical mixture ($\dot{O}H$ adduct radicals to phenol).
[9]) 34...44% e^--transfer.
[10]) Dose rate dependence of yields and competition kinetics.
[11]) Relative to $2k(C_6H_5\dot{C}HOH + C_6H_5\dot{C}HOH) = 1.0 \cdot 10^9\,M^{-1}\,s^{-1}$.

Reaction Radical generation Method	Solvent	T[K]	Rate data	Ref./ add. ref.
$C_6H_5\dot{C}HOH$ + 4-toluenediazonium ($CH_3C_6H_4N_2^+$) $\longrightarrow$ products				
γ-rad. of benzylalcohol + N_2O + H_2O				80 Pac 1
[10]	H_2O	RT	$k = 2.9 \cdot 10^5\ M^{-1}\ s^{-1}$ [11]	
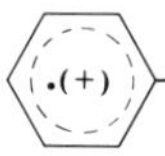 — OCH_3 + Fe^{2+} $\longrightarrow$ $C_6H_5OCH_3$ + Fe^{3+}				
Pulse rad. of anisole + N_2O + H_2O				76 Hol 1
KAS	H_2O, pH = 1	RT	$k = 1.0(2) \cdot 10^9\ M^{-1}\ s^{-1}$	
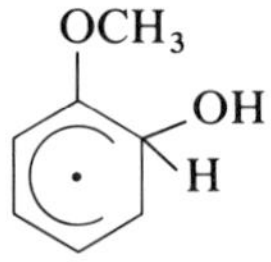 + $Fe(CN)_6^{3-}$ $\longrightarrow$ products $\dot{K}$				
Pulse rad. of anisole + N_2O + H_2O				79 Ste 2
KAS	H_2O	298	$k = 2.41(26) \cdot 10^9\ M^{-1}\ s^{-1}$	
$\dot{K}$ + anthraquinone-2-sulfonate $\longrightarrow$ products				
Pulse rad. of anisole + N_2O + H_2O				79 Ste 2
KAS	H_2O	298	$k \leqslant 1 \cdot 10^7\ M^{-1}\ s^{-1}$	
$\dot{K}$ + *p*-benzoquinone $\longrightarrow$ products				
Pulse rad. of anisole + N_2O + H_2O				79 Ste 2
KAS	H_2O	298	$k = 1.15(27) \cdot 10^9\ M^{-1}\ s^{-1}$	
$\dot{K}$ + 2,5-dimethylbenzoquinone $\longrightarrow$ products				
Pulse rad. of anisole + N_2O + H_2O				79 Ste 2
KAS	H_2O	298	$k = 1.10(12) \cdot 10^8\ M^{-1}\ s^{-1}$	
$\dot{K}$ + 2,6-dimethylbenzoquinone $\longrightarrow$ products				
Pulse rad. of anisole + N_2O + H_2O				79 Ste 2
KAS	H_2O	298	$k = 8.7(8) \cdot 10^7\ M^{-1}\ s^{-1}$	
$\dot{K}$ + 2-methylbenzoquinone $\longrightarrow$ products				
Pulse rad. of anisole + N_2O + H_2O				79 Ste 2
KAS	H_2O	298	$k = 3.51(26) \cdot 10^8\ M^{-1}\ s^{-1}$	
$\dot{K}$ + 4-nitrobenzoate $\longrightarrow$ products				
Pulse rad. of anisole + N_2O + H_2O				79 Ste 2
KAS	H_2O	298	$k \leqslant 2 \cdot 10^7\ M^{-1}\ s^{-1}$	
$\dot{K}$ + tetramethylbenzoquinone [12] $\longrightarrow$ products				
γ- and pulse rad. of anisole + N_2O + H_2O				79 Ste 2
PR, KAS	H_2O	298	$k = 6 \cdot 10^5 \ldots 10^7\ M^{-1}\ s^{-1}$	

[10]) Dose rate dependence of yields and competition kinetics.
[11]) Relative to $2k(C_6H_5\dot{C}HOH + C_6H_5\dot{C}HOH) = 1.0 \cdot 10^9\ M^{-1}\ s^{-1}$.
[12]) Duroquinone.

Reaction Radical generation Method	Solvent	T[K]	Rate data	Ref./ add. ref.
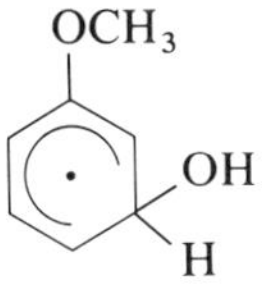 $+ Fe(CN)_6^{3-} \longrightarrow$ products L̇				
γ- and pulse rad. of anisole + N_2O + H_2O PR, KAS	H_2O	298	$k \approx 2 \cdot 10^7\ M^{-1}s^{-1}$	79 Ste 2
L̇ + 1,4-benzoquinone ⟶ products				
γ-rad. of anisole + N_2O + H_2O PR	H_2O	298	$k \leqslant 8 \cdot 10^5\ M^{-1}s^{-1}$	79 Ste 2
L̇ + 2,5-dimethylbenzoquinone ⟶ products				
γ-rad. of anisole + N_2O + H_2O PR	H_2O	298	$k \leqslant 2 \cdot 10^5\ M^{-1}s^{-1}$	79 Ste 2
L̇ + 2-methylbenzoquinone ⟶ products				
γ-rad. of anisole + N_2O + H_2O PR	H_2O	298	$k \leqslant 2 \cdot 10^5\ M^{-1}s^{-1}$	79 Ste 2
L̇ + tetramethylbenzoquinone [12]) ⟶ products				
γ-rad. of anisole + N_2O + H_2O PR	H_2O	298	$k \leqslant 1 \cdot 10^5\ M^{-1}s^{-1}$	79 Ste 2
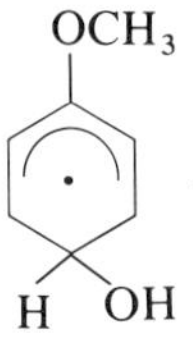 $+ Fe(CN)_6^{3-} \longrightarrow$ products Ṁ				
Pulse rad. of anisole + N_2O + H_2O KAS	H_2O	298	$k = 2.41(26) \cdot 10^9\ M^{-1}s^{-1}$	79 Ste 2
Ṁ + anthraquinone-2-sulfonate ⟶ products				
Pulse rad. of anisole + N_2O + H_2O KAS	H_2O	298	$k \leqslant 1 \cdot 10^7\ M^{-1}s^{-1}$	79 Ste 2
Ṁ + 1,4-benzoquinone ⟶ products				
Pulse rad. of anisole + N_2O + H_2O KAS	H_2O	298	$k = 4.36(100) \cdot 10^9\ M^{-1}s^{-1}$	79 Ste 2
Ṁ + 2,5-dimethylbenzoquinone ⟶ products				
Pulse rad. of anisole + N_2O + H_2O KAS	H_2O	298	$k = 2.03(6) \cdot 10^9\ M^{-1}s^{-1}$	79 Ste 2
Ṁ + 2,6-dimethylbenzoquinone ⟶ products				
Pulse rad. of anisole + N_2O + H_2O KAS	H_2O	298	$k = 1.87(4) \cdot 10^9\ M^{-1}s^{-1}$	79 Ste 2

[12]) Duroquinone.

Reaction Radical generation Method	Solvent	T[K]	Rate data	Ref./ add. ref.
$\dot{M}$ *) + 2-methylbenzoquinone ⟶ products				
Pulse rad. of anisole + N_2O + H_2O				79 Ste 2
KAS	H_2O	298	$k = 2.31(49) \cdot 10^9\ M^{-1}s^{-1}$	
$\dot{M}$ + 4-nitrobenzoate ⟶ products				
Pulse rad. of anisole + N_2O + H_2O				79 Ste 2
KAS	H_2O	298	$k \leqslant 2 \cdot 10^7\ M^{-1}s^{-1}$	
$\dot{M}$ + tetramethylbenzoquinone [12]) ⟶ products				
Pulse rad. of anisole + N_2O + H_2O				79 Ste 2
KAS	H_2O	298	$k \approx 5 \cdot 10^7\ M^{-1}s^{-1}$	
$C_6H_5\dot{C}(O^-)CH_3 + O_2$ ⟶ products [13])				
Pulse rad. of acetophenone + t-butanol + O_2 + H_2O				71 Wil 2
KAS	H_2O, pH = 12	RT	$k = 2.3 \cdot 10^9\ M^{-1}s^{-1}$	
$C_6H_5\dot{C}(O^-)CH_3$ + benzophenone ⟶ $C_6H_5COCH_3 + (C_6H_5)_2\dot{C}O^-$				
Pulse rad. of acetophenone + t-butanol + N_2 + H_2O				73 Ada 1
KAS	H_2O, pH = 12	RT	$k = 7.8 \cdot 10^8\ M^{-1}s^{-1}$	
$C_6H_5\dot{C}(O^-)CH_3$ + 4-nitroacetophenone(PNAP) ⟶ $C_6H_5COCH_3$ + PNAP$^{\cdot-}$				
Pulse rad. of acetophenone + t-butanol + N_2 + H_2O				73 Ada 1
KAS	H_2O, pH = 11	RT	$k = 5.2(5) \cdot 10^9\ M^{-1}s^{-1}$	
$C_6H_5\dot{C}(O^-)CH_3$ + 2,2,6,6-tetramethyl-4-oxo-piperidinyloxy(TAN) ⟶ products [13])				
Pulse rad. of acetophenone + t-butanol + N_2 + H_2O				71 Wil 2
KAS	H_2O, pH = 12	RT	$k = 4.6 \cdot 10^8\ M^{-1}s^{-1}$	
$(C_6H_5CH_2CO_2H)^{\cdot-}$ [14]) + cytochrome(III)-c ⟶ products				
Pulse rad. of $C_6H_5CH_2COOH$ + N_2 + H_2O				75 Sim 1
KAS	H_2O	RT	$k = 1.8 \cdot 10^9\ M^{-1}s^{-1}$	
$C_6H_5\dot{C}OHCH_3$ + cytochrome(III)-c(Fe(III)) ⟶ products				
Pulse rad. of acetophenone + N_2 + H_2O				75 Sim 1
KAS	H_2O	RT	$k = 8 \cdot 10^8\ M^{-1}s^{-1}$	
$C_6H_5\dot{C}OHCH_3$ + N-ethylmaleimide ⟶ $C_6H_5COCH_3 + H^+$ + products				
Pulse rad. of acetophenone + t-butanol + N_2 + H_2O				72 Hay 1
KAS	H_2O	RT	$k = 3.8 \cdot 10^9\ M^{-1}s^{-1}$	
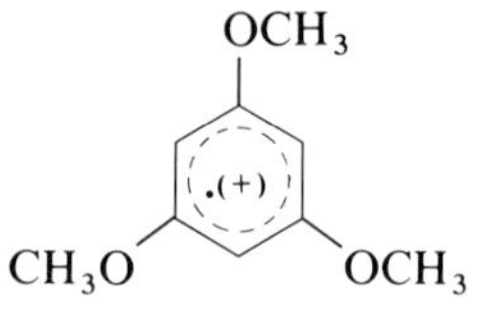 $\dot{N}$ + diethyldisulfide ⟶ 1,3,5-trimethoxybenzene + $(C_2H_5SSC_2H_5)^{\cdot+}$				
Pulse rad. of 1,3,5-trimethoxybenzene + N_2O + H_2O				76 Bon 1
KAS	H_2O	RT	$k = 2.1 \cdot 10^9\ M^{-1}s^{-1}$	

*) For $\dot{M}$, see p. 359.
[12]) Duroquinone.
[13]) e^--transfer and/or addition.
[14]) Electron adduct to benzylic acid, located likely on aromatic ring.

Reaction Radical generation Method	Solvent	T[K]	Rate data	Ref./ add. ref.
Ṅ*) + dimethyldisulfide ⟶ 1,3,5-trimethoxybenzene + $(CH_3SSCH_3)^{\dot{+}}$				
Pulse rad. of 1,3,5-trimethoxybenzene + N_2O + H_2O				76 Bon 1
KAS	H_2O	RT	$k = 2.2 \cdot 10^9\ M^{-1}s^{-1}$	

4.2.1.2.3 Radicals containing C, H, N, and other atoms

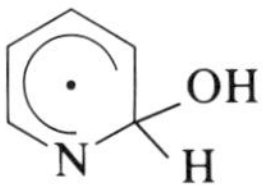

+ $IrCl_6^{2-}$ ⟶ products

Pulse rad. of pyridine + N_2O + H_2O				80 Sel 1
KAS	H_2O	RT	$k = 5.7 \cdot 10^8\ M^{-1}s^{-1}$	

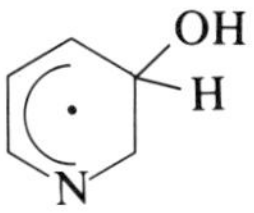

+ $Fe(CN)_6^{3-}$ ⟶ products

Ȯ

γ- and pulse rad. of pyridine + N_2O + H_2O				80 Sel 1
PR, KAS	H_2O	RT	$k = 10^4 \ldots 5 \cdot 10^6\ M^{-1}s^{-1}$	
Ȯ + $IrCl_6^{2-}$ ⟶ products				
Pulse rad. of pyridine + N_2O + H_2O				80 Sel 1
KAS	H_2O	RT	$k = 5.7 \cdot 10^8\ M^{-1}s^{-1}$	

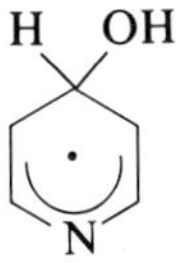

+ $Fe(CN)_6^{3-}$ ⟶ products

Ṗ

γ-rad. of pyridine + N_2O + H_2O				80 Sel 1
PR	H_2O	RT	$k \leq 5 \cdot 10^4\ M^{-1}s^{-1}$	
Ṗ + $IrCl_6^{2-}$ ⟶ products				
γ-rad. of pyridine + N_2O + H_2O				80 Sel 1
PR	H_2O	RT	$k \leq 5 \cdot 10^4\ M^{-1}s^{-1}$	

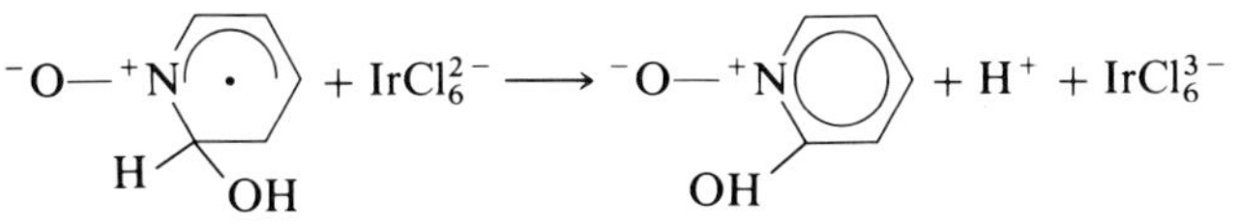

Pulse rad. of pyridine N-oxide + N_2O + H_2O				80 Net 2
KAS	H_2O, pH = 7	RT	$k = 3.3 \cdot 10^7\ M^{-1}s^{-1}$ (±20%)	

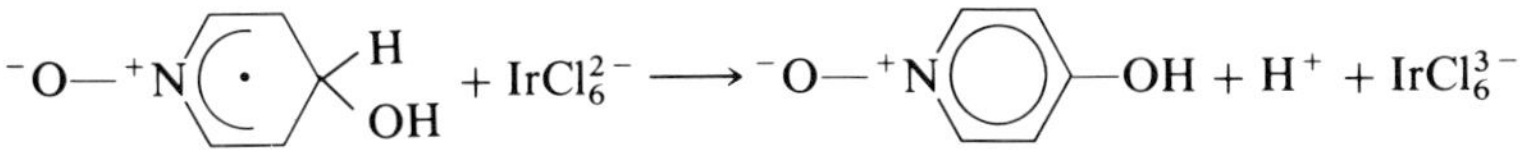

Pulse rad. of pyridine N-oxide + N_2O + H_2O				80 Net 2
KAS	H_2O, pH = 7	RT	$k = 1.3 \cdot 10^8\ M^{-1}s^{-1}$ (±20%)	

*) For Ṅ, see p. 360.

Reaction Radical generation Method	Solvent	T[K]	Rate data	Ref./ add. ref.
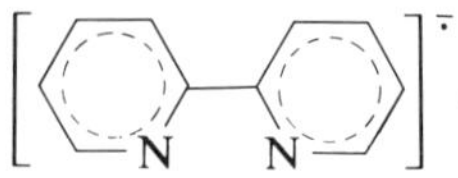 $+$ cytochrome(III)-c(Fe(III)) $\longrightarrow$ products				
Pulse rad. of 2,2'-bipyridine $+ N_2 + H_2O$				75 Sim 1
KAS	H_2O	RT	$k = 5.7 \cdot 10^8\ M^{-1}s^{-1}$	
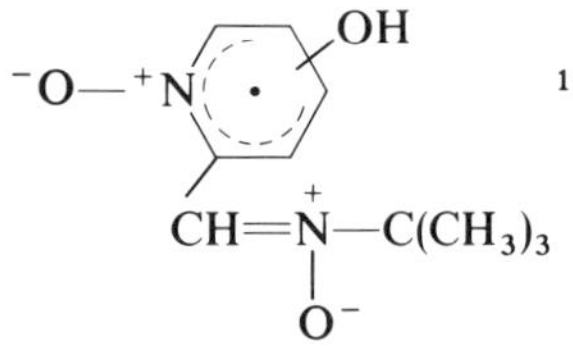 [15]) $+ IrCl_6^{2-} \longrightarrow IrCl_6^{3-}$ + products [15]) $+ H^+$				
Pulse rad. of α-2-pyridyl 1-oxide N-*t*-butyl nitrone $+ N_2O + H_2O$				80 Net 2
KAS	H_2O	RT	$k \approx 5 \cdot 10^7\ M^{-1}s^{-1}$	
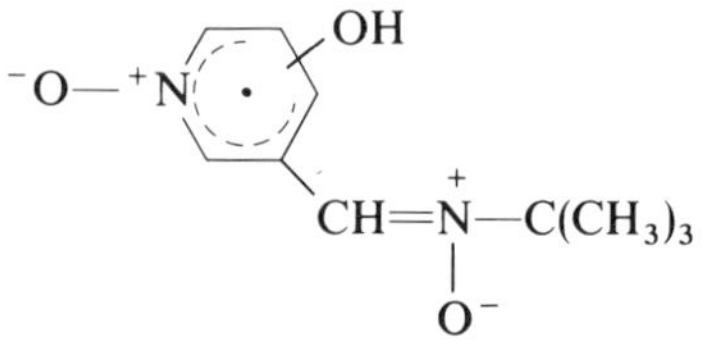 [15]) $+ IrCl_6^{2-} \longrightarrow IrCl_6^{3-} + H^+$ + products [15])				
Pulse rad. of α-3-pyridyl 1-oxide N-*t*-butyl nitrone $+ N_2O + H_2O$				80 Net 2
KAS	H_2O	RT	$k \approx 5 \cdot 10^7\ M^{-1}s^{-1}$	
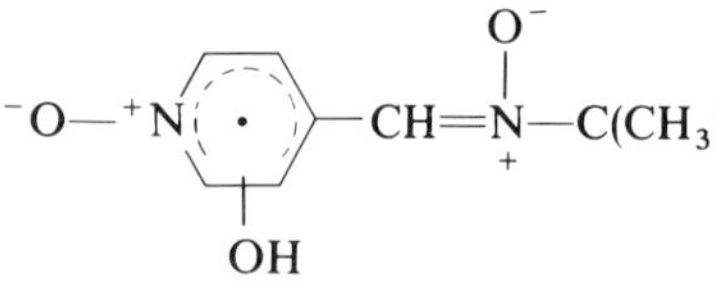 [15]) $+ IrCl_6^{2-} \longrightarrow IrCl_6^{3-} + H^+$ + products [15])				
Pulse rad. of α-4-pyridyl 1-oxide N-*t*-butyl nitrone $+ N_2O + H_2O$				80 Net 2
KAS	H_2O	RT	$k \approx 5 \cdot 10^7\ M^{-1}s^{-1}$	
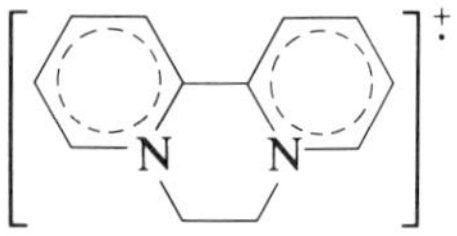 $\dot{Q}^+$ [16]) + cytochrome-c(Fe(III)) $\longrightarrow [\ldots]^{2+} + (\text{cyt}\ldots)^{\dot{-}}$				
Pulse rad. of 1,1'-ethylene-2,2'-bipyridinium + formate $+ H_2O$				75 Lan 1/
KAS	H_2O, pH $= 4\ldots8$	RT	$k = 1.4 \cdot 10^8\ M^{-1}s^{-1}$	74 Ste 1, 75 Mac 1
$\dot{Q}^+$ [16]) $+ O_2 \longrightarrow [\ldots]^{2+} + \dot{O}_2^-$				
Pulse rad. of 1,1'-ethylene-2,2'-bipyridinium + formate $+ H_2O$				78 Far 1
KAS	H_2O	RT	$k = 4.7(3) \cdot 10^8\ M^{-1}s^{-1}$	
$\dot{Q}^+$ [17]) + flavin adenine dinucleotide(FAD) products				
Pulse rad. of 1,1'-ethylene-2,2'-bipyridinium + *t*-butanol $+ N_2 + H_2O$				76 And 1
KAS	H_2O	RT	$k = 1.85 \cdot 10^8\ M^{-1}s^{-1}$	

[15]) Mixture of $\dot{O}H$ adducts at various positions of pyridine ring, and of hydroxylated products.
[16]) Radical from $\dot{C}O_2^- +$ 1,1'-ethylene-2,2'-bipyridinium reaction.
[17]) Electron adduct to 1,1'-ethylene-2,2'-bipyridinium.

Reaction Radical generation Method	Solvent	T[K]	Rate data	Ref./ add. ref.

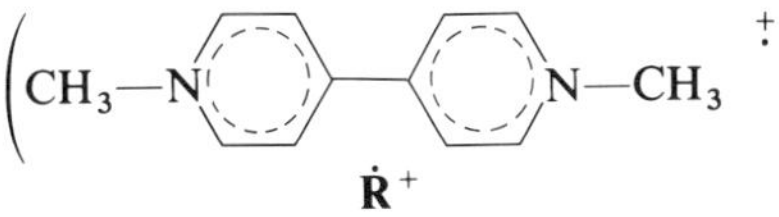

$\dot{R}^+$

[18]) + cytochrome-c(Fe(III)) ⟶ (cytochrome-c)$^{\dot{-}}$ + (1,1′-dimethyl...)$^{2+}$

Pulse rad. of 1,1′-dimethyl-4,4′-bipyridinium [19]) + formate + N_2O				75 Lan 1/
KAS	H_2O, pH = 4...8	RT	$k = 2.1 \cdot 10^8\ M^{-1}s^{-1}$	74 Ste 1, 75 Mac 1, 75 Sim 1

$\dot{R}^+$ + cytochrome(III)-c(Fe(III)) ⟶ products

Pulse rad. of 1,1′-dimethyl-4,4′-bipyridinium [19]) + N_2 + H_2O				75 Sim 1/
KAS	H_2O, pH = 7	RT	$k = 2.2 \cdot 10^9\ M^{-1}s^{-1}$	74 Ste 1,
	pH = 10.8		$3.4 \cdot 10^8$	74 Mac 1, 75 Lan 1

$\dot{R}^+ + O_2 \longrightarrow [...]^{2+}$ [19]) + $\dot{O}_2^-$

Phot. of air-saturated aqueous solutions of 1,1′-dimethyl-4,4′-bipyridinium [19])				77 Pat 1/
KAS	H_2O	RT	$k = 6.0 \cdot 10^8\ M^{-1}s^{-1}$	78 Far 1, 73 Far 1

$\dot{R}^+ + O_2 \longrightarrow [...]^{2+}$ [19]) + $\dot{O}_2^-$

Pulse rad. of 1,1′-dimethyl-4,4′-bipyridinium [19]) + formate + H_2O				78 Far 1/
KAS	H_2O	RT	$k = 8.0(3) \cdot 10^8\ M^{-1}s^{-1}$	73 Far 1, 77 Pat 1

$\dot{R}^+$ + flavin adenine dinucleotide(FAD) ⟶ products

Pulse rad. of 1,1′-dimethyl-4,4′-bipyridinium [19]) + t-butanol + N_2 + H_2O				76 And 1
KAS	H_2O	RT	$k = 1.7 \cdot 10^9\ M^{-1}s^{-1}$	

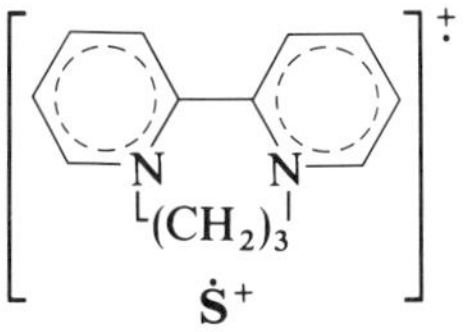

$\dot{S}^+$

[20]) + cytochrome-c(Fe(III)) ⟶ $[...]^{2+}$ + (cyt...)$^{\dot{-}}$

Pulse rad. of 1,1′-trimethylene-2,2′-bipyridinium + formate + H_2O				75 Lan 1/
KAS	H_2O, pH = 4...8	RT	$k = 1.9 \cdot 10^8\ M^{-1}s^{-1}$	74 Ste 1, 75 Mac 1

$\dot{S}^+$ [20]) + $O_2 \longrightarrow [...]^{2+} + \dot{O}_2^-$

Pulse rad. of 1,1′-trimethylene-2,2′-bipyridinium + formate + H_2O				78 Far 1
KAS	H_2O	RT	$k = 8.4(6) \cdot 10^8\ M^{-1}s^{-1}$	

$\dot{S}$ + flavin adenine dinucleotide(FAD) ⟶ products

Pulse rad. of 1,1′-trimethylene-2,2′-bipyridinium + t-butanol + N_2 + H_2O				76 And 1
KAS	H_2O, pH = 7	RT	$k = 3.0 \cdot 10^9\ M^{-1}s^{-1}$	
	pH = 11.5		$1.19 \cdot 10^8$	

[18]) Radical from 1,1′-dimethyl-4,4′-bipyridinium + $\dot{C}O_2^-$ reaction.
[19]) Methylviologen, paraquat.
[20]) Radical from $\dot{C}O_2^-$ + 1,1′-trimethylene-2,2′-bipyridinium reaction.

Reaction Radical generation Method	Solvent	T[K]	Rate data	Ref./ add. ref.
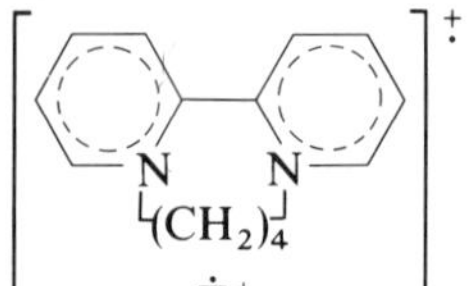 $\dot{T}^+$ [21]) $+ O_2 \longrightarrow [\ldots]^{2+} + \dot{O}_2^-$				
Pulse rad. of 1,1′-tetramethylene-2,2′-bipyridinium + formate + H_2O				78 Far 1
KAS	H_2O	RT	$k = 9.6(4) \cdot 10^8\ M^{-1}\,s^{-1}$	
$\dot{T}^+$ + flavin adenine dinucleotide(FAD) ⟶ products				
Pulse rad. of 1,1′-tetramethylene-2,2′-bipyridinium + t-butanol + N_2 + H_2O				76 And 1
KAS	H_2O	RT	$k = 2.7 \cdot 10^9\ M^{-1}\,s^{-1}$	
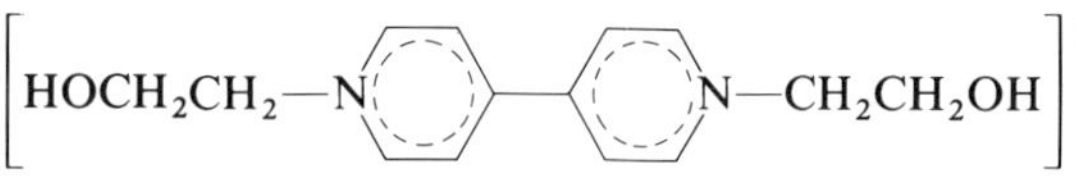 [22]) $+ O_2 \longrightarrow [\ldots]^{2+} + \dot{O}_2^-$				
Pulse rad. of 1,1′-bis-2-hydroxyethyl-4,4′-bipyridinium + formate + H_2O				78 Far 1
KAS	H_2O	RT	$k = 3.6(2) \cdot 10^8\ M^{-1}\,s^{-1}$	
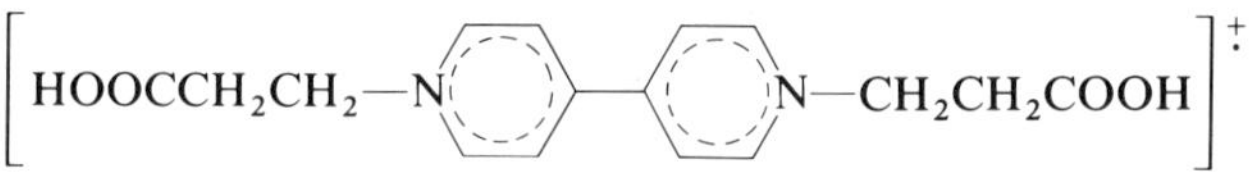 + flavin adenine dinucleotide(FAD) ⟶ products				
Pulse rad. of 1,1′-dicarboxyethyl-4,4′-bipyridinium + t-butanol + N_2 + H_2O				76 And 1
KAS	H_2O	RT	$k = 6.4 \cdot 10^8\ M^{-1}\,s^{-1}$	
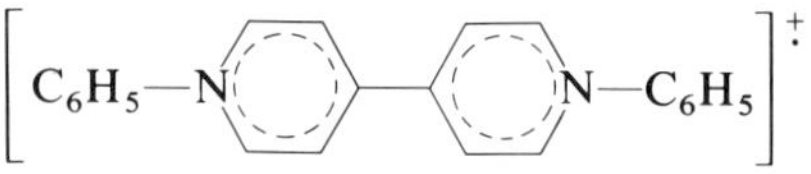 $+ O_2 \longrightarrow [\ldots]^{2+} + \dot{O}_2^-$				
Pulse rad. of 1,1′-diphenyl-4,4′-bipyridinium + formate + H_2O				78 Far 1
KAS	H_2O	RT	$k = 2.2(1) \cdot 10^7\ M^{-1}\,s^{-1}$ [23])	
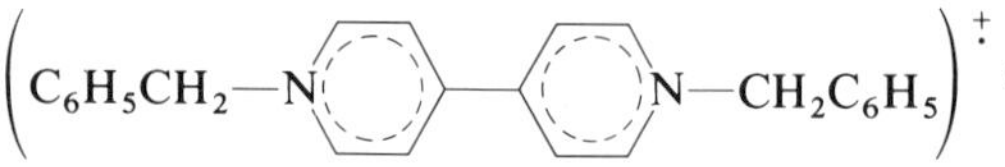 $\dot{U}^+$ [24]) + cytochrome-c(Fe(III)) ⟶ (cytochrome-c)$^{\dot{-}}$ + $(1,1'\text{-di}\ldots)^{2+}$				
Pulse rad. of 1,1′-dibenzyl-4,4′-bipyridinium [25]) + formate + H_2O				75 Lan 1/ 74 Ste 1, 75 Mac 1, 75 Sim 1
KAS	H_2O, pH = 4…8	RT	$k = 1.8 \cdot 10^8\ M^{-1}\,s^{-1}$	
$\dot{U}^+$ + cytochrome(III)-c(Fe(III)) ⟶ products				
Pulse rad. of 1,1′-dibenzyl-4,4′-bipyridinium [25]) + N_2 + H_2O				75 Sim 1/ 75 Lan 1, 74 Ste 1, 75 Mac 1
KAS	H_2O	RT	$k = 4.3 \cdot 10^8\ M^{-1}\,s^{-1}$	
$\dot{U}^+$ + anthraquinone-2-sulfonate ⟶ products				
Pulse rad. of 1,1′-dibenzyl-4,4′-bipyridinium [25]) + t-butanol + N_2 + H_2O				76 War 1
KAS	H_2O	RT	$k > 3 \cdot 10^9\ M^{-1}\,s^{-1}$	

[21]) Radical from $\dot{C}O_2^-$ + 1,1′-tetramethylene-2,2′-bipyridinium reaction.
[22]) Radical from $\dot{C}O_2^-$ + 1,1′-bis-2-hydroxyethyl-4,4′-bipyridinium reaction.
[23]) Rate constant listed in the reference paper as preliminary value.
[24]) Radical from $\dot{C}O_2^-$ + 1,1′-dibenzyl-4,4′-dipyridinium reaction.
[25]) Benzylviologen.

Reaction Radical generation Method	Solvent	T[K]	Rate data	Ref./ add. ref.
$\dot{U}^+$ *) + flavin adenine dinucleotide(FAD) $\longrightarrow$ products				
Pulse rad. of 1,1′-dibenzyl-4,4′-bipyridinium [25]) + t-butanol + N_2 + H_2O				76 And 1
KAS	H_2O, pH = 7	RT	$k = 3.65 \cdot 10^8\ M^{-1}s^{-1}$	
	pH = 11		$7.4 \cdot 10^7$	
$\dot{U}^+$ + 1-(2-hydroxy-3-methoxypropyl)-2-nitroimidazole $\longrightarrow$ products				
Pulse rad. of 1,1′-dibenzyl-4,4′-bipyridinium [25]) + t-butanol + N_2 + H_2O				76 War 1
KAS	H_2O	RT	$k = 4.8(4) \cdot 10^8\ M^{-1}s^{-1}$	
$\dot{U}^+$ + tetramethylbenzoquinone [26]) $\longrightarrow$ products				
Pulse rad. of 1,1′-dibenzyl-4,4′-bipyridinium [25]) + t-butanol + N_2 + H_2O				76 War 1
KAS	H_2O	RT	$k = 3.2(3) \cdot 10^9\ M^{-1}s^{-1}$	

4.2.1.3 Radicals with undefined stoichiometry and structure

The rate constants listed in this section refer to radicals of undefined stoichiometry. The latter result essentially from reactions of $\dot{O}H$ radicals, $\dot{H}$ atoms and hydrated electrons, e_{aq}^-, with organic substrates and are generally written as (compound-$\dot{O}H$), (compound-$\dot{H}$) or (compound)$^{\dot{-}}$. The variety of radicals resulting from these reactions essentially reflects the properties of $\dot{O}H$, $\dot{H}$ and e_{aq}^-. The $\dot{O}H$ radicals for example undergo electrophilic addition, abstraction and electron transfer (oxidation) reactions. They are generally highly reactive and thus rather unselective. If an $\dot{O}H$ radical reacts for example with a C_4-compound with each carbon carrying one or more H-atoms hydrogen abstraction usually leads to the formation of four different radicals with the radical site at either C_1, C_2, C_3 or C_4. The relative yields of the latter may significantly differ and depend on the nature of the ruptured C—H bond and also on the influence of functional groups. Similar considerations apply also for $\dot{H}$ atom reaction except that direct electron transfer is usually a reduction process. The radical resulting from the reaction of e_{aq}^- in most cases is likely to be the radical anion formed by electron addition to the functional group with the highest electron affinity. A structural and stoichiometric uncertainty often results however from spin and charge delocalization, and from H_2O, OH^- or H^+ addition or elimination. The latter, which is also observed for radicals formed from $\dot{O}H$ and $\dot{H}$ reactions, may frequently in fact occur on about the same time scale as the electron transfer to the other substrate of interest. In several cases the identity of radicals from these $\dot{O}H$, $\dot{H}$ and e_{aq}^- reactions has been established unambiguously by time-resolved ESR-measurements or other ingeniously designed investigations. Electron transfer rate constants referring to such species are of course listed in the appropriate section of well defined radicals. It can be expected that many more species will be unambiguously characterized in the future, and it is quite possible that for some of these, their assignment as carbon-centered radicals will be inappropriate. The present listing of all these undefined radicals in this section is based on an assumed finite probability of at least partial spin localization on a particular carbon atom or on the carbon skeleton of the radicals.

(Acetonitrile-$\dot{O}H$) [1]) + Fe^{2+} $\longrightarrow$ Fe^{3+} + products				
$Fe(II)/H_2O_2$ in $CH_3CN + H_2O$				73 Wal 1
PR	H_2O	RT	$k = 1.2 \cdot 10^6\ M^{-1}s^{-1}$ [2])	
(Acetylasparagine-$\dot{O}H$) [3]) + 2-methyl-1,4-naphthoquinone $\longrightarrow$ ...semiquinone + products [4])				
Pulse rad. of acetylasparagine + H_2O + N_2O				73 Rao 3
KAS	H_2O, pH = 12.5	RT	$k = 1.5(2) \cdot 10^9\ M^{-1}s^{-1}$	
(Acetyldiglycine-$\dot{O}H$) [5]) + 2-methyl-1,4-naphthoquinone $\longrightarrow$ ...semiquinone + products [6])				
Pulse rad. of acetyldiglycine + H_2O + N_2O				73 Rao 1
KAS	H_2O, pH = 12.3	RT	$k = 3.8(4) \cdot 10^9\ M^{-1}s^{-1}$	

*) For $\dot{U}$, see p. 364.
[25]) Benzylviologen.
[26]) Duroquinone.
[1]) Reducing radical formed from $\dot{O}H + CH_3CN$ reaction.
[2]) Calc. on the basis of various assumptions.
[3]) Radicals from acetylasparagine + $\dot{O}H$ reaction.
[4]) 39% e^--transfer.
[5]) Radicals from acetyldiglycine + $\dot{O}H$ reaction.
[6]) 55% e^--transfer.

Reaction Radical generation Method	Solvent	T[K]	Rate data	Ref./ add. ref.
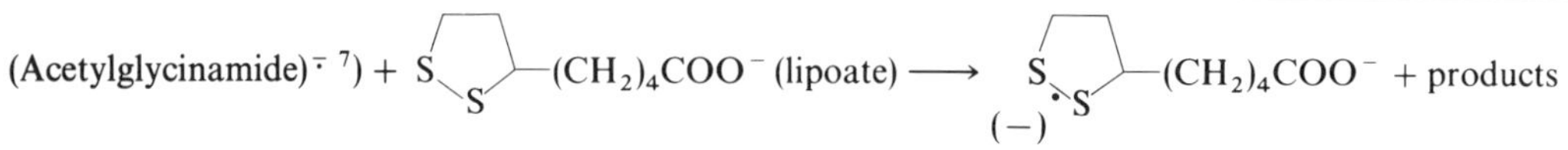 (Acetylglycinamide)$^{\overline{\cdot}}$ 7) + S S —(CH$_2$)$_4$COO$^-$ (lipoate) ⟶ S S —(CH$_2$)$_4$COO$^-$ + products (−)				
Pulse rad. of acetylglycinamide + t-butanol + N_2 + H_2O				75 Far 1,
KAS	H_2O, pH = 10	RT	$k = 5.8 \cdot 10^8\ M^{-1}s^{-1}$	75 Far 3
(Acetylglycylglycinamide)$^{\overline{\cdot}}$ [8]) + 2-methyl-1,4-naphthoquinone ⟶ ...semiquinone + products				
Pulse rad. of acetylglycylglycinamide + H_2O				73 Rao 1
KAS	H_2O	RT	$k = 2.7(3) \cdot 10^9\ M^{-1}s^{-1}$	
(Acetylserinamide-$\dot{O}$H) [9]) + 2-methyl-1,4-naphthoquinone ⟶ ...semiquinone + products				
Pulse rad. of acetylserinamide + H_2O + N_2O				73 Rao 3
KAS	H_2O, pH = 6	RT	$k = 1.5(2) \cdot 10^9\ M^{-1}s^{-1}$ [10])	
	pH = 11		$1.9(2) \cdot 10^9$ [11])	
(Acetyltrialanine-$\dot{O}$H) [12]) + 2-methyl-1,4-naphthoquinone ⟶ ...semiquinone + products				
Pulse rad. of acetyltrialanine + H_2O + N_2O				73 Rao 1
KAS	H_2O, pH = 6.9	RT	$k = 2.1(2) \cdot 10^9\ M^{-1}s^{-1}$ [13])	
	pH = 12.3		$2.6(3) \cdot 10^9$ [14])	
(Acetyltriglycine-$\dot{O}$H) [15]) + 2-methyl-1,4-naphthoquinone ⟶ ...semiquinone + products [16])				
Pulse rad. of acetyltriglycine + H_2O + N_2O				73 Rao 1
KAS	H_2O, pH = 12.5	RT	$k = 3.7(4) \cdot 10^9\ M^{-1}s^{-1}$	
(Acetyltrisarcosine-$\dot{O}$H) [17]) + 2-methyl-1,4-naphthoquinone ⟶ ...semiquinone + products [18])				
Pulse rad. of acetyltrisarcosine + H_2O + N_2O				73 Rao 3
KAS	H_2O, pH = 12.5	RT	$k = 1.3(1) \cdot 10^9\ M^{-1}s^{-1}$	
(Acrylate)$^{\overline{\cdot}}$ [19]) + *anti*-5-nitro-2-furaldoxime [20]) ⟶ products [21])				
Pulse rad. of acrylate + H_2O				73 Gre 1
KAS	H_2O	RT	$k = 4.0 \cdot 10^9\ M^{-1}s^{-1}$	
(Adenine)$^{\overline{\cdot}}$ [22]) + O_2 ⟶ $\dot{O}_2^-$ + products				
Pulse rad. of adenine + H_2O				71 Wil 2
KAS	H_2O	RT	$k = 3.6 \cdot 10^9\ M^{-1}s^{-1}$	
(Adenine)$^{\overline{\cdot}}$ [22]) + benzophenone ⟶ $C_6H_5(\dot{C}O^-)CH_3$ + products				
Pulse rad. of adenine + t-butanol + N_2 + H_2O				72 Ada 1
KAS	H_2O, pH = 12	RT	$k = 2.7 \cdot 10^9\ M^{-1}s^{-1}$	

7) Radicals from acetylglycinamide + e_{aq}^- reaction (radical presumed to be of $>\dot{C}$—O^- type).
8) Radicals from acetylglycylglycinamide + e_{aq}^- reaction.
9) Radicals from acetylserinamide + $\dot{O}$H reaction.
10) 52% e^--transfer.
11) 68% e^--transfer.
12) Radicals from acetyltrialamine + $\dot{O}$H reaction.
13) 18% e^--transfer.
14) 47% e^--transfer.
15) Radicals from acetyltriglycine + $\dot{O}$H reaction.
16) 50% e^--transfer.
17) Radicals from acetyltrisarcosine + $\dot{O}$H reaction.
18) 39% e^--transfer.
19) Radicals from e_{aq}^- + acrylate reaction.
20) Nifuroxime.
21) 65% e^--transfer.
22) Radicals from adenine + e_{aq}^- reaction.

Reaction Radical generation Method	Solvent	T[K]	Rate data	Ref./ add. ref.
(Adenine)$^{\bar{\cdot}}$ [22]) + 6-carboxyuracil [23]) ⟶ products				
Pulse rad. of adenine + t-butanol + N_2 + H_2O				72 Ada 1
KAS	H_2O	RT	$k = 3.5 \cdot 10^9\ M^{-1}s^{-1}$	
(Adenine)$^{\bar{\cdot}}$ [22]) + 8-methoxypsoralen(8-MOP) ⟶ (8-MOP)$^{\bar{\cdot}}$ + products				
Pulse rad. of adenine + t-butanol + H_2O				78 Red 1
KAS	H_2O	RT	$k = 5.5 \cdot 10^9\ M^{-1}s^{-1}$	
(Adenine)$^{\bar{\cdot}}$ [22]) + 2-methyl-1,4-naphthoquinone ⟶ 2-methyl...semiquinone + products				
Pulse rad. of adenine + t-butanol + N_2 + H_2O				72 Ada 1
KAS	H_2O	RT	$k = 4.2 \cdot 10^9\ M^{-1}s^{-1}$	
(Adenine)$^{\bar{\cdot}}$ [22]) + 4-nitroacetophenone(PNAP) ⟶ (PNAP)$^{\bar{\cdot}}$ + products				
Pulse rad. of adenine + t-butanol + N_2 + H_2O				72 Ada 1
KAS	H_2O, pH = 7	RT	$k = 5.5 \cdot 10^9\ M^{-1}s^{-1}$	
	pH = 12		$4.2 \cdot 10^9$	
(Adenine)$^{\bar{\cdot}}$ [22]) + *anti*-5-nitro-2-furaldoxime ⟶ products				
Pulse rad. of adenine + t-butanol + N_2 + H_2O				73 Gre 1
KAS	H_2O	RT	$k = 6.9 \cdot 10^9\ M^{-1}s^{-1}$	
(Adenine)$^{\bar{\cdot}}$ [22]) + 2,2,6,6-tetramethyl-4-oxo-1-piperidinyloxy(TAN) ⟶ products [24])				
Pulse rad. of adenine + t-butanol + N_2 + H_2O				71 Wil 2
KAS	H_2O	RT	$k = 1.6 \cdot 10^9\ M^{-1}s^{-1}$	
(Adenine-$\dot{O}$H) [24a]) + *anti*-5-nitro-2-furaldoxime ⟶ products [25])				
Pulse rad. of adenine + N_2O + H_2O				73 Gre 1
KAS	H_2O	RT	$k = 2.5 \cdot 10^9\ M^{-1}s^{-1}$	
(Adenine-$\dot{O}$H) [24a]) + nor-pseudopelletierine-N-oxy ⟶ products [26])				
Pulse rad. of adenine + N_2O + H_2O				72 Bru 1
KAS	H_2O	RT	$k = (5.7...6.7) \cdot 10^8\ M^{-1}s^{-1}$	
(Adenosine)$^{\bar{\cdot}}$ [27]) + 5-bromouracil ⟶ Br^- + products				
Pulse rad. of adenosine + t-butanol + N_2 + H_2O				72 Ada 2
PR, KAS (competition kinetics)	H_2O	RT	$k = 3.5 \cdot 10^8\ M^{-1}s^{-1}$	
(Adenosine)$^{\bar{\cdot}}$ [27]) + 6-carboxyuracil [23]) ⟶ products				
Pulse rad. of adenosine + t-butanol + N_2 + H_2O				72 Ada 1
KAS	H_2O	RT	$k = 2.8 \cdot 10^9\ M^{-1}s^{-1}$	
(Adenosine)$^{\bar{\cdot}}$ [27]) + 2-methyl-1,4-naphthoquinone ⟶ 2-methyl...semiquinone + products				
Pulse rad. of adenosine + t-butanol + N_2 + H_2O				72 Ada 1
KAS	H_2O	RT	$k = 3.4 \cdot 10^9\ M^{-1}s^{-1}$	
(Adenosine)$^{\bar{\cdot}}$ [27]) + 4-nitroacetophenone(PNAP) ⟶ (PNAP)$^{\bar{\cdot}}$ + products				
Pulse rad. of adenosine + t-butanol + N_2 + H_2O				72 Ada 1
KAS	H_2O	RT	$k = 5.4 \cdot 10^9\ M^{-1}s^{-1}$	

[22]) Radicals from adenine + e_{aq}^- reaction.
[23]) Orotic acid.
[24]) e^--transfer and/or addition.
[24a]) Radicals from adenine + $\dot{O}$H reaction.
[25]) 8% e^--transfer.
[26]) Addition and/or e^--transfer.
[27]) Radicals from adenosine + e_{aq}^- reaction.

Reaction Radical generation Method	Solvent	T[K]	Rate data	Ref./ add. ref.
(Adenosine)$^{\cdot -}$ [27]) + *anti*-5-nitro-2-furaldoxime ⟶ products				
Pulse rad. of adenosine + *t*-butanol + N_2 + H_2O				73 Gre 1
KAS	H_2O	RT	$k = 4.1 \cdot 10^9\ M^{-1}s^{-1}$	
(Adenosine)$^{\cdot -}$ [27]) + 2,2,6,6-tetramethyl-4-oxo-1-piperidinyloxy(TAN) ⟶ products [28])				
Pulse rad. of adenosine + *t*-butanol + N_2 + H_2O				71 Wil 2
KAS	H_2O	RT	$k = 1.5 \cdot 10^9\ M^{-1}s^{-1}$	
(Adenosine-$\dot{O}$H) [28a]) + *anti*-5-nitro-2-furaldoxime ⟶ products [29])				
Pulse rad. of adenosine + N_2O + H_2O				73 Gre 1
KAS	H_2O	RT	$k = 2.2 \cdot 10^9\ M^{-1}s^{-1}$	
(Adenosine-5′-monophosphate)$^{\cdot -}$ [30]) + 6-carboxyuracil [31]) ⟶ products				
Pulse rad. of adenosine-5′-monophosphate + *t*-butanol + N_2 + H_2O				72 Ada 1
KAS	H_2O	RT	$k = 1.8 \cdot 10^9\ M^{-1}s^{-1}$	
(Adenosine-5′-monophosphate)$^{\cdot -}$ [30]) + 2-methyl-1,4-naphthoquinone ⟶ 2-methyl...semiquinone + products				
Pulse rad. of adenosine-5′-monophosphate + *t*-butanol + N_2 + H_2O				72 Ada 1
KAS	H_2O	RT	$k = 2.9 \cdot 10^9\ M^{-1}s^{-1}$	
(Adenosine-5′-monophosphate)$^{\cdot -}$ [30]) + 4-nitroacetophenone(PNAP) ⟶ (PNAP)$^{\cdot -}$ + products				
Pulse rad. of adenosine-5′-monophosphate + *t*-butanol + N_2 + H_2O				72 Ada 1
KAS	H_2O	RT	$k = 4.6 \cdot 10^9\ M^{-1}s^{-1}$	
(Adenosine-5′-monophosphate)$^{\cdot -}$ [30]) + *anti*-5-nitro-2-furaldoxime ⟶ products				
Pulse rad. of adenosine-5′-monophosphate + *t*-butanol + N_2 + H_2O				73 Gre 1
KAS	H_2O	RT	$k = 3.6 \cdot 10^9\ M^{-1}s^{-1}$	
(Adenosine-5′-monophosphate-$\dot{O}$H) [31a]) + *anti*-5-nitro-2-furaldoxime products [29])				
Pulse rad. of adenosine-5′-monophosphate + N_2O + H_2O				73 Gre 1
KAS	H_2O	RT	$k = 2.0 \cdot 10^9\ M^{-1}s^{-1}$	
(Alanine)$^{\cdot -}$ [32]) + *anti*-5-nitro-2-furaldoxime ⟶ products				
Pulse rad. of alanine + *t*-butanol + N_2 + H_2O				73 Gre 1
KAS	H_2O	RT	$k = 5.0 \cdot 10^9\ M^{-1}s^{-1}$	
(Alanine anhydride)$^{\cdot -}$ [33]) + acetophenone ($C_6H_5COCH_3$) ⟶ products				
Pulse rad. of alanine anhydride + *t*-butanol + N_2 + H_2O				71 Hay 1
KAS	H_2O, pH = 5.2	RT	$k = 2.0(3) \cdot 10^9\ M^{-1}s^{-1}$	
	pH = 12.2		$1.5(4) \cdot 10^9$	
(Alanine anhydride)$^{\cdot -}$ [33]) + benzophenone ($(C_6H_5)_2CO$) ⟶ products				
Pulse rad. of alanine anhydride + *t*-butanol + N_2 + H_2O				71 Hay 1
KAS	H_2O, pH = 5.2	RT	$k = 1.6(2) \cdot 10^9\ M^{-1}s^{-1}$	
	pH = 12.2		$1.9(3) \cdot 10^9$	

[27]) Radicals from adenosine + e_{aq}^- reaction.
[28]) e^--transfer and/or addition.
[28a]) Radicals from adenosine + $\dot{O}$H reaction.
[29]) 10% e^--transfer.
[30]) Radicals from adenosine-5′-monophosphate + e_{aq}^- reaction.
[31]) Orotic acid.
[31a]) Radicals from adenosine-5′-monophosphate + $\dot{O}$H reaction.
[32]) Radicals from alanine + e_{aq}^- reaction.
[33]) Radicals from alanine anhydride + e_{aq}^- reaction; most likely $CH(CH_3)\dot{C}O^-NHCH(CH_3)CONH$.

Reaction Radical generation Method	Solvent	T[K]	Rate data	Ref./ add. ref.
(Alanine anhydride)$^{-\cdot}$ [33]) + cystamine ⟶ products				
Pulse rad. of alanine anhydride + t-butanol + N_2 + H_2O				71 Hay 1
KAS	H_2O, pH = 5.1 and 11.4	RT	$k = 1.1(2) \cdot 10^8\ M^{-1}\,s^{-1}$	
(Alanine anhydride)$^{-\cdot}$ [33]) + cysteine ⟶ products				
Pulse rad. of alanine anhydride + t-butanol + N_2 + H_2O				71 Hay 1
KAS	H_2O	RT	$k = 1.4(3) \cdot 10^8\ M^{-1}\,s^{-1}$	
(Alanine anhydride)$^{-\cdot}$ [34]) + 2-methyl-1,4-naphthoquinone ⟶ ...semiquinone + products				
Pulse rad. of alanine anhydride + H_2O				73 Rao 3
KAS	H_2O	RT	$k = 4.6(5) \cdot 10^9\ M^{-1}\,s^{-1}$	
(Alanine anhydride-$\dot{O}H$) [35]) + 2-methyl-1,4-naphthoquinone ⟶ ...semiquinone + products [36])				
Pulse rad. of alanine anhydride + H_2O + N_2O				73 Rao 3
KAS	H_2O, pH = 10.9	RT	$k = 3.1(3) \cdot 10^9\ M^{-1}\,s^{-1}$	
(Aniline-$\dot{O}H$) [37]) + 2-methyl-1,4-naphthoquinone ⟶ products [38])				
Pulse rad. of aniline + N_2O + H_2O				73 Rao 3
KAS	H_2O	RT	$k = 4.0 \cdot 10^9\ M^{-1}\,s^{-1}$	
(Benzoic acid)$^{-\cdot}$ [39]) + N-ethylmaleimide ⟶ products				
Pulse rad. of benzoic acid + t-butanol + N_2 + H_2O				72 Hay 1
KAS	H_2O	RT	$k = 3.2 \cdot 10^9\ M^{-1}\,s^{-1}$	
(Benzoic acid-$\dot{O}H$) [40]) + $Fe(CN)_6^{3-}$ ⟶ $Fe(CN)_6^{4-}$ + products				
Pulse rad. of benzoic acid + N_2O + H_2O				75 Kle 1
KAS	H_2O	RT	$k = 2.0(10) \cdot 10^5\ M^{-1}\,s^{-1}$	
(4-Bromoacetophenone)$^{-\cdot}$ [41]) + 4-nitroacetophenone(PNAP) ⟶ products + (PNAP)$^{-\cdot}$				
Pulse rad. of 4-bromoacetophenone + t-butanol + N_2 + H_2O				73 Ada 1
KAS	H_2O	RT	$k = 4.8(5) \cdot 10^9\ M^{-1}\,s^{-1}$	

(2′-Bromo-2′-deoxyuridine)$^{-\cdot}$ [41a]) ⟶ 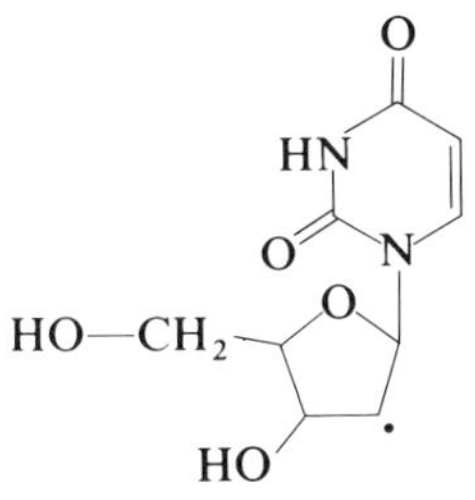 + Br^- [41b])

Pulse rad. of 2′-bromo-2′-deoxyuridine + H_2O + t-butanol				81 His 1
Time resolved Cond.	H_2O	RT	$k \approx 10^5\ s^{-1}$	

[33]) Radicals from alanine anhydride + e_{aq}^- reaction; most likely $CH(CH_3)\dot{C}O^-NHCH(CH_3)CONH$.
[34]) Electron adduct to alanine anhydride.
[35]) Radicals from alanine anhydride + $\dot{O}H$ reaction.
[36]) 75% e^--transfer.
[37]) Radicals from aniline + $\dot{O}H$ reaction.
[38]) 26% e^--transfer.
[39]) Radicals from benzoic acid + e_{aq}^- reaction.
[40]) Radicals from benzoic acid + $\dot{O}H$ reaction.
[41]) Radicals from 4-bromoacetophenone + e_{aq}^- reaction.
[41a]) Electron adduct to nucleobase.
[41b]) Intramolecular e^--transfer to sugar-bound bromine.

Reaction Radical generation Method	Solvent	T[K]	Rate data	Ref./ add. ref.
(2,3-Butanediol-$\dot{O}H$)[42]) + tetranitromethane ($C(NO_2)_4$) ⟶ $C(NO_2)_3^-$ + NO_2 + H^+ + products				
Pulse rad. of 2,3-butanediol + N_2O + H_2O				73 Asm 1
KAS	H_2O	RT	$k = 3.3 \cdot 10^9\ M^{-1}\,s^{-1}$	
(6-Carboxyuracil)$^{\cdot-}$ [43]) + N-ethylmaleimide ⟶ products				
Pulse rad. of 6-carboxyuracil[44]) + t-butanol + N_2 + H_2O				72 Hay 1
KAS	H_2O, pH = 3.2	RT	$k = 1.3 \cdot 10^9\ M^{-1}\,s^{-1}$ [45])	
	pH = 6.8		$2.4 \cdot 10^9$	
(6-Carboxyuracil)$^{\cdot-}$ [43]) + *anti*-5-nitro-2-furaldoxime ⟶ products				
Pulse rad. of 6-carboxyuracil[44]) + t-butanol + N_2 + H_2O				73 Gre 1
KAS	H_2O	RT	$k = 6.5 \cdot 10^9\ M^{-1}\,s^{-1}$	
(6-Carboxyuracil-$\dot{O}H$)[46]) + 2-methyl-1,4-naphthoquinone ⟶ 2-methyl...semiquinone + products				
Pulse rad. of 6-carboxyuracil[44]) + N_2O + H_2O				73 Hay 3
KAS	H_2O, pH = 6.8	RT	$k = 6.0 \cdot 10^9\ M^{-1}\,s^{-1}$ [47])	
	pH = 10.9		$8.5 \cdot 10^{8}$ [48])	
(4-Chloroacetophenone)$^{\cdot-}$ [49]) + 4-nitroacetophenone(PNAP) ⟶ PNAP$^{\cdot-}$ + products				
Pulse rad. of 4-chloroacetophenone + t-butanol + N_2 + H_2O				73 Ada 1
KAS	H_2O	RT	$k = 5.1(5) \cdot 10^9\ M^{-1}\,s^{-1}$	
(Chlorouracil)$^{\cdot-}$ [50]) + 4-nitroacetophenone(PNAP) ⟶ PNAP$^{\cdot-}$ + products				
Pulse rad. of chlorouracil + t-butanol + N_2 + H_2O				76 Bur 1
KAS	H_2O	RT	$k = 5.3 \cdot 10^9\ M^{-1}\,s^{-1}$	
(Chlorouracil-$\dot{H}$)[51]) + 4-nitroacetophenone(PNAP) ⟶ PNAP$^{\cdot-}$ + products				
Pulse rad. of chlorouracil + t-butanol + N_2 + H_2O				76 Bur 1
KAS	H_2O, acid pH	RT	$k = 3.3 \cdot 10^9\ M^{-1}\,s^{-1}$	
(Crotonate)$^{\cdot-}$ [52]) + *anti*-5-nitro-2-furaldoxime[53]) ⟶ products[54])				
Pulse rad. of crotonate + H_2O				73 Gre 1
KAS	H_2O	RT	$k = 1.4 \cdot 10^9\ M^{-1}\,s^{-1}$	
(Cytidine)$^{\cdot-}$ [55]) + 6-carboxyuracil[56]) ⟶ products				
Pulse rad. of cytidine + t-butanol + N_2 + H_2O				72 Ada 1
KAS	H_2O	RT	$k = 1.3 \cdot 10^9\ M^{-1}\,s^{-1}$	
(Cytidine)$^{\cdot-}$ [55]) + 4-nitroacetophenone(PNAP) ⟶ PNAP$^{\cdot-}$ + products				
Pulse rad. of cytidine + t-butanol + N_2 + H_2O				72 Ada 1
KAS	H_2O	RT	$k = 4.0 \cdot 10^9\ M^{-1}\,s^{-1}$	

[42]) Radicals from 2,3-butanediol + $\dot{O}H$ reaction (likely to be $CH_3\dot{C}OHCHOHCH_3$ and/or $CH_3\dot{C}HCOCH_3$).
[43]) Radicals from 6-carboxyuracil + e_{aq}^- reaction.
[44]) Orotic acid.
[45]) Electron transfer from protonated form of 6-carboxyuracil + e_{aq}^- reaction product.
[46]) Radicals from 6-carboxyuracil + $\dot{O}H$ reaction.
[47]) 13% e^--transfer.
[48]) 50% e^--transfer.
[49]) Radicals from 4-chloroacetophenone + e_{aq}^- reaction, possibly e_{aq}^- adduct.
[50]) Radicals from chlorouracil + e_{aq}^- reaction.
[51]) Protonated product radical of chlorouracil + e_{aq}^- reaction.
[52]) Radicals from crotonate + e_{aq}^- reaction.
[53]) Nifuroxime.
[54]) 25% e^--transfer.
[55]) Radicals from cytidine + e_{aq}^- reaction.
[56]) Orotic acid.

Reaction Radical generation Method	Solvent	T[K]	Rate data	Ref./ add. ref.
(Cytidine-5′-monophosphate)$^{-\cdot}$ [57]) + 6-carboxyuracil [56]) ⟶ products				
Pulse rad. of cytidine-5′-monophosphate + *t*-butanol + N_2 + H_2O				72 Ada 1
KAS	H_2O	RT	$k = 6.5 \cdot 10^8\ M^{-1}\,s^{-1}$	
(Cytidine-5′-monophosphate)$^{-\cdot}$ [57]) + 4-nitroacetophenone(PNAP) ⟶ PNAP$^{-\cdot}$ + products				
Pulse rad. of cytidine-5′-monophosphate + *t*-butanol + N_2 + H_2O				72 Ada 1
KAS	H_2O	RT	$k = 3.4 \cdot 10^9\ M^{-1}\,s^{-1}$	
(Cytidine-5′-monophosphate-ȮH) [58]) + 2-methyl-1,4-naphthoquinone ⟶ 2-methyl…semiquinone + products [59])				
Pulse rad. of cytidine-5′-monophosphate + N_2O + H_2O				72 Sim 1
KAS	H_2O	RT	$k = 2.0 \cdot 10^9\ M^{-1}\,s^{-1}$	
(Cytosine)$^{-\cdot}$ [60]) + benzophenone ⟶ $(C_6H_5)_2\dot{C}O^-$ + products				
Pulse rad. of cytosine + *t*-butanol + N_2 + H_2O				72 Ada 1
KAS	H_2O, pH = 12	RT	$k = 3.4 \cdot 10^9\ M^{-1}\,s^{-1}$	
(Cytosine)$^{-\cdot}$ [60]) + 1,4-benzoquinone ⟶ 1,4-benzosemiquinone + products				
Pulse rad. of cytosine + *t*-butanol + N_2 + H_2O				74 Rao 1
KAS	H_2O	RT	$k = 4.2 \cdot 10^9\ M^{-1}\,s^{-1}$	
(Cytosine)$^{-\cdot}$ [60]) + 6-carboxyuracil [56]) ⟶ products				
Pulse rad. of cytosine + *t*-butanol + N_2 + H_2O				72 Ada 1
KAS	H_2O	RT	$k = 2.6 \cdot 10^9\ M^{-1}\,s^{-1}$	
(Cytosine)$^{-\cdot}$ [60]) + fluorescein ⟶ products				
Pulse rad. of cytosine + *t*-butanol + N_2 + H_2O				73 Rao 2
KAS	H_2O, pH = 9.2	RT	$k = 1.5 \cdot 10^9\ M^{-1}\,s^{-1}$	
(Cytosine)$^{-\cdot}$ [60]) + 8-methoxypsoralen(8-MOP) ⟶ (8-MOP)$^{-\cdot}$ + products				
Pulse rad. of cytosine + *t*-butanol + H_2O				78 Red 1
KAS	H_2O	RT	$k = 3.9 \cdot 10^9\ M^{-1}\,s^{-1}$	
(Cytosine)$^{-\cdot}$ [60]) + 2-methyl-1,4-naphthoquinone ⟶ 2-methyl…semiquinone + products				
Pulse rad. of cytosine + *t*-butanol + N_2 + H_2O				72 Ada 1/
KAS	H_2O	RT	$k = 4.0 \cdot 10^9\ M^{-1}\,s^{-1}$	74 Rao 1
(Cytosine)$^{-\cdot}$ [60]) + 4-nitroacetophenone(PNAP) ⟶ PNAP$^{-\cdot}$ + products				
Pulse rad. of cytosine + *t*-butanol + N_2 + H_2O				72 Ada 1
KAS	H_2O, pH = 7	RT	$k = 5.3 \cdot 10^9\ M^{-1}\,s^{-1}$	
	pH = 12		$5.0 \cdot 10^9$	
(Cytosine)$^{-\cdot}$ [60]) + *anti*-5-nitro-2-furaldoxime ⟶ products				
Pulse rad. of cytosine + *t*-butanol + N_2 + H_2O				73 Gre 1
KAS	H_2O	RT	$k = 7.6 \cdot 10^9\ M^{-1}\,s^{-1}$	
(Cytosine-Ḣ) [61]) + 4-nitroacetophenone(PNAP) ⟶ PNAP$^{-\cdot}$ + products				
Pulse rad. of cytosine + *t*-butanol + H_2O				79 His 1
KAS	H_2O, acid pH	RT	$k = 5.3(3) \cdot 10^9\ M^{-1}\,s^{-1}$	

[56]) Orotic acid.
[57]) Radicals from cytidine-5′-monophosphate + e_{aq}^- reaction.
[58]) Radicals from cytidine-5′-monophosphate + ȮH reaction.
[59]) 37% e^--transfer.
[60]) Radicals from cytosine + e_{aq}^- reaction.
[61]) Protonated form of radical anion formed in cytosine + e_{aq}^- reaction.

Reaction Radical generation Method	Solvent	T[K]	Rate data	Ref./ add. ref.
(Cytosine-$\dot{O}$H) [62]) + $Fe(CN)_6^{3-}$ ⟶ $Fe(CN)_6^{4-}$ + products				
Pulse rad. of cytosine + N_2O + H_2O				73 Rao 1
KAS	H_2O	RT	$k = 4.29 \cdot 10^9 M^{-1} s^{-1}$	
(Cytosine-$\dot{O}$H) [62]) + hemin-c ⟶ products				
Pulse rad. of cytosine + N_2O + H_2O				75 Gof 1
KAS	H_2O	RT	$k = 1.1 \cdot 10^9 M^{-1} s^{-1}$	
(Cytosine-$\dot{O}$H) [62]) + anthraquinone-2,6-disulfonate ⟶ anthrasemiquinone... + products [63])				
Pulse rad. of cytosine + N_2O + H_2O				74 Rao 1/
KAS	H_2O	RT	$k = 2.2 \cdot 10^9 M^{-1} s^{-1}$	73 Rao 1
(Cytosine-$\dot{O}$H) [62]) + anthraquinone-2-sulfonate ⟶ anthrasemiquinone... + products				
Pulse rad. of cytosine + N_2O + H_2O				73 Rao 1
KAS	H_2O	RT	$k = 1.48 \cdot 10^9 M^{-1} s^{-1}$	
(Cytosine-$\dot{O}$H) [62]) + 1,4-benzoquinone ⟶ 1,4-benzosemiquinone + products				
Pulse rad. of cytosine + N_2O + H_2O				73 Sim 1,
KAS	H_2O	RT	$k = 5.0 \cdot 10^9 M^{-1} s^{-1}$ [64]) $7.2 \cdot 10^9$ [65])	73 Rao 1
(Cytosine-$\dot{O}$H) [62]) + 4-hydroxy-2,2,6,6-tetramethylpiperidino-1-oxy ⟶ products [66])				
Pulse rad. of cytosine + N_2O + H_2O				72 Bru 1
KAS	H_2O	RT	$k = 3.1(2) \cdot 10^8 M^{-1} s^{-1}$	
(Cytosine-$\dot{O}$H) [62]) + indigo disulfonate ⟶ products [67])				
Pulse rad. of cytosine + N_2O + H_2O				73 Rao 2
KAS	H_2O	RT	$k = (1.9 \ldots 2.0) \cdot 10^9 M^{-1} s^{-1}$	
(Cytosine-$\dot{O}$H) [62]) + indigo tetrasulfonate ⟶ products [68])				
Pulse rad. of cytosine + N_2O + H_2O				73 Rao 2
KAS	H_2O	RT	$k = 2.0 \cdot 10^9 M^{-1} s^{-1}$	
(Cytosine-$\dot{O}$H) [62]) + indophenol ⟶ products [69])				
Pulse rad. of cytosine + N_2O + H_2O				73 Rao 2
KAS	H_2O, pH = 9	RT	$k = 2.2 \cdot 10^9 M^{-1} s^{-1}$	
(Cytosine-$\dot{O}$H) [62]) + 8-methoxypsoralen ⟶ products [70])				
Pulse rad. of cytosine + N_2O + H_2O				78 Red 1
KAS	H_2O	RT	$k = 1.0 \cdot 10^9 M^{-1} s^{-1}$	
(Cytosine-$\dot{O}$H) [62]) + methylene blue ⟶ products [71])				
Pulse rad. of cytosine + N_2O + H_2O				73 Rao 2
KAS	H_2O	RT	$k = 2.3 \cdot 10^9 M^{-1} s^{-1}$	

[62]) Radicals from cytosine + $\dot{O}$H reaction.
[63]) 40% e^--transfer.
[64]) 75% e^--transfer [73 Sim 1].
[65]) From [73 Rao 1].
[66]) Addition and possibly some e^--transfer.
[67]) 81% e^--transfer at pH = 7; 55% e^--transfer at pH = 9.
[68]) 84% e^--transfer.
[69]) 88% e^--transfer.
[70]) 24% e^--transfer.
[71]) 85% e^--transfer.

Reaction Radical generation Method	Solvent	T[K]	Rate data	Ref./ add. ref.
(Cytosine-$\dot{O}$H)[62]) + 2-methyl-1,4-naphthoquinone ⟶ 2-methyl…semiquinone + products				
Pulse rad. of cytosine + N_2O + H_2O				73 Hay 3,
KAS	H_2O, pH = 2.9	RT	$k = 3.0 \cdot 10^9\ M^{-1}s^{-1}$ [72])	73 Rao 1,
	pH = 7		$4.9 \cdot 10^9$ [73])	74 Rao 1
			$3.8 \cdot 10^9$ [74])	
			$2.7 \cdot 10^9$ [75])	
(Cytosine-$\dot{O}$H)[62]) + *anti*-5-nitro-2-furaldoxime ⟶ products[76])				
Pulse rad. of cytosine + N_2O + H_2O				73 Gre 1
KAS	H_2O	RT	$k = 6 \cdot 10^8\ M^{-1}s^{-1}$	
(Cytosine-$\dot{O}$H)[62]) + nor-pseudopelletierine-N-oxy ⟶ products[66])				
Pulse rad. of cytosine + N_2O + H_2O				72 Bru 1
KAS	H_2O	RT	$k = 7.0(2) \cdot 10^8\ M^{-1}s^{-1}$	
(Cytosine-$\dot{O}$H)[62]) + riboflavin ⟶ products				
Pulse rad. of cytosine + N_2O + H_2O				73 Rao 1
KAS	H_2O	RT	$k = 1.61 \cdot 10^9\ M^{-1}s^{-1}$	
(Cytosine-$\dot{O}$H)[62]) + safranine T ⟶ products[77])				
Pulse rad. of cytosine + N_2O + H_2O				73 Rao 2
KAS	H_2O	RT	$k = 1.2 \cdot 10^9\ M^{-1}s^{-1}$	
(Cytosine-$\dot{O}$H)[62]) + 2,2,6,6-tetramethyl-4-oxo-1-piperidinyloxy(TAN) ⟶ products[66])				
Pulse rad. of cytosine + N_2O + H_2O				72 Bru 1
KAS	H_2O	RT	$k = 3.7(2) \cdot 10^8\ M^{-1}s^{-1}$	
(Cytosine-$\dot{O}$H)[62]) + thionine ⟶ products[68])				
Pulse rad. of cytosine + N_2O + H_2O				73 Rao 2
KAS	H_2O, pH = 8	RT	$k = 2.7 \cdot 10^9\ M^{-1}s^{-1}$	
(Deoxyadenosine)$^{\dot{-}}$[78]) + 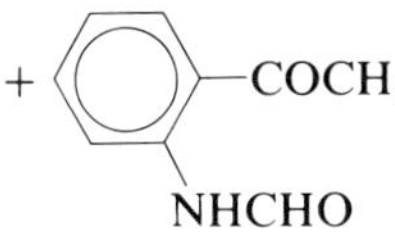 —$COCH_2CH(NH_2)COOH$[79]) ⟶ products				
Pulse rad. of deoxyadenosine + *t*-butanol + H_2O				77 Pil 1
KAS	H_2O	RT	$k = 2.5 \cdot 10^9\ M^{-1}s^{-1}$ (±20%)	
(Deoxyadenosine-5′-monophosphate)$^{\dot{-}}$[80]) + 6-carboxyuracil[81]) ⟶ products				
Pulse rad. of deoxyadenosine-5′-monophosphate + *t*-butanol + N_2 + H_2O				72 Ada 1
KAS	H_2O	RT	$k = 1.9 \cdot 10^9\ M^{-1}s^{-1}$	

[62]) Radicals from cytosine + $\dot{O}$H reaction.
[66]) Addition and possibly some e^--transfer.
[68]) 84% e^--transfer.
[72]) 14% e^--transfer [73 Hay 3].
[73]) 60% e^--transfer [73 Hay 3].
[74]) From [73 Rao 1].
[75]) 80% e^--transfer [74 Rao 1].
[76]) 10% e^--transfer.
[77]) 17% e^--transfer.
[78]) Radicals from deoxyadenosine + e^-_{aq} reaction.
[79]) N′-formylkynurenine; 2-amino-4-[2-(formylamino)phenyl]-4-oxobutanoic acid.
[80]) Radicals from deoxyadenosine-5′-monophosphate + e^-_{aq} reaction.
[81]) Orotic acid.

Reaction Radical generation Method	Solvent	T[K]	Rate data	Ref./ add. ref.
(Deoxyadenosine-5′-monophosphate)$^{\bar{\cdot}}$ [80]) + 2-methyl-1,4-naphthoquinone ⟶ 2-methyl... semiquinone + products				
Pulse rad. of deoxyadenosine-5′-monophosphate + t-butanol + N_2 + H_2O				72 Ada 1
KAS	H_2O	RT	$k = 2.9 \cdot 10^9\ M^{-1}s^{-1}$	
(Deoxyadenosine-5′-monophosphate)$^{\bar{\cdot}}$ [80]) + 4-nitroacetophenone(PNAP) ⟶ (PNAP)$^{\bar{\cdot}}$ + products				
Pulse rad. of deoxyadenosine-5′-monophosphate + t-butanol + N_2 + H_2O				72 Ada 1
KAS	H_2O	RT	$k = 4.3 \cdot 10^9\ M^{-1}s^{-1}$	
(Deoxycytidine)$^{\bar{\cdot}}$ [82]) + 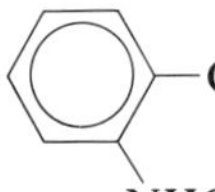 —$COCH_2CH(NH_2)COOH$ [79]) ⟶ products NHCHO				
Pulse rad. of deoxycytidine + t-butanol + H_2O				77 Pil 1
KAS	H_2O	RT	$k = 1.7 \cdot 10^9\ M^{-1}s^{-1}$ (±20%)	
(2′-Deoxycytidine-Ḣ) [83]) + 4-nitroacetophenone(PNAP) ⟶ PNAP$^{\bar{\cdot}}$ + products				
Pulse rad. of 2′-deoxycytidine + t-butanol + H_2O				79 His 1
KAS	H_2O	RT	$k = 3.5(3) \cdot 10^9\ M^{-1}s^{-1}$	
(Deoxyguanosine monophosphate-ȮH$^+$) [84]) + cysteamine(cySH, cyS$^-$) ⟶ cyṠ + products				
Pulse rad. of deoxyguanosine monophosphate + N_2O + H_2O				74 Wil 3
KAS	H_2O	RT	$k = 1.7 \cdot 10^8\ M^{-1}s^{-1}$	
(Deoxyguanosine monophosphate-ȮH$^+$) [84]) + promethazine(PZ) ⟶ PZ$^{\dot{+}}$ + products				
Pulse rad. of deoxyguanosine monophosphate + N_2O + H_2O				74 Wil 3
KAS	H_2O	RT	$k = 2.8 \cdot 10^9\ M^{-1}s^{-1}$	
(Deoxyribose-ȮH) [85]) + 1,4-benzoquinone ⟶ 1,4-benzosemiquinone + products				
Pulse rad. of deoxyribose + H_2O + N_2O				73 Sim 1
KAS	H_2O	RT	$k = 2.7 \cdot 10^9\ M^{-1}s^{-1}$	
(Deoxyribose-ȮH) [85]) + 2-methyl-1,4-naphthoquinone ⟶ ... semiquinone + products [86])				
Pulse rad. of deoxyribose + H_2O + N_2O				72 Sim 1,
KAS	H_2O	RT	$k = 4.0 \cdot 10^9\ M^{-1}s^{-1}$ $2.1(2) \cdot 10^9$ [87])	73 Rao 3
(Deoxyribose-ȮH) [85]) + 2,2,6,6-tetramethyl-4-oxo-1-piperidinyloxy(TAN) ⟶ products				
Pulse rad. of deoxyribose + N_2O + H_2O				71 Wil 2
Competition kinetics	H_2O	RT	$k = 3.9 \cdot 10^8\ M^{-1}s^{-1}$ [88])	
(Deoxyribose-ȮH) [85]) + 4-nitroacetophenone(PNAP) ⟶ PNAP$^{\bar{\cdot}}$ + products				
Pulse rad. of deoxyribose + N_2O + H_2O				75 Whi 2
KAS	H_2O	RT	$k = 1.0(3) \cdot 10^9\ M^{-1}s^{-1}$	

[79]) N′-formylkynurenine; 2-amino-4-[2-(formylamino)phenyl]-4-oxobutanoic acid.
[80]) Radicals from deoxyadenosine-5′-monophosphate + e^-_{aq} reaction.
[82]) Radicals from deoxycytidine + e^-_{aq} reaction.
[83]) Protonated form of radical anions formed in 2′-deoxycytidine + e^-_{aq} reaction.
[84]) Radical cations formed from deoxyguanosine monophosphate + ȮH reaction.
[85]) Radicals from deoxyribose + ȮH reaction.
[86]) 81% e^--transfer.
[87]) From [73 Rao 3].
[88]) Relative to $k(\dot{R} + Fe(CN)_6^{3-}) = 2.8 \cdot 10^9\ M^{-1}s^{-1}$.

Reaction Radical generation Method	Solvent	T[K]	Rate data	Ref./ add. ref.
(Deoxyribose-$\dot{O}H$)[85] + *anti*-5-nitro-2-furaldoxime[89] ⟶ products[90]				
Pulse rad. of deoxyribose + N_2O + H_2O				73 Gre 1
KAS	H_2O	RT	$k = 1.0 \cdot 10^9\ M^{-1}s^{-1}$	
(3,6-Diaminoacridine)$^{\dot{-}}$[91] + 1,1′-dimethyl-4,4′-bipyridinium(MV^{2+})[92] ⟶ $MV^{\dot{+}}$ + products				
Pulse rad. of 3,6-diaminoacridine + *t*-butanol + H_2O				81 Nen 1
KAS	H_2O	RT	$k = 4.2 \cdot 10^{10}\ M^{-1}s^{-1}$	
(N,N-Diethylnicotinamide)$^{\dot{-}}$[93] + 2-methyl-1,4-naphthoquinone ⟶ products				
Pulse rad. of N,N-diethylnicotinamide + *t*-butanol + N_2 + H_2O				73 Rao 3
KAS	H_2O	RT	$k = 4.1 \cdot 10^9\ M^{-1}s^{-1}$	
(Diglycine)$^{\cdot}$[94] + lipoate(—S—S—) ⟶ (lipoate)$^{\dot{-}}$(—S$\overset{(-)}{\therefore}$S—) + products				
Pulse rad. of diglycine + *t*-butanol + H_2O				75 Far 3
KAS	H_2O	RT	$k = 4.0 \cdot 10^6\ M^{-1}s^{-1}$	
(Diglycine)$^{\cdot}$[95] + lipoate(—S—S—) ⟶ (lipoate)$^{\dot{-}}$(—S$\overset{(-)}{\therefore}$S—) + products				
Pulse rad. of chloroacetyldiglycine + *t*-butanol + H_2O				75 Far 3
KAS	H_2O	RT	$k = 5 \cdot 10^6\ M^{-1}s^{-1}$	
(Diglycine-$\dot{O}H$)[96] + 2-methyl-1,4-naphthoquinone ⟶ ...semiquinone + products[97]				
Pulse rad. of diglycine + H_2O + N_2O				73 Rao 3
KAS	H_2O	RT	$k = 1.2(1) \cdot 10^9\ M^{-1}s^{-1}$	
(Dihydrothymine-$\dot{O}H$)[1] + 1,4-benzoquinone ⟶ 1,4-benzosemiquinone + products[2]				
Pulse rad. of dihydrothymine + N_2O + H_2O				73 Sim 1,
KAS	H_2O	RT	$k = (4.0...5.4) \cdot 10^9\ M^{-1}s^{-1}$	74 Sim 2,
				74 Rao 1
(Dihydrothymine-$\dot{O}H$)[1] + 2,6-dichloroindophenol ⟶ products[3]				
Pulse rad. of dihydrothymine + N_2O + H_2O				73 Rao 2
KAS	H_2O	RT	$k = 3.4 \cdot 10^9\ M^{-1}s^{-1}$	
(Dihydrothymine-$\dot{O}H$)[1] + indigo disulfonate ⟶ products[4]				
Pulse rad. of dihydrothymine + N_2O + H_2O				73 Rao 2
KAS	H_2O	RT	$k = 3.2 \cdot 10^9\ M^{-1}s^{-1}$	
(Dihydrothymine-$\dot{O}H$)[1] + indigo tetrasulfonate ⟶ products[5]				
Pulse rad. of dihydrothymine + N_2O + H_2O				73 Rao 2
KAS	H_2O	RT	$k = 2.7 \cdot 10^9\ M^{-1}s^{-1}$	

[85]) Radicals from deoxyribose + $\dot{O}H$ reaction.
[89]) Nifuroxime.
[90]) 70% e^--transfer.
[91]) Proflavin.
[92]) Methylviologen.
[93]) Radicals from N,N-diethylnicotinamide + e^-_{aq} reaction.
[94]) Deaminated radical from diglycine + e^-_{aq} reaction.
[95]) Dechlorinated radical from chloroacetyldiglycine + e^-_{aq} reaction.
[96]) Radicals from diglycine + $\dot{O}H$ reaction.
[97]) 19% e^--transfer at pH = 6.6; 47% e^--transfer at pH = 11.0.
[1]) Radicals from dihydrothymine + $\dot{O}H$ reaction.
[2]) 53...80% e^--transfer.
[3]) 82% e^--transfer.
[4]) 77% e^--transfer.
[5]) 78% e^--transfer.

Reaction Radical generation Method	Solvent	T[K]	Rate data	Ref./ add. ref.
(Dihydrothymine-$\dot{O}$H) [1]) + methylene blue ⟶ products [6])				
Pulse rad. of dihydrothymine + N_2O + H_2O				73 Rao 2
KAS	H_2O	RT	$k = 2.7 \cdot 10^9 M^{-1} s^{-1}$	
(Dihydrothymine-$\dot{O}$H) [1]) + 2-methyl-1,4-naphthoquinone ⟶ 2-methyl...semiquinone + products				
Pulse rad. of dihydrothymine + N_2O + H_2O				73 Hay 3,
KAS	H_2O	RT	$k = 7.0 \cdot 10^9 M^{-1} s^{-1}$ [7]) $5.1 \cdot 10^9$ [8])	74 Rao 1
(Dihydrothymine-$\dot{O}$H) [1]) + phenosafranine ⟶ products [9])				
Pulse rad. of dihydrothymine + N_2O + H_2O				73 Rao 2
KAS	H_2O	RT	$k = 1.2 \cdot 10^9 M^{-1} s^{-1}$	
(Dihydrothymine-$\dot{O}$H) [1]) + safranine T ⟶ products [10])				
Pulse rad. of dihydrothymine + N_2O + H_2O				73 Rao 2
KAS	H_2O	RT	$k = 9 \cdot 10^8 M^{-1} s^{-1}$	
(Dihydrouracil-$\dot{O}$H) [11]) + $Fe(CN)_6^{3-}$ ⟶ $Fe(CN)_6^{4-}$ + products				
Pulse rad. of dihydrouracil + N_2O + H_2O				75 Hay 1
KAS	H_2O	RT	$k = 2 \cdot 10^9 M^{-1} s^{-1}$	
(Dihydrouracil-$\dot{O}$H) [11]) + hemin-c ⟶ products				
Pulse rad. of dihydrouracil + N_2O + H_2O				75 Gof 1
KAS	H_2O	RT	$k = 1.6 \cdot 10^9 M^{-1} s^{-1}$	
(Dihydrouracil-$\dot{O}$H) [11]) + 1,4-benzoquinone ⟶ 1,4-benzosemiquinone + products [12])				
Pulse rad. of dihydrouracil + N_2O + H_2O				74 Rao 1
KAS	H_2O	RT	$k = 4.5 \cdot 10^9 M^{-1} s^{-1}$	
(Dihydrouracil-$\dot{O}$H) [11]) + 2-methyl-1,4-naphthoquinone ⟶ 2-methyl...semiquinone + products [13])				
Pulse rad. of dihydrouracil + N_2O + H_2O				73 Hay 3,
KAS	H_2O	RT	$k = (4.6...4.8) \cdot 10^9 M^{-1} s^{-1}$	74 Rao 1
(Dihydrouracil-$\dot{O}$H) [11]) + *anti*-5-nitro-2-furaldoxime ⟶ products [14])				
Pulse rad. of dihydrouracil + N_2O + H_2O				73 Gre 1
KAS	H_2O	RT	$k = 2.4 \cdot 10^9 M^{-1} s^{-1}$	
(2,7-Dimethyl-3,6-diaminoacridine [15]))$^{\dot{-}}$ [16]) + $EuCl_3$ ⟶ Eu(II) + products				
Pulse rad. of acridine yellow + *t*-butanol + H_2O				81 Mic 1
KAS	H_2O, pH = 3.5	292	$k < 4 \cdot 10^6 M^{-1} s^{-1}$	
(2,7-Dimethyl-3,6-diaminoacridine [15]))$^{\dot{-}}$ [16]) + Eu(III)-EDTA ⟶ Eu(II) + products				
Pulse rad. of acridine yellow + *t*-butanol + H_2O				81 Mic 1,
KAS	H_2O, pH = 4.9	292	$k = 2.2 \cdot 10^8 M^{-1} s^{-1}$	81 Nen 1

[1]) Radicals from dihydrothymine + $\dot{O}$H reaction.
[6]) 81% e^--transfer.
[7]) 56% e^--transfer [73 Hay 3].
[8]) 70% e^--transfer [74 Rao 1].
[9]) 57% e^--transfer.
[10]) 22% e^--transfer.
[11]) Radicals from dihydrouracil + $\dot{O}$H reaction.
[12]) 75% e^--transfer.
[13]) 66...75% e^--transfer.
[14]) 20% e^--transfer.
[15]) Acridine yellow.
[16]) Radicals from 2,7-dimethyl-3,6-diaminoacridine + e^-_{aq} reaction.

Reaction Radical generation Method	Solvent	T[K]	Rate data	Ref./ add. ref.
(2,7-Dimethyl-3,6-diaminoacridine [15]))$^{\cdot-}$ [16]) + Eu(III)-salycilate ⟶ Eu(II) + products				
Pulse rad. of acridine yellow + t-butanol + H_2O				81 Mic 1,
KAS	H_2O, pH = 6.7	292	$k = 1.1 \cdot 10^8 M^{-1} s^{-1}$	81 Nen 1
(2,7-Dimethyl-3,6-diaminoacridine [15]))$^{\cdot-}$ [16]) + V(III) + salycilate ⟶ V(II) + products				
Pulse rad. of acridine yellow + t-butanol + H_2O				81 Mic 1,
KAS	H_2O, pH = 6.8	292	$k = 4 \cdot 10^7 M^{-1} s^{-1}$	81 Nen 1
(2,7-Dimethyl-3,6-diaminoacridine)$^{\cdot-}$ [16]) + 1,1'-dimethyl-4,4'-bipyridinium(MV^{2+}) [17]) ⟶ $MV^{\cdot+}$ + products				
Pulse rad. of 2,7-dimethyl-3,6-diaminoacridine + t-butanol + H_2O				81 Nen 1,
KAS	H_2O	RT	$k = 1.4 \cdot 10^7 M^{-1} s^{-1}$	81 Mic 1
(Dimethyl fumerate)$^{\cdot-}$ [18]) + N-ethylmaleimide ⟶ products				
Pulse rad. of dimethyl fumerate + t-butanol + N_2 + H_2O				73 Hay 2,
KAS	H_2O, pH ≈ 6	RT	$k = 2.2 \cdot 10^9 M^{-1} s^{-1}$	67 Ada 1
	pH = 11		$5.3 \cdot 10^9$ [19])	
(1,3-Dimethyluracil)$^{\cdot-}$ [20]) + 1,4-benzoquinone ⟶ products [21])				
Pulse rad. of 1,3-dimethyluracil + t-butanol + N_2 + H_2O				74 Rao 1
KAS	H_2O, pH = 5.4	RT	$k = 2.9 \cdot 10^9 M^{-1} s^{-1}$	
(1,3-Dimethyluracil)$^{\cdot-}$ [20]) + 2-methyl-1,4-napthoquinone ⟶ 2-methyl...semiquinone + products [21])				
Pulse rad. of 1,3-dimethyluracil + t-butanol + N_2 + H_2O				74 Rao 1
KAS	H_2O, pH = 5.4	RT	$k = 4.1 \cdot 10^9 M^{-1} s^{-1}$	
(1,3-Dimethyluracil-ȮH) [21a]) + 1,4-benzoquinone ⟶ 1,4-benzosemiquinone + products [22])				
Pulse rad. of 1,3-dimethyluracil + N_2O + H_2O				74 Rao 1
KAS	H_2O	RT	$k = 3.2 \cdot 10^9 M^{-1} s^{-1}$	
(1,3-Dimethyluracil-ȮH) [21a]) + methylene blue ⟶ products [23])				
Pulse rad. of 1,3-dimethyluracil + N_2O + H_2O				74 Rao 1
KAS	H_2O	RT	$k = 3.0 \cdot 10^9 M^{-1} s^{-1}$	
(1,3-Dimethyluracil-ȮH) [21a]) + 2-methyl-1,4-naphthoquinone ⟶ 2-methyl...semiquinone + products [24])				
Pulse rad. of 1,3-dimethyluracil + N_2O + H_2O				74 Rao 1
KAS	H_2O, pH = 7	RT	$k = 2.7 \cdot 10^9 M^{-1} s^{-1}$	
	pH = 10.8		$2.3 \cdot 10^9$	
(Dodecylsulfate)$^{\cdot}$ [25]) + 1,4-benzoquinone ⟶ 1,4-benzosemiquinone + products				
Pulse rad. of sodium dodecylsulfate				79 Alm 1
KAS	H_2O	RT	$k = 1.2 \cdot 10^9 M^{-1} s^{-1}$	
(Ethylene glycol-ȮH) [26]) + tetranitromethane($C(NO_2)_4$) ⟶ $C(NO_2)_3^-$ + NO_2 + H^+ + products				
Pulse rad. of ethylene glycol + N_2O + H_2O				73 Asm 1
KAS	H_2O	RT	$k = 1.7 \cdot 10^9 M^{-1} s^{-1}$	

[15]) Acridine yellow.
[16]) Radicals from 2,7-dimethyl-3,6-diaminoacridine + e_{aq}^- reaction.
[17]) Methylviologen.
[18]) Radicals from dimethyl fumerate + e_{aq}^- reaction.
[19]) From [67 Ada 1].
[20]) Radicals from 1,3-dimethyluracil + e_{aq}^- reaction.
[21]) Reaction does not appear to be simple e^--transfer.
[21a]) Radicals from 1,3-dimethyluracil + ȮH reaction.
[22]) 60% e^--transfer.
[23]) 35% e^--transfer.
[24]) 30% e^--transfer.
[25]) Radicals from sodium dodecylsulfate.
[26]) Radicals from $(CH_2OH)_2$ + ȮH reaction (likely to be ĊHOHCH$_2$OH and/or ĊH$_2$CHO).

Reaction Radical generation Method	Solvent	T[K]	Rate data	Ref./ add. ref.
(Ethylene oxide)$_n^{\cdot}$ [27]) + 4-nitrosodimethylaniline ⟶ products				
Pulse rad. of ethylene oxide oligomer and polymer + H_2O				72 Beh 1
KAS	H_2O	298	$k = 9 \cdot 10^8$(n = 1)... $2 \cdot 10^8$ (n ≈ 900) $M^{-1} s^{-1}$	
(Ethylene oxide)$_n^{\cdot}$ [27]) + tetranitromethane($C(NO_2)_4$) ⟶ $C(NO_2)_3^-$ + NO_2 + H^+ + products				
Pulse rad. of ethylene oxide oligomer and polymer + H_2O				72 Beh 1
KAS	H_2O	298	$k = 3.2 \cdot 10^9$(n = 1)... $1.7 \cdot 10^9$(n ≈ 900) $M^{-1} s^{-1}$	
(*meso*-Erythritol-$\dot{O}$H) [28]) + hydroquinone ⟶ $^-OC_6H_4\dot{O}$ + products				
Pulse rad. of *meso*-erythritol + N_2O + H_2O				79 Ste 1
KAS	H_2O, pH ≈ 11.5	RT	$k = 1.3(3) \cdot 10^9\ M^{-1} s^{-1}$	
(N-Ethylmaleamate)$^{\bar{\cdot}}$ [29]) + O_2 ⟶ $\dot{O}_2^-$ + products				
Pulse rad. of N-ethylmaleamate + *t*-butanol + N_2 + H_2O				72 Hay 1
KAS	H_2O, pH = 11	RT	$k = 1.5(2) \cdot 10^9\ M^{-1} s^{-1}$	
(N-Ethylmaleamate)$^{\bar{\cdot}}$ [30]) + N-ethylmaleimide ⟶ products				
Pulse rad. of N-ethylmaleamate + *t*-butanol + N_2 + H_2O				72 Hay 1
KAS	H_2O, pH = 6	RT	$k = 1.8(3) \cdot 10^9\ M^{-1} s^{-1}$	
(N-Ethylmaleimide)$^{\bar{\cdot}}$ [31]) + O_2 ⟶ $\dot{O}_2^-$ + products				
Pulse rad. of N-ethylmaleimide + *t*-butanol + N_2 + H_2O				72 Hay 1
KAS	H_2O, pH = 5.1	RT	$k = 2.0(2) \cdot 10^9\ M^{-1} s^{-1}$	
(N-Ethylmaleimide)$^{\bar{\cdot}}$ [31]) + anthraquinone ⟶ anthrasemiquinone + products				
Pulse rad. of N-ethylmaleimide + *t*-butanol + N_2 + H_2O				73 Rao 1
KAS	H_2O	RT	$k = 8.3 \cdot 10^8\ M^{-1} s^{-1}$	
(N-Ethylmaleimide)$^{\bar{\cdot}}$ [31]) + anthraquinone-2,6-disulfonate ⟶ anthrasemiquinone... + products				
Pulse rad. of N-ethylmaleimide + *t*-butanol + N_2 + H_2O				73 Rao 1
KAS	H_2O	RT	$k = 4.6 \cdot 10^8\ M^{-1} s^{-1}$	
(N-Ethylmaleimide)$^{\bar{\cdot}}$ [31]) + anthraquinone-2-sulfonate ⟶ anthrasemiquinone... + products				
Pulse rad. of N-ethylmaleimide + *t*-butanol + N_2 + H_2O				72 Hay 1,
KAS	H_2O, pH = 4.9 pH = 6	RT	$k = 1.3(2) \cdot 10^9\ M^{-1} s^{-1}$ $8.2 \cdot 10^8$ [32])	73 Rao 1
(N-Ethylmaleimide)$^{\bar{\cdot}}$ [31]) + 1,4-benzoquinone ⟶ 1,4-benzosemiquinone + products				
Pulse rad. of N-ethylmaleimide + *t*-butanol + N_2 + H_2O				73 Rao 1
KAS	H_2O	RT	$k = 5.5 \cdot 10^9\ M^{-1} s^{-1}$	
(N-Ethylmaleimide)$^{\bar{\cdot}}$ [31]) + 2,5-dimethylbenzoquinone ⟶ 2,5-dimethylbenzosemiquinone + products				
Pulse rad. of N-ethylmaleimide + *t*-butanol + N_2 + H_2O				73 Rao 1
KAS	H_2O	RT	$k = 4.4 \cdot 10^9\ M^{-1} s^{-1}$	
(N-Ethylmaleimide)$^{\bar{\cdot}}$ [31]) + 2-methyl-1,4-naphthoquinone ⟶ 2-methyl...semiquinone + products				
Pulse rad. of N-ethylmaleimide + *t*-butanol + N_2 + H_2O				72 Hay 1,
KAS	H_2O, pH = 5.2 pH = 6	RT	$k = 4.3(3) \cdot 10^9\ M^{-1} s^{-1}$ $3.0 \cdot 10^9$ [32])	73 Rao 1

[27]) Radicals from $\dot{O}$H reaction with ethylene oxide monomer, oligomers and polymers.
[28]) Radicals from *meso*-erithritol + $\dot{O}$H reaction.
[29]) Radicals from N-ethylmaleamate + e_{aq}^- reaction.
[30]) Protonated radicals from N-ethylmaleate + e_{aq}^- reaction.
[31]) Radicals from N-ethylmaleimide + e_{aq}^- reaction.
[32]) From [73 Rao 1].

Reaction Radical generation Method	Solvent	T[K]	Rate data	Ref./ add. ref.
(4-Fluoroacetophenone)$^{\cdot-}$ [33]) + 4-nitroacetophenone(PNAP) ⟶ (PNAP)$^{\cdot-}$ + products				
Pulse rad. of p-fluoroacetophenone + *t*-butanol + N_2 + H_2O				73 Ada 1
KAS	H_2O, pH = 11	RT	$k = 4.9(5) \cdot 10^9\ M^{-1}\,s^{-1}$	
(Glucose-$\dot{O}$H) [34]) + 4-chloronitrobenzene ⟶ products				
Pulse rad. of glucose + H_2O + N_2O				77 Bia 1
KAS	H_2O	RT	$k = 4.0 \cdot 10^8\ M^{-1}\,s^{-1}$	
(Glucose-$\dot{O}$H) [34]) + 3,5-dinitrobenzonitrile ⟶ (3,5-dinitrobenzonitrile)$^{\cdot-}$ [35]) + products				
Pulse rad. of glucose + H_2O + N_2O				77 Bia 1
KAS	H_2O	RT	$k = 1.0 \cdot 10^9\ M^{-1}\,s^{-1}$	
(Glucose-$\dot{O}$H) [34]) + (1-(2,4-dinitrophenyl)pyridinium)$^+$ ⟶ (1-(dinitrophenyl)pyridinium)$^{\pm\cdot}$ [35]) + products				
Pulse rad. of glucose + H_2O + N_2O				77 Bia 1
KAS	H_2O	RT	$k = 1.0 \cdot 10^9\ M^{-1}\,s^{-1}$	
(Glucose-$\dot{O}$H) [34]) + hydroquinone ⟶ $^-OC_6H_4\dot{O}$ + products				
Pulse rad. of glucose + N_2O + H_2O				79 Ste 1
KAS	H_2O, pH ≈ 11.5	RT	$k = 7.1(9) \cdot 10^8\ M^{-1}\,s^{-1}$	
(Glucose-$\dot{O}$H) [34]) + 4-nitroacetophenone(PNAP) ⟶ PNAP$^{\cdot-}$ + products				
Pulse rad. of glucose + N_2O + H_2O				77 Bia 1
KAS	H_2O	RT	$k = 9 \cdot 10^8\ M^{-1}\,s^{-1}$	
(Glucose-$\dot{O}$H) [34]) + 2,2,6,6-tetramethyl-4-hydroxy-1-piperidinyloxy(TMPN) ⟶ products				
Pulse rad. of glucose + N_2O + H_2O				76 Asm 1
Cond. (time-resolved)	H_2O	RT	$k = 4.9(5) \cdot 10^7\ M^{-1}\,s^{-1}$	
(Glucose-$\dot{O}$H) [34]) + 2,2,6,6-tetramethyl-4-oxo-1-piperidinyloxy(TAN) ⟶ products				
Pulse rad. of glucose + N_2O + H_2O				76 Asm 1/
Cond. (time-resolved)	H_2O, pH = 3…5	RT	$k = 5.9(6) \cdot 10^7\ M^{-1}\,s^{-1}$ [36])	71 Wil 2
(Glucose-$\dot{O}$H) [34]) + 2,2,5,5-tetramethyl-1-pyrrolidinyloxy-3-carboxamide ⟶ products [37])				
Pulse rad. of glucose + N_2O + H_2O				76 Nig 1
Cond. (time-resolved)	H_2O, pH = 3…5	RT	$k = 5.1 \cdot 10^7\ M^{-1}\,s^{-1}$	
(Glucose-$\dot{O}$H) [34]) + 2,2,5,5-tetramethyl-3-pyrroline-1-yloxy-3-carboxamide ⟶ products [37])				
Pulse rad. of glucose + N_2O + H_2O				76 Nig 1
Cond. (time-resolved)	H_2O, pH = 3…5	RT	$k = 4.3 \cdot 10^7\ M^{-1}\,s^{-1}$	
(Glucose-$\dot{O}$H) [34]) + tetranitromethane ($C(NO_2)_4$) ⟶ $C(NO_2)_3^-$ + NO_2 + H^+ + products				
Pulse rad. of glucose + N_2O + H_2O				65 Rab 1
KAS	H_2O	RT	$k = 2.6 \cdot 10^9\ M^{-1}\,s^{-1}$	
(Glucose-$\dot{O}$H) [34]) + trinitrobenzenesulfonate ion ⟶ products				
Pulse rad. of glucose + N_2O + H_2O				77 Bia 1
KAS	H_2O	RT	$k = 1.1 \cdot 10^9\ M^{-1}\,s^{-1}$	

33) Radicals from 4-fluoroacetophenone + e_{aq}^- reaction.
34) Radicals from glucose + $\dot{O}$H reaction.
35) Electron transfer to nitro groups.
36) Includes some addition reaction.
37) e^--transfer and possibly some addition reactions.

Reaction Radical generation Method	Solvent	T[K]	Rate data	Ref./ add. ref.
(Glycerol-$\dot{O}$H) [38]) + tetranitromethane ($C(NO_2)_4$) ⟶ $C(NO_2)_3^-$ + NO_2 + H^+ + products				
Pulse rad. of glycerol + N_2O + H_2O				64 Asm 1/
KAS	H_2O	RT	$k = 2.4(3) \cdot 10^9\ M^{-1}s^{-1}$	65 Asm 1
(Glycerol)$^\cdot$ [39]) + tetranitromethane ($C(NO_2)_4$) ⟶ $C(NO_2)_3^-$ + NO_2 + H^+ + products				
Pulse rad. of glycerol + N_2O + H_2O				65 Asm 1/
KAS	H_2O with	RT		64 Asm 1
	1 wt% glycerol		$k = 2.25 \cdot 10^9\ M^{-1}s^{-1}$	
	24		$1.75 \cdot 10^9$	
	35		$1.45 \cdot 10^9$	
	42		$1.2 \cdot 10^9$	
	50		$0.9 \cdot 10^9$	
	60		$0.55 \cdot 10^9$	
	69		$0.45 \cdot 10^9$	
(Glycine)$^{\dot{-}}$ [40]) + Fe(III)cytochrome c ⟶ products + Fe(II)cytochrome c				
Pulse rad. of glycine anhydride + H_2O				78 Sim 1
KAS	H_2O	RT	$k = 8 \cdot 10^8\ M^{-1}s^{-1}$	
(Glycine)$^{\dot{-}}$ [40]) + *anti*-5-nitro-2-furaldoxime ⟶ products				
Pulse rad. of glycine + *t*-butanol + N_2 + H_2O				73 Gre 1
KAS	H_2O	RT	$k = 3.1 \cdot 10^9\ M^{-1}s^{-1}$	
(Glycine)$^\cdot$ [41]) + lipoate(—S—S—) ⟶ (lipoate)$^{\dot{-}}$(—S$\overset{(-)}{\therefore}$S—) + products				
Pulse rad. of chloroacetyl glycine + *t*-butanol + H_2O				75 Far 3
KAS	H_2O	RT	$k = 5 \cdot 10^6\ M^{-1}s^{-1}$	
(Glycine anhydride)$^{\dot{-}}$ [42]) + acetophenone($C_6H_5COCH_3$) ⟶ products				
Pulse rad. of glycine anhydride + *t*-butanol + N_2 + H_2O				71 Hay 1
KAS	H_2O, pH = 5.2	RT	$k = 2.3(3) \cdot 10^9\ M^{-1}s^{-1}$	
	pH = 12.3		$2.0(3) \cdot 10^9$	
(Glycine anhydride)$^{\dot{-}}$ [42]) + benzophenone ($(C_6H_5)_2CO$) ⟶ products				
Pulse rad. of glycine anhydride + *t*-butanol + N_2 + H_2O				71 Hay 1
KAS	H_2O, pH = 5.5	RT	$k = 2.2(3) \cdot 10^9\ M^{-1}s^{-1}$	
	pH = 12.3		$2.5(4) \cdot 10^9$	
(Glycine anhydride)$^{\dot{-}}$ [42]) + cystamine ⟶ products				
Pulse rad. of glycine anhydride + *t*-butanol + N_2 + H_2O				71 Hay 1
KAS	H_2O, pH = 5.7 and 11.0	RT	$k = 1.2(3) \cdot 10^8\ M^{-1}s^{-1}$	
(Glycine anhydride)$^{\dot{-}}$ [42]) + cysteine ⟶ products				
Pulse rad. of glycine anhydride + *t*-butanol + N_2 + H_2O				71 Hay 1
KAS	H_2O	RT	$k = 2.1(4) \cdot 10^8\ M^{-1}s^{-1}$	
(Glycine anhydride)$^{\dot{-}}$ [42]) + glutathione (RSH) ⟶ products				
Pulse rad. of glycine anhydride + *t*-butanol + N_2 + H_2O				71 Hay 1
KAS	H_2O	RT	$k = 1.8(3) \cdot 10^8\ M^{-1}s^{-1}$	

[38]) Radicals from glycerol + $\dot{O}$H reaction (likely to be $CH_2OHCHOH\dot{C}HOH$, $CH_2OH\dot{C}OHCH_2OH$ and/or $CH_2OH\dot{C}HCHO$, $\dot{C}H_2COCH_2OH$).
[39]) Radicals from glycerol + $\dot{O}$H reaction and direct energy absorption by glycerol.
[40]) Radicals from glycine + e_{aq}^- reaction.
[41]) Dechlorinated radical from chloroacetyl glycine + e_{aq}^- reaction.
[42]) Radicals from glycine anhydride + e_{aq}^- reaction; most likely $CH_2\dot{C}O^-NHCH_2CONH$ (ring).

Reaction Radical generation Method	Solvent	T[K]	Rate data	Ref./ add. ref.
(Glycine anhydride)$^{\overline{\cdot}}$ [42]) + glutathione disulfide (RSSR) ⟶ products				
Pulse rad. of glycine anhydride + t-butanol + N_2 + H_2O				71 Hay 1
KAS	H_2O	RT	$k = 4.0(4)\cdot 10^7\ M^{-1}s^{-1}$	
(Glycine anhydride)$^{\overline{\cdot}}$ [43]) + 2-methyl-1,4-naphthoquinone ⟶ ... semiquinone + products				
Pulse rad. of glycine anhydride + H_2O				73 Rao 3
KAS	H_2O	RT	$k = 4.9(5)\cdot 10^9\ M^{-1}s^{-1}$	
(Glycine anhydride)$^{\overline{\cdot}}$ [42]) + 1-mercaptopropionic acid ($HSCH_2CH_2COO^-$) ⟶ products				
Pulse rad. of glycine anhydride + t-butanol + N_2 + H_2O				71 Hay 1
KAS	H_2O, pH = 5.3 and 7.4	RT	$k = 3.0(3)\cdot 10^8\ M^{-1}s^{-1}$	
(Glycine anhydride-ȮH) [44]) + 1,4-benzoquinone ⟶ 1,4-benzosemiquinone + products				
Pulse rad. of glycine anhydride + H_2O + N_2O				73 Hay 1
KAS	H_2O, pH = 10.5	RT	$k = 2.2\cdot 10^9\ M^{-1}s^{-1}$	
(Glycine anhydride-ȮH) [44]) + 2-methyl-1,4-naphthoquinone ⟶ ... semiquinone + products [45])				
Pulse rad. of glycine anhydride + H_2O + N_2O				73 Rao 3/
KAS	H_2O, pH = 10.9	RT	$k = 4.0(4)\cdot 10^9\ M^{-1}s^{-1}$ [46])	72 Sim 1
(Glycol amide-ȮH) [47]) + 2-methyl-1,4-naphthoquinone ⟶ 2-methyl ... semiquinone + products [48])				
Pulse rad. of glycol amide + N_2O + H_2O				73 Rao 3
KAS	H_2O	RT	$k = 2.3\cdot 10^9\ M^{-1}s^{-1}$	
(Glycylglycinamide-ȮH) [49]) + 2-methyl-1,4-naphthoquinone ⟶ ... semiquinone + products				
Pulse rad. of glycylglycinamide + H_2O + N_2O				73 Rao 1
KAS	H_2O, pH = 6.8 pH = 11.0	RT	$k = 8.5(9)\cdot 10^8\ M^{-1}s^{-1}$ [50]) $8.5(9)\cdot 10^9$ [51])	
(Glycylsarcosine-ȮH) [52]) + 2-methyl-1,4-naphthoquinone ⟶ ... semiquinone + products [53])				
Pulse rad. of glycylsarcosine + H_2O + N_2O				73 Rao 3
KAS	H_2O, pH = 10.9	RT	$k = 1.0(1)\cdot 10^9\ M^{-1}s^{-1}$	
(Guanine-ȮH) [54]) + 4-hydroxy-2,2,6,6-tetramethylpiperidino-1-oxy ⟶ products [55])				
Pulse rad. of guanine + N_2O + H_2O				72 Bru 1
KAS	H_2O	RT	$k = 1.3(2)\cdot 10^8\ M^{-1}s^{-1}$	
(Guanine-ȮH) [54]) + nor-pseudopelletierine-N-oxy ⟶ products [55])				
Pulse rad. of guanine + N_2O + H_2O				72 Bru 1
KAS	H_2O	RT	$k = 5.3(3)\cdot 10^8\ M^{-1}s^{-1}$	
(Guanine-ȮH) [54]) + 2,2,6,6-tetramethyl-4-oxo-1-piperidinyloxy ⟶ products [55])				
Pulse rad. of guanine + N_2O + H_2O				72 Bru 1
KAS	H_2O	RT	$k = 1.7(2)\cdot 10^8\ M^{-1}s^{-1}$	

[42]) Radicals from glycine anhydride + e^-_{aq} reaction; most likely $CH_2\dot{C}O^-NHCH_2CONH$ (ring).
[43]) Radicals from glycine anhydride + e^-_{aq} reaction.
[44]) Radicals from glycine anhydride + ȮH reaction.
[45]) 88% e^--transfer.
[46]) No e^--transfer at pH = 6.7.
[47]) Radicals from glycol amide + ȮH reaction.
[48]) 48% e^--transfer.
[49]) Radicals from glycylglycinamide + ȮH reaction.
[50]) 25% e^--transfer.
[51]) 45% e^--transfer.
[52]) Radicals from glycylsarcosine + ȮH reaction.
[53]) 46% e^--transfer.
[54]) Radicals from guanine + ȮH reaction.
[55]) Addition and/or e^--transfer.

Reaction / Radical generation / Method	Solvent	T[K]	Rate data	Ref./ add. ref.
(Histidine)$^{\bar{\cdot}}$ [56]) + 2-methyl-1,4-naphthoquinone ⟶ 2-methyl...semiquinone + products				
Pulse rad. of histidine + *t*-butanol + N_2 + H_2O				73 Rao 3
KAS	H_2O	RT	$k = 1.2 \cdot 10^9\, M^{-1} s^{-1}$	
(Histidine)$^{\bar{\cdot}}$ [56]) + *anti*-5-nitro-2-furaldoxime ⟶ products				
Pulse rad. of histidine + *t*-butanol + N_2 + H_2O				73 Gre 1
KAS	H_2O	RT	$k = 3.5 \cdot 10^9\, M^{-1} s^{-1}$	
(Histidine-$\dot{O}$H) [57]) + 1,4-benzoquinone ⟶ 1,4-benzosemiquinone + products [58])				
Pulse rad. of histidine + N_2O + H_2O				73 Sim 1
KAS	H_2O	RT	$k = 1.8 \cdot 10^9\, M^{-1} s^{-1}$	
(Histidine-$\dot{O}$H) [57]) + 2,6-dichloroindophenol ⟶ products				
Pulse rad. of histidine + N_2O + H_2O				73 Rao 2
KAS	H_2O	RT	$k = 1.1 \cdot 10^9\, M^{-1} s^{-1}$	
(Histidine-$\dot{O}$H) [57]) + indigo disulfonate ⟶ products				
Pulse rad. of histidine + N_2O + H_2O				73 Rao 2
KAS	H_2O	RT	$k = (1.1 \ldots 1.3) \cdot 10^9\, M^{-1} s^{-1}$	
(Histidine-$\dot{O}$H) [57]) + indophenol ⟶ products				
Pulse rad. of histidine + N_2O + H_2O				73 Rao 2
KAS	H_2O, pH = 9	RT	$k = (1.0 \ldots 1.6) \cdot 10^9\, M^{-1} s^{-1}$	
(Histidine-$\dot{O}$H) [57]) + 2-methyl-1,4-naphthoquinone ⟶ 2-methyl...semiquinone + products [59])				
Pulse rad. of histidine + N_2O + H_2O				73 Rao 3
KAS	H_2O	RT	$k = 1.2 \cdot 10^9\, M^{-1} s^{-1}$	
(Histidine-$\dot{O}$H) [57]) + *anti*-5-nitro-2-furaldoxime ⟶ products [60])				
Pulse rad. of histidine + N_2O + H_2O				73 Gre 1
KAS	H_2O	RT	$k = 2.1 \cdot 10^9\, M^{-1} s^{-1}$	
(Imidazole)$^{\bar{\cdot}}$ [61]) + 2-methyl-1,4-naphthoquinone ⟶ 2-methyl...semiquinone + products				
Pulse rad. of imidazole + *t*-butanol + N_2 + H_2O				73 Rao 3
KAS	H_2O	RT	$k = 1.2 \cdot 10^9\, M^{-1} s^{-1}$	
(Imidazole-$\dot{O}$H) [62]) + 2-methyl-1,4-naphthoquinone ⟶ 2-methyl...semiquinone + products [63])				
Pulse rad. of imidazole + N_2O + H_2O				73 Rao 3
KAS	H_2O	RT	$k = 1.6 \cdot 10^9\, M^{-1} s^{-1}$	
(Indole-$\dot{O}$H) [64]) + 2-methyl-1,4-naphthoquinone ⟶ 2-methyl...semiquinone + products [65])				
Pulse rad. of indole + N_2O + H_2O				73 Rao 3
KAS	H_2O	RT	$k = 2.9 \cdot 10^9\, M^{-1} s^{-1}$	
(Isocytosine-$\dot{O}$H) [66]) + 2-methyl-1,4-naphthoquinone ⟶ 2-methyl...semiquinone + products				
Pulse rad. of isocytosine + N_2O + H_2O				73 Hay 3
KAS	H_2O, pH = 3.2	RT	$k = 3.0 \cdot 10^9\, M^{-1} s^{-1}$ [65])	
	pH = 7		$4.8 \cdot 10^9$ [67])	

[56]) Radicals from histidine + e_{aq}^- reaction.
[57]) Radicals from histidine + $\dot{O}$H reaction.
[58]) 93% e^--transfer.
[59]) 88% e^--transfer.
[60]) 20% e^--transfer.
[61]) Radicals from imidazole + e_{aq}^- reaction.
[62]) Radicals from imidazole + $\dot{O}$H reaction.
[63]) 84% e^--transfer.
[64]) Radicals from indole + $\dot{O}$H reaction.
[65]) 20% e^--transfer.
[66]) Radicals from isocytosine + $\dot{O}$H reaction.
[67]) 70% e^--transfer.

Reaction Radical generation Method	Solvent	T[K]	Rate data	Ref./ add. ref.
(Isonicotinamide)$^{\bar{\cdot}}$ [68]) + $Co(III)(NH_3)_6^{3+}$ ⟶ products				
Pulse rad. of isonicotinamide + H_2O				74 Coh 1
KAS	H_2O	RT	$k = 7.5 \cdot 10^6\ M^{-1}\,s^{-1}$	
(Isonicotinamide)$^{\bar{\cdot}}$ [68]) + $Co(III)(NH_3)_5(H_2O)^{3+}$ ⟶ products				
Pulse rad. of isonicotinamide + H_2O				74 Coh 1
KAS	H_2O, pH = 3.9	RT	$k = 1.6 \cdot 10^8\ M^{-1}\,s^{-1}$	
(Isonicotinamide)$^{\bar{\cdot}}$ [68]) + $Co(III)(NH_3)_5$(benzoate) ⟶ products				
Pulse rad. of isonicotinamide + H_2O				74 Coh 1
KAS	H_2O	RT	$k = 1.8 \cdot 10^7\ M^{-1}\,s^{-1}$	
(Isonicotinamide)$^{\bar{\cdot}}$ [68]) + $Co(III)(NH_3)_5Br^{2+}$ ⟶ products				
Pulse rad. of isonicotinamide + H_2O				74 Coh 1
KAS	H_2O	RT	$k = 7.0 \cdot 10^8\ M^{-1}\,s^{-1}$	
(Isonicotinamide)$^{\bar{\cdot}}$ [68]) + $Co(III)(NH_3)Cl^{2+}$ ⟶ products				
Pulse rad. of isonicotinamide + H_2O				74 Coh 1
KAS	H_2O	RT	$k = 2.3 \cdot 10^8\ M^{-1}\,s^{-1}$	
(Isonicotinamide)$^{\bar{\cdot}}$ [69]) + $Co(III)(NH_3)_5$(isonicotinamide) ⟶ products				
Pulse rad. of isonicotinamide + H_2O				74 Coh 1
KAS	H_2O, pH = 1	RT	$k = 3(1) \cdot 10^6\ M^{-1}\,s^{-1}$	
	pH = 6.4...6.7		$1.6 \cdot 10^8$	
(Isonicotinamide)$^{\bar{\cdot}}$ [68]) + $Co(III)(NH_3)_5$(nicotinamide) ⟶ products				
Pulse rad. of isonicotinamide + H_2O				74 Coh 1
KAS	H_2O	RT	$k = 3.3 \cdot 10^8\ M^{-1}\,s^{-1}$	
(Isonicotinamide)$^{\bar{\cdot}}$ [68]) + $Co(III)(NH_3)_5$(pyridine) ⟶ products				
Pulse rad. of isonicotinamide + H_2O				74 Coh 1
KAS	H_2O	RT	$k = 1.8 \cdot 10^8\ M^{-1}\,s^{-1}$	
(Lumazine)$^{\bar{\cdot}}$ [70]) + anthraquinone-2,6-disulfonate ⟶ products				
Pulse rad. of lumazine + *t*-butanol + H_2O				75 Moo 2
KAS	H_2O	RT	$k \approx 1.2 \cdot 10^9\ M^{-1}\,s^{-1}$	
(Lumiflavin-3-acetate)$^{\bar{\cdot}}$ [71]) + O_2 ⟶ products				
Pulse rad. of lumiflavin-3-acetate + *t*-butanol + H_2O				75 Far 2
KAS	H_2O, pH = 10	RT	$k = 2.5(3) \cdot 10^8\ M^{-1}\,s^{-1}$	
(Methionine-$\dot{O}$H) [72]) + *anti*-5-nitro-2-furaldoxime ⟶ products [73])				
Pulse rad. of methionine + *t*-butanol + N_2 + H_2O				73 Gre 1/
KAS	H_2O	RT	$k = 3.7 \cdot 10^9\ M^{-1}\,s^{-1}$	81 Hil 1, 83 Hil 1
(N-Methyl-3-carbamido-pyridinium)$^{\cdot}$ [74]) + cytochrome-c ⟶ products				
Pulse rad. of N-methyl-3-carbamido pyridinium + formate + H_2O				75 Lan 1
KAS	H_2O	RT	$k = 9.4 \cdot 10^8\ M^{-1}\,s^{-1}$	

[68]) Radicals from isonicotinamide + e^-_{aq} reaction.
[69]) Radicals from isonicotinamide + e^-_{aq} reaction; at pH = 1 from $\dot{H}$ reaction.
[70]) Radicals from lumazine + e^-_{aq} reaction.
[71]) Radicals from lumiflavin-3-acetate + e^-_{aq} reaction.
[72]) Radicals from methionine + $\dot{O}H$ reaction.
[73]) 30% e^--transfer (likely by $CH_3SCH_2CH_2\dot{C}HNH_2$) [81 Hil 1].
[74]) Radicals from N-methyl-3-carbamido-pyridinium + $\dot{C}O_2^-$ reaction.

Reaction Radical generation Method	Solvent	T[K]	Rate data	Ref./ add. ref.
(N-Methyl-4-carbamido-pyridinium)˙ [75]) + cytochrome-c ⟶ products				
Pulse rad. of N-methyl-4-carbamido pyridinium + formate + H_2O				75 Lan 1
KAS	H_2O	RT	$k = 7.2 \cdot 10^8\ M^{-1}\ s^{-1}$	
(5-Methylcytosine-H)˙ [76]) + 4-nitroacetophenone(PNAP) ⟶ $PNAP^{\cdot-}$ + products				
Pulse rad. of 5-methylcytosine + *t*-butanol + H_2O				79 His 1
KAS	H_2O	RT	$k = 5.3(3) \cdot 10^9\ M^{-1}\ s^{-1}$	
(1-Methylcytosine-ȮH) [77]) + 2-methyl-1,4-naphthoquinone ⟶ 2-methyl...semiquinone + products [78])				
Pulse rad. of 1-methylcytosine + N_2O + H_2O				73 Hay 3
KAS	H_2O	RT	$k = 4.6 \cdot 10^9\ M^{-1}\ s^{-1}$	
(3-Methylcytosine-ȮH) [79]) + 2-methyl-1,4-naphthoquinone ⟶ 2-methyl...semiquinone + products [80])				
Pulse rad. of 3-methylcytosine + N_2O + H_2O				73 Hay 3
KAS	H_2O, pH = 6.9	RT	$k = 2.7 \cdot 10^9\ M^{-1}\ s^{-1}$	
	pH = 9.3		$1.2 \cdot 10^9$	
(5-Methylcytosine-ȮH) [81]) + 2-methyl-1,4-naphthoquinone ⟶ 2-methyl...semiquinone + products [82])				
Pulse rad. of 5-methylcytosine + N_2O + H_2O				73 Hay 3
KAS	H_2O	RT	$k = 2.0 \cdot 10^9\ M^{-1}\ s^{-1}$	
$(1\text{-Methylnicotinamide})^{\cdot-}$ [83]) + cytochrome(III)-c ⟶ products				
Pulse rad. of 1-methylnicotinamide + N_2 + H_2O				75 Sim 1
KAS	H_2O	RT	$k = 1.4 \cdot 10^9\ M^{-1}\ s^{-1}$	
$(1\text{-Methylnicotinamide})^{\cdot-}$ [83]) + 1,4-benzoquinone ⟶ products				
Pulse rad. of 1-methylnicotinamide + *t*-butanol + N_2 + H_2O				74 Bru 1
KAS	H_2O	RT	$k = 5.2 \cdot 10^9\ M^{-1}\ s^{-1}$	
$(1\text{-Methylnicotinamide})^{\cdot-}$ [83]) + 3-benzoylpyridine ⟶ products				
Pulse rad. of 1-methylnicotinamide + *t*-butanol + N_2 + H_2O				74 Bru 1
KAS	H_2O	RT	$k = 2.6 \cdot 10^9\ M^{-1}\ s^{-1}$	
$(1\text{-Methylnicotinamide})^{\cdot-}$ [83]) + flavin adenine dinucleotide(FAD) ⟶ products				
Pulse rad. of 1-methylnicotinamide + *t*-butanol + N_2 + H_2O				76 And 1
KAS	H_2O	RT	$k = 3.6 \cdot 10^9\ M^{-1}\ s^{-1}$	
$(1\text{-Methylnicotinamide})^{\cdot-}$ [83]) + 2-methyl-1,4-naphthoquinone ⟶ products				
Pulse rad. of 1-methylnicotinamide + *t*-butanol + N_2 + H_2O				73 Rao 3
KAS	H_2O, pH = 6.8 and 10.9	RT	$k = 4.9 \cdot 10^9\ M^{-1}\ s^{-1}$	
(1-Methyluracil-ȮH) [84]) + 1,4-benzoquinone ⟶ 1,4-benzosemiquinone + products				
Pulse rad. of 1-methyluracil + N_2O + H_2O				74 Rao 1
KAS	H_2O	RT	$k = 4.0 \cdot 10^9\ M^{-1}\ s^{-1}$	

[75]) Radicals from N-methyl-4-carbamido-pyridinium + $\dot{C}O_2^-$ reaction.
[76]) Protonated form of radical anions formed in 5-methylcytosine + e_{aq}^- reaction.
[77]) Radicals from 1-methylcytosine + ȮH reaction.
[78]) 50% e^--transfer.
[79]) Radicals from 3-methylcytosine + ȮH reaction.
[80]) 12...14% e^--transfer.
[81]) Radicals from 5-methylcytosine + ȮH reaction.
[82]) 26% e^--transfer.
[83]) Radicals from 1-methylnicotinamide + e_{aq}^- reaction.
[84]) Radicals from 1-methyluracil + ȮH reaction.

Reaction Radical generation Method	Solvent	T[K]	Rate data	Ref./ add. ref.
(1-Methyluracil-$\dot{O}H$)[84]) + methylene blue ⟶ products				
Pulse rad. of 1-methyluracil + N_2O + H_2O				74 Rao 1
KAS	H_2O, pH = 10.8	RT	$k = 4.7 \cdot 10^9\ M^{-1}s^{-1}$	
(1-Methyluracil-$\dot{O}H$)[84]) + 2-methyl-1,4-naphthoquinone ⟶ 2-methyl...semiquinone + products				
Pulse rad. of 1-methyluracil + N_2O + H_2O				73 Hay 3
KAS	H_2O, pH = 7	RT	$k = 3.7 \cdot 10^9\ M^{-1}s^{-1}$ [85])	
	pH = 11		$4.1 \cdot 10^9$ [86])	
(3-Methyluracil-$\dot{O}H$)[87]) + 2-methyl-1,4-naphthoquinone ⟶ 2-methyl...semiquinone + products				
Pulse rad. of 3-methyluracil + N_2O + H_2O				73 Hay 3
KAS	H_2O, pH = 7	RT	$k = 2.7 \cdot 10^9\ M^{-1}s^{-1}$ [88])	
	pH = 11		$3.5 \cdot 10^9$ [89])	
(6-Methyluracil-$\dot{O}H$)[90]) + 2-methyl-1,4-naphthoquinone ⟶ 2-methyl...semiquinone + products				
Pulse rad. of 6-methyluracil + N_2O + H_2O				73 Hay 3
KAS	H_2O, pH = 7	RT	$k = 3.9 \cdot 10^9\ M^{-1}s^{-1}$ [91])	
	pH = 10.5		$6.8 \cdot 10^9$ [92])	
$NAD^{\cdot}$ [93]) + cytochrome(III)-c ⟶ products				
Pulse rad. of NAD^+ + N_2 + H_2O				75 Sim 1/
KAS	H_2O	RT	$k = 7.4 \cdot 10^8\ M^{-1}s^{-1}$	75 Lan 1
$NAD^{\cdot}$ [94]) + cytochrome-c ⟶ products				
Pulse rad. of NAD^+ + formate + H_2O				75 Lan 1/
KAS	H_2O, pH = 4...8	RT	$k = 7.7 \cdot 10^8\ M^{-1}s^{-1}$	75 Sim 1
$NAD^{\cdot}$ [93]) + O_2 ⟶ $\dot{O}_2^-$ + products				
Pulse rad. of NAD^+ + *t*-butanol + N_2 + H_2O				74 Bru 1
KAS	H_2O	RT	$k = 2.0 \cdot 10^9\ M^{-1}s^{-1}$	
$NAD^{\cdot}$ [93]) + anthraquinone ⟶ products				
Pulse rad. of NAD^+ + *t*-butanol + N_2 + H_2O				73 Rao 1
KAS	H_2O	RT	$k = 4.1 \cdot 10^8\ M^{-1}s^{-1}$	
$NAD^{\cdot}$ [93]) + anthraquinone-2,6-disulfonate ⟶ products				
Pulse rad. of NAD^+ + *t*-butanol + N_2 + H_2O				73 Rao 1
KAS	H_2O	RT	$k = 9.6 \cdot 10^8\ M^{-1}s^{-1}$	
$NAD^{\cdot}$ [93]) + 1,4-benzoquinone ⟶ products				
Pulse rad. of NAD^+ + *t*-butanol + N_2 + H_2O				73 Rao 1,
KAS	H_2O	RT	$k = (3.6...4.4) \cdot 10^9\ M^{-1}s^{-1}$	74 Bru 1, 71 Wil 1
$NAD^{\cdot}$ [93]) + flavin adenine dinucleotide(FAD) ⟶ products				
Pulse rad. of NAD^+ + *t*-butanol + N_2 + H_2O				76 And 1
KAS	H_2O	RT	$k = 1.0 \cdot 10^9\ M^{-1}s^{-1}$	

[84]) Radicals from 1-methyluracil + $\dot{O}H$ reaction.
[85]) 19% e^--transfer.
[86]) 55% e^--transfer.
[87]) Radicals from 3-methyluracil + $\dot{O}H$ reaction.
[88]) 31% e^--transfer.
[89]) 79% e^--transfer.
[90]) Radicals from 6-methyluracil + $\dot{O}H$ reaction.
[91]) 69% e^--transfer.
[92]) 34% e^--transfer.
[93]) Radicals from nicotinamide adenine dinucleotide + e_{aq}^- reaction.
[94]) Radicals from nicotinamide adenine dinucleotide + $\dot{C}O_2^-$ reaction.

Reaction Radical generation Method	Solvent	T[K]	Rate data	Ref./ add. ref.
NAD$^{\cdot}$ [93]) + 2-methyl-1,4-naphthoquinone ⟶ products				
Pulse rad. of NAD$^+$ + t-butanol + N_2 + H_2O				73 Rao 1,
KAS	H_2O	RT	$k = 3.1 \cdot 10^9\ M^{-1}s^{-1}$	73 Rao 3
NAD$^{\cdot}$ [93]) + *anti*-5-nitro-2-furaldoxime ⟶ products				
Pulse rad. of NAD$^+$ + t-butanol + N_2 + H_2O				76 Bia 1
KAS	H_2O	RT	$k = 3.1 \cdot 10^9\ M^{-1}s^{-1}$	
NAD$^{\cdot}$ [93]) + riboflavin ⟶ products				
Pulse rad. of NAD$^+$ + t-butanol + N_2 + H_2O				73 Rao 1/
KAS	H_2O	RT	$k = 1.0 \cdot 10^9\ M^{-1}s^{-1}$	74 Bru 1
NAD$^{\cdot}$ [93]) + 2,2,6,6-tetramethyl-4-oxo-1-piperidinyloxy(TAN) ⟶ products				
Pulse rad. of NAD$^+$ + t-butanol + N_2 + H_2O				71 Wil 2
KAS	H_2O	RT	$k = 1.5 \cdot 10^8\ M^{-1}s^{-1}$	
(Nicotinamide)$^{\cdot-}$ [96]) + Co(III)$(NH_3)_6^{3+}$ ⟶ products				
Pulse rad. of nicotinamide + H_2O				74 Coh 1
KAS	H_2O	RT	$k = 1.1 \cdot 10^8\ M^{-1}s^{-1}$	
(Nicotinamide)$^{\cdot-}$ [96]) + Co(III)$(NH_3)_5(H_2O)^{3+}$ ⟶ products				
Pulse rad. of nicotinamide + H_2O				74 Coh 1
KAS	H_2O, pH = 3.9	RT	$k = 1.3 \cdot 10^9\ M^{-1}s^{-1}$	
(Nicotinamide)$^{\cdot-}$ [96]) + Co(III)$(NH_3)_5$(benzoate) ⟶ products				
Pulse rad. of nicotinamide + H_2O				74 Coh 1
KAS	H_2O	RT	$k = 7.0 \cdot 10^8\ M^{-1}s^{-1}$	
(Nicotinamide)$^{\cdot-}$ [96]) + Co(III)$(NH_3)_5Br^{2+}$ ⟶ products				
Pulse rad. of nicotinamide + H_2O				74 Coh 1
KAS	H_2O	RT	$k = 2.4 \cdot 10^9\ M^{-1}s^{-1}$	
(Nicotinamide)$^{\cdot-}$ [96]) + Co(III)$(NH_3)_5Cl^{2+}$ ⟶ products				
Pulse rad. of nicotinamide + H_2O				74 Coh 1
KAS	H_2O	RT	$k = 2.5 \cdot 10^9\ M^{-1}s^{-1}$	
(Nicotinamide)$^{\cdot-}$ [97]) + Co(III)$(NH_3)_5$(isonicotinamide) ⟶ products				
Pulse rad. of nicotinamide + H_2O				74 Coh 1
KAS	H_2O, pH = 1	RT	$k = 1.7 \cdot 10^8\ M^{-1}s^{-1}$	
	pH = 6.4...6.7		$1.7 \cdot 10^9$	
(Nicotinamide)$^{\cdot-}$ [97]) + Co(III)$(NH_3)_5$(nicotinamide) ⟶ products				
Pulse rad. of nicotinamide + H_2O				74 Coh 1
KAS	H_2O, pH = 1	RT	$k = 3.5 \cdot 10^8\ M^{-1}s^{-1}$	
	pH = 6.4...6.7		$2.1 \cdot 10^9$	
(Nicotinamide)$^{\cdot-}$ [97]) + Co(III)$(NH_3)_5$(pyridine) ⟶ products				
Pulse rad. of nicotinamide + H_2O				74 Coh 1
KAS	H_2O, pH = 1	RT	$k = 2.5 \cdot 10^8\ M^{-1}s^{-1}$	
	pH = 6.4...6.7		$2.0 \cdot 10^9$	
(Nicotinamide)$^{\cdot-}$ [96]) + anthraquinone ⟶ products				
Pulse rad. of nicotinamide + H_2O				73 Rao 1
KAS	H_2O	RT	$k = 3.5 \cdot 10^9\ M^{-1}s^{-1}$	

[93]) Radicals from nicotinamide adenine dinucleotide + e_{aq}^- reaction.
[96]) Radicals from nicotinamide + e_{aq}^- reaction.
[97]) Radicals from nicotinamide + e_{aq}^- reaction; at pH = 1 from $\dot{H}$ atom reaction.

Reaction Radical generation Method	Solvent	T[K]	Rate data	Ref./ add. ref.
(Nicotinamide)$^{\cdot-}$ [96]) + anthraquinone-2-sulfonate ⟶ products				
Pulse rad. of nicotinamide + H_2O				73 Rao 1
KAS	H_2O	RT	$k = 3.8 \cdot 10^9\ M^{-1}s^{-1}$	
(Nicotinamide)$^{\cdot-}$ [96]) + 1,4-benzoquinone ⟶ products				
Pulse rad. of nicotinamide + H_2O				74 Bru 1,
KAS	H_2O	RT	$k = 5.0 \cdot 10^9\ M^{-1}s^{-1}$ $7.0 \cdot 10^9$ [98])	73 Rao 1
(Nicotinamide)$^{\cdot-}$ [96]) + 2,5-dimethylbenzoquinone ⟶ products				
Pulse rad. of nicotinamide + H_2O				73 Rao 1
KAS	H_2O	RT	$k = 7.25 \cdot 10^9\ M^{-1}s^{-1}$	
(Nicotinamide)$^{\cdot-}$ [96]) + eosin Y ⟶ products				
Pulse rad. of nicotinamide + H_2O				74 Bru 1
KAS	H_2O	RT	$k = 2.5 \cdot 10^9\ M^{-1}s^{-1}$	
(Nicotinamide)$^{\cdot-}$ [96]) + 2-methyl-1,4-naphthoquinone ⟶ products				
Pulse rad. of nicotinamide + H_2O				73 Rao 1,
KAS	H_2O, pH = 7...10.9	RT	$k = (5.1 \dots 5.4) \cdot 10^9\ M^{-1}s^{-1}$	73 Rao 3
(Nicotinamide)$^{\cdot-}$ [96]) + 1,4-naphthoquinone-2-sulfonate ⟶ products				
Pulse rad. of nicotinamide + H_2O				73 Rao 1
KAS	H_2O	RT	$k = 5.45 \cdot 10^9\ M^{-1}s^{-1}$	
(Nicotinic acid)$^{\cdot-}$ [99]) + 1,4-benzoquinone ⟶ products				
Pulse rad. of nicotinic acid + H_2O				74 Bru 1
KAS	H_2O	RT	$k = 5.2 \cdot 10^9\ M^{-1}s^{-1}$	
(Nicotinic acid)$^{\cdot-}$ [99]) + eosin Y ⟶ products				
Pulse rad. of nicotinic acid + H_2O				74 Bru 1
KAS	H_2O	RT	$k = 3.0 \cdot 10^9\ M^{-1}s^{-1}$	
(Nicotinic acid)$^{\cdot-}$ [100]) + isonicotinic acid ⟶ products				
Pulse rad. of nicotinic acid + H_2O				74 Net 1
KAS	H_2O, $H_0 = -1$	RT	$k = 2.0 \cdot 10^9\ M^{-1}s^{-1}$	
(Nicotinic acid)$^{\cdot-}$ [99]) + 2-methyl-1,4-naphthoquinone ⟶ 2-methyl...semiquinone + products				
Pulse rad. of nicotinic acid + H_2O				73 Rao 3
KAS	H_2O	RT	$k = 4.4 \cdot 10^9\ M^{-1}s^{-1}$	
(Phenylalanine)$^{\cdot-}$ [1]) + *anti*-5-nitro-2-furaldoxime ⟶ products				
Pulse rad. of phenylalanine + *t*-butanol + N_2 + H_2O				73 Gre 1
KAS	H_2O	RT	$k = 3.7 \cdot 10^9\ M^{-1}s^{-1}$	
(Phenylalanine-ȮH) [2]) + *anti*-5-nitro-2-furaldoxime ⟶ products [3])				
Pulse rad. of phenylalanine + N_2O + H_2O				73 Gre 1
KAS	H_2O	RT	$k = 3.0 \cdot 10^9\ M^{-1}s^{-1}$	

[96]) Radicals from nicotinamide + e_{aq}^- reaction.
[98]) From [73 Rao 1].
[99]) Radicals from nicotinic acid + e_{aq}^- reaction.
[100]) One-electron reduced form of nicotinic acid.
[1]) Radicals from phenylalanine + e_{aq}^- reaction.
[2]) Radicals from phenylalanine + ȮH reaction.
[3]) 8% e^--transfer.

Reaction Radical generation Method	Solvent	T[K]	Rate data	Ref./ add. ref.
(Phenylphosphate-$\dot{O}H$)[4]) + Fe^{2+} $\longrightarrow$ Fe^{3+} + products				
Pulse rad. of $C_6H_5OPO_3H_2$ + H_2O				79 Gra 1
KAS	H_2O, pH < 1.7	RT	$k = 1.2(1) \cdot 10^9\ M^{-1}\ s^{-1}$	
(1,2-Propanediol-$\dot{O}H$)[5]) + tetranitromethane ($C(NO_2)_4$) $\longrightarrow$ $C(NO_2)_3^-$ + NO_2 + H^+ + products				
Pulse rad. of 1,2-propanediol + N_2O + H_2O				73 Asm 1
KAS	H_2O	RT	$k = 3.2 \cdot 10^9\ M^{-1}\ s^{-1}$	
(Pterin)$^{\dot{-}}$ [6]) + anthraquinone-2,6-disulfonate $\longrightarrow$ products				
Pulse rad. of pterin + *t*-butanol + H_2O				76 Moo 1
KAS	H_2O	RT	$k \approx 10^9\ M^{-1}\ s^{-1}$	
(Pyrazine)$^{\dot{-}}$ [7]) + O_2 $\longrightarrow$ products				
Pulse rad. of pyrazine + *t*-butanol + H_2O				74 Moo 1
KAS	H_2O	RT	$k = 1.9 \cdot 10^9\ M^{-1}\ s^{-1}$	
(Quinoxaline)$^{\dot{-}}$ [8]) + O_2 $\longrightarrow$ products				
Pulse rad. of quinoxaline + *t*-butanol + H_2O				74 Moo 1
KAS	H_2O, pH = 10.9 ... 13.6	RT	$k = 3.7 \cdot 10^8\ M^{-1}\ s^{-1}$	
(Riboflavin)$^{\dot{-}}$ [9]) + 2,6-dichloroindophenol $\longrightarrow$ products				
Pulse rad. of riboflavin + *t*-butanol + N_2 + H_2O				73 Rao 2
KAS	H_2O, pH = 10.8	RT	$k = 6.2 \cdot 10^8\ M^{-1}\ s^{-1}$	
(Riboflavin)$^{\dot{-}}$ [9]) + indigo disulfonate $\longrightarrow$ products				
Pulse rad. of riboflavin + *t*-butanol + N_2 + H_2O				73 Rao 2
KAS	H_2O, pH = 10.8	RT	$k = 5.0 \cdot 10^8\ M^{-1}\ s^{-1}$	
(Riboflavin)$^{\dot{-}}$ [9]) + methylene blue $\longrightarrow$ products				
Pulse rad. of riboflavin + *t*-butanol + N_2 + H_2O				73 Rao 2
KAS	H_2O, pH = 10.8	RT	$k = 6.2 \cdot 10^8\ M^{-1}\ s^{-1}$	
(Riboflavin)$^{\dot{-}}$ [9]) + tetramethylbenzoquinone [10]) $\longrightarrow$ products				
Pulse rad. of riboflavin + *t*-butanol + N_2 + H_2O				75 Mei 1
KAS	H_2O	RT	$k = 2.5 \cdot 10^8\ M^{-1}\ s^{-1}$	
(Riboflavin)$^{\dot{-}}$ [9]) + toluidine blue $\longrightarrow$ products				
Pulse rad. of riboflavin + *t*-butanol + N_2 + H_2O				73 Rao 2
KAS	H_2O, pH = 10.8	RT	$k = 6.0 \cdot 10^8\ M^{-1}\ s^{-1}$	
(Ribose-$\dot{O}H$)[11]) + N-ethylmaleimide $\longrightarrow$ products [12])				
Pulse rad. of ribose + H_2O + N_2O				72 Hay 1
KAS	H_2O	RT	$k = 2.1 \cdot 10^9\ M^{-1}\ s^{-1}$	
(Ribose-$\dot{O}H$)[11]) + hydroquinone $\longrightarrow$ $^-OC_6H_4\dot{O}$ + products				
Pulse rad. of ribose + N_2O + H_2O				79 Ste 1
KAS	H_2O, pH ≈ 11.5	RT	$k = 9.6(10) \cdot 10^8\ M^{-1}\ s^{-1}$	

[4]) Radicals from $C_6H_5OPO_3H_2$ + $\dot{O}H$ reaction in acid solution.
[5]) Radicals from 1,2-propanediol + $\dot{O}H$ reaction (likely to be $CH_2OH\dot{C}OHCH_3$, $\dot{C}HOHCHOHCH_3$ and/or $\dot{C}H_2COCH_3$, $CHO\dot{C}HCH_3$).
[6]) Radicals and deprotonated radicals from pterin + e_{aq}^- reaction.
[7]) Radicals from pyrazine + e_{aq}^- reaction.
[8]) Radicals from quinoxaline + e_{aq}^- reaction.
[9]) Radicals from riboflavin + e_{aq}^- reaction.
[10]) Duroquinone.
[11]) Radicals from ribose + $\dot{O}H$ reaction.
[12]) 30% e^--transfer.

Reaction Radical generation Method	Solvent	T[K]	Rate data	Ref./ add. ref.
(Ribose-$\dot{O}H$)[11]) + 2-methyl-1,4-naphthoquinone ⟶ ... semiquinone + products[13])				
Pulse rad. of ribose + H_2O + N_2O				73 Rao 3/
KAS	H_2O	RT	$k = 1.4(1) \cdot 10^9 M^{-1} s^{-1}$	72 Sim 1
(Ribose-$\dot{O}H$)[11]) + *anti*-5-nitro-2-furaldoxime[14]) ⟶ products[15])				
Pulse rad. of ribose + N_2O + H_2O				73 Gre 1
KAS	H_2O	RT	$k = 7.5 \cdot 10^8 M^{-1} s^{-1}$	
(Ribose phosphate-$\dot{O}H$)[16]) + *anti*-5-nitro-2-furaldoxime[14]) ⟶ products[17])				
Pulse rad. of ribose phosphate + N_2O + H_2O				73 Gre 1
KAS	H_2O	RT	$k = 1.7 \cdot 10^8 M^{-1} s^{-1}$	
(Sarcosine anhydride)$^{\dot{-}}$ [18]) + acetophenone ($C_6H_5COCH_3$) ⟶ products				
Pulse rad. of sarcosine anhydride + *t*-butanol + N_2 + H_2O				71 Hay 1
KAS	H_2O, pH = 5.2	RT	$k = 2.0(2) \cdot 10^9 M^{-1} s^{-1}$	
	pH = 12.4		$2.1(2) \cdot 10^9$	
(Sarcosine anhydride)$^{\dot{-}}$ [18]) + benzophenone ($(C_6H_5)_2CO$) ⟶ products				
Pulse rad. of sarcosine anhydride + *t*-butanol + N_2 + H_2O				71 Hay 1
KAS	H_2O, pH = 5.2	RT	$k = 2.3(2) \cdot 10^9 M^{-1} s^{-1}$	
	pH = 12.2		$2.4(2) \cdot 10^9 M^{-1} s^{-1}$	
(Sarcosine anhydride)$^{\dot{-}}$ [18]) + cysteine ⟶ products				
Pulse rad. of sarcosine anhydride + *t*-butanol + N_2 + H_2O				71 Hay 1
KAS	H_2O	RT	$k = 1.5(3) \cdot 10^8 M^{-1} s^{-1}$	
(Sarcosine anhydride)$^{\dot{-}}$ [19]) + 2-methyl-1,4-naphthoquinone ⟶ ... semiquinone + products				
Pulse rad. of sarcosine anhydride + H_2O				73 Rao 3
KAS	H_2O	RT	$k = 4.8(5) \cdot 10^9 M^{-1} s^{-1}$	
(Sorbitol-$\dot{O}H$)[20]) + hydroquinone ⟶ $^{-}OC_6H_4\dot{O}$ + products				
Pulse rad. of sorbitol + N_2O + H_2O				79 Ste 1
KAS	H_2O, pH ≈ 11.5	RT	$k = 9.8(13) \cdot 10^8 M^{-1} s^{-1}$	
(Sucrose)$^{\cdot}$ [21]) + tetranitromethane ($C(NO_2)_4$) —— $C(NO_2)_3^-$ + NO_2 + H^+ + products				
Pulse rad. of sucrose + N_2O + H_2O				65 Asm 1
KAS	H_2O with	RT		
	2 wt % sucrose		$k = 6.0 \cdot 10^8 M^{-1} s^{-1}$	
	20		$4.4 \cdot 10^8$	
	35		$3.35 \cdot 10^8$	
	40		$3.05 \cdot 10^8$	
	50		$2.45 \cdot 10^8$	
	60		$1.15 \cdot 10^8$	

[11]) Radicals from ribose + $\dot{O}H$ reaction.
[13]) 60% e^--transfer.
[14]) Nifuroxime.
[15]) 50% e^--transfer.
[16]) Radicals from ribose phosphate + $\dot{O}H$ reaction.
[17]) 10% e^--transfer.
[18]) Radicals from sarcosine anhydride + e_{aq}^- reaction, most likely $CH_2\dot{C}O^-N(CH_3)CH_2CON(CH_3)$.
[19]) Radicals from sarcosine anhydride + e_{aq}^- reaction.
[20]) Radicals from sorbitol + $\dot{O}H$ reaction.
[21]) Radicals from sucrose + $\dot{O}H$ reaction and direct energy absorption by sucrose.

Reaction Radical generation Method	Solvent	T[K]	Rate data	Ref./ add. ref.
(Sucrose-$\dot{O}H$) 22) + tetranitromethane ($C(NO_2)_4$) ⟶ $C(NO_2)_3^-$ + NO_2 + H^+ + products				
Pulse rad. of sucrose + N_2O + H_2O				64 Asm 1,
KAS	H_2O	RT	$k = 7.0(10) \cdot 10^8\ M^{-1}s^{-1}$ $8.5(10) \cdot 10^8$ 23)	65 Rab 1
(Thymidine)$^{\dot{-}}$ 24) + 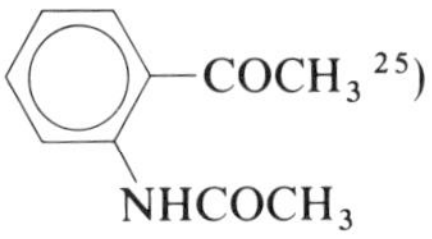 25) ⟶ products				
Pulse rad. of thymidine + t-butanol + H_2O				77 Pil 1
KAS	H_2O	RT	$k = 7.3 \cdot 10^9\ M^{-1}s^{-1} (\pm 10\%)$	
(Thymidine)$^{\dot{-}}$ 24) + 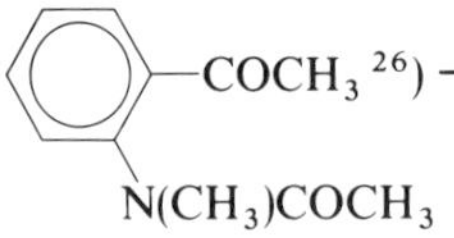 26) ⟶ products				
Pulse rad. of thymidine + t-butanol + H_2O				77 Pil 1
KAS	H_2O	RT	$k = 4.7 \cdot 10^9\ M^{-1}s^{-1} (\pm 10\%)$	
(Thymidine)$^{\dot{-}}$ 24) + 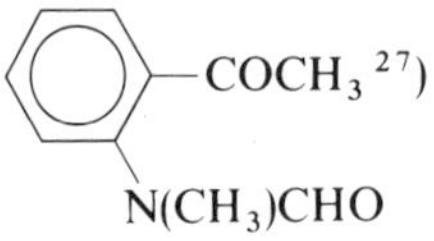 27) ⟶ products				
Pulse rad. of thymidine + t-butanol + H_2O				77 Pil 1
KAS	H_2O	RT	$k = 2.8 \cdot 10^9\ M^{-1}s^{-1} (\pm 10\%)$	
(Thymidine)$^{\dot{-}}$ 24) + 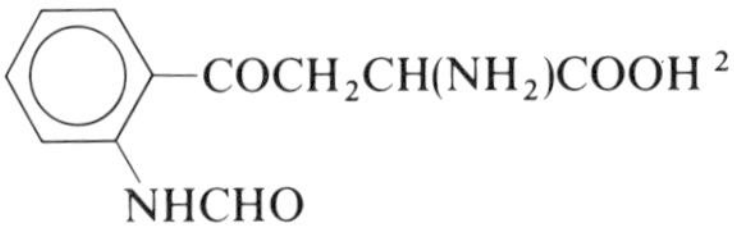 28) ⟶ products				
Pulse rad. of thymidine + t-butanol + H_2O				77 Pil 1
KAS	H_2O	RT	$k = 2.3 \cdot 10^9\ M^{-1}s^{-1} (\pm 10\%)$	
(Thymidine)$^{\dot{-}}$ 24) + 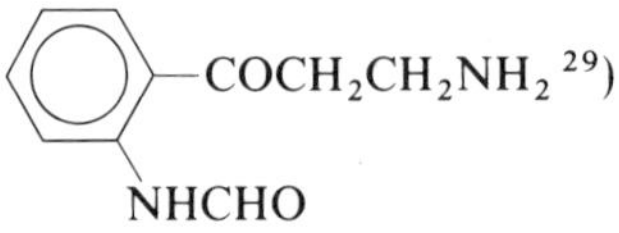 29) ⟶ products				
Pulse rad. of thymidine + t-butanol + H_2O				77 Pil 1
KAS	H_2O	RT	$k = 5.2 \cdot 10^9\ M^{-1}s^{-1} (\pm 10\%)$	
(Thymidine)$^{\dot{-}}$ 24) + benzophenone ⟶ $(C_6H_5)_2\dot{C}O^-$ + products				
Pulse rad. of thymidine + t-butanol + N_2 + H_2O				72 Ada 1
KAS	H_2O, pH = 12	RT	$k = 2.6 \cdot 10^9\ M^{-1}s^{-1}$	

Figure text for the structures above:

---

22) Radicals from sucrose + $\dot{O}H$ reaction.
23) From [65 Rab 1].
24) Radicals from thymidine + e_{aq}^- reaction.
25) N-(2-acetylphenyl)acetamide.
26) N-(2-acetylphenyl)-N-methylacetamide.
27) N-(2-acetylphenyl)-N-methylformamide.
28) N'-formylkynurenine; 2-amino-4-[2-(formylamino)phenyl]-4-oxobutanoic acid.
29) N-[2-(3-amino-1-oxopropyl)phenyl]formamide.

Reaction Radical generation Method	Solvent	T[K]	Rate data	Ref./ add. ref.
(Thymidine)$^{\cdot-}$ [24]) + 5-bromodeoxyuridine ⟶ products				
Pulse rad. of thymidine + t-butanol + N_2 + H_2O				72 Ada 2
PR, KAS	H_2O	RT	$k = 3.2 \cdot 10^8\ M^{-1}s^{-1}$ [30])	
(Thymidine)$^{\cdot-}$ [24]) + 5-bromouracil ⟶ products				
Pulse rad. of thymidine + t-butanol + N_2 + H_2O				72 Ada 2
PR, KAS	H_2O	RT	$k = 7.2 \cdot 10^8\ M^{-1}s^{-1}$ [30])	
(Thymidine)$^{\cdot-}$ [24]) + 6-carboxyuracil [31]) ⟶ products				
Pulse rad. of thymidine + t-butanol + N_2 + H_2O				72 Ada 1
KAS	H_2O	RT	$k = 9 \cdot 10^8\ M^{-1}s^{-1}$	
(Thymidine)$^{\cdot-}$ [24]) + 8-methoxypsoralen products				
Pulse rad. of thymidine + t-butanol + H_2O				78 Red 1
KAS	H_2O	RT	$k = 2.3 \cdot 10^9\ M^{-1}s^{-1}$	
(Thymidine)$^{\cdot-}$ [24]) + 2-methyl-1,4-naphthoquinone ⟶ 2-methyl...semiquinone + products				
Pulse rad. of thymidine + t-butanol + N_2 + H_2O				72 Ada 1
KAS	H_2O	RT	$k = 3.2 \cdot 10^9\ M^{-1}s^{-1}$	
(Thymidine)$^{\cdot-}$ [24]) + 4-nitroacetophenone(PNAP) ⟶ PNAP$^{\cdot-}$ + products				
Pulse rad. of thymidine + t-butanol + N_2 + H_2O				72 Ada 1,
KAS	H_2O, pH = 7	RT	$k = 3.7 \cdot 10^9\ M^{-1}s^{-1}$	75 Whi 2
	pH = 6.5(5)		$4.8(2) \cdot 10^9$ [32])	
	pH = 12		$4.1 \cdot 10^9$	
(Thymidine-5'-monophosphate)$^{\cdot-}$ [33]) + benzophenone ⟶ products				
Pulse rad. of thymidine-5'-monophosphate + t-butanol + N_2 + H_2O				72 Ada 1
KAS	H_2O, pH = 12	RT	$k = 1.8 \cdot 10^9\ M^{-1}s^{-1}$	
(Thymidine-5'-monophosphate)$^{\cdot-}$ [33]) + 6-carboxyuracil [31]) ⟶ products				
Pulse rad. of thymidine-5'-monophosphate + t-butanol + N_2 + H_2O				72 Ada 1
KAS	H_2O	RT	$k = 7 \cdot 10^8\ M^{-1}s^{-1}$	
(Thymidine-5'-monophosphate)$^{\cdot-}$ [33]) + 2-methyl-1,4-naphthoquinone ⟶ 2-methyl...semiquinone + products				
Pulse rad. of thymidine-5'-monophosphate + t-butanol + N_2 + H_2O				72 Ada 1
KAS	H_2O	RT	$k = 3.9(10) \cdot 10^9\ M^{-1}s^{-1}$	
(Thymidine-5'-monophosphate)$^{\cdot-}$ [33]) + 4-nitroacetophenone(PNAP) ⟶ PNAP$^{\cdot-}$ + products				
Pulse rad. of thymidine-5'-monophosphate + t-butanol + N_2 + H_2O				75 Whi 2,
KAS	H_2O, pH = 6.5(5)	RT	$k = 5.0(3) \cdot 10^9\ M^{-1}s^{-1}$	72 Ada 1
	pH = 7		$4.2 \cdot 10^9$ [34])	
	pH = 12		$3.8 \cdot 10^9$ [34])	

[24]) Radicals from thymidine + e_{aq}^- reaction.
[30]) Based on k[(thymidine)$^{\cdot-}$ + PNAP] = $3.7 \cdot 10^9\ M^{-1}s^{-1}$.
[31]) Orotic acid.
[32]) From [75 Whi 2].
[33]) Radicals from thymidine-5'-monophosphate + e_{aq}^- reaction.
[34]) From [72 Ada 1].

Reaction Radical generation Method	Solvent	T[K]	Rate data	Ref./ add. ref.
(Thymidine-5'-monophosphate-$\dot{O}H$)[35] + 2-methyl-1,4-naphthoquinone ⟶ 2-methyl…semiquinone + products				
Pulse rad. of thymidine-5'-monophosphate + N_2O + H_2O				73 Hay 3
KAS	H_2O, pH = 7	RT	$k = 5.0 \cdot 10^9\ M^{-1}\ s^{-1}$ [36]	
	pH = 11		$1.0 \cdot 10^9$ [37]	
(Thymidine-5'-monophosphate-$\dot{O}H$)[35] + 4-nitroacetophenone(PNAP) ⟶ $PNAP^{\dot{-}}$ + products[38]				
Pulse rad. of thymidine-5'-monophosphate + N_2O + H_2O				75 Whi 2
KAS	H_2O	RT	$k = 1.3(3) \cdot 10^9\ M^{-1}\ s^{-1}$	
(Thymidine-5'-monophosphate-$\dot{O}H$)[35] + nor-pseudopelletierine-N-oxy ⟶ products[39]				
Rad. of thymidine-5'-monophosphate + N_2O + H_2O				71 Rob 1
KAS	H_2O	RT	$k = 3.7(4) \cdot 10^8\ M^{-1}\ s^{-1}$	
(Thymidine-$\dot{O}H$)[39a] + 8-methoxypsoralen ⟶ products[40]				
Pulse rad. of thymidine + N_2O + H_2O				78 Red 1
KAS	H_2O	RT	$k = 1.3 \cdot 10^9\ M^{-1}\ s^{-1}$	
$(Thymine)^{\dot{-}}$[41] + 2-amino-5-nitrothiazole ⟶ products				
Pulse rad. of thymine + t-butanol + N_2 + H_2O				76 Gre 1
KAS	H_2O	RT	$k = 2.0 \cdot 10^9\ M^{-1}\ s^{-1}$	
$(Thymine)^{\dot{-}}$[41] + benzophenone ⟶ $(C_6H_5)_2\dot{C}O^-$ + products				
Pulse rad. of thymine + t-butanol + N_2 + H_2O				72 Ada 1
KAS	H_2O, pH = 12	RT	$k = 3.75 \cdot 10^9\ M^{-1}\ s^{-1}$	
$(Thymine)^{\dot{-}}$[41] + 1,4-benzoquinone ⟶ 1,4-benzosemiquinone + products				
Pulse rad. of thymine + t-butanol + N_2 + H_2O				74 Rao 1,
KAS	H_2O, pH = 5.4	RT	$k = 4.8 \cdot 10^9\ M^{-1}\ s^{-1}$ [42]	71 Wil 1
	neutral solution		$6.0 \cdot 10^9$ [43]	
$(Thymine)^{\dot{-}}$[41] + 5-bromodeoxyuridine ⟶ products				
Pulse rad. of thymine + t-butanol + N_2 + H_2O				72 Ada 2
PR, KAS	H_2O	RT	$k = 8.4 \cdot 10^8\ M^{-1}\ s^{-1}$ [44]	
$(Thymine)^{\dot{-}}$[41] + 2-bromo-5-nitrothiazole ⟶ products				
Pulse rad. of thymine + t-butanol + N_2 + H_2O				76 Gre 1
KAS	H_2O	RT	$k = 4.9 \cdot 10^9\ M^{-1}\ s^{-1}$	
$(Thymine)^{\dot{-}}$[41] + 5-bromouracil ⟶ products				
Pulse rad. of thymine + t-butanol + N_2 + H_2O				72 Ada 2
PR, KAS	H_2O	RT	$k = 1.1 \cdot 10^9\ M^{-1}\ s^{-1}$ [44]	
$(Thymine)^{\dot{-}}$[41] + 6-carboxyuracil[45] ⟶ products				
Pulse rad. of thymine + t-butanol + N_2 + H_2O				72 Ada 1
KAS	H_2O	RT	$k = 1.5 \cdot 10^9\ M^{-1}\ s^{-1}$	

35) Radicals from thymidine-5'-monophosphate + $\dot{O}H$ reaction.
36) 22% e^--transfer.
37) 40% e^--transfer.
38) 7% e^--transfer.
39) Addition and possibly some e^--transfer.
39a) Radicals from thymidine + $\dot{O}H$ reaction.
40) 30% e^--transfer.
41) Radicals from thymine + e^-_{aq} reaction.
42) 70% e^--transfer [74 Rao 1].
43) From [71 Wil 1].
44) Relative to k(thymine$^{\dot{-}}$ + PNAP) = $4.8 \cdot 10^9\ M^{-1}\ s^{-1}$.
45) Orotic acid.

Reaction Radical generation Method	Solvent	T[K]	Rate data	Ref./ add. ref.
$(Thymine)^{\overline{\cdot}}$ 41) + crystal violet ⟶ products				
Pulse rad. of thymine + t-butanol + N_2 + H_2O				73 Rao 2
KAS	H_2O, pH = 5.4	RT	$k = 2.4 \cdot 10^9\ M^{-1}\,s^{-1}$	
$(Thymine)^{\overline{\cdot}}$ 41) + diazenedicarboxylic acid bis-dimethylamide ⟶ products				
Pulse rad. of thymine + t-butanol + N_2 + H_2O				75 Whi 1
KAS	H_2O	RT	$k = 5.0(1) \cdot 10^9\ M^{-1}\,s^{-1}$	
$(Thymine)^{\overline{\cdot}}$ 41) + dimethylfumarate ⟶ products				
Pulse rad. of thymine + t-butanol + N_2 + H_2O				73 Hay 2
KAS	H_2O	RT	$k = 5 \cdot 10^9\ M^{-1}\,s^{-1}$	
$(Thymine)^{\overline{\cdot}}$ 41) + 1-(2-hydroxyethyl)-2-methyl-5-nitroimidazole (metronidazole) ⟶ products				
Pulse rad. of thymine + t-butanol + N_2 + H_2O				74 Wil 2, 75 Whi 3
KAS	H_2O	RT	$k \approx 2.0 \cdot 10^9\ M^{-1}\,s^{-1}$ $3.1(3) \cdot 10^9$ 46)	
$(Thymine)^{\overline{\cdot}}$ 41) + 8-methoxypsoralen ⟶ products				
Pulse rad. of thymine + t-butanol + H_2O				78 Red 1
KAS	H_2O	RT	$k = 3.1 \cdot 10^9\ M^{-1}\,s^{-1}$	
$(Thymine)^{\overline{\cdot}}$ 41) + 2-methyl-1,4-naphthoquinone ⟶ 2-methyl...semiquinone + products				
Pulse rad. of thymine + t-butanol + N_2 + H_2O				74 Rao 1, 72 Ada 1
KAS	H_2O, pH = 5.4 pH = 7 pH = 9.4	RT	$k = 4.6 \cdot 10^9\ M^{-1}\,s^{-1}$ 47) $4.0 \cdot 10^9$ 48) $4.1 \cdot 10^9$ 49)	
$(Thymine)^{\overline{\cdot}}$ 41) + 2-methyl-5-nitroimidazol ⟶ products				
Pulse rad. of thymine + t-butanol + N_2 + H_2O				76 Gre 1
KAS	H_2O	RT	$k = 4.0 \cdot 10^9\ M^{-1}\,s^{-1}$	
$(Thymine)^{\overline{\cdot}}$ 41) + 4-nitroacetophenone(PNAP) ⟶ $PNAP^{\overline{\cdot}}$ + products				
Pulse rad. of thymine + t-butanol + N_2 + H_2O				72 Ada 1, 75 Whi 2
KAS	H_2O, pH = 7 pH = 6.5(5) pH = 12	RT	$k = 4.8 \cdot 10^9\ M^{-1}\,s^{-1}$ $5.5 \cdot 10^9$ 50) $5.0 \cdot 10^9$	
$(Thymine)^{\overline{\cdot}}$ 41) + *anti*-5-nitro-2-furaldoxime ⟶ products				
Pulse rad. of thymine + t-butanol + N_2 + H_2O				73 Gre 1, 76 Gre 1
KAS	H_2O	RT	$k = 6.4 \cdot 10^9\ M^{-1}\,s^{-1}$ $5.5 \cdot 10^9$ 51)	
$(Thymine)^{\overline{\cdot}}$ 41) + 2-nitroimidazole ⟶ products				
Pulse rad. of thymine + t-butanol + N_2 + H_2O				76 Gre 1
KAS	H_2O	RT	$k = 4.0 \cdot 10^9\ M^{-1}\,s^{-1}$	
$(Thymine)^{\overline{\cdot}}$ 41) + 4-nitroimidazole ⟶ products				
Pulse rad. of thymine + t-butanol + N_2 + H_2O				76 Gre 1
KAS	H_2O	RT	$k = 4.0 \cdot 10^9\ M^{-1}\,s^{-1}$	

41) Radicals from thymine + e_{aq}^- reaction.
46) From [75 Whi 3].
47) 60% e^--transfer [74 Rao 1].
48) From [72 Ada 1].
49) From [74 Rao 1].
50) From [75 Whi 2].
51) From [76 Gre 1].

Reaction Radical generation Method	Solvent	T[K]	Rate data	Ref./ add. ref.
(Thymine)$^{\bar{\cdot}}$ [41]) + 1-(2-nitro-1-imidazolyl)-3-methoxy-2-propanol (misonidazole) ⟶ products				
Pulse rad. of thymine + t-butanol + N_2 + H_2O				75 Whi 3
KAS	H_2O	RT	$k = 3.1(3) \cdot 10^9\ M^{-1}s^{-1}$	
(Thymine)$^{\bar{\cdot}}$ [41]) + 2-nitropyrrole ⟶ products				
Pulse rad. of thymine + t-butanol + N_2 + H_2O				76 Gre 1
KAS	H_2O	RT	$k = 3.0 \cdot 10^9\ M^{-1}s^{-1}$	
(Thymine)$^{\bar{\cdot}}$ [41]) + 3-nitropyrrole ⟶ products				
Pulse rad. of thymine + t-butanol + N_2 + H_2O				76 Gre 1
KAS	H_2O	RT	$k = 3.5 \cdot 10^9\ M^{-1}s^{-1}$	
(Thymine)$^{\bar{\cdot}}$ [41]) + 2-nitrothiophene products				
Pulse rad. of thymine + t-butanol + N_2 + H_2O				76 Gre 1
KAS	H_2O	RT	$k = 3.5 \cdot 10^9\ M^{-1}s^{-1}$	
(Thymine)$^{\bar{\cdot}}$ [41]) + 3-nitrothiophene ⟶ products				
Pulse rad. of thymine + t-butanol + N_2 + H_2O				76 Gre 1
KAS	H_2O	RT	$k = 2.5 \cdot 10^9\ M^{-1}s^{-1}$	
(Thymine)$^{\bar{\cdot}}$ [41]) + 2,2,6,6-tetramethyl-4-oxo-1-piperidinyloxy(TAN) ⟶ products [52])				
Pulse rad. of thymine + t-butanol + N_2 + H_2O				71 Wil 2
PR, KAS	H_2O	RT	$k = 1.0 \cdot 10^9\ M^{-1}s^{-1}$	
(Thymine-$\dot{O}$H) [53]) + 1,4-benzoquinone ⟶ products [54])				
Pulse rad. of thymine + N_2O + H_2O				74 Rao 1
KAS	H_2O	RT	$k = 3.8 \cdot 10^9\ M^{-1}s^{-1}$	
(Thymine-$\dot{O}$H) [53]) + crystal violet ⟶ products				
Pulse rad. of thymine + N_2O + H_2O				73 Rao 2
KAS	H_2O, pH = 10.8	RT	$k = 1.3 \cdot 10^9\ M^{-1}s^{-1}$	
(Thymine-$\dot{O}$H) [53]) + fluorescein ⟶ products				
Pulse rad. of thymine + N_2O + H_2O				73 Rao 2
KAS	H_2O, pH = 10.8	RT	$k = 5.7 \cdot 10^8\ M^{-1}s^{-1}$	
(Thymine-$\dot{O}$H) [53]) + 4-hydroxy-2,2,6,6-tetramethylpiperidino-1-oxy ⟶ products [55])				
Pulse rad. of thymine + N_2O + H_2O				72 Bru 1, 71 Emm 1
KAS	H_2O	RT	$k = 2.3(2) \cdot 10^8\ M^{-1}s^{-1}$ $2.6 \cdot 10^8$ [56])	
(Thymine-$\dot{O}$H) [53]) + indophenol ⟶ products				
Pulse rad. of thymine + N_2O + H_2O				73 Rao 2
KAS	H_2O, pH = 10.8	RT	$k = 2.0 \cdot 10^9\ M^{-1}s^{-1}$	
(Thymine-$\dot{O}$H) [53]) + 8-methoxypsoralen ⟶ products [57])				
Pulse rad. of thymine + N_2O + H_2O				78 Red 1
KAS	H_2O	RT	$k = 1.0 \cdot 10^9\ M^{-1}s^{-1}$	

[41]) Radicals from thymine + e_{aq}^- reaction.
[52]) Electron transfer and/or addition.
[53]) Radicals from thymine + $\dot{O}$H reaction.
[54]) 40% e^--transfer.
[55]) Addition and/or e^--transfer.
[56]) From [71 Emm 1].
[57]) 23% e^--transfer.

Reaction Radical generation Method	Solvent	T[K]	Rate data	Ref./ add. ref.
(Thymine-$\dot{O}$H) [53]) + methylene blue ⟶ products				
Pulse rad. of thymine + N_2O + H_2O				74 Rao 1
KAS	H_2O	RT	$k = 4.0 \cdot 10^9\ M^{-1}\ s^{-1}$	
(Thymine-$\dot{O}$H) [53]) + 2-methyl-1,4-naphthoquinone ⟶ products				
Pulse radiolysis of thymine + N_2O + H_2O				74 Rao 1,
KAS	H_2O, pH = 7 pH = 10.8...11	RT	$k = 3.9 \cdot 10^9\ M^{-1}\ s^{-1}$ [58]) $5.2 \cdot 10^9$ [59])	73 Hay 3
(Thymine-$\dot{O}$H) [53]) + 4-nitroacetophenone ⟶ products				
Pulse rad. of thymine + N_2O + H_2O				75 Whi 2
KAS	H_2O	RT	$k \leqslant 10^7\ M^{-1}\ s^{-1}$	
(Thymine-$\dot{O}$H) [53]) + nor-pseudopelletierine-N-oxy ⟶ products [55])				
Pulse rad. of thymine + N_2O + H_2O				71 Rob 1/
KAS	H_2O	RT	$k = 6.2(6) \cdot 10^8\ M^{-1}\ s^{-1}$	71 Emm 1, 72 Bru 1
(Thymine-$\dot{O}$H) [53]) + 2,2,6,6-tetramethyl-4-oxo-1-piperidinyloxy(TAN) ⟶ products [55])				
Pulse rad. of thymine + N_2O + H_2O				72 Bru 1,
KAS	H_2O	RT	$k = 3.2(2) \cdot 10^8\ M^{-1}\ s^{-1}$ $3.5 \cdot 10^8$ [56])	71 Emm 1
(Triglycine)$^{\dot{-}}$ [60]) + lipoic acid ⟶ products				
Pulse rad. of triglycine + t-butanol + N_2 + H_2O				75 Far 1
KAS	H_2O	RT	$k = 10^8 \ldots 10^9\ M^{-1}\ s^{-1}$	
(Triglycine)$^{\cdot}$ [61]) [62]) + lipoate (—S—S—) (lipoate)$^{\dot{-}}$ (—S$\overset{(-)}{\therefore}$S—) + products				
Pulse rad. of triglycine + t-butanol + H_2O				75 Far 3
KAS	H_2O	RT	$k = 2.5 \cdot 10^7\ M^{-1}\ s^{-1}$ [61]) $5 \cdot 10^6$ [62])	
(Triglycine-Cu(II))$^{\dot{-}}$ [63]) ⟶ Cu(I)-triglycine [64])				
Pulse rad. of Cu(II)-triglycine + t-butanol + N_2 + H_2O				76 Far 1
KAS	H_2O, pH = 9	RT	$k = 1.7(3) \cdot 10^4\ s^{-1}$	
(Triglycine-$\dot{O}$H) [65]) + 9,10-anthraquinone-2,6-disulfonate ion ⟶ ...semiquinone + products				
Pulse rad. of triglycine + H_2O + N_2O				73 Rao 1
KAS	H_2O, pH = 10	RT	$k = 1.8(2) \cdot 10^9\ M^{-1}\ s^{-1}$	
(Triglycine-$\dot{O}$H) [65]) + 9,10-anthraquinone-2-sulfonate ion ⟶ ...semiquinone + products				
Pulse rad. of triglycine + H_2O + N_2O				73 Rao 1
KAS	H_2O, pH = 10	RT	$k = 1.4(1) \cdot 10^9\ M^{-1}\ s^{-1}$	
(Triglycine-$\dot{O}$H) [65]) + 1,4-benzoquinone ⟶ 1,4-benzosemiquinone + products				
Pulse rad. of triglycine + H_2O + N_2O				73 Rao 1
KAS	H_2O, pH = 10	RT	$k = 2.5(2) \cdot 10^9\ M^{-1}\ s^{-1}$	

[53]) Radicals from thymine + $\dot{O}$H reaction.
[55]) Addition and/or e^--transfer.
[56]) From [71 Emm 1].
[58]) 12% e^--transfer [73 Hay 3].
[59]) 41...45% e^--transfer [74 Rao 1, 73 Hay 3].
[60]) Radicals from triglycine + e_{aq}^- reaction.
[61]) Electron adduct to carboyl group presumed.
[62]) Deaminated radical from triglycine + e_{aq}^- reaction.
[63]) Radicals from triglycine-Cu(II) + e_{aq}^- reaction, e^--addition to carbonyl group of peptide presumed.
[64]) *Intra*-complex e^--transfer.
[65]) Radicals from triglycine + $\dot{O}$H reaction.

Reaction Radical generation Method	Solvent	T[K]	Rate data	Ref./ add. ref.
(Triglycine-$\dot{O}H$)[65]) + 2-hydroxy-1,4-naphthoquinone ⟶ ...semiquinone + products				
Pulse rad. of triglycine + H_2O + N_2O				73 Rao 1
KAS	H_2O, pH = 10	RT	$k = 1.9 \cdot 10^9\ M^{-1}s^{-1}$	
(Triglycine-$\dot{O}H$)[65]) + 2-methyl-1,4-naphthoquinone ⟶ ...semiquinone + products[66])				
Pulse rad. of triglycine + H_2O + N_2O				73 Rao 3,
KAS	H_2O	RT	$k = 1.8(2) \cdot 10^9\ M^{-1}s^{-1}$	73 Rao 1
(Triglycine-$\dot{O}H$)[65]) + 1,4-naphthoquinone-2-sulfonate ion ⟶ ...semiquinone + products				
Pulse rad. of triglycine + H_2O + N_2O				73 Rao 1
KAS	H_2O, pH = 10	RT	$k = 2.0(2) \cdot 10^9\ M^{-1}s^{-1}$	
(Tryptophane amide)$^{\dot{-}}$[67]) + 2-methyl-1,4-naphthoquinone ⟶ 2-methyl...semiquinone + products[68])				
Pulse rad. of tryptophane amide + *t*-butanol + N_2 + H_2O				73 Rao 3
KAS	H_2O	RT	$k = 2.0 \cdot 10^9\ M^{-1}s^{-1}$	
(Tryptophane-$\dot{O}H$)[69]) + 2-methyl-1,4-naphthoquinone ⟶ 2-methyl...semiquinone + products[70])				
Pulse rad. of tryptophane + N_2O + H_2O				73 Rao 3
KAS	H_2O	RT	$k = 2.8 \cdot 10^9\ M^{-1}s^{-1}$	
(Tryptophane-$\dot{O}H$)[69]) + *anti*-5-nitro-2-furaldoxime ⟶ products[70])				
Pulse rad. of tryptophane + N_2O + H_2O				73 Gre 1
KAS	H_2O	RT	$k = 2.9 \cdot 10^9\ M^{-1}s^{-1}$	
(Tryptophane-$\dot{O}H$)$^+$[71]) + promethazine(PZ) ⟶ $PZ^{\dot{+}}$ + products				
Pulse rad. of tryptophane + N_2O + H_2O				79 Asm 1
KAS	H_2O	RT	$k = 7 \cdot 10^8\ M^{-1}s^{-1}$	
(Tyrosine)$^{\dot{-}}$[72]) + *anti*-5-nitro-2-furaldoxime ⟶ products				
Pulse rad. of tyrosine + *t*-butanol + N_2 + H_2O				73 Gre 1
KAS	H_2O	RT	$k = 5.0 \cdot 10^9\ M^{-1}s^{-1}$	
(Tyrosine amide)$^{\dot{-}}$[73]) + 2-methyl-1,4-naphthoquinone ⟶ products[74])				
Pulse rad. of tyrosine amide + *t*-butanol + N_2 + H_2O				73 Rao 3
KAS	H_2O	RT	$k = 1.5 \cdot 10^9\ M^{-1}s^{-1}$	
(Tyrosine-$\dot{O}H$)[75]) + 2-methyl-1,4-naphthoquinone ⟶ products[76])				
Pulse rad. of tyrosine + N_2O + H_2O				73 Rao 3
KAS	H_2O, pH = 7.7...11.2	RT	$k \approx 4.0 \cdot 10^9\ M^{-1}s^{-1}$	
(Tyrosine-$\dot{O}H$)[75]) + *anti*-5-nitro-2-furaldoxime ⟶ products[76a])				
Pulse rad. of tyrosine + N_2O + H_2O				73 Gre 1
KAS	H_2O	RT	$k = 3.0 \cdot 10^9\ M^{-1}s^{-1}$	

[65]) Radicals from triglycine + $\dot{O}H$ reaction.
[66]) 11% e^--transfer at pH = 7 [73 Rao 3]; 77% e^--transfer at pH = 12 [73 Rao 1].
[67]) Radicals from tryptophane amide + e^-_{aq} reaction.
[68]) 70% e^--transfer.
[69]) Radicals from tryptophane + $\dot{O}H$ reaction.
[70]) 20% e^--transfer.
[71]) Radical cations formed from $\dot{O}H$ + tryptophane reaction.
[72]) Radicals from tyrosine + e^-_{aq} reaction.
[73]) Radicals from tyrosine amide + e^-_{aq} reaction.
[74]) 38% e^--transfer.
[75]) Radicals from tyrosine + $\dot{O}H$ reaction.
[76]) 32...37% e^--transfer.
[76a]) 12% e^--transfer.

Reaction Radical generation Method	Solvent	T[K]	Rate data	Ref./ add. ref.
Trp˙—TyrOH [77]) ⟶ TrpH—TyrȮ [78])				
Pulse rad. of N_3^- + *L*-tryptophyl-*L*-tyrosine + N_2O + H_2O				79 Pru 1
KAS	H_2O	RT	$k = 7.3 \cdot 10^4\,s^{-1}$	
TyrOH—Trp˙ [79]) ⟶ TyrȮ—TrpH [78])				
Pulse rad. of N_3^- + *L*-tyrosyl-*L*-tryptophane + N_2O + H_2O				79 Pru 1
KAS	H_2O	RT	$k = 5.4 \cdot 10^4\,s^{-1}$	
(Uracil)$^{\dot{-}}$ [80]) + benzophenone ⟶ $(C_6H_5)_2\dot{C}O^-$ + products				
Pulse rad. of uracil + *t*-butanol + N_2 + H_2O				72 Ada 1
KAS	H_2O, pH = 12	RT	$k = 3.0 \cdot 10^9\,M^{-1}s^{-1}$	
(Uracil)$^{\dot{-}}$ [80]) + 1,4-benzoquinone ⟶ 1,4-benzosemiquinone + products				
Pulse rad. of uracil + *t*-butanol + N_2 + H_2O				74 Rao 1
KAS	H_2O, pH = 5.4	RT	$k = 2.8 \cdot 10^9\,M^{-1}s^{-1}$ [81])	
	pH = 8		$3.0 \cdot 10^9$	
(Uracil)$^{\dot{-}}$ [80]) + 5-bromouracil ⟶ products				
Pulse rad. of uracil + *t*-butanol + N_2 + H_2O				72 Ada 2
PR, KAS	H_2O	RT	$k = 6.8 \cdot 10^8\,M^{-1}s^{-1}$ [82])	
(Uracil)$^{\dot{-}}$ [80]) + 6-carboxyuracil [83]) ⟶ products				
Pulse rad. of uracil + *t*-butanol + N_2 + H_2O				72 Ada 1
KAS	H_2O	RT	$k = 1.75 \cdot 10^9\,M^{-1}s^{-1}$	
(Uracil)$^{\dot{-}}$ [80]) + 8-methoxypsoralen ⟶ products				
Pulse rad. of uracil + *t*-butanol + H_2O				78 Red 1
KAS	H_2O	RT	$k = 3.6 \cdot 10^9\,M^{-1}s^{-1}$	
(Uracil)$^{\dot{-}}$ [80]) + 2-methyl-1,4-naphthoquinone ⟶ 2-methyl...semiquinone + products				
Pulse rad. of uracil + *t*-butanol + N_2 + H_2O				74 Rao 1,
KAS	H_2O, pH = 5.4	RT	$k = 2.9 \cdot 10^9\,M^{-1}s^{-1}$ [84])	72 Ada 1
	pH = 7		$3.6 \cdot 10^9$ [85])	
(Uracil)$^{\dot{-}}$ [80]) + 4-nitroacetophenone(PNAP) ⟶ PNAP$^{\dot{-}}$ + products				
Pulse rad. of uracil + *t*-butanol + N_2 + H_2O				72 Ada 1
KAS	H_2O, pH = 7	RT	$k = 5.2 \cdot 10^9\,M^{-1}s^{-1}$	
	pH = 12		$5.5 \cdot 10^9$	
(Uracil)$^{\dot{-}}$ [80]) + *anti*-5-nitro-2-furaldoxime ⟶ products				
Pulse rad. of uracil + *t*-butanol + N_2 + H_2O				73 Gre 1
KAS	H_2O	RT	$k = 6.0 \cdot 10^9\,M^{-1}s^{-1}$	
(Uracil-ȮH) [86]) + 1,4-benzoquinone ⟶ products [87])				
Pulse rad. of uracil + N_2O + H_2O				74 Rao 1
KAS	H_2O	RT	$k = 4.0 \cdot 10^9\,M^{-1}s^{-1}$	

[77]) Radicals from $\dot{N}_3$ + L-tryptophyl-L-tyrosine (TrpH-TyrOH) reaction.
[78]) *Intra*-molecular e^--transfer.
[79]) Radicals from $\dot{N}_3$ + L-tyrosyl-L-tryptophane (TyrOH-TrpH) reaction.
[80]) Radicals from uracil + e_{aq}^- reaction.
[81]) 65% e^--transfer.
[82]) Relative to k(uracil$^{\dot{-}}$ + PNAP) = $5.2 \cdot 10^9\,M^{-1}s^{-1}$.
[83]) Orotic acid.
[84]) 55% e^--transfer [74 Rao 1].
[85]) From [72 Ada 1].
[86]) Radicals from uracil + ȮH reaction.
[87]) 60% e^--transfer.

Reaction Radical generation Method	Solvent	T[K]	Rate data	Ref./ add. ref.
(Uracil-$\dot{O}$H)[86]) + crystal violet ⟶ products				
Pulse rad. of uracil + N_2O + H_2O				73 Rao 2
KAS	H_2O, pH = 10.8	RT	$k = 1.6 \cdot 10^9\ M^{-1}\,s^{-1}$	
(Uracil-$\dot{O}$H)[86]) + fluorescein ⟶ products				
Pulse rad. of uracil + N_2O + H_2O				73 Rao 2
KAS	H_2O, pH = 10.8	RT	$k = 6.0 \cdot 10^8\ M^{-1}\,s^{-1}$	
(Uracil-$\dot{O}$H)[86]) + 8-methoxypsoralen ⟶ products[88])				
Pulse rad. of uracil + N_2O + H_2O				78 Red 1
KAS	H_2O	RT	$k = 9 \cdot 10^8\ M^{-1}\,s^{-1}$	
(Uracil-$\dot{O}$H)[86]) + methylene blue ⟶ products[89])				
Pulse rad. of uracil + N_2O + H_2O				74 Rao 1
KAS	H_2O, pH = 10.8	RT	$k = 4.7 \cdot 10^9\ M^{-1}\,s^{-1}$	
(Uracil-$\dot{O}$H)[86]) + 2-methyl-1,4-naphthoquinone ⟶ products				
Pulse rad. of uracil + N_2O + H_2O				76 Mic 1,
KAS	H_2O	RT	$k = 4.2 \cdot 10^9\ M^{-1}\,s^{-1}$ [90])	74 Rao 1, 72 Sim 1
(Uridine)$^{\dot{-}}$ [91]) + 6-carboxyuracil[92]) ⟶ products				
Pulse rad. of uridine + t-butanol + N_2 + H_2O				72 Ada 1
KAS	H_2O	RT	$k = 7.5 \cdot 10^8\ M^{-1}\,s^{-1}$	
(Uridine)$^{\dot{-}}$ [91]) + 4-nitroacetophenone(PNAP) ⟶ PNAP$^{\dot{-}}$ + products				
Pulse rad. of uridine + t-butanol + N_2 + H_2O				72 Ada 1
KAS	H_2O	RT	$k = 4.1 \cdot 10^9\ M^{-1}\,s^{-1}$	
(Uridine)$^{\dot{-}}$ [91]) + *anti*-5-nitro-2-furaldoxime ⟶ products				
Pulse rad. of uridine + t-butanol + N_2 + H_2O				73 Gre 1
KAS	H_2O	RT	$k = 3.6 \cdot 10^9\ M^{-1}\,s^{-1}$	
(Uridine-$\dot{O}$H)[93]) + *anti*-5-nitro-2-furaldoxime ⟶ products[94])				
Pulse rad. of uridine + N_2O + H_2O				73 Gre 1
KAS	H_2O	RT	$k = 2.0 \cdot 10^9\ M^{-1}\,s^{-1}$	
(Uridine-5′-monophosphate)$^{\dot{-}}$ [95]) + 6-carboxyuracil[92]) ⟶ products				
Pulse rad. of uridine-5′-monophosphate + t-butanol + N_2 + H_2O				72 Ada 1
KAS	H_2O	RT	$k = 7 \cdot 10^8\ M^{-1}\,s^{-1}$	
(Uridine-5′-monophosphate)$^{\dot{-}}$ [95]) + 4-nitroacetophenone(PNAP) ⟶ PNAP$^{\dot{-}}$ + products				
Pulse rad. of uridine-5′-monophosphate + t-butanol + N_2 + H_2O				72 Ada 1
KAS	H_2O	RT	$k = 3.5 \cdot 10^9\ M^{-1}\,s^{-1}$	
(Uridine-5′-monophosphate)$^{\dot{-}}$ [95]) + *anti*-5-nitro-2-furaldoxime ⟶ products				
Pulse rad. of uridine-5′-monophosphate + t-butanol + N_2 + H_2O				73 Gre 1
KAS	H_2O	RT	$k = 3.3 \cdot 10^9\ M^{-1}\,s^{-1}$	

[86]) Radicals from uracil + $\dot{O}$H reaction.
[88]) 25% e^--transfer.
[89]) 80% e^--transfer.
[90]) <28% e^--transfer at pH = 7; 62% e^--transfer at pH = 11.
[91]) Radicals from uridine + e^-_{aq} reaction.
[92]) Orotic acid.
[93]) Radicals from uridine + $\dot{O}$H reaction.
[94]) 10% e^--transfer.
[95]) Radicals from uridine-5′-monophosphate + e^-_{aq} reaction.

Reaction Radical generation Method	Solvent	T[K]	Rate data	Ref./ add. ref.
(Uridine-5′-monophosphate-$\dot{O}H$)[96]) + 2-methyl-1,4-naphthoquinone ⟶ 2-methyl...semiquinone + products				
Pulse rad. of uridine-5′-monophosphate + N_2O + H_2O				73 Hay 3
KAS	H_2O, pH = 7	RT	$k = 3.1 \cdot 10^9\,M^{-1}s^{-1}$ [97])	
	pH = 11		$1.0 \cdot 10^9$ [98])	
(Uridine-5′-monophosphate-$\dot{O}H$)[96]) + *anti*-5-nitro-2-furaldoxime ⟶ products [97])				
Pulse rad. of uridine-5′-monophosphate + N_2O + H_2O				73 Gre 1
KAS	H_2O	RT	$k = 1.0 \cdot 10^9\,M^{-1}s^{-1}$	
(Xylitol-$\dot{O}H$)[99]) + hydroquinone ⟶ $^{-}OC_6H_4\dot{O}$ + products				
Pulse rad. of xylitol + N_2O + H_2O				79 Ste 1
KAS	H_2O, pH ≈ 11.5	RT	$k = 1.2(1) \cdot 10^9\,M^{-1}s^{-1}$	

4.2.2 Reactions in nonaqueous solutions

4.2.2.1 Aliphatic radicals and radicals derived from other non-aromatic compounds

4.2.2.1.1 Radicals containing only C and H atoms

4.2.2.1.1.1 Neutral radicals

Reaction Radical generation Method	Solvent	T[K]	Rate data	Ref./ add. ref.
$\dot{C}H_3 + Cu(II)(OOCCH_3)_2 \longrightarrow CH_3OOCCH_3 + Cu(I)OOCCH_3$ [1])				
Catalytic decomp. of CH_3OOCH_3				72 Jen 1
PR by glc	CH_3COOH/ CH_3CN mixt.	RT	$k = 1.5 \cdot 10^6\,M^{-1}s^{-1}$ [2])	
$CH_3CH_2\dot{C}H_2$ + Cu(II)acetate complexes $\xrightarrow{e}$ Cu(I)... + products				
Cu(II) catalyzed decomp. of $C_3H_7OOC_3H_7$				65 Koc 1
PR by glc	CH_3COOH/H_2O mixt.	330	$k_e = 4.36 \cdot 10^7\,M^{-1}s^{-1}$ [3]) $\log[A/M^{-1}s^{-1}] = 8.3$ $E_a = 28.05\,kJ\,mol^{-1}$	
$CH_3CH_2\dot{C}H_2 + Cu(II) \xrightarrow{e} Cu(I)$ + products + *n*-butyraldehyde $\xrightarrow{H}$ C_3H_8 + *n*-butyraldehyde($-\dot{H}$)				
Catalytic decomp. of *n*-valerylperoxide (*A*) and 2-methylbutyrylperoxide (*B*)				65 Koc 1, 65 Koc 2
PR, glc	CH_3COOH/H_2O (≈2:1 V/V)	330	$k_e/k_H = 5.56 \cdot 10^3$ (*A*) $= 6.25 \cdot 10^3$ (*B*)	
$CH_3CH_2\dot{C}H_2 + Cu(II) \xrightarrow{e} Cu(I)$ + products + *n*-valeraldehyde $\xrightarrow{H}$ C_3H_8 + *n*-valeraldehyde($-\dot{H}$)				
Catalytic decomp. of *n*-butyrylperoxide				65 Koc 1, 65 Koc 2
PR, glc	CH_3COOH/H_2O (≈2:1 V/V)	330	$k_e/k_H = 6.67 \cdot 10^3$	

96) Radicals from uridine-5′-monophosphate + $\dot{O}H$ reaction.
97) 10% e^--transfer.
98) 50% e^--transfer.
99) Radicals from xylitol + $\dot{O}H$ reaction.
1) Reaction presumed to be electron transfer.
2) Based on $k(n\text{-}C_4H_9 + Cu(II)(OOCCH_3)_2) = 3.1 \cdot 10^6\,M^{-1}s^{-1}$.
3) Based on gas-phase rate constants k_a for H-atom abstraction by $CH_3CH_2\dot{C}H_2$ [59 Ker 1] and $k_a/k_e = 1.8 \cdot 10^{-4}$ [65 Koc 1].

Reaction				
Radical generation				Ref./
Method	Solvent	T[K]	Rate data	add. ref.

$(CH_3)_2\dot{C}H + Cu(II)acetate\ complexes \xrightarrow{e} Cu(I)\dots + products$

Cu(II) catalyzed decomp. of $(CH_3)_2CHOOCH(CH_3)_2$				65 Koc 1
PR by glc	CH_3COOH/H_2O mixt.	330	$k_e = 5.0 \cdot 10^7\ M^{-1}s^{-1}$ [4]) $\log[A/M^{-1}s^{-1}] = 8.3$ $E_a = 26.4\ kJ\,mol^{-1}$	

$(CH_3)_2\dot{C}H + Cu(II)(NCCH_3)_4^{2+} \longrightarrow CH_3CH{=}CH_2 + H^+ + Cu(I)(NCCH_3)_4^+$

Cu(II) catalyzed decomp. of $(CH_3)_2CHOOCH(CH_3)_2$				68 Koc 1
PR by glc	CH_3CN/CH_3COOH (1:1.5)	298.5	$k = 5.0 \cdot 10^6\ M^{-1}s^{-1}$ [5])	

$(CH_3)_2\dot{C}H + Cu(II) \xrightarrow{e} Cu(I) + products$
    $+ isobutyraldehyde \xrightarrow{H} (CH_3)_2CH_2 + isobutyraldehyde(-\dot{H})$

Catalytic decomp. of n-valerylperoxide				65 Koc 1,
PR, glc	CH_3COOH/H_2O ($\approx$2:1 V/V)	330	$k_e/k_H = 3.57 \cdot 10^3$	65 Koc 2

$CH_3CH_2CH_2\dot{C}H_2 + Cu(II)acetate\ complexes \longrightarrow Cu(I)\dots + products$

Cu(II) catalyzed decomp. of $C_4H_9OOC_4H_9$				65 Koc 1
PR by glc	CH_3COOH/H_2O mixt.	330	$k_e = 1.1 \cdot 10^8\ M^{-1}s^{-1}$ [6]) $\log[A/M^{-1}s^{-1}] = 7.9$ $E_a = 22.6\ kJ\,mol^{-1}$	

$CH_3CH_2CH_2\dot{C}H_2 + Cu(II)(\alpha,\alpha'\text{-bipyridine})^{2+} \longrightarrow CH_3CH_2CH{=}CH_2 + H^+ + Cu(I)(\alpha,\alpha'\text{-bipyridine})^+$

Cu(II) catalyzed decomp. of n-C_4H_9OO-n-C_4H_9				68 Koc 1
PR by glc	CH_3CN/CH_3COOH (1:1.5)	298.5	$k = 1.7 \cdot 10^7\ M^{-1}s^{-1}$ [7])	

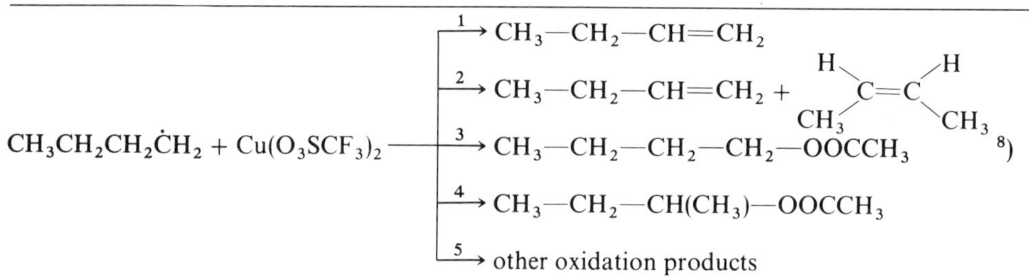

$CH_3CH_2CH_2\dot{C}H_2 + Cu(O_3SCF_3)_2 \longrightarrow$

1 → $CH_3{-}CH_2{-}CH{=}CH_2$

2 → $CH_3{-}CH_2{-}CH{=}CH_2 +$ $\underset{CH_3}{\overset{H}{C}}{=}\underset{CH_3}{\overset{H}{C}}$

3 → $CH_3{-}CH_2{-}CH_2{-}CH_2{-}OOCCH_3$ 8)

4 → $CH_3{-}CH_2{-}CH(CH_3){-}OOCCH_3$

5 → other oxidation products

Catalytic decomp. of n-C_4H_9OO-n-C_4H_9				72 Jen 1
PR by glc	CH_3COOH/CH_3CN mixt.	RT	$k_1 = 1.0 \cdot 10^7\ M^{-1}s^{-1}$ [9]) $k_2 = 2.0 \cdot 10^6\ M^{-1}s^{-1}$ [9]) $k_3 = 8.0 \cdot 10^6\ M^{-1}s^{-1}$ [9]) $k_4 = 4.0 \cdot 10^6\ M^{-1}s^{-1}$ [9]) $k_5 = 2.3 \cdot 10^7\ M^{-1}s^{-1}$ [9])	

$CH_3CH_2CH_2\dot{C}H_2 + Cu(II)(NCCH_3)_4^{2+} \longrightarrow CH_3CH_2CH{=}CH_2 + H^+ + Cu(I)(NCCH_3)_4^+$

Cu(II) catalyzed decomp. of n-C_4H_9OO-n-C_4H_9				68 Koc 1
PR by glc	CH_3CN/CH_3COOH (1:1.5)	298.5	$k = 3.1 \cdot 10^6\ M^{-1}s^{-1}$ [7])	

---

[4]) Based on gas-phase rate constants k_a for H-atom abstraction by $(CH_3)_2\dot{C}H$ [59 Ker 2] and $k_a/k_e = 2.8 \cdot 10^{-4}$ [65 Koc 1].
[5]) Assuming $k = 5.0 \cdot 10^3\ M^{-1}s^{-1}$ for competing reaction $(CH_3)_2\dot{C}H + (CH_3)_2CHCHO \longrightarrow (CH_3)_2CH_2 + (CH_3)_2CH\dot{C}O$.
[6]) Based on gas-phase rate constants k_a for H-atom abstraction by $C_4H_9^{\cdot}$ [60 Ker 1] and $k_a/k_e = 2.0 \cdot 10^{-4}$ [65 Koc 1].
[7]) Assuming $k = 1 \cdot 10^4\ M^{-1}s^{-1}$ for competing reaction $CH_3(CH_2)_2\dot{C}H_2 + (CH_3)_2CHCHO \longrightarrow CH_3(CH_2)_2CH_3 +$ $(CH_3)_2CH\dot{C}O$.
[8]) Reactions presumed to be electron transfer.
[9]) Based on $k(n\text{-}C_4H_9^{\cdot} + C_4H_7CHO) = 1 \cdot 10^4\ M^{-1}s^{-1}$.

Reaction Radical generation Method	Solvent	T[K]	Rate data	Ref./ add. ref.
$CH_3CH_2CH_2\dot{C}H_2 + Cu(II) \xrightarrow{e} Cu(I) + \text{products}$ $+ \text{acetic acid} \xrightarrow{H} n\text{-}C_4H_{10} + \text{acetic acid}(-\dot{H})$				
Catalytic decomp. of n-valerylperoxide PR, glc	CH_3COOH/H_2O ($\approx$2:1 V/V)	330	$k_e/k_H = 4.35 \cdot 10^5$	65 Koc 1, 65 Koc 2
$CH_3CH_2CH_2\dot{C}H_2 + Cu(II) \xrightarrow{e} Cu(I) + \text{products}$ $+ \text{benzyl alcohol} \xrightarrow{H} n\text{-}C_4H_{10} + \text{benzyl alcohol}(-\dot{H})$				
Catalytic decomp. of n-valerylperoxide PR, glc	CH_3COOH/H_2O ($\approx$2:1 V/V)	330	$k_e/k_H = 3.33 \cdot 10^4$	65 Koc 1, 65 Koc 2
$CH_3CH_2CH_2\dot{C}H_2 + Cu(II) \xrightarrow{e} Cu(I) + \text{products}$ $+ \text{isobutyraldehyde} \xrightarrow{H} n\text{-}C_4H_{10} + \text{isobutyraldehyde}(-\dot{H})$				
Catalytic decomp. of n-valerylperoxide PR, glc	CH_3COOH/H_2O ($\approx$2:1 V/V)	330	$k_e/k_H = 2.17 \cdot 10^3$	65 Koc 1, 65 Koc 2
	glacial acetic acid	330	$k_e/k_H = 1.01 \cdot 10^3$	
$CH_3CH_2CH_2\dot{C}H_2 + Cu(II) \xrightarrow{e} Cu(I) + \text{products}$ $+ n\text{-butyraldehyde} \xrightarrow{H} n\text{-}C_4H_{10} + n\text{-butyraldehyde}(-\dot{H})$				
Catalytic decomp. of n-valerylperoxide PR, glc	CH_3COOH/H_2O ($\approx$2:1 V/V)	330	$k_e/k_H = 5.88 \cdot 10^3$	65 Koc 1, 65 Koc 2
	glacial acetic acid	330	$k_e/k_H = 2.27 \cdot 10^3$	
$CH_3CH_2CH_2\dot{C}H_2 + Cu(II) \xrightarrow{e} Cu(I) + \text{products}$ $+ \text{diallylether} \xrightarrow{H} n\text{-}C_4H_{10} + \text{diallylether}(-\dot{H})$				
Catalytic decomp. of n-valerylperoxide PR, glc	CH_3COOH/H_2O ($\approx$2:1 V/V)	330	$k_e/k_H = 1.11 \cdot 10^4$	65 Koc 1, 65 Koc 2
$CH_3CH_2CH_2\dot{C}H_2 + Cu(II) \xrightarrow{e} Cu(I) + \text{products}$ $+ \text{dibenzylether} \xrightarrow{H} n\text{-}C_4H_{10} + \text{dibenzylether}(-\dot{H})$				
Catalytic decomp. of n-valerylperoxide PR, glc	CH_3COOH/H_2O ($\approx$2:1 V/V)	330	$k_e/k_H = 3.33 \cdot 10^3$	65 Koc 1, 65 Koc 2
	glacial acetic acid	330	$k_e/k_H = 4.17 \cdot 10^3$	
$CH_3CH_2CH_2\dot{C}H_2 + Cu(II) \xrightarrow{e} Cu(I) + \text{products}$ $+ \text{dichloroacetic acid} \xrightarrow{H} n\text{-}C_4H_{10} + \text{dichloroacetic acid}(-\dot{H})$				
Catalytic decomp. of n-valerylperoxide PR, glc	CH_3COOH/H_2O ($\approx$2:1 V/V)	330	$k_e/k_H = 3.23 \cdot 10^3$	65 Koc 1, 65 Koc 2
	glacial acetic acid	330	$k_e/k_H = 2.94 \cdot 10^3$	
$CH_3CH_2CH_2\dot{C}H_2 + Cu(II) \xrightarrow{e} Cu(I) + \text{products}$ $+ \text{2-ethylbutyraldehyde} \xrightarrow{H} n\text{-}C_4H_{10} + \text{2-ethylbutyraldehyde}(-\dot{H})$				
Catalytic decomp. of n-valerylperoxide PR, glc	CH_3COOH/H_2O ($\approx$2:1 V/V)	330	$k_e/k_H = 3.23 \cdot 10^3$	65 Koc 1, 65 Koc 2

Reaction Radical generation Method	Solvent	T[K]	Rate data	Ref./ add. ref.
$CH_3CH_2CH_2\dot{C}H_2 + Cu(II) \xrightarrow{e} Cu(I)$ + products + phenylacetaldehyde $\xrightarrow{H}$ n-C_4H_{10} + phenylacetaldehyde($-\dot{H}$)				
Catalytic decomp. of n-valerylperoxide				65 Koc 1,
PR, glc	CH_3COOH/H_2O ($\approx$2:1 V/V)	330	$k_e/k_H = 1.49 \cdot 10^3$	65 Koc 2
$CH_3CH_2CH_2\dot{C}H_2 + Cu(II) \xrightarrow{e} Cu(I)$ + products + pivaldehyde $\xrightarrow{H}$ n-C_4H_{10} + pivaldehyde($-\dot{H}$)				
Catalytic decomp. of n-valerylperoxide				65 Koc 1,
PR, glc	CH_3COOH/H_2O ($\approx$2:1 V/V)	330	$k_e/k_H = 7.69 \cdot 10^3$	65 Koc 2
$CH_3CH_2CH_2\dot{C}H_2 + Cu(II) \xrightarrow{e} Cu(I)$ + products + trichloromethane $\xrightarrow{H}$ n-$C_4H_{10} + \dot{C}Cl_3$				
Catalytic decomp. of n-valerylperoxide				65 Koc 1,
PR, glc	CH_3COOH/H_2O ($\approx$2:1 V/V)	330	$k_e/k_H = 1.43 \cdot 10^3$	65 Koc 2
$CH_3CH_2CH_2\dot{C}H_2 + Cu(II) \xrightarrow{e} Cu(I)$ + products + isovaleraldehyde $\xrightarrow{H}$ n-C_4H_{10} + isovaleraldehyde($-\dot{H}$)				
Catalytic decomp. of n-valerylperoxide				65 Koc 1,
PR, glc	CH_3COOH/H_2O ($\approx$2:1 V/V)	330	$k_e/k_H = 5.0 \cdot 10^3$	65 Koc 2
$CH_3CH_2CH_2\dot{C}H_2 + Cu(II) \xrightarrow{e} Cu(I)$ + products + n-valeraldehyde $\xrightarrow{H}$ n-C_4H_{10} + n-valeraldehyde($-\dot{H}$)				
Catalytic decomp. of isovalerylperoxide				65 Koc 1,
PR, glc	CH_3COOH/H_2O ($\approx$2:1 V/V)	330	$k_e/k_H = 5.0 \cdot 10^3$	65 Koc 2
$CH_3CH_2\dot{C}HCH_3$ + Cu(II)acetate complexes $\xrightarrow{e}$ Cu(I)... + products				
Cu(II) catalyzed decomp. of ROOR, (R = $CH_3CH_2CHCH_3$)				65 Koc 1
PR by glc	CH_3COOH/H_2O mixt.	330	$k_e = 7.6 \cdot 10^7\ M^{-1}s^{-1}$ [10]) $\log[A/M^{-1}s^{-1}] = 7.7$ $E_a = 20.5\ kJ\ mol^{-1}$	
$CH_3CH_2\dot{C}HCH_3 + Cu(II) \xrightarrow{e} Cu(I)$ + products + 2-methylbutyraldehyde $\xrightarrow{H}$ n-C_4H_{10} + 2-methylbutyraldehyde($-\dot{H}$)				
Catalytic decomp. of n-butyrylperoxide (A) and decanoylperoxide (B)				65 Koc 1,
PR, glc	CH_3COOH/H_2O ($\approx$2:1 V/V)	330	$k_e/k_H = 2.56 \cdot 10^3$ (A) $= 2.04 \cdot 10^3$ (B)	65 Koc 2
$(CH_3)_2CH\dot{C}H_2$ + Cu(II)acetate complexes $\xrightarrow{e}$ Cu(I)... + products				
Cu(II) catalyzed decomp. of $(CH_3)_2CHCH_2OOCH_2CH(CH_3)_2$				65 Koc 1
PR by glc	CH_3COOH/H_2O mixt.	330	$k_e = 5.0 \cdot 10^7\ M^{-1}s^{-1}$ [11]) $\log[A/M^{-1}s^{-1}] = 8.7$ $E_a = 27.2\ kJ\ mol^{-1}$	
$(CH_3)_2CH\dot{C}H_2 + Cu(II) \xrightarrow{e} Cu(I)$ + products + 2-methylbutyraldehyde $\longrightarrow$ $(CH_3)_2CHCH_3$ + 2-methylbutyraldehyde($-\dot{H}$)				
Catalytic decomp. of isovalerylperoxide				65 Koc 1,
PR, glc	CH_3COOH/H_2O ($\approx$2:1 V/V)	330	$k_e/k_H = 1.27 \cdot 10^3$	65 Koc 2

[10]) Based on gas-phase rate constants k_a for H-atom abstraction by $CH_3CH_2\dot{C}HCH_3$ [56 Gru 1] and $k_a/k_e = 3.9 \cdot 10^{-4}$ [65 Koc 1].

[11]) Based on gas-phase rate constants k_a for H-atom abstraction by $(CH_3)_2CH\dot{C}H_2$ [60 Met 1] and $k_a/k_e = 5.3 \cdot 10^{-4}$ [65 Koc 1].

Reaction Radical generation Method	Solvent	T[K]	Rate data	Ref./ add. ref.
$(CH_3)_2CH\dot{C}H_2 + Cu(II) \xrightarrow{e} Cu(I) + \text{products}$ $+ \text{isovaleraldehyde} \xrightarrow{H} (CH_3)_2CHCH_3 + \text{isovaleraldehyde}(-\dot{H})$				
Catalytic decomp. of *n*-valerylperoxide				65 Koc 1,
PR, glc	CH_3COOH/H_2O (≈2:1 V/V)	330	$k_e/k_H = 1.64 \cdot 10^3$	65 Koc 2
$(CH_3)_2CH\dot{C}H_2 + Cu(II) \xrightarrow{e} Cu(I) + \text{products}$ $+ n\text{-valeraldehyde} \xrightarrow{H} (CH_3)_2CHCH_3 + n\text{-valeraldehyde}(-\dot{H})$				
Catalytic decomp. of isovalerylperoxide				65 Koc 1,
PR, glc	CH_3COOH/H_2O (≈2:1 V/V)	330	$k_e/k_H = 1.89 \cdot 10^3$	65 Koc 2
$(CH_3)_3\dot{C} + Cu(II)\text{acetate complexes} \xrightarrow{e} Cu(I)\ldots + \text{products}$				
Cu(II) catalyzed decomp. of $(CH_3)_3COOC(CH_3)_3$				65 Koc 1
PR by glc	CH_3COOH/H_2O mixt.	330	$k_e = 5.5 \cdot 10^8\ M^{-1}s^{-1}$ [12] $\log[A/M^{-1}s^{-1}] = 7.5$ $E_a = 18.0\ kJ\ mol^{-1}$	
$(CH_3)_3\dot{C} + Cu(II) \xrightarrow{e} Cu(I) + \text{products}$ $+ \text{pivaldehyde} \xrightarrow{H} (CH_3)_3CH + \text{pivaldehyde}(-\dot{H})$				
Catalytic decomp. of *n*-valerylperoxide				65 Koc 1,
PR, glc	CH_3COOH/H_2O (≈2:1 V/V)	330	$k_e/k_H \geq 1.09 \cdot 10^4$	65 Koc 2
$(CH_3)_3C\dot{C}H_2 + Cu(II)(\alpha,\alpha'\text{-bipyridine})^{2+} \longrightarrow Cu(I)(\alpha,\alpha'\text{-bipyridine})^+ + \text{products}$				
Cu(II) catalyzed decomp. of $(CH_3)_3CCH_2OOCH_2C(CH_3)_3$				68 Koc 1
PR by glc	CH_3CN/CH_3COOH (1:1.5)	298.5	$k = 2.5 \cdot 10^4\ M^{-1}s^{-1}$ [13]	
$(CH_3)_3C\dot{C}H_2 + Cu(II)(NCCH_3)_4^{2+} \longrightarrow Cu(I)(NCCH_3)_4^+ + \text{products}$				
Cu(II) catalyzed decomp. of $(CH_3)_3CCH_2OOCH_2C(CH_3)_3$				68 Koc 1
PR by glc	CH_3CN/CH_3COOH (1:1.5)	298.5	$k = 4.5 \cdot 10^5\ M^{-1}s^{-1}$ [13]	
$CH_2{=}CH{-}(CH_2)_3{-}\dot{C}H_2 + Cu(II)(OOCCH_3)_2 \longrightarrow$ $CH_2{=}CH{-}(CH_2)_2{-}CH{=}CH_2 + Cu(I)OOCCH_3 + HOOCCH_3$ [14]				
Catalytic decomp. of di-5-hexenylperoxide				72 Jen 1
PR by glc	60% CH_3COOH/ 40% CH_3CN	RT	$k = 1.2 \cdot 10^6\ M^{-1}s^{-1}$ [15]	
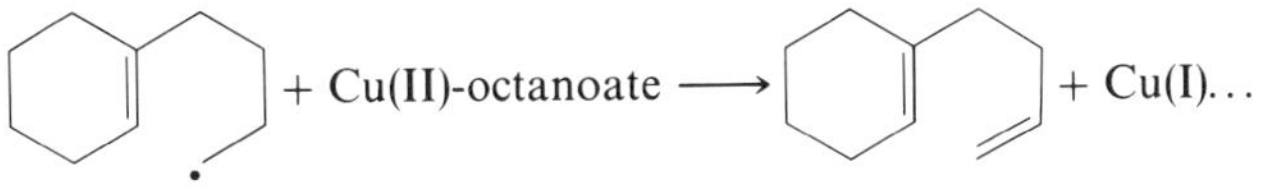				
Thermal decomp. of cupric octanoate or bis-5-(cyclohex-1-enyl)pentanoylperoxide [16] in benzene sol.				70 Str 1, 72 Bec 2
PR by glc	C_6H_6	353	$k = 5 \cdot 10^7\ M^{-1}s^{-1}$ $\approx 4 \cdot 10^8$ [16]	

[12]) Based on gas-phase rate constant k_a for H-atom abstraction by $(CH_3)_3\dot{C}$ [54 Bir 1] and $k_a/k_e = 9.2 \cdot 10^{-5}$.
[13]) Assuming $k = 1 \cdot 10^4\ M^{-1}s^{-1}$ for competing reaction $(CH_3)_3C\dot{C}H_2 + (CH_3)_2CHCHO \longrightarrow (CH_3)_3CCH_3 + (CH_3)_2CH\dot{C}O$.
[14]) Reaction presumed to be electron transfer.
[15]) Relative to $k(CH_2{=}CH(CH_2)_3\dot{C}H_2 \longrightarrow c\text{-}C_5H_9\dot{C}H_2) = 1 \cdot 10^5\ s^{-1}$.
[16]) From [72 Bec 2].

Reaction Radical generation Method	Solvent	T[K]	Rate data	Ref./ add. ref.
4.2.2.1.1.2 Anionic radicals				
(Septapreno-β-carotene)$^{\cdot-}$ + chlorophyll a ⟶ (chlorophyll a)$^{\cdot-}$ + septapreno-β-carotene [17])				
Pulse rad. of septapreno-β-carotene [17]) + n-hexane				78 Laf 1
KAS	n-C_6H_{14}	RT	$k = 7.0 \cdot 10^9\ M^{-1}s^{-1}$ [18])	
(Septapreno-β-carotene)$^{\cdot-}$ + chlorophyll b ⟶ (chlorophyll b)$^{\cdot-}$ + septapreno-β-carotene [17])				
Pulse rad. of septapreno-β-carotene [17]) + n-hexane				78 Laf 1
KAS	n-C_6H_{14}	RT	$k = 8.0 \cdot 10^{10}\ M^{-1}s^{-1}$ [18])	
(15,15′-*cis*-β-Carotene)$^{\cdot-}$ + chlorophyll a ⟶ (chlorophyll a)$^{\cdot-}$ + 15,15′-*cis*-β-carotene [19])				
Pulse rad. of 15,15′-*cis*-β-carotene [19]) + n-hexane				78 Laf 1
KAS	n-C_6H_{14}	RT	$k = 8.7 \cdot 10^9\ M^{-1}s^{-1}$ [18])	
(15,15′-*cis*-β-Carotene)$^{\cdot-}$ + chlorophyll b ⟶ (chlorophyll b)$^{\cdot-}$ + 15,15′-*cis*-β-carotene [19])				
Pulse rad. of 15,15′-*cis*-β-carotene [19]) + n-hexane				78 Laf 1
KAS	n-C_6H_{14}	RT	$k = 1.45 \cdot 10^{10}\ M^{-1}s^{-1}$ [18])	
(All-*trans*-β-carotene)$^{\cdot-}$ + chlorophyll a ⟶ (chlorophyll a)$^{\cdot-}$ + all-*trans*-β-carotene [20])				
Pulse rad. of β-carotene [20]) + n-hexane				76 Laf 1/
KAS	n-C_6H_{14}	RT	$k = 8.5(10) \cdot 10^9\ M^{-1}s^{-1}$	78 Laf 1
(All-*trans*-β-carotene)$^{\cdot-}$ + chlorophyll b ⟶ (chlorophyll b)$^{\cdot-}$ + all-*trans*-β-carotene [20])				
Pulse rad. of all-*trans*-β-carotene [20]) + n-hexane				78 Laf 1
KAS	n-C_6H_{14}	RT	$k = 1.75 \cdot 10^{10}\ M^{-1}s^{-1}$ [18])	
(All-*trans*-β-carotene)$^{\cdot-}$ + copper pheophytin a ⟶ (copper pheophytin a)$^{\cdot-}$ + *trans*-β-carotene [20])				
Pulse rad. of *trans*-β-carotene [20]) + n-hexane				79 McV 1
KAS	n-C_6H_{14}	RT	$k = 2.98 \cdot 10^{10}\ M^{-1}s^{-1}$	
(All-*trans*-β-carotene)$^{\cdot-}$ + copper pheophytin b ⟶ (copper pheophytin b)$^{\cdot-}$ + *trans*-β-carotene [20])				
Pulse rad. of *trans*-β-carotene [20]) + n-hexane				79 McV 1
KAS	n-C_6H_{14}	RT	$k = 3.2 \cdot 10^9\ M^{-1}s^{-1}$	
(All-*trans*-β-carotene)$^{\cdot-}$ + etioporphyrine ⟶ (etioporphyrine)$^{\cdot-}$ + *trans*-β-carotene [20])				
Pulse rad. of *trans*-β-carotene [20]) + n-hexane				79 McV 1
KAS	n-C_6H_{14}	RT	$k = 1.04 \cdot 10^{10}\ M^{-1}s^{-1}$	

---

[17])

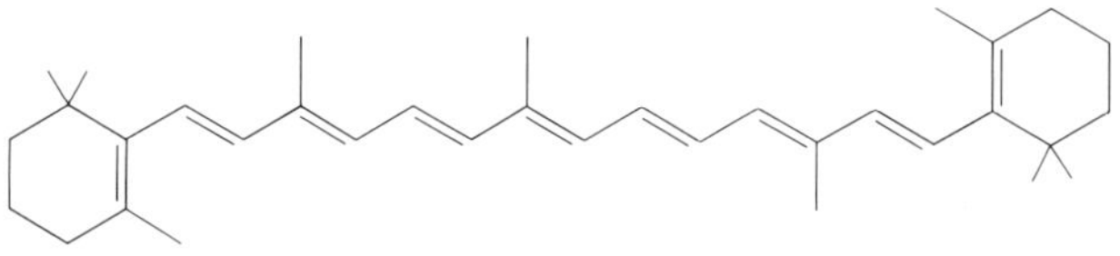

[18]) Based on assumption of monomeric chlorophyll.

[19])

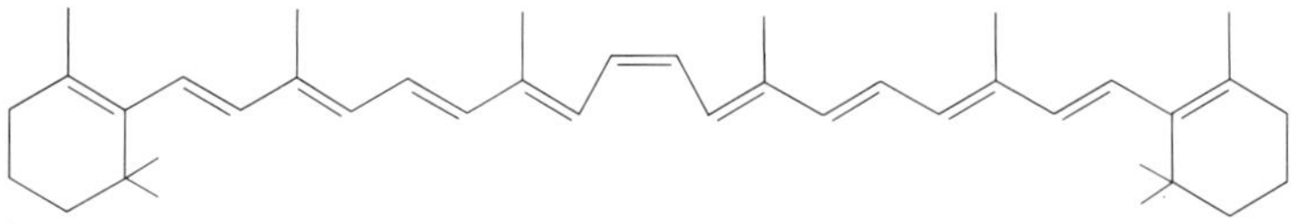

[20]) β-carotene = β-carotin = provitamine A = $C_{40}H_{56}$

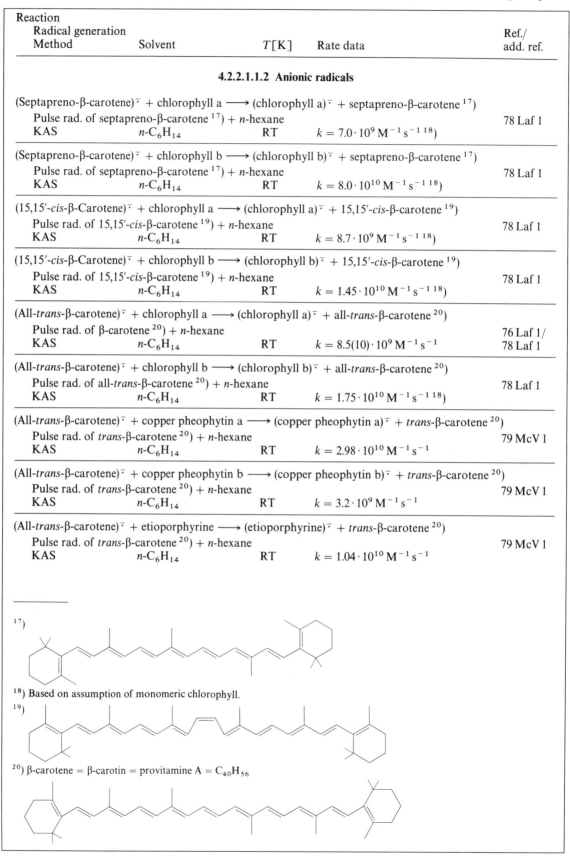

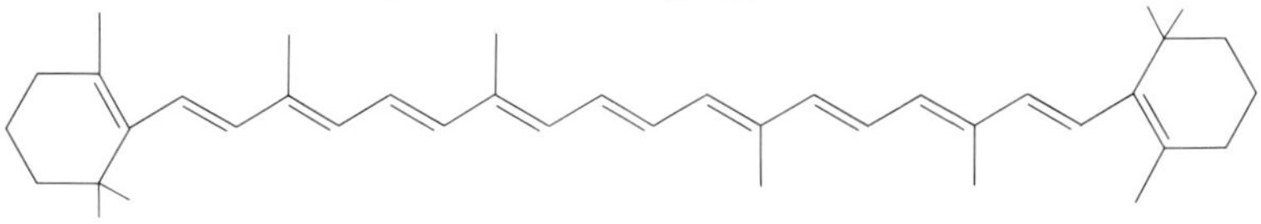

Reaction Radical generation Method	Solvent	T[K]	Rate data	Ref./ add. ref.
(All-*trans*-β-carotene)$^{\cdot-}$ + mesoporphyrine ⟶ (mesoporphyrine)$^{\cdot-}$ + *trans*-β-carotene [20])				
Pulse rad. of *trans*-β-carotene [20]) + *n*-hexane				79 McV 1
KAS	*n*-C_6H_{14}	RT	$k = 1.20 \cdot 10^{10}\ M^{-1} s^{-1}$	
(All-*trans*-β-carotene)$^{\cdot-}$ + pheophytin a ⟶ (pheophytin a)$^{\cdot-}$ + *trans*-β-carotene [20])				
Pulse rad. of *trans*-β-carotene [20]) + *n*-hexane				79 McV 1
KAS	*n*-C_6H_{14}	RT	$k = 1.93 \cdot 10^{10}\ M^{-1} s^{-1}$	
(All-*trans*-β-carotene)$^{\cdot-}$ + pheophytin b ⟶ (pheophytin b)$^{\cdot-}$ + *trans*-β-carotene [20])				
Pulse rad. of *trans*-β-carotene [20]) + *n*-hexane				79 McV 1
KAS	*n*-C_6H_{14}	RT	$k = 2.45 \cdot 10^{10}\ M^{-1} s^{-1}$	
(*trans*-Lycopene)$^{\cdot-}$ + chlorophyll a ⟶ (chlorophyll a)$^{\cdot-}$ + lycopene [21])				
Pulse rad. of lycopene [21]) + *n*-hexane				76 Laf 1/
KAS	*n*-C_6H_{14}	RT	$k = 7.0(10) \cdot 10^{9}\ M^{-1} s^{-1}$	78 Laf 1
(*trans*-Lycopene)$^{\cdot-}$ + copper pheophytin a ⟶ (copper pheophytin a)$^{\cdot-}$ + *trans*-lycopene [21])				
Pulse rad. of *trans*-lycopene [21]) + *n*-hexane				79 McV 1
KAS	*n*-C_6H_{14}	RT	$k = 2.00 \cdot 10^{10}\ M^{-1} s^{-1}$	
(*trans*-Lycopene)$^{\cdot-}$ + copper pheophytin b ⟶ (copper pheophytin b)$^{\cdot-}$ + *trans*-lycopene [21])				
Pulse rad. of *trans*-lycopene [21]) + *n*-hexane				79 McV 1
KAS	*n*-C_6H_{14}	RT	$k = 2.6 \cdot 10^{9}\ M^{-1} s^{-1}$	
(*trans*-Lycopene)$^{\cdot-}$ + etioporphyrine ⟶ (etioporphyrine)$^{\cdot-}$ + *trans*-lycopene [21])				
Pulse rad. of *trans*-lycopene [21]) + *n*-hexane				79 McV 1
KAS	*n*-C_6H_{14}	RT	$k = 1.2 \cdot 10^{9}\ M^{-1} s^{-1}$	
(*trans*-Lycopene)$^{\cdot-}$ + mesoporphyrine ⟶ (mesoporphyrine)$^{\cdot-}$ + *trans*-lycopene [21])				
Pulse rad. of *trans*-lycopene [21]) + *n*-hexane				79 McV 1
KAS	*n*-C_6H_{14}	RT	$k = 2.3 \cdot 10^{9}\ M^{-1} s^{-1}$	
(*trans*-Lycopene)$^{\cdot-}$ + pheophytin a ⟶ (pheophytin a)$^{\cdot-}$ + *trans*-lycopene [21])				
Pulse rad. of *trans*-lycopene [21]) + *n*-hexane				79 McV 1
KAS	*n*-C_6H_{14}	RT	$k = 9.9 \cdot 10^{9}\ M^{-1} s^{-1}$	
(*trans*-Lycopene)$^{\cdot-}$ + pheophytin b ⟶ (pheophytin b)$^{\cdot-}$ + *trans*-lycopene [21])				
Pulse rad. of *trans*-lycopene [21]) + *n*-hexane				79 McV 1
KAS	*n*-C_6H_{14}	RT	$k = 1.49 \cdot 10^{10}\ M^{-1} s^{-1}$	

[20]) β-carotene = β-carotin = provitamine A = $C_{40}H_{56}$

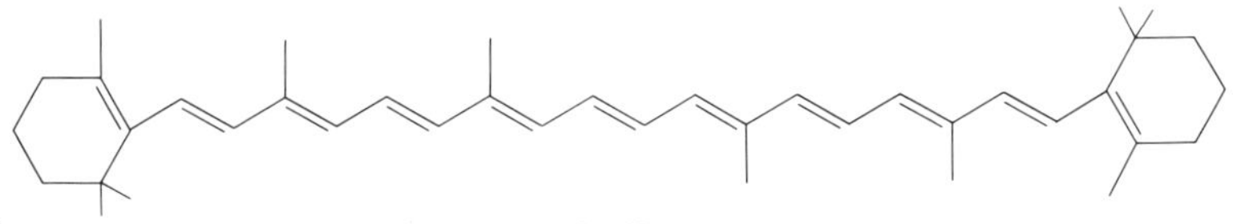

[21]) Lycopene = lycopin = neolycopene = $C_{40}H_{56}$

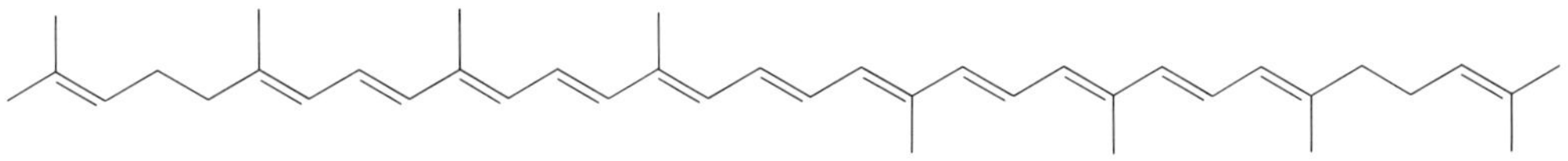

Reaction Radical generation Method	Solvent	T[K]	Rate data	Ref./ add. ref.
(7,7′-Dihydro-β-carotene)$^{\cdot-}$ + chlorophyll a ⟶ (chlorophyll a)$^{\cdot-}$ + 7,7′-dihydro-β-carotene [22]				
Pulse rad. of 7,7′-dihydro-β-carotene [22] + n-hexane				78 Laf 1
KAS	n-C_6H_{14}	RT	$k = 8.0 \cdot 10^9\ M^{-1}\ s^{-1}$ [23]	
(7,7′-Dihydro-β-carotene)$^{\cdot-}$ + chlorophyll b ⟶ (chlorophyll b)$^{\cdot-}$ + 7,7′-dihydro-β-carotene [22]				
Pulse rad. of 7,7′-dihydro-β-carotene [22] + n-hexane				78 Laf 1
KAS	n-C_6H_{14}	RT	$k = 2.5 \cdot 10^{10}\ M^{-1}\ s^{-1}$ [23]	
(Decapreno-β-carotene)$^{\cdot-}$ + chlorophyll a ⟶ (chlorophyll a)$^{\cdot-}$ + decapreno-β-carotene [24]				
Pulse rad. of decapreno-β-carotene [24] + n-hexane				78 Laf 1
KAS	n-C_6H_{14}	RT	$k = 5.4 \cdot 10^9\ M^{-1}\ s^{-1}$ [23]	
(Decapreno-β-carotene)$^{\cdot-}$ + chlorophyll b ⟶ (chlorophyll b)$^{\cdot-}$ + decapreno-β-carotene [24]				
Pulse rad. of decapreno-β-carotene [24] + n-hexane				78 Laf 1
KAS	n-C_6H_{14}	RT	$k = 1.0 \cdot 10^{10}\ M^{-1}\ s^{-1}$ [23]	
(Decapreno-β-carotene)$^{\cdot-}$ + copper pheophytin a ⟶ (copper pheophytin a)$^{\cdot-}$ + decapreno-β-carotene [24]				
Pulse rad. of decapreno-β-carotene [24] + n-hexane				79 McV 1
KAS	n-C_6H_{14}	RT	$k = 2.00 \cdot 10^{10}\ M^{-1}\ s^{-1}$	
(Decapreno-β-carotene)$^{\cdot-}$ + copper pheophytin b ⟶ (copper pheophytin b)$^{\cdot-}$ + decapreno-β-carotene [24]				
Pulse rad. of decapreno-β-carotene [24] + n-hexane				79 McV 1
KAS	n-C_6H_{14}	RT	$k = 2.00 \cdot 10^{10}\ M^{-1}\ s^{-1}$	
(Decapreno-β-carotene)$^{\cdot-}$ + etioporphyrine ⟶ (etioporphyrine)$^{\cdot-}$ + decapreno-β-carotene [24]				
Pulse rad. of decapreno-β-carotene [24] + n-hexane				79 McV 1
KAS	n-C_6H_{14}	RT	$k = 2.4 \cdot 10^9\ M^{-1}\ s^{-1}$	
(Decapreno-β-carotene)$^{\cdot-}$ + mesoporphyrine ⟶ (mesoporphyrine)$^{\cdot-}$ + decapreno-β-carotene [24]				
Pulse rad. of decapreno-β-carotene [24] + n-hexane				79 McV 1
KAS	n-C_6H_{14}	RT	$k = 2.3 \cdot 10^9\ M^{-1}\ s^{-1}$	
(Decapreno-β-carotene)$^{\cdot-}$ + pheophytin a ⟶ (pheophytin a)$^{\cdot-}$ + decapreno-β-carotene [24]				
Pulse rad. of decapreno-β-carotene [24] + n-hexane				79 McV 1
KAS	n-C_6H_{14}	RT	$k = 6.1 \cdot 10^9\ M^{-1}\ s^{-1}$	
(Decapreno-β-carotene)$^{\cdot-}$ + pheophytin b ⟶ (pheophytin b)$^{\cdot-}$ + decapreno-β-carotene [24]				
Pulse rad. of decapreno-β-carotene [24] + n-hexane				79 McV 1
KAS	n-C_6H_{14}	RT	$k = 1.14 \cdot 10^{10}\ M^{-1}\ s^{-1}$	

[22]

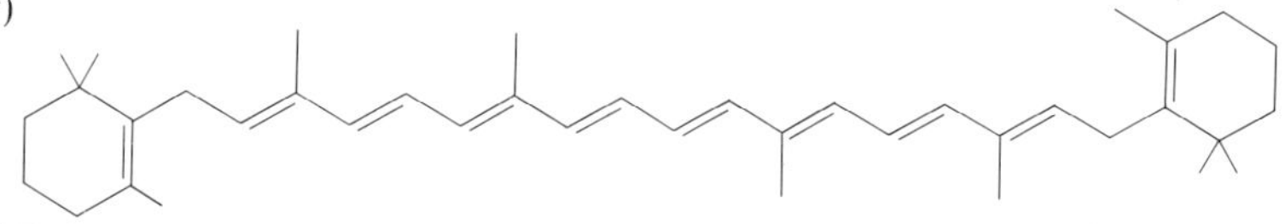

[23] Based on the assumption of monomeric chlorophyll.

[24]

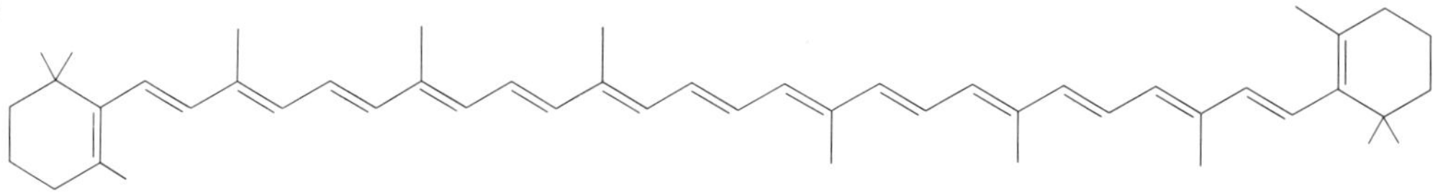

Reaction Radical generation Method	Solvent	T[K]	Rate data	Ref./ add. ref.
4.2.2.1.1.3 Cationic radicals				
$(c\text{-}C_5H_{10})^{\overset{+}{\cdot}}$ + pyrene ⟶ $(\text{pyrene})^{\overset{+}{\cdot}}$ + $c\text{-}C_5H_{10}$ [25])				
Pulse rad. of $c\text{-}C_5H_{10}$				73 Zad 1
KAS	$c\text{-}C_5H_{10}$	RT	$k = 3 \cdot 10^{10}\,M^{-1}\,s^{-1}$	
$(c\text{-}C_6H_{12})^{\overset{+}{\cdot}}$ + diphenyl mercury$\{(C_6H_5)_2Hg\}$ ⟶ products [26])				
Pulse rad. of $c\text{-}C_6H_{12}$				81 War 1
[27])	$c\text{-}C_6H_{12}$	RT	$k = 3.0 \cdot 10^{11}\,M^{-1}\,s^{-1}$	
$(c\text{-}C_6H_{12})^{\overset{+}{\cdot}}$ + anthracene ⟶ $(\text{anthracene})^{\overset{+}{\cdot}}$ + $c\text{-}C_6H_{12}$ [25])				
Pulse rad. of $c\text{-}C_6H_{12}$				74 Bre 1
KAS	$c\text{-}C_6H_{12}$	RT	$k \geqslant 1.0 \cdot 10^{12}\,M^{-1}\,s^{-1}$	
$(c\text{-}C_6H_{12})^{\overset{+}{\cdot}}$ + benzene (C_6H_6) ⟶ products [26])				
Pulse rad. of $c\text{-}C_6H_{12}$				81 Bax 1/
Cond. [28])	$c\text{-}C_6H_{12}$	RT	$k = 2.4 \cdot 10^{11}\,M^{-1}\,s^{-1}$	81 War 1,
				76 War 1
$(c\text{-}C_6H_{12})^{\overset{+}{\cdot}}$ + benzene (C_6H_6) ⟶ $c\text{-}C_6H_{12}$ + $(C_6H_6)^{\overset{+}{\cdot}}$ [25])				
Pulse rad. of $c\text{-}C_6H_{12}$				76 War 2
[27])	$c\text{-}C_6H_{12}$	296	$k = 1.9(3) \cdot 10^{11}\,M^{-1}\,s^{-1}$	
		318	$2.0(3) \cdot 10^{11}$	
		344	$2.1(3) \cdot 10^{11}$	
			$E_a < 4\,kJ\,mol^{-1}$	
$(c\text{-}C_6H_{12})^{\overset{+}{\cdot}}$ + benzophenone$\{(C_6H_5)_2CO\}$ ⟶ products [25])				
Pulse rad. of $c\text{-}C_6H_{12}$				74 Bre 1
KAS	$c\text{-}C_6H_{12}$	RT	$k \geqslant 2.5 \cdot 10^{12}\,M^{-1}\,s^{-1}$	
$(c\text{-}C_6H_{12})^{\overset{+}{\cdot}}$ + biphenyl ⟶ $(\text{biphenyl})^{\overset{+}{\cdot}}$ + $c\text{-}C_6H_{12}$ [25])				
Pulse rad. of $c\text{-}C_6H_{12}$				76 War 2/
[27])	$c\text{-}C_6H_{12}$	296	$k = 1.3(2) \cdot 10^{11}\,M^{-1}\,s^{-1}$	73 Hum 1
KAS	$c\text{-}C_6H_{12}$	RT	$k \geqslant 2.0 \cdot 10^{12}\,M^{-1}\,s^{-1}$	74 Bre 1
$(c\text{-}C_6H_{12})^{\overset{+}{\cdot}}$ + decalin ⟶ products [26])				
Pulse rad. of $c\text{-}C_6H_{12}$				81 Bax 1/
Cond. [28])	$c\text{-}C_6H_{12}$	RT	$k = 1.85 \cdot 10^{11}\,M^{-1}\,s^{-1}$	81 War 1
$(c\text{-}C_6H_{12})^{\overset{+}{\cdot}}$ + diethylether $(C_2H_5OC_2H_5)$ ⟶ products [26])				
Pulse rad. of $c\text{-}C_6H_{12}$				81 Bax 1/
Cond. [28])	$c\text{-}C_6H_{12}$	RT	$k = 1.3 \cdot 10^{11}\,M^{-1}\,s^{-1}$	81 War 1
$(c\text{-}C_6H_{12})^{\overset{+}{\cdot}}$ + N,N-dimethylaniline(DMA) ⟶ $(\text{DMA})^{\overset{+}{\cdot}}$ + $c\text{-}C_6H_{12}$ [25])				
Pulse rad. of $c\text{-}C_6H_{12}$				76 War 2
[27])	$c\text{-}C_6H_{12}$	298	$k = 2.9(4) \cdot 10^{11}\,M^{-1}\,s^{-1}$	
		315	$3.2(4) \cdot 10^{11}$	
		341	$3.5(4) \cdot 10^{11}$	
			$E_a < 4\,kJ\,mol^{-1}$	

[25]) Hole reaction.
[26]) e^-- and/or H^+-transfer; hole reaction.
[27]) Time-resolved microwave absorption.
[28]) Time-resolved dc cond. method.

Reaction Radical generation Method	Solvent	T[K]	Rate data	Ref./ add. ref.
$(c\text{-C}_6\text{H}_{12})^{\dot{+}}$ + hexene-2 ⟶ (hexene-2)$^{\dot{+}}$ + $c\text{-C}_6\text{H}_{12}$ [29]				
Pulse rad. of $c\text{-C}_6\text{H}_{12}$ KAS	$c\text{-C}_6\text{H}_{12}$	RT	$k = 3.0(13) \cdot 10^{10}\,\text{M}^{-1}\,\text{s}^{-1}$	81 Meh 1
$(c\text{-C}_6\text{H}_{12})^{\dot{+}}$ + c-hexene ⟶ (c-hexene)$^{\dot{+}}$ + $c\text{-C}_6\text{H}_{12}$ [29]				
Pulse rad. of $c\text{-C}_6\text{H}_{12}$ KAS	$c\text{-C}_6\text{H}_{12}$	RT	$k = 4.0(30) \cdot 10^{10}\,\text{M}^{-1}\,\text{s}^{-1}$	81 Meh 1
$(c\text{-C}_6\text{H}_{12})^{\dot{+}}$ + methanol (CH_3OH) ⟶ products [26]				
Pulse rad. of $c\text{-C}_6\text{H}_{12}$ Cond. [28]	$c\text{-C}_6\text{H}_{12}$	RT	$k = 1.65 \cdot 10^{11}\,\text{M}^{-1}\,\text{s}^{-1}$	81 Bax 1/ 81 War 1
$(c\text{-C}_6\text{H}_{12})^{\dot{+}}$ + naphthalene ⟶ (naphthalene)$^{\dot{+}}$ + $c\text{-C}_6\text{H}_{12}$ [25]				
Pulse rad. of $c\text{-C}_6\text{H}_{12}$ KAS	$c\text{-C}_6\text{H}_{12}$	RT	$k = 2.3(5) \cdot 10^{10}\,\text{M}^{-1}\,\text{s}^{-1}$	74 Bre 1
$(c\text{-C}_6\text{H}_{12})^{\dot{+}}$ + phenanthrene ⟶ (phenanthrene)$^{\dot{+}}$ + $c\text{-C}_6\text{H}_{12}$ [25]				
Pulse rad. of $c\text{-C}_6\text{H}_{12}$ KAS	$c\text{-C}_6\text{H}_{12}$	RT	$k \geqslant 2.0 \cdot 10^{12}\,\text{M}^{-1}\,\text{s}^{-1}$	74 Bre 1
$(c\text{-C}_6\text{H}_{12})^{\dot{+}}$ + pyrene ⟶ (pyrene)$^{\dot{+}}$ + $c\text{-C}_6\text{H}_{12}$ [25]				
Pulse rad. of $c\text{-C}_6\text{H}_{12}$ KAS	$c\text{-C}_6\text{H}_{12}$	RT	$k = 5 \cdot 10^{11}\,\text{M}^{-1}\,\text{s}^{-1}$ $4 \cdot 10^{11}$ [30]	75 Zad 1, 72 Bec 1/ 73 Zad 1
$(c\text{-C}_6\text{H}_{12})^{\dot{+}}$ + tetramethylethylene ⟶ (tetramethylethylene)$^{\dot{+}}$ + $c\text{-C}_6\text{H}_{12}$ [29]				
Pulse rad. of $c\text{-C}_6\text{H}_{12}$ KAS	$c\text{-C}_6\text{H}_{12}$	RT	$k = 9(3) \cdot 10^{10}\,\text{M}^{-1}\,\text{s}^{-1}$	81 Meh 1
$(c\text{-C}_6\text{H}_{12})^{\dot{+}}$ + N,N,N′,N′-tetramethyl-p-phenylene diamine(TMPD) ⟶ $c\text{-C}_6\text{H}_{12}$ + (TMPD)$^{\dot{+}}$ [25]				
Pulse rad. of $c\text{-C}_6\text{H}_{12}$ KAS	$c\text{-C}_6\text{H}_{12}$	RT	$k = 3.5 \cdot 10^{11}\,\text{M}^{-1}\,\text{s}^{-1}$ [31]	73 Zad 1
$(c\text{-C}_6\text{H}_{12})^{\dot{+}}$ + tetraphenylsilane{$(\text{C}_6\text{H}_5)_4\text{Si}$} ⟶ products [26]				
Pulse rad. of $c\text{-C}_6\text{H}_{12}$ [27]	$c\text{-C}_6\text{H}_{12}$	RT	$k = 3.3 \cdot 10^{11}\,\text{M}^{-1}\,\text{s}^{-1}$	81 War 1
$(c\text{-C}_6\text{H}_{12})^{\dot{+}}$ + triethylamine{$(\text{C}_2\text{H}_5)_3\text{N}$} ⟶ products [26]				
Pulse rad. of $c\text{-C}_6\text{H}_{12}$ Cond. [28]	$c\text{-C}_6\text{H}_{12}$	RT	$k = 1.3 \cdot 10^{11}\,\text{M}^{-1}\,\text{s}^{-1}$	81 Bax 1/ 81 War 1
$(c\text{-C}_6\text{H}_{12})^{\dot{+}}$ + 2,2,4-trimethylpentane ⟶ $c\text{-C}_6\text{H}_{12}$ + (2,2,4-tri…)$^{\dot{+}}$ [32]				
Pulse rad. of $c\text{-C}_6\text{H}_{12}$ KAS	$c\text{-C}_6\text{H}_{12}$	RT	$k = 4.3 \cdot 10^{9}\,\text{M}^{-1}\,\text{s}^{-1}$	69 Cap 1
$(c\text{-C}_6\text{H}_{12})^{\dot{+}}$ + triphenylchloromethane $(\text{C}_6\text{H}_5)_3\text{CCl}$ ⟶ $c\text{-C}_6\text{H}_{12}$ + $((\text{C}_6\text{H}_5)_3\text{CCl})^{\dot{+}}$ [25]				
Pulse rad. of $c\text{-C}_6\text{H}_{12}$ [27]	$c\text{-C}_6\text{H}_{12}$	293	$k = 2.5(2) \cdot 10^{11}\,\text{M}^{-1}\,\text{s}^{-1}$	79 Zad 1/ 75 Dav 1, 69 Cap 1

[25] Hole reaction.
[26] e^-- and/or H^+-transfer; hole reaction.
[27] Time-resolved microwave absorption.
[28] Time-resolved dc cond. method.
[29] Mobile hole reaction.
[30] From [72 Bec 1].
[31] Lower limit.
[32] Possibly hole reaction.

Reaction Radical generation Method	Solvent	T[K]	Rate data	Ref./ add. ref.
$(n\text{-}C_6H_{14})^{+\cdot}$ + pyrene ⟶ (pyrene)$^{+\cdot}$ + $n\text{-}C_6H_{14}$ [32]				
Pulse rad. of $n\text{-}C_6H_{14}$				73 Zad 1
KAS	$n\text{-}C_6H_{14}$	RT	$k = 4 \cdot 10^{10}\,M^{-1}\,s^{-1}$	
$(n\text{-}C_6H_{14})^{+\cdot}$ + N,N,N′,N′-tetramethyl-*p*-phenylene diamine(TMPD) ⟶ (TMPD)$^{+\cdot}$ + $n\text{-}C_6H_{14}$ [32]				
Pulse rad. of $n\text{-}C_6H_{14}$				73 Zad 1
KAS	$n\text{-}C_6H_{14}$	RT	$k \approx 2 \cdot 10^{10}\,M^{-1}\,s^{-1}$	
$(C_6H_{11}CH_3)^{+\cdot}$ + pyrene ⟶ (pyrene)$^{+\cdot}$ + $C_6H_{11}CH_3$ [25]				
Pulse rad. of methyl-*c*-hexane				73 Zad 1
KAS	methyl-*c*-hexane	RT	$k = 2 \cdot 10^{11}\,M^{-1}\,s^{-1}$	
$(n\text{-}C_7H_{16})^{+\cdot}$ + heptene-1 ⟶ (heptene-1)$^{+\cdot}$ + $n\text{-}C_7H_{16}$ [29]				
Pulse rad. of *n*-heptane				81 Meh 1
KAS	$n\text{-}C_7H_{16}$	RT	$k = 1.9(3) \cdot 10^{10}\,M^{-1}\,s^{-1}$	
$(n\text{-}C_7H_{16})^{+\cdot}$ + *c*-hexene ⟶ (*c*-hexene)$^{+\cdot}$ + $n\text{-}C_7H_{16}$ [29]				
Pulse rad. of *n*-heptane				81 Meh 1
KAS	$n\text{-}C_7H_{16}$	RT	$k = 8(2) \cdot 10^{10}\,M^{-1}\,s^{-1}$	
$(n\text{-}C_7H_{16})^{+\cdot}$ + tetramethylethylene (tetramethylethylene)$^{+\cdot}$ + $n\text{-}C_7H_{16}$ [29]				
Pulse rad. of *n*-heptane				81 Meh 1
KAS	$n\text{-}C_7H_{16}$	RT	$k = 1.8(3) \cdot 10^{11}\,M^{-1}\,s^{-1}$	
$(c\text{-}C_8H_{16})^{+\cdot}$ + pyrene ⟶ (pyrene)$^{+\cdot}$ + $c\text{-}C_8H_{16}$ [32]				
Pulse rad. of *c*-octane				73 Zad 1
KAS	$c\text{-}C_8H_{16}$	RT	$k = 8 \cdot 10^{9}\,M^{-1}\,s^{-1}$	
$(CH_3C(CH_3)_2CH_2CH(CH_3)CH_3)^{+\cdot}$ + pyrene ⟶ (pyrene)$^{+\cdot}$ + $CH_3C(CH_3)_2CH_2CH(CH_3)CH_3$ [25]				
Pulse rad. of 2,2,4-trimethylpentane				73 Zad 1
KAS	2,2,4-trimethylpentane	RT	$k = 2 \cdot 10^{10}\,M^{-1}\,s^{-1}$	
$(CH_3C(CH_3)_2CH_2CH(CH_3)CH_3)^{+\cdot}$ + N,N,N′,N′-tetramethyl-*p*-phenylene diamine(TMPD) ⟶ TMPD$^{+\cdot}$ + $CH_3C(CH_3)_2CH_2CH(CH_3)CH_3$ [25]				
Pulse rad. of 2,2,4-trimethylpentane				73 Zad 1
KAS	2,2,4-trimethylpentane	RT	$k = 2.0(5) \cdot 10^{10}\,M^{-1}\,s^{-1}$	
(Septapreno-β-carotene)$^{+\cdot}$ + chlorophyll a ⟶ (chlorophyll a)$^{+\cdot}$ + septapreno-β-carotene [33]				
Pulse rad. of septapreno-β-carotene [33] + *n*-hexane				78 Laf 1
KAS	$n\text{-}C_6H_{14}$	RT	$k = 1.07 \cdot 10^{10}\,M^{-1}\,s^{-1}$ [34]	
(Septapreno-β-carotene)$^{+\cdot}$ + chlorophyll b ⟶ (chlorophyll b)$^{+\cdot}$ + septapreno-β-carotene [33]				
Pulse rad. of septapreno-β-carotene [33] + *n*-hexane				78 Laf 1
KAS	$n\text{-}C_6H_{14}$	RT	$k = 6 \cdot 10^{9}\,M^{-1}\,s^{-1}$ [34]	

[25] Hole reaction.
[29] Mobile hole reaction.
[32] Possibly hole reaction.
[33]

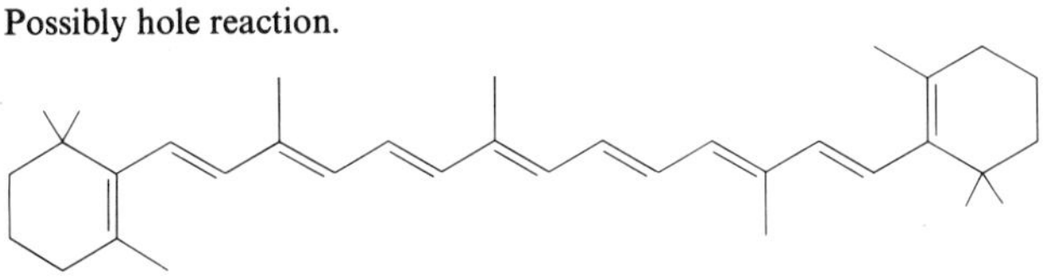

[34] Based on assumption of monomeric chlorophyll.

Reaction Radical generation Method	Solvent	T[K]	Rate data	Ref./ add. ref.
(15,15′-*cis*-β-Carotene)$^{+\cdot}$ + chlorophyll a ⟶ (chlorophyll a)$^{+\cdot}$ + 15,15′-*cis*-β-carotene [35])				
Pulse rad. of 15,15′-*cis*-β-carotene [35]) + *n*-hexane				78 Laf 1
KAS	n-C_6H_{14}	RT	$k = 1.18 \cdot 10^{10}\,M^{-1}\,s^{-1}$ [34])	
(15,15′-*cis*-β-Carotene)$^{+\cdot}$ + chlorophyll b ⟶ (chlorophyll b)$^{+\cdot}$ + 15,15′-*cis*-β-carotene [35])				
Pulse rad. of 15,15′-*cis*-β-carotene [35]) + *n*-hexane				78 Laf 1
KAS	n-C_6H_{14}	RT	$k < 1 \cdot 10^{8}\,M^{-1}\,s^{-1}$ [34])	
(All-*trans*-β-carotene)$^{+\cdot}$ + chlorophyll a ⟶ (chlorophyll a)$^{+\cdot}$ + *trans*-β-carotene [36])				
Pulse rad. of β-carotene [36]) + *n*-hexane				76 Laf 1/
KAS	n-C_6H_{14}	RT	$k = 6.0(10) \cdot 10^{9}\,M^{-1}\,s^{-1}$	78 Laf 1
(All-*trans*-β-carotene)$^{+\cdot}$ + chlorophyll b ⟶ (chlorophyll b)$^{+\cdot}$ + all-*trans*-β-carotene [36])				
Pulse rad. of all-*trans*-β-carotene [36]) + *n*-hexane				78 Laf 1
KAS	n-C_6H_{14}	RT	$k < 1 \cdot 10^{8}\,M^{-1}\,s^{-1}$ [34])	
(All-*trans*-β-carotene)$^{+\cdot}$ + copper pheophytin a ⟶ (copper pheophytin a)$^{+\cdot}$ + *trans*-β-carotene [36])				
Pulse rad. of *trans*-β-carotene [36]) + *n*-hexane				79 McV 1
KAS	n-C_6H_{14}	RT	$k < 1 \cdot 10^{8}\,M^{-1}\,s^{-1}$	
(All-*trans*-β-carotene)$^{+\cdot}$ + copper pheophytin b ⟶ (copper pheophytin b)$^{+\cdot}$ + *trans*-β-carotene [36])				
Pulse rad. of *trans*-β-carotene [36]) + *n*-hexane				79 McV 1
KAS	n-C_6H_{14}	RT	$k < 3 \cdot 10^{8}\,M^{-1}\,s^{-1}$	
(All-*trans*-β-carotene)$^{+\cdot}$ + etioporphyrine ⟶ (etioporphyrine)$^{+\cdot}$ + *trans*-β-carotene [36])				
Pulse rad. of *trans*-β-carotene [36]) + *n*-hexane				79 McV 1
KAS	n-C_6H_{14}	RT	$k < 1 \cdot 10^{8}\,M^{-1}\,s^{-1}$	
(All-*trans*-β-carotene)$^{+\cdot}$ + mesoporphyrine – (mesoporphyrine)$^{+\cdot}$ + *trans*-β-carotene [36])				
Pulse rad. of *trans*-β-carotene [36]) + *n*-hexane				79 McV 1
KAS	n-C_6H_{14}	RT	$k < 3 \cdot 10^{8}\,M^{-1}\,s^{-1}$	
(All-*trans*-β-carotene)$^{+\cdot}$ + pheophytin a ⟶ (pheophytin a)$^{+\cdot}$ + *trans*-β-carotene [36])				
Pulse rad. of *trans*-β-carotene [36]) + *n*-hexane				79 McV 1
KAS	n-C_6H_{14}	RT	$k < 1 \cdot 10^{8}\,M^{-1}\,s^{-1}$	
(All-*trans*-β-carotene)$^{+\cdot}$ + pheophytin b ⟶ (pheophytin b)$^{+\cdot}$ + *trans*-β-carotene [36])				
Pulse rad. of *trans*-β-carotene [36]) + *n*-hexane				79 McV 1
KAS	n-C_6H_{14}	RT	$k < 1 \cdot 10^{8}\,M^{-1}\,s^{-1}$	

[34]) Based on assumption of monomeric chlorophyll.

[35])

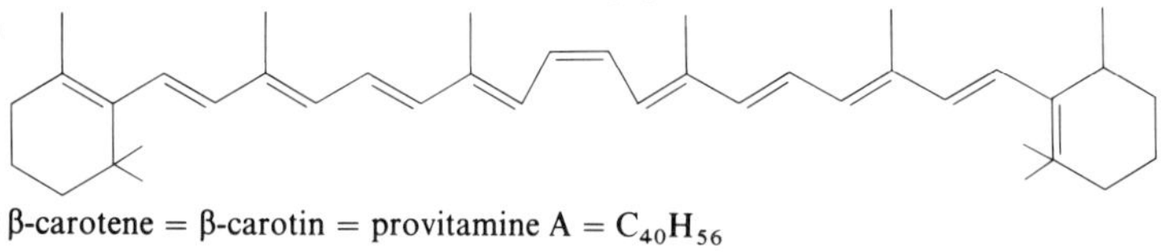

[36]) β-carotene = β-carotin = provitamine A = $C_{40}H_{56}$

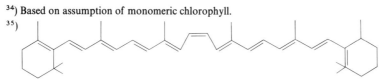

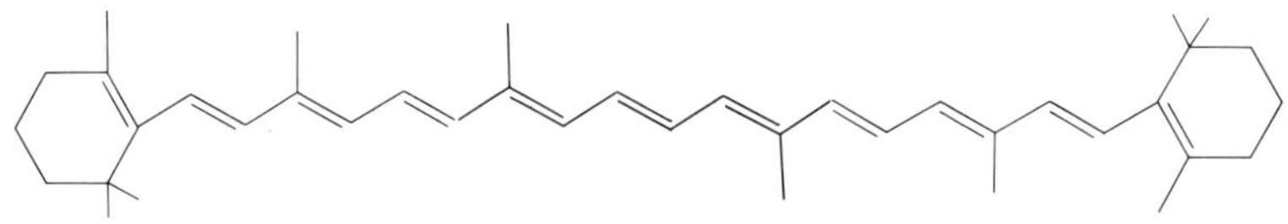

Reaction Radical generation Method	Solvent	T[K]	Rate data	Ref./ add. ref.
(*trans*-Lycopene)$^{\cdot+}$ + chlorophyll a ⟶ (chlorophyll a)$^{\cdot+}$ + *trans*-lycopene [37])				
Pulse rad. of lycopene [37]) + n-hexane				76 Laf 1/
KAS	n-C_6H_{14}	RT	$k = 1.7(3) \cdot 10^9\ M^{-1}s^{-1}$	78 Laf 1
(*trans*-Lycopene)$^{\cdot+}$ + copper pheophytin a ⟶ (copper pheophytin a)$^{\cdot+}$ + *trans*-lycopene [37])				
Pulse rad. of *trans*-lycopene [37]) + n-hexane				79 McV 1
KAS	n-C_6H_{14}	RT	$k < 1 \cdot 10^8\ M^{-1}s^{-1}$	
(*trans*-Lycopene)$^{\cdot+}$ + copper pheophytin b ⟶ (copper pheophytin b)$^{\cdot+}$ + *trans*-lycopene [37])				
Pulse rad. of *trans*-lycopene [37]) + n-hexane				79 McV 1
KAS	n-C_6H_{14}	RT	$k = 6 \cdot 10^8\ M^{-1}s^{-1}$	
(*trans*-Lycopene)$^{\cdot+}$ + etioporphyrine ⟶ (etioporphyrine)$^{\cdot+}$ + *trans*-lycopene [37])				
Pulse rad. of *trans*-lycopene [37]) + n-hexane				79 McV 1
KAS	n-C_6H_{14}	RT	$k = 6 \cdot 10^8\ M^{-1}s^{-1}$	
(*trans*-Lycopene)$^{\cdot+}$ + mesoporphyrine ⟶ (mesoporphyrine)$^{\cdot+}$ + *trans*-lycopene [37])				
Pulse rad. of *trans*-lycopene [37]) + n-hexane				79 McV 1
KAS	n-C_6H_{14}	RT	$k < 1 \cdot 10^8\ M^{-1}s^{-1}$	
(*trans*-Lycopene)$^{\cdot+}$ + pheophytin a ⟶ (pheophytin a)$^{\cdot+}$ + *trans*-lycopene [37])				
Pulse rad. of *trans*-lycopene [37]) + n-hexane				79 McV 1
KAS	n-C_6H_{14}	RT	$k = 1.3 \cdot 10^9\ M^{-1}s^{-1}$	
(*trans*-Lycopene)$^{\cdot+}$ + pheophytin b ⟶ (pheophytin b)$^{\cdot+}$ + *trans*-lycopene [37])				
Pulse rad. of *trans*-lycopene [37]) + n-hexane				79 McV 1
KAS	n-C_6H_{14}	RT	$k < 1 \cdot 10^8\ M^{-1}s^{-1}$	
(7,7′-Dihydro-β-carotene)$^{\cdot+}$ + chlorophyll a ⟶ (chlorophyll a)$^{\cdot+}$ + 7,7′-dihydro-β-carotene [38])				
Pulse rad. of 7,7′-dihydro-β-carotene [38]) + n-hexane				78 Laf 1
KAS	n-C_6H_{14}	RT	$k = 5.4 \cdot 10^9\ M^{-1}s^{-1}$ [39])	
(7,7′-Dihydro-β-carotene)$^{\cdot+}$ + chlorophyll b ⟶ (chlorophyll b)$^{\cdot+}$ + 7,7′-dihydro-β-carotene [38])				
Pulse rad. of 7,7′-dihydro-β-carotene [38]) + n-hexane				78 Laf 1
KAS	n-C_6H_{14}	RT	$k = 1.0 \cdot 10^{10}\ M^{-1}s^{-1}$ [39])	
(Decapreno-β-carotene)$^{\cdot+}$ + chlorophyll a ⟶ (chlorophyll a)$^{\cdot+}$ + decapreno-β-carotene [40])				
Pulse rad. of decapreno-β-carotene [40]) + n-hexane				78 Laf 1
KAS	n-C_6H_{14}	RT	$k = 4.7 \cdot 10^9\ M^{-1}s^{-1}$ [39])	

[37]) Lycopene = lycopin = neolycopene = $C_{40}H_{56}$

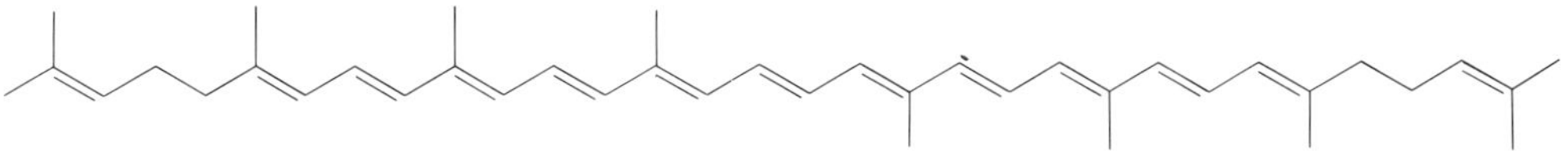

[38])

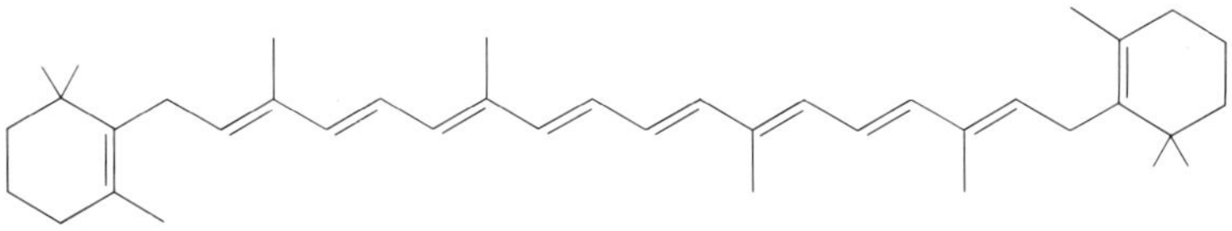

[39]) Based on assumption of monomeric chlorophyll.

[40])

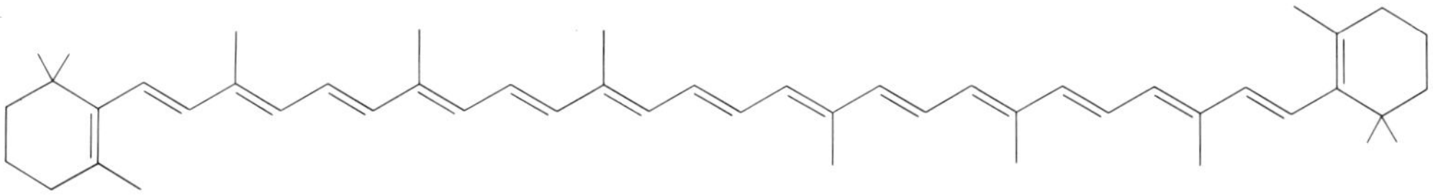

Reaction Radical generation Method	Solvent	T[K]	Rate data	Ref./ add. ref.
(Decapreno-β-carotene)$^{+\cdot}$ + chlorophyll b ⟶ (chlorophyll b)$^{+\cdot}$ + decapreno-β-carotene [40]				
Pulse rad. of decapreno-β-carotene [40] + n-hexane				78 Laf 1
KAS	n-C_6H_{14}	RT	$k < 1 \cdot 10^8\ M^{-1}s^{-1}$ [39]	
(Decapreno-β-carotene)$^{+\cdot}$ + copper pheophytin a ⟶ (copper pheophytin a)$^{+\cdot}$ + decapreno-β-carotene [40]				
Pulse rad. of decapreno-β-carotene [40] + n-hexane				79 McV 1
KAS	n-C_6H_{14}	RT	$k = 6 \cdot 10^8\ M^{-1}s^{-1}$	
(Decapreno-β-carotene)$^{+\cdot}$ + copper pheophytin b ⟶ (copper pheophytin b)$^{+\cdot}$ + decapreno-β-carotene [40]				
Pulse rad. of decapreno-β-carotene [40] + n-hexane				79 McV 1
KAS	n-C_6H_{14}	RT	$k = 4 \cdot 10^8\ M^{-1}s^{-1}$	
(Decapreno-β-carotene)$^{+\cdot}$ + etioporphyrine ⟶ (etioporphyrine)$^{+\cdot}$ + decapreno-β-carotene [40]				
Pulse rad. of decapreno-β-carotene [40] + n-hexane				79 McV 1
KAS	n-C_6H_{14}	RT	$k < 1 \cdot 10^8\ M^{-1}s^{-1}$	
(Decapreno-β-carotene)$^{+\cdot}$ + mesoporphyrine ⟶ (mesoporphyrine)$^{+\cdot}$ + decapreno-β-carotene [40]				
Pulse rad. of decapreno-β-carotene [40] + n-hexane				79 McV 1
KAS	n-C_6H_{14}	RT	$k < 1 \cdot 10^8\ M^{-1}s^{-1}$	
(Decapreno-β-carotene)$^{+\cdot}$ + pheophytin a ⟶ (pheophytin a)$^{+\cdot}$ + decapreno-β-carotene [40]				
Pulse rad. of decapreno-β-carotene [40] + n-hexane				79 McV 1
KAS	n-C_6H_{14}	RT	$k = 4 \cdot 10^8\ M^{-1}s^{-1}$	
(Decapreno-β-carotene)$^{+\cdot}$ + pheophytin b ⟶ (pheophytin b)$^{+\cdot}$ + decapreno-β-carotene [40]				
Pulse rad. of decapreno-β-carotene [40] + n-hexane				79 McV 1
KAS	n-C_6H_{14}	RT	$k < 1 \cdot 10^8\ M^{-1}s^{-1}$	

4.2.2.1.2 Radicals containing only C, H, and Cl atoms

Reaction Radical generation Method	Solvent	T[K]	Rate data	Ref./ add. ref.
$(CCl_4)^{+\cdot}$ + benzene (C_6H_6) ⟶ CCl_4 + $(C_6H_6)^{+\cdot}$				
Pulse rad. of CCl_4				79 Meh 1
KAS	CCl_4	293(2)	$k = 6.5 \cdot 10^9\ M^{-1}s^{-1}$	
$(CCl_4)^{+\cdot}$ + 1-chlorobutane (n-C_4H_9Cl) ⟶ CCl_4 + (n-$C_4H_9Cl)^{+\cdot}$				
Pulse rad. of CCl_4				79 Meh 1
KAS	CCl_4	293(2)	$k = 8 \cdot 10^9\ M^{-1}s^{-1}$	
$(CCl_4)^{+\cdot}$ + 2-chloropropane ($(CH_3)_2CHCl$) ⟶ CCl_4 + $((CH_3)_2CHCl)^{+\cdot}$				
Pulse rad. of CCl_4				79 Meh 1
KAS	CCl_4	293(2)	$k = 2 \cdot 10^9\ M^{-1}s^{-1}$	
$(CCl_4)^{+\cdot}$ + dichloroethane ($C_2H_4Cl_2$) [41] ⟶ CCl_4 + $(C_2H_4Cl_2)^{+\cdot}$				
Pulse rad. of CCl_4				79 Meh 1
KAS	CCl_4	293(2)	$k = 5 \cdot 10^8\ M^{-1}s^{-1}$	

[39] Based on assumption of monomeric chlorophyll.

[40]

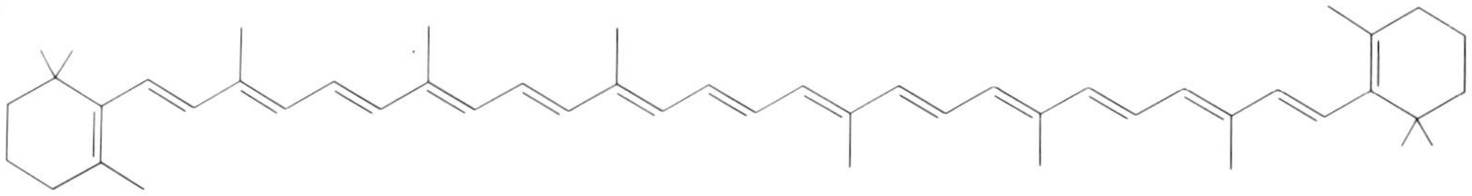

[41] Not specified whether 1,1- or 1,2-dichloroethane.

Reaction Radical generation Method	Solvent	T[K]	Rate data	Ref./ add. ref.
$(CCl_4)^{+\cdot}$ + dichloromethane (CH_2Cl_2) ⟶ CCl_4 + $(CH_2Cl_2)^{+\cdot}$				
Pulse rad. of CCl_4 KAS	CCl_4	293(2)	$k = 1 \cdot 10^8\ M^{-1}s^{-1}$	79 Meh 1
$(CCl_4)^{+\cdot}$ + *n*-heptane (n-C_7H_{16}) ⟶ CCl_4 + (n-$C_7H_{16})^{+\cdot}$				
Pulse rad. of CCl_4 KAS	CCl_4	293(2)	$k = 7 \cdot 10^9\ M^{-1}s^{-1}$	79 Meh 1
$(CCl_4)^{+\cdot}$ + heptene-1 ⟶ CCl_4 + (heptene-1)$^{+\cdot}$				
Pulse rad. of CCl_4 KAS	CCl_4	293(2)	$k = 9 \cdot 10^9\ M^{-1}s^{-1}$	79 Meh 1
$(CCl_4)^{+\cdot}$ + *c*-hexane (c-C_6H_{12}) ⟶ CCl_4 + (c-$C_6H_{12})^{+\cdot}$				
Pulse rad. of CCl_4 KAS	CCl_4	293(2)	$k = 1 \cdot 10^{10}\ M^{-1}s^{-1}$	79 Meh 1
$(CCl_4)^{+\cdot}$ + hexene-1 ⟶ CCl_4 + (hexene-1)$^{+\cdot}$				
Pulse rad. of CCl_4 KAS	CCl_4	293(2)	$k = 9 \cdot 10^9\ M^{-1}s^{-1}$	79 Meh 1
$(CCl_4)^{+\cdot}$ + hexene-2 ⟶ CCl_4 + (hexene-2)$^{+\cdot}$				
Pulse rad. of CCl_4 KAS	CCl_4	293(2)	$k = 1 \cdot 10^{10}\ M^{-1}s^{-1}$	79 Meh 1
$(CCl_4)^{+\cdot}$ + *c*-hexene (c-C_6H_{10}) ⟶ CCl_4 + (c-$C_6H_{10})^{+\cdot}$				
Pulse rad. of CCl_4 KAS	CCl_4	293(2)	$k = 7 \cdot 10^9\ M^{-1}s^{-1}$	79 Meh 1
$(CCl_4)^{+\cdot}$ + 3-methylcyclohexene ⟶ CCl_4 + (3-methylcyclohexene)$^{+\cdot}$				
Pulse rad. of CCl_4 KAS	CCl_4	293(2)	$k = 9 \cdot 10^9\ M^{-1}s^{-1}$	79 Meh 1
$(CCl_4)^{+\cdot}$ + 3-methylheptane ⟶ CCl_4 + (3-methylheptane)$^{+\cdot}$				
Pulse rad. of CCl_4 KAS	CCl_4	293(2)	$k = 1 \cdot 10^{10}\ M^{-1}s^{-1}$	79 Meh 1
$(CCl_4)^{+\cdot}$ + 2-methylhexene-1 ⟶ CCl_4 + (2-methylhexene-1)$^{+\cdot}$				
Pulse rad. of CCl_4 KAS	CCl_4	293(2)	$k = 6 \cdot 10^9\ M^{-1}s^{-1}$	79 Meh 1
$(CCl_4)^{+\cdot}$ + *c*-pentane (c-C_5H_{10}) ⟶ CCl_4 + (c-$C_5H_{10})^{+\cdot}$				
Pulse rad. of CCl_4 KAS	CCl_4	293(2)	$k = 9 \cdot 10^9\ M^{-1}s^{-1}$	79 Meh 1
$(CCl_4)^{+\cdot}$ + toluene ($C_6H_5CH_3$) ⟶ CCl_4 + $(C_6H_5CH_3)^{+\cdot}$				
Pulse rad. of CCl_4 KAS	CCl_4	293(2)	$k = 6.5 \cdot 10^9\ M^{-1}s^{-1}$	79 Meh 1
$(CCl_4)^{+\cdot}$ + trichloromethane ($CHCl_3$) ⟶ CCl_4 + $(CHCl_3)^{+\cdot}$				
Pulse rad. of CCl_4 KAS	CCl_4	293(2)	$k = 3.5 \cdot 10^8\ M^{-1}s^{-1}$	79 Meh 1
$(ClCH_2CH_2Cl)^{+\cdot}$ + Cd(II)-tetraphenylporphyrine ⟶ products [42]				
Pulse rad. of 1,2-dichloroethane KAS	$ClCH_2CH_2Cl$ + 1% pyridine	RT	$k = 3.4(5) \cdot 10^8\ M^{-1}s^{-1}$	81 Net 1

[42]) Oxidation at ligand.

Reaction Radical generation Method	Solvent	T[K]	Rate data	Ref./ add. ref.
$(ClCH_2CH_2Cl)^{+\cdot}$ + Co(II)-tetraphenylporphyrine ⟶ Co(III)-tetraphenylporphyrine + $ClCH_2CH_2Cl$				
Pulse rad. of 1,2-dichloroethane				81 Net 1
KAS	$ClCH_2CH_2Cl$ + 1% pyridine	RT	$k = 5(1) \cdot 10^8\ M^{-1}s^{-1}$	
$(ClCH_2CH_2Cl)^{+\cdot}$ + Cu(II)-tetraphenylporphyrine ⟶ products [42]				
Pulse rad. of 1,2-dichloroethane				81 Net 1
KAS	$ClCH_2CH_2Cl$ + 1% pyridine	RT	$k = 8(1) \cdot 10^7\ M^{-1}s^{-1}$	
$(ClCH_2CH_2Cl)^{+\cdot}$ + $(C_6H_5CH_2)_2Hg$ ⇝ $(C_6H_5CH_2)^+$ + products				
Pulse rad. of $ClCH_2CH_2Cl$				78 Dor 1
KAS	$ClCH_2CH_2Cl$	297	$k = 1.3 \cdot 10^{10}\ M^{-1}s^{-1}$ [43]	
$(ClCH_2CH_2Cl)^{+\cdot}$ + Mg(II)-tetraphenylporphyrine ⟶ products [42]				
Pulse rad. of 1,2-dichloroethane				81 Net 1
KAS	$ClCH_2CH_2Cl$ + 1% pyridine	RT	$k = 6(1) \cdot 10^8\ M^{-1}s^{-1}$	
$(ClCH_2CH_2Cl)^{+\cdot}$ + Mn(III)$(OOCCH_3)$tetraphenylporphyrine ⟶ products [42]				
Pulse rad. of 1,2-dichloroethane				81 Net 1
KAS	$ClCH_2CH_2Cl$ + 1% pyridine	RT	$k = 4(1) \cdot 10^8\ M^{-1}s^{-1}$	
$(ClCH_2CH_2Cl)^{+\cdot}$ + Ni(II)-tetraphenylporphyrine ⟶ products [44]				
Pulse rad. of 1,2-dichloroethane				81 Net 1
KAS	$ClCH_2CH_2Cl$ + 1% pyridine	RT	$k \approx 5 \cdot 10^8\ M^{-1}s^{-1}$	
$(ClCH_2CH_2Cl)^{+\cdot}$ + Pb(II)-tetraphenylporphyrine ⟶ products [42]				
Pulse rad. of 1,2-dichloroethane				81 Net 1
KAS	$ClCH_2CH_2Cl$ + 1% pyridine	RT	$k \approx 5 \cdot 10^8\ M^{-1}s^{-1}$	
$(ClCH_2CH_2Cl)^{+\cdot}$ + V(IV)O-tetraphenylporphyrine ⟶ products [42]				
Pulse rad. of 1,2-dichloroethane				81 Net 1
KAS	$ClCH_2CH_2Cl$ + 1% pyridine	RT	$k = 5(1) \cdot 10^8\ M^{-1}s^{-1}$	
$(ClCH_2CH_2Cl)^{+\cdot}$ + Zn(II)-tetraphenylporphyrine ⟶ products [42]				
Pulse rad. of 1,2-dichloroethane				81 Net 1
KAS	$ClCH_2CH_2Cl$ + 1% pyridine	RT	$k = 7(1) \cdot 10^8\ M^{-1}s^{-1}$	
$(ClCH_2CH_2Cl)^{+\cdot}$ + chlorophyll a ⟶ (chlorophyll a)$^{+\cdot}$ + $ClCH_2CH_2Cl$				
Pulse rad. of 1,2-dichloroethane				81 Net 1
KAS	$ClCH_2CH_2Cl$ + 1% pyridine	RT	$k = 2.3(5) \cdot 10^9\ M^{-1}s^{-1}$	
$(ClCH_2CH_2Cl)^{+\cdot}$ + biphenyl ⟶ $ClCH_2CH_2Cl$ + (biphenyl)$^{+\cdot}$ [45]				
Pulse rad. of $ClCH_2CH_2Cl$				70 Sha 1
KAS	$ClCH_2CH_2Cl$	298	$k > 3 \cdot 10^{10}\ M^{-1}s^{-1}$	

[42]) Oxidation at ligand.
[43]) k is determined via $(C_6H_5CH_2)^+$ formation kinetics. There is some evidence that this rate constant refers to an intermediate process and may not be attributable to initial step.
[44]) Oxidation at metal center (in absence of pyridine oxidation at ligand).
[45]) Electron jump mechanism.

Reaction Radical generation Method	Solvent	T[K]	Rate data	Ref./ add. ref.
$(ClCH_2CH_2Cl)^{+\cdot}$ + diphenylbromomethane$\{(C_6H_5)_2CHBr\}$ $\rightsquigarrow$ $(C_6H_5)_2CH^+$ + products				
Pulse rad. of $ClCH_2CH_2Cl$				78 Dor 1/ 79 Wan 1
KAS	$ClCH_2CH_2Cl$	297	$k = 1.6\cdot10^{10}\ M^{-1}s^{-1}$ [46]	
$\dot{R}^{+}$ [47] + diphenylbromomethane$\{(C_6H_5)_2CHBr\}$ $\longrightarrow$ $(C_6H_5)_2CH^+$ + Br + R				
Pulse rad. of 1,2-dichloroethane				79 Wan 1/ 78 Dor 1
—	$ClCH_2CH_2Cl$	RT	$k = 1.2\cdot10^{10}\ M^{-1}s^{-1}$	
$(ClCH_2CH_2Cl)^{+\cdot}$ + p-terphenyl $\longrightarrow$ $ClCH_2CH_2Cl$ + $(p\text{-terphenyl})^{+\cdot}$ [45]				
Pulse rad. of $ClCH_2CH_2Cl$				70 Sha 1
KAS	$ClCH_2CH_2Cl$	242	$k > 1\cdot10^{10}\ M^{-1}s^{-1}$	
$(ClCH_2CH_2Cl)^{+\cdot}$ + triphenylbromomethane$\{(C_6H_5)_3CBr\}$ $\rightsquigarrow$ $(C_6H_5)_3C^+$ + products				
Pulse rad. of $ClCH_2CH_2Cl$				78 Dor 1
KAS	$ClCH_2CH_2Cl$	297	$k = 8.4\cdot10^{9}\ M^{-1}s^{-1}$ [48]	
$(ClCH_2CH_2Cl)^{+\cdot}$ + triphenylchloromethane$\{(C_6H_5)_3CCl\}$ $\rightsquigarrow$ $(C_6H_5)_3C^+$ + products				
Pulse rad. of $ClCH_2CH_2Cl$				78 Dor 1
KAS	$ClCH_2CH_2Cl$	297	$k = 4.0\cdot10^{8}\ M^{-1}s^{-1}$ [48]	
$(ClCH_2CH_2Cl)^{+\cdot}$ + triphenylmethanol$\{(C_6H_5)_3COH\}$ $\rightsquigarrow$ $(C_6H_5)_3C^+$ + products				
Pulse rad. of $ClCH_2CH_2Cl$				78 Dor 1
KAS	$ClCH_2CH_2Cl$	297	$k = 5.7\cdot10^{8}\ M^{-1}s^{-1}$ [48]	
$(n\text{-}C_4H_9Cl)^{+\cdot}$ + 1,3-c-hexadiene $\longrightarrow$ (1,3-c-hexadiene)$^{+\cdot}$ + $n\text{-}C_4H_9Cl$				
Pulse rad. of n-butylchloride				82 Meh 1
KAS	$n\text{-}C_4H_9Cl$	RT	$k = 1.0(3)\cdot10^{10}\ M^{-1}s^{-1}$	
$(n\text{-}C_4H_9Cl)^{+\cdot}$ + 1,4-c-hexadiene $\longrightarrow$ (1,4-c-hexadiene)$^{+\cdot}$ + $n\text{-}C_4H_9Cl$				
Pulse rad. of n-butylchloride				82 Meh 1
KAS	$n\text{-}C_4H_9Cl$	RT	$k = 1.0(3)\cdot10^{10}\ M^{-1}s^{-1}$	
$(n\text{-}C_4H_9Cl)^{+\cdot}$ + isoprene $\longrightarrow$ (isoprene)$^{+\cdot}$ + $n\text{-}C_4H_9Cl$				
Pulse rad. of n-butylchloride				82 Meh 1
KAS	$n\text{-}C_4H_9Cl$	RT	$k = 5(2)\cdot10^{9}\ M^{-1}s^{-1}$	
$(n\text{-}C_4H_9Cl)^{+\cdot}$ + 1,5-c-octadiene $\longrightarrow$ (1,5-c-octadiene)$^{+\cdot}$ + $n\text{-}C_4H_9Cl$				
Pulse rad. of n-butylchloride				82 Meh 1
KAS	$c\text{-}C_4H_9Cl$	RT	$k = 1.0(3)\cdot10^{10}\ M^{-1}s^{-1}$	
$(n\text{-}C_4H_9Cl)^{+\cdot}$ + cis-1,3-pentadiene $\longrightarrow$ (cis-1,3-pentadiene)$^{+\cdot}$ + $n\text{-}C_4H_9Cl$				
Pulse rad. of n-butylchloride				82 Meh 1
KAS	$n\text{-}C_4H_9Cl$	RT	$k = 9(3)\cdot10^{9}\ M^{-1}s^{-1}$	

[45]) Electron jump mechanism.
[46]) k is determined via $(C_6H_5)_2CH^+$ formation kinetics. There is some evidence that this rate constant refers to an intermediate process and may not be attributable to initial step.
[47]) Radical cations from irr. of 1,2-dichloroethane: $(CH_2ClCH_2Cl)^{+\cdot}$ and/or $(CH_2CHCl)^{+\cdot}$.
[48]) k is determined via $(C_6H_5)_3C^+$ formation kinetics. There is some evidence that this rate constant refers to an intermediate process and may not be attributable to initial step.

Reaction Radical generation Method	Solvent	T[K]	Rate data	Ref./ add. ref.
4.2.2.1.3 Radicals containing only C, H, and O atoms				
4.2.2.1.3.1 Neutral radicals				
$\dot{C}H_2OH$ + (5,7,7,12,14,14-hexamethyl-1,4,8,11-tetraazacyclotetradeca-4,11,-diene-N,N′,N″,N‴)-copper(II) ion ⟶ Cu(I)...				
Flash phot., via excited state of CuN_4...complex KAS	CH_3OH	RT	$k = 2.2 \cdot 10^4\ M^{-1}s^{-1}$	77 Fer 1
$\dot{C}H_2OH$ + carbontetrachloride (CCl_4) ⟶ HCHO + HCl + $\dot{C}Cl_3$				
Phot. of di-*t*-butylperoxide + CH_3OH KESR	CH_3OH	233 218... 290	$k = (2.5...3.0) \cdot 10^5\ M^{-1}s^{-1}$ $\log[A/M^{-1}s^{-1}] = 8.146$ $E_a = 11.9\ kJ\ mol^{-1}$	79 Pau 1
$\dot{C}H_2OH$ + hexachloroethane $\xrightarrow{a}$ HCHO + HCl + $CCl_3\dot{C}Cl_2$ $CH_3\dot{C}HOH$ + hexachloroethane $\xrightarrow{b}$ CH_3CHO + HCl + $CCl_3\dot{C}Cl_2$ $(CH_3)_2\dot{C}OH$ + hexachloroethane $\xrightarrow{c}$ $(CH_3)_2CO$ + HCl + $CCl_3\dot{C}Cl_2$				
γ-rad. of CH_3OH, C_2H_5OH or $(CH_3)_2CHOH$ PR, glc	CH_3OH or C_2H_5OH or $(CH_3)_2CHOH$	RT [48a]	$k_a:k_b:k_c = 1:3:14$ [48b]	78 Saw 1/ 75 Joh 1, 67 Sed 1, 69 Sim 1
$\dot{C}H_2OH$ + retinal Schiff's base [49] (R) + CH_3OH ⟶ $\dot{R}^-$ + CH_2O + $CH_3OH_2^+$				
Pulse rad. of $CH_3OH + N_2O$ KAS	CH_3OH	RT	$k = 9 \cdot 10^7\ M^{-1}s^{-1}$	81 Rag 1
$\dot{C}H_2OH$ + tetranitromethane{$C(NO_2)_4$} ⟶ $C(NO_2)_3^-$ + NO_2 + H^+ + CH_2O				
Pulse rad. of CH_3OH KAS	CH_3OH	RT	$k = 7.0(4) \cdot 10^9\ M^{-1}s^{-1}$	72 Cha 2
$CH_3\dot{C}HOH$ + hexachloroethane ⟶ CH_3CHO + HCl + $CCl_3\dot{C}Cl_2$			See $\dot{C}H_2OH + C_2Cl_6$ reaction	
$CH_3\dot{C}HOH$ + tetranitromethane{$C(NO_2)_4$} ⟶ $C(NO_2)_3^-$ + H^+ + NO_2 + CH_3CHO				
Pulse rad. of C_2H_5OH KAS	C_2H_5OH	RT	$k = 4.0(4) \cdot 10^9\ M^{-1}s^{-1}$	72 Cha 2
$CH_2OH\dot{C}HOH$ + tetranitromethane{$C(NO_2)_4$} ⟶ $C(NO_2)_3^-$ + H^+ + CH_2OHCHO + NO_2				
Pulse rad. of ethylene glycol KAS	CH_2OHCH_2OH	RT	$k = 5.3(3) \cdot 10^8\ M^{-1}s^{-1}$	72 Cha 2
$(CH_3)_2\dot{C}OH$ + chlorophyll a ⟶ (chlorophyll a)$^{\dot{-}}$ + H^+ + $(CH_3)_2CO$				
Pulse rad. of 2-propanol KAS	2-propanol	RT	$k = 7(3) \cdot 10^7\ M^{-1}s^{-1}$	79 Net 2

[48a]) Temp. not given, presumed to be RT.
[48b]) Based on $2k(\dot{C}H_2OH + \dot{C}H_2OH) = 2.7 \cdot 10^9\ M^{-1}s^{-1}$ [75 Joh 1], $2k(CH_3\dot{C}HOH + CH_3\dot{C}HOH) = 2.0 \cdot 10^9\ M^{-1}s^{-1}$ [67 Sed 1], and $2k(CH_3)_2\dot{C}OH + (CH_3)_2\dot{C}OH = 1.4 \cdot 10^9\ M^{-1}s^{-1}$ [69 Sim 1].
[49])

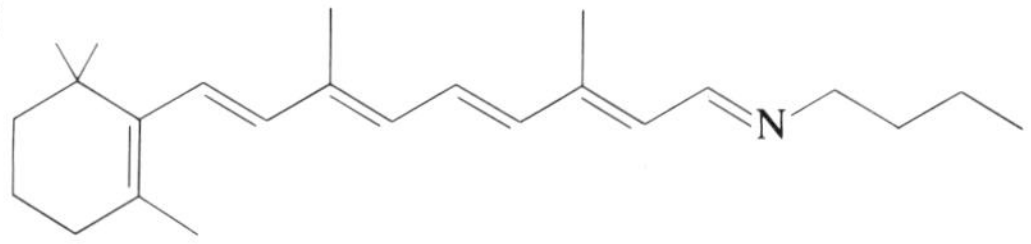

Reaction Radical generation Method	Solvent	T[K]	Rate data	Ref./ add. ref.
$(CH_3)_2\dot{C}OH$ + vitamin B12r(cobal(II) amin) $\longrightarrow$ B12s(Co(I)) + $(CH_3)_2CO + H^+$				
Flash phot. of 2-propanol				79 End 1
KAS	2-propanol (90%)/ H_2O (10%)	RT	$k = 4(2) \cdot 10^9\ M^{-1}s^{-1}$	
$(CH_3)_2\dot{C}OH$ + Zn-tetraphenylporphyrine(ZnTPP) $\longrightarrow$ $(ZnTPP)^{\dot{-}} + H^+ + (CH_3)_2CO$				
Pulse rad. of 2-propanol				79 Net 2
KAS	2-propanol	RT	$k \approx 10^7\ M^{-1}s^{-1}$	
$(CH_3)_2\dot{C}OH$ + *anti*-azobenzene(A_a) $(CH_3)_2CO + A_aH^{\cdot}$ [50])				
Pulse rad. of 2-propanol				77 Net 1
KAS	2-propanol	RT	$k = 3.0 \cdot 10^7\ M^{-1}s^{-1}$	
$(CH_3)_2\dot{C}OH$ + hexachloroethane $\longrightarrow$ $(CH_3)_2CO + HCl + CCl_3\dot{C}Cl_2$			See $\dot{C}H_2OH + C_2Cl_6$ reaction	
$(CH_3)_2\dot{C}OH$ + tetranitromethane$\{C(NO_2)_4\}$ $\longrightarrow$ $C(NO_2)_3^- + H^+ + NO_2 + (CH_3)_2CO$				
Pulse rad. of 2-propanol				71 Asm 1
KAS	2-propanol	RT	$k = 7.0(3) \cdot 10^9\ M^{-1}s^{-1}$	
$(CH_3)_2\dot{C}OH$ + *meso*-tetraphenylporphyrine(H_2TPP) $\longrightarrow$ $(H_2TPP)^{\dot{-}} + H^+ + (CH_3)_2CO$				
Pulse rad. of 2-propanol				79 Net 2
KAS	2-propanol	RT	$k \approx 1 \cdot 10^8\ M^{-1}s^{-1}$	
$(CH_3)_2\dot{C}OH$ + trichloroacetic acid (CCl_3COOH) $\longrightarrow$ $(CH_3)_2CO + HCl + \dot{C}Cl_2COOH$				
Phot. of $(CH_3)_2CO$ in 2-propanol				78 Ays 1
KESR	2-propanol	293	$k = 6.1(16) \cdot 10^6\ M^{-1}s^{-1}$	
$C_2H_5\dot{C}(CH_3)OH$ + 4-chloroacetophenone $\xrightarrow{a}$ $C_2H_5COCH_3$ + 4-$ClC_6H_4\dot{C}(OH)CH_3$ [50a]) + acetophenone $\xrightarrow{b}$ $C_2H_5COCH_3 + C_6H_5\dot{C}(OH)CH_3$ [50a])				
Thermal decomp. of di-*t*-butylperoxide				63 Huy 1
PR, glc	2-butanol	398	$k_a/k_b = 3.01(23)$	
$C_2H_5\dot{C}(CH_3)OH$ + 2,4-dimethylacetophenone $\xrightarrow{a}$ $C_2H_5COCH_3$ + 2,4-$(CH_3)_2C_6H_3\dot{C}(OH)CH_3$ [50a]) + acetophenone $\xrightarrow{b}$ $C_2H_5COCH_3 + C_6H_5\dot{C}(OH)CH_3$ [50a])				
Thermal decomp. of di-*t*-butylperoxide				63 Huy 1
PR, glc	2-butanol	398	$k_a/k_b = 0.18(10)$	
$C_2H_5\dot{C}(CH_3)OH$ + 4-methoxyacetophenone $\xrightarrow{a}$ $C_2H_5COCH_3$ + 4-$CH_3OC_6H_4\dot{C}(OH)CH_3$ [50a]) + acetophenone $\xrightarrow{b}$ $C_2H_5COCH_3 + C_6H_5\dot{C}(OH)CH_3$ [50a])				
Thermal decomp. of di-*t*-butylperoxide				63 Huy 1
PR, glc	2-butanol	398	$k_a/k_b = 0.17(1)$	
$C_2H_5\dot{C}(CH_3)OH$ + 3-methylacetophenone $\xrightarrow{a}$ $C_2H_5COCH_3$ + 3-$CH_3C_6H_4\dot{C}(OH)CH_3$ [50a]) + acetophenone $\xrightarrow{b}$ $C_2H_5COCH_3 + C_6H_5\dot{C}(OH)CH_3$ [50a])				
Thermal decomp. of di-*t*-butylperoxide				63 Huy 1
PR, glc	2-butanol	398	$k_a/k_b = 0.85(1)$	
$C_2H_5\dot{C}(CH_3)OH$ + 4-methylacetophenone $\xrightarrow{a}$ $C_2H_5COCH_3$ + 4-$CH_3C_6H_4\dot{C}(OH)CH_3$ [50a]) + acetophenone $\xrightarrow{b}$ $C_2H_5COCH_3 + C_6H_5\dot{C}(OH)CH_3$ [50a])				
Thermal decomp. of di-*t*-butylperoxide				63 Huy 1
PR, glc	2-butanol	398	$k_a/k_b = 0.59(6)$	

[50]) Electron transfer and subsequent protonation of $A_a^{\dot{-}}$.
[50a]) Reaction likely to occur via e^--transfer mechanism.

Reaction Radical generation Method	Solvent	T[K]	Rate data	Ref./ add. ref.
$C_2H_5\dot{C}(CH_3)OH$ + 3-(trifluoromethyl)acetophenone $\xrightarrow{a}$ $C_2H_5COCH_3$ + 3-$CF_3C_6H_4\dot{C}(OH)CH_3$ [50a]) + acetophenone $\xrightarrow{b}$ $C_2H_5COCH_3$ + $C_6H_5\dot{C}(OH)CH_3$ [50a])				
Thermal decomp. of di-*t*-butylperoxide				63 Huy 1
PR, glc	2-butanol	398	$k_a/k_b = 4.85(50)$	
$C_2H_5\dot{C}(CH_3)OH$ + 2,4,6-trimethylacetophenone $\longrightarrow$ $C_2H_5COCH_3$ + 2,4,6-$(CH_3)_3C_6H_2\dot{C}(OH)CH_3$ [50a]) + acetophenone $\longrightarrow$ $C_2H_5COCH_3$ + $C_6H_5\dot{C}(OH)CH_3$ [50a])				
Thermal decomp. of di-*t*-butylperoxide				63 Huy 1
PR, glc	2-butanol	398	$k_a/k_b = 0.05(3)$	
$C_6H_{13}\dot{C}HCH_2CH_2COCH_3$ + Ce(IV)$(OOCCH_3)_4$ $\xrightarrow{e}$ Ce(III)... + products + CH_3COCH_3 $\xrightarrow{a}$ $CH_3CO\dot{C}H_2$ + products				
Ox. of CH_3COCH_3 by Ce(IV)-acetate and $CH_3CO\dot{C}H_2$ addition to $C_6H_{13}CH{=}CH_2$ in glacial acetic acid				71 Hei 1
PR	glacial acetic acid/ 10% $NaOOCCH_3$	340 318	$k_e/k_a = 5.4 \cdot 10^2$ $k_e/k_a = 6.3 \cdot 10^2$ $E_a = 39.8\,kJ\,mol^{-1}$ [51])	
$C_6H_{13}\dot{C}HCH_2CH_2COCH_3$ + Cu(II)$(OOCCH_3)_2$ $\xrightarrow{a}$ Cu(I)... + products + Mn(III)$(OOCCH_3)_3$ $\xrightarrow{b}$ Mn(II)... + products				
Metal acetate induced oxidation of CH_3COCH_3 and $CH_3CO\dot{C}H_2$ addition to $C_6H_{13}CHCH_2$ in glacial acetic acid				71 Hei 1
PR	glacial acetic acid/ 10% $NaOOCCH_3$	298	$k_b/k_a = 350$ $E_a = 52.3\,kJ\,mol^{-1}$ [52])	
$C_6H_{13}\dot{C}HCH_2CH_2COCH_3$ + Mn(III)$(OOCCH_3)_3$ $\xrightarrow{e}$ Mn(II)... + products + CH_3COCH_3 $\xrightarrow{a}$ $CH_3CO\dot{C}H_2$ + products				
Ox. of CH_3COCH_3 by Mn(III)-acetate and $CH_3CO\dot{C}H_2$ addition to $C_6H_{13}CH{=}CH_2$ in glacial acetic acid				71 Hei 1
PR	glacial acetic acid/ 10% $NaOOCCH_3$	343 318	$k_e/k_a = 45.4$ $k_e/k_a = 38.2$ $E_a = 52.4\,kJ\,mol^{-1}$ [51])	
4.2.2.1.3.2 Anionic radicals				
$(CH_3)_2\dot{C}O^-$ + chlorophyll a $\longrightarrow$ (chlorophyll a)$^{\dot{-}}$ + $(CH_3)_2CO$				
Pulse rad. of $(CH_3)_2CHONa$ + 2-propanol				79 Net 1
KAS	2-propanol [53])	RT	$k = 6(1) \cdot 10^8\,M^{-1}s^{-1}$	
$(CH_3)_2\dot{C}O^-$ + Na-tetraphenylporphyrine(Na_2TPP) $\longrightarrow$ $(Na_2TPP)^{\dot{-}}$ + $(CH_3)_2CO$				
Pulse rad. of $(CH_3)_2CHONa$ + 2-propanol				79 Net 2
KAS	2-propanol [53])	RT	$k = 2.0(4) \cdot 10^8\,M^{-1}s^{-1}$	
$(CH_3)_2\dot{C}O^-$ + Zn-tetraphenylporphyrine(ZnTPP) $\rightarrow$ $(ZnTPP)^{\dot{-}}$ + $(CH_3)_2CO$				
Pulse rad. of $(CH_3)_2CHONa$ + 2-propanol				79 Net 2
KAS	2-propanol [53])	RT	$k = 6(1) \cdot 10^8\,M^{-1}s^{-1}$	
$(CH_3)_2\dot{C}O^-$ + acenaphthylene $\longrightarrow$ (acenaphthylene)$^{\dot{-}}$ + $(CH_3)_2CO$ [54])				
Pulse rad. of 2-propanol + $(CH_3)_2CHONa$				78 Lev 1
KAS	2-propanol	295(1)	$k = 1.4 \cdot 10^9\,M^{-1}s^{-1}$	

[50a]) Reaction likely to occur via e^--transfer mechanism.
[51]) Calc. from $\Delta E_a = -6.3\,kJ\,mol^{-1}$ and assuming $E_a = 46.1\,kJ\,mol^{-1}$ for H-atom abstraction process.
[52]) For both processes. Based on $E_a = 46.1\,kJ\,mol^{-1}$ for H-atom abstraction process $C_6H_{13}\dot{C}HCH_2CH_2COCH_3$ + CH_3COCH_3.
[53]) 10^{-1} M $(CH_3)_2CHONa$ added.
[54]) k[(acenaphthylene)$^{\dot{-}}$ + $(CH_3)_2CHOH$ $\longrightarrow$ (acenaphthylene − H)$^{\cdot}$ + $(CH_3)_2CHO^-$] $= 4 \cdot 10^2\,s^{-1}$.

Reaction Radical generation Method	Solvent	T[K]	Rate data	Ref./ add. ref.
$(CH_3)_2\dot{C}O^- +$ anthracene $\longrightarrow$ (anthracene)$^{\dot{-}}$ $+ (CH_3)_2CO$				
Pulse rad. of acetone				73 Rob 1
KAS	CH_3COCH_3	298	$k = 4.9 \cdot 10^9\,M^{-1}s^{-1}$	
$(CH_3)_2\dot{C}O^- +$ azobenzene(A)[55] $\longrightarrow (CH_3)_2CO + \dot{A}^-$[56]				
Pulse rad. of 2-propanol + 0.04 M $(CH_3)_2CHONa$				77 Net 1
KAS	2-propanol	RT	$k = 2 \cdot 10^9\,M^{-1}s^{-1}$	
$(CH_3)_2\dot{C}O^- +$ azulene $\longrightarrow$ (azulene)$^{\dot{-}}$ $+ (CH_3)_2CO$[57]				
Pulse rad. of 2-propanol + $(CH_3)_2CHONa$				78 Lev 1
KAS	2-propanol	295(1)	$k = 1.5 \cdot 10^9\,M^{-1}s^{-1}$	
$(CH_3)_2\dot{C}O^- +$ benzo(a)pyrene $\longrightarrow$ (benzo(a)pyrene)$^{\dot{-}}$ $+ (CH_3)_2CO$[58]				
Pulse rad. of 2-propanol + $(CH_3)_2CHONa$				78 Lev 1
KAS	2-propanol	295(1)	$k = 7.0 \cdot 10^8\,M^{-1}s^{-1}$	
$(CH_3)_2\dot{C}O^- +$ chrysene $\longrightarrow$ (chrysene)$^{\dot{-}}$ $+ (CH_3)_2CO$[59]				
Pulse rad. of 2-propanol + $(CH_3)_2CHONa$				78 Lev 1
KAS	2-propanol	295(1)	$k < 10^8\,M^{-1}s^{-1}$	
$(CH_3)_2\dot{C}O^- +$ 1,2,3,4-dibenzanthrazene $\longrightarrow$ (1,2,3,4-...)$^{\dot{-}}$ $+ (CH_3)_2CO$[60]				
Pulse rad. of 2-propanol + $(CH_3)_2CHONa$				78 Lev 1
KAS	2-propanol	295(1)	$k = 4.0 \cdot 10^8\,M^{-1}s^{-1}$	
$(CH_3)_2\dot{C}O^- +$ 1,2,5,6-dibenzanthracene $\longrightarrow$ (1,2,5,6-...)$^{\dot{-}}$ $+ (CH_3)_2CO$[61]				
Pulse rad. of 2-propanol + $(CH_3)_2CHONa$				78 Lev 1
KAS	2-propanol	295(1)	$k = 1.5 \cdot 10^8\,M^{-1}s^{-1}$	
$(CH_3)_2\dot{C}O^- +$ fluoranthene $\longrightarrow$ (fluoranthene)$^{\dot{-}}$ $+ (CH_3)_2CO$[62]				
Pulse rad. of 2-propanol + $(CH_3)_2CHONa$				78 Lev 1
KAS	2-propanol	295(1)	$k = 7.2 \cdot 10^8\,M^{-1}s^{-1}$	
$(CH_3)_2\dot{C}O^- +$ fluorenone $\longrightarrow$ products				
Pulse rad. of acetone + $C_2H_5ONa + C_2H_5OH$				67 Ada 1
KAS	C_2H_5OH[63]	RT	$k = 2.0 \cdot 10^9\,M^{-1}s^{-1}$	
$(CH_3)_2\dot{C}O^- +$ perylene $\longrightarrow$ (perylene)$^{\dot{-}}$ $+ (CH_3)_2CO$[64]				
Pulse rad. of 2-propanol + $(CH_3)_2CHONa$				78 Lev 1
KAS	2-propanol	295(1)	$k = 1.4 \cdot 10^9\,M^{-1}s^{-1}$	
$(CH_3)_2\dot{C}O^- +$ pyrene $\longrightarrow$ (pyrene)$^{\dot{-}}$ $+ (CH_3)_2CO$[65]				
Pulse rad. of 2-propanol + $(CH_3)_2CHONa$				78 Lev 1
KAS	2-propanol	295(1)	$k = 1.1 \cdot 10^8\,M^{-1}s^{-1}$	
Pulse rad. of acetone				73 Rob 1
KAS	CH_3COCH_3	298	$k = 1.3 \cdot 10^{10}\,M^{-1}s^{-1}$	

[55]) *Syn* and *anti* form.
[56]) Only *anti* form.
[57]) k[(azulene)$^{\dot{-}}$ $+ (CH_3)_2CHOH \longrightarrow$ (azulene − H)$^{\cdot}$ $+ (CH_3)_2CHO^-$] $= 1.5\,s^{-1}$.
[58]) k[(benzo(a)pyrene)$^{\dot{-}}$ $+ (CH_3)_2CHOH \longrightarrow$ (benzo(a)pyrene − H)$^{\cdot}$ $+ (CH_3)_2CHO^-$] $= 2.5 \cdot 10^3\,s^{-1}$.
[59]) k[(chrysene)$^{\dot{-}}$ $+ (CH_3)_2CHOH \longrightarrow$ (chrysene − H)$^{\cdot}$ $+ (CH_3)_2CHO^-$] $= 2.0 \cdot 10^4\,s^{-1}$.
[60]) k[(1,2,3,4-dibenzanthrazene)$^{\dot{-}}$ $+ (CH_3)_2CHOH \longrightarrow$ (1,2,3,4-dibenzanthrazene − H)$^{\cdot}$ $+ (CH_3)_2CHO^-$] $= 3.2 \cdot 10^3\,s^{-1}$.
[61]) k[(1,2,5,6-dibenzanthracene)$^{\dot{-}}$ $+ (CH_3)_2CHOH \longrightarrow$ (1,2,5,6-dibenzanthracene − H)$^{\cdot}$ $+ (CH_3)_2CHO^-$] $= 2 \cdot 10^2\,s^{-1}$.
[62]) k[(fluoranthene)$^{\dot{-}}$ $+ (CH_3)_2CHOH \longrightarrow$ (fluoranthene − H)$^{\cdot}$ $+ (CH_3)_2CHO^-$] $\approx 7\,s^{-1}$.
[63]) 10^{-2} M C_2H_5ONa added.
[64]) k[(perylene)$^{\dot{-}}$ $+ (CH_3)_2CHOH \longrightarrow$ (perylene − H)$^{\cdot}$ $+ (CH_3)_2CHO^-$] $= 20\,s^{-1}$.
[65]) k[(pyrene)$^{\dot{-}}$ $+ (CH_3)_2CHOH \longrightarrow$ (pyrene − H)$^{\cdot}$ $+ (CH_3)_2CHO^-$] $= 1.0 \cdot 10^4\,s^{-1}$.

Reaction Radical generation Method	Solvent	T[K]	Rate data	Ref./ add. ref.
$(CH_3)_2\dot{C}O^- + $ *cis-* and *trans*-stilbene $\longrightarrow$ (*cis-* and *trans*-stilbene)$^{\dot{-}}$ $+ (CH_3)_2CO$				
Pulse rad. of 2-propanol + $(CH_3)_2CHONa$				78 Lev 1
KAS	2-propanol	295(1)	$k < 10^7\ M^{-1}s^{-1}$	
$(CH_3)_2\dot{C}O^- +$ tetracene $\longrightarrow$ (tetracene)$^{\dot{-}}$ $+ (CH_3)_2CO$ [66])				
Pulse rad. of 2-propanol + $(CH_3)_2CHONa$				78 Lev 1
KAS	2-propanol	295(1)	$k = 1.2\cdot 10^9\ M^{-1}s^{-1}$	
$(CH_3)_2\dot{C}O^- +$ tetranitromethane$\{C(NO_2)_4\}$ $\longrightarrow C(NO_2)_3^- + NO_2 + (CH_3)_2CO$				
Pulse rad. of acetone				72 Cha 1
KAS	CH_3COCH_3	RT	$k = 1.2(2)\cdot 10^{10}\ M^{-1}s^{-1}$	
$(CH_3)_2\dot{C}O^- +$ trichloromethane($CHCl_3$) $\longrightarrow$ products				
Pulse rad. of acetone				73 Rob 1
KAS	CH_3COCH_3	298	$k = 3.3\cdot 10^8\ M^{-1}s^{-1}$	
$(CH_3)_2\dot{C}O^- +$ triphenylene $\longrightarrow$ (triphenylene)$^{\dot{-}}$ $+ (CH_3)_2CO$ [67])				
Pulse rad. of 2-propanol + $(CH_3)_2CHONa$				78 Lev 1
KAS	2-propanol	295(1)	$k < 10^8\ M^{-1}s^{-1}$	
(Dimethylfumerate)$^{\dot{-}}$ [68]) $+ O_2 \longrightarrow \dot{O}_2^- +$ dimethylfumerate				
Pulse rad. of dimethylfumerate + $C_2H_5ONa + C_2H_5OH$				67 Ada 1
KAS	C_2H_5OH [69])	RT	$k = 1.5\cdot 10^9\ M^{-1}s^{-1}$	
(β-*apo*-8′-Carotenal)$^{\dot{-}}$ + copper pheophytin a $\longrightarrow$ (copper pheophytin a)$^{\dot{-}}$ + β-*apo*-8′-carotenal [70])				
Pulse rad. of β-*apo*-8′-carotenal [70]) + *n*-hexane				79 McV 1
KAS	*n*-C_6H_{14}	RT	$k = 1.57\cdot 10^{10}\ M^{-1}s^{-1}$	
(β-*apo*-8′-Carotenal)$^{\dot{-}}$ + copper pheophytin b $\longrightarrow$ (copper pheophytin b)$^{\dot{-}}$ + β-*apo*-8′-carotenal [70])				
Pulse rad. of β-*apo*-8′-carotenal [70]) + *n*-hexane				79 McV 1
KAS	*n*-C_6H_{14}	RT	$k = 1.06\cdot 10^{10}\ M^{-1}s^{-1}$	
(β-*apo*-8′-Carotenal)$^{\dot{-}}$ + etioporphyrine $\longrightarrow$ (etioporphyrine)$^{\dot{-}}$ + β-*apo*-8′-carotenal [70])				
Pulse rad. of β-*apo*-8′-carotenal [70]) + *n*-hexane				79 McV 1
KAS	*n*-C_6H_{14}	RT	$k = 1.13\cdot 10^{10}\ M^{-1}s^{-1}$	
(β-*apo*-8′-Carotenal)$^{\dot{-}}$ + mesoporphyrine $\longrightarrow$ (mesoporphyrine)$^{\dot{-}}$ + β-*apo*-8′-carotenal [70])				
Pulse rad. of β-*apo*-8′-carotenal [70]) + *n*-hexane				79 McV 1
KAS	*n*-C_6H_{14}	RT	$k = 3.00\cdot 10^{10}\ M^{-1}s^{-1}$	
(β-*apo*-8′-Carotenal)$^{\dot{-}}$ + pheophytin a $\longrightarrow$ (pheophytin a)$^{\dot{-}}$ + β-*apo*-8′-carotenal [70])				
Pulse rad. of β-*apo*-8′-carotenal [70]) + *n*-hexane				79 McV 1
KAS	*n*-C_6H_{14}	RT	$k = 7.8\cdot 10^9\ M^{-1}s^{-1}$	
(β-*apo*-8′-Carotenal)$^{\dot{-}}$ + pheophytin b $\longrightarrow$ (pheophytin b)$^{\dot{-}}$ + β-*apo*-8′-carotenal [70])				
Pulse rad. of β-*apo*-8′-carotenal [70]) + *n*-hexane				79 McV 1
KAS	*n*-C_6H_{14}	RT	$k = 9.9\cdot 10^9\ M^{-1}s^{-1}$	

[66]) k[(tetracene)$^{\dot{-}}$ $+ (CH_3)_2CHOH \longrightarrow$ (tetracene − H)$^{\cdot}$ $+ (CH_3)_2CHO^-] = 4\cdot 10^2\ s^{-1}$.
[67]) k[(triphenylene)$^{\dot{-}}$ $+ (CH_3)_2CHOH \longrightarrow$ (triphenylene − H)$^{\cdot}$ $+ (CH_3)_2CHO^-] = 3.0\cdot 10^5\ s^{-1}$.
[68]) Electron adduct to dimethylfumerate ($^-OOCC(CH_3){=}C(CH_3)COO^-$).
[69]) 10^{-2} M C_2H_5ONa added.
[70])

Reaction Radical generation Method	Solvent	T[K]	Rate data	Ref./ add. ref.
4.2.2.1.3.3 Cationic radicals				
$[(CH_3)_2CO]^{+\cdot} + Br^- \longrightarrow \dot{Br} + (CH_3)_2CO$				
Pulse rad. of acetone				73 Rob 1
KAS	CH_3COCH_3	298	$k = 4.2 \cdot 10^{10}\ M^{-1}s^{-1}$	
$[(CH_3)_2CO]^{+\cdot} + H_2O \longrightarrow$ products				
Pulse rad. of acetone				73 Rob 1
KAS	CH_3COCH_3	298	$k = 7.0 \cdot 10^{8}\ M^{-1}s^{-1}$	
$[(CH_3)_2CO]^{+\cdot} + SCN^- \longrightarrow (SCN)^{\cdot} + (CH_3)_2CO$				
Pulse rad. of acetone				73 Rob 1
KAS	CH_3COCH_3	298	$k = 1.1 \cdot 10^{11}\ M^{-1}s^{-1}$	
$[(CH_3)_2CO]^{+\cdot}$ + pyrene $\longrightarrow$ (pyrene)$^{+\cdot}$ + $(CH_3)_2CO$				
Pulse rad. of acetone				72 Rod 1/
KAS	CH_3COCH_3	303	$k = 5.1 \cdot 10^{10}\ M^{-1}s^{-1}$	73 Rob 1
	+ 2 Vol% CH_3NO_2	198	$0.66 \cdot 10^{10}$	
$[(CH_3)_2CO]^{+\cdot}$ + pyrene $\longrightarrow$ (pyrene)$^{+\cdot}$ + $(CH_3)_2CO$				
Pulse rad. of acetone				73 Rob 1/
KAS	CH_3COCH_3	298	$k = 5.1 \cdot 10^{10}\ M^{-1}s^{-1}$	72 Rod 1
(β-*apo*-8′-Carotenal)$^{+\cdot}$ + copper pheophytin a $\longrightarrow$ (copper pheophytin a)$^{+\cdot}$ + β-*apo*-8′-carotenal [70]				
Pulse rad. of β-*apo*-8′-carotenal [70] + *n*-hexane				79 McV 1
KAS	*n*-C_6H_{14}	RT	$k = 1.08 \cdot 10^{10}\ M^{-1}s^{-1}$	
(β-*apo*-8′-Carotenal)$^{+\cdot}$ + copper pheophytin b $\longrightarrow$ (copper pheophytin b)$^{+\cdot}$ + β-*apo*-8′-carotenal [70]				
Pulse rad. of β-*apo*-8′-carotenal [70] + *n*-hexane				79 McV 1
KAS	*n*-C_6H_{14}	RT	$k = 2.4 \cdot 10^{9}\ M^{-1}s^{-1}$	
(β-*apo*-8′-Carotenal)$^{+\cdot}$ + etioporphyrine $\longrightarrow$ (etioporphyrine)$^{+\cdot}$ + β-*apo*-8′-carotenal [70]				
Pulse rad. of β-*apo*-8′-carotenal [70] + *n*-hexane				79 McV 1
KAS	*n*-C_6H_{14}	RT	$k = 1.07 \cdot 10^{10}\ M^{-1}s^{-1}$	
(β-*apo*-8′-Carotenal)$^{+\cdot}$ + mesoporphyrine $\longrightarrow$ (mesoporphyrine)$^{+\cdot}$ + β-*apo*-8′-carotenal [70]				
Pulse rad. of β-*apo*-8′-carotenal [70] + *n*-hexane				79 McV 1
KAS	*n*-C_6H_{14}	RT	$k = 1.50 \cdot 10^{10}\ M^{-1}s^{-1}$	
(β-*apo*-8′-Carotenal)$^{+\cdot}$ + pheophytin a $\longrightarrow$ (pheophytin a)$^{+\cdot}$ + β-*apo*-8′-carotenal [70]				
Pulse rad. of β-*apo*-8′-carotenal [70] + *n*-hexane				79 McV 1
KAS	*n*-C_6H_{14}	RT	$k = 1.17 \cdot 10^{10}\ M^{-1}s^{-1}$	
(β-*apo*-8′-Carotenal)$^{+\cdot}$ + pheophytin b $\longrightarrow$ (pheophytin b)$^{+\cdot}$ + β-*apo*-8′-carotenal [70]				
Pulse rad. of β-*apo*-8′-carotenal [70] + *n*-hexane				79 McV 1
KAS	*n*-C_6H_{14}	RT	$k = 7.8 \cdot 10^{9}\ M^{-1}s^{-1}$	

[70])

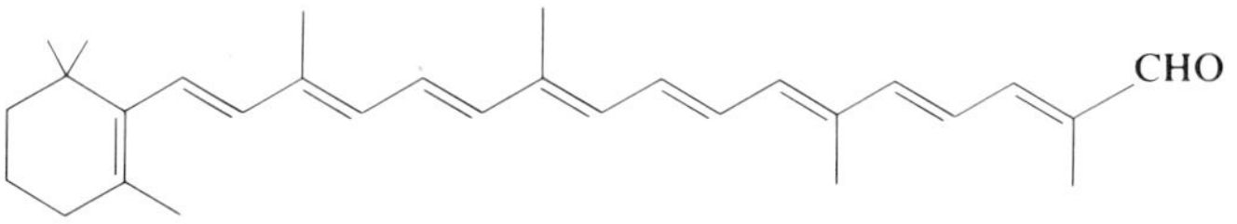

Reaction Radical generation Method	Solvent	T[K]	Rate data	Ref./ add. ref.
4.2.2.1.4 Radicals containing only C, H, and N atoms				
4.2.2.1.4.1 Neutral radicals				
$\dot{C}H_2NHCH_3$ + benzil ($C_6H_5COCOC_6H_5$) ⟶ $C_6H_5CO\dot{C}OHC_6H_5$ + products				
Laser phot. of $(CH_3)_2NH$ + benzil containing soln.; initial step: photoreduction of benzil				81 Sca 1
KAS	$C_6H_5COCH_3/H_2O$ (90:10%)	295	$k = 1.0 \cdot 10^9\ M^{-1}s^{-1}$	
$\dot{C}H_2N(CH_3)_2$ + benzil ($C_6H_5COCOC_6H_5$) ⟶ $C_6H_5CO\dot{C}OHC_6H_5$ + products				
Laser phot. of benzil + $(CH_3)_3N$ containing solutions; initial step: photoreduction of benzil				81 Sca 1
KAS	$C_6H_5COCH_3/H_2O$ (90:10%)	295	$k = 1.2 \cdot 10^9\ M^{-1}s^{-1}$	
$\dot{C}H_2N(CH_3)_2$ + benzil ($C_6H_5COCOC_6H_5$) ⟶ $C_6H_5CO\dot{C}OHC_6H_5$ + products				
Laser phot. of $(CH_3)_3N$ + di-*t*-butylperoxide				81 Sca 1
KAS	$(CH_3)_3COOC(CH_3)_3$	295	$k = 6.0 \cdot 10^8\ M^{-1}s^{-1}$	
(Pyrrolidine)˙ [71]) + benzil ($C_6H_5COCOC_6H_5$) ⟶ $C_6H_5CO\dot{C}OHC_6H_5$ + products				
Laser phot. of pyrrolidine + di-*t*-butylperoxide				81 Sca 1
KAS	$(CH_3)_3COOC(CH_3)_3$	295	$k = 3.3 \cdot 10^9\ M^{-1}s^{-1}$	
$CH_3\dot{C}HN(C_2H_5)_2$ + benzil ($C_6H_5COCOC_6H_5$) ⟶ $C_6H_5CO\dot{C}OHC_6H_5$ + products				
Laser phot. of $(C_2H_5)_3N$ + benzil + H_2O + $C_6H_5COCH_3$; initial step: photoreduction of benzophenone				81 Sca 1
KAS	$C_6H_5COCH_3/H_2O$ (90:10%)	295	$k = 2.7 \cdot 10^9\ M^{-1}s^{-1}$	
Laser phot. of $(C_2H_5)_3N$ + benzil containing soln.; initial step: photoreduction of benzil				81 Sca 1
KAS	$C_6H_5COCH_3/H_2O$	295	$k = 1.6 \cdot 10^9\ M^{-1}s^{-1}$	
	(90:10%)	256... 345	$\log[A/M^{-1}s^{-1}] = 11.4$ $E_a = 12.4\ kJ\ mol^{-1}$	
	+ 0.003 M NaOH		$k = 1.8 \cdot 10^9\ M^{-1}s^{-1}$	
$CH_3\dot{C}HN(C_2H_5)_2$ + benzil ($C_6H_5COCOC_6H_5$) ⟶ $CH_2{=}CHN(C_2H_5)_2$ + $C_6H_5CO\dot{C}OHC_6H_5$				
Laser phot. of $(C_2H_5)_3N$ + DTBP				81 Sca 1
KAS	$(CH_3)_3COOC(CH_3)_3$	295	$k = 1.8 \cdot 10^9\ M^{-1}s^{-1}$	
		237... 347	$\log[A/M^{-1}s^{-1}] = 10.4$ $E_a = 6.5\ kJ\ mol^{-1}$	
$CH_3\dot{C}HN(C_2H_5)_2$ + benzil ($C_6H_5COCOC_6H_5$) ⟶ $C_6H_5COCOHC_6H_5$ + $C_2H_5NHCH{=}CH_2$ or $C_2H_5N{=}CH{-}CH_3$ or $(C_2H_5NH{=}CHCH_3)^+$				
Laser phot. of $(C_2H_5)_3N$ + DTBP				81 Sca 1
KAS	$(CH_3)_3COOC(CH_3)_3$	295	$k = 3.3 \cdot 10^9\ M^{-1}s^{-1}$	

[71]) Radicals from $(CH_3)_3C\dot{O}$ + pyrolidine reaction, likely to be

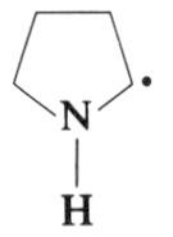
and/or
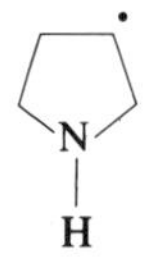

Reaction Radical generation Method	Solvent	T[K]	Rate data	Ref./ add. ref.

4.2.2.1.4.2 Anionic radicals

$(CH_3CN)^{\cdot -}$ [72]) + benzophenone $((C_6H_5)_2CO) \longrightarrow (C_6H_5)_2\dot{C}O^- + CH_3CN$

| Pulse rad. of acetonitrile | | | | 77 Bel 1 |
| KAS | CH_3CN | RT | $k = 5.5 \cdot 10^{10} \, M^{-1} s^{-1}$ | |

$(CH_3CN)^{\cdot -}$ [72]) + biphenyl $\longrightarrow$ (biphenyl)$^{\cdot -}$ + CH_3CN

| Pulse rad. of acetonitrile | | | | 77 Bel 1 |
| KAS | CH_3CN | RT | $k = 3.3 \cdot 10^{10} \, M^{-1} s^{-1}$ | |

$(CH_3CN)^{\cdot -}$ [72]) + carbontetrachloride $(CCl_4) \longrightarrow \dot{C}Cl_3 + Cl^- + CH_3CN$

| Pulse rad. of acetonitrile | | | | 77 Bel 1 |
| KAS | CH_3CN | RT | $k = 6.6 \cdot 10^{10} \, M^{-1} s^{-1}$ | |

$(CH_3CN)^{\cdot -}$ [72]) + pyrene $\longrightarrow$ (pyrene)$^{\cdot -}$ + CH_3CN

| Pulse rad. of acetonitrile | | | | 77 Bel 1 |
| KAS | CH_3CN | RT | $k = 3.9 \cdot 10^{10} \, M^{-1} s^{-1}$ | |

$(CH_3CN)^{\cdot -}$ [72]) + trans-stilbene $\longrightarrow$ (trans-stilbene)$^{\cdot -}$ + CH_3CN

| Pulse rad. of acetonitrile | | | | 77 Bel 1 |
| KAS | CH_3CN | RT | $k = 3.3 \cdot 10^{10} \, M^{-1} s^{-1}$ | |

$(CH_3CN)^{\cdot -}$ [72]) + tetracyanobenzene $\longrightarrow$ (tetracyanobenzene)$^{\cdot -}$ + CH_3CN

| Pulse rad. of acetonitrile | | | | 77 Bel 1 |
| KAS | CH_3CN | RT | $k = 6.1 \cdot 10^{10} \, M^{-1} s^{-1}$ | |

$(CH_3CN)^{\cdot -}$ [72]) + trichloroethylene $(CHCl{=}CCl_2) \longrightarrow$ products

| Pulse rad. of acetonitrile | | | | 77 Bel 1 |
| KAS | CH_3CN | RT | $k = 3.3 \cdot 10^{10} \, M^{-1} s^{-1}$ | |

4.2.2.2 Aromatic radicals and radicals derived from compounds containing aromatic and heterocyclic constituents

4.2.2.2.1 Radicals containing only C and H atoms

4.2.2.2.1.1 Neutral radicals

$C_6H_5CH_2\dot{C}H_2 + Cu(II)(\alpha,\alpha'\text{-bipyridine})^{2+} \xrightarrow{e} C_6H_5CH{=}CH_2 + H^+ + Cu(I)(\alpha,\alpha'\text{-bipyridine})^+$
$\qquad + Cu(II)(\alpha,\alpha'\text{-bipyridine})^{2+} + CH_3COOH \xrightarrow{s}$
$\qquad\qquad\qquad\qquad C_6H_5CH_2CH_2OOCCH_3 + H^+ + Cu(I)(\alpha,\alpha'\text{-bipyridine})^+$

| Cu(II) catalyzed decomp. of $C_6H_5CH_2CH_2OOCH_2CH_2C_6H_5$ | | | | 68 Koc 1 |
| PR by glc | CH_3CN/CH_3COOH (1:1.5) | 298.5 | $k_e = 1.4 \cdot 10^7 \, M^{-1} s^{-1}$ [1]) $k_s = 8.3 \cdot 10^5 \, M^{-1} s^{-1}$ [1]) | |

$C_6H_5CH_2\dot{C}H_2 + Cu(II)(NCCH_3)_4^{2+} \xrightarrow{e} C_6H_5CH{=}CH_2 + H^+ + Cu(I)(NCCH_3)_4^+$
$\qquad + Cu(II)(NCCH_3)_4^{2+} + CH_3COOH \xrightarrow{s} C_6H_5CH_2CH_2OOCCH_3 + H^+ + Cu(I)(NCCH_3)_4^+$

| Cu(II) catalyzed decomp. of $C_6H_5CH_2CH_2OOCH_2CH_2C_6H_5$ | | | | 68 Koc 1 |
| PR by glc | CH_3CN/CH_3COOH (1:1.5) | 298.5 | $k_e = 1.6 \cdot 10^6 \, M^{-1} s^{-1}$ [1]) $k_s = 5 \cdot 10^4 \, M^{-1} s^{-1}$ [1]) | |

---

[72]) Electron adduct to CH_3CN.
[1]) Assuming $k = 1 \cdot 10^4 \, M^{-1} s^{-1}$ for competing reaction $C_6H_5CH_2\dot{C}H_2 + (CH_3)_2CHCHO \longrightarrow C_6H_5CH_2CH_3 + (CH_3)_2CH\dot{C}O$.

Reaction Radical generation Method	Solvent	T[K]	Rate data	Ref./ add. ref.
$4\text{-}CH_3C_6H_4CH_2\dot{C}H_2 + Cu(II)(NCCH_3)_4^{2+} \xrightarrow{e} 4\text{-}CH_3C_6H_4CH{=}CH_2 + H^+ + Cu(I)(NCCH_3)_4^+$ $+ Cu(II)(NCCH_3)_4^{2+} + CH_3COOH \xrightarrow{s}$ $4\text{-}CH_3C_6H_4CH_2CH_2OOCCH_3 + H^+ + Cu(I)(NCCH_3)_4^+$				
Cu(II) catalyzed decomp. of $4\text{-}CH_3C_6H_4CH_2CH_2OOCH_2CH_2C_6H_4CH_3$				68 Koc 1
PR by glc	CH_3CN/CH_3COOH (1:1.5)	298.5	$k_e = 1.0 \cdot 10^6\ M^{-1}s^{-1}$ [2]) $k_s = 8.1 \cdot 10^5\ M^{-1}s^{-1}$ [2])	
4.2.2.2.1.2 Anionic radicals				
$(Naphthalene)^{\dot{-}}$ + bromobenzene (C_6H_5Br) ⟶ products				
Stopped flow expt., naphthalene + THF on Na mirror				76 Ban 1
KAS	THF	293	$k = 6.4 \cdot 10^3\ M^{-1}s^{-1}$	
$(Naphthalene)^{\dot{-}}$ + 1-bromobutane ($n\text{-}C_4H_9Br$) ⟶ naphthalene + $n\text{-}C_4H_9^{\cdot}$ + Br^-				
Pulse rad. of naphthalene + THF				73 Boc 1
KAS	THF	298	$k = 3.3(6) \cdot 10^7\ M^{-1}s^{-1}$	
$(Naphthalene)^{\dot{-}}$ + 1-bromohexane ($n\text{-}C_6H_{13}Br$) ⟶ naphthalene + $n\text{-}C_6H_{13}^{\cdot}$ + Br^-				
Stopped flow expt., naphthalene + THF on metal mirror				78 Ban 1
KAS	THF	RT	$k = 1.5 \cdot 10^5\ M^{-1}s^{-1}$	
$(Naphthalene)^{\dot{-}}$ + chlorobenzene (C_6H_5Cl) ⟶ products				
Stopped flow expt., naphthalene + THF on Na mirror				76 Ban 1
KAS	THF	293	$k = 6.0 \cdot 10^2\ M^{-1}s^{-1}$	
$(Naphthalene)^{\dot{-}}$ + 1-chlorohexane ($n\text{-}C_6H_{13}Cl$) ⟶ naphthalene + $n\text{-}C_6H_{13}^{\cdot}$ + Cl^-				
Stopped flow expt., naphthalene + THF on metal mirror				78 Ban 1
KAS	THF	RT	$k = 4.0 \cdot 10^2\ M^{-1}s^{-1}$	
$(Naphthalene)^{\dot{-}}$ + 1-chlorohexane ($n\text{-}C_6H_{13}Cl$) ⟶ naphthalene + $n\text{-}C_6H_{13}^{\cdot}$ + Cl^-				
Reduct. of naphthalene by metallic Na in stopped flow expt.				75 Ban 1
KAS	THF/tetraglyme (1:1)	293	$k = 1.20 \cdot 10^3\ M^{-1}s^{-1}$	
$(Naphthalene)^{\dot{-}}$ + fluorobenzene (C_6H_5F) ⟶ products				
Stopped flow expt., naphthalene + THF on Na mirror				76 Ban 1
KAS	THF	293	$k = 19.4\ M^{-1}s^{-1}$	
$(Naphthalene)^{\dot{-}}$ + 1-iodobutane ($n\text{-}C_4H_9I$) ⟶ naphthalene + $n\text{-}C_4H_9^{\cdot}$ + I^-				
Pulse rad. of naphthalene + THF				73 Boc 1
KAS	THF	298	$k = 7.4(11) \cdot 10^9\ M^{-1}s^{-1}$	
$(Naphthalene)^{\dot{-}}$ + 1-iodohexane ($n\text{-}C_6H_{13}I$) ⟶ naphthalene + $n\text{-}C_6H_{13}^{\cdot}$ + I^-				
Stopped flow expt., naphthalene + THF on metal mirror				78 Ban 1
KAS	THF	RT	$k = 4.4 \cdot 10^7\ M^{-1}s^{-1}$	
$(Naphthalene)^{\dot{-}}/Cs^+$ + 5-hexenylfluoride ⟶ products				
Reduct. of naphthalene by Cs				74 Gar 1
KAS	1,2-dimethoxyethane	298	$k < 5 \cdot 10^{-7}\ M^{-1}s^{-1}$	
$(Naphthalene)^{\dot{-}}/K^+$ + 5-hexenylfluoride(RF) ⟶ naphthalene + K^+ + $\dot{R}$ + F^-				
Reduct. of naphthalene by K				74 Gar 1
KAS	1,2-dimethoxyethane	298	$k = 2 \cdot 10^{-6}\ M^{-1}s^{-1}$	

[2]) Assuming $k = 1 \cdot 10^4\ M^{-1}s^{-1}$ for competing reaction $4\text{-}CH_3C_6H_4CH_2\dot{C}H_2 + (CH_3)_2CHCHO \longrightarrow 4\text{-}CH_3C_6H_4C_2H_5 + (CH_3)_2CH\dot{C}O$.

Reaction Radical generation Method	Solvent	T[K]	Rate data	Ref./ add. ref.
(Naphthalene)$^{\dot{-}}$/Li$^+$ + 1-fluorohexane (n-$C_6H_{13}F$) ⟶ naphthalene + Li$^+$ + n-$C_6H_{13}^{\cdot}$ + F$^-$				
Reduct. of naphthalene by Li				74 Gar 1
KAS	1,2-dimethoxyethane	298	$k = 7.5(10) \cdot 10^{-3}\,M^{-1}s^{-1}$	
(Naphthalene)$^{\dot{-}}$/Li$^+$ + 1-fluorooctane (n-$C_8H_{17}F$) ⟶ naphthalene + Li$^+$ + n-$C_8H_{17}^{\cdot}$ + F$^-$				
Reduct. of naphthalene by Li				74 Gar 1
KAS	1,2-dimethoxyethane	298	$k = 1.22(8) \cdot 10^{-2}\,M^{-1}s^{-1}$	
(Naphthalene)$^{\dot{-}}$/Li$^+$ + 5-hexenylfluoride(RF) ⟶ naphthalene + Li$^+$ + $\dot{R}$ + F$^-$				
Reduct. of naphthalene by Li				74 Gar 1
KAS	1,2-dimethoxyethane	298	$k = 2.7(2) \cdot 10^{-2}\,M^{-1}s^{-1}$	
(Naphthalene)$^{\dot{-}}$/Na$^+$ + 1-bromobutane ⟶ naphthalene + Na$^+$ + Br$^-$ + n-$C_4H_9^{\cdot}$				
Pulse rad. of naphthalene + Na$^+$ + THF				73 Boc 1
KAS	THF	298	$k < 4 \cdot 10^5\,M^{-1}s^{-1}$	
(Naphthalene)$^{\dot{-}}$/Na$^+$[3]) + 1-bromohexane (n-$C_6H_{13}Br$) ⟶ naphthalene + n-$C_6H_{13}^{\cdot}$ + Br$^-$				
Reduct. of naphthalene by metallic Na$^+$ in stopped flow expt.				75 Ban 1
KAS	THF	293	$k = 7.2 \cdot 10^4\,M^{-1}s^{-1}$	
(Naphthalene)$^{\dot{-}}$/Na$^+$[4]) + 1-chlorohexane (n-$C_6H_{13}Cl$) ⟶ naphthalene + n-$C_6H_{13}^{\cdot}$ + Cl$^-$				
Reduct. of naphthalene by metallic Na in stopped flow expt.				75 Ban 1
KAS	1,2-dimethoxyethane	293	$k = 1.08 \cdot 10^3\,M^{-1}s^{-1}$	
(Naphthalene)$^{\dot{-}}$/Na$^+$[3]) + 1-chlorohexane (n-$C_6H_{13}Cl$) ⟶ naphthalene + n-$C_6H_{13}^{\cdot}$ + Cl$^-$				
Reduct. of naphthalene by metallic Na in stopped flow expt.				75 Ban 1
KAS	THF	323	$k = 4.56 \cdot 10^2\,M^{-1}s^{-1}$	
		313	$3.84 \cdot 10^2$	
		303	$2.37 \cdot 10^2$	
		293	$2.00 \cdot 10^2$	
		283	$1.64 \cdot 10^2$	
		272	$1.87 \cdot 10^2$	
		260	$6.98 \cdot 10^2$	
	+0.025 M Na$^+$	292	$2.00 \cdot 10^2$	
	+0.05 M Na$^+$		$2.14 \cdot 10^2$	
	+0.1 M Na$^+$		$1.91 \cdot 10^2$	
(Naphthalene)$^{\dot{-}}$/Na$^+$ + 1-fluorohexane (n-$C_6H_{13}F$) ⟶ naphthalene + Na$^+$ + n-$C_6H_{13}^{\cdot}$ + F$^-$				
Reduct. of naphthalene on Na mirror				75 Gar 1
KAS	THF	279	$k = 2.4 \cdot 10^{-5}\,M^{-1}s^{-1}$	
		298	$1.9 \cdot 10^{-4}$	
		308	$3.6 \cdot 10^{-4}$	
		319	$9.2 \cdot 10^{-4}$	
			$\Delta H^{\ddagger} = 62.8\,kJ\,mol^{-1}$	
			$\Delta S^{\ddagger} = -105\,J\,mol^{-1}K^{-1}$	
KAS	THF/dicyclohexyl-18-crown-6 mixt.	298	$k = 3.4 \cdot 10^{-5}\,M^{-1}s^{-1}$	75 Gar 1
KAS	THF/tetraglyme mixt.	298	$k = 1.2 \cdot 10^{-4}\,M^{-1}s^{-1}$	75 Gar 1
(continued)				

[3]) Tight (naphthalene)$^{\dot{-}}$/Na$^+$ ion pair.
[4]) Loose (naphthalene)$^{\dot{-}}$/Na$^+$ ion pair.

Reaction Radical generation Method	Solvent	T[K]	Rate data	Ref./ add. ref.
(Naphthalene)$^{\bar{\cdot}}$/Na$^+$ + 1-fluorohexane(n-$C_6H_{13}F$) ⟶ naphthalene + Na$^+$ + n-$C_6H_{13}^{\cdot}$ + F^- (continued)				
Reduct. of naphthalene on Na mirror				
KAS	2-methyl-tetrahydrofuran	298.00(5) 308.00(5) 319.00(5)	$k = 3.4\cdot10^{-4}\,M^{-1}\,s^{-1}$ $7.0\cdot10^{-4}$ $1.7\cdot10^{-3}$ $\Delta H^{\ddagger} = 58.6\,kJ\,mol^{-1}$ $\Delta S^{\ddagger} = -117.6\,J\,mol^{-1}\,K^{-1}$	75 Gar 1
KAS	1,2-dimethoxyethane	298 298.00(5) 308.00(5) 319.00(5)	$k = 2.2(5)\cdot10^{-4}\,M^{-1}\,s^{-1}$ $2.5\cdot10^{-4}$ [5]) $5.0\cdot10^{-4}$ [5]) $1.3\cdot10^{-3}$ [5]) $\Delta H^{\ddagger} = 54.4\,kJ\,mol^{-1}$ [5]) $\Delta S^{\ddagger} = -117.6\,J\,mol^{-1}\,K^{-1}$ [5])	74 Gar 1, 75 Gar 1
Reduct. of naphthalene on Na mirror in 1,2-dimethoxyethane				69 Gar 1
Not given	1,2-dimethoxyethane	298	$k = 4.3(11)\cdot10^{-4}$	
Reduct. of naphthalene on Na mirror				75 Gar 1
KAS	1,2-dimethoxyethane/dicyclohexyl-18-crown-6 mixt.	299 308 318	$k = 4.7\cdot10^{-5}\,M^{-1}\,s^{-1}$ $1.1\cdot10^{-4}$ $4.2\cdot10^{-4}$ $\Delta H^{\ddagger} = 87.9\,kJ\,mol^{-1}$ $\Delta S^{\ddagger} = -29.4\,J\,mol^{-1}\,K^{-1}$	
(Naphthalene)$^{\bar{\cdot}}$/Na$^+$ + 1-fluorooctane (n-$C_8H_{17}F$) ⟶ naphthalene + Na$^+$ + n-$C_8H_{17}^{\cdot}$ + F^-				
Reduct. of naphthalene on Na mirror				74 Gar 1,
KAS	1,2-dimethoxyethane	298	$k = 2.2(4)\cdot10^{-4}\,M^{-1}\,s^{-1}$ $4.3(11)\cdot10^{-4}$ [6])	69 Gar 1
(Naphthalene)$^{\bar{\cdot}}$/Na$^+$ + 5-hexenylfluoride ⟶ naphthalene + Na$^+$ + F^- + $CH_2{=}CH{-}(CH_2)_3\dot{C}H_2$				
Reduct. of naphthalene on Na mirror				74 Gar 1,
KAS	1,2-dimethoxyethane	298 308 318	$k = 2.8(5)\cdot10^{-4}\,M^{-1}\,s^{-1}$ $5.7(16)\cdot10^{-4}$ [6]) $9(3)\cdot10^{-4}$ $19(1)\cdot10^{-4}$	69 Gar 1
	THF	298	$k = 2.1\cdot10^{-4}\,M^{-1}\,s^{-1}$	
(Naphthalene)$^{\bar{\cdot}}$/Na$^+$ + 1-iodobutane ⟶ naphthalene + Na$^+$ + I^- + n-$C_4H_9^{\cdot}$				
Pulse rad. of naphthalene + Na$^+$ + THF				73 Boc 1
KAS	THF	298	$k = 9.3(5)\cdot10^{7}\,M^{-1}\,s^{-1}$	
(Naphthalene)$^{\bar{\cdot}}$/Na$^+$ + N-methyl-N-phenyl-4-toluenesulfonamide $\xrightarrow{a}$ products + N,N-di-n-butyl-4-toluenesulfonamide $\xrightarrow{b}$ products				
Mixing of Na and naphthalene in dimethoxyethane				70 Clo 1
PR by glc	dimethoxyethane	298	$k_a/k_b = 1.26(6)$ [7])	
(Naphthalene)$^{\bar{\cdot}}$/Na$^+$ + tetrahydrofurfurylfluoride(RF) ⟶ naphthalene + Na$^+$ + $\dot{R}$ + F^-				
Reduct. of naphthalene on Na mirror				74 Gar 1
KAS	1,2-dimethoxyethane	298	$k = 4.0\cdot10^{-4}\,M^{-1}\,s^{-1}$	
(Biphenyl)$^{\bar{\cdot}}$ + O_2 ⟶ biphenyl + $\dot{O}_2^-$				
Pulse rad. of biphenyl + c-C_6H_{12}				71 Ric 1
KAS	c-C_6H_{12}	296	$k = 2.3\cdot10^{10}\,M^{-1}\,s^{-1}$	

[5]) From [75 Gar 1].
[6]) From [69 Gar 1].
[7]) Rate determining steps assumed to be e^--transfer.

Reaction Radical generation Method	Solvent	T[K]	Rate data	Ref./ add. ref.
(Biphenyl)$^{\cdot-}$ + SF_6 ⟶ biphenyl + products				
Pulse rad. of biphenyl + c-C_6H_{12}				71 Ric 1
KAS	c-C_6H_{12}	296	$k = 7.5 \cdot 10^9\ M^{-1}s^{-1}$	
(Biphenyl)$^{\cdot-}$ + anthracene ⟶ biphenyl + (anthracene)$^{\cdot-}$				
Pulse rad. of biphenyl + 2-propanol				67 Ara 1
KAS	2-propanol	298	$k = 6.4(20) \cdot 10^9\ M^{-1}s^{-1}$	
(Biphenyl)$^{\cdot-}$ + 1-bromobutane (n-C_4H_9Br) ⟶ biphenyl + n-$C_4H_9^{\cdot}$ + Br^-				
Pulse rad. of biphenyl + THF				73 Boc 1
KAS	THF	298	$k = 3.4(6) \cdot 10^7\ M^{-1}s^{-1}$	
(Biphenyl)$^{\cdot-}$ + 1-chlorohexane (n-$C_6H_{13}Cl$) ⟶ biphenyl + n-$C_6H_{13}^{\cdot}$ + Cl^-				
Stopped flow expt., biphenyl + naphthalene + THF on Na mirror				76 Ban 1
KAS	THF	293	$k = 3.2(2) \cdot 10^4\ M^{-1}s^{-1}$	
(Biphenyl)$^{\cdot-}$ + diphenylchloromethane ($(C_6H_5)_2CHCl$) ⟶ biphenyl + $(C_6H_5)_2\dot{C}H$ + Cl^-				
Pulse rad. of biphenyl + 2-methyltetrahydrofuran				81 Tak 1
KAS	2-methyl- tetrahydrofuran	RT	$k = 7.0 \cdot 10^9\ M^{-1}s^{-1}$	
(Biphenyl)$^{\cdot-}$ + 1-iodobutane ⟶ biphenyl + I^- + n-$C_4H_9^{\cdot}$				
Pulse rad. of biphenyl + THF				73 Boc 1
KAS	THF	298	$k = 9.6(14) \cdot 10^9\ M^{-1}s^{-1}$	
(Biphenyl)$^{\cdot-}$ + naphthalene ⟶ biphenyl + (naphthalene)$^{\cdot-}$				
Pulse rad. of biphenyl + 2-propanol				67 Ara 1
KAS	2-propanol	298	$k = 2.6(8) \cdot 10^8\ M^{-1}s^{-1}$	
(Biphenyl)$^{\cdot-}$ + phenanthrene ⟶ biphenyl + (phenanthrene)$^{\cdot-}$				
Pulse rad. of biphenyl + 2-propanol				67 Ara 1
KAS	2-propanol	298	$k = 6.0(30) \cdot 10^8\ M^{-1}s^{-1}$	
(Biphenyl)$^{\cdot-}$ + phenylchloromethane ($C_6H_5CH_2Cl$) ⟶ biphenyl + $C_6H_5CH_2^{\cdot}$ + Cl^-				
Pulse rad. of biphenyl + c-C_6H_{12}				71 Ric 1
KAS	c-C_6H_{12}	296	$k = 1.0 \cdot 10^{10}\ M^{-1}s^{-1}$	
(Biphenyl)$^{\cdot-}$ + pyrene ⟶ biphenyl + (pyrene)$^{\cdot-}$				
Pulse rad. of biphenyl + c-C_6H_{12}				71 Ric 1/
KAS	c-C_6H_{12}	296	$k = 3.2 \cdot 10^{10}\ M^{-1}s^{-1}$	72 Rae 1
(Biphenyl)$^{\cdot-}$ + pyrene ⟶ biphenyl + (pyrene)$^{\cdot-}$				
Flash phot.; photoionization of (pyrene)$^{\cdot-}$ and subsequent e^--capture by biphenyl [8])				72 Rae 1/
KAS	THF	RT	$k = 4.8(5) \cdot 10^{10}\ M^{-1}s^{-1}$	71 Ric 1, 71 Fis 1
(Biphenyl)$^{\cdot-}$ + pyrene ⟶ biphenyl + (pyrene)$^{\cdot-}$				
Flash phot. of biphenyl + THF				71 Fis 1/
KAS	THF	RT	$k = (2.0 \ldots 2.7) \cdot 10^{10}\ M^{-1}s^{-1}$	72 Rae 1, 71 Ric 1
(Biphenyl)$^{\cdot-}$ + pyrene ⟶ biphenyl + (pyrene)$^{\cdot-}$				
Pulse rad. of biphenyl + 2-propanol				67 Ara 1
KAS	2-propanol	298	$k = 5.0(18) \cdot 10^9\ M^{-1}s^{-1}$	

[8]) (Pyrene)$^{\cdot-}$ was produced via reduct. of pyrene on Na mirror in THF.

Reaction Radical generation Method	Solvent	T[K]	Rate data	Ref./ add. ref.
(Biphenyl)$^{\cdot-}$ + p-terphenyl $\longrightarrow$ biphenyl + (p-terphenyl)$^{\cdot-}$				
Pulse rad. of biphenyl + 2-propanol				67 Ara 1
KAS	2-propanol	298	$k = 3.2(7) \cdot 10^9\ M^{-1} s^{-1}$	
(Biphenyl)$^{\cdot-}$/Li^+ + 5-hexenylfluoride(RF) $\longrightarrow$ biphenyl + Li^+ + $\dot{R}$ + F^-				
Reduct. of biphenyl by Li				74 Gar 1
KAS	1,2-dimethoxyethane	298	$k = 1.24(2) \cdot 10^{-2}\ M^{-1} s^{-1}$	
(Biphenyl)$^{\cdot-}$/Na^+ + 1-bromobutane $\longrightarrow$ biphenyl + Na^+ + Br^- + n-$C_4H_9^{\cdot}$				
Pulse rad. of biphenyl + Na^+ + THF				73 Boc 1
KAS	THF	298	$k = 1.3(1) \cdot 10^6\ M^{-1} s^{-1}$	
(Biphenyl)$^{\cdot-}$/Na^+ + 5-hexenylfluoride(RF) $\longrightarrow$ biphenyl + Na^+ + $\dot{R}$ + F^-				
Reduct. of biphenyl on Na mirror				74 Gar 1
KAS	1,2-dimethoxyethane	298	$k = 1.4 \cdot 10^{-4}\ M^{-1} s^{-1}$	
(Biphenyl)$^{\cdot-}$/Na^+ + 1-iodobutane $\longrightarrow$ biphenyl + Na^+ + I^- + n-$C_4H_9^{\cdot}$				
Pulse rad. of biphenyl + Na^+ + THF				73 Boc 1
KAS	THF	298	$k = 4.3(2) \cdot 10^8\ M^{-1} s^{-1}$ $E_a \geqslant 8.8\ kJ\ mol^{-1}$ $\Delta H^{\ddagger} = 6.9\ kJ\ mol^{-1}$ [9] $\Delta S^{\ddagger} = -8.9\ J\ mol^{-1} K^{-1}$ [9]	
(Biphenyl)$^{\cdot-}$/Na^+ + N-methyl-N-phenyl-4-toluenesulfonamide $\xrightarrow{a}$ products + N,N-di-n-butyl-4-toluenesulfonamide $\xrightarrow{b}$ products				
Mixing of Na and biphenyl in dimethoxyethane				70 Clo 1
PR by glc	dimethoxyethane	298	$k_a/k_b = 1.31(6)$ [10]	
(Biphenyl)$^{\cdot-}$/Na^+ + pyrene $\longrightarrow$ biphenyl + (pyrene)$^{\cdot-}$/Na^+				
Flash phot.; photoionization of (pyrene$^{\cdot-}$, Na^+) and e^--capture by biphenyl [11]				72 Rae 1/
KAS	THF	RT	$k = 6 \cdot 10^9\ M^{-1} s^{-1}$	71 Ric 1,
	THP		$k = 5 \cdot 10^9\ M^{-1} s^{-1}$	71 Fis 1
(Biphenyl)$^{\cdot-}$/Na^+ + pyrene $\longrightarrow$ biphenyl + (pyrene)$^{\cdot-}$/Na^+				
Pulse rad. of biphenyl + Na^+ + THF				71 Fis 1/
KAS	THF	RT	$k = 7 \cdot 10^9\ M^{-1} s^{-1}$	72 Rae 1,
				71 Ric 1
(Anthracene)$^{\cdot-}$ + bromobenzene (C_6H_5Br) $\longrightarrow$ products				
Stopped flow expt., anthracene + naphthalene + THF on Na mirror				76 Ban 1
KAS	THF	293	$k = 1.54\ M^{-1} s^{-1}$	
(Anthracene)$^{\cdot-}$ + 1-bromohexane (n-$C_6H_{13}Br$) $\longrightarrow$ anthracene + n-$C_6H_{13}^{\cdot}$ + Br				
Stopped flow expt., anthracene + THF on metal mirror				78 Ban 1
KAS	THF	RT	$k = 6.6 \cdot 10^2\ M^{-1} s^{-1}$ [12]	
(Anthracene)$^{\cdot-}$ + 2-bromopropane $\longrightarrow$ anthracene + $(CH_3)_2\dot{C}H$ + Br^-				
Stopped flow expt., anthracene + THF on metal mirror				78 Ban 1
KAS	THF	RT	$k = 2.6 \cdot 10^3\ M^{-1} s^{-1}$ [13]	

[9]) For loose ion pairs.
[10]) Rate determining steps assumed to be possibly e^--transfer.
[11]) (Pyrene$^{\cdot-}$/Na^+) was produced by reduction of pyrene on Na mirror in THF or THP.
[12]) k(anthracene$^{\cdot-}$ + n-$C_6H_{13}Br$)/k(anthracene − H^- + n-$C_6H_{13}Br$) = 0.24.
[13]) k(anthracene$^{\cdot-}$ + 2-C_3H_7Br)/k(anthracene − H^- + 2-C_3H_7Br) = 41.5.

Reaction Radical generation Method	Solvent	T[K]	Rate data	Ref./ add. ref.
(Anthracene)$^{\cdot-}$ + 1-chlorohexane (n-$C_6H_{13}Cl$) anthracene + n-$C_6H_{13}^{\cdot}$ + Cl^-				
Stopped flow expt., anthracene + THF on metal mirror				78 Ban 1
KAS	THF	RT	$k = 1.3 \cdot 10^{-1}\,M^{-1}s^{-1}$ [14]	
(Anthracene)$^{\cdot-}$ + diphenylchloromethane{$(C_6H_5)_2CHCl$} ⟶ anthracene + $(C_6H_5)_2\dot{C}H$ + Cl^-				
Pulse rad. of anthracene + 2-methyltetrahydrofuran				81 Tak 1
KAS	2-methyl-tetrahydrofuran	RT	$k \approx 6 \cdot 10^{7}\,M^{-1}s^{-1}$ [15]	
(Anthracene)$^{\cdot-}$ + iodobenzene (C_6H_5I) ⟶ products				
Stopped flow expt., anthracene + naphthalene + THF on Na mirror				76 Ban 1
KAS	THF	293	$k = 4.0 \cdot 10^{4}\,M^{-1}s^{-1}$	
(Anthracene)$^{\cdot-}$ + 1-iodohexane (n-$C_6H_{13}I$) ⟶ anthracene + n-$C_6H_{13}^{\cdot}$ + I^-				
Stopped flow expt., anthracene + THF on metal mirror				78 Ban 1,
KAS	THF	RT	$k = 4.4 \cdot 10^{4}\,M^{-1}s^{-1}$ [16] $2.2 \cdot 10^{4}$ [17]	75 Ban 1
(Anthracene)$^{\cdot-}$/Na^+ + 1-bromobutane (n-C_4H_9Br) ⟶ anthracene + n-$C_4H_9^{\cdot}$ + Br^- + Na^+				
Stopped flow expt., anthracene + THF on Na mirror				78 Ban 1
KAS	THF	273	$k = 2.42 \cdot 10^{2}\,M^{-1}s^{-1}$	
(Anthracene)$^{\cdot-}$/Na^+ + 2-bromobutane ⟶ anthracene + $CH_3\dot{C}HCH_2CH_3$ + Br^- + Na^+				
Stopped flow expt., anthracene + THF on Na mirror				78 Ban 1
KAS	THF	273	$k = 9.38 \cdot 10^{2}\,M^{-1}s^{-1}$	
(Anthracene)$^{\cdot-}$/Na^+ + 1-bromohexane (n-$C_6H_{13}Br$) ⟶ anthracene + n-$C_6H_{13}^{\cdot}$ + Br^- + Na^+				
Reduct. of anthracene by metallic Na in stopped flow expt.				75 Ban 1
KAS	1,2-dimethoxyethane	293	$k = 3.3 \cdot 10^{2}\,M^{-1}s^{-1}$	
	THF	293	$k = 3.3 \cdot 10^{2}\,M^{-1}s^{-1}$	
(Anthracene)$^{\cdot-}$/Na^+ + 2-bromo-2-methylpropane ⟶ anthracene + $(CH_3)_3\dot{C}$ + Br^- + Na^+				
Stopped flow expt., anthracene + THF on Na mirror				78 Ban 1
KAS	THF	273	$k = 2.402 \cdot 10^{3}\,M^{-1}s^{-1}$	
(Anthracene)$^{\cdot-}$/Na^+ + 1-chlorohexane (n-$C_6H_{13}Cl$) ⟶ anthracene + n-$C_6H_{13}^{\cdot}$ + Na^+ + Cl^-				
Reduct. of anthracene by metallic Na in stopped flow expt.				75 Ban 1
KAS	1,2-dimethoxyethane	293	$k = 6.6 \cdot 10^{-2}\,M^{-1}s^{-1}$	
(Anthracene)$^{\cdot-}$/Na^+ + 4-chlorobenzene-N-methyl-sulfonanilide [18] [19] $\xrightarrow{a}$ products + benzene-N-methyl-sulfonanilide [18] [20] $\xrightarrow{b}$ products				
Na + anthracene in THF				78 Qua 1
PR by glc	THF	298	$k_a/k_b = 1.2$ [21]	

[14]) k(anthracene$^{\cdot-}$ + n-$C_6H_{13}Cl$)/k(anthracene − H^- + n-$C_6H_{13}Cl$) = 0.021.
[15]) Numerical value evaluated from graphical plot.
[16]) k(anthracene$^{\cdot-}$ + n-$C_6H_{13}I$)/k(anthracene − H^- + n-$C_6H_{13}I$) = 2.0.
[17]) From [75 Ban 1].
[18]) 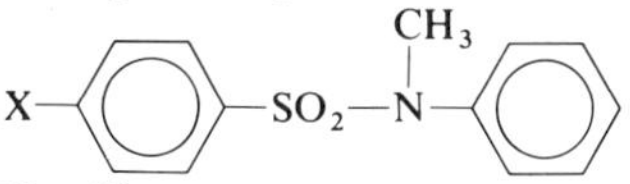
[19]) X = Cl.
[20]) X = H.
[21]) Rate determining steps assumed to be e^--transfer.

Reaction Radical generation Method	Solvent	T[K]	Rate data	Ref./ add. ref.
(Anthracene)$^{\dot{-}}$/Na$^+$ + 4-cyanobenzene-N-methyl-sulfonanilide [18] [22] $\xrightarrow{a}$ products + benzene-N-methyl-sulfonanilide [18] [20] $\xrightarrow{b}$ products				
Na + anthracene in THF				78 Qua 1
PR by glc	THF	298	$k_a/k_b = 0.60$ [21]	
(Anthracene)$^{\dot{-}}$/Na$^+$ + 4-(dimethylamino)benzene-N-methyl-sulfonanilide [18] [23] $\xrightarrow{a}$ products + benzene-N-methyl-sulfonanilide [18] [20] $\xrightarrow{b}$ products				
Na + anthracene in THF				78 Qua 1
PR by glc	THF	298	$k_a/k_b = 0.024$ [21]	
(Anthracene)$^{\dot{-}}$/Na$^+$ + 4-fluorobenzene-N-methyl-sulfonanilide [18] [24] $\xrightarrow{a}$ products + benzene-N-methyl-sulfonanilide [18] [20] $\xrightarrow{b}$ products				
Na + anthracene in THF				78 Qua 1
PR by glc	THF	298	$k_a/k_b = 0.92$ [21]	
(Anthracene)$^{\dot{-}}$/Na$^+$ + 4-(methanesulfinyl)benzene-N-methyl-sulfonanilide [18] [25] $\xrightarrow{a}$ products + benzene-N-methyl-sulfonanilide [18] [20] $\xrightarrow{b}$ products				
Na + anthracene in THF				78 Qua 1
PR by glc	THF	298	$k_a/k_b = 0.63$ [21]	
(Anthracene)$^{\dot{-}}$/Na$^+$ + 4-methoxybenzene-N-methyl-sulfonanilide [18] [26] $\xrightarrow{a}$ products + benzene-N-methyl-sulfonanilide [18] [20] $\xrightarrow{b}$ products				
Na + anthracene in THF				78 Qua 1
PR by glc	THF	298	$k_a/k_b = 0.19$ [21]	
(Anthracene)$^{\dot{-}}$/Na$^+$ + 4-methylthiobenzene-N-methyl-sulfonanilide [18] [27] $\xrightarrow{a}$ products + benzene-N-methyl-sulfonanilide [18] [20] $\xrightarrow{b}$ products				
Na + anthracene in THF				78 Qua 1
PR by glc	THF	298	$k_a/k_b = 1.04$ [21]	
(Anthracene)$^{\dot{-}}$/Na$^+$ + 4-nitrobenzene-N-methyl-sulfonanilide [18] [28] $\xrightarrow{a}$ products + benzene-N-methyl-sulfonanilide [18] [20] $\xrightarrow{b}$ products				
Na + anthracene in THF				78 Qua 1
PR by glc	THF	298	$k_a/k_b = 0.078$ [21]	
(Anthracene)$^{\dot{-}}$/Na$^+$ + toluene-4-(N-ethyl-sulfonanilide) [29] $\xrightarrow{a}$ products + benzene-N-methyl-sulfonanilide [18] [20] $\xrightarrow{b}$ products				
Na + anthracene in THF				78 Qua 1
PR by glc	THF	298	$k_a/k_b = 0.45$ [21]	
(Anthracene)$^{\dot{-}}$/Na$^+$ + toluene-4-(N-methyl-sulfonanilide) [18] [30] $\xrightarrow{a}$ products + benzene-N-methyl-sulfonanilide [18] [20] $\xrightarrow{b}$ products				
Na + anthracene in THF				78 Qua 1
PR by glc	THF	298	$k_a/k_b = 0.54$ [21]	

[18]
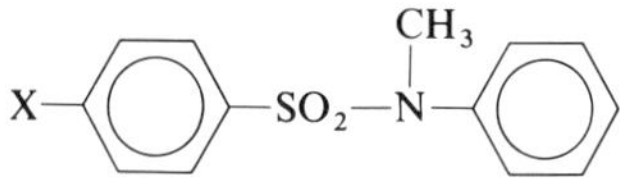

[20] X = H.
[21] Rate determining steps assumed to be e^--transfer.
[22] X = CN.
[23] X = $(CH_3)_2N$.
[24] X = F.
[25] X = CH_3SO.
[26] X = CH_3O.
[27] X = CH_3S.
[28] X = NO_2.
[29]
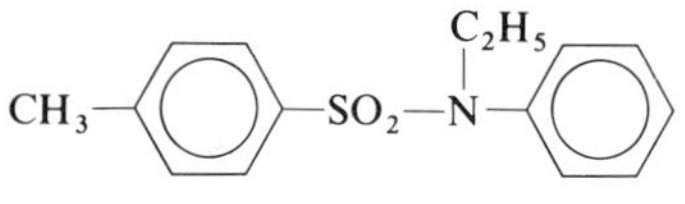

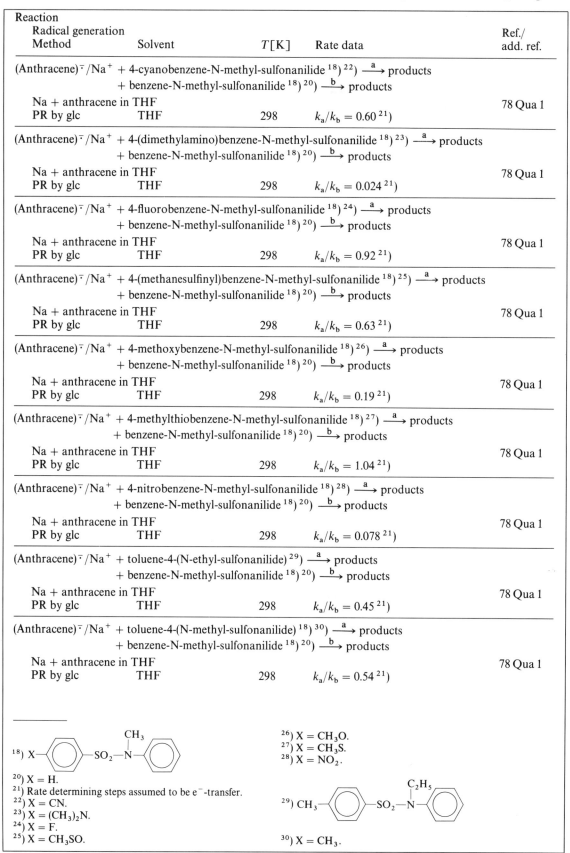

[30] X = CH_3.

Reaction Radical generation Method	Solvent	T[K]	Rate data	Ref./ add. ref.
(Phenanthrene)$^{\cdot-}$ + diphenylchloromethane{$(C_6H_5)_2CHCl$} ⟶ phenanthrene + $(C_6H_5)_2\dot{C}H$ + Cl^-				
Pulse rad. of phenanthrene + 2-methyltetrahydrofuran				81 Tak 1
KAS	2-methyl- tetrahydrofuran	RT	$k \approx 6\cdot10^9\,M^{-1}s^{-1}$ [31])	
(*cis*-Stilbene)$^{\cdot-}$ + perylene ⟶ *cis*-stilbene + (perylene)$^{\cdot-}$				
Flash phot. of Na-perylenide + THF + stilbene				77 Wan 1
KAS	THF	RT	$k = 1.5\cdot10^{10}\,M^{-1}s^{-1}$	
(*trans*-Stilbene)$^{\cdot-}$ + SF_6 ⟶ *trans*-stilbene + $SF_6^{\cdot-}$				
Pulse rad. of *trans*-stilbene + *c*-hexane				78 Rob 1
KAS	*c*-hexane	RT	$k = 3.9(4)\cdot10^9\,M^{-1}s^{-1}$	
(*trans*-Stilbene)$^{\cdot-}$ + perylene ⟶ *trans*-stilbene + (perylene)$^{\cdot-}$				
Flash phot. of Na-perylenide + stilbene + THF				77 Wan 1
KAS	THF	RT	$k = 1.8\cdot10^{10}\,M^{-1}s^{-1}$	
(*trans*-1,2-Diphenylethylene)$^{\cdot-}$ + diphenylchloromethane{$(C_6H_5)_2CHCl$} ⟶ *trans*-1,2-diphenylethylene + $(C_6H_5)_2\dot{C}H$ + Cl^-				
Pulse rad. of *trans*-1,2-diphenylethylene + 2-methyltetrahydrofuran				81 Tak 1
KAS	2-methyl- tetrahydrofuran	RT	$k \approx 1\cdot10^9\,M^{-1}s^{-1}$ [31])	
(Fluoranthene)$^{\cdot-}$ + 1-bromohexane ⟶ fluoranthene [32]) + *n*-$C_6H_{13}^{\cdot}$ + Br^-				
Stopped flow expt., fluoranthene [32]) + naphthalene + THF on Na mirror				76 Ban 1
KAS	THF	RT	$k = 1.6(6)\,M^{-1}s^{-1}$	
(Fluoranthene)$^{\cdot-}$ + 1-iodohexane ⟶ fluoranthene [32]) + *n*-$C_6H_{13}^{\cdot}$ + I^-				
Stopped flow expt., fluoranthene [32]) + naphthalene + THF on Na mirror				76 Ban 1
KAS	THF	RT	$k = 3.6(2)\cdot10^2\,M^{-1}s^{-1}$	
(Pyrene)$^{\cdot-}$ + 1-bromohexane (*n*-$C_6H_{13}Br$) ⟶ pyrene + *n*-$C_6H_{13}^{\cdot}$ + Br^-				
Stopped flow expt., pyrene + naphthalene + THF on Na mirror				76 Ban 1
KAS	THF	293	$k = 6.4(4)\cdot10^3\,M^{-1}s^{-1}$	
(Pyrene)$^{\cdot-}$ + 1-chlorohexane (*n*-$C_6H_{13}^{\cdot}Cl$) ⟶ pyrene + *n*-$C_6H_{13}^{\cdot}$ + Cl^-				
Stopped flow expt., pyrene + naphthalene + THF on Na mirror				76 Ban 1
KAS	THF	293	$k = 7.0(8)\,M^{-1}s^{-1}$	
(Pyrene)$^{\cdot-}$ + diphenylchloromethane{$(C_6H_5)_2CHCl$} ⟶ pyrene + $(C_6H_5)_2\dot{C}H$ + Cl^-				
Pulse rad. of pyrene + 2-methyltetrahydrofuran				81 Tak 1
KAS	2-methyl- tetrahydrofuran	RT	$k \approx 2.4\cdot10^8\,M^{-1}s^{-1}$ [31])	
(*o*-Terphenyl)$^{\cdot-}$ + pyrene ⟶ *o*-terphenyl + (pyrene)$^{\cdot-}$				
Pulse rad. of *o*-terphenyl [33]) + 2-propanol				67 Ara 1
KAS	2-propanol	298	$k = 4.0(18)\cdot10^9\,M^{-1}s^{-1}$	

[31]) Numerical value evaluated from graphical plot.

[32]) 1,2-benzacenaphthene

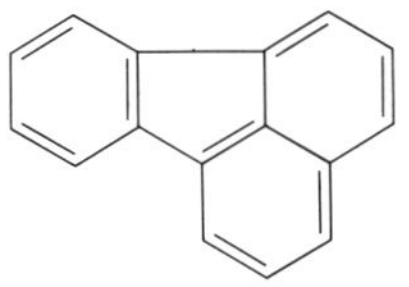

[33]) 1,2-Diphenylbenzene.

Reaction Radical generation Method	Solvent	$T[K]$	Rate data	Ref./ add. ref.
$(m\text{-Terphenyl})^{\bar{\cdot}} + \text{pyrene} \longrightarrow m\text{-terphenyl} + (\text{pyrene})^{\bar{\cdot}}$				
Pulse rad. of m-terphenyl [34]) + 2-propanol KAS	2-propanol	298	$k = 3.5(12) \cdot 10^9 \, \text{M}^{-1}\text{s}^{-1}$	67 Ara 1
$(p\text{-Terphenyl})^{\bar{\cdot}} + \text{anthracene} \longrightarrow p\text{-terphenyl} + (\text{anthracene})^{\bar{\cdot}}$				
Pulse rad. of p-terphenyl [35]) + 2-propanol KAS	2-propanol	298	$k = 5.5(9) \cdot 10^9 \, \text{M}^{-1}\text{s}^{-1}$	67 Ara 1
$(p\text{-Terphenyl})^{\bar{\cdot}} + \text{pyrene} \longrightarrow p\text{-terphenyl} + (\text{pyrene})^{\bar{\cdot}}$				
Pulse rad. of p-terphenyl [35]) + 2-propanol KAS	2-propanol	298	$k = 3.6(11) \cdot 10^9 \, \text{M}^{-1}\text{s}^{-1}$	67 Ara 1
$(\text{Perylene})^{\bar{\cdot}} + 1\text{-bromohexane} \, (n\text{-C}_6\text{H}_{13}\text{Br}) \longrightarrow \text{perylene} + n\text{-C}_6\text{H}_{13}^{\cdot} + \text{Br}^-$				
Stopped flow expt.; perylene + naphthalene + THF on Na mirror KAS	THF	293	$k = 5.8(2) \cdot 10^{-1} \, \text{M}^{-1}\text{s}^{-1}$	76 Ban 1
$(\text{Perylene})^{\bar{\cdot}} + \text{ethyliodide} \, (\text{C}_2\text{H}_5\text{I}) \longrightarrow \text{perylene} + \text{C}_2\text{H}_5^{\cdot} + \text{I}^-$				
Electrochem. reduct. of perylene [36])	N,N-dimethyl-formamide	293	$k = 3.7 \cdot 10^3 \, \text{M}^{-1}\text{s}^{-1}$	79 Par 1
$(\text{Perylene})^{\bar{\cdot}} + \text{oxalic acid} \{(\text{COOH})_2\} \longrightarrow \text{products} \, [37])$				
Electrochem. reduct. of perylene [36])	N,N-dimethyl-formamide	293	$k = 1.5 \cdot 10^4 \, \text{M}^{-1}\text{s}^{-1}$	79 Par 1

4.2.2.2.1.3 Cationic radicals

$(\alpha\text{-Methylstyrene})^{\dot{+}} + \text{C}_2\text{H}_5\text{OH} \xrightarrow{a} \text{products} \, [38])$ $(\alpha\text{-methylstyrene})_2^{\dot{+}} + \text{C}_2\text{H}_5\text{OH} \xrightarrow{b} \text{products} \, [38])$				
Pulse rad. of $\text{CH}_2\text{ClCH}_2\text{Cl} + \alpha$-methylstyrene KAS	$\text{CH}_2\text{ClCH}_2\text{Cl}$	RT	$k_a = 3.5(7) \cdot 10^7 \, \text{M}^{-1}\text{s}^{-1}$ $k_b = 8.5(10) \cdot 10^7 \, \text{M}^{-1}\text{s}^{-1}$	77 Hay 1
$(\text{Naphthalene})^{\dot{+}} + \text{Eu(II)} \longrightarrow \text{naphthalene} + \text{Eu(III)}$				
Flash phot. KAS	CH_3CN	RT	$k = 3.9 \cdot 10^9 \, \text{M}^{-1}\text{s}^{-1}$	78 Lev 2
$(\text{Biphenyl})^{\dot{+}} + \text{pyrene} \longrightarrow \text{biphenyl} + (\text{pyrene})^{\dot{+}}$				
Pulse rad. of biphenyl + $\text{CH}_2\text{ClCH}_2\text{Cl}$ KAS	$\text{CH}_2\text{ClCH}_2\text{Cl}$	RT	$k = 9.9(10) \cdot 10^9 \, \text{M}^{-1}\text{s}^{-1}$	70 Sha 1
$(\text{Biphenyl})^{\dot{+}} + p\text{-terphenyl} \longrightarrow \text{biphenyl} + (p\text{-terphenyl})^{\dot{+}}$				
Pulse rad. of biphenyl + $\text{CH}_2\text{ClCH}_2\text{Cl}$ KAS	$\text{CH}_2\text{ClCH}_2\text{Cl}$	RT	$k = 5.1(10) \cdot 10^9 \, \text{M}^{-1}\text{s}^{-1}$	70 Sha 1
$(trans\text{-Stilbene})^{\dot{+}} + 1,2,4\text{-trimethoxybenzene} \longrightarrow (1,2,4\text{-trimethoxybenzene})^{\dot{+}} + trans\text{-stilbene}$				
Photooxidation through singlet excited 9,10-dicyanoanthracene KAS	CH_3CN	RT	$k = 1.2(1) \cdot 10^{10} \, \text{M}^{-1}\text{s}^{-1}$	80 Spa 1

[34]) 1,3-Diphenylbenzene.
[35]) 1,4-Diphenylbenzene.
[36]) Second harmonic ac voltametry.
[37]) Reaction may involve e^--transfer.
[38]) e^-- and/or H^+-transfer.

Reaction Radical generation Method	Solvent	T[K]	Rate data	Ref./ add. ref.
(1,1-Diphenylethylene)$^{+\cdot}$ + C_2H_5OH ⟶ products[37])				
Pulse rad. of 1,1-diphenylethylene + CH_2Cl_2				78 Hay 1
KAS	CH_2Cl_2	RT	$k = 1 \cdot 10^8\ M^{-1}s^{-1}$	
(Tetracene)$^{+\cdot}$ + Eu(II) ⟶ tetracene + Eu(III)				
Flash phot. of tetracene[39]) + CH_3CN				78 Lev 2
KAS	CH_3CN	RT	$k = 1.6 \cdot 10^7\ M^{-1}s^{-1}$	
(p-Terphenyl)$^{+\cdot}$ + anthracene ⟶ (anthracene)$^{+\cdot}$ + p-terphenyl				
Pulse rad. of p-terphenyl[40]) + CH_2ClCH_2Cl				70 Sha 1
KAS	CH_2ClCH_2Cl	RT	$k = 8.1(8) \cdot 10^9\ M^{-1}s^{-1}$	
(Perylene)$^{+\cdot}$ + Eu(II) ⟶ perylene + Eu(III)				
Flash phot. of perylene + CH_3CN				78 Lev 2
KAS	CH_3CN	RT	$k = 2.8 \cdot 10^6\ M^{-1}s^{-1}$	
(Perylene)$^{+\cdot}$ + I^- ⟶ perylene + $\dot{I}$				
Biphotonic photoionization of perylene + CH_3CN				77 Eva 1
KAS	CH_3CN	RT	$k = 2.1 \cdot 10^{10}\ M^{-1}s^{-1}$	
Biphotonic photoionization of perylene + CH_3OH				77 Eva 1
KAS	CH_3OH	RT	$k = 4.7 \cdot 10^9\ M^{-1}s^{-1}$	
(Coronene)$^{+\cdot}$ + Eu(II) ⟶ Eu(III) + coronene				
Flash phot. of coronene + CH_3CN				78 Lev 2
KAS	CH_3CN	RT	$k = 1.3 \cdot 10^9\ M^{-1}s^{-1}$	
(9,10-Diphenylanthracene)$^{+\cdot}$ + Br^- ⟶ 9,10-di... + $\dot{Br}$				
Stopped flow expt.; electrochem. generation in 9,10-diphenylanthracene + CH_3CN				78 Eva 1
KAS	CH_3CN	298	$k = 6.91(85) \cdot 10^5\ M^{-1}s^{-1}$	
(9,10-Diphenylanthracene)$^{+\cdot}$ + CN^- ⟶ 9,10-di... + $CN^{\cdot}$				
Stopped flow expt.; electrochem. generation in 9,10-diphenylanthracene + CH_3CN				78 Eva 1
KAS	CH_3CN	298	$k = 6.3(12) \cdot 10^6\ M^{-1}s^{-1}$	
2(9,10-Diphenylanthracene)$^{+\cdot}$ + H_2S ⟶ 2(9,10-di...) + S + $2H^+$				
Stopped flow expt.; electrochem. generation in 9,10-diphenylanthracene + CH_3CN				78 Eva 1
KAS	CH_3CN	298	$k = 6.6(12)\ M^{-1}s^{-1}$	
(9,10-Diphenylanthracene)$^{+\cdot}$ + H_2S ⟶ 9,10-DPA + $H_2\dot{S}^+$				
Electrochem. ox. of DPA				76 Eva 1
[41])	CH_3CN	298.0(2)	$k = 6.6(12)\ M^{-1}s^{-1}$	
(9,10-Diphenylanthracene)$^{+\cdot}$ + I^- ⟶ 9,10-di... + $\dot{I}$				
Stopped flow expt.; electrochem. generation in 9,10-diphenylanthracene + CH_3CN				78 Eva 1
KAS	CH_3CN	298	$k = 1.49(55) \cdot 10^7\ M^{-1}s^{-1}$	
(9,10-Diphenylanthracene)$^{+\cdot}$ + SCN^- ⟶ 9,10-di... + $SCN^{\cdot}$				
Stopped flow expt.; electrochem. generation in 9,10-diphenylanthracene + CH_3CN				78 Eva 1
KAS	CH_3CN	298	$k = 3.35(23) \cdot 10^6\ M^{-1}s^{-1}$	

[37]) Reaction may involve e^--transfer.
[39]) Naphthacene, 2,3-benzanthracene.
[40]) 1,4-Diphenylbenzene.
[41]) Stopped flow KAS; electrochem. methods.

Reaction Radical generation Method	Solvent	T[K]	Rate data	Ref./ add. ref.

4.2.2.2.2 Radicals containing C, H, L, and other atoms

4.2.2.2.2.1 Neutral radicals

$CH_3\dot{C}OHC_6H_5$ + benzil ($C_6H_5COCOC_6H_5$) $\longrightarrow$ $C_6H_5CO\dot{C}OHC_6H_5 + CH_3COC_6H_5$

Laser phot. of acetophenone				81 Sca 1
KAS	$C_6H_5COCH_3/H_2O$ (90:10%)	295	$k \approx 8 \cdot 10^7\ M^{-1}s^{-1}$ [42]	

$4\text{-}CH_3OC_6H_4CH_2\dot{C}H_2$ + Cu(II)(α,α'-bipyridine)$^{2+}$ $\xrightarrow{e}$ $4\text{-}CH_3OC_6H_4CH{=}CH_2 + H^+$ + Cu(I)(α,α'-bipyridine)$^+$

+ Cu(II)(α,α'-bipyridine)$^{2+}$ + CH_3COOH $\xrightarrow{s}$ $4\text{-}CH_3OC_6H_4CH_2CH_2OOCCH_3 + H^+$ + Cu(I)(α,α'-bipyridine)$^+$

Cu(II) catalyzed decomp. of $4\text{-}CH_3OC_6H_4CH_2CH_2OOCH_2CH_2C_6H_4OCH_3$				68 Koc 1
PR by glc	CH_3CN/CH_3COOH (1:1.5)	298.5	$k_e = 3.0 \cdot 10^6\ M^{-1}s^{-1}$ [43] $k_s = 1.3 \cdot 10^7\ M^{-1}s^{-1}$ [43]	

$4\text{-}CH_3OC_6H_4CH_2\dot{C}H_2 + Cu(II)(NCCH_3)_4^{2+} \xrightarrow{e} 4\text{-}CH_3OC_6H_4CH{=}CH_2 + H^+ + Cu(I)(NCCH_3)_4^+$

$+ Cu(II)(NCCH_3)_4^{2+} + CH_3COOH \xrightarrow{s} 4\text{-}CH_3OC_6H_4CH_2CH_2OOCCH_3 + H^+ + Cu(I)(NCCH_3)_4^+$

Cu(II) catalyzed decomp. of $4\text{-}CH_3OC_6H_4CH_2CH_2OOCH_2CH_2C_6H_4OCH_3$				68 Koc 1
PR by glc	CH_3CN/CH_3COOH (1:1.5)	298.5	$k_e = 2.1 \cdot 10^4\ M^{-1}s^{-1}$ [43] $k_s = 1.6 \cdot 10^6\ M^{-1}s^{-1}$ [43]	

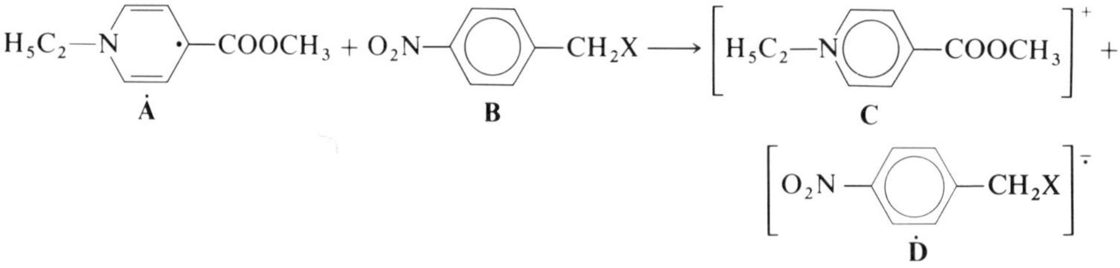

X = Br

Reduct. of 1-ethyl-4-carbomethoxypyridinium iodide by Zn				71 Moh 1
KAS	1,2-dimethoxyethane	298	$k = 1.9 \cdot 10^4\ M^{-1}s^{-1}$	
	2-methyltetra-hydrofuran		$k = 2.0 \cdot 10^3\ M^{-1}s^{-1}$	

Ȧ + B ⟶ C + Ḋ X = Cl

Reduct. of 1-ethyl-4-carbomethoxypyridinium iodide by Zn				68 Kos 1/ 71 Moh 1
KAS	CH_3CN	298	$k = 2.4 \cdot 10^4\ M^{-1}s^{-1}$	
	dimethylformamide		$1.2 \cdot 10^4$	
	CH_3COCH_3		$4.5 \cdot 10^2$	
	CH_2Cl_2		$7.5 \cdot 10^1$	
	1,2-dimethoxyethane		8.3	
	2-methyltetra-hydrofuran		1.62	

Ȧ + B ⟶ C + Ḋ X = F

Reduct. of 1-ethyl-4-carbomethoxypyridinium iodide by Zn				71 Moh 2
KAS	C_2H_5OH	298	$k = 4.8\ M^{-1}s^{-1}$	
	2-propanol		0.3	
	CH_3CN		$1.8 \cdot 10^{-2}$	

[42]) Error limits possibly ±300%.

[43]) Assuming $k = 1 \cdot 10^4\ M^{-1}s^{-1}$ for competing reaction $4\text{-}CH_3OC_6H_4CH_2\dot{C}H_2 + (CH_3)_2CHCHO \longrightarrow 4\text{-}CH_3OC_6H_4C_2H_5 + (CH_3)_2CH\dot{C}O$.

Reaction Radical generation Method	Solvent	T[K]	Rate data	Ref./ add. ref.
$(C_6H_5)_2\dot{C}OH + Ag^+ \longrightarrow$ products				
Flash phot. of benzilic acid + Ag^+ + acetone				79 Kem 1
KAS	$(CH_3)_2CO$	RT	$k = 3.71(37) \cdot 10^6\ M^{-1}s^{-1}$	
$(C_6H_5)_2\dot{C}OH + Cu^{2+} \longrightarrow$ products				
Flash phot. of benzilic acid + Cu^{2+} + acetone				79 Kem 1
KAS	$(CH_3)_2CO$	RT	$k = 1.83(5) \cdot 10^6\ M^{-1}s^{-1}$	
$(C_6H_5)_2\dot{C}OH + Fe^{3+} \longrightarrow$ products				
Flash phot. of benzilic acid + Fe^{3+} + acetone				79 Kem 1
KAS	$(CH_3)_2CO$	RT	$k = 4.91(26) \cdot 10^7\ M^{-1}s^{-1}$	
$(C_6H_5)_2\dot{C}OH + HgCl_2 \longrightarrow$ products				
Flash phot. of benzilic acid + $HgCl_2$ + acetone				79 Kem 1
KAS	$(CH_3)_2CO$	RT	$k = 6.50(50) \cdot 10^4\ M^{-1}s^{-1}$	
$(C_6H_5)_2\dot{C}OH$ + 1,1′-azobis-(N,N-dimethylformamide)[44]) $\longrightarrow$ (1,1′-azobis-(N,N-...))$^{\cdot-}$ + H^+ + $(C_6H_5)_2CO$				
Laser phot. of benzilic acid + uranyl nitrate + acetone				80 Kem 1
KAS	CH_3COCH_3	293	$k = 2.75(16) \cdot 10^7\ M^{-1}s^{-1}$	
$(C_6H_5)_2\dot{C}OH$ + benzoic acid (C_6H_5COOH) $\longrightarrow$ products				
Laser phot. of benzilic acid + uranyl nitrate + acetone				80 Kem 1
KAS	CH_3COCH_3	293	$k = 3.3(6) \cdot 10^3\ M^{-1}s^{-1}$	
$(C_6H_5)_2\dot{C}OH$ + 1,4-benzoquinone $\longrightarrow$ (1,4-benzosemiquinone) + $(C_6H_5)_2CO$				
Laser phot. of benzilic acid + uranyl nitrate + acetone				80 Kem 1
KAS	CH_3COCH_3	293	$k = 2.63(12) \cdot 10^8\ M^{-1}s^{-1}$	
$(C_6H_5)_2\dot{C}OH$ + 3-chlorobenzoic acid $\longrightarrow$ products				
Laser phot. of benzilic acid + uranyl nitrate + acetone				80 Kem 1
KAS	CH_3COCH_3	293	$k = 3.00(17) \cdot 10^4\ M^{-1}s^{-1}$	
$(C_6H_5)_2\dot{C}OH$ + 4-chlorobenzoic acid $\longrightarrow$ products				
Laser phot. of benzilic acid + uranyl nitrate + acetone				80 Kem 1
KAS	CH_3COCH_3	293	$k = 2.26(24) \cdot 10^4\ M^{-1}s^{-1}$	
$(C_6H_5)_2\dot{C}OH$ + 2,5-dimethyl-1,4-benzoquinone $\longrightarrow$ (2,5-dimethyl-... semiquinone) + $(C_6H_5)_2CO$				
Laser phot. of benzilic acid + uranyl nitrate + acetone				80 Kem 1
KAS	CH_3COCH_3	293	$k = 2.00(11) \cdot 10^8\ M^{-1}s^{-1}$	
$(C_6H_5)_2\dot{C}OH$ + 1,2-dinitrobenzene $\longrightarrow$ (1,2-dinitrobenzene)$^{\cdot-}$ + H^+ + $(C_6H_5)_2CO$				
Laser phot. of benzilic acid + uranyl nitrate + acetone				80 Kem 1
KAS	CH_3COCH_3	293	$k = 4.36(13) \cdot 10^6\ M^{-1}s^{-1}$	
$(C_6H_5)_2\dot{C}OH$ + 1,3-dinitrobenzene $\longrightarrow$ (1,3-dinitrobenzene)$^{\cdot-}$ + H^+ + $(C_6H_5)_2CO$				
Laser phot. of benzilic acid + uranyl nitrate + acetone				80 Kem 1,
KAS	CH_3COCH_3	293	$k = 1.28(12) \cdot 10^6\ M^{-1}s^{-1}$	80 Kem 2
Laser phot. of benzilic acid + uranyl nitrate + acetone + H_2O				80 Kem 1,
KAS	CH_3COCH_3/H_2O (75:25 V/V)	293	$k = 4.19(37) \cdot 10^6\ M^{-1}s^{-1}$	80 Kem 2

[44]) "Diamide".

Reaction Radical generation Method	Solvent	T[K]	Rate data	Ref./ add. ref.
$(C_6H_5)_2\dot{C}OH$ + 1,4-dinitrobenzene ⟶ (1,4-dinitrobenzene)$^{\bar{\cdot}}$ + H^+ + $(C_6H_5)_2CO$				
Laser phot. of benzilic acid + uranyl nitrate + acetone				80 Kem 1
KAS	CH_3COCH_3	293	$k = 9.04(4) \cdot 10^6\ M^{-1}\,s^{-1}$	
Laser phot. of benzilic acid + uranyl nitrate + acetone + H_2O				80 Kem 1
KAS	CH_3COCH_3/H_2O (75:25 V/V)	293	$k = 3.78(41) \cdot 10^7\ M^{-1}\,s^{-1}$	
$(C_6H_5)_2\dot{C}OH$ + 2,4-dinitrobenzoic acid ⟶ (2,4-dinitrobenzoic acid)$^{\bar{\cdot}}$ + H^+ + $(C_6H_5)_2CO$				
Laser phot. of benzilic acid + uranyl nitrate + acetone				80 Kem 1
KAS	CH_3COCH_3	293	$k = 3.79(13) \cdot 10^6\ M^{-1}\,s^{-1}$	
$(C_6H_5)_2\dot{C}OH$ + 2,5-dinitrobenzoic acid ⟶ (2,5-dinitrobenzoic acid)$^{\bar{\cdot}}$ + H^+ + $(C_6H_5)_2CO$				
Laser phot. of benzilic acid + uranyl nitrate + acetone				80 Kem 1
KAS	CH_3COCH_3	293	$k = 1.04(2) \cdot 10^7\ M^{-1}\,s^{-1}$	
$(C_6H_5)_2\dot{C}OH$ + 3,4-dinitrobenzoic acid ⟶ (3,4-dinitrobenzoic acid)$^{\bar{\cdot}}$ + H^+ + $(C_6H_5)_2CO$				
Laser phot. of benzilic acid + uranyl nitrate + acetone				80 Kem 1
KAS	CH_3COCH_3	293	$k = 6.49(20) \cdot 10^6\ M^{-1}\,s^{-1}$	
Laser phot. of benzilic acid + uranyl nitrate + acetone + H_2O				80 Kem 1
KAS	CH_3COCH_3/H_2O (75:25 V/V)	293	$k = 3.71(24) \cdot 10^7\ M^{-1}\,s^{-1}$	
$(C_6H_5)_2\dot{C}OH$ + 3,5-dinitrobenzoic acid ⟶ (3,5-dinitrobenzoic acid)$^{\bar{\cdot}}$ + H^+ + $(C_6H_5)_2CO$				
Laser phot. of benzilic acid + uranyl nitrate + acetone				80 Kem 1
KAS	CH_3COCH_3	293	$k = 3.76(4) \cdot 10^6\ M^{-1}\,s^{-1}$	
$(C_6H_5)_2\dot{C}OH$ + 4-fluorobenzoic acid ⟶ products				
Laser phot. of benzilic acid + uranyl nitrate + acetone				80 Kem 1
KAS	CH_3COCH_3	293	$k = 6.7(2) \cdot 10^3\ M^{-1}\,s^{-1}$	
$(C_6H_5)_2\dot{C}OH$ + 1-fluoro-2,4-dinitrobenzene ⟶ (1-fluoro-2,4-...)$^{\bar{\cdot}}$ + H^+ + $(C_6H_5)_2CO$				
Laser phot. of benzilic acid + uranyl nitrate + acetone				80 Kem 1
KAS	CH_3COCH_3	293	$k = 4.22(10) \cdot 10^6\ M^{-1}\,s^{-1}$	
$(C_6H_5)_2\dot{C}OH$ + 5-hydroxy-1,4-naphthoquinone [45]) ⟶ (5-hydroxy-... semiquinone) + $(C_6H_5)_2CO$				
Laser phot. of benzilic acid + uranyl nitrate + acetone				80 Kem 1
KAS	CH_3COCH_3	293	$k = 3.18(20) \cdot 10^8\ M^{-1}\,s^{-1}$	
$(C_6H_5)_2\dot{C}OH$ + 2-methyl-1,4-naphthoquinone [46]) ⟶ (2-methyl-... semiquinone) + $(C_6H_5)_2CO$				
Laser phot. of benzilic acid + uranyl nitrate + acetone				80 Kem 1
KAS	CH_3COCH_3	293	$k = 7.22(20) \cdot 10^7\ M^{-1}\,s^{-1}$	
$(C_6H_5)_2\dot{C}OH$ + methyl-3-nitrobenzoate ⟶ (methyl-3-nitrobenzoate)$^{\bar{\cdot}}$ + H^+ + $(C_6H_5)_2CO$				
Laser phot. of benzilic acid + uranyl nitrate + acetone				80 Kem 1
KAS	CH_3COCH_3	293	$k = 7.12(14) \cdot 10^5\ M^{-1}\,s^{-1}$	
$(C_6H_5)_2\dot{C}OH$ + methyl-4-nitrobenzoate ⟶ (methyl-4-nitrobenzoate)$^{\bar{\cdot}}$ + H^+ + $(C_6H_5)_2CO$				
Laser phot. of benzilic acid + uranyl nitrate + acetone				80 Kem 1
KAS	CH_3COCH_3	293	$k = 1.21(4) \cdot 10^6\ M^{-1}\,s^{-1}$	
$(C_6H_5)_2\dot{C}OH$ + 1,2-naphthoquinone ⟶ (1,2-naphthosemiquinone) + $(C_6H_5)_2CO$				
Laser phot. of benzilic acid + uranyl nitrate + acetone				80 Kem 1
KAS	CH_3COCH_3	293	$k = 2.53(21) \cdot 10^8\ M^{-1}\,s^{-1}$	

[45]) "Juglone".
[46]) "Menadiane".

Reaction Radical generation Method	Solvent	T[K]	Rate data	Ref./ add. ref.
$(C_6H_5)_2\dot{C}OH$ + 3-nitroacetophenone ⟶ (3-nitroacetophenone)$^{\cdot -}$ + H^+ + $(C_6H_5)_2CO$				
Laser phot. of benzilic acid + uranyl nitrate + acetone				80 Kem 1
KAS	CH_3COCH_3	293	$k = 5.48(13) \cdot 10^5\ M^{-1}\,s^{-1}$	
$(C_6H_5)_2\dot{C}OH$ + 4-nitroacetophenone ⟶ (4-nitroacetophenone)$^{\cdot -}$ + H^+ + $(C_6H_5)_2CO$				
Laser phot. of benzilic acid + uranyl nitrate + acetone				80 Kem 1
KAS	CH_3COCH_3	293	$k = 1.61(5) \cdot 10^6\ M^{-1}\,s^{-1}$	
$(C_6H_5)_2\dot{C}OH$ + 3-nitrobenzaldehyde ⟶ (3-nitrobenzaldehyde)$^{\cdot -}$ + H^+ + $(C_6H_5)_2CO$				
Laser phot. of benzilic acid + uranyl nitrate + acetone				80 Kem 1
KAS	CH_3COCH_3	293	$k = 5.72(19) \cdot 10^5\ M^{-1}\,s^{-1}$	
$(C_6H_5)_2\dot{C}OH$ + 4-nitrobenzaldehyde ⟶ (4-nitrobenzaldehyde)$^{\cdot -}$ + H^+ + $(C_6H_5)_2CO$				
Laser phot. of benzilic acid + uranyl nitrate + acetone				80 Kem 1
KAS	CH_3COCH_3	293	$k = 1.75(7) \cdot 10^6\ M^{-1}\,s^{-1}$	
$(C_6H_5)_2\dot{C}OH$ + 4-nitrobenzamide ⟶ (4-nitrobenzamide)$^{\cdot -}$ + H^+ + $(C_6H_5)_2CO$				
Laser phot. of benzilic acid + uranyl nitrate + acetone				80 Kem 1
KAS	CH_3COCH_3	293	$k = 1.28(3) \cdot 10^6\ M^{-1}\,s^{-1}$	
$(C_6H_5)_2\dot{C}OH$ + nitrobenzene ⟶ $C_6H_5\dot{N}O_2^-$ + $(C_6H_5)_2CO$ + H^+				
Laser phot. of benzilic acid + uranyl nitrate + acetone				80 Kem 1
KAS	CH_3COCH_3	293	$k = 2.62(18) \cdot 10^5\ M^{-1}\,s^{-1}$	
Laser phot. of benzilic acid + uranyl nitrate + acetone + H_2O				80 Kem 1
KAS	CH_3COCH_3/H_2O (75:25 V/V)	293	$k = 3.50(25) \cdot 10^5\ M^{-1}\,s^{-1}$	
$(C_6H_5)_2\dot{C}OH$ + 2-nitrobenzoic acid ⟶ (2-nitrobenzoic acid)$^{\cdot -}$ + H^+ + $(C_6H_5)_2CO$				
Laser phot. of benzilic acid + uranyl nitrate + acetone				80 Kem 1,
KAS	CH_3COCH_3	293	$k = 1.00(12) \cdot 10^6\ M^{-1}\,s^{-1}$	80 Kem 2
$(C_6H_5)_2\dot{C}OH$ + 3-nitrobenzoic acid ⟶ (3-nitrobenzoic acid)$^{\cdot -}$ + H^+ + $(C_6H_5)_2CO$				
Laser phot. of benzilic acid + uranyl nitrate + acetone				80 Kem 1
KAS	CH_3COCH_3	293	$k = 6.01(12) \cdot 10^5\ M^{-1}\,s^{-1}$	
Laser phot. of benzilic acid + uranyl nitrate + acetone + H_2O				80 Kem 1
KAS	CH_3COCH_3/H_2O (75:25 V/V)	293	$k = 1.10(8) \cdot 10^6\ M^{-1}\,s^{-1}$	
$(C_6H_5)_2\dot{C}OH$ + 4-nitrobenzoic acid ⟶ (4-nitrobenzoic acid)$^{\cdot -}$ + H^+ + $(C_6H_5)_2CO$				
Laser phot. of benzilic acid + uranyl nitrate + acetone				80 Kem 1
KAS	CH_3COCH_3	293	$k = 1.13(9) \cdot 10^6\ M^{-1}\,s^{-1}$	
$(C_6H_5)_2\dot{C}OH$ + 3-nitrobenzonitrile ⟶ (3-nitrobenzonitrile)$^{\cdot -}$ + H^+ + $(C_6H_5)_2CO$				
Laser phot. of benzilic acid + uranyl nitrate + acetone				80 Kem 1
KAS	CH_3COCH_3	293	$k = 7.36(49) \cdot 10^5\ M^{-1}\,s^{-1}$	
$(C_6H_5)_2\dot{C}OH$ + 4-nitrobenzonitrile ⟶ (4-nitrobenzonitrile)$^{\cdot -}$ + H^+ + $(C_6H_5)_2CO$				
Laser phot. of benzilic acid + uranyl nitrate + acetone				80 Kem 1
KAS	CH_3COCH_3	293	$k = 4.38(11) \cdot 10^6\ M^{-1}\,s^{-1}$	
$(C_6H_5)_2\dot{C}OH$ + 4-nitrobenzylbromide ⟶ (4-nitrobenzylbromide)$^{\cdot -}$ + H^+ + $(C_6H_5)_2CO$				
Laser phot. of benzilic acid + uranyl nitrate + acetone				80 Kem 1
KAS	CH_3COCH_3	293	$k = 9.12(5) \cdot 10^5\ M^{-1}\,s^{-1}$	

Reaction Radical generation Method	Solvent	T[K]	Rate data	Ref./ add. ref.
$(C_6H_5)_2\dot{C}OH$ + *anti*-5-nitro-2-furaldoxime[47]) ⟶ (*anti*-5-nitro-...)$^{\cdot -}$ + H^+ + $(C_6H_5)_2CO$				
Laser phot. of benzilic acid + uranyl nitrate + acetone				80 Kem 1
KAS	CH_3COCH_3	293	$k = 2.16(9) \cdot 10^7\ M^{-1}s^{-1}$	
$(C_6H_5)_2\dot{C}OH$ + 5-nitro-2-furoic acid ⟶ (5-nitro-2-furoic acid)$^{\cdot -}$ + H^+ + $(C_6H_5)_2CO$				
Laser phot. of benzilic acid + uranyl nitrate + acetone				80 Kem 1
KAS	CH_3COCH_3	293	$k = 1.16(9) \cdot 10^7\ M^{-1}s^{-1}$	
$(C_6H_5)_2\dot{C}OH$ + 3-nitrotoluene ⟶ (3-nitrotoluene)$^{\cdot -}$ + H^+ + $(C_6H_5)_2CO$				
Laser phot. of benzilic acid + uranyl nitrate + acetone				80 Kem 1
KAS	CH_3COCH_3	293	$k = 2.88(27) \cdot 10^5\ M^{-1}s^{-1}$	
$(C_6H_5)_2\dot{C}OH$ + 9,10-phenanthraquinone ⟶ (9,10-phenanthrasemiquinone) + $(C_6H_5)_2CO$				
Laser phot. of benzilic acid + uranyl nitrate + acetone				80 Kem 1
KAS	CH_3COCH_3	293	$k = 2.28(10) \cdot 10^8\ M^{-1}s^{-1}$	
$(C_6H_5)_2\dot{C}OH$ + phenyl-1,4-benzoquinone ⟶ (phenyl-1,4-benzosemiquinone) + $(C_6H_5)_2CO$				
Laser phot. of benzilic acid + uranyl nitrate + acetone				80 Kem 1
KAS	CH_3COCH_3	293	$k = 2.57(12) \cdot 10^8\ M^{-1}s^{-1}$	
$(C_6H_5)_2\dot{C}OH$ + tetrachlorophthalic anhydride ⟶ (tetra...)$^{\cdot -}$ + H^+ + $(C_6H_5)_2CO$				
Laser phot. of benzilic acid + uranyl nitrate + acetone				80 Kem 1
KAS	CH_3COCH_3	293	$k = 1.36(34) \cdot 10^6\ M^{-1}s^{-1}$	
$(C_6H_5)_2\dot{C}OH$ + 2,3,5,6-tetramethyl-1,4-benzoquinone[48]) ⟶ (2,3,5,6-tetra... semiquinone) + $(C_6H_5)_2CO$				
Laser phot. of benzilic acid + uranyl nitrate + acetone				80 Kem 1
KAS	CH_3COCH_3	293	$k = 3.27(11) \cdot 10^7\ M^{-1}s^{-1}$	
$(C_6H_5)_2\dot{C}OH$ + 3-trifluoromethylbenzoic acid ⟶ products				
Laser phot. of benzilic acid + uranyl nitrate + acetone				80 Kem 1
KAS	CH_3COCH_3	293	$k = 4.06(42) \cdot 10^4\ M^{-1}s^{-1}$	
4.2.2.2.2.2 Anionic radicals				
$(C_6F_6)^{\cdot -}$ + O_2 ⟶ C_6F_6 + $\dot{O}_2^-$				
Pulse rad. of C_6F_6				81 End 1,
[49])	C_6F_6	RT	$k = 7 \cdot 10^9\ M^{-1}s^{-1}$	81 War 1
$(C_6F_6)^{\cdot -}$ + SF_6 ⟶ products				
Pulse rad. of C_6F_6				81 End 1,
[49])	C_6F_6	RT	$k < 10^8\ M^{-1}s^{-1}$	81 War 1
$(C_6F_6)^{\cdot -}$ + dibromomethane (CH_2Br_2) ⟶ Br^- + $\dot{C}H_2Br$ + C_6F_6				
Pulse rad. of C_6F_6				81 End 1,
[49])	C_6F_6	RT	$k < 10^8\ M^{-1}s^{-1}$	81 War 1
$(C_6F_6)^{\cdot -}$ + tetrabromomethane (CBr_4) ⟶ C_6F_6 + Br^- + $\dot{C}Br_3$				
Pulse rad. of C_6F_6				81 End 1,
[49])	C_6F_6	RT	$k = 1.5 \cdot 10^{11}\ M^{-1}s^{-1}$	81 War 1

[47]) "Nifuroxime".
[48]) "Duroquinone".
[49]) Time-resolved microwave absorption.

Reaction Radical generation Method	Solvent	T[K]	Rate data	Ref./ add. ref.
$(C_6F_6)^{\bar{\cdot}}$ + tetrachloromethane (CCl_4) ⟶ $C_6F_6 + Cl^- + \dot{C}Cl_3$				
Pulse rad. of C_6F_6 [49]	C_6F_6	RT	$k = 1.6 \cdot 10^9\ M^{-1}s^{-1}$	81 End 1, 81 War 1
$(C_6F_6)^{\bar{\cdot}}$ + tetracyanoethylene(TCNE) ⟶ C_6F_6 + $(TCNE)^{\bar{\cdot}}$				
Pulse rad. of C_6F_6 [49]	C_6F_6	RT	$k > 1.4 \cdot 10^{11}\ M^{-1}s^{-1}$	81 End 1, 81 War 1
$(C_6F_6)^{\bar{\cdot}}$ + tribromomethane ($CHBr_3$) ⟶ $C_6F_6 + Br^- + \dot{C}HBr_2$				
Pulse rad. of C_6F_6 [49]	C_6F_6	RT	$k = 9 \cdot 10^9\ M^{-1}s^{-1}$	81 End 1, 81 War 1
$(Fluorenone)^{\bar{\cdot}}$ [50] + O_2 ⟶ $\dot{O}_2^-$ + fluorenone				
Pulse rad. of fluorenone + $C_2H_5ONa + N_2O + C_2H_5OH$				67 Ada 1
KAS	$C_2H_5OH + 10^{-2}\ M$ C_2H_5ONa	RT	$k = 9.8 \cdot 10^8\ M^{-1}s^{-1}$	
Fluorescein, semireduced ($\dot{F}^{3-}$) + rhodamine B ⟶ $(\text{rhodamine B})^{\bar{\cdot}} + F^{2-}$				
Phot. of fluorescein [51] + leucofluorescein in CH_3CN				74 Kru 1
KAS	CH_3CN	RT	$k = (1.0\ldots1.3) \cdot 10^9\ M^{-1}s^{-1}$ [52]	

4.2.2.2.2.3 Cationic radicals

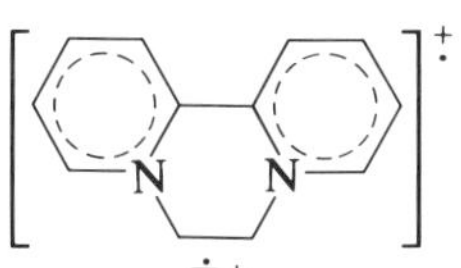

$\dot{E}^+$

+ Cu(II) ⟶ $[\ldots]^{2+}$ + Cu(I)

Method	Solvent	T[K]	Rate data	Ref./ add. ref.
Stopped flow expt. with (6,7-dihydrodipyrido[1,2-a:1′,2′-c]pyrazinium dibromide)				77 Eva 2
KAS	CH_3OH	193	$k = 9.9(2) \cdot 10^4\ M^{-1}s^{-1}$	
$\dot{E} + O_2$ ⟶ $[\ldots]^{2+} + \dot{O}_2^-$				
Stopped flow expt. with (6,7-dihydrodipyrido[1,2-a:1′,2′-c]pyrazinium dibromide)				77 Eva 2
KAS	C_2H_5OH	193	$k = 4.4(4) \cdot 10^3\ M^{-1}s^{-1}$	
		173… 203	$E_a = 35.5(40)\ kJ\ mol^{-1}$	

[49]) Time-resolved microwave absorption.

[50]) Likely to be 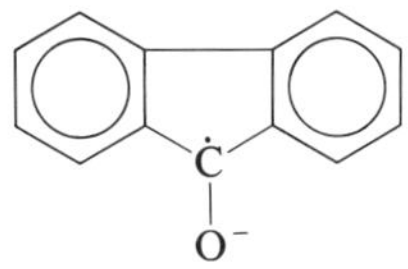

[51]) 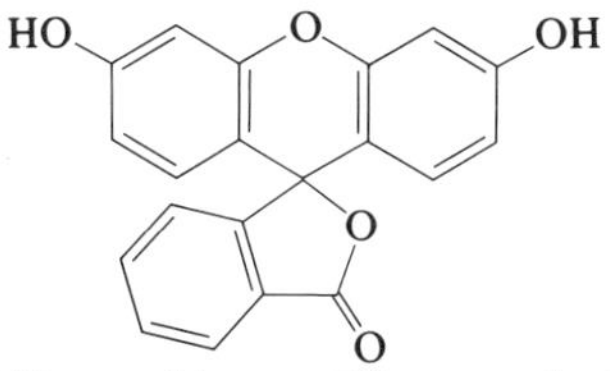
, radical possibly 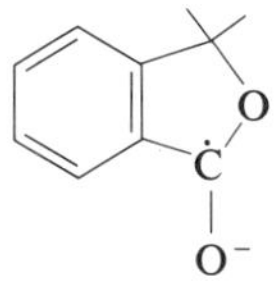

[52]) Measured by two different methods.

Reaction Radical generation Method	Solvent	T[K]	Rate data	Ref./ add. ref.

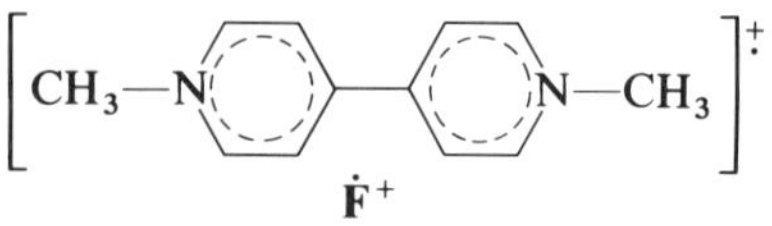 $+ O_2 \longrightarrow \dot{O}_2^- + [\ldots]^{2+}$

$\dot{F}^+$

Phot. of air sat. solutions of 1,1′-dimethyl-4,4′-bipyridinium [53]) — 77 Pat 1

KAS	CH_3OH	RT	$k = 3.3 \cdot 10^6\ M^{-1}s^{-1}$	

$\dot{F}^+ + O_2 \longrightarrow \dot{O}_2^- + [\ldots]^{2+}$

Phot. of air sat. solutions of 1,1′-dimethyl-4,4′-bipyridinium [53]) — 77 Pat 1

KAS	$C_2H_5OH + 5\%\ H_2O$	RT	$k = 1.3 \cdot 10^6\ M^{-1}s^{-1}$
	n-$C_3H_7OH + 5\%\ H_2O$	RT	$k = 3.0 \cdot 10^6\ M^{-1}s^{-1}$
	2-propanol $+ 5\%\ H_2O$	RT	$k = 8.7 \cdot 10^5\ M^{-1}s^{-1}$

$\dot{F}^+$ + 1,4-benzoquinone ⟶ (1,4-benzosemiquinone)$^{\dot{-}}$ + $[\ldots]^{2+}$

Flash phot. of chlorophyll a + 1,1′-dimethyl-4,4′-bipyridinium [53]) + C_2H_5OH — 80 Dar 1

KAS	C_2H_5OH (90%)	RT	$k = 1.3(1) \cdot 10^5\ M^{-1}s^{-1}$

$\dot{F}^+$ + 2,3,5-trimethyl-1,4-benzoquinone ⟶ (2,3,5-…-semiquinone)$^{\dot{-}}$ + $[\ldots]^{2+}$

Flash phot. of chlorophyll a + 1,1′-dimethyl-4,4′-bipyridinium [53]) + C_2H_5OH — 80 Dar 1

KAS	C_2H_5OH (90%)	RT	$k = 9.3(7) \cdot 10^5\ M^{-1}s^{-1}$

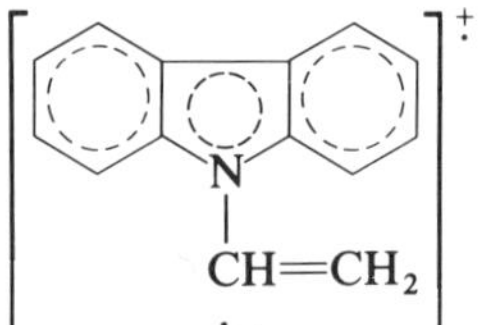 + dimethylphenylamine ⟶ $[C_6H_5N(CH_3)_2]^{\dot{+}}$ + N-vinylcarbazole

$\dot{G}^+$

Pulse rad. of N-vinylcarbazole + nitrobenzene — 80 Was 1

KAS	$C_6H_5NO_2$	RT	$k = 3.9 \cdot 10^9\ M^{-1}s^{-1}$

$\dot{G}^+$ + diphenylamine ⟶ $[(C_6H_5)_2NH]^{\dot{+}}$ + N-vinylcarbazole

Pulse rad. of N-vinylcarbazole + nitrobenzene — 80 Was 1

KAS	$C_6H_5NO_2$	RT	$k = 2.0 \cdot 10^9\ M^{-1}s^{-1}$

$\dot{G}^+$ + triphenylamine ⟶ $[(C_6H_5)_3N]^{\dot{+}}$ + N-vinylcarbazole

Pulse rad. of N-vinylcarbazole + nitrobenzene — 80 Was 1

KAS	$C_6H_5NO_2$	RT	$k = 1.3 \cdot 10^9\ M^{-1}s^{-1}$

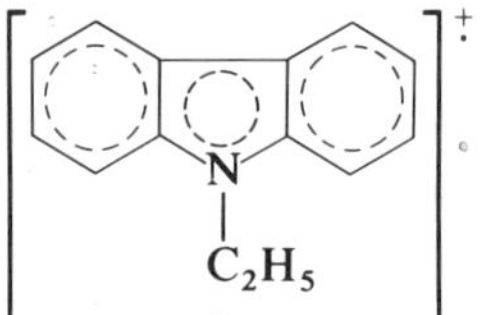 + dimethylphenylamine ⟶ $[C_6H_5N(CH_3)_2]^{\dot{+}}$ + N-ethylcarbazole

$\dot{H}^+$

Pulse rad. of N-ethylcarbazole + nitrobenzene — 80 Was 1

KAS	$C_6H_5NO_2$	RT	$k = 3.6 \cdot 10^9\ M^{-1}s^{-1}$

$\dot{H}^+$ + diphenylamine ⟶ $[(C_6H_5)_2NH]^{\dot{+}}$ + N-ethylcarbazole

Pulse rad. of N-ethylcarbazole + nitrobenzene — 80 Was 1

KAS	$C_6H_5NO_2$	RT	$k = 1.8 \cdot 10^9\ M^{-1}s^{-1}$

[53]) Paraquat, methylviologen.

Reaction Radical generation Method	Solvent	T[K]	Rate data	Ref./ add. ref.
$\dot{H}^+$ *) + triphenylamine ⟶ $[(C_6H_5)_3N]^{\dot{+}}$ + N-ethylcarbazole				
Pulse rad. of N-ethylcarbazole + nitrobenzene				80 Was 1
KAS	$C_6H_5NO_2$	RT	$k = 1.3 \cdot 10^9\ M^{-1}s^{-1}$	
(Zn-tetraphenylporphyrine)$^{\dot{+}}$ + chlorophyll a ⟶ (chlorophyll a)$^{\dot{+}}$ + ZnTPP				
Pulse rad. of Zn-tetraphenylporphyrine + 1,2-dichloroethane				80 Lev 1
KAS	CH_2ClCH_2Cl	RT	$k \approx 4 \cdot 10^9\ M^{-1}s^{-1}$	

*) For $\dot{H}$, see p. 440.

References for 4.2

53 Bax 1 Baxendale, J.H., Smithies, D.: J. Chem. Soc. **1953**, 779.
54 Bir 1 Birrell, R.N., Trotman-Dickenson, A.F.: J. Chem. Phys. **22** (1954) 678.
56 Gru 1 Gruver, J.T., Calver, J.G.: J. Am. Chem. Soc. **78** (1956) 5208.
59 Ker 1 Kerr, J.A., Trotman-Dickenson, A.F.: Trans. Faraday Soc. **55** (1959) 572.
59 Ker 2 Kerr, J.A., Trotman-Dickenson, A.F.: Trans. Faraday Soc. **55** (1959) 921.
60 Ker 1 Kerr, J.A., Trotman-Dickenson, A.F.: J. Chem. Soc. **1960**, 1611.
60 Met 1 Metcalfe, E.L., Trotman-Dickenson, A.F.: J. Chem. Soc. **1960**, 5072.
63 Huy 1 Huyser, E.S., Neckers, D.C.: J. Am. Chem. Soc. **85** (1963) 3641.
64 Asm 1 Asmus, K.-D., Henglein, A., Ebert, M., Keene, J.P.: Ber. Bunsenges. Phys. Chem. **68** (1964) 657.
65 Asm 1 Asmus, K.-D.: Dissertation, Techn. Univ. Berlin, **1965**, D 83.
65 Kee 1 Keene, J.P., Land, E.J., Swallow, A.J.: "Pulse Radiolysis", Ebert, M., Keene, J.P., Swallow, A.J., Baxendale, J.H. (eds.), New York: Academic Press, **1965**, p. 227.
65 Koc 1 Kochi, J.K., Subramanian, R.V.: Inorg. Chem. **4** (1965) 1527.
65 Koc 2 Kochi, J.K., Subramanian, R.V.: J. Am. Chem. Soc. **87** (1965) 4855.
65 Rab 1 Rabani, J., Mulac, W.A., Matheson, M.S.: J. Phys. Chem. **69** (1965) 53.
66 Asm 1 Asmus, K.-D., Wigger, A., Henglein, A.: Ber. Bunsenges. Phys. Chem. **70** (1966) 862.
66 Asm 2 Asmus, K.-D., Beck, G., Henglein, A., Wigger, A.: Ber. Bunsenges. Phys. Chem. **70** (1966) 869.
67 Ada 1 Adams, G.E., Michael, B.D., Richards, J.T.: Nature (London) **215** (1967) 1248.
67 Ara 1 Arai, S., Grev, D.A., Dorfman, L.M.: J. Chem. Phys. **46** (1967) 2572.
67 Chr 1 Chrysochoos, J., Ovadia, J., Grossweiner, L.I.: J. Phys. Chem. **71** (1967) 1629.
67 Pru 1 Pruetz, W., Land, E.J.: Biophysik **3** (1967) 349.
67 Sed 1 Seddon, W.A., Allen, A.O.: J. Phys. Chem. **71** (1967) 1914.
68 Ada 1 Adams, G.E., Michael, B.D., Willson, R.L.: Adv. Chem. Ser. **81** (1968) 289.
68 Cor 1 Cordier, P., Grossweiner, L.I.: J. Phys. Chem. **72** (1968) 2018.
68 Koc 1 Kochi, J.K., Bemis, A., Jenkins, C.L.: J. Am. Chem. Soc. **90** (1968) 4616.
68 Kos 1 Kosower, E.M., Mohammad, M.: J. Am. Chem. Soc. **90** (1968) 3271.
68 Lan 1 Land, E.J., Swallow, A.J.: Biochim. Biophys. Acta **162** (1968) 327.
68 Lil 1 Lilie, J., Beck, G., Henglein, A.: Ber. Bunsenges. Phys. Chem. **72** (1968) 529.
69 Ada 1 Adams, G.E., Willson, R.L.: Trans. Faraday Soc. **65** (1969) 2981.
69 Bax 1 Baxendale, J.H., Khan, A.A.: Int. J. Radiat. Phys. Chem. **1** (1969) 11.
69 Bux 1 Buxton, G.V., Dainton, F.S., Kalecinski, J.: Int. J. Radiat. Phys. Chem. **1** (1969) 87.
69 Cap 1 Capellos, C., Allen, A.O.: J. Phys. Chem. **73** (1969) 3264.
69 Chr 1 Christensen, H.C., Gustafsson, R.: Nukleonik **12** (1969) 49.
69 Gar 1 Garst, J.F., Barton, F.E.: Tetrahedron Lett. **1969**, 587.
69 Gru 1 Gruenbein, W., Henglein, A.: Ber. Bunsenges. Phys. Chem. **73** (1969) 376.
69 Lan 1 Land, E.J., Swallow, A.J.: Biochemistry **8** (1969) 2117.
69 Sim 1 Simic, M., Neta, P., Hayon, E.: J. Phys. Chem. **73** (1969) 3794.
69 Zim 1 Zimbrick, J.D., Ward, J.F., Myers, L.S. Jr.: Int. J. Radiat. Biol. Relat. Stud. Phys., Chem. Med. **16** (1969) 505.
70 Bar 1 Barker, G.C., Fowles, P., Stringer, B.: Trans. Faraday Soc. **66** (1970) 1509.
70 Bur 1 Burchill, C.E., Ginns, I.S.: Can. J. Chem. **48** (1970) 2628.
70 Cha 1 Chambers, K.W., Collinson, E., Dainton, F.S.: Trans. Faraday Soc. **66** (1970) 142.
70 Clo 1 Closson, W.D., Ji, S., Schulenberg, S.: J. Am. Chem. Soc. **92** (1970) 650.
70 Foj 1 Fojtik, A., Czapski, G., Henglein, A.: J. Phys. Chem. **74** (1970) 3204.
70 Pru 1 Pruetz, W.A., Land, E.J.: J. Phys. Chem. **74** (1970) 2107.
70 Sha 1 Shank, N.E., Dorfman, L.M.: J. Chem. Phys. **52** (1970) 4441.

70 Str 1 Struble, D.L., Beckwith, A.L.J., Gream, G.E.: Tetrahedron Lett. **1970**, 4795.
70 Wil 1 Willson, R.L.: Chem. Commun. **1970**, 1425.
70 Wil 2 Willson, R.L.: Chem. Commun. **1970**, 1005.
71 Asm 1 Asmus, K.-D., Chaudhri, S.A., Nazhat, N.B., Schmidt, W.F.: Trans. Faraday Soc. **67** (1971) 2607.
71 Bur 1 Burchill, C.E., Jones, P.W.: Can. J. Chem. **49** (1971) 4005.
71 Emm 1 Emmerson, P.T., Fielden, E.M., Johansen, I.: Int. J. Radiat. Biol. **19** (1971) 229.
71 Fie 1 Fielden, E.M., Roberts, P.B.: Int. J. Radiat. Biol. Relat. Stud. Phys., Chem. Med. **20** (1971) 355.
71 Fis 1 Fischer, M., Raemme, G., Claesson, S., Szwarc, M.: Chem. Phys. Lett. **9** (1971) 306.
71 Hay 1 Hayon, E., Simic, M.: J. Am. Chem. Soc. **93** (1971) 6781.
71 Hei 1 Heiba, E.J., Dessau, R.M.: J. Am. Chem. Soc. **93** (1971) 524.
71 Koe 1 Koester, R., Asmus, K.-D.: Z. Naturforsch. **26b** (1971) 1104.
71 Lan 1 Land, E.J., Swallow, A.J.: Arch. Biochem. Biophys. **145** (1971) 365.
71 Moh 1 Mohammad, M., Kosower, E.M.: J. Am. Chem. Soc. **93** (1971) 2709.
71 Moh 2 Mohammad, M., Kosower, E.M.: J. Am. Chem. Soc. **93** (1971) 2713.
71 Ric 1 Richards, J.T., Thomas, J.K.: Chem. Phys. Lett. **10** (1971) 317.
71 Rob 1 Roberts, P.M., Fielden, E.M.: Int. J. Radiat. Biol. **20** (1971) 363.
71 Sto 1 Stockhausen, K., Henglein, A.: Ber. Bunsenges. Phys. Chem. **75** (1971) 833.
71 Wil 1 Willson, R.L.: Trans. Faraday Soc. **67** (1971) 3020.
71 Wil 2 Willson, R.L.: Trans. Faraday Soc. **67** (1971) 3008.
72 Ada 1 Adams, G.E., Greenstock, C.L., Hemmen, J.J.v., Willson, R.L.: Radiat. Res. **49** (1972) 85.
72 Ada 2 Adams, G.E., Willson, R.L.: Int. J. Radiat. Biol. **22** (1972) 589.
72 Bec 1 Beck, G., Thomas, J.K.: J. Phys. Chem. **76** (1972) 3856.
72 Bec 2 Beckwith, A.L.J., Gream, G.E., Struble, D.L.: Australian J. Chem. **25** (1972) 1081.
72 Beh 1 Behzadi, A., Bargwardt, V., Schnabel, W.: Chem. Zvesti **26** (1972) 242.
72 Bru 1 Brustad, T., Bugge, H., Jones, W.B.G., Wold, E.: Int. J. Radiat. Biol. **22** (1972) 115.
72 Bur 1 Burchill, C.E., Wollner, G.P.: Can. J. Chem. **50** (1972) 1751.
72 Cha 1 Chaudhri, S.A., Asmus, K.-D.: J. Phys. Chem. **76** (1972) 26.
72 Coh 1 Cohen, H., Meyerstein, D.: J. Am. Chem. Soc. **94** (1972) 6944.
72 Hay 1 Hayon, E., Simic, M.: Radiat. Res. **50** (1972) 464.
72 Hof 1 Hoffman, M.Z., Simic, M.: J. Am. Chem. Soc. **94** (1972) 1757.
72 Hul 1 Hulme, B.E., Land, E.J., Phillips, G.O.: J. Chem. Soc., Faraday Trans. I **68** (1972) 1992.
72 Jen 1 Jenkins, C.L., Kochi, J.K.: J. Am. Chem. Soc. **94** (1972) 843.
72 Mar 1 Martin, J.E., Hart, E.J., Adamson, A.W., Gafney, H., Halpern, J.: J. Am. Chem. Soc. **94** (1972) 9238.
72 Nel 1 Nelson, D.A., Hayon, E.: J. Phys. Chem. **76** (1972) 3200.
72 Rae 1 Raemme, G., Fisher, M., Claesson, S., Szwarc, M.: Proc. Roy. Soc. (London) Ser. A **327** (1972) 467.
72 Rod 1 Rodgers, M.A.J.: J. Chem. Soc. Faraday Trans. I **68** (1972) 1278.
72 Sim 1 Simic, M., Hayon, E.: Int. J. Radiat. Biol. Relat. Stud. Phys., Chem. Med. **22** (1972) 507.
72 Ste 1 Stelter, L.H.: Ph.D. Thesis, Univ. of Toledo, Ohio, **1972**, 115 p.
73 Ada 1 Adams, G.E., Willson, R.L.: J. Chem. Soc., Faraday Trans. I **69** (1973) 719.
73 Asm 1 Asmus, K.-D., Möckel, H., Henglein, A.: J. Phys. Chem. **77** (1973) 1218.
73 Beh 1 Behar, D., Samuni, A., Fessenden, R.W.: J. Phys. Chem. **77** (1973) 2055.
73 Boc 1 Bockrath, B., Dorfman, L.M.: J. Phys. Chem. **77** (1973) 2618.
73 Bux 1 Buxton, G.V., Sellers, R.M.: J. Chem. Soc., Faraday Trans. I **69** (1973) 555.
73 Ell 1 Ellis, J.D., Green, M., Sykes, A.G., Buxton, G.V., Sellers, R.M.: J. Chem. Soc., Dalton Trans. **1973**, 1724.
73 Far 1 Farrington, J.A., Ebert, M., Land, E.J., Fletcher, K.: Biochim. Biophys. Acta **314** (1973) 372.
73 Far 2 Faraggi, M., Leopold, J.G.: Biochem. Biopyhs. Res. Commun. **50** (1973) 413.
73 Gil 1 Gilbert, B.C., Norman, R.O.C., Sealy, R.C.: J. Chem. Soc. Perkin Trans. II **1973**, 2174.
73 Gre 1 Greenstock, C.L., Dunlop, I.: Radiat. Res. **56** (1973) 428.
73 Gre 2 Greenstock, C.L., Dunlop, I.: J. Am. Chem. Soc. **95** (1973) 6917.
73 Gre 3 Greenstock, C.L., Dunlop, I., Neta, P.: J. Phys. Chem. **77** (1973) 1187.
73 Hay 1 Hayon, E., Simic, M.: J. Am. Chem. Soc. **95** (1973) 6681.
73 Hay 2 Hayon, E., Simic, M.: J. Am. Chem. Soc. **95** (1973) 2433.
73 Hay 3 Hayon, E., Simic, M.: J. Am. Chem. Soc. **95** (1973) 1029.
73 Hof 1 Hoffman, M.Z., Simic, M.: Inorg. Chem. **12** (1973) 2471.
73 Hum 1 Hummel, A., Luthjens, L.H.: J. Chem. Phys. **59** (1973) 654.
73 Naz 1 Nazhat, N.B., Asmus, K.-D.: J. Phys. Chem. **77** (1973) 614.
73 Pat 1 Patel, K.B., Willson, R.L.: J. Chem. Soc., Faraday Trans. I **69** (1973) 814.
73 Rao 1 Rao, P.S., Hayon, E.: Nature (London) **243** (1973) 344.
73 Rao 2 Rao, P.S., Hayon, E.: J. Phys. Chem. **77** (1973) 2753.
73 Rao 3 Rao, P.S., Hayon, E.: Biochem. Biophys. Acta **292** (1973) 516.
73 Rob 1 Robinson, A.J., Rodgers, M.A.J.: J. Chem. Soc., Faraday Trans. I **69** (1973) 2036.
73 Sch 1 Schoeneshoefer, M.: Int. J. Radiat. Phys. Chem. **5** (1973) 375.

73 Sim 1 Simic, M., Hayon, E.: Biochem. Biopyhs. Res. Commun. **50** (1973) 364.
73 Wal 1 Walling, C., El-Taliawi, G.M.: J. Am. Chem. Soc. **95** (1973) 844.
73 Wil 1 Willson, R.L.: Trans. Biochem. Soc. **1** (1973) 929.
73 Zad 1 Zador, E., Warman, J.M., Hummel, A.: Chem. Phys. Lett. **23** (1973) 363.
74 Bla 1 Blackburn, R., Erkol, A.Y., Phillips, G.O.: J. Chem. Soc., Faraday Trans. I **70** (1974) 1693.
74 Bre 1 Brede, O., Helmstreit, W., Mehnert, R.: Chem. Phys. Lett. **28** (1974) 43.
74 Bru 1 Bruehlmann, U., Hayon, E.: J. Am. Chem. Soc. **96** (1974) 6169.
74 But 1 Butler, J., Jayson, G.G., Swallow, A.J.: J. Chem. Soc., Faraday Trans. I **70** (1974) 1394.
74 Coh 1 Cohen, H., Meyerstein, D.: Israel J. Chem. **12** (1974) 1049.
74 Gar 1 Garst, J.F., Barton, F.E.: J. Am. Chem. Soc. **96** (1974) 523.
74 Gil 1 Gilbert, B.C., Norman, R.O.C., Sealy, R.C.: J. Chem. Soc., Perkin Trans. II **1974**, 824.
74 Gil 2 Gilbert, B.C., Norman, R.O.C., Sealy, R.C.: J. Chem. Soc., Perkin Trans. II **1974**, 1435.
74 Har 1 Harel, Y., Meyerstein, D.: J. Am. Chem. Soc. **96** (1974) 2720.
74 Kel 1 Kelm, M., Lilie, J., Henglein, A., Janata, E.: J. Phys. Chem. **78** (1974) 882.
74 Kru 1 Krüger, U., Memming, R.: Ber. Bunsenges. Phys. Chem. **78** (1974) 684.
74 Lil 1 Lilie, J., Heckel, E., Lamb, R.C.: J. Am. Chem. Soc. **96** (1974) 5543.
74 Mic 1 Micic, O.I., Cercek, B.: J. Phys. Chem. **78** (1974) 285.
74 Moo 1 Moorthy, P.N., Hayon, E.: J. Phys. Chem. **78** (1974) 2615.
74 Mul 1 Mulazzani, Q.G., Ward, M.D., Semerano, G., Emmi, S.S., Giordani, P.: Int. J. Radiat. Phys. Chem. **6** (1974) 187.
74 Mul 2 Mulazzani, Q.G., Emmi, S., Roffi, G., Hoffman, M.Z.: Quaderni dell'Area di Ricerca dell'Emilia-Romagna, C.N.R., Vol. 5, Lab. di Fotochimica e Radiazioni d'Alta Energia, Rapporto annuale 1973, Bologna: Consiglio Nazionale delle Ricerche, **1974**, p. 111.
74 Net 1 Neta, P., Patterson, L.K.: J. Phys. Chem. **78** (1974) 2211.
74 Rao 1 Rao, P.S., Hayon, E.: J. Am. Chem. Soc. **96** (1974) 1295.
74 Sha 1 Shafferman, A., Stein, G.: Science **183** (1974) 428.
74 Sim 1 Simic, M., Lilie, J.: J. Am. Chem. Soc. **96** (1974) 291.
74 Sim 2 Simic, M., Hayon, E.: FEBS Lett. **44** (1974) 334.
74 Ste 1 Steckhan, E., Kuwana, T.: Ber. Bunsenges. Phys. Chem. **78** (1974) 253.
74 Wil 1 Willson, R.L., Gilbert, B.C., Marshall, P.D.R., Norman, R.O.C.: Int. J. Radiat. Biol. Relat. Stud. Phys., Chem. Med. **26** (1974) 427.
74 Wil 2 Willson, R.L., Cramp, W.A., Ings, R.M.J.: Int. J. Radiat. Biol. **26** (1974) 557.
74 Wil 3 Willson, R.L., Wardman, P., Asmus, K.-D.: Nature (London) **252** (1974) 323.
75 Arm 1 Armor, J.N., Hoffman, M.Z.: Inorg. Chem. **14** (1975) 444.
75 Ays 1 Ayscough, P.B., Elliot, A.J., Salmon, G.A.: Int. J. Radiat. Biol. Relat. Stud. Phys., Chem. Med. **27** (1975) 603.
75 Ban 1 Bank, S., Juckett, D.A.: J. Am. Chem. Soc. **97** (1975) 567.
75 Bre 1 Brede, O., Helmstreit, W., Mehnert, R.: Z. Phys. Chem. (Leipzig) **256** (1975) 513.
75 Dav 1 Davids, E.L., Warman, J.M., Hummel, A.: J. Chem. Soc., Faraday Trans. I **71** (1975) 1252.
75 Eri 1 Eriksen, T.E.: Radiochem. Radioanal. Lett. **22** (1975) 33.
75 Far 1 Faraggi, M., Redpath, J.L., Tal, Y.: Radiat. Res. **64** (1975) 452.
75 Far 2 Faraggi, M., Hummerich, P., Pecht, I.: FEBS Lett. **51** (1975) 47.
75 Far 3 Faraggi, M., in: "Fast Processes in Radiation Chemistry and Biology", G.E. Adams, E.M. Fielden and B.D. Michael, (eds.), London: Wiley, **1975**, p. 285.
75 Fuj 1 Fujita, S., Horii, H., Mori, T., Taniguchi, S.: Bull. Chem. Soc. Jpn. **48** (1975) 3067.
75 Gar 1 Garst, J.F., Roberts, R.D., Abels, B.N.: J. Am. Chem. Soc. **97** (1975) 4925.
75 Gof 1 Goff, H., Simic, M.G.: Biochim. Biophys. Acta **392** (1975) 201.
75 Hay 1 Haysom, H.R., Phillips, J.M., Richards, J.T., Scholes, G., Willson, R.L., in: "Fast Processes in Radiation Chemistry and Biology", G.E. Adams, E.M. Fielden and B.D. Michael, (eds.), London: Wiley, **1975**, p. 241.
75 Hof 1 Hoffman, M.Z., Hayon, E.: J. Phys. Chem. **79** (1975) 1362.
75 Joh 1 Johnson, D.W., Salmon, G.A.: J. Chem. Soc., Faraday Trans. I **71** (1975) 583.
75 Kel 1 Kelm, M., Lilie, J., Henglein, A.: J. Chem. Soc., Faraday Trans. I **5** (1975) 1132.
75 Kle 1 Klein, G.W., Bhatia, K., Madhavan, V., Schuler, R.H.: J. Phys. Chem. **79** (1975) 1767.
75 Lan 1 Land, E.J., Swallow, A.J.: Ber. Bunsenges. Phys. Chem. **79** (1975) 436.
75 Mac 1 Mackey, L., Steckhan, E., Kuwana, T.: Ber. Bunsenges. Phys. Chem. **79** (1975) 587.
75 Mei 1 Meisel, D., Neta, P.: J. Phys. Chem. **79** (1975) 2459.
75 Moo 1 Moorthy, P.N., Hayon, E.: J. Am. Chem. Soc. **97** (1975) 3345.
75 Moo 2 Moorthy, P.N., Hayon, E.: J. Phys. Chem. **79** (1975) 1059.
75 Moo 3 Moorthy, P.N., Hayon, E.: J. Am. Chem. Soc. **97** (1975) 2048.
75 Rao 1 Rao, P.S., Hayon, E.: J. Am. Chem. Soc. **97** (1975) 2986.
75 Sim 1 Simic, M.G., Taub, I.A., Tocci, J., Hurwitz, P.A.: Biochem. Biophys. Res. Commun. **62** (1975) 161.
75 Sto 1 Storer, D.K., Waltz, W.L., Brodovitch, J.C., Eager, R.L.: Int. J. Radiat. Phys. Chem. **7** (1975) 693.

75 Wal 1	Walrant, P., Santus, R., Grossweiner, L.I.: Photochem. Photobiol. **22** (1975) 63.
75 Whi 1	Whillans, D.W., Neta, P.: Radiat. Res. **64** (1975) 416.
75 Whi 2	Whillans, D.W., Adams, G.E.: Int. J. Radiat. Biol. Relat. Stud. Phys., Chem. Med. **28** (1975) 501.
75 Whi 3	Whillans, D.W., Adams, G.E., Neta, P.: Radiat. Res. **62** (1975) 407.
75 Wil 1	Wilting, J., Van Buuren, K.J.H., Braams, R., Van Gelder, B.F.: Biochim. Biophys. Acta **376** (1975) 285.
75 Wil 2	Willson, R.L., Slater, T.F., in: "Fast Processes in Radiation Chemistry and Biology", G.E. Adams, E.M. Fielden, B.D. Michael, (eds.), London: Wiley, **1975**, p. 147.
75 Zad 1	Zador, E., Warman, J.M., Hummel, A.: J. Chem. Phys. **62** (1975) 3897.
76 And 1	Anderson, R.F.: Ber. Bunsenges. Phys. Chem. **80** (1976) 969.
76 Asm 1	Asmus, K.-D., Nigam, S., Willson, R.L.: Int. J. Radiat. Biol. Relat. Stud. Phys., Chem. Med. **29** (1976) 211.
76 Ban 1	Bank, S., Juckett, D.A.: J. Am. Chem. Soc. **98** (1976) 7742.
76 Bia 1	Biaglow, J.E., Nygaard, O.F., Greenstock, C.L.: Biochem. Pharmac. **25** (1976) 393.
76 Bon 1	Bonifačić, M., Asmus, K.-D.: J. Phys. Chem. **80** (1976) 2426.
76 Bre 1	Breitenkamp, M., Henglein, A., Lilie, J.: Ber. Bunsenges. Phys. Chem. **80** (1976) 973.
76 Bur 1	Burr, J.G., Wagner, B.O., Schulte-Frohlinde, D.: Int. J. Radiat. Biol. **29** (1976) 433.
76 Bux 1	Buxton, G.V., Sellers, R.M., McCracken, D.R.: J. Chem. Soc., Faraday Trans. I **72** (1976) 1464.
76 Eva 1	Evans, J.F., Blount, H.N.: J. Phys. Chem. **80** (1976) 1011.
76 Far 1	Faraggi, M., Leopold, J.G.: Radiat. Res. **65** (1976) 238.
76 Fra 1	Frank, A.J., Graetzel, M., Henglein, A., Janata, E.: Ber. Bunsenges. Phys. Chem. **80** (1976) 294.
76 Fra 2	Frank, A.J., Graetzel, M., Henglein, A., Janata, E.: Ber. Bunsenges. Phys. Chem. **80** (1976) 547.
76 Fuj 1	Fujita, S., Horii, H., Mori, T., Taniguchi, S.: Bull. Chem. Soc. Jpn. **49** (1976) 1250.
76 Gre 1	Greenstock, C.L., Ruddock, G.W., Neta, P.: Radiat. Res. **66** (1976) 472.
76 Hol 1	Holeman, J., Sehested, K.: J. Phys. Chem. **80** (1976) 1642.
76 Ila 1	Ilan, Y., Rabani, J.: Int. J. Radiat. Phys. Chem. **8** (1976) 609.
76 Ila 2	Ilan, Y.A., Rabani, J., Czapski, G.: Biochim. Biophys. Acta **446** (1976) 277.
76 Jun 1	Jungbluth, H., Beyrich, J., Asmus, K.-D.: J. Phys. Chem. **80** (1976) 1049.
76 Laf 1	Lafferty, J., Land, E.J., Truscott, T.G.: J. Chem. Soc., Chem. Comm. **1976**, 70.
76 Mic 1	Michaels, H.B., Rasburn, E.J., Hunt, J.W.: Radiat. Res. **65** (1976) 250.
76 Moo 1	Moorthy, P.N., Hayon, E.: J. Org. Chem. **41** (1976) 1607.
76 Net 1	Neta, P., Meisel, D.: J. Phys. Chem. **80** (1976) 519.
76 Net 2	Neta, P., Simic, M.G., Hoffman, M.Z.: J. Phys. Chem. **80** (1976) 2018.
76 Net 3	Neta, P.: Radiat. Res. **68** (1976) 422.
76 Nig 1	Nigam, S., Asmus, K.-D., Willson, R.L.: J. Chem. Soc., Faraday Trans. I **72** (1976) 2324.
76 Sek 1	Seki, H., Ilan, Y.A., Ilan, Y., Stein, G.: Biochim. Biophys. Acta **440** (1976) 573.
76 Tai 1	Tait, A.M., Hoffman, M.Z., Hayon, E.: J. Am. Chem. Soc. **98** (1976) 86.
76 Tai 2	Tait, A.M., Hoffman, M.Z., Hayon, E.: Int. J. Radiat. Phys. Chem. **8** (1976) 691.
76 Tai 3	Tait, A.M., Hoffman, M.Z., Hayon, E.: Inorg. Chem. **15** (1976) 934.
76 Tof 1	Toffel, P., Henglein, A.: Ber. Bunsenges. Phys. Chem. **80** (1976) 525.
76 War 1	Wardman, P., Clarke, E.D.: J. Chem. Soc., Faraday Trans. I **72** (1976) 1377.
76 War 2	Warman, J.M., Infelta, P.P., deHaas, M.P., Hummel, A.: Chem. Phys. Lett. **43** (1976) 321.
77 Bel 1	Bell, I.P., Rodgers, M.A.J.: J. Chem. Soc., Faraday Trans. I **73** (1977) 315.
77 Ber 1	Berdnikov, V.M., Zhuravleva, O.S., Terent'eva, L.A.: Bull. Acad. Sci. USSR, Div. Chem. Sci. **26** (1977) 2050; Transl. from Izv. Akad. Nauk SSSR, Ser. Khim. **26** (1977) 2214.
77 Bet 1	Bettelheim, A., Faraggi, M.: Radiat. Res. **72** (1977) 71.
77 Bia 1	Biaglow, J.E., Jacobson, B., Greenstock, C.L., Raleigh, J.: Mol. Pharmacol. **13** (1977) 269.
77 Che 1	Cheney, R.P., Simic, M.G., Hoffman, M.Z., Taub, I.A., Asmus, K.-D.: Inorg. Chem. **16** (1977) 2187.
77 Coh 1	Cohen, H., Meyerstein, D.: J. Chem. Soc., Dalton Trans. **1977**, 1056.
77 Dor 1	Dorfman, L.M., Wang, Y., Wang, H.Y., Sujdak, R.J.: Faraday Disc., Chem. Soc. **63** (1977) 149.
77 Eva 1	Evans, T.R., Hurysz, L.F.: Tetrahedron Lett. **1977**, 3103.
77 Eva 2	Evans, A.G., Alford, R.E., Rees, N.H.: J. Chem. Soc., Perkin Trans. II **1977**, 445.
77 Fer 1	Ferraudi, G.J., Endicott, J.F.: Inorg. Chem. **16** (1977) 2762.
77 Hay 1	Hayashi, K., Irie, M., Lindenau, D., Schnabel, W.: Eur. Polym. J. **13** (1977) 925.
77 Lin 1	van der Linde, H.J.: Radiat. Phys. Chem. **10** (1977) 199.
77 Moo 1	Moorthy, P.N., Hayon, E.: J. Org. Chem. **42** (1977) 879.
77 Net 1	Neta, P., Levanon, H.: J. Phys. Chem. **81** (1977) 2288.
77 Pat 1	Patterson, L.K., Small, R.D., Jr., Scaiano, J.C.: Radiat. Res. **72** (1977) 218.
77 Pil 1	Pileni, M.P., Santus, R., Land, E.J.: Int. J. Radiat. Biol. **32** (1977) 23.
77 Rab 1	Rabani, J., Mulac, W.A., Matheson, M.S.: J. Phys. Chem. **81** (1977) 99.
77 Sek 1	Seki, H., Imamura, M., Illian, X.A., Ilan, Y., Stein, G.: J. Radiat. Res. **18** (1977) 6.
77 Sha 1	Shafferman, A., Stein, G.: Biochim. Biophys. Acta **462** (1977) 161.
77 Sim 1	Simic, M.G., Hoffman, M.Z., Brezniak, N.V.: J. Am. Chem. Soc. **99** (1977) 2166.
77 Wan 1	Wang, H.C., Levin, G., Szwarc, M.: J. Am. Chem. Soc. **99** (1977) 2642.

78 Ays 1 Ayscough, P.B., Lambert, G.: J. Chem. Soc., Faraday Trans. I **74** (1978) 2481.
78 Ban 1 Bank, S., Frost Bank, J.: Am. Chem. Soc., Symp. Series, No 69, "Organic Free Radicals", W.A. Pryor, (ed.) **1978**, p. 343.
78 Bux 1 Buxton, G.V., Green, J.C.: J. Chem. Soc., Faraday Trans. I **74** (1978) 697.
78 Dor 1 Dorfman, L.M., Wang, Y., Wang, H.-Y., Sujdak, R.J.: Farad. Disc., Chem. Soc. **63** (1978) 149.
78 Eva 1 Evans, J.F., Blount, H.N.: J. Am. Chem. Soc. **100** (1978) 4191.
78 Eve 1 Evers, E.L., Jayson, G.G., Swallow, A.J.: J. Chem. Soc., Faraday Trans. I **74** (1978) 418.
78 Far 1 Farrington, J.A., Ebert, M., Land, E.J.: J. Chem. Soc., Faraday Trans. I **74** (1978) 665.
78 Fav 1 Favaudon, V., Ferradini, C., Pucheault, J., Gilles, L., Le Gall, J.: Biochem. Biophys. Res. Commun. **84** (1978) 435.
78 Fuj 1 Fujita, S., Horii, H., Mori, T., Taniguchi, S.: J. Phys. Chem. **82** (1978) 1693.
78 Hay 1 Hayashi, K., Irie, M., Lindenau, D., Schnabel, W.: Radiat. Phys. Chem. **11** (1978) 139.
78 Ila 1 Ilan, Y., Ilan, Y.A., Czapski, G.: Biochim. Biophys. Acta **503** (1978) 399.
78 Jon 1 Jonah, C.D., Matheson, M.S., Meisel, D.: J. Am. Chem. Soc. **100** (1978) 1449.
78 Laf 1 Lafferty, J., Truscott, T.G., Land, E.J.: J. Chem. Soc., Faraday Trans. I **74** (1978) 2760.
78 Lev 1 Levanon, H., Neta, P., Trozzolo, A.M.: Chem. Phys. Lett. **54** (1978) 181.
78 Lev 2 Levin, G.: J. Phys. Chem. **82** (1978) 1584.
78 Mul 1 Mulazzani, Q.G., Emmi, S., Fuochi, P.G., Hoffman, M.Z., Venturi, M.: J. Am. Chem. Soc. **100** (1978) 981.
78 Qua 1 Quaal, K.S., Ji, S., Kim, Y.M., Closson, W.D., Zubieta, J.A.: J. Org. Chem. **43** (1978) 1311.
78 Red 1 Redpath, J.L., Ihara, J., Patterson, L.K.: Int. J. Radiat. Biol. **33** (1978) 309.
78 Rob 1 Robinson, E.A., Salmon, G.A.: J. Phys. Chem. **82** (1978) 382.
78 Saw 1 Sawai, T., Ohara, N., Shimokawa, T.: Bull. Chem. Soc. Jpn. **51** (1978) 1300.
78 Sim 1 Simic, M.G., Taub, I.A.: Biophys. J. **24** (1978) 285.
78 Sut 1 Sutton, H.C., Seddon, W.A., Sopchyshyn, F.C.: Can. J. Chem. **56** (1978) 1961.
78 Tau 1 Tausch-Treml, R., Henglein, A., Lilie, J.: Ber. Bunsenges. Phys. Chem. **82** (1978) 1335.
78 Wie 1 Wieghardt, K., Cohen, H., Meyerstein, D.: Ber. Bunsenges. Phys. Chem. **82** (1978) 388.
78 Wie 2 Wieghardt, K., Cohen, H., Meyerstein, D.: Angew. Chem., Int. Ed. Engl. **17** (1978) 608.
79 Alm 1 Almgren, M., Grieser, F., Thomas, J.K.: J. Chem. Soc., Faraday Trans. I **75** (1979) 1674.
79 And 1 Anderson, R.F.: Radiat. Phys. Chem. **13** (1979) 155.
79 Asm 1 Asmus, K.-D., Bahnemann, D., Mönig, J., Searle, A., Willson, R.L., in: "Radiation Biology and Chemistry" H.E. Edwards, S. Navaratnam, B.J. Parsons and G.O. Philligas (eds.), Amsterdam: Elsevier, **1979**, p. 39.
79 End 1 Endicott, J.F., Netzel, T.L.: J. Am. Chem. Soc. **101** (1979) 4000.
79 Gra 1 Grabner, G., Sacher, M., Getoff, N.: Radiat. Res. **77** (1979) 69.
79 Har 1 Harrington, P.C., Wilkins, R.G.: J. Biol. Chem. **254** (1979) 7505.
79 Hil 1 Hiller, K.-O.: Ph.D. Thesis, Techn. Univ. Berlin, D83, **1979**.
79 His 1 Hissung, A., v. Sonntag, C.: Int. J. Radiat. Biol. **35** (1979) 449.
79 Hof 1 Hoffman, M.Z., Kimmel, D.W., Simic, M.G.: Inorg. Chem. **18** (1979) 2479.
79 Hou 1 Houee-Levin, C., Gardes-Albert, M., Ferradini, C., Pucheault, J.: Biochem. Biophys. Res. Commun. **91** (1979) 1196.
79 Ila 1 Ilan, Y., Shafferman, A., Feinberg, B.A., Lau, Y.-K.: Biochim. Biophys. Acta **548** (1979) 565.
79 Jan 1 Janovsky, I.: Radiochem. Radioanal. Lett. **39** (1979) 337.
79 Kem 1 Kemp, T.J., Martins, L.J.A.: J. Chem. Soc., Chem. Commun. **1979**, 227.
79 Lee 1 van Leeuwen, J.W., Tromp, J., Nauta, H.: Biochim. Biophys. Acta **577** (1979) 394.
79 McV 1 McVie, J., Sinclair, R.S., Tait, D., Truscott, T.G.: J. Chem. Soc., Faraday Trans. I **75** (1979) 2869.
79 Meh 1 Mehnert, R., Brede, O., Bös, J., Naumann, W.: Ber. Bunsenges. Phys. Chem. **83** (1979) 992.
79 Mul 1 Mulazzani, Q.G., Emmi, S., Fuochi, P.G., Venturi, M., Hoffman, M.Z., Simic, M.G.: J. Phys. Chem. **83** (1979) 1582.
79 Net 1 Neta, P.: J. Phys. Chem. **83** (1979) 3096.
79 Net 2 Neta, P., Scherz, A., Levanon, H.: J. Am. Chem. Soc. **101** (1979) 3624.
79 Par 1 Parker, V.D.: Pure Appl. Chem. **51** (1979) 1021.
79 Pau 1 Paul, H.: Int. J. Chem. Kinet. **11** (1979) 495.
79 Pru 1 Prütz, W.A., Land, E.J.: Int. J. Radiat. Biol. **36** (1979) 513.
79 Shi 1 Shieh, J.J., Sellers, R.M., Hoffman, M.Z., Taub, I.A., in: "Radiation Biology and Chemistry", H.E. Edwards, S. Navaratnam, B.J. Parsons and G.O. Philligas (eds.), Amsterdam: Elsevier, **1979**, p. 179.
79 Sim 1 Simic, M.G., Hoffman, M.Z., Cheney, R.P., Mulazzani, Q.G.: J. Phys. Chem. **83** (1979) 439.
79 Ste 1 Steenken, S.: J. Phys. Chem. **83** (1979) 595.
79 Ste 2 Steenken, S., Raghavan, N.V.: J. Phys. Chem. **83** (1979) 3101.
79 Suk 1 Sukhov, N.L., Ershov, B.G.: High Energy Chem. **13** (1979) 45; Transl. from Khim. Vys. Energ. **13** (1979) 55.
79 Tam 1 Tamminga, J.J., van den Ende, C.A.M., Warman, J.M., Hummel, A.: Recl. Trav. Chim. Pays-Bas **98** (1979) 305.

79 Wan 1 Wang, Y., Tria, J.J., Dorfman, L.M.: J. Phys. Chem. **83** (1979) 1946.
79 Zad 1 Zador, E., Warman, J.M., Hummel, A.: J. Chem. Soc., Faraday Trans. I **75** (1979) 914.
80 Bie 1 Bielski, B.H.J., Shine, G.G., Bajuk, S.: J. Phys. Chem. **84** (1980) 830.
80 Bra 1 Brault, D., Bizet, C., Morliere, P., Rougee, M., Land, E.J., Santus, R., Swallow, A.J.: J. Am. Chem. Soc. **102** (1980) 1015.
80 But 1 Butler, J., Henglein, A.: Radiat. Phys. Chem. **15** (1980) 603.
80 Dar 1 Darwent, J.R., Kalyanasundaram, K., Porter, Sir George: Proc. R. Soc. London Ser. A **373** (1980) 179.
80 Ell 1 Elliot, A.J., Wiekinson, F., Armstrong, D.A.: Int. J. Radiat. Biol. **38** (1980) 1.
80 Eri 1 Eriksen, T.E., Sjöberg, L.: Radiat. Phys. Chem. **16** (1980) 213.
80 Fer 1 Ferraudi, G., Patterson, L.K.: J. Chem. Soc., Dalton Trans. **1980**, 476.
80 Hou 1 Houee-Levin, C., Gardes-Albert, M., Ferradini, C., Pucheault, J.: Radiat. Res. **83** (1980) 270.
80 Kem 1 Kemp, T.J., Martins, L.J.A.: J. Chem. Soc., Perkin Trans. II **1980**, 1708.
80 Kem 2 Kemp, T.J., Martins, L.J.A., in: "Techniques and Applications of Fast Reactions in Solution", eds. W.J. Gettins and E. Wyn-Jones, NATO Advanced Study Institute Series, Series C, D. Reidel, The Hague, 1979, p. 549.
80 Lev 1 Levanon, H., Neta, P.: Chem. Phys. Lett. **70** (1980) 100.
80 Mad 1 Madhavan, V., Schuler, R.H.: Radiat. Phys. Chem. **16** (1980) 139.
80 Nel 1 Nelsen, S.F., Parmelee, W.P., Göbl, M., Hiller, K.-O., Veltwisch, D., Asmus, K.-D.: J. Am. Chem. Soc. **102** (1980) 5606.
80 Net 1 Neta, P., Behar, D.: J. Am. Chem. Soc. **102** (1980) 4798.
80 Net 2 Neta, P., Steenken, S., Janzen, E.G., Shetty, R.V.: J. Phys. Chem. **84** (1980) 532.
80 Pac 1 Packer, J.E., Heighway, C.H., Miller, H.M., Dobson, B.D.: Australian J. Chem. **33** (1980) 965.
80 Rag 1 Raghavan, N.V., Steenken, S.: J. Am. Chem. Soc. **102** (1980) 3495.
80 Sel 1 Selvarajan, N., Raghavan, N.V.: J. Phys. Chem. **84** (1980) 2548.
80 Spa 1 Spada, L.T., Foote, C.S.: J. Am. Chem. Soc. **102** (1980) 391.
80 Tep 1 Teplý, J., Janovský, I., Mehnert, R., Brede, O.: Radiat. Phys. Chem. **15** (1980) 169.
80 Ven 1 Venturi, M., Emmi, S., Fouchi, P.G., Mulazzani, Q.G.: J. Phys. Chem. **84** (1980) 2160.
80 Was 1 Washio, M., Tagawa, S., Tabata, Y.: J. Phys. Chem. **84** (1980) 2876.
81 Bah 1 Bahnemann, D., Asmus, K.-D., Willson, R.L.: J. Chem. Soc. Perkin Trans. II **1981**, 890.
81 Bax 1 Baxendale, J.H., Sharpe, P.H.G.: Unpublished data cited in J.M. Warman: "The dynamics of electrons and ions in non-polar liquids" NATO Advanced Study Institute, Capri, Italy, 7–18 Sept., **1981**, in "The Study of Fast Processes and Transient Species by Electron Pulse Radiolysis" J.H. Baxendale, F. Busi, (eds.), Dordrecht: D. Reidel Publ. Co., **1982**.
81 Bha 1 Bhattacharyya, S.N., Saha, N.C., Neta, P.: J. Phys. Chem. **85** (1981) 300.
81 Bra 1 Brault, D., Neta, P.: J. Am. Chem. Soc. **103** (1981) 2705.
81 End 1 van den Ende, C.A.M., Nyikos, L., Sowada, U., Warman, J.M., Hummel, A.: 7th Intern. Conf. on Conduction and Breakdown in Dielectric Liquids. Proceedings. W.F. Schmidt (ed.), Institute of Electrical and Electronics Engineers, Inc., N.Y. (USA). Electrical Insulation Society **1981**, p. 50.
81 Fuj 1 Fujita, S., Steenken, S.: J. Am. Chem. Soc. **103** (1981) 2540.
81 Gol 1 Golding, B.T., Kemp, T.J., Sheena, H.H.: J. Chem. Res. (S) **1981**, 34; ibid. (M) **1981**, 334.
81 Hil 1 Hiller, K.-O., Masloch, B., Göbl, M., Asmus, K.-D.: J. Am. Chem. Soc. **103** (1981) 2734.
81 Hil 2 Hiller, K.-O., Asmus, K.-D.: Int. J. Radiat. Biol. **40** (1981) 597.
81 His 1 Hissung, A., Isildar, M., von Sonntag, C., Witzel, H.: Int. J. Radiat. Biol. **39** (1981) 185.
81 Kum 1 Kumar, A., Neta, P.: J. Phys. Chem. **85** (1981) 2830.
81 Meh 1 Mehnert, R., Brede, O., Cserép, Gy.: Radiochem. Radioanal. Lett. **47** (1981) 173.
81 Mic 1 Micic, O.I., Nenadovic, M.T.: J. Chem. Soc., Faraday Trans. I **77** (1981) 919.
81 Mul 1 Mulazzani, Q.G., Emmi, S., Hoffman, M.Z., Venturi, M.: J. Am. Chem. Soc. **103** (1981) 3362.
81 Nat 1 Natarajan, P., Raghaven, N.V.: J. Phys. Chem. **85** (1981) 188.
81 Nen 1 Nenadović, M.T., Mičić, O.I., Kosanić, M.M.: Radiat. Phys. Chem. **17** (1981) 159.
81 Net 1 Neta, P.: J. Phys. Chem. **85** (1981) 3678.
81 Net 2 Neta, P., Grebel, V., Levanon, H.: J. Phys. Chem. **85** (1981) 2117.
81 Pac 1 Packer, J.E., Mönig, J., Dobson, B.D.: Australian J. Chem. **34** (1981) 1433.
81 Rag 1 Raghavan, N.V., Das, P.K., Bobrowski, K.: J. Am. Chem. Soc. **103** (1981) 4569.
81 Sca 1 Scaiano, J.C.: J. Phys. Chem. **85** (1981) 2851.
81 Ser 1 Serpone, N., Jamieson, M.A., Emmi, S.S., Fuochi, P.G., Mulazzani, Q.G., Hoffman, M.Z.: J. Am. Chem. Soc. **103** (1981) 1091.
81 Tak 1 Takamuku, S., Kigawa, H., Toki, S., Tsumori, K., Sakurai, H.: Bull, Chem. Soc. Jpn. **54** (1981) 3688.
81 War 1 Warman, J.M.: "The dynamics of electrons and ions in non-polar liquids" NATO Advanced Study Institute, Capri, Italy, 7–18 Sept., **1981**, in: "The Study of Fast Processes and Transient Species by Electron Pulse Radiolysis" J.H. Baxendale, F. Busi, (eds.), Dordrecht: D. Reidel Publ. Co., **1982**.
82 Meh 1 Mehnert, R., Brede, O., Cserép, Gy.: Ber. Bunsenges. Phys. Chem. **86** (1982) 1123.
82 Ste 1 Steenken, S., Neta, P.: J. Phys. Chem. **86** (1982) 3661.
83 Hil 1 Hiller, K.-O., Asmus, K.-D.: J. Phys. Chem. **87** (1983) 3682.